2017 NORTH CAROLINA AGRICULTURAL CHEMICALS MANUAL

PESTICIDE USE AND SAFETY • APPLICATION EQUIPMENT • SPECIMEN IDENTIFICATION • FERTILIZER USE • INSECT CONTROL • WEED CONTROL • GROWTH REGULATORS • ANIMAL CONTROL• DISEASE CONTROL

This book is also available on the Extension Resource Catalog at content.ces.ncsu.edu/north-carolina-agricultural-chemicals-manual. For more information on the book or Web version, email extension_publications@ncsu.edu.

© College of Agriculture and Life Sciences, NC State University, Raleigh, NC

USEFUL PHONE NUMBERS

Carolinas Poison Center
1-800-222-1222
www.ncpoisoncenter.org
P.O. Box 32861, Charlotte, NC 28232-2861
Carolinas Poison Center can give advice on diagnosis and treatment of human illness resulting from toxic substances.

Chemical Transportation Emergency Center (CHEMTREC)
1-800-262-8200
www.chemtrec.com
CHEMTREC offers 24-hour information and help to aid in responding to emergencies involving hazardous chemicals.

National Pesticide Information Center
1-800-858-7378
npic.orst.edu
NPTN provides information by phone about pesticides Monday through Friday, 8 a.m. to 6 p.m., Central Time.

Pesticide Disposal Assistance Program
1-919-733-3556 (Raleigh, N.C.)
www.ncagr.gov/SPCAP/pesticides/PDAP/
PDAP gives information in disposal of unwanted pesticides.

MISUSE OF PESTICIDES

It is illegal to use any pesticide in a manner not permitted by its labeling. All recommendations for pesticide use included in this manual were legal at the time of publication, but the status of registration and use patterns are subject to change by actions of state and federal regulatory agencies.

2017 North Carolina Agricultural Chemicals Manual
Table of Contents — PESTICIDE USE AND SAFETY INFORMATION

I — PESTICIDE USE AND SAFETY INFORMATION

Restricted Use Pesticides

Wayne Buhler, Pesticide Education Specialist

Because of their potential to cause adverse effects on human health and the environment, many very hazardous pesticides are classified **RESTRICTED USE** by the U.S. Environmental Protection Agency (EPA). For an updated list of these pesticides, visit https://www.epa.gov/pesticide-worker-safety/restricted-use-products-rup-report and click on the link for "RUP Updated List." The label states **RESTRICTED USE PESTICIDE**, as indicated below.

Restricted Use Pesticide

For retail sale to and use only by certified applicators or by persons under their direct supervision and only for those uses covered by the certified applicator's certification.

Pesticide formulations labeled RESTRICTED USE PESTICIDE can only be sold in North Carolina by licensed dealers and purchased or used by licensed commercial applicators, public operators, certified or licensed structural pest control applicators, and certified private pesticide applicators, or by persons working under their direct supervision.

Pesticide Restricted by North Carolina Regulation — Arsenic Trioxide

All pesticide formulations containing the active ingredient arsenic trioxide shall not be used or stored inside or in the immediate vicinity of any building used as a human dwelling.

The pesticide dealer shall secure the signature of the purchaser attesting to the fact that the purchaser has knowledge of the use restrictions on arsenic trioxide.

Any licensed pesticide applicator under the authority of Section 143-452 or 106-65.25 as amended of the General Statutes of North Carolina may use such formulations containing arsenic trioxide in or around human dwellings provided such use is performed in a manner consistent with the product label and access to the pesticide by children is limited.

Local Need — 24 (c) Registrations in North Carolina

The North Carolina Department of Agriculture and Consumer Services (NCDA&CS) has registered a number of pesticides as Special Local Need products in North Carolina (Table 1-1). The state-approved labels have been submitted to EPA for review. These products can be used for the purposes listed in this manual and as detailed on the supplemental labels. It is important to remember, however, that the products listed here are for reference only, and any information given on the product labeling takes precedence over information given in this manual.

Application of a pesticide can begin as soon as the state has registered the Special Local Need. However, the applicator must have within his or her possession a copy of the supplemental labeling, including directions for use. Within 10 days, the state must notify EPA of what is registered, including a copy of the label. Within 90 days after the date of issuance of the Special Local Need registration by the state, EPA can disapprove the registration. If EPA does nothing within the 90 days, the registration automatically becomes a full federal registration for use only within the state.

If EPA should disapprove a 24 (c) label within the specified 90-day period, it is still legal for the grower to sell the agricultural commodity treated with the pesticide during the time the state label was effective because during that time, the application was legal. However, the commodity is subject to regulations governing tolerance, and residues must not exceed tolerance limits established for the pesticide.

Special Local Need - Active 24 (c) Registrations in North Carolina

The following table is provided for reference only and is subject to change at any time. If you have questions about a specific North Carolina 24 (c) registration, please contact Lee Davis, N.C. Department of Agriculture and Consumer Services Pesticide Registration Manager, (919) 733-3556 or send e-mail to: lee.davis@ncagr.gov.

Table 1-1. Special Local Need—24 (c) Registrations in North Carolina

Product Name	Use	SLN Number
Reflex	Nutsedge in pine seedling plantations	NC-950010
Witchaway Permviro Systems	Witchweed in various sites	NC-980005
Goal 2XL herbicide	Witchweed control in corn (allows planting of small grains within 10 months of last application)	NC-020001
Goal 2XL Herbicide	Witchweed in noncropland (USDA/NCDA witchweed project)	NC-020002
Select 2EC herbicide	Annual and perennial weeds in kenaf	NC-020006
Actigard 50WG plant activator	Suppression of Tomato Spotted Wilt Virus (TSWV) in flue-cured tobacco	NC-020007
Orthene 97 insecticide	Thrips in peanuts (in furrow application)	NC-030003
Orthene 97 insecticide	Slash pine flower thrips, coneworms, coneborers, and seedbugs in southern pine seed orchards	NC-030004
Stinger herbicide	Broadleaf weeds in strawberries	NC-030005
Orthene 97 insecticide	Thrips on kenaf	NC-050003
Curfew fumigant	Parasitic nematodes and mole crickets in golf course turf	NC-050004
Intrepid 2F insecticide	Lepidoptera larvae on sweet potatoes	NC-060003
Brigade 2EC insecticide/miticide	Allows a different timing of application (pre-plant) at a higher rate	NC-070002
Select Max herbicide	Annual and perennial grasses in clary sage	NC-070003
Manzate Pro-Stick fungicide	Diseases of tobacco	NC-080002
Penncozeb 75DF dry flowable fungicide	Diseases on tobacco	NC-080003
Tenacity herbicide	Certain winter annual broadleaf weeds in DOT wildflower beds	NC-080005

Table 1-1. Special Local Need—24 (c) Registrations in North Carolina

Product Name	Use	SLN Number
Lorsban Advanced insecticide	Reduce PHI from 120 days to 60 days	NC-090001
Lorsban Advanced insecticide	Allows a higher application rate to be applied to tobacco	NC-090004
Dual Magnum herbicide	Reduce PHI to 60 days when used on tomatoes at 1.67 pints or fewer per acre	NC-100001
Admire Pro Systemic protectant	Reduces PHI to 60 days when used on sweet potatoes	NC-100003
Perm-Up 3.2 EC insecticide	Control regeneration weevils in conifer nurseries	NC-110001
Quadris Flowable fungicide	Control target spot on tobacco seedlings grown in greenhouses	NC-110003
Ridomil Gold SL	Allows for application in tobacco transplant water	NC-110004
Dual Magnum herbicide	Control pigweed and yellow nutsedge in various vegetable and berry crops	NC-110005
Select Max herbicide	Allows for use without an adjuvant when applied to certain vegetables	NC-110007
Corvus herbicide	Control glyphosate-resistant Palmer amaranth and other broadleaf weeds and grasses in field corn	NC-120001
Cotoran 4L herbicide	Control certain summer annual broadleaf weeds in DOT wildflower beds	NC-120002
Milestone herbicide	Control herbaceous broadleaf weeds and woody plants in forests and grazed areas in and around these sites	NC-120003
Gramoxone SL 2.0 herbicide	Suppression and/or control of Palmer amaranth in peanuts	NC-120004
Reflex herbicide	Control weeds in transplanted tomatoes and transplanted peppers	NC-120006
Linex 4L herbicide	Control cutleaf evening primrose and certain other weeds in clary sage	NC-120005
Milestone herbicide	Control or suppression of blackberry (briars), vines, and susceptible broadleaf weeds in newly planted longleaf pine	NC-120010
Prowl H2O herbicide	Control annual grass weeds and certain broadleaf weeds as they germinate in biofuel crops	NC-120011
Avipel Hopper Box (Dry) corn seed treatment	Protect field and sweet corn seed against consumption by black birds and crows	NC-130001
Avipel Liquid corn seed treatment	Protect field and sweet corn seed against consumption by black birds and crows	NC-130002
Vapam HL Soil fumigant	Establishes the use of a revised buffer zone table when the product is applied to peanuts using shank injection to raised beds.	NC-130003
Sectagon 42 Agricultural fumigant	Establishes the use of a revised buffer zone table when the product is applied to peanuts using shank injection to raised beds	NC-130004
Metam CLR 42%	Establishes the use of a revised buffer zone table when the product is applied to peanuts using shank injection to raised beds	NC-130005
Fyfanon ULV AG insecticide	Control of Spotted Wing Drosophila in blueberries	NC-130006
Malathion 8 Flowable Ag insecticide	Control of Spotted Wing Drosophila in blueberries	NC-130007
Malathion 8 Flowable Ag insecticide	Control of Spotted Wing Drosophila in caneberries	NC-130008
Sandea herbicide	Reduce PHI from 30 days to 21 days when used on cucumbers	NC-140001
Bean Guard / Allegiance Fungicide	Seed treatment for control of seed and soilborne damping off diseases of clary sage	NC-140002
Sterilizing Gas 8	Control American foulbrood and other pests	NC-140003
Pyrimax 3.2L Herbicide	Control annual broadleaf and grass weeds in DOT wildflower beds	NC-150001
Reflex Herbicide	Control Palmer amaranth and other weeds in pumpkins and watermelons	NC-150002
Sonar Genesis Aquatic Herbicide	Control hydrilla on portions of the Eno River in Durham and Orange counties	NC-150003
Parazone 3SL Herbicide	Control weeds in clary sage	NC-150005
Fierce Herbicide	Allows over-the-top applications to winter wheat	NC-150004
Armezon	Control weeds in miscanthus grown for nonfood and nonfeed biomass production	NC-150006
Valor SX Herbicide	Control weeds in clary sage	NC-150007
Sivanto Prime Insecticide	Control insects in Christmas trees	NC-160001
Profume Fumigant	Fumigation of non-edible commodities and for use under quarantine/regulatory treatment schedules	NC-160002

The Safe Use of Pesticides

Wayne Buhler, Pesticide Education Specialist

General Safety Instructions

- Use pesticides only when needed.
- Always ask the advice of an expert on problems of pests and pesticides.
- Use the correct pesticide for the problem.
- Know any hazards that the pesticide might present.
- Read and follow the label.
- Commercial pest control operators, farmers, and other applicators of organophosphate and carbamate pesticides should contact their physician at the beginning of the season. At this time, you should inform the physician of the types of pesticides you will be using. The physician will determine the level of enzymes in your blood that may be affected by the use of certain pesticides. While discussing the pesticides to be used, review the signs and symptoms of pesticide poisoning.
- Know what to do in the event of an accident. *Plan ahead.* Call your physician or 911 immediately in the event of an accident.
- Have your physician's phone number programmed into your phone. In an emergency, time is extremely important.
- Take time to explain the safe pesticide use to employees.
- Check your application equipment for leaks or clogged lines, nozzles, and strainers.
- Calibrate equipment frequently for proper output, using water.
- Check respirator for cleanliness, clean filter, and proper fit.
- Check gloves and other protective clothing for holes and cleanliness before each use.
- Make sure plenty of clean water, detergent, towels, and a clean change of clothing are available.
- Do not permit delivery of pesticides unless a responsible representative is on hand to receive and properly store them.
- Make sure that people have been warned and livestock and pets that may be exposed have been removed from the area to be treated.
- Notify beekeepers who maintain beehives in the vicinity of a pesticide application.
- Cover food and water containers.
- Never eat, drink, or smoke when handling pesticides.
- Wash your hands before eating, smoking, or drinking.
- Make sure the time intervals between date of application and reentry, harvest, slaughter, or milking will comply with those given on the label.
- Rinse pesticide containers before recycling or disposal. (Put rinsate in sprayer tank.)

Select the Appropriate Product

Pesticides are classified according to their hazard potential to humans, animals, and the environment. Hazard potential to humans is based on animal test results and/or use experience. Each pesticide is identified with one of the following signal words: **"DANGER," "WARNING,"** or **"CAUTION."**

"DANGER" on the label tells you the product has the highest toxicity to humans. The word **"POISON"** (in red) and the skull and crossbones are also required on the label of highly toxic materials. If **"DANGER"** is printed on the label without the word **"POISON"** and the skull and crossbones symbol, the pesticide can cause severe skin injury or irreversible eye damage.

"WARNING" identifies products with moderate toxicity.

"CAUTION" identifies those pesticides that are least toxic to humans.

Table 1-2 further defines pesticide label signal words.

Table 1-2. Pesticide Signal Words

Signal Word on Container	Toxicity to Humans	Amount Required to Kill an Adult if Swallowed
CAUTION	Slightly toxic	An ounce to a pint
WARNING	Moderately toxic	A teaspoon to an ounce
DANGER*	Very toxic	A taste to a teaspoon

*DANGER **with** the word "poison" and the skull and crossbones symbol means the pesticide is very toxic if swallowed or inhaled. **Without** the word "poison" and the skull and crossbones symbol, DANGER usually means that a pesticide has a high potential as a skin or eye irritant.

Contrary to common belief, the signal word **DOES NOT** tell you how well a pesticide will control a pest. For example, **"DANGER"** means the pesticide can be more dangerous to *you*, not that it is more toxic to pests than a product labeled **"CAUTION."**

Because of the risks involved in handling, many of the very hazardous pesticides have been **"RESTRICTED"** and can be bought and applied only by or under the direct supervision of certified or licensed individuals. These pesticides bear the words **"RESTRICTED USE PESTICIDE"** on the label.

Follow Label Directions

The label tells you how to use the pesticide properly and safely. Use a pesticide only on crops, animals, or other sites as the label directs. **Use the recommended rate, and apply the pesticide at the time and in the manner stated.**

Obey all precautions for using a pesticide safely, such as "Keep out of reach of children;" "Keep away from pets;" "Do not use near fire, sparks, or flame;" "Do not inhale, ingest, or allow to get on skin;" "Do not store near food, feed, seed, or animals;" "Do not contaminate water supplies;" and any other warnings on the label.

Wear Protective Clothing Called for on the Label

While people generally realize the danger in getting pesticides in the mouth or eyes or breathing gaseous fumes, they are frequently unaware of how harmful many pesticides are when absorbed through the skin. Anytime you handle or apply pesticides, wear at least a long-sleeved shirt and long-legged trousers made of a closely woven fabric, socks, and liquid-proof shoes. Wear personal protective equipment listed on the label.

With most **DANGER**- and **WARNING**-labeled pesticides, a respirator covering the nose and mouth and goggles or a face shield protecting the eyes are necessary. Again, the label will tell you the kind of protective equipment you need.

Use Proper Application Equipment

You cannot apply a pesticide properly or safely unless you have the correct equipment. Small jobs around the home with less hazardous pesticides may be done with simple equipment such as a pump-up sprayer, hose attachment sprayer, granular applicator, or hand duster. Larger jobs and those where more hazardous materials are used often require specially designed equipment. Contact your county Cooperative Extension Service agent, pesticide dealer, or equipment dealer if you need further advice on proper equipment for applying pesticides.

Know Emergency First-Aid Procedures

In case of suspected poisoning — stomach cramps, dizziness, vomiting, heavy sweating — follow the label's first-aid advice and **IMMEDIATELY** call a doctor or take the person to a hospital. Take the pesticide label with you because the doctor needs it to prescribe the proper treatment.

If you spill a pesticide on yourself, remove any contaminated clothing immediately. Wash skin immediately with soap and water. **DO NOT USE AN ABRASIVE OR PETROLEUM-BASED CLEANER**, as this would allow the pesticide to penetrate your skin more easily. If an individual is exposed to pesticide vapors, get him to fresh air. Start artificial respiration if the person has stopped breathing. If a pesticide is splashed in the eyes or mouth, rinse it out with large quantities of clean water for at least 15 minutes. If swallowed, read the label to see if you should induce vomiting (this could be harmful, depending on the pesticide). Never give anything by mouth to an unconscious individual.

The doctor may want to call a poison control center for specific treatment of pesticide poisoning. (The number of the Carolinas Poison Center is **1-800-222-1222**.)

Hazard and Toxicity of Pesticides

Do not depend on toxicity values as the *only* factor to be considered regarding the hazards of a chemical to people or other animals. Pesticide users should be concerned with the *hazard(s)* associated with exposure to the chemical and not the *toxicity* of the material itself. Hazard and toxicity are not synonymous.

Toxicity is the inherent capacity of a substance to produce injury or death.

Hazard is a function of two primary variables, *toxicity* and *exposure*, and is the potential threat that injury will result from the use of a substance in a given formulation or quantity. Some hazards do not involve toxicity to people or other animals. For example, sulfur, oils, and numerous other chemicals are considered relatively non-hazardous to animals, but may pose considerable hazard to some plants (phytotoxicity).

A pesticide may be extremely toxic but present little hazard to the applicator or others when used:

- in a very dilute formulation;
- in a formulation that is not readily absorbed through the skin or readily inhaled;
- under conditions to which humans are not exposed;
- only by experienced applicators who are properly equipped to handle the chemical safely.

However, a chemical may exhibit a relatively low mammalian *toxicity* but present a *hazard* because it is normally used in a concentrated form, which may be readily absorbed or inhaled.

Pesticide Toxicity to People

Most pesticides are harmful to people if they are handled or applied in an unsafe way. A pesticide may harm a person if it is:

- swallowed (oral toxicity);
- breathed (inhalation toxicity);
- allowed to get on the skin or in the eye (dermal toxicity).

Children may be poisoned if a pesticide is left where they can eat it, play in it, or drink it. Pesticides should not be stored in unlabeled containers, such as soft drink bottles. A few pesticides give off harmful vapors that must not be breathed. Some applicators are poisoned when they allow pesticides to contact their skin. Oil-based pesticides (such as emulsifiable concentrates) penetrate the skin in greater quantity and more quickly than dusts, granules, or wettable powders. Sun-burnt or hot, sweaty skin with cuts or abrasions allows more rapid penetration.

Pesticide Hazards to the Environment

A pesticide may not affect people the same way it does the environment. Some pesticides may not harm the environment, even though they are moderately to highly hazardous to people. And some pesticides that are only slightly hazardous to people may cause greater environmental damage. A restricted use pesticide may be hazardous to people, the environment, or both.

A given pesticide may be hazardous in the air (particle or vapor drift), soil, or water. It may leave harmful residues in food or injure non-target plants and animals, such as fish, bees, birds, other wildlife, and domestic animals.

Some pesticides are potentially more harmful to the environment because they last for a long time once they are applied. Others may accumulate in the body and cause poisoning. Most uses for persistent, accumulative pesticides (such as DDT) have been cancelled in the U.S. Some uses of other persistent pesticides are now restricted.

Wildlife Exposure: Managing the Risk

Wildlife may contact residues of pesticides applied to forests, aquatic habitats, farmland, rights-of-way, turf, and gardens. Pesticide poisonings to wildlife may be caused by runoff to surface water during rainfall, spray drift, foraging on pesticide-treated vegetation or insects, or consumption of pesticide-treated granules, baits, or seeds. Also, secondary poisoning occurs when an animal eats prey species that contain pesticide residues.

Fortunately, not all pesticides have detrimental effects on all wildlife, nor do pesticide residues necessarily have serious consequences for wildlife. Before using pesticides, get advice from wildlife, conservation, and pesticide professionals at universities and state and federal agencies on the choice and proper use of pesticides and alternative pest control strategies. Also, consider strategies to improve wildlife habitats.

Implementing the following suggestions will benefit wildlife while allowing you to control pests. As you look over the suggestions, keep in mind that you must also be in compliance with all pesticide product labels. North Carolina Pesticide Law of 1971 Article 52 G.S. 143, and 2 NCAC 9 L regulations.

Be Careful Around Natural Areas

- All wildlife need natural areas in which to feed, rest, reproduce, raise young, and take shelter. Create habitats by encouraging and promoting the growth of native vegetation. This also reduces the need for mowing.
- Plant disease- and insect-resistant trees and shrubs, thereby reducing the need for pesticides.
- Always store pesticides and wildlife feed separately, and do not feed wildlife near pesticide storage and mixing areas.

Wildlife Benefit When You Understand and Follow Pesticide Labels

- Keep wildlife habitats in mind when reading pesticide labels.
- Compare labels and select highly specific products that pose less risk to nontarget species. Read the label carefully, and use the lowest effective rate.
- Calibrate equipment carefully to assure that the pesticide is applied at labeled rates.
- Get the EPA *Endangered Species Protection* county bulletin from the Web at http://www.epa.gov/espp (click on "Bulletins Live! Two") or by calling 1-800-447-3813. Follow the limitations in the bulletin. Also, check regulation 2 NCAC 9L .2200 on pesticide use limitations to protect the Carolina

heelsplitter freshwater mussel.

- Take heed of the label. The environmental wildlife precautions on labels are based on scientific and regulatory actions. They must be followed. It's the law, it's good business, and it's the right thing to do.

- Consult the NCDA&CS Structural Pest Control and Pesticides Division or your county Cooperative Extension center for label clarification or to determine potential pesticidal impacts on wildlife. Consult natural resource agencies, natural heritage programs, and the Nature Conservancy for additional information about wildlife, native vegetation, and endangered species.

Be Alert for Wildlife Before and During Pesticide Application

- Avoid spraying near areas frequented by wildlife, especially flocks of birds, or, if possible, reduce the application rates.

- Homeowners should search for bird and mammal nests before spraying fruit or ornamental trees, shrubs, or lawns, and then avoid applications near those areas.

- Use mechanical, cultural, and biological pest control tactics when available and practical. For example, tillage, crop rotation, pest-resistant plants, natural predators, and trapping can help control pests.

- Scouting and pest identification are critical components of wise pesticide use. To save money and reduce impacts on wildlife, apply pesticides only when pests are present at unacceptable levels.

- Remember to guard against pesticide drift and runoff. Apply pesticides under low, directional wind conditions, and use adjuvants when appropriate. Buffer zones of unsprayed crops or grass strips adjacent to important habitats will help protect wildlife.

- Do not apply pesticides when heavy rain is imminent. Surface runoff may move some pesticides into ponds, streams, and wetlands inhabited by wildlife. In urban areas, such runoff may flow into storm drains leading directly to streams and rivers.

- Multiple pesticide applications may have cumulative effects, especially during breeding season. Reduce application frequency when possible, and target each application to the specific site of the pest instead of making applications over entire fields or lawns. Spot treating weeds and insects in lawns and gardens will reduce the amount of pesticide applied.

- Where practical, do not apply pesticides in and around field edges and corners, fencerows, set-aside acreage, nesting sites, vegetation near streams and wetlands, and areas dedicated to wildlife except to spot treat state-listed noxious weeds. Especially sensitive areas include endangered species habitats, native plant communities, and sinkholes.

- Check the label for instructions on incorporating or watering pesticide granules into the soil. These techniques allow the pesticide to reach the target pests more readily, and foraging birds are less likely to ingest them.

- Never spray leftover pesticides or wash equipment near wetlands, rivers, streams, creeks, potholes, ponds, marshes, sinkholes, wildlife habitats, or drains leading to these areas. Dispose of leftover pesticide as specified on the label.

- For rules and guidelines on protecting honey bees, see section on Bee Protection below and in Chapter 5, Reducing the Risk of Pesticides Poisoning to Honey Bees.

- For additional information, contact your county Cooperative Extension center or call the NCDA&CS Structural Pest Control and Pesticides Division at 919-733-3556. You may

also be interested in the Extension publication AG-463, *Pesticides and Wildlife* and the AG-463 fact sheet series on using integrated pest management on specific crops, including soybeans, tobacco, cotton, and peanuts. Publications are available through county Extension centers.

This section was written by Henry Wade, Environmental Programs Manager, NCDA&CS, Structural Pest Control and Pesticides Division.

Hazardous Chemicals Right-to-Know Act

The Hazardous Chemicals Right-To-Know Act (N.C.G.S. 95-173 *et seq.*) was adopted by the N.C. General Assembly in 1985. The purposes of this act are (1) to see that firefighters have all the information they need to respond to chemical emergencies and (2) to ensure that citizens have access to sufficient information to make informed judgments about hazards in their communities.

Public and private employers who normally use or store at least 55 gallons or 500 pounds of any hazardous chemical must comply with this law. Although the full requirements of this act do not apply to farms with 10 or fewer full-time employees, the employer must tell fire departments whom to contact in case of emergency.

If you receive a Safety Data Sheet (SDS) with a product you purchase, you will know the material has been classified as hazardous. The SDS gives health-related information, emergency and first-aid procedures, and other information needed to use, store, and dispose of the chemical properly.

North Carolina Worker Protection Standard Regulations

The Worker Protection Standard (WPS) requires the agricultural employer to reduce the risks of pesticide exposure for farmworkers by providing them with specific pesticide safety training, personal protective equipment, application notification, and a means to mitigate pesticide exposure through emergency assistance. The primary requirements for compliance are listed below.

- Provide annual pesticide safety training to both workers and handlers.

- Provide label-specified personal protective equipment to the appropriate employees, including respirators, fit tests, and training

- Ensure no entry into treated area or the application exclusion zone (AEZ), which is an area up to 100 feet surrounding application equipment, during pesticide application on farms, forests, and nurseries.

- Post warnings, oral notification, or both to inform employees of pesticide applications, restricted entry intervals (see below), and restricted areas.

- Provide the following safety information at a central location and at sites where decontamination supplies are located (if the decontamination supplies are at a permanent site).

  1. Emergency medical information (name, telephone numbers, and address of nearest medical facility);

  2. SDS sheets and pesticide-specific information (location of the area to be treated, product name, EPA registration number, active ingredient(s) of pesticide, time and date the pesticide is scheduled to be applied and time and date the application was completed, and the restricted-entry interval for the pesticide). Each day of application must be recorded as a separate application record. Recorded information must be retained for a period of 2 years and made available to

any employee, treating medical professional, or designated representative who requests it.

3. The WPS pesticide safety information poster. This no longer needs to be displayed as a poster as long as the following information is included: 7 concepts for preventing pesticides from entering your body; instructions for employees to seek medical help immediately if they have been poisoned, injured, or made ill by pesticides; name, address and phone of state or tribal pesticide regulatory authority; and name, telephone and address of nearby medical facility.

- Provide ample supplies at designated decontamination stations (single-use towels, soap, water, and clean change of clothing).

Restricted Entry Intervals

The restricted entry interval (REI) is the time after the end of a pesticide application when entry into the treated area is prohibited. The agricultural employer must not allow or direct any worker to enter or remain in the treated area before the REI has expired except under the conditions of *early entry.*

Early entry provisions of the Worker Protection Standard (WPS) allow trained and label-specified early entry PPE-equipped workers with a minimum age of 18 to enter a treated area during the REI to perform short-term activities with "limited contact," such as moving irrigation equipment or opening ventilation systems.

The following conditions must be met for "limited contact" early entry activities.

1. Worker's contact with the treated surfaces is minimal and limited to the feet, lower legs, hands, and forearms.

2. Pesticide product does not have a statement in the labeling requiring double notification.

3. PPE for early entry conforms with the label requirements or includes at least coveralls, chemical resistant gloves, socks, chemical resistant footwear, and eyewear, if eyewear is required by the label.

4. No hand labor, such as hoeing, picking, or pruning is performed.

5. The time in the area under REI for any worker does not exceed 8 hours in a 24-hour period.

6. Workers do not enter the area during the first 4 hours after application and not until applicable ventilation criteria and label specified inhalation exposure levels are reached.

7. Agricultural employer must give an oral or written notification of the specifics of the early entry. Notification must be given in a language that workers understand.

Preharvest Intervals

The preharvest interval is the time in days that must pass between the last application of a pesticide and harvesting a food or feed crop. The interval varies with the pesticide, depending largely on its persistence (how long the pesticide lasts) on or in the crop as well as on the pesticide's toxicity.

For example, one insecticide label for peach tree borer on peaches states: "Make only one application per season. Do not apply within 14 days before harvest." Fourteen days is the preharvest interval, and it means peaches cannot be harvested from an orchard where this pesticide has been applied until at least 14 days have passed. If harvested sooner, the peaches cannot be legally sold.

If you do not obey the preharvest interval, your crop can be seized and destroyed, and you can be fined. Most **importantly, if crops are harvested and consumed before the preharvest interval has passed, people or animals may be poisoned.**

Aerial Application Limitations

If you expect to have your crops sprayed by an aerial applicator, in order to obey N.C. Pesticide Regulations, you must carefully *plan the location of these crops.* Certain areas are restricted, and aerial pesticide applications cannot be made to these areas unless certain rules are followed.

DO NOT SPRAY (or otherwise allow pesticides to be deposited):

1. In any congested area unless permission is granted by appropriate authorities;

2. Within 300 feet of the premises of schools, hospitals, nursing homes, churches, or any building used for business or social activities if either the premises or the building is occupied by people;

3. Within 100 feet of a residence;

4. On a right-of-way of a public road or within 25 feet of the road, whichever is the greater distance;

5. In or near any body of water, if the pesticide is labeled as toxic or harmful to aquatic life, unless such aquatic life is the target of the pesticide.

Farmers can prevent potential problems by not planting crops near those areas just described and by always following directions on the pesticide label concerning the spraying of crops, especially near sensitive and restricted areas. The Bee Protection section below also reviews the limitations on aerial application.

Bee Protection

Anyone who hires an aerial applicator to apply a pesticide **labeled as toxic to bees shall notify all beekeepers with registered apiaries** located within 1 mile of the area to be treated. This notification must not take place less than 48 hours nor more than 10 days before the application.

A list of beekeepers with registered apiaries may be obtained in one of the following ways.

1. The NCDA&CS Plant Protection Section mails a list of registered apiaries to aerial applicators licensed in North Carolina quarterly. They will also provide the list on request (call 919-233-8214). This list will have the names of any beekeepers who have registered apiaries located within the required 1 mile from the target area.

2. The NCDA&CS Pesticide Section will mail a list of registered apiaries to farmers who have been identified by the beekeeper on the Apiary Registration Form as having farms within 1 mile of the applicable apiary. The failure of a beekeeper to list a farmer on the Apiary Registration Form does not relieve the farmer of responsibility for notifying the beekeeper of an aerial application of a bee-toxic pesticide within 1 mile of the registered apiary.

The list of registered apiaries is mailed on the first day of each quarter. Revised lists will be issued on the first day of each successive quarter. The lists of revised registered apiary locations will become effective on the fifth day of the first month of that quarter. The registration period will be effective for the calendar year and applies only to the listed apiary locations. Moving an apiary to a new site does not provide protection under the law, unless the new site is also registered.

The farmer can notify the beekeeper digitally, orally, or in writing of the approximate time the pesticide application will be made and the type of pesticide to be used. Digital communication may take place thru email or by texting. Oral notification can be by

PESTICIDE USE, SAFETY

telephone or in person to the beekeeper or the alternate person designated on the apiary registration list. Acceptable written notification is by mail or by notice left at the beekeeper's residence or at an alternate location designated on the apiary registration list.

NCDA&CS has initiated a free voluntary beehive registration program in coordination with FieldWatch. Growers may register at http://www.fieldwatch.com as an applicator and view maps of the area surrounding their farms to help identify local beekeepers. Contact information is available for beekeepers, and this tool may be used to enhance communication with beekeepers to reduce the adverse effects of pesticide to bees. Growers may also see additional flags or signs to designate the location of apiaries.

Beekeepers who wish to voluntarily map and register their apiaries for free should first go to FieldWatch.com, then complete the registration process thru BeeCheck. Apiaries registered thru Plant Industry will automatically be mapped and registered on BeeCheck.

Growers should be aware of upcoming label changes to 75 active ingredients that have been classified as acutely toxic to bees (products with an LD 50 of <11 micrograms per bee). New labels may place restrictions on pesticide applications when contract pollinators are on site. There are also application restrictions for neonicotinoid products being applied to crops and ornamentals that are blooming and attractive to bees.

Protecting Surface and Groundwater

Groundwater makes up 96 percent of the world's total water resources. Ninety percent of rural residents and 50 percent of the people in the United States depend on groundwater for drinking water. Although only very small amounts of pesticides have been detected in North Carolina's groundwater, we must reduce the likelihood of pesticides entering groundwater and surface water in order to avoid future water quality problems.

What can pesticide users do to prevent groundwater and surface water contamination?

Follow Label Directions Exactly

Pesticide labels provide valuable information concerning the pesticide's potential dangers of contaminating water and the environment. When applying a pesticide, the timing and placement instructions on the label must be followed correctly to ensure the pesticide is applied properly. Applying a pesticide when heavy rains are predicted could lead to water contamination. Likewise, placing a pesticide on top of the soil, when it should be incorporated, not only minimizes pest control but could lead to unnecessary runoff to surface water and perhaps groundwater.

Use Integrated Pest Management Practices

Cultural practices, such as crop rotation and cover crops, not only reduce pest populations, they also maintain and improve good soil and water conditions. Careful pest monitoring will also ensure that pesticides are used only when needed.

Prevent Spills and Back Siphoning

Pesticides spilled near wells, sinkholes, surface waters, or anywhere else can move into surface water and groundwater. Avoid mixing and loading pesticides near wells and other water sources. Use a long hose from the water source to the sprayer so if any spills occur, they will be farther away from the clean water supply. If a spill occurs, be sure to clean it up and move the contaminated soil to a place where it will not seep into the water or otherwise harm the environment. If contaminated soil cannot be applied to a labeled site that does not exceed the rate of application described on the product label, it must be disposed of at an approved waste disposal facility. When using water from a

hose to dilute pesticides in a spray tank, do not allow the hose to be submerged in the spray tank, which can lead to a backflow situation where pesticides may be siphoned back into the water supply. The Chemigation section of this chapter also stresses regulations required to prevent the backflow of pesticides into water supplies.

Dispose of Wastes Properly

It's illegal and dangerous to dispose of pesticides improperly. If pesticides, their containers, or other hazardous materials are discarded where they can contaminate the water supply or environment, you (and your family) could drink pesticide-contaminated water. This contamination could move into your neighbors' or livestock's water supply as well as affect wildlife and conditions of soil and air. Don't take it for granted; many pesticides don't just disappear. Your responsibility for these hazardous materials includes proper disposal.

The guidelines for disposing of such materials can be found first on the pesticide label. Follow these instructions for disposal carefully. (See the sections in this chapter on Disposal of Pesticides and Disposal of Empty Pesticide Containers.)

Surface Water Protection

If there is more water in the soil than the soil can absorb, water (with pesticides in it) may flow into the groundwater or run off into streams, rivers, and lakes. Prolonged heavy rains and too much irrigation will also produce excess surface water. Pay attention to weather forecasts, maintain proper irrigation scheduling, and use strip crops to restrict potential surface water problems. The chemigation regulations in this manual are also designed to reduce both surface water and groundwater contamination.

Land Characteristics

Geology plays a key role in protecting groundwater and surface water. If groundwater is within a few feet of the soil surface, pesticides are much more likely to reach groundwater. If pesticides are applied to an area that drains into a sinkhole, irrigation or moderate rainfall may carry some of the pesticides directly into groundwater. You must select pesticides carefully when either groundwater is close to the soil surface or soil permeability is great (see Evaluating the Potential for Groundwater Contamination section).

N.C. Well Construction Standards — and in some cases more stringent local regulations relating to well location, casing, grouting, and other requirements — help ensure that groundwater is not contaminated. For example, N.C. regulations require that well casings be at least 12 inches above the soil surface and that casings be cemented at least 20 feet below the soil surface.

Soil Characteristics

Soil texture (sand, silt, clay), soil permeability, and soil organic matter all play a major part in pesticide movement. Soils containing large amounts of organic matter and clay, for example, will hold (absorb) some pesticides before they reach groundwater. But pesticides are more likely to move into groundwater through sandy soils — low in organic matter and clay — and loose, porous soils. Table 1-4 gives the relative leaching potential ratings for Southeastern U.S. soils.

Pesticide Characteristics

Some pesticides move into the soil more easily than others. Those with high water solubility are more likely to seep into the soil than those pesticides with extremely low water solubility.

Table 1-3 gives the relative mobility (movement) of certain pesticides, listed by common and brand names, in soils. Other sources of groundwater contamination include abandoned and uncontrolled waste sites, landfills, holding pits and ponds, leaking storage tanks, and septic tanks.

Chemigation

Applying pesticides to land, crops, or plants through an irrigation system is called chemigation. A limited number of pesticides are cleared for chemigation application. **Chemigation is only legal when the pesticide label has directions for such uses.**

The North Carolina Pesticide Board has adopted chemigation regulations to protect water resources from pesticide pollution by reducing the potential for backsiphoning or direct injection of pesticides into water sources. Farmers, greenhouse operators, nurserymen, golf course operators, turf growers, and others must comply with these regulations.

The types of irrigation equipment covered by these regulations include, but are not limited to, drip or trickle, center pivot, lateral move, traveler gun, and solid set systems. The regulations do not apply to hand-held hose-end sprayers that are constructed so that an interruption in water flow automatically prevents any backflow to the water supply. Protected water resources include, but are not limited to, private ponds, lakes, rivers, streams, canals, wells, and public water systems.

The following antipollution devices must be installed and maintained on an irrigation system used to apply any pesticide. Safety devices must meet the following qualifications.

Automatic low pressure drain—located on the bottom of the horizontal irrigation pipeline between the discharge side of the irrigation pump and the inlet side of the double check valves. This device shall be level and have an orifice size at least 3/16 the diameter of the irrigation pipe. The top of the drain shall not exceed beyond the inside surface of the bottom of the irrigation pipeline and shall be at least 2 inches above grade. The drain shall discharge at least 20 feet from any water supply. Furthermore, the discharge must be controlled to prevent it from reentering the water supply. In the event that the mainline check valves leak slowly, solution will drain away from rather than flow into the water supply.

Inspection port—located between the irrigation pump discharge side and the inlet side of the mainline check valves. The inspection port may be part of the vacuum relief valve. The purpose of the inspection port is to allow an individual to observe whether or not the mainline check valves are leaking.

Vacuum relief valve—located on the top of the horizontal irrigation pipeline between the discharge side of the irrigation pump and the inlet side of the double check valves. The orifice size of the valve shall be 3/16 of the diameter of the irrigation pipe. The purpose of the vacuum relief valve is to allow air into the pipeline when the water flow stops, preventing the creation of a vacuum that could lead to backsiphoning.

Double check valves—located between the irrigation pump discharge (or pressure side) and the point of pesticide injection into the irrigation pipeline. These valves must be within 10 degrees of horizontal. Double check valves prevent solution from draining or backsiphoning into the irrigation water source and polluting groundwater or surface water. Check valves must have positive closing action and a watertight seal. Note: For irrigation systems that contain media filters, refer to the section below entitled Chemigation Systems that Contain Media Filters.

Check valve—located on the pesticide line between the point of pesticide injection into the irrigation system and the pesticide injection unit. The check valve stops the flow of water from the irrigation system into the chemical supply tank. It should be constructed of chemically resistant materials. The check valve should always be flushed with clean water after injecting a chemical to prevent the deposition of chemical precipitates. Whenever a Dosatron, DosmaticPlus, or a similar metering device is used, the check valve needs to be positioned near the metering device on its outlet side. If this metering device is on a bypass line, the check valve must be located on the main irrigation line immediately downstream of the bypass line.

Flow interruption device, solenoid valve—located in the pesticide supply line between the pesticide injection unit and the pesticide supply tank or container. The solenoid valve provides a positive shut off on the chemical injection line. This prevents both chemical and water from flowing in either direction if the chemical pump is stopped. Because this valve will be subjected to different chemicals, it must be compatible with the chemicals being injected. The valve should be inspected often to assure that it is performing properly. A solenoid valve is not required with the Dosatron, DosmaticPlus, or a similar hydraulic injection device.

Functional systems interlock—The irrigation pump and the chemical injection pump must be interlocked or connected so that if the irrigation pump stops, the chemical injection pump will stop. The functional systems interlock ensures that a pesticide is applied with water through the irrigation system. The Dosatron, DosmaticPlus or a similar hydraulic injection device does not require the functional systems interlock.

Illegal Techniques

1. Some pesticide product labels prohibit application of the product by any irrigation system. Others prohibit applications through certain specific irrigation systems.

2. It is illegal to inject a pesticide into an irrigation system on the suction (or inlet) side of the irrigation pump.

3. It is illegal to connect an irrigation system directly to a public water system when applying any pesticide.

NOTE: When a public water system is used, the water must first be discharged into a reservoir tank. An air gap at least twice the inside diameter of the fill pipe must exist between the end of the fill pipe from the public water system and the top rim of the reservoir tank.

Chemigation Systems that Contain Media Filters

Some chemigators are using chemigation systems that have one or more sand-containing media filters. Surface water, and in some cases, groundwater, flows through these filters to remove debris that would clog the small orifices of the emitters on drip irrigation systems. **The injection of pesticides into the irrigation line must be on the outlet side of all media filters.** This prevents pesticide from passing through the media filters and contaminating the debris, which will be discharged into the environment whenever the media filters are backflushed.

Any water dumping or open dumping of pesticide dilutions is an illegal discharge of a pesticide in violation of NC Pesticide Board rule 2 NCAC 9L .0604. Further, pesticide product labels have enforceable language on the illegal disposal of pesticides.

Additionally, a check valve is required between the outlet side of all media filters and the point of pesticide injection into the irrigation line. If the injection system has bypass piping, a check valve would be positioned between the outlet side of all media filters and the inlet side of the bypass on the irrigation pipeline.

The purpose of the check valve mentioned above is to reduce the risk of media filter contamination if a backsiphonage occurs. Systems operating without this safeguard could dispose of pesticides unintentionally during a backflush cycle. This would be illegal. The chemigator can be fined for any illegal disposals.

Any chemigation system that is not in compliance with pesticide regulations will be issued a stop-use order. This order can only be released when a follow-up inspection indicates that the appropriate antipollution devices have been installed.

Hand-Held Hose-End Sprayers

Hand-held hose-end sprayers are allowed on the outlet side of a water hose. This device must contain a check valve that will prevent any backsiphoning from the pesticide reservoir into the water hose. The use of devices connected to a faucet or spigot that siphon pesticide from a reservoir or container is not permitted in North Carolina.

System Inspections

One of the requirements of the regulations is that the system operator must inspect the antisiphon devices and the functional systems interlock during periods of chemigation to ensure that they are functioning properly. If components of the system are defective, they must be repaired or replaced before any chemigation is employed with a pesticide.

Representatives of the NCDA&CS, Pesticide Section, may inspect an irrigation system used for chemigation at any time. If the system is not in compliance with the regulations, a stop-use order will be issued by the department, and the system must be inspected again by the departmental representative before the stop-use order can be removed.

For Additional Information

For a copy of regulation 2 NCAC 9L .2000 — Chemigation or the chemigation and fertigation brochure, visit the NCDA&CS Web site at http://www.ncagr.com/SPCAP/pesticides/sitemap.htm, or call the Structural Pest Control and Pesticides Division at 919-733-3556.

The chemigation section was written by Henry Wade, Environmental Programs Manager, NCDA&CS, Structural Pest Control and Pesticides Division.

Evaluating the Potential for Groundwater Contamination

When estimating the groundwater contamination potential (GWCP) index for a pesticide at a given site, the characteristics of the soil at the site must also be evaluated. Soil properties are as important as a pesticide's chemical properties in determining mobility and risk to groundwater. See Table 1-4. Soil Leaching Potential (SLP) Indices. The GWCP index for a pesticide on a given soil is the mean of the PLP index and the SLP index, i.e.; GWCP = (PLP + SLP)/2. More detailed formulas are given at the end of the following table.

Table 1-3. Relative Pesticide Leaching Potential (PLP) Indices and Ratings for Commonly Used Pesticides

The PLP value will change with changes in application rate. Values in this table are calculated using average rates.*
KEY: Very High (VH) = 90 to 100, High (H) = 70 to 89, Moderate (M) = 50 to 69, Low (L) = 30 to 49, Very Low (VL) = 0 to 29

Common Name	Brand Name	Application Method	PLP	Rating
Herbicides				
2,4-D	Weedone	foliage	45	L
2,4-DB	Butyrac 200	foliage	41	L
acetochlor	Harness	soil	53	M
acifluorfen	Blazer	foliage	42	L
alachlor	Lasso	soil	55	M
ametryn	Evik	foliage	45	L
amitrole	Amizole	foliage	51	M
AMS	Ammate	foliage	74	H
arsenic acid	Hy-Yield	foliage	39	L
asulam	Asulox	foliage	47	L
atrazine	AAtrex	soil	56	M
benefin	Balan	soil	31	L
bensulfuron	Londax	foliage	20	VL
bensulide	Prefar	soil	57	M
bentazon	Basagran	foliage	48	M
bispyribac	Velocity	foliage	45	L
bromacil	Hyvar	soil	85	H
bromoxynil	Buctril	foliage	30	L
butachlor	Machete	soil	36	L
butylate	Sutan	soil	48	L
cacodylic	Rad-E-Cate	foliage	19	VL
carfentrazone	Aim	foliage	26	VL
carfentrazone	Aim	soil	23	VL
chlorimuron	Classic	soil	19	VL
chlorsulfuron	Glean	foliage	28	VL
clethodim	Envoy	foliage	40	L
clodinafop-prop	Discover	foliage	1	VL
clomazone	Command	soil	42	L
clopyralid	Stinger	foliage	46	L
cloransulam-me	First-Rate	soil	34	L
cloransulam-me	First-Rate	foliage	25	VL
cyanazine	Bladex	soil	41	L
cyanazine	Bladex	foliage	37	L
cycloate	Ro-Neet	soil	47	L
cyhalofop-bu	Clincher	foliage	1	VL
DCPA	Dacthal	soil	47	L
desmedipham	Betanex	foliage	27	VL
dicamba	Banvel	foliage	49	L
dichlobenil	Carson	soil	55	M
dichlorprop	DP-Amine	foliage	44	L
diclofop	Hoelon	foliage	50	M
diclosulam	Strongarm	soil	33	L
difenzoquat	Avenge	foliage	0	VL
diflufenzopyr	Distinct	foliage	21	VL
dimethenamid	Frontier	soil	46	L
diphenamid	Enide	soil	54	M
dimethipin	Harvaid	foliage	70	H
diquat	Diquat	foliage	15	VL
dithiopyr	Dimension	foliage	45	L
diuron	Karmex	soil	65	M
DSMA	Ansar	foliage	39	L
endothall	Aquathol	aquatic	64	M
EPTC	Double Play	soil	53	M
ethalfluralin	Sonalan	soil	25	VL
ethamesulfuron	Muster	foliage	28	VL
ethofumesate	Nortron	soil	52	M
fenarimol	Rubigan	foliage	44	L
fenoxaprop	Whip	foliage	24	VL
fluazifop	Fusilade	foliage	30	L
flufenacet	Define	soil	43	L
flumetsulam	Broadstrike	soil	50	M
flumiclorac	Resource	foliage	14	VL
flumioxazin	Valor	soil	45	L
flumioxazin	Valor	foliage	29	VL
fluometuron	Cotoran	soil	62	M
fluridone	Sonar	aquatic	34	L
fluroglycofen	Compete	foliage	18	VL
flroxypyr	Starane u	foliage	45	L
fluthicet-me	Action	foliage	17	VL
fomesafen	Reflex	foliage	49	L
formasulfron	Option	foliage	37	L
fosamine	Krenite	foliage	16	VL
glufosinate	Ignite	foliage	8	VL
glyphosate	Roundup	foliage	20	VL
halosulfuron-methyl	Permit	foliage	23	VL
haloxyfop	Galant	foliage	48	L
hexazinone	Velpar	soil	73	H
imazamethabenz	Assert	foliage	47	L
imazamox	Raptor	foliage	43	L
imazapic	Cadre	foliage	49	L
imazapyr	Arsenal	foliage	65	M
imazaquin	Scepter	foliage	44	L
imazethapyr	Pursuit	foliage	52	M

PESTICIDE USE, SAFETY

Table 1-3. Relative Pesticide Leaching Potential (PLP) Indices and Ratings for Commonly Used Pesticides

The PLP value will change with changes in application rate. Values in this table are calculated using average rates.*
KEY: Very High (VH) = 90 to 100, High (H) = 70 to 89, Moderate (M) = 50 to 69, Low (L) = 30 to 49, Very Low (VL) = 0 to 29

Common Name	Brand Name	Application Method	PLP	Rating
imazethapyr	Pursuit	soil	49	L
ioxynil	Totril	foliage	22	VL
isoxaflutole	Balance	soil	23	VL
isoxoben	Gallery	soil	47	L
lactofen	Cobra	foliage	25	VL
linuron	Lorox	foliage	42	L
linuron	Lorox	soil	48	L
MAA	MAA	foliage	37	L
MAMA	MAMA	foliage	37	L
MCPA	Chiptox	foliage	45	L
MCPB	Thistrol	foliage	51	M
mecoprop	Mecomec	foliage	58	M
mesotrione	Lexor	soil	39	L
metolachlor	Dual	soil	63	M
metribuzin	Sencor	soil	52	M
metsulfuron methyl	Ally	foliage	16	VL
molinate	Ordram	soil	52	M
MSMA	Daconate	foliage	35	L
napropamide	Devrinol	soil	52	M
naptalam	Alanap	foliage	63	M
nicosulfuron	Accent	foliage	30	L
norflurazon	Zorial	soil	51	M
oryzalin	Surflan	soil	41	L
oxadiazon	Ronstar	soil	39	L
oxyfluorofen	Goal	soil	24	VL
paraquat	Gramoxone	foliage	6	VL
pebulate	Tillam	soil	45	L
pendimethalin	Prowl	soil	20	VL
phenmedipham	Spin-Aid	foliage	21	VL
picloram	Tordon	soil	68	M
pinoxaden	Axial	foliage	8	VL
primisulfuron	Beacon	foliage	29	VL
prodiamine	Barricade	foliage	28	VL
prometon	Pramitol	soil	95	VH
prometryn	Caparol	soil	51	M
prometryn	Caparol	foliage	42	L
pronamide	Kerb	soil	48	L
propachlor	Ramrod	soil	50	M
propanil	Stam	foliage	27	VL
propham	Chemhoe	soil	50	M
prosulfuron	Peak	foliage	27	VL
pyrazon	Pyramin	foliage	51	M
pyridate	Tough	foliage	28	VL
pyrithiobac	Staple	soil	41	L
quinclorac	Facet	foliage	47	L
quizalofop	Assure	foliage	31	L

Table 1-3. Relative Pesticide Leaching Potential (PLP) Indices and Ratings for Commonly Used Pesticides

The PLP value will change with changes in application rate. Values in this table are calculated using average rates.*
KEY: Very High (VH) = 90 to 100, High (H) = 70 to 89, Moderate (M) = 50 to 69, Low (L) = 30 to 49, Very Low (VL) = 0 to 29

Common Name	Brand Name	Application Method	PLP	Rating
rimsulfuron	Matrix	soil	41	L
rimsulfuron	Matrix	foliage	14	VL
sethoxydim	Poast	foliage	29	VL
siduron	Tupersan	soil	63	M
simazine	Princep	soil	62	M
sulfentrazone	Authority	soil	54	M
sulfometuron	Oust	foliage	37	L
sulfometuron	Oust	soil	42	L
sulfosulfuron	Certainty	foliage	23	VL
TCA	Varitox	soil	94	VH
tebuthiuron	Spike	soil	78	H
terbacil	Sinbar	foliage	51	M
terbutryn	Igran	foliage	34	L
thiazopyr	Visor	soil	57	M
thifensulfuron methyl	Pinnacle	foliage	20	VL
thiobencarb	Saturn	soil	45	L
topramezone	Impact	foliage	1	VL
tralkoxydim	Achieve	foliage	43	L
triallate	Avadex-2	soil	38	L
triasulfuron	Amber	foliage	21	VL
tribenuron	Express	foliage	18	VL
triclopyr	Garlon	foliage	54	M
trifluralin	Treflan	soil	25	VL
trifloxysulfuron	Envoke	foliage	21	VL
triflusulfuron	Debut	foliage	12	VL
vernolate	Vernam	soil	44	L

Growth Regulators, Defoliants, Desiccants

chlormequat	Cycocel	foliage	4	VL
clofencet	Genesis	foliage	62	M
dimethipin	Harvade	foliage	72	H
ethephon	Super Boll	foliage	71	H
fumetralin	Prime+	foliage	22	VL
mefluidide	Embark	foliage	25	VL
mepiquat	PIX	foliage	4	VL
MH	Royal MH	foliage	49	L
pyraflufen	Ecopart	foliage	1	VL
sodium chlorate	Defol	foliage	80	H
thidiazuron	Dropp	foliage	30	L
tribufos	DEF	foliage	25	VL

Fungicides, Biocides

azoxystrobin	Heritage	foliage	42	L
benomyl	Benlate	foliage	43	L
captan	Captan	foliage	54	M
carboxin	Evershield V	seed	13	VL
chlorothalonil	Bravo	foliage	33	L

Table 1-3. Relative Pesticide Leaching Potential (PLP) Indices and Ratings for Commonly Used Pesticides

The PLP value will change with changes in application rate. Values in this table are calculated using average rates.*
KEY: Very High (VH) = 90 to 100, High (H) = 70 to 89, Moderate (M) = 50 to 69, Low (L) = 30 to 49, Very Low (VL) = 0 to 29

Common Name	Brand Name	Application Method	PLP	Rating
copper hydroxide		foliage	41	L
DCNA	Botran	foliage	29	VL
dimethomorph	Acrobat	foliage	37	L
dodine	Syllit	foliage	1	VL
ethoprop	Mocap	soil	60	M
etridiazole	Terrazole	soil	17	VL
fenarimol	Rubigan	foliage	59	M
fenbutatin oxide	Vendex	foliage	38	L
fenhexamid	Elevate	foliage	14	VL
flutolanil	Moncut	foliage	39	L
fosethyl-Al	Alliette	foliage	9	VL
Iprodione	Rovral	foliage	27	VL
mancozeb	Dithane	foliage	38	L
maneb	Manzate	foliage	40	L
metalaxyl	Ridomil	soil	59	M
metiram	Polyram	foliage	4	VL
myclobutanil	Nova	foliage	32	L
oxamyl	Vydate	foliage	46	L
oxythioquinox	Morestan	foliage	24	VL
PCNB	Terrachlor	seed	28	VL
penconazole	Topaz	foliage	28	VL
propamocarb	Previcur	foliage	17	VL
propiconazole	Tilt	foliage	30	L
tebuconazole	Folicure	foliage	27	VL
terbufos	Counter	seed	31	L
thiabendazole	Mertect	foliage	37	L
thiophanate	Topsin	foliage	18	VL
thiophanate-methyl	Cercorbin	foliage	41	L
trifloxystrobin	Stratego	foliage	8	VL
triflumizole	Procure	foliage	27	VL
triphenyltin acetate	Fentin Acetate	foliage	20	VL
triphenyltin chloride	Fentin Chloride	foliage	20	VL
triphenyltin hydroxide	Super Tin	foliage	20	VL
vinclozolin	Ronilan	foliage	18	VL
zineb	Dithane	foliage	35	L
ziram	Zirex	foliage	41	L
Insecticides, Acaricides, Miticides, Nematicides				
acephate	Orthene	foliage	52	M
acetamiprid	Assail	foliage	27	VL
aldicarb	Temik	soil	67	M
aldoxycarb	Standak	foliage	72	H
amitraz	Mitac	foliage	11	VL
azinphosmethyl	Guthion	foliage	22	VL
bendiocarb	Turcam	seed	26	VL
bifenthrin	Biflex	foliage	0	VL

Table 1-3. Relative Pesticide Leaching Potential (PLP) Indices and Ratings for Commonly Used Pesticides

The PLP value will change with changes in application rate. Values in this table are calculated using average rates.*
KEY: Very High (VH) = 90 to 100, High (H) = 70 to 89, Moderate (M) = 50 to 69, Low (L) = 30 to 49, Very Low (VL) = 0 to 29

Common Name	Brand Name	Application Method	PLP	Rating
carbaryl	Sevin	foliage	37	L
carbofuran	Furadan	foliage	54	M
carbofuran	Furadan	soil	73	H
carbosulfan	Advantage	foliage	11	VL
chlordimeform	Galecron	foliage	6	VL
chlorfenvinphos	Birlane	soil	40	L
chlorfenvinphos	Birlane	foliage	35	L
chlorobenzilate	Folbex	foliage	26	VL
chlorpyrifos	Lorsban	foliage	27	VL
chlorpyrifos	Lorsban	soil	30	L
clofentezine	Ovation	foliage	1	VL
cyfluthrin	Baythroid	foliage	0	VL
cyhalothrin	Karate	foliage	11	VL
cypermethrin	Ammo	foliage	0	VL
cyromazine	Trigard	foliage	39	L
deltamethrin	Decis	foliage	1	VL
diazinon	Diazinon	foliage	41	L
dicofol	Kelthane	foliage	33	L
dicrotophos	Bidrin	foliage	39	L
dietholate	Eradicane-Extra	soil	26	VL
diflubenzuron	Dimilin	foliage	0	VL
dimethoate	Dimethoate	foliage	47	L
disulfoton	Di-syston	foliage	33	L
endosulfan	Thiodan	foliage	18	VL
esfenvalerate	Asana	foliage	2	VL
ethion	Ethion	foliage	34	L
ethoprop	Mocap	soil	61	M
fenamiphos	Nemacur	soil	58	M
fenoxycarb	Logic	soil	0	VL
fenpropathrin	Danitol	foliage	0	VL
fenthion	Baytex	foliage	27	VL
fenvalerate	Asana XL	foliage	9	VL
flucynthrinate	Pay-Off	foliage	0	VL
fluvalinate	Spur	foliage	0	VL
fonofos	Dyfonate	soil	29	VL
formetanate	Carzol	foliage	28	VL
hexythiazox	Hexygon	foliage	9	VL
hydramethylnon	Amdro	foliage	0	VL
imidacloprid	Timax	foliage	1	VL
indoxacarb	Steward	foliage	4	VL
lambda-cyhalothrin	Karate	foliage	0	VL
lindane	Lindane	foliage	55	M
malathion	Cythion	foliage	13	VL
methamidophos	Monitor	foliage	53	M
methidathion	Supracide	foliage	27	VL
methomyl	Lannate	foliage	43	L

Table 1-3. Relative Pesticide Leaching Potential (PLP) Indices and Ratings for Commonly Used Pesticides

The PLP value will change with changes in application rate. Values in this table are calculated using average rates.*
KEY: Very High (VH) = 90 to 100, High (H) = 70 to 89, Moderate (M) = 50 to 69, Low (L) = 30 to 49, Very Low (VL) = 0 to 29

Common Name	Brand Name	Application Method	PLP	Rating
methoxychlor	Marlate	foliage	7	VL
methoxyfenozide	Intrepid	foliage	20	VL
methyl parathion	Penncap-M	foliage	11	VL
mevinphos	Phosdrin	foliage	30	L
monocrotophos	Azodrin	foliage	61	M
naled	Dibrom	foliage	25	VL
oxamyl	Vydate	foliage	46	L
oxydemetonmethyl	Metasystox-R	foliage	49	L
parathion	Ethyl-Parathion	foliage	19	VL
PCNB	Terraclor	soil	37	L
permethrin	Ambush	foliage	0	VL
phorate	Thimet	soil	48	L
phosmet	Imidan	foliage	37	L
phosphamidon	Dimecron	foliage	53	M
profenophos	Curacron	foliage	15	VL
propargite	Comite	foliage	39	L
propoxur	Baygon	foliage	49	L
pymetrozine	Fulfill	foliage	7	VL
spinosad	Tracer	foliage	11	VL
sulprophos	Bolstar	foliage	27	VL
temephos	Abate	aquatic	0	VL
terbufos	Counter	soil	31	L
thiamethoxam	Platinum	seed	54	M
thiamethoxam	Centric	foliage	12	VL
thiodicarb	Larvin	foliage	24	VL
tralomethrin	Scout	foliage	0	VL
trichlorfon	Dylox	foliage	38	L

Table 1-3. Relative Pesticide Leaching Potential (PLP) Indices and Ratings for Commonly Used Pesticides

The PLP value will change with changes in application rate. Values in this table are calculated using average rates.*
KEY: Very High (VH) = 90 to 100, High (H) = 70 to 89, Moderate (M) = 50 to 69, Low (L) = 30 to 49, Very Low (VL) = 0 to 29

Common Name	Brand Name	Application Method	PLP	Rating
trimethacarb	Landin	soil	38	L
zetamethrin	Fury	foliage	1	VL
Fumigants				
1,3-dichloropropene	Telone-2	soil	79	H
chloropicrin	Larvacide	soil	65	M
dazomet	Basamid	soil	92	VH
metam sodium	Vapam	soil	94	VH
methyl bromide	Brom-O-Gas	soil	100	VH
Molluscicides				
metaldehyde	Deadline Bullets	soil	41	L

*Formulae used to determine the values in Table 1-3 include the following:

$$PLP_{value} = \frac{(\text{Application rate of kg. ai/ha.})(\text{fraction hitting the soil}) (T1/2)}{Koc}$$

Where $T\frac{1}{2}$ = half life of the parent compound under field conditions

Koc = soil organic carbon binding value

$$PLP_{INDEX} = (Log\ PLP_{value})(14.3) + 57$$

Table 1-4. Relative Soil Leaching Potential (SLP) Indices and Ratings for Soils in the Southeastern U.S.

The SLP index will change slightly for the site where the soil series is located. Values in this table are calculated using the profile description of the original site where the soil was named. KEY: Very High (VH) = 90 to 100, High (H) = 70 to 89, Moderate (M) = 50 to 69, Low (L) = 30 to 49, Very Low (VL) = 0 to 29

Soil	SLP	Rating	Soil	SLP	Rating	Soil	SLP	Rating	Soil	SLP	Rating
Alamance	69	M	Corolla	95	VH	Johnston	22	VL	Plummer	79	H
Alpin	84	H	Coxville	36	L	Kalmia	77	H	Ponzer	08	VL
Altavista	73	H	Craven	66	M	Kenansville	84	H	Portsmouth	26	VL
Appling	62	M	Creedmoor	64	M	Kureb	100	VH	Pungo	00	VL
Arapahoe	39	L	Croatan	14	VL	Lakeland	84	H	Rains	45	L
Argent	58	M	Cullowhee	77	H	Leaf	35	L	Rimini	84	H
Augusta	66	M	Dare	09	VL	Lenoir	32	L	Rion	64	M
Autryville	68	M	Davidson	55	M	Leon	84	H	Roanoke	57	M
Aycock	68	M	Deloss	29	VL	Liddell	47	L	Roper	16	VL

Table 1-4. Relative Soil Leaching Potential (SLP) Indices and Ratings for Soils in the Southeastern U.S.

The SLP index will change slightly for the site where the soil series is located. Values in this table are calculated using the profile description of the original site where the soil was named. **KEY: Very High (VH) = 90 to 100, High (H) = 70 to 89, Moderate (M) = 50 to 69, Low (L) = 30 to 49, Very Low (VL) = 0 to 29**

Soil	SLP	Rating	Soil	SLP	Rating	Soil	SLP	Rating	Soil	SLP	Rating
Ballahack	20	VL	Dogue	67	M	Lignum	63	M	Rosman	55	M
Barclay	77	H	Dorovan	10	VL	Louisburg	81	H	Rumford	84	H
Bayboro	04	VL	Dothan	66	M	Lumbee	53	M	Seabrook	84	H
Baymeade	87	H	Dragston	80	H	Lynchburg	40	L	Stallings	58	M
Belhaven	10	VL	Duckston	98	VH	Lynn Haven	61	M	State	73	H
Bibb	75	H	Dunbar	35	L	Madison	68	M	Stockade	35	L
Blaney	73	H	Duplin	62	M	Mandarin	89	H	Tarboro	90	VH
Blanton	82	H	Durham	72	H	Marlboro	63	M	Tate	46	L
Bojac	82	H	Echaw	84	H	Marvyn	44	L	Tatum	64	M
Bonneau	70	H	Edneyville	54	M	Masada	37	L	Toisnot	68	M
Braddock	64	M	Emporia	67	M	Mayodan	63	M	Tomahawk	86	H
Bragg	64	M	Enon	77	H	Mccoll	24	VL	Tomotley	52	M
Brookman	07	VL	Evard	70	H	Mecklenburg	69	M	Torhunta	29	VL
Buncombe	87	H	Exum	69	M	Meggett	29	VL	Vance	66	M
Butters	83	H	Faceville	58	M	Munden	81	H	Varina	62	M
Byars	21	VL	Fanin	70	H	Nahunta	41	L	Vaucluse	76	H
Cainhoy	89	H	Foreston	81	H	Nankin	63	M	Wagram	71	H
Candor	79	H	Fork	72	H	Nason	41	L	Wahee	43	L
Cape Fear	13	VL	Fuquay	73	H	Nimmo	78	H	Wakulla	85	H
Caroline	64	M	Gaston	63	M	Nixonton	82	H	Wando	87	H
Cecil	60	M	Georgeville	59	M	Norfolk	67	M	Wasda	18	VL
Centenary	85	H	Gilead	62	M	Ocilla	53	M	Watauga	46	L
Chandler	47	L	Goldsboro	70	H	Onslow	70	H	Wedowee	64	M
Charleston	83	H	Goldston	78	H	Orangeburg	67	M	Weeksville	34	L
Chastain	40	L	Grantham	44	L	Ousley	84	H	Wehadkee	48	L
Chester	44	L	Grifton	61	M	Pacolet	63	M	White Store	61	M
Chewacla	47	L	Gritney	34	L	Pactolus	85	H	Wickham	75	H
Chipley	61	M	Hayesville	40	L	Pamlico	53	M	Wilbanks	43	L
Chowan	35	L	Helena	66	M	Pantego	12	VL	Wilkes	82	H
Clifton	40	L	Herndon	63	M	Pasquotank	52	M	Winnsboro	82	H
Colvard	88	H	Hullete	62	M	Paxville	40	L	Winton	72	H
Conaby	23	VL	Hyde	17	VL	Pender	69	M	Woodington	58	M
Conetoe	85	H	Invershiel	70	H	Perquimans	44	L	Worsham	43	L
Congaree	28	VL	Johns	78	H	Pinkston	73	H	Yaupon	72	H
									Yonges	57	M

SAMPLE CALCULATION FOR GWCP INDEX: Acetochlor (PLP index = 55 M) applied to an Alamance soil (SLP index = 69 M).
GWCP index = 55 + 69 / 2 = 62 M.

Table 1-5. Groundwater Contamination Potential (GWCP) Risk of Pesticide-Soil Combinations

Obtain numbers for PLP and SLP for your soil and pesticide from Tables 1-3 and 1-4, respectively.

Pesticide Leaching Potentional (PLP) Rating	Soil Leaching Potential (SLP) Rating				
	0–29 Very Low	30–49 Low	50–69 Moderate	70–89 High	90–100 Very High
0–29 Very Low	Very Low Risk	Very Low Risk	Low Risk	Low Risk	Moderate Risk
30–49 Low	Very Low Risk	Low Risk	Low Risk	Moderate Risk	Moderate Risk
50–69 Moderate	Low Risk	Low Risk	Moderate Risk	Moderate Risk	High Risk
70–89 High	Low Risk	Moderate Risk	Moderate Risk	High Risk	High Risk
90–100 Very High	Moderate Risk	Moderate Risk	High Risk	High Risk	Very High Risk

Proper Pesticide Storage

Safe and proper storage can extend the shelf (storage) life of your pesticides, keep the containers in good condition, and keep the labels clean and legible.

Only rules for storing pesticides on the farm and in household situations are presented here. If you store restricted use pesticides in commercial storage facilities, you must meet additional requirements as outlined in Regulation 2 NCAC 9L.1903–.1913. Details on these requirements can be obtained at your county Cooperative Extension center or by calling the NCDA&CS Pesticide Section at 919-733-3556.

Household and Farm Situation Storage Rule (simplified)

1. This rule applies to all pesticides.

2. Store pesticides to prevent leaking and to aid inspection.

3. Do not store formulated pesticides in unlabeled containers. The following minimum information must be shown clearly and prominently on any containers of formulated pesticides:

    a. common chemical name (such as carbaryl for the product Sevin)

    b. percentage of each active ingredient

    c. EPA registration number

    d. signal word (DANGER, WARNING, CAUTION)

    e. use classification (restricted use or general use).

4. Do not store pesticides (formulated products or dilutions) in any food, feed, beverage, or medicine container that has previously been used for such purposes or that is specifically designed to contain only those products.

5. Do not store pesticides in a way that could contaminate foods, feeds, beverages, eating utensils, tobacco, or tobacco products, or otherwise result in accidental ingestion by people or domestic animals. In addition, pesticides should not be stored in such a way that could contaminate other pesticides, seeds, or fertilizers.

6. Store pesticides based on the following:

    a. storage recommendations, if any, on their labels, and

    b. labels on all other products, including nonpesticide products held in the same storage area.

7. When unattended, store pesticides to prevent unauthorized access.

8. Store pesticides in an area that is dry (does not accumulate water) and well-ventilated.

9. Pesticide storage areas should be free of combustible materials—such as gasoline, kerosene, or petroleum solvents other than those associated with pesticide application—and debris like waste paper, rags, or used cardboard boxes that may provide an ignition source. They must also be separated from other operations that present a fire hazard, such as welding or burning. Take appropriate care to reduce fire hazards when providing supplemental heating to storage areas during the winter.

Disposal of Pesticides

Reduce the Need for Disposal of Unwanted Pesticides

Because of the expenses, environmental hazards, and legal responsibilities associated with the disposal of pesticides and other hazardous waste, the best solution is to minimize or eliminate the need for disposal altogether. Careful planning of spray programs is essential in order to avoid purchasing more pesticides than will actually be needed for a particular application or season. However, in spite of careful planning, farmers (and homeowners) will occasionally need to dispose of unwanted pesticides. In addition, there are old, obsolete, or banned pesticide products in storage throughout the state that require disposal. Some of these materials are in containers that are structurally unsound and have incomplete or no labeling. The following information will aid in accomplishing disposal in the safest, most economical, and environmentally acceptable manner possible.

Donation of Excess Pesticides

If no longer needed, unopened containers of recently purchased pesticides may possibly be returned to the local dealer or manufacturer/formulator for a refund. If this cannot be arranged, the pesticide may be sold or donated to a neighbor or to someone who can and will use it properly. If donation is a possibility, there are two important factors that must be taken into consideration.

1. A pesticide designated **RESTRICTED USE** on the label must not be given to someone who is not a certified applicator.

2. A pesticide that has been banned or one for which all uses have been cancelled should not be donated to another person.

NCDA&CS Disposal Systems Program for Excess Pesticides

If donation of unwanted pesticides is not a possibility, two other options are available. The NCDA&CS has a Pesticide Disposal Assistance Program for farmers and homeowners. This service is provided on an individual request basis with no fees charged. Specific requests for assistance should be made to Derrick Bell, NCDA&CS Pesticide Waste Specialist, (919) 733-7366, ext. 327 or 328.

Commercial Hazardous Waste Disposal

In some situations the NCDA&CS may not be able to provide the disposal assistance that is needed. Commercial disposal companies, permitted by the U.S. Environmental Protection Agency (EPA) through the Resource Conservation Recovery Act (RCRA) program are available to dispose of pesticide wastes.

While all pesticides are hazardous substances, not all are legally classified as RCRA hazardous wastes. However, because pesticides cannot be legally disposed of at county landfills, a permitted disposal facility is an option that may have to be used.

The NCDA&CS will furnish a listing of hazardous waste disposal companies upon request. Anyone contemplating this option is advised to contact several firms, compare costs, and ask for and check references. All paper work (manifests) generated by the disposal should be retained permanently.

One must not assume that disposal of a pesticide (which is also a RCRA hazardous waste) at an EPA permitted facility eliminates all further legal responsibility for that product. The person who generates a hazardous waste (farmer/homeowner) is legally and financially responsible for that material for as long as it remains in existence. Therefore, even though destructive disposal (incineration) may cost more than nondestructive disposal (landfilling), there are worthwhile, long-term benefits for using the incineration method.

In summary, the disposal of some pesticides via commercial disposal companies involves RCRA regulations. There are stringent requirements for storage, transport, and disposal of hazardous waste and severe penalties for failure to comply with the regulations. For information regarding hazardous waste management and disposal contact:

Division of Waste Management, Hazardous Waste Section, N.C. Department of Environment and Natural Resources, P.O. Box 27687, Raleigh, NC 27611-7687.

Disposal of Excess Spray Solution from Tank and Equipment Rinses

As already indicated, proper planning and careful calculations should eliminate the need for disposal of large quantities of excess spray solutions. Small quantities should be sprayed out along field borders or on the row ends. Care must be taken not to exceed the labeled application rate.

Tank and equipment rinses should also be applied along field borders or on the row ends. If decontamination solutions, cleaners, detergents, ammonia, chlorine bleaches, etc., are used to remove residues, adequate dilution may be necessary to prevent soil and plant injury.

Disposal of Empty Pesticide Containers

Metal, Plastic, or Glass Containers of Liquid Formulations (5 Gallons or Fewer)

Before pesticide containers can be accepted for recycling or disposed of properly, they must be rinsed by one of the following methods.

Pressure rinsing

1. Drain the container into the spray tank for 30 to 60 seconds after the last amount starts to drip.

2. Insert tip of the pressure nozzle through the side of the pesticide container near its base.

3. While holding the container so the opening can drain into the spray tank, spray the inside of the container for at least 30 seconds.

4. Drain all rinse water into the spray tank.

Triple rinsing

1. Drain the container into the spray tank for 30 to 60 seconds after the last amount starts to drip.

2. Fill the containers 1/3 full with water, cap, and shake thoroughly. Empty this rinse water into the spray tank.

3. Repeat the above rinse procedure at least two more times, adding each amount of rinsate to the spray tank.

4. Punch holes in the bottom and sides of metal and plastic containers. The holes will prevent the containers from being reused and will indicate that they are indeed empty.

Properly-rinsed plastic pesticide containers can be delivered to a container recycling collection site. Nearly every North Carolina county has one or more of these sites. Call your county Extension center for directions to the nearest collection site. Properly rinsed metal or glass pesticide containers can be taken to a county solid waste collection system or to a county landfill.

Metal or Plastic Drums (30- to 55-Gallon)

1. Rinse and drain these containers into the spray tank.

2. Attempt to return drums to dealer or distributor (recycling).

3. If you are not successful, attempt to donate drums to a drum reconditioner (contact Pesticide Disposal Specialist, NCDA&CS, Derrick Bell, 919-733-7366 ext. 327 for information on drum reconditioners operating in the state).

4. If steps 2 or 3 above fail, attempt to dispose of rinsed drums in the county sanitary landfill. This may not be possible in some cases because of the difficulty for landfill personnel to verify that the drums have been properly rinsed.

5. If the above procedures fail, contact the Solid and Hazardous Waste Management Branch, 919-733-2178, for assistance with pesticide drum disposal.

Containers of Non-Liquid Pesticides

1. Shake container into the applicator tank until all the pesticide has been removed.

2. Tear open the container to make sure it is completely empty.

3. If the pesticide is a wettable powder and the container can be triple rinsed with water, do so. Add the rinsate to the tank.

4. Puncture, crush, or otherwise render the container incapable of being reused and then place in the solid waste collection system or carry to a county sanitary landfill facility.

5. If large containers can be returned for recycling, do so.

Controlling Pesticide Drift

Once discharged from the application equipment, pesticides may drift through the air and injure susceptible plants or sensitive animal life before the pesticides actually reach the target area. Movement through the air may be by spray drift, vapor, or dust.

If herbicides are allowed to drift, they can often cause extensive damage to susceptible crops. Drift of other pesticides—although not as likely to injure nearby crops—can damage livestock, bees, fish, and people, in addition to leaving illegal residues on crops.

Spray drift is the movement of airborne spray particles beyond the target. The amount of spray drift is influenced by the (1) droplet size, (2) amount of wind, and (3) height from which the spray is released.

When using ground equipment certain precautions can reduce spray drift.

1. Select nozzles that produce a minimum proportion of small droplets in relation to overall droplet size. The droplet size is influenced by pounds of pressure per square inch (psi) and the nozzle's design (opening size). Lower pressures and larger openings, for example, tend to produce larger droplets and less drift. Standard fan nozzles operate at 30 to 40 psi. Whirl chamber nozzles and low-pressure fan nozzles work at 15 to 20 psi. Select the largest nozzle opening that will provide enough gallons per acre for penetration, uniform coverage, mixing with the pesticide, and effective control.

2. Spray when wind velocity is at a minimum, when temperatures are moderate, and preferably before adjacent susceptible crops emerge. Where a susceptible crop has emerged, spray when the wind is moving away from the crop or consider leaving an untreated strip along the edge of the field.

3. Adjust the boom as close as possible to the target without losing uniform distribution of the pesticide. To bring the boom closer to the target, space the nozzles closer together or use wide-angle nozzles. The boom's height above the ground affects how long it takes a spray droplet to reach the ground. Wind velocities are usually lower closer to the ground.

Special nozzles that produce large droplets have been developed to control this drift from aerial application. Also, invert emulsions and drift control agents can be used.

Vapor drift occurs when vapor or fumes move from the area of application. Vapor drift may damage susceptible crops or may simply reduce the effectiveness of the herbicide or pesticide.

Ester formulations of 2,4-D may volatilize in hot weather and drift to susceptible crops. Therefore, select amine and sodium salts of 2,4-D that are less prone to volatilize and continue to take steps to prevent spray particle drift.

Fumigants like methyl bromide and chloropicrin drift after they turn into a gas. Make sure the covers or buildings where these chemicals are applied are airtight and application equipment lines and tanks do not leak.

17

Mixing and Loading Pesticides

People often get into trouble when mixing and loading pesticides. Some common problems are discussed briefly here.

Tank Mixes

Tank mixes (combinations of two or more pesticides in the spray tank at time of application) may fall into one of two categories.

- Instructions provided for such use on one or more labels of registered products.
- Tank mixes that are recommended by Cooperative Extension or are common agricultural practices.

Tank mixes recommended on labels are obviously consistent with the label.

Tank mixes will not be deemed "use inconsistent with the label" if:

- The products in the mix are applied at a dosage rate not to exceed the label instructions for use of any product in the mix used singly for the same set of pests on the same crop; and
- The label on one or more of the products does not explicitly instruct against such mixture.

The only mixtures proven effective and safe are those specified on product labels. The user applies all other mixtures at his or her own risk with respect to effects on crops and application equipment, applicator safety, environmental effects, and preharvest interval tolerances.

Compatibility

Some pesticides will not mix (are not compatible) with other pesticides or with liquid fertilizers in spray-tank mixtures. For example, wettable sulphur cannot be mixed with Lorsban or Morestan. Some herbicides are not compatible with liquid fertilizers and herbicide oils. Any time you plan to mix two or more pesticides, first make sure they are compatible. Follow the specific directions on the label to test for compatibility if you still have questions. Also remember that a pesticide may mix physically with another pesticide, but the activity of one or both may be altered (based on their chemical or biological incompatibility).

Adjuvants

Adjuvants are inert ingredients that are added to pesticide formulations or tank mixes to increase the effectiveness of the pesticide's active ingredients. Adjuvants may be wetting agents, emulsifiers, spreaders, stickers, penetrants, drift reduction agents, thickeners, buffers, and compatibility agents just to name a few. Adjuvants should not be used unless needed. Often, the pesticide formulation already contains the adjuvants needed for the application.

Many herbicide and other pesticide formulations require that a surfactant, penetrating agent, or other adjuvant be added to the spray tank to increase the pesticide's effectiveness. Read the label to find out what should or should not be added to the pesticide formulation to give you the best possible control.

Formulation Sequence

If you use more than one pesticide formulation (WP, WDG, DF, L, EC) in a spray tank, there is a proper order for adding them.

1. Add a small amount of water or other liquid carrier to the spray tank.
2. Dry materials go into the spray before liquid chemicals. If a wettable powder (WP) is used, put it in first as follows: Make a slurry with the wettable powder by adding a small amount of water to it until it forms a gravy-like consistency. Slowly add this slurry to the tank with the spray tank agitator (mixer) running.
3. Dry flowables (DF) or water-dispersible granules (WDG) go in second. Flowables should be premixed (1 part flowable to 1 part water) and poured slowly into the tank.
4. Liquid flowables (F or L) should be added third. Exception: When using Furadan 4F, this material should be put in last. Liquids should also be premixed (1 part liquid chemical to 2 parts water or liquid fertilizer) before blending in the tank. Many labels will give you the proper pesticide mixing sequence.
5. Emulsifiable concentrates (EC), should be combined last.

Safety Warning

When mixing and loading pesticides, you usually work with concentrated pesticides. Accident reports have shown that the danger of being poisoned may be greatest at this time. Wear the appropriate gloves and other protective clothing to avoid getting a pesticide on you or your clothing.

If a pesticide gets into your eyes, immediately rinse them with plenty of clean water; continue for at least 15 minutes. (The label on some pesticides calls for longer flushing times). If pesticides get on your skin, wash them off with water. Remove contaminated clothing and wash it separately from the family laundry before wearing any of it again. Clothing saturated with a highly hazardous pesticide (labeled DANGER) should be disposed of the same way you would discard the pesticide. Remember: Injuries from most pesticide accidents can be prevented if you know what to do and do it FAST. Take the person to a doctor if you suspect pesticide poisoning.

Cleaning Sprayer Systems

Most people are aware that spraying sensitive crops with a sprayer that has been used earlier to apply certain pesticides can lead to crop damage. Most pesticides can be washed out of sprayers. Dicamba (Banvel), 2,4-D, 2,4-DB, and MCPP are more difficult to wash out, however, and many crops are very sensitive to these herbicides. So it is best to have a separate sprayer to apply these herbicides. Certain crops are very sensitive to sulfonylurea herbicides (Classic, Canopy, Gemini, Glean, Harmony), but these can easily be washed out of sprayers if the proper procedure is used.

Before applying a pesticide with a sprayer that was previously used for some other pesticide, always wash out the sprayer thoroughly. It is best to wash out the sprayer immediately after use. Some pesticide labels give instructions on how to properly clean that pesticide out of the sprayer. If the label does not contain this information, that does not imply that residues of that particular pesticide in a sprayer will not harm other crops.

Pesticide Record-Keeping Requirements

The U.S. Department of Agriculture (USDA), through the 1990 Farm Bill, and the North Carolina Pesticide Board, through the N.C. Pesticide Law of 1971 (NCPL), require that applicators, dealers, and agricultural employers record certain pesticide information. The NCDA&CS, Structural Pest Control and Pesticides Division administers and enforces record keeping provisions. Records must be kept for certain lengths of time and made available to representatives of the NCDA&CS and USDA upon request.

Table 1-6 summarizes the pesticide application records required by federal and state regulations for certain pesticides and for compliance with the Worker Protection Standard (WPS). Readers are encouraged to read the actual laws and regulations for more detailed information. Dealers, certified applicators, licensed aerial and ground applicators, and agricultural employers all have some responsibilities for recording pesticide information.

Agricultural employers who hire pesticide handlers, workers, or both must display application information as required by the WPS. This information must be posted at a central location accessible by employees prior to application and kept for 30 days after the expiration of the restricted-entry interval (REI). Under state law, the time when the application was completed must be recorded. This and all other required record-keeping items must be maintained for a period of two years after the REI expires.

Pesticide dealers in North Carolina are required to keep sales records for all restricted-use pesticides. The 10 elements that are required for each restricted-use pesticide sale are:

1. Date of sale;
2. Initials of sales clerk;
3. Name of certified or licensed applicator;
4. Certification or license number from card;
5. Expiration date as shown on the card;
6. Product brand name;
7. EPA registration number;
8. Number of individual containers;
9. Size of individual containers;
10. Total quantity sold.

Table 1-6. Summary of Pesticide Record-Keeping Requirements for Growers and Applicators

Required Items	USDA Requirements for Private & Commercial Applicators (Restricted Use Pesticides)	NCPL[1] Requirements for Commercial Applicators & Public Operators (Restricted Use Pesticides)	NCPL[1] Requirements for Aerial Applicators (All Pesticides)	Federal & State Requirements for Agricultural Employers (WPS*) (Agricultural Use Pesticides)
Brand name/product name	✓	✓	✓	✓
EPA registration number	✓	✓	✓	✓
Total amount of pesticide used	✓	✓ amount/unit of measure (e.g. acre)	✓ amount of formulated product or active ingredient/acre PLUS amount of tank mix/acre	—
Date of application	✓	✓	✓	✓
Time of application		✓ Time completed	✓ Time completed	✓ Start time and completed
Description/location of treated area	✓	✓	✓	✓
Crop, commodity, or stored product	✓	✓	✓	✓
Size of area treated	✓	✓	✓	—
Name and address of property owner or operator	—	✓	✓	—
Name of applicator	✓	✓	✓	—
Name of licensee	✓ or name of supervisor	✓	✓ name of contractor and signature of record keeper	—
Certification number	✓	—	—	—
Active ingredients	—	—	—	✓
Restricted entry interval	—	—	—	✓
Record must be:	completed within 14 days of application and kept 2 years (commercial applicator, only, must furnish records to customer within 30 days)	kept 3 years	completed within 72 hours after application and kept 3 years	displayed, along with SDS, within 24 hours after end of application, posted for 30 days after the Restricted-Entry Interval expires and kept for 2 years after the REI expires.

For these records, each day of application must be recorded as a separate application.

*The federal Worker Protection Standard has been adopted by reference by the N.C. Pesticide Board. This standard requires that pesticide information be posted in a central location on an agricultural establishment.

[1] NCPL = North Carolina Pesticide Law

Source: North Carolina Department of Agriculture and Consumer Services

Record-keeping forms for the USDA restricted-use pesticide regulation and the Worker Protection Standard are available online at http://pesticidesafety.ces.ncsu.edu/.

II — CHEMICAL APPLICATION EQUIPMENT

Chemical Application Equipment

G. T. Roberson, Extension Specialist, Biological and Agricultural Engineering

Introduction

The objective of a chemical application program can be summed up with a single sentence: *Put the right product on the right target at the right time.* To accomplish this objective, good management practices and proper equipment are necessary. In selecting agricultural chemical application equipment, follow these guiding principles:

1. Considering acreage, crops, and labor, select the equipment that will best fit your farming operation.

2. For medium to small farms, select smaller units. Large equipment is impractical on small farms with fields of irregular shape, leading to frequent breakdown and poor application.

3. If feasible, use separate spray equipment for certain types of weed-control chemicals to prevent accidental contamination and crop damage.

4. Remember that some chemical application equipment is designed for specific applications, and some components are not suitable for all jobs. For example, some spray nozzles recommended for insecticide application may not be suitable for herbicides.

5. Select a machine that is convenient to set up, hitch, and operate. Convenient features can help a busy operator complete an application on time.

6. Keep in mind that high-clearance, self-propelled sprayers often cannot be justified on small farms; consider hiring a custom operator if this type of machine is needed.

7. Use the proper pressure for spraying; avoid excessively high pressures. Check the chemical manufacturer's label for recommendations. You should also check the nozzle manufacturer's charts to ensure you have the correct droplet size for the pressure and application.

8. Check all adjustments for any type of application equipment carefully. Refer to your operator's manual for recommended settings. **Remember to CALIBRATE!**

9. Always wear the proper protective clothing and use other recommended personal protective devices or practices when handling or using agricultural chemicals.

Types of Equipment

Dusters

Dusters are no longer used for field crops. Sprayers and granular applicators are generally preferred. However, small hand-held dusters and backpack dusters are still quite popular for use in gardens and around the home.

Small dusters for home or garden use are typically hand-held and hand-powered. Large dusters can be backpack units powered by small gasoline engines or battery-powered motors. They can hold from a few ounces to several pounds of material. The fan or air pump generates the necessary airflow to move the dust through the discharge tube to the nozzle. Aim the nozzle at the proper target and watch the application rate closely. It is easy to apply too much material in one small area, which can be hazardous.

Granular Applicators

Granular applicators are used for a wide range of jobs, from applying small volumes of pesticides at planting to applying fertilizer or lime using band or broadcast methods.

Granular applicators typically consist of a hopper, a metering system, drop tubes, and a diffuser. The metering system can work one of two ways: gravity flow or positive displacement. In the gravity flow type, the size of an orifice or opening is adjusted to control how the product flows out of the hopper. An agitator in the bottom of the hopper ensures smooth delivery. In the positive displacement type, the metering unit delivers a fixed volume of material for every revolution it makes. The drive mechanism is often connected to a ground wheel, so material is delivered whenever the wheel turns and the orifice or gate is open. Some units are driven by PTO, hydraulic drives, or small electric motors rather than ground drives. Positive displacement systems are generally more accurate.

The metering device must function properly. Clean it regularly to remove caked material or other obstructions. Check the condition of components and replace them if they appear worn. Badly worn components may result in large application errors. Check the tubes and diffusers for leaks or blockage. Drop tubes or hoses should allow the material to fall freely without collecting in the tube. Keep the tubes as nearly vertical as possible to minimize blockage.

Broadcast Spreaders

Broadcast spreaders are widely used to apply fertilizer, lime, or amendments on lawns, gardens, or fields. These machines may be small hand-held or cart-mounted units for home or garden use, or they may be three-point hitch, trailer-mounted, or truck-mounted units for field use. Spreaders for home or garden use can be simple drop spreaders with a series of holes along the underside of the hopper to meter and spread the material, or they may be spinner spreaders, which use one or two rotating disks or spinners to spread the product across the swath. Larger field spreaders can be single-spinner, twin-spinner, or air-boom designs.

A large spinner spreader typically consists of a hopper, a drag chain or belt, a discharge gate, a chute, and one or two spinners. Some smaller units do not use the drag chain but rely on gravity flow through an orifice and an agitator to achieve the desired discharge rate. Larger broadcast spreaders can have a drag chain or belt driven by a ground wheel, hydraulic drive, or by PTO.

Spinner spreaders should produce a pattern that is heavy in the center and tapers to the edges. Desirable patterns for a spinner spreader are the triangle, oval, and flat top patterns shown in Figure 2-1. Proper spacing of the swaths in the field is critical to apply product correctly. Swath spacing should be the width across the pattern where each side delivers 50 percent of the rate (Figure 2-1). If the swath spacing is too wide, some areas will not receive enough product between the passes of the spreader. If the swath spacing it too close, some areas will receive too much product. Be sure to check the spread pattern and maintain proper swath spacing in the field. The effects of improper swath spacing or a bad spread pattern for products such as fertilizer will often show up in the field as stripped or uneven crop development. It is very important to recognize the difference between the width of the spread pattern and the swath width. Figure 2-1 illustrates the spread pattern width is the maximum width the product is distributed, where swath width is the spacing of the applicator's passes through the field.

Air boom spreaders use a high volume air stream to suspend the product particles and convey them through tubes to diffusers spaced along the boom. The product is metered into an air chamber where the air stream catches the material and divides it into the tubes running to the diffusers. The product is uniformly distributed along the width of the boom with a very slight taper on the outside edges. As with the spinner spreader, proper swath spacing is critical. Because the air boom pattern has little taper, precise swath spacing must be maintained.

Smooth delivery of material is important. Check the discharge mechanism for blockage or wear. Check the drive mechanism to make sure it is functioning properly. Slipping wheels, worn belts, and worn chains can seriously affect performance and should be repaired. Check the spinners for holes in the bottom or in the vanes. Check for caked material on the vanes as well. Pay close attention to the speed of the spinner; excessively high or low speeds can cause improper application patterns. On air-boom spreaders, be sure to check the air chamber and tubes for blockages and leaks. Refer to your operator's manual for correct settings and adjustments on all machines.

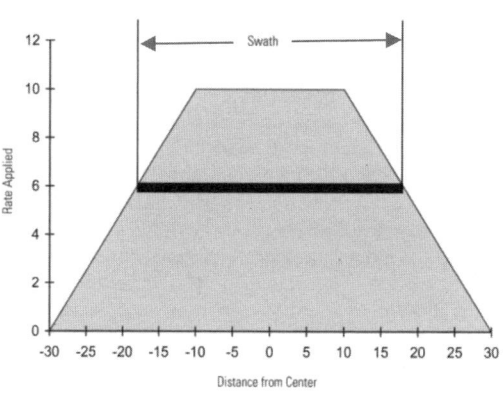

Figure 2-1. Desirable patterns for broadcast spinner spreader are the triangle, oval, and flat top, as shown above.

Sprayers

Most field sprayers in use today are hydraulic, which means the spray pressure is built up by the action of the pump on the spray mixture. Sprayers range from low- to high-pressure and may be mounted, pull-type, or self-propelled models. There are also electrostatic and air-directed sprayers. Sprayer designs also include boom and boomless units to match a wide range of applications.

The basic components are the tank, pump, agitator, hoses, valves and fittings, and nozzles. Some modern units also incorporate electronic control systems to enhance accuracy and performance. When purchasing a new sprayer or when replacing components, choose carefully to ensure quality.

Two types of sprayer designs are currently available: tank mix and injection. The tank mix design consists of a large tank in which water and chemicals are mixed. Tank mix sprayers are less expensive and easy to operate. Injection sprayers have a large tank that contains only water. An additional tank or tanks are used to hold the chemical. For products that are difficult to clean, you may want to have dedicated chemical tanks. While more costly to purchase, injection sprayers have the advantage of reduced risk since you never mix more than needed. The water tank is never contaminated with chemicals, so cleanup is simpler. Electronic controls are used to measure the flow of water and inject the proper quantity of chemical into the water stream before it is delivered to the nozzles. This design reduces contamination potential and eliminates the need for tank agitation.

Tanks are usually made of stainless steel, aluminum, fiberglass, or plastic and should include a large splash-proof filler. They should be resistant to rust and all chemicals placed in them. Check with your chemical supplier regarding compatibility.

A suction line connection and a drain should be connected at the bottom of the tank. Bottom suction makes the pump self-priming. When possible, use only one tank for the sprayer; multiple tanks are more difficult to agitate properly. The injection sprayer can use multiple tanks without problem since agitation is not required.

Pumps are available in brass, aluminum, cast iron, and corrosion-resistant alloys (Table 2-1, at end of chapter). Select a pump material that will not be corroded by the chemical you are using. The pump seal material (for example, viton or buna-N) should be compatible with the chemical.

Pump capacity is measured in gallons per minute (GPM) or gallons per hour (GPH). Select a pump with a capacity at least 50 percent larger than the sum of the nozzle outputs and the agitator. Oversizing the pump compensates for wear and keeps the system running longer. Pumps are usually attached directly to the tractor PTO. They should be allowed to "float" as they rotate, but should not be so loose that they twist the hoses. Secure the pump with a length of chain, leaving some slack rather than bolting the pump down rigidly. Hydraulically driven pumps can be mounted on the sprayer frame. Characteristics of various types of pumps are listed in Table 2-1.

Jet agitators are usually used in sprayer tanks. Place jet agitators in the bottom of the tank and connect to the high-pressure side of the control valve. This arrangement allows full flow to the agitator as well as full pressure control for the nozzles. Check with your sprayer supplier for proper size of agitator nozzles and flow rates. The return line from the pressure regulator is not adequate for agitation.

Every field sprayer should have a sprayer control unit. The control can be manual or electronic. The control should allow the operator to select the section of the boom or nozzles to use and regulate pressure. The control should also have a high accuracy pressure gauge. Measuring and controlling pressure accurately is essential for most sprayer designs. Never shortchange the quality of the pressure gauge.

Electronic control units for sprayers can enhance performance by improving application accuracy. These controls may include pressure sensors, flow rate sensors, or a combination of the two. These controls rely on solenoid or ball valves to control flow to the system and a throttling valve or bypass valve to regulate system pressure. These valves must be checked periodically to ensure proper performance. Some controls also include some type of ground speed sensor, such as GNSS (Global Navigation Satellite System), radar or wheel sensors. By maintaining accurate ground speed, the sprayer can compensate for changes in speed and maintain better accuracy.

Electronic or automatic controls can improve accuracy and performance of a sprayer; however, they are **no substitute for proper calibration.**

Hoses should be oil resistant and durable. They should have a test pressure twice the sprayer operating pressure. Suction hoses should be strong enough to prevent collapsing. The hoses should be large enough for proper flow while keeping pressure loss to a minimum (Table 2-2, at end of chapter). If a hose is too small, the pressure at the nozzles will be much less than that measured at the control valve, causing under-application. Suction hoses should be at least as large as the pump's intake port.

Sprayers should have a suction strainer to protect the pump. Surface area for a suction strainer should be 2 square inches per GPM. Mesh size is determined by the type of chemical used. Use strainers to protect valves and nozzles.

Strainers used on the pressure side should have an area of at least 1 square inch per GPM.

Nozzles are perhaps the most important part of the sprayer. Unfortunately, no one nozzle can cover every type of application. Nozzles vary according to capacity (gallons per minute or GPM), spray pattern angle, and shape of spray pattern (Table 2-3, at end of chapter). Nozzles are usually equipped with strainers to protect them from abrasive particles. Even with protection, nozzles will wear. Check nozzles regularly and replace them if wear is observed.

Electrostatic sprayers offer better plant coverage and more efficient use of chemicals. Electrostatic sprayers use a high-voltage electrical charge to create an electrostatic field between the plant and the spray droplet. The plant and the spray droplet have opposite charges causing them to attract each other. The droplets cling to any exposed plant surface, top or bottom. More uniform coverage and potential for reduced spray volume make electrostatic spraying a cost-effective method of chemical application.

Air-directed sprayers fall into three groups: airblast, air boom, and air nozzle. Airblast sprayers have been widely used for years in vineyards and orchards. A nozzle or nozzle cluster sprays the droplets into the discharge of a high-volume fan. The airflow spreads the droplets onto the crop. Air boom sprayers are similar to conventional hydraulic sprayers. In addition to the sprayer boom and nozzles, an air tube is added with air ports or nozzles over the sprayer nozzles. The air blast from the air tube blows the spray down into the crop canopy, improving penetration and reducing drift. Air nozzles fall into two categories: pressure and induction. Air pressure nozzles use compressed air to atomize liquid and create the spray pattern rather than water pressure. Because droplet size and pattern are controlled by air pressure, there is no need to have different nozzles for each application rate. Air induction nozzles draw air into the spray stream to control drop formation. The drops formed are larger air-filled drops that are less likely to drift.

Drift control is an important consideration for any type of sprayer. Drift can be caused by wind at or near the ground or by high nozzle pressures. The ground speed of the sprayer can contribute to drift. Drift can also be caused by evaporation of the liquid on hot days. Often a combination of the three is present. Drift contaminates other crops or the surrounding woods and streams. Electrostatic, air boom, and air nozzle sprayers all have drift-reducing potential. Conventional sprayers can also take advantage of drift reducing technology by incorporating spray shields or drift-reducing nozzles.

Backpack Sprayers
Backpack sprayers consist of a tank, a pump, and a spray wand with one or more nozzles. Some sprayers have a pressure-regulating valve or a pressure gauge to help the user maintain

desired pressure. Small size, portability, and ease of use make the backpack sprayer a valuable tool for many users. Backpack sprayers are best suited for small acreage, spot spraying, hard-to-reach areas, and other areas where a larger sprayer is impractical.

Most backpack sprayers use hand pumps; however, some units have a small battery or engine-powered pumping system. Regardless, consistent and uniform pressure control is essential to ensure proper application.

Hand-operated sprayers should have a comfortably located, reversible handle (to allow for left- or right-hand use). The shoulder straps should distribute the load evenly across the shoulders. Consider a hip belt to help carry the weight of the larger units. The wand should be comfortable and allow for easy use of the trigger. The sprayer should also have removable screens to protect the pump and nozzles. These should be cleaned regularly. Finally, the sprayer should have a stable base to hold it upright for filling and mixing.

A portable containment and mixing system for backpack sprayers has been developed for use in pickup trucks and utility vehicles. This system secures the backpack to prevent tipping over and provides a means to shake the unit for mixing. Further, hand pumps and meters are provided to allow accurate mixing of doses for the sprayer.

Wick Applicators
A convenient method to apply contact herbicides is the wick applicator. In this design, a wick (rope, sponge, etc.) is soaked with a solution. Herbicide is applied to the plant target by brushing against the plant. The application rate is controlled by adjusting the chemical solution, the rate it is soaked by the wick, or the speed of travel over the target. Ropes are commonly used as wicks, which explains the name "rope wick applicator." Wick applicators offer the ability to apply chemical only to a selected target, e.g., weeds and grasses taller than the crop. Reduced chemical usage compared to broadcast spraying is possible.

GNSS and Variable Rate Technology
Improved accuracy, site-specific application rates, and higher profitability are among the advantages offered by Global Navigation Satellite Systems (GNSS) and variable rate applicators. GNSS refers to all satellite navigation systems world-wide, which also includes the Global Positioning System (GPS) maintained by the U.S. Department of Defense. GPS is a system of navigational satellites established by the military and made available for private use. GNSS signals alone are not accurate enough to guide or control field equipment. Differential correction signals must be used to achieve the necessary accuracy. Depending on your location, you may wish to subscribe to a commercial signal service or use one of the free correction signals, such as the Coast Guard Beacon or Wide Area Augmentation System (WAAS). Real Time Kinematic (RTK) correction is the most advanced correction service available and offers accuracy to within a few inches. RTK correction can be achieved by operating your own base station, using a dealer base station network, or using the North Carolina Real Time Network (NCRTN).

To use a variable rate system, a machine such as a tractor, spreader truck, sprayer, or aircraft is equipped with a DGPS receiver. The operator uses the receiver and an on board computer to guide the equipment across the field. The ability to establish and maintain uniformly spaced swaths is an advantage with GPS guidance systems. Once the swaths are established, a light bar, screen display, or automatic steering system is used to help the equipment operator stay on course. With the ability to determine position in a field accurately, the grower can take advantage of variable rate control systems and soil, weed, or yield maps prepared for the field. Information from these maps is used to determine the proper amount of material—fertilizer, lime, or pesticides—to apply to each area within the field. Therefore, only the areas that need or will benefit from higher application rates receive those rates, while other areas receive more moderate rates.

Control systems can be selected to manage single or multiple products. A single product (also called single channel) controller can be used for granular applicators, broadcast spreaders, or sprayers. Single channel controls for sprayers are used on tank mix sprayer systems. Multiple channel controllers can also be used on broadcast spreaders and sprayers. On broadcast spreaders, multiple channel controllers are used to control rates on multiple hopper spreaders to blend products on the go in the field as called for by prescription maps, crop sensors, or commands from the spreader operator. On sprayers, multiple channel controls can be used to inject pesticides into a constant stream of water before the spray is applied through the nozzles. This allows variable rates of pesticides to be applied as indicted by prescription maps, crop sensors, or operator commands. Another advantage of the computerized variable rate control systems is the ability to record a map of the product as it is applied to the field. Application record maps can be valuable assets in a complete management plan.

Cleaning Equipment

Sprayers, granular applicators, broadcast spreaders, and other application systems should be properly cleaned after use. Some chemicals are corrosive and may damage the components if left in the equipment. Other materials may contaminate the next application.

Rinsate in sprays can often be applied over the crop if it has been diluted properly. Check with your chemical supplier or county Cooperative Extension center for suggestions. Most sprayers can be equipped with an in-field rinse system for a few hundred dollars. The system consists of a rinse tank, control valves, and hoses to connect it to the sprayer circuit. A rinse nozzle must be installed inside the tank.

Wash the equipment on an approved wash pad or site. Protect components from rust and corrosion after washing with a light oil or other suitable material.

Calibrating Chemical Application Equipment

Purpose

To determine if the proper amount of chemical is being applied, the operator must measure the output of the application equipment. This technique is known as *calibration*.

Calibration not only ensures accuracy, a critical factor with regard to many chemicals, but it can also save time and money and benefit the environment.

Getting Started

Careful and accurate control of ground speed is important for any type of chemical application procedure. From large self-propelled sprayers and spreaders to small walk-behind or backpack units, precise ground speed is a key for success. Ground speed can be determined by one of two methods. The first method requires a test course and stopwatch. For this procedure, measure a suitable test course in the field and record the time it takes to cover the course with the equipment. The course should be between 100 and 300 feet long. Drive or walk the course at least twice (once in each direction) and average the times for greater accuracy. Calculate the speed with Equation 1 or refer to Table 2-6 at the end of the chapter.

Equation 1. Ground Speed (MPH) = $\dfrac{Distance \ x \ 60}{Seconds \ x \ 88}$

The second method is to use a true ground speed indicator such as a tractor-mounted radar or GNSS receiver. Do not rely on transmission speed charts and engine tachometers. They are not accurate enough for calibration.

Calibrating a Sprayer

Preparing to Calibrate

For calibration to be successful, several items need to be taken care of before going to the field. Calibration will not be worthwhile if the equipment is not properly prepared. Whenever possible, calibration should be performed using water only. If you must calibrate using spray mixture, do so on a site listed on the chemical label and with wind speeds less than 5 MPH. Follow the steps outlined below to prepare spraying equipment for calibration.

1. Inspect the sprayer. Be sure all components are in good working order and undamaged. On backpack sprayers, pay particular attention to the pump, control wand, strainers, and hoses. On boom sprayers, pay attention to the pump, control valves, strainers, and hoses. On airblast sprayers, be sure to

inspect the fan and air tubes or deflectors as well. Be sure there are no obstructions or leaks in the sprayer.

2. Check the label of the product or products to be applied and record the following:

   - *Application Rate*, gallons per acre (GPA)
   - *Nozzle Type*, droplet size and shape of pattern
   - *Nozzle Pressure,* pounds per square inch (PSI)
   - *Type of Application*, broadcast, band, or directed

3. Next, determine some information about the sprayer and how it is to be operated. This includes:

   - *Type of Sprayer:* backpack, boom, or airblast. The type of sprayer may suggest the type of calibration procedure to use.
   - *Nozzle Spacing (inches):* For broadcast applications, nozzle spacing is the distance between nozzles.
   - *Nozzle Spray Width (inches):* For broadcast applications, nozzle spray width is the same as nozzle spacing—the distance between nozzles. For band applications, use the width of the sprayed band if the treated area in the band is specified on the chemical label; use nozzle spacing if the total area is specified. For directed spray applications, use the row spacing divided by the number of nozzles per row. Some directed spray applications use more than one type or size of nozzle per row. In this case, the nozzles on each row are added together and treated as one. Spray width would be the row spacing.
   - In most cases, a backpack sprayer uses a single nozzle. Some sprayers use mini-booms or multiple nozzles. The spray width is the effective width of the area sprayed, being sure to account for overlap. If you are using a sweeping motion from side to side, be sure to use the full width sprayed as you walk forward. If you are spraying on foliage in a row, use the row spacing. Dyes are available to blend with the spray to show what has been covered.
   - *Spray Swath (feet):* The width covered by all the nozzles on the boom of a sprayer. For airblast or other boomless sprayers, it is the effective width covered in one pass through the field.
   - *Ground Speed, miles per hour (MPH).* When using a backpack sprayer, walk a comfortable pace that is easy to maintain. Slow walking speeds will take longer to complete the task, while high speeds may be tiresome. Choose a safe, comfortable speed that will enable you to

finish the job in a timely manner. On tractor-mounted sprayers, select a ground speed appropriate for the crop and type of sprayer used. Slow speeds will take longer to complete the task, while high speeds may be difficult to control and unsafe. Choose a safe, controllable speed that will enable you to finish the job in a timely manner. Ground speed can be determined from Equation 1 (page 23) or Table 2-6.

4. The *discharge rate, gallons per minute (GPM),* required for the nozzles must be calculated in order to choose the right nozzle size. Discharge rate depends on the application rate; ground speed; and nozzle spacing, spray width, or spray swath.

For applications using nozzle spacing or nozzle spray width (inches), use Equation 2 or Table 2-9 at the end of the chapter.

Equation 2. Discharge Rate =

$$\frac{Application\ Rate\ x\ Ground\ Speed\ x\ Nozzle\ Spray\ Width}{5,940}$$

For applications using the spray swath (feet):

Equation 3. Discharge Rate =

$$\frac{Application\ Rate\ x\ Ground\ Speed\ x\ Spray\ Swath}{495}$$

5. Choose an appropriate nozzle or nozzles from the manufacturer's charts and install on the sprayer. Check each nozzle to be sure it is clean and that the proper strainer is installed with it. Be sure to check the droplet size recommendation on the chemical label when you choose your nozzle. Spray nozzles are classified by the droplet size produced at the selected operating pressure as well as flow rate. *Do not confuse the flow rate color codes with the droplet size color codes.*

6. Fill the tank half full of water and adjust the nozzle pressure to the recommended setting. Measure the discharge rate for the nozzle. This can be done by using a flow meter or by using a collection cup and stopwatch. The flow meter should read in gallons per minute (GPM). If you are using the collection cup and stopwatch method, the following equation is helpful to convert ounces collected and collection time, in seconds, into gallons per minute.

Equation 4. Discharge Rate =

$$\frac{Ounces\ Collected\ x\ 60}{Collection\ Time\ x\ 128}$$

7. Whenever possible, calibrate with water instead of spray solution. Do not calibrate with spray solution unless required by the chemical label. Follow all recommendations on the label. If the spray solution has a density different than water, the rate can be corrected using the procedure shown in Calibration Variables.

8. On boom sprayers or sprayers with multiple nozzles, average the discharge rates of all the nozzles on the sprayer. Reject any nozzle that has a bad pattern or that has a discharge rate 10 percent more or less than the overall average. Install a new nozzle to replace the rejected one and measure its output. Calculate a new average and recheck the nozzles compared to the new average. Again, reject any nozzle that is 10 percent more or less than the average or has a bad pattern. When finished, select a nozzle that is closest to the average to use later as your quick check nozzle.

9. On backpack sprayers or sprayers with a single nozzle, compare the discharge rate of the nozzle on the sprayer to the manufacturer's tables for that nozzle size. Reject any nozzle that has a bad pattern or that has a discharge rate 10 percent

more or less than the advertised rate. Install a new nozzle to replace the rejected one and measure its output.

Once the sprayer has been properly prepared for calibration, select a calibration method. When calibrating a sprayer, changes are often necessary to achieve the application rates needed. The sprayer operator needs to understand the changes that can be made to adjust the rate and the limits of each adjustment. The adjustments and the recommended approach are:

- *Pressure:* if the error in application rate is less than 10 percent, adjust the pressure.
- *Ground speed:* if the error is greater than 10 percent but less than 25 percent, change the ground speed of the sprayer.
- *Nozzle size:* if the error is greater than 25 percent, change nozzle size.

The goal is to have application rate errors less than 5 percent.

Calibration Methods

There are four methods commonly used to calibrate a sprayer:
- basic
- nozzle
- 128th acre
- area

The basic, nozzle, and 128th acre methods are time-based methods that require using a stopwatch or watch with a second hand to ensure accuracy. The area method is based on spraying a test course measured in the field. Each method offers certain advantages. Some are easier to use with certain types of sprayers. For example, the basic and area methods can be used with any type of sprayer. The 128th acre and nozzle methods work well for boom and backpack sprayers. Choose a method you are comfortable with and use it whenever calibration is needed.

Basic Method

1. Accurate ground speed is very important to good calibration with the basic method. For tractor-mounted sprayers, set the tractor for the desired ground speed and run the course at least twice. For backpack sprayers, walk the course and measure the time required. Walk across the course at least twice. Average the times required for the course distance and determine ground speed from Equation 1 or Table 2-6 at the end of the chapter.

2. Calculate the application rate based on the average discharge rate measured for the nozzles, the ground speed over the test course, and the nozzle spacing, nozzle spray width, or spray swath on the sprayer.

When using nozzle spacing or nozzle spray width measured in inches, use the following equation:

Equation 5. Application Rate =

$$\frac{5,940\ x\ Discharge\ Rate}{Ground\ Speed\ x\ Nozzle\ Spray\ Width}$$

For spray swath applications measured in feet:

Equation 6. Application Rate =

$$\frac{495\ x\ Discharge\ Rate}{Ground\ Speed\ x\ Spray\ Swath}$$

3. Compare the application rate calculated to the rate required. If the rates are not the same, choose the appropriate adjustment and reset the sprayer.

4. Recheck the system if necessary. Once you have the accuracy you want, calibration is complete.

Nozzle Method

1. Accurate ground speed is very important to good calibration with the nozzle method. For tractor-mounted sprayers, set the tractor for the desired ground speed and run the course at

least twice. For backpack sprayers, walk the course and measure the time required. Walk across the course at least twice. Average the times required for the course distance and determine ground speed from Equation 1 or Table 2-6 at the end of the chapter.

2. Calculate the nozzle discharge rate based on the application rate required, the ground speed over the test course, and the nozzle spacing, spray width, or spray swath of the sprayer. For nozzle spacing or spray width measured in inches use the following equations (or refer to Table 2-9 at end of chapter):

Equation 7. Discharge Rate =

$$\frac{Application\ Rate\ x\ Speed\ x\ Spray\ Width}{5,940}$$

For spray swath measured in feet:

Equation 8. Discharge Rate =

$$\frac{Application\ Rate\ x\ Speed\ x\ Spray\ Width}{495}$$

Set the sprayer and determine the average nozzle rate.

3. Compare the rate calculated to the average rate from the nozzles. If the two don't match, choose the appropriate adjustment and reset the system.

4. Recheck the system if necessary. Once you have the accuracy you want, calibration is complete.

128th Acre Method

1. The distance for one nozzle to cover 128th of an acre must be calculated. The nozzle spacing or spray width in inches is used to determine the spray distance. Spray distance is measured in feet. On backpack sprayers, be sure to measure the full width sprayed as you walk forward. Use Equation 9 or Table 2-14 at the end of the chapter.

Equation 9. Spray Distance =

$$\frac{4,084}{Spray\ Width}$$

2. Measure the spray distance on a test course in the field. Check the ground speed as you travel across the course. Be sure to maintain an accurate and consistent speed. Travel the course at least twice and average the time to cover the course.

3. For backpack sprayers, collect the output from the nozzle for the time measured in step 2. For tractor-mounted sprayers, park the sprayer, select the nozzle closest to the average, and collect the output for the time determined in step 4. Ounces collected will equal application rate in GPA.

4. Compare the application rate measured for the nozzle to the rate determined in step 3. If the rates are not the same, choose the appropriate adjustment and reset the system.

5. Recheck the system if necessary. Once you have the accuracy you want, calibration is complete.

Area Method

1. Determine the distance that can be sprayed by one tank using the full spray swath measured in feet.

Equation 10. Tank Spray Distance (ft) =

$$\frac{Tank\ Volume\ (gal)\ x\ 43,560}{Application\ Rate\ (GPA)\ x\ Swath\ (ft)}$$

2. Lay out a test course that is at least 10 percent of the tank spray distance from Step 1. Fill the sprayer tank with water only, mark the level in the tank, set the sprayer as

recommended, and spray the water out on the course. Be sure to maintain an accurate and consistent speed.

3. After spraying the test course, carefully measure the volume of water required to refill the tank to the original level. Calculate the application rate as shown:

Equation 11. Application Rate (GPA) =

$$\frac{Volume\ Sprayed\ (gal)\ x\ 43,560}{Test\ Course\ Distance\ (ft)\ x\ Swath\ (ft)}$$

4. Compare the application rate measured to the rate required. If the rates are not the same, choose the appropriate adjustment method and reset the sprayer.

5. Recheck the system. Once you have the accuracy you want, calibration is complete.

Calibrating a Granular Applicator

Preparing to Calibrate

Granular application calibration is usually done with the chemical to be applied. It is difficult to find a blank material that matches the granular product. Extra care should be taken in handling this product. Minimize worker exposure and take precautions against spills during calibration.

To prepare for calibration, follow these steps:

1. Before calibrating, carefully inspect the equipment to ensure that all components are in proper working order. Check the hopper, the metering rotor, the orifice, and the drop tubes. Be sure there are no leaks or obstructions.

2. Determine the type of application required for the product:

- Broadcast: treats the entire area (includes band applications based on broadcast rates).
- Band: treats only the area under the band.
- Row: treats along the length of the row.

3. Determine the application rate needed:

- Broadcast: pounds per acre.
- Band: pounds per acre of treated band width.
- Row: pounds per acre or pounds per 1,000 feet of row length.

4. What type of drive system does the applicator use?

- Independent: uses PTO, hydraulic, or electric motor drive.
- Ground Drive: uses ground driven wheel.

4. Regardless of how the application rate is expressed or type of application, calibration is easier if the rate is expressed in terms of pounds per foot of row length. Use one of the following steps to determine the correct row rate in pounds per foot.

For broadcast and row applications (Application Rate = lb/ac):

Equation 12. Row Rate, lb/ft =

$$\frac{Application\ Rate\ x\ Row\ Width\ (ft)}{43,560}$$

For banded applications (Application Rate = lb/ac of Treated Band):

Equation 13. Row Rate, lb/ft =

$$\frac{Application\ Rate\ x\ Band\ Width\ (ft)}{43,560}$$

For directed (row) applications (Application Rate = lb per 1,000 ft):

Equation 14. Row Rate, lb/ft =

$$\frac{Application\ Rate}{1,000}$$

CHEMICAL APPLICATION EQUIPMENT

6. Choose a calibration distance to work with and measure a test course of this distance in the field you will be working in. Choose an area that is representative of field conditions. The calibration distance should be at least 50 feet but not more than 500 feet. Longer distances are generally more accurate.

7. Calculate the weight of material that should be collected for the calibration distance chosen.

Equation 15. Weight Collected =

Row Rate x Calibration Distance

8. Select a ground speed appropriate for the crop and type of equipment used. Slow speeds take longer to finish the task, while high speeds may be inefficient and unsafe. Consult your equipment manual for a recommended speed. Even ground-driven application equipment can be sensitive to changes in speed. Maintaining an accurate and consistent speed is very important. Choose a safe, controllable speed that will enable you to complete the job in a timely and efficient manner.

9. Set your equipment according to recommendations from the equipment or chemical manufacturer. Most equipment manufacturers and chemical manufacturers provide rate charts to determine the correct orifice setting or rotor speed for each applicator. Fill the hopper at least half full to represent average capacity for calibration.

10. Attach a suitable collection container to each outlet on the applicator. You should be able to collect all material discharged from the applicator. Locate a scale capable of weighing the samples collected in calibration. Some samples may be very small, so a low-capacity scale may be needed. An accurate scale for measurement is very important.

Calibration Methods

Two methods for calibrating granular applicators are commonly used. The first is the distance method. This method is preferred by many operators because it applies to any type of granular machine and is easy to perform. The second method is the time method. This method is similar to sprayer calibration and can be used for applicators driven by PTO, hydraulic, or electric motors.

Distance Method

1. On the test course selected in the field, collect the output from the applicator in a container as you travel the course and weigh the material collected. Record the time required to travel the course also. Run the course twice, once in each direction, and average the results for both weight and time.

2. Determine the weight of the product that should be collected for the calibration distance.

Equation 16. Weight Collected (lb) =

Row Rate (lb/ft) x Calibration Distance (ft)

Time Method

1. On the test course selected in the field, record the time required to travel the course. Run the course twice, once in each direction, and average the results. Accurate ground speed is very important to good calibration with the time method.

2. With the equipment parked, set the orifice control as recommended and run the applicator for the time measured to run the calibration distance. Collect and weigh the output of the applicator for this time measurement.

3. Determine the weight of the product that should be collected for the calibration distance.

Equation 17. Weight Collected (lb) =

Row Rate (lb/ft) x Calibration Distance (ft)

4. Compare the weight of the product actually collected during the time it took to cover the calibration distance to the weight expected for the calibration distance. If the rates differ by more than 10 percent, adjust the orifice, rotor speed, or ground speed and repeat. Bear in mind, speed adjustments are not effective for ground-driven equipment.

5. Repeat the procedure until the error is less than 10 percent.

Calibrating a Broadcast Spreader

1. Carefully inspect all machine components. Repair or replace any elements that are not in good working order.

2. Determine the type of drive system that is being used: ground drive or independent PTO. This may help determine the method of calibration.

3. Determine the application rate and the bulk density of the product to be applied.

4. Determine the spreader pattern and swath of the spreader. Check the pattern to ensure uniformity. To check the pattern, place collection pans across the path of the spreader. For drop spreaders, be sure to place a pan under each outlet. For centrifugal and pendulum spreaders, space the pans uniformly with one in the center and an equal number on each side. The pattern should be the same on each side of the center and should taper smoothly as you go to the outer edge. The swath would be set as the width from side to side where a pan holds 50 percent of the maximum amount collected in the center pan.

5. Fill the hopper half full to simulate average conditions.

6. Set the ground speed of the spreader.

7. Set the spreader according to the manufacturer's recommendations and begin calibration.

Calibration Methods

There are two common methods used to calibrate broadcast spreaders. The first method is the discharge method. To use this procedure, collect and measure the total discharge from the spreader as it runs across a test course. The second method, the pan method, is used on centrifugal and pendulum spreaders. The pattern test pans used to determine pattern shape and swath are used to determine the application rate.

Discharge Method

1. Determine the test distance to use. Longer distances may give better accuracy but may be difficult to manage. A distance of 300 to 400 feet is usually adequate. Use shorter distances if necessary to avoid collecting more material than you can reasonably handle or weigh.

2. Set the ground speed. Be sure to maintain a constant ground speed at all times.

3. If using a ground drive spreader, attach a collection bin to the discharge chute or under the outlets and collect all the material discharged from the spreader as it runs across the test distance. If using an independent drive spreader, record the time required to run the test course. Park the spreader at a convenient location and measure the discharge from the spreader for the time measured on the test distance. The course should be run twice and the times averaged for better accuracy.

4. Calculate the application rate (pounds per acre):

Equation 18. Application Rate, lb/A =

$$\frac{Weight\ Collected\ (lb)\ x\ 43,560}{Distance\ (ft)\ x\ Swath\ (ft)}$$

5. Compare the application rate measured to the rate required. Adjust and repeat as necessary.

Pan Method

1. Place pans in the field across the swath to be spread. Pans should be uniformly spaced to cover the full swath. One pan should be at the center of the swath with equal numbers of pans on each side. Use enough pans, 11 or more, to get a good measurement.

2. Make three passes with the spreader using the driving pattern to be used in the field. One pass should be directly over the center pan and the other passes at the recommended distance, lane spacing, to the left and right of the center pass.

3. Combine the material collected in the pans and determine the weight or volume collected. Divide by the number of pans used to determine the average weight or volume per pan.

4. Calculate the application rate.

If you are measuring the weight in the pans in grams:

Equation 19. Application Rate, lb/A =

$$\frac{13{,}829 \; x \; Weight \; (grams)}{Pan \; Area \; (square \; inches)}$$

If you are measuring the volume in the pans in cubic centimeters (cc):

Equation 20. Application Rate, lb/A =

$$\frac{13{,}829 \; x \; Bulk \; Density \; (lb \; per \; cubic \; ft) \; x \; Volume \; (cc)}{Pan \; Area \; (square \; inches) \; x \; 62.4}$$

5. Compare the rate measured to the rate required.

Calibration Variables

Several factors can affect proper calibration. The ground speed of any type of PTO-powered machine can make a difference. On the other hand, ground-driven machines are usually only slightly affected by changes in ground speed. If using dry or granular material, product density will affect the discharge rate and may change the pattern for broadcast spreaders. For liquids, calibration can be affected by pressure, nozzle size, density, and viscosity of the liquid, and application type: band or broadcast. The following adjustments may help in adjusting these variables.

Speed

For PTO-powered equipment or other equipment in which the discharge rate is independent of ground speed, Equation 21 is useful.

Equation 21. New Application Rate =

$$Old \; Application \; Rate \; x \; \frac{Old \; Speed}{New \; Speed}$$

For ground-driven equipment, there should be little or no change in application rate when speed is changed.

Pressure

For liquids in sprayers, the discharge rate changes in proportion to the square root of the ratio of the pressures.

Equation 22. New Discharge Rate =

$$Old \; Discharge \; Rate = \sqrt{\frac{New \; pressure}{Old \; Pressure}}$$

Density

For liquids in sprayers, the discharge rate changes if the specific gravity (S.G.) of the liquid changes. Use water for calibration and adjust as shown below. Calibrate with spray solution only if recommended by the supplier.

Equation 23. Water Discharge Rate =

$$Spray \; Discharge \; Rate \; x \; \sqrt{S.G.} \; of \; Spray \; Solution$$

Band Application Versus Broadcast Application

Some pesticide application recommendations are based on area of cropland covered. Other recommendations are based on area of land treated in the band covered. Check the label for the product you are using to see how it is listed.

Broadcast application is based on area of cropland covered. Nozzle spacing is the distance between nozzles. Band applications in which the area of covered cropland is used for calibration and those applications in which multiple nozzles per row are used are both treated like broadcast applications. Divide the row spacing by the number of nozzles used per row to get a nozzle spacing for calibration.

For band applications in which area of treated land—not cropland covered—is specified, use the width of the band at the ground as the spacing for calibration.

Determining Upper and Lower Limits

Upper and lower limits provide a range of acceptable error. To set these limits for a given sample size, use the equations below. First, however, you must decide upon the degree of accuracy you wish to achieve. Select a percent error: 2 percent, 5 percent, 10 percent, or any other level of accuracy.

Equation 24. Upper Limit =

$$Target \; Rate \; x \; (1 + \frac{Percent \; Error}{100\%})$$

Equation 25. Lower Limit =

$$Target \; Rate \; x \; (1 - \frac{Percent \; Error}{100\%})$$

Useful Tables and Data

Table 2-1. Pump Types and Characteristics

Pump Type	Available Construction Material	Typical Maximum Pressure (PSI)	Capacity (GPM)	Effect of Abrasives	Repairable
Gear	Bronze	80	3 to 7	Severe	No
Roller[1] (4 to 8)	Cast iron, Ni-resist, or bronze (nylon, Teflon, polypropylene, or rubber rollers)	75 to 150	6 to 68	Moderate	Yes
Centrifugal	Cast iron	50	40[2]	Moderate	Yes
Diaphragm	Cast iron or aluminum	75 to 850	5 to 62	Practically none	Yes
Piston	Cast iron	400 to 600	6 to 24	Moderately little	Yes

[1] Roller pumps are available with rubber rollers in lieu of nylon rollers. The rubber rollers have the longest life when used with abrasive materials, but the maximum pressure is limited to 75 psi. Pumps equipped with rubber rollers cost more.

[2] Capacity for a typical 1.25-inch suction size centrifugal pump. Flow is much greater than at low pressures.

Table 2-2. Normal Permissible Hose Flow Rates

Maximum Flow, GPM	2	4	8	12	20	40
Hose Size (inches)	0.375	0.5	0.625	0.75	1	1.25

Table 2-3. Nozzle Types and Materials

Nozzle Type	Nozzle Description
Flat fan	Typically used for broadcast applications.
Even	Typically used for band applications. Pattern is uniform across width. Do not overlap.
Whirl chamber	Wide-angle hollow cone used in place of flat fan. Minimum clogging and low drift.
Flooding	Wide spray patterns used for herbicides and fertilizers. Overlapping usually required.
Boomless	Wide swath of 30 feet or more. May be single or cluster nozzles. Coverage not as uniform as boom nozzles and more affected by wind.
Hollow cone	Very uniform distribution and better coverage of crop foliage.
Solid cone	Concentrates spray material on plants.
Off-center	Flat or cone pattern, used on end of boom to increase width of swath.
Double-outlet	Wider pattern angle and width.
Nozzle Material	**Material Description**
Brass	Most common material used, relatively inexpensive. High wear rate with abrasive materials
Aluminum	Resistant to corrosives. High wear rate with abrasives.
Monel	Same as aluminum.
Plastic	Low cost, corrosion resistant, used for nonabrasive materials.
Stainless steel	Noncorrosive but relatively expensive. Low wear rate with abrasives.
Ceramic	Suitable for use with abrasives at high pressures. Long life.

CHEMICAL APPLICATION EQUIPMENT

Table 2-4. Weights and Measures

Weights

28.35 grams = 1 ounce
16 ounces = 1 pound = 453.6 grams
1 kilogram = 1,000 grams = 2.205 pounds
1 gallon water = 8.34 pounds = 3.8 kilograms
1 cubic foot water = 62.4 pounds = 28.3 kilograms
1 kilogram water = 33.81 ounces = 2.205 pounds
1 ton = 2,000 pounds = 907 kilograms
1 metric ton = 1,000 kilograms = 2,205 pounds

Length and Land Measure

1 inch = 2.54 centimeters
100 centimeters = 1 meter = 3.281 feet
16.5 feet = 5.5 yards = 1 rod
66 feet = 4 rods = 1 chain
1 mile = 5,280 feet = 1.609 kilometers
272.5 square feet = 30.25 square yards = 1 square rod
4,356 square feet = 16 square rods = 1 square chain
43,560 square feet = 160 square rods = 1 acre
43,560 square feet = 10 square chains = 1 acre
1 acre = 0.4047 hectare = 4,047 square meters
1 hectare = 2.471 acres = 10,000 square meters

Volume and Liquid Measure

3 teaspoons = 1 tablespoon = 14.8 cubic centimeters
2 tablespoons = 1 fluid ounce = 29.6 cubic centimeters
8 fluid ounces = 16 tablespoons = 1 cup = 236.6 cubic centimeters
2 cups = 32 tablespoons = 1 pint = 473.1 cubic centimeters
2 pints = 64 tablespoons = 1 quart = 946.2 cubic centimeters
1 liter = 1,000 cc = 0.2642 gallons = 33.81 ounces
4 quarts = 256 tablespoons = 1 gallon = 3,785 cubic centimeters
1 gallon = 128 fluid ounces = 231 cubic inches = 3,785 cubic centimeters
1 bushel = 1.244 cubic feet = 9.309 liquid gallons = 35.2 liters
1 cubic centimeter = 1 milliliter

Table 2-5. Length of Row Required for 1 Acre

Row Spacing (inches)	Length or Distance
24	7,260 yards = 21,780 feet
30	5,808 yards = 17,424 feet
36	4,840 yards = 14,520 feet
42	4,149 yards = 12,447 feet
48	3,630 yards = 10,890 feet
54	3,227 yards = 9,860 feet
60	2,904 yards = 8,712 feet

Table 2-6. Travel Speed Chart

Miles per Hour	Time Required to Travel (in seconds)		
	100 feet	200 feet	300 feet
1	68	136	205
2	34	68	102
3	23	45	68
4	17	34	51
5	14	27	41
6	11	23	34
7	10	20	29
8	9	17	26
9	8	15	23
10	7	14	21

Table 2-7. Dilutions for Liquids and Dusts

Equivalent Quantities of Liquid Materials When Mixed by Parts

Water	1 to 100	1 to 200	1 to 400	1 to 800	1 to 1,600
100 gal	1 gal	2 qt	1 qt	1 pt	1 cup
50 gal	2 qt	1 qt	1 pt	1 cup	1/2 cup
5 gal	0.4 pt	0.2 pt	2 tbsp	5 tsp	2-1/2 tsp
1 gal	1.28 oz	0.64 oz	2 tsp	1 tsp	1/2 tsp

Example: If a recommendation calls for 1 part of the chemical to 800 parts of water, it would take 5 tsp in 5 gal of water to give 5 gal of a mixture of 1 to 800.

Equivalent Quantities of Dry Materials (Wettable Powders) for Various Quantities of Water

Water	Quantity of Material					
100 gal	1 lb	2 lb	3 lb	4 lb	5 lb	6 lb
50 gal	8 oz	1 lb	1.5 lb	2 lb	2.5 lb	3 lb
5 gal	3 tbsp	1.5 oz	2.5 oz	3.25 oz	4 oz	5 oz
1 gal	2 tsp	1 tbsp	1.5 tbsp	2 tbsp	5 tsp	1 oz

Example: If a recommendation calls for a mixture of 4 lb of a wettable powder to 100 gal of water, it would take 3.25 oz (approximately 6.5 tsp) to 5 gal of water to give 5 gal of the spray mixture of the same strength.

Note: Wettable pesticide materials vary considerably in density. Therefore, the teaspoonful (tsp) and tablespoonful (tbsp) measurements in this table are not exact dosage by weight but are within the bounds of safety and efficiency for mixing small amounts of spray.

CHEMICAL APPLICATION EQUIPMENT

Table 2-8. Chemical Formulations

Pounds of Active Ingredients Per Gallon, Pounds Per Pint of Liquid, and the Number of Pints for Various Per Acre Rates

Pounds of Active Ingredients In 1 Gallon of Commercial Products	Pounds of Active Ingredients per Pint	Pints of Commercial Product Needed for Each Acre to Give the Following Pounds of Active Ingredient per Acre					
		0.25	0.50	0.75	1	1.5	2
1.00	0.125	2.00	4.00	6.00	8.00	12.0	16.00
2.00	0.250	1.00	2.00	3.00	4.00	6.0	8.00
2.64	0.330	0.75	1.50	2.25	3.00	4.5	6.00
3.00	0.380	0.67	1.33	2.00	2.67	4.0	5.33
3.45	0.420	0.60	1.20	1.08	2.40	3.6	4.80
4.00	0.500	0.50	1.00	1.50	2.00	3.0	4.00
6.00	0.750	0.33	0.67	1.00	1.33	2.0	2.67

Available Commercial Materials in Pounds Active Ingredients Per Gallon Necessary to Make Various Percentage Concentration Solution

Pounds of Active Ingredients In 1 Gallon of Commercial Products	Pounds of Active Ingredients per Pint	Liquid Ounces of Commercial Product per 1 Gallon of Solution to Make:				
		0.5%	1%	2%	5%	10%
1.00	0.125	5.34	10.68	21.35	53.38	106.80
2.00	0.250	2.68	5.36	10.72	26.80	53.40
2.64	0.330	2.02	4.05	8.10	20.25	40.44
3.00	0.380	1.78	3.56	7.12	17.80	35.58
3.34	0.420	1.59	3.18	6.36	15.90	31.96
4.00	0.500	1.33	2.67	5.33	13.34	26.69
6.00	0.750	0.89	1.78	3.56	8.90	17.79

Based on 8.34 pounds per gallon (weight of water).

CHEMICAL APPLICATION EQUIPMENT

Table 2-9. Required Nozzle Discharge Rates (Gallons per Minute) for Various Application Rates, Nozzle Spacings, and Travel Speeds

Travel Speed (MPH)	Nozzle Spacing (inches)	Application Rate (gallons per acre)							
		5	7.5	10	12.5	15	20	25	35
3	6	0.0152	0.0227	0.0303	0.0379	0.0455	0.0606	0.0758	0.106
	8	0.0202	0.0303	0.0404	0.0505	0.0606	0.0808	0.101	0.141
	10	0.0253	0.0379	0.0505	0.0631	0.0758	0.101	0.126	0.177
	12	0.0303	0.0455	0.0606	0.0758	0.0909	0.121	0.152	0.212
	14	0.0354	0.053	0.0707	0.0884	0.106	0.141	0.177	0.247
	16	0.0404	0.0606	0.0808	0.101	0.121	0.162	0.202	0.283
	18	0.0455	0.0682	0.0909	0.114	0.136	0.182	0.227	0.318
	20	0.0505	0.0758	0.101	0.126	0.152	0.202	0.253	0.354
	21	0.053	0.0795	0.106	0.133	0.159	0.212	0.265	0.371
	22	0.0556	0.0833	0.111	0.139	0.167	0.222	0.278	0.389
	24	0.0606	0.0909	0.121	0.152	0.182	0.242	0.303	0.424
	30	0.0758	0.114	0.152	0.189	0.227	0.303	0.379	0.53
	36	0.0909	0.136	0.182	0.227	0.273	0.364	0.455	0.636
	42	0.106	0.159	0.212	0.265	0.318	0.424	0.53	0.742
	48	0.121	0.182	0.242	0.303	0.364	0.485	0.606	0.848
	60	0.152	0.227	0.303	0.379	0.455	0.606	0.758	1.06
4	6	0.0202	0.0303	0.0404	0.0505	0.0606	0.0808	0.101	0.141
	8	0.0269	0.0404	0.0539	0.0673	0.0808	0.108	0.135	0.189
	10	0.0337	0.0505	0.0673	0.0842	0.101	0.135	0.168	0.236
	12	0.0404	0.0606	0.0808	0.101	0.121	0.162	0.202	0.283
	14	0.0471	0.0707	0.0943	0.118	0.141	0.189	0.236	0.330
	16	0.0539	0.0808	0.108	0.135	0.162	0.215	0.269	0.377
	18	0.0606	0.0909	0.121	0.152	0.182	0.242	0.303	0.424
	20	0.0673	0.101	0.135	0.168	0.202	0.269	0.337	0.471
	21	0.0707	0.106	0.141	0.177	0.212	0.283	0.354	0.495
	22	0.0741	0.111	0.148	0.185	0.222	0.296	0.370	0.519
	24	0.0808	0.121	0.162	0.202	0.242	0.323	0.404	0.566
	30	0.101	0.152	0.202	0.253	0.303	0.404	0.505	0.707
	36	0.121	0.182	0.242	0.303	0.364	0.485	0.606	0.848
	42	0.141	0.212	0.283	0.354	0.424	0.566	0.707	0.990
	48	0.162	0.242	0.323	0.404	0.485	0.646	0.808	1.130
	60	0.202	0.303	0.404	0.505	0.606	0.808	1.010	1.410
5	6	0.0253	0.0379	0.0505	0.0631	0.0758	0.101	0.126	0.177
	8	0.0337	0.0505	0.0673	0.0842	0.101	0.135	0.168	0.236
	10	0.0421	0.0631	0.0842	0.105	0.126	0.168	0.210	0.295
	12	0.0505	0.0758	0.101	0.126	0.152	0.202	0.253	0.354
	14	0.0589	0.0884	0.118	0.147	0.177	0.236	0.295	0.412
	16	0.0673	0.101	0.135	0.168	0.202	0.269	0.337	0.471
	18	0.0758	0.114	0.152	0.189	0.227	0.303	0.379	0.530
	20	0.0842	0.126	0.168	0.210	0.253	0.337	0.421	0.589
	21	0.0884	0.133	0.177	0.221	0.265	0.354	0.442	0.619
	22	0.0926	0.139	0.185	0.231	0.278	0.370	0.463	0.648
	24	0.101	0.152	0.202	0.253	0.303	0.404	0.505	0.707
	30	0.126	0.189	0.253	0.316	0.379	0.505	0.631	0.88
	36	0.152	0.227	0.303	0.379	0.455	0.606	0.758	1.060
	42	0.177	0.265	0.354	0.442	0.530	0.707	0.884	1.240
	48	0.202	0.303	0.404	0.505	0.606	0.808	1.010	1.410
	60	0.253	0.379	0.505	0.631	0.758	1.010	1.260	1.770

CHEMICAL APPLICATION EQUIPMENT

Table 2-10. Time (in Seconds) Needed To Collect 1 Pint (0.125 Gallon) of Liquid from a Single Nozzle at Various Application Rates, Nozzle Spacings, and Travel Speeds

Travel Speed (MPH)	Nozzle Spacing (inches)	Application Rate (gallons per acre)							
		5	7.5	10	12.5	15	20	25	35
3	6	495	330	247	198	165	124	99	70.7
	8	371	247	186	148	124	92.8	74.3	53
	10	297	198	148	119	99	74.3	59.4	42.4
	12	247	165	124	99	82.5	61.9	49.5	35.4
	14	212	141	106	84.9	70.7	53	42.4	30.3
	16	186	124	92.8	74.3	61.9	46.4	37.1	26.5
	18	165	110	82.5	66	55	41.3	33	23.6
	20	148	99	74.3	59.4	49.5	37.1	29.7	21.2
	21	141	94.3	70.7	56.6	47.1	35.4	28.3	20.2
	22	135	90	67.5	54	45	33.8	27	19.3
	24	124	82.5	61.9	49.5	41.3	30.9	24.8	17.7
	30	99	66	49.5	39.6	33	24.8	19.8	14.1
	36	82.5	55	41.3	33	27.5	20.6	16.5	11.8
	42	70.7	47.1	35.4	28.3	23.6	17.7	14.1	10.1
	48	61.9	41.3	30.9	24.8	20.6	15.5	12.4	8.84
	60	49.5	33	24.8	19.8	16.5	12.4	9.9	7.07
4	6	371	247	186	148	124	92.8	74.3	53
	8	278	186	139	111	92.8	69.6	55.7	39.8
	10	223	148	111	89.1	74.3	55.7	44.5	31.8
	12	186	124	92.8	74.3	61.9	46.4	37.1	26.5
	14	159	106	79.6	63.6	53	39.8	31.8	22.7
	16	139	92.8	69.6	55.7	46.4	34.8	27.8	19.9
	18	124	82.5	61.9	49.5	41.3	30.9	24.8	17.7
	20	111	74.3	55.7	44.5	37.1	27.8	22.3	15.9
	21	106	70.7	53	42.4	35.4	26.5	21.2	15.2
	22	101	67.5	50.6	40.5	33.8	25.3	20.3	14.5
	24	92.8	61.9	46.4	37.1	30.9	23.2	18.6	13.3
	30	74.3	49.5	37.1	29.7	24.8	18.6	14.8	10.6
	36	61.9	41.3	30.9	24.8	20.6	15.5	12.4	8.84
	42	53	35.4	26.5	21.2	17.7	13.3	10.6	7.58
	48	46.4	30.9	23.2	18.6	15.5	11.6	9.28	6.63
	60	37.1	24.8	18.6	14.8	12.4	9.28	7.72	5.3
5	6	297	198	148	119	99	74.3	59.4	42.4
	8	223	148	111	89.1	74.3	55.7	44.5	31.8
	10	178	119	89.1	71.3	59.4	44.5	35.6	25.5
	12	148	99	74.3	59.4	49.5	37.1	29.7	21.2
	14	127	84.9	63.6	50.9	42.4	31.8	25.5	18.2
	16	111	74	55.7	44.5	37.1	27.8	22.3	15.9
	18	99	66	49.5	39.6	33	24.8	19.8	14.1
	20	89.1	59.4	44.5	35.6	29.7	22.3	17.8	12.7
	21	84.9	56.6	42.4	33.9	28.3	21.2	17	12.1
	22	81	54	40.5	32.4	27	20.3	16.2	11.6
	24	74.3	49.5	37.1	29.7	24.8	18.6	14.8	10.6
	30	59.4	39.6	29.7	23.8	19.8	14.8	11.9	8.49
	36	49.5	33	24.8	19.8	16.5	12.4	9.9	7.07
	42	42.4	28.3	21.2	17	14.1	10.6	8.49	6.06
	48	37.1	24.8	18.6	14.8	12.4	9.28	7.42	5.3
	60	29.7	19.8	14.8	11.9	9.9	7.42	5.94	4.24

Table 2-11. Fumigant Application Rate Table

Application Rate (gal/acre)	Quantity per 100 Feet at Given Bandwidth												
	8-Inch Width		12-Inch Width		24-Inch Width		30-Inch Width		36-Inch Width		42-Inch Width		48-Inch Width	
	oz	cc	oz	cc	oz	cc	oz	cc	oz	cc	oz	cc	oz	cc
1	0.2	6	0.3	9	0.6	17	0.7	22	0.9	26	1.0	30	1.2	35
3	0.6	17	0.9	26	1.7	52	2.2	65	2.6	78	3.1	91	3.5	104
5	1.0	29	1.5	44	2.9	87	3.7	109	4.4	130	5.1	152	5.9	174
7	1.4	41	2.1	61	4.1	122	5.1	152	6.2	183	7.2	213	8.2	243
9	1.8	52	2.6	78	5.3	157	6.1	196	7.9	235	9.3	274	10.6	313
12	2.4	70	3.6	104	7.1	209	8.8	261	10.6	313	12.3	365	14.1	418
15	2.9	87	4.4	131	8.8	261	11.1	326	13.2	391	15.4	457	17.6	522
20	3.9	116	5.9	174	11.8	348	14.7	435	17.6	522	20.6	609	23.5	696

Table 2-12. Fertilizer Application Rate Table

Rate (lb/acre)	Pounds per 100 Feet of Row				
	24-inch Rows	30-inch Rows	36-inch Rows	42-inch Rows	48-inch Rows
100	0.5	0.6	0.7	0.8	0.9
200	0.9	1.1	1.4	1.6	1.8
300	1.4	1.7	2.1	2.4	2.8
400	1.8	2.2	2.8	3.2	3.7
500	2.3	2.8	3.4	4.0	4.6
600	2.8	3.4	4.1	4.8	5.5
700	3.2	3.9	4.8	5.6	6.4
800	3.7	4.5	5.5	6.4	7.3
900	4.1	5.1	6.2	7.2	8.8
1,000	4.6	5.6	6.9	8.0	9.2
1,100	5.1	6.2	7.6	8.8	10.1
1,200	5.5	6.7	8.3	9.6	11.0
1,300	6.0	7.3	9.0	10.4	11.9
1,400	6.4	8.0	9.6	11.2	12.6
1,500	6.9	8.6	10.3	12.1	13.8

Table 2-13. Granular Application Rate Table

Rate (lb/acre)	Grams per 100 Feet of Row				
	24-in. rows	30-in. rows	36-in. rows	42-in. rows	48-in. rows
5	10	13	16	18	21
6	13	16	19	22	25
7	15	18	22	26	29
8	17	21	25	29	33
9	19	23	28	33	38
10	21	26	31	36	42
12	25	31	37	44	50
15	31	39	47	55	62
20	42	52	62	73	83
25	52	65	78	91	104
30	63	78	93	109	125
40	83	104	125	146	167
50	104	130	156	182	208
60	125	156	187	218	250
75	156	195	234	273	312

Table 2-14. Distance To Cover for Each Nozzle To Spray 1/128TH Acre

Average Nozzle Spacing (inches)	Distance to Cover (feet)
6	681
8	510
10	408
12	340
14	292
16	255
18	227
20	204
22	186
24	170
30	136
36	113
38	107
40	102
42	97
48	85

Note: Each ounce per nozzle represents 1 gallon per acre application rate.

CHEMICAL APPLICATION EQUIPMENT

III — HOW TO SEND SPECIMENS FOR DISEASE, INSECT, AND WEED IDENTIFICATION

SENDING SPECIMENS

How to Send Specimens for Disease, Insect, and Weed Identification

B. Shew, Extension Plant Pathology

Plant Disease and Insect Clinic (PDIC)

Fees for Disease Diagnosis and Insect Identification Samples

Turf Sample Fees

- $120 out-of-state, all turf samples
- $60 in-state, golf courses
- $30 in-state, other turf, submitted by individual or business
- $20 in-state, other turf, submitted by Cooperative Extension agent with online submission form

Non-Turf Fees (all others)

- $75 out-of-state
- $30 in-state, submitted by individual or business
- $20 in-state, submitted by Cooperative Extension agent or NCDA&CS specialist with online submission form
- No charge for image-only samples submitted online. If a follow-up physical sample is necessary, the fees outlined above apply.

Please include a check made payable to NCSU with your sample.

Contact Information

Hours: 8 a.m. to 4:30 p.m., Monday through Friday

Plant Disease/Insect Clinic
State Courier #: 53-61-21
Campus Box 7211, Room 1227 Gardner Hall
100 Derieux Place
North Carolina State University
Raleigh, NC 27695-7211
http://www.ncsu.edu/pdic

General Information: 919-515-3619; plantclinic@ces.ncsu.edu

Insects: Matt Bertone, 919-515-9530; matt_bertone@ncsu.edu

Turf: Lee Butler, 919-513-3878; lee_butler@ncsu.edu

Sending Samples and Images Directly to the PDIC: Our database allows users to create an account, create a sample record, and upload sample information and images for diagnosis. For information and instructions on how to create an account, click the "new user" button at the top of the PDIC home page, http://www.ncsu.edu/pdic. Once you enter your sample online, you can track sample progress, review past reports and check status of your invoices. Log into the database by clicking the "login to database" button on the PDIC homepage.

We send reports as soon as the diagnosis is ready.

What to Sample and How to Ship

Visit our website at http://www.ncsu.edu/pdic and go to the "How to Submit a Sample" tab for instructions and an illustrated guide. Here are some things to remember:

- *Dead plants tell no tales* — Most plants that are totally dead, dry, or rotten are useless for diagnosis. Collect plants that show a range of the symptoms but are not yet dead. Likewise, collect live but diseased limbs and branches from trees or large shrubs.

- *More is better* — We may miss the main concern if you send only one plant. For bedding plants, young vegetables, and field crops, collect several plants with each type of problem.

- *Getting to the root of the problem* — Many plant problems are related to the roots and soil. **Submit whole plants, including roots and soil whenever possible.** Dig (don't pull) plants up to keep roots intact. For large plants, submit a portion of the root ball and a pint of soil.

- *A GOOD picture is worth a thousand words* — Images help us to better understand the situation. Use our database to upload one or more images of the problem as it appears in the field, greenhouse, or landscape. Make sure images are in focus and clearly illustrate the problem, symptoms, or insect of concern from different angles and distances. Do not attempt to upload more than 10 MB of image files at one time.

- *A place for everything* — Keep soil *off* the foliage and *on* the roots. This keeps roots from drying out and the foliage free of secondary organisms. Put the roots and soil in a plastic bag and tie at the main stem. Loosely wrap foliage in newspaper, then pull the bag over the rest of the plant and tie again to keep foliage from drying out. Make sure foliage is dry for packaging.

- *Details, details* — The more you tell us about the situation on the PDIC form, the better. Please give complete information, including a client name and location, percent of planting affected, variety or cultivar, and insect collection information (if applicable).

- *History 101* - List all fertilizer, fungicide, herbicide, and insecticide applications made in the last 30 days. For field and vegetable crops, give the rotation history and planting date. For orchards and landscapes, indicate when the trees or other plants were transplanted or established.

- *Fresher is better* — Mail or deliver samples as soon as possible. On hot days or in very cold weather, store samples in an insulated cooler until you can send them. Avoid mailing on Fridays since many samples will degrade before delivery on Monday. Samples can be held over the weekend in a refrigerator or cooler. Succulent plants, fruits, vegetables, and mushrooms rot quickly and should be sent overnight express.

- *Fragile, handle with care* — Ship all samples in a crush-proof box. Do not send your sample in an envelope.

- *Be cool* — Never leave any sample or insect specimen in the sun or a closed car for even a few minutes; "cooked" samples may be impossible to diagnose.

- *We don't work for peanuts (or shredded paper)* — Please don't use foam peanuts to pad samples. Crumpled newspaper is cheaper, works well, and is easier for us to handle.

- *Noxious weeds* — Send no more than 1 quart of soil with samples from counties regulated under state and/or federal measures to prevent movement of noxious weeds. http://www.ncagr.gov/plantindustry/plant/weed/weedprog.htm

Disease Sample Submission and Instructions

Turf Grasses

1. To diagnose turfgrass problems, we need **at least a 6-inch x 6-inch piece of the turf**, including the root system and soil. If using a golf course cup cutter, please send at least two plugs.

2. Collect samples from the border between healthy and diseased turf so that two-thirds of the sample is diseased and one-third is healthy.

3. Wrap the soil and roots in two to three layers of aluminum foil to **ensure that no soil leaks out during transit.**

4. **DO NOT store or transport the samples in plastic bags.** Instead, place the samples in a cardboard box and stuff it with newspaper or other packing material to hold the samples in place.

5. Download and print the turfgrass sample submission form from the clinic Web site: https://turfpathology.plantpath.ncsu.edu/diagnostics-lab/how-to-submit-a-sample/

6. Please fill out the sample submission form completely and describe the symptoms you are observing as accurately as possible. All of the information requested on the form is needed to make an accurate diagnosis.

7. **List all fertilizer, fungicide, herbicide, and insecticide applications made in the last 30 days.** Also, list any major cultural practices (aerification, topdressing, etc.) conducted in the last 30 days. These practices have a major impact on disease and insect development and provide valuable clues that will help us make an accurate diagnosis.

Field, Garden, or Landscape: Small Grains, Most Field Crops, Vegetable Plants, Small Shrubs, Perennials, and Annuals

1. Select plants with a range of symptoms but do not choose dead plants.

2. **Dig (do not pull) up several whole plants, leaving roots/soil intact.**

3. Wrap soil and roots in a plastic bag and tie the bag closed at the main stem.

4. Package and ship as described in general guidelines above.

Mushrooms, Fruits, and Vegetables

1. Wrap in dry newspaper, and then place in a plastic bag. Please ship overnight in a strong box with dry newspaper padding.

Small Plants in Greenhouse or Nursery

1. Send two to four entire plants. Leave in the pot if possible. For very large plants, remove most of the media and enclose root system with remaining media in a plastic bag.

2. For seedlings: send at least half the flat or up to a dozen plugs if the problem is scattered. Newspaper padding can help keep soil off leaves.

3. Label each bag with your code if the bag represents different types of samples (Good, Bad or A, B, C, etc.)

4. Package and ship as described in the general guidelines above.

Large Shrubs and Small Trees

1. Dig a generous shovel full of the small fine roots.

2. Seal these in a plastic bag along with 1 quart of soil.

3. In a separate plastic bag, include several branches showing symptoms. **Where cankers or diebacks are observed, we need stem or branch sections where live wood changes to dead wood.**

4. For trees and shrubs that are going to be removed, please also send the bottom 8 to 10 inches of the main trunk (cut below the soil line if possible, including some of the larger roots).

Large Trees

Decline of large trees often cannot be diagnosed by a clinic sample alone. However, with a good sample we can often identify leaf spots, vascular wilts, wood decays, root rots, and cankers. Images are invaluable aids to diagnosis and should be included with the sample if at all possible.

- *Cankers, swellings, diebacks* — Cut several affected branches to include several inches of healthy wood attached below the canker or dieback. Package in a plastic bag.

- *Vascular wilt diseases* — Find a wilting or dying but NOT dead branch, peel back the bark and look for dark tan, black, or greenish streaks in the sapwood. Send several branches with these symptoms.

- *Leaf spots* — Send several small branches with leaves attached. Seal loosely in a plastic bag. Don't add water to any sample.

- *Wood rots* — These can be identified ONLY by the presence of mushrooms, conks, or other fungal fruiting structures. Wrap mushrooms or conks loosely in several layers of newspaper; place the wrapped specimen in a plastic bag, add more newspaper padding, and ship in a crush-proof box. Fleshy mushrooms rot quickly. Please ship overnight and take photos as a back-up. Badly decayed samples are useless for identification.

Harvested Root Crops (sweet potato, potato, onion, etc.)

1. Collect six to eight roots/tubers showing the symptoms of concern.

2. Wrap each in newspaper separately or place in individual paper bags (do not use plastic bags).

3. **Package and ship in a crush-proof box with lots of newspaper padding.**

Houseplants or Interiorscape Plants

These plants typically have very few disease problems due to the dry air in most buildings. Most problems are related to improper moisture, temperature, nutrition, or light conditions. Please screen houseplants for these problems and avoid sending such samples to the clinic. When a disease or insect problem is suspected on a valuable specimen plant:

1. Send the whole plant if practical;

2. For plants too large to send, please send at least 2 cups of soil along with a generous handful of the smaller roots;

3. Cut several stems showing the symptoms of concern;

4. Package and ship as described in the general guidelines.

Insect Identification

- *Most insects* — Cockroaches, termites, bugs, beetles, flies, wasps, ants, maggots, spiders, etc. should be sent in alcohol, 70% (or higher) concentration.

- *Mites, scales, aphids, and thrips* — Send alive on some of the affected foliage/stems, collected as you would a plant specimen. Place in a plastic bag when collected. If there are numerous individuals, save some in a vial of alcohol.

- *Butterflies and moths* — Freeze specimens or kill them with fumes of ethyl acetate (commonly found in nail polish remover) and package them lightly in tissue paper in a crush-proof box.

- *Caterpillars* — Send alive on some of the host plant in a plastic bag. If you cannot send alive on the host plant, please note the plant and send the caterpillar in alcohol. For larger caterpillars, it is advisable to drop the caterpillar into boiling water before putting in the alcohol (70% or higher concentration).

- *Grubs* — Send alive in a pint or two of soil enclosed in a plastic bag.

When in doubt, put specimens in alcohol, 70% or higher concentration. Collect several specimens if possible. Send insect specimens to the Plant Disease and Insect Clinic at the address given above.

Please *do not* wrap insects in tissue or cellophane, and then put them in an envelope. Also, don't stick specimens to paper with tape or place them in an empty or overcrowded vial; they may decay or get damaged in these situations. Insects that arrive badly damaged and/or without data (see below) will require more time to identify or may be too poor in quality to diagnose at all.

Important: Data submitted with insects should be complete and accurate. When filling out the clinic forms, please see that the following four pieces of information are provided with all specimens:

- *Date specimen collected* — Date actually found, not date received by agent or other second party.

- *Town and county where specimen was collected* — Location where the insect was actually found, which may not be the agent's or property owner's address. If not found within a town, give nearest town and distance and direction from it, or GPS coordinates.

- *Name of collector* — Name the specific person who actually captured or collected the insect. The agent or other second party should not be listed as the collector.

- *Where were the specimens, or on what were they feeding when collected?* — If a plant, specifically name the host plant, e.g., oak, tobacco, marigold, etc. If not on a plant, name spot they were found, e.g., windowsill, closet, in a box of dog food, etc.

We strongly advise sending in insect specimens with the information above. This information is very important for our diagnoses and records. Your sample identification and report may be delayed if we have to reach you to ask follow-up questions.

Human and animal pests

We do not accept nor search through blood, tissue, fecal, or other such samples. Potential organisms must be isolated and sent in 70% or higher concentration alcohol. For guidelines on submitting insects and other arthropods associated with human/animal health, please refer to this document:

http://www.cals.ncsu.edu/plantpath/extension/clinic/ Submit/PDIC Procedures for Submitting Health Samples.pdf

Precautions

Some identifications and diagnoses require extensive studies and are based on the material and information submitted. Time devoted to individual specimens must necessarily be limited, and the samples usually represent only a small percentage of the crop or problem. Reports reflect considered opinion and best judgment but may not always be statements of absolute fact or define the major problems affecting the crop or site.

Plant and Weed Identification

General plant identification services are provided by the NCSU Herbarium. Weed identification services (turfgrass weeds, cropland, and non-cropland), are provided by the Department of Crop Science. All forms can be found at: http://www.cals.ncsu.edu/plantbiology/ncsc/identification.htm

A complete plant or a specimen containing leaves, stem, roots, and flowers or fruit is absolutely necessary to get a definite identification. Please prepare specimens properly and pack well. It is impossible to identify a plant when only a leaf or stem is sent. Crushed specimens from poor packing or specimens that are rotted from being tightly packed when green are especially hard to identify.

Soil Testing, Nematode Assay, Plant Tissue Nutrient Testing, Waste Analysis, and Solution Analysis

Information and forms are available in each North Carolina Cooperative Extension county center, by calling the NCDA&CS labs, and on the Web at

http://www.ncagr.gov/agronomi/index.htm

For a list of fees for various NCDA&CS services, see

http://www.ncagr.com/agronomi/fees.htm

NCDA&CS Agronomic Division,
Dr. Colleen M. Hudak-Wise, Director

Mailing Address:

1040 Mail Service Center, Raleigh NC 27699-1040

Physical Address:

4300 Reedy Creek Rd., Raleigh NC 27607-6465

Phone: 919-733-2655; FAX: 919-733-2837

IV — FERTILIZER USE

FERTILIZER USE

Lime and Fertilizer Suggestions for Field, Pasture, and Hay Crops

M. Castillo, C. R. Crozier, E. J. Dunphy, K. Edmisten, L. Fisher, R. W. Heiniger,

D. L. Jordan, J. M. Luginbuhl, P. Mueller, D. L. Osmond, and A. Post, Crop and Soil Sciences; D. H. Hardy, NCDA&CS

For best results, apply lime and nutrients (except nitrogen and boron) based on soil test information. If this information is unavailable, use these suggestions as general guidelines. Ranges in rates are given for P_2O_5 and K_2O. Sandy soils generally respond to higher rates of K_2O than clayey ones, but the opposite is true for P_2O_5. Also, attend to lime needs. For row crops (including corn, cotton, sorghum, soybean, and tobacco), experimental data across North Carolina demonstrate that starter P is not necessary on soils with very high P levels (P-index >100). On soils with P indexes between 50 and 100, starter band P may be useful with no-till management or when soils are cool and wet.

A liming program is essential to manage acidity in North Carolina soils. Base lime rates and frequency of applications on a routine soil sampling and testing program. In the absence of soil tests, most fields should receive 1 ton of lime per acre every 3 to 4 years. Liming more frequently at lesser rates may be necessary on sandy, light-colored soils where fertility changes more rapidly than in clay or organic soils, especially in years with excessive rainfall. Although liming is imperfect, over liming can cause micronutrient deficiencies, particularly manganese deficiency in small grain and soybeans grown on sandy soils. Dolomitic lime contains magnesium and is generally preferred over calcitic lime, especially on sandy soils. Lack of lime is the dominant soil fertility problem in North Carolina, even though liming is known to be a highly economical practice on either a short- or long-term basis.

Nitrogen rates are not based on soil testing. Use the RYE (Realistic Yield Expectation) for the soil and the appropriate nitrogen factor (NF) to establish the nitrogen rate. A range of NFs is found in Table 4-1A. The appropriate NF for each crop is determined by the soil mapping unit. Nitrogen rates based on RYE and NF are provided for each county by soil series at: http://nutrients.soil.ncsu.edu/yields/

Table 4-1A. Lime and Fertilizer Suggestions — Field Crops

Area of State or Soil Type	Optimum pH	Plant Nutrient Suggestions When Soil Test is Unavailable[1]			Remarks
		N factor	P_2O_5 lb/acre	K_2O lb/acre	
Corn Grain					
Mineral soils	6.0	1.0 to 1.25 lb/bu	10 to 20	80 to 100	Banded starter fertilizer with 20 to 30 pounds per acre of both N and P_2O_5 is recommended under no-till management or on cool, wet soils, especially when planting early. Starter P not likely to benefit on soils with very high initial P levels (P-index >100). Apply 1/4 to 1/3 of the N at planting. Sidedress remaining N when plants are 15 to 24 inches high. Under irrigation increase N rate by 10% to 15%. Starter band P if using no-till management or when soils are cool and wet. On deep sandy soils, apply K just before planting or use a split application of K at planting and at sidedress. Mineral soils that are sandy or greater than 18 inches to clay should receive 20 pounds S per acre. Test organic soils to determine copper needs.
Mineral-organic soils	5.5				
Organic soils	5.0				
Corn Silage					
Mineral soils	6.0	10 to 12 lb/ton	10 to 20	100 to 120	
Mineral-organic soils	5.5				
Organic soils	5.0				
Cotton					
Mineral soils	6.2	0.06 to 0.12 lb/lb lint	10 to 20	50 to 70	Apply 20 to 30 pounds N per acre at or before planting. Apply remaining N 2 to 3 weeks after first square. Starter P not likely to benefit on soils with very high initial P levels (P-index >100). After peanuts or soybeans, reduce total N by 25 to 30 pounds per acre. Apply 0.5 pound per acre B at planting or as foliar spray at first bloom. On deep sands (more than 18 inches to clay), use higher end of pounds per RYE range and apply 1/2 the remaining N at early square and 1/2 just prior to bloom. Deep sands should also receive 20 to 25 pounds S per acre.
Mineral-organic soils	5.5	0.05 to 0.09			
Organic soils	5.0	0.03 to 0.06 lb/lb lint			
Peanut					
Coastal plain	6.0	0	10 to 20	0	To minimize Ca deficiency risk, reduce unnecessary application of other cations. Apply K based on soil test recommendations and incorporate. During the growing season, at early to mid-flowering (late June-early July), apply gypsum to all Virginia market types. Use soil testing to determine Ca need for small-seeded runner market types. For larger-seeded "jumbo" runners, use half the rate as for Virginia market types. See *Peanut Information* series (http://www.peanuts.ncsu.edu/) for more information on gypsum products and rates. Apply Mg only as recommended based on soil test. Apply 0.5 pound B per acre (liquid or dry) at peak flower. Apply *Bradyrhizobia* inoculant to seed or in the seed furrow regardless of previous rotation history to ensure peanut is capable of fixing N. Inoculation is especially important on land where peanut has not been grown recently. Peanut taproot should have 10 or more nodules approximately 45 days after planting or supplemental N is needed. Apply 600 pounds/acre of ammonium sulfate (i.e., 125 pounds of N/acre) as soon as possible if roots are not nodulating effectively and foliage is N deficient. Zinc toxicity may occur in fields with soil test zinc indices greater than 250 or at lower indices with pH less than 6.0. Apply Mn at 0.5 to 1.0 pound Mn per acre when deficiency symptoms appear, often associated with high pH. See *Peanut Information* series for selection of appropriate B and Mn products if needed.
Small Grain, Grain					
Mineral soils	6.0	Wheat, Rye: 1.7 to 2.4 lb/bu	10 to 20	80 to 100	Apply 15 to 20 pounds N at planting (sometimes can be skipped if after soybean or peanut). If tiller density at spring green up (Feekes GS-3) is low, split N topdress (Feb., March). If tiller density is high, apply all N just before jointing. Use a tissue test at GS-5 to find the optimum nitrogen rate for wheat. Test organic soils to determine copper needs. Mineral soils that are sandy or greater than 18 inches to clay should receive 20 pounds S per acre. Sensitive to manganese deficiency when soil pH is greater than 6.2.
Mineral-organic soils	5.5				
Organic soils	5.0	Barley, Triticale: 1.4 to 1.6 lb/bu			
		Oats: 1.0 to 1.3 lb/bu	10 to 20	60 to 80	

FERTILIZER USE

Table 4-1A. Lime and Fertilizer Suggestions — Field Crops

Area of State or Soil Type	Optimum pH	Plant Nutrient Suggestions When Soil Test is Unavailable[1]			Remarks
		N factor	P$_2$O$_5$ lb/acre	K$_2$O lb/acre	
Sorghum, Grain					
Mineral soils	6.0	1.5 to 2.0 lb/bu	10 to 20	50 to 70	Apply 20% to 25% of the N before planting. Apply remainder as a topdressing. If used for silage, increase N and K$_2$O by 40 pounds per acre. Sensitive to manganese deficiency when soil pH is greater than 6.2.
Mineral-organic soils	5.5				
Organic soils	5.0				
Soybean					
Mineral soils	6.0	0	10 to 20	50 to 70	Fertilizer may be applied on preceding crop. Inoculate when planting in new land. If soil pH is low and lime is not applied, use 0.5 ounce of sodium molybdate per acre seed treatment to facilitate N fixation (not a recommended substitute for lime). Test organic soils to determine copper needs. Soybean is sensitive to manganese deficiency when soil pH is greater than 6.2.
Mineral-organic soils	5.5				
Organic soils	5.0				
Tobacco, Burley Greenhouse or Outdoor Float System - Float system: 5 to 7 days after seeding					
	5.5 to 6.0 (water)	See Remarks	See Remarks	See Remarks	75 to 100 ppm N from 20-10-20 or similar ratio fertilizer. Choose fertilizers with no more than 0.2% boron. Avoid fertilizers with 50% or more of N from urea. Test source water before using, and solutions during the season for nutrient levels, alkalinity, and conductivity.
Tobacco, Burley Greenhouse or Outdoor Float System - 4 weeks after seeding					
	5.5 to 6.0 (water)	See Remarks	See Remarks	See Remarks	75 to 100 ppm N from 20-10-20 or similar ratio fertilizer. Test solutions for nutrient concentrations.
Tobacco, Burley Field - Planting					
	6.0 (5.5 if history of Black Shank)	See Remarks	10 to 20	40 to 200	Apply 80 to 100 pounds N per acre under conditions where no manure or legume is involved. Avoid excess chloride by using only fertilizers formulated for tobacco.
Tobacco, Burley Field - Sidedressing					
	6.0 (5.5 if history of Black Shank)	See Remarks	0	0	Apply 100 to 150 pounds N per acre 2 to 3 weeks after transplanting. It may contain up to 75% ammonium-N. For most soils, total N application (planting + sidedressing) of 180 to 200 pounds per acre is adequate for optimum yield in the mountains and 250 pounds is adequate in the piedmont.
Tobacco, Flue-Cured Greenhouse - Float system: 5 to 7 days after seeding					
Coastal plain and piedmont	5.5 to 6.0 (water)	See Remarks	See Remarks	See Remarks	100 to 150 ppm N from 20-10-20 or similar ratio fertilizer. Choose fertilizer with no more than 0.2% boron. Avoid fertilizers with 50% or more of N from urea. Test source water before using, and solutions during the season for nutrient levels, alkalinity, and conductivity.
Tobacco, Flue-Cured Greenhouse - Float system: 4 weeks after seeding					
Coastal plain and piedmont	5.5 to 6.0 (water)	See Remarks	See Remarks	See Remarks	100 ppm N from 20-10-20 or similar ratio fertilizer or ammonium nitrate. Test solutions for nutrient concentrations.
Tobacco, Flue-Cured Field - Planting					
Coastal plain and piedmont	6.0	See Remarks and *Flue-Cured Tobacco Infor-mation*, AG-187	10 to 20	90 to 110	Apply 35 to 40 pounds N per acre from tobacco fertilizer containing up to 75% ammonium-N. If needed, apply P$_2$O$_5$ at or within 7 days after transplanting. Piedmont soils are more likely to require fertilizer P than those in the coastal plain.
Tobacco, Flue-Cured Field - Sidedressing					
0 to 10 in. to clay		See Remarks	0	See Remarks	Apply 20 pounds N per acre 2 to 3 weeks after transplanting. N source may contain up to 75% ammonium-N. N-P-K ratios of 1-0-0 or 1-0-1 are sufficient for most tobacco soils if 90 to 110 pounds of K$_2$O per acre were supplied by the base fertilizer at planting. Use plant tissue analysis to identify nutrient deficiencies that need to be corrected.
11 to 15 in. to clay		See Remarks	0	See Remarks	Apply 30 pounds N per acre under conditions described in remarks for 0 to 10 inches to clay. Use plant tissue analysis to identify nutrient deficiencies that need to be corrected.
Over 15 in. to clay		See Remarks and *Flue-Cured Tobacco Information*, AG-187	0	See Remarks	Apply 40 pounds N per acre under conditions described in remarks for 0 to 10 inches to clay. Use plant tissue analysis to identify nutrient deficiencies that need to be corrected.
Tobacco, Flue-Cured Field - Adjustment for leaching					
		See Remarks and *Flue-Cured Tobacco Information*, AG-187			Replace N and K that are lost by leaching as early as conditions permit. Adjustments are normally not needed on soils with clay less than 10 inches from the surface. Magnesium and sulfur leaching may also be a concern on deep sandy soils. Use plant tissue analysis as a guide for determining nutrient deficiencies that need to be corrected.

[1] Suggested rates may overestimate or underestimate actual needs. Take soil samples to assure accurate nutrient requirements.

FERTILIZER USE

Table 4-1B. Lime and Fertilizer Suggestions — Pasture and Hay Crops

Commodity	Purpose	Area of State	Optimum pH	Plant Nutrient Suggestions When Soil Test is Unavailable[1]			Remarks
				N Based on RYE	P$_2$O$_5$ lb/acre	K$_2$O lb/acre	
Alfalfa	Seeding	All	6.5	See Remarks	90	150	Apply 20 pounds N per acre before planting. Apply 3 pounds boron per acre. Use inoculant at seeding.
	Annual maintenance	All	6.5	0	45	130	Apply first topdressing in spring of second year, following seeding. Each annual topdressing should include 2 pounds boron per acre.
Ladino Clover/Grass Mixture (>30% clover)	Seeding	All	6.5	See Remarks	90	150	Apply 20 pounds N per acre before planting.
	Annual maintenance	All	6.5	0	45	150	Apply first topdressing in spring of second year, following seeding. When legumes make up 30% of stands, yield will be similar to pure grass receiving 150 to 200 pounds of N per acre.
Bluegrass/ White Clover Mixture (>30% clover)	Annual maintenance	Mountain	6.0	0	40	90	
Tall Fescue Orchardgrass, Timothy, Prairiegrass	Seeding	All	6.5	See Remarks	45	70	Apply 40 to 60 pounds N per acre at planting.
	Annual maintenance	All	6.0	40 to 50 lb/ton[2]	40	90	Use higher N rate for maximum production. Apply 1/2 N in February to March and 1/2 N in August to September; or 1/3 N in February, 1/3 N in April, and 1/3 N in August.
Bermudagrass, hybrid and improved seed cultivars	Sprigging	All	6.0	See Remarks	45	90	Apply 60 to 80 pounds N per acre before planting. If sprigged in March, apply 1/2 N in May and 1/2 N in July first year. May apply additional 60 to 80 pounds N/acre if complete soil covered before Sept. 1. If planted with a companion legume, lime to soil pH 6.5.
	Annual maintenance	All	6.0	40 to 50 lb/ton[2]	40	180	Apply N in three or four applications, but not later than Sept. 1. If planted with a companion legume, lime to soil pH 6.5.
Sudan Hybrid Sorghum — Sudan hybrids Pearl Millet Ryegrass Small Grain	Seeding	All	6.0	See Remarks	40	90	Apply 50 to 60 pounds N per acre planting. Apply N in two to four applications following each harvest except the last one.
	Topdressing	All	6.0	40 to 60 lb/ton[2]	0	0	Apply N in 2 to 4 applications following each harvest except the last one.
Red Clover, Fescue, Orchardgrass Mixture (>30% clover)	Seeding	Piedmont	6.5	See Remarks	90	150	Apply 20 pounds N per acre before planting.
	Annual maintenance	Piedmont	6.5	0	25 to 35	60 to 80	Apply in second and third year if stand is adequate.
Switchgrass Flaccidgrass Caucasian Bluestem Eastern Gammagrass	Seeding	All	5.5 to 6.0	See Remarks	20	80	Apply 40 pounds N per acre after plants are 6 to 10 inches high to minimize grassy weed competition.
	Maintenance	All	5.5 to 6.0	Switchgrass: 30 to 40 lb/ton All others: 35 to 45 lb/ton	20	80	Apply 1/2 N in May, 1/2 N in July, or 1/3 each in May, June, and July.

[1] Suggested rates may overestimate or underestimate actual needs. Take soil samples to assure accurate nutrient requirements.

[2] Rate is for tons of dry matter produced. Recognize that while hay and silage crops remove most or all of the above ground vegetation and the nutrients contained therein, grazing animals excrete 90% or more of the nutrients they consume from feed. Therefore, consider how grazing animals are managed to uniformly distribute the excreta over the pasture. To minimize N loss to the environment, N requirements for pastures should be reduced by 25% to 50% compared to what would be used for hay-silage production; yields may or may not be reduced.

FERTILIZER USE

Fertilizer Suggestions for Tree Fruit

M. L. Parker, Horticultural Science; and D. H. Hardy, NCDA&CS

Table 4-2. Fertilizer Suggestions for Tree Fruit

Purpose	Material	Amount	Precautions and Remarks
Apples			
Preplant	Lime and P fertilizer	Depends on soil test.	Prepare soil as deep as possible before planting. Take soil samples at least 12 inches deep (preferably 0 to 8 inches and 8 to 16 inches) for lime and phosphorus recommendation. Apply one-half of total, adjusted for the depth of incorporation, and plow down; apply other half and work in well.
Improve and/or maintain growth of young trees	10-10-10 or its equivalent	1 lb for each year of age until tree begins bearing. Then as recommended from leaf analysis.	Apply before rainfall or irrigate after application before buds swell in the spring.
To raise boron level of tree	Solubor	1 lb/100 gal of spray at first cover	If leaf analysis shows a deficiency, use additional cover sprays as recommended from leaf analysis to reduce "cork spot." Dry years and large fruit may enhance the incidence of "cork spot."
Growth and fruit development	Nitrogen	1.25 lb of actual N for trees producing 10 to 15 bu of apples	Annual terminal growth should be about 12 inches. Use observations of growth, crop size, and fruit condition plus the leaf analysis to determine the yearly application.
Increase foliar level	Potassium	Apply according to leaf analysis. Rate dependent upon soil analysis.	Leaf analysis is a good indicator of the need for a soil application.
Increase calcium level of tree	Gypsum (CaSO$_4$)	15 to 50 lb/tree with a 6- to 10-ft radius.	Apply only as needed by low soil or tissue calcium. One application will usually last 3 to 5 years.
	Calcium nitrate	Apply in late fall or early spring at rate to supply recommended nitrogen.	Applied as soil applications to increase calcium supply and reduce "Bitter Pit."
Foliar application	Calcium nitrate	3 lb/100 gal for sprays two weeks apart and ending 2 to 3 weeks before harvest.	Apply to reduce the incidence of "Bitter Pit." Excessive use of CaNO$_3$ may result in excessive tree vigor, which may actually worsen Bitter Pit. Both soil and foliar applications may be needed on large-fruited varieties. Leaf analysis may be beneficial.
	Calcium chloride	Same time as calcium nitrate. 2 lb/100 gal water.	Apply to reduce incidence of "Bitter Pit." DO NOT apply when temperature is 85 degrees F or above.
Peaches			
Preplant	Lime and Phosphorus	Depends on soil test.	Apply dolomitic lime necessary to raise soil pH above 6.0. Apply phosphorus to raise levels to desired range as indicated by soil test.
Tree growth, first year	N, P$_2$O$_5$, & K$_2$O	5 lb/acre of each per application.	Broadcast 0.5 pound of 10-10-10 around trees after growth starts in spring (April). Repeat every 4 to 6 weeks until August on sandy soils. On heavier soils, apply 0.5 pound of 10-10-10 one month after planting and 0.5 pound of 10-10-10 in May.
Tree growth, second year	N, P$_2$O$_5$, and K$_2$O	10 lb/acre of each per application.	Double amounts used first year. Make first application before growth starts in March and repeat in May. In sandy soils or if leaching is severe, an additional application may be made in July.
Tree growth, third year	N, P$_2$O$_5$, and K$_2$O	30 lb/acre of each	Make first application of 15 pound of each before growth starts and repeat in 6 to 8 weeks. If leaching is severe, repeat in July.
Growth and fruit development of young, bearing trees	N, P$_2$O$_5$, & K$_2$O	70 lb/acre of each (Determine by soil and foliar analysis.)	Broadcast under trees 40 pounds/acre each of N, P$_2$O$_5$, and K$_2$O (for example 400 pounds/acre 10-10-10) before growth starts. Add 30 pounds/acre each of N, P$_2$O$_5$, and K$_2$O after fruit set. If soil phosphorus test is high, omit P in second application. If leaching is severe, apply 20 to 30 pounds N/acre after harvest.
Growth and fruit development of mature trees	N, P$_2$O$_5$, and K$_2$O	70 lb/acre of each (Determine by soil and foliar analysis.)	Broadcast under trees 40 pounds/acre each of N, P$_2$O$_5$, and K$_2$O (for example, 400 pounds/acre 10-10-10) before growth starts. Add 30 pounds/acre of N, P$_2$O$_5$, and K$_2$O after fruit set. If soil test phosphorus is high, omit P in second application. If leaching is severe, apply 30 pounds N/acre after harvest.
Increase boron level of tree	Solubar		Apply boron to producing trees by spraying foliage with Solubor or equivalent product at a rate of 0.75 pounds Solubor/100 gallons of water once each year.
Soil pH maintenance	Dolomitic lime		Maintain soil pH above 6.0.

FERTILIZER USE

Fertilizer Suggestions for Small Fruit

J. Pattison, B. Cline, and G. E. Fernandez, Horticultural Science; and K. Hicks, NCDA&CS

For best results, fertilize using soil and tissue test information. If unavailable, use the general suggestions below.

Table 4-3. Fertilizer Suggestions for Small Fruit

Purpose	Material	Amount	Precautions and Remarks
Blackberries			
Preplant	Lime, P_2O_5, and K_2O	Apply based on a recent soil test report.	Blackberries can be grown on a variety of soil types. Regardless of the soil type, however, organic matter additions, pH adjustments, and incorporation of phosphorus (P) and potassium (K) should be completed before planting to optimize productivity. Take a soil test three to six months prior to planting to ensure that soil amendments are added appropriately.
Growth first year	N	20 to 50 lb/acre	Apply N in a split application: 1-2 weeks after an early spring planting and again 30 days later. Optimally, N can be portioned out through the drip irrigation system on a weekly basis.
	P_2O_5 and K_2O	30 to 60 lb/acre	Assuming pre-plant P and K were applied according to soil test recommendations, additional P and K during the first year should not be needed. If this is not the case, apply P and K in 4-inch bands around but not closer than 6 inches from stems. Optimally, portion out P and K through the drip irrigation system on a weekly basis during the growing season.
Growth second year	N	35 to 60 lb/acre (up to 65 lb/acre on sandy soil)	Apply N in single or split application. For single application, apply in March. Spread fertilizer uniformly in 4-inch bands around but not closer than 6 inches from stems. For split application, apply N between the spring flush of growth and harvest with the greater portion applied as spring growth begins. Optimally, N can be portioned out through the drip irrigation system on a weekly basis. Only apply N after harvest if tissue analysis indicates it is low. Use primocane leaf tissue analysis 10-14 days post-harvest to optimize N fertilization. Monitor growth and adjust N appropriately to achieve optimal growth rate; check excessive vegetative growth with a reduction in N rate.
	P_2O_5 and K_2O	30 to 60 lb/acre	P and K can be applied in the fall if this is most convenient. Optimally, apply P and K during the growing season as per N recommendations. Adjust P and K rates as needed according to tissue analysis.
Growth third year and mature planting	N	60 to 80 lb/acre	Apply N, P and K according to recommendations for second year. Use primocane leaf tissue analysis 10-14 days post-harvest to optimize and manage fertilization program.
	P_2O_5 and K_2O	30 to 60 lb/acre	
Blueberries			
Site modification one year to six months prior to planting	Raised beds, Sulfur, organic additions of peat moss or pine bark, P_2O_5, and K_2O	Apply based on a recent soil test report	Most NC soils require modification (raised beds, organic additions) to grow blueberries successfully, with the exception of certain high-organic, acid soils in the coastal plain. Blueberries require an acid soil with an organic matter content (HM%) above 3%, with good aeration and constant moisture. A soil pH of 5.0 or less is needed for highbush and southern highbush blueberries. Rabbiteye blueberries will tolerate slightly higher pH, up to 5.3. Sulfur at a rate of 1 pound per 100 square feet can be used to lower pH one point (for example, from 6.0 down to 5.0). Use 2 pounds per 100 square feet on heavier soils. Where needed, apply sulfur one year in advance and re-test prior to planting. Do not over-apply sulfur.
Growth first year	N, P_2O_5, and K_2O	40 to 80 lb 14-28-14/acre per application	Apply after first flush of growth and repeat every 4 to 6 weeks until mid-August. Extend application interval during dry periods until rainfall has totaled 4 inches Based on 1,360 plants per acre.
Growth second year	N, P_2O_5, and K_2O	Double first year amount at first application only	Use same schedule and amounts after first application as first year.
Growth and fruit development of bearing plants	N, P_2O_5, and K_2O	150 to 250 lb 14-28-14/acre	Apply 2/3 of this amount before bloom and 1/3 4 to 6 weeks later (early May). Apply 10 to 30 pounds additional N/acre immediately after harvest if more vigorous growth is desired. Apply 50 pounds per acre of diammonium phosphate (18-48-0) in mid-August to maintain plant vigor if P_2O_5 is low or leaching has been severe.

Grapes, Bunch

Same as muscadine schedule. Petiole analysis can show which nutrients are limiting. Collect petioles opposite the first or second flower/fruit cluster at full bloom to veraiso. Contact your county Cooperative Extension agent for further information.

Purpose	Material	Amount	Precautions and Remarks
Grapes, Muscadine			
Preplant	Lime, P_2O_5, and K_2O	Apply based on a recent soil test report.	Any pH adjustments and incorporation of phosphorus (P) and potassium (K) should be completed before planting to optimize productivity. Take a soil test three to six months prior to planting to ensure that soil amendments are added appropriately.
First year	N, P_2O_5, and K_2O	0.25 lb 10-10-10 per vine per application	Apply after growth starts (late April to early May) and repeat in June and July (but no later than mid-July as winter injury may occur). Broadcast in a circle at least 18 inches from the trunk.
Second year	N, P_2O_5, and K_2O	0.5 lb 10-10-10 per vine per application	Apply in March and again in May and early July. To minimize the potential for winter cold injury, Piedmont and foothills growers should omit the July fertilizer application. Do not put fertilizer closer than 21 inches from the trunk.
Third year	N, P_2O_5, and K_2O	0.75 lb 10-10-10 per vine per application	Apply in March, in May and again in late June. Piedmont and foothills growers should omit the late June fertilizer application. Do not put fertilizer closer than 21 inches from the trunk.
Mature vines	N, P_2O_5, and K_2O	200 lb 10-10-10/acre per application	Apply in March (near bud break), and again in late May. If more vigorous growth is desired, add an additional 20 pounds N per acre in late June (from 200 pounds 10-10-10). Omit this last application in the piedmont and foothills. In Eastern NC, an alternative fertilizer to 10-10-10 that shows promise involves the application of 6-6-18 in March and mid-to-late May at the rate of 333 pounds/acre per application (instead of 10-10-10 at 200 pounds/acre per application). A final application of calcium nitrate is applied in late June at 133 pounds /acre (provides 20 pounds N per acre). Use leaf tissue analysis to monitor nutrient uptake and fine-tune fertilization program. Tissue samples should be collected in early to mid-June. Collect a double fist full of mature leaves located opposite fruit clusters on fruiting shoots. Detach the petioles from the leaves before placing the leaf blades in a paper bag. Send samples to the Agronomic Division, NC Dept. of Agriculture and Consumer Services (see Chapter 3).
	B	1 lb Solubor/acre	For mature vineyards, a common recommendation has been to apply 1 pound of Solubor (20% boron) annually with 100 gallons of water per acre just before bloom. Boron deficiency is more likely on sandy soils with high pH. Excessive boron causes injury; do not exceed boron recommendations.

Table 4-3. Fertilizer Suggestions for Small Fruit

Purpose	Material	Amount	Precautions and Remarks
Raspberries			
Preplant	Lime, P_2O_5, and K_2O	Apply based on a recent soil test report.	Test before planting and apply P_2O_5, and K_2O and lime according to soil test.
Growth first year	N, P_2O_5, and K_2O	250 to 500 lb 10-10-10/acre	Fertilize 30 to 60 days after planting. Apply fertilizer in a band at the side of the row but not closer than 6 inches from stems. If using a drip system, the nutrients can be added via the drip system. Portion out the fertilizer at the recommended rates weekly or as needed.
Growth second year	N, P_2O_5, and K_2O	350 to 500 lb 10-10-10/acre	Apply fertilizer in a band at the side of the row but not closer than 6 inches from stems. If using a drip system, the nutrients can be added via the drip system. Portion out the fertilizer at the recommended rates weekly or as needed.
Growth third year and mature planting	N, P_2O_5, and K_2O	500 to 800 lb 10-10-10/acre	Apply fertilizer in a band at the side of the row but not closer than 6 inches from stems. If using a drip system, the nutrients can be added via the drip system. Portion out the fertilizer at the recommended rates weekly or as needed.
Strawberries, Matted-row			
Growth of new planting	N	30 to 40 lb N/acre	Apply in May and repeat in August or September on sandy soils. An additional 20 to 30 pounds N per acre may be applied in January.
	P_2O_5, K_2O, and lime	Depends on soil test	Test before planting and apply P_2O_5, K_2O, and lime based on soil test.
Growth and fruit development	N	30 to 40 lb N/acre	Apply in August or September and in sandy soils again in January.
	N, P_2O_5, and K_2O	300 to 400 lb 10-10-10	Apply after harvest. If soil test for P and K are high, 30 to 40 pounds of N may be used rather than 10-10-10.
Strawberries, Plasticulture			
Preplant (fall)	N	60 lb/acre	Broadcast and incorporate before bedding. Calcium ammonium nitrate (CAN) or a complete fertilizer (if P and K recommended) may be used.
	P_2O_5, K_2O, and lime	Refer to soil test. If not available, apply 60 lb P_2O_5 and 120 lb K_2O/acre.	Soil test 3-6 months prior to planting. Apply lime at least 3 months before planting. Broadcast and incorporate recommended nutrients before bedding.
Preharvest (spring)	N	1/2 to 1 lb/acre/day (3.5 to 7 lb/acre/week) based upon petiole nitrate test	Begin biweekly tissue testing when plants begin growing in the spring. Adjust rate or omit applications depending on tissue test interpretation. Weekly injection of fertilizer is preferred; however, biweekly applications are a common practice. N fertilizer needs to be greenhouse-grade to ensure solubility and avoid clogging the drip tape emitters [i.e., Calcium Nitrate, Potassium Nitrate, Urea Ammonium Nitrate (UAN)].
	Other nutrients	Depends on tissue test	Biweekly tissue tests will indicate need. If B is needed, apply at 1/8 pound B per acre. If tissue tests indicate the need for multiple nutrient supplements, do not assume that all fertilizer components will be compatible for a single injection. Perform a jar test with a small batch to check for mixture solubility and compatibility.

Lime and Fertilizer Suggestions for Lawns

Charles Peacock, Grady Miller, and Matt Martin, Crop and Soil Sciences

Suggested Establishment Fertilization

Collect a soil sample for NCDA&CS analysis and follow lime and fertilizer recommendations. If the soil has not been tested, incorporate 75 pounds of ground limestone, except for centipedegrass, and 15 pounds of 0-14-14 fertilizer (or equivalent) per 1,000 square feet into the soil to a depth of 4 to 8 inches before seeding. At seeding, apply 1 pound nitrogen per 1,000 square feet from a turf-grade fertilizer in which one-fourth to one-half of the nitrogen is slowly available (e.g., 12-4-8 or 16-4-8). Use half of this fertilization rate when establishing centipedegrass. For more information, see *Carolina Lawns* or the lawn maintenance calendar for your specific grass. These can be found at http://www.turffiles.ncsu.edu. You can also request copies at your county North Carolina Cooperative Extension center.

Table 4-4A. Suggested Maintenance Fertilization for Coastal Plain[1]

Lawn Grass Type	Monthly Nitrogen Application Rate per 1,000 Square Feet[2]												Total lb N/ 1,000 sq ft/yr
	Jan	Feb	March	April	May	June	July	Aug[3]	Sept	Oct	Nov	Dec	
Bermudagrass													
Basic				1		1		1					3
High				1	1	1	1	1	1				6
Centipedegrass[3]													
Basic					1								1
High					1			1					2
Fescue, Tall													
Basic		0.5							1		0.5		2
High		1	0.5						1	1	0.5		4
St. Augustinegrass													
Basic					1			1					2
High				0.5	1	0.5	1	0.5	0.5				4
Zoysiagrass (Emerald and Meyer cultivars)													
Basic					1			1					2
High				1		1		1					3
Zoysiagrass (other cultivars)													
Basic					1			1					2
High				1	0.5	1	0.5	1					4

[1] All rates are per 1,000 square feet. Multiply by 43.5 to convert to an acre basis. Follow table suggestions in the absence of soil test recommendations to the contrary. With the exception of centipedegrass, use a complete (N-P-K) turf-grade fertilizer in which 1/4 to 1/2 of the nitrogen is slowly available and that has a 3-1-2 or 4-1-2 analysis (e.g., 12-4-8, 16-4-8).

[2] In the absence of soil test recommendations, apply about 1 pound potassium per 1,000 square feet using 1.6 pounds of muriate of potash (0-0-60), 5 pounds of sul-po-mag (0-0-22), or 2 pounds of potassium sulfate (0-0-50) to bermudagrass, centipedegrass, and St. Augustinegrass.

[3] Centipedegrass should be fertilized very lightly after establishment. Fertilize established centipedegrass using a low-phosphorus, high-potassium fertilizer with an analysis approaching 1-1-2 or 1-1-3. Fertilizers absent of phosphorus are preferred if soils supporting centipedegrass exhibit moderate to high levels of phosphorus.

FERTILIZER USE

Table 4-4B. Suggested Maintenance Fertilization for Central Piedmont[1]

Lawn Grass Type	Monthly Nitrogen Application Rate per 1,000 Square Feet[2]												Total lb N/ 1,000 sq ft/yr
	Jan	Feb	March	April	May	June	July	Aug[3]	Sept	Oct	Nov	Dec	
Bermudagrass													
Basic					1		1		1				3
High				1	1	1	1	1	1				6
Centipedegrass[3]													
Basic					1								1
High					1			1					2
Fescue, Tall													
Basic		0.5							1		0.5		2
High		1	0.5						1	1	0.5		4
Kentucky Bluegrass													
Basic		0.5							1		0.5		2
High		1	0.5						1	1	0.5		4
Kentucky Bluegrass-Fine Fescue Mix													
Basic		0.5							1		0.5		2
High		1	0.5						1	1	0.5		4
Kentucky Bluegrass-Tall Fescue Mix													
Basic		0.5							1		0.5		2
High		1	0.5						1	1	0.5		4
Kentucky Bluegrass-Tall Fescue-Fine Fescue Mix													
Basic		0.5							1		0.5		2
High		1	0.5						1	1	0.5		4
Kentucky Bluegrass-Perennial Ryegrass Mix													
Basic		1							1		1		3
High		1	0.5						1	1	0.5		4
St. Augustinegrass													
Basic					1				1				2
High					1		1		1				3
Zoysiagrass (Emerald and Meyer cultivars)													
Basic					1								1
High				1			1						2
Zoysiagrass (other cultivars)													
Basic					1		1						2
High				1		1		1					3

[1] All rates are per 1,000 square feet. Multiply by 43.5 to convert to an acre basis. Follow table suggestions in the absence of soil test recommendations to the contrary. With the exception of centipedegrass, use a complete (N-P-K) turf-grade fertilizer in which 1/4 to 1/2 of the nitrogen is slowly available and that has a 3-1-2 or 4-1-2 analysis (e.g., 12-4-8, 16-4-8).

[2] In the absence of soil test recommendations, apply about 1 pound potassium per 1,000 square feet using 1.6 pounds of muriate of potash (0-0-60), 5 pounds of sul-po-mag (0-0-22), or 2 pounds of potassium sulfate (0-0-50) to bermudagrass, centipedegrass, St. Augustinegrass, and zoysiagrass.

[3] Centipedegrass should be fertilized very lightly after establishment. Fertilize established centipedegrass using a low-phosphorus, high-potassium fertilizer with an analysis approaching 1-1-2 or 1-1-3. Fertilizers absent of phosphorus are preferred if soils supporting centipedegrass exhibit moderate to high levels of phosphorus.

FERTILIZER USE

Table 4-4C. Suggested Maintenance Fertilization for the Mountains[1]

Lawn Grass Type	Monthly Nitrogen Application Rate per 1,000 Square Feet[2]												Total lb N/ 1,000 sq ft/yr
	Jan	Feb	March	April	May	June	July	Aug	Sept	Oct	Nov	Dec	
Bermudagrass													
Basic					1		1						2
High					1	1	1	1					4
Fescue, Tall													
Basic			0.5					1		0.5			2
High			1					1		1			3
Kentucky Bluegrass													
Basic			1					1					2
High			1					1					3
Kentucky Bluegrass-Fine Fescue Mix													
Basic			1					1					2
High			1					1		1			3
Kentucky Bluegrass-Tall Fescue Mix													
Basic			1					1					2
High			1					1		1			3
Kentucky Bluegrass-Tall Fescue- Fine Fescue Mix													
Basic			1					1					2
High			1					1		1			3
Kentucky Bluegrass- Perennial Ryegrass Mix													
Basic			1					1		0.5			2.5
High			1					1	1	0.5			3.5
Zoysiagrass (other cultivars)													
Basic					0.5			0.5					1
High					1		1						2

[1] All rates are per 1,000 square feet. Multiply by 43.5 to convert to an acre basis. Follow table suggestions in the absence of soil test recommendations to the contrary. With the exception of centipedegrass, use a complete (N-P-K) turf-grade fertilizer in which 1/4 to 1/2 of the nitrogen is slowly available and that has a 3-1-2 or 4-1-2 analysis (e.g., 12-4-8, 16-4-8).

[2] In the absence of soil test recommendations, apply about 1 pound potassium per 1,000 square feet using 1.6 pounds of muriate of potash (0-0-60), 5 pounds of sul-po-mag (0-0-22), or 2 pounds of potassium sulfate (0-0-50) to bermudagrass, St. Augustinegrass, and zoysiagrass.

FERTILIZER USE

Fertilizer Suggestions for Ornamental Plants in Landscapes

B. Fair, Horticultural Science

The fertilizer suggestions given in the table are intended as a general guide. They are not a replacement for soil analyses from samples taken on a regular basis to assure maintenance of good nutrient availability and soil pH. It is always a good idea to collect samples for a foliar analysis as well to determine if the plant itself is functioning properly.

Table 4-5. Fertilizer Suggestions for Ornamental Plants

Kind of Plant	Ratio	Type	Amount to Use	When to apply	Remarks
Trees	3-1-1 or 3-1-2	Slow Release	2-4 lbs. N/1,000 sq.ft of root zone (at least 70% of dripzone) Do not exceed 6 lbs.N/1,000 sq. ft. per year	Apply slow-release split application between spring budbreak and leaf color in fall. Fall application does not necessarily predispose a plant to winter injury or promote additional growth.	Refer to ANSI A300 (Part 2) - 2011 Soil Management a. Modification, b. Fertilization, c. Drainage. Before applying fertilizer to any woody plant, you should outline your goals for fertilizing the plant(s). Acceptable reasons for fertilizing are to increase growth, flowering/fruiting, increase vigor/vitality, balance root and shoot growth, and to address a visible deficiency. Fertilizing with too much nitrogen can lead to increased herbivory or susceptibility to insects and some disease.
		Quick Release	1-2 lbs.N/1,000 sq ft root zone	With sandy soils and areas with heavy rainfall or constant irrigation, multiple applications will be necessary. Split total dose evenly over these applications.	
Palms	3-1-3	15-5-15 including micronutrients, calcium and mangnesium	1-1.5 lbs. N/1,000 sq ft in root zone (maybe 2-3 times out from the dripline on some palm species)	Do not fertilize during drought periods. Do fertilize when soil is moist.	
Shrubs	3-1-1 or 3-1-2	Slow Release	2-4 lbs. N/1,000 sq.ft of root zone (at least 70% of dripzone) Do not exceed 6 lbs.N/1,000 sq. ft. per year	Only use quick release when a sufficient response cannot be accomplished with slow-release products.	Best application method is sub-surface liquid injection. Use broadcast dry granules only where no turf is present and only for nitrogen application. Be sure to irrigate or apply prior to a rain event. Phosphorous will bind to soil before plant roots are able to obtain sufficient quantities. Other methods include sub-surface dry drill-hole application, surface liquid application, foliar application, tree injection or implants. The last few methods should be used only when other options are impractical or under special circumstances. Always use products in accordance with manufacturer's recommendations.
		Quick Release	1-2 lbs.N/1,000 sq ft root zone		
Flowers	1-2-2 or 1-2-1 or 1-1-1	Slow release Quick release	4-6 lbs. N/1,000 sq ft	Incorporate half slow release granular forms before planting; apply second half 6 weeks after planting. If using liquid quick release form, apply every 1-4 weeks throughout growing season, using 1 qt. per sq ft	Refer to Horticultural Information Leaflet 551, *Bed Preparation and Fertilizer Recommendations for Bedding Plants in the Landscape.*

FERTILIZER USE

Fertilizer Suggestions for Nursery Crops

A. V. LeBude, Horticultural Science, and K. Hicks, NCDA&CS

Table 4-6. Fertilizer Suggestions for Nursery Crops

Kind of Production	Amount to Use[1]	Remarks
Deciduous Tree Seedling Beds	Year 1: 50 lb nitrogen (N) per acre or 18 oz N per 1,000 sq ft of bed. Year 2: Beds receive 100lb N per acre.	Year 1: Surface apply after first true leaves appear. Base applications of other elements upon a soil test.
Field Production	Year 1: 50 lb of nitrogen per acre incorporated as preplant. Year 2: 0.5 to 1 oz N per plant. Year 3: 1 to 2 oz N per plant. Do not exceed 100 to 200 lb N per acre. For liquid fertilizer applications through drip irrigation, reduce N application by 1/2 rate of annual field grade fertilizer. Apply equal rates of N during several irrigation events.	Year 1: Incorporate nutrients into the soil before planting. For field grade fertilizers, apply 0.67 of total amount before bud swell, 0.33 in June. Base applications of other elements upon a soil test. For more information, see http://www.ces.ncsu.edu/depts/hort/nursery/cultural/cultural_docs/field-bmps/building_nursery_soils.pdf
Container Production	All essential elements must be provided when soilless mixes are used to produce nursery crops in containers. Controlled release fertilizers offer a consistent and reliable source of nutrients through the growing season. Formulations of nutrients and their release over time differ by fertilizer company and the optimal product chosen depends upon the species being grown as well as the location of the nursery in the state. Irrigation water quality can contribute significant amounts of some nutrients, such as calcium, magnesium, and iron. Test water quality of irrigation supplies at least once a year (e.g., mid-summer) to determine if nutrient adjustments are required. Diagnosis of whole production system nutritional problems requires analysis of foliar nutrient content (sample uppermost fully expanded leaves), collection of approximately 8 ounces of leachate solution (collect from containers approximately 30 minutes after irrigation), and an irrigation water sample (collect from the irrigation head). Current research suggests that dolomitic limestone rates depend upon the calcium and magnesium content in irrigation water. As a result, limestone may not be required if provided by irrigation, however if dolomitic limestone is needed add only 2 to 5 pounds per cubic yard of pine bark and sand potting mixes. Minor element supplements included in either NPK controlled release fertilizer products or from separate minor element packages are necessary in all pine bark mixes and should be incorporated if possible. Potting mixes containing composts, however, generally do not require dolomitic limestone or minor element supplements. Leachates, irrigation water, and foliar samples can be analyzed for $5 each by the Agronomic Division, NCDA&CS, 4300 Reedy Creek Road, Raleigh, NC 27607-6465.	

[1] Rates may change with irrigation and soil type.

Lime and Fertilizer Suggestions for Vegetable Crops

J. R. Schultheis and J. M. Davis, Horticultural Science;
C. R. Crozier and D. L. Osmond, Crop and Soil Sciences

Important Notes:

1. Consult the Southeastern U.S. Vegetable Crop Handbook: http://www.thegrower.com/south-east-vegetable-guide/ for numerous crop management suggestions.

2. Since optimum fertilizer management practices vary widely due to the specific vegetable production system, soil type, weather, and previous management, plant tissue analysis may be needed to fine-tune decisions. Plant tissue samples can be analyzed for essential plant nutrients at a nominal fee by the N.C. Department of Agriculture and Consumer Services, Agronomic Services Lab, 1040 Mail Service Center, Raleigh, NC 27699-1040; (919) 733-2655. When using a private carrier to deliver the samples or dropping them off in person, use the physical address of 4300 Reedy Creek Road, Raleigh, NC 27607-6465. Consult the new plant tissue analysis guide for instructions for each specific crop http://www.ncagr.gov/agronomi/pdffiles/plantguide.pdf.

3. For most vegetables grown on light-textured soils, apply the total recommended P_2O_5 and K_2O together with 25 to 50 percent of the recommended nitrogen before planting. The remaining nitrogen can be sidedressed with a fertilizer containing nitrogen only. Sidedressing or topdressing potash (K_2O) is recommended only on extremely light sandy soils with very low cation exchange capacities.

4. It may be desirable to build up the phosphorus and potassium levels in infertile loam and silt loam soils more rapidly than provided by these recommendations. In such instances, add an additional 40 to 50 pounds of P_2O_5 and K_2O, respectively, to the recommendations listed in the table for soils testing low in phosphorus and potassium. Apply the additional amounts as a broadcast and plow down or broadcast and disk-in application.

5. In the absence of soil tests, use recommendations listed under medium phosphorus and medium potassium levels on light-textured soils that have been in intensive vegetable production.

6. For Piedmont growers producing vegetables on clay loam soils: Reduce the recommended nitrogen and potassium rates by 20 percent and increase the phosphorus rate by 25 percent of the rates indicated in this table.

Table 4-7. Lime and Fertilizer Suggestions for Vegetable Crops

Crop	Desirable pH	Nitrogen (N) lb/acre	Soil Phosphorus Level Low	Med	High	Very High	Soil Potassium Level Low	Med	High	Very High	Total Amount of Nutrient Recommended and Suggested Methods of Application
			P_2O_5 lb/acre				K_2O lb/acre				
Asparagus	6.5										
Growing crowns		100	200	100	50	0	200	150	50	0	Total recommended.
		50	200	100	50	0	100	75	50	0	Broadcast and disk in.
		50	0	0	0	0	100	75	0	0	Sidedress after cutting.
New Planting Crowns and direct seeding		50	200	100	50	0	200	100	100	0	Total recommended.
		0	200	100	50	0	100	75	50	0	Broadcast and plow down.
		50	0	0	0	0	100	25	50	0	Sidedress at first cultivation.
Cutting Bed or Nonhybrids		100	150	100	50	0	200	150	100	0	Total recommended.
		50	150	100	50	0	100	150	100	0	Broadcast and disk in.
		50	0	0	0	0	100	75	50	0	Sidedress at first cultivation.
New hybrids		100	200	150	100	0	300	225	150	0	Total recommended.
		50	200	150	100	0	150	100	75	0	Broadcast before cutting season.
		50	0	0	0	0	100	125	75	0	Sidedress after cutting.
		Apply 2 pounds boron (B) per acre every 3 years on most soils.									
Bean, Lima Single crop	6 to 6.5	70 to 110	120	80	40	20	160	120	80	20	Total recommended.
		25 to 50	80	40	20	0	120	80	60	0	Broadcast and disk-in.
		20	40	40	20	20	40	40	20	20	Band-place with planter.
		25 to 40	0	0	0	0	0	0	0	0	Sidedress 3 to 5 weeks after emergence.
Bean, Snap	6 to 6.5	40 to 80	80	60	40	20	80	60	40	20	Total recommended.
		20 to 40	40	40	0	0	40	40	0	0	Broadcast and disk-in.
		20 to 40	40	20	40	20	40	20	40	20	Band-place with planter.
Beet	6 to 6.5	75 to 100	150	100	50	0	150	100	50	0	Total recommended.
		50	150	100	50	0	150	100	50	0	Broadcast and disk-in.
		25 to 50	0	0	0	0	0	0	0	0	Sidedress 4 to 6 weeks after planting.
Broccoli	6 to 6.5	125 to 175	200	100	50	0	200	100	50	0	Total recommended.
		50 to 100	150	100	50	0	150	100	50	0	Broadcast and disk-in.
		50	50	0	0	0	50	0	0	0	Sidedress 2 to 3 weeks after planting.
		25	0	0	0	0	0	0	0	0	Sidedress every 2 to 3 weeks after first sidedressing.
		Apply 2 pounds boron (B) per acre with broadcast fertilizer.									

FERTILIZER USE

Table 4-7. Lime and Fertilizer Suggestions for Vegetable Crops

Crop	Desirable pH	Nitrogen (N) lb/acre	Recommended Nutrients Based on Soil Tests								Total Amount of Nutrient Recommended and Suggested Methods of Application
			Soil Phosphorus Level				Soil Potassium Level				
			Low	Med	High	Very High	Low	Med	High	Very High	
			P$_2$O$_5$ lb/acre				K$_2$O lb/acre				
Brussel Sprout, Cabbage, and Cauliflower	6 to 6.5	100 to 175	200	100	50	0	200	100	50	0	Total recommended.
		50 to 75	200	100	50	0	200	100	50	0	Broadcast and disk-in.
		25 to 50	0	0	0	0	0	0	0	0	Sidedress 2 to 3 weeks after planting.
		25 to 50	0	0	0	0	0	0	0	0	Sidedress if needed, according to weather.
		Apply 2 to 3 pounds boron (B) per acre and molybdenum (mo) per acre as 0.5 pound sodium molybdate per acre with broadcast fertilizer.									
Carrot	6 to 6.5	50 to 80	150	100	50	0	150	100	50	0	Total recommended.
		50	150	100	50	0	150	100	50	0	Broadcast and disk-in.
		25 to 30	0	0	0	0	0	0	0	0	Sidedress if needed.
		Apply 1 to 2 pounds boron (B) per acre with broadcast fertilizer.									
Celery	6 to 6.5	75 to 100	250	150	100	0	250	150	100	0	Total recommended.
		50	250	150	100	0	250	150	100	0	Broadcast and disk-in or drill deep.
		25 to 50	0	0	0	0	0	0	0	0	Sidedress 2 to 3 weeks after planting.
		Apply 2 to 3 pounds boron (B) per acre with broadcast fertilizer.									
Corn, Sweet	6 to 6.5	110 to 155	160	120	80	20	160	120	80	20	Total recommended.
		40 to 60	120	100	60	0	120	100	60	0	Broadcast before planting.
		20	40	20	20	20	40	20	20	20	Band-place with planter.
		50 to 75	0	0	0	0	0	0	0	0	Sidedress when corn is 12 to 18 in. tall.
		Apply 1 to 2 pounds boron (B) per acre with broadcast fertilizer. NOTE: On very light sandy soils, sidedress 40 pounds N per acre when corn is 6 in. tall and another 40 pounds N per acre when corn is 12 to 18 in. tall.									
Cucumber	6 to 6.5	80 to 160	150	100	50	25	200	150	100	25	Total recommended.
		40 to 100	125	75	25	0	175	125	75	0	Broadcast and disk-in.
		20 to 30	25	25	25	25	25	25	25	25	Band-place with planter 7 to 14 days after planting.
		20 to 30	0	0	0	0	0	0	0	0	Sidedress when vines begin to run, or apply in irrigation water.
		Drip fertilization: See "cucumber" in specific recommendations in the current Southeastern Vegetable Crop Handbook.									
Eggplant Bareground	6 to 6.5	100 to 200	250	150	100	0	250	150	100	0	Total recommended.
		50 to 100	250	150	100	0	250	150	100	0	Broadcast and disk-in.
		25 to 50	0	0	0	0	0	0	0	0	Sidedress 3 to 4 weeks after planting.
		25 to 50	0	0	0	0	0	0	0	0	Sidedress 6 to 8 weeks after planting.
		Apply 1 to 2 pounds boron (B) per acre with broadcast fertilizer..									
Plasticulture		145	250	150	100	0	240	170	100	0	Total recommended.
		50	250	150	100	0	100	100	100	0	Broadcast and disk in.
		95	0	0	0	0	140	70	0	0	Fertigate.
		Apply 1 to 2 pounds boron (B) per acre with broadcast fertilizer. Drip fertilization: See "eggplant" in specific recommendations in the current Southeastern Vegetable Crop Handbook.									
Endive, Escarole, Leaf Lettuce	6 to 6.5	75 to 125	200	150	100	0	200	150	100	0	Total recommended.
		50 to 75	200	150	100	0	200	150	100	0	Broadcast and disk-in.
		25 to 50	0	0	0	0	0	0	0	0	Sidedress 3 to 5 weeks after planting.
Iceberg Lettuce	6 to 6.5	85 to 175	200	150	100	0	200	150	100	0	Total recommended.
		60 to 80	200	150	100	0	200	150	100	0	Broadcast and disk-in.
		25 to 30	0	0	0	0	0	0	0	0	Sidedress 3 times beginning 2 weeks after planting.
Leafy Greens, Collard, Kale, Mustard	6 to 6.5	75 to 80	150	100	50	0	150	100	50	0	Total recommended.
		50	150	100	50	0	150	100	50	0	Broadcast and disk-in.
		25 to 30	0	0	0	0	0	0	0	0	Sidedress, if needed.
		Apply 1 to 2 pounds boron (B) per acre with broadcast fertilizer.									
Leek	6 to 6.5	75 to 125	200	150	100	0	200	150	100	0	Total recommended.
		50 to 75	200	150	100	0	200	150	100	0	Broadcast and disk-in.
		25 to 50	0	0	0	0	0	0	0	0	Sidedress 3 to 4 weeks after planting if needed.
		Apply 1 to 2 pounds boron (B) per acre with broadcast fertilizer.									

Table 4-7. Lime and Fertilizer Suggestions for Vegetable Crops

Crop	Desirable pH	Nitrogen (N) lb/acre	Recommended Nutrients Based on Soil Tests								Total Amount of Nutrient Recommended and Suggested Methods of Application
			Soil Phosphorus Level				Soil Potassium Level				
			Low	Med	High	Very High	Low	Med	High	Very High	
			P₂O₅ lb/acre				K₂O lb/acre				
Cantaloupes and Mixed Melons Bareground	6 to 6.5	75 to 115	150	100	50	25	200	150	100	25	Total recommended.
		25 to 50	125	75	25	0	175	125	75	0	Broadcast and disk-in.
		25	25	25	25	25	25	25	25	25	Band-place with planter.
		25 to 40	0	0	0	0	0	0	0	0	Sidedress when vines start to run.
		Apply 1 to 2 pounds boron (B) per acre with broadcast fertilizer.									
Cantaloupes and Mixed Melons Plasticulture	6 to 6.5	75 to 150	150	100	50	25	200	150	100	25	Total recommended.
		25	150	100	50	25	100	75	50	25	Broadcast and disk in.
		50 to 100	0	0	0	0	100	75	50	0	Fertigate.
		Apply 1 to 2 pounds boron (B) per acre with broadcast fertilizer. Drip fertilization: See "muskmelon" in specific recommendations in the current Southeastern Vegetable Crop Handbook.									
Okra	6 to 6.5	100 to 200	250	150	100	0	250	150	100	0	Total recommended.
		50 to 100	250	150	100	0	250	150	100	0	Broadcast and disk-in.
		25 to 50	0	0	0	0	0	0	0	0	Sidedress 3 to 4 weeks after planting.
		25 to 50	0	0	0	0	0	0	0	0	Sidedress 6 to 8 weeks after planting.
		Apply 1 to 2 pounds boron (B) per acre with broadcast fertilizer. NOTE: Where plastic mulches are being used, broadcast 50 to 100 pounds nitrogen (N) per acre with recommended P₂O₅ and K₂O and disk incorporate prior to laying mulch. Drip fertilization: See "okra" in specific recommendations in the current Southeastern Vegetable Crop Handbook.									
Onion, Bulb	6 to 6.5	75 to 125	200	100	50	0	200	100	50	0	Total recommended.
		50 to 75	200	100	50	0	200	100	50	0	Broadcast and disk-in.
		25 to 50	0	0	0	0	0	0	0	0	Sidedress 4 to 5 weeks after planting.
Onion, Green		150 to 175	200	100	50	0	200	100	50	0	Total recommended.
		50 to 75	200	100	50	0	200	100	50	0	Broadcast and disk-in.
		50	0	0	0	0	0	0	0	0	Sidedress 4 to 5 weeks after planting.
		50	0	0	0	0	0	0	0	0	Sidedress 3 to 4 weeks before harvest.
		Apply 1 to 2 pounds boron (B) and 20 pounds sulfur (S) per acre with broadcast fertilizer.									
Parsley	6 to 6.5	100 to 175	200	150	100	0	200	150	100	0	Total recommended.
		50 to 75	200	150	100	0	200	150	100	0	Broadcast and disk-in.
		25 to 50	0	0	0	0	0	0	0	0	Sidedress after first cutting.
		25 to 50	0	0	0	0	0	0	0	0	Sidedress after each additional cutting.
Parsnip	6 to 6.5	50 to 100	150	100	50	0	150	100	50	0	Total recommended.
		25 to 50	150	100	50	0	150	100	50	0	Broadcast and disk-in.
		25 to 50	0	0	0	0	0	0	0	0	Sidedress 4 to 5 weeks after planting.
		Apply 1 to 2 pounds boron (B) per acre with broadcast fertilizer.									
Pea, English Spring plowed	5.8 to 6.5	40 to 60	120	80	40	0	120	80	40	0	Total recommended. Broadcast and disk-in before seeding.
Pea, Southern	5.8 to 6.5	16	96	48	0	0	96	48	0	0.	Broadcast and disk-in.
Pepper Bareground	6 to 6.5	100 to 130	200	150	100	0	200	150	100	0	Total recommended.
		50	200	150	100	0	200	150	100	0	Broadcast and disk-in.
		25 to 50	0	0	0	0	0	0	0	0	Sidedress after first fruit set.
		25 to 30	0	0	0	0	0	0	0	0	Sidedress later in season if needed.
Pepper Plasticulture		100 to 185	200	150	100	0	365	300	235	0	Total recommended
		50	200	150	100	0	100	100	100	0	Broadcast and disk in.
		50 to 135	0	0	0	0	265	200	135	0	Fertigage
		Drip fertilization: See "pepper" in specific commodity recommendations in Plasticulture for Commercial Vegetables (AG-489).									
Potato, Irish Loams and silt loams	5.8 to 6.2	100 to 150	110	90	70	50	200	150	50	50	Total recommended.
		85 to 135	60	40	20	0	200	150	50	50	Broadcast and disk in.
		15	50	50	50	50	0	0	0	0	Band-place with planter at planting.
Sandy loams and loamy sands		150	200	150	100	50	300	200	100	50	Total recommended.
		50	200	150	100	50	300	200	100	50	Broadcast and disk in.
		100	0	0	0	0	0	0	0	0	Sidedress 4 to 5 weeks after planting.
Pumpkin and Squash (Winter) Bareground	6 to 6.5	80 to 90	150	100	50	0	200	150	100	0	Total recommended.
		40 to 50	150	100	50	0	200	150	100	0	Broadcast and disk in.
		40 to 45	0	0	0	0	0	0	0	0	Sidedress when vines begin to run.

FERTILIZER USE

Table 4-7. Lime and Fertilizer Suggestions for Vegetable Crops

Crop	Desirable pH	Nitrogen (N) lb/acre	Soil Phosphorus Level P_2O_5 lb/acre				Soil Potassium Level K_2O lb/acre				Total Amount of Nutrient Recommended and Suggested Methods of Application
			Low	Med	High	Very High	Low	Med	High	Very High	
Pumpkin and Squash (Winter) Plasticulture		80 to150	150	100	50	0	200	150	100	0	Total recommended.
		25 to50	150	100	50	0	100	75	50	0	Disk in row
		55 to 100	0	0	0	0	100	75	50	0	Sidedress when vines begin to run.
Radish	6 to 6.5	50	150	100	50	0	150	100	50	0	Total recommended. Broadcast and disk-in or drill deep.
		Apply 1 to 2 pounds boron (B) per acre with broadcast fertilizer.									
Rutabaga and Turnip	6 to 6.5	50 to 75	150	100	50	0	150	100	50	0	Total recommended.
		25 to 50	150	100	50	0	150	100	50	0	Broadcast and disk in.
		25 to 50	0	0	0	0	0	0	0	0	Sidedress when plants are 4 to 6 in. tall.
		Apply 1 to 2 pounds boron (B) per acre with broadcast fertilizer.									
Spinach Fall	6 to 6.5	75 to 125	200	150	100	0	200	150	100	0	Total recommended.
		50 to 75	200	150	100	0	200	150	100	0	Broadcast and disk in.
		25 to 50	0	0	0	0	0	0	0	0	Sidedress or topdress.
Overwinter		80 to 120	0	0	0	0	0	0	0	0	Total recommended for spring application on overwintered crop.
		50 to 80	0	0	0	0	0	0	0	0	Apply in late February.
		30 to 40	0	0	0	0	0	0	0	0	Apply in late March.
Squash, Summer	6 to 6.5	100 to 130	150	100	50	0	200	150	100	0	Total recommended.
		25 to 50	150	100	50	0	150	100	50	0	Broadcast and disk in.
		50	0	0	0	0	0	0	0	0	Sidedress when vines start to run.
		25 to 30	0	0	0	0	0	0	0	0	Apply through irrigation system.
		Apply 1 to 2 pounds boron (B) per acre with broadcast fertilizer. Drip/trickle fertilization: See "summer squash" in specific recommendations in the current Southeastern Vegetable Crop Handbook.									
Sweet Potato	5.8 to 6.2	50 to 80	200	100	50	0	300	200	150	0	Total recommended.
		0	150	60	30	0	150	50	30	0	Broadcast and disk in.
		50 to 80	0	0	0	0	150	150	120	120	Sidedress 21 to 28 days after planting.
		Add 0.5 pounds of actual boron (B) per acre 40 to 80 days after planting.									
Tomato Bareground for sandy loams and loamy sands	6 to 6.5	80 to 90	200	150	100	0	300	200	100	0	Total recommended.
		40 to 45	200	150	100	0	300	200	100	0	Broadcast and disk in.
		40 to 45	0	0	0	0	0	0	0	0	Sidedress when first fruits are set and as needed.
TOMATO Bareground for loams and clay	6 to 6.5	75 to 80	200	150	100	0	250	150	100	0	Total recommended.
		50	200	150	100	0	250	150	100	0	Broadcast and plow down.
		25 to 30	0	0	0	0	0	0	0	0	Sidedress when first fruits are set and as needed.
		Apply 1 to 2 pounds boron (B) per acre with broadcast fertilizer.									
Tomato Plasticulture		130 to 210	200	150	100	0	420	325	275	0	Total recommended.
		50	200	150	100	0	295	220	125	0	Broadcast and disk in.
		80 to 160	0	0	0	0	295	220	125	0	Fertigate
		Apply 1 to 2 pounds boron (B) per acre with broadcast fertilizer. Drop fertilization: See "tomato" in specific recommendations in the current Southeastern Vegetable Crop Handbook.									
Watermelon Nonirrigated	6 to 6.5	75 to 90	150	100	50	0	200	150	100	0	Total recommended.
		50	150	100	50	0	200	150	100	0	Broadcast and disk in.
		25 to 40	0	0	0	0	0	0	0	0	Topdress when vines start to run.
Irrigated		100 to 150	150	100	50	0	200	150	100	0	Total recommended.
		50	150	100	50	0	150	150	100	0	Broadcast and disk in.
		25 to 50	0	0	0	0	0	0	0	0	Topdress when vines start to run.
		25 to 50	0	0	0	0	0	0	0	0	Topdress at first fruit set.
Plasticulture		125 to 150	150	100	50	0	200	150	100	0	Total recommended.
		25 to 50	150	100	50	0	100	75	50	0	Disk in row.
		100	0	0	0	0	100	75	50	0	Fertigation
		NOTE: Excessive rates of N may increase hollow heart in seedless watermelons. Drip fertilization: See "watermelon" in specific recommendations in the current Southeastern Vegetable Crop Handbook.									

FERTILIZER USE

Fertilizer Rules and Regulations

D. A. Crouse, C. R. Crozier, and D. L. Osmond, Crop and Soil Sciences; D. Turner, NCDA&CS

Fertilizer, lime and landplaster are regulated by Plant Industry- NCDA&CS and rules and regulations in their entirety are found at http://www.ncagr.gov/plantindustry/pubs.htm.

Chlorine Guarantees for Tobacco Fertilizer

The maximum chlorine (Cl) guarantees permitted for tobacco plantbed fertilizer shall be:

1. For fertilizers with a nitrogen (N) guarantee up to and including 6 percent, 0.5 percent chlorine (Cl).

2. For fertilizers with a nitrogen (N) guarantee above 6 percent, 1 percent chlorine (Cl).

The maximum chlorine (Cl) guarantees permitted on field crop tobacco fertilizer shall be:

1. For fertilizer with a nitrogen (N) guarantee up to and including 4 percent, a maximum chlorine (Cl) guarantee of 2 percent.

2. For fertilizer with a nitrogen (N) guarantee greater than 4 percent, a maximum percent chlorine (Cl) guarantee not more than one-half of the respective total nitrogen (N) guarantee.

The maximum chlorine (Cl) permitted in tobacco top-dressers shall be 2 percent.

Size Standards for Agricultural Liming Material

Agricultural liming material shall conform to the following minimum screening standards:

1. Ninety percent must pass through U.S. Standard 20-mesh screen, with a tolerance of 5 percent.

2. For dolomitic limestone, 35 percent must pass through U.S. Standard 100-mesh screen; for calcitic limestone, 25 percent must pass through U.S. Standard 100-mesh screen, with a tolerance of 5 percent.

Additional Criteria for Agricultural Liming Material

1. A product must contain at least 6 percent magnesium (Mg) from magnesium carbonate to be classified as a dolomitic limestone.

2. There is no minimum calcium carbonate equivalent (CCE) requirement for limestone sold in North Carolina. However, the product must be labeled to show the amount necessary to equal that required from a liming material having a 90 percent CCE. Lime recommendations in North Carolina are based on 90 percent CCE. For example, a product having a CCE of 50 percent would be labeled "3,600 pounds of this material equals 1 ton of standard agricultural liming material."

3. Pelleted lime must slake down when it comes in contact with moisture, thereby meeting the size standards for agricultural liming materials given above.

Nutrient Content of Fertilizer Materials

C. R. Crozier and D. L. Osmond, Crop and Soil Sciences; R. Sherman, Biological and Agricultural Engineering; and A. V. LeBude, Horticultural Science

Although fertilizer materials may contain two or more nutrients, usually only one of the nutrients is commonly associated with the material. The content of that nutrient is generally well-known, but the contents of other associated nutrients are less well-known. Information on the contents of lesser-known nutrients in fertilizer materials exists but is often not readily available. Information presented below gives the nutrients and their contents that are normally found in fertilizer materials. In the case of micronutrient fertilizers, only the principle micronutrient is considered relevant. This information should be considered only as a guide, since actual nutrient contents may vary slightly, depending on the source of the fertilizer material. Additional information about the nutrient content of fertilizer and organic materials can be found in the N.C. Cooperative Extension Service publications AG-439-18, AG-439-19, and AG-467 or on the Internet at http://www.bae.ncsu.edu/bae/programs/extension/publicat/wqwm/ag473_14.html.

Table 4-8. Composition of Selected Micronutrient Fertilizers

Element	Fertilizer	Chemical Formula	Elemental Percentage (%)
Mn	Manganese sulfate	$MnSO_4 \cdot 3H_2O$	27
	Manganous oxide	MnO	41 to 68
	Manganese chloride	$MnCl_2$	7
	Manganese oxide	MnO_2	62 to 70
Mo	Sodium molybdate	$Na_2MoO_4 \cdot H_2O$	38
Cu	Copper sulfate	$CuSO_4 \cdot 5H_2O$	25
Zn	Zinc sulfate	$ZnSO_4 \cdot H_2O$	22 to 36
	Zinc oxide	ZnO	78
Fe	Ferrous sulfate	$FeSO_4 \cdot 7H_2O$	20
	Ferric sulfate	$Fe_2(SO_4)_3 \cdot 4H_2O$	23
B	Borax	$Na_2B_4O_7 \cdot 10H_2O$	11
	Solubor	$Na_2B_8O_{13} \cdot 4H_2O$	20

FERTILIZER USE

Table 4-9. Composition of Selected Organic Fertilizer Materials

Nutrient Source	Analysis (%)[1]			Relative Rate of Nutrient Release
	N	P_2O_5	K_2O	
Animal tankage	7	9	0	Medium
Bone meal Raw steamed	 3 2	 22 28	 0 0	 Very slow Slow
Castor pomace	5	1	1	Slow
Cotton seed meal	7	2	2	Slow
Dried blood	12	3	0	Fast
Feather meal	14	0.3	0.1	Fast
Hardwood ashes[2]	0.5	1	5	Slow
Linseed meal	5	2	2	Slow
Municipal yard and leaf compost[4]	0.5	0.4	0.8	Slow
Sheep wool pellets	5.8	0.9	2.4	Fast/Slow
Softwood ashes[2]	0.5	2	4	Slow
Tobacco stems	2	1	7	Slow
Turkey litter compost[3]	1.5	3.5	1	Slow
Food waste vermicompost	2.11	0.79	1.47	Slow
Animal manure vermicompost	1.67	3.04	0.55	Slow

[1] Percentages of nutrient elements may vary depending upon source. Values given are approximations.

[2] Plant-available N less than or equal to 0.25%.

[3] Compost products must be stabilized before use, or they may cause nutrient depletion.

[4] Municipal yard and leaf waste composts may contain measurable concentrations of metals, such as lead and zinc. A complete analysis of these soil amendments should be conducted before use.

Table 4-10. Composition of Typical N Solutions[1]

Contents (%)		No-Pressure Nonammonia Solutions[2]						Low Pressure Ammonia Solution	Aqua-Ammonia
Total N	16.0	19.0	21.0	28.0	30.0	32.0	37.0	20.0
NH_3	0	0	0	0	0	0	16.6	24.4
NH_4NO_3	45.8	54.3	60.0	40.0	42.2	44.3	66.8	0
Urea	0	0	0	30.0	32.7	35.4	0	0
Water	54.2	45.7	40.0	30.0	25.1	20.3	16.6	75.6
NO_3-N	8.0	9.5	10.5	7.0	7.4	7.7	11.7	0
NH_4-N	8.0	9.5	10.5	7.0	7.4	7.8	25.4[3]	20.0[3]
Urea-N	0	0	0	13.8	15.2	16.4	0	0
		Salt-out Temperatures (F)[4]						
		41	41	0	15	32	56	-58

[1] Values may vary depending on product.

[2] Several companies provide N solutions containing 3.5% to 5.0% sulfur.

[3] Includes N present as NH_3.

[4] Proprietary additives can alter salt-out temperatures; this should be verified with the dealer.

Table 4-11. Composition of Selected Fertilizer Materials[1]

Fertilizer Material	Chemical Formula	Nutrient Percentage (%)							
		NO_3-N	NH_4-N	Total N	P_2O_5	K_2O	Ca[4]	Mg	S
Ammonium nitrate	NH_4NO_3	17	17	34					
Monoammonium phosphate	$NH_4H_2PO_4$		11	11	48		1		2
Diammonium phosphate	$(NH_4)_2HPO_4$		16 to 18	16 to 18	46 to 48				
Ammonium sulfate	$(NH_4)_2SO_4$		21	21					24
Ammonium thiosulfate	$(NH_4)_2S_2O_3$			12					26
Anhydrous ammonia	NH_3		82	82					
Urea-Form[2]			38	38					
Calcium nitrate	$Ca(NO_3)_2$	15		15			19	1	
Nitrate of soda potash	$NaNO_3 \cdot KNO_3$	15		15		14			
Sodium nitrate	$NaNO_3$	16		16					
Urea	$CO(NH_2)_2 \cdot H_2O$		45 to 46	45 to 46					
Single superphosphate	$Ca(H_2PO_4)_2 + CaSO_4$				18 to 20		18 to 21		12
Triple superphosphate	$Ca(H_2PO_4)_2 \cdot H_2O$				42 to 50		12 to 14		1
Basic slag[3]	$5CaO \cdot P_2O_5 \cdot SiO$				2 to 17		3 to 33	3	
Potassium chloride	KCl					60 to 62			
Potassium nitrate	KNO_3	13		13		44			
Potassium sulfate	K_2SO_4					50 to 53		1	18
Potassium-magnesium sulfate	$K_2SO_4 \cdot 2MgSO_4$					22		11	23
Epsom salt	$MgSO_4 \cdot 7H_2O$						2	10	14
Gypsum	$CaSO_4 \cdot 2H_2O$						23		18.5

[1] Values may vary depending on product.

[2] Slow release N source.

[3] Lime value — about 0.67 agricultural limestone.

[4] Evaluate gypsum products marked as Ca sources based on guaranteed fertilizer analysis. Pure calcium sulfiate is 29 percent calcium. Landplasters typically contain 70 to 85 percent calcium sulfate (21 to 25 percent calcium), while phosphogypsum and other products may contain 50 percent or less calcium sulfate (15 percent calcium). See Table 4-1 for rate recommendations for peanut.

FERTILIZER USE

Table 4-12. Composition of Selected Specialty, Alternative, or Fertilizer Efficiency Enhancer Materials

Not intended to be all inclusive, and inclusion does not imply endorsement. See product label for specific instructions. Some product evaluations at http://extension.agron.iastate.edu/compendium/index.aspx, "Compendium of Research Reports on Use of Non-Traditional Materials for Crop Production."

Product (Source)	Ingredient(s)	Primary Uses	Typical rate	Nutrient supply at typical rate (lb/A)	Comments
AG-TEK™ BIO-D 4200 (Global Green Products)	Protein synthesized with aspartic acid, 4-0-0	in furrow and sidedress	1-2 qt/A	If 2 qt/A N-0.2	Product literature states this is not a fertilizer replacement and must be used with a normal soil fertility program with full fertilizer rates, mentions enhanced rooting and solution chemistry benefits.
Agrotain Dry, ADII (Koch)	NBPT (urease inhibitor)	Mix with urea or UAN solutions			Duration of urease inhibition shorter at higher temperatures and perhaps organic matter-dependent. Affects urea component of mixed N sources. Potential to reduce volatilization losses, especially in dry high pH situations.
Agrotain Plus (Koch)	NBPT+DCD (urease + nitrification inhibitors)	Mix with UAN solutions			Duration of urease and nitrification inhibition shorter at higher temperatures and perhaps organic matter-dependent. Affects urea and ammonium components of mixed N sources. Potential to reduce volatilization and leaching losses, especially in dry high pH situations or in sandy soils with heavy rainfall after application.
AVAIL (SFP)	Maleic itaconic copolymer 0-0-0	Broadcast or band with P fertilizer			Indirect mechanism, intended to increase soil CEC and influence soil cation interactions
Bio-forge (Stoller)	diformyl urea + potassium hydroxide (2-0-7)	In furrow	1 pt/A	N-0.025 K_2O-0.09	
ESN (Agrium)	Urea (plastic-encapsulated) 44-0-0	broadcast		Each 1 lb 44-0-0: N-0.44	Temperature-dependent slow release, duration shorter at higher temperatures. Potential to reduce volatilization and leaching losses, especially in dry high pH situations or in sandy soils with heavy rainfall after application.
Grasshopper (Grasshopper Fertilizer)	Multiple products (e.g. 30-8-10)	Foliar (also other products)	10 lb/A	10 lb of 30-8-10:N-3, P_2O_5-0.8, K_2O-1, B-0.005, Cu-0.005, Fe-0.05, Mn-0.005, Zn-0.01, Mo-0.0001	Multiple products with highly soluble ingredients and many with chelated micronutrients. Product list includes slow release N provided as methylene urea source.
GYP-Soil	Calcium sulfate (FGD source)	broadcast			Considered equivalent sources of Ca and S as conventional gypsum when rate adjusted based on composition. Potential effects on soil physical properties and subsoil aluminum toxicity.
Instinct (Dow)	Nitrapyrin (nitrification inhibitor)	Mix with UAN solutions			Duration of nitrification inhibition lower at higher temperatures and lower in soils with high organic matter content. Affects urea and ammonium components of mixed N sources. Potential to reduce leaching losses, especially in sandy soils with heavy rainfall after application.
Invigor 8 (Genesis Ag)	0-37-37 derived from potassium acid phosphate	Seed treatment	1 lb per 20 units corn (approx. 1 lb/40 A)	0.01 lb P_2O_5, 0.01 lb K_2O	
MicroPack (Agriculture Solutions)	11-8-5 + B, Cu, Fe, Mn, Zn, Co, Mo	Foliar	2 qt/A, possibly 2-3 times	N-0.55, P_2O_5-0.4, K_2O-0.25, B-0.001, Cu-0.0025, Fe-0.005, Mn-0.0025, Zn-0.0025, Co-0.000025, Mo-0.000025	
Nachurs Solutions Multiple NPKS and micronutrient blends	Multiple products (e.g. HKW18 =3-18-18)	In furrow (IF), foliar	IF:2-6 gpa Foliar:1-3 gpa	2 gpa of 3-18-18: N-0.7, P_2O_5-4.2, K_2O-4.2	Multiple products with highly soluble ingredients formulated to minimize salt index.
Nutrisphere (SFP)	Maleic itaconic copolymer 0-0-0	Band or broadcast with N fertilizers			Indirect mechanism, intended to increase soil CEC and influence soil cation interactions
N Zone (AgXplore International)	Adjuvant, glycols & poly-saccharides 0-0-0	Mix with urea, UAN, liquid manures			Marketed as penetrant aid for N sources. Most likely effect is with foliar application.
Optimize (Novozymes)	Rhizobial inoculants plus LCO promoter	Seed trt			Not intended to replace in-furrow inoculation on land not previously planted to soybean.
Pentilex (Biovante)	0-37-37	Seed trt	1-1.25gal /1000 lb	If 1 gal on 35k seed/A: P_2O_5-0.007, K_2O-0.007	
Quick-Sol	Sodium silicate	Soil, foliar	14-20 oz/A		Soluble silicon source, of potential benefit to some crops.
Riser (Loveland Products)	7-17-3 + Cu, Fe, Mn, Zn	In furrow	2-5 gpa	If 3 gpa: N-2.2, P_2O_5-5.4, K_2O-1, Cu-0.02, Fe-0.06, Mn-0.02, Zn-0.3	
Soygro [Soygro (Pty) Ltd]	Microbial inoculants				Unspecified on public website, but could include rhizobial and non-nodulating species.
TakeOff Sulfone (Verdesian)	5-20-15+14%S+0.5%Fe	Soil or foliar application	0.5-4 lb/A	If 2.5 lb/A: N-0.125, P_2O_5-0.5, K_2O-0.375, Fe-0.01	If foliar, do not exceed product concentration of 1.25% by volume in water.

Solubility of Selected Fertilizer Materials

To be available to plants, at least some of a nutrient must be slightly soluble in the soil solution. The amount of substance that will dissolve at a given temperature in water is known as its solubility. Solubility of most chemicals is slightly higher at higher temperatures; that of others, especially ammonium and potassium nitrates, increases rapidly with temperature. The presence of other substances in the solution may either increase or decrease solubility. The solubility of selected pure fertilizer materials in water at 32 degrees F is shown below.

Table 4-13. Solubility of Selected Fertilizer Materials

Fertilizer Material	Chemical Formula	Solubility (lb/100 gal)	Salt Index (relative effect on the soil solution)
Ammonia	NH_3	750	47.1
Ammonium nitrate	NH_4NO_3	983	104.7
Ammonium sulfate	$(NH_4)_2SO_4$	592	69.0
Borax	$Na_2B_4O_7 \cdot 10H_2O$	25	
Calcium carbonate (limestone)	$CaCO_3$	0.050	4.7
Calcium metaphosphate	$Ca(PO_3)_2$	0.008	
Calcium nitrate	$Ca(NO_3)_2 \cdot 4H_2O$	1,117	52.5
Calcium sulfate	$CaSO_4 \cdot 2H_2O$	2	8.1
Copper sulfate	$CuSO_4 \cdot 5H_2O$	267	
Diammonium phosphate	$(NH_4)_2HPO_4$	209	29.9
Dicalcium phosphate	$CaHPO_4 \cdot 2H_2O$	0.168	8
Magnesia	MgO	0.005	1.4
Magnesium sulfate	$MgSO_4 \cdot 7H_2O$	580	44
Manganese sulfate	$MnSO_4 \cdot 4H_2O$	875	
Monoammonium phosphate	$NH_4H_2PO_4$	358	34.2
Monocalcium phosphate	$CaH_4(PO_4)_2 \cdot H_2O$	See below[1]	15.4
Potassium chloride	KCl	233	116.3
Potassium-magnesium sulfate	$K_2SO_4 \cdot 2MgSO_4$	200	
Potassium nitrate	KNO_3	108	73.6
Potassium sulfate	K_2SO_4	67	46.1
Sodium nitrate	$NaNO_3$	608	100.0
Urea	$CO(NH_2)_2$	559	75.4
Zinc sulfate	$ZnSO_4 \cdot 6H_2O$	584	

[1] Material decomposes with a small amount of water and is soluble in a large amount. The solubility varies with the conditions.

FERTILIZER USE

Mixing Herbicides with Nitrogen Solutions or Fluid Fertilizers

C. R. Crozier and W. Everman, Crop and Soil Sciences

Tank-mixing and applying herbicides with nonpressure nitrogen solutions or fluid fertilizers may offer savings in labor and time by eliminating at least one trip over the field. Effectiveness of some postemergence-directed herbicides for corn and sorghum is increased by applying them in nitrogen solution.

Herbicide labels will indicate if the product can be mixed with sprayable fertilizer. Herbicides may not always mix evenly throughout a sprayable fluid fertilizer, or the components may separate too quickly to make their combined use practical. Therefore, test every batch of fertilizer for compatibility before adding the herbicide. Batches can vary in pH, salt content, and salt concentration. Even these minor differences may affect compatibility. Sulfur-containing nitrogen solutions can be particularly troublesome.

The information presented below is intended as a guide only. The term "compatibility" as used here refers to chemical and physical compatibility and is not intended to supplant label directions.

A guide for making your own compatibility test follows. Also, herbicide labels give information on compatibility, tank-mix combinations, and procedures for testing compatibility with sprayable fluid fertilizer. Be sure to read the label!

Precautions

To help ensure successful results from application of herbicides and nonpressure nitrogen solutions or fluid fertilizers:

1. Always check compatibility by making a small scale test (directions given below) before mixing in field.

2. Use a compatibility agent if indicated on the herbicide label or if your small-scale test indicates the need.

3. Vigorously agitate tank contents while mixing and applying. The spray application equipment should use a high-capacity (eight-roller, internal gear, or centrifugal) pump. Spray tanks should be equipped with hydraulic jet agitators mounted in the bottom of the tank. Bypass lines usually do not provide adequate agitation. Jet agitators should be attached to a separate line from the pump.

4. Do not connect agitators to by-pass lines. For simple hydraulic jet agitators, a flow rate of 6 gallons per minute for every 100 gallons of tank capacity is sufficient. If volume-booster nozzles are used for agitation, the flow rate can be reduced to 2 to 3 gallons per minute for every 100 gallons of tank capacity. If a liquid-fertilizer type applicator is used with a metering pump, agitation must also be supplied. The best method is to use a separate power takeoff pump for extra circulation.

5. If a wettable powder or dry flowable is used, make a slurry with water and add slowly to the tank. Add wettable powder first, dry flowables second, flowables third, and liquids last.

6. A number of dry flowable herbicides now come packaged in water-soluble film packets. These packets usually will not dissolve in nitrogen solution or fluid fertilizer. When using nitrogen solution or fluid fertilizer as the carrier for a product packed in these film packets, slurry the herbicide in clean water before adding to the spray tank.

7. If a flowable product is used, premix one part flowable with one part water and add diluted mixture slowly to tank. The fluid fertilizer may be substituted for the water after compatibility has been checked.

8. Premix liquid products with two parts water or the fertilizer carrier before adding into the tank.

9. Reduce drift by choosing an appropriate nozzle size to put out the desired amount of solution without causing excessive pressure.

10. *Caution:* Take extra care in applying herbicide-fluid fertilizer mixtures to ensure that the correct herbicide rate is applied, that the herbicide is distributed uniformly, and that all directions concerning application of the herbicide are followed. Sprayer output will not be the same using fluid fertilizer as when using water as the carrier. Recalibrate sprayers for the fertilizer carrier. Do not apply herbicide-fluid fertilizer mixes overtop any crop except small grains, as injury will result.

How to Test Compatibility of Herbicides with Fluid Fertilizers

C. R. Crozier and W. Everman, Crop and Soil Sciences

Follow the compatibility test procedures on the herbicide label. If not on the label, follow the directions listed below.

1. Put 1 pint of fluid fertilizer in each of two 1-quart jars.

2. Following the adjacent table, add 0.25 to 0.375 teaspoon of a compatibility agent to one jar and shake for 5 or 10 seconds to mix. One-fourth teaspoon is equivalent to 2 pints/100 gallons of fluid fertilizer. Mark the jar "with" to indicate the compatibility agent has been added.

3. Next, add the proper amount of herbicide to each jar, according to the table. For herbicides used in small quantities per acre, one will need a greater volume of fluid fertilizer for the compatibility test. Adjust the herbicide/fertilizer ratio as follows:

    1 ounce dry flowable per acre = 0.7 teaspoon in 1 gallon
    1 fluid ounce liquid per acre = 0.25 teaspoon in 1 gallon

    If more than one herbicide is to be used in the mixture, add them separately with the wettable powders or dry flowables first, flowables second, and liquids last. Shake the jar gently for 5 to 10 seconds after each addition.

4. Let the jars stand for 5 minutes, then check for formation of large flakes, sludge, gels, or other precipitates, or to see if the herbicide remains as small, oily particles in the solution.

5. Allow both jars to stand for 30 minutes, checking them periodically. An emulsifiable concentrate normally will go to the top after standing, whereas wettable powders or dry flowables will either settle to the bottom of the jar or float to the top, depending on the density of the fertilizer and the herbicide. If the separate layers of fluid fertilizers and additives (herbicide and compatibility agent) can be resuspended by shaking, commercial application is possible.

6. If incompatibility of any form occurs in the jar with the compatibility agent added, do not mix the fluid fertilizer and herbicide in the same spray tank. If incompatibility occurs only in the jar without the compatibility agent, use of a compatibility agent is recommended.

Table 4-14. Amounts of Herbicide to Use to Test Compatibility[1]

Application Rate Product Per Acre	Add to 1 Pint of Fluid Fertilizer
Wettable powders or dry herbicides	
1 pound	1.4 teaspoon
2 pound	2.9 teaspoon
3 pound	4.3 teaspoon
4 pound	5.8 teaspoon
5 pound	7.2 teaspoon
Emulsifiable concentrates, flowables, liquids, or solutions	
1 pint	0.5 teaspoon
1 quart	0.9 teaspoon
2 quart	1.9 teaspoon
3 quart	2.9 teaspoon
1 gallon	3.8 teaspoon
5 quart	4.8 teaspoon

[1] This compatibility test is designed for 25 gallons of spray per acre with the maximum labeled rate of herbicide. For changes in spray volume or herbicide concentration, make the appropriate proportional change in the ingredients in the test. Regardless of spray volume, the amount of compatibility agents should be equal to 2 to 3 pints (0.25 teaspoon equals 2 pint; 0.375 teaspoon equals 3 pints) per 100 gallons of fertilizer.

FERTILIZER USE

Fertilizer Placement

C. R. Crozier and D. L. Osmond, Crop and Soil Sciences, and D. H. Hardy, NCDA&CS

Proper fertilizer placement can be as important as what kind of fertilizer you use. The first step, though, is a soil test to determine phosphorus and potassium needs and the use of realistic yield expectations to determine nitrogen. Once fertilizer needs are determined, fertilizer placement should be assessed.

Starter P (along with N) does not always lead to yield increases, but it can be an effective tool to enhance early season growth and reduce risks of losses associated with billbugs, competitive weeds, and summer droughts (see SoilFacts: Starter Phosphorus Fertilizer and Additives in North Carolina Soils: Use, Placement, and Plant Response, http://www.soil.ncsu.edu/publications/Soilfacts/AG-439-75.pdf). Use of starter fertilizers is more important in no-till since soil warming is delayed, and the cooler temperatures can reduce the rate of crop growth. For fields that already have very high soil test P levels (P-index >100), numerous North Carolina tests have shown there is no advantage to applying additional P in a starter band. Even most mineral soils testing greater than a 50 P-Index generally do not need starter P.

Fertilizer placed in contact with or too close to seeds and young plants can cause salt injury, resulting in poor stands and slow starts. Nonuniform broadcasting of fertilizer with shallow mixing just before planting may give streaks through the field due to delayed germination or seedling injury. Salt injury is most severe in dry weather or following light rains that dissolve the fertilizer salts and leave highly concentrated salt solutions in the root zone. Nitrogen and potassium salts account for most of this injury.

To reduce salt injury risk, a side-band placement of starter fertilizer at planting is generally preferred over application directly in contact with the seed, commonly referred to as "pop-up" placement. Good results are often achieved for seeded crops like corn, cotton, sorghum, and soybeans using a 2 by 2-inch placement, placing the fertilizer 2 inches to the side and 2 inches below the planted seed. The risk of salt injury is also related to the amount of salt applied. Generally, a maximum rate of 80 pounds per acre of nitrogen alone, K_2O alone, or a combination of nitrogen plus K_2O is suggested for 2 x 2 band placement. When a greater rate is necessary, make a split application; broadcast part and apply the remainder in the row at seeding. If side placement is not possible or practical and starter fertilizer is placed directly in the seed furrow, the maximum rate should be much lower; 5 gpa of typical starter solution products contain less than 20 pounds N plus K2O per acre and is the maximum rate that should be used. An alternative method is to broadcast fertilizer prior to planting, with thorough mixing into the soil preferred, except for conservation tillage systems. The lack of soil mixing is an additional reason that banded starters are advantageous with no-till management.

For tobacco, place the band 3 to 5 inches from the transplants. This distance reduces the chance of placing plants in fertilizer bands. If side placement equipment is not available, place the fertilizer deep in the row so it will be 3 to 5 inches below the roots of the transplants.

Seedlings of small grain or plants such as clovers, grasses, and alfalfa respond very early to phosphate; consequently, it can be advantageous to place phosphate close to the seed as is done with the conventional grain drill.

Livestock & Poultry Manure Production Rates and Nutrient Content

D. A. Crouse, C. R. Crozier, and T. J. Smyth, Crop and Soil Sciences; and K. Hicks, NCDA&CS

The use of livestock and poultry manure as a fertilizer supplement in crop production has come full circle. Before the advent of inexpensive inorganic fertilizers, farmers routinely used manure to supply essential plant nutrients. Today, because of rising costs of commercial fertilizers and increasing emphasis on good manure management practices to protect water quality renewed interest has been focused on optimizing the nutrient benefits of manures.

Manure production and characteristics are influenced by on-farm management practices. For example, manure from open housing systems or from open storage ponds is diluted by rainfall. Drying and storing solid manures can increase ammonia loss and reduce the nitrogen content and fertilizer value. Microbial digestion in a waste treatment lagoon reduces total nitrogen by 50 to 85% and converts as much as 90% of the phosphorus to forms that settle into the sludge rather than remaining in lagoon liquid. All of the practices ultimately affect the total amount of recoverable nutrients that can be used in crop production.

Manure Production (Volume and Weights)

The total manure produced in volume or weight is related to the animal production system, the number of animals in that system, and the manure collection and treatment system. In liquid systems, the production of manure is often reported in gallons per animal per year or gallons per 1,000 birds capacity per year (Table 4-15). For dry or solid systems, the production is reported in tons per animal per year or tons per 1,000 birds capacity per year (Table 4-15).

Manure Nutrient Content

The manure nutrient content varies with animal age, diet, and waste (manure) management system. Nutrient data reported in Table 4-16 are statewide averages and may not reflect the actual nutrient content on any individual farm. Because of the variability of end products in the numerous different animal production and waste management systems manure must be sampled and analyzed within 60 days of application to land to determine its actual nutrient content. Manure samples can be analyzed for 11 essential plant nutrients for a nominal fee by the N.C. Department of Agriculture and Consumer Services. The Waste/Compost Analysis Laboratory (http://ncagr.gov/agronomi/uyrwaste.htm) provides details on collecting and submitting manure samples for waste analysis).

Plant Availability Coefficients

Two factors are important in the plant availability of nutrients in manure: mineralization potential and application method. Since the nutrients in manure are primarily in organic compounds, they must undergo microbial mineralization in order to become plant available. Once in the plant-available form, some nutrients are at risk of loss to the environment. For example, nitrogen is susceptible to gaseous losses due to volatilization of ammonia (a mineralized form of N). As a result, the availability of nitrogen for plant uptake is dependent on the mineralization potential of the manure. Further, the application method plays an important role in the availability of nutrients. Subsurface placement such as soil incorporation tends to reduce volatile losses whereas surface placement such as irrigation tends to increase volatile losses. Most nutrients including phosphorus and potassium are not subject to volatile losses, meaning the application method has minimal impact on their availability. Plant availability coefficients (Table 4-17) are used to estimate the nutrients that would be available to the first crop, taking into consideration both the mineralization potential and the application method.

Calculating On-Farm Availability of Manure Nutrients

Land application of manures is an integral part of the overall soil-fertility management strategy on many farms. For farm nutrient budgeting, an assessment of the nutrients available for plant utilization must be completed. Using Tables 4-15 through 4-17, the calculation is simple:

$$Total\ Plant\ Available\ Nutrients = \frac{No.\ of\ Animals}{Year} \times \frac{Weight\ or\ Volume}{Animal} \times \frac{Total\ Nutrients}{Weight\ or\ Volume} \times Availability\ Coefficient$$

For example, consider a 900-sow farrow to finish operation where the lagoon effluent is soil incorporated following application but before planting of a grain crop.

$$Total\ Plant\ Available\ N\ (PAN) = No.\ of\ Sows \times \frac{gallons}{sows/year} \times \frac{Total\ N}{1,000\ gal} \times \frac{X\ lbs\ PAN}{lbs\ Total\ N}$$

$$Total\ Plant\ Available\ N\ (PAN) = 900 \times \frac{10,478\ gallons}{sows/year} \times \frac{3.6\ Total\ N}{1,000\ gal} \times \frac{0.6\ lbs\ PAN}{lbs\ Total\ N}$$

$$Total\ Plant\ Available\ N\ (PAN) = 20,369\ lbs\ of\ N\ available\ for\ the\ grain\ crop\ in\ the\ first\ year$$

This same calculation can also be performed for other nutrients such as P, Zn, and Cu.

Management Considerations

Always apply manure as close to the period of maximum plant demand for nutrients as possible. Base manure application rates on the available portion of the nutrients and not the total concentration. Do not apply more than the receiver crop needs since excessive amounts not only waste valuable nutrients but may result in surface and/or groundwater pollution. Nutrient management planning guidelines (http://nutrients.soil.ncsu.edu) should be used to determine if application rates should be based on N or P content. Monitoring of Zn and Cu may be needed to avoid accumulation in soils to toxic levels. Use soil testing (http://ncagr.gov/agronomi/sthome.htm) to predict nutrient and lime requirements and proper application rates of the manure. Use plant analysis (http://ncagr.gov/agronomi/uyrplant.htm) to monitor the crop nutritional status (actual uptake of soil nutrients) and effectiveness of the nutrient management program.

Additional manure management information is available in at http://www.soil.ncsu.edu/publications/extension.htm

Table 4-15. Manure Volume and Weights in Typical North Carolina Animal Production Systems

Animal Production System	NCDA&CS Waste Code	Accumulated Manure
		gallons/animal/year[1]
Anaerobic Lagoon Liquid - Swine	**ALS (except Farrow-to-Wean)**	
Farrow-to-Wean (per sow)	ALF	3,203
Farrow-to-Feeder (per sow)		3,861
Farrow-to-Finish (per sow)		10,478
Wean-to-Feeder (per pig)		191
Wean-to-Finish (per pig)		776
Feeder-to-Finish (per pig)		927
Anaerobic Lagoon Sludge – Swine	**ASS**	
Farrow-to-Wean (per sow)		433
Farrow-to-Feeder (per sow)		522
Farrow-to-Finish (per sow)		1,417
Wean-to-Feeder (per pig)		30
Wean-to-Finish (per pig)		26.3
Feeder-to-Finish (per pig)		135
Dairy – Slurry	**LSD**	
Calf		1,876
Heifer		5,535
Milk Cow		7,749
		gallons/1,000 bird capacity/year[1,2]
Anaerobic Lagoon Liquid – Poultry	**ALP**	
Pullet (non-laying)		9,110
Pullet (laying)		22,201
Layer		25,373

FERTILIZER USE

Table 4-15. Manure Volume and Weights in Typical North Carolina Animal Production Systems

Animal Production System	NCDA&CS Waste Code	Accumulated Manure
Anaerobic Lagoon Sludge – Poultry	**ASP**	
Pullet (non-laying)		1,659
Pullet (laying)		4,147
Layer		4,739
		tons/1,000 bird capacity/year[2]
Poultry Litter – Breeders	**HBB**	**24**
Poultry Litter – Broilers	**HLB**	
Whole House		7.2
Cake		4.0
Poultry Litter – Broiler Pullets	**HBP**	**7.2**
Poultry Litter – Layers	**HLL**	**24**
Poultry Litter – Layer Pullets	**HLP**	**24**
Poultry Litter – Turkeys	**HLT**	
Poult		5.3
Hen		17
Tom		25
Breeder		37
		tons/animal/year
Dairy – Scraped	**SSD**	
Calf		4.1
Heifer		12
Milk Cow		17
Beef – Scraped	**SSB**	
Stocker		1.5
Feeder		2.2
Brood Cow		3.0
Horse – Scraped	**SSH**	**9.1**

[1] To convert gallons to acre-inches, divide gallons by 27,154.

[2] Capacity is based on the maximum number of birds on the farm at any time, which is a measure of the size of the flock and not the total annual production on the farm.

Table 4-16. Total Nitrogen (N), Phosphorus (P AS P_2O_5), and Potassium (K AS K_2O) from Manure Sources

Production System	NCDA&CS Waste Code	N	P_2O_5	K_2O
		pounds of total nutrient per 1,000 gallons[1]		
Anaerobic Lagoon Liquid - Swine	**ALS**			
Boar		3.6	1.4	8.3
Farrow-to-Wean	ALF	2.4	0.9	4.1
Farrow-to-Feeder		3.6	1.4	8.3
Farrow-to-Finish		3.6	1.4	8.3
Wean-to-Feeder		3.6	1.4	8.3
Wean-to-Finish		3.6	1.4	8.3
Feeder-to-Finish		3.6	1.4	8.3
Anaerobic Lagoon Sludge – Swine	ASS	20.4	30.6	7.5
Anaerobic Lagoon Liquid - Poultry	ALP	3.1	1.0	13.8
Anaerobic Lagoon Sludge - Poultry	ASP	24.4	38.1	10.3
Dairy - Slurry	LSD	16.7	9.1	15.4
		pounds of nutrient per ton		
Dairy - Scraped	SSD	11.2	7.0	9.8
Horse - Scraped	SSH	9.3	7.0	9.8
Beef - Scraped	SSB	13.0	8.3	13.6
Poultry Litter - Breeders	HBB	47.6	44.7	39.5
Poultry Litter - Broilers	HLB	57.8	40.0	48.6
Poultry Litter – Broiler Pullets	HBP	57.8	40.0	48.6
Poultry Litter – Layers	HLL	47.6	44.7	39.5

FERTILIZER USE

Table 4-16. Total Nitrogen (N), Phosphorus (P AS P$_2$O$_5$), and Potassium (K AS K$_2$O) from Manure Sources

Production System	NCDA&CS Waste Code	N	P$_2$O$_5$	K$_2$O
Poultry Litter – Layer Pullets	HLP	47.6	44.7	39.5
Poultry Litter - Turkeys	HLT	54.0	48.2	33.8

1 To convert gallons to acre-inches, divide gallons by 27,154.

Table 4-17. First-Year Nutrient Availability Coefficients for Nitrogen (N), Phosphorus (P) and Potassium (K) from Manure Sources

Production System	NCDA&CS Waste Code	N Broadcast or Irrigated	N Incorporated or Injected	P Broadcast or Irrigated	P Incorporated or Injected	K Broadcast or Irrigated	K Incorporated or Injected
Aanerobic Lagoon Liquids							
Swine	ALS	0.5	0.6	1.0	1.0	1.0	1.0
Swine Farrow to Wean	ALF	0.5	0.6	1.0	1.0	1.0	1.0
Poultry	ALP	0.5	0.6	1.0	1.0	1.0	1.0
Other	ALO	0.5	0.6	1.0	1.0	1.0	1.0
Anaerobic Lagoon Sludges							
Swine	ASS	0.5	0.6	1.0	1.0	1.0	1.0
Poultry	ASP	0.5	0.6	1.0	1.0	1.0	1.0
Other	ASO	0.5	0.6	1.0	1.0	1.0	1.0
Slurries							
Beef	LSB	0.4	0.6	1.0	1.0	1.0	1.0
Dairy	LSD	0.4	0.6	1.0	1.0	1.0	1.0
Swine	LSS	0.4	0.6	1.0	1.0	1.0	1.0
Other	LSO	0.4	0.6	1.0	1.0	1.0	1.0
Scraped or Stockpiled Manure							
Beef	SSB	0.4	0.6	1.0	1.0	1.0	1.0
Dairy	SSD	0.4	0.6	1.0	1.0	1.0	1.0
Horse	SSH	0.4	0.6	1.0	1.0	1.0	1.0
Swine	SSS	0.4	0.6	1.0	1.0	1.0	1.0
Other	SSO	0.4	0.6	1.0	1.0	1.0	1.0
Poultry Litters							
Breeders	HBB	0.5	0.6	1.0	1.0	1.0	1.0
Broilers	HLB	0.5	0.6	1.0	1.0	1.0	1.0
Broiler Pullets	HBP	0.5	0.6	1.0	1.0	1.0	1.0
Layers	HLL	0.5	0.6	1.0	1.0	1.0	1.0
Layer Pullets	HLP	0.5	0.6	1.0	1.0	1.0	1.0
Turkeys	HLT	0.5	0.6	1.0	1.0	1.0	1.0
Other	HLO	0.5	0.6	1.0	1.0	1.0	1.0

FERTILIZER USE

Beneficial Use of Municipal Biosolids

J. G. White, Crop and Soil Sciences

Biosolids are the nutrient-rich, largely organic residuals of wastewater (sewage) treatment, i.e., sludge that has been generated at a water resource recovery facility (WRRF, a.k.a., wastewater treatment plant) and treated to meet US-EPA standards for land application. These standards include limitations on pathogens, potentially toxic metals, and disease-vector attractants. Biosolids can be recycled via land application to sustainably improve and maintain productive soils and foster plant growth. Most biosolids contain sufficient plant nutrients (e.g., nitrogen [N], phosphorus [P], potassium [K]) to have fertilizer value, and some have liming value (agricultural lime equivalent). Previous concerns regarding the levels of potentially toxic metals in biosolids have been alleviated by current US-EPA standards. However, most biosolids do contain small quantities of pharmaceuticals, personal care products, steroids, hormones, and flame retardants. These might pose risks to crops, the livestock or humans that consume them, and/or to the environment. Currently, these substances in biosolids are not regulated, and any potentially adverse or advantageous effects are not well characterized.

Biosolids are classified as either Class B, Class A, or Class A – Exceptional Quality (EQ) based on US-EPA regulatory reduction requirements for pathogens, vector attractants, and potentially toxic metals. Class B biosolids meet the least restrictive limits for concentrations of potentially toxic metals and have been treated to reduce pathogens to levels considered protective of human health. Land application of Class B biosolids requires site-use and access-restriction permits. Class A biosolids are further treated to reduce pathogens below detectable limits. Class A – Exceptional Quality biosolids meet stricter metal concentration requirements and may be applied by the general public in reasonable quantities without permit. Physically, North Carolina biosolids typically fall into one of four categories depending on their dry matter content and consistency. Class B slurries have very low solids content (2 – 5%). Class B cakes range from about 15 to 25% solids. Class A alkaline-treated biosolids range from ~35 to 80% solids, and Class A heat-treated biosolids have from 90 to 98% solids and are typically pelleted. Biosolids are also used in the production of commercial bulk compost.

Anyone who wants to distribute and land apply biosolids must comply with all relevant federal and state regulations. In some cases, a permit may be required and obtained from the North Carolina Department of Environmental Quality (NCDEQ). Typically, any WRRF wishing to apply biosolids to its own land or to private farmland must obtain a permit. To obtain a permit, sufficient data must be submitted to show that the biosolids meet the federal and state limits for pathogens, potentially toxic metals, and vector attraction, and that the proposed site is suitable to receive biosolids. Some of the factors considered in site selection include topography; soil type; distance from streams, wells, and property lines; flood hazard; and depth to bedrock and the water table. If the site is suitable, then application rates are calculated based on biosolids composition, crop nutrient requirements, previous history of biosolids application, and soil pH. The primary purposes of these criteria are to: 1) protect ground and surface water from contamination by biosolids nitrogen and phosphorus, and 2) foster crop nutrient-use efficiency. If a property owner wishes to apply certain biosolids to his or her land, he or she must obtain a permit. The preparation of permit applications is the responsibility of the municipality, industry, and land owner. The process is somewhat involved, and we do not attempt to outline it here. Information on permitting may be obtained from any NCDEQ regional office or at the NCDEQ Division of Water Resources - Water Quality Permitting Section - Non-Discharge Permitting Branch (http://deq.nc.gov/about/divisions/water-resources/water-resources-permits/wastewater-branch/non-discharge-permitting).

Many biosolids are treated with alkaline materials to raise sludge pH to reduce pathogens and vector attraction ("lime stabilized"). These biosolids typically have liming value that should be identified before use. If these biosolids are land-applied based on nutrient content (e.g., N, P) without regard to their agricultural lime equivalent, over-liming may occur and negatively affect plant growth.

Since most biosolids produced in North Carolina are suitable for use on agricultural land, we encourage interested farmers to contact their county Cooperative Extension Service Center (https://www.ces.ncsu.edu/local-county-center/) and/or their NCDEQ Regional Office (http://deq.nc.gov/contact/regional-offices) about possible sources of biosolids in their area. Table 4-18 shows representative average characteristics of some biosolids available in North Carolina. The user should obtain actual, current analyses of any product that is to be land-applied in order to determine application rates that optimize plant nutrient use efficiency, moderate soil pH, and protect water quality.

Table 4-18. Plant Nutrient* and Metal Contents of Some Municipal Biosolids from or Available in North Carolina[1]

Biosolids Source	Solids	Total Nitrogen*	Plant-available Nitrogen[2]	P₂O₅*	K₂O*	Ca*	Mg*	Cu*	Zn*	Ni*	Cr	Pb	Cd	Mo*	As
	%														

Biosolids Source	Solids (%)	Total Nitrogen*	Plant-available Nitrogen[2]	P$_2$O$_5$*	K$_2$O*	Ca*	Mg*	Cu*	Zn*	Ni*	Cr	Pb	Cd	Mo*	As
		Constituents in pounds per dry ton of biosolids													
Cape Fear Public Utility Authority (Wilmington): Class B[3]	15	76	24	76	3	38	4	0.46	1.4	0.04	0.08	0.05	0.004	0.02	0.02
Cary: "Enviro Gems" pellets Class A Exceptional Quality[3]	90	135	40	149	13	23	12	0.46	1.2	0.03	0.05	0.03	0.004	0.02	0.01
Charlotte: Class B[3]															
Irwin Creek WWTP[4]	14	122	40	83	4	46	8	0.6	1.5	0.08	0.24	0.15	0.02	0.03	0.01
McAlpine Creek WWTP	17	110	36	133	4	36	14	0.42	1.5	0.06	0.18	0.08	0.004	0.02	0.01
McDowell Creek WWTP	14	131	47	163	7	48	12	0.32	1.3	0.03	0.12	0.04	0.004	0.01	0.01
Elizabeth City: Class B[3]	2	122	26	55	24	330	8	0.26	0.64	0.02	NA	0.03	0.002	0.01	0.01
Goldsboro: compost Class A Exceptional Quality[3]		45	7	8	22	31	10	0.18	0.28	0.04	NA	0.01	0.001	0.003	0.02
Milwaukee Metro Sewage District: "Milorganite" Class A Exceptional Quality[3]	95	100	32	92	8	42	14	0.5	1	0.06	0.46	0.09	0.002	0.02	0.01
Orange County: Class B[3]	3	222	47	206	21	42	11	0.54	1.6	0.03	0.04	0.01	0.001	0.01	0.004

Table 4-18. Plant Nutrient* and Metal Contents of Some Municipal Biosolids from or Available in North Carolina[1]

Biosolids Source	Solids	Total Nitrogen*	Plant-available Nitrogen[2]	P$_2$O$_5$*	K$_2$O*	Ca*	Mg*	Cu*	Zn*	Ni*	Cr	Pb	Cd	Mo*	As
Raleigh (Neuse River Resource Recovery Facility): "Raleigh Plus" Class A Exceptional Quality[3]	60	27	8	7	10	743	6	0.01	0.2	0.02	0.02	0.02	0.002	0.004	0.01
Raleigh (Neuse River Resource Recovery Facility): Class B[3]	3	144	56	275	180	20	6	0.5	1	0.05	0.05	0.04	0.01	0.03	0.03
Rocky Mount: Tar River Regional WWTP	4	15.4	3.5	4	0.5	4	0.4	0.03	0.04	0.06	NA	0.01	0.001	0.002	0.003
Winston-Salem Archie Elledge WWTP Pellets: Class A	93	79	23	73	4	39	8	0.56	3	0.05	0.1	0.08	0.004	0.02	0.01

* Essential plant nutrient

[1] Results vary between samples. This table represents average or median values. Data are from 2015 Annual Residual Sampling Summaries submitted to NC-DEQ Division of Water Quality except: Goldsboro Compost, which is from prior year data, and Milorganite, which is from www.milorganite.com.

[2] Plant-available nitrogen (PAN) estimated for the year of application. PAN depends on waste processing and land application methods and N forms in the biosolids. PAN is some fraction of total N = organic N + inorganic N (ammonium + nitrate + nitrite). Values here are for surface application; if injected or incorporated, PAN will be somewhat greater. For current methods used to estimate PAN, see the North Carolina Department of Agriculture and Consumer Services Waste and Compost Analysis Guide (http://www.ncagr.gov/agronomi/pdffiles/wasteguide.pdf).

[3] Class B biosolids meet minimum US-EPA land-application standards for pathogens, potentially toxic metals, and vector-attraction; Class A biosolids meet stricter pathogen standards; Class A Exceptional Quality biosolids are Class A biosolids with very low metal content.

[4] WWTP, Waste water treatment plant, a.k.a. water resource recovery facility.

Certified Organic Farm Management Alternatives

C. R. Crozier and S.C. Reberg-Horton, Crop and Soil Sciences

Although crop nutritional requirements are the same for organic and conventional farms, organic producers need to be more creative due to the limitations on allowable inputs. See the "North Carolina Organic Grain Production Guide" http://www.cefs.ncsu.edu/resources/organicgrainfinal.pdf for additional information. Since use of some soil amendments is limited to cases of nutrient deficiency, organic producers should maintain soil testing and plant tissue analysis records documenting specific nutrient deficiencies that need correction. With tissue testing, the appropriate plant part must be collected at the proper growth stage as specified by laboratory guidelines. See a detailed plant tissue analysis guide at www.ncagr.gov/agronomi/pdffiles/plantguide.pdf or contact your local county Extension Center for more details.

Certain inputs are allowable on organic production systems, if applied according to guidelines. These include many (not all) natural and certain synthetic materials. Current USDA regulations can be found at http://www.ams.usda.gov/AMSv1.0/nop. Find the section on "Organic Standards" and select "National List and Petitioned Substances," then find "National List and Petitions" and select "Current National List (eCFR)" to see a detailed listing of allowed and prohibited substances. See a list of commercially available materials that have been reviewed by the Organic Materials Review Institute (OMRI) at http://www.omri.org/omri-lists. In all cases, input use should be considered prohibited unless included in the farm plan and confirmed by the certifying authority prior to application.

Critical aspects of soil fertility management include pH, major nutrients (N, P, K,S, Ca, Mg), and micronutrients (especially B, Cu, Mn, Zn; but also Fe, Mo, Cl, etc.). A summary of soil fertility parameters and organic management options is given in Table 4-19 below.

Specific guidelines must be followed when applying composts and manures in organic production systems. Materials must be applied at agronomic rates in compliance with any applicable nutrient management guidelines (http://nutrients.soil.ncsu.edu/), and which avoid excess nutrients. Raw animal manures must either be: 1) composted according to specific criteria specified in the USDA National List, 2) applied to land used for a crop not intended for human consumption, 3) incorporated into the soil at least 90 days prior to the harvest of an edible product not contacting soil or soil particles, or 4) incorporated into the soil at least 120 days prior to the harvest of an edible product that does contact soil or soil particles. The guidelines for compost production for organic agriculture state that the initial C:N ratio must be between 25:1 and 40:1; and a temperature between 131 and 170 degrees Fahrenheit must be achieved and maintained for at least 3 days for in-vessel or static aerated pile systems, or for at least 15 days during which there are at least 5 turnings for windrow systems. Composts not meeting these criteria must be applied based on other raw manure criteria, which also apply to animal waste lagoon liquids and solids, and stockpiled poultry litter. Human or industrial wastes such as lime stabilized sludges or composted biosolids are not allowed. Ashes of manures may not be used, but ashes from other untreated plant and animal materials may be applied if not combined with any prohibited substances. Avoid over-reliance on animal manures, since this could lead to accumulation of excess P, Cu, and Zn in soils. Sporadic use of manures in conjunction with more frequent use of legume cover crops, green manures or other N sources is an excellent way to supply several plant nutrients in appropriate amounts.

FERTILIZER USE

Table 4-19. Soil Fertility Parameters and Management Options

Parameter	Problem Documentation	Supply Options[1]	Not Allowed
pH	Soil test	Standard calcitic or dolomitic agricultural ground limestone; pH can be lowered by adding elemental sulfur.	Hydrated or burnt lime [$Ca(OH)_2$, CaO], industrial wastes, slags, lime-stabilized biosolids
Major Nutrients			
Nitrogen (N)	Tissue analysis	Legumes, manures[3], animal by-products (blood, fish), plant by-products (cotton, apple, fermentation wastes), mined sodium nitrate ($NaNO_3$)[3]	Synthetic fertilizers, sewage sludges, municipal waste composts
Phosphorus (P)	Soil test, tissue analysis	Manures[3], rock phosphate, animal by-products (bone meal; fish, shrimp, and oyster scraps; leather)	Processed rock phosphates
Potassium (K)	Soil test, tissue analysis	Manures[3], plant by-products (ash, dried seaweed), greensand, sulfate of potash (K_2SO_4)[4], possibly muriate of potash (KCl)[3,4]	KCl if excess chloride
Sulfur (S)	Tissue analysis	Manures[3], plant by-products (cotton motes, peanut meal), elemental sulfur[4], gypsum ($CaSO_4$), Epsom salt ($MgSO_4$)[4], sulfate of potash (K_2SO_4)[4]	Synthetic fertilizers
Calcium (Ca)	Soil test, tissue analysis	Standard calcitic or dolomitic agricultural ground limestone, gypsum ($CaSO_4$), bone meal, ash	$Ca(OH)_2$, CaO, calcium nitrate [$Ca(NO_3)_2$]
Magnesium (Mg)	Soil test, tissue analysis	Standard dolomitic agricultural ground limestone, Epsom salts ($MgSO_4$)[4], sulfate of potash magnesium, bone meal, plant by-products (cottonseed meal, wood ash)	Synthetic fertilizers
Micronutrients[2]			
Boron (B)	Tissue analysis	Manures, animal and plant by-products, soluble boron fertilizers[4]	
Copper (Cu)	Soil test, tissue analysis	Manures, animal and plant by-products, sulfates & oxides[4]	chlorides
Manganese (Mn)	Soil test, tissue analysis	Manures, animal and plant by-products, sulfates & oxides[4]	chlorides
Zinc (Zn)	Soil test, tissue analysis	Manures, animal and plant by-products, sulfates & oxides[4]	chlorides
Co, Fe, Mo, Se	Tissue analysis[5]	Manures, animal and plant by-products, sulfates, carbonates, oxides, or silicates[4]	Chlorides, nitrates

[1] Inputs must be on the National Organic Program or the OMRI-approved source list and approved by the certifying agents.

[2] Avoid over-application of micronutrients since toxicities can occur.

[3] Note restrictions.

[4] Documentation of nutrient deficiency required.

[5] Deficiencies of Co, Mo, and Se are not common in North Carolina, and these elements are not included in routine tissue analysis performed by the North Carolina Department of Agriculture and Consumer Services, although Mo can be included upon request and for an additional fee. Consult a Cooperative Agricultural Extension office for information regarding private agricultural laboratories.

FERTILIZER USE

V — INSECT CONTROL

Relative Toxicity of Pesticides to Honey Bees

DAVID R. TARPY, Professor and Extension Apiculturist

Most pesticides are at least somewhat toxic to honey bees and other pollinators; however, the degree of toxicity varies considerably from product to product. Insecticides are generally the most likely to cause a bee kill; herbicides, fungicides, and defoliants present relatively minor danger to bees if used according to label directions. Pesticides are listed by active ingredients (AI) and common and/or brand name(s) where appropriate.

Table 5-1A. Relative Toxicity of Pesticides to Honey Bees

Group 1 — Highly Toxic. Severe bee losses may be expected **IF** the following pesticides are used when bees are present, or the product is applied near beehives, or within a day after application to foraging bees in the pesticide application area.

abemectin (Zephyr)	ethion	milbemectin (Milbemycin A4 + Milbemycin A3)
acephate (Orthene)	etofenprox	momfluorothrin
aldicarb (Temik)	famoxadone (Famoxate)	monocrotophos (Azodrin)
alpha-cypermethrin	famphur (Famophos)	naled (Dibrom)
arsenicals	fenamiphos	oxamyl
avermectin	fenazaquin	parathion
azadirachtin	fenitrothion (Sumithion)	pentachloronitrobenzene (PCNB)
azinphos-methyl (Guthion)	fenoxaprop-ethyl	permethrin (Ambush, Pounce)
bensulide	fenpropathrin (Danitol, Dasanit)	phenothrin
beta-cyfluthrin	fenthion (Baytex)	phorate
bifenthrin (Brigade, Discipline)	fenvalerate	phosmet (Imidan)
carbaryl (Sevin 80 S)	fipronil	phosphamidon
carbofuran	fluvalinate	pirimiphos-methyl
chlorethoxyfos	foramsulfuron	prallethrin (ETOC)
chlorfenapyr	fosthiazate	profenofos
chlorpyrifos (Dursban, Lorsban)	gamma-cyhalothrin (Prolex)	propoxur (Baygon)
clothianidin (Poncho 600)	imidacloprid	pyrazophos (Afugan)
cyantraniliprole	imiprothrin	pyrethrins
cyfluthrin (Baythroid)	indoxacarb (Steward, Avaunt)	pyridaben
cypermethrin (Ammo)	isofenphos	resmethrin (Synthrin)
cyphenothrin	lambda-cyhalothrin (Karate)	rotenone
deltamethrin (Decis)	lindane	spinetoram
d-phenothrin (Sumithrin)	LPOS (Sulfotine, RAID TVK)	spinosad (XDE-105, Tracer)
diazinon (Spectracide)	malathion (Cythion)	sulfoxaflor
dichlorvos (DDVP, Vapona)	methamidophos (Monitor, Tamaron)	tefluthrin
dicrotophos (Bidrin)	methidathion (Supracide)	tetrachlorvinphos
dimethoate (Cygon, DE-FEND)	methiocarb (Mesurol)	tetramethrin
dinotefuran	methomyl (Lannate, Nudrin)	thiamethoxam
diuron	methoprene	tralomethrin
D-trans-allethrin	methyl parathion	tolfenpyrad
emamectin benzoate	mexacarbate (Zectran)	zetamethrin (Mustang max)
esfenvalerate (Asona)		zineb

Group 2 — Moderately Toxic. These pesticides can be used in the vicinity of bees if dosage, timing, and method of application are correct; but these products should never be applied directly on bees in the field or at the colony location (apiaries).

acetamiprid	disulfoton	pirimicarb (Pirimor)
allethrin, D-trans	demeton	profenofox (Curacron)
arsenic acid	disulfoton	propamocarb (Carbamult)propamocarb
azadirachtin	endosulfan (Thiodan)endothion	hydrochloride (Banol)
benomyl	ethoprop (Mocap)	pyrethrum
bifenazate (Floramite)	fenarimol	ronnel
bromoxynil octanoate	fluazinam	sethoxydim
aluminum phosphide (Phostoxin)	fonofos (Dyfonate)	ppiroxamine
Bacillius thuringiensis (Di-Beta)	formetanate (Carzol)	sumithrin (Anvillollo)
Biothion	mesosulfuron methyl	tartar emetic
carbophenothion	oxamyl (Vydate)	temephos (Abate)
coumaphos (Co-Ral)	oxydemeton-methyl (Metasystox-R)	terbufos (Counter)
crotoxyphos	paraquat dichloride	thiacloprid (Calypso, YRC-2894)
crotoxyphos (Ciodrin)	phorate (Thimet)	thiazopyr (MANDATE, VISOR)
demeton	phosalone (Zolone)	thiodicarb (Larvin)

Table 5-1A. Relative Toxicity of Pesticides to Honey Bees

Group 3 — Relatively Non-Toxic. These pesticides can be used around bees with a minimum risk of injury.

Acaricides, Diseases, IGRs, and Insecticides

aldoxycarb (Standak)	cryolite (Kyocide)	Nosema locustae (Canning)
allethrin	cymiazole (Apitol)	oxythioquinox (Morestan)
amitraz	cyromazine (Trigard)	polynactins
azadirachtin (Margosan-O)	dibromochloropropane (Nemagon)	propargite (Comite, Omite)
Bacillus thuringiensis (Biotrol)	dicofol (Dicofol, Griffin dicofol)	pymetrozine (Fulfill, Endeavor)
Bacillus thuringiensis (Dipel)	diflubenzuron (Dimilin)	pyrethrum
Bacillus thuringiensis (Thuricide)	dinobuton (Dessin)	pyriproxyfen
B.t. kurstaki (Dipel 4L)	dioxathion (Delnav)	rotenone
B.t. kurstaki (Javelin)	ethion (Ethiol)	ryania
B.t. tenebrionis	Heliothis virus	tetradifon (Tedion)
chlorobenzilate (Acaraben)	metaflumizone	tetraflubenzuron (CME)
chlorobenzilate (Folbex)	methoxychlor (Marlate)	trichlorfon (Dylox)
clofentizine (Apollo)	multimethylalkenols (Stirrup)	Z-11-hexadecanol (tomato pinworm pheromone)
	nicotine	

Fungicides

acibenzolar-S-methyl (Actigard)	cyclanilide (Finish)	mancozeb
anilazine (Dyrene)	cymoyanil (Curzate 60DF)	maneb (Manzate)
anilazine (Kemate)	cyprodinil (Vangard WP)	metiram (Polyram)nabam (Parzate)
azoxystrobin (Heritage)	dazomet (Mylone)	polyphase P-100 (Troyson)
benomyl (Benlate)	dicholone	prochloraz
bordeaux mixture	dimethomorph (Acrobat MZ)	prochloraz/carbendazin (Sportac)
captafol (Difolatan)	diniconazole (Spotless)	sulfur
captan (Orthocide)	dinocap (Karathane)	thiram
chloropicrin	dithianon (Thynon)	thiramlmethoxychlor (Atasan)
copper 8-quinolinate	dodine (Cyprex)	trifloxystrobin (Flint, Stratego, Compass)
copper hydroxide	fenaminosulf (Lesan)	triforine (Funginex)
copper oxychloride sulfate	fenhexamid (Elevate 50 WDG)	triphenyltin hydroxide (Du-Ter)
copper sulfate—monohydrated	fluazinam (Omega 500F)	ziram (Zerlate)
cuprous oxide	folpet (Phaltan)	zoxamide (Zoxium 80W)
	glyodin (Glyoxide)	

Herbicides, Defoliants, Desiccants, and PGRs

2,3,6-TBA (Trysben)	diquat	naptalam (Alanap)
2,4,5-T	diuron (Karmex)	nitrofen (TOK)
2,4-D (2, 4-D)	EPTC (Eptam)	norflurazon (Zorial)
2,4-DB (Butoxon)	etephon (Ethrel)	ovasyn
2,4-DB (Butyrac)	ethalfluralin (Sonalan)	paraquat
acetochlor	EXD (Herbisan)	pendimethalin
alachlor (Lasso)	flufenacet (Axiom DF)	phenmedipham (Betanal)
amitrole	fluometuron (Cotoran)	picloram (Tordon)
ammonium sulfate	flumioxazin (Valor WDG)	prohexadione calcium (Apogee PGR, Baseline)
atrazine (AAtrex)	fluridone (BRAKE, Sonar)	prometryn (Caparol)
benomyl (Benlate)	fluroxypyr (Starane EC)	pronamide (Kerb)
bensulide	fluthiacet-methyl (Action)	propanil (Stam F-34)
bentazon (Basagran)	foramsulfron (Option)	propazine (Milogard)
bromacil (Hyvar)	glyphosate (Roundup)	propham (Ban-Hoe, IPC)
butifos (DEF)	hydrogen cyanamide (Dormex)	PT807-HCl (Ecolyst)
chlorbromuron (Maloran)	imadagylin (Arsenal)	quinchlorac (FACET)
chloroxuron (Tenoran)	imazamox (Raptor)	simazine (Princep)
clodinafop-propergyl (Discover)	isoxaflutole (Balance)	sodium chlorate (KNOCK 'UM OFF)
clofencet (Genesis)	linuron (Lorox)	terbacil (Sinbar)
cloproxydim (Select)	MCPA (Mapica)	terbutryn
cloransulam-methyl (First-Rate)	metaldehyde propazine (Milogard)	terbutryne (Igran)
cyanazine (Bladex)	methazole (Probe)	thiadiazuron (DROPP)
cyhalofop-butyl (Clincher)	metribuzin (Lexone)	tralkoxydim (Achieve 40DG)
dalapon	metribuzin (Sencor)	tribufos (DEF)
dicamba (Banvel)	mesotrione (Callisto)	tribuphos (6EC)
dichlobenil (Casoron)	metolachlor	tribuphos (Folex)
diflufenzopyr (Distinct)	monuron	tributyl phosphorotrithioite (Folex)
		trifluralin

INSECT CONTROL

Table 5-1B. Pesticide Use Inside and Around Honey Bee Hives — Formulations for use by the general public, unless otherwise noted

Pests	Chemical (Brand)	Formulation	Precautions and Remarks (Always follow product label directions for handling, product application, and disposal)
Tracheal Mite	menthol (Mite-A-Thol)	Crystalline granules	Both products generate vapors that kill tracheal mites. Apply onto inner cover/top super according to label directions. Best if used when ambient temperatures are above 70 degrees F for menthol and 50 degrees F for formic acid. Use gloves when handling crystals or gel packets.
	formic acid (Mite-Away Quick Strips)	Various delivery methods	
Varroa Mite	tau-fluvalinate (Apistan)	Plastic strip; pesticide-impregnated	Strips contain contact poison to kill mites. Use protective gloves when handling strips. Hang strips in brood-chamber according to label directions. Caution should be used, as mites have evolved a resistance to this particular chemical, and it may not be effective in many instances.
	formic acid (Mite-Away Quick Strips)	Various delivery methods	Product generates vapors to kill mites. Kills mites in sealed brood cells. Treat colonies according to label directions.
	coumaphos (Check-Mite+)	Plastic strip; pesticide-impregnated	For varroa mites, product should be used **only** when fluvalinate-resistance has been confirmed by NCDA Bee Inspectors. Caution should be exercised, as mites have evolved a resistance to this particular chemical and may not be effective in many instances.
	amitraz (Apivar)	Plastic strip; pesticide-impregnated	Strips contain active ingredient to kill mites upon contact. Use protective gloves when handling strips.
	thymol (ApiLife VAR or Apiguard)	Pesticide-impregnated vermiculite tablets or gel	Essential oils volatilize to kill mites outside of brood cells.
	sucrose octonoate (Sucrocide)	Liquid; mix with water	Spray all adult bees with fine mist; must be completely wetted to kill mites.
Small Hive Beetle (adults)	coumaphos (Check-Mite+)	Plastic strip; pesticide-impregnated	Use protective gloves when handling strips. Attach to cardboard or other material as specified on label direction and place strip-side down on bottom board to kill adult beetles. Application for varroa mites (see above) is **not** simultaneously effective for SHB.
(pupae)	permethrin (GardStar)	Liquid; mix with water	For ground treatment around hive(s) only. Kills larvae/pupae during soil-inhabiting phase of beetle life cycle. Mix and apply to soil according to label directions.
Wax Moth	paradichlorobenzene (Para-Moth)	Crystalline granules	Use to prevent infestation of stored hive equipment (drawn-comb) only. Do not use in hives containing honey bees. Use protective gloves when handling crystals. Store product in sealed container when not in use.

Always follow label directions, which require the removal of honey from beehives prior to most pesticide treatments.

Reducing the Risk of Pesticide Poisoning to Honey Bees

Pesticide poisoning of honey bees can usually be kept to a minimum if the pesticide applicators and the beekeepers take several precautions.

Precautions for the Pesticide Applicator

1. Always read and follow any warning statements regarding honey bees on the pesticide label.

2. If more than one product gives good control of the target pest, select a pesticide from Group 2 or 3 instead of Group 1 from the preceding "Relative Toxicity of Pesticides to Honey Bees."

3. Avoid applying any bee toxic pesticides on blooming plants that attract bees. Keep pesticide drift from nearby blooming weeds that are attracting bees.

4. Time of pesticide application is very important. Apply pesticides that are toxic to bees in the late afternoon (after 3 p.m.) or in the evening if at all possible. Most honey bees have stopped foraging and have returned to their hives by 3 p.m. This allows maximum time for the pesticide to decompose before the bees come into contact with it the next day.

5. Select the safest formulation of the pesticide that is available for the intended use. "Drifting" of the pesticide from the target pest and/or crop to areas frequented by bees should be minimized and formulation selection is the key to this problem.

   a. "Dusts" almost always drift more than other pesticide formulations and are generally more dangerous to bees than are sprays or granular applications.

   b. Spray formulations are usually safer to bees than are dusts, but there are differences among the spray formulation types. Generally, water-soluble formulations are safer than are emulsifiable-formulations, and fine sprays are less dangerous than are coarse sprays. Sprays of undiluted technical pesticide (ULV) may be more dangerous than diluted sprays.

   c. **Granular applications generally are the least likely to drift and accidentally kill bees.** Consider a granular formulation if it is suitable for destroying the target pest.

   d. Microencapsulated pesticides present a very distinct and serious threat to honey bees. The particle size of this pesticide formulation is very similar to that of pollen, and adult honey bees may carry this pesticide back to the hive where it will be combined with stored pollen. This pesticide will not kill the adult bees that collected it, but the microencapsulated pesticide will kill the brood (immature) stages of the bees and the young adult (nurse) bees that feed the brood when it is later fed to those bees. Bees have little protection against these products.

6. The mode of pesticide application is also important, particularly from a drifting standpoint. Aerial applications are generally more dangerous than applications by ground equipment. If a pesticide application is being made by air, it is the contractor's responsibility to notify any beekeepers that have *registered* apiaries (one or more hives of bees) within 1/2 mile of the area to be aerially sprayed. These regulations are defined in the N.C. Pesticide Laws, and the person responsible for the notification is the person who contracts for the aerial application.

7. Never apply any pesticide directly over a beehive.

8. Notify beekeepers who have beehives near an area to be treated with a pesticide so that they may attempt to protect their bees.

9. Follow proper precautions in disposing of unused pesticides and pesticide containers. Be particularly careful not to contaminate water with pesticides, as the water may be collected by bees and result in bee kills.

Precautions for the Beekeeper

1. If your bees are located in any area where pesticides are commonly used, then identify yourself as a beekeeper to your neighbors who may use pesticides.

2. Identify your apiaries with your name and address or telephone number if the apiary is not associated with your residence so that you may be notified if pesticides are to be used by a neighboring individual.

3. Explain the importance of your bees in the pollination of crops being grown on nearby fields to those growers so that they may consider the value of the bees in pollination before applying any pesticides that may kill the pollinating insects.

4. Be aware of the precautions that apply to the pesticide applicator (above) so that you can serve as a resource in providing solutions to reducing bee kills.

5. Do not place apiaries in areas used to grow crops that require heavy and frequent usage of pesticides.

6. Register your apiary locations with the N.C. Department of Agriculture if aerial applications of pesticides are used in your apiary locations.

7. As a very last resort, move your beehives if possible when bee-toxic pesticides are being applied near your apiary. Covering the hives (e.g., with wet burlap) is usually not possible for large apiaries and can cause bees to overheat or suffocate.

INSECT CONTROL

Insect Control in Field Corn

D. D. Reisig, Entomology Extension

Table 5-2. Insect Control in Field Corn

Insecticide, Mode of Action Code, and Formulation	Per Acre		Acres/gal (lb)	Preharvest Interval (PHI) (Days)	Precautions and Remarks
	Amount	Active (lb)			
Annual White Grub — At Planting Seed Treatments/ In Furrow					
bifenthrin, MOA 3 (Capture) LFR	3.4 to 13.6 oz	0.047 to 0.062	38 to 9.4	30	Provides control alone, without addition of seed treatment
clothianidin, MOA 4A (Poncho) 600 FS	—	0.025 mg per kernel	—	—	0.5 and 1.25 mg per kernel rate can provide improved control under high pest pressure or slow grow off conditions.
thiamethoxam, MOA 4A (Cruiser) 5 FS	—	0.025 mg per kernel	—	—	0.5 and 1.25 mg per kernel rate can provide improved control under high pest pressure or slow grow off conditions.
Billbug — At Planting Seed Treatments					
clothianidin, MOA 4A (Poncho) 600 FS	—	1.25 mg per kernel	—	—	Must be special-ordered from a seedsman. In most situations, these products will provide adequate control. Corn planted near previous year's corn, corn planted mid-April, and corn near good overwintering habitats are most at risk. In these situations, these products will not provide adequate control.
thiamethoxam, MOA 4A (Cruiser) 5 FS	—	1.25 mg per kernel	—	—	
Brown Stink Bug					
beta-cyfluthrin, MOA 3 (Baythroid XL) 1.0 EC	2.8 fl oz	0.022	45.7	21	Seedling injury mainly occurs in no-till. On larger plants, apply to stages just prior to tasseling. Use ground application only at 15+ gallons volume per acre. Results may be poor to mediocre depending on application. Applications can be effective up to, or less than, one week after treatment.
bifenthrin (Brigade, Discipline, Sniper, and others) 2 EC	6.4 fl oz	0.10	20	30	
bifenthrin, MOA 3 + zeta-cypermethrin, MOA 3 (Hero) 1.24 EC	10.3 fl oz	0.1	12.4	60 (forage) 30 (grain and stover)	
bifenthrin, MOA 3 + zeta-cypermethrin, MOA 3 (Steed) 1.5 EC	4.7 fl oz	0.055	27.2	60 (forage) 30 (grain and stover)	
cyfluthrin, MOA 3 (Tombstone) 1.0 EC	2.8 fl oz	0.044	45.7	21	
lambda-cyhalothrin, MOA 3 (Karate, Lambda-cyhalothrin, Silencer) 1.0 EC	3.84 fl oz	0.03	33.3	21	
(Warrior II and Karate Z) 2.08 CS	1.92 fl oz	0.03	66.7	21	
zeta-cypermethrin, MOA 3 (Mustang Maxx) 0.8 EC	4.0 fl oz	0.025	32	30	
zeta-cypermethrin, MOA 3 + bifenthrin, MOA 3 (Hero) 1.24 EC	10.3 fl oz	0.033 + 0.066	12.4	30	
Corn Leaf Aphid					
pyrethroids, MOA 3 and pyrethroid combinations	(see brown stink bug above for rates)	—	—	—	
Corn Earworm — In Whorl					
Bt transgenic corn, MOA 11 (Agrisure Viptera and Leptera)	—	—	—	—	This is transgenic corn seed. Observe the refuge specifications on the label. Corn earworm is not a yield-limiting pest in timely-planted corn.
chlorantraniliprole, MOA 28 (Prevathon) 0.43 SC	14 to 20 fl oz	0.047 to 0.067	9.1 to 6.4	14	Corn earworm is not a yield-limiting pest in timely-planted corn.
Cutworm — Postemergence					
beta-cyfluthrin, MOA 3 (Baythroid XL) 1.0 EC	1.6 to 2.8 fl oz	0.017 to 0.022	80 to 45.7	21	Best to direct spray to the plant base and use at least 15 gallons volume per acre by ground. Pyrethroids are suggested for organic soils. Use higher rates for heavier infestations or aerial application. Do not feed Lorsban-treated corn until 35 days post-treatment.
bifenthrin (Brigade, Discipline, Sniper, and others) 2 EC	2.1 to 6.4 fl oz	0.033 to 0.10	61 to 20	30	
bifenthrin, MOA 3 + zeta-cypermethrin, MOA 3 (Hero) 1.24 EC	2.6 to 6.1 fl oz	0.25 to 0.06	49.2 to 21	60 (forage) 30 (grain and stover)	
bifenthrin, MOA 3 + zeta-cypermethrin, MOA 3 (Steed) 1.5 EC	2.5 to 3.5 fl oz	0.029 to 0.041	51.2 to 36.6	60 (forage) 30 (grain and stover)	
Bt transgenic corn, MOA 11 (Agrisure Viptera, Herculex, Leptra, PowerCore, Optimum Intrasect, SmartStax)	See remarks	—	—	NA	This is transgenic corn seed. Plants will express Bt endotoxin. Observe the refuge specifications on the label.
chlorpyrifos, MOA 1B (Lorsban) 4 E	2 pt	1	4	14 (silage) 35 (grain)	
cyfluthrin, MOA 3 (Tombstone) 1.0 EC	0.8 to 1.6 fl oz	0.013 to 0.025	160 to 80	21	
esfenvalerate, MOA 3 (Asana XL) 0.66 EC	5.8 to 9.6 fl oz	0.03 to 0.05	22.1 to 13.3	21	
gamma-cyhalothrin, MOA 3 (Declare) 1.25 EC	0.77 to 1.28 fl oz	0.0075 to 0.0125	166.2 to 100	21	

Table 5-2. Insect Control in Field Corn

Insecticide, Mode of Action Code, and Formulation	Per Acre Amount	Per Acre Active (lb)	Acres/gal (lb)	Preharvest Interval (PHI) (Days)	Precautions and Remarks
Cutworm — Postemergence (continued)					
lambda-cyhalothrin, MOA 3 (Karate, Lambda-cyhalothrin, Silencer) 1.0 EC	1.9 to 3.2 fl oz	0.015 to 0.025	67.4 to 40	21	
(Warrior II and Karate Z) 2.08 CS	1 to 1.6 fl oz	0.015 to 0.025	128 to 80	21	
methoxyfenozide, MOA 18A (Intrepid) 2F	4 to 8 fl oz	0.06 to 0.12	32 to 16	21	
zeta-cypermethrin, MOA 3 (Mustang Maxx) 0.8 EC	1.3 to 2.8 fl oz	0.008 to 0.0175	98.5 to 45.7	30	
European Corn Borer					
beta-cyfluthrin, MOA 3 (Baythroid XL) 1.0 EC	1.6 to 2.8 oz	0.017 to 0.022	80 to 45.7	21	Must be applied before borers enter stalk. Apply by ground only and into plant whorls with at least 25 gallons water per acre. Use 30 psi or less. A surfactant may improve whorl penetration.
bifenthrin, MOA 3 (Brigade, Discipline, Sniper, and others) 2 EC	2.1 to 6.4 oz	0.033 to 0.10	61 to 20	30	
bifenthrin, MOA 3 + zeta-cypermethrin, MOA 3 (Hero) 1.24 EC	4.0 to 10.3 oz	0.4 to 0.10	32 to 12.4	60 (forage) 30 (grain and stover)	
bifenthrin, MOA 3 + zeta-cypermethrin, MOA 3 (Steed) 1.5 EC	3.5 to 4.7 oz	0.041 to 0.055	36.6 to 27.2	60 (forage) 30 (grain and stover)	
Bt transgenic corn, MOA 11 (Agrisure Viptera, Genuity VT Double/Triple PRO,, Herculex, Leptra, Optimum Intrasect, PowerCore, SmartStax)	See remarks	—	—	NA	This is transgenic corn seed. Plants will express Bt endotoxin. Observe the refuge specifications on the label.
chlorantraniliprole, MOA 28 (Prevathon) 0.43 SC	14 to 20 fl oz	0.047 to 0.067	9.1 to 6.4	14	
chlorpyrifos, MOA 1B (Lorsban) 15 G	6.5 lb	1	0.154	35	Apply by air or ground. Will handle whorl infestations, but effectiveness decreases with stalk boring. Rainfall soon after enhances control.
(Lorsban) 4 E	2 pt	1	4	35	
cyfluthrin, MOA 3 (Tombstone) 1.0 EC	1.6 to 2.8 fl oz	0.025 to 0.044	80 to 45.7	21	Must be applied before borers enter stalk. Apply by ground only and into plant whorls with at least 25 gallons water per acre. Use 30 psi or less. A surfactant may improve whorl penetration.
esfenvalerate, MOA 3 (Asana XL) 0.66 EC	9.6 fl oz	0.05	13.3	21	
gamma-cyhalothrin, MOA 3 (Declare) 1.25 EC	1.02 to 1.54 fl oz	0.01 to 0.015	125.5 to 83.1	21	
lambda-cyhalothrin, MOA 3 (Karate, Lambda-cyhalothrin, Silencer) 1.0 EC	2.6 to 3.8 fl oz	0.02 to 0.03	49.2 to 33.7	21	
(Warrior II and Karate Z) 2.08 CS	1.28 to 1.92 fl oz	0.02 to 0.03	100 to 66.7	21	
methoxyfenozide, MOA 18A (Intrepid) 2F	4 to 8 fl oz	0.06 to 0.12	32 to 16	21	
spinosad, MOA 5 (Blackhawk) 4 SC	1.67 to 3.3 fl oz	0.038 to 0.075	76.6 to 38.8	28	
zeta-cypermethrin, MOA3 (Mustang Maxx) 0.8 EC	2.7 to 4.0 fl oz	0.017 to 0.025	47.4 to 32	30	
Fall Armyworm — In Whorl					
Bt transgenic corn, MOA 11 (Agrisure Viptera, Genuity VT Double/Triple PRO, Leptra, PowerCore, SmartStax)	—	See remarks	—	—	This is transgenic corn seed. Oberve the refuge specifications on the label.
chlorantraniliprole, MOA 28 (Prevathon) 0.43 SC	14 to 20 fl oz	0.047 to 0.067	9.1 to 6.4	14	Use a minimum of 15 gallons per acre by ground for whorl treatment (not by air). Low pressure spray and addition of surfactant may help liquid to penetrate into whorl. Application to large caterpillars may not give satisfactory results.
Grasshopper					
bifenthrin, MOA 3 (Brigade, Discipline, Sniper, and others) 2 EC	2.1 to 6.4 fl oz	0.033 to 0.10	61 to 20	30	Apply by air or ground uniformly over foliage as a broadcast treatment. Early morning treatment preferred. Use higher rates for heavy infestation. Grasshoppers are often confined to field margins.
chlorpyrifos, MOA 1B (Lorsban) 4 E	0.5 to 1 pt	0.25 to 0.5	16 to 8	21	
pyrethroids, MOA 3 and pyrethroid combinations	(see European corn borer above for rates)	—	—	—	
Sod Webworm, Chinch Bug					
bifenthrin, MOA 3 (Brigade, Discipline, Sniper, and others) 2 EC	2.1 to 6.4 fl oz	0.033 to 0.1	61 to 20	30	Apply to base of seedlings as a directed spray or over the row. Seldom an economic problem. Use higher rates for chinch bugs. Drop nozzles at 15 gallons per acre or above will give better results.
pyrethroids, MOA 3 and pyrethroid combinations	(see European corn borer above for rates)	—	—	—	
clothianidin, MOA 4A (Poncho) 600 FS	—	0.25 to 1.25 mg per kernel	—	—	1250 rate must be special-ordered from a seedsman.
thiamethoxam, MOA 4A (Cruiser) 5 FS	—	0.5 to 1.25 mg per kernel	—	—	

77

INSECT CONTROL

Table 5-2. Insect Control in Field Corn

Insecticide, Mode of Action Code, and Formulation	Per Acre		Acres/gal (lb)	Preharvest Interval (PHI) (Days)	Precautions and Remarks
	Amount	Active (lb)			
Sod Webworm, Chinch Bug (continued)					
carbaryl, MOA 1A (Sevin XLR Plus) 4 EC	2 pt	1	4	14	Apply to base of seedlings as directed spray or over the row. Seldom an economic problem. Use higher rates for chinch bugs. Drop nozzles at 15 gallons/acre or above will give better results.
chlorpyrifos, MOA 1B (Lorsban) 4 E	1 pt	0.5	8	21	
Sugarcane Beetle — At Planting Treatments					
clothianidin, MOA 4A (Poncho) 600 FS	—	1.25 mg per kernel	—	—	This seed treatment, combined with an in-furrow insecticidal granular or liquid application will still provide only fair control. 1250 rate must be special-ordered from a seedsman.
clothianidin, MOA 4A + in-furrow insecticide, MOA 1B (Poncho 500) + (various, e.g., chlorpyrifos (Lorsban), phosphorothioic acid + bifenthrin (SmartChoice), tebupirimphos + cyfluthrin (Aztec), terbofos (Counter), etc.))	—	—	—	—	See recommendations for seed treatment above. Granular insecticide alone or 500 rate of seed treatment alone will not provide adequate control without granular insecticide. Expect only fair control.
True Armyworm — In Whorl and on Foliage					
bifenthrin, MOA 3 (Brigade, Discipline, Sniper, and others) 2 EC	2.1 to 6.4 oz	0.04 to 0.16	61 to 20	30	Apply into plant whorls where worms are located and use a minimum of 15 gallons water per acre. Treat when worms are small. Aerial application is satisfactory when caterpillars are not in whorl (post-tassel). Armyworm problems are usually confined to no-till planted corn seedlings. Consult county agent for scouting information.
chlorantraniliprole, MOA 28 (Prevathon) 0.43 SC	14 to 20 fl oz	0.047 to 0.067	9.1 to 6.4	14	
chlorpyrifos, MOA 1B (Lorsban) 4 EC	2 pt	1	4	35	
methomyl, MOA 1A (Lannate) 2.4 LV (Lannate) 90 SP	0.75 to 1.5 pt 0.25 to 0.5 lb	0.23 to 0.45 0.23 to 0.45	10.7 to 5.3 4 to 2	3 (forage) 21 (fodder)	
pyrethroids, MOA 3 and pyrethroid combinations	(see European corn borer above for rates)	—	—	—	
spinosad, MOA 6 (Blackhawk) 4 SC	1.67 to 3.3 fl oz	0.038 to 0.075	76.6 to 38.8	7 (forage or seed) 28 (grain)	
Western or Northern Corn Rootworm — At Planting, Seed Treatments					
clothianidin, MOA 4A (Poncho) 600 FS	—	1.25 mg/kernel	—	—	Must be special-ordered from a seedsman. Rootworms mainly a problem in Piedmont and mountain regions where corn is not rotated.
Bt transgenic corn, MOA 11 (Agrisure, Herculex XTRA, Genuity VT Triple Pro, SmartStax)	—	See remarks	—	—	This transgenic corn is designed to prevent root injury from rootworm larvae. Usually only needed in corn following corn. Observe the refuge specifications on the label.
chlorpyrifos, MOA 1B (Lorsban) 15 G	8 oz/1,000 ft	*	—	—	Rootworms are presently a problem in non-rotated corn in the Piedmont and mountains. Apply granules 6- to 7-inch band over the open seed furrow and in front of the planter press wheel at planting time. Consult product label for incorporation instructions. Terbufos may be applied directly into the seed furrow. Do not apply phorate into seed furrow as seedling injury may occur. Terbufos may interact with Beacon herbicide and injure plants. Consult label.
phorate, MOA 1B (Thimet) 20 G	6 oz/1,000 ft	—	—	—	
tefluthrin, MOA 1A (Force) 3.0 G (Force) CS	4 to 5 oz/1,000 ft 0.46 to 0.57 oz/1,000 ft	*	—	—	
terbufos, MOA 1B (Counter) 20 G	6 oz/1,000 ft	*	—	—	
Wireworm — At Planting Treatments					
bifenthrin, MOA 3 (Capture) LFR	3.4 to 13.6 oz	0.047 to 0.062	—	—	Apply as an in-furrow spray, microstream, or t-band.
clothianidin, MOA 4A (Poncho) 600 FS	0.5 to 1.25 mg/kernel	—	—	—	1250 rate must be special-ordered from a seedsman.
thiamethoxam, MOA 4A (Cruiser) 5 FS	0.5 to 1.25 mg/kernel	—	—	—	
thiamethoxam, MOA 4A + chlorantraniliprole, MOA 28 (Lumivia)	—	0.25 mg + 0.25 mg per kernel	—	—	
phorate, MOA 1B (Thimet) 20G	6 oz/1,000 ft	—	—	—	Apply only in T-band over open furrows. Results may be poor if approximately 50% fails to fall with the seed (into seed furrows); however, in-furrow application may reduce stand.
tefluthrin, MOA 1A (Force) 3.0 G (Force) CS	4 to 5 oz/1,000 ft 0.46 to 0.57 oz/1,000 ft	*	—	—	T-band or in-furrow. If T-banded, some granules must fall with seed for wireworm control. Wireworm control is improved when used in-furrow. Terbufos may interact with Beacon herbicide when used in-furrow.
terbufos, MOA 1A (Counter) 20 G	6 oz/1,000 ft	*	—	—	

* For 30-inch or wider row spacings.

PRECAUTIONS: Always use pesticides according to label directions. Be mindful of reducing the impact of pesticides on wildlife and groundwater. See Extension publication AG-463-2, *Pesticides and Wildlife—Corn*, for further information.

Insect Control in Grain Sorghum

D. D Reisig, Entomology Extension

Table 5-3. Insect Control in Grain Sorghum

Insecticide, Mode of Action Code, and Formulation	Per Acre Amount	Active (lb)	Acres/gal (lb)	Preharvest Interval (PHI) (Days)	Precautions and Remarks
Aphid (including sugarcane aphid) — At Planting					
clothianidin, MOA 4A (Poncho) 600 FS	5.1 to 6.4 oz/cwt	See label	—	—	Follow label instructions for mixing.
clothianidin, MOA 4A + *Bacillus firmus* (for nematodes) (Poncho/VOTiVO)	6.13 fl oz/cwt	See label	—	—	
imidacloprid, MOA 4A (Gaucho) 480 FS (Gaucho) 600 FS	8 fl oz/cwt 6.4 fl oz/cwt	See label	—	45 (forage)	
thiamethoxam, MOA 4A (Cruiser) 5 FS	5.1 to 7.6 fl oz	See label	—	45 (forage)	
Aphid (excluding sugarcane aphid) — Foliar					
beta-cyfluthrin, MOA 3 (Baythroid XL) 1.0 EC	1.6 to 2.8 oz	0.017 to 0.022	80 to 45.7	21	Ground application with at least 15 gallons water per acre is preferred. Aerial application should use at least 5 gallons water per acre. At least 300 aphids per plant are necessary to justify treatment.
chlorpyrifos, MOA 1B (Lorsban) 75 WG	0.5 to 1 pt	0.25 to 0.5	16 to 8	28	
chlorpyrifos, MOA 1B + lambda-cyhalothrin, MOA 3 (Cobalt Advanced) 75 WG	11 to 38 fl oz	See label	11.6 to 3.4	30 to 60 (See label)	
cyfluthrin, MOA 3 (Tombstone) 1.0 EC	1.3 to 2.8 oz	0.2 to 0.044	98.5 to 45.7	14	
dimethoate, MOA 1B (Dimethoate) 4 EC	0.5 to 1 pt	0.25 to 0.5	16 to 8	28	
lambda-cyhalothrin, MOA 3 (Karate, Lambda-cyhalothrin, Silencer) 1.0 EC (Warrior II and Karate Z) 2.08 CS	2.56 to 3.84 fl oz 1.28 to 1.92 fl oz	0.02 to 0.03 0.02 to 0.3	50 to 33.3 100 to 66.7	30 30	
zeta-cypermethrin, MOA 3 (Mustang Maxx) 0.8 EC	3.2 to 4.0 fl oz	0.02 to 0.25	40 to 32	14 (grain) 45 (forage)	
Aphid (sugarcane aphid only) — Foliar					
flupyradifurone, MOA 4D (Sivanto) 200 SL	4 to 7 fl oz	0.052 to 0.091	32 to 18.3	21 (grain) 7 (forage)	A maximum of 28 ounces per acre can be used in a season.
sulfoxaflor, MOA 4C (Transform) 50 WG	0.75 to 1.5 oz	0.024 to 0.047	171 to 85	14 (grain) 7 (forage)	Note that this use is provided only under a Section 18 emergency exemption set to expire Nov. 30, 2016. This exemption is expected to be granted again in 2017, but check the label for confirmation. A maximum of 3 ounces per acre can be used in a season.
Chinch Bug — At Planting					
clothianidin, MOA 4A (Poncho) 600 FS	5.1 to 6.4 oz/100 lb seed	See label	—	—	Follow label instructions for mixing.
imidacloprid, MOA 4A (Gaucho) 480 FS (Gaucho) 600 FS	8 fl oz/cwt 6.4 fl oz/cwt	See label	—	45 (forage)	
imidacloprid, MOA 4A (Gaucho) 480 FS (Gaucho) 600 FS	8 fl oz/cwt 6.4 fl oz/cwt	See label	—	45 (forage)	
thiamethoxam, MOA 4A (Cruiser) 5 FS	7.6 fl oz	See label	—	45 (forage)	
Chinch Bug — Foliar					
carbaryl, MOA 1A (Sevin XLR Plus) 4 EC	3 pt	1.5	2.7	21	Apply to base of plants where insects congregate. Begin applications when insects migrate from small grains or grass weeds to sorghum. Expect fair control from pyrethroids (MOA 3).
chlorpyrifos, MOA 1B (Lorsban) 75 WG	0.67 to 1.33 lbs	0.5 to 1.0	1.5 to 0.75	28	
pyrethroids, MOA 3 and pyrethroid combinations	(use highest labeled rates)	See label	—	—	
Corn Earworm/Webworm — In Heads					
Bacillus thuringiensis, MOA 11B2 (Various)	—	—	—	0	Best when larvae are small.
beta-cyfluthrin, MOA 3 (Baythroid XL) 1.0 EC	1.3 to 2.8 fl oz	0.01 to 0.022	98.5 to 45.7	14	Ground application with at least 15 gallons water per acre is preferred. Aerial application should use at least 5 gallons water per acre. Use higher rates by air for serious infestation. Threshold is one medium to large earworm or armyworm per head or three webworms per head. See label for Asana. Entrust is OMRI listed.
carbaryl, MOA 1A (Sevin XLR Plus) 4 EC	3 pt	1.5	2.7	21	
chlorantraniliprole, MOA 28 (Prevathon) 0.43 SC	14 to 20 oz	0.047 to 0.067	9.1 to 6.4	14	
cyfluthrin, MOA 3 (Tombstone) 1.0 EC	1.3 to 2.8 fl oz	0.02 to 0.044	98.5 to 45.7	14	
esfenvalerate, MOA 3 (Asana XL) 0.66 EC	5.8 to 9.6 fl oz	0.03 to 0.05	22 to 13.3	21	
lambda-cyhalothrin, MOA 3 (Karate, Lambda-cyhalothrin, Silencer) 1.0 EC (Warrior II and Karate Z) 2.08 C	2.6 to 3.8 fl oz 1.28 to 1.92 fl oz	0.02 to 0.03 0.02 to 0.03	49.2 to 33.7 100 to 66.7	30 30	
methomyl, MOA 1A (Lannate) 2.4 LV (Lannate) 90 SP	0.75 to 1.5 pt 0.25 to 0.5 lb	0.23 to 0.45 0.23 to 0.45	10.7 to 5.3 4 to 2	14 14	
spinosad, MOA 5 (Blackhawk) 4 SC (Entrust) 80 WP	1.7 to 3.0 oz 1 to 2 oz	0.039 to 0.068 0.05 to 0.01	75.3 to 42.7 16 to 8	21 (grain) 3 (forage)	
zeta-cypermethrin, MOA 3 (Mustang Maxx) 0.8 EC	1.8 to 4.0 oz	0.011 to 0.025	71.1 to 32	14 (grain) 45 (forage)	

INSECT CONTROL

Table 5-3. Insect Control in Grain Sorghum

Insecticide, Mode of Action Code, and Formulation	Per Acre		Acres/gal (lb)	Preharvest Interval (PHI) (Days)	Precautions and Remarks
	Amount	Active (lb)			
Fall Armyworm					
chlorantraniliprole, MOA 28 (Prevathon) 0.43 SC	14 to 20 oz	0.047 to 0.067	9.1 to 6.4	14	Difficult to control—ground application only with high volume. Direct spray into whorls. Treat at 80% infestation (1 worm per plant) or 40% infestation (multiple worms per plant). Treat when worms are small. Addition of surfactant and application when dew is on plant may be helpful. Entrust is OMRI listed.
chlorpyrifos, MOA 1B (Lorsban) 75 WG	0.67 to 1.33 oz	0.5 to 1	191 to 96.2	See label	
methomyl, MOA 1A (Lannate) 2.4 LV (Lannate) 90 SP	0.75 to 1.5 pt 0.25 to 0.5 lb	0.23 to 0.45 0.23 to 0.45	10.7 to 5.3 4 to 2	14 14	
spinosad, MOA 5 (Blackhawk) 4 SC (Entrust) 80 WP	1.7 to 3.0 oz 1 to 2 oz	0.039 to 0.068 0.05 to 0.01	75.3 to 42.7 16 to 8	21 (grain) 3 (forage)	

Insect Control in Small Grains

D. D. Reisig, Entomology Extension

Table 5-4. Insect Control in Small Grains

Insecticide, Mode of Action Code, and Formulation	Per Acre		Acres/gal (lb)	Preharvest Interval (PHI) (Days)	Precautions and Remarks
	Amount	Active (lb)			
Aphid — At Planting					
imidacloprid, MOA 4A (Gaucho) 480 FS (Gaucho) 600 FS (Gaucho) XT	1 to 3 fl oz/cwt 0.8 to 2.4 fl oz/cwt 3.5 fl oz/cwt	See label	—	45 (forage)	Early season protection against aphids. Has shown barley yellow dwarf suppression. Most effective on early planted grains. Acknowledge plant-back restrictions. See Hessian fly section.
thiamethoxam, MOA 4A (Cruiser) 5 F)	0.75 to 1.33 fl oz/cwt	See label	—	45 (forage)	
Aphid — Foliar					
beta-cyfluthrin, MOA 3 (Baythroid XL) 1.0 EC	1.8 to 2.4 fl oz	0.014 to 0.019	71.1 to 53.3	7 (forage) 30 (harvest)	
cyfluthrin, MOA 3 (Tombstone) 1.0 EC	1.8 to 2.4 fl oz	0.028 to 0.038	71.1 to 53.3	30	
dimethoate, MOA 1B (Dimethoate) 4 EC	0.5 to 0.75 pt	0.25 to 0.37	16 to 10.7	35	Will not reduce barley yellow dwarf virus infection. Consult county extension agent for scouting and threshold suggestions. Keep lambda-cyhalothrin away from waterways.
lambda-cyhalothrin, MOA 3 (Karate, Lambda-cyhalothrin, Silencer) 1.0 EC (Warrior II and Karate Z) 2.08 C	2.56 fl oz 1.28 fl oz	0.02 0.03	50 100	30 30	
zeta-cypermethrin, MOA 3 (Mustang Maxx) 0.8 EC	3.2 to 4.0 oz	0.02 to 0.025	40 to 32	14	
Cereal Leaf Beetle					
beta-cyfluthrin, MOA 3 (Baythroid XL) 1.0 EC	1.0 to 1.8 fl oz	0.008 to 0.014	128 to 71.1	7 (forage) 30 (harvest)	Use where beetle eggs/larvae are above threshold. Application of insecticide with topdress fertilizer for preventative control is not advised. Lower rates should only be used where population densities are above threshold, but moderate.
carbaryl*, MOA 1A (Sevin XLR Plus) 4 EC	1 pt	0.5	8	21	
chlorpyrifos, MOA 1B + lambda-cyhalothrin, MOA 3 (Cobalt Advanced) 2.63 EC	11 to 25 fl oz	See label	11.6 to 2.3	30	
cyfluthrin, MOA 3 (Tombstone) 1.0 EC	1.0 to 1.8 fl oz	0.016 to 0.028	128 to 71.1	30	
gamma-cyhalothrin, MOA 3 (Declare) 1.25 EC	1.02 to 1.54 oz	0.01 to 0.015	125.5 to 83.1	30	
lambda-cyhalothrin, MOA 3 (Karate, Lambda-cyhalothrin, Silencer) 1.0 EC (Warrior II and Karate Z) 2.08	2.56 fl oz 1.92 fl oz	0.02 0.03	50 66.7	30 30	
methomyl, MOA 1A (Lannate) 2.4 LV (Lannate) 90 SP	1 to 2 pt 0.25 to 0.5 lb	0.22 to 0.45 0.22 to 0.45	8 to 4	7 7	
zeta-cypermethrin, MOA 3 (Mustang Maxx) 0.8 EC	1.6 to 4.0 oz	0.011 to 0.025	80 to 32	14	

INSECT CONTROL

Table 5-4. Insect Control in Small Grains

Insecticide, Mode of Action Code, and Formulation	Per Acre		Acres/gal (lb)	Preharvest Interval (PHI) (Days)	Precautions and Remarks
	Amount	Active (lb)			
Hessian Fly— Fall Generation					
imidacloprid, MOA 4A (Gaucho) 600 FS (Gaucho) XT (Rancona Crest)	1.2 to 2.4 fl oz/cwt 3.5 fl oz/cwt 5.0 to 8.3 fl oz/cwt	See label	—	45 (forage)	Early season protection against Hessian fly. Seed usually treated by seedsman. Acknowledge plant-back restriction.
thiamethoxam, MOA 4A (Cruiser) 5 FS	0.75 to 1.33 oz/cwt	See label	—	45 (forage)	
Hessian Fly— Fall and Late Winter Generations					
beta-cyfluthrin, MOA 3 (Baythroid XL) 1.0 EC	2.4 fl oz	0.019	53.3	3 (forage) 30 (harvest)	Apply to fields with high egg count in fall; preferable at or before the 2-3 leaf stage. In spring, apply to infested fields as flies emerge. Use high rates for heavy infestations. Recent NCSU experiments suggest that a resistant variety or seed treatment are far superior to foliar sprays as rescue treatments.
cyfluthrin, MOA 3 (Tombstone) 1.0 EC	2.4 fl oz	0.038	53.3	30	
lambda-cyhalothrin, MOA 3 (Karate, Lambda-cyhalothrin, Silencer) 1.0 EC (Warrior II and Karate Z) 2.08 EC	3.8 fl oz 1.92 fl oz	0.03 0.03	33.7 66.7	30 30	
zeta-cypermethrin, MOA 3 (Mustang Maxx) 0.8 EC	4 fl oz	0.025	32	14	
True Armyworm — Spring					
beta-cyfluthrin, MOA 3 (Baythroid XL) 1.0 EC	1.8 to 2.4 fl oz	0.013 to 0.019	71.1 to 53.3	3 (forage) 30 (harvest)	Apply by air or ground when armyworms are at 2 per square foot or greater. Use higher rates when caterpillars are very numerous. High volume (3 to 5 gallons per acre) may be beneficial in thickly planted wheat. Poor performance may result when temperatures are cool or when rainfall washes residues from plants. Best to apply when conditions are warm (60 degrees F plus) and armyworms are active. Carbaryl may stimulate aphid populations. Entrust is OMRI listed.
carbaryl, MOA 1A (Sevin XLR Plus) 4 EC	1.5 pt	0.75	5.3	21	
cyfluthrin, MOA 3 (Tombstone) 1.0 EC	1.8 to 2.4 fl oz	0.028 to 0.038	71.1 to 53.3	30	
gamma-cyhalothrin, MOA 3 (Declare) 1.25 EC	1.02 to 1.54 oz	0.01 to 0.015	125.5 to 83.1	30	
lambda-cyhalothrin, MOA 3 (Karate, Lambda-cyhalothrin, Silencer) 1.0 EC (Warrior II and Karate Z) 2.08	2.6 to 3.8 fl oz 1.28 to 1.92 fl oz	0.02 to 0.03 0.02 to 0.03	49.2 to 33.7 100 to 66.7	30 30	
methomyl, MOA 1A (Lannate) 2.4 LV (Lannate) 90 SP	1.5 pt 0.5 lb	0.45 0.45	5.3 2	7 7	
spinosad, MOA 5 (Blackhawk) 4 SC (Entrust) 80 WP	1.1 to 3.0 oz 1 to 2 oz	0.026 to 0.068 0.05 to 0.01	116.4 to 42.7 16 to 8	3 (forage) 21 (harvest)	
zeta-cypermethrin, MOA 3 (Mustang Maxx) 0.8 EC	1.6 to 4.0 oz	0.011 to 0.025	80 to 32	14	
Wireworm — At Planting					
imidacloprid, MOA 4A (Gaucho) 480 FS (Gaucho) 600 FS (Gaucho) XT (Rancona Crest)	1 fl oz/cwt 0.8 fl oz/cwt 3.5 fl oz/cwt 8.3 fl oz/cwt	See label	—	45 (forage)	See remarks under Aphids. Seed treatments must be applied by seedsman.
thiamethoxam, MOA 4A (Cruiser) 5 FS	0.75 fl oz/cwt	See label	—	45 (forage)	

CAUTION: Always use pesticides according to label directions. Be mindful of reducing the impact of pesticides on wildlife and groundwater. See Extension publication AG-463-6, *Pesticides and Wildlife—Small Grains*, for further information.

INSECT CONTROL

Insect Control on Cotton

D. D. Reisig, Entomology Extension

NOTE: Use the Mode of Action (MOA) codes following each insecticide to combat the development of insecticide resistance. Active ingredients sharing the same letter/number have the same mode of action.

Table 5-5A. Insect Control on Cotton

Insect Insecticide, Mode of Action (MOA), and Formulation	Per Acre		Acres/gal (lb)	Pre-harvest Interval (days)	Precautions and Remarks
	Amount	Active (lb)			
Beet Armyworm					
emamectin benzoate, MOA 6 (Denim) 0.16 EC	6 to 8 oz	0.0075 to 0.01	21.3 to 16	21	Bollgard II, TwinLink, WideStrike and WideStrike 3 varieties show high resistance to beet armyworm damage, unless larvae move to cotton from late burned-down weed hosts (see Bollworm/Budworm section for Bt cotton notes).
indoxacarb, MOA 22 (Steward) 1.25 SC	9.2 to 11.3 oz	0.09 to 0.11	14 to 11.5	14	
methoxyfenozide, MOA 18A (Intrepid) 2F	4.0 oz	0.06	33	14	Refer to labels for seasonal total active ingredient restrictions for all products.
methoxyfenozide, MOA 18A + spinetoram, MOA 5 (Intrepid Edge) 3F	4.0 to 8.0 oz	0.094 to 0.188	32 to 16	28	
spinosad, MOA 5 (Blackhawk) 4 SC	2.14 to 2.9 oz	0.067 to 0.089	60 to 45	28	
chlorantraniliprole, MOA 28 (Prevathon) 0.43 SC	14 to 27 oz	0.047 to 0.09	9.1 to 4.8	21	
Bollworm[a] /Tobacco Budworm					
Bollgard II, MOA 11B2 (varous varieties)					CryIA(c) and Cry2A(b) alleles in Bollgard II produce two *Bacillus thuringiensis* (*Bt*) delta endotoxins. Moderate activity against bollworm and high activity against other pest caterpillar species on cotton except cutworms. No activity against insects other than caterpillars.
TwinLink, MOA 11B2 (various varieties)					Cry1Ab and Cry2Ae alleles in TwinLink produce two *Bacillus thuringiensis* (*Bt*) delta endotoxins. Moderate activity against bollworm and high activity against other pest caterpillar species on cotton except cutworms. No activity against insects other than caterpillars.
WideStrike, MOA 11B2 (various varieties)					CryIA(c) and CryIF alleles in WideStrike produce two *Bacillus thuringiensis* (*Bt*) delta endotoxins. Low to moderate activity against bollworm and high activity against other pest caterpillar species on cotton except cutworms. No activity against insects other than caterpillars.
WideStrike 3, MOA 11B2 (various varieties)					CryIA(c), CryIF and Vip3A alleles in WideStrike 3 produce two *Bacillus thuringiensis* (*Bt*) delta endotoxins and one exotoxin. High activity against all pest caterpillar species on cotton except cutworms. No activity against insects other than caterpillars.
beta-cyfluthrin (MOA 3) Baythroid XL 1.0 EC	1.6 to 2.6 oz	0.013 to 0.021	77 to 47.6	0	High pressure (50 to 70 psi) and low volume (6 to 10 gpa) generally advised for pyrethroids against caterpillars.
bifenthrin, MOA 3 (Brigade, Fanfare, Discipline, Sniper and others) 2 EC	2.6 to 6.4 oz	0.04 to 0.1	50 to 20	14	
bifenthrin, MOA 3 + zeta-cypermethrin, MOA 3 (Hero) 1.24 EC	5.2 to 10.3 oz	0.05 to 0.1	24.8 to 12.4	14	
bifenthrin, MOA 3 + zeta-cypermethrin, MOA 3 (Steed) 1.5 EC	3.5 to 4.7 oz	0.041 to 0.055	36.6 to 27.3	14	
bifenthrin, MOA 3 + imidacloprid, MOA 4 (Swagger) 1 F	7.6 to 15.4 oz	0.06 to 0.12	16.7 to 8.3	14	
chlorpyrifos, MOA 1B + zeta-cypermethrin, MOA 3 (Stallion) 3.03	8.25 to 11.75	0.2 to 0.275	15.5 to 10.9	14	
chlorantraniliprole, MOA 28 + lambda-cyhalothrin MOA 3 (Besiege) 1.25	6.5 to 12.5 oz	0.063 to 0.12	19.8 to 10.4	14	
cyfluthrin MOA 3 (Tombstone) 1.0EC	1.6 to 2.6 oz	0.025 to 0.041	40 to 25	0	
cypermethrin, MOA 3 (Ammo) 2.5 EC	2 to 5 oz	0.04 to 0.1	62.5 to 25	14	
deltamethrin, MOA 3 (Decis)1.5 EC	1.6 to 2.6 oz	0.019 to 0.03	79 to 50	14	
esfenvalerate, MOA 3 (Asana XL) 0.66 EC	5.8 to 9.6 oz	0.03 to 0.05	22 to 13.2	21	
gamma-cyhalothrin, MOA 3 (Prolex, Declare) 1.25 EC	1.28 to 2.05 oz	0.0125 to 0.02	100 to 62.5	21	
lambda-cyhalothrin, MOA 3 (Karate Z) 2.08 CS (Karate, Silencer) 1.0 EC	1.6 to 3.2 oz 2.56 to 5.12 oz	0.025 0.04	83 to 40 52 to 2525	21 21	
zetamethrin, MOA 3 (Mustang Maxx) 0.8 EC	2.64 to 3.6 oz	0.0165 to 0.0225	48.5 to 35.5	14	
chlorantraniliprole, MOA 28 (Prevathon) 0.43 SC	14 to 27 oz	0.047 to 0.09	9.1 to 4.8	21	
indoxacarb, MOA 22 (Steward) 1.25 SC	9.2 to 11.3 oz	0.09 to 0.11	13.9 to 11.4	14	Steward must be applied to early stage larvae for effective control. Use lower rate for *Bt* cottons.
methoxyfenozide, MOA 18A + spinetoram, MOA 5 (Intrepid Edge) 3F	6.0 to 8.0 oz	0.140 to 0.188	21.3 to 16	28	

Table 5-5A. Insect Control on Cotton

Insect Insecticide, Mode of Action (MOA), and Formulation	Per Acre		Acres/gal (lb)	Pre- harvest Interval (days)	Precautions and Remarks
	Amount	Active (lb)			
Bollworm[a] /Tobacco Budworm (continued)					
spinosad, MOA 5 (Blackhawk) 4 SC	2.4 to 3.2 oz	0.054 to 0.073	74 to 55	28	
Cotton Aphid					
acetamiprid, MOA 4A (Assail) 70 WP	0.6 to 1.1 oz	0.025 to 0.05	28 to 14	28	Due to a high potential for cotton aphid resistance to insecticides and because of the routine presence of significant levels of predators, parasites and pathogens that limit cotton aphid build-ups, treat for cotton aphids only as a last resort.
clothianidin, MOA 4A (Belay) 2.13 WDG	3 to 4 oz	0.05 to 0.067	42.6 to 31.8	21	
flonicamid, MOA 9C (Carbine) 50 WG	1.4 to 2.8 oz	0.044 to 0.089	22.7 to 11.2	30	In 2012, cotton aphid resistance to the chloronicotinoid insecticide class was confirmed in North Carolina. Try to limit the use of this class of insecticides, especially for stink bugs. All insecticides in this section are nicotinoids except for Carbine and Transform.
imidacloprid, MOA 4A (Trimax Pro, Admire Pro, other generics) 4.0 F	1 to 1.5 oz	0.03 to 0.047	128 to 85	14	
sulfoxaflor MOA 4C (Transform) 50 WG	0.75 to 1.0 oz	0.023 to 0.031	21.3 to 16	14	
thiamethoxam, MOA 4A (Centric) 40 WG	1.25 to 2.5 oz	0.03 to 0.06	13.3 to 8	21	
Cotton Aphid and Bollworm					
imidacloprid, MOA 4A + cyfluthrin, MOA 3 (Leverage 360) 3 SE	2.8 to 3.2 oz	0.065 to 0.075	46 to 40	14	May be used with threshold levels of both cotton aphids and bollworms. The pyrethroid in these materials also provides control of stink bugs.
imidacloprid, MOA 4A + bifenthrin, MOA 3 (Brigadier) 2 EC	5.1 to 7.7 oz	0.08 to 0.12	25.1 to 16.6	14	
thiamethoxam, MOA 4A + lambda-cyhalothrin, MOA 3 (Endigo ZC) 2.06 SE	4.5 to 5.5 oz	0.072 to 0.089	28.4 to 23.3	21	
bifenthrin, MOA 3 + imidacloprid, MOA 4 (Swagger) 1 F	10.2 to 15.4 oz	0.08 to 0.12	12.5 to 8.3	14	
European Corn Borer					
Bollgard II, MOA 11B2 (varous varieties)					CryIA(c) and Cry2A(b) alleles in Bollgard II produce two *Bacillus thuringiensis (Bt)* delta endotoxins. Moderate activity against bollworm and high activity against other pest caterpillar species on cotton except cutworms. No activity against insects other than caterpillars.
TwinLink, MOA 11B2 (various varieties)					Cry1Ab and Cry2Ae alleles in TwinLink produce two *Bacillus thuringiensis (Bt)* delta endotoxins. Moderate activity against bollworm and high activity against other pest caterpillar species on cotton except cutworms. No activity against insects other than caterpillars.
WideStrike, MOA 11B2 (various varieties)					CryIA(c) and CryIF alleles in WideStrike produce two *Bacillus thuringiensis (Bt)* delta endotoxins. Low to moderate activity against bollworm and high activity against other pest caterpillar species on cotton except cutworms. No activity against insects other than caterpillars.
bifenthrin, MOA 3 (Brigade, Fanfare, Declare, Discipline, Sniper and others) 2 EC	3.2 oz	0.05	40	14	European corn borers are generally more of a problem in rank, non-Bt cotton. Other bollworm materials may provide some control.
beta-cyfluthrin, MOA 3 (Baythroid XL) 1.0 EC	1.6 to 2.6 oz	0.013 to 0.021	77 to 47.6	0	
cypermethrin, MOA 3 (Ammo) 2.5 EC	3.1 to 4.1 oz	0.06 to 0.08	41.7 to 31.3	14	
lamda-cyhalothrin, MOA 3 (Karate Z) 2.08 CS (Karate, Silencer) 1 EC	1.6 oz 3.2 to 5.12 oz	0.025 0.025 to 0.04	80 40 to 25	21	
Zeta-cypermethrin, MOA 3 (Mustang Maxx) 0.8 EC	2.9 to 3.55 oz	0.018 to 0.025 oz	44.4 to 32	14	
Fall Armyworm					
chlorpyrofos, MOA 1B (Lorsban) 4 E	1 to 2 pt	0.5 to 1	8 to 4	14	Various rates and combinations may be recommended, depending upon cotton phenology and the age distribution and population levels of larvae. Pyrethroids keep some fall armyworms from hatching. Bollgard II and WideStrike varieties show high resistance to fall armyworm damage.
emamectin benzoate, MOA 6 (Denim) 0.16 EC	8 to 12 oz	0.01 to 0.015	16 to 10.7	21	
indoxacarb, MOA 22 (Steward) 1.25 SC	9.2 to 11.3 oz	0.09 to 0.11	14 to 11.5	14	
lambda-cyhalothrin, MOA 3 +chlorantraniliprole, MOA 28 (Besiege) 1.25 ZC	6.5 to 12.5 oz	0.063 to 0.12	19.8 to 10.4	14	
methomyl, MOA 1A (Lannate) 2.4 LV (Lannate) 90 SP	1.5 pt 0.5 lb	0.45 0.45	5.3 2	15 15	
methoxyfenozide, MOA 1BA (Intrepid) 2F	4 to 10 oz	0.06 to 0.16	33 to 12.5	14	
methoxyfenozide, MOA 18A + spinetoram, MOA 5 (Intrepid Edge) 3F	6.0 to 8.0 oz	0.140 to 0.188	21.3 to 16	28	
novaluron, MOA 15 (Diamond) 0.83 EC	6 to 12 oz	0.04 to 0.08	21.3 to 10.7	30	
spinosad, MOA 5 (Blackhawk) 4 SC	2.14 to 2.9 oz	0.067 to 0.089	60 to 45	28	

83

INSECT CONTROL

Table 5-5A. Insect Control on Cotton

Insect Insecticide, Mode of Action (MOA), and Formulation	Per Acre		Acres/gal (lb)	Pre-harvest Interval (days)	Precautions and Remarks
	Amount	Active (lb)			
Fall Armyworm (continued)					
chlorantraniliprole, MOA 28 (Prevathon) 0.43 SC	14 to 27 oz	0.047 to 0.09	4.8 to 27	21	
Plant Bug					
acephate, MOA 1B (Orthene and other brands) 75 S 90 S 97 ST	0.3 to 1.3 lb 0.25 to 1 lb 0.25 to 1 lb	0.25 to 1 0.225 to 0.9 0.24 to 0.97	3.3 to 0.77 4 to 1 4 to 1	21 21 21	Prebloom treatment not recommended if square retention is in excess of 80%. If square retention is less than 80%, confirmation of threshold levels of plant bugs should be met prior to treatment.

Postbloom treatment more likely in low-spray environment, such as with Bt cottons. Neonicotinoids (MOA 4A) tend to be less effective mid to late season, but control can be erratic, as they will sometimes work season-long. Some populations appear to be developing resistance to bifenthrin. **Rotating insecticide modes of action is critical for long-term management of this insect.** Nearly any insecticide can be improved by an immediate follow-up insecticide spray within three days of the initial spray.

Fields adjacent to corn, potatoes, weedy areas, ditch banks, and other sources of plant bugs may be at higher risk of plant bug injury.

Likelihood of damage levels of plant bugs on cotton generally higher in northeastern North Carolina counties.

Bidrin is toxic to humans. Be sure to follow label directions and observe 6-day reentry interval. |
acetamiprid, MOA 4A (Assail) 70 WP	1.1 oz	0.5	14	28	
chlorpyrifos, MOA 1B (Lorsban) 4 EC	6.1 oz	0.19	21	14	
clothianidin, MOA 4A (Belay) 2.13 SC	3 to 4 oz	0.05 to 0.067	42.6 to 31.8	21	
dicrotophos, MOA 1B (Bidrin) 8 EC	4 to 8 oz	0.25 to 0.5	32 to 16	10	
dicrotophos, MOA 1B + bifenthrin MOA 3 (Bidrin XP II) 5 EC	8 to 12 oz	0.313 to 0.54	16 to 9.3	30	
flonicamid, MOA 9C (Carbine) 50 WG	1.7 to 2.8 oz	0.054 to 0.089	75.3 to 45.7	30	
imidacloprid, MOA 4A (Admire Pro, Trimax Pro, other generics) 4 F	1.3 to 1.7 oz	0.047 to 0.061	98 to 75	14	
methomyl, MOA 1A (Lannate) 2.4 LV (Lannate) 90 SP	12 oz 0.25 lb	0.225 0.225	10.7 4	15 15	
novaluron, MOA 15 (Diamond) 0.83 EC	9 to 12 oz	0.06 to 0.08	14 to 11	30	
oxamyl, MOA 1A (Vydate) 3.77 C LV	8.5 oz	0.25	15	21	
sulfoxaflor, MOA 4C (Transform) 50 WG + acephate	1.5 to 2.25 oz	0.047 to 0.071	10.7 to 7.1	14	Sulfoxaflor is generally ineffective when used alone and should be paired with acephate as a rotational product to avoid developing resistance.
thiamethoxam, MOA 4A (Centric) 40 WP	2 oz	0.047	8	21	
Soybean Looper					
lambda-cyhalothrin, MOA 3 + chlorantraniliprole, MOA 28 (Besiege) 1.25 ZC	10.0 to 12.5 oz	0.098 to 0.12	12.8 to 10.4	14	Bollgard II, TwinLink, WideStrike, and WideStrike 3 varieties show high resistance to looper damage.
emamectin benzoate, MOA 6 (Denim) 0.16 EC	6 to 12 oz	0.01 to 0.015	10.6 to 16	21	
indoxacarb, MOA 22 (Steward) 1.25 SC	6.7 to 9.2 oz	0.065 to 0.09	19 to 14	14	
methoxyfenozide, MOA 18A (Intrepid) 2 F	4 to 10 oz	0.098 to 0.16	33 to 12.5	14	
methoxyfenozide, MOA 18A + spinetoram, MOA 5 (Intrepid Edge) 3F	4.0 to 8.0 oz	0.094 to 0.188	32 to 16	28	
spinosad, MOA 5 (Blackhawk) 4 SC	2.4 to 3.2	0.054 to 0.073	74 to 54	28	
chlorantraniliprole, MOA 28 (Prevathon) 0.43 SC	20 to 29 oz	0.067 to 0.097	6.4 to 4.4	21	
Spider Mite					
abemectin, MOA 6 (Zephyr, Abemectin) 0.15 EC	8 to 16 oz	0.01 to 0.019	15 to 7.9	20	Control often unnecessary because of beneficial arthropods and fungi. Apply with 20-plus gallons of water (applies to all chemicals).
bifenthrin, MOA 3 (Brigade, Fanfare, Sniper, Declare, Discipline and others) 2 EC	3.8 oz	0.06	33	14	
dicofol, MOA UNC (Dicofol) 4 E	0.8 to 1.6 qt	0.8 to 1.6	5 to 2.5	14	
entoxazole, MOA 10B (Zeal) 72 WP	0.66 to 1 oz	0.03 to 0.045	45 to 30	28	
fenpropathrin, MOA 3 (Danitol) 2.4 EC	10.7 to 16 oz	0.2 to 0.3	12 to 8	21	
fenpyroximate, MOA 21A (Portal, Fujimite) 0.4 E	12 to 16 oz	0.037 to 0.05	10.8 to 8	14	Use 1.5 to 2X the amount of product if applied by aircraft.
propargite, MOA 12C (Comite) 6.55 L	1 qt	1.6	4	14	
spiromesifen, MOA 23 (Oberon) 2 SC	6 to 16 oz	0.094 to 0.25	21.3 to 8	30	Use 6 ounces only in early season to control low populations.
Stink Bug					
acephate, MOA 1B (Orthene) 75 S (Orthene and others) 97 S	1 lb 0.75 lb	0.75 0.75	1.3 1	21	Stink bugs may be more prevalent on unsprayed or less sprayed Bt cottons. Bidrin and methyl parathion are more effective against brown stink bugs than other products, though extremely toxic to humans. Be sure to follow label directions and observe 6-day reentry interval.
dicrotophos, MOA 1B (Bidrin) 8 EC	4 to 8 oz	0.25 to 0.5	32 to 16	10	**Be sure to observe the 6-day reentry interval with Bidrin.**

Table 5-5A. Insect Control on Cotton

Insect Insecticide, Mode of Action (MOA), and Formulation	Per Acre		Acres/gal (lb)	Pre-harvest Interval (days)	Precautions and Remarks
	Amount	Active (lb)			
Stink Bug (continued)					
dicrotophos, MOA 1B + bifenthrin, MOA 3 (Bidrin XP II) 5EC	8.0 to 12.8 oz	0.313 to 0.54	16 to 9.3	30	Product contains 4.0 pounds dicrotophos and 1.0 pound bifenthrin per gallon. Toxic to humans; be sure to follow label directions and observe 6-day reentry interval.
oxamyl, MOA 1A (Vydate) 3.77 SL	17 oz	0.5	7.5	21	
pyrethroids, MOA 3 and pyrethroid combinations	(see bollworms above for rates)		—	—	Pyrethroids provide good to excellent control green and brown marmorated stink bugs, but are **less effective against brown stink bugs**. Bifenthrin is more effective than other pyrethroids against brown stink bugs.
Thrips (at planting treatment)					
imidacloprid, MOA 4A (Gaucho Grande 600 FS, Acceleron-I)	—	0.375 mg/seed	—	—	Each of the following seed treatment lacks persistence beyond approximately 3 weeks and may require a supplemental foliar treatment for thrips control. Supplemental sprays are less likely in late planted (after May 20) cotton. Note that resistance to neonicotinoids (imidacloprid and thiamethoxam) has been confirmed in tobacco thrips throught the state. Variable control should be expected.
thiamethoxam, MOA 4A (Cruiser) 5 FS	—	0.34 mg/seed	—	—	
abamectin, MOA 6, + thiamethoxam MOA 4A (Avicta Duo 500FS, Avicta Complete, Acceleron-N)	—	0.15 abamectin + 0.375 thiamethoxam mg/seed	—	—	
imidacloprid, MOA 4A + thiodicarb, MOA 1A (Aeris) 48DS	—	0.375 imidacloprid + 0.375 thiodicarb mg/seed	—	—	
imidacloprid (MOA 4A) + clothianidin (MOA 4A) + thiodicarb (MOA 1A) + *Bacillus firmus* (biological (Aeris/Poncho/VOTiVO)	—	0.375 imidacloprid + 0.375 thiodicarb + 0.424 clothianidin mg/seed + 2 x 10^9 cfu/ml *B. fermis* units	—	—	
imidacloprid, MOA 4A (Admire Pro) 4.6F (Wrangler) 4.0F	7.4 to 9.2 8.5 to 10.5	0.27 to 0.33 0.27 to 0.33	17.3 to 13.9 15.1 to 12.2	—	Apply liquid into open furrow directly onto seed before furrow closure with either a seed firmer or with a drop tube (i.e., with a #55 oriface) producing a narrow jet of finished product into the center of the open furrow. Works best in combination with another at-planting treatment, such as a seed treatment. Note that resistance to imidacloprid has been confirmed in tobacco thrips throught the state. Variable control should be expected.
Thrips (post-emergence)					
acephate, MOA 1B (Orthene) 75 S (Orthene) 90 S (Orthene) 97 S (Orthene) 97 ST[b]	3 to 4 oz 0.2 lb 2.5 to 3 oz 6 oz	0.14 to 0.19 0.18 0.15 to 0.18 0.375	5.3 to 4 5 6.4 to 5.3 2.67	21	Not suggested to replace at-plant insecticides in conventional cotton. With the high thrips populations often found in North Carolina, consider at least 0.25 pound a.i. per acre the standard rate for Orthene. Pyrethroids do not provide adequate thrips control on cotton.
dicrotophos, MOA 1B (Bidrin) 8 EC	4 oz	0.25	32	10	
dimethoate, MOA 1B (Dimethoate) 4 EC	8 oz	0.25	16	10	
methamidophos, MOA 1B (Monitor) 4 EC	6.4 oz	0.2	20	50	
spinetoram, MOA 5 (Radiant) 1 SC	1.5 to 3 oz	0.01 to 0.02	85 to 43	28	Provides improved control of western flower thrips, as well as good control of tobacco thrips. Use higher rates for improved control.

[a] Lowest labeled rates for bollworms and budworms
[b] 2 (ee) state local need label for higher rates
NOTE: Upper or lower rate ranges do not indicate equivalent activity.

Cotton Insect Resistance Management

D. D. Reisig, Entomology Department

Resistance occurs when some insects in a population survive a chemical treatment and are therefore able to pass on an inherited gene(s) for this survival to their offspring. Because these offspring are better able to survive the insecticide than those that are not resistant, the resistant individuals increase their numbers faster in the presence of the insecticide. After several generations, the resistant insects can outnumber the susceptible ones, and the insecticide becomes ineffective. Because the alleles that allow insects to survive an insecticide are often initially present in very few individuals out of a very large population of susceptible insects, resistance development may take years. Five to 20 years would be a common range for effectiveness of many insecticides.

Insects vary greatly in their ability to develop resistance to insecticides. For example, cotton aphids have been able to develop resistance to various classes of chemicals rapidly, while the boll weevil remains susceptible to several organophosphate insecticides after more than 50 years of exposure.

Insects develop resistance to insecticides in several ways. Some are able to break down (metabolize) insecticides, while others are able eliminate the toxins. Some can sequester insecticides (move them to a less harmful place in or on the body), and still others can avoid the toxin (behavioral resistance). The above are examples of different modes of action (MOA). Unfortunately, once an insect develops resistance to one insecticide, in most cases the insect is also resistant to others in the same class or group of insecticides sharing the same mode of action. For example, if tobacco budworms are resistant to the pyrethroid Baythroid, they are also resistant to the pyrethroid Karate. To make matters worse, some insects may be resistant to several classes of insecticides, such as is presently the case with plant bugs in the Midsouth. In North Carolina some populations of cotton aphids (chloronicotinoid class) and corn earworms (pyrethroid class) have developed resistance to these chemical classes that were initially very effective.

As you can see from the table below, many different kinds of possible insecticide resistance have been identified. Most have complicated, hard-to-remember names. To make it easy to recognize different classes or modes of actions that can lead to resistance development, each chemical has been identified with a number, and occasionally subdivided with a letter. Products sharing the same number or letter and number combination have the same mode of action (for additional detail see: http://pested.okstate.edu/pdf/insecticide 20moa.pdf).

One major strategy in managing resistance is to avoid using products with the same mode of action (sharing the same number in the table) in the same year. Also, tank mixing insecticides with different modes of action may delay resistance development. Additionally, if only a single class of insecticides is listed for control of an insect (e.g., Assail, Centric, and Trimax Pro – all chloronicotinoids – for cotton aphids), one should try to either limit insecticide use to a single spray or try to avoid treatment. One final strategy in minimizing insect resistance to insecticides is to avoid unneeded treatments by following recommended thresholds.

Listed below are the economically important cotton pests found in North Carolina, followed by the chemical and brand names and mode of action.

Table 5-5B. Cotton Insecticide Modes of Action (MOA); Insecticide Resistance Action Committee Designations

Insect	Chemical Name (Brand Name)	Mode of Action
Beet Armyworm	clorantraniliprole (Prevathon)	2B
	emabectin benzoate (Denim)	6
	fluendiamide (Belt)	28
	indoxacarb (Steward)	22
	methoxyfenozide (Intrepid)	18A
	spinosad (Blackhawk)	5
Bollworm/Tobacco Budworm	*Bacillus thuringiensis* var. *kurstaki* (Bollgard II and Widestrike: delta endotoxin expressed by various varieties)	11B2
	beta-cyfluthrin (Baythroid XL)	3
	bifenthrin (Brigade, Capture, Discipline, others)	3
	bifenthrin + zeta-cypermethrin (Hero, Steed)	3
	bifenthrin + imidacloprid (Swagger)	3 + 4A
	chlorantraniliprole (Prevathon)	28
	cypermethrin (Ammo)	3
	esfenvalerate (Asana XL)	3
	flubendiamide (Belt)	28
	gamma-cyhalothrin (Prolex, Declare)	3
	lambda-cyhalothrin (Karate, Karate Z)	3
	lambda-cyhalothrin + clorantraniliprole (Besiege)	3 + 28
	zetamethrin (Mustang Max)	3
	spinosad (Blackhawk)	5
	indoxacarb (Steward)	22
	methomyl (Lannate)	1A
Cotton Aphid	acetamiprid (Assail)	4A
	flonicamid (Carbine)	9C
	imidacloprid (Trimax Pro)	4A
	thiamethoxam (Centric)	4A
	sulfoxaflor (Transform)	4C
Cotton Aphid & Bollworm	imidacloprid + bifenthrin(Swagger, Brigadier)	4A + 3
	imidacloprid + cyfluthrin (Leverage 360)	4A + 3
	thiamethoxam + lambda-cyhalothrin (Endigo)	4A + 3
European Corn Borer	Bollgard II (various varieties)	11B2
	Widestrike (various varieties)	11B2
	beta-cyfluthrin (Baythroid XL)	3
	bifenthrin (Brigade, Fanfare, Discipline, Sniper and others)	3
	chlorantraniliprole (Prevathon) cyfluthrin (Baythroid)	28
	cypermethrin (Ammo)	3
	esfenvalerate (Asana XL)	3

Table 5-5B. Cotton Insecticide Modes of Action (MOA); Insecticide Resistance Action Committee Designations

Insect	Chemical Name (Brand Name)	Mode of Action
European Corn Borer (continued)	gamma-cyhalothrin (Prolex)	3
	lambda-cyhalothrin (Karate, Karate Z)	3
	lambda-cyhalothrin + clorantraniliprole (Besiege)	3
	zetamethrin (Mustang Max)	3 + 28
	spinosad (Blackhawk)	3
	indoxacarb (Steward)	22
	methomyl (Lannate)	1A
Fall Armyworm	chlorantraniliprole (Prevathon)	28
	chlopyrofos (Lorsban)	1B
	indoxacarb (Steward)	22
	methomyl (Lannate)	1A
	methoxyfenozide (Intrepid)	18A
	novaluron (Diamond)	15
	spinosad (Tracer)	5
	thiodicarb (Larvin)	1A
Plant Bug	acephate (Orthene, and others)	1B
	acetamiprid (Assail)	4A
	chlopyrofos (Lorsban)	1B
	clothianidin (Belay)	4A
	dicrotophos (Bidrin)	1B
	flonicamid (Carbine)	9C
	imidacloprid (Trimax Pro, Admire Pro)	4A
	methomyl (Lannate)	1A
	methyl parathion (Methyl parathion, Penncal-M)	1B
	oxamyl (Vydate)	1A
	sulfoxaflor (Transform)	4C
	thiamethoxam (Centric)	4A
Soybean & Cabbage Looper	emamectin benzoate (Denim)	6
	lambda-cyhalothrin+ clorantraniliprole (Besiege)	3 + 28
	indoxacarb (Steward)	22
	methoxyfenozide (Intrepid)	18A
	spinosad (Blackhawk)	5
Spider Mite	abemectin (Zephyr, Abemectin)	6
	bifenthrin (Brigade, Capture, Discipline, Sniper and others)	3
	dicofol (Dicofol)	UNC*
	entoxazole (Zeal)	10B
	fenpyroximate (Portal)	21A
	fenpropathrin (Danitol)	3
	propargate (Comite)	12C
	spiromesfen (Oberon)	23
Stink Bug	acephate (Orthene, and others)	1B
	clothianidin (Belay)	4A
	dicrotophos (Bidrin)	18
	dicrotophos + bifenthrin (Bidrin XP II)	18 + 3
	oxamyl (Vydate)	1A
	pyrethroids	3
THRIPS (At-Planting)	imidacloprid (Gaucho Grande, Acceleron I)	4A
	thiamethoxam (Cruiser)	4A
	thiamethoxam + abamectin (Avicta Duo,Acceleron N))	4A + 6
	imidacloprid + thiodicarb (Aeris)	4A + 1A
	imidacloprid + clothianidin + thiodicarb (Aeris/Poncho/VOTiVO)	4A + 1A
THRIPS (Postemergence)	acephate (Orthene, and others)	1B
	dicrotophos (Bidrin)	1B
	dimethoate (Dimethoate)	1B
	methamidophos (Monitor)	1B
	spinetoram (Radiant)	5

*UNC: Compound with unknown mode of action.

87

Insect Control on Peanuts

R. L. Brandenburg, Entomology Extension and Research

Table 5-6. Insect Control on Peanuts

Insecticide and Formulation	Amount of Formulation Per Acre	Precautions and Remarks
Seasonal Control of Thrips and Leafhoppers		
Thrips at Planting		
acephate (Orthene 97) (generics available)	0.75 to 1 lb	Apply as in-furrow spray in 3 to 5 gallons of water per acre. State (24c) label must be in possession at time of application.
phorate (Thimet) (generics available)	5.0 lb of 20% granules	
fluopyram + imidacloprid (Velum Total)	14-18 fl oz	Application rate appropriate for in-furrow spray during planting directed on or below seed, or chemigation into root-zone through low-pressure drip or trickle irrigation.
Imidacloprid (Admire Pro)	7.0-10.5 fl oz	In furrow spray during planting, directed on or below seed.
Thiamethoxam + Mefenoxam + Fludioxonil + azoxystrobin (Cruiser Maxx Peanuts)	treated peanut seed	Suppression only
Thrips foliar postermergence		
acephate (Orthene) 97 (generics available)	0.375 to 0.75 lb	Do not feed or graze livestock on treated vines. Apply 10 to 50 gallons spray solution per acre to foliage. Do not apply more than 4.125 pounds per acre (4 pounds a.i. per acre) per season.
Beta-cyfluthrin (Baythroid XL)	2.8 oz	
Bifenthrin (Brigade)	2.1-6.4 fl oz	Pre-harvest interval of 14 days.
Control of Specific Pests		
Beet Armyworm		
Bacillus thuringiensis (Xentari)	0.5 to 2 lb	Apply to small caterpillars. Use highest rate for larger worms or high populations; 0 day harvest restriction.
methomyl (Lannate LV)	0.75 to 3 pt	Apply broadcast in sufficient water for good coverage when worms are small. Do not apply within 21 days of harvest. See fall armyworm for additional restrictions.
Methoxyfenozide + spinetoram (Intrepid Edge)	4-8 fl oz	Application rate varies with timing. Lower rates appropriate for light infestations, smaller larvae and/or small plants.
indoxacarb (Steward)	9.2 to 11.3 oz	Do not apply more than 45 ounces per acre per crop. 14-day preharvest interval.
spinosad (Blackhawk) Bifenthrin (Brigade) Chlorantraniliprole (Prevathon)	1.5 to 3 fl oz 2.1-6.4 fl oz 14.0-20.0 fl oz/A	Do not apply more than 9 fluid ounces per season or make more than three applications. 3-day preharvest interval. Pre-harvest interval of 14 days. Make no more than 4 applications per crop per year.
Corn Earworm, Southern Armyworm, Green Cloverworm, Velvetbean Caterpillar		
acephate (Orthene) 97 (generics available)	0.75 to 1 lb	Do not feed or graze livestock on acephate-treated vines. Do not apply within 14 days of harvest (digging).
Bacillus thuringiensis (Dipel DF) (Dipel ES) (Xentari)	0.5 to 2 lb 1 to 2 pt 0.5 to 2 lb	For velvetbean caterpillar control only. Apply to small caterpillars and use highest rate for larger worms and/or high populations; 0 day harvest restriction. Xentari also controls southern armyworm.
esfenvalerate (Asana XL)	2.9 to 5.8 oz	Do not feed Asana-treated vines or graze livestock on treated plants.
fenpropathrin (Danitol) 2.4 EC	10.67 to 16 fl oz	Do not exceed 2.67 pints per acre per season. Use 10 to 50 gallons per acre by ground and 5 to 10 gallons per acre by air. Repeat no more often than every 7 days. Do not apply within 14 days of digging and do not feed or graze vines within 14 days of last application.
indoxacarb (Steward)	9.2 to 11.3 oz	Do not apply more than 45 ounces per acre per crop. 14 day preharvest interval. For corn earworm.
lambda-cyhalothrin (Karate Z)	1.28 to 1.92 oz	Do not feed or graze livestock on Karate-treated plants.
methomyl (Lannate LV)	0.75 to 3 pt	Apply to foliage when four or more worms are present per foot of row and preferably when worms are small. Do not apply methomyl within 21 days of harvest. Do not feed methomyl-treated vines to livestock. Use minimum of 3 gallons of water for aerial application.
Methoxyfenozide + spinetoram (Intrepid Edge)	4-8 fl oz	Application rate varies with timing. Lower rates appropriate for light infestations, smaller larvae and/or small plants.
spinosad (Blackhawk) Bifenthrin (Brigade) Chlorantraniliprole+ lambda-cyhalothrin (Besiege) Chlorantraniliprole (Prevathon)	2 to 3 fl oz 2.1-6.4 fl oz 6.0-10.0 fl oz/A 14.0-20.0 fl oz/A	Do not apply more than 9 fluid ounces per season or make more than three applications. 3-day preharvest interval. Pre-harvest interval of 14 days. Pre-harvest interval 14 days. Do not exceed a total of 31 fluid ounces of Besiege per acre per year. Make no more than 4 applications per crop per year.
Cutworm		
chlorpyrifos (Lorsban) 15 G	1.33 lb	Apply in 16- to 18-inch band over row when infestation is first seen. May be applied by air. Do not graze or feed immature crop to livestock.
esfenvalerate (Asana XL)	5.8 to 9.6 oz	Do not feed treated vines to livestock.
indoxacarb (Steward)	9.2 to 11.3 oz	Do not apply more than 45 ounces per acre per crop. 14 day preharvest interval.
lambda-cyhalothrin (Karate Z)	0.96 to 1.6 oz	Do not use treated vines or hay for animal feed.
methomyl (Lannate LV) Bifenthrin (Brigade) Chlorantraniliprole + lambda-cyhalothrin (Besiege)	1.5 to 3 pt 2.1-6.4 fl oz 5.0-8.0 fl oz/A	Do not apply within 21 days of harvest. Do not feed treated vines to livestock. Pre-harvest interval of 14 days. Pre-harvest interval 14 days. Do not exceed a total of 31 fluid ounces of Besiege per acre per year.

Table 5-6. Insect Control on Peanuts

Insecticide and Formulation	Amount of Formulation Per Acre	Precautions and Remarks
Fall Armyworm		
acephate (Orthene) 97 (generics available)	0.75 to 1 lb	Do not apply within 14 days of harvest (digging). Do not feed or graze livestock on vines treated with acephate. Apply 10 to 50 gallons spray solution per acre. Do not apply more than 4.13 pounds per acre (4 pounds a.i. per acre per season).
fenpropathrin (Danitol) 2.4 EC	10 2/3 to 16 fl oz	Do not exceed 2.67 pints per acre per season. Repeat no more often than every 7 days. Do not apply within 14 days of digging and do not feed or graze vines within 14 days of last application.
indoxacarb (Steward)	9.2 to 11.3 oz	Do not apply more than 45 ounces per acre per crop. 14 day preharvest interval.
lambda-cyhalothrin (Karate Z)	1.28 to 1.92 oz	
methomyl (Lannate LV)	0.75 to 1.5 pt	Effective against all sizes of worms. Use minimum of 3 gallons of water for aerial application. Do not apply within 21 days of harvest. Do not feed methomyl-treated vines to livestock.
Methoxyfenozide + spinetoram (Intrepid Edge)	4-8 fl oz	Application rate varies with timing. Lower rates appropriate for light infestations, smaller larvae and/or small plants.
spinosad (Blackhawk) Bifenthrin (Brigade) Chlorantraniliprole+lambda-cyhalothrin (Besiege) Chlorantraniliprole (Prevathon)	2 to 3 fl oz 2.1-6.4 fl oz 6.0-10.0 fl oz/A 14.0-20.0 fl oz/A	Do not apply more than 9 fluid ounces per season or make more than three applications. 3-day preharvest interval. Pre-harvest interval of 14 days. Pre-harvest interval 14 days. Do not exceed a total of 31 fluid ounces of Besiege per acre per year. Make no more than 4 applications per crop per year
Leafhoppers		
acephate (Orthene) 97 (generics available)	0.75 to 1 lb	See remarks under Thrips.
esfenvalerate (Asana XL)	2.9 to 5.8 oz	Do not feed livestock Asana-treated vines or graze livestock on treated plants.
fenpropathrin (Danitol) 2.4 EC	6 to 10.67 oz	Do not exceed 2 2/3 pints per acre per season. Repeat no more often than every 7 days. Do not apply within 14 days of digging and do not feed or graze vines within 14 days of last application.
Chlorantraniliprole+lambda-cyhalothrin (Besiege)	6.0-10.0 fl oz/A	
lambda-cyhalothrin (Karate Z)	0.96 to 1.6 oz	Do not use treated vines or hay for animal feed.
methomyl (Lannate LV) Bifenthrin (Brigade)	0.75 to 3 pt 2.1-6.4 fl oz	Do not apply within 21 days of harvest. Do not use treated vines as feed. Pre-harvest interval of 14 days.
Lesser Cornstalk Borer		
chlorpyrifos (Lorsban, Pilot) 15 G (generics available) Chlorantraniliprole+lambda-cyhalothrin (Besiege) Chlorantraniliprole (Prevathon)	7 to 14 lb 10.0 fl oz/A 14-20.0 fl oz/A	Pre-harvest interval 14 days. Do not exceed a total of 31 fluid ounces of Besiege per acre per year. See label
Southern Corn Rootworm		
chlorpyrifos (Lorsban, Pilot) 15 G (generics available)	13.3 lb	Apply in a 16- to 18-inch band over the row just before pegging.
phorate (Thimet) 20 G (generics available)	10 lb	
Spider Mite		
propargite (Comite) 73 L	2 pt	Apply in at least 25 gallons of water per acre. Spider mite outbreaks are less likely to develop if foliar insecticides are not used during July and August and copper fungicides are used for Cercospora leafspot. Do not apply propargite within 14 days of harvest.
fenpropathrin (Danitol) 2.4 EC	10.67 to 16 fl oz	Do not exceed 2.67 pints (42 2/3 fluid ounces) per acre per season. Use 10 to 50 gallons per acre by ground and 5 to 10 gallons per acre by air. Repeat no more often than every 7 days. Do not apply within 14 days of digging and do not feed or graze vines within 14 days of last application.
Bifenthrin (Brigade)	5.1-6.4 fl oz.	Pre-harvest interval of 14 days.

See N.C. Cooperative Extension Service publication AG-463-5, *Pesticides and Wildlife—Peanuts*, for additional information on minimizing pesticide impact on wildlife.

INSECT CONTROL

Insect Control in Soybeans

D. D. Reisig, Entomology Extension

Table 5-7. Insect Control on Soybeans

Insect Insecticide and Formulation	per Acre		Acres/gal. (lb)	Preharvest Interval (PHI) (Days)	Precautions and Remarks
	Amount of Formulation	Active Ingredient			
Bean Leaf Beetle					
acephate, MOA 1B (Orthene) 97 S	0.75 to 1 lb	0.75 to 1	1.25 to 1	14	Treat when defoliation reaches threshold levels or buildup is obvious. Threshold is 30% prebloom defoliation or 15% defoliation 2 weeks prior to bloom through podfill. Pod skinning by this insect can be a concern in soybeans grown for seed. Selected pyrethroids will suppress bean leaf beetle. Tolerance can quickly develop if chemistries are not rotated. **In the premixed products listed, the effective chemistries are in MOA's 3 and 1B.**
acetamiprid, MOA 4A + bifenthrin, MOA 3 (Justice) 1.8 EC	5 fl oz	See label	25.6	30	
beta-cyfluthrin, MOA 3 (Baythroid XL) 1.0 EC	2.8 fl oz	0.022	45.7	45	
bifenthrin, MOA 3 (Brigade, Discipline, Sniper, and others) 2 EC	4 to 6.4 fl oz	0.062 to 0.10	32 to 20	30	
chlorantraniliprole, MOA 28 + lambda-cyhalothrin, MOA 3 (Besiege) 1.25 SC	5 to 8 fl oz	See label	25.6 to 16	21	
chlorpyrifos, MOA 1B + gamma-cyhalothrin, MOA 3 (Cobalt Advanced) 2.63 EC	19 to 24 fl oz	See label	6.7 to 5.3	30	
chlorpyrifos, MOA 1B (Lorsban) 4 E	1 pt	0.5	8	14	
cyfluthrin, MOA 3 (Tombstone) 2 E	1.6 to 2.8 fl oz	0.025 to 0.04	80 to 45.7	45	
diflubenzuron, MOA 15 + lambda-cyhalothrin, MOA 3 (DoubleTake) 3 SC	4 fl oz	See label	32	30	
imidacloprid, MOA 4A + cyfluthrin, MOA 3 (Leverage 360) 3.0 SE	2.8 fl oz	See label	45.7	45	
lambda-cyhalothrin, MOA 3 (Warrior, Lambda-cyhalothrin, Silencer) 1.0 EC (Karate Z and Warrior II) 2.08 CS	1.92 to 3.2 fl oz 0.96 to 1.6 fl oz	0.015 to 0.025 0.015 to 0.025	66.7 to 40 133.3 to 80	30 30	
lambda-cyhalothrin, MOA 3 + thiamethoxam, MOA 4A (Endigo ZC) 2.06 SE	4 to 4.5 fl oz	See label	32 to 28.4	30	
Beet Armyworm					
chlorantraniliprole, MOA 28 (Prevathon) 0.43 SC	14 to 20 oz	0.047 to 0.067	9.1 to 64	1	Ground application only for larger caterpillars. Control of large armyworms is difficult. Chlorantraniliprole, indoxacarb, methoxyfenozide, and spinosad are the superior products.
chlorantraniliprole, MOA 28 + lambda-cyhalothrin, MOA 3 (Besiege) 1.25 SC	9 fl oz	0.04 + 0.02	14.2	21	
indoxacarb, MOA 22 (Steward) 1.25 EC	5.6 to 11.3 fl oz	0.06 to 0.11	22.9 to 11.3	21	
methomyl, MOA 1A (Lannate) 2.4 LV (Lannate) 90 SP	1.5 pt 0.5 lb	0.45 0.45	5.3 2	14 14	
methoxyfenozide, MOA 18A (Intrepid) 2 F	4 to 8 fl oz	0.06 to 0.12	32 to 16	14 (grain) 7 (hay)	
methoxyfenozide, MOA 18A + spinetoram, MOA 5 (Intrepid Edge) 3F	4.0 to 6.4 oz	See label	32 to 20	28	
spinosad, MOA 5 (Blackhawk) 4 SC	1.7 to 2.2 fl oz	0.04 to 0.05	75.3 to 58.2	28	
Corn Earworm					
beta-cyfluthrin, MOA 3 (Baythroid XL) 1.0 EC	1.6 to 2.8 fl oz	0.0125 to 0.022	80 to 45.7	45	Treat when earworm numbers exceed threshold as determined by scouting. Be sure worms are present and 3/8 to 1/2 inch in size when treatment is applied. Use low rates for light infestations. **Use higher rates by air.** Some tolerance to pyrethroids has been found in NC corn earworm populations. Distinguishing between corn earworm and tobacco budworm is important as pyrethroids provide poor control of tobacco budworm. A pyrethroid combined with 0.5 pound/acre acephate will improve control in situations where tobacco budworm is present and/or resistance of corn earworm to pyrethroids is present. Effective products with a pyrethroid-alternative chemistry include chlorantraniliprole, indoxacarb, spinetoram, and spinosad. Go to Web page http://www.ces.ncsu.edu/wp-content/uploads/2013/02/CEW-calculator-v0.005.html for an online threshold calculator. At $10.00 per bushel, the plant compensates due to the low caterpillar levels needed to reach threshold at $10.00 and above.
chlorantraniliprole, MOA 28 (Prevathon) 0.43 SC	14 to 20 oz	0.047 to 0.067	9.1 to 6.4	1	
chlorantraniliprole, MOA 28 + lambda-cyhalothrin, MOA 3 (Besiege) 1.25 SC	6 to 9 fl oz	See label	21.3 to 14.2	21	
chlorpyrifos, MOA 1B (Lorsban) 4 EC	1.5 to 2 pt	0.75 to 1	5.3 to 4	14	
cyfluthrin, MOA 3 (Tombstone) 2 E	1.6 to 2.8 fl oz	0.025 to 0.04	80 to 45.7	45	
esfenvalerate, MOA 3 (Asana XL) 0.66 EC	5.8 to 9.6 fl oz	0.03 to 0.05	22.1 to 13.3	21	
chlorpyrifos, MOA 1B + gamma-cyhalothrin, MOA 3 (Cobalt Advanced) 2.63 EC	16 to 38 fl oz	See label	8 to 3.4	30	
gamma-cyhalothrin, MOA 3 (Declare) 1.25 EC	1.54 fl oz	0.015	83.1	21	
imidacloprid, MOA 4A +beta-cyfluthrin, MOA 3 (Leverage 360) 3.0 SE	2.8 fl oz	0.04 + 0.02	45.7	15	
indoxacarb, MOA 22 (Steward) 1.25 EC	5.6 to 11.3 fl oz	0.06 to 0.11	22.9 to 11.3	21	
lambda-cyhalothrin, MOA 3 (Warrior, Lambda-cyhalothrin, Silencer) 1.0 EC (Karate Z and Warrior II) 2.08 CS	1.92 to 3.2 fl oz 0.96 to 1.6 fl oz	0.015 to 0.025 0.015 to 0.025	66.7 to 40 133.3 to 80	30 30	
lambda-cyhalothrin, MOA 3 + thiamethoxam, MOA 4A (Endigo ZC) 2.06 SE	3.5 to 4 fl oz	See label	36.6 to 32	30	
methoxyfenozide, MOA 18A + spinetoram, MOA 5 (Intrepid Edge) 3F	4.0 to 6.4 oz	See label	32 to 20	28	
spinosad, MOA 5 (Blackhawk) 4 SC	1.7 to 2.2 fl oz	0.04 to 0.05	75.3 to 58.2	28	

INSECT CONTROL

Table 5-7. Insect Control on Soybeans

Insect Insecticide and Formulation	per Acre		Acres/gal. (lb)	Preharvest Interval (PHI) (Days)	Precautions and Remarks
	Amount of Formulation	Active Ingredient			
Corn Earworm (continued)					
zeta-cypermethrin, MOA 3 (Mustang Max) 0.8 EC	2.8 to 4 fl oz	0.0175 to 0.025	45.7 to 32	21	
zeta-cypermethrin, MOA 3 + bifenthrin, MOA 3 (Hero) 1.24 EC	10.3 fl oz	0.033 + 0.066	12.4	30	
Grasshopper					
acephate, MOA 1B (Orthene 97)	0.25 to 0.5 lb	0.25 to 0.5	4 to 2	14	Apply by air or ground uniformly over foliage as a broadcast treatment. Early morning treatment is preferred. Use higher rates for heavy infestations. Diflubenzuron is not effective to control adult grasshoppers. See label for additional instructions and suggestions.
chlorpyrifos, MOA 1B + gamma-cyhalothrin, MOA 3 (Cobalt Advanced) 2.63 EC	10 to 13 fl oz	See label	12.8 to 9.8	30	
diflubenzuron, MOA 15 (Dimilin) 2L, 25W	2 fl oz 0.25 lb.	0.06 0.06	64 8	21	
Green Cloverworm					
Bacillus thuringiensis, MOA 11B2 (Various)	—	—	—	0	Treat when defoliation reaches threshold. This insect is seldom an economic pest. See label of specific *B.t.* products. Thresholds are listed under bean leaf beetle. Pyrethroid insecticides are labeled for green cloverworm control at, or below, the rates shown under "Corn Earworm."
beta-cyfluthrin, MOA 3 (Baythroid XL) 1.0 EC	1.6 to 2.8 fl oz	0.0125 to 0.022	80 to 45.7	45	
chlorantraniliprole, MOA 28 (Prevathon) 0.43 SC	14 to 20 fl oz	0.047 to 0.067	9.1 to 6.4	1	
chlorantraniliprole, MOA 28 + lambda-cyhalothrin, MOA 3 (Besiege) 1.25 SC	5 to 8 fl oz	See label	25.6 to 16	21	
cyfluthrin, MOA 3 (Tombstone) 2E	1.6 to 2.8 fl oz	0.025 to 0.04	80 to 45.7	45	
esfenvalerate, MOA 3 (Asana XL) 0.66 EC	5.8 to 9.6 fl oz	0.03 to 0.05	22.1 to 13.3	21	
gamma-cyhalothrin, MOA 3 (Declare) 1.25 EC	1.54 fl oz	0.015	83.1	21	
indoxacarb, MOA 22 (Steward) 1.25 EC	8 to 11.3 fl oz	0.08 to 0.11	16 to 11.3	21	
lambda-cyhalothrin, MOA 3 (Warrior, Lambda-cyhalothrin, Silencer) 1.0 EC (Karate Z and Warrior II) 2.08 CS	1.92 to 3.2 fl oz 0.96 to 1.6 fl oz	0.015 to 0.025 0.015 to 0.025	66.7 to 40 133.3 to 80	30 30	
lambda-cyhalothrin, MOA 3 + thiamethoxam, MOA 4A (Endigo ZC) 2.06 SE	3.5 to 4 fl oz	See label	36.6 to 32	30	
methoxyfenozide, MOA 18A + spinetoram, MOA 5 (Intrepid Edge) 3F	4.0 to 6.4 oz	See label	32 to 20	28	
spinosad, MOA 5 (Blackhawk) 4 SC	1.1 to 2.2 fl oz	0.025 to 0.05	116.4 to 58.2	28	
zeta-cypermethrin, MOA 3 (Mustang Maxx) 0.8 EC	2.8 to 4 fl oz	0.0175 to 0.025	45.7 to 32	21	
zeta-cypermethrin, MOA 3 + bifenthrin, MOA 3 (Hero) 1.24 EC	10.3 fl oz	0.033 + 0.066	12.4	30	
Kudzu Bug					
acephate, MOA 1B (Orthene) 97 S	1 lb	1	1	14	
bifenthrin, MOA 3 (Brigade, Discipline, Sniper, and others) 2 EC	4 to 6.4 fl oz	0.062 to 0.10	32 to 20	30	
bifenthrin, MOA 3 + acetamiprid, MOA 4a (Justice) 1.8 EC	5 fl oz	See label	25.6	30	
bifenthrin, MOA 3 + imidacloprid, MOA 4A (Brigadier) 2 E (Swagger) 1 F	6.1 fl oz 12.2 fl oz	See label See label	21 10.5	7 18	
gamma-cyhalothrin, MOA 3 (Declare) 1.25 EC	1.54 fl oz	0.015	83.1	21	
lambda-cyhalothrin, MOA 3 (Karate, Lambda-cyhalothrin, Silencer) 1.0 EC (Karate Z, Warrior II) 2.08 CS	3.84 fl oz 1.92 fl oz	0.03 0.03	33.3 66.7	30 30	
lambda-cyhalothrin, MOA 3 + thiamethoxam, MOA 4A (Endigo ZC) 2.06 SE	3.5 to 4.5 fl oz	See label	36.6 to 28.4	30	
zeta-cypermethrin, MOA 3 (Mustang Maxx) 0.8 EC	4 fl oz	0.025	32	21	
zeta-cypermethrin, MOA 3 + bifenthrin, MOA 3 (Hero) 1.24 EC	6.4 to 10.3 fl oz	See label	20 to 12.4	30	
Soybean Looper					
chlorantraniliprole, MOA 28 (Prevathon) 0.43 SC	14 to 20 fl oz	0.047 to 0.067	9.1 to 6.4	1	Treat when thresholds are reached or when buildup is obvious. Threshold is 15% defoliation in soybeans two weeks prior to flowering, but can be increased to 20% during R6 when growing conditions are ideal. Ground application is superior. **Resistance is occurring in this species and has been documented in the Blacklands for MOA 3 and 18A;** insecticides work best on small caterpillars. The most consistent insecticides in Blacklands are those containing MOA 5 and MOA 22.
chlorantraniliprole, MOA 28 + lambda-cyhalothrin, MOA 3 (Besiege) 1.25 SC	5 to 9 fl oz	See label	25.6 to 14.2	21	
chlorpyrifos, MOA 1B + gamma-cyhalothrin, MOA 3 (Cobalt Advanced) 2.63 EC	20 to 38 fl oz	See label	6.4 to 3.4	30	
indoxacarb, MOA 22 (Steward) 1.25 EC	5.6 to 11.3 fl oz	0.06 to 0.11	22.9 to 11.3	21	
methoxyfenozide, MOA 18A (Intrepid) 2F	4 to 8 fl oz	0.06 to 0.12	32 to 16	7 (hay) 14 (grain)	

91

Table 5-7. Insect Control on Soybeans

Insect Insecticide and Formulation	per Acre		Acres/gal. (lb)	Preharvest Interval (PHI) (Days)	Precautions and Remarks
	Amount of Formulation	Active Ingredient			
Soybean Looper (continued)					
methoxyfenozide, MOA 18A + spinetoram, MOA 5 (Intrepid Edge) 3F	4.0 to 6.4 oz	See label	32 to 20	28	
spinetoram, MOA 5 (Radiant) 1 SC	2 to 4 fl oz	0.016 to 0.12	64 to 32	7 (hay) 14 (grain)	
spinosad, MOA 5 (Blackhawk) 4 SC	1.1 to 2.2 fl oz	0.025 to 0.05	116.4 to 58.2	28	
spinosad, MOA 5 + gamma-cyhalothrin, MOA 3 (Consero)	2 to 3 fl oz	See label	64 to 42.7	See label	
Spider Mite					
bifenthrin, MOA 3 (Brigade, Discipline, Sniper, and others) 2 EC	5.12 to 6.4 fl oz	0.08 to 0.10	25 to 20	18	Miticides registered on soybean often provide erratic control. Two applications may be needed for high populations. The only true miticidal product listed is extoxazole, which has activity on the immatures.
chlorpyrifos, MOA 1B (Lorsban) 4E	1 to 2 pints	0.5 to 1	8 to 4	28	
etoxazole, MOA 11A1 (Zeal) SC	2 to 6 fl oz	0.045 to 0.135	64 to 21		
Stink Bug (Brown, Brown Marmorated, Green, and Southern Green)					
acephate, MOA 1B (Orthene) 97 S	0.5 to 1 lb	0.5 to 1	2 to 1	14	Treat when bug numbers exceed one per row foot or 5 per 15 sweeps (narrow-row beans). Acephate and the highest rates of pyrethroids are preferred for brown stink bug, with bifenthrin the preferred pyrethroid. Stink bugs are often late-season pests so be aware of the preharvest interval of insecticides. Reduce thresholds by 50% in soybeans intended for seed use. **In the premixed products listed, the effective chemistries are in MOA's 3 and 1B.** Go to Web page http://www.ces.ncsu.edu/plymouth/ent/soysbthresholdcalc.html for an online threshold calculator. At $10.00 per bushel, the plant compensates due to the low stink bug levels needed to reach threshold at $10.00 and above.
bifenthrin, MOA 3 (Brigade, Discipline, Sniper, and others) 2 EC	2.1 to 6.4 fl oz	0.033 to 0.10	61 to 20	30	
chlorpyrifos, MOA 1B + gamma-cyhalothrin, MOA 3 (Cobalt Advanced) 2.63 EC	20 to 38 fl oz	See label	6.4 to 3.4	30	
cyfluthrin, MOA 3 (Tombstone) 2E	1.6 to 2.8 fl oz	0.025 to 0.04	80 to 45.7	45	
diflubenzuron, MOA 15 + lambda-cyhalothrin, MOA 3 (DoubleTake) 3 SC	4 fl oz	See label	32	30	
gamma-cyhalothrin, MOA 3 (Declare) 1.25 EC	1.54 fl oz	0.015	83.1	21	
imidacloprid, MOA 4A + cyfluthrin, MOA 3 (Leverage 360) 3.0 SE	2.8 fl oz	See label	45.7	45	
lambda-cyhalothrin, MOA 3 (Warrior, Lambda-cyhalothrin, Silencer) 1.0 EC (Karate Z and Warrior II) 2.08 CS	1.92 to 3.2 fl oz 0.96 to 1.6 fl oz	0.015 to 0.025 0.015 to 0.025	66.7 to 40 133.3 to 80	30 30	
lambda-cyhalothrin, MOA 3 + thiamethoxam, MOA 4A (Endigo ZC) 2.06 SE	4 to 4.5 fl oz	See label	32 to 28.4	30	
zeta-cypermethrin, MOA 3 (Mustang Maxx) 0.8 EC	4 fl oz	0.025	32	21	
zeta-cypermethrin, MOA 3 + bifenthrin, MOA 3 (Hero) 1.24 EC	10.3 fl oz	0.033 + 0.066	12.4	21	
Velvetbean Caterpillar					
Bacillus thuringiensis, MOA 11B2 (various)	—	—	—	0	See specific labels for use rates.
pyrethroids, MOA 3	—	—	—	—	
chlorantraniliprole, MOA 28 (Prevathon) 0.43 SC	14 to 20 fl oz	0.047 to 0.067	9.1 to 6.4	1	
chlorantraniliprole, MOA 28 + lambda-cyhalothrin, MOA 3 (Besiege) 1.25 SC	5 to 9 fl oz	See label	25.6 to 14.2	21	
diflubenzuron, MOA 15 (Dimilin) 2L	2 to 4 fl oz	0.06 to 0.125	64 to 32	21	
methoxyfenozide, MOA 18A (Intrepid) 2F	4 to 8 fl oz	0.06 to 0.12	32 to 16	7 (hay) 14 (grain)	
methoxyfenozide, MOA 18A + spinetoram, MOA 5 (Intrepid Edge) 3F	4.0 to 6.4 oz	See label	32 to 20	28	
spinetoram, MOA 5 (Radiant) 1 SC	2 to 4 fl oz	0.016 to 0.12	64 to 32	7 (hay) 14 (grain)	
spinosad, MOA 5 (Blackhawk) 4 SC	1.1 to 2.2 fl oz	0.025 to 0.05	116.4 to 58.2	28	
Grape Colaspis, Blister Beetle, Japanese Beetle, Mexican Bean Beetle, Spotted Cucumber Beetle, Three Cornered Alfalfa Hopper					
acephate, MOA 1B (Orthene) 97 S	0.75 to 1 lb	0.75 to 1	1.25 to 1	14	These insects are rarely pests; exercise care in determining if a problem exists. Do not spray Mexican bean beetle when many eggs and pupae are present; wait 4 to 5 days. Thrips have never been demonstrated to reduce soybean yields in North Carolina. Three cornered alfalfa hopper girdle mainstems when plants are below 10 inches tall and petioles when plants are larger. Treatments for three-cornered alfalfa hopper only impact yield when applied to seedling soybeans.
pyrethroids, MOA 3 combinations	(see corn earworm above for rates)	—	—	—	

CAUTION: Always use pesticides according to label directions. Be mindful of reducing the impact of pesticides on wildlife and groundwater. See Extension publication AG-463-1, *Pesticides and Wildlife—Soybeans*, for further information.

INSECT CONTROL

Insect Control on Flue-Cured and Burley Tobacco

H. J. Burrack, Entomology Extension

The Insect Resistance Action Committee (**IRAC**) has grouped insecticides sharing the same mode of action (**MOA**) into categories. The categories are listed following insecticide and formulation names. To minimize the likelihood of resistance development, avoid successive treatment with insecticides having the same MOA. The Organic Materials Registry Institute (**OMRI**) lists products acceptable for use in organic production. These products are identified in the Precautions and Remarks section.

Sanitation is important in controlling greenhouse pests. Keep all trash, equipment, etc., out of and away from the greenhouse. Growing plants other than tobacco can introduce difficult-to-control pests. Leaving the empty greenhouses open during cold periods and closed during the summer can help reduce insect pests.

In general, information is provided for the most commonly used formulations of active ingredients available in multiple forumations. Carefully check the label of the product you plan to use in the event that it differs from those listed. The label is the law! Pesticides residues remain of concern for purchasers. Growers are encouraged to discuss insecticide options with their purchasers before treating to reduce potential residue concerns.

Flue-Cured and Burley Tobacco — Greenhouse

Table 5-8A. Insect Control on Flue-Cured and Burley Tobacco in Greenhouses

Insecticide, Formulation[1] and IRAC Group	Amount of Formulation Per 1,000 sq ft	Restricted Entry Interval (REI) (hours)	Preharvest Interval (PHI) (days)	Precautions and Remarks[3]
Green peach aphid				
acephate, IRAC 1B (Orthene) 97 PE	3/4 Tbsp (3/4 lb/acre)	24	3	There are many formulations of acephate. Apply in 3 gallons water per 1,000 sq ft. Even and thorough coverage is necessary for good control. Do not use more than 4 1/8 lb/acre Orthene (4 lb AI/acre). This includes greenhouse, transplant water, soil, and foliar applications.
imidacloprid, IRAC 4A (Admire Pro)	Rate per 1,000 plants 0.5 fl oz	12	14	**Only apply imidacloprid to control aphids in the greenhouse if tobacco will be transplanted within a week.** This application replaces tray drench applications for field control of aphids and flea beetles described below. There are many other formulations of imidacloprid.
thiamethoxam, IRAC 4A (Platinum) 75 SG (Platinum) SC	Rate per 1,000 plants 0.17 to 0.43 oz 0.5 to 1.3 fl oz	12	None given	**Only apply thiamethoxam to control aphids in the greenhouse if tobacco will be transplanted within a week.** This application replaces tray drench applications for field control of aphids and flea beetles described below.
Tobacco flea beetle				
acephate, IRAC 1B (Orthene) 97 PE	3/4 tbsp (3/4 lb/acre)	24	3	There are many formulations of acephate. Apply in 3 gallons water per 1,000 square feet. Even and thorough coverage is necessary for good control. Do not use more than 4 1/8 lb/acre Orthene (4 lb AI/acre). This includes greenhouse, transplant water, soil, and foliar applications.
imidacloprid, IRAC 4A (Admire Pro)	Rate per 1,000 plants 0.5 fl oz	12	14	**Only apply imidacloprid to control aphids in the greenhouse if tobacco will be transplanted within a week.** This application replaces tray drench applications for field control of aphids and flea beetles described below. There are many other formulations of imidacloprid.
thiamethoxam, IRAC 4A (Platinum) 75 SG (Platinum) SC	Rate per 1,000 plants 0.27 to 0.43 oz 0.5 to 1.3 fl oz	12	None given	**Only apply thiamethoxam to control aphids in the greenhouse if tobacco will be transplanted within a week.** This application replaces tray drench applications for field control of aphids and flea beetles described below.
Slugs or snails				
hydrated or air-slaked lime		—	—	Apply lime in a band 3 to 4 inches wide around margins of beds.
metaldehyde bait (Deadline Bullets)	0.2 to 0.6 lb		12	At dusk scatter bait around margins of beds and in walkways and open spaces. TO AVOID PLANT INJURY, DO NOT PUT BAIT ON PLANTS.
Iron phosphate bait (Sluggo)	0.5 to 1 lb		0	**OMRI** listed. TO AVOID PLANT INJURY, DO NOT PUT BAIT ON PLANTS.

[1] Some insecticides are available in several formulations. Those listed are generally the most commonly used or available. Other formulations may or may not be suitable for use on tobacco or a specific pest. Check labels carefully.

Flue-Cured and Burley Tobacco — Field

Table 5-8B. Insect Control on Flue-Cured and Burley Tobacco in the Field

Insecticide, Formulation[1] and IRAC Group	Amount of Formulation Per Acre	Restricted Entry Interval (REI) (hours)	Preharvest Interval (PHI) (days)	Precautions and Remarks
Green peach aphid				

Aphids are primarily pretopping pests. Greenhouse or transplant treatments may provide control through topping, and additional foliar treatments are not typically needed. Post topping, aphids are most common on suckers or regrowth. Sucker management via contact materials or hand removal is often sufficient to control post topping aphid populations. The threshold for green peach aphids in the field is 10% of plants scouted with 50 or more aphids on the upper leaves.
Aphid control in organic production should be initiated upon first aphid appearance, and treatment should continue on 7-10 day intervals until topping. **Data on specific organic aphid controls are limited.** Organic tobacco with aphid populations should be topped as early as feasible. Post topping sucker control is very important for aphid control in organic tobacco.

Insecticide, Formulation[1] and IRAC Group	Amount of Formulation Per Acre	Restricted Entry Interval (REI) (hours)	Preharvest Interval (PHI) (days)	Precautions and Remarks
acephate, IRAC 1B (Orthene) 97	0.75 lb	24 If significant foliar contact will occur, gloves must be worn for 14 days after treatment.	3	**TRANSPLANT WATER APPLICATION.** Apply in a minimum of 100 gallons of transplant water/acre. To avoid plant injury, do not exceed 0.75 pound a.i. acephate per acre. **SUPPRESSION ONLY,** but may not provide suppression through topping. Continue to scout plants post transplant. Do not use more than 4 1/8 lb/acre Orthene (4 lb AI/acre). This includes greenhouse, transplant water, soil, and foliar applications.
acephate, IRAC 1B (Orthene) 97 PE	0.5 lb	24 If significant foliar contact will occur, gloves must be worn for 14 days after treatment.	3	**FIELD FOLIAR APPLICATIONS.** Use at least 25 gallons per acre at 60 PSI. Using hollow cone or small solid cone nozzles cover entire plant with spray. If control 4 days after treatment is not adequate, choose another MOA for subsequent applications. Do not use more than 4 1/8 lb/acre Orthene (4 lb AI/acre). This includes greenhouse, transplant water, soil, and foliar applications.
imidacloprid, IRAC 4A (Admire Pro)	**Rate per 1,000 plants** 0.6 fl oz	12	14	**TRANSPLANT WATER APPLICATION.** Rate is per 1,000 plants and should be converted for transplant water applications based on plant population. Proper calibration of application equipment is essential for effective transplant water applications. A metered or pressurized application system is recommended. Several concentrations of imidacloprid (1.6F, 2F, 4F, and 4.6F) are available. Carefully read the label to determine the correct rate for target pests.
imidacloprid, IRAC 4A (Admire Pro)	**Rate per 1,000 plants** 0.5 to 0.6 fl oz	12	14	**GREENHOUSE TRAY DRENCH APPLICATION.** Rate is per 1,000 plants. Apply no more than 5 days before transplanting. Immediately after application, wash the material off the plants onto the potting soil. The lowest label rate is sufficient for aphid and flea beetle management. See below for recommendations for areas with high incidence of Tomato Spotted Wilt Virus (TSWV). Several concentrations of imidacloprid (1.6F, 2F, 4F, and 4.6F) are available. Carefully read the label to determine the correct rate for target pests.
imidacloprid, IRAC 4A (Admire Pro) (several products) 2F	0.7-1.4 fl oz 1.6 to 3.2 fl oz	12	14	**FIELD FOLIAR APPLICATION.** Avoid using only Group 4A insecticides as foliar field applications for aphids on plants which were treated in the greenhouse with imidacloprid or thiamethoxam. Several concentrations of imidacloprid (1.6F, 2F, 4F, and 4.6F) are available. Carefully read the label to determine the correct rate for target pests.
thiamethoxam, IRAC 4A (Platinum) 75 SG (Platinum) SC	**Rate per 1,000 plants** 0.17 oz 0.5 fl oz	12	None given	**TRANSPLANT WATER APPLICATION.** Use lower label rate for aphids. Rate is per 1,000 plants and should be converted for transplant water applications based on plant population. Proper calibration of application equipment is essential for effective transplant water applications. A metered or pressurized application system is recommended.
thiamethoxam, IRAC 4A (Platinum) 75 SG (Platinum) SC	**Rate per 1,000 plants** 0.17 oz 0.5 fl oz	12	None given	**GREENHOUSE TRAY DRENCH APPLICATION.** Use lower label rate for aphids. Rate is per 1,000 plants. Apply no more than 5 days before transplant. Immediately after application, wash the material off the plants onto the potting soil OR apply in transplant water.
thiamethoxam, IRAC 4A (Actara) 25 WDG	2 to 3 oz	12	14	**FIELD FOLIAR APPLICATION.** **Make only one application of thiamethoxam per season. Thiamethoxam is also the active ingredient in Platinum.**
acetamiprid, IRAC 4A (Assail) 30 SG	1.5 to 4 oz	12	7	**FIELD FOLIAR APPLICATION.** Make no more than four applications of acetamiprid per season, and do not apply more than once every seven days. Avoid using only Group 4A insecticides as foliar field applications for aphids on plants which were treated in the greenhouse with imidacloprid or thiamethoxam.
pymetrozine, IRAC 9B (Fulfill) 50 WG	2.75 oz	12	14	**FIELD FOLIAR APPLICATION.** Make no more than two applications of pymetrozine per year.
lambda-cyhalothrin, IRAC 3A (Warrior) (Karate Xeon)	2.5 to 3.0 oz 0.96 to 1.92 fl oz	24	40	**FIELD FOLIAR APPLICATION.** NOTE LONG PREHARVEST INTERVAL.
Pyrethrins IRAC 3 (Pyganic) 1.4 EC (Pyganic) 5.0 EC	16 to 64 fl oz 4.5 to 18 fl oz	12	0	**FIELD FOLIAR APPLICATION.** Pyganic should be buffered to pH 5.5 to 7. **OMRI** listed. Limited data.
sorbitol octanoale (SucraShield)	0.8 to 1.0% v/v	48	0	**FIELD FOLIAR APPLICATION.** **OMRI** listed. Limited data.
Azadirachtin, IRAC UN (Aza Direct)	1-2 pt	4	0	**FIELD FOLIAR APPLICATION.** Optimal pH range 5.6 - 6.5. **OMRI** listed. Limited data.
rosemary and peppermint oil (Ecotec)	2-4 pt	0	0	**FIELD FOLIAR APPLICATION.** **OMRI** listed. Limited data.
petroleum oil (Saf-T-Side)	1-2 gal	4	0	**FIELD FOLIAR APPLICATION.** **OMRI** listed. Limited data.

INSECT CONTROL

94

Table 5-8B. Insect Control on Flue-Cured and Burley Tobacco in the Field

Insecticide, Formulation[1] and IRAC Group	Amount of Formulation Per Acre	Restricted Entry Interval (REI) (hours)	Preharvest Interval (PHI) (days)	Precautions and Remarks
Tobacco flea beetle				
Greenhouse or transplant treatments may provide control through topping, and additional foliar treatments are not typically needed. The threshold for foliar treatments on small, recently planted tobacco is 4 beetles per plant. Flea beetles populations may increase near harvest and require management if populations exceed 60 beetles per fully grown plant. Good coverage is required for effective flea beetle control in large plants. Use appropriate equipment and sufficient water volume to achieve coverage from the base to the top of the plant.				
acephate, IRAC 1B (Orthene) 97	0.75 lb	24 If significant foliar contact will occur, gloves must be worn for 14 days after treatment.	3	**TRANSPLANT WATER APPLICATION.** Apply in a minimum of 100 gallons of transplant water/acre. To avoid plant injury, do not exceed 0.75 pound a.i. acephate per acre. Do not use more than 4 1/8 lb/acre Orthene (4 lb AI/acre). This includes greenhouse, transplant water, soil, and foliar applications.
acephate, IRAC 1B (Orthene) 97 PE	0.5 lb	24 If significant foliar contact will occur, gloves must be worn for 14 days after treatment.	3	**FIELD FOLIAR APPLICATIONS.** Use at least 25 gallons per acre at 60 PSI. Using hollow cone or small solid cone nozzles cover entire plant with spray. If control 4 days after treatment is not adequate, choose another MOA for subsequent applications. Do not use more than 4 1/8 lb/acre Orthene (4 lb AI/acre). This includes greenhouse, transplant water, soil, and foliar applications.
imidacloprid, IRAC 4A (Admire Pro)	Rate per 1,000 plants 0.6 fl oz	12	14	**TRANSPLANT WATER APPLICATION.** Proper calibration of application equipment is essential for effective transplant water applications. A metered or pressurized application system is recommended. Several concentrations of imidacloprid (1.6F, 2F, 4F, and 4.6F) are available. Carefully read the label to determine the correct rate for target pests.
imidacloprid, IRAC 4A (Admire Pro)	Rate per 1,000 plants 0.5 to 0.6 fl oz	12	14	**GREENHOUSE TRAY DRENCH APPLICATION.** Rate is per 1,000 plants. Apply no more than 5 days before transplanting. Immediately after application, wash the material off the plants onto the potting soil. The lowest label rate is sufficient for aphid and flea beetle management. See below for recommendations for areas with high incidence of Tomato Spotted Wilt Virus (TSWV). Several concentrations of imidacloprid (1.6F, 2F, 4F, and 4.6F) are available. Carefully read the label to determine the correct rate for target pests.
imidacloprid, IRAC 4A (Admire Pro) (several products) 2F	0.7-1.4 fl oz 1.6 to 3.2 fl oz	12	14	**FIELD FOLIAR APPLICATION.** Avoid using only Group 4A insecticides as foliar field applications for aphids on plants which were treated in the greenhouse with imidacloprid or thiamethoxam. Several concentrations of imidacloprid (2F, 4F, and 4.6F) are available. Carefully read the label to determine the correct rate for target pests.
thiamethoxam, IRAC 4A (Platinum) 75 SG (Platinum) SC	Rate per 1,000 plants 0.27 oz 0.8 fl oz	12	None given	**TRANSPLANT WATER APPLICATION.** Use lower label rate for aphids. Rate is per 1,000 plants and should be converted for transplant water applications based on plant population. Proper calibration of application equipment is essential for effective transplant water applications. A metered or pressurized application system is recommended.
thiamethoxam, IRAC 4A (Platinum) 75 SG (Platinum) SC	Rate per 1,000 plants 0.27 oz 0.8 fl oz	12	None given	**GREENHOUSE TRAY DRENCH APPLICATION.** Use lower label rate for aphids. Rate is per 1,000 plants. Apply no more than 5 days before transplant. Immediately after application, wash the material off the plants onto the potting soil OR apply in transplant water.
acetamiprid, IRAC 4A (Assail) 30 SG	2.5 to 4 oz	12	7	Make no more than four applications of acetamiprid per season, and do not apply more than once every seven days. Avoid using only Group 4A materials for season long control of insects with more than 1 generation. Following treatments of Group 4A materials, rotate to a different MOA before making additional applications of a Group 4A material.
methomyl, IRAC 1A (Lannate) 2.4 LV	1.5 pt	48	5 flue-cured; 14 burley	Some tobacco purchasers may have concerns about methomyl residue. Discuss the use of this material with potential purchasers before making applications.
lambda-cyhalothrin, IRAC 3A (Warrior) 1CS (Karate Xeon)	2.5 to 3.0 oz 0.96 to 1.92 fl oz	24	40	NOTE LONG PREHARVEST INTERVAL.
thiamethoxam, IRAC 4A (Actara) 25 WDG	2 to 3 oz	12	14	**Make only one application of thiamethoxam per season. Thiamethoxam is also the active ingredient in Platinum.**
Armyworm				
Armyworms are typically most common late in the growing season. Preventative treatment is not recommended.				
chlorantraniliprole, IRAC 28 (Coragen)	3.5 to 7 fl oz	4	1	**FIELD FOLIAR APPLICATION.** Make no more than 4 applications per season (with at least 3 days between applications), and apply no more than 15.4 fl oz per season.
emamectin benzoate, IRAC 6 (Denim) 0.16EC	8 to 12 oz	48	14	Apply by ground only in a minimum of 20 gallons per acre of finished spray.
lambda-cyhalothrin, IRAC 3A (Warrior) (Karate Xeon)	2.5 to 3.0 oz 0.96 to 1.92 fl oz	24	40	NOTE LONG PREHARVEST INTERVAL.
Budworm				
The threshold for tobacco budworm is 10% infested plants. This threshold is very conservative, and budworms should not be treated unless infestations exceed 10%. Coverage is important for budworm management. Use 1 to 3 full cone nozzles 6 to 12 inches above bud and a minimum of 25 gallons water per acre.				
acephate, IRAC 1B (Orthene) 97 PE	0.75 lb	24	3	There are many formulations of acepate. Do not use more than 4 1/8 lb/acre Orthene (4 lb AI/acre). This includes greenhouse, transplant water, soil, and foliar applications. Other materials provide more

INSECT CONTROL

Table 5-8B. Insect Control on Flue-Cured and Burley Tobacco in the Field

Insecticide, Formulation[1] and IRAC Group	Amount of Formulation Per Acre	Restricted Entry Interval (REI) (hours)	Preharvest Interval (PHI) (days)	Precautions and Remarks
Budworm (continued)				
Bacillus thuringiensis, IRAC 11 DIPel DF	0.5 to 1 lb	4	0	There are many *B.t.* formulations, including Agree, Biobit, Condor, Crymax, Deliver, Dipel, Javelin, and Lepinox. Highest labeled rates are generally needed for budworm control. DiPel DF and many other B.t. formulations are **OMRI** listed, but not all Bt formulations are **ORMI** listed. Carefully read the label to determine if a material is acceptable for use on organically certified plants.
chlorantraniliprole, IRAC 28 (Coragen)	5.0 to 7.5 fl oz	4	1	**TRANSPLANT WATER APPLICATION.** Transplant applications of Coragen may suppress tobacco budworm populations for 4 to 7 weeks. Proper calibration of application equipment is essential for effective transplant water applications. A metered or pressurized application system is recommended. Apply no more than 15.4 fluid ounces of Coragen or more than 0.2 pound chlorantraniliprole per acre per crop, which includes applications of Coragen, Beseige, and Endigo.
chlorantraniliprole, IRAC 28 (Coragen)	3.5 to 7.5 fl oz	4	1	**FIELD FOLIAR APPLICATION.** Make no more than 4 applications per season (with at least 3 days between applications), and apply no more than 15.4 fluid ounces of Coragen or more than 0.2 pound chlorantraniliprole per acre per crop, which includes applications of Coragen, Beseige, and Endigo.
emamectin benzoate, IRAC 6 (Denim) 0.16EC	8 to 12 oz	48	14	Apply by ground only in a minimum of 20 gallons per acre of finished spray.
lambda-cyhalothrin, IRAC 3A (Warrior) 1CS (Karate Xeon)	2.5 to 3.0 oz 0.96 to 1.92 fl oz	24 24	40 40	To avoid build-up of resistance, rotate use of this product with other insecticides. NOTE THE LONG PREHARVEST USE RESTRICTION.
lambda-cyhalothrin + chloratraniliprole IRAC 3 + 28 (Beseige)	5.0 to 9.0 fl oz	24	40	NOTE THE LONG PREHARVEST USE RESTRICTION. Apply no more than 0.2 pound chlorantraniliprole per acre per crop, which includes applications of Coragen, Beseige, and Endigo.
lambda-cyhalothrin + thiamethoxam, IRAC 3 + 4A (Endigo) ZC	4.0 to 4.5 fl oz	24	40	NOTE THE LONG PREHARVEST USE RESTRICTION. Apply no more than 0.2 pound chlorantraniliprole per acre per crop, which includes applications of Coragen, Beseige, and Endigo.
spinosad, IRAC 5 (Tracer) 4 L (Blackhawk)	1.4 to 2 fl oz 1.6 to 3.2 oz	4	3	Tracer is a liquid formulation; Blackhawk is a solid formulation. While spinosad is a naturally derived active ingredient, neither Tracer nor Blackhawk is **OMRI** listed.
Cutworm				
Preventative insecticide applications are not recommended for cutworms because they are infrequent pests and rescue materials are effective. Scout fields in the first 4 weeks following transplant for cutworm injury and treat if 10% of plants are clipped. Cutworm treatments should be applied in a directed spray over rows in the late afternoon or at dusk, when cutworms are most likely to be active.				
acephate, IRAC 1B (Orthene) 97 PE	0.75 lb	24	3	There are many formulations of acephate. Do not use more than 4 1/8 lb/acre Orthene (4 lb AI/acre). This includes greenhouse, transplant water, soil, and foliar applications. Other materials provide more
chlorantraniliprole, IRAC 28 (Coragen)	3.5 to 7 fl oz	4	1	Make no more than 4 applications per season (with at least 3 days between applications). Apply no more than 15.4 fluid ounces of Coragen or more than 0.2 pound chlorantraniliprole per acre per crop, which includes applications of Coragen, Beseige, and Endigo.
lambda-cyhalothrin, IRAC 3A (Warrior) (Karate Xeon)	2.5 to 3 oz 0.96 to 1.92 fl oz	24	40	NOTE LONG PREHARVEST INTERVAL.
lambda-cyhalothrin + chloratraniliprole IRAC 3 + 28 (Beseige)	5.0 to 9.0 fl oz	24	40	NOTE LONG PREHARVEST USE RESTRICTION. Apply no more than 15.4 fluid ounces of Coragen or more than 0.2 pound chlorantraniliprole per acre per crop, which includes applications of Coragen, Beseige, and Endigo.
Grasshopper				
acephate, IRAC 1B (Orthene) 97	0.25 to 0.5 lb	24	3	Nymphs (young) are more easily controlled than adults. There are many formulations of acephate. Do not use more than 4 1/8 lb/acre Orthene (4 lb AI/acre). This includes greenhouse, transplant water, soil, and foliar applications.
Hornworm				
Treat for hornworms when five or more larvae longer than 1 inch and without cocoons are found per 50 plants. Hornworm larvae with cocoons should be considered 1/5 of a larvae when counting. If treatment is necessary during harvesting, be certain to follow all labeled preharvest intervals.				
acephate, IRAC 1B (Orthene) 97 PE	0.5 lb	24	3	There are many formulations of acephate. Do not use more than 4 1/8 lb/acre Orthene (4 lb AI/acre). This includes greenhouse, transplant water, soil, and foliar applications.
Bacillus thuringiensis, IRAC 11 DIPel DF	0.5 to 1 lb	4	0	There are many *B.t.* formulations, including Agree, Biobit, Condor, Crymax, Deliver, Dipel, Javelin, and Lepinox. Highest labeled rates are generally needed for budworm control. DiPel DF and many but not all Bt formulations are **ORMI** listed. Carefully read the label to determine if a material is acceptable for use on organically certified plants.
chlorantraniliprole, IRAC 28 (Coragen)	3.5 to 7.5 fl oz	4	1	**FIELD FOLIAR APPLICATION.** Because they are not frequent pests before topping, transplant water applications of Coragen for hornworms alone are not recommended. Make no more than 4 applications per season (with at least 3 days between applications). Apply no more than 15.4 fluid ounces of Coragen or more than 0.2 pound chlorantraniliprole per acre per crop, which includes applications of Coragen, Beseige, and Endigo. Lower label rates of Coragen are likely sufficient for hornworms.
emamectin benzoate, IRAC 6 (Denim) 0.16EC	8 oz	48	14	
lambda-cyhalothrin + chloratraniliprole IRAC 3 + 28 (Beseige)	5.0 to 9.0 fl oz	24	40	NOTE THE LONG PREHARVEST USE RESTRICTION. Apply no more than 0.2 pound chlorantraniliprole per acre per crop, which includes applications of Coragen, Beseige, and Endigo.

Table 5-8B. Insect Control on Flue-Cured and Burley Tobacco in the Field

Insecticide, Formulation[1] and IRAC Group	Amount of Formulation Per Acre	Restricted Entry Interval (REI) (hours)	Preharvest Interval (PHI) (days)	Precautions and Remarks
Hornworm (continued)				
lambda-cyhalothrin + thiamethoxam, IRAC 3 + 4A (Endigo) ZC	4.0 to 4.5 fl oz	24	40	NOTE THE LONG PREHARVEST USE RESTRICTION. Apply no more than 0.2 pound chlorantraniliprole per acre per crop, which includes applications of Coragen, Beseige, and Endigo.
spinosad, IRAC 5 (Tracer) 4 L (Blackhawk)	1.4 to 2 fl oz 1.6 to 3.2 oz	4	3	Tracer is a liquid formulation; Blackhawk is a solid formulation. While spinosad is a naturally derived active ingredient, neither Tracer nor Blackhawk is OMRI listed.
Japanese beetle				
Infestations may be spotty within fields and do not typically require treatment.				
acephate, IRAC 1B (Orthene) 97	0.75 lb	24	3	There are many formulations of acephate. Do not use more than 4 1/8 lb/acre Orthene (4 lb AI/acre). This includes greenhouse, transplant water, soil, and foliar applications.
lambda-cyhalothrin + chloratraniliprole IRAC 3 + 28 (Besiege)	5.0 to 9.0 fl oz	24	40	NOTE THE LONG PREHARVEST USE RESTRICTION. Apply no more than 0.2 pound chlorantraniliprole per acre per crop, which includes applications of Coragen, Beseige, and Endigo.
lambda-cyhalothrin + thiamethoxam, IRAC 3 + 4A (Endigo) ZC	4.0 to 4.5 fl oz	24	40	NOTE THE LONG PREHARVEST USE RESTRICTION. Apply no more than 0.2 pound chlorantraniliprole per acre per crop, which includes applications of Coragen, Beseige, and Endigo.
imidacloprid, IRAC 4A (Admire Pro) (several products) 2F	1.4 fl oz 3.2 fl oz	12	14	**FIELD FOLIAR APPLICATION.** Avoid using only Group 4A materials for season long control of insects with more than 1 generation. Following treatments of Group 4A materials, rotate to a different MOA before making additional applications of a Group 4A material.
thiamethoxam, IRAC 4A (Actara) 25 WDG	2 to 3 oz	12	14	**Make only one application of thiamethoxam per season. Thiamethoxam is also the active ingredient in Platinum.**
Slug				
Slugs are only potential pests in the greenhouse and shortly following transplant. They do not present a risk to larger plants.				
Iron phosphate bait (Sluggo)	20 to 44 lb	0	—	**OMRI** listed. TO AVOID PLANT INJURY, DO NOT PUT BAIT ON PLANTS.
metaldehyde bait (Deadline Bullets)	12 to 40 lb	12	—	Apply at dusk to soil surface between rows and around margins of field. DO NOT PUT BAIT ON PLANTS.
Stink bug				
Stink bugs rarely cause economic damage to tobacco and rarely require treatment.				
acephate, MOA 1B (Orthene) 97	0.75 lb	24	3	There are many formulations of acephate. Do not use more than 4 1/8 lb/acre Orthene (4 lb AI/acre). This includes greenhouse, transplant water, soil, and foliar applications.
bifenthrin, IRAC 3 (Capture LFR)	3.4 to 6.8 fl oz	12	Do not apply after Layby	**FIELD FOLIAR APPLICATION.** NOTE THE LONG PREHARVEST USE RESTRICTION.
bifenthrin + imidacloprid, IRAC 3, 4A (Brigadier) 2SC	6.4 fl oz	12	Do not apply after Layby	**FIELD FOLIAR APPLICATION.** NOTE THE LONG PREHARVEST USE RESTRICTION.
lambda-cyhalothrin, IRAC 3A (Warrior) 1CS (Karate Xeon)	2.5 to 3 oz 0.96 to 1.92 fl oz	24 24	40 40	To avoid build-up of resistance, rotate use of this product with other modes of action. NOTE THE LONG PREHARVEST USE RESTRICTION.
lambda-cyhalothrin + chloratraniliprole IRAC 3 + 28 (Besiege)	5.0 to 9.0 fl oz	24	40	NOTE THE LONG PREHARVEST USE RESTRICTION. Apply no more than 0.2 pound chlorantraniliprole per acre per crop, which includes applications of Coragen, Beseige, and Endigo.
Tomato spotted wilt virus (TSWV) suppression				
imidacloprid, IRAC 4A (Admire Pro)	**Rate per 1,000 plants** 0.8 fl oz	12	14	**TRANSPLANT WATER APPLICATION.** Rate is per 1,000 plants and should be converted for transplant water applications based on plant population. Proper calibration of application equipment is essential for effective transplant water applications. A metered or pressurized application system is recommended. Several concentrations of imidacloprid (1.6F, 2F, 4F, and 4.6F) are available. Carefully read the label to determine the correct rate for target pests.
imidacloprid, IRAC 4A (Admire Pro)	**Rate per 1,000 plants** 0.8 fl oz	12	14	**GREENHOUSE TRAY DRENCH APPLICATION.** Rate is per 1,000 plants. Apply no more than 5 days before transplanting. Immediately after application, wash the material off the plants onto the potting soil. Several concentrations of imidacloprid (1.6F, 2F, 4F, and 4.6F) are available. Carefully read the label to determine the correct rate for target pests.
thiamethoxam, IRAC 4A (Platinum) 75 SG (Platinum) SC	**Rate per 1,000 plants** 0.27 oz 0.8 fl oz	12	None given	**TRANSPLANT WATER APPLICATION.** Rate is per 1,000 plants and should be converted for transplant water applications based on plant population. Proper calibration of application equipment is essential for effective transplant water applications. A metered or pressurized application system is recommended.
thiamethoxam, IRAC 4A (Platinum) 75 SG (Platinum) SC	**Rate per 1,000 plants** 0.27 oz 0.8 fl oz	12	None given	**GREENHOUSE TRAY DRENCH APPLICATION.** Use lower label rate for aphids. Rate is per 1,000 plants. Apply no more than 5 days before transplant. Immediately after application, wash the material off the plants onto the potting soil OR apply in transplant water.

INSECT CONTROL

Table 5-8B. Insect Control on Flue-Cured and Burley Tobacco in the Field

Insecticide, Formulation[1] and IRAC Group	Amount of Formulation Per Acre	Restricted Entry Interval (REI) (hours)	Preharvest Interval (PHI) (days)	Precautions and Remarks
Vegetable weevil				
acephate, IRAC 1B (Orthene) 97	0.5 to 0.75 lb	24	3	Treat plants in late afternoon for best control. Spray a band over center of row using a good volume of water. Do not use more than 4 1/8 lb/acre Orthene (4 lb AI/acre). This includes greenhouse, transplant water, soil, and foliar applications.
lambda-cyhalothrin, IRAC 3A (Warrior) 1CS (Karate Xeon)	2.5 to 3 oz 0.96 to 1.92 fl oz	24 24	40 40	NOTE THE LONG PREHARVEST USE RESTRICTION.
lambda-cyhalothrin + chloratraniliprole IRAC 3 + 28 (Besiege)	5.0 to 9.0 fl oz	24	40	NOTE THE LONG PREHARVEST USE RESTRICTION. Apply no more than 0.2 pound chlorantraniliprole per acre per crop, which includes applications of Coragen, Beseige, and Endigo.
Wireworm				

Wireworm treatments should be applied pretransplant in fields with a history of significant damage. If fields do not have a history of wireworm injury, greenhouse tray drench or transplant water treatments of imidacloprid or thiamethoxam will also suppress wireworm damage if they are present.

Insecticide, Formulation[1] and IRAC Group	Amount of Formulation Per Acre	Restricted Entry Interval (REI) (hours)	Preharvest Interval (PHI) (days)	Precautions and Remarks
bifenthrin + imidacloprid, IRAC 3, 4A (Brigadier 2SC)	6.4 fl oz	12	Do not apply after Layby	Use as described above for transplant water treatments for imidacloprid. Brigadier is not intended for greenhouse use. Data on wireworm control are limited.
bifenthrin, IRAC 3 (Capture LFR)	3.4 to 6.8 fl oz	12	Do not apply after Layby	Apply as a pretransplant soil treatment and incorporate into 4 inches of soil OR apply in transplant water at 3.4 to 6.8 fluid ounces per acre. Data on wireworm control are limited.
chlorpyrifos, IRAC 1B (Lorsban) Advanced (Lorsban) 15 G	2 pt 13.5 to 20 lb	24	NA	**Preplant soil application only.** Apply 15G formulation broadcast 1 week before setting. Incorporate immediately. Apply liquid formulation 24-48 hours before bedding in at least 10 gallons water per acre. Data on wireworm control are limited.
ethoprop, IRAC 1B (Mocap) 15 G	13-40 lb (broadcast) 3.2 lb per 100 row ft (banded)	48	NA	**Preplant soil application only.** Rates depend on application timing and target pests.
imidacloprid, IRAC 4A (Admire Pro)	**Rate per 1,000 plants** 1.2 fl oz	12	14	**GREENHOUSE TRAY DRENCH APPLICATION.** Rate is per 1,000 plants. Apply no more than 5 days before transplanting. Immediately after application, wash the material off the plants onto the potting soil. Several concentrations of imidacloprid (1.6F, 2F, 4F, and 4.6F) are available. Carefully read the label to determine the correct rate for target pests. Data on wireworm control are limited.
thiamethoxam, IRAC 4A (Platinum) 75 SG (Platinum) SC	**Rate per 1,000 plants** 0.43 oz 1.3 fl oz	12	None given	**GREENHOUSE TRAY DRENCH APPLICATION.** Use lower label rate for aphids. Rate is per 1,000 plants. Apply no more than 5 days before transplant. Immediately after application, wash the material off the plants onto the potting soil OR apply in transplant water. Data on wireworm control are limited.

[1] Some insecticides are available in several formulations. Those listed are generally the most commonly used or are readily available. Other formulations may or may not be suitable for use on tobacco or a specific pest. Check labels carefully.

[2] Many soil-applied insecticides can injure plants under certain conditions. Some soil-applied insecticides are very soluble and pose a threat to surface and groundwater; check labels carefully for warnings.

[3] Tobacco purchasers are concerned about pesticide residues in cured leaf. Use caution in making applications of methomyl, acephate, Group 3 (pyrethriod) insecticides. Select other materials when available.

More production information is available at http://www.tobacco.ces.ncsu.edu.

Insect Control for Commercial Vegetables

J. F. Walgenbach, Entomology Extension, and G. G. Kennedy, Entomology Research

Read the pesticide label before application. High pressure (200 psi) and high volume (50 gallons per acre) aid in vegetable insect control. Ground sprays with airblast sprayers or sprayers with hollow cone drop nozzles are suggested. Incorporate several methods of control for best results. In recent years, the number of generic products has increased significantly. For brevity, these generic products typically are not listed within each section. The trade names listed are intended to aid in identification of products and are neither intended to promote use of specific trade names nor to discourage use of generic products. A list of active ingredients and generic brand names appears in a separate table at the end of this section.

Insecticides are placed into IRAC MOA classes based on their mode of action (insecticides in the same MOA class have the same mode of action). Effective insecticide resistance management involves the use of alternations, rotations, or sequences of different insecticide MOA classes. To prevent the development of resistance, it is important not to apply insecticides with the same MOA to successive generations of the same insect.

Table 5-9. Insect Control for Commercial Vegetables

CROP Insect	Insecticide, Mode of Action Code, and Formulation	Amount of Formulation Per Acre	Restricted Entry Interval (REI)	Pre harvest Interval (PHI) (Days)	Precautions and Remarks
Asparagus					
Aphid	dimethoate 400, MOA 1B	1 pt	48 hrs	180	Do not exceed 5 pints per acre per year.
	malathion, MOA 1B (various) 57 EC	2 pt	12 hrs	1	Aphid colonies appear by early September. The use of carbamates may result in aphid buildup.
	pymetrozine, MOA 9B (Fulfill) 50 WDG	2.75 oz	12 hrs	—	For aphid control on ferns after harvest.
Asparagus beetle, Japanese beetle, Grasshopper	carbaryl, MOA 1A (Sevin) 50 WP (Sevin) 80 S (Sevin) XLR Plus	2 to 4 lb 1.25 to 2.5 lb 1 to 2 qt	12 hrs	1	Low rate to be used on seedlings or spears. Do not apply more often than once every 3 days. With established beetle populations, three consecutive weekly sprays are required. Manage beetles and grasshoppers in the fall.
	dimethoate 400, MOA 1B	1 pt	48 hrs	180	Do not exceed 5 pints per acre per year.
	malathion, MOA 1B (various) 57 EC	2 pt	12 hrs	1	Apply as needed.
	methomyl, MOA 1A (Lannate) 2.4 LV	1.5 pt	48 hrs	1	Leave a row on edge of field near overwintering sites of asparagus beetles fern out. This will attract and hold beetles for that directed insecticide spray (trap and destroy).
	pyrethroid, MOA 3				See table 5-9B for a list of registered pyrethroids and pre-harvest intervals.
	spinetoram, MOA 5 (Radiant) 1 SC	4 to 8 fl oz	4 hrs	60	For asparagus beetle only. This use is only for asparagus ferns; do not apply within 60 days of spear harvest.
Beet armyworm, Cutworm, Yellow-striped armyworm	*Bacillus thuringiensis*, MOA 11A (Dipel) DF	0.5 to1 lb	4 hrs	0	
	chlorantraniliprole, MOA 28 (Coragen) 1.67SC	3.5 to 5 fl oz	4 hrs	1	
	cyantraniliprole, MOA 28 (Exirel) 0.83EC	7 to 13.5 fl oz	12 hr	1	Do not make applcations within 25 ft of water sources.
	methomyl, MOA 1A (Lannate) 2.4 LV (Lannate) 90 SP	1.5 to 3 pt 0.5 to 1 lb	48 hrs	1	
	spinetoram, MOA 5 (Radiant) 1 SC	4 to 8 fl oz	4 hrs	60	This use is only for asparagus ferns; do not apply within 60 days of spear harvest.
	spinosad MOA 5 (Entrust 2SC)	4 to 6 fl oz	4 hrs	60	This use is only for asparagus ferns; do not apply within 60 days of spear harvest. OMRI approved.
Beans (Snap, Lima, Pole)					
Aphid	acetamiprid MOA 4A (Assail) 30SG	2.5 to 5.3 oz	12 hrs	7	
	dimethoate 4 EC, MOA 1B	0.5 to 1 pt	48 hrs	0	On foliage as needed. Re-entry interval of 48 hours
	imidacloprid, Soil treatment (Admire Pro) 4.6 F (various) 2F	7 to 10.5 fl oz 16 to 24 fl oz	12 hrs	21	See label for soil application instructions. Also controls leafhoppers and thrips
	Foliar treatment Admire Pro 4.6 F (various) 1.6 F	1.2 fl oz 3.5 fl oz	12 hrs	7	
	pyrethroid, MOA 3		12 hrs		See table 5-9B for a list of registered pyrethroids and pre-harvest intervals.
	Spirotetramat, MOA 23 (Movento) 2 SC	4 to 5 fl oz	24 hrs	1 (succulent) 7 (dried)	
Thrips	acephate, MOA 1B (Orthene) 97 PE	0.5 to 1 lb	24 hrs	14	Lima beans may be treated and harvested the same day. Do not apply more than 2 pounds a.i. per acre per season.
	acetamiprid MOA 4A (Assail) 30SG	2.5 to 5.3 oz	12 hrs	7	

Table 5-9. Insect Control for Commercial Vegetables

CROP Insect	Insecticide, Mode of Action Code, and Formulation	Amount of Formulation Per Acre	Restricted Entry Interval (REI)	Pre harvest Interval (PHI) (Days)	Precautions and Remarks
Beans (Snap, Lima, Pole) (continued)					
Thrips (continued)	pyrethroid, MOA 3		12 hrs		See table 5-9B for a list of registered pyrethroids and pre-harvest intervals.
	methomyl, MOA 1A (Lannate) 90 SP (Lannate) 2.4 LV	0.5 lb 1.5 pt	48 hrs	1	
	novaluron MOA 15 (Rimon) 0.83 EC	12 fl oz	12 hrs	1	Effective against immature thrips only.
	spinetoram, MOA 5 (Radiant) 1 SC	5 to 6 fl oz	4 hrs	3 (succulent); 28 (dried)	Do not apply more than 28 fluid ounces per acre per season on succulent beans or more than 12 fluid ounces on dried beans.
	Spinosad, MOA 5 (Blackhawk)	2.5 to 3.3 oz	4 hrs	3 (succulent); 28 (dried	Do not apply more than 20 ounces per acre per season on succulent beans or more than 8.3 ounces on dried beans.
Corn earworm, European corn borer, Lesser cornstalk borer, Looper	chlorantraniliprole, MOA 28 (Coragen) 1.67 SC	3.5 to 5 fl oz	4 hrs	1	
	flubendiamide, MOA 28 (Belt) 4 SC	2 to 3 fl oz	12 hrs	1 (succulent) 14 (dried)	1-day PHI for podded and succulent, 14 for dry beans.
	novaluron MOA 15 (Rimon) 0.83 EC	6 to 12 fl oz	12 hrs	1	
	spinetoram, MOA 5 (Radiant) 1 SC	4.5 to 6 fl oz	4 hrs	3 (succulent); 28 (dried)	Do not apply more than 28 fluid ounces per acre per season on succulent beans or more than 12 fluid ounces on dried beans.
	Spinosad, MOA 5 (Blackhawk)	1.7 to 3.3 oz	4 hrs	3 (succulent); 28 (dried	Do not apply more than 20 ounces per acre per season on succulent beans or more than 8.3 ounces on dried beans.
	pyrethroid, MOA 3		12 hrs		See table 5-9B for a list of registered pyrethroids and pre-harvest intervals.
Cowpea curculio	pyrethroid, MOA 3				See table 5-9B for a list of registered pyrethroids and pre-harvest intervals. Control may be poor in areas where resistant populations occur, primarily in the Gulf Coast areas.
Cucumber beetle, Bean leaf beetle, Japanese beetle, Cutworm	carbaryl, MOA 1A (Sevin) 50 WP 80 S XLR Plus	4 lb 2.5 lb 1 qt	12 hrs	3 (succulent) 21 (dried)	
	pyrethroid, MOA 3		12 hrs		See table 5-9B for a list of registered pyrethroids and pre-harvest intervals.
Grasshopper	pyrethroid, MOA 3		12 hrs		See table 5-9B for a list of registered pyrethroids and pre-harvest intervals.
Leafminer	cryomazine, MOA 17 (Trigard) 75 WP	2.66 oz	12 hrs	7	
	naled, MOA 1B (Dibrom) 8 EC	1 pt	48 hrs	1	Re-entry interval is 48 hours
	spinetoram, MOA 5 (Radiant) 1 SC	4 to 8 fl oz	4 hrs	3 (succulent); 28 (dried)	Do not apply more than 28 fluid ounces per acre per season on succulent beans or more than 12 fluid ounces on dried beans.
	Spinosad, MOA 5 (Blackhawk)	2.5 to 3.3 oz	4 hrs	3 (succulent); 28 (dried)	Do not apply more than 20 ounces per acre per season on succulent beans or more than 8.3 ounces on dried beans.
Lygus bug	Pyrethroid, MOA 3		12 hrs		See table 5-9B for a list of registered pyrethroids and pre-harvest intervals.
	carbaryl, MOA 1A (Sevin) 50 WP 80 S XLR Plus	3 lb 1.875 lb 1.5 qt	12 hrs	3 (succulent) 21 (dried)	On foliage when pods begin to form.
	dimethoate, MOA 1B (Dimethoate) 4 EC	1 pt	48 hrs	7	Do not apply if bees are visiting area to be treated when crops or weeds are in bloom.
Mexican bean beetle	acetamiprid MOA 4A (Assail) 30SG	2.5 to 5.3 oz	12 hrs	7	
	pyrethroid, MOA 3				See table 5-9B for a list of registered pyrethroids and pre-harvest intervals.
	carbaryl, MOA 1A (Sevin) 50 WP (Sevin) 80 S (Sevin) XLR Plus	1 to 2 lb 0.625 to 1.25 lb 1 qt	12 hrs	3 (succulent) 21 (dry)	On foliage as needed. Use low rate on young plants.
	dimethoate, MOA 1B (Dimethoate) 4 EC	1 pt	48 hrs	0	48-hour re-entry interval.
	novaluron MOA 15 (Rimon) 0.83 EC	9 to 12 oz	12 hrs	1	Controls immature stages only.
	phorate, MOA 1B (Thimet) 20 G	4.9 to 9.4 oz/1,000 ft row	48 hrs	60	Drill granules to the side of seed at planting. Avoid contact with seed.

Table 5-9. Insect Control for Commercial Vegetables

CROP Insect	Insecticide, Mode of Action Code, and Formulation	Amount of Formulation Per Acre	Restricted Entry Interval (REI)	Pre harvest Interval (PHI) (Days)	Precautions and Remarks
Beans (Snap, Lima, Pole) (continued)					
Potato leafhopper	acetamiprid MOA 4A (Assail) 30SG	2.5 to 5.3 oz	12 hrs	7	
	carbaryl, MOA 1A (Sevin) 50 WP (Sevin) 80 S (Sevin) XLR Plus	4 lb 2.5 lb 1 qt	12 hrs	3 (succulent) 21 (dry)	On foliage as needed.
	dimethoate 4 EC, MOA 1B	0.5 to 1 pt	48 hrs	7	
	methomyl, MOA 1A (Lannate) 90 SP (Lannate) 2.4 L	0.5 lb 1.5 to 3 pt	48 hrs	1 1 to 3	Do not graze before 3 days or use for hay before 7 days.
	phorate, MOA 1B (Thimet) 20 G	4.9 to 9.4 oz/ 1,000 ft row	48 hrs	60	Drill granules to the side of seed at planting. Avoid contact with seed.
	pyrethroid, MOA 3		12 hrs		See table 5-9B for a list of registered pyrethroids and their re entry and pre-harvest intervals.
Seedcorn maggot, Wireworm	Use seed pretreated with insecticide for seedcorn maggot control.				Seed can be purchased pretreated. Pretreated seed will not control wireworms.
	bifenthrin MOA 3 (Empower) 1.15G	3.5 to 8.7 lb	9 days	9	Apply preplant broadcast incorporated in the top 1 to 3 inches of soil.
	chlorpyrifos MOA 1B (Lorsban) 4E	2 pts	24 hrs		Can be applied preplant broadcast incorporated in the too p1 to 3 inches of soil, or at planting as a T-band application. For at planting application, apply 1.8 fluid ounces per 1,000 feet of row at 30-inch row spacing. Apply the spray in a 3 to 5 inch wide band over the row behind the planting shoe and in front of the press wheel to achieve shallow incorporation. Do not make more than one application per year or apply more than 1 pound ai per acre.
	phorate, MOA 1B (Thimet) 20 G	4.9 to 9.4 oz/ 1,000 ft row	12 hrs	60	Drill granules to the side of seed at planting. Avoid contact with seed.
Spider mite	pyrethroid, MOA 3		12 hrs		See table 5-9B for a list of registered pyrethroids and pre-harvest intervals.
	bifenazate MOA 20D (Acramite) 4 SC	16 to 24 fl oz	12 hrs	3	
Stink bug, Kudzu bug	pyrethroid, MOA 3		12 hrs		See table 5-9B for a list of registered pyrethroids and pre-harvest intervals.
	naled, MOA 1B (Dibrom) 8 EC	1.5 pt/100 gal water	48 hrs	1	
Whiteflies	acetamiprid MOA 4A (Assail) 30 SG	4.0 to 5.3 oz	12 hrs	7	
	buprofezin, MOA 16 (Courier) 40 SC	9 to 13.6 fl oz	12 hrs	14	For use on snap beans only.
	imidacloprid, MOA 4A Soil treatment (Admire Pro) 4.6 F (various) 2 F	7 to 10.5 fl oz 16 to 24 fl oz	12 hrs	21	See label for soil application instructions.
	Foliar treatment (Admire Pro) 4.6 F (various) 1.6 F	1.2 fl oz 3.5 fl oz	12 hrs	7	
	spirotetramat, MOA 23 (Movento)	4 to 5 fl oz	24 hrs	1 (succulent) 7 (dry)	PHI is 1 day for succulent beans and 7 days for dry beans.
Beet					
Aphid	flonicamid, MOA 29 (Beleaf) 50SG	2 to 2.8 pz	12 hrs	7	
	imidacloprid, MOA 4A Soil treatment (Admire Pro) 4.6 F (various) 2 F	4.4 to 10.5 fl oz 10 to 24 fl oz	12 hrs	21	See label for soil application instructions. Will also control flea beetle.
	flupyradifurone, MOA 4D (Sivanto) 200 SL	7.0 to 10.5 fl oz	4 hrs	1	
	Foliar treatment (Admire Pro) 4.6 F (various) 1.6 F	1.2 fl oz 3.5 fl oz	12 hrs	7	
	thiamethoxam, MOA 4A (Platinum) 75 SG	1.7 to 2.17 oz	12 hrs		Platinum may be applied to direct-seeded crops in-furrow at seed or transplant depth, post seeding or transplant as a drench, or through drip irrigation. Do not exceed 12 ounces per acre per season of Platinum. Check label for plant-back restrictions for a number of crops.
	(Actara) 25 WDG	1.5 to 3 oz	12 hrs	7	

Table 5-9. Insect Control for Commercial Vegetables

CROP Insect	Insecticide, Mode of Action Code, and Formulation	Amount of Formulation Per Acre	Restricted Entry Interval (REI)	Pre harvest Interval (PHI) (Days)	Precautions and Remarks
Beet (continued)					
Armyworm, Beet webworm	chlorantraniliprole MOA 28 (Coragen) 1.67 SC	3.5 to 5 fl oz	4 hrs	1	
	methoxyfenozide MOA 18 (Intrepid) 2F	6 to 16 fl oz	4 hrs	7	
	spinetoram, MOA 5 (Radiant) 1 SC	6 to 8 fl oz	4 hrs	7	Do not apply more than 32 fluid ounces per acre per season.
	Spinosad, MOA 5 (Blackhawk)	1.7 to 3.3 oz	4 hrs	3	
Blister beetle, Flea beetle	carbaryl, MOA 1A (Sevin) 50 WP 80 S XLR	3 lb 1.875 lb 1 qt	12 hrs	7	
	pyrethroid, MOA 3		12 hrs		See table 5-9B for a list of registered pyrethroids and pre-harvest intervals.
Leafminer	spinetoram, MOA 5 (Radiant) 1 SC	6 to 10 fl oz	4 hrs	7	Control will be improved with addition of a spray adjuvant.
Broccoli, Brussels Sprouts, Cabbage, Cauliflower, Kohlrabi					
Aphid	Where whitefly resistance is an issue (or any other insect with a high potential for resistance to Group 4A MOA insecticides), a foliar-applied Group 4A insecticide program and a soil-applied Group 4A program should not be used in the same season. Also, if using a foliar-applied program, avoid using a block of more than three consecutive applications of any products belonging to Group 4A insecticides.				
	acetamiprid, MOA 4A (Assail) 30 SG	2 to 3 oz	12 hrs	7	
	clothianidin, MOA 4A (Belay) 50WD	4.8 to 6.4 oz (soil) 1.6 to 2.1 oz (foliar)	12 hrs	21 (soil) 7 (foliar)	Soil application at planting only.
	dimethoate 4 EC, MOA 1B	0.5 to 1 pt	48 hrs	7	
	flonicamid, MOA 29 (Beleaf) 50SG	2 to 2.8 oz	12 hrs	0	
	flupyradifurone, MOA 4D (Sivanto) 200 SL	7.0 to 12.0 fl oz	4 hrs	1	
	imidacloprid, MOA 4A Soil treatment (Admire Pro) 4.6 F (various) 2 F	4.4 to 10.5 fl oz 10 to 24 fl oz	12 hrs	21	Do not follow soil applications of Admire with foliar applications of any neonicotinoid insecticide. Use only one application method. See label for soil application instructions. Imidacloprid also controls whiteflies.
	Foliar treatment (Admire Pro) 4.6 F (various) 1.6 F	1.3 fl oz 3.75 fl oz	12 hrs	7	Imidacloprid also controls whiteflies. Not effective against flea beetle.
	pymetrozine, MOA 9B (Fulfill) 50 WDG	2.75 oz	12 hrs	7	
	spirotetramat, MOA 23 (Movento) 2 SC	4 to 5 fl oz	24 hrs	1	Do not exceed 10 fluid ounces per season. Requires surfactant.
	thiamethoxam MOA 4A Soil treatment (Platinum) 75SG Foliar treatment (Actara) 25WDG	1.66 to 3.67 oz 1.5 to 3.0 oz	12 hrs	30 0	Platinum may be applied to direct-seeded crops in-furrow at seed or transplant depth, postseeding or transplant as a drench, or through drip irrigation. Do not exceed 3.67 ounces per acre per season. Thiamethoxam also controls whiteflies and certain thrips species.
Diamondback moth, Cabbage looper, Imported cabbageworm, Corn earworm, Cross-striped cabbageworm, Cabbage webworm, Amryworms	Insecticide-resistant populations, widespread in Southeastern U.S., may not be controlled with some registered insecticides. To manage resistance, avoid transplants from Georgia and Florida and avoid repeated use of the same materials for extended periods of time. Repeated use of pyrethroid insecticides destroys natural enemies and often aggravates diamondback moth problems. Do not allow populations to increase to large densities before initiating treatments.				
	Bacillus thuringiensis, MOA 11A (Dipel) 2X (Dipel) 4 L (Javelin) WG (Xentari) WDG	8 oz 1 to 2 qt 0.5 to 1 lb 0.5 to 1 lb	4 hrs	0	On foliage every 7 days. On summer or fall plantings, during periods when eggs and larvae are present. This usually occurs when true leaves appear; on other plantings, it may occur later. A spreader-sticker will be helpful. **Not effective against Cabbage Webworm**
	chlorantraniliprole, MOA 28 (Coragen) 1.67 SC	3.5 to 5 fl oz	4 hrs	3	Foliar or soil application. See label for soil application instructions.
	cyantraniliprole, MOA 28 (Verimark) 1,67SC (Exirel) 0.83SE	5 to 10 fl oz 7 to 17 fl oz	12 hrs 12 hrs	NA 1	Verimark is for soil application only. Apply at planting only. See label for application options. Exirel is for foliar application only. Use higher rates for cabbage looper.
	emamectin benzoate, MOA 6 (Proclaim) 5 WDG	3.2 to 4.8 oz	12 hrs	7	
	flubendiamide, MOA 28 (Belt) 4SC	2 to 2.4 fl oz	12 hrs	1	
	indoxacarb, MOA 22 (Avaunt) 30 WDG	2.5 to 3.5 oz	12 hrs	3	Add a wetting agent to improve spray. Do not apply more than 14 ounces (0.26 pound a.i.) per acre per crop. The minimum interval between sprays is 3 days.

Table 5-9. Insect Control for Commercial Vegetables

CROP Insect	Insecticide, Mode of Action Code, and Formulation	Amount of Formulation Per Acre	Restricted Entry Interval (REI)	Pre harvest Interval (PHI) (Days)	Precautions and Remarks
Broccoli, Brussels Sprouts, Cabbage, Cauliflower, Kohlrabi (continued)					
Diamondback moth, Cabbage looper, Imported cabbageworm, Corn earworm, Cross-striped cabbageworm, Cabbage webworm, Amryworm (continued)	Novaluron, MOA 15 (Rimon) 0.83 EC	6 to 12 fl oz	12 hrs	7	Use lower rates when targeting eggs or small larvae, and use higher rates when larvae are large. Make no more than three applications, or 24 fluid ounces per acre per season.
	Spinetoram, MOA 5 (Radiant) 1 SC	5 to 10 fl oz	4 hrs	1	
	pyrethroid, MOA 3		12 hrs		See table 5-9B for a list of registered pyrethroids and pre-harvest intervals. Not for use where diamondback moth is a concern.
Flea beetle	acetamiprid, MOA 4A (Assail) 30 SG	2 to 3 oz	12 hrs	7	
	Clothianidin, MOA 4A (Belay) 50WDG	4.8 to 6.4 oz (soil) 1.6 to 2.1 oz (foliar)	12 hrs	7 (foliar)	Soil applications may only be made at planting.
	cyantraniliprole, MOA 28 (Verimark) 1,67SC	6.75 to 13.5 fl oz	4 hrs	1	Verimark is for at planting soil application only. See label for application options.
	(Exirel) 0.83SE	13.5 to 20.5 fl oz	12 hrs	1	Exirel is for foliar application only.
	Dinotefuran, MOA 4A Foliar treatment (Venom) 70 SG (Scorpion) 35SL	1 to 4 oz 2 to 7 fl oz	12 hrs	1	See lable for soil application options.
	Soil treatment (Venom) 70 SG (Scorpion) 35SL	5 to 6 oz 9 to 10.5 fl oz		21	
	Dimethoate 4 EC, MOA 1B	0.5 to 1 pt	48 hrs	7	
	Pyrethroid MOA 3		12 hrs		See table 5-9B for a list of registered pyrethroids and pre-harvest intervals.
Harlequin bug, Stink bug	clothianidin, MOA 4A (Belay) 50WDG	4.8 to 6.4 oz (soil) 1.6 to 2.1 oz (foliar)	12 hrs	NA 7 (foliar)	Soil application at planting only.
	Dinotefuran, MOA 4A (Venom) 70 SG (Scorpion) 35 SL	3 to 4 oz 2 to 7 fl oz	12 hrs	1	Do not exceed 6 ounces of Venom per season.
	Pyrethroid, MOA 3		12 hrs		See table 5-9B for a list of registered pyrethroids and pre-harvest intervals.
Root maggot	chlorpyrifos, MOA 1B (Lorsban) 4 EC (Lorsban) 75 WG	2 pt/100 gal 1.33 lb	24 hrs	—	Directed spray to transplants: Spray the base of the plant immediately after transplanting, using a minimum of 40 gallons per acre.
	Chlorpyrifos, MOA 1B (Lorsban) 4 EC	1.6 to 2.75 oz/ 1,000 ft row	24 hrs	—	Direct seeded: Apply in a 4-inch wide band behind planter shoe and in front of press wheel for shallow incorporation.
	(Lorsban) 15 G	4.6 to 9.2 oz/ 1,000 ft row	24 hrs	—	Direct seeded: Place across seed row in 4-inch band behind planter shoe and in front of press wheel.
	Diazinon, MOA 1B (Diazinon 50 W) 50 WP	0.25 to 0.5 lb/ 50 gal	4 days	—	Transplant water: Apply in transplant water or drench water at 4 to 6 ounces per plant at transplanting.
Thrips	dimethoate 4 EC, MOA 1B	0.5 to 1 pt	48 hrs	7	
	imidacloprid, MOA 4A (Admire Pro) 4.6F (various) 2F (various) 1.6 F	1.3 fl oz 3.0 fl oz 3.75 fl oz	12 hrs	7	Check label for rates for other formulations. Foliar applications only.
	Methomyl, MOA 1A (Lannate) 2.4 LV	1.5 fl oz	48 hrs	1	
	novaluron, MOA 15 (Rimon) 0.83 EC	6 to 12 fl oz	12 hrs	7	Make no more than three applications, or 24 fluid ounces, per acre per season.
	Spinetoram, MOA 5 (Radiant) 1 SC	6 to 10 fl oz	4 hrs	1	
Whitefly	acetamiprid, MOA 4A (Assail) 30 SG	2.5 to 4.0 oz	12 hrs	7	Use s spreader stick to improve control.
	Dinotefuran, MOA 4A Foliar treatment (Venom) 70 SG (Scorpion) 35SL	1 to 4 oz 2 to 7 fl oz	12 hrs	1	Do not follow soil applications with foliar applications of any neonicotinoid insecticide. Use only one application method. Do not apply more than 6 ounces per acre per season using foliar applications, or 12 ounces per acre per season using soil applications. Soil applications may be applied by: a narrow band below or above the seed line at planting; a post-seeding or transplant drench with sufficient water to ensure incorporation to the root zone; or through drip irrigation.
	Soil treatment (Venom) 70 SG (Scorpion) 35SL	5 to 6 oz 9 to 10.5 fl oz		21	
	flupyradifurone, MOA 4D (Sivanto) 200 SL	10.5 to 14.0 oz	4 hrs	1	

INSECT CONTROL

Table 5-9. Insect Control for Commercial Vegetables

CROP Insect	Insecticide, Mode of Action Code, and Formulation	Amount of Formulation Per Acre	Restricted Entry Interval (REI)	Pre harvest Interval (PHI) (Days)	Precautions and Remarks
Broccoli, Brussels Sprouts, Cabbage, Cauliflower, Kohlrabi (continued)					
Whitefly (continued)	Spiromesifen, MOA 23 (Oberon) 2 SC	7 to 8.5 fl oz	12 hrs	7	Do not exceed 25.5 fluid ounces per acre per season.
	Spirotetramat, MOA 23 (Movento) 2 SC	4 to 5 fl oz	24 hrs	1	Do not exceed 10 fluid ounces per season. Requires surfactant.
	Pyriproxyfen, MOA 7 (Knack) 0.86EC	8 to 10 fl oz	12 hrs	7	Only treat whole fields, and do not any crop other than those that Knack is registered within 30 days after the last application.
Cantaloupe, Muskmelon					

Insecticide applications in cucurbits should be made in late evening to protect pollinating insects. Refer to the section of this chapter on Reducing the Risk of Pesticide Poisoning to Honey Bees for more information about protecting pollinators.

CROP Insect	Insecticide, Mode of Action Code, and Formulation	Amount of Formulation Per Acre	Restricted Entry Interval (REI)	Pre harvest Interval (PHI) (Days)	Precautions and Remarks
Aphid	Where whitefly resistance is an issue (or any other insect with a high potential for resistance to Group 4A MOA insecticides), a folai applied Group 4A insecticide program and a soil-applied Group 4A program shoud not be used in the same season. Also, if using a foliar-applied program, avoid using a block of more than three consecutive applications of any products belonging to Group 4A insecticides.				
	acetamiprid MOA 4A (Assail) 30SG	2.5 to 4.0 oz	12 hrs	0	Do not exceed 0.5 pound per acre per season.
	Clothianidin, MOA 4A (Belay) 50 WDG	4.8 to 6.4 oz (soil) 1.6 to 2.1 oz (foliar)	12 hrs	7 (foliar)	Soil application at planting only.
	Dimethoate, MOA 1B (Dimethoate E267) 2E (Dimethoate E267) 2.67E	2 pt 1.5 pt	48 hrs	3	
	flonicamid, MOA 29 (Beleaf) 50 SG	2 to 2.8 oz	12 hrs	0	
	flupyradifurone, MOA 4D (Sivanto) 200 SL	10.5 to 12.0 fl oz	4 hrs	1	
	imidacloprid, MOA 4A (Admire Pro) 4.6 F (various) 2 F	7 to 10.5 fl oz 16 to 24 fl oz	12 hrs	21	Must be applied to the soil. May be applied preplant; at planting; as a post-seeding drench, transplant water drench, or hill drench; subsurface side-dress or by chemigation using low-pressure drip, or trickle irrigation. See label for information on approved application methods. Will also control cucumber beetles and whiteflies.
	Pymetrozine, MOA 9B (Fulfill) 50 WDG	2.75 oz	12 hrs	0	Apply before aphids reach damaging levels. Do not exceed 5.5 ounces per acre per season.
	Thiamethoxam, MOA 4A (Platinum) 75 SG (Actara) 25WDG	1.66 to 3.67 oz 1.5 to 3 oz	12 hrs	30 0	Platinum is for soil application and may be applied to direct-seeded crops in-furrow at seed or transplant depth, post seeding or transplant as a drench, or through drip irrigation. Do not exceed 8 ounces per acre per season of Platinum. Check label for plant-back restrictions for a number of crops. Actara is for foliar application only.
Armyworm, Cabbage looper	*Bacillus thuringiensis*, MOA 11A (Crymax) WDG, (Dipel) 2X (Xentari) WDG	0.5 to 1.5 lb 8 oz 0.5 to 1 lb	4 hrs	0	On foliage as needed.
	pyrethroid, MOA 3		12 hrs		See table 5-9B for a list of registered pyrethroids and pre-harvest intervals. Not recommended for armyworm.
	chlorantraniliprole, MOA 28 (Coragen) 1.67 SC	3.5 to 5 fl oz	4 hrs	1	Coragen may be used for foliar or drip chemigation.
	cyantraniliprole, MOA 28 (Verimark) 1,67SC (Exirel) 0.83SE	5 to 13.5 fl oz 7 to 17 fl oz	4 hrs 12 hrs	1 1	Verimark is for soil application only. It may be applied to the soil at planting at 6.75 to 13.5 ounces, or via drip chemigation at 5 to 10 fluid ounces. Do not make more than two soil or chemigation applications per season. See label for application options. Exirel is for foliar application only. Use higher rates for cabbage looper.
	flubendiamide, MOA 28 (Belt) 4SC	1.5 fl oz	12 hrs	1	
	indoxacarb, MOA 22 (Avaunt) 30WDG		12 hrs		
	methoxyfenozide, MOA 18 (Intrepid) 2 F	4 to 10 fl oz	4 hrs	3	Use higher rates against large larvae.
	novaluron, MOA 15 (Rimon) 0.83EC	9 to 12 fl oz	12 hrs	1	
	spinetoram, MOA 5 (Radiant) 1 SC	5 to 10 fl oz	4 hrs	3	
Cucumber beetle	acetamiprid MOA 4A (Assail) 30SG	2.5 to 5.3 oz	12 hrs	0	Do not exceed 0.5 pound per acre per season.
	pyrethroid, MOA 3		12 hrs		See table 5-9B for a list of registered pyrethroids and pre-harvest intervals.

Table 5-9. Insect Control for Commercial Vegetables

CROP Insect	Insecticide, Mode of Action Code, and Formulation	Amount of Formulation Per Acre	Restricted Entry Interval (REI)	Pre harvest Interval (PHI) (Days)	Precautions and Remarks
Cantaloupe, Muskmelon (continued)					
Insecticide applications in cucurbits should be made in late evening to protect pollinating insects. Refer to the section of this chapter on Reducing the Risk of Pesticide Poisoning to Honey Bees for more information about protecting pollinators.					
Cucumber beetle (continued)	carbaryl MOA 1A (Sevin) 50 WP (Sevin) 80 S (Sevin) XLR Plus	2 lb 1.25 lb 1 qt	12 hrs	3	
	clothianidin, MOA 4A (Belay) 50 WDG	4.8 to 6.4 oz (soil) 1.6 to 2.1 oz (foliar)	12 hrs	21 (foliar)	Soil application at planting only.
	Dinotefuran, MOA 4A Foliar treatment (Venom) 70 SG (Scorpion) 35SL	1 to 4 oz 2 to 7 fl oz	12 hrs	1	Do not make both a soil and foliar application, use one or the other. At planting applications are most effective against cucumber beetle.
	Soil treatment (Venom) 70 SG (Scorpion) 35SL	5 to 6 oz 9 to 10.5 fl oz		21	
	imidacloprid, MOA 4A (Admire Pro) 4.6 F (various) 2F	7 to 10.5 fl oz 16 to 24 fl oz	12 hrs	21	Must be applied to the soil. See label for informaiton on approved application methods. Will also control aphids and whiteflies.
Leafminer	abamectin, MOA 6 (Agri-mek) 0.7 SC	1.75 to 3.5 fl oz	12 hrs	7	Do not use more than six applications per season.
	cyromazine, MOA 17 (Trigard) 75 WS	2.7 oz	12 hrs	0	
	chlorantraniliprole, MOA 28 (Coragen) 1.67 SC	2 to 3.5 fl oz	4 hrs	1	For foliar or drip chemigation. Drip chemigation must be applied uniformly to the root zone. See label for instructions.
	spinetoram, MOA 5 (Radiant) 1 SC	5 to 10 fl oz	4 hrs	3	
	dimethoate 4 EC, MOA 1B	1 pt	48 hrs	3	
Pickleworm, Melonworm	pyrethroid, MOA 3		12 hrs		See table 5-9B for a list of registered pyrethroids and pre-harvest intervals.
	carbaryl, MOA 1A (Sevin) 50 WP (Sevin) 80 S (Sevin) XLR Plus	2 lb 1.25 lb 1 qt	12 hrs	3	On foliage when worms appear in blossoms. Repeat as needed. Protect pollinators. Rarely a problem before July.
	chlorantraniliprole, MOA 28 (Coragen) 1.67 SC	2 to 3.5 fl oz	4 hrs	1	For foliar or drip chemigation. Drip chemigation must be applied uniformly to the root zone. See label for instructions.
	cyantraniliprole, MOA 28 (Verimark) 1,67SC	5 to 13.5 fl oz	4 hrs	1	Verimark is for soil application only. It may be applied to the soil at planting at 6.75 to 13.5 ounces, or via drip chemigation at 5 to 10 fluid ounces. Do not make more than two soil or chemigation applications per season. See label for application options. Exirel is for foliar application only.
	(Exirel) 0.83SE	7 to 13.5 fl oz	12 hrs	1	
	flubendiamide, MOA 28 (Belt) 4SC	1.5 fl oz	12 hrs	1	
	methoxyfenozide, MOA 18 (Intrepid) 2 F	4 to 10 fl oz	4 hrs	3	
	spinetoram, MOA 5 (Radiant) 1 SC	5 to 10 fl oz	4 hrs	3	
Spider mite	abamectin, MOA 6 (Agri-mek) 0.7 SC	1.75 to 3.4 fl oz	12 hrs	7	
	bifenazate, MOA 20D (Acramite) 50 WS	0.75 to 1.0 lb	12 hrs	3	Do not make more than one application per season.
	etoxazole, MOA 10B (Zeal) 72 WSP	2 to 3 oz	12 hrs	7	Does not kill adults
	fenpyroximate MOA 21 (Portal) 0.4EC	2 pt	12 hrs	3	Do not make more than two applications per season.
	spiromesifen, MOA 23 (Oberon) 2 SC	7 to 8.5 fl oz	12 hrs	7	
Thrips	dimethoate 4EC, MOA 1B	1 pt	48 hrs	3	On foliage as needed.
	Dinotefuran, MOA 4A Foliar treatment (Venom) 70 SG (Scorpion) 35SL	1 to 4 oz 2 to 7 fl oz	12 hrs	1	Do not follow soil applications of Venom with foliar applications of any neonicotinoid insecticide. Use only one application method. Do not apply more than 6 ounces per acre per season using foliar applications, or 12 ounces per acre per season using soil applications. Soil applications may be applied by: a narrow band below or above the seed line at planting; a post-seeding or transplant drench with sufficient water to ensure incorporation to the root zone; or through drip irrigation.
	Soil treatment (Venom) 70 SG (Scorpion) 35SL	5 to 6 oz 9 to 10.5 fl oz		21	
	spinetoram, MOA 5 (Radiant) 1 SC	6 to 10 fl oz	4 hrs	3	

105

INSECT CONTROL

Table 5-9. Insect Control for Commercial Vegetables

CROP Insect	Insecticide, Mode of Action Code, and Formulation	Amount of Formulation Per Acre	Restricted Entry Interval (REI)	Pre harvest Interval (PHI) (Days)	Precautions and Remarks
Cantaloupe, Muskmelon (continued)					
	Insecticide applications in cucurbits should be made in late evening to protect pollinating insects. Refer to the section of this chapter on Reducing the Risk of Pesticide Poisoning to Honey Bees for more information about protecting pollinators.				
Whitefly	acetamiprid, MOA 4A (Assail)	1.1 to 2.3 oz	12 hrs	0	
	buprofezin, MOA 16 (Courier) 40 SC	9 to 13.6 oz	12 hrs	7	Use sufficient water to ensure good coverage. Do not apply more than twice per crop cycle.
	cyantraniliprole, MOA 28 (Verimark) 1,67SC	10 fl oz	4 hrs	1	Verimark is for soil application only. It may be applied to the soil at planting at 6.75 to 13.5 ounces, or via drip chemigation at 5 to 10 fluid ounces. See label for application options.
	(Exirel) 0.83SE	13.5 to 20.5 fl oz	12 hrs	1	Exirel is for foliar application only. Use an adjuvant for best results.
	Dinotefuran, MOA 4A Foliar treatment (Venom) 70 SG (Scorpion) 35SL	1 to 4 oz 2 to 7 fl oz	12 hrs	1	Do not follow soil applications with foliar applications. Use only one application method. Do not apply more than 6 ounces per acre per season using foliar applications, or 12 ounces per acre per season using soil applications. Soil applications may be applied by: a narrow band below or above the seed line at planting; a post-seeding or transplant drench with sufficient water to ensure incorporation to the root zone; or through drip irrigation.
	Soil treatment (Venom) 70 SG (Scorpion) 35SL	5 to 6 oz 9 to 10.5 fl oz		21	
	flupyradifurone, MOA 4D (Sivanto) 200 SL	10.5 to 14.0 fl oz	4 hrs	1	
	imidacloprid, MOA 4A (Admire Pro) 4.6 F (various) 2 F	7 to 10.5 oz 16 to 24 fl oz	12 hrs	21	Must be applied to the soil. May be applied preplant; at planting; as a post-seeding drench or hill drench; subsurface sidedress; or by chemigation using low pressure drip or trickle irrigation. See label for information on approved application methods. Will also control aphids and cucumber beetles.
	pyriproxyfen, MOA 7C (Knack) 0.86 EC	8 to 10 oz	12 hrs	7	Do not make more than two applications per season, and do not make applications closer than 14 days apart.
	spiromesifen, MOA 23 (Oberon) 2 SG	7 to 8.5 oz	12 hrs	7	Appy against adults, before nymphs are present. Do not exceed 3 applications per season.
	thiamethoxam, MOA 4A (Platinum) 75 SG	1.66 to 3.67 fl oz	12 hrs	30	Platinum is for soil application and may be applied to direct-seeded crops in-furrow at seed or transplant depth, postseeding or transplant as a drench, or through drip irrigation. Do not exceed 11 ounces per acre per season of Platinum. Check label for plant-back restrictions for a number of crops.
	(Actara) 25WDG	3 to 5.5 oz		0	Actara is for foliar application.
Wireworm	diazinon, MOA 1B (Diazinon) AG 500	3 to 4 qt	3 days	—	Broadcast on soil before planting and thoroughly work into upper 6 inches.
Carrot					
Aphid, Leafhopper	imidacloprid, MOA 4A Soil treatment (Admire Pro) 4.6 F (various) 2 F	4.4 to 10.5 fl oz 10 to 24 fl oz	12 hrs	21	Must be applied to the soil. May be applied via chemigation into the root zone through low-pressure drip, trickle, micro-sprinkler, or equivalent equipment; in-furrow spray or shanked-in 1 to 2 inches below seed depth during planting; or in a narrow band (2 inches or fewer) 1 to 2 inches directly below the eventual seed row in a bedding operation 14 or fewer days before planting. Higher rates provide longer lasting control. See label for information on approved application methods and rate per 100 row feet for different row spacings.
	Foliar treatment (Admire Pro) 4.6 F (various) 1.6 F	1.2 fl oz 3.5 fl oz	12 hrs	7	
	thiamethoxam, MOA 4A (Platinum) 75 SG	1.66 to 3.67 oz	12 hrs	30	Platinum may be applied to direct-seeded crops in-furrow at seeding, immediately after seeding with sufficient water to ensure incorporation into the root zone, or through trickle irrigation.
	(Actara) 25 WDG	1.5 to 3 oz	12 hrs	7	Actara is applied to foliage. Do not exceed 4 ounces Actara per acre per season.
	flonicamid, MOA 29 (Beleaf) 50SG	2 to 2.8 fl oz	12 hrs	3	
	flupyradifurone, MOA 4D (Sivanto) 200 SL	7.0 to 10.5 fl oz	4 hrs	7	
Armyworm, Parsleyworm	pyrethroid, MOA 3		12 hrs		See table 5-9B for a list of registered pyrethroids and pre-harvest intervals.
	carbaryl, MOA 1A (Sevin) 80 S (Sevin) XLR Plus	1.25 lb 1 qt	12 hrs	7	On foliage as needed.
	chlorantraniliprole, MOA 28 (Coragen) 1.67 SC	3.5 to 5 fl oz	4 hrs	1	Coragen may be used for foliar or drip chemigation.
	methomyl, MOA 1A (Lannate) 2.4 LV (Lannate) 90 SP	0.75 to 1.5 pt 0.25 to 0.5 lb	48 hrs	1	
	methoxyfenozide, MOA 18 (Intrepid) 2 F	4 to 10 fl oz	4 hrs	1	Use higher rates against large larvae.
	spinetoram, MOA 5 (Radiant) 1 SC	6 to 8 fl oz	4 hrs	3	Radiant will not control leafhoppers. Do not make more than 4 applications per year.

INSECT CONTROL

Table 5-9. Insect Control for Commercial Vegetables

CROP Insect	Insecticide, Mode of Action Code, and Formulation	Amount of Formulation Per Acre	Restricted Entry Interval (REI)	Pre harvest Interval (PHI) (Days)	Precautions and Remarks
Carrot (continued)					
Leafminer	spinetoram, MOA 5 (Radiant) 1 SC	6 to 8 fl oz	4 hrs	3	
Wireworm	diazinon, MOA 1B (Diazinon) (AG 500)	4 qt	3 days	—	Broadcast and incorporate preplant.
Celery					
Aphid, Leafhopper, Flea beetle	imidacloprid, MOA 4A (Admire Pro) 4.6 F (various) 2 F	7 to 10.5 fl oz 16 to 24 fl oz	12 hrs	21	Apply via chemigation into the root zone, as an in-furrow spray at planting on/or below the seed, or as a post-seeding or transplant drench.
	flonicamid, MOA 29 (Beleaf) 50SG	2 to 2.8 oz	12 hrs	0	
	flupyradifurone, MOA 4D (Sivanto) 200 SL	10.5 to 12.0 fl oz	4 hrs	1	Will not control flea beetle
	spirotetramat, MOA 23 (Movento) 2SC	4 to 5 fl oz	24 hrs	3	Do not exceed 10 fluid ounces per season. Not for flea beetle. Requires surfactant. Will not control flea beetle
	Tolfenpyrad, MOA 21A (Torac) 1.29 EC	17 to 21 fl oz	12 hrs	1	
Armyworm, Corn earworm, Looper	chlorantraniliprole, MOA 28 (Coragen) 1.67 SC	3.5 to 5 fl oz	4 hrs	1	Foliar or drip chemigation. Drip chemigation must be applied uniformly to the root zone. See label for instructions.
	emamectin benzoate, MOA 6 (Proclaim) 5 WDG	2.4 to 4.8 oz	12 hrs	7	Do not make more than two sequential applications without rotating to another product with a different mode of action.
	methomyl, MOA 1A (Lannate) 2.4 LV	3 pt	48 hrs	7	Methomyl may induce leafminer infestations.
	methoxyfenozide, MOA 18 (Intrepid) 2 F	4 to 10 fl oz	4 hrs	7	For early season applications only to young crop and small plants. For mid- to late-season applications and to heavier infestations and under conditions in which thorough coverage is more difficult. Do not apply more than 16 fluid ounces per application, and do not exceed 64 fluid ounces per season. See Rotational Crop Restrictions on label.
	pyrethroid, MOA 3		12 hrs		See table 5-9B for registered pyrethroids and pre-harvest intervals.
	spinetoram, MOA 5 (Radiant) 1 SC	5 to 10 fl oz	4 hrs	1	Use higher rates for armyworms.
Leafminer	abamectin, MOA 6 (Agri-Mek) 0.15EC	1.75 to 3.5 fl oz	12 hrs	7	
	chlorantraniliprole, MOA 28 (Coragen) 1.67 SC	5 to 7.5 fl oz	4 hrs	1	Foliar or drip chemigation. Drip chemigation must be applied uniformly to the root zone. See label for instructions.
	cryomazine, MOA 17 (Trigard 75WP)	2.66 oz	12 hrs	7	
	spinetoram, MOA 5 (Radiant) 1 SC	6 to 10 fl oz	4 hrs	1	
Collard, Kale, Mustard Greens					
Aphid	acetamiprid, MOA 4A (Assail) 30 SG	2 to 3 oz	12 hrs	7	
	clothianidin, (Belay) 50 WDG	4.8 to 6.4 oz (soil) 1.6 to 2.1 oz (foliar)	12 hrs	7 (foliar)	Soil application at planting only.
	flonicamid, MOA 9C (Beleaf) 50SG	2 to 2.8 fl oz	12 hrs	0	
	flupyradifurone, MOA 4D (Sivanto) 200 SL	10.5 to 12.0 fl oz	4 hrs	1	
	imidacloprid, MOA 4A Soil treatment (Admire Pro) 4.6 F (various) 2 F	4.4 to 10.5 fl oz 10 to 24 fl oz	12 hrs	21	See label for soil application instructions. Admire Pro will also control flea beetle.
	Foliar treatment (Admire Pro) 4.6 F (various) 1.6 F	3.8 fl oz	12 hrs	7	
	pymetrozine, MOA 9B (Fulfill) 50 WDG	2.75 oz	12 hrs	7	
	spirotetramat, MOA 23 (Movento) 2SC	4 to 5 fl oz	24 hrs	1	Do not exceed 10 fluid ounces per season. Requires surfactant.

107

Table 5-9. Insect Control for Commercial Vegetables

CROP Insect	Insecticide, Mode of Action Code, and Formulation	Amount of Formulation Per Acre	Restricted Entry Interval (REI)	Pre harvest Interval (PHI) (Days)	Precautions and Remarks
Collard, Kale, Mustard Greens (continued)					
Diamondback moth, Caterpillars, including Cabbage looper, Imported cabbageworm, Cross-striped cabbageworm, Cabbage webworm, Armyworm	Insecticide-resistant populations may not be controlled with some registered insecticides. To manage resistance, avoid transplants from Georgia and Florida, and avoid the repeated use of the same materials for extended periods of time. Use of pyrethroid insecticides destroys natural enemies and aggravates diamondback moth problems. Do not allow populati-ons to increase to large densities before treatments are initiated.				
	Bacillus thuringiensis, MOA 11A (Crymax) WDG (Dipel) 2 X, DF (Dipel) (Xentari) WDG	0.5 to 1.5 lb 8 oz 1 pt 0.5 to 1 lb	4 hrs	0	Use a spreader/sticker. Do not apply insecticides with the same mode of action more than twice to any generation of diamondback moth. After two applications, rotate to an insecticide with a different mode of action.
	chlorantraniliprole, MOA 28 (Coragen) 1.67 SC	3.5 to 4 fl oz	4 hrs	3	Foliar or drip chemigation. Drip chemigation must be applied uniformly to the root zone. See label for instructions.
	emamectin benzoate, MOA 6 (Proclaim) 5 WDG	2.4 to 4.8 oz	12 hrs	14	
	flubendiamide, MOA 28 (Belt) 4SC	2 to 2.4 fl oz	12 hrs	1	
	indoxacarb, MOA 22 (Avaunt) 30 WDG	3.5 oz	12 hrs	3	Do not apply Avaunt more than twice to any generation of diamondback moth. After two applications, rotate to an insecticide with a different mode of action. Do not make more than 6 applications (4 in GA), or exceed 14 ounces per season per crop.
	spinetoram, MOA 5 (Radiant) 1 SC	5 to 10 fl oz	4 hrs	1	
Flea beetle	carbaryl, MOA 1A (Sevin) 50 WP (Sevin) 80 S (Sevin) XLR	3 lb 1.875 lb 1 qt	12 hrs	14	
	acetamiprid, MOA 4A (Assail) 30SG	2 to 4 oz	12 hrs	7	
	pyrethroid, MOA 3		12 hrs		See table 5-9B for a list of registered pyrethroids and pre-harvest intervals.
Grasshopper	pyrethroid, MOA 3		12 hrs		See table 5-9B for a list of registered pyrethroids and pre-harvest intervals. May flare diamond back moth populations.
Harlequin bug, Stink bug	acetamiprid, MOA 4A (Assail) 30 SG	3 to 4 oz	12 hrs	7	
	clothianidin, MOA 4A (Belay) 50 WDG	4.8 to 6.4 oz (soil); 1.6 to 2.1 oz (foliar)	12 hrs	7 (foliar)	Soil application at planting only.
	pyrethroid, MOA 3		12 hrs		See table 5-9B for a list of registered pyrethroids and pre-harvest intervals.
	thiamethoxam, MOA 4A (Actara) 25WDG	3 to 5.5 oz	12 hrs	7	
Root maggot	chlorpyrifos, MOA 1B (Lorsban) 4 EC (Lorsban) 75WDG	1.6 to 2.75 fl oz 1.1 to 1.8/ 1,000 ft row	24 hrs	—	For directed-seeded crops, apply as a 4-inch band over the row after planting. For transplanted crops, apply as a directed spray immediately after transplanting.
Whitefly	acetamiprid, MOA 4A (Assail) 30 SG	2.5 to 4.0 oz	12 hrs	7	Apply against adults, before nymphs are present. Use a spreader stick to improve control.
	flupyradifurone, MOA 4D (Sivanto) 200 SL	10.5 to 14.0 fl oz	4 hrs	1	Do not make more than 3 applications or apply more than 28 fluid ounces per season.
	pyriproxyfen, MOA 7C (Knack) 0.86 EC	8 to 10 fl oz	12 hrs	7	Do not apply Knack more than twice per season or exceed 0.134 pound per acre per season.
	spiromesifen, MOA 23 (Oberon) 2 SC	7 to 8.5 fl oz	12 hrs	7	Do not make more than 3 applications or apply more than 25.5 fluid ounces per season.
	spirotetramat, MOA 23 (Movento) 2 SC	4 to 5 fl oz	24 hrs	1	Do not exceed 10 fluid ounces per season. Requires surfactant.
Corn, Sweet					
Corn earworm, Fall armyworm, European corn borer	transgenic sweet corn varieties expressing *Bt* protein				Highly effective against European corn borer. Effectiveness against corn earworm will vary among BT traits; additional insecticide applications may be required to prevent damage to the ear tips of some varieties.
	pyrethroid, MOA 3		12 hrs		Check label for variety limitations and grazing restrictions. Apply as needed until first tassel shoots appear in whorl. To protect ears, spray when tassel shoots first appear, 3 days later, then every 2 to 3 days for 5 applications. Following the fifth application, apply at 2- to 3-days until harvest. Corn tasseling after July 1 may require daily applications from first silk through 60% dry silk followed by applications at 2-day intervals until harvest to ensure worm-free ears. Corn earworms and fall armyworms present in the late whorl stage must be controlled before tassel emergence to prevent migration to ears.
	chlorantraniliprole MOA 28 (Coragen) 1.67 SC	3.5 to 5 fl oz	4 hrs	1	
	flubendiamide MOA 28	2.0 to 3.0 oz	12 hrs	1	

Table 5-9. Insect Control for Commercial Vegetables

CROP Insect	Insecticide, Mode of Action Code, and Formulation	Amount of Formulation Per Acre	Restricted Entry Interval (REI)	Pre harvest Interval (PHI) (Days)	Precautions and Remarks
	(Belt) 4 SC				
Corn, Sweet (continued)					
Corn earworm, Fall armyworm, European corn borer (continued)	methomyl, MOA 1A (Lannate) 90 SP (Lannate) 2.4 LV	4 to 6 oz 0.75 to 1.5 pt	48 hrs	0	Do not use methomyl for European corn borer control.
	indoxacarb, MOA 22 (Avaunt) 30 WDG	2.5 to 3.5 oz	12 hrs	3	For control of fall armyworm and European corn borer in whorl stage only. Do not apply more than 14 ounces Avaunt (0.26 lb a.i.) per acre per crop. Minimum interval between sprays is 3 days. Make no more than 4 applications per season.
	spinetoram, MOA 5 (Radiant) 1 SC	3 to 6 fl oz	4 hrs	1	Do not apply more than 36 ounces per acre per year.
	Spinosad, MOA 5 (Blackhawk)	1.7 to 3.3 oz	4 hrs		
Cutworm	pyrethroid, MOA 3		12 hrs		See table 5-9B for a list of registered pyrethroids and pre-harvest intervals.
Flea beetle, Grasshopper, Japanese beetle, Rootworm beetle	pyrethroid, MOA 3		12 hrs		See table 5-9B for a list of registered pyrethroids and pre-harvest intervals.
Sap beetle	pyrethroid, MOA 3		12 hrs		See table 5-9B for a list of registered pyrethroids and pre-harvest intervals.
	carbaryl, MOA 1A (Sevin) 50 WP (Sevin) 80 S (Sevin) XLR Plus	2 lb 1.25 lb 1 qt	12 hrs	2	Infestations usually associated with prior ear damage. Populations build on overmature and damaged fruit and vegetables. Sanitation is important.
Southern corn billbug, Rootworm, Wireworm	Seed treatments: clothianidin, MOA 4A (Poncho 600) imidacloprid, MOA 4A (Gaucho 600)	1.13 fl oz per 80,000 seeds 4 to 8 oz per cwt seed		—	Seed treatments are applied by commercial seed treaters only. Not for use in hopper bins, slurry mixes, or any other type of on-farm treatment.
	pyrethroid, MOA 3		12 hrs		See table 5-9B for a list of registered pyrethroids and pre-harvest intervals.
	chlorpyrifos, MOA 1B (Lorsban) 4 E	4 pt	24 hrs	0	Preplant incorporation treatment. For postemergence treatment use 2 to 3 pints.
	terbufos, MOA 1B (Counter) 15 G	Banded: 6.5 to 13 lb (40 in. row spacing) OR 8 to 16 oz/1,000 ft row In-Furrow: 6.5 lb (40 in. row) OR 8 oz/10 ft row		— —	Place granules in a 7-inch band over the row directly behind the planter shoe in front of press wheel. Place granules directly in the seed furrow behind the planter shoe. Rotation is advised.
Stink bug	pyrethroids, MOA 3				See table 5-9B for a list of registered pyrethroids and pre-harvest intervals.
	methomyl, MOA 1A (Lannate) 90SP	0.5 lb	48 hrs	0	Re-entry interval is 48 hours.
Cucumber					

Insecticide applications in cucurbits should be made in late evening to protect pollinating insects. Refer to the section of this chapter on Reducing the Risk of Pesticide Poisoning to Honey Bees for more information about protecting pollinators.

CROP Insect	Insecticide, Mode of Action Code, and Formulation	Amount of Formulation Per Acre	Restricted Entry Interval (REI)	Pre harvest Interval (PHI) (Days)	Precautions and Remarks
Aphid	acetamiprid MOA 4A (Assail) 30SG	2.5 to 4.0 oz	12 hrs	0	Do not exceed 0.5 pound per acre per season.
	clothianidin, MOA 4A (Belay) 50 WDG	4.8 to 6.4 oz (soil); 1.6 to 2.1 oz (foliar)	12 hrs	7 (foliar)	Soil application at plant only.
	flonicamid, MOA 29 (Beleaf) 50 SG	2 to 2.8 oz	12 hrs	0	
	flupyradifurone, MOA 4D (Sivanto) 200 SL	7.0 to 12.0 fl oz	4 hrs	1	
	imidacloprid, MOA 4A (Admire Pro) 4.6 F (various) 2 F	7 to 10.5 fl oz 10 to 24 fl oz	12 hrs	21	Must be applied to the soil. May be applied preplant; at planting; as a post-seeding drench, transplant water drench, or hill drench; subsurface side-dress; or by chemigation using low-pressure drip or trickle irrigation. See label for information on approved application methods. Will also control cucumber beetles and whiteflies.
	pymetrozine, MOA 9B (Fulfill) 50 WDG	2.75 oz	12 hrs	0	Apply before aphids reach damaging levels. Do not exceed 5.5 ounces per acre per season.

INSECT CONTROL

Table 5-9. Insect Control for Commercial Vegetables

CROP Insect	Insecticide, Mode of Action Code, and Formulation	Amount of Formulation Per Acre	Restricted Entry Interval (REI)	Pre harvest Interval (PHI) (Days)	Precautions and Remarks
Cucumber (continued)					
Insecticide applications in cucurbits should be made in late evening to protect pollinating insects. Refer to the section of this chapter on Reducing the Risk of Pesticide Poisoning to Honey Bees for more information about protecting pollinators.					
Aphid (continued)	thiamethoxam, MOA 4A (Platinum) 75 SG	1.66 to 3.67 oz	12 hrs	30	Platinum may be applied to direct-seeded crops in-furrow seeding or transplant depth, post seeding or transplant as a drench, or through drip irrigation. Do not exceed 11 ounces per acre per season of Platinum. Check label for plant-back restrictions for a number of crops.
	(Actara) 25 WDG	1.5 to 3 oz	12 hrs	0	
	cyantraniliprole, MOA 28 (Verimark), 1,67SC	6.75 to 10 fl oz	4 hrs	1	May be applied to the soil at planting and/or via drip irrigation. Do not make more than two drip chemigation applications per crop per season, or one if an application is made at planting.
Cucumber beetle, Flea beetle	acetamiprid MOA 4A (Assail) 30SG	2.5 to 5.3 oz	12 hrs	0	Do not exceed 0.5 pound per acre per season.
	pyrethroid, MOA 3		12 hrs		See table 5-9B for a list of registered pyrethroids and pre-harvest intervals.
	carbaryl, MOA 1A (Sevin) 50 WP (Sevin) 80 S (Sevin) XLR Plus	2 lb 1.25 lb 1 qt	12 hrs	0	On foliage as needed. Beetles are most destructive to seedlings. They also spread bacterial wilt disease.
	clothianidin, MOA 4A (Belay) 50 WDG	4.8 to 6.4oz (soil); 1.6 to 2.1fl oz (foliar)	12 hrs	7 (foliar)	Soil application at plant only.
	Dinotefuran, MOA 4A Foliar treatment (Venom) 70 SG (Scorpion) 35SL	1 to 4 oz 2 to 7 fl oz	12 hrs	1	Foliar applications should not be made after plants have started to bloom. Do not follow soil applications with foliar applications on any neonicotinoid insecticide. Use only one application method. Do not apply more than 6 ounces per acre per season using foliar applications, or 12 ounces per acre per season using soil applications. Soil application may be applied by: 1) a narrow band below or above the seed line at planting; 2) a post-seeding or transplant drench with sufficient water to ensure incorporation to the root zone; or 3) drip irrigation.
	Soil treatment (Venom) 70 SG (Scorpion) 35SL	5 to 6 oz 9 to 10.5 fl oz		21	
	imidacloprid, MOA 4A (Admire Pro) 4.6 F (various) 2 F	7 to 10.5 fl oz 16 to 24 fl oz	12 hrs	21	See application method under Aphid.
Cutworm	pyrethroid, MOA 3		12 hrs		See table 5-9B for a list of registered pyrethroids and pre-harvest intervals.
Leafminer	abamectin, MOA 6 (Agri-Mek) 0.7 SC	1.75 to 3.5 fl oz	12 hrs	7	
	chlorantraniliprole, MOA 28 (Coragen) 1.67 SC	5 to 7.5 fl oz	4 hrs	1	Soil, foliar, or drip chemigation. See label for application instructions.
	cyromazine, MOA 17 (Trigard) 75 WS	2.7 oz	12 hrs	0	Do not make more than six applications per season.
	spinetoram, MOA 5 (Radiant) 1 SC	6 to 8 fl oz	4 hrs	1	
Pickleworm, Melon worm, Cabbage looper	chlorantraniliprole, MOA 28(Coragen) 1.67 SC	3.5 to 5 fl oz	4 hrs	1	Foliar or drip chemigation. Drip chemigation must be applied uniformly to the root zone. See label for instructions.
	flubendiamide, MOA 28 (Belt) 4SC	1.5 fl oz	12 hrs	1	Do not apply more than 4.5 fluid ounces per crop per season.
	indoxacarb, MOA 22 (Avaunt) 30 WDG	2.5 to 6 oz	12 hrs	3	
	pyrethroid, MOA 3		12 hrs		See table 5-9B for a list of registered pyrethroids and pre-harvest intervals.
	chlorantraniliprole, MOA 28 (Coragen) 1.67 SC	3.5 to 5 fl oz	4 hrs	1	Foliar or drip chemigation. Drip chemigation must be applied uniformly to the root zone. See label for instructions.
	spinetoram, MOA 5 (Radiant) 1 SC	5 to 10 fl oz	4 hrs	1	
Spider mite	abamectin, MOA 6 (Agri-mek) 0.7 SC	1.75 to 3.5 fl oz	12 hrs	7	No more than two applications.
	bifenazate MOA 20D (Acramite) 50 WS	0.75 to 1 lb	12 hrs	3	Do not make more than one application per season.
	extoxazole, MOA 10B (Zeal) 72 WSP	2 to 3 oz	12 hrs	7	
	fenpyrroximate, MOA 21A (Portal) 4EC	2 pts	12 hrs	1	Do not apply within 75 feet of fish-bearing waters. Do not make more than two applications per crop per season, and allow 14 days between applications.
	spiromesifen, MOA 23 (Oberon) 2 SG	7 to 8.5 fl oz	12 hrs	7	

110

Table 5-9. Insect Control for Commercial Vegetables

CROP Insect	Insecticide, Mode of Action Code, and Formulation	Amount of Formulation Per Acre	Restricted Entry Interval (REI)	Pre harvest Interval (PHI) (Days)	Precautions and Remarks
Cucumber (continued)					
Insecticide applications in cucurbits should be made in late evening to protect pollinating insects. Refer to the section of this chapter on Reducing the Risk of Pesticide Poisoning to Honey Bees for more information about protecting pollinators.					
Thrips	dinotefuran, MOA 4A (Venom) 70 SG (Scorpion) 35SL	1 to 4 oz 2 to 7 fl oz	12 hrs	1	Foliar applications should not be made after plants have started to bloom.Do not follow soil applications with foliar applications on any neonicotinoid insecticide. Use only one application method. Do not apply more than 6 ounces per acre per season using foliar applications, or 12 ounces per acre per season using soil applications. Soil application may be applied by: 1) a narrow band below or above the seed line at planting; 2) a post-seeding or transplant drench with sufficient water to ensure incorporation to the root zone; or 3) drip irrigation.
	methomyl, MOA 1A (Lannate) 2.4 LV	1.5 pt	48 hrs	1	
	spinetoram, MOA 5 (Radiant) 1 SC	6 to 10 fl oz	4 hrs	1	
Whitefly	acetamiprid MOA 4A (Assail) 30SG	2.5 to 5.3 oz	12 hrs	0	Do not exceed 0.5 pound per acre per season.
	buprofezin, MOA 16 (Courier) 40 SC	9 to 13.6 fl oz	12 hrs	7	Use sufficient water to ensure good coverage. Do not apply more than twice per crop cycle.
	chlorantraniliprole, MOA 28 (Coragen) 1.67 SC	5 to 7.5 fl oz	4 hrs	1	For foliar or drip chemigation. Drip chemigation must be applied uniformly to the root zone. See label for instructions.
	cyantraniliprole, MOA 28 (Verimark) 1,67SC	5 to 10 fl oz	4 hrs	1	Verimark is for soil application. Applications may be made at planting or via drip chemigation. See label for application options.
	(Exirel) 0.83SE	13.5 to 20.5 fl oz	12 hrs	1	Exirel is for foliar application only
	dinotefuran, MOA 4A (Venom) 70 SG (Scorpion) 35SL	1 to 4 oz 2 to 7 fl oz	12 hrs	1	See comments under thrips for application instructions and restrictions.
	flupyradifurone, MOA 4D (Sivanto) 200 SL	10.5 to 14.0 fl oz	4 hrs	1	
	imidacloprid, MOA 4A (Admire Pro) 4.6 F (various) 2 F	7 to 10.5 fl oz 16 to 24 fl oz	12 hrs	21	See comments under Aphids.
	pyriproxyfen, MOA 7C Knack) 0.86 EC	8 to 10 fl oz	12 hrs	7	Do not make more than two applications per season, and do not make applications closer than 14 days apart.
	spiromesifen, MOA 23 (Oberon), 2 SC	7 to 8.5 fl oz	12 hrs	7	Do not make more than 3 applications per season.
	thiamethoxam, MOA 4A (Platinum) 75 SG	1.66 to 3.67 oz	12 hrs	30	Platinum is for soil applications. See comments under Aphids.
	(Actara) 25WDG	3 to 5.5 oz		0	Actara is for foliar applications.
Eggplant					
Aphid	Where whitefly resistance is an issue (or any other insect with a high potential for resistance to Group 4A MOA insecticides), avoid making foliar applications of Group 4A insecticides when a soil-applied Group 4A program is used; i.e., do not make both foliar and soil applications of Group 4A insecticides. Also, if using a foliar-applied program, avoid using a block of more than three consecutive applications of any products belonging to Group 4A insecticides.				
	acetamiprid, MOA 4A (Assail) 30 SG	2 to 4 oz	12 hrs	7	Thoroughly cover foliage to effectvely control aphids. Do not apply more than once every 7 days, and do not exceed a total of 7 ounces per season.
	clothianidin, (Belay) 50 WDG	4.8 to 6.4 oz (Soil) 1.6 to 2.1 oz (Foliar)	12 hrs	7 (Foliar)	Soil application at planting only.
	flonicamid, MOA 29 (Beleaf) 50 SG	2 to 4.8 oz	12 hrs	0	
	flupyradifurone, MOA 4D (Sivanto) 200 SL	7.0 to 12.0 fl oz	4 hrs	1	
	imidacloprid, MOA 4A Soil treatment (Admire Pro) 4.6 F (various) 2 F	7 to 10.5 fl oz 16 to 24 fl oz	12 hrs	21	See label for soil application instructions. For short-term protection of transplants at planting, apply Admire Pro (0.44 ounces per 10,000 plants) not more than 7 days before transplanting by 1) uniformly spraying on transplants, followed immediately by sufficient overhead irrigation to wash product into potting media; or 2) injection into overhead irrigation system with adequate volume to thoroughly saturate soil media.
	Foliar treatment (Admire Pro) 4.6 F (various) 1.6 F	1.3 to 2.2 fl oz 3.75 fl oz	12 hrs	0	
	pymetrozine, MOA 9B (Fulfill) 50 WDG	2.75 oz	12 hrs	14	Apply before aphids reach damaging levels. Do not exceed 5.5 ounces per acre per season.
	spirotetramat, MOA 23 (Movento) 2 SC	4 to 5 fl oz	24 hrs	1	Do not exceed 10 fluid ounces per season. Requires surfactant.

111

Table 5-9. Insect Control for Commercial Vegetables

CROP Insect	Insecticide, Mode of Action Code, and Formulation	Amount of Formulation Per Acre	Restricted Entry Interval (REI)	Pre harvest Interval (PHI) (Days)	Precautions and Remarks
Eggplant (continued)					
Aphid (continued)	thiamethoxam, MOA 4A Soil treatment (Platinum) 75 SG	1.66 to 3.67 oz	12 hrs	30	Platinum may be applied to direct-seeded crops in-furrow at seed or transplant depth, postseeding or transplant as a drench, or through drip irrigation. Do not exceed 8 ounces per acre per season. Check label for plant-back restrictions for a number of plants.
	Foliar treatment (Actara) 25 WDG	2 to 3 oz	12 hrs	0	Actara is for foliar application.
Blister beetle	pyrethroid, MOA 3		12 hrs		See table 5-9B for a list of registered pyrethroids and pre-harvest intervals.
Colorado potato beetle	Resistance to many insecticides is widespread in Colorado potato beetle. To reduce risk of resistance, scout fields and apply insecticides only when needed to prevent damage to the crop. Crop rotation will help prevent damaging Colorado potato beetle infestations. If control failures or reduced levels of control occur with a particular insecticide, do NOT make a second application of the same insecticide at the same or higher rate. If an additional insecticide application is necessary, a different insecticide representing a different MOA class should be used. Do NOT use insecticides belonging to the same class 2 years in a row for Colorado potato beetle control.				
	abamectin, MOA 6 (Agri-Mek) 0.7 SC	1.75 to 3.5 fl oz	12 hrs	7	Apply when adults and small larvae are present but before large larvae appear. For resistance management, use the higher rate.
	acetamiprid, MOA 4A (Assail) 30 SG	2 to 4 oz	12 hrs	7	Do not apply more than once every 7 days, and do not exceed 7 ounces of formulation per season.
	pyrethroid, MOA 3		12 hrs		See table 5-9B for a list of registered pyrethroids and pre-harvest intervals.
	chlorantraniliprole, MOA 28 (Coragen) 1.67 SC	3.5 to 5 fl oz	4 hrs	1	Foliar or drip chemigation. Drip chemigation must be applied uniformly to the root zone. See label for instructions.
	clothianidin, (Belay) 50WDG	4.8 to 6.4 oz (Soil) 1.6 to 2.1 oz (Foliar)	12 hrs	7 (Foliar)	Soil application at planting only.
	Dinotefuran, MOA 4A Foliar treatment (Venom) 70 SG (Scorpion) 35SL Soil treatment (Venom) 70 SG (Scorpion) 35SL	1 to 4 oz 2 to 7 fl oz 5 to 6 oz 9 to 10.5 fl oz	12 hrs	1 21	Do not follow soil applications with foliar applications on any neonicotinoid insecticide. Use only one application method. Do not apply more than 6 ounces per acre per season using foliar applications, or 12 ounces per acre per season using soil applications. Soil application may be applied by: 1) a narrow band below or above the seed line at planting; 2) a post-seeding or transplant drench with sufficient water to ensure incorporation to the root zone; or 3) drip irrigation.
	imidacloprid, MOA 4A Soil treatment (Admire Pro) 4.6 F (various) 2 F	7 to 10.5 fl oz 16 to 24 fl oz	12 hrs	21	See application methods under Aphids, Thrips.
	Foliar treatment (Admire Pro) 4.6 F (various) 1.6 F	1.3 fl oz 3.75 fl oz	12 hrs	0	
	novaluron, MOA 15 (Rimon) 0.83 EC	9 to 12 fl oz	12 hrs	1	
	spinetoram, MOA 5 (Radiant) 1 SC	5 to 10 fl oz	4 hrs	1	
	thiamethoxam, MOA 4A (Platinum) 75 SG	1.66 to 3.67 oz	12 hrs	30	See application methods under Aphids.
	(Actara) 25 WDG	2 to 3 oz	12 hrs	0	
Eggplant lace bug	imidacloprid, MOA 4A Foliar treatment (Admire Pro) 4.6 F (various) 1.6 F	1.3 to 2.2 fl oz 3.8 to 6.2 fl oz	12 hrs	0	
	malathion, MOA 1B (various brands) 57 EC	3 pt	12 hrs	3	
Flea beetle	pyrethroid, MOA 3		12 hrs		See table 5-9B for a list of registered pyrethroids and pre-harvest intervals.
	carbaryl, MOA 1A (Sevin) 50 WP (Sevin) 80 S (Sevin) XLR Plus	2 lb 1.25 lb 1 lb	12 hrs	3	
	clothianidin, MOA 4A (Belay) 50WDG	4.6 to 6.8 oz (soil); 1.6 to 2.1 fl oz (foliar)	12 hrs	7 (foliar)	Soil application at planting only.
	cyantraniliprole, MOA 28 (Verimark) 1,67SC	6.75 to 13.5 fl oz	4 hrs	1	Verimark for soil application only. Apply at planting or via drip chemigation. See label for application options.
	Dinotefuran, MOA 4A Foliar treatment (Venom) 70 SG (Scorpion) 35SL Soil treatment (Venom) 70 SG (Scorpion) 35SL	1 to 4 oz 2 to 7 fl oz 5 to 6 oz 9 to 10.5 fl oz	12 hrs	1 21	Do not follow soil applications with foliar applications on any neonicotinoid insecticide. Use only one application method. Do not apply more than 6 ounces per acre per season using foliar applications, or 12 ounces per acre per season using soil applications. Soil application may be applied by: 1) a narrow band below or above the seed line at planting; 2) a post-seeding or transplant drench with sufficient water to ensure incorporation to the root zone; or 3) drip irrigation.

Table 5-9. Insect Control for Commercial Vegetables

CROP Insect	Insecticide, Mode of Action Code, and Formulation	Amount of Formulation Per Acre	Restricted Entry Interval (REI)	Pre harvest Interval (PHI) (Days)	Precautions and Remarks
Eggplant (continued)					
Flea Beetle (continued)	thiamethoxam, MOA 4A (Platinum) 75 SG	1.66 to 3.67 oz	12 hrs	30	See application methods under Aphids.
	(Actara) 25 WDG	2 to 3 oz	12 hrs	0	
Hornworm, European corn borer, Beet army worm, Corn earworm	chlorantraniliprole, MOA 28 (Coragen) 1.67 SC	3.5 to 4 fl oz	4 hrs	1	Foliar or drip chemigation. Drip chemigation must be applied uniformly to the root zone. See label for instructions.
	cyantraniliprole, MOA 28 (Verimark) 1,67SC	5 to 10 fl oz	4 hrs	1	Verimark is for soil application only. Applications made at planting and/or via drip chemigation. See label for application options. Exirel is for foliar application only.
	(Exirel) 0.83SE	7 to 13.5 fl oz	12 hrs	1	
	flubendiamide, MOA 28 (Belt) 4SC	1.5 fl oz	12 hrs	1	
	indoxacarb, MOA 22 (Avaunt) 30 WDG	2.5 to 3.5 oz	12 hrs	3	Do not apply more than 14 ounces per acre per season.
	methomyl, MOA 1A (Lannate) 2.4 LV	1.5 to 3 pt	48 hrs	5	
	methoxyfenozide, MOA 18 (Intrepid) 2 F	4 to 16 fl oz	4 hrs	1	Apply at rates of 4 to 8 fluid ounces early in season when plants are small. Apply at rates of 8 to 16 ounces to large plants or when infestations are heavy. During periods of continuous moth flights, retreatments at 7 to 14 days may be required. Do not apply more than 16 fluid ounces per application or 64 fluid ounces of Intrepid 2F per acre per season.
	spinetoram, MOA 5 (Radiant) 1 SC	5 to 10 fl oz	4 hrs	1	
	pyrethroid, MOA 3		12 hrs		See table 5-9B for a list of registered pyrethroids and pre-harvest intervals.
Leafminer	abamectin, MOA 6 (Agri-Mek) 0.15 EC	8 to 16 fl oz	12 hrs	7	Use low rates for low to moderate infestations, and high rates for severe infestations
	chlorantraniliprole, MOA 28 (Coragen) 1.67 SC	5 to 7.5 fl oz	4 hrs	1	Foliar, soil, or drip chemigation. Drip chemigation must be applied uniformly to the root zone. See label for application instructions.
	oxamyl, MOA 1A (Vydate) 2 L	1 to 2 qt	48 hrs	7	
	spinetoram, MOA 5 (Radiant) 1 SC	5 to 10 fl oz	4 hrs	1	
Stink bug, leaffooted bug	pyrethroid MOA 3		12 hrs		See table 5-9B for a list of registered pyrethroids and preharvest intervals.
	Dinotefuran, MOA 4A Foliar treatment (Venom) 70 SG (Scorpion) 35SL	1 to 4 oz 2 to 7 fl oz	12 hrs	1	
	Soil treatment (Venom) 70 SG (Scorpion) 35SL	5 to 6 oz 9 to 10.5 fl oz		21	
	Thiamethoxam, MOA 4A (Actara) 25 WDG	3 to 5.5 oz	12 hrs	0	Do not exceed 11 ounces Actara per acre per season.
Spider mite	abamectin, MOA 6 (Agri-Mek) 0.7 SC	1.75 to 3.5 fl oz	12 hrs	7	Use low rates for low to moderate infestations, and high rates for severe infestations.
	acequinocyl, MOA 20B (Kanemite) 15SC	31 fl oz	12 hrs	1	
	bifenazate, MOA 20D (Acramite) 50 WS	0.75 to 1.0 lb	12 hrs	3	Do not make more than one application per season.
	etoxazole, MOA 10B (Zeal)	2 to 3 oz	12 hrs	7	Do not make more than one Zeal application per season.
	fenpyroximate MOA 21 (Portal) 0.4EC	2 pts	12 hrs	3	Do not make more than two applications per season.
	hexakis, MOA 12B (Vendex) 50 WP	2 to 3 lb	48 hrs	3	
	spiromesifen, MOA 23 (Oberon) 2 SG	7 to 8.5 fl oz	12 hrs	7	
Thrips	Dinotefuran, MOA 4A Foliar treatment (Venom) 70 SG (Scorpion) 35SL	1 to 4 oz 2 to 7 fl oz	12 hrs	1	See Whitefly for application instructions. Soil applications are more effective against thrips than foliar applications are.
	Soil treatment (Venom) 70 SG (Scorpion) 35SL	5 to 6 oz 9 to 10.5 fl oz		21	
	imidacloprid, MOA 4A Admire Pro 4.6 F (various) 2 F	7 to 10.5 fl oz 16 to 24 fl oz	12 hrs	21	See Aphids for application instructions.

INSECT CONTROL

Table 5-9. Insect Control for Commercial Vegetables

CROP Insect	Insecticide, Mode of Action Code, and Formulation	Amount of Formulation Per Acre	Restricted Entry Interval (REI)	Pre harvest Interval (PHI) (Days)	Precautions and Remarks
Eggplant (continued)					
Thrips (continued)	methomyl, MOA 1A (Lannate) 2.4 LV	1.5 to 3 pt	48 hrs	3	
	spinetoram, MOA 5 (Radiant) 1 SC	6 to 10 fl oz	4 hrs	1	
Whitefly	acetamiprid, MOA 4A (Assail) 30 SG	2.5 to 4 oz	12 hrs	7	Begin applications when significant populations of adults appear. Do not wait until heavy populations have become established. Do not apply more than once every 7 days, and do not exceed 4 applications per season. Do not apply more than 7 ounces per season.
	chlorantraniliprole, MOA 28 (Coragen) 1.67 SC	5 to 7.5 oz	12 hrs	1	For foliar or drip chemigation. Drip chemigation must be applied uniformly to the root zone. See label for instructions.
	cyantraniliprole, MOA 28 (Verimark) 1,67SC	6.75 to 13.5 fl oz	4 hrs	1	Verimark for soil application only. Apply at planting or via drip chemigation. See label for application options.
	(Exirel) 0.83SE	13.5 to 20.5 fl oz	12 hrs	1	Exirel for foliar application only.
	Dinotefuran, MOA 4A Foliar treatment (Venom) 70 SG (Scorpion) 35SL	1 to 4 oz 2 to 7 fl oz	12 hrs	1	Use only one application method (foliar or soil) of Group 4A insecticides. Soil applications may be applied in a narrow band on the plant row in bedding operations, as a post-seeding or transplant drench, as a side-dress after planting and incorporated 1 or more inches, or through a drip irrigation system.
	Soil treatment (Venom) 70 SG (Scorpion) 35SL	5 to 6 oz 9 to 10.5 fl oz		21	
	flupyradifurone, MOA 4D (Sivanto) 200 SL	10.5 to 14.0 fl oz	4 hrs	1	
	imidacloprid, MOA 4A (Admire Pro) 4.6 F (various) 2 F	7 to 10.5 fl oz 16 to 24 fl oz	12 hrs	21	Do not follow soil applications with applications of other neonicotinoid insecticides (Assail or Venom). See Aphids for application methods and restrictions.
	pyriproxyfen, MOA 7C (Knack) 0.86 EC	8 to 10 fl oz	12 hrs	14	Knack prevents eggs from hatching. It does not kill whitefly adults. Applications should begin when 3 to 5 adults per leaf are present. Do not make more than 2 applications per season, and do not apply a second application within 14 days of the first application. Do not exceed 20 fluid ounces of Knack per acre per season. Check label for plant-back restrictions.
	spirotetramat, MOA 23 (Movento) 2SC	4 to 5 fl oz	24 hrs	1	Do not exceed 10 fl oz per season. Requires surfactant.
	spiromesifen, MOA 23 (Oberon) 2 SC	7 to 8.5 fl oz	12 hrs	7	Do not exceed 3 applications or 25.5 fluid ounces per season.
	thiamethoxam, MOA 4A (Platinum) 75 SG	1.66 to 3.67 oz	12 hrs	30	Platinum is for soil applications and may be applied to direct-seeded crops in furrow at seed or transplant depth, at postseeding or transplant as a drench, or through drip irrigation. Do not exceed 11 ounces per acre per season. Check label for plant-back restrictions for a number of plants.
	(Actara) 25WDG	3 to 5.5		0	Actara is for foliar application.
Lettuce					
Aphid	acetamiprid, MOA 4A (Assail) 30 SG	2 to 4 oz	12 hrs	7	Do not apply more than once every 7 days, and do not exceed 4 applications per season.
	clothianidin, MOA 4A (Belay) 2.13 SC	4.8 to 6.8 oz (soil); 1.6 to 2.1 oz (foliar)	12 hrs	7 (foliar)	Soil application at planting only.
	dimethoate 4 EC, MOA 1B	0.5 pt	48 hrs	14	
	flonicamid, MOA 29 (Beleaf) 50 SG	2 to 2.8 oz	12 hrs	0	
	flupyradifurone, MOA 4D (Sivanto) 200 SL	10.5 to 12.0 fl oz	4 hrs	1	
	imidacloprid, MOA 4A Soil treatment (Admire Pro) 4.6 F (various) 2 F	4.4 to 10.5 fl oz 10 to 24 fl oz	12 hrs	21	Do not follow soil applications with foliar applications of any neonicotinoid insecticide. See label for soil application instructions.
	Foliar treatment (Admire Pro) 4.6 F (various) 1.6 F	1.3 fl oz 3.8 fl oz	12 hrs	7	
	pymetrozine, MOA 9B (Fulfill) 50 WDG	2.75 oz	12 hrs	7	Apply before aphids reach damaging levels. Do not exceed 5.5 ounces per acre per season.
	spirotetramat, MOA 23 (Movento) 2SC	4 to 5 fl oz	24 hrs	3	Do not exceed 10 fluid ounces per season. Requires surfactant.
	thiamethoxam, MOA 4A (Platinum) 75 SG	1.66 to 3.67 oz	12 hrs	30	Do not follow applications of Platinum with foliar applications of any neonicotinoid insecticide. Platinum may be applied to direct-seeded crops in-furrow at the seeding or transplant depth, or as a narrow surface band above the seedling and followed by irrigation. Post seeding, it may be applied as a transplant or through drip irrigation. Actara is applied as a foliar spray.
	(Actara) 25 WDG	1.5 to 3 oz	12 hrs	7	
	Tolfenpyrad, MOA 21A (Torac) 1.29 EC	17 to 21 fl oz	12 hrs	1	Do not apply until at least 14 days after plant emergence or after transplanting to allow time for root establishment.

Table 5-9. Insect Control for Commercial Vegetables

CROP Insect	Insecticide, Mode of Action Code, and Formulation	Amount of Formulation Per Acre	Restricted Entry Interval (REI)	Pre harvest Interval (PHI) (Days)	Precautions and Remarks
Lettuce (continued)					
Armyworm	chlorantraniliprole, MOA 28 (Coragen) 1.67 SC	3.5 to 5 fl oz	4 hrs	1	Foliar or drip chemigation.
	cyantraniliprole, MOA 28 (Verimark) 1,67SC	5 to 13.5 fl oz	4 hrs	1	Verimark is for soil application only. Applications made at planting and/or via drip chemigation. See label for application options.
	(Exirel) 0.83SE	7 to 13.5 fl oz	12 hrs	1	Exirel is for foliar application only.
	emamectin benzoate, MOA 6 (Proclaim) 5 WDG	2.4 to 4.8 oz	12 hrs	7	Do not make more than two sequential applications without rotating to another product with a different mode of action.
	flubendiamide, MOA 28 (Belt) 4SC	1.5 fl oz	12 hrs	1	
	indoxacarb, MOA 22 (Avaunt) 30 WDG	3.5 oz	12 hrs	3	For control of low numbers of beet armyworm and not for corrective treatments of higher numbers of larvae. Do not apply more than 14 ounces of Avaunt (0.26 lb a.i.) per acre per crop. The minimum interval between sprays is 3 days.
	methoxyfenozide, MOA 18 (Intrepid) 2 F	4 to 10 oz	4 hrs	1	Use low rates for early-season applications when plants are small. For mid- and late-season applications, use 10 to 16 ounces.
	spinetoram, MOA 5 (Radiant) 1 SC	4 to 8 fl oz	4 hrs	1	
Cabbage looper, Corn earworm, Tobacco budworm	*Bacillus thuringiensis*, MOA 11A (Crymax) WDG (Dipel) DF	0.5 to 1.5 lb 8 oz	4 hrs	0	
	chlorantraniliprole, MOA 28 (Coragen) 1.67 SC	3.5 to 5 fl oz	4 hrs	1	Foliar or drip chemigation.
	cyantraniliprole, MOA 28 (Verimark) 1,67SC	5 to 13.5 fl oz	4 hrs	1	Verimark is for soil application only. Applications made at planting and/or via drip chemigation. Use higher rates (>10 fluid ounces) where cabbage looper is a concern. See label for application options.
	(Exirel) 0.83SE	7 to 17 fl oz	12 hrs	1	Exirel is for foliar application only. Use higher rates (>13.5 fluid ounces) for Cabbage.
	emamectin benzoate, MOA 6 (Proclaim) 5 WDG	3.2 to 4.8 oz	12 hrs	7	Do not make more than two sequential applications without rotating to another product with a different mode of action.
	flubendiamide, MOA 28 (Belt) 4SC	1.5 fl oz	12 hrs	1	
	indoxacarb, MOA 22 (Avaunt) 30 WDG	2.5 to 3.5 lb	12 hrs	3	Do not apply more than 14 ounces of Avaunt (0.26 lb a.i.) per acre per crop. The minimum interval between sprays is 3 days.
	pyrethroid, MOA 3		12 hrs		See table 5-9B for registered pyrethroids and pre-harvest intervals.
	methomyl, MOA 1A (Lannate) 2.4 LV	1.5 to 3 pt	48 hrs	7 to 10	
	methoxyfenozide, MOA 18 (Intrepid) 2 F	4 to 10 fl oz	4 hrs	1	Low rates for early-season applications to young or small plants. For mid- and late-season applications, use 6 to 10 ounces.
	spinetoram, MOA 5 (Radiant) 1 SC	5 to 10 fl oz	4 hrs	1	
Leafhopper	dinotefuran, MOA 4A (Venom) 70 SG	1 to 3 oz (foliar) 5 to 6 oz (soil)	12 hrs	7 21	Do not follow soil applications with foliar applications of any neonicotinoid insecticide. Use only one application method. Do not apply more than 6 ounces per acre (foliar) or 12 ounces per acre (soil). Soil applications may be applied by: 1. Narrow band below or above the seed line at planting; 2. post seeding or transplant drench with sufficient water to ensure incorporation; or 3. drip irrigation.
	dimethoate 4 EC, MOA 1B	0.5 pt	48 hrs	14	14-day interval for leaf lettuce.
	flupyradifurone, MOA 4D (Sivanto) 200 SL	7.0 to 10.5 fl oz	4 hrs	1	
	imidacloprid, MOA 4A (various) 1.6 F	3.75 fl oz	12 hrs	7	There is a 12-month plant-back restriction for a number of crops. Check label for restrictions.
	pyrethroid, MOA 3		12 hrs		See table 5-9B for registered pyrethroids and pre-harvest intervals.
	thiamethoxam, MOA 4A (Actara) 25 WDG	1.5 to 3 oz	12 hrs	7	
	Tolfenpyrad, MOA 21A (Torac) 1.29 EC	14 to 21 fl oz	12 hrs	1	Do not apply until at least 14 days after plant emergence or after transplanting to allow time for root establishment.
Melon (See Cantaloupe)					
Mustard Greens (See Collard, Kale, Mustard Greens)					
Okra					
Aphid	imidacloprid, MOA 4A Soil treatment (Admire Pro) 4.6 F (various) 2 F	7 to 10.5 fl oz 16 to 24 fl oz	12 hrs	21	See label for soil treatment instructions.
	Foliar treatment (Admire Pro) 4.6 F (various) 1.6 F	1.3 to 2.2 fl oz 3.8 fl oz	12 hrs	0	

Table 5-9. Insect Control for Commercial Vegetables

CROP Insect	Insecticide, Mode of Action Code, and Formulation	Amount of Formulation Per Acre	Restricted Entry Interval (REI)	Pre harvest Interval (PHI) (Days)	Precautions and Remarks
Okra (continued)					
Aphid (continued)	flonicamid, MOA 29 (Beleaf) 50 SG	2 to 2.8 oz	12 hrs	0	
	flupyradifurone, MOA 4D (Sivanto) 200 SL	7.0 to 12 fl oz	4 hrs	1	
	spirotetramat, MOA 23 (Movento) 2SC	4 to 5 fl oz	24 hrs	3	Do not exceed 10 fluid ounces per season. Not for flea beetle. Requires surfactant.
	malathion, MOA 1B (various brands) 8 F (various brands) 25 WP	1.5 pt 6 lb	12 hrs	1	
Blister beetle, Flea beetle, Japanese beetle	carbaryl, MOA 1A (Sevin) 50 WP (Sevin) 80 S (Sevin) XLR Plus	4 lb 2.5 lb 2 qt	12 hrs	3	On foliage as needed.
	pyrethroid, MOA 3		12 hrs		See table 5-9B for a list of registered pyrethroids and pre-harvest intervals.
Corn earworm, Tobacco budworm, European corn borer	carbaryl, MOA 1A (Sevin) 50 WP (Sevin) 80 S (Sevin) XLR Plus	4 lb 2.5 lb 2 qt	12 hrs	3	On foliage as needed.
	chlorantraniliprole, MOA 28 (Coragen) 1.67 SC	2 to 3.5 fl oz	4 hrs	1	Foliar or drip chemigation. Drip chemigation must be applied uniformly to the root zone. See label for instructions.
	cyantraniliprole, MOA 28 (Verimark) 1,67SC	5 to 10 fl oz	4 hrs	1	Verimark is for soil application only. Applications made at planting and/or via drip chemigation. See label for application options.
	(Exirel) 0.83SE	7 to 17 fl oz	12 hrs	1	Exirel is for foliar application only. Rates >13.5 for loopers only.
	flubendiamide, MOA 28 (Belt) 4SC	1.5 fl oz	12 hrs	1	
	methoxyfenozide, MOA 18 (Intrepid) 2 F	8 to 16 fl oz	4 hrs	1	
	novaluron, MOA 15 (Rimon) 0.83 EC	9 to 12 fl oz	12 hrs	1	
	spinetoram, MOA 5 (Radiant) 1 SC	5 to 10 fl oz	4 hrs	1	For corn earworm only.
	pyrethroid, MOA 3		12 hrs		See table 5-9B for a list of registered pyrethroids and pre-harvest intervals.
Spider mites	bifenazate, MOA 20D (Acramite) 50 WP	0.75 to 1 lb	12 hrs	3	Do not make more than one application per season.
	fenpyroximate MOA 21 (Portal) 0.4EC	2 pt	12 hrs	3	Do not make more than two applications per season.
Stink bug, leaffooted bug	pyrethroid, MOA 3		12 hrs		See table 5-9B for a list of registered pyrethroids and pre-harvest intervals.
Whitefly	buprofezin, MOA 16 (Courier) 40 SC	9 to 13.6 fl oz	12 hrs	1	
	chlorantraniliprole, MOA 28 (Coragen) 1.67 SC	2 to 3.5 fl oz	4 hrs	1	Foliar or drip chemigation. Drip chemigation must be applied uniformly to the root zone. See label for instructions.
	cyantraniliprole, MOA 28 (Verimark) 1,67SC	6.75 to 13.5 fl oz	4 hrs	1	Apply Verimark to at planting and/or later via drip irrigation or soil injection. See label for application options.
	(Exirel) 0.83SE	13.5 to 20.5 fl oz	12 hrs	1	Exirel is for foliar application.
	flupyradifurone, MOA 4D (Sivanto) 200 SL	10.5 to 14.0 fl oz	4 hrs	1	
	imidacloprid, MOA 4A Soil treatment (Admire Pro) 4.6 F (various) 2 F	7 to 14 fl oz 16 to 32 fl oz	12 hrs	21	See label for soil application instructions.
	Foliar treatment (Admire Pro) 4.6 F (various) 1.6 F	1.3 to 2.2 fl oz 3.8 oz	12 hrs	0	
	pyriproxyfen, MOA 7C (Knack) 0.86 EC	8 to 10 fl oz	12 hrs	1	Do not make more than two applications per season.
	spirotetramat, MOA 23 (Movento) 2SC	4 to 5 fl oz	24 hrs	3	Do not exceed 10 fluid ounces per season. Not for flea beetle. Requires surfactant.
Onion					
Armyworm, Cutworm	methoxyfenozide MOA 18 (Intrepid) 2F	4 to 8 fl oz 8 to 12 fl oz	4 hrs	1	Green onion only. Use lower rates in early season on small plants; use higher rates in late season and heavy infestations.
	pyrethroid, MOA 3		12 hrs		See table 5-9B for a list of registered pyrethroids and pre-harvest intervals.
	spinetoram, MOA 5 (Radiant) 1 SC	5 to 10 fl oz	4 hrs	1	

Table 5-9. Insect Control for Commercial Vegetables

CROP Insect	Insecticide, Mode of Action Code, and Formulation	Amount of Formulation Per Acre	Restricted Entry Interval (REI)	Pre harvest Interval (PHI) (Days)	Precautions and Remarks
Onion (continued)					
Leafminer	cryomazine, MOA 17 (Trigard) 75 WS	2.66 oz	12 hrs	7	
	spinetoram, MOA 5 (Radiant) 1 SC	6 to 8 fl oz	4 hrs	1	
Onion maggot, Seed corn maggot	Onion seed pre-treated with cyromazine (Trigard) can be used to control onion and seed corn maggot.				
	chlorpyrifos, MOA 1B (Lorsban) 4 E	32 fl oz	24 hrs		Apply as in-furrow drench at planting. Use a minimumof 40 gal per acre and incorporate to a depth of 1 to 2 inches. Do not make more than one application per year.
	diazinon, MOA 1B (Diazinon) (AG 500)	2 to 4 qt	3 days		Furrow application; drench the seed furrow at planting time. Apply as a furrow treatment at time of planting. Use separate hoppers for seed and chemical.
	pyrethroid, MOA 3		12 hrs		See table 5-9B for a list of registered pyrethroids and pre-harvest intervals.
Thrips	acetamiprid MOA 4A (Assail) 70 WP	2.1 to 3.4 oz	12 hrs	7	
	methomyl, MOA 1A (Lannate) 2.4 LV	1.5 pt	48 hrs	7	
	spinetoram, MOA 5 (Radiant) 1 SC	6 to 8 fl oz	4 hrs	1	
	pyrethroid, MOA 3		12 hrs		See table 5-9B for a list of registered pyrethroids and pre-harvest intervals.
Pea, English and Snow Pea (Succulent and dried)					
Aphid	acetamiprid MOA 4A (Assail) 70 WP	1 to 2.3 oz	12 hrs	7	Also controls leafhoppers. Succulent peas only.
	pyrethroid, MOA 3		12 hrs		See table 5-9B for a list of registered pyrethroids and pre-harvest intervals.
	dimethoate, MOA 1B (Dimethoate) 400 (4E)	0.33 pt	48 hrs	0	Do not make more than one application per season, and do not feed or graze if a mobile viner is used, or for 21 days if a stationary viner is used. Re-entry interval is 48 hours.
	flupyradifurone, MOA 4D (Sivanto) 200 SL	7.0 to 10.5 fl oz	4 hrs	7	Will also control leafhopper
	imidacloprid, MOA 4A Soil treatment (Admire Pro) 4.6 F (various) 2 F	7 to 10.5 fl oz 16 to 24 fl oz	12 hrs	21	See label for soil application instructions..
	Foliar treatment (Admire Pro) 4.6 F (various) 1.6 F	1.2 fl oz 3.5 fl oz	12 hrs	7	
Armyworm, Cloverworm, Cutworm, Looper	chlorantraniliprole MOA 28 (Coragen) 1.67 SC	3.5 to 5 fl oz	4 hrs	1	
	pyrethroid, MOA 3		12 hrs		See table 5-9B for a list of registered pyrethroids and pre-harvest intervals.
	spinetoram, MOA 5 (Radiant) 1 SC	4 to 8 fl oz	4 hrs	3 (succulent); 28 (dried)	Not for cutworm.
	spinosad, MOA 5 (Blackhawk)	2.2 to 3.3 oz	4 hrs	3 (succulent); 28 (dried)	
Leafhopper, Lygus bug, Stink bug	dimethoate, MOA 1B (Dimethoate) 400 (4E)	0.33 to 1 pt	48 hrs	See label	Do not make more than one application per season. Do not feed or graze if a mobile viner is used, or for 21 days if a stationary viner is used.
	methomyl, MOA 1A (Lannate) 2.4 LV	1.5 to 3 pt	48 hrs	3	Apply to foliage as needed.
	pyrethroid, MOA 3		12 hrs		See table 5-9B for registered pyrethroids and pre-harvest intervals.
Seedcorn maggot	See **Beans** for control				
Pea (Cowpea, southernpeas)					
Aphid, Thrips	acetamiprid MOA 4A (Assail) 70 WP	1 to 2.3 oz	12 hrs	7	Succulent peas only.
	pyrethroid, MOA 3		12 hrs		See table 5-9B for registered pyrethroids and pre-harvest intervals.
	flupyradifurone, MOA 4D (Sivanto) 200 SL	7.0 to 10.5 fl oz	4 hrs	7	Will not control thrips.
	imidacloprid, MOA 4A Soil treatment (Admire Pro) 4.6 F (various) 2 F	7 to 10.5 fl oz 16 to 24 fl oz	12 hrs	21	See label for soil application instructions.
	Foliar treatment (Admire Pro) 4.6 F (various) 1.6 F	1.3 fl oz 3.5 fl oz	12 hrs	7	
	spinetoram, MOA 5 (Radiant) 1 SC	5 to 8 fl oz	4 hrs	3 (succulent); 28 (dried)	Radiant is not effective against aphids.

Table 5-9. Insect Control for Commercial Vegetables

CROP Insect	Insecticide, Mode of Action Code, and Formulation	Amount of Formulation Per Acre	Restricted Entry Interval (REI)	Pre harvest Interval (PHI) (Days)	Precautions and Remarks
Pea (Cowpea, southernpeas) (continued)					
Aphid, Thrips (continued)	spinosad, MOA 5 (Blackhawk)	2.2 to 3.3 oz	4 hrs	3 (succulent); 28 (dried)	Blackhawk is not effective against aphids.
Bean leaf beetle	carbaryl, MOA 1A (Sevin) 4 L (Sevin) 80 S	0.5 to 1 qt 0.625 to 1.25 lb	12 hrs	3	Do not feed treated foliage to livestock.
	pyrethroid, MOA 3		12 hrs		See table 5-9B for a list of registered pyrethroids and pre-harvest intervals.
Corn earworm, Loopers, European corn borer, Armyworm	chlorantraniliprole MOA 28 (Coragen) 1.67 SC	3.5 to 5 fl oz	4 hrs	1	
	methoxyfenozide, MOA 18 (Intrepid) 2 F	4 to 16 fl oz	4 hrs	7	Use lower rates on smaller plants and higher rates for mid- to late season applications, against corn earworm. Do not apply more than 16 fluid ounces (0.25 pound a.i.) per acre per season.
	spinetoram, MOA 5 (Radiant) 1 SC	3 to 6 fl oz	4 hrs	3 (succulent) 28 (dried)	Do not apply more than 12 fluid ounces (0.188 a.i.) per acre per season.
	pyrethroid, MOA 3		12 hrs		See table 5-9B for a list of registered pyrethroids and pre-harvest intervals.
	methomyl, MOA 1A (Lannate) 90SP	0.5 to 1 lb	48 hrs	1	Re-entry interval is 48 hr.
Cowpea curculio	Pyrethroids, MOA 3		12 hrs		See table 5-9B for a list of registered pyrethroids and pre-harvest intervals. Control may be poor in areas where resistant populations occur, primarily in parts of Alabama and Georgia. In areas where resistance is a problem, pyrethroid insecticides should be used at the highest labeled rate and synergized by tank-mixing with 1 pint piperonyl butoxide synergist per acre. In fields where resistance is a problem, applications every 3 to 5 days may be necessary to maintain control of the cowpea curculio population.
	methomyl, MOA 1A (Lannate) 90 SP	0.5 to 1 lb	48 hrs	1	Re-entry interval is 48 hours. Not effective against resistant cowpea curculio populations.
Stink bug	pyrethroid, MOA 3		12 hrs		See table 5-9B for a list of registered pyrethroids and pre-harvest intervals. Control may be poor in areas where resistant populations occur, primarily in the Gulf Coast areas.
	methomyl, MOA 1A (Lannate) 90SP	0.5 to 1 lb	48 hrs	1	Re-entry interval is 48 hours.
Leafminer	spinetoram, MOA 5 (Radiant) 1 SC	5 to 8 fl oz	4 hrs	3 (succulent); 28 (dried)	
	Spinosad, MOA 5 (Blackhawk)	2.5 to 3.3 oz	4 hrs	3 (succulent); 28 (dried)	
Pepper					
Aphid, Flea beetle	acetamiprid, MOA 4A (Assail) 70 WP	0.8 to 1.2 oz	12 hrs	7	Do not apply more than once every 7 days, and do not exceed 4 applications per season.
	clothianidin, MOA 4A (Belay) 50WDG	4.8 to 6.4 oz (soil) 1.6 to 2.1oz (foliar)	12 hrs	7	Soil application at planting only.
	cyantraniliprole, MOA 28 (Verimark)	6.75 to 13.5 fl oz	4 hr	1	Apply to soil at planting, as a transplant tray drench, in transplant water or hill drench. After planting may be applied via drip irrigation.
	Dinotefuran, MOA 4A Foliar treatment (Venom) 70 SG (Scorpion) 35SL	1 to 4 oz 2 to 7 fl oz	12 hrs	1	Do not follow soil applications with foliar applications. Use only one application method. Do not apply more than 6 ouces per acre per season using foliar applications, or 12 ounces per acre per season using soil applications. Soil applications may be appled by 1) a narrow band below or above the seed line at planting; 2) a post-seeding or transplant drench with sufficient water to ensure incorporation to the root zone; or 3) drip irrigation. For flea beetle control only.
	Soil treatment (Venom) 70 SG (Scorpion) 35SL	5 to 6 oz 9 to 10.5 fl oz		21	
	flonicamid, MOA 29 (Beleaf) 50 SG	2 to 4.8 oz	12 hrs	0	Will not control flea beetle.
	flupyradifurone, MOA 4D (Sivanto) 200 SL	7.0 to 12.0 fl oz	4 hrs	1	
	imidacloprid, MOA 4A Soil treatment (Admire Pro) 4.6 F (various) 2 F	7 to 14 fl oz 16 to 32 fl oz	12 hrs	21	Where whitefly resistance is a concern, do not follow soil applications with foliar applications of any neonicotinoid. See label for soil application instructions. For short-term protection of transplants at planting, apply Admire Pro (0.44 oz/10,000 plants) not more than 7 days before transplanting by 1) uniformly spraying on transplants, followed immediately by sufficient overhead irrigation to wsh product into potting media; or 2) injection into overhead irrigation system using adequate volume to thoroughly saturate soil media.
	Foliar treatment (Admire Pro) 4.6 F (various) 1.6 F	1.3 fl oz 3.8 fl oz	12 hrs	0	
	oxamyl, MOA 1A (Vydate) 2 L	1 to 2 qt	48 hrs	7	
	pymetrozine, MOA 9B (Fulfill) 50 WDG	2.75 oz	12 hrs	0	Apply before aphids reach damaging levels. Do not exceed 5.5 ounces per acre per season. Not for flea beetle.
	spirotetramat, MOA 23 (Movento) 2SC	4 to 5 fl oz	24 hrs	1	Do not exceed 10 fluid ounces per season. Requires surfactant. Will not control flea beetle.

Table 5-9. Insect Control for Commercial Vegetables

CROP Insect	Insecticide, Mode of Action Code, and Formulation	Amount of Formulation Per Acre	Restricted Entry Interval (REI)	Pre harvest Interval (PHI) (Days)	Precautions and Remarks
Pepper (continued)					
Aphid, Flea beetle (continued)	thiamethoxam, MOA 4A Soil treatment (Platinum) 75 SG	1.66 to 3.67 oz	12 hrs	30	Platinum may be applied to direct-seeded crops in-furrow seeding or transplant depth, post seeding or transplant as a drench, or through drip irrigation. Actara is applied as a foliar spray. Do not exceed 11 ounces per acre per season of Platinum or Actara. Check label for plant-back restrictions for a number of crops.
	Foliar treatment (Actara) 25 WDG	2 to 4 oz	12 hrs	0	
Armyworm, Corn earworm, Looper, Hornworm	*Bacillus thuringiensis*, MOA 11A (Dipel) DF (Xentari) WDG	0.5 to 1.5 lb 0.5 to 1 lb	4 hrs	0	
	chlorantraniliprole, MOA 28 (Coragen) 1.67 SC	2 to 3.5 fl oz	4 hrs	1	Foliar or drip chemigation. Drip chemigation must be applied uniformly to the root zone. See label for instructions.
	cyantraniliprole, MOA 28 (Verimark) 1,67SC	5 to 10 fl oz	4 hrs	1	Verimark is for soil application only. Applications made at planting and/or via drip chemigation. See label for application options.
	(Exirel) 0.83SE	7 to 13.5 fl oz	12 hrs	1	Exirel is for foliar application only.
	emamectin benzoate, MOA 6 (Proclaim) 5 WDG	2.4 to 4.8 oz	12 hrs	7	Apply when larvae are first observed. Additional applications may be necessary to maintain control.
	flubendiamide, MOA 28 (Belt) 4 SC	1.5 fl oz	12 hrs	1	
	indoxacarb, MOA 22 (Avaunt) 30 WDG	2.5 to 3.5 oz	12 hrs	3	Use only higher rate for control of armyworm and corn earworm. Do not apply more than 14 ounces of Avaunt (0.26 pound a.i. per acre per crop). Minimum interval between sprays is 5 days.
	methoxyfenozide, MOA 18 (Intrepid) 2 F	4 to 16 fl oz	4 hrs	1	Apply at rates of 4 to 8 fluid ounces early in season when plants are small. Apply at rates of 8 to 16 ounces to large plants or when infestations are heavy. During periods of continuous moth flights re-treatments at 7 to 14 days may be required. Do not apply more than 16 fluid ounces per application or 64 fluid ounces of Intrepid per acre per season.
	novaluron, MOA 15 (Rimon) 0.83 EC	9 to 12 fl oz	12 hrs	1	The use of a surfactant/adjuvant with Rimon is prohibited on pepper.
	spinetoram, MOA 5 (Radiant) 1 SC	5 to 10 fl oz	4 hrs	1	
	pyrethroid, MOA 3		12 hrs		See table 5-9B for a list of registered pyrethroids and pre-harvest intervals.
Blister beetle, Stink bug, Leaffooted bug	pyrethroid, MOA 3		12 hrs		See table 5-9B for a list of registered pyrethroids and pre-harvest intervals.
	Dinotefuran, MOA 4A Foliar treatment (Venom) 70 SG (Scorpion) 35SL	1 to 4 oz 2 to 7 fl oz	12 hrs	1	Do not combine foliar applications with soil applications, or vice versa. Use only one application method.
	Soil treatment (Venom) 70 SG (Scorpion) 35SL	5 to 6 oz 9 to 10.5 fl oz		21	
	thiamethoxam, MOA 4A (Actara) 25WDG	3 to 5.5 oz	12 hrs	0	
European corn borer	For all insecticides, begin applications at first fruit set when European corn borer moths are flying, as indicated by light trap catches. Applications should be made at 5- to 7-day intervals as long as moths continue to fly or egg masses are present on the plants.				
	acephate, MOA 1B (Orthene) 97 PE	0.75 to 1 lb	24 hrs	7	For use on bell-type peppers only. Do not apply more than 2 pounds a.i. per acre per season.
	chlorantraniliprole, MOA 28 (Coragen) 1.67 SC	2 to 3.5 fl oz	4 hrs	1	Foliar or drip chemigation. Drip chemigation must be applied uniformly to the root zone. See label for instructions.
	cyantraniliprole, MOA 28 (Verimark) 1.67SC (Exirel) 0.83SE	10 oz 7 to 13.5 fl oz	4 hr 12 hrs	1 1	Verimark should be applied via drip irrigation or soil injection only. Exirel is for foliar application only.
	Emamectin benzoate (Proclaim) 5SG	2.4 to 4.8 oz	12 hrs	7	Do not allow animals to graze in treated areas.
	Flubendiamide, MOA 28 (Belt) 4SC	1.5 fl oz	12 hrs	1	
	methomyl, MOA 1A (Lannate) 2.4 LV	1.5 pt	48 hrs	3	
	pyrethroid, MOA 3		12 hrs		See table 5-9B for registered pyrethroids and pre-harvest intervals.
Leafminer	abamectin, MOA 6 (Agri-Mek) 0.7 SC	1.75 to 3.5 fl oz	12 hrs	7	
	cyromazine, MOA 17 (Trigard) 75 WP	2.66 oz	12 hrs	0	
	dimethoate 4 EC, MOA 1B	0.5 pt	48 hrs	0	Re-entry interval is 48 hr.
	spinetoram, MOA 5 (Radiant) 1 SC	6 to 10 fl oz	4 hrs	1	

119

Table 5-9. Insect Control for Commercial Vegetables

CROP Insect	Insecticide, Mode of Action Code, and Formulation	Amount of Formulation Per Acre	Restricted Entry Interval (REI)	Pre harvest Interval (PHI) (Days)	Precautions and Remarks
Pepper (continued)					
Pepper maggot	acephate, MOA 1B (Orthene) 97 PE	0.75 to 1 lb	24 hrs	7	See comments under European corn borer.
	Dimethoate 4 EC, MOA 1B	0.5 to 0.67 pt	48 hrs	0	
	pyrethroid, MOA 3		12 hrs		See table 5-9B for registered pyrethroids and pre-harvest intervals.
Pepper weevil	acetamiprid, MOA 4A (Assail) 30 SG	4 oz	12 hrs	7	
	oxamyl, MOA 1A (Vydate) 2 L	2 to 4 pt	48 hrs	7	
	thiamethoxam, MOA 4A (Actara) 25 WP	3 to 4 oz	12 hrs	0	Do not exceed 8 oz of Actara per acre per season.
	Pyrethroid, MOA 3		12 hrs		See table 5-9B for registered pyrethroids and pre-harvest intervals.
Broad mite	abamectin, MOA 6 (Agri-Mek) 0.7 SC	1.75 to 3.5 fl oz	12 hrs	7	On foliage as needed.
	Fenpyroximate MOA 21 (Portal) 0.4EC	2 pt	12 hrs	3	Do not make more than two applications per season.
	Spiromesifen, MOA 23 (Oberon) 2 SG	7 to 8.5 oz	12 hrs	7	Do not exceed 3 applications per season.
Thrips	Dinotefuran, MOA 4A Foliar treatment (Venom) 70 SG (Scorpion) 35SL	1 to 4 oz 2 to 7 fl oz	12 hrs	1	See label for application instructions and restrictions.
	Soil treatment (Venom) 70 SG (Scorpion) 35SL	5 to 6 oz 9 to 10.5 fl oz		21	
	Imidacloprid, MOA 4A (Admire Pro) 4.6 F (various) 2 F	7 to 14 fl oz 16 to 32 fl oz	12 hrs	21	See Aphids for application instructions. Treating transplants before setting in the field, followed by drip irrigation may suppress incidence of tomato spotted virus. Imidacloprid is ineffective against western flower thrips.
	Methomyl, MOA 1A (Lannate) 2.4 LV	1.5 pt	48 hrs	3	
	spinetoram, MOA 5 (Radiant) 1 SC	6 to 10 fl oz	4 hrs	1	Do not exceed 29 fluid ounces per acre per season. Control of thrips may be improved by adding a spray adjuvant. See label for instructions.
Potato, Irish					
Aphid	acetamiprid, MOA 4A (Assail) 30 SG	1.5 to 4 oz	12 hrs	7	Do not make more than 4 applications per season. Thorough coverage is important. Assail belongs to the same class of insecticides (neonicotinoid) as Admire Pro, Provado, Actara, and Platinum and Colorado protato beetle populations have the potential to become resistant to this class.
	Clothianadin MOA 4A Belay 50 WDG	1.0 to 1.5 oz	12 hrs	7	Apply Belay 50 WDG as foliar spray when populations reach a threshold level. Do not apply more than 3 applications. Belay belongs to the same class of insecticides (neonicotinoid) as Admire Pro, Provado, Actara, and Platinum and Colorado protato beetle populations have the potential to become resistant to this class.
	Flonicamid, MOA 29 (Beleaf) 50 SG	2 to 2.8 oz	12 hrs	7	
	flupyradifurone, MOA 4D (Sivanto) 200 SL	7.0 to 12.0 fl oz	4 hrs	1	
	dimethoate 4 EC, MOA 1B	0.5 to 1 pt	48 hrs	0	Do not apply more than 2 pints total per year.
	Imidacloprid, MOA 4A (Admire Pro) 4.6F (various) 1.6 F	1.2 fl oz 3.75 fl oz	12 hrs	7	To minimize selection for resistance in Colorado potato beetle, do not use acetamiprid, 120midacloprid, or thiamethoxam for aphid control if either of these compounds was applied to the crop for control of Colorado potato beetle. See comments on insecticide rotation under Colorado potato beetle.
	Pymetrozine, MOA 9B (Fulfill) 50 WDG	2.75 oz	12 hrs	14	Allow at least 7 days between applications. Do not exceed a total of 5.5 ounces (0.17 lb a.i.) per acre per season.
	Thiamethoxam, MOA 4A (Actara) 25 WDG	3 oz	12 hrs	14	To minimize selection for resistance in Colorado potato beetle, do not use imidacloprid or thiamethoxam for aphid control if either of these compounds was applied to the crop for control of Colorado potato beetle.

Table 5-9. Insect Control for Commercial Vegetables

CROP Insect	Insecticide, Mode of Action Code, and Formulation	Amount of Formulation Per Acre	Restricted Entry Interval (REI)	Pre harvest Interval (PHI) (Days)	Precautions and Remarks
Potato, Irish (continued)					
Colorado potato beetle	Colorado potato beetle populations in most commercial potato-growing areas have developed resistance to many insecticides. As a result, insecticides that are effective in some areas, or were effective in the past, may no longer provide control in particular areas. Colorado potato beetle readily develops resistance to insecticides. The following practices help to reduce the risk of resistance developing: CROP ROTATION AND INSECTICIDE ROTATION (the use of insecticides representing different modes of action IRAC MoA class number in different years and against different generations of potato beetle within a year)) are essential if insecticide resistance is to be managed and the risks of control failures due to resistance minimized. If control failures or reduced levels of control are observed with a particular insecticide, do NOT make a second application of the same insecticide at the same or higher rate. If an additional insecticide application is necessary, a different insecticide representing a different IRAC MoA class number should be used. Because potato beetle adults will move between adjacent and nearby fields from one year to the next, It is important to maintain the same rotation schedule of insecticide MOA classes in adjacent fields ar groups of nearby fields. SCOUT FIELDS: All insecticide applications to the potato crop, regardless of the target insect pest, have the potential to increase the resistance of the Colorado potato beetle to insecticides. Unnecessary insecticide applications should be avoided by scouting fields for insect pests and applying insecticides only when potentially damaging insect populations are present. SPOT TREATMENTS: Because overwintered potato beetles invade rotated fields from sources outside the field, potato beetle infestations in rotated fields occur first along field edges early in the season. Limiting insecticide applications to infested portions of the field will provide effective control and reduce costs. Growers are advised to keep accurate records on which insecticides have been applied to their potato crop for control of Colorado potato beetle and on how effective those insecticides were at controlling infestations. This will make choosing an insecticide and maintaining insecticide rotations easier. Monitoring the insecticide resistance status of local populations will also make insecticide selection easier.				
	Abamectin, MOA 6 (Agri-Mek) 0.7 SC	1.75 to 3.5 fl oz	12 hrs	14	Apply when adults and/or small larvae are present but before large larvae appear. Do not exceed two applications per season. Apply in at least 20 gallons water per acre.
	Acetamiprid, MOA 4A (Assail) 70 WP	0.6 to 1.7 oz	12 hrs	7	Apply when most of the egg masses have hatched and many small but few large larvae are present. An additional application should be used only if defoliation increases. Allow at least 7 days between foliar applications. To minimize selection for resistance, do not use foliar applications of any IRAC MOA class 4A insecticides if any IRAC MOA class 4A insecticides were applied to the crop as soil or seed piece treatments. See comments on insect rotation under Colorado potato beetle.
	chlorantraniliprole, MOA 28 (Coragen) 1.67	3.5 to 5 oz	4 hrs	14	Do not apply more than 15.4 ounces Coragen per acre per crop season.
	clothianadin MOA 4A Belay 50 WDG	1.9 to 2.8 fl oz	12 hrs	7	Apply Belay 50 WDG as foliar spray Apply when adults and/or small larvae are present but before large larvae appear. Do not apply more than 3 applications. Belay belongs to the same class of insecticides (neonicotinoid) as Admire Pro, Provado, Actara, and Platinum and Colorado protato beetle populations have the potential to become resistant to this class.
	cyantraniliprole, MOA 28 (Verimark) 1.67SC	6.75 to 13.5 fl oz	4 hr	NA	Apply in-furrow at planting. Do not apply any other MOA Group 28 insecticide for Coloarado potato beetle control following an at-plant application for cyantraniliprole. When applied at 10-13.5 fluid ounces per acre will provide control of Eurpean conrn borer in most years, except possibly in very early planted potatoes.
	dinotefuran, MOA 4A (Venom) 70 SG	1 to 1.5 oz (foliar) 6.5 to 7.5 oz (soil)	12 hrs	7	Soil treatment for preplant, preemergence, or at ground crack only application only. To minimize selection for resistance, do not use foliar applications of any IRAC MOA class 4A insecticides if any IRAC MOA class 4A insecticides were applied to the crop as soil or seed piece treatments. See comments on insecticide rotation under Colorado potato beetle.
	imidacloprid seed piece treatment, MOA 4A (Genesis) 240 g/L	0.4 to 0.6 fl oz/100 lb of seed tubers			See label for specific instructions. For early planted potatoes control may be marginal because of the prolonged time between application and Colorado potato beetle emergence. Limit use to locations where Colorado potato beetles were a problem in the same or adjacent fields during the previous year. Do not apply other IRAC MOA class 4A insecticides to a field if seed pieces were treated with Genesis. See product label for restrictions on rotational crops.
	imidacloprid, MOA 4A Soil treatment (Admire Pro) 4.6 F (various) 2.0 F	0.74 fl oz/ 1,000 ft row	12 hrs	—	Admire Pro applied in-furrow at planting time may provide season-long control. However, for early planted potatoes control may be marginal due to the prolonged time between application and Colorado potato beetle emergence. Use only in potato fields that have a history of potato beetle infestations. If potatoes are rotated to a field adjacent to one planted in potato last year, a barrier treatment may be effective. (See Vegetable IPM Insect Note #45.) Admire Pro may also be applied as a seed treatment. Check label for instructions regarding this use. Check label for restrictions on planting crops following Admire Pro treated potatoes. There have been reports of low levels of resistance to imidacloprid. To minimize selection for resistance, do not use foliar applications of any IRAC MoA class 4A insecticides if any of these compounds were applied to the crop as soil or seed piece treatments. See comments on insecticide rotation under Colorado potato beetle.
	Foliar treatment (Admire Pro) 4.6 (various) 1.6 F	1.3 fl oz 3.75 fl oz	12 hrs	7	Apply when most of the egg masses have hatched and most larvae are small (1/8 to 3/16 in.). An additional application should be made only if defoliation increases. Allow at least 7 days between foliar applications. Do not exceed 5.6 fluid ounces of Admire Pro per field per acre per season. Regardless of formulation, do NOT make more than a total of 0.31 pound imidacloprid per season. Foliar applications of imidacloprid should not be applied If soil applicaton was used. There have been reports of resistance to imidacloprid. To minimize selection for resistance, do not use foliar applications of any IRAC MOA class 4A insecticides if any of these compounds were applied to the crop as soil or seed piece treatments. See comments on insecticide rotation under Colorado potato beetle.

Table 5-9. Insect Control for Commercial Vegetables

CROP Insect	Insecticide, Mode of Action Code, and Formulation	Amount of Formulation Per Acre	Restricted Entry Interval (REI)	Pre harvest Interval (PHI) (Days)	Precautions and Remarks
Potato, Irish (continued)					
Colorado potato beetle (continued)	imidacloprid cyfluthrin premix, MOA 4A and 3 (Leverage) 2.7 SE	3 to 3.75 fl oz		7	Apply when most of the egg masses have hatched and most larvae are small (1/8 to 3/16 inch). An additional application should be made only if defoliation increases. Leverage will control European corn borer if application coincides with egg hatch and presence of small corn borer larvae. Leverage should not be used in fields treated with Admire Pro. There have been reports of low levels of resistance to imidacloprid. To minimize selection for resistance, do not use foliar applications of any IRAC MOA class 4A insecticides if any of these compounds were applied to the crop as soil or seed piece treatments. See comments on insecticide rotation under Colorado potato beetle.
	novaluron, MOA 15 (Rimon) 0.83 EC	9 to 12 fl oz	12 hrs	14	
	spinosad, MOA 5 (Blackhawk) 36WG	1.7 to 3.3 oz		3	Apply when most egg masses have hatched and both small and large larvae are present. Thorough coverage is important. Do not apply more than a total of 0.33 pound a.i. (14.4 ounces of Blackhawk or 21 ounces of Radiant) per crop. Do not apply in consecutive generations of Colorado potato beetle and do not make more than two applications per single generation of Colorado potato beetle. Do not make successive applications less than 7 days apart. To minimize the potential for resistance, do NOT use spinosad or spinetoram if it either product was applied to a potato crop in the field or an adjacent field within the last year.
	spinetoram, MOA 5 (Radiant) 1 SC	6 to 8 fl oz	4 hrs	7	
	thiamethoxam seed piece treatment, MOA 4A (Cruiser) 5 FS	0.11 to 0.16 fl oz/100 lb			See label for specific instructions. For early planted potatoes control may be marginal because of the prolonged time between application and Colorado potato beetle emergence. Limit use to locations where Colorado potato beetles were a problem in the same or adjacent fields during the previous year. To minimize selection for resistance, do not use foliar applications of any IRAC MOA class 4A insecticides if any of these compounds were applied to the crop as soil or seed piece treatments. See comments on insecticide rotation under Colorado potato beetle.
	thiamethoxam, MOA 4A (Platinum) 75 SG	1.66 to 2.67 oz	12 hrs	7	Platinum applied in-furrow at planting time may provide season-long control. For early planted potatoes control may be marginal because of the prolonged time between application and Colorado potato beetle emergence. Limit use to locations where Colorado potato beetles were a problem in the same or adjacent fields in the previous year. To minimize selection for resistance, do not use foliar applications of any IRAC MOA class 4A insecticides if any of these compounds were applied to the crop as soil or seed piece treatments. See comments on insecticide rotation under Colorado potato beetle. See product label for restrictions on rotational crops.
	(Actara) 25 WDG	3 oz	12 hrs	7	Actara is applied as foliar spray. Apply when most of the eggs have hatched and most of the larvae are small (1/8 to 3/16 inch). An additional application should be made only if defoliation increases. Allow at least 7 days between applications. Do not make more than 2 applications of Actara per crop per season. To minimize selection for resistance, do not use foliar applications of any IRAC MOA class 4A insecticides if any of these compounds were applied to the crop as soil or seed piece treatments. See label for rotational restrictions.
	thiamethoxam, MOA 4A F chlorantraniliprole, MOA 28 Premx (Voliarm Flexi)	4 oz		14	Voliam Flexi is applied as a foliar spray. Apply when most of the eggs have hatched and most of the larvae are small (1/8 to 3/16 inch.). An additional application should be made only if defoliation increases. Allow at least 7 days between applications. To minimize selection for resistance, do not use foliar applications of any IRAC MOA class 4A insecticides if any of these compounds were applied to the crop as soil or seed piece treatments. Do not exceed 8 ounces of Voliam Flexi. See label for rotational restrictions Volam Flexi can be expected to provide control of European corn borer if application is timed correctly (see European corn borer for correct timing.
European corn borer	The Atlantic variety of potato is very tolerant of injury by European corn borer larvae. Consequently, control is not recommended on Atlantic unless more than 30 percent of the stems are infested. Control on all other varieties is recommended when infestations reach 20 percent infested stems. Application timing is critical. Scout for eggs and treat when eggs hatch or at the first sign of larvae entering petioles. Several days of cool wet weather will kill larvae and may eliminate the need for insecticide applications. If this occurs, flag additional egg masses and apply insecticide at hatch.				
	pyrethroid, MOA 3		12 hrs		Apply when threshold is reached (usually during the first half of May). A second application may be needed if the percentage of infested stems increases substantially 7 to 10 days after the first application. Ground applications are usually more effective than aerial applications. See table 5-9B for a list of registered pyrethroids and pre-harvest intervals.
	chlorantraniliprole, MOA 28 (Coragen) 1.67	3.5 to 5 oz	4 hrs	14	Do not apply more than 15.4 ounces Coragen per acre per crop season.
	thiamethoxam, MOA 4A chlorantraniliprole MOA 28 Premix (Voliam Flexi)	4 oz	12 hrs	14	Voliam Flexi is applied as a foliar spray. Apply when most of the eggs have hatched and most of the larvae are small (1/8 to 3/16 inch.). An additional application should be made only if defoliation increases. Allow at least 7 days between applications. To minimize selection for resistance, do not use foliar applications of any IRAC MOA class 4A insecticides if any of these compounds were applied to the crop as soil or seed piece treatments. Do not exceed 8 ounces of Voliam Flexi. See label for rotational restrictions Volam Flexi can be expected to provide control of Colorado potato beetle if application is timed correctly (see Colorado potato beetle section for correct timing.

Table 5-9. Insect Control for Commercial Vegetables

CROP Insect	Insecticide, Mode of Action Code, and Formulation	Amount of Formulation Per Acre	Restricted Entry Interval (REI)	Pre harvest Interval (PHI) (Days)	Precautions and Remarks
Potato, Irish (continued)					
European corn borer (continued)	indoxacarb, MOA 22 (Avaunt) 30 WDG	3.5 to 6.0 oz	12 hrs	7	Apply when threshold is reached (usually during the first half of May). A second application may be needed if the percentage of infested stems increases substantially 7 to 10 days after the first application. Ground applications are usually more effective than aerial applications. Do not apply more than 24 ounces of Avaunt per acre per crop.
	spinetoram, MOA 5 (Radiant) 1 SC	6 to 8 fl oz	4 hrs	7	Do not apply more than a total of 0.25 pound a.i. (32 fluid ounces product) per crop.
Flea beetle	imidacloprid, MOA 4A Soil treatment (Admire Pro) 4.6 F (various) 2.0 F	0.74 fl oz/ 1,000 ft row	12 hrs	—	imidacloprid applied in-furrow at planting time may provide season-long control. However, for early planted potatoes control may be marginal due to the prolonged time between application and crop emergence. Check label for restrictions on planting crops following Admire Pro treated potatoes. See comments on resistance in Colorado potato beetle to imidacloprid. To minimize selection for resistance, do not use foliar applications of any IRAC MoA class 4A insecticides if any of these compounds were applied to the crop as soil or seed piece treatments.
	Foliar treatment (Admire Pro) 4.6 (various) 1.6 F	1.3 fl oz 3.75 fl oz	12 hrs	7	See comments for imidacloprid resistance in Colorado potato beetle.
	thiamethoxam seed piece treatment, MOA 4A (Cruiser) 5 FS	0.11 to 0.16 fl oz/100 lb	12 hrs		See label for specific instructions. For early planted potatoes control may be marginal because of the prolonged time between application and flea beetle emergence. Limit use to locations where Colorado potato beetles were a problem in the same or adjacent fields during the previous year. To minimize selection for resistance, do not use foliar applications of any IRAC MOA class 4A insecticides if any of these compounds were applied to the crop as soil or seed piece treatments. See comments on insecticide rotation under Colorado potato beetle.
	thiamethoxam, MOA 4A (Platinum) 2 SC	5 to 8 fl oz	12 hrs	7	Platinum applied in-furrow at planting time may provide season-long control. However, for early planted potatoes control may be marginal due to the prolonged time between application and crop emergence. See product label for restrictions on rotational crops.
	(Actara) 25 WDG	3 oz	12 hrs	7	Actara is applied as foliar spray.
	Thiamethxsam MOA 4A chlorantraniliprole moa 28 (Volium Flexi)	4 fl oz		14	Do not exceed a total of 8.0 fluid ounces per acre Volium Flexi or 0.094 lb ai/acre of thiamethoxam-containing products or 0.2 pound ai/acre of chlorantraniliprole-containing products per growing season.
	pyrethroid, MOA 3		12 hrs		See table 5-9B for a list of registered pyrethroids and pre-harvest intervals.
Leafhopper	carbaryl, MOA 1A (Sevin) 50 WP (Sevin) 80 S (Sevin) XLR Plus	1 to 2 lb 0.625 to 1.25 lb 1 pt	12 hrs	7	On foliage when leafhoppers first appear. Repeat every 10 days as needed. Often a problem in the mountains.
	dimethoate 4 EC, MOA 1B	0.5 to 1 pt	48 hrs	0	Do not apply more than 2 pints total per acre per year.
	methomyl, MOA 1A (Lannate) 2.4 LV	1.5 pt	48 hrs	6	
	pyrethroid, MOA				See table 5-9B for a list of registered pyrethroids and pre-harvest intervals.
Leafminer	dimethoate 4 EC, MOA 1B	0.5 to 1 pt	48 hrs	0	Do not apply more than 2 pints total per acre per year.
	chlorantraniliprole, MOA 28 (Coragen) 1.67 SC	3.5 to 5 fl oz	4 hrs	14	
Blister beetle, Leaffooted bug, Plant bug, Stink bug, Vegetable weevil	carbaryl, MOA 1A (Sevin) 50 WP (Sevin) XLR Plus	2 to 4 lb 1 to 2 qt	12 hrs	7	On foliage as needed.
	pyrethroid, MOA 3		12 hrs		See table 5-9B for a list of registered pyrethroids and pre-harvest intervals.
Potato tuberworm	chlorantraniliprole, MOA 28 (Coragen) 1.67 SC	3.5 to 5 fl oz	4 hrs	14	Do not exceed 4 applications per acre per crop. Do not apply more than 15.4 ounces Coragen per acre per crop season. Minimum interval between applications is 5 days.
	methomyl, MOA 1A (Lannate) 2.4 LV	1.5 to 3 pt	48 hrs	6	Prevent late-season injury by keeping potatoes covered with soil. To prevent damage in storage, practice sanitation.
	pyrethroid, MOA 3		12 hrs		See table 5-9B for a list of registered pyrethroids and pre-harvest intervals.
Thrips	dimethoate 4 EC, MOA 1B	0.5 pt	48 hrs	0	
	spinetoram, MOA 5 (Radiant) 1 SC	6 to 8 fl oz	4 hrs	7	
	spinosad, MOA 5 (Blackhawk) 36WG	2.25 to 3.5 oz	4 hrs	3	Control may be improved by addition of an adjuvant to the spray mixture.
Wireworm	Planting in fields previously in corn, soybean, or fallow may increase risk of wireworm.				
	bifenthrin, MOA 3 (Capture LFR)	25.5 fl oz			In furrow at planting.

123

Table 5-9. Insect Control for Commercial Vegetables

CROP Insect	Insecticide, Mode of Action Code, and Formulation	Amount of Formulation Per Acre	Restricted Entry Interval (REI)	Pre harvest Interval (PHI) (Days)	Precautions and Remarks
Potato, Irish (continued)					
Wireworm (continued)	clothianidin (Belay) 50 WDG	6 oz oz	12 hrs		In-furrow at planting.
	ethoprop, MOA 1B (Mocap) 15 G	1.4 lb per 1,000 row ft	48 hrs	90	In-furrow at planting.
	fipronil, MOA 2B (Regent) 4 SC	3.2 fl oz	0 hrs	90	In-furrow at planting. Do NOT use T-banding over the top of a closed furrow.
	phorate, MOA 1B (Thimet) 20 G	Row Treatment: 10 to 20 oz (38 in. row spacing)	12 hrs	90	Can contribute to insecticide-resistance problems with Colorado potato beetle.
Pumpkin, Squash					
Insecticide applications in cucurbits should be made in late evening to protect pollinating insects. Refer to the section of this chapter on Reducing the Risk of Pesticide Poisoning to Honey Bees for more information about protecting pollinators.					
Aphid	Where whitefly resistance is an issue (or any other insect with a high potential for resistance to Group 4A MOA insecticides), avoid making foliar applications of Group 4A insecticides when a soil-applied Group 4A program is used; i.e., do not make both foliar and soil applications of Group 4A insecticides. Also, if using a foliar-applied program, avoid using a block of more than three consecutive applications of any products belonging to Group 4A.				
	acetamiprid, MOA 4A (Assail) 30SG	2.5 to 4 oz	12 hrs	0	
	clothianidin, MOA 4A (Belay) 50WDG	4.8 to 6.4 oz (Soil) 1.6 to 2.1 oz (foliar)	12 hrs	At planting 7	Soil applications may only be applied at planting. Will also control cucumber beetle. Do not apply Belay during bloom or if bees are actively foraging.
	cyantraniliprole, MOA 28 (Verimark) 1,67SC	10 to 13.5 fl oz	4 hrs	1	Applied to the soil at planting or later via drip irrigation system. See lable for application options.
	flonicamid, MOA 29 (Beleaf) 50 SG	2 to 2.8 oz	12 hrs	0	
	flupyradifurone, MOA 4D (Sivanto) 200 SL	7.0 to 12.0 fl oz	4 hrs	1	
	imidacloprid, MOA 4A (Admire Pro) 4.6 F (various) 2 F	7 to 10.5 fl oz 16 to 24 fl oz	12 hrs	21	Must be applied to the soil. May be applied preplant; at planting; as a post-seeding drench, transplant water drench, or hill drench; subsurface sidedress or by chemigation using low-pressure drip or trickle irrigation. See label for approved application methods. Will also control whitefly and cucumber beetles.
	pymetrozine, MOA 9B (Fulfill) 50 WDG	2.75 oz	12 hrs	14	Apply before populations reach damaging levels. Do not exceed 5.5 ounces per acre per season.
	thiamethoxam, MOA 4A Soil treatment (Platinum) 75 SG	1.66 to 3.67 oz	12 hrs	30	Platinum may be applied to direct-seeded crops in-furrow seeding or transplant depth, post seeding or transplant as a drench, or through drip irrigation. Do not exceed 11 ounces per acre per season of Platinum. Check label for plant-back restrictions for a number of crops.
	Foliar treatment (Actara) 25 WDG	1.5 to 3.0 oz	12 hrs	0	
Armyworm	chlorantraniliprole, MOA 28(Coragen) 1.67 SC	2 to 3.5 fl oz	4 hrs	1	Foliar or drip chemigation. Drip chemigation must be applied uniformly to the root zone. See label for instructions.
	flubendiamide, MOA 28 (Belt) 4SC	1.5 fl oz	12 hrs	1	
	methoxyfenozide, MOA 18 (Intrepid) 2 F	4 to 10 fl oz	4 hrs	3	Do not exceed 4 applications per season, and do not reapply in less than 7 days.
	spinetoram, MOA 5 (Radiant) 1 SC	5 to 10 fl oz	4 hrs	3	
Cucumber beetle	pyrethroid, MOA				See table 5-9B for a list of registered pyrethroids and pre-harvest intervals.
	carbaryl, MOA 1A (Sevin) 50 WP (Sevin) 80 S (Sevin) XLR Plus	2 lb 1.25 lb 1 qt	12 hrs	3	Phytotoxicity may occur following application of carbaryl during hot, humid weather.
	acetamiprid, MOA 4A (Assail) 30SG	2.5 to 5.3 oz	12 hrs	0	
	clothianidin, MOA 4A (Belay) 50WDG	4.8 to 6.8oz (soil); 1.6 to 2.1oz (foliar)	12 hrs	At Planting 7	See application instructions and precautionary bee statement under aphids.
	Dinotefuran, MOA 4A Foliar treatment (Venom) 70 SG (Scorpion) 35SL Soil treatment (Venom) 70 SG (Scorpion) 35SL	1 to 4 oz 2 to 7 fl oz 5 to 6 oz 9 to 10.5 fl oz	12 hrs	1 21	Do not follow soil applications with foliar applications. Use only one application method. Do not apply more than 6 ounces per acre per season using foliar applications, or 12 ounces per acre per season using soil applications. Soil applications may be appled by 1) a narrow band below or above the seed line at planting; 2) a post-seeding or transplant drench with sufficient water to ensure incorporation to the root zone; or 3) drip irrigation.
	imidacloprid, MOA 4A (Admire Pro) 4.6 F (various) 2 F	7 to 10.5 fl oz 16 to 24 fl oz	12 hrs	21	See application methods under Aphid.

Table 5-9. Insect Control for Commercial Vegetables

CROP Insect	Insecticide, Mode of Action Code, and Formulation	Amount of Formulation Per Acre	Restricted Entry Interval (REI)	Pre harvest Interval (PHI) (Days)	Precautions and Remarks
Pumpkin, Squash (continued)					
Insecticide applications in cucurbits should be made in late evening to protect pollinating insects. Refer to the section of this chapter on Reducing the Risk of Pesticide Poisoning to Honey Bees for more information about protecting pollinators.					
Cutworm	chlorantraniliprole, MOA 28 (Coragen) 1.67 SC	3.5 to 5 fl oz	4 hrs	1	Foliar or drip chemigation. Drip chemigation must be applied uniformly to the root zone. See label for instructions.
	pyrethroid, MOA 3		12 hrs		See table 5-9B for a list of registered pyrethroids and pre-harvest intervals.
Corn earworm, Looper, Pickleworm, Melonworm	chlorantraniliprole, MOA 28 (Coragen) 1.67 SC	3.5 to 5 fl oz	4 hrs	1	Foliar or drip chemigation. Drip chemigation must be applied uniformly to the root zone. See label for instructions.
	flubendiamide, MOA 28 (Belt) 4SC	1.5 fl oz	12 hrs	1	
	indoxacarb, MOA 22 (Avaunt) 30 WDG	2.5 to 6 oz	12 hrs	3	
	methoxyfenozide, MOA 18 (Intrepid) 2 F	4 to 10 fl oz	4 hrs	3	Do not exceed 4 applications per season, and do not reapply in less than 7 days.
	pyrethroid, MOA 3		12 hrs		See table 5-9B for a list of registered pyrethroids and pre-harvest intervals.
	spinetoram, MOA 5 (Radiant) 1 SC	5 to 10 fl oz	4 hrs	3	
Spider mite	abamectin, MOA 6 (Agri-Mek) 0.7 SC	1.75 to 3.5 fl oz	12 hrs	7	
	bifenazate, MOA 20D (Acramite) 50 WS	0.75 to 1.0 lb	12 hrs	3	Do not make more than one application per season.
	etoxazole, MOA 10B (Zeal)	2 to 3 oz	12 hrs	7	Do not make more than one Zeal application per season.
	spiromesifen, MOA 23 (Oberon) 2 SG	7 to 8.5 oz	12 hrs	7	Do not exceed 3 applications per season.
Squash bug	acetamiprid, MOA 4A (Assail) 30 SG	5.3 oz	12 hrs	0	Assail is most effective against newly laid eggs and nymphs.
	clothianidin, MOA 4A (Belay) 50SDG	4.8 to 6.8oz (soil); 1.6 to 2.1oz (foliar)	12 hrs	At planting 7	See application instructions and precautionary bee statement above under aphid.
	dinotefuran, MOA 4A (Venom) 70 SG (Scorpion) 35 SL	3 to 4 oz 2 to 7 fl oz	12 hrs	1	Do not exceed 6 ounces Venom per acre per season.
	pyrethroid, MOA 3		12 hrs		See table 5-9B for a list of registered pyrethroids and pre-harvest intervals.
Squash vine borer	acetamiprid, MOA 4A (Assail) 30 SG	5.3 oz	12 hrs	0	
	chlorantraniliprole, MOA 28 (Coragen) 1.67 SC	3.5 to 5 fl oz	4 hrs	1	Foliar or drip chemigation. Drip chemigation must be applied uniformly to the root zone. See label for instructions.
	flubendiamide, MOA 28 (Belt) 4SC	1.5 fl oz	12 hrs	1	
	pyrethroid, MOA 3		12 hrs		See table 5-9B for a list of registered pyrethroids and pre-harvest intervals.
Thrips	dinotefuran, MOA 4A (Venom) 70 SG (Scorpion) 35 SL	1 to 4 oz 2 to 7 fl oz	12 hrs	1	
	spinetoram, MOA 5 (Radiant) 1 SC	6 to 10 fl oz	4 hrs	3	
Whitefly	acetamiprid, MOA 4A (Assail) 30 SG	5.3 oz	12 hrs	0	
	buprofezin, MOA 16 (Courier) 40 WP	9 to 13.6 oz	12 hrs	7	Use sufficient water to ensure good coverage. Do not apply more than twice per crop cycle or 4 applications per year total.
	chlorantraniliprole, MOA 28 (Coragen) 1.67 SC	5 to 7.5 fl oz	4 hrs	1	Foliar or drip chemigation. Drip chemigation must be applied uniformly to the root zone. See label for instructions.
	cyantraniliprole, MOA 28 (Verimark) 1,67SC (Exirel) 0.83SE	6.75 to 13.5 fl oz 13.5 to 20.5 fl oz	4 hrs 12 hrs	1 1	Apply Verimark at planting and/or later via drip irrigation or soil injection. See label for application options. Exirel is for foliar application.
	Dinotefuran, MOA 4A Foliar treatment (Venom) 70 SG (Scorpion) 35SL	1 to 4 oz 2 to 7 fl oz	12 hrs	1	See comments under cucumber beetle for application instructions and restrictions.
	Soil treatment (Venom) 70 SG (Scorpion) 35SL	5 to 6 oz 9 to 10.5 fl oz		21	

125

INSECT CONTROL

Table 5-9. Insect Control for Commercial Vegetables

CROP Insect	Insecticide, Mode of Action Code, and Formulation	Amount of Formulation Per Acre	Restricted Entry Interval (REI)	Pre harvest Interval (PHI) (Days)	Precautions and Remarks
Pumpkin, Squash (continued)					
Insecticide applications in cucurbits should be made in late evening to protect pollinating insects. Refer to the section of this chapter on Reducing the Risk of Pesticide Poisoning to Honey Bees for more information about protecting pollinators.					
Whitefly (continued)	flupyradifurone, MOA 4D (Sivanto) 200 SL	10.5 to 14.0 fl oz	4 hrs	1	
	imidacloprid, MOA 4A (Admire Pro) 4.6 F (various) 2 F	7 to 10.5 fl oz 16 to 24 fl oz	12 hrs	21	Admire Pro must be applied to the soil. May be applied preplant; at planting; as a post-seeding drench, transplant water drench, or hill drench; subsurface sidedress or by chemigation using low-pressure drip or trickle irrigation. See label for information on approved application methods. Will also control aphids and cucumber beetle.
	pyriproxifen, MOA 7C (Knack) 0.86 EC	8 to 10 fl oz	12 hrs	7	Do not make more than two applications per season, and do not make applications closer than 14 days apart.
	spiromesifen, MOA 23 (Oberon) 2 SC	7 to 8.5 fl oz	12 hrs	7	
	thiamethoxam, MOA 4A (Platinum) 75 SG	1.66 to 3.67 oz	12 hrs	30	Platinum may be applied to direct-seeded crops in-furrow at seed or transplant depth, postseeding or transplant as a drench, or through drip irrigation. Do not exceed 11 ounces per acre per season. Check label for plant-back restrictions for a number of crops.
	(Actara) 25 WDG	1.5 to 3.0 oz	12 hrs	0	
Radish					
Aphid, Flea beetle, Leafminer	pyrethroid, MOA 3		12 hrs		See table 5-9B for a list of registered pyrethroids and pre-harvest intervals.
	flupyradifurone, MOA 4D (Sivanto) 200 SL	7.0 to 10.5 fl oz	4 hrs	7	Will not control flea beetle or leafminer.
	Foliar treatment - imidacloprid (Admire Pro) 4.6 F (various) 1.6 F	1.2 fl oz 3.5 fl oz	12 hrs	7	Will not control leafminer.
	flonicamid, MOA 29 (Beleaf) 50SG	2 to 2.8 oz	12 hrs	0	
	thiamethoxam, MOA 4A (Platinum) 75SG (Actara) 25WDG	1.7 to 2.17 oz 1.5 to 3 oz	12 hrs	30 7	See label for soil application instructions.
Root maggot, Wireworm	chlorpyrifos, MOA 1B (Lorsban) 4E	1 fl oz/1,000 linear ft	24 hrs	—	Water-based drench in-furrow planting. Use a minimum of 40 gal of water per acre.
	diazinon, MOA 1B (AG 500) 50 WP	3 to 4 qt 6 to 8 lb	3 days		Broadcast just before planting and immediately incorporate into the upper 4 to 8 inches of soil.
Spinach					
Aphid	acetamiprid, MOA 4A (Assail) 30SG	2 to 4 oz	12 hrs	7	Do not apply more than once every 7 days, and do not exceed 5 applications per season.
	clothianidin, MOA 4A (Belay) 50 WDG	4.8 to 6.0 oz (soil) 1.6 to 2.1 fl oz (foliar)	12 hrs	7	Soil application at planting only.
	cyantraniliprole, MOA 28 (Verimark) 1,67SC	6.75 to 10 fl oz	4 hrs	1	Soil applications made at planting only. See label for application options.
	flonicamid, MOA 9C (Beleaf) 50 SG	2 to 2.8	12 hrs	0	
	flupyradifurone, MOA 4D (Sivanto) 200 SL	10.5 to 12.0 fl oz	4 hrs	1	
	imidacloprid, MOA 4A Soil treatment (Admire Pro) 4.6 F (various) 2 F	4.4 to 10.5 fl oz 10 to 24 fl oz	12 hrs	21	Do not follow soil applications with foliar applications of any neonicotinoid insecticides. See label for soil application instructions.
	Foliar treatment (Admire Pro) 4.6 F (various) 1.6 F	1.2 fl oz 3.8 fl oz	12 hrs	7	
	pymetrozine, MOA 9B (Fulfill) 50 WDG	2.75 oz	12 hrs	7	Apply before aphids reach damaging levels. Use sufficient water to ensure good coverage.
	spirotetramat, MOA 23 (Movento) 2SC	4 to 5 fl oz	24 hrs	3	Do not exceed 10 fluid ounces per season. Requires surfactant.
	thiamethoxam, MOA 4A Soil treatment (Platinum) 75SG	1.7 to 2.17 oz	12 hrs	30	See label for soil application instructions.
	Foliar treatment (Actara) 25WDG	1.5 to 3 oz	12 hr	7	
	tolfenpyrad, MOA 21A (Torac) 1.29 EC	17 to 21 fl oz	12 hrs	1	

Table 5-9. Insect Control for Commercial Vegetables

CROP Insect	Insecticide, Mode of Action Code, and Formulation	Amount of Formulation Per Acre	Restricted Entry Interval (REI)	Pre harvest Interval (PHI) (Days)	Precautions and Remarks
Spinach (continued)					
Leafminer	chlorantraniliprole, MOA 28 (Coragen) 1.67 SC	5 to 7.5 fl oz	4 hrs	1	Foliar or drip chemigation. Drip chemigation must be applied uniformly to the root zone. See label for instructions.
	cryomazine, MOA 17 (Trigard) 75 WP	2.66 oz	12 hrs	7	
	spinetoram, MOA 5 (Radiant) 1 SC	6 to 10 fl oz	4 hrs	1	Spray adjuvants may enhance efficacy against leafminers. See label for information on adjuvants.
Armyworm, Beet webworm, Corn earworm, Cutworm, Looper	chlorantraniliprole, MOA 28 (Coragen) 1.67 SC	3.5 to 5 fl oz	4 hrs	3	
	emamectin benzoate, MOA 6 (Proclaim) 5 SG	2.4 to 4.8 oz	12 hrs	7	
	flubendiamide, MOA 28 (Belt) 4SC	1.5 fl oz	12 hrs	1	
	indoxacarb, MOA 22 (Avaunt) 30 SG	2.5 to 3.5 oz	12 hrs	3	
	methomyl, MOA 1A (Lannate) 90 SP (Lannate) 2.4 LV	0.5 lb 1.5 pt	48 hrs	7	Air temperature should be well above 32 degrees F. Do not apply to seedlings less than 3 in. in diameter.
	methoxyfenozide, MOA 18 (Intrepid) 2 F	4 to 10 fl oz	4 hrs	1	Use low rates for early-season applications to young or small plants and 6 to 10 oz for mid- to late-season applications.
	spinetoram, MOA 5 (Radiant) 1 SC	5 to 10 fl oz	4 hrs	1	
	pyrethroid, MOA 3		12 hrs		See table 5-9B for a list of registered pyrethroids and pre-harvest intervals.
Squash (see Pumpkin, Squash)					
Sweetpotato					
Aphids, Leafhopper, Whitefly	Aphids, leafhoppers, and whiteflies are rarely a problem.				
	acetamiprid, MOA 4A (Assail) 30SG	1.5 to 4 oz	12 hrs	7	Do not make more than 4 applications per season. Do not apply more frequently than once every 7 days. Use 2.5 to 4 ounces for aphids.
	clothianidin, MOA 4A (Belay) 2.13 SC	9 to 12 oz (soil)	12 hrs	21	Soil application as an in-furrow or sidedress application. For sidedress applications, immediately cover with soil.
	flonicamid, MOA 29 (Beleaf) 50 SG	2 to 2.8 oz	12 hrs	7	
	flupyradifurone, MOA 4D (Sivanto) 200 SL	7.0 to 14.0 fl oz	4 hrs	1	For aphids and leafhopper use 7.0 to 10.5 fluid ounces, for whitefly use 10.5 to 14.0 fluid ounces.
	imidacloprid, MOA 4A (Admire Pro) 4.6 F (various) 1.6 F	1.2 fl oz 3.5 fl oz	12 hrs	7	Two applications may be needed to control heavy populations. Allow 5 to 7 days between applications.
	pymetrozine, MOA 9B (Fulfill) 50 WDG	2.75 to 5.5 oz	12 hrs	14	
	spirotetramat MOA 23 (Movento) 2 SC	4 to 5 fl oz	24 hrs	7	Will not control leafhopper. Requires surfactant.
	thiamethoxam, MOA 4A (Actara) 25 WDG	3 oz		14	Two applications of Actara may be needed to control heavy populations. Allow 7 to 10 days between applications. Do not exceed a total of 6 ounces of Actara per crop per season.
Armyworm, Looper, Corn earworm, Hornworm	chlorantraniliprole, MOA 28 (Coragen) 1.67 SC	3.5 to 5 fl oz	4 hrs	1	Foliar application only on sweetpotato.
	methoxyfenozide, MOA 18 (Intrepid) 2 F	6 to 10 fl oz	4 hrs	7	Damaging earworm infestations may occur in August or September. If significant infestations are present on foliage during harvest, larvae may feed on exposed root. Do not make more than 3 applications or apply more than 30 fl oz of Intrepid per acre per season.
	novaluron, MOA 15 (Rimon) 0.83 EC	9 to 12 fl oz	12 hrs	14	Do not make more than 2 applications per crop per season.
	spinetoram, MOA 5 (Radiant) 1 SC	6 to 8 fl oz	4 hrs	7	
Cucumber beetle (adults), Japanese beetle (adults), Tortoise beetle	Cucumber beetle larvae (diabrotica) are a serious pest of sweetpotato in LA and MS. Controlling adult cucumber beetles in areas with a history of diabrotica damage can reduce damage to roots. Foliage feeding by beetles rarely causes economic loss, and control is not warranted unless defoliation is severe.				
	pyrethroid, MOA 3		12 hrs		See table 5-9B for registered pyrethroids and pre-harvest intervals.
	carbaryl, MOA 1A (Sevin) 50 WP (Sevin) 80 S, WSB (Sevin) XLR Plus	4 lb 2.5 lb 2 qt	12 hrs	7	Treat for tortoise beetles only if significant defoliation is observed. Tortoise beetles are frequently present but rarely reach levels requiring treatment.
	spinetoram, MOA 5 (Radiant) 1 SC	6 to 8 fl oz	4 hrs	7	

127

Table 5-9. Insect Control for Commercial Vegetables

CROP Insect	Insecticide, Mode of Action Code, and Formulation	Amount of Formulation Per Acre	Restricted Entry Interval (REI)	Pre harvest Interval (PHI) (Days)	Precautions and Remarks
Sweetpotato (continued)					
Flea beetle, Wireworm, White grub	bifenthrin, MOA 3 (various) 2 EC	9.6 to 19.2 fl oz		21	Apply as broadcast, preplant application to the soil and incorporate 4 to 6 inches prior to bed formation. This use has been demonstrated to control overwintered wireworm populations and reduce damage to roots at harvest. Chlorpyrifos will not control whitefringed beetle or other grubs that attack sweetpotato. Research has shown that best control is achieved when chlorpyrifos is applied as a preplant application incorporated 4 to 6 inches deep prior to bed formation, followed by 1 or more soil-directed, incorporations of bifenthrin during routine cultivation. Bifenthrin should be directed onto each side of the bed from the drill to the middle of the furrow and incorporated with cultivating equipmet set to throw soil toward the drill. The objective is to provide a barrier of treated soil that covers the bed and furrows.Foliar sprays of various insecticides that target adults to prevent egg laying have not been shown to provide any reduction in damage to roots by wireworm larvae at harvest.
	chlorpyrifos, MOA 1B (Lorsban) 15 G (Lorsban) 4 E (Lorsban Advanced)	13.5 lb 4 pt 4 pt	24 hrs	125 (60 in NC for Lorsban Advanced only)	
	Imicacloprid (Admire Pro) 4.6SC	10.5 fl oz or 0.75 fl oz per 1,000 ft	3 days	60 days (NC Only) 125 days elsewhere	
Fruit fly (vinegar fly)	pyrethrins, MOA 3(Pyrenone)	1 gal/100,000 cu ft	12 hrs	—	Postharvest application in storage. Apply as a space fog with a mechanical or thermal generator. Do not make more than 10 applications.
Sweetpotato weevil	pyrethroid, MOA 3		12 hrs		See table 5-9B for registered pyrethroids and pre-harvest intervals.
	phosmet, MOA 1B (Imidan) 70 W	1.33 lb	5 days	7	
Thrips	spinetoram, MOA 5 (Radiant) 1 SC	6 to 8 fl oz	4 hrs	7	
Whitefringed beetle	phosmet, MOA 1B (Imidan) 70 W	1.33 lb	5 days	7	Do not make more than five applications per season. Whitefringed beetle adults are active in July and August. Do not plant in fields with a recent history of whitefringed beetles.
Tomato					
Aphid, Flea beetle	acetamiprid, MOA 4A (Assail) 30 SG	2 to 4 oz	12 hrs	7	Do not apply more than once every 7 days, and do not exceed 5 applications per season.
	clothianidin, MOA 4A (Belay) 50 WDG	4.8 to 6.4 oz (soil) 1.6 to 2.1 oz (foliar)	12 hrs	7	Soil applications at planting only.
	cyantraniliprole, MOA 28 (Verimark) 1.67SC	6.75 to 13.5 fl oz	4	1	Soil applications at planting will control aphids and flea beetles. See lable for application options.
	dimethoate 4 EC, MOA 1B	0.5 to 1 pt	48 hrs	7	Do not exceed rate with dimethoate as leaf injury may result.
	flonicamid, MOA 29 (Beleaf) 50 SG	2 to 4.8 oz	12 hrs	0	Will not control flea beetle.
	flupyradifurone, MOA 4D (Sivanto) 200 SL	7.0 to 10.5 fl oz	4 hrs	1	Will not control flea beetle.
	imidacloprid, MOA 4A Soil treatment (Admire Pro) 4.6 F (various) 2 F	7 to 10.5 fl oz 16 to 24 fl oz	12 hrs	21	For short-term protection at planting. Admire Pro may also be applied to transplants in the planthouse not more than 7 days before planting at the rate of 0.44 (4.6 F formulation) or 1 ounce (2 F formulation) per 10,000 plants. See label for soil application instructions.
	Foliar treatment (Admire Pro) 4.6 F (various) 1.6 F	1.2 fl oz 3.75 fl oz	12 hrs	0	
	pymetrozine, MOA 9B (Fulfill) 50 WDG	2.75 oz	12 hrs	0	For aphids only.
	spirotetramat, MOA 23 (Movento) 2SC	4 to 5 fl oz	24 hrs	1	Do not exceed 10 fl oz per season. Requires surfactant.
	thiamethoxam, MOA 4A (Platinum) 75 SG	1.66 to 3.67 oz	12 hrs	30	Platinum may be applied to direct-seeded crops in-furrow seeding or transplant depth, post seeding or transplant as a drench, or through drip irrigation. Do not exceed 11 ounces per acre per season. Check label for plant-back restrictions for a number of crops.
	(Actara) 25 WDG	2 to 3 oz	12 hrs	0	Actara is for foliar applications.
Armyworm	*Bacillus thuringiensis*, MOA 11A (Crymax) WDG (Dipel) 2X (Xentari)	0.5 to 1.5 lb 0.5 to 1 lb 0.5 to 1 lb	4 hrs	0	Start applications when larvae are small, and continue at 5- to 7-day intervals during periods of infestation.
	chlorantraniliprole, MOA 28 (Coragen) 1.67 SC	3.5 to 4 oz	4 hrs	1	Foliar or drip chemigation. Drip chemigation must be applied uniformly to the root zone. See label for instructions.
	emamectin benzoate, MOA 6 (Proclaim) 5 WDG	2.4 to 4.8 oz	12 hrs	7	Apply when larvae are first observed.
	flubendiamide, MOA 28 (Belt) 4SC	1.5 fl oz	12 hrs	1	
	indoxacarb, MOA 22 (Avaunt) 30 DG	3.5 oz	12 hrs	3	Do not apply more than 14 ounces of Avaunt (0.26 a.i.) per acre per crop. The minimum interval between sprays is 5 days.
	methoxyfenozide, MOA 18 (Intrepid) 2 F	4 to 10 fl oz	4 hrs	1	Use low rates for early-season applications to young or small plants and 6 to 10 ounces for mid- and late-season applications.
	novaluron, MOA 15 (Rimon) 0.83 EC	9-12 fl oz	12 hrs	1	Do not make more than 3 applications per season.
	spinetoram, MOA 5 (Radiant) 1 SC	5 to 10 fl oz	4 hrs	1	

Table 5-9. Insect Control for Commercial Vegetables

CROP Insect	Insecticide, Mode of Action Code, and Formulation	Amount of Formulation Per Acre	Restricted Entry Interval (REI)	Pre harvest Interval (PHI) (Days)	Precautions and Remarks
Tomato (continued)					
Colorado potato beetle	acetamiprid, MOA 4A (Assail) 30 SG	1.5 to 2.5 oz	12 hrs	7	
	chlorantraniliprole, MOA 28 (Coragen) 1.67 SC	3.5 to 5 fl oz	4 hrs	1	Foliar or drip chemigation. Drip chemigation must be applied uniformly to the root zone. See label for instructions.
	cyantraniliprole, MOA 28 (Verimark) 1,67SC (Exirel) 0.83SE	5 to 10 fl oz 7 to 13.5 fl oz	4 hrs 12 hrs	1 1	Apply Verimark to soil via drip irrigation or soil injection. Exirel is for foliar application.
	imidacloprid, MOA 4A Soil treatment (Admire Pro) 4.6 F (various) 2 F	7 fl oz 16 fl oz	12 hrs	21	Use Admire Pro for soil or transplant drench treatment and 1.6 F formulation for foliar applications.
	Foliar treatment (Admire Pro) 4.6 F (various) 1.6 F	1.2 fl oz 3.75 fl oz	12 hrs	0	
	spinetoram, MOA 5 (Radiant)1SC	5 to 10 fl oz	4 hrs	1	
	thiamethoxam, MOA 4A (Platinum) 75 SG	1.66 to 3.67 oz	12 hrs	30	Platinum may be applied to direct-seeded crops in-furrow seeding or transplant depth, post seeding or transplant as a drench, or through drip irrigation. Do not exceed 11 oz per acre per season of Platinum. Check label for plant-back restrictions for a number of crops.
	(Actara) 25 WDG	2 to 3 oz	12 hrs	0	Actara is for foliar applications.
Cabbage looper, Hornworm, Tomato fruitworm, Pinworm	*Bacillus thuringiensis*, MOA 11A (Dipel) DF, MOA (Crymax) WDG	0.5 to 1 lb 0.5 to 1.5 lb	4 hrs	0	
	pyrethroid, MOA				See table 5-9B for a list of registered pyrethroids and pre-harvest intervals.
	chlorantraniliprole, MOA 28 (Coragen) 1.67 SC	3.5 to 5 fl oz	4 hrs	1	Foliar or drip chemigation. Drip chemigation must be applied uniformly to the root zone. See label for instructions.
	cyantraniliprole, MOA 28 (Verimark) 1,67SC	5 to 10 fl oz	4 hrs	1	Verimark is for soil application only. Applications made at planting and/or via drip chemigation after planting. See label for application options.
	(Exirel) 0.83SE	7 to 13.5 fl oz	12 hrs	1	Exirel is for foliar application only.
	emamectin benzoate, MOA 6 (Proclaim) 5 WDG	2.4 to 4.8 oz	12 hrs	7	
	flubendiamide, MOA 28 (Belt) 4 SC	1.5 fl oz	12 hrs	1	
	indoxacarb, MOA 22 (Avaunt) 30 WDG	2.5 to 3.5 oz	12 hrs	3	Do not apply more than 14 ounces of Avaunt (0.26 pound a.i.) per acre per crop. The minimum interval between sprays is 5 days.
	methomyl, MOA 1A (Lannate) 2.4 LV	1.5 to 3 pt	48 hrs	1	Methomyl may induce leafminer infestation.
	methoxyfenozide, MOA 18 (Intrepid) 2 F	4 to 10 fl oz	4 hrs	1	Use low rates for early-season applications to young or small plants and 6 to 10 ounces for mid- and late-season applications. Intrepid provides suppression of pinworm only.
	novaluron, MOA 15 (Rimon) 0.83 EC	9 to12 fl oz	12 hrs	1	Do not make more than 3 applications per season.
	spinetoram, MOA 5 (Radiant) 1 SC	5 to 10 fl oz	4 hrs	1	
Cutworm	pyrethroid, MOA 3		12 hrs		See table 5-9B for a list of registered pyrethroids and pre-harvest intervals.
Leafminer	abamectin, MOA 6 (Agri-Mek) 0.7 SC	1.75 to 3.5 fl oz	12 hrs	7	Do not exceed 48 fluid ounces per acre per season, or more than two sequential applications.
	chlorantraniliprole, MOA 28 (Coragen) 1.67 SC	5 to 7.5 fl oz	4 hrs	1	Foliar or soil chemigation. Drip chemigation must be applied uniformly to the root zone. See label for soil application instructions.
	cryomazine, MOA 17 (Trigard) 75 WP	2.66 oz	12 hrs	0	See label for plant-back restrictions.
	spinetoram, MOA 5 (Radiant) 1 SC	6 to 8 fl oz	4 hrs	1	Do not exceed 29 fl oz per acre per season.
Spider mite	abamectin, MOA 6 (Agri-Mek) 0.7 SC	1.75 to 3.5 fl oz	12 hrs	7	Do not exceed 48 fluid ounces per acre per season, or more than two sequential applications.
	acequinocyl, MOA 29 (Kanemite) 15SC	31 fl oz	12 hrs	1	The use of a surfactant/adjuvant with Kanemite on tomatoes is prohibited.
	bifenazate, MOA 20D (Acramite) 50 WS	0.75 to 1.0 lb	12 hrs	3	Do not make more than one application per season.
	cyflumetofen, MOA 25 (Nealta) 1.67 SC	13.7 fl oz	12 hrs	3	Do not make more than one application before using an effective miticide with a different mode of action.
	fenpyroximate MOA 21 (Portal) 0.4EC	2 pts	12 hrs	3	Do not make more than two applications per season.
	spiromesifen, MOA 23 (Oberon) 2 SG	7 to 8.5 fl oz	12 hrs	7	Do not exceed 3 applications per season.

INSECT CONTROL

Table 5-9. Insect Control for Commercial Vegetables

CROP Insect	Insecticide, Mode of Action Code, and Formulation	Amount of Formulation Per Acre	Restricted Entry Interval (REI)	Pre harvest Interval (PHI) (Days)	Precautions and Remarks
Tomato (continued)					
Stink bug	pyrethroid, MOA 3		12 hrs		See table 5-9B for a list of registered pyrethroids and pre-harvest intervals.
	Dinotefuran, MOA 4A Foliar treatment (Venom) 70 SG (Scorpion) 35SL	1 to 4 oz 2 to 7 fl oz	12 hrs	1	
	Soil treatment (Venom) 70 SG (Scorpion) 35SL	5 to 6 oz 9 to 10.5 fl oz		21	
	thiamethoxam, MOA 4A(Actara) 25 WDG	3 to 5.5 oz	12 hrs	0	Do not exceed 11 ounces Actara per acre per season.
Thrips	dimethoate 4 EC, MOA 1B	0.5 to 1 pt	48 hrs	7	
	Dinotefuran, MOA 4A Foliar treatment (Venom) 70 SG (Scorpion) 35SL	1 to 4 oz 2 to 7 fl oz	12 hrs	1	See comments under Whitefly for application instructions and restrictions.
	Soil treatment (Venom) 70 SG (Scorpion) 35SL	5 to 6 oz 9 to 10.5 fl oz		21	
	methomyl, MOA 1A (Lannate) 2.4 LV	1.5 to 3 pt	48 hrs	1	On foliage as needed.
	novaluron, MOA 15 (Rimon) 0.83 EC	9 to12 fl oz	12 hrs	1	Do not make more than 3 applications per season.
	spinetoram, MOA 5 (Radiant) 1 SC	6 to 10 fl oz	4 hrs	1	Will control thrips on foliage, not in flowers.
Whitefly	For resistance management of whiteflies, do not follow a soil application of a neonicotinoid (MOA group 4A) with a foliar application of any neonicotinoid.				
	acetamiprid, MOA 4A (Assail) 30 SG	2.5 to 4 oz	12 hrs	7	Do not apply more than once every 7 days, and do not exceed 5 applications per season.
	buprofezin, MOA 16 (Courier) 40 SC	9 to 13.6 fl oz	12 hrs	1	Use sufficient water to ensure good coverage. Do not apply more than twice per crop cycle, and allow 28 days between applications.
	chlorantraniliprole, MOA 28 (Coragen) 1.67 SC	5 to 7.5 fl oz	4 hrs	1	Foliar or soil application. Drip chemigation must be applied uniformly to the root zone. See label for soil application instructions.
	cyantraniliprole, MOA 28 (Verimark) 1,67SC	6.75 to 13.5 fl oz	4 hrs	1	Apply Verimark to at planting and/or later via drip irrigation or soil injection. See label for application options.
	(Exirel) 0.83SE	13.5 to 20.5 fl oz	12 hrs	1	Exirel is for foliar application..
	dinotefuran MOA 4A Soil treatment (Venom) 70 SG (Scorpion) 35 SL	5 to 6 oz 9 to 10.5 fl oz	12 hrs	21	Soil applications of Venom or Scorpion may be made in a narrow band under the plant row as a post transplant drench, as a soil incorporated sidedress after plants are established, or in drip irrigation water. See label for instructions.
	Foliar treatment (Venom) 70 SG (Scorpion) 35 SL	1 to 4 oz 2 to 7 fl oz		1	
	imidacloprid, MOA 4A (Admire Pro) 4.6 F (various) 2 F	16 to 24 fl oz 7 to 10.5 fl oz	12 hrs	21	Apply through a drip irrigation system or as a transplant drench with sufficient water to reach root zone. As a sidedress, apply 2 to 4 inches to the side of the row and incorporate 1 or more in. Residual activity will increase with increasing rates applied. Use higher rate for late-season or continuous infestations. Trickle irrigation applications will also control aphids and stinkbugs.
	pyriproxyfen, MOA 7C (Knack) 0.86EC	8 to 10 fl oz	12 hrs	1	Do not apply more than two applications per growing season, and do not make applications closer than 14 days.
	spiromesifen, MOA 23 (Oberon) 2 SC	7 to 8.5 fl oz	12 hrs	7	Do not make more than 3 applications per season.
	spirotetramat, MOA 23 (Movento) 2SC	4 to 5 fl oz	24 hrs	1	Do not exceed 10 fluid ounces per season. Requires surfactant.
	thiamethoxam, MOA 4A (Platinum) 75 SG	1.66 to 3.67 oz	12 hrs	30	Platinum may be applied to direct-seeded crops in-furrow seeding or transplant depth, post seeding or transplant as a drench, or through drip irrigation. Do not exceed 11 ounces per acre per season of Platinum. Check label for plant-back restrictions for a number of crops.Actara is for foliar applications.
	(Actara) 25 WDG	3 to 5.5 oz	12 hrs	0	
Wireworm	diazinon, MOA 1B (Diazinon) AG 500 or 50 WP	2 to 4 qt	48 hrs	—	Broadcast before planting and incorporate. Wireworms may be a problem in fields previously in pasture, corn, or soybean.
Turnip					
Aphid, Flea beetle	clothianidin, MOA 4A (Belay) 50 WDG	4.8 to 6.4 oz (soil) 1.6 to 2.1 oz (foliar)	12 hrs	7 (Foliar)	Soil application as in in-furrow, side dress application, seed or transplant drench, or chemigation. See label for application instructions.
	cyantraniliprole, MOA 28 (Verimark) 1,67SC	6.75 to 13.5 fl oz	4 hrs	4	Soil applications made at planting only. See label for application options.

Table 5-9. Insect Control for Commercial Vegetables

CROP Insect	Insecticide, Mode of Action Code, and Formulation	Amount of Formulation Per Acre	Restricted Entry Interval (REI)	Pre harvest Interval (PHI) (Days)	Precautions and Remarks
Turnip (continued)					
Aphid, Flea beetle (continued)	dimethoate 4 EC, MOA 1B	0.5 pt	48 hrs	14	
	flonicamid, MOA 29 (Beleaf) 50SG	2 to 2.8 oz	12 hrs	0	
	flupyradifurone, MOA 4D (Sivanto) 200 SL	7.0 to 10.5 fl oz	4 hrs	7	Will not control flea beetle
	imidacloprid, MOA 4A Soil treatment (Admire Pro) 4.6 F (various) 2 F	4.4 to 10.5 fl oz 10 to 24 fl oz	12 hrs	21	See label for soil application instructions.
	Foliar treatment (Admire Pro) 4.6 F (various) 1.6 F	1.2 fl oz 3.8 fl oz	12 hrs	7	
	pymetrozine, MOA 9B (Fulfill) 50 WDG	2.75 oz	12 hrs	7	Will not control flea beetle.
	thiamethoxam, MOA 4A (Platinum) 75 SG	1.7 to 4.01 oz	12 hrs	Apply at plant	Platinum is for soil application and Actara for foliar application..
	(Actara) 25 WDG	1.5 to 3 oz	12 hrs	7	
Harlequin bug, Vegetable weevil, Yellow margined leaf beetle	clothianidin, MOA 4A (Belay) 50 WDG	4.8 to 6.0 oz (soil) 1.6 to 2.1 oz (foliar)	12 hrs	7 (Foliar)	Soil application as in in-furrow, side dress application, seed or transplant drench, or chemigation. See label for application instructions.
	Imidacloprid, MOA 4A Soil treatment (Admire Pro) 4.6F (Various) 2F	4.4 to 10.5 fl oz 10 to 24 fl oz	12 hrs	21	Soil applications of imidacloprid will not control harlequin bug past 20 days after application.
	Foliar treatment (Admire Pro) 4.6F (Various) 2F	1.2 fl oz 2.8 fl oz		7	
	thiamethoxam, MOA 4A (Platinum) 75 SG (Actara) 25 WDG	1.7 to 4.01 oz 1.5 to 3 oz	12 hrs	Apply at plant 7	Platinum is for soil application and Actara for foliar application. .
	pyrethroid, MOA 3		12 hrs		See table 5-9B for a list of registered pyrethroids and pre-harvest intervals.
Cabbage looper, Diamondback moth	Insecticide-resistant populations, widespread in the Southeast, may not be controlled with some registered insecticides. To manage resistance, avoid transplants from Georgia and Florida, and avoid the repeated use of the same materials for extended periods of time. Repeated use of pyrethroid insecticides often aggravates diamondback moth problems. Do not allow populations to increase to large densities before treatments are initiated.				
	Bacillus thuringiensis, MOA 11A (Crymax) WDG (Dipel) 2 X (Dipel) 4 L (Xentari) WDG	0.5 to 1.5 lb 8 oz 1 to 2 pt 0.5 to 1 lb	4 hrs	0	On foliage every 7 days as needed.
	chlorantraniliprole, MOA 28 (Coragen)	3.5 to 5.0 fl oz	4 hrs	1	For turnip greens or root turnips.
	cyantraniliprole, MOA 28 (Verimark) 1,67SC (Exirel) 0.83SE	5 to 10 fl oz 7 to 13.5 fl oz	4 hrs 12 hrs	1 1	Verimark and Exirel are for greens only, not root turnips Verimark is for soil application only. Applications made at planting and/or later via drip chemigation. See label for application options. Exirel is for foliar application only.
	emamectin benzoate, MOA 6 (Proclaim) 5 WDG	2.4 to 4.8 oz	12 hrs	14	For turnip greens only.
	flubendiamide, MOA 28 (Belt) 4 SC	2 to 2.4 fl oz	12 hrs	8	For turnip greens only
	indoxacarb, MOA 22 (Avaunt) 30 WDG	2.5 to 3.5 oz	12 hrs	3	Avaunt may be applied only to turnip greens, not root turnips.
	spinetoram, MOA 5 (Radiant) 1 SC	3 to 6 fl oz	4 hrs	1	
Root maggot	chlorpyrifos, MOA 1B (Lorsban) 4 E (Lorsban) 75 WDG	1 to 2 pt 1.1 to 1.8 oz/1,000 ft row	24 hrs	21	Irrigation or rainfall after application will enhance activity.
Watermelon					
Insecticide applications in cucurbits should be made in late evening to protect pollinating insects. Refer to the section of this chapter on Reducing the Risk of Pesticide Poisoning to Honey Bees for more information about protecting pollinators.					
Aphid	acetamiprid, MOA 4A (Assail) 30 SG	2.5 to 4 oz	12 hrs	0	
	clothianidin, MOA 4A (Belay) 50WDG	4.8 to 6.4oz (soil); 1.6 to 2.1oz (foliar)	12 hrs	At planting 7 (foliar)	Soil application at planting only.
	dimethoate, MOA 1B 2 E 2.67 E	2 pt 1.5 pt	48 hrs	3	

Table 5-9. Insect Control for Commercial Vegetables

CROP Insect	Insecticide, Mode of Action Code, and Formulation	Amount of Formulation Per Acre	Restricted Entry Interval (REI)	Pre harvest Interval (PHI) (Days)	Precautions and Remarks
Watermelon (continued)					
Insecticide applications in cucurbits should be made in late evening to protect pollinating insects. Refer to the section of this chapter on Reducing the Risk of Pesticide Poisoning to Honey Bees for more information about protecting pollinators.					
Aphid (continued)	flonicamid, MOA 29 (Beleaf) 50 SG	2 to 2.8 oz	12 hrs	0	
	flupyradifurone, MOA 4D (Sivanto) 200 SL	10.5 to 12.0 fl oz	4 hrs	1	
	imidacloprid, MOA 4A (Admire Pro) 4.6 F (various) 2 F	7 to 10.5 fl oz 16 to 24 fl oz	12 hrs	21	Admire Pro must be applied to soil. May apply preplant; at planting; as a post-seeding drench, transplant water drench, or hill drench; subsurface sidedress or by chemigation using low-pressure drip or trickle irrigation. See label for information on approved application method. Will also control cucumber beetles and whiteflies.
	pymetrozine, MOA 9B (Fulfill) 50 WDG	2.75 oz	12 hrs	0	Apply before populations reach damaging levels. Do not exceed 5.5 oz per acre per season.
	thiamethoxam, MOA 4A (Platinum) 75 SG	1.66 to 3.67 oz	12 hrs	30	Platinum may be applied to direct-seeded crops in-furrow seeding or transplant depth, post seeding or transplant as a drench, or through drip irrigation. Do not exceed 11 oz per acre per season of Platinum. Check label for plant-back restrictions for a number of crops.
	(Actara) 25 WDG	1.5 to 3 oz	12 hrs	0	Actara is for foliar applications.
Armyworm, Cabbage looper	Bacillus thuringiensis, MOA 11A (Xantari) DF (Dipel) DF	0.5 to 2 lb 0.5 to 2 lb	4 hrs	0	On foliage as needed.
	chlorantraniliprole, MOA 28 (Coragen) 1.67 SC	3.5 to 5 fl oz	4 hrs	1	Foliar or drip chemigation. Drip chemigation must be applied uniformly to the root zone. See label for instructions.
	cyantraniliprole, MOA 28 (Verimark) 1,67SC	5 to 10 fl oz	4 hrs	1	Verimark is for soil application only. Applications made at planting and/or later via drip chemigation. See label for application options. Verimark is for foliar application only.
	(Exirel) 0.83SE	7 to 13.5 fl oz	12 hrs	1	
	flubendiamide, MOA 28 (Belt) 4 SC	1.5 fl oz	12 hrs	1	
	methoxyfenozide, MOA 18 (Intrepid) 2 F	4 to 10 fl oz	4 hrs	3	Use higher rates against large larvae.
	pyrethroid, MOA				See table 5-9B for a list of registered pyrethroids and pre-harvest intervals.
	spinetoram, MOA 5 (Radiant) 1 SC	5 to 10 fl oz	4 hrs	3	
Cucumber beetle	acetamiprid, MOA 4A (Assail) 30 SG	2.5 to 5.3 oz	12 hrs	0	
	Dinotefuran, MOA 4A Foliar treatment (Venom) 70 SG (Scorpion) 35SL	1 to 4 oz 2 to 7 fl oz	12 hrs	1	Dinotefuran may be applied foliarly or to the soil. See labels for soil application instructions.
	Soil treatment (Venom) 70 SG (Scorpion) 35SL	5 to 6 oz 9 to 10.5 fl oz		21	
	clothianidin, MOA 4A (Belay) 50WDG	4.8 to 6.4 oz (soil); 1.6 to 2.1oz (foliar)	12 hrs	At planting 21 (foliar)	Soil application at planting only.
	imidacloprid, MOA 4A (Admire Pro) 4.6 F (various) 2 F	7 to 10.5 fl oz 16 to 24 fl oz	12 hrs	21	Admire Pro must be applied to the soil. May be applied preplant; at planting; as a post-seeding drench, transplant water drench, or hill drench; subsurface sidedress or by chemigation using low-pressure drip or trickle irrigation. See label for information on approved application method. Will also control aphids and whiteflies.
	pyrethroid, MOA 3		12 hrs		See table 5-9B for registered pyrethroids and pre-harvest intervals.
	thiamethoxam, MOA 4A Soil Treatment (Platinum) 75 SG	1.66 to 3.67 oz	12 hrs	30	
	Foliar treatment (Actara) 25 WDG	3 to 5.5 oz	12 hrs	0	
Cutworm	pyrethroid, MOA 3		12 hrs		See table 5-9B for registered pyrethroids and pre-harvest intervals.
Thrips	dimethoate, MOA 1B (various brands and formulations)	See label	48 hrs	3	
	dinotefuran, MOA 4A (Venom) 70 SG (Scorpion) 35 SL	1 to 4 oz 2 to 7 fl oz	12 hrs	1	
	spinetoram, MOA 5 (Radiant) 1 SC	6 to 10 fl oz	4 hrs	3	

Table 5-9. Insect Control for Commercial Vegetables

CROP Insect	Insecticide, Mode of Action Code, and Formulation	Amount of Formulation Per Acre	Restricted Entry Interval (REI)	Pre harvest Interval (PHI) (Days)	Precautions and Remarks
Watermelon (continued)					
Insecticide applications in cucurbits should be made in late evening to protect pollinating insects. Refer to the section of this chapter on Reducing the Risk of Pesticide Poisoning to Honey Bees for more information about protecting pollinators.					
Leafminer	abamectin, MOA 6 (Agri-Mek) 0.7 SC	1.75 to 3.5 fl oz	12 hrs	7	
	chlorantraniliprole, MOA 28 (Coragen) 1.67 SC	3.5 to 5 fl oz	4 hrs	1	Foliar or drip chemigation. Drip chemigation must be applied uniformly to the root zone. See label for instructions.
	cyromazine, MOA 17 (Trigard) 75 WP	2.66 oz	12 hrs	0	
	spinetoram, MOA 5 (Radiant) 1 SC	8 fl oz	4 hrs	3	
Spider mite	abamectin, MOA 6 (Agri-Mek) 0.7 SC	1.75 to 3.5 fl oz	12 hrs	7	Do not feed or graze vines.
	bifenazate, MOA 20D (Acramite) 50 WS	0.75 to 1.0 lb	12 hrs	3	Do not make more than one application per season.
	etoxazole, MOA 10B (Zeal) 72 WSP	2 to 3 oz	12 hrs	7	
	fenpyroximate MOA 21 (Portal) 0.4EC	2 pts	12 hrs	3	Do not make more than two applications per season.
	spiromesifen, MOA 23 (Oberon) 2 SG	7 to 8.5 fl oz	12 hrs	7	Do not exceed 3 applications per season.
Squash bug, Leaffooted bug	acetamiprid, MOA 4A (Assail) 30 SG	5.3 oz	12 hrs	0	
	dinotefuran MOA 4A (Venom) 70 SG (Scorpion) 35 SL	1 to 4 oz 2 to 7 fl oz	12 hrs	1	Foliar use only.
	pyrethroid, MOA 3		12 hrs		See table 5-9B for a list of registered pyrethroids and pre-harvest intervals.
Whitefly	acetamiprid, MOA 4A (Assail) 30 SG	2.5 to 5.3 oz	12 hrs	0	
	buprofezin, MOA 16 (Courier) 40 SC	9 to 12.5 fl oz	12 hrs	1	Allow at least 7 days between applications.
	cyantraniliprole, MOA 28 (Verimark) 1,67SC	6.75 to 13.5 fl oz	4 hrs	1	Apply Verimark to at planting and/or later via drip irrigation or soil injection. See label for application options.
	(Exirel) 0.83SE	13.5 to 20.5 fl oz	12 hrs	1	Exirel is for foliar application.
	flupyradifurone, MOA 4D (Sivanto) 200 SL	10.5 to 14.0 fl oz	4 hrs	1	
	pyriproxyfen, MOA 7D (Knack) 0.86EC	8 to 10 fl oz	12 hrs	1	Do not make more than two applications per growing season.
	Dinotefuran, MOA 4A Foliar treatment (Venom) 70 SG (Scorpion) 35SL	1 to 4 oz 2 to 7 fl oz	12 hrs	1	Do not follow soil applications with foliar applications of any neonicotinoid insecticide. Use only one application method. Do not apply more than 6 ounces per acre per season using soil applications. Soil applications may be applied by a narrow band below or above the seed line at planting, by a post-seeding or transplant drench with sufficient water to ensure incorporation into the soil, or by drip irrigation.
	Soil treatment (Venom) 70 SG (Scorpion) 35SL	5 to 6 oz 9 to 10.5 fl oz		21	
	imidacloprid, MOA 4A (Admire Pro) 4.6 F (various) 2 F	7 to 10.5 fl oz 16 to 24 fl oz	12 hrs	21	Do not follow soil applications with foliar applications of any neonicotinoid insecticides. Must be applied to the soil. Do not use a foliar application of any neonicotinoid insecticide if using Admire Pro. May apply preplant; at planting; as a post-seeding drench, transplant water drench, or hill drench; subsurface sidedress or by chemigation using low-pressure drip or trickle irrigation. See label for information on approved application method. Will also control aphids and cucumber beetles.
	spiromesifen, MOA 23 (Oberon) 70 SC	7 to 8.5 fl oz	12 hrs	7	
	thiamethoxam, MOA 4A (Platinum) 75 SG	1.66 to 3.67 oz	12 hrs	30	Apply Platinum to direct-seeded crops in-furrow at seed or transplant depth, postseeding or transplant as a drench, or through drip irrigation. Do not exceed 11 ounces per acre per season. Check label for plant-back restrictions for a number of crops.
	(Actara) 25 WDG	3 to 5.5 oz	12 hrs	0	Actara is for foliar applications. Do not use a foliar application of any neonicotinoid insecticide if using Admire Pro.
Wireworm	diazinon, MOA 1B (Diazinon) AG 500	3 to 4 qt	3 days	—	Broadcast on soil before planting and thoroughly work into upper 6 inches.

133

INSECT CONTROL

Relative Effectiveness of Insecticides and Miticides for Insect and Mite Control on Vegetables

Table 5-9A. Relative Effectiveness of Insecticides and Miticides for Insect and Mite Control on Vegetables

J. F. Walgenbach, Entomology Extension, and G. G. Kennedy, Entomology Research

Not all insecticides listed are registered on all vegetable crops. Refer to label before applying to a specific crop. Ratings are based on a consensus of vegetable entomologists in the southeastern United States. Table continued on following page.

(E = very effective; G = effective; F = somewhat effective; I = ineffective or insufficient data)

Chemical class (IRAC)	Common name	Example Product	Flea Beetle	Colorado potato beetle*	Cucumber beetles	Corn earworm*	European corn borer	Fall armyworm	Cabbage looper	Imported cabbageworm	Diamondback moth*	Squash vine borer
1A	carbaryl	Sevin	E	F	G	F	G	F	F	G	F	F
	methomyl	Lannate	F	I	I	G	G	G	G	G	G	I
	oxamyl	Vydate	F	F	F	I	I	I	I	I	I	I
1B	malathion	Malathion	G	F	G	F	F	F	F	G	F	F
	chlorpyrifos	Lorsban	I	I	I	F	F	F	F	G	F	I
	acephate	Orthene	I	I	I	F	E	G	F	G	I	I
	diazinon	Diazinon	I	I	I	I	I	I	I	I	I	I
	dimethoate	Dimethoate	G	I	F	I	I	I	I	I	I	I
3	permethrin	Pounce	G	F	G	G	G	F	G	E	F	E
	alpha cypermethrin	Fastac	E	F	E	G	E	G	G	E	F	E
	zeta cypermethrin	Mustang Max	E	F	E	G	E	G	G	E	F	E
	cyfluthrin	Baythroid/renounce	G	F	G	G	G	F	G	E	F	E
	lambda cyhalothrin	Karate	E	F	E	G	E	G	G	E	F	E
	esfenvalerate	Asana XL	G	G	G	G	G	F	G	E	F	G
	gamma cyhalothrin	Proaxis	E	F	E	G	E	G	G	E	F	E
	fenpropathrin	Danitol	G	I	G	G	G	F	F	E	F	G
	bifenthrin	Brigade	E	F	E	G	G	F	F	E	F	E
4A	imidacloprid	Admire	F	G	E	I	I	I	I	I	I	I
	acetamiprid	Assail	G	E	G	I	I	I	I	I	I	F
	clothianidin	Belay	E	E	G	I	I	I	I	I	I	I
	thiamethoxam	Platinum/Actara	E	G	G	I	I	I	I	I	I	I
	dinotefuran	Venom/Scorpion	E	E	G	I	I	I	I	I	I	I
4D	flupyradifurone	Savinto	I	I	I	I	I	I	I	I	I	I
5	spinosad	Blackhawk/Entrust	I	E	I	G	G	G	G	E	G	G
	spinetoram	Radiant	I	E	I	E	E	G	G	E	G	G
6	emamectin benzoate	Proclaim	I	I	I	G	G	G	E	E	E	G
	abamectin	AgriMek	I	E	I	I	I	I	I	I	I	I
7C	pyriproxyfen	Knack/Distance	I	I	I	I	I	I	I	I	I	I
9B	pymetrozine	Fulfill	I	I	I	I	I	I	I	I	I	I
29	flonicamid	Beleaf	I	I	I	I	I	I	I	I	I	I
10	etoxazole	Zeal	I	I	I	I	I	I	I	I	I	I
11	Bt	Dipel, various	I	I	I	F	F	F	G	E	G	F
15	novaluron	Rimon	I	E	I	E	E	E	G	E	F	G
16	buprofezin	Courier	I	I	I	I	I	I	I	I	I	I
17	cyromazine	Trigard	I	G	I	I	I	I	I	I	I	I
18	methoxyfenozide	Intrepid	I	I	I	G	G	E	E	E	F	G
20B	acequinocyl	Kanemite	I	I	I	I	I	I	I	I	I	I
21A	fenpyroximate	Portal	I	I	I	I	I	I	I	I	I	I
	tolfenpyrad	Torac	G	I	I	F	F	F	F	G	I	I
22A	indoxacarb	Avaunt	F	G	F	E	G	G	E	E	G	G
23	spiromesifen	Oberon	I	I	I	I	I	I	I	I	I	I
	spirotetramat	Movento	I	I	I	I	I	I	I	I	I	I
25	cyflumetofen	Nealta	I	I	I	I	I	I	I	I	I	I
28	chlorantraniliprole	Coragen	I	E	I	E	E	E	E	E	E	G
	cyantraniliprole	Verimark/Exirel	G	E	I	E	E	E	E	E	E	G
	flubendiamide	Belt	I	G	I	E	E	G	E	E	E	G
20D	bifenazate	Acramite	I	I	I	I	I	I	I	I	I	I

Table 5-9A. Relative Effectiveness of Insecticides and Miticides for Insect and Mite Control on Vegetables (continued)

Chemical class (IRAC)	Common name	Example Product	Beet armyworm*	Stinkbugs/ Harlequin bug	Squash bug	Aphids*	Thrips	Western Flower Thrips*	Leafminer	Maggots	Whiteflies*	Cutworms	Wireworms	White grubs	Spider mites*
1A	carbaryl	Sevin	I	I	I	I	F	I	I	I	I	F	I	I	I
	methomyl	Lannate	F	G	G	F	E	G	F	I	F	I	I	I	I
	oxamyl	Vydate	I	F	F	G	G	F	I	I	F	I	I	I	==
1B	malathion	Malathion	I	F	F	F	F	I	I	F	I	F	I	I	I
	chlorpyrifos	Lorsban	I	I	I	I	F	I	I	E	I	G	G	G	I
	acephate	Orthene	I	I	I	G	G	I	F	I	F	G	I	I	I
	diazinon	Diazinon	I		I	I	I	I	I	G	I	F	G	F	I
	dimethoate	Dimethoate	I	G	F	E	E	F	G	I	I	I	I	I	I
3	permethrin	Pounce	I	F	G	F	F	I	F	I	I	G	I	I	I
	zeta cypermethrin	Mustang Max	I	G	E	F	F	I	F	I	I	E	I	I	I
	cyfluthrin	Baythroid/Renounce	I	G	E	F	F	I	F	I	I	E	I	I	I
	lambda cyhalothrin	Karate	I	G	E	F	F	I	F	I	I	E	I	I	I
	esfenvalerate	Asana XL	I	F	G	F	F	I	F	I	I	G	I	I	I
	gamma cyhalothrin	Proaxis	I	E	E	F	F	I	F	I	I	E	I	I	I
	fenpropathrin	Danitol	I	E	E	F	F	I	F	I	F	G	I	I	F
	bifenthrin	Brigade	I	E	E	F	G	I	F	F	F	E	G	F	F
4A	imidacloprid	Admire	I	F	G	E	G	I	I	G	G	I	F	G	I
	acetamiprid	Assail	I	F	F	E	G	I	I	I	G	I	I	I	I
	clothianidin	Belay	I	I	I	G	I	I	F	G	I	I	F	G	I
	thiamethoxam	Platinum/Actara	I	G	G	E	F	I	F	G	G	I	F	F	I
	dinotefuran	Venom/Scorpion	I	G	G	F	G	I	F	I	G	I	I	I	I
4D	flupyradifurone	Savinto	I	I	I	E	I	I	I	I	G	I	I	E	
5	spinosad	Blackhawk/Entrust	G	I	I	I	G	G	E	I	I	F	I	I	I
	spinetoram	Radiant	G	I	I	I	E	G	E	I	I	F	I	I	I
6	emamectin benzoate	Proclaim	E	I	I	I	I	I	F	I	I	F	I	I	I
	abamectin	AgriMek	I	I	I	I	G	F	E	I	I	I	I	I	E
7C	pyriproxyfen	Knack/Distance	I	I	I	I	I	I	I	I	G	I	I	I	I
9B	pymetrozine	Fulfill	I	I	I	E	I	I	I	I	F	I	I	I	I
29	flonicamid	Beleaf	I	I	I	E	I	I	I	I	I	I	I	I	I
10	etoxazole	Zeal	I	I	I	I	I	I	I	I	I	I	I	I	G
11	Bt	Dipel, various	F	I	I	I	I	I	I	I	I	I	I	I	I
15	novaluron	Rimon	E	F	F	I	G	G	G	I	G	I	I	I	I
16	buprofezin	Courier	I	I	I	I	I	I	I	I	G	I	I	I	I
17	cyromazine	Trigard	I	I	I	I	I	I	E	I	I	I	I	I	I
18	methoxyfenozide	Intrepid	E	I	I	I	I	I	I	I	I	I	I	I	I
20B	acequinocyl	Kanemite	I	I	I	I	I	I	I	I	I	I	I	I	E
21A	fenpyroximate	Portal	I	I	I	I	I	I	I	I	I	I	I	I	G
	tolfenpyrad	Torac	F	I	G	G	F	I	I	I	F	I	I	I	I
22	indoxacarb	Avaunt	E	I	I	I	I	I	F	I	I	F	I	I	I
23	spiromesifen	Oberon	I	I	I	I	I	I	I	I	G	I	I	I	G
	spirotetramat	Movento	I	I	I	E	I	I	I	I	G	I	I	I	I
25	cyflumetofen	Nealta	I	I	I	I	I	I	I	I	I	I	I	I	G
28	chlorantraniliprole	Coragen	E	I	I	I	F	I	E	I	G	I	I	I	I
	cyantraniliprole	Verimark/Exirel	E	I	I	G	F	F	E	I	G	I	I	I	I
	flubendiamide	Belt	E	I	I	I	I	I	F	I	I	I	I	I	I
20D	bifenazate	Acramite	I	I	I	I	I	I	I	I	I	I	I	I	E

* Denotes that insecticide-resistant populations may occur in some areas and can affect the performance of insecticides.

135

INSECT CONTROL

Preharvest Intervals for Pyrethroid Insecticides in Vegetable Crops

Table 5-9B. Preharvest Intervals (in Days) for Pyrethroid Insecticides in Vegetable Crops

See Table 5-9A to compare relative efficacy of these products against specific insect pests. Read the pesticide label for specific rates and application instructions.

		Common Name/Example Product (Restrictricted Entry Interval – REI)										
		alpha cypermethrin Fastac (12 hrs)	beta cyfluthrin Baythroid XL (12 hrs)	bifenthrin Brigade (12 hrs)	cypermethrin Various brands (12 hrs)	cyfluthrin Tombstone (12 hrs)	esfenvalerate Asana XL (12 hrs)	fenpropathrin Danitol (24 hrs)	gamma cyhalothrin Proaxis (24 hrs)	lambda cyhalothrin Karate/Warrior (24 hrs)	permethrin Pounce (12 hrs)	zeta cypermethrin Mustang Max (12 hrs)
	Asparagus	NR	NR	NR	NR	NR	NR	NR	NR	NR	1	NR
Bulb Vegetables	Onions, Green	NR	NR	NR	7	NR	NR	NR	NR	NR	NR	7
	Onions, Dry Bulb	NR	NR	NR	7	NR	NR	NR	14	14	1	7
Brassica Leafy Vegetables	Broccoli, Brussels Sprout, Cabbage, Cauliflower, Kohlrabi	1	0	7	1	0	3	7	1	1	1	1
	Collard, Mustard Green	1	0	7	1	0	7	NR	NR	NR	1	1
Cereal Corn	Sweet Corn	3	0	1	NR	0	1	NR	1	1	1	3
Cucurbits	Cantaloupe, Watermelon	1	0	3	NR	0	3	7	NR	1	0	1
	Cucumber, Pumpkin, Summer Squash, Winter Squash	1	0	3	NR	0	3	7	NR	1	0	1
Fruiting Vegetables	Eggplant, Pepper	1	7	7	NR	0	7	3	5	5	3	1
	Tomato	1	0	1	NR	7	1	3	5	5	0	1
	Okra	1	NR	7	NR	NR	NR	NR	NR	NR	NR	1
Legumes	Edible-podded	1	NR	3	NR	NR	3	NR	7	7	NR	1
	Succulent Shelled Pea and Bean	1		3	NR		3	7	7	7	NR	1
	Dried Shelled Pea and Bean	21	7	14	NR	7	21	NR	21	21	NR	21
Leafy Vegetables, Except Brassica	Head and Leaf Lettuce	1	0	7	5^C	0	7^A	NR	1	1	1	1
	Spinach	1	0	40	NR	0	NR	NR	NR	NR	1	1
	Celery	1	0	NR	NR	0	NR	NR	NR	NR	3	1
Root and Tuber Vegetables	Beet, Carrot, Radish. Turnip	1	0	21	NR	0	7	NR	NR	NR	1	1
	Potato	1	0	21	NR	0	NR	NR	NR	7	14	1
	Sweetpotato	1	0	21	NR	0	NR	NR	NR	7	NR	1

NR Not registered
[A] Head lettuce only

Insect Control for Greenhouse Vegetables

J. F. Walgenbach, Entomology Extension and G. G. Kennedy, Entomology Research

Sound cultural practices, such as sanitation and insect-free transplants, help prevent insect establishment and subsequent damage. Separate plant production houses, use of yellow sticky traps, and timely sprays will help prevent whitefly buildup. Use of *Encarsia* parasites for whitefly and other biological control agents in conjunction with use of pesticides is encouraged. Unless a pesticide label specifically states that a product cannot be used in a greenhouse vegetable crop, the product can be used on those crops for which it is registered. However, pesticides behave differently in the field and the greenhouse, and for many products, information is not available on greenhouse crop phytotoxicity and residue retention. If unsure of the safety of a product to a crop, apply to a small area before treating the entire crop.

Table 5-10. Insect Control for Greenhouse Vegetables

CROP Insect	Insecticide and Formulation	Amount of Formulation	Re Entry Interval	Pre Harvest Interval (PHI) (Days)	Precautions and Remarks
Cucumber					
Aphid	flonicamid, MOA 29 (Beleaf) 50SG	0.065 to 0.1 oz per 1000 sq ft	12 hrs	0	May be applied either to the soil as a drench or drip irrigation for preventive control, or sprayed onto plans as a rescue treatment.
	imidacloprid, MOA 4A (Admire Pro) 4.6 F	0.6 fl oz/1,000 plants	12 hrs	0	Apply in a minimum of 21 gallons water using soil drenches, micro-irrigation, or drip irrigation. Do not apply to immature plants as phytotoxicity may occur. Make only one application per crop per season.
	insecticidal soap (M-Pede) 49 EC	2 tbsp/gal water	12 hrs	0	
Cabbage looper	*Bacillus thuringiensis*, MOA 11 (various)	0.5 to 1 lb OR 3 pt/100 gal water	4 hrs	—	
	spinosad, MOA 5 (Entrust) SC	3 fl oz/100 gal	4 hrs	1	Do not make more than two consecutive applications. OMRI approved.
Spider mite	insecticidal soap (M-pede) 49 EC	2 tbsp/gal water	12 hrs		Use predatory mites.
	mineral oil (TriTek)	1 to 2 gal/100 gal	4 hrs	0	Begin applications when mite populations are low, and repeat at weekly intervals.
	Fenpyroximate, MOA 21A (Akari) 5SC	1 to 2 pts per 100 gal	12 hrs	7	
Whitefly, Leafminer	acetamiprid, MOA 4A (Assail) 30 SG	0.1 oz per 1000 sq ft	12	0	
	flonicamid, MOA 20 (Beleaf) 30SG	0.065 to 0.1 oz per 1000 sq ft	12 hrs	0	
	imidacloprid, MOA 4A (Admire Pro) 4.6 F	0.6 fl oz/1,000 plants	12 hrs	0	Apply in a minimum of 21 gallons water using soil drenches, micro-irrigation, or drip irrigation. Do not apply to immature plants as phytotoxicity may occur. Make only one application per crop per season.
	insecticidal soap (M-Pede) 49 EC	2 tbsp/gal water	12 hrs	0	May be used alone or in combination. Acts as an exciter.
	Beauveria bassiana (Mycotrol WP)	0.25 lb/20 gal water		0	Apply when whiteflies observed. Repeat in 4- to 5-day intervals.
Lettuce					
Aphid, Leafminer, Whitefly	Fulymetrozine, MOA 9B (Fulfill) 50 WG	0.063 oz per 1000 sq ft	12 hrs	0	Will not control leafminer.
	pyrethrins, MOA 3 (Pyganic) 5EC	0.25 to 0.5 fl oz per gal water	12 hrs	0	May be used alone or tank mixed with a companion insecticide (see label for details).
	malathion, MOA 1B (various) 57 EC / 25 WP	1 qt/100 gal water / 4 lb/100 gal water	24 hrs	14 / 14	
	insecticidal soap (M-Pede) 49 EC	2 tbsp/gal water	12 hrs	0	May be used alone or in combination. Acts as an exciter. Insecticidal soaps can cause phytotoxicity under high temperatures or slow drying conditions. If unsure, apply to a small area before treating the entire crop.
	Beauveria bassiana (Mycotrol WP)	0.25 lb/20 gal water	4 hrs	0	Under high aphid or whitely pressure, apply at 2 to 5 day intervals.
Cabbage looper	*Bacillus thuringiensis*, MOA 11 (Javelin) WG	0.5 to 1.25/100 gal water	4 hrs	0	
	spinosad, MOA 5 Entrust SC	3 fl oz/100 gal	4 hrs	1	Do not make more than two consecutive applications.
Slugs	iron phosphate (Sluggo)	0.5 to 1 lb/1,000 sq ft	4 hr	1	Scatter the bait around the perimeter of the greenhouse to provide a protective barrier. If slugs are within the crop, then scatter the bait on the ground around the plants. Do not make more than 3 applicationsi within 21 days. Sluggo will control slugs and snails, while Bug-N-Sluggo will also control earwigs, cutworms, sowbugs and pillbugs. Both are OMRI approved.
	iron phosphate + spinosad (Bug-N-Sluggo	0.5 to 1 lb/1,000 sq ft	4 hr	1	
Spider mite	insecticidal soap (M-Pede)49 EC	2 tbsp/gal water	12 hrs	0	
	mineral oil (TriTek)	1 to 2 gal/100 gal	4 hrs	0	Begin applications when mite populations are low, and repeat at weekly intervals.

INSECT CONTROL

Table 5-10. Insect Control for Greenhouse Vegetables

CROP Insect	Insecticide and Formulation	Amount of Formulation	Re Entry Interval	Pre Harvest Interval (PHI) (Days)	Precautions and Remarks
Tomato, Pepper					
Aphid	flonicamid, MOA 20 (Beleaf) 50SG	0.1 oz per 1000 sq ft	12 hrs	0	May be applied to the soil as a drench or drip irrigation for preventive control, or as a spray for rescue treatments. Will also control whiteflies.
	imidacloprid, MOA 4A (Admire Pro) 4.6 F	0.6 fl oz/1,000 plants	12 hrs	0	Apply in a minimum of 21 gallons water using soil drenches, micro-irrigation, or drip irrigation. Do not apply to immature plants as phytotoxicity may occur. Make only one application per crop per season. Also controls whiteflies.
	malathion, MOA 1B (various) 10 A 57 EC 25 WP	 1 lb/50,000 cu ft 1 qt/100 gal water 4 lb/100 gal water	12 hrs	 15 hr 1 1	
	insecticidal soap (M-Pede) 49 EC	2 tbsp/gal water	12 hrs	0	May be used alone or in combination. Acts as an exciter.
	Beauveria bassiana (Mycotrol WP)	0.25 lb/20 gal water		0	Apply when whiteflies are observed. Repeat in 4-to 5-day intervals.
Armyworm, Fruitworm, Cabbage looper, Cutworm, Pinworm	*Bacillus thuringiensis* , MOA 11 (Javelin) WG (Agree) WP (Dipel) DF Xentari DF	 0.5 lb to 1.25 lb/100 gal water 1 to 2 lb 0.5 to 1.25 0.5 to 1.5	4 hrs	0	
	Chlorfenapyr MOA 13 (Pylon) 2SC	6.5 to 13 fl oz/100 gal water or per acrea area	12 hrs	0	For use on tomatoes more than 1 inch in diameter at maturity. Do not make more than two applications at 5 to 10 day intervals before rotating to an insecticide with a different mode of action.
	cyantraniliprole, MOA 28 (Exirel) SE	7 to 13.5 fl oz per acre, or per 100 gal	12 hrs	1	
	spinosad, MOA 5 Entrust SC	3 fl oz/100 gal	4 hrs	1	Do not make more than two consecutive applications. Do not apply to seedling tomatoes or peppers grown for transplants.
Leafminer	Cyantraniliprole, MOA 28 (Exirel) SE	13.5 to 20.5 fl oz per acre, or per 100 gal	12 hrs	1	
	diazinon, MOA 1B (Diazinon, Spectracide) (AG 500) 50 WP	4 to 8 oz/100 gal water	48 hrs	3	Keep ventilators closed for 2 hours or overnight. Plant injury may result if labeling directions are not followed. For use by members of N.C. Greenhouse Vegetable Growers Association only.
	spinosad, MOA 5 (Entrust) SC	10 fl oz/100 gal	4 hrs	1	Do not apply to seedlings grown for transplants.
Slug	metaldehyde (various) bait	Follow label directions	12 hr		Apply to soil surface around plants. Do not contaminate fruit.
	Iron phosphate (Sluggo)	½ teaspoon per 9-inch pot		0	
Spider mite, broad mite, rust mite	acequinocyl, MOA 20B (Kanemite) 15SC	31 fl oz/100 gal'	12 hr	1	
	Bifenazate (Floramite) SC,	4 to 8 fl oz/100 gal water (1/4 to 1/2 tsp/gal)	12 hr	3	For use on tomatoes more than 1 inch in diameter at maturity. Not registered on pepper. Not for rust mite
	mineral oil (TriTek)	1 to 2 gal/100 gal		0	Begin applications when mite populations are low, and repeat at weekly intervals.
	Chlorfenapyr, MOA 13 (Pylon) 2SC	9.8 to 13 fl oz/100 gal water or per acrea area		0	For use on tomatoes more than1 inch in diameter at maturity. Do not make more than two applications at 5 to 10 day intervals before rotating to an insecticide with a different mode of action.
	fenpyroximate, MOA 21A (Akari) 5SC	1 to 2 pts per 100 gal	12 hrs	1	
	insecticidal soap (M-Pede) 49 EC	2 tbsp/gal water	12 hrs	0	
Thrips, including western flower	*Beauveria bassiana* (Mycotrol WP)	0.25 lb/20 gal water		0	Use screens on intake vents. Apply when whiteflies observed. Repeat in 4- to 5-day intervals.
	Chlorfenapyr, MOA 13 (Pylon) 2SC	9.8 to 13 fl oz/100 gal water or per acrea area		0	For use on tomatoes more than1 inch in diameter at maturity. Do not make more than two applications at 5 to 10 day intervals before rotating to an insecticide with a different mode of action.
	cyantraniliprole, MOA 28 (Exirel) SE	13.5 to 20.5 fl oz per acre, or per 100 gal	12 hrs	1	For foliage-feeding thrips only, not those in flowers.
	spinosad, MOA 5 (Entrust) SC	5.5 fl oz/100 gal	4 hrs	1	Do not make more than two consecutive applications, and do not apply more than 6 times in a 12 month period against thrips. Do not apply to seedlings grown for transplants.
Whitefly	imidacloprid, MOA 4A (Admire Pro) 4.6 F	0.6 fl oz/1,000 plants	12 hrs	0	Apply in a minimum of 21 gallons water using soil drenches, micro-irrigation, or drip irrigation. Do not apply to immature plants as phytotoxicity may occur. Make only one application per crop per season. Also controls aphids.
	cyantraniliprole, MOA 28 (Exirel) 0.83SE	13.5 to 20.5 fl oz/100 gal water or per acre area	12 hrs	1	
	flonicamid, MOA 29 (Beleaf) 50SG	0.1 oz per 1,000 sq ft	12 hrs	0	For use on tomato only.

Table 5-10. Insect Control for Greenhouse Vegetables

CROP Insect	Insecticide and Formulation	Amount of Formulation	Re Entry Interval	Pre Harvest Interval (PHI) (Days)	Precautions and Remarks
Tomato, Pepper (continued)					
Whitefly (continued)	insecticidal soap (M-Pede) 49 EC	2 tbsp/gal water	12 hrs	0	
	pyrethrins and PBO, MOA 3 (Pyganic) 5EC	0.25 to 0.5 fl oz per gal	12 hrs	0	May be used alone or tank mixed with a companion insecticide. (See label for details.)
	Beauveria bassiana (Mycotrol WP)	0.25 lb/20 gal water	4 hrs	0	Apply when whiteflies are observed. Repeat in 4- to 5-day intervals.
	buprofezin, MOA 16 (Talus) 40SC	9 to 13.6 oz/100 gal water or per acre area	12 hrs	1	Insect growth regulator that affects immature stages of whiteflies. Will not kill adults. For use on tomatoes only.
	pyriproxyfen, MOA 7C (Distance) 0.86EC	6 fl oz/100 gal water	12 hrs	<1	Insect growth regulator that affects immature stages of whiteflies. Will not kill adults. Do not use on tomatoes more than 1 inch in diameter. Do not apply on non-bell pepper.

Insect Control for Livestock and Poultry

W. Watson, Entomology Extension

Table 5-11A. Insect Control for Cattle

Insect Insecticide and Formulation	Amount of Formulation to Use in Water	Dosage per Animal	Minimum Interval (Days) Between Application and Harvest	Precautions and Remarks
Cattle Grub—(a) Beef and non-lactating dairy animals				
				Make all grub treatment after heel fly season ends but before Oct. 1.
coumaphos (CoRal) 11.6 EC	15 oz/4 gal	See label	14 freshening	Apply using high pressure spray (250 to 350 psi).
doramectin (Dectomax) injectable	—	1 cc/110 lb	35	Not for female dairy cattle over 20 months of age.
ivermectin injectable pour-on bolus	— — —	1 cc/110 lb 1 ml/22 lb See label	49 48 —	Not for female dairy cattle of breeding age. For calves older than 12 weeks of age.
moxidectin (Cydectin) 0.5 PO	—	5 ml/110 lb	0	Not for use on lactating dairy cattle.
Cattle Grub—(b) Dairy animals (also beef and non-lactating dairy animals)				
eprinomectin (Eprinex) pour-on	—	1 ml/22 lb	0	
Horn Fly—(a) Dairy and beef animals				
coumaphos (CoRal) 1 D	—	3 to 6 tbsp	0	Repeat as necessary.
5.8 EC 11.6 EC	5 oz/4 gal 2.5 oz/4 gal	—	0 14 freshening	Not for use on lactating dairy cattle. Follow all label directions.
cyfluthrin (CyLence) 1 PO	—	—	0	Follow label instructions.
diflubenzeron bolus (Vigilante)	—	0.5 to 2 boluses according to weight	—	Controls fly larvae in manure. All cattle in herd should be treated for best results.
diflubenzeron oral larvicide (Clarify)	—	—	—	In feed according to label.
eprinomectin (Eprinex) pour-on	—	1 ml/22 lb	0	Effective control for 7 days only.
methoprene bolus (Inhibitor)	—	0.5 to 1 bolus according to weight	—	
methoprene mineral mix	—	—	0	Daily in feed according to label.
moxidectin (Cydectin) 0.5 PO	—	5 ml/100 lb	0	Not for use on lactating dairy cattle.
permethrin EC or PO	—	—	0	See label for rate and application directions.
permethrin + diflubenzuron	—	3 ml/110 lb		See label for rate and application directions.
pyrethrins 0.1 OS + synergist	—	1 to 2 oz	0	Oil sprays will harm skin if not applied properly. Apply oil solutions daily as a mist.
spinosad (Elector) 2.5 pour-on spray	Ready to use 10 oz/5 gal water	4 ml/100 lb Spray to runoff	No withholding for milk	Do not use more than once each week. Do not make more than 5 consecutive applications.
tetrachlorvinphos (Rabon) 3 D oral larvacide	— —	2 oz —	—	Daily in feed according to label.

139

Table 5-11A. Insect Control for Cattle

Insect Insecticide and Formulation	Amount of Formulation to Use in Water	Dosage per Animal	Minimum Interval (Days) Between Application and Harvest	Precautions and Remarks
Horn Fly—(a) Dairy and beef animals (continued)				
SELF-APPLICATING DEVICES coumaphos (CoRall) 1 D 1 OS 5.8 EC 11.6 EC permethrin tetrachlorvinphos (Rabon) 3 D tetrachlorvinphos+dichlorvos (RaVap) 23 EC	 — — — — — — 5 oz/ 1 gal oil	 — — — — — — —	0	For dairy and beef animals. These devices aid in face fly and louse control. Follow all label instructions. Inspect and charge oilers and dust bags weekly as needed.
EAR TAGS abamectin (XP820) beta-cyfluthrin (CyGuard) coumaphos + diazinon (CoRal Plus, Corathon) cyfluthrin (Cutter Gold, CyLence Ultra) cypermethrin (Python, Magnum) diazinon (40%) (Patriot) diazinon (20%) (Optimizer) diazinon + chlorpyrifos (Warrior) lambda-cyhalothrin (Saber) permethrin (GardStar) pirimiphos-methyl (Dominator)	—	2/head		These devices give season-long fly control. Some tags are not for use on lactating dairy cattle. Some tags are restricted from use on calves under the age of 3 months. Use according to label. Other ear tags are available. Contact Entomology Department, N.C. State University, for current tag list.
Horn Fly—(b) Beef animals				
lambda-cyhalothrin Aim Capsule		1 capsule (600 lb)		Smart Vet applicator required
gamma cyhalothrin (Stangard) pour-on		10 ml < 600 lb 15 ml > 600 lb		Do not apply more than once in 2 weeks or more than 4 times in 6 months.
ivermectin PO bolus	 — —	 1 ml/22 lb —	 48 —	Not for female dairy cattle of breeding age. Controls horn flies for up to 28 days. Bolus for calves older than 12 weeks of age.
spinosad (Elector) 2.5 pour-on spray	 Ready to use 10 oz/5 gal water	 4 ml/100 lb Spray to runoff	Do not apply within 2 days of slaughter	Do not use more than once each week. Do not make more than 5 consecutive applications.
tetrachlorvinphos (Rabon) 50 WP	5 oz/5 gal	2 to 4 qt	0	
SELF-APPLICATING DEVICES tetrachlorvinphos+dichlorvos (RaVap) 23 EC	5 oz/1 gal oil	—	0	For beef only. These devices aid in face fly and louse control.
Lice—(a) Dairy and beef animals				
coumaphos (CoRal) 1 D	—	3 to 6 tbsp		
5.8 EC 11.6 EC	2.5 oz/4 gal 1.25 oz/4 gal	— —	0	Spray thoroughly—wet to skin.
cyfluthrin (CyLence) 1 PO	—	—	—	Follow label instructions.
eprinomectin (Eprinex) pour-on	—	1 ml/22 lb	0	Follow label instructions.
permethrin EC PO permethrin plus diflubenzuron (Cleanup)	See label	 — —	0	Follow label instructions. Spray entire animal, second treatment at 14 to 21 days. Pyrethroid and IGR blend to control all louse life stages. Follow label instructions.
spinosad (Elector) 2.5 pour-on spray	 Ready to use 10 oz/5 gal water	 4 ml/100 lb Spray to runoff	No withholding for milk	Second application in 45 to 60 days. Do not use more than once each week. Do not make more than 5 consecutive applications.
tetrachlorvinphos (Rabon) 3 D	—	2 oz	0	
Lice—(b) Beef animals				
gamma cyhalothrin (Stangard) pour-on		10 ml < 600 lb 15 ml > 600 lb		Do not apply more than once in 2 wks or more than 4 times in 6 months.
coumaphos 5.8 EC 11.6 EC	 5 oz/4 gal 2.5 oz/4 gal	 — —	 0 14 freshening	Spray—wet to skin.
doramectin (Dectomax) injectable	—	1 cc/110 lb	35	Not for female dairy cattle over 20 months of age.
ivermectin injectable pour-on bolus	 — — —	 1 cc/110 lb 1 ml/22 lb —	 49 48 —	Not for female dairy cattle of breeding age. Injection ineffective for control of biting lice. Pour-on controls both biting and sucking lice. Bolus for calves older than 12 weeks of age.
lambda-cyhalothrin (Saber) 1 PO	—	—	0	Follow label instructions.
lambda-cyhalothrin Aim Capsule		1 capsule (600 lb)		Smart Vet applicator required
moxidectin (Cydectin) 0.5 PO	—	5 ml/110 lb	0	Not for lactating dairy cattle.
spinosad (Elector) 2.5 pour-on spray	 Ready to use 10 oz/5 gal water	 4 ml/110 lb Spray to runoff	Do not apply within 2 days of slaughter	Make second application in 45 to 60 days. Do not use more than once each week. Do not make more than 5 consecutive applications.

Table 5-11A. Insect Control for Cattle

Insect Insecticide and Formulation	Amount of Formulation to Use in Water	Dosage per Animal	Minimum Interval (Days) Between Application and Harvest	Precautions and Remarks
Lice—(b) Beef animals (continued)				
tetrachlorvinphos (Rabon) 50 WP	5 oz/5 gal	2 to 4 oz	0	Spray thoroughly.
tetrachlorvinphos+dichlorvos (RaVap) 23 EC	See label	—	0	Do not treat more often than every 10 days. Spray entire animal.
Note: Self-applicating devices under horn fly aid in louse control.				
Face Fly				
lambda-cyhalothrin Aim Capsule		1 capsule (600 lb)		Smart Vet applicator required
cyfluthrin (CyLence) 1 PO	See label	—	—	Follow label instructions.
permethrin EC PO	 See label See label	 — —	0	Follow label instructions.
diflubenzeron oral larvicide (Clarify)	—	—	—	In feed according to label.
EAR TAGS abamectin (XP820) beta cyfluthrin (Cyguard) cyfluthrin (Cutter Gold, CyLence Ultra) coumaphos+diazinon (Corathon) cypermethrin (Python, Magnum) diazinon+chlorpyifos (Warrior) diazinon (40%) (Patriot) fenvalerate (Ectrin) lambda-cyhalothrin (Saber) permethrin (GardStar) pirimiphos-methyl (Dominator)		2/head	 0 0	These devices give season-long fly control or aid in the control of face flies. Some tags are not for use on lactating dairy cattle. Use according to label. Other ear tags are available. Contact Entomology Department, N.C. State University, for current tag list.
Note: Self-applicating devices under horn fly aid in face fly control.				
Mange				
amitraz (Taktic) 12.5 EC	1 pt/50 gal	—	0	
doramectin (Dectomax) injectable	—	1 cc/110 lb	35	Not for female dairy cattle over 20 months of age.
eprinomectin (Eprinex) pour-on	—	1 ml/22 lb	0	Follow label instructions.
ivermectin injectable pour-on bolus	 — — —	 1 cc/110 lb 1 ml/22 lb —	 49 48 —	Not for female dairy cattle of breeding age. Injection ineffective for control of biting lice. Pour-on controls both biting and sucking lice. Bolus for calves older than 12 weeks of age.
moxidectin (Cydectin) 0.5 PO	—	5 ml/110 lb	0	Not for lactating dairy cattle.
permethrin EC or PO	See label	—	0	Follow label instructions. Spray entire animal, second treatment at 14 to 21 days.
Maggots in Wounds				
chlorpyrifos spray bomb	See label	—	—	
coumaphos (CoRal) 5.8 EC or 11.6 EC	See label	—	—	
pyrethrin + PBO + Dinpropyl isocinchomeronate	See label	—	—	
Stable Fly, Horse Fly, Deer Fly				
pyrethrins 0.1 OS plus synergist				May give protection for short periods.
Mosquitoes; Dairy and beef animals				
permethrin			0	
Ticks—Dairy and beef animals				
coumaphos (CoRal) 5.8 EC 11.6 EC	 10 oz/4 gal 5 oz/4 gal	 — —	 0	Not for use on lactating dairy animals. Spray animals thoroughly.
permethrin	See label	—	0	
amitraz (Taktic)	See label	—	—	
tetrachlorvinphos (Rabon) 50 WP	4 lb/50 gal	0.5 to 1 gal	—	Do not treat lactating dairy animals. Treat about every 3 weeks during periods of heavy tick activity. Spray animals thoroughly.
tetrachlorvinphos+dichlorvos (Rabon +Vapona, RaVap)	1 qt/50 gal	—	0	Spray animals completely.
House Fly, Lesser House Fly, Stable Fly, Other Filth Flies—Premises: beef and dairy				
bifenthrin (ActiShield) 7.9L	See label	0.33 to 1 fl oz/1,000 sq ft	—	May be applied as crack and crevice treatment while animals are present.
chlorpyrifos (Durashield), 20 CS	See label	—	—	Restricted use insecticide.
cyfluthrin (Tempo, Countdown) 20 WP or 2 L	See label	—	—	Do not apply when animals are present.

141

Table 5-11A. Insect Control for Cattle

Insect Insecticide and Formulation	Amount of Formulation to Use in Water	Dosage per Animal	Minimum Interval (Days) Between Application and Harvest	Precautions and Remarks
House Fly, Lesser House Fly, Stable Fly, Other Filth Flies—Premises: beef and dairy (continued)				
Deltamethrin (Annihilator polyzone)	0.25-1.5 oz/gal	1 pt/10,666-64,000 sq ft	—	Do not apply when animals are present
dichlorvos (Vapona) 1 OS or 0.3 OS	—	—	—	Fog, mist, or surface spray.
fenvalerate (Ectrin) 10 WDL	2 oz/6 gal	—	—	Do not treat animals. Remove livestock before spraying building surfaces.
gamma-cyhalothrin (StandGuard) 5.9 MC	See label	—	—	
lambda-cyhalothrin (Grenade, Oxyfly) 9.7 ER	See label	—	—	
permethrin 25 WP or EC	See label	—	—	
pyrethrins 0.1 OS + synergist	—	—	—	Fog or mist.
spinosad (Elector) 2.5 spray	20 oz/5 gal water	See label	Lactating and non-lactating cattle may be present when applied	Do not use more than once each week. Do not make more than 5 consecutive applications.
tetrachlorvinphos (Rabon) 50 WP	4 lb/25 gal	0.5 to 1 gal/500 sq ft	—	
tetrachlorvinphos+dichlorvos (RaVap) 23 EC	5 oz/1 gal	1 gal/500 to 1,000 sq ft	—	Surface treatment only. DO NOT use as a space spray.
LARVACIDE cyromazine (Neporex) 2 SG	See label	Spray or dry application: 1 lb/200 sq ft	21	For larval control in manure or animal bedding only.
BAIT MIXTURES dichlorvos (Vapona) imidacloprid (QuickBayt) cyantraniliprole (Zyrox) methomyl (Golden Malrin, Apache) nithiazine (QuikStrike) strip spinosad (Elector Bait) Beauveria bassiana (balEnce Bait)				Do not apply baits in areas accessible to animals. Labeled for organic farming.

Table 5-11B. Insect Control for Sheep and Goats

Insecticide and Formulation	Amount of Formulation to Use in Water	Dosage per Animal	Minimum Interval (Days) Between Application and Harvest	Precautions and Remarks
Lice and Sheep Ked				
fenvalerate (Ectrin) 10 WDL	see label	—	—	Do not apply more often than every 14 days as needed.
permethrin (Boss) 5 PO (DeLice, Expar) 1 PO	— —	1.5 ml/50 lb 7.5 ml/50 lb	—	
Blow Fly, other maggots in wounds				
chlorpyrifos spray bomb	—	—	—	Apply directly to wounds and surrounding areas. DO NOT treat lactating dairy goats. DO NOT use on sick animals.

Table 5-11C. Insect Control for Swine

Insecticide and Formulation	Amount of Formulation to Use in Water	Dosage per Animal	Minimum Interval (Days) Between Application and Harvest	Precautions and Remarks
Cockroaches, Spiders				
chlorpyrifos (Durashield) 20 CS	See label	—	—	Chlorpyrifos products are restricted use insecticides. Do apply when animals are present. Restricted use insecticide. Do not apply when animals are present.
cyfluthrin (Tempo) 20 WP or 2 L	See label	—	—	
pyriproxyfen (Pyri-Shield) 1.3 EC	1 fl oz/gal	1 gal/1,000 to 1,500 sq ft		An insect growth regulator best used in tank mixes with other insecticides.
House Fly, Stable Fly—Premises				
bifenthrin (ActShield) 7.9 L	See label	0.33 to 1 fl oz/1,000 sq ft	—	May be applied as crack and crevice treatment while animals are present.
chlorpyrifos (Durashield) 20 CS	See label	—	Hogs may not be present when applied. Allow to dry before introducing animals.	Restricted use insecticide.

Table 5-11C. Insect Control for Swine

Insecticide and Formulation	Amount of Formulation to Use in Water	Dosage per Animal	Minimum Interval (Days) Between Application and Harvest	Precautions and Remarks
House Fly, Stable Fly—Premises (continued)				
cyromazine (Neporex) 2 G	See label	Spray or dry application: 1 lb/200 sq ft	21	For larval control only in manure or animal bedding.
deltamethrin (Annihilator polyzone)	0.25-1.5 oz/gal	1 pt/10,666-64,000 sq ft	—	Do not apply when animals are present
esfenvalerate (Valorshield) 35 WP	—	—	—	Follow label directions.
gamma-cyhalothrin (StandGuard) 5.9 MC	See label	—	—	
lambda-cyhalothrin (OxyFly) 97 ER	—	—	—	
pyriproxyfen (PyriShield) 1.3 EC	1 fl oz/gal	1 gal/1,000 to 1,500 sq ft		This slow-acting insect growth regulator is most effective when tank mixed with other insecticides.
spinosad (Elector) 2.5 spray	20 oz/5 gal water	Spray to runoff; 1 gal per 500 to 1,000 sq ft	Hogs may not be present when applied. Allow to dry before introducing animals.	Do not spray more than once each week. Do not make more than 5 consecutive spray applications.
Beauveria bassiana (balEnce)	See label	See label	—	Labeled for organic farming.
Lice				
coumaphos (CoRal) 1 D 1 D	— —	1 oz to shoulders and back 2 oz/30 sq ft of bedding	0	Do not treat 10 days before or after shipping, weaning, or exposure to disease. Allow at least 10 days between applications.
fenvalerate (Ectrin) 10 WDL	2 oz/3 gal	1 pt		Spray with particular attention to neck and ears. Re-treat in 30 days (if necessary).
ivermectin injectable pre mix	—	1 cc/75 lb 300 g/ton	18 5	Continually feed for 7 days. For feeder pigs and finish hogs ONLY.
permethrin		—	5	Spray entire animal until thoroughly wet.
phosmet (Prolate/Lintox 11.75%)			1	Retreat in 14 days.
tetrachlorvinphos (Rabon) 50 WP	7 oz/5 gal	1 to 2 qt	0	
Mange Mite				
amitraz (Taktic) 12 EC 2 PO	760 ml/50 gal		3 7	Follow label instructions.
doramectin (Dectomax) injectable	—	1 cc/75 lb	24	
fenvalerate (Ectrin)	—	—	—	Spray entire animal until thoroughly wet.
ivermectin injectable pre mix (Ivomec only)	— —	1 cc/75 lb 300 g/ton	18 5	Continually feed for 7 days. For feeder pigs and finishing hogs ONLY.
permethrin EC 10 PO (Swine Guard)	—	— 3 ml/100 lb	5	Spray entire animal until thoroughly wet. See label for correct rates and treatment intervals.
phosmet (Prolate/Lintox 11.75%)	2 qt in 50 gal		1 to harvest	Retreat in 14 days
Maggots in Wounds				
See CATTLE				
House Fly				
tetrachlorvinphos (Rabon oral larvacide)				See label.
Also see CATTLE		—	—	Treat according to label.

Table 5-11D. Insect Control for Horses

Insecticide and Formulation	Amount of Formulation to Use in Water	Precautions and Remarks
Bot		
ivermectin (Zimecterin, Eqvalan)		Follow all instructions.
Horse Fly, Deer Fly, Mosquito		
For materials and control suggestions see CATTLE section.		

Table 5-11D. Insect Control for Horses

Insecticide and Formulation	Amount of Formulation to Use in Water	Precautions and Remarks
House Fly, Stable Fly—Premises		
bifenthrin (ActiShield) 7.9 L	See label	May be applied as crack and crevice treatment while animals are present.
chlorpyrifos (Durashield) 20 CS	See label	Restricted use insecticide.
Cyromazine (Neporex) 2G	See label	Spray or dry application to stall bedding or muck pile.
(Solitude IGR) 2.1		In feed to control fly larvae in manure.
esfenvalerate (ValorShield) 35 WP	See label	
gamma-cyhalothrin (StandGuard) 5.9 MC	See label	
lambda-cyhalothrin		
pyriproxyfen (Pyri-Shield)	1 fl oz/gal	This slow-acting insect growth regulator is most effective when used with other insecticides.
spinosad (Elector) 2.5 spray	20 oz/5 gal water	Spray to runoff; 1 gallon per 500 to 1,000 squre feet. Horses may be present when applied. Do not use more than once each week. Do not make more than 5 consecutive applications.
Beauveria bassiana (balEnce)	See label	Organic labeling.
Horn Fly, Face Fly, House Fly, Stable Fly		
coumaphos (CoRal)		Dust or spray. Follow label instructions.
cypermethrin (Tri-Tec 14)		Follow label instructions.
dichlorvos (Vapona) + pyrethrin + piperonyl butoxide		Follow label instructions.
permethrin (Ectiban, Atroban, Tech- Trol, Tech-Trol 12, Permectrin II)		Follow label instructions.
permethrin + piperonyl butoxide (Poridon) (Flysect-7)		Pour on for fly control. Spray.
pyrethrin + piperonyl butoxide		Follow label instructions.
tetrachlorvinphos (Rabon oral larvacide)		In feed, mixed, or topdressed for control of fly larvae in manure.
AUTOMATIC SPRAY SYSTEMS resmethrin; natural pyrethrins + piperonyl buxide		Follow label instructions.
BAIT MIXTURES dichlorvos (Vapona), imidacloprid (QuickBayt), methomyl (Golden Malrin, Apache), nithiazine (Quik Strike) Strip, spinosad (Elector Bait)		Do not apply baits in areas accessible to animals.

Table 5-11E. Insect Control for Poultry

Insecticide and Formulation	Amount of Formulation in Water	Dosage	Precautions and Remarks
Chicken Mite			
permethrin	See label	—	Provide easy to clean roosts and nests with few hiding places. Apply sprays thoroughly to roosts and cracks in surrounding areas. Repeat application as necessary. Follow labels carefully. Treatment of birds as for northern mite also helps.
Northern Fowl Mite, Lice			
permethrin	—	1 gal spray/100 birds	
permethrin (Poultry Mite Tags)	—	2 tags/bird	Follow label directions.
tetrachlorvinphos (Rabon) 50 WP	6.5 oz/5 gal	1 gal/100 birds or 1 to 2 gal/1,000 sq ft of litter	Direct on birds. Thorough coverage and feather penetration is essential. Follow labels carefully. Use 100 to 125 psi for good penetration. Apply premises spray as necessary to reduce NFM/lice dislodged from birds.
3 D	—	1 lb/300 birds or 1 lb/100 sq ft of litter	
tetrachlorvinphos+ dichlorovos (RaVap) 23 EC	5 oz/1 gal	1 gal/100 birds; 1 to 2 gal/1,000 sq ft of litter	Direct on birds. Thorough coverage and feather penetration is essential. Follow labels carefully. Use 100 to 125 psi for good penetration. Apply premises spray as necessary to reduce NFM/lice dislodged from birds.
House Fly, Lesser House Fly, Stable Fly, Other Filth Flies—Premises			
bifenthrin (ActiShield) 7.9 L	See label	0.33 to 1 fl oz/1,000 sq ft	May be applied as crack and crevice treatment while animals are present.
chlorpyrifos (Durashield) 20 CS	See label		Restricted use insecticide. Surface treatment only. DO NOT use as a space spray.
cyfluthrin (Tempo, Countdown, Optem) 20 WP or 2 L	See label	—	Remove birds from building prior to treatment of interior surfaces.
deltamethrin (Annihilator polyzone)	0.25-1.5 oz/gal	1 pt/10,666-64,000 sq ft	Remove birds from building prior to treatment of interior surfaces.
dichlorvos (Vapona) 40 EC	—		Fog, mist, or surface spray. See label.
esfenvalerate (ValorShield) 35 WP	See label		

Table 5-11E. Insect Control for Poultry

Insecticide and Formulation	Amount of Formulation in Water	Dosage	Precautions and Remarks
House Fly, Lesser House Fly, Stable Fly, Other Filth Flies—Premises (continued)			
gamma-cyhalothrin (StandGuard) 5.9 MC	See label		
lambda-cyhalothrin (Grenade, OxyFly) 9.7 ER	See label	—	See cyfluthrin.
permethrin	See label	—	
pyrethrins 0.1 OS + synergist	See label	—	Fog or mist.
spinosad (Elector PSP) 44.2 spray	See label		
(Elector) 2.56 spray	20 oz/5 gal water	Spray to runoff; 1 gal per 500 to 1,000 sq ft	Poultry may not be present when applied. Allow to dry before introducing animals. Do not use more than once each week. Do not make more than 5 consecutive applications.
Beauveria bassiana (balEnce) spray	—	—	Apply as directed. Organic labeling.
tetrachlorvinphos (Rabon) 50 WP	4 lb/25 gal	0.5 to 1 gal/500 sq ft	
tetrachlorvinphos+ dichlorvos (RaVap) 23EC	5 to 10 oz/1 gal	1 gal/500 to1,000 sq ft	
tetrachlorvinphos+ dichlorvos (RaVap) 23 EC	5 oz/1 gal		Apply larvicide as spot treatment.
tetrachlorvinphos (Rabon) 50 WP	4 lb/25 gal		Apply larvicide as spot treatment.
LARVICIDES cyromazine (Neporex) 2 G			For use in all poultry.
(Flyzine, Larvadex) premix	See label	1 lb/ton of feed	Approved as a manure treatment for broiler breeders and caged layers only. Feed continuously for 4 to 6 weeks.
(Larvadex) 2 SL		Spray or dry application: 1 lb/200 sq ft	For use as manure spray for broiler breeders and caged layers.
pyriproxyfen (Pyri-Shield) 1.3 EC	1 fl oz/gal	1 gal/1,000 to 1,500 sq ft	This slow-acting insect growth regulator is most effective when used in combination with other insecticides.
BAIT MIXTURES cyantraniliprole (Zyrox) dichlorvos (Vapona) imidacloprid (QuickBayt) methomyl (Golden Malrin, Apache) nithiazine (Quik Strike) bait strip spinosad (Elector Bait) Beauveria bassiana (balEnce Bait)	—		Do not apply baits in areas accessible to poultry. Use as directed.
Scaly-Leg Mite			
crude petroleum oil	Undiluted	Dip shanks	
Chigger			
permethrin	—	See label	Apply day before poultry is put on range. Repeat in 2 to 3 weeks.
Stick-Tight Flea			
permethrin	—	See label	May be applied to birds.
pyriproxyfen (Pyri-Shield) 1.3 EC	—	—	Use in tank mix with permethrin as premise treatment.
Vaseline	—	Rub into areas of head where pest is attached	Keep dogs and other animals out of poultry areas. Yards, nesting, and roosting areas should be cleaned frequently.
Bed Bug, Fowl Tick			
bifenthrin (Actishield) 7.9 L	See label	0.33 to 1 fl oz/1,000 sq ft	May be applied as crack and crevice treatment while birds are present.
cyfluthrin (Tempo, Countdown) 20 WP or 2 L	See label	—	Remove birds prior to treatment.
dichlorvos (Vapona) 40 EC	—	—	Use according to label.
lambda-cyhalothrin (Grenade) 9.7 ER	See label	—	
permethrin	—	—	
Darkling Beetle (Lesser Mealworm)			
bifenthrin (ActiShield) 7.9 L	See label	0.33 to 1 fl oz/1,000 sq ft	May be applied as crack and crevice treatment while poultry are present.
carbaryl (Sevin) 80 WSP 43 SL	— —	— —	Limited to building exteriors; see label.
cyfluthrin (Tempo, Countdown, Optem) 20 WP or 2 L	See label		Remove birds prior to treatment.
chlorpyrifos (Durashield) 20 CS	See label		Restricted use insecticide.
gamma-cyhalothrin (StandGuard) 5.9 MC	See label		

Table 5-11E. Insect Control for Poultry

Insecticide and Formulation	Amount of Formulation in Water	Dosage	Precautions and Remarks
Darkling Beetle (Lesser Mealworm) (continued)			
imidacloprid (Credo, Dominion, Exile DB) 428 CS	3 fl oz/0.5 to 2 gal water	1 gal/1,000 sq ft	
lambda-cyhalothrin (Grenade) 9.7 ER	See label		Remove birds prior to treatment.
(OxyFly) 9.7 R	—	—	
permethrin	—	—	
pyrifproxyfen (Pyri-Shield) 1.3 EC	1 fl oz/gal	1 gal/1,000 to 1,500 sq ft	This slow-acting insect growth regulator is most effective when used in combination with other insecticides.
spinosad (Elector PSP) 44.2 spray	See label		
tetrachlorvinphos (Rabon) 50 WP 3 D	4 lb/50 gal —	1 to 2 gal/1,000 sq ft 1 lb/100 sq ft	Do not treat houses with birds 6 weeks old or less.
tetrachlorvinphos+ dichlorvos (RaVap) 23 EC	5 to 10 oz/1 gal	1 gal/500 to1,000 sq ft	
zetacypermethrin (Zetagard LBT) Granular	50 Lbs/house	See Label	6 wks withdrawel period before slaughter
Imported Fire Ants			
See COMMUNITY PEST CONTROL			
Rodents			
See ANIMAL DAMAGE CONTROL chapter— Rodenticides			

Community Pest Control

M. Waldvogel, Entomology Extension and M. Reiskind, Entomology Research

NOTE: Insecticides recommended for use by Certified Applicators only. For rodents, see Animal Control, Chapter 9.

Table 5-12A. Community Pest Control — Mosquito Adults[1]

Read pesticide labels carefully. Some pesticide products are not approved for application to edible plants. Avoid spraying when pollinators are actively foraging.
KEY: Dv 0.9 = 90% of the spray volume droplets are smaller than value given VMD = Volume Median Diameter; *um* = micrometer

TYPE OF APPLICATION Insecticide and Formulation	Mixing Instructions and Application Equipment	Application Rate at 10 mph	Droplet Size Requirements on Label (*um*)	Precautions and Remarks
Ground Application				
bifenthrin 7.9L	0.33 to 1.0 fl oz/gal water in backpack or hydraulic sprayer			Apply at a rate of 1 gallons per1000 square feet for thorough coverage of lawns and/or ornamentals.
garlic oil% (ATSB concentrate)	38 fl oz/ga water in a backpack or hydraulic sprayer			Apply at a rate of 15 ounces per 100 linear feet to vegetation 1 to 5 feet above the ground wetting both surfaces of foliage to the point of runoff. Do not apply with handheld or truck-mounted cold ULV or thermal foggers or by aircraft.
malathion 96.5% concentrate (Fyfanon ULV)	Use undiluted on aerosol ULV sprayer.	2 to 4.3 fl oz	VMD < 30 *um* Dv 0.9 < 50 *um*	Do not spray when wind speed is more than 5 mph.
	Dilute 3.9 to 5.2 gal to 100 gal with No. 2 fuel or diesel oil; use in thermal fog sprayer.			Avoid direct application to vehicles; these insecticides may damage paint. Apply when air temperatures are cool and wind speed is 3 mph or less. Toxic to fish, aquatic invertebrates, and wildlife.
naled (Dibrom) 87.4% concentrate	10 fl oz to 10 gal No. 2 fuel or diesel oil; use in thermal fog sprayer.	80 gal/hr	VMD < 40 *um* Dv 0.9 < 75 *um*	Toxic to fish, aquatic invertebrates, and wildlife. Restricted Use Pesticide.
	Dilute 0.5 gal to 5 gal with soybean oil or HAN; use in ULV sprayer.	6 to 12 fl oz/min	VMD < 40 *um* Dv 0.9 < 75 *um*	Do not directly apply to water or to areas where runoff into water is likely to occur.
permethrin 10% to 57% concentrate	Apply undiluted or mix with refined mineral or soybean oil.	0.31 to 15 oz/min depending on dilution	VMD = 150 to 300 *um*	Permethrin 57% is not for use in residential misting systems. Do not allow drift onto cropland, poultry ranges or potable water supplies. Do not use on crops used for foor or forage.
(Permanone) 10% EC	Dilute 1:20 with water (6.5 fl oz/ 1 gal of water).			Treat surfaces using course wet spray. Spray to runoff.
permethrin (20%) and piperonyl butoxide (20%) (Aqua-Reslin)	Dilute 1 gal with 2 to 12 gal water	2.1 to 9 oz/min depending on dilution	VMD < 30 *um* Dv 0.9 < 50 *um*	Dilute with water only. Toxic to fish and aquatic invertebrates.
permethrin and piperonyl botoxide (Permanone 31-66, Biomist 4+12 ULV)	Dilute 1 gal to 2.4 gal with light weight oil; use in ULV sprayer.	0.5 to 3 fl oz/min	VMD < 30 *um* Dv 0.9 < 50 *um*	
prallithrin (1%) and sumithrin (5%) and piperonyl butoxide (5%) (Duet)	Apply undiluted in aerosol ULV sprayer	2.5-7.5 oz/min	VMD = 8 to 30 *um* Dv 0.9 < 50*um*	Do not allow drift onto pastureland, rangeland, or potable water supplies.
pyrethrins (5%) and piperonyl butoxide (25%) (Aquahalt)	Apply undiluted or diluted with water and applied as an ultra low volume (ULV)	0.27 to 0.76 oz/ac	VMD < 30 *um* Dv 0.9 < 50*um*	Do not apply more than 0.2 pounds of pyrethrins/acre/year or 1.0 pound PBO/acre/year in any treated area
resmethrin (18%) + piperonyl butoxide (54%) (Scourge)	Dilute 0.67 gal with 1 gal of light mineral oil; use in ULV sprayer.	4.5 to 9 fl oz/min	VMD < 30 *um* Dv 0.9 < 50*um*	Restricted-Use Pesticide. Can be applied ULV or diluted with refined soybean oil, light mineral oil of 54 second viscosity or other suitable solvent or diluent.
sumithrin and piperonyl butoxide (Anvil 10+10 ULV or 2+2 ULV)	Use undiluted or dilute 10+10 formulation with light mineral oil.	1.3 to 18.6 oz/min	VMD < 30 *um* Dv 0.9 < 50*um*	
Fixed Wing Aerial Application				
malathion 96.5% concentrate (Fyfanon ULV)	Use undiluted.	2.6 to 3 fl oz/acre	VMD < 60 *um* Dv 0.9 < 100*um*	Toxic to fish, aquatic invertebrates, and wildlife. Do not directly apply to water or to areas where runoff into water is likely to occur. Do not retreat a site more than 3 times in any one week except in emergencies.
naled (Dibrom) 87.4% concentrate	Use undiluted.	0.5 to 1 fl oz/acre	VMD = 60 *um* Dv 0.9 < 115*um*	Toxic to fish, aquatic invertebrates, and wildlife. Do not directly apply to water, except when necessary to target areas where adult mosquitoes are present or to areas where runoff into water is likely to occur. Not for use in or around homes.
	Dilute 50 to 100 fl oz to 100 gal with No. 2 fuel oil or diesel oil.	1 gal/acre	VMD = 60 *um* Dv 0.9 < 115*um*	Toxic to fish, aquatic invertebrates, and wildlife. Do not directly apply to water, except when necessary to target areas where adult mosquitoes are present or to areas where runoff into water is likely to occur. Not for use in or around homes.
permethrin (20%) and piperonyl butoxide (20%) (Aqua-Reslin)	Dilute 1 gal with 2 to 12 gal water	2.1 to 9 oz/min depending on dilution	VMD < 60 *um* Dv 0.9 < 100*um*	Dilute with water only. Toxic to fish and aquatic invertebrates.
resmethrin (18%) + piperonyl butoxide (54%) (Scourge)	Dilute 0.67 gal with 1 gal of light mineral oil; use in ULV sprayer.	4.5 to 9 fl oz/min	VMD < 60 *um* Dv 0.9 < 100 *um*	Restricted-Use Pesticide. Can be applied ULV or diluted with refined soybean oil, light mineral oil of 54 second viscosity or other suitable solvent or diluent.
prallithrin (1%) and sumithrin (5%) and piperonyl butoxide (5%) (Duet)	Apply undiluted on aerosol ULV sprayer	0.41 - 1.24 oz/ac	VMD = < 60 *um*	Do not allow drift onto pastureland, rangeland, or potable water supplies.
pyrethrins (5%) and piperonyl butoxide (25%) (Aquahalt)	Apply undiluted or diluted with water and applied as an ultra low volume (ULV)	0.27 to 0.76 oz/ac	VMD < 60 *um* Dv 0.9 < 80*um*	Do not apply more than 0.2 pounds of pyrethrins/acre/year or 1.0 pound PBO/acre/year in any treated area

INSECT CONTROL

Table 5-12A. Community Pest Control — Mosquito Adults[1]

Read pesticide labels carefully. Some pesticide products are not approved for application to edible plants. Avoid spraying when pollinators are actively foraging.
KEY: Dv 0.9 = 90% of the spray volume droplets are smaller than value given VMD = Volume Median Diameter; *um* = micrometer

TYPE OF APPLICATION Insecticide and Formulation	Mixing Instructions and Application Equipment	Application Rate at 10 mph	Droplet Size Requirements on Label (*um*)	Precautions and Remarks
Fixed Wing Aerial Application (continued)				
sumithrin and piperonyl butoxide (Anvil 10+10)	Use undiluted.	3.8 to 5.7 fl oz/acre	VMD < 60 *um* Dv 0.9 < 80 *um*	

[1] Avoid direct applications to flowering plants when pollinators are active. Do not allow drift onto adjoining non-target areas. When treating residential properties. Cover or remove pet food and water sources, grills, swimming pools and children's toys.

Table 5-12B. Community Pest Control — Mosquito Immatures and Other Pests

PEST Insecticide and Formulation	Mixing Instructions and Application Equipment	Application Rate Per Acre	Precautions and Remarks
Mosquito—Immatures			
Bacillus thuringiensis, var.*israelensis* (Bactimos, Teknar, Vectobac) 50 WP 2 WP 14.3% aqueous conc. 15% aqueous conc. 1.2% aqueous conc. 0.8% aqueous conc.	Dilute with sufficient water to obtain uniform coverage.	6 to 12 oz 4 to 16 oz 0.5 to 3 pt 0.5 to 3 pt 0.25 to 2 pt 0.5 to 2 pt	Only effective against larvae. Can be applied to all breeding habitats, including potable water supplies.
Bacillus thuringiensis, var. *israelensis* (Bactimos) briquets 10%			Use one briquet per 100 square feet of surface area regardless of depth.
(Bactimos, Teknar, Vectobac) granules 0.2% pellets 0.4%	Ready to use	—	Apply 4 to 10 pounds per acre with aircraft or ground equipment.
methoprene (Altosid) 20% EC (Altosid) briquet 2.1%, 8.6% pellet 4.2% (Altosid) granule 0.27%, 1.5%	3 to 4 fl oz/gal water Ready to use	1 gal —	Apply when larvae are in 3rd and 4th instar. Methoprene will not kill pupae or adults. Water less than 2 feet; 1 briquet per 100 square feet; deeper or flowing water; 1 briquet per 10 cubic feet. 2.5- to 10-pound pellet per acre; use high rate in breeding sites with high organic content.
Spinosad (Natular XRG) 2.5% granule	Ready to use	5 to 20 lb	
monomolecular surface film (Agnique MMF)	—	0.2 to 0.5 gal	Use in conjunction with indicator oil to avoid over treatment.
proprietary mosquito control oils (GB-1111, etc.)	—	1 to 5 gal	Dosage depends on amount of floatage and vegetation in water.
temephos (Abate) 43% EC 1 G 2 G 5 G	0.5 to 1.5 fl oz/gal water — — —	1 gal 5 to 10 lb 2.5 to 5 lb 1 to 2 lb	
Midge ("fuzzy bills")			
temephos (Abate) 1 G 2 G 5 G	— — —	5 to 10 lb 2.5 to 5 lb 2 lb	Double recommended rates for water high in organic content.
methoprene 20% EC (Strike) 4.25% pellet (Strike)	4 to 5 oz/ 1 million gal wastewater —	— 5 to 10 lb/acre	For use in wastewater treatment facilities. Uniformly apply at the influent side over a 24-hr period. Apply to natural and manmade aquatic habitats. High rate recommended for wastewater.
Spinosad (Natular XRG) 2.5% granule	Ready to use	5 to 20 lb	
Tick			
acetamiprid-permethrin (Transport)	—	Apply 0.11% concentration of active ingredient to cover 1,000 sq. ft.	Do not apply more than 0.11% finished dilution per 1,000 square feet.
carbaryl (Sevin) 50 WP	0.1 lb/10 gal water	870 gal	Keep children and pets off treated areas until they have dried.
chlorpyrifos (Dursban) 2 E, M, 4E	8 to 16 fl oz/40 to 100 gal water	40 to 100 gal	Golf courses, rights-of-way and industrial sites only
cyfluthrin (Tempo) 24% EC 20% WP	5.9 fl oz/40 to 100 gal water 7.7 oz	40 to 100 gal	
bifenthrin (Talstar) 0.2% G 7.9% L	Ready to use 1 fl. oz/100 gal water	100-200 lb/acre	Do not allow public use of area during treatment 1 gallon per 1,000 square feet.

Table 5-12B. Community Pest Control — Mosquito Immatures and Other Pests

PEST Insecticide and Formulation	Mixing Instructions and Application Equipment	Application Rate Per Acre	Precautions and Remarks
Tick (continued)			
Imidacloprid-cyfluthrin (Temprid)	0.075% - 0.15% fl oz/gal water		Apply at rate not to cause drip/run-off from site
Rosemary Oil, Geraniol, Wintergreen (Essentria IC3)	1 to 8 ounces of Essentria IC3 per gallon of water	43 gal	2 gallons per 1,000 square feet
Imported Fire Ants			
acetamiprid (Transport)		Apply 0.11% concentration of active ingredient to cover 1,000 sq. ft.	Do not apply more than 0.11% finished dilution per 1,000 square feet.
avermectin (Ascend, Black Flag Fire Ant Ender) 0.011% B	—	1 lb	For use on turf, lawns, and other noncrop areas, such as parks and golf courses. Apply when soil temperature is greater than 60 degrees F. Apply after dew or rainfall has dried for maximum effectiveness.
fenoxycarb (Award, Logic) 1.0% B	—	1 to 1.5 lb	Uniformly distribute 1 to 3 tablespoons around the edge of each mound. For broadcast applications, apply 1 to 1.5 pounds per acre. May be used on pastures and grazed areas on horse farms if horses are not intended for human consumption.
fipronil (Topchoice) 0.0143%	— —	87 lb	For use on home lawns, golf courses, commercial and recreational turf, and sod farms. One application of 87 pounds of product/acre per year. Restricted-Use Pesticide.
hydramethylnon (Amdro) 0.73% B	—	1 to 1.5 lb	Broadcast uniformly on pasture and range grass, lawns, turf, and nonagricultural lands. Or distribute 5 level tbsp 3 to 4 feet around base of each mound (do not exceed 1.5 pounds per acre).
hydramethylnon 0.365% + S-Methoprene 0.25% (Extinguish Plus) B	—	1.5 lb	Broadcast uniformly on pasture and range grass, lawns, turf, and nonagricultural lands. Or distribute 2 to 5 level tablespoons 3 to 4 feet around base of each mound (do not exceed 1.5 pounds per acre).
indoxacarb (Advion) 0.045%	—	1.5 lb	For use in outdoor areas on noncroplands.
Metaflumizone (Siesta)(0.0653% B	-	1.5 lb	Broadcast uniformly on target area or use 2 to 4 level tablespoons 3 to 4 feet around base of each mound (do not exceed 1.5 pounds per acre).
methoprene (Extinquish) 0.5% B	—	1 to 1.5 lb	For use on crop and noncroplands, such as parks, zoos, sports fields, and school grounds.
pyriproxyfen (Distance) 0.5% B	—	1 to 1.5 lb	For use in outdoor areas on noncroplands.
spinosad (Conserve) 0.15% B	-	4 lbs	May require two applications per year. (OMRI certified)

For treatment of individual ant mounds with liquid insecticides, refer to the section on insect control for home lawns.

149

Industrial and Household Pests

M. Waldvogel and P. Alder, Entomology Extension

For Use by Licensed Pest Control Operators

Space limitations preclude listing all pesticide formulations and trade names. Other products or formulations may be used. Some products may contain a mixture of active ingredients. Read the product label for specific information about the active ingredients, application rates, and detailed instructions on use—particularly on permitted sites for application.

Mention of pesticides in this section does not imply that chemicals are or should be the first or only means of pest control. Nonchemical methods, including exclusion and sanitation, are important to long-term pest management.

Table 5-13. Industrial and Household Pests—For use by licensed pest control operators only

Pesticide	Boric acid (Niban, Perma-Dust, Intice)	Diatomaceous earth (Mother Earth D)	Silica gel (Drione, Tri-Die, Cimex Dust)	Sodium Tetraborate (Advance 388B, Cymex)	Methomyl (Apache, Flytek)	Propoxur (PT 2, Invader)	Propoxur (PT 2, Invader)	Acephate (Orthene)	Dichlorvos (Nuvan)	Dichlorvos (Nuvan)
Chemical Class[1]	Inorganic	Inorganic	Inorganic	Inorganic	Carbamate	Carbamate	Carbamate	Organophosphate	Organophosphate	Organophosphate
Formulation[2]	Bait, Dust	Dust[3]	Dust[3]	Bait, Dust	Bait	Bait	Sprayable	Sprayable (outdoor use only)	Strip	Sprayable
Pests										
ANTS	X	X	X	X		X	X	X		X
BED BUGS	X	X	X	X					X	X
BEES			X					X		X
BOOKLICE	X	X	X							
BUGS (TRUE)[4]	X		X				X	X	X	
CARPET BEETLES		X		X					X	X
CENTIPEDES	X	X	X				X	X		
CLOTHES MOTHS		X							X	X
CLOVER MITES	X	X	X				X	X		
COCKROACHES	X	X	X			X	X	X	X	X
CRICKETS	X	X	X			X	X	X		X
EARWIGS	X	X	X					X	X	
FLEAS		X	X	X				X		X
FLIES	X	X	X		X				X	X
HORNETS/WASPS			X					X		X
LADY BEETLES		X	X							X
MILLIPEDES	X	X	X				X	X		X
MOSQUITOES										X
STORED PRODUCT PESTS	X	X	X							
SCORPIONS		X	X							
SILVERFISH	X	X	X		X		X	X	X	X
SPIDERS		X					X	X	X	X
SOWBUGS		X					X	X		X
SPRINGTAILS		X	X				X	X		X
TICKS			X				X			

[1] Alternating uses of insecticides in different chemical classes can help reduce the likelihood of the pests developing resistance to one group or class of compounds.

[2] **Formulations:**
  Aerosol includes Crack & Crevice.
  Bait may be granular, gel, or station.
  Sprayable may be concentrate or powder, some RTU formulations.

[3] Some formulations of diatomaceous earth and silica gel contain pyrethrins as a flushing agent.

[4] True bugs includes boxelder bugs, stink bugs, kudzu bugs, and similar occasional invaders.

Table 5-13 (continued). Industrial and Household Pests—For use by licensed pest control operators only

Pesticide	Allethrin (PT565 Plus XLO)	Bifenthrin (Bifen, Talstar)	Cyfluthrin (Tempo Ultra, Ultrashield CS)	Cypermethrin (Demon, Cynoff)	Deltamethrin (DeltaDust, DeltaGuard, Suspend)	Deltamethrin (DeltaDust, DeltaGuard, Suspend)	Esfenvalerate (Onslaught)	Etofenprox (Zenprox)	Fenvalerate (Pyrid)	Lambda-cyhalothrin (Demand, 228L)	Lambda-cyhalothrin (Demand, 228L)	Permethrin (Flee, Dragnet, Prelude)	Phenothrin (Bedlam, Nyguard Plus[5])	Prallethrin (ULD Spy-300, Altocirrus Fog)	Pyrethrins and pyrethrum (Kicker, Pyrenone)	Pyrethrins and pyrethrum (Kicker, Pyrenone)	Pyrethrins and pyrethrum (Kicker, Pyrenone)	Sumithrin (Bedlam) Nyguard Plus[5]	Tetramethrin (CB Stinger)
Chemical Class	Pyrethrids[1]																		
Formulation[2]	S	S, G	S	S	D	G	S	S	S	S	G	S	S	S	A[3]	S[3]	D[3]	S	S
Pests																			
ANTS	X	X	X	X	X	X	X	X	X	X	X	X	X	X	X	X	X	X	
BED BUGS	X		X		X		X	X	X	X		X		X		X	X		
BEES		X	X	X	X		X	X				X		X				X	X
BOOKLICE	X		X		X		X					X	X			X			
BUGS (TRUE)[4]		X	X	X			X			X		X	X	X				X	
CARPET BEETLES	X		X		X		X	X	X	X		X		X	X	X			
CENTIPEDES	X	X	X	X	X	X	X	X	X	X		X	X	X				X	
CLOTHES MOTHS	X		X		X		X							X					
CLOVER MITES	X		X	X	X						X	X		X	X	X	X		
COCKROACHES	X	X	X	X	X	X	X	X	X	X		X		X	X	X	X	X	
CRICKETS	X	X	X	X	X		X	X	X	X	X	X		X	X	X			
EARWIGS		X	X	X				X		X	X	X		X	X	X			
FLEAS	X	X	X	X	X		X	X	X	X		X		X	X	X		X	
FLIES/GNATS	X	X	X	X			X	X		X		X		X	X	X			
HORNETS/WASPS	X	X	X	X			X	X	X	X				X	X	X			X
LADY BEETLES														X	X				
MILLIPEDES	X	X	X	X		X	X	X	X	X	X	X							
MOSQUITOES	X	X	X	X			X	X	X	X				X	X	X			
STORED PRODUCT														X					
PESTS	X		X				X		X	X			X	X	X	X			
SCORPIONS		X	X	X			X				X	X	X		X	X			
SILVERFISH	X	X	X	X			X	X	X	X	X	X	X	X	X	X			
SPIDERS	X	X	X	X		X	X	X	X	X		X	X	X	X	X			
SOWBUGS		X	X	X	X	X	X	X		X		X	X	X	X	X			
SPRINGTAILS		X	X	X	X	X	X	X	X	X		X	X	X	X	X			
TICKS	X	X	X	X	X	X	X	X	X	X		X	X	X	X	X			

[1] Alternating uses of insecticides in different chemical classes can help reduce the likelihood of the pest developing resistance to one class or group of compounds. Many pyrethroids can be tank-mixed with piperonyl butoxide products to enhance insecticidal activity.

[2] **KEY TO FORMULATION SYMBOLS:**
A = aerosol
B = bait (granular or station)
D = dust
G = granular
S = sprayable (concentrate or powder, some RTU formulations)

[3] Some formulations of pyrethrins contain piperonyl butoxide as a synergist.

[4] True bugs includes boxelder bugs, stink bugs, kudzu bugs, and similar occasional invaders.

[5] Nyguard Plus contains the IGR pyriproxyfen

INSECT CONTROL

Table 5-13 (continued). Industrial and Household Pests — For use by licensed pest control operators only

Pesticide	Hydroprene (Gencor)[3]	Fenoxycarb (Altosid, Pre-Strike)[3]	Methoprene (Altosid, Kabat, Pharorid, Precor, Vigren)[3]	Pyriproxyfen (Archer, Ultracide)V[3]	Acetamiprid+Bifenthrin (Transport)	Dinotefuran (Advance), Alpine)	Imidacloprid (FlyBait, Maxforce, Premise, Temprid[8])	Thiamethoxam (Optiguard)[9]	Clothianidin (Maxforce Impact)	Abamectin (Ascend, Avert, Advance)	Aluminum phosphide (Phostoxin)[5]	Chlorfenapyr (Phantom)[6]	Cyanotraniliprole (Zyrox)	d-Limonene (ProCitra-DL)	Fipronil (Maxforce F, TopChoice, Termidor[7])	2-Phenyl Propionate (EcoPCO EC)	Hydramethylnon (Amdro, Siege, MaxForce)	Indoxacarb (Advion, Arilion)	Mint oil (Victor)	Sulfuryl floride (Vikane, Profume, Zythor)[9]
Chemical class[1]	Insect Growth Regulators					Neonicotinoids				Other Classes										
Formulation[2]	A,S	S	B	A,S	A,S	B,S	B,D,S	B	B,S	B	F	S	B	S	B,G,S	A,S	B	B,S	A	F
Pests																				
ANTS	X		X		X	X	X	X[9]		X		X			X	X	X	X	X	
BED BUGS				X	X	X						X				X				X
BEES					X	X											X		X	
BOOKLICE					X	X														
BUGS (TRUE)[4]					X	X			X				X			X				
CARPET BEETLES					X	X										X				X
CENTIPEDES					X	X			X							X				
CLOTHES MOTHS					X	X										X				X
CLOVER MITES					X							X								
COCKROACHES	X				X	X	X	X	X	X		X		X	X	X	X	X	X	X
CRICKETS	X				X	X		X				X		X	X	X		X	X	
EARWIGS					X	X		X				X		X	X	X			X	
FLEAS	X			X	X	X	X					X		X	X	X			X	
FLIES/GNATS					X	X		X				X	X	X	X	X			X	
HORNETS/WASPS					X	X						X[10]		X	X	X			X	
LADY BEETLES					X	X			X					X	X	X			X	
MILLIPEDES					X	X			X					X	X	X		X	X	
MOSQUITOES	X	X	X		X									X		X				
STORED																				
PRODUCT PESTS	X			X	X	X			X		X	X				X				X
SCORPIONS					X	X						X								
SILVERFISH					X	X			X			X			X	X			X	
SPIDERS					X	X						X		X	X	X				
SOWBUGS					X	X						X							X	
SPRINGTAILS					X	X						X							X	
TICKS				X	X							X			X					

[1] Alternating uses of insecticides in different chemical classes can help reduce the likelihood of the pest developing resistance to one class or group of compounds.

[2] **KEY TO FORMULATION SYMBOLS:**

    A = Aerosol (includes Crack & Crevice)   B = Bait (granular, gel or station)

    D = Dust   F = Fumigant

    G = Granular   S = sprayable (concentrate or powder, some RTU formulations)

[3] IGR products are not typically effective against adult stage of pests; use with an adulticide to provide quicker control of pest population.

[4] True bugs includes boxelder bugs, stink bugs, kudzu bugs, and similar occasional invaders

[5] Requires an F-Phase Structural Pest Control License.

[6] Chlorfenapyr labeled for indoor use only for these pests or limited spot treatment outdoors.

[7] Termidor liquid formulations are labeled for outdoor use only; use other insecticide products indoors.

[8] Temprid contains both imidacloprid and cyfluthrin.

[9] Optigard not for use against pharaoh ants or carpenter ants.

[10] Phantom is not a knockdown insecticide for pests such as wasps.

ORNAMENTALS

Arthropod Management for Ornamental Plants Grown in Greenhouses

S. D. Frank, Entomology Extension

Successful pest management programs use a combination of appropriate pest control tactics. Always follow label precautions when handling or applying pesticides. Make chemical control part of an integrated pest management program that includes monitoring and pest identification along with appropriate cultural, physical, horticultural, and biological controls.

Responsible pesticide use includes resistance management. A system has been developed by the inter-company Insecticide Resistance Action Committee (IRAC; www.irac-online.org) to help you rotate chemicals correctly. Pesticides have been assigned an IRAC classification number based on their mode of action. To rotate properly, choose a product with a different IRAC number for each successive application directed against the same pest. Follow resistance management instructions on the label.

The information in this chart is not a substitute for the label. Pesticide labels and restrictions change frequently. Read and understand all label information before using any pesticide. Do not use pesticides for uses other than those on the label. Check county and state regulations for any local restrictions on the use of products listed here before using them.

Table 5-14. Arthropod Management for Ornamental Plants Grown in Greenhouses

Permitted application sites: G = greenhouse, L = landscape, N = Nursery. (Trade names listed are common examples of products that contain the active ingredient, not an endorsement of a particular product.)

Insect or Mite	Pesticide common name (Trade name)	Minimum Hours Between Application and Reentry	IRAC Mode of Action Group	Permitted application sites
Aphid	abamectin (Avid)	12 hr	6	G, L, N
	acephate (Orthene)	24 hr	1B	G, L, N
	acetamiprid (TriStar)	12 hr	4A	G, L, N
	azadirachtin (Azatin)	4 hr	18B	G, L, N
	Beauveria bassiana (Botanigard/Naturalis)	4 hr	M	G, L, N
	bifenthrin (Talstar)	12 hr	3	follow label
	cyantraniliprole (Mainspring)	4 hr	28	G
	cyfluthrin (Decathlon)	12 hr	3A	G, L, N
	dinotefuran (Safari)	12 hr	4A	G, L, N
	flonicamid (Aria)	12 hr	9B	G, L, N
	fluvalinate (Maverik)	12 hr	3A	G, L, N
	horticultural oil (various)	4 hr		G, L, N
	imidacloprid (Marathon II)	12 hr	4A	G, N
	insecticidal soaps	12 hr		G, N, L
	kinoprene (Enstar II)	4 hr	7A	G
	neem oil (Various)	4 hr	UN	G, L, N
	permethrin (Astro, others)	12 hr	3	follow label
	pymetrozine (Endeavor)	12 hr	9B	G, L, N
	pyrethrins (various)	12 hr	3A	G, L, N
	pyrifluquinazon (Rycar)	12 hr	UN	G
	spinetoram + sulfoxaflor (XXpire)	12 hr	4C + 5	G, L, N
	spirotetramat (Kontos)	24 hr foliar (see exception for drench application)	23	G, N
	thiamethoxam (Flagship)	12 hr	4A	G, L, N
	tolfenpyrad (Hachi-Hachi)	12 hr	21A	G
Broad Mite	abamectin (Avid)	12 hr	6	G, L, N
	chlorfenapyr (Pylon)	12 hr	13	G
	fenpyroximate (Akari)	12 hr	21A	G, N
	pyridaben (Sanmite)	12 hr	21A	G, L, N
	spiromesifen (Judo)	12 hr	23	G, N
Caterpillar	acephate (Orthene)	24 hr	1B	G, L, N
	acetamiprid (Tri-Star)	12 hr	4A	G, L, N
	azadirachtin (Azatin)	4 hr	18B	G, L, N
	Bacillus thuringiensis var. kurstaki	4 hr	11B2	follow label
	Beauveria bassiana	12 hr		follow label
	bifenthrin (Talstar)	12 hr	3	follow label
	chlorfenapyr (Pylon)	12 hr	13	G
	cyantraniliprole (Mainspring)	4 hr	28	G
	cyfluthrin (Decathlon)	12 hr	3A	G, L, N
	diflubenzuron (Adept)	12 hr	15	G

Table 5-14. Arthropod Management for Ornamental Plants Grown in Greenhouses

Permitted application sites: G = greenhouse, L = landscape, N = Nursery. (Trade names listed are common examples of products that contain the active ingredient, not an endorsement of a particular product.)

Insect or Mite	Pesticide common name (Trade name)	Minimum Hours Between Application and Reentry	IRAC Mode of Action Group	Permitted application sites
Caterpillar (continued)	fluvalinate (Mavrik)	12 hr	3A	G, L, N
	insecticidal soaps	12 hr		G, L, N
	novaluron (Pedestal)	12 hr	15	G, N
	permethrin (Astro, others)	12 hr	3	Follow label
	pyridalyl (Overture)	12 hr	UN	G
	spinetoram + sulfoxaflor (XXpire)	12 hr	4C + 5	G, L, N
	spinosad (Conserve)	4 hr	5	G, L, N
	tolfenpyrad (Hachi-Hachi)	12 hr	21A	G
Cyclamen Mite	abamectin (Avid)	12 hr	6	G, L, N
	chlorfenapyr (Pylon)	12 hr	13	G
	fenpyroximate (Akari)	12 hr	21A	G, N
	pyridaben (Sanmite)	12 hr	21A	G, L, N
	spiromesifen (Judo)	12 hr	23	G, N
Fungus Gnat Adults	bifenthrin (Talstar)	12 hr	3	Follow label
	cyfluthrin (Decathlon)	12 hr	3A	G, L, N
	fluvalinate (Mavrik)	12 hr	3A	G, L, N
	insecticidal soaps	12 hr		G, L, N
	permethrin (Astro, others)	12 hr	3	Follow label
Fungus Gnat Larvae	acetamiprid (Tri-Star)	12 hr	4A	G, L, N
	azadirachtin (Azatin)	4 hr	18B	G, L, N
	Bacillus thuringiensis var. israelensis	4 hr	11A1	Follow label
	chlorfenapyr (Pylon)	12 hr	13	G
	cyromazine (Citation)	12 hr	17	G, L, N
	diflubenzuron (Adept)	12 hr	15	G
	kinoprene (Enstar II)	4 hr	7A	G
	pyriproxyfen (Distance)	12 hr	7C	G, L, N
	Steinernema feltiae (various; beneficial nematode)	0 hr	Biological	G, L, N
Leafminer	abamectin (Avid)	12 hr	6	G, L, N
	acephate (Orthene)	24 hr	1B	G, L, N
	acetamiprid (Tri-Star)	12 hr	4A	G, L, N
	azadirachtin (Azatin)	4 hr	18B	G, L, N
	cyromazine (Citation)	12 hr	17	G, L, N
	dinotefuran (Safari)	12 hr	4A	G, L, N
	fenoxycarb (Preclude)	12 hr	7B	G
	imidacloprid (Marathon II, others)	12 hr	4A	Follow label
	spinosad (Conserve)	4 hr	5	G, L, N
	thiamethoxam (Flagship)	12 hr	4A	G, L, N
Mealybug	acephate (Orthene)	24 hr	1B	G, L, N
	acetamiprid (Tri-Star)	12 hr	4A	G, L, N
	azadirachtin (Azatin)	4 hr	18B	G, L, N
	Beuveria bassiana	12 hr		Follow label
	bifenthrin (Talstar)	12 hr	3	Follow label
	buprofezin (Talus)	12 hr	16	G, N
	cyfluthrin (Decathlon)	12 hr	3A	G, L, N
	dinotefuran (Safari)	12 hr	4A	G, L, N
	flonicamid (Aria)	12 hr	9B	G, L, N
	horticultural oil (various)	4 hr		G, L, N
	imidacloprid (Marathon II, others)	12 hr	4A	Follow label
	insecticidal soaps	12 hr		G, L, N
	kinoprene (Enstar II)	4 hr	7A	G
	neem oil (Various)	4 hr	UN	G, L, N
	permethrin (Astro, others)	12 hr	3	Follow label
	pyrifluquinazon (Rycar)	12 hr	UN	G

Table 5-14. Arthropod Management for Ornamental Plants Grown in Greenhouses

Permitted application sites: G = greenhouse, L = landscape, N = Nursery. (Trade names listed are common examples of products that contain the active ingredient, not an endorsement of a particular product.)

Insect or Mite	Pesticide common name (Trade name)	Minimum Hours Between Application and Reentry	IRAC Mode of Action Group	Permitted application sites
Mealybug (continued)	spinetoram + sulfoxaflor (XXpire)	12 hr	4C + 5	G, L, N
	spirotetramat (Kontos)	24 hr foliar (see exception for drench application)	23	G, N
	thiamethoxam (Flagship)	12 hr	4A	G, L, N
Scale (Armored) check label to be sure it lists scale to be treated	acephate (Orthene)	24 hr	1B	G, L, N
	acetamiprid (Tri-Star)	12 hr	4A	G, L, N
	bifenthrin (Talstar)	12 hr	3	Follow label
	buprofezin (Talus)	12 hr	16	G, N
	dinotefuran (Safari)	12 hr	4A	G, L, N
	horticultural oil (various)	4 hr		G, L, N
	kinoprene (Enstar II)	4 hr	7A	G
	thiamethoxam (Flagship)	12 hr	4A	G, L, N
Scale (Soft) check label to be sure it lists scale to be treated	acephate (Orthene)	24 hr	1B	G, L, N
	acetamiprid (Tri-Star)	12 hr	4A	G, L, N
	bifenthrin (Talstar)	12 hr	3	Follow label
	buprofezin (Talus)	12 hr	16	G, N
	cyantraniliprole (Mainspring)	4 hr	28	G
	dinotefuran (Safari)	12 hr	4A	G, L, N
	horticultural oil (various)	4 hr		G, L, N
	imidacloprid (Marathon II, others)	12 hr	4A	Follow label
	kinoprene (Enstar II)	4 hr	7A	G
	neem oil (Various)	4 hr	UN	G, L, N
	pyriproxyfen (Distance)	12 hr	7C	G, L, N
	thiamethoxam (Flagship)	12 hr	4A	G, N
Shorefly	acephate (Orthene)	24 hr	1B	G, L, N
	azadirachtin (Azatin)	4 hr	18B	G, L, N
	bifenthrin (Talstar)	12 hr	3	Follow label
	diflubenzuron (Adept)	12 hr	15	G
	imidacloprid (Marathon II, others)	12 hr	4A	Follow label
	kinoprene (Enstar II)	4 hr	7A	G
	pyriproxyfen (Distance)	12 hr	7C	G, L, N
	spinetoram + sulfoxaflor (XXpire)	12 hr	4C + 5	G, L, N
Slugs	iron phosphate (bait)	Follow label	UN	Follow label
	metaldehyde (bait)	Follow label	UN	Follow label
	methiocarb (bait)	Follow label	1A	Follow label
Spider Mites	abamectin (Avid)	12 hr	6	G, L, N
	acequinocyl (Shuttle)	12 hr	20B	G, N
	bifenazate (Floramite)	12 hr	UN	G, L, N
	chlorfenapyr (Pylon)	12 hr	13	G
	clofentezine (Ovation)	12 hr	10A	G, N
	cyflumetofen (Sultan)	12 hr	25	G, L, N
	etoxazole (TetraSan)	12 hr	10B	G, L, N
	fenazaquin (Magus)	12 hr	21A	G, L, N
	fenpyroximate (Akari)	12 hr	21A	G, N
	hexythiazox (Hexygon)	12 hr	10B	G, L, N
	horticultural oil (various)	4 hr		Follow label
	insecticidal soaps	12 hr		Follow label
	pyridaben (Sanmite)	12 hr	21A	G, L, N
	spiromesifen (Judo)	12 hr	23	G, N
Thrips	abamectin (Avid)	12 hr	6	G, L, N
	acephate (Orthene)	24 hr	1B	G, L, N
	acetamiprid (Tri-Star)	12 hr	4A	G, L, N
	azadirachtin (Azatin)	4 hr	18B	G, L, N

INSECT CONTROL

Table 5-14. Arthropod Management for Ornamental Plants Grown in Greenhouses

Permitted application sites: G = greenhouse, L = landscape, N = Nursery. (Trade names listed are common examples of products that contain the active ingredient, not an endorsement of a particular product.)

Insect or Mite	Pesticide common name (Trade name)	Minimum Hours Between Application and Reentry	IRAC Mode of Action Group	Permitted application sites
Thrips (continued)	*Beuveria bassiana*	12 hr		Follow label
	bifenthrin (Talstar)	12 hr	3	Follow label
	chlorfenapyr (Pylon)	12 hr	13	G
	cyantraniliprole (Mainspring)	4 hr	28	G
	cyfluthrin (Decathlon)	12 hr	3A	G, L, N
	flonicamid (Aria)	12 hr	9B	G, L, N
	fluvalinate (Mavrik)	12 hr	3A	G, L, N
	horticultural oil (various)	4 hr		Follow label
	kinoprene (Enstar II)	4 hr	7A	G
	novaluron (Pedestal)	12 hr	5	G, N
	pyrethrins (various)	12 hr	3A	G, L, N
	pyridalyl (Overture)	12 hr	UN	G
	spinetoram + sulfoxaflor (XXpire)	12 hr	4C + 5	G, L, N
	spinosad (Conserve)	4 hr	5	G, L, N
	tolfenpyrad (Hachi-Hachi)	12 hr	21A	G
Whitefly	abamectin (Avid)	12 hr	6	G, L, N
	acephate (Orthene)	24 hr	1B	G, L, N
	acetamiprid (Tri-Star)	12 hr	4A	G, L, N
	azadirachtin (Azatin)	4 hr	18B	G, L, N
	Beuveria bassiana	12 hr		Follow label
	bifenthrin (Talstar)	12 hr	3	Follow label
	buprofezin (Talus)	12 hr	16	G, N
	cyantraniliprole (Mainspring)	4 hr	28	G
	cyfluthrin (Decathlon)	12 hr	3A	G, L, N
	dinotefuran (Safari)	12 hr	4A	G, L, N
	fenazaquin (Magus)	12 hr	21A	G, L, N
	fenoxycarb (Preclude)	12 hr	7B	G
	flonicamid (Aria)	12 hr	9B	G, L, N
	fluvalinate (Mavrik)	12 hr	3A	G, L, N
	horticultural oil (various)	4 hr		G, L, N
	imidacloprid (Marathon II, others)	12 hr	4A	Follow label
	insecticidal soaps	12 hr		G, L, N
	kinoprene (Enstar II)	4 hr	7A	G
	neem oil (Various)	4 hr	UN	G, L, N
	novaluron (Pedestal)	12 hr	5	G, N
	permethrin (Astro, others)	12 hr	3	Follow label
	pyridaben (Sanmite)	12 hr	21A	G, L, N
	pyriproxyfen (Distance)	12 hr	7C	G, L, N
	pyrifluquinazon (Rycar)	12 hr	UN	G
	spinetoram + sulfoxaflor (XXpire)	12 hr	4C + 5	G, L, N
	spirotetramat (Kontos)	24 hr foliar (see exception for drench application)	23	G, N
	thiamethoxam (Flagship)	12 hr	4A	G, N
	tolfenpyrad (Hachi-Hachi)	12 hr	21A	G

Arthropod Management for Ornamental Plants Grown in Nurseries or Landscapes

S. D. Frank, Entomology Extension

Successful pest management programs use a combination of appropriate pest control tactics. Always follow label precautions when handling or applying pesticides. Make chemical control part of an integrated pest management program that includes monitoring and pest identification along with appropriate cultural, physical, horticultural, and biological controls.

Responsible pesticide use includes resistance management. A system has been developed by the Insecticide Resistance Action Committee (IRAC; www.irac-online.org) to help you rotate chemicals correctly. Pesticides have been assigned an IRAC classification number based on their mode of action. To rotate properly, choose a product with a different IRAC number for each successive application directed against the same pest. Follow resistance management instructions on the label.

The information in this chart is not a substitute for the label. Pesticide labels and restrictions change frequently. The label will provide the most updated information. Read and understand all label information before using any pesticide. Do not use pesticides for uses other than those on the label. Check county and state regulations for any local restrictions on the use of products listed here before using them.

Table 5-15. Arthropod Management for Ornamental Plants Grown in Nurseries or Landscapes

Permitted application sites: G = greenhouse, L = landscape, N = Nursery. (Trade names listed are common examples of products that contain the active ingredient, not an endorsement of a particular product.)

Insect or Mite	Pesticide common name (Trade name)	Minimum Hours Between Application and Reentry	IRAC Mode of Action Group	Permitted application sites
Adelgid	acetamprid (TriStar)	12 hr	4A	G, L, N
	chlorantraniliprole (Acelepryn)	4 hr	28	L
	dinotefuran (Safari)	12 hr	4A	G, L, N
	horticultural oil (various)	4 hr		G, L, N
	imidacloprid (Merit, Marathon, others)	12 hr	4A	Follow label
	insecticidal soap (various)	12 hr		G, L, N
	spirotetramat (Kontos)	24 hr foliar (see exception for drench application)	23	G, N
	thiamethoxam (Flagship)	12 hr	4A	G, N
Aphid	abamectin (Avid)	12 hr	6	G, L, N
	acephate (Orthene)	24 hr	1B	G, L, N
	acetamiprid (TriStar)	12 hr	4A	G, L, N
	azadirachtin (Azatin)	4 hr	18B	G, L, N
	bifenthrin + imidacloprid (Allectus)	12 hr	3 + 4A	L
	bifenthrin + clothianidin (Aloft)	12 hr	4 + 4A	L
	Beauveria bassiana (BotaniGard)	4 hr		G, L, N
	carbaryl (Sevin)	12 hr	1A	L, N
	clothianidin (Celero, Arena)	12 hr	4A	Follow label
	cyfluthrin (Decathlon)	12 hr	3	G, N
	fluvalinate (Mavrik)	12 hr	3	G, L
	horticultural oil (various)	4 hr		G, L, N
	imidacloprid (Merit, Marathon)	12 hr	4A	Follow label
	neem oil (Triact) 70	4 hr	18B	G, L, N
	permethrin (Astro, Perm-up, others)	12 hr	3	Follow label
	pymetrozine (Endeavor)	12 hr	9B	G, L, N
	pyrethrins (various)	12 hr	3A	G, L, N
	insecticidal soap (various)	12 hr Follow label directions		G, L, N
	spinetoram + sulfoxaflor (XXpire)	12 hr	4C + 5	G, L, N
	spirotetramat (Kontos)	24 hr foliar (see exception for drench application)	23	G, N
	thiamethoxam (Flagship)	12 hr	4A	G, N
Armored Scale (such as Juniper scale, Oystershell scale, Pine needle scale, Tea scale, Euonymus scale, White peach scale)	acephate (Orthene)	24 hr	1B	G, L, N
	acetamiprid (TriStar)	12 hr	4A	G, L, N
	bifenthrin (Talstar)	12 hr	3	Follow label
	buprofezin (Talus)	12 hr	16	G, L, N
	carbaryl (Sevin)	Follow label directions	1A	L, N
	dinotefuran (Safari)	12 hr	4A	G, L, N
	horticultural oil (various)	4 hr		G, L, N
	insecticidal soap (various)	Follow label directions 12 hr		G, L, N
	neem oil (Triact) 70	4 hr	18B	G, L, N
	pyriproxyfen (Distance)	12 hr	7C	G, L, N
	spinetoram + sulfoxaflor (XXpire)	12 hr	4C + 5	G, L, N

157

INSECT CONTROL

Table 5-15. Arthropod Management for Ornamental Plants Grown in Nurseries or Landscapes

Permitted application sites: G = greenhouse, L = landscape, N = Nursery. (Trade names listed are common examples of products that contain the active ingredient, not an endorsement of a particular product.)

Insect or Mite	Pesticide common name (Trade name)	Minimum Hours Between Application and Reentry	IRAC Mode of Action Group	Permitted application sites
Asian Ambrosia Beetle	permethrin (Astro, Perm-up, Permethrin Pro)	12 hr	3	Follow label
Bagworm	acephate (Orthene)	24 hr	1B	G, L, N
	acetamiprid (Tri-Star)	12 hr	4A	G, L, N
	azadirachtin (Azatin)	4 hr	18B	G, L, N
	bifenthrin + imidacloprid (Allectus)	12 hr	3 + 4A	L
	bifenthrin + clothianidin (Aloft)	12 hr	4 + 4A	L
	Bacillus thuringiensis kurstaki (BiobitHP, DiPel, or Foray)	4 hr	11B2	G, L, N
	bifenthrin (Talstar)	Follow label directions	3	G, L, N
	carbaryl (Sevin)	Follow label directions	1A	L, N
	chlorantraniliprole (Acelepryn)	4 hr	28	L
	fluvalinate (Mavrik)	Follow label directions	3	G, L
	indoxacarb (Provaunt)	12 hr	22	L
	novaluron (Pedestal)	12 hr	15	G, N
	spinetoram + sulfoxaflor (XXpire)	12 hr	4C + 5	G, L, N
	spinosad (Conserve SC)	4 hr	5	G, N
Bark Beetles	permethrin (Astro, Perm-up, others	12 hr	3	Follow label
	bifenthrin (Onyx, Talstar)	Follow label directions	3	Follow label
Black Vine Weevil	acephate (Orthene)	Follow label directions	1A	G, L, N
	Beauveria bassiana (BotaniGard)	4 hr		G, L, N
	bifenthrin (Onyx, Talstar)	Follow label directions	3	Follow label
	cyfluthrin + imidacloprid (Discus)	12 hr	3 + 4A	N
	dinotefuran (Safari)	12 hr	4A	G, L, N
	fluvalinate (Mavrik)	Follow label directions	3	G, L
	imidacloprid (Merit, Marathon, others)	12 hr	4A	Follow label
Borers (Clearwing, flatheaded, and roundheaded borers are included in this section. Make sure label specifically lists the type of borer you are trying to control.)	azadirachtin (Azatin)	4 hr	18B	G, L, N
	chlorantraniliprole (Acelepryn)	4 hr	28	L
	cyfluthrin + imidacloprid (Discus)	12 hr	3 + 4A	N
	dinotefuran (Safari)	12 hr	4A	G, L, N
	imidacloprid (Merit, Marathon II, others)	12 hr	4A	Follow label
	bifenthrin (Onyx, Talstar)	Follow local regulations for landscape reentry	3	Follow label
	permethrin (Astro, Perm-up, Permethrin Pro)	12 hr	3	Follow label
Caterpillars (such as armyworm, budworm, eastern tent caterpillar, fall webworm, orangestriped oakworm, leafrollers)	acephate (Orthene)	24 hr	1B	G, L, N
	acetamiprid (Tri-Star)	12 hr	4A	G, L, N
	azadirachtin (Azatin)	4 hr	18B	G, L, N
	Bacillus thuringiensis kurstaki (DiPel)	4 hr	11B2	G, L, N
	bifenthrin (Onyx, Talstar)	Follow label directions	3	Follow label
	bifenthrin + imidacloprid (Allectus)	12 hr	3 + 4A	L
	bifenthrin + clothianidin (Aloft)	12 hr	4 + 4A	L
	carbaryl (Sevin)	12 hr	1A	L, N
	chlorantraniliprole (Acelepryn)	4 hr	28	L
	indoxacarb (Provaunt)	12 hr	22	L
	insecticidal soap (various)	Follow label directions		G, L, N
	novaluron (Pedestal)	12 hr	15	G, N
	permethrin (Astro, Perm-up, Permethrin Pro)	12 hr	3	Follow label
	spinetoram + sulfoxaflor (XXpire)	12 hr	4C + 5	G, L, N
	spinosad (Conserve SC)	4 hr	5	G, N
	tebufenozide (Confirm)	4 hr	18A	L, N
Cricket	bifenthrin (Onyx, Talstar)	12 hr	3	Follow label
	cyfluthrin (Decathlon)	Follow label directions	3	G, N
	pyrethrins (Pyrenone)	Follow label directions	3	Follow label
	insecticidal soap (various)	Follow label directions		G, L, N

Table 5-15. Arthropod Management for Ornamental Plants Grown in Nurseries or Landscapes

Permitted application sites: G = greenhouse, L = landscape, N = Nursery. (Trade names listed are common examples of products that contain the active ingredient, not an endorsement of a particular product.)

Insect or Mite	Pesticide common name (Trade name)	Minimum Hours Between Application and Reentry	IRAC Mode of Action Group	Permitted application sites
Eriophyid Mite	abamectin (Avid)	12 hr	6	G, L, N
	horticultural oil (various)	4 hr		G, L, N
	spiromesifen (Judo, Forbid)	12 hr	23	G, N
False Spider Mites (such as privet mite)	acequinocyl (Shuttle)	12 hr	20B	G, N
	bifenazate (Floramite)	12 hr	Un	G, N, L
	etoxazole (TetraSan)	12 hr	10B	G, N, L
	horticultural oil (various)	4 hr		G, N, L
	insecticidal soaps	12 hr		G, N, L
	spiromesifen (Judo, Forbid)	12 hr	23	follow label
Fungus Gnat Adults	bifenthrin (Talstar)	12 hr	3	follow label
	cyfluthrin (Decathlon)	12 hr	3A	G, L, N
	fluvalinate (Mavrik)	12 hr	3A	G, L, N
	insecticidal soaps	12 hr		G, L, N
	permethrin (Astro, others)	12 hr	3	Follow label
Fungus Gnat Larvae	acetamiprid (Tri-Star)	12 hr	4A	G, L, N
	azadirachtin (Azatin)	4 hr	18B	G, L, N
	Bacillus thuringiensis var. israelensis	4 hr	11A1	Gollow label
	chlorfenapyr (Pylon)	12 hr	13	G
	cyromazine (Citation)	12 hr	17	G, L, N
	diflubenzuron (Adept)	12 hr	15	G
	kinoprene (Enstar II)	4 hr	7A	G
	pyriproxyfen (Distance)	12 hr	7C	G, L, N
	Steinernema feltiae (various; beneficial nematode)	0 hr	Biological	G, L, N
Grasshopper	bifenthrin (Onyx, Talstar)	12 hr	3	Follow label
	carbaryl (Sevin) 5 bait	Follow label directions	1A	Follow label
	cyfluthrin (Decathlon)	Follow label directions	3	G, N
	insecticidal soap (various)	12 hr		G, L, N
Japanese Beetle (Adult) and other leaf-feeding scarab beetles	acetamiprid (Tri-Star)	12 hr	4A	G, L, N
	acephate (Orthene)	Follow label directions	1A	G, L, N
	azadirachtin (Azatin XL)	4 hr	18B	G, L, N
	bifenthrin (Onyx, Talstar)	Follow label directions	3	Follow label
	bifenthrin (Talstar, Onyx)	12 hr	3	Follow label
	bifenthrin + imidacloprid (Allectus)	12 hr	3 + 4A	L
	bifenthrin + clothianidin (Aloft)	12 hr	4 + 4A	L
	carbaryl (Sevin)	Follow label directions	3	L, N
	chlorantraniliprole (Acelepryn)	4 hr	28	L
	clothianidin (Arena)		4A	L
	cyfluthrin + imidacloprid (Discus)	12 hr	3 + 4A	N
	cyfluthrin (Decathlon) 20 WP	Follow label directions	3	G, N
	dinotefuran (Safari)	12 hr	4A	G, L, N
	imidacloprid (Merit, Marathon II, others)	12 hr	4A	Follow label
	permethrin (Astro, Perm-up, Permethrin Pro)	12 hr	3	Follow label
	spinetoram + sulfoxaflor (XXpire)	12 hr	4C + 5	G, L, N
	thiamethoxam (Flagship)	12 hr	4A	G, N
Lacebugs	acephate (Orthene)	Follow label directions	1A	G, L, N
	Beauveria bassina (BotaniGard)	4 hr		G, L, N
	bifenthrin (Talstar, Onyx)	12 hr	3	Follow label
	bifenthrin + imidacloprid (Allectus)	12 hr	3 + 4A	L
	bifenthrin + clothianidin (Aloft)	12 hr	4 + 4A	L
	carbaryl (Sevin)	12 hr	1A	L, N
	chlorantraniliprole (Acelepryn)	4 hr	28	L
	cyfluthrin + imidacloprid (Discus)	12 hr	3 + 4A	N
	dinotefuran (Safari)	12 hr	4A	G, L, N
	imidacloprid (Merit, Marathon, others)	12 hr	4A	Follow label

INSECT CONTROL

Table 5-15. Arthropod Management for Ornamental Plants Grown in Nurseries or Landscapes

Permitted application sites: G = greenhouse, L = landscape, N = Nursery. (Trade names listed are common examples of products that contain the active ingredient, not an endorsement of a particular product.)

Insect or Mite	Pesticide common name (Trade name)	Minimum Hours Between Application and Reentry	IRAC Mode of Action Group	Permitted application sites
Lacebugs (continued)	permethrin (Astro, Perm-up, Permethrin Pro)	12 hr	3	Follow label
	spinetoram + sulfoxaflor (XXpire)	12 hr	4C + 5	G, L, N
	soap (Olympic Insecticidal)	Follow label directions 12 hr		Follow label
	thiamethoxam (Flagship)	12 hr	4A	G, N
Leaf Beetles (such as cucmber beetle, elm leaf beetle, willow leaf beetle, and flea beetles including Altica spp.)	acephate (Orthene)	12 hr	1A	G, L, N
	acetamiprid (TriStar)	12 hr	4A	G, L, N
	bifenthrin (Onyx, Talstar)	12 hr	3	Follow label
	bifenthrin + imidacloprid (Allectus)	12 hr	3 + 4A	L
	bifenthrin + clothianidin (Aloft)	12 hr	4 + 4A	L
	carbaryl (Sevin)	12 hr	3	L, N
	chlorantraniliprole (Acelepryn)	4 hr	28	L
	cyfluthrin + imidacloprid (Discus)	12 hr	3 + 4A	N
	dinotefuran (Safari)	12 hr	4A	G, L, N
	imidacloprid (Merit, Marathon II, others)	12 hr	4A	Follow label
	spinetoram + sulfoxaflor (XXpire)	12 hr	4C + 5	G, L, N
	spinosad (Conserve SC)	4 hr	5	G, N
	thiamethoxam (Flagship)	12 hr	4A	G, N
Leafhoppers (such as potato leafhopper and sharpshooters)	acephate (Orthene)	Follow label directions	1A	G, L, N
	acetamiprid (TriStar)	12 hr	4A	G, L, N
	bifenthrin (Onyx, Talstar)	Follow label directions	3	Follow label
	bifenthrin + imidacloprid (Allectus)	12 hr	3 + 4A	L
	bifenthrin + clothianidin (Aloft)	12 hr	4 + 4A	L
	carbaryl (Sevin)	Follow label directions	1A	L, N
	clothianidin (Arena)	12 hr	4A	L
	cyfluthrin (Decathlon)	Follow label directions	3	G, N
	cyfluthrin + imidacloprid (Discus)	12 hr	3 + 4A	N
	dinotefuran (Safari)	12 hr	4A	G, L, N
	fluvalinate (Mavrik)	Follow label directions	3	G, L
	imidacloprid (Merit, Marathon II, others)	12 hr	4A	Follow label
	neem oil (Triact) 90 EC	4 hr	18B	G, L, N
	permethrin (Astro, Perm-up, Permethrin Pro)	12 hr	3	Follow label
	thiamethoxam (Flagship)	12 hr	4A	G, N
	insecticidal soap	Follow label directions		G, L, N
	spirotetramat (Kontos)	24 hr foliar (see exception for drench application)	23	G, N
Leafminers (such as boxwood leafminer, holly leafminer, birch leafminer) Note this includes dipterous, lepidopterous, and coleopterus leafminers. Make sure leafminer to be treated is listed on label.	abamectin (Avid)	Follow label directions	6	G, L, N
	acephate (Orthene)	Follow label directions	1A	G, L, N
	acetamiprid (TriStar)	24 hr	4A	G, L, N
	azadirachtin (Azatin XL)	12 hr	18B	G, L, N
	bifenthrin (Onyx, Talstar)	Follow label directions	3	Follow label
	chlorantraniliprole (Acelepryn SC)	4 hr	28	L
	clothianidin (Arena)	12 hr	4A	L
	cyfluthrin + imidacloprid (Discus)	12 hr	3 + 4A	N
	dinotefuran (Safari)	12 hr	4A	G, L, N
	imidacloprid (Merit, Marathon, others)	12 hr	4A	Follow label
	permethrin (Astro, Perm-up, Permethrin Pro)	12 hr	3	Follow label
	pyriproxyfen (Distance)	12 hr	7C	G, L, N
	spinosad (Conserve SC)	4 hr	5	G, N
Mealybugs	acephate (Orthene)	12 hr	1A	G, L, N
	acetamiprid (TriStar)	24 hr	4A	G, L, N
	Beauveria bassina (BotaniGard)	4 hr		G, L, N
	bifenthrin (Onyx, Talstar)	Follow label directions	3	Follow label
	buprofezin (Talus)	12 hr	16	G, N
	carbaryl (Sevin)	Follow label directions	1A	L, N

160

Table 5-15. Arthropod Management for Ornamental Plants Grown in Nurseries or Landscapes

Permitted application sites: G = greenhouse, L = landscape, N = Nursery. (Trade names listed are common examples of products that contain the active ingredient, not an endorsement of a particular product.)

Insect or Mite	Pesticide common name (Trade name)	Minimum Hours Between Application and Reentry	IRAC Mode of Action Group	Permitted application sites
Mealybugs (continued)	cyfluthrin (Decathlon) 20 WP	Follow label directions	3	G, N
	clothianidin (Arena, Celero)		4A	L
	cyfluthrin + imidacloprid (Discus)	12 hr	3 + 4A	N
	dinotefuran (Safari)	12 hr	4A	G, L, N
	fluvalinate (Mavrik) 22.3 F	Follow label directions	3	G, L
	imidacloprid (Merit, Marathon, others)	12 hr	4A	Follow label
	neem oil (Triact)	4 hr	18B	G, L, N
	permethrin (Astro, Perm-up, Permethrin Pro)	12 hr	3	Follow label
	insecticidal soap (various)	Follow label directions 12 hr		G, L, N
	horticultural oil (various)	4 hr		G, L, N
	spinetoram + sulfoxaflor (XXpire)	12 hr	4C + 5	G, L, N
	spirotetramat (Kontos)	24 hr foliar (see exception for drench application)	23	G, N
	thiamethoxam (Flagship)	12 hr	4A	G, N
Pillbug	bifenthrin (Onyx, Talstar)	12 hr	3	Follow label
	cyfluthrin (Decathlon) 20 WP	Follow label directions	3	G, N
Plantbugs	bifenthrin (Onyx, Talstar)	Follow label directions	3	Follow label
	cyfluthrin (Decathlon)	Follow label directions	3	G, N
	permethrin (Astro, others)	12 hr	3	Follow label
	spinetoram + sulfoxaflor (XXpire)	12 hr	4C + 5	G, L, N
	thiamethoxam (Flagship)	12 hr	4A	G, N
	insecticidal soap (various)	Follow label directions 12 hr		G, L, N
Psyllid	acephate (Orthene)	Follow label directions	1A	G, L, N
	acetamiprid (TriStar)	24 hr	4A	G, L, N
	azadirachtin (Azatin XL)	12 hr	18B	G, L, N
	Beauveria bassiana (BotaniGard)	4 hr		G, L, N
	dinotefuran (Safari)	12 hr	4A	G, L, N
	imidacloprid (Merit, Marathon, others)	12 hr	4A	Follow label
	insecticidal soap (various)	12 hr		G, L, N
	neem oil (Triact)	4 hr	18B	G, L, N
	spinosad (Conserve SC)	4 hr	5	G, N
	thiamethoxam (Flagship)	12 hr	4A	G, N
Sawfly	acephate (Orthene)	Follow label directions	1A	G, L, N
	acetamiprid (TriStar)	24 hr	4A	G, L, N
	carbaryl (Sevin)	Follow label directions	1A	L, N
	chlorantraniliprole (Acelepryn SC)	4 hr	28	L
	cyfluthrin (Decathlon) 20WP	Follow label directions	3	G, N
	cyfluthrin + imidacloprid (Discus)	12 hr	3 + 4A	N
	imidacloprid (Merit, Marathon, others)	12 hr	4A	Follow label
	indoxacarb (Provaunt)	12 hr	22	L
	insecticidal soap (various)	12 hr		G, L, N
	spinetoram + sulfoxaflor (XXpire)	12 hr	4C + 5	G, L, N
	spinosad (Conserve SC)	4 hr	5	G, N
Slug, Snail	iron phosphate (bait)	follow label	UN	Follow label
	metaldehyde + carbaryl (Sevin) bait	Follow label directions	Follow label	Follow Label
	methiocarb (Mesurol)	24 hr	1A	Follow label
Soft Scale (such as fletcher scale, cottony maple scale, wax scale)	acetamiprid (Tri-Star)	12 hr	4A	G, L, N
	buprofezin (Talus)	12 hr	16	G, N
	cyfluthrin + imidacloprid (Discus)	12 hr	3 + 4A	N
	dinotefuran (Safari)	12 hr	4A	G, L, N
	horticultural oil (various)	4 hr		G, L, N
	imidacloprid (Merit, Marathon, others)	Follow label directions	4A	Follow label
	pyriproxyfen (Distance)	12 hr	7C	G, L, N

161

Table 5-15. Arthropod Management for Ornamental Plants Grown in Nurseries or Landscapes

Permitted application sites: G = greenhouse, L = landscape, N = Nursery. (Trade names listed are common examples of products that contain the active ingredient, not an endorsement of a particular product.)

Insect or Mite	Pesticide common name (Trade name)	Minimum Hours Between Application and Reentry	IRAC Mode of Action Group	Permitted application sites
Soft Scale (such as fletcher scale, cottony maple scale, wax scale) (continued)	spinetoram + sulfoxaflor (XXpire)	12 hr	4C + 5	G, L, N
	thiamethoxam (Flagship)	12 hr	4A	G, N
Sowbug	cyfluthrin (Decathlon)	Follow label directions	3	G, N
Spider Mite (such as twospotted, southern red, and spruce spider mite)	abamectin (Avid)	12 hr	6	G, L, N
	acequinocyl (Shuttle)	12 hr	20B	G, N
	bifenazate (Floramite)	12 hr	Un	G, L, N
	clofentezine (Ovation)	12 hr	10A	G, N
	cyflumetofen (Sultan)	12 hr	25	G, L, N
	etoxazole (TetraSan)	12 hr	10B	G, N, L
	fenazaquin (Magus)	12 hr	21A	G, L, N
	fenpyroximate (Akari)	12 hr	21A	G, N
	hexythiazox (Hexygon)	12 hr	10B	G, L, N
	horticultural oil (various)	4 hr		follow label
	insecticidal soaps	12 hr		follow label
	pyridaben (Sanmite)	12 hr	21A	G, L, N
	spiromesifen (Judo, Forbid)	12 hr	23	Follow label
Spittlebug	acephate (Orthene)	12 hr	1A	G, L, N
	cyfluthrin (Decathlon)	Follow label directions	11B2	G, N
	horticultural oil (various)	4 hr		follow label
	insecticidal soaps	12 hr		follow label
Thrips	abamectin (Avid)	12 hr	6	G, L, N
	acephate (Orthene)	24 hr	1B	G, L, N
	acetamiprid (Tri-Star)	12 hr	4A	G, L, N
	azadirachtin (Azatin)	4 hr	18B	G, L, N
	Beauveria bassina (BotaniGard)	4 hr		G, L, N
	bifenthrin (Onyx, Talstar)	Follow label directions	3	Follow label
	cyfluthrin (Decathlon)	12 hr	3A	G, L, N
	flonicamid (Aria)	12 hr	9B	G, L, N
	fluvalinate (Mavrik)	12 hr	3A	G, L, N
	horticultural oil (various)	4 hr		Follow label
	novaluron (Pedestal)	12 hr	5	G, N
	spinetoram + sulfoxaflor (XXpire)	12 hr	4C + 5	G, L, N
	spinosad (Conserve SC)	4 hr	4	G, N
Twig Borer	bifenthrin (Onyx, Talstar)	12 hr	3	Follow label
Whitefly	abamectin (Avid)	12 hr	6	G, L, N
	acephate (Orthene)	12 hr	1A	G, L, N
	acetamiprid (TriStar)	12 hr	4A	G, L, N
	azadirachtin (Azatin)	4 hr	18B	G, L, N
	Beauveria bassina (BotaniGard)	4 hr		G, L, N
	bifenthrin (Onyx, Talstar)	12 hr	3	Follow label
	buprofezin (Talus)	12 hr	16	G, N
	cyfluthrin (Decathlon)	Follow label directions	3	G, N
	dinotefuran (Safari)	Follow label directions	4A	G, L, N
	fenazaquin (Magus)	12 hr	21A	G, L, N
	fluvalinate (Mavrik)	Follow label directions	3	G, L
	flonicamid (Aria)	12 hr	9B	G, L, N
	horticultural oil (various)	4 hr		G, L, N
	imidacloprid (Merit, Marathon, others)	12 hr	4A	Follow label
	insecticidal soap (various)	Follow label directions 12 hr		G, L, N
	neem oil (Triact)	4 hr	18B	G, L, N
	novaluron (Pedestal)	12 hr	5	G, N
	permethrin (Astro, others)	12 hr	3	Follow label
	pyridaben (Sanmite)	12 hr	21A	G, L, N

Table 5-15. Arthropod Management for Ornamental Plants Grown in Nurseries or Landscapes

Permitted application sites: G = greenhouse, L = landscape, N = Nursery. (Trade names listed are common examples of products that contain the active ingredient, not an endorsement of a particular product.)

Insect or Mite	Pesticide common name (Trade name)	Minimum Hours Between Application and Reentry	IRAC Mode of Action Group	Permitted application sites
Whitefly (continued)	pyriproxyfen (Distance) 11.2 EC	12 hr	7C	G, L, N
	spinetoram + sulfoxaflor (XXpire)	12 hr	4C + 5	G, L, N
	spirotetramat (Kontos)	24 hr foliar (see exception for drench application)	23	G, N
	thiamethoxam (Flagship)	12 hr	4A	G, N
White Grubs (in containers or landscape plants (not turf) such as oriental and Japanese beetle)	*Beauveria bassina* (BotaniGard)	4 hr		G, L, N
	chlorantraniliprole (Acelepryn)	4 hr	28	L
	clothianidin (Arena)	12 hr	4A	L
	dinotefuran (Safari)	12 hr	4A	G, L, N
	imidacloprid (Merit, Marathon, others)	12 hr	4A	Follow label
	thiamethoxam (Flagship)	12 hr	4A	G, N

Arthropod Control on Christmas Trees

J. R. Sidebottom, Entomology Forestry

Table 5-16. Arthropod Control on Christmas Trees

** N.C. label

Insect or Mite Insecticide and Formulations	Amount of Formulation per Gallon of Spray	Amount per 100 Gallons of Water	Minimum Interval (Hours) Between Application and Reentry	Precautions and Remarks
Adelgids (Balsam Woolly Adelgid, Cooley, Eastern Spruce Gall)				
bifenthrin (Talstar Nursery Flowable)		20 to 40 oz/acre	12	Will also control twig aphids and spider mites but not rust mites.
bifenthrin 25% (Sniper)		3.9 to 12.8 oz/acre	12	Will also control twig aphid and spider mites but not rust mites.
bifenthrin (OnyxPro)		1.8 to 14.4 oz/100 gal	12	
chlorpyrifos (Lorsban 4E or Nufos 4E)		1 qt/acre	24	Do not treat plants under extreme heat or drought stress. Control is achieved only when eggs and crawlers are not present.
dinotefuran (Safari)		4 to 8 oz/100 gal	12	Do not apply more than 2.7 pounds per acre.
esfenvalerate (Asana XL or Adjourn)		5.8 to 9.6 oz/100 gal	12	Use full rate to control balsam woolly adelgid.
imidacloprid (Couraze 1.6F, Pasada 1.6F, or Provado 1.6F)		4 to 8 oz/acre OR 2 oz/100 gal	12	Adding a spray adjuvant may improve coverage. Do not apply more than 40 ounces per acre per year.
lambda-cyhalothrin (Lambda-T, Silencer or Warrior)		2.58 to 5.12 oz/acre	24	Maximum use 1.92 pints/acre/year
petroleum oil (Damoil)		2 to 4 gal/100 gal dormant use 1 to 3 gal/100 gal summer use	4 hr	
spirotetramat (Movento)		5 to 10 oz/acre	24	Maximum use 20 ounces/acre/year. Use adjuvant to increase penetration.
Ants				
bifenthrin (Talstar Nursery Flowable)		5 to 10 oz/acre	12	
carbaryl (Sevin SL)		1 qt/acre	12	
chlorpyrifos (Lorsban 4E)		1 qt/acre	24	Do not treat plants under extreme heat or drought stress.

INSECT CONTROL

Table 5-16. Arthropod Control on Christmas Trees

** N.C. label

Insect or Mite Insecticide and Formulations	Amount of Formulation per Gallon of Spray	Amount per 100 Gallons of Water	Minimum Interval (Hours) Between Application and Reentry	Precautions and Remarks
Aphid (including Balsam Twig Aphid and Cinara Aphid)				
abamectin (Avid 0.15 EC)		8 oz/100 gal	12	Do not apply more than 16 ounces or less than 8 ounces per acre. To suppress aphids, spray must contact young immatures.
azadirachtin (Aza-Direct)		1 to 2 pts/acre	4	Under extremely heavy pest pressure up to 3.5 pints may be used.
Beaveria bassiana (Naturalis T&O)	0.3 to 1 oz/gal	30 to 100 oz/100 gal	4	Spray immediately after mixing.
bifenthrin (Talstar Nursery Flowable)		5 to 40 oz/acre	12	
bifenthrin 25% (Sinper)		3.9 to 12.8 oz/acre	12	Will also control twig aphid and spider mites but not rust mites.
bifenthrin (OnyxPro)		1.8 to 14.4 oz/100 gal	12	
carbaryl (Chipco Sevin SL)		1 qt/acre	12	
chlorpyrifos (Lorsban 4E or Nufos 4E)		1 qt/acre	24	Do not treat plants under extreme heat or drought stress.
cinnamaldehyde (Cinnamite)	0.85 oz/gal	85 oz/100 gal	4	
dimethoate (Dimethoate 400 or Clean Crop)		1 to 1 1/2 pt/acre	10 days	
disulfoton (Di-Syston 15 G)		1 tsp/tree OR 20 to 30 lb/acre	48 where rainfall exceeds 25 in./year	Spread the granules in the root zone of the trees at the dripline and work into the soil or water thoroughly within 48 hours of application. Not for use in bare-ground plantations.
esfenvalerate (Asana XL or Adjourn)		5.8 to 9.6 oz/100 gal	12	
flupyradifurone (Sivanto Prime)		7 to 14 oz/acre	4	Not for use in bare-ground plantations.
imidacloprid (Couraze 1.6F, Pasada 1.6F, or Provado 1.6 F)		4 to 8 oz/acre or 2 oz/100 gal	12	Adding a spray adjuvant may improve control. Do not apply more than 40 ounces per acre per year.
lambda-cyhalothrin (Lambda-T, Silencer or Warrior)		2.58 to 5.12 oz/acre	24	Maximum use 1.92 pints/acre/year
petroleum oil (Damoil)		2 to 4 gal/100 gal dormant use 1 to 3 gal/100 gal summer use	4 hr	
pymetrozine (Endeavor)		Up to 10 oz/acre	12	
spirotetramat (Movento)		5 to 10 oz/acre	24	Maximum use 20 ounces/acre/year. Use adjuvant to increase penetration.
thiamethoxam (Flagship 25WP)		2 to 4 oz/100 gal or 4 to 8 oz/acre	12	Maximum use 8 ounces/acre/year
Bagworm				
azadirachtin (Aza-Direct)		1 to 2 pt/acre	4	Under extremely heavy pest pressure up to 3.5 pints may be used.
bifenthrin (Talstar Nursery Flowable)		5 to 10 oz/acre	12	
carbaryl (Sevin SL)		1 qt/acre	12	
diflourorbenzamide (Dimilin 4L)		1 to 8 oz/acre	12	Apply to early instars in mid- to late June.
dimethoate (Dimethoate 400 or Clean Crop)		1 to 1 1/2 pt/acre	10 days	
lambda-cyhalothrin (Lambda-T, Silencer or Warrior)		2.58 to 5.12 oz/acre	24	Maximum use 1.92 pints/acre/year
spinosad (Conserve SC)		4 to 16 oz/acre	4	
tebufenozide (Confirm)		4 to 8 oz/acre	4	Apply to early instar larvae; foliage development should be minimum of 20%. Do not apply more than 16 ounces per acre per year.
Elongate Hemlock Scale and Cryptomeria Scale				
bifenthrin 25% (Sniper)		3.9 to 12.8 oz/acre	12	Will also control twig aphid and spider mites but not rust mites.
buprofezin (Talus 70 DF)		14 oz/acre	12	
buprofezin (Talus 40 SC)		21.5 oz/acre	12	
dimethoate (Dimethoate 400 orClean Crop)		1 to 1 1/2 pt/acre	10 days	
esfenvalerate (Asana XL)		5.8 to 9.6 oz/100 gal	12	Best results when mixed with dimethoate.
dinotefuran (Safari)		4 to 8 oz/100 gal	12	Do not apply more than 2.7 pounds per acre.
Talus		14 oz/100 gal	12	
European Pine Shoot Moth				
azadirachtin (Aza-Direct)		1 to 2 pts/acre	4	Under extremely heavy pest pressure up to 3.5 pints may be used.
chlorpyrifos (Lorsban 4E or Nufos 4E)		1 qt/acre	24	Do not treat plants under extreme heat or drought stress.
dimethoate (Dimethoate 400 or Clean Crop)		1 to 1 1/2 pt/acre	10 days	
phosmet (Imidan 70-WSB or Gowan)		1.3 to 1.5 lb/acre	24	

Table 5-16. Arthropod Control on Christmas Trees

** N.C. label

Insect or Mite Insecticide and Formulations	Amount of Formulation per Gallon of Spray	Amount per 100 Gallons of Water	Minimum Interval (Hours) Between Application and Reentry	Precautions and Remarks
Gypsy Moth				
azadirachtin (Aza-Direct)		1 to 2 pts/acre	4	Under extremely heavy pest pressure up to 3.5 pints may be used.
bifenthrin (Talstar Nursery Flowable)		10 to 20 oz/acre	12	
chlorpyrifos (Lorsban 4E or Nufos 4E)		1 qt/acre	24	Do not treat plants under extreme heat or drought stress.
diflourorbenzamide (Dimilin 4L)		0.5 to 2 oz/acre	12	Apply to early instar and prior to full leaf expansion.
lambda-cyhalothrin (Lambda-T, Silencer or Warrior)		2.58 to 5.12 oz/acre	24	Maximum use 1.92 pints/acre/year
phosmet (Imidan 70-WSB or Gowan)		1.3 to 1.5 lb/acre	24	
spinosad (Conserve SC)		4 to 16 oz/acre	4	
terbufenozide (Confirm)		4 to 8 oz/acre	4	Apply to early instar larvae after each foliage flush at approximately 25% foliage expansion. Allow at least 6 hours between application and rainfall to assure thorough spray drying.
Midge (Douglas fir needle midge, pine needle midge)				
chlorpyrifos (Lorsban 4E)		1 qt/acre	24	Do not treat plants under extreme heat or drought stress.
esfenvalerate (Asana XL)		5.8 to 9.6 oz/100 gal	12	
Nantucket Pine Tip Moth				
azadirachtin (Aza-Direct)		1 to 2 pts/acre	4	Under extremely heavy pest pressure up to 3.5 pints may be used.
carbaryl (Sevin SL)		1 qt/acre	12	
diflourorbenzamide (Dimilin 4L)		1 to 2 oz/acre	12	Apply when second generation instars are present or 70% of first generation pupal cases are empty.
dimethoate (Dimethoate 400 or Clean Crop)		1 to 1 1/2 pt/acre	10 days	
**disulfoton (Di-Syston) 15 G	1 tsp/tree; 20 to 30 lb/acre		48	Spread granules in the root zone of the trees at the dripline and work into the soil or water thoroughly within 48 hours of application. Not for use in bare-ground plantations.
esfenvalerate (Asana XL or Adjourn)		5.8 to 9.6 fl oz	12	Apply as needed for control. Spray sufficient gallonage to obtain good coverage of entire tree.
permethrin (Permethrin 3.2 EC or Pounce)		4 to 8 oz/100 gal	12	Begin applications when adults appear. Repeat applications may be made in 5- to 7-day intervals as needed.
phosmet (Imidan 70-WSB or Gowan)		1.3 to 1.5 lb/acre	24	
tebufenozide (Confirm)		8 oz/acre	4	Apply to early instar larvae after each foliage flush at approximately 25% foliage expansion. Allow at least 6 hours between application and rainfall to assure thorough spray drying.
esfenvalerate (Asana XL)		5.8 to 9.6 oz/100 gal	12	
Pine Chafer				
esfenvalerate (Asana XL or Adjourn)		5.8 to 9.6 oz/100 gal	12	
lambda-cyhalothrin (Lambda-T, Silencer or Warrior)		2.58 to 5.12 oz/acre	24	Maximum use 1.92 pints/acre/year
Rosette Bud Mite				
dimethoate (various brands)		1.3 pt/100 gal	10 days	
spirotetramat (Movento)		5 to 10 oz/acre	24	Maximum use 20 ounces/acre/year. Use adjuvant to increase penetration.
Rust Mites				
abamectin (Ardent 0.15EC)		4 oz/100 gal	12	
chloropyridan (Sanmite)		4 oz/100 gal or 10.7 oz/acre	12	
chlorpyrifos (Lorsban 4E)		1 qt/acre	24	Do not treat plants under extreme heat or drought stress.
dimethoate (Dimethoate 400 or Clean Crop)		1 to 1.5 pt/acre	10 days	
petroleum oil (Damoil)		2 to 4 gal/100 gal dormant use 1 to 3 gal/100 gal summer use	4 hr	
spirodiclofen (Envidor 2SC)		18 to 24.7 oz/acre	24	Make only one application per season.
Sawflies (Redheaded pine, red pine, European pine)				
carbaryl (Sevin SL)		1 qt/acre	12	
chlorpyriphos (Nufos 4E)		1 qt/acre	24	
diflourorbenzamide (Dimilin 4L)		2 to 4 oz/acre	12	Apply to early instars.
esfenvalerate (Asana XL or Adjourn)		5.8 to 9.6 oz/100 gal	12	
imidacloprid (Couraze 1.6F, Pasada 1.6F, or Provado 1.6 F)		4 to 8 oz/acre or 2 oz/100 gal	12	Adding a spray adjuvant may improve control. Do not apply more than 40 ounces per acre per year.

INSECT CONTROL

Table 5-16. Arthropod Control on Christmas Trees

** N.C. label

Insect or Mite Insecticide and Formulations	Amount of Formulation per Gallon of Spray	Amount per 100 Gallons of Water	Minimum Interval (Hours) Between Application and Reentry	Precautions and Remarks
Sawflies (Redheaded pine, red pine, European pine) (continued)				
lambda-cyhalothrin (Lambda-T, Silencer or Warrior)		2.58 to 5.12 oz/acre	24	Maximum use 1.92 pints/acre/year
malathion (Malation 8 or Gowan)	2 tbsp/gal	1 pt/100 gal	12	
phosmet (Imidan 70-WSB or Gowan)		1.3 to 1.5 lb/acre	24	
Scale (Pine needle, pine tortoise, spruce bud, black pine, stripped pine; see also Elongate Hemlock and Cryptomeria Scale)				
azadirachtin (Aza-Direct)		1 to 2 pts/acre	4	Under extremely heavy pest pressure up to 3.5 pints may be used.
chlorpyrifos (Lorsban 4E)		1 qt/acre	24	Do not treat plants under extreme heat and drought stress. Apply when scale crawlers are active.
dinotefuran (Safari)		4 to 8 oz/100 gal	12	Do not apply more than 2.7 pounds per acre.
lambda-cyhalothrin (Lambda-T or Silencer)		2.58 to 5.12 oz/acre	24	Maximum use 1.92 pints/acre/year
petroleum oil (Damoil)		2 to 4 gal/100 gal dormant use 1 to 3 gal/100 gal summer use	4	
thiamethoxam (Flagship 25WP)		2 to 4 oz/100 gal or 4 to 8 oz/acre	12	Maximum use 8 ounces/acre/year
Seed Bugs				
permethrin (Permethrin 3.2 EC)		30 oz/acre	12	
lambda-cyhalothrin (Lambda-T, Silencer or Warrior)		2.58 to 5.12 oz/acre	24	Maximum use 1.92 pints/acre/year
Spider Mite (Spruce spider mites)				
abamectin (Avid 0.15 EC)		4 to 8 oz/100 gal	12	Do not apply more than 16 ounces or less than 8 ounces per acre.
abamectin (Ardent 0.15EC)		4 oz/100 gal	12	
Beauveria bassiana (Naturalis T&O)	0.3 to 1 oz/gal	30 to 100 oz/100 gal	4	Spray immediately after mixing
bifenazate (Floramite)		2 to 8 oz/100 gal	12	Add an adjuvant like Silwet L-77 or Slygard 309 to the Floramite solution.
bifenthrin (Talstar Nursery Flowable)		5 to 40 oz/acre	12	
bifenthrin 25% (Sniper)		3.9 to 12.8 oz/acre	12	Will also control twig aphid and spider mites but not rust mites.
bifenthrin (OnyxPro)		1.8 to 14.4 oz/100 gal	12	
chloropyridan (Sanmite)		4 oz/100 gal or 10.7 oz/acre	12	
chlorpyrifos (Lorsban 4E or Nufos 4E)		1 qt/acre	24	Do not treat plants under extreme heat or drought stress. If eggs are present, reapply in 7 to 10 days to control newly hatched nymphs.
clofentezine (Apollo SC)	—	4 to 8 oz/acre	12	Most effective when applied at first sign of mite activity and mite eggs.
cinnamaldehyde (Cinnamite)	2 tbsp/gal	85 oz/100 gal	4	
cyflumetofen (Sultan)		13.7 oz/100 gal	12	Do not make more than 2 applications per year. Use at least 100 gallons of water per acre and get thorough coverage. Do not tank mix with insect or plant growth regulators or cabamate, organophosphate, or pyrethroid insecticides.
dimethoate (Dimethoate 400 or Clean Crop)		1 to 1 1/2 pt/acre	48	
disulfoton (Di-Syston 15G)		1 tsp/tree20 to 30 lb/acre	48 where rainfall exceeds 25 in/year	Spread the granules in the root zone of the trees at the dripline and work into the soil or water thoroughly within 48 hours of application. Not for use in bare-ground plantations.
etoxazole (TetraSan 5 WDG)		8 to 16 oz/100 gal	12	TetraSan kills mite eggs and nymphs but not adult mite. Treated adults will not produce viable eggs.
fenpyroximate (Akari 5SC)		16 to 24 oz/100 gal	12	
hexathiazox (Savey) 50 WP	3 to 6 oz/acre	2 oz/100 gal	12	Do not make more than one application per year.
propargite (Ornamite-CR)		3 to 7.5 lb/acre	7 days	Make no more than three applications per year. Compatibility restrictions.
spirodichlofen (Envidor 2SC)		18 to 24.7 oz/acre	24	Make only one application per season.
spiromesifen (Judo)		1 to 4 oz/100 gal	12	Do not apply more than four times per season.
Spittlebug				
chlorpyrifos (Lorsban 4F)		1 qt/acre	24	Do not treat plants under extreme heat or drought stress.
esfenvalerate (Asana XL)		5.8 to 9.6 oz/100 gal	12	
lambda-cyhalothrin (Lambda-T, Silencer or Warrior)		2.58 to 5.12 oz/acre	24	Maximum use 1.92 pints/acre/year

Table 5-16. Arthropod Control on Christmas Trees

** N.C. label

Insect or Mite Insecticide and Formulations	Amount of Formulation per Gallon of Spray	Amount per 100 Gallons of Water	Minimum Interval (Hours) Between Application and Reentry	Precautions and Remarks
Spruce Needle Miner				
chlorpyrifos (Lorsban 4E)		1 qt/acre	24	Do not treat plants under extreme heat or drought stress.
Weevils (pales, northern pine, pitch eating, root collar, white pine				
azadirachtin (Aza-Direct)		1 to 2 pts/acre	4	Under extremely heavy pest pressure up to 3.5 pints may be used.
chlorpyrifos (Lorsban 4E or Nufos 4E)	2 tbsp/gal	3 qt/100 gal	24	Apply as a cut stump drench.
diflourobenzamide (Dimilin 4L)		4 to 8 oz/acre	12	Treat prior to egg deposition.
esfenvalerate (Asana XL or Adjourn)		5.8 to 9.6 oz/100 gal	12	
phosmet (Imidan 70-WSP or (Gowan)		1.3 to 1.5 lb/acre	24	
White Grubs				
chlorpyrifos (Lorsban 4E)		1 qt/acre	24	Incorporate into the soil if possible.
imidacloprid (Admire Pro)		7 to 14 oz/acre	12	Maximum per season: 14 ounces/acre
thiamethoxam (Flagship 25WG)		8 oz/acre	12	Apply from adult flight through peak hatch of targeted species.
Zimmerman Pine Moth				
azadirachtin (Aza-Direct)		1 to 2 pts/acre	4	Under extremely heavy pest pressure up to 3.5 pints may be used.
dimethoate (Dimethoate 400 or Clean Crop)		1 to 1 1/2 pt/acre	10 days	

Commercial Turf Insect Control

R. L. Brandenburg, Entomology Extension

Table 5-17. Insect Control in Commercial Turf

Pest Insecticide and Formulation	Amount per 1,000 sq ft	Precautions and Remarks
Annual Bluegrass Weevil		
		Recommmendations available at www.turffiles.ncsu.edu early 2017.
Ant (also see Imported Fire Ant)		
bifenthrin[1] (Menace, Talstar, others) F, GC; G form also available	0.5 to 1 fl oz	Use GC formulation for golf courses.
carbaryl[1] (Sevin) 80 WSP	1 to 1.5 oz	
chlorpyrifos[1] (Dursban) 2E, 4E, 2G, 50 WP, Pro eta-Cypermethrin, bifenthrin, and imadacloprid (Triple Crown)	See label 20-35 fl oz/acre	For use on golf courses. See new label.
chlothianidin + bifenthrin (Aloft) GC SC LC SC GC G LC G	0.27 to 0.54 fl oz 0.27 to 0.54 fl oz 1.8 to 3.6 lb 1.8 to 3.6 lb	
cyfluthrin (Tempo 2)	0.143 fl oz	Home lawns only.
cypermethrin[1] (Demon) TC	See label	
deltamethrin (Deltagard) G	2 to 3 lb/1,000 ft	
fipronil 0.0143 G (Top Choice, Taurus G)	2 lb	
hydramethylnon[1] (Maxforce G, Amdro)	See label	
lambda-cyhalothrin[1] (Battle, Scimitar, Cyonara)	See label	Do not make applications within 20 feet of any body of water. No reentry until spray has dried.

167

Table 5-17. Insect Control in Commercial Turf

Pest Insecticide and Formulation	Amount per 1,000 sq ft	Precautions and Remarks
Bee and Wasp (Burrowing)		
carbaryl[1] (Sevin) 80 WSP	1.5 oz	
pyrethroids[1] (Advanced Garden, Battle, Deltagard, Menace, Scimitar, Talstar, Tempo)	See label	
Billbug		
bifenthrin[1] (Menace, Talstar, others) F, GC; G form also available	0.25 to 0.5 fl oz	Use GC formulation for golf courses.
chlorantraniliprole (Acelepryn)	0.184 to 0.46 fl oz	
chlorpyrifos[1] (Dursban) 50 WSP, Pro	See label	For use on golf courses; check new label.
clothianidin (Arena) .5G 50 WDG	 14 to 22 oz 0.15 to 0.22 oz	
chlothianidin + bifenthrin (Aloft) GC SC LC SC GC G LC G	 0.27 to 0.44 fl oz 0.27 to 0.54 fl oz 1.8 to 3.6 lb 1.8 to 3.6 lb	
deltamethrin (Deltagard) G	2 to 3 lb/1,000 ft	
imidacloprid[1] (Merit) 75 WSP	3 to 4 level tsp	Make application prior to egg hatch.
lambda-cyhalothrin[1] (Battle, Scimitar, Cyonara)	See label	Observe restrictions near water.
zeta-Cypermethrin, bifenthrin, and imadacloprid (Triple Crown)	10-20 fl oz/acre	
thiamethoxam (Meridian) 0.33 G 25 WG Dinotefuran (Zylam) 20SG	 60 to 80 lb/acre 12.7 to 17 oz/acre 1 oz per 1,000 sf	Optimum control when applied from peak flight of adults to peak of egg hatch. Also suppresses mole crickets and chinch bugs.
Chinch Bug		
acephate[1] (Orthene T, T&O) 75 S	1.2 to 2.4 oz	
chlorantraniliprole (Acelepryn)	0.184 to 0.46 fl oz	Suppression.
bifenthrin[1] (Menace, Talstar, others) F, GC; G form also available	0.25 to 0.5 fl oz	Use GC formulation for golf courses.
carbaryl[1] (Sevin) 80 WSP	2.5 to 3 oz	
chlothianidin (Arena) .5G 50 WDG	 1.4 to 1.8 lb 0.2 to 0.3 oz	
chlothianidin + bifenthrin (Aloft) GC SC LC SC GC G LC G	 0.27 to 0.44 fl oz 0.27 to 0.54 fl oz 1.8 to 3.6 lb 1.8 to 3.6 lb	
cypermethrin (Demon) TC	0.33 to 0.65 fl oz	
chlorpyrifos[1] (Dursban), 2E, 4E, 50 WP, Pro	See label	For use on golf courses; check new label.
cyfluthrin (Tempo 2)	0.2 fl oz	Home lawns only.
deltamethrin (Deltagard) G	2 to 3 lb/1,000 ft	
lambda-cyhalothrin[1] (Battle, Scimitar, Cyonara)	See label	Do not make applications within 20 feet of any body of water. No reentry until spray has dried.
zeta-Cypermethrin, bifenthrin, and imadacloprid (Triple Crown)	20-35 fl oz/acre	
permethrin[1] (Astro) Dinotefuran (Zylam) 20SG	0.4 to 0.8 fl oz 1 oz per 1000 sf	For suppression.
Cutworm, Armyworm		
acephate[1] (Orthene T, T&O)	1.2 to 2.4 oz	Commercial and residential turf only.
azadirachtin[1] (Neemix, Turplex)	See label	
bifenthrin[1] (Menace, Talstar, others) F, GC; G form also available	0.18 to 0.25 fl oz	Use GC formulation for golf courses.
Bt products, various labels	See label	
carbaryl[1] (Sevin) 80 WSP and baits	0.75 to 1.5 oz	Treat in late afternoon. Apply in adequate water for good coverage but do not flood or water in. Do not cut grass for 1 to 3 days after treatment.
chlorantraniliprole (Acelepryn)	0.046 to 0.092 fl oz	
chlorpyrifos[1] (Dursban) 4 E, 2 ES, 50 WP, Pro	See label	For use on golf courses; check new label.
chlothianidin (Arena) .5G 50 WDG	 1.4 to 1.8 lb 0.2 to 0.3 oz	Cutworms only.

Table 5-17. Insect Control in Commercial Turf

Pest Insecticide and Formulation	Amount per 1,000 sq ft	Precautions and Remarks
Cutworm, Armyworm (continued)		
chlothianidin + bifenthrin (Aloft) GC SC LC SC GC G LC G	 0.27 to 0.54 fl oz 0.27 to 0.54 fl oz 1.8 to 3.6 lb 1.8 to 3.6 lb	
cyfluthrin[1] (Tempo 2)	0.143 fl oz	Home lawns only.
deltamethrin (Deltagard) G	2 to 3 lb/1,000 ft	
entomogenous nematodes[1]	See label	Read and follow special application instructions. Effective only against small cutworms.
indoxacarb (Provuant) SC	0.0625 to 0.25 fl oz	Not labeled for use on sod farms.
lambda-cyhalothrin[1] (Battle, Scimitar, Cyonara)	See label	Do not make applications within 20 feet of any body of water. No reentry until spray has dried.
spinosad A + D (Conserve) SC	1.25 fl oz	Rate varies with size and species.
trichlorfon (Dylox, Proxol) 80 SP Dinotefuran (Zylam) 20SG	1.5 to 3 oz 1 oz per 1,000 sf	
Earthworm		
		Usually not a problem. No effective controls available.
Fall Armyworm		
acephate[1] (Orthene, T, T&O)	0.5 to 1.2 oz	Water in immediately after application.
chlorantraniliprole (Acelepryn)	0.046 to 0.092 fl oz	
chlorpyrifos[1] (Dursban) 4 E, 2 E, 50WP, Pro	See label	For use on golf courses; check new label.
indoxacarb (Provuant) SC	0.0625 to 0.25 fl oz	Not labeled for use on sod farms.
pyrethroids[1] (Advanced Garden, Battle, Deltagard, Menace, Scimitar, Talstar, Tempo, Cyonara)	See label	
spinosad A + D (Conserve SC)	1.25 fl oz	Rate varies with size and species.
Grasshopper		
acephate[1] (Orthene T, T&O)	0.5 oz	Do not mow turfgrass for at least 24 hours after application.
deltamethrin (Deltagard) G	2 to 3 lb/1,000 ft	
lambda-cyhalothrin[1] (Battle, Scimitar, Cyonara)	See label	Do not make applications within 20 feet of any body of water. No reentry until spray has dried.
Ground Pearl		
		No effective control—practice good management.
Imported Fire Ant (See http://www.ncagr.gov/plantindustry/plant/entomology/documents/ncifaquarantine.pdf for latest quarantine areas.)		
acephate[1] (Lesco-Fate) (Orthene, T, T&O) 75 S	See label 1 to 2 tsp/mound	Distribute uniformly over mound. For best results apply in early morning or late afternoon.
hydramethylnon[1] (Amdro) 0.88% bait (Maxforce G)	— See label	Uniformly broadcast 1 to 1.5 pound of bait per acre with ground equipment on pastures, range grasses, lawns, and nonagricultural lands. Or distribute uniformly 5 level tablespoons of bait 3 to 4 feet around base of each mound. Do not exceed 1.5 pounds per acre.
avermectin/B[1] (Affirm) 0.011% bait	5 to 7 tbsp/ mound	Distribute uniformly 5 to 7 level tablespoons of bait 3 to 4 feet around base of each mound.
bifenthrin[1] (Menace, Talstar, others) F; G form also available	—	Follow label directions.
chlorpyrifos[1] (Dursban) 4 E	See label	
chlothianidin + bifenthrin (Aloft) GC SC LC SC GC G LC G	See label 0.27 to 0.44 fl oz 0.27 to 0.54 fl oz 1.8 to 3.6 lb 1.8 to 3.6 lb	
deltamethrin (Deltagard) G	2 to 3 lb/	
fenoxycarb (Award) [1] B	1 to 3 level tbsp 1 to 1.5 lb/acre	Single mound treatment. Apply uniformly with ground equipment.
Fipronil + bifenthrin + lamda-cyhalothrin (Taurus Trio G)	2 lb	Apply as a broadcast. Irrigate prior to treatment.
spinosad (Justice bait)	See label	
spinosad A + D (Conserve SC)	0.1 fl oz/gal/mound	Dilute 0.1 fluid ounce in 1 gallon water. Use 1-2 gallons per mound.
fipronil (Topchoice, Fipronil, others) 0.0143	2 lb	Apply as a broadcast.
imidacloprid + bifenthrin (Allectus, Atera)	See label	Rate varies with pest. Different formulations for different sites.
indoxacarb (Advion) bait	1.5 lb/acre	Bait formulation.
lambda-cyhalothrin[1] (Battle, Scimitar, Cyonara)	See label	
Metaflumizone (Siesta) bait	1.0 to 1.5 lbs/acre 2 to 4 tbsp/mound	Do not exceed 4 applications in a one-year period.
methoprene (Extinguish) 0.5 % bait	1.5 lb/acre	Mound or broadcast.
methoprene + hydromethylnon (Extinguish Plus)	1.5 lb/acre	
pyriproxyfen (Distance Fire Ant Bait)	See label	Mound or broadcast.

169

Table 5-17. Insect Control in Commercial Turf

Pest Insecticide and Formulation	Amount per 1,000 sq ft	Precautions and Remarks
Leafhopper, Spittlebug		
acephate[1] (Orthene, T, T&O) 75 S	1 oz	
bifenthrin[1] (Menace, Talstar, others) F, GC; G form also available	0.25 to 0.5 fl oz	Use GC formulation for golf courses.
carbaryl[1] (Sevin) 80 WSP	0.75 to 1.5 oz	
chlorpyrifos[1] (Dursban) 4 E, 50 WSP, Pro	See label	For use on golf courses; check new label.
deltamethrin (Deltagard) G	2 to 3 lb	
Millipede		
bifenthrin[1] (Menace, Talstar, others) F, GC; G form also available	0.25 to 0.5 fl oz	Use GC formulation for golf courses.
carbaryl[1] (Sevimol) (Sevin) 80 WSP	 1.5 to 3 oz 0.75 to 1.5 oz	
chlorpyrifos[1] (Dursban) 2 E, Pro	See label	For use on golf courses; check new label.
cypermethrin (Demon) TC	See label	
lambda-cyhalothrin[1] (Battle, Scimitar, Cyonara)	See label	Do not make applications within 20 ft of any body of water. No reentry until spray has dried.
Mole Cricket		
acephate[1] (Orthene T, T&O, Lesco-Fate)	1 to 1.9 oz	Water soil before application. Do not water in.
bifenthrin[1] (Menace, Talstar, others) F, GC; G form also available	0.5 to 1 fl oz	Use GC formulation for golf course.
carbaryl[1] (Sevin) baits	See label	
Zeta-Cypermethrin, bifenthrin, and imadacloprid (Triple Crown)	20-35 fl oz/acre	
cyfluthrin[1] (Tempo 2, Tempo Ultra)	0.2 fl oz	Home lawn use only.
deltamethrin (Deltagard) G	2 to 3 lb	
entomogenous nematodes[1]	See label	Various formulations now available. Adequate soil moisture critical for good control.
fipronil (Chipco Choice, others) 0.1 G (Top Choice, Fipronil, others) 0.0143	 12.5-25 lb/A 2 lb	Use slit placement equipment. Apply as a broadcast.
imidacloprid (Merit) 75 WP 0.5G	 4 level tsp 1.8 lb	Apply while crickets are less than ½ inch long (June, early July).
indoxacarb (Advion) Insect G	50 to 200 lb/acre	Not for use on sod farms. DO NOT water in after application.
indoxacarb (Provaunt)	0.275 oz	Two applications 2-4 weeks apart work best, following egg hatch.
lambda-cyhalothrin[1] (Battle, Scimitar, Cyonara)	See label	Do not make applications within 20 feet of any body of water. No reentry until spray has dried.
propoxur (Baygon) B Dinotefuran (Zylam) 20SG	See label 1 oz per 1,000 sf	Apply at egg hatch
Slug, Snail		
Mesurol 2 B	1 lb	Apply late in afternoon.
Metaldehyde	See label	
Sod Webworm		
acephate[1] (Lesco-Fate, Orthene T, T&O) (Precise 4G)	 0.5 to 1 oz 2.8 lb	Home lawns only. Irrigate immediately.
azadirachtin[1] (Azatrol, Neemix, Turplex)	0.5 fl oz	
Bacillus thuringiensis, various brands	1 to 2 lb/acre	
bifenthrin[1] (Menace, Talstar, others) F, GC; G form also available	0.18 to 0.25 fl oz	Use GC formulation for golf courses.
carbaryl[1] (Sevin) 80 WSP	2.5 to 3 oz	
chlorantraniliprole (Acelepryn)	0.046 to 0.092 fl oz	
chlorpyrifos[1] (Dursban) 4 E, 2 E, 5 G, Pro	See label	For use on golf courses; check new label.
clothianidin (Arena) .5G 50 WDG	 14 to 22 oz 0.15 to 0.22 oz	
chlothianidin + bifenthrin (Aloft) GC SC LC SC GC G LC G	 0.27 to 0.54 fl oz 0.27 to 0.54 fl oz 1.8 to 3.6 lb 1.8 to 3.6 lb	
cyfluthrin[1] (Tempo 2, Tempo Ultra)	0.143 fl oz	Irrigate immediately after application. Do not apply to newly seeded stands or bentgrass.
deltamethrin (Deltagard) G	2 to 3 lb	
indoxacarb (Provaunt) SC	0.0625 to 0.25 fl oz	Not labeled for use on sod farms.

Table 5-17. Insect Control in Commercial Turf

Pest Insecticide and Formulation	Amount per 1,000 sq ft	Precautions and Remarks
Sod Webworm (continued)		
lambda-cyhalothrin[1] (Cyonara, Scimitar, Battle)	See label	Do not make applications within 20 feet of any body of water. No reentry until spray has dried.
permethrin[1] (Astro)	0.4 to 0.8 fl oz	
spinosad A + D (Conserve) SC	1.25 fl oz	Rate varies with size and species.
trichlorfon[1] (Dylox, Proxol) 80 SP Dinotefuran (Zylam) 20SG	1.5 to 3 oz 1 oz per 1,000 sf	
Sowbug, Pillbug		
bifenthrin[1] (Talstar) F, GC G form also available	0.25 to 0.5 fl oz	Use GC formulation for golf courses.
carbaryl[1] (Sevin) 80 WSP	0.75 to 1.5 oz	
cypermethrin[1] (Demon) TC	See label	
deltamethrin (Deltagard) G	2 to 3 lb	
lambda-cyhalothrin[1] (Battle, Cyonara, Scimitar)	See label	Do not make applications within 20 feet of any body of water. No reentry until spray has dried.
White Grub (May beetle, chafers, green June beetle, and others)		
chlorantraniliprole (Acelepryn) zeta-Cypermethrin, bifenthrin, and imadacloprid (Triple Crown)	0.184 to 0.367 fl oz 20-35 fl oz/acre	Optimal control when applied at egg hatch. Use higher rates later in summer.
clothianidin (Arena) .5G 50 WDG	14 to 22 oz 0.15 to 0.22 oz	Mole cricket suppression.
chlothianidin + bifenthrin (Aloft) GC SC LC SC GC G LC G	0.27 to 0.54 fl oz 0.27 to 0.54 fl oz 1.8 to 3.6 lb 1.8 to 3.6 lb	
imidacloprid[1] (Merit) 75 WP	3 to 4 level tsp	Make application prior to egg hatch. (Offers some suppression of caterpillars.)
thiamethoxam (Meridian) 0.33 G 25 WG	60 to 80 lb/acre 12.7 to 17 oz/acre	Optimum control when applied from peak flight of adults to peak of egg hatch. Also suppresses mole crickets and chinch bugs.
trichlorfon (Dylox, Proxol) 80 SP Dinotefuran (Zylam) 20SG	3.75 oz 1 oz per 1,000 sf	Can be used with some success as a rescue treatment in August and September. Apply at egg hatch.
B.t. subspecies galleriae (grubGoneG)	100-150 lbs per acre	
White Grub, Green June Beetle (only)		
carbaryl[1] (Sevin) 80 WSP	1 to 1.5 oz	
chlorantraniliprole (Acelepryn)	0.184 to 0.367 fl oz	Optimal control when applied at egg hatch. Use higher rates later in summer.
chlorpyrifos[1] (Dursban) 50 WSP, Pro	See label	For use on golf courses; see new label.
clothianidin (Arena) .5G 50 WDG	14 to 22 oz 0.15 to 0.22 oz	Mole cricket suppression.
chlothianidin + bifenthrin (Aloft) GC SC LC SC GC G LC G	0.27 to 0.54 fl oz 0.27 to 0.54 fl oz 1.8 to 3.6 lb 1.8 to 3.6 lb	
B.t. subspecies galleriae (grubGoneG)	100-150 lbs per acre	
imidacloprid[1] (Merit) 75 WP	3 to 4 level tsp	Make application prior to egg hatch. Do not use on sod farms. Offers some suppression of caterpillars.
thiamethoxam (Meridian) 0.33 G 25 WG Dinotefuran (Zylam) 20SG	60 to 80 lb/acre 12.7 to 17 oz/acre 1 oz per 1,000 sf	Optimum control when applied from peak flight of adults to peak of egg hatch. Also suppresses mole crickets and chinch bugs. Apply at egg hatch
White Grub (Japanese beetle)		
B.t. subspecies galleriae (grubGoneG) Zeta-Cypermethrin, bifenthrin, and imidacloprid (Triple Crown)	100-150 lbs per acre 20-35 fl oz/acre	
carbaryl[1] (Sevin) 80 WSP	3 oz	
chlorantraniliprole (Acelepryn)	0.184 to 0.367 fl oz	Optimal control when applied at egg hatch. Use higher rates later in summer.
chlothianidin + bifenthrin (Aloft) GC SC LC SC GC G LC G	0.27 to 0.54 fl oz 0.27 to 0.54 fl oz 1.8 to 3.6 lb 1.8 to 3.6 lb	
clothianidin (Arena) .5G 50 WDG	14 to 22 oz 0.15 to 0.22 oz	Mole cricket suppression.
imidacloprid[1] (Merit) 75 WP	3 to 4 level tsp	Make application prior to egg hatch. (Offers some suppression of caterpillars.)

INSECT CONTROL

171

Table 5-17. Insect Control in Commercial Turf

Pest Insecticide and Formulation	Amount per 1,000 sq ft	Precautions and Remarks
White Grub (Japanese beetle) (continued)		
thiamethoxam (Meridian) 0.33 G 25 WG	60 to 80 lb/acre 12.7 to 17 oz/acre	Optimum control when applied from peak flight of adults to peak of egg hatch. Also suppresses mole crickets and chinch bugs.
trichlorfon[1] (Dylox, Proxol) 80 SP Dinotefuran (Zylam) 20SG	3.75 oz 1 oz per 1,000 sf	Can be used with some success as a rescue treatment in August and September. Apply at egg hatch.

[1] Several tradenames available. Check label for active ingredient. Always follow label instructions.

Insect Control for Wood and Wood Products

M. G. Waldvogel and P. Alder, Entomology Extension

Space limitations preclude listing all pesticide formulations and trade names. Other products or formulations may be used—but only those products labeled for the intended use. Products labeled for outdoor use only should never be applied indoors. Some insecticides listed here are designated for professional use only; others may have different formulations for professionals and the general public. Read the product label for specific information about the active ingredient, application rates, and detailed instructions on use.

Mention of pesticides in this section does not imply that chemicals are or should be the first or only means of pest control. Nonchemical methods, including exclusion, proper sanitation/maintenance, and moisture reduction, are critical to controlling wood-destroying pests.

Table 5-18. Insect Control for Wood and Wood Products

Insect Insecticide	Formulation[1]	Use[2]	Precautions and Remarks
Carpenter Ant—(a) Indoors			
abamectin (Advance)	Bait	P	Apply as directed on label.
acetarmiprid+bifenthrin (Transport Mikron)	Sprayable	P	Apply as directed on label
allethrin (Ortho)	Aerosol	G	Apply as directed on label.
avermectin (Advance)	Bait	P	Apply as directed on label.
bifenthrin (Ortho) (Talstar)	Aerosol Sprayable	G G,P	Apply as directed on label.
boric acid (Niban, MptherEarth G)	Bait	P	Apply as directed on label
chlorfenapyr (Phantom)	Sprayable	P	Apply as directed on label.
cyfluthrin (Bayer Advanced) (Tempo)	Sprayable	G P	Apply as directed on label.
cypermethrin (Cynoff)	Dust	P	Apply dust formulation directly to galleries
deltamethrin (Bayer Advanced) (Suspend)	Sprayable	G, P G P	Apply as directed on label.
dinotefuran (Alpine)	Foam & Spray	P	Apply as directed on label
Esfenvalerate (Onslaught)	Sprayable	P	Apply as directed on label.
fipronil (Combat) (Maxforce)	Bait	G P	Bait where you see ant activity. Apply as directed on label.
hydramethylnon (Combat)	Bait	P	Bait where you see ant activity. Apply as directed on label.
imidacloprid+cyfluthrin (Temprid SC)	Sprayable	P	Apply as directed on label.
indoxacarb (Advion, Arilon)	Bait (gel) Sprayable	P P	Bait where you see ant activity. Apply as directed on label. Apply as directed on label
lambda-cyhalothrin (Demand) (Spectracide)	Sprayable Spray and foam	P G	Apply as directed on label.
permethrin (Dragnet, Masterline)	Sprayable	P	Apply as directed on label.
sodium borate (Boracare, Borathor) (Spectracide, Terminate)	Sprayable, Dust	P G	Apply as directed on label.
thiamethoxam (Optigard)	Sprayable	P	Apply as foam to wall voids or infested wood.

Table 5-18. Insect Control for Wood and Wood Products

Insect Insecticide	Formulation[1]	Use[2]	Precautions and Remarks
Carpenter Ant—(b) outdoors			
acetarmiprid+bifenthrin (Transport Mikron)	Sprayable	P	Apply outdoors only as pinstream, spot, crack and crevice, or perimeter spray.
abamectin (Advance)	Bait	P	Place bait around perimeter.
bifenthrin (Ortho) (Bifen, Talstar)	Sprayable	G P	Spray or inject into wood.
boric acid (Perma-Dust Niban)	Aerosol, Bait	P	Place bait granules around perimeter.
chlorfenapyr (Phantom)	Sprayable	P	Exterior use limited to spot (2 square feet) and crack and crevice treatments at points of entry.
cyfluthrin (Bayer Advanced) (Tempo)	Sprayable	G P	Treat into and around the nest, then seal holes.
cypermethrin (Cynoff)	Dust	P	Apply dust formulation directly to galleries
cypermethrin (Demon TC)	Sprayable	P	Treat into and around the nest.
deltamethrin (Suspend SC)	Sprayable	P	Treat into and around the nest.
dinotefuran (Alpine)	Foam & Spray	P	Apply as directed on label (apply to damaged shrubs, tree stumps, fences, etc.)
Esfenvalerate (Onslaught)	Sprayable	P	Apply as directed on label.
fipronil (Maxforce, Termidor, Taurus)	Bait, Granular, Powder	P	Apply bait granules in ant foraging areas. Water area after applying granules.
hydramethylnon (Maxforce)	Bait	P	Apply granules along perimeter of building or nest. (Maxforce is for professional use.)
imidacloprid+cyfluthrin (Temprid SC)	Sprayable	P	Apply as directed on label.
Indoxacarb (Arilon, Advion)	Bait (Granular/gel) Sprayable	P	Apply as directed on label
cyhalothrin (Demand) (Spectracide)	Sprayable	P G	Apply as directed on label.
permethrin (Dragnet, Masterline)	Sprayable	P G	Apply as crack and crevice or spot treatment or paint onto surface. Application by drilling and injecting is also permitted.
sodium borate (Boracare, Borathor) (Spectracide)	Sprayable	P G	Spray, brush on, or inject into wood. For long-term protection, apply a water repellent stain to exterior wood surfaces 2 to 3 weeks after treatment.
Carpenter Bee			
carbaryl (Sevin)	Dust, Sprayable	G	Apply liquid as a coarse surface spray and into gallery entrance. Puff into and around entrance holes, using dust applicator. Seal with wood plugs, putty, or stainless steel wool.
bifenthrin (Ortho) (Bifen, Talstar)	Sprayable	G P	Apply as a coarse surface spray and into entrance hole. Seal entrance hole. Spectracide is for the general public.
boric acid (Perma-Dust PT 240)	Aerosol	P	Inject into entrance hole or tunnels with wood injector nozzle. Seal entrance hole.
chlorfenapyr (Phantom)	Sprayable	P	Apply as directed on label.
cyfluthrin (Bayer Advanced) (Tempo)	Sprayable	G P	Apply liquid as a surface spray and into entrance hole. Seal entrance hole.
lambda-cyhalothrin (Demand) (Spectracide)	Sprayable	P G	Spray or inject into wood. Seal holes in wood before injecting. Avoid runoff.
deltamethrin (Bayer Advanced) (Suspend SC)	Sprayable, Dust	G P	Apply liquid as a coarse surface spray and into gallery entrance. Puff into and around entrance holes, using dust applicator. Seal with wood plugs, putty, or stainless steel or copper wool.
imidacloprid (Premise)	Foam	P	Apply to galleries as directed on label.
imidacloprid+cyfluthrin (Temprid SC)	Sprayable	P	Apply as directed on label.
permethrin (Dragnet, Masterline) (Permethrin 3.2)	Sprayable	P G	Spray or inject into wood. Seal holes in wood before injecting. Avoid runoff.
sodium borate (Boracare, Borathor) (Spectracide)	Sprayable Dust	P G	Apply dust formulation directly to galleries.
cypermethrin (Cynoff)	Dust	P	Apply dust formulation directly to galleries
Old House Borer			
aluminum phosphide (Phostoxin)	Fumigant	P	For infested furniture, stacked lumber, other wood products. Apply under gas-tight tarpaulins or in sealed chamber. Requires an F-Phase N.C. Structural Pest Control License.
bifenthrin (Ortho) (Bifen, Talstar)	Sprayable	G P	

Table 5-18. Insect Control for Wood and Wood Products

Insect Insecticide	Formulation[1]	Use[2]	Precautions and Remarks
Old House Borer (continued)			
cyfluthrin (Bayer Advanced) (Tempo)	Sprayable	G P	Coarse spray, brush on, or inject into wood. Avoid excessive runoff.
cypermethrin (Demon TC)	Sprayable	P	
deltamethrin (Suspend SC)	Sprayable	P	
imidacloprid+cyfluthrin (Temprid SC)	Sprayable	P	Apply as directed on label.
permethrin (Dragnet, Masterline) (Permethrin 3.2)	Sprayable	P G	
sodium borate (Boracare, Timbor) (Spectracide)	Sprayable Dust	P G	Spray, brush on, or inject into wood. For permanent protection, a water repellent should be applied to exterior surfaces 2 to 3 weeks after treatment.
sulfuryl fluoride (Vikane, Zythor)	Fumigant	P	Apply under gas-tight tarpaulins only. Hold for 20-24 hours at temperature above 60 degrees F. Requires an F-Phase N.C. Structural Pest Control License.
Powderpost Beetle			
aluminum phosphide (Phostoxin)	Fumigant	P	For infested furniture, stacked lumber, other wood products. Apply under gas-tight tarpaulin or in a sealed chamber. Requires an F-Phase N.C. Structural Pest Control License.
bifenthrin (Ortho) (Bifen, Talstar)	Sprayable	G P	Coarse spray, brush on, or inject into wood. Avoid excessive runoff
chlorfenapyr (Phantom)	Sprayable	P	
cyfluthrin (Bayer Advanced) (Tempo)	Sprayable	G P	
cypermethrin (Demon TC)	Sprayable	P	
deltamethrin (Suspend SC)	Sprayable	P	
imidacloprid+cyfluthrin(Temprid SC)	Sprayable	P	Apply as directed on label.
lambda-cyhalothrin (Demand) (Spectracide)	Sprayable	P G	Apply as directed on label.
permethrin (Dragnet, Masterline)	Sprayable	P G	
sodium borate (Boracare, Timbor) (Spectracide)	Sprayable Dust	P G	For long-term protection, apply a water repellent to exterior surfaces 2 to 3 weeks after treatment.
sulfuryl fluoride (Vikane, Zythor)	Fumigant	P	For infested furniture, stacked lumber, other wood products. Apply under gas-tight tarpaulin. Hold for 20-24 hours at a temperature above 60 degrees F. Requires an F-Phase N.C. Structural Pest Control License.
Termite—Drywood Species (Wood Treatment)			
acetarmiprid+bifenthrin (Transport)[3]	Sprayable	P	Coarse spray or drill and inject wood.
aluminum phosphide (Phostoxin)	Fumigant	P	Apply under gas-tight tarpaulins or in sealed chamber.
bifenthrin (Ortho) (Bifen, Talstar)	Sprayable	G P	Coarse spray or inject into wood.
cyfluthrin (Bayer Advanced) (Tempo)	Sprayable	G P	Coarse surface spray or inject wood.
lambda-cyhalothrin (Demand) (Spectracide)	Sprayable	P G	Apply as directed on label. Localized treatments. Spectracide is not recommended as a sole protection against termites.
cypermethrin (Demon TC)	Sprayable	P	Coarse spray or inject into wood for localized infestations.
fipronil (Termidor, Taurus)	Sprayable, Foam, Dry	P	Coarse surface spray or inject wood.
dinotefuran (Alpine)	Foam, Spray	P	Apply as directed on label (can be used on infested shrubs, fence posts, utility poles, etc.)
imidacloprid (Dominion, Premise)	Sprayable, Foam	P	Drill and inject spray or foam into voids.
imidacloprid+cyfluthrin (Temprid SC)	Sprayable	P	Apply as directed on label.
methyl bromide (Meth-O-Gas Q)	Fumigant	P	Apply under gas-tight tarpaulins only. Regulatory use only.
permethrin (Dragnet, Masterline)	Sprayable	P G	Coarse spray on wood for localized infestation.
sodium borate (Boracare, Timbor) (Spectracide)	Sprayable	P G	Coarse surface spray or inject wood.
sulfuryl fluoride (Vikane, Zythor)	Fumigant	P	Apply under gas-tight tarpaulins only. Hold for 20-24 hours at temperature above 60 degrees F. Requires an F-Phase N.C. Structural Pest Control License.
thiamethoxam (Optiguard)	Sprayable	P	Coarse spray or drill and inject wood.

Table 5-18. Insect Control for Wood and Wood Products

Insect Insecticide	Formulation[1]	Use[2]	Precautions and Remarks
Termite—Subterranean Species (a) (Wood treatment)			
acetarmiprid+bifenthrin (Transport)[3]	Sprayable	P	
bifenthrin (Bifen, Talstar) (Ortho)	Sprayable	P G	For use only in voids or channels in damaged wood or to cracks and spaces between wooden members of structures.
boric acid (Perma-Dust PT 240)	Aerosol	P	Coarse surface spray or inject wood.
chlorantraniliprole (Altriset)	Spraying	P	Coarse spray around or inject into infested poles, trees and stumps (Outdoors)
chlorfenapyr (Phantom)	Sprayable	P	Coarse spray or inject into wood.
cyfluthrin (Tempo) (Bayer Advanced)	Sprayable	P G	Coarse spray, brush on, or inject into wood. Avoid excessive runoff.
lambda-cyhalothrin (Demand) (Spectracide)	Sprayable	P G	Apply as directed on label. Localized treatments. Spectracide is not recommended as a sole protection against termites.
deltamethrin (Suspend SC)	Sprayable	P	Coarse surface spray or inject wood.
diflubenzuron (Exterra, Advance)	Bait	P	Above-ground stations used in conjunction with in-ground baiting systems.
dinotefuran (Alpine)	Foam and Spray	P	Apply as directed on label (can be used on infested shrubs, fence posts, utility poles, etc.)
fipronil (Termidor, Taurus)	Sprayable, Foam	P	Coarse spray or inject into wood.
imidacloprid (Premise)	Sprayable, Gel, Foam	P	Gel and foam formulations may be injected into voids or damaged wood.
imidacloprid+cyfluthrin (Temprid SC)	Sprayable	P	Apply as directed on label.
noviflumuron (Recruit IV AG)	Bait	P	Available only as part of the Sentricon in-ground system (see below).
permethrin (Dragnet, Masterline)	Sprayable	P	Coarse spray, brush on, or inject into wood. Avoid excessive runoff.
sodium borate (Boracare, Penetreat) (Spectracide)	Sprayable Dust	P G	Spray, brush on, or inject into wood. For long-term protection, apply a water repellent to exterior wood surfaces 2 to 3 weeks after treatment. Not a replacement for a soil treatment.
sulfluramid (FirstLine)	Bait	P	Above-ground stations used in conjunction with in-ground system.
Termite—Subterranean Species (b) Soil treatment			
Acetarmiprid+bifenthrin (Transport)[3]	Sprayable	P	Dig trenches 6 inches wide and at least 4 inches deep along the foundation. Never trench below the top of the footing. Depending upon the depth of footer, rodding may be needed. Dilutions and rates of applications vary among specific products. Vertical barriers usually require about 4 gallons of spray per 10 linear feet for each foot of depth along a foundation. Follow label restrictions on treatment in crawlspaces containing wells or cisterns. Follow instructions if "excavation and backfill" is permitted. Exercise extreme caution when treating crawlspaces. Wear appropriate protective equipment as specified on product label. General (broadcast) treatments of crawlspace soil for termites are prohibited, except as noted on the label. **NOTE:** Most termite infestations require treatment by a W-phase licensed structural pest control operator. Requirements for termite treatments are outlined in 2NCAC 34:.0503, .0505. Apply Premise or Bayer Advanced granules to trenches as a spot treatment. Bayer Advanced for the general public is available only in granular formations.
bifenthrin (Bifen, Talstar) (Ortho)	Sprayable	P G	
chlorfenapyr (Phantom)	Sprayable	P	
chlorantraniliprole (Altriset)	Spraying	P	
cyfluthrin (Bayer Advanced) (Tempo)	Sprayable	G P	
lambda-cyhalothrin (Demand) (Spectracide)	Sprayable	P G	Apply as directed on label. Localized treatments.
fipronil (Termidor, Taurus, Ultrathor)	Sprayable	P	
imidacloprid (Premise) (Bayer Advanced)	Sprayable, Granular	P G	
imidacloprid+cyfluthrin (Temprid SC)	Sprayable	P	
indoxacarb (Arilon)	Sprayable	P	Use for spot or local treatment only (Arilon is not intended as sole protection against termites)
permethrin (Dragnet FT, MasterLine)	Sprayable	P G	
diflubenzuron (Advance, Exterra)	Bait	P	Termite monitoring and baiting program. Available only through manufacturer-authorized pest control companies.
hexaflumuron (Shatter)	Bait	P	Termite monitoring and baiting program. Available only through manufacturer-authorized pest control companies
novaluron (Trelona CTB)	Bait	P	Termite monitoring and baiting program. Available only through manufacturer-authorized pest control companies.
noviflumuron (Recruit HD)	Bait	P	Termite monitoring and baiting program. Available only through manufacturer-authorized pest control companies.
sulfluramid (FirstLine) (Terminate)	Bait	P G	FirstLine is for professional use only. Terminate is for the general public. Terminate is not intended as the sole protection against termite infestation.

[1]   Formulation designations: Aerosol = injectable or spray; Dust = dry application; Fumigant = gas in pressurized cylinder or pellets; Foam = Injectable foam; Sprayable = liquid concentrate or wettable powder for mixing with water or in a ready-to-use form

[2]   Use designations: P = Professional applicator (licensed in structural pest control); G = General public use

*   Several trade names available. Check label for active ingredient. Always follow label instructions.

INSECT CONTROL

INSECT CONTROL FOR HOME USE

Insect Control for the Home Vegetable Garden

J. F. Walgenbach, Entomology Extension

Homeowner products are numerous and names change frequently. Insecticides listed below are identified by the active ingredient. Brand names for homeowner products identify the active ingredient; always check the "active ingredients" portion of the product label to determine if the product is appropriate for your needs. Refer to the product label for rates and pre harvest intervals.

Table 5-19. Insect Control for the Home Vegetable Garden

Commodity Insect	Insecticide Active ingredient	Minimum Interval (Days) Between Last Application and Harvest	Precautions and Remarks
Asparagus			
Asparagus beetle, Japanese beetle, grasshopper, and aphid	Carbaryl	1	Carbaryl will not control aphids.
	permethrin	3	
Bean			
Aphid	malathion	1	
	bifenthrin	3	
	cyfluthrin	7	
	insecticidal soap	0	
Corn earworm, Mexican bean beetle, bean leaf beetle, flea beetle, Japanese beetle, and cucumber beetle, potato leafhopper, fleahopper, lygus, and stink bug	carbaryl	3	
	spinosad	3	Will not control Japanese beetle, cucumber beetle or stink bug.
	bifenthrin	3	
	cyfluthrin	7	
	Lambda-cyhalothrin	7	21-day preharvest interval for dried beans.
Spider mite	bifenthrin	3	
	malathion	1	
	insecticidal soap	0	Apply treatment at first sign of mites and speckled plants.
Whitefly	Beauveria bassiana	0	
	insecticidal soap	0	
Beet			
Flea beetle, beet webworm, and blister beetle	carbaryl	3 (14)	On foliage as needed. Fourteen days if tops used; 3 days if tops not used.
Broccoli, Cabbage, Cauliflower, Brussels Sprouts, Rutabaga			
Aphid	bifenthrin	7	
	cyfluthrin	3	
	malathion	7	
	insecticidal soap	0	
Cabbage looper, imported cabbageworm, diamondback moth, and cutworm	Bacillus thuringiensis	0	Start control program when worms are small and treat foliage every 5 to 7 days.
	carbaryl	3	On foliage as needed. Will not control cabbage looper. Carbaryl is suggested for cutworm.
	bifenthrin	7	
	esfenvalerate	3	
	lambda-cyhalothrin	1	
	spinosad	1	
Flea beetle and thrips	carbaryl	3	
	malathion	7	
	spinosad	1	For thrips only.
Harlequin bug	bifenthrin	7	On foliage as needed.
	lambda-cyhalothrin	1	On foliage as needed.
	malathion	7	On foliage as needed.
Cantaloupe			
Aphid and thrips	cyfluthrin	0	
	Esfenvalerate	3	
	malathion	1	
	insecticidal soap	0	On foliage as needed.

Table 5-19. Insect Control for the Home Vegetable Garden

Commodity Insect	Insecticide Active ingredient	Minimum Interval (Days) Between Last Application and Harvest	Precautions and Remarks
Cantaloupe (continued)			
Cucumber beetle (spotted and striped), pickleworm, squash bug, and squash vine borer	bifenthrin	3	
	esfenvalerate	3	
	cyfluthrin	0	
Spider mite	insecticidal soap	0	On foliage as needed.
Carrot			
Armyworm, leafminer, and leafhopper	*Bacillus thuringiensis*	0	*B.t.* will not control leafhoppers.
	carbaryl	0	On foliage as needed.
	cyfluthrin	0	
Celery			
Aphid, flea beetle, leafminer, and flea hopper	malathion	7	On foliage as needed.
	permethrin	3	On foliage as needed.
Collard			
Aphid and flea beetle	bifenthrin	7	
	malathion	7	On foliage as needed.
	insecticidal soap	0	On foliage as needed.
Cabbage looper, diamondback moth, and imported cabbageworm	*Bacillus thuringiensis*	0	Begin foliage treatments early and repeat as necessary. Include a spreader/sticker.
	spinosad	1	
Harlequin bug	malathion	7	
	bifenthrin	7	
	lambda-cyhalothrin	1	
	cyfluthrin	0	
Corn (Sweet)			
Corn earworm, sap beetle, flea beetle, and Japanese beetle	bifenthrin	1	
	esfenvalerate	1	
	carbaryl	2	
Corn earworm, European corn borer, and fall armyworm	*Bacillus thuringiensis*	0	Consult specific label. *B.t.* is effective while worms are feeding on the foliage.
	cyfluthrin	0	
	esfenvalerate	1	
	lambda-cyhalothrin	1	
	permethrin	3	
	spinosad	1	
Cucumber			
Cucumber beetle (spotted and striped), and pickleworm	bifenthrin	3	
	esfenvalerate	3	
	cyfluthrin	0	
Spider mite	insecticidal soap	0	On foliage as needed.
Whitefly	insecticidal soap	0	On foliage as needed.
	Beauveria bassiana	0	
Eggplant			
Aphid, flea beetle, whitefly, lace bug	bifenthrin	7	
	lambda-dyhalothrin	5	
	malathion	3	On foliage as needed.
Colorado potato beetle, hornworm, and corn earworm	*Bacillus thuringiensis* var. *tennebrionus*	0	For Colorado potato beetle only. Treat when small larvae are present. Not effective against adults or large larvae.
	spinosad	1	
spider mite	insecticidal soap	0	On foliage as needed.

177

INSECT CONTROL

Table 5-19. Insect Control for the Home Vegetable Garden

Commodity Insect	Insecticide Active ingredient	Minimum Interval (Days) Between Last Application and Harvest	Precautions and Remarks
Lettuce			
Aphid, leafhopper	bifenthrin	7	
	lambda-cyhalothrin	1	
	malathion	14 leaf, 7 head	On foliage as needed.
	insecticidal soap	0	On foliage as needed.
Cabbage looper, corn earworm, and leafhopper	Bacillus thuringiensis	0	On foliage as needed.
	spinosad	1	On foliage as needed.
	lambda-cyhalothrin	1	
Mustard Greens			
Aphid, Flea beetle	bifenthrin	7	
	malathion	7	On foliage as needed.
	insecticidal soap	0	On foliage as needed.
Cabbage looper, diamondback moth, and imported cabbageworm	Bacillus thuringiensis	0	Begin foliage treatments early and repeat as necessary.
	spinosad	1	
Okra			
Aphid and leafminer	bifenthrin	7	
	malathion		
Corn earworm, European corn borer, flea beetle, and stink bug	spinosad	1	
	bifenthrin	7	
	cyfluthrin	1	
	esfenvalerate	1	
	permethrin	1	
Onion			
Onion thrips	lambda-cyhalothrin	14	
	malathion	3 (Green)	
	insecticidal soap	0	
Peas			
Aphid and leafminer	insecticidal soap	0	
Pepper			
Aphid and thrips	esfenvalerate	1	
	malathion	3	
	insecticidal soap	0	
European corn borer, flea beetle, tomato fruitworm, hornworm, and stink bug	carbaryl	3	Will not control stink bug
	Bifenthrin	7	Excellent control of stink bug
	cyfluthrin	7	
	esfenvalerate	1	
	permethrin	3	
	spinosad	1	Will not control stink bug
Potato, Irish			
Aphid	cyfluthrin	0	
	esfenvalerate	0	
European corn borer, potato tuberworm	Bacillus thuringiensis	0	
	carbaryl	0	Apply when eggs begin to hatch and every 5 days as needed.
	esfenvalerate	1	
	permethrin	3	
Potato leafhopper, potato flea beetle, Colorado potato beetle, and blister beetle	Imidacloprid	21	Apply to the soil immediately at planting for long-term control.
	Bacillus thuringiensis var. san diego var. tennebrionus	0	For Colorado potato beetle only. Treat when small larvae are present. Not effective against adults or large larvae.
	carbaryl	0	On foliage as needed. Treat when most Colorado potato beetle eggs have hatched.

Table 5-19. Insect Control for the Home Vegetable Garden

Commodity Insect	Insecticide Active ingredient	Minimum Interval (Days) Between Last Application and Harvest	Precautions and Remarks
Pumpkin—See SQUASH AND PUMPKIN			
Radish			
Aphid	malathion	7	On foliage as needed.
Flea beetle and imported cabbageworm	cyfluthrin	0	
Spinach			
Aphid, thrips, and leafminer	permethrin	1	
	malation	7	
	insecticidal soap	0	On foliage as needed.
Corn earworm and loopers	*Bacillus thuringiensis*	0	
	permethrin	1	
	spinosad	1	
Squash and Pumpkin			
Aphid	bifenthrin	3	
	malathion	1	
	insecticidal soap	0	
Cucumber beetle (spotted and striped), flea beetle, and leafhopper	esfenvalerate bifenthrin	3 3	
Pickleworm	esfenvalerate	3	
	spinosad	3	
Squash bug	bifenthrin	3	
Tomato			
Aphid, fleabeetle	bifenthrin	1	
	malathion	1	
	insecticidal soap	0	
Cutworm (surface type)	esfenvalerate	1	
Colorado potato beetle	*Bacillus thuringiensis var. san diego* var. *tennebrionus*	0	For Colorado potato beetle only. Treat when small larvae are present. Not effective against adults or large larvae.
	spinosad	1	
Spider mite	insecticidal soap	0	On foliage as needed.
Stink bug	cyfluthrin	7	Do not make more than 6 applications per season.
	lambda-cyhalothrin	5	
	malathion	1	
	permethrin	7	Do not apply on cherry tomatoes or varieties less than 1 inch in diameter.
Thrips	Spinosad	1	
	insecticidal soap	0	
Tomato fruitworm, cabbage looper, tobacco hornworm	*Bacillus thuringiensis*	0	Treat weekly, if necessary. Begin when fruits are 0.5 inch in diameter. Fruitworms are most serious after August 1.
	carbaryl	3	
	cyfluthrin	7	Do not make more than 6 applications per season.
	esfenvalerate	1	
	lambda-cyhalothrin	5	
	permethrin	7	Do not apply on cherry tomatoes or varieties less than 1 inch in diameter.
	spinosad	1	
Whitefly	*Beauveria bassiana*	0	Apply when whiteflies observed. Repeat in 4- to 5-day intervals.
	malathion	1	
	pyrethrum products	0	
	insecticidal soap	0	

INSECT CONTROL

Table 5-19. Insect Control for the Home Vegetable Garden

Commodity Insect	Insecticide Active ingredient	Minimum Interval (Days) Between Last Application and Harvest	Precautions and Remarks
Turnip, Turnip Greens			
Aphid, flea beetle	bifenthrin	7	
	malathion	7	On foliage as needed.
	insecticidal soap	0	
Cabbage looper, diamondback moth, imported cabbageworm	*Bacillus thuringiensis*	0	On foliage as needed.
	spinosad	1	
Harlequin bug	Gamma-cyhalothrin	1	On foliage as needed.
Watermelon			
Aphid	bifenthrin	3	
	malathion	1	
	insecticidal soap	0	On foliage as needed.
Cucumber beetle (spotted and striped)	bifenthrin	3	
	esfenvalerate	3	
	malathion	1	
Spider mite	bifenthrin	3	
	malathion	1	
	insecticidal soap	0	
Thrips	Spinosad	3	
	malathion	1	
	insecticidal soap	0	

Control of Household Pests

(Products for Use by the General Public)

P. Alder and M. G. Waldvogel, Entomology Extension

Mention of pesticides in this section does not imply that chemicals are or should be the first or only means of control. Nonchemical methods, including exclusion and sanitation, are important to long-term pest management.

Space limitations preclude listing all pesticide formulations and trade names. Other appropriate products or formulations may be used.

Never use products that are not labeled for the intended use. Products labeled for outdoor use only should never be applied indoors. Read the product label for specific pest information about the active ingredient, application rates, and detailed instructions on the product's use.

NOTE: The insecticides listed below are identified by the common name. The brand names of most consumer insecticide products do not identify the specific chemical used, and the formulation and/or its contents may be changed by the manufacturer. Always check the "Active ingredients" portion of the product label to determine if the product is appropriate for your needs.

Table 5-20. Control of Household Pests–Products for Use by the General Public

Insecticide	Formulation	Precautions and Remarks
Ant (a) Indoors *(For information on carpenter ants, see Insect Control for Wood and Wood Products)*		
avermectin (Raid)	Bait Station	Place bait stations in areas where ants are active. Keep of out of reach of children and pets. Use dust formulations only in inaccessible areas.
bifenthrin (Ortho)	Liquid, Aerosol Spray	
borax/boric acid (Terro)	Bait, Dust, Bait Station	Treat ant-traveled areas. Re-treat as effectiveness diminishes. Some products are not suitable for use in residential kitchens or commercial food/feed preparation sites. Read the product label carefully. Remove food from storage areas before treating.
cyfluthrin (Bayer)	Liquid, Aerosol Spray	
cypermethrin (Raid)	Aerosol Spray	Apply products as directed on the label
deltamethrin (Black Flag, Raid, Terro)	Aerosol, Liquid	
dinotefuran (Hot Shot, Spectracide)	Bait	
hydramethylnon (Amdro, Combat)	Bait Station	
diatomaceous earth (PermaGuard)	Dust	
borax (Terro)	Bait	
fipronil (Combat)	Bait	
lambda-cyhalothrin (Spectracide, Hot Shot)	Liquid, Aerosol Spray	
imiprothrin (Raid, Black Flag)	Aerosol Spray	Imoprothrin is usually formulated with other pesticides in these products.
indoxacarb (Hot Shot, Spectracide)	Bait Station	
lemongrass oil (Hot Shot)	Liquid, Aerosol Spray	
mint oil (Victor)	Aerosol Spray	
d-phenothrin (Raid)	Aerosol Spray	
prallethrin (Raid, Hot Shot, Bengal)	Aerosol Spray	
pyrethrins, pyrethrum (Hot Shot, Black Flag)	Aerosol Spray	
sodium o-phenylphenate (Bayer)	Liquid	
thiamethoxam (Raid)	Bait	
Ant (b) Outdoors *(Also see "Ant" and "Imported Fire Ant" under Home Lawns table.)*		
bifenthrin (Ortho)	Granular, Liquid	Apply granular bait around nest. Place bait stations in areas where ants are active. Treat nest and surrounding area. May be applied along building perimeter.
borax (Terro)	Bait	
cyfluthrin (Bayer)	Granular, Liquid	Apply chemicals as directed on the label.
deltamethrin (Black Flag, Raid, Terro)	Liquid	
diatomaceous earth (PermaGuard)	Dust	
dinotefuran (Hot Shot, Spectracide)	Bait	
fipronil (Combat)	Bait	
lambda-cyhalothrin (Spectracide, Hot Shot)	Liquid, Granular, Aerosol Spray	
hydramethylnon (Amdro, Combat)	Bait	
indoxacarb (Spectracide, Hot Shot)	Bait Station	
Bed Bug		
cyflutrhin (Bayer)	Liquid	
deltamethrin (Black Flag, Enforcer, Terro)	Aerosol Spray, Liquid	
diatomaceous earth (PermaGuard)	Dust	
dichlorvos (Hot Shot)	Pest Strip	
d-phenothrin (Raid)	Aerosol Spray	
lambda-cyhalothrin (Hot Shot)	Liquid	
N-octyl bicycloheptene dicarboximide (Raid)	Aerosol Spray	
permethrin (Hot Shot)	Liquid	

INSECT CONTROL

Table 5-20. Control of Household Pests–Products for Use by the General Public

Insecticide	Formulation	Precautions and Remarks
Bed Bug (continued)		
pyrethrins (Enforcer)	Fogger	
phenoxybenzl (Enforcer)	Aerosol Spray	
pralletrhin (Hot Shot)	Aerosol Spray	
silicon dioxide (Hot Shot)	Dust	
Bee (a) Indoors		
deltamethrin (Raid)	Liquid	Apply only for sporadic invaders. If bees are found frequently, locate and remove the nest.
diatomaceous earth (PermaGuard)	Dust	
d-phenothrin (Raid)	Aerosol Spray	Apply products as directed on the label.
pyrethrins, pyrethrum	Aerosol Spray, Liquid	
Bee (b) Outdoors For carpenter bees, see section *Insect control for Wood and Wood Products*		
bifenthrin (Ortho)	Liquid	Apply after dark when insects have returned to nest. Some materials available in pressurized cans that propel an insecticide stream up to 10 feet. Re-treatment may be necessary.
carbaryl (Sevin)	Dust, Liquid, Powder	
cyfluthrin (Bayer)	Liquid	Apply products as directed on the label.
deltamethrin (Black Flag)	Liquid	
lambda-cyhalothrin (Spectracide)	Liquid	
d-phenothrin (Raid)	Aerosol Spray	
Booklouse (psocid) (Indoors and outdoors)		
bifenthrin (Ortho)	Liquid	Apply as a barrier spray along foundation and entry points (doors and windows). Read labels to determine which products are suitable for indoor use. Clean up moisture problems, which may attract insects indoors. Excess moisture may impede product effectiveness.
cyfluthrin (Bayer)	Liquid	
diatomaceous earth (PermaGuard)	Dust	
deltamethrin (Black Flag)	Liquid	
mint oil (Victor)	Aerosol Spray	
pyrethrins, pyrethrum	Aerosol Spray	
Boxelder Bug (Outdoors)		
bifenthrin (Ortho)	Liquid	Harmless insects become nuisances when searching indoors for hibernation sites in the fall. Treat door thresholds, window ledges, and other areas where the insects congregate or may gain entry.
cyfluthrin (Bayer)	Liquid	
deltamethrin (Black Flag, Raid)	Liquid	
lambda-cyhalothrin (Spectracide)	Liquid	
d-phenothrin (Raid)	Aerosol Spray	
Brown Dog Tick (a) Indoors		
cyfluthrin (Bayer)	Aerosol Spray, Liquid	
deltamethrin (Raid, Terro)	Liquid	
diatomaceous earth (PermaGuard)	Dust	
d-phenothrin (Raid)	Raid	
permethrin (Hot Shot)	Aerosol Spray,	
tetramethrin (Raid)	Aerosol Spray	
gamma-cyhalothrin (Spectracide)	Liquid	
deltamethrin (Spectracide)	Aerosol Spray, Liquid	
pyrethrins, pyrethrum	Aerosol Spray	
imiprothrin (Black Flag)	Aerosol Spray	
Brown Dog Tick (b) Outdoors and under buildings		
bifenthrin (Ortho)	Granules	
cyfluthrin (Bayer)	Liquid, Granule	
deltamethrin (Raid)	Liquid	
diatomaceous earth (PermaGuard)	Dust	
eugenol (Bioganic, Raid)	Aerosol Spray, Dust	
lambda-cyhalothrin (Spectracide)	Liquid, Granule	
deltamethrin (Spectracide, Black Flag)	Aerosol Spray, Liquid	
Carpet Beetle (a) Nonfabric areas and infested areas of carpets only		
cyfluthrin (Bayer)	Liquid	
cypermethrin (Raid)	Aerosol Spray	
diatomaceous earth (PermaGuard)	Dust	
deltamethrin (Black Flag)	Aerosol Spray	
d-phenothrin (Raid)	Aerosol Spray	

INSECT CONTROL

Table 5-20. Control of Household Pests–Products for Use by the General Public

Insecticide	Formulation	Precautions and Remarks
Carpet Beetle (a) Nonfabric areas and infested areas of carpets only (continued)		
lamma-cyhalothrin (Spectracide, Hot Shot)	Liquid	
pyrethrins, pyrethrum	Aerosol Spray	
bifenthrin (Ortho)	Aerosol Spray, Liquid	
deltamethrin (Raid)	Liquid	
Carpet Beetle (b) On fabric		
diatomaceous earth (PermaGuard)	Dust	
pyrethrins, pyrethrum	Aerosol Spray, Liquid	
Centipede (a) Indoors		
cyfluthrin (Bayer)	Liquid	
deltamethrin (Black Flag, Raid)	Liquid, Aerosol Spray	
diatomaceous earth (PermaGuard)	Dust	
lambda–cyhalothrin (Spectracide)	Liquid,	
imoprothrin (Raid, Black Flag)	Aerosol Spray, Liquid	
deltamethrin (Raid, Black Flag)	Aerosol Spray, Liquid	
pyrethrins, pyrethrum	Aerosol Spray	
Centipede (b) Outdoors		
bifenthrin (Ortho)	Granules, Liquid	Treat infested areas around building foundations, vents, and similar access points. Barrier sprays of 12 to 18 inches along perimeter may be effective.
cyfluthrin (Bayer)	Granules, Liquid	
deltamethrin (Raid, Black Flag)	Liquid, Aerosol Spray	
diatomaceous earth (PermaGuard)	Dust	
lambda-cyhalothrin (Spectracide, Hot Shot)	Liquid	
deltamethrin (Black Flag, Raid)	Aerosol, Liquid	
Chigger (Red bug) Outdoors		
bifenthrin (Ortho)	Granular, Liquid	Apply to grass, bushes, and weeds in the infested areas. Thoroughly saturate soil, but avoid runoff into ponds, lakes, or other bodies of water. Repeat as needed. Apply labeled repellent products to shoes, ankles, and legs before entering suspected chigger-infested areas.
cyfluthrin (Bayer)	Liquid	
gamma-cyhalothrin (Spectracide)	Liquid, Granule	
lamda-cyhalothrin (Spectracide)	Liquid	
deltamethrin (Black Flag)	Liquid	
Clothes Moth (a) Nonfabric areas and infested areas of carpet only, See Carpet Beetle		
CLothes Moth (b) On fabric, See Carpet Beetle		
Clothes Moth (c) In storage areas		
dichlorvos (Pest Strip) Ortho No-Pest Strip	Strip	Hang on strip in clothes closets or storage chests up to 1,000 cubic feet in capacity. Not for use in occupied rooms or in closets in occupied rooms. Follow label instructions carefully.
paradichlorobenzene (PDB)napthalene	Crystals or similar solid	Effective repellents on clean fabric in airtight enclosures. Avoid contact with plastic buttons and zippers.
Clover Mite (a) Indoors		
cyfluthrin (Bayer)	Aerosol Spray, Liquid	
deltamethrin (Raid Black Flag)	Liquid, Aerosol Spray	
diatomaceous earth (PermaGuard)	Dust	
lambda-cyhalothrin (Spectracide)	Liquid	
pyrethrins, pyrethrum	Aerosol Spray	
Clover Mite (b) Outdoors		
bifenthrin (Ortho)	Granular	Treat around points of entry, such as foundations, vents, windows, and doors. Maintain a 12-inch wide vegetation-free zone along foundation. Spray 1 to 2 feet high along the foundation wall and a 3- to 5-feet barrier on the grass or landscaped areas around the foundation. Water immediately after applying granules. Apply products as directed on the label.
cyfluthrin (Bayer)	Granular, Liquid	
deltamethrin (Raid, Black Flag)	Liquid, Aerosol Spray	
diatomaceous earth (PermaGuard)	Dust	
lambda-cyhalothrin (Spectracide)	Liquid	
Cockroach (a) Indoors		
avermectin (Raid)	Bait Station	Apply sprays along baseboards, under sinks, in cabinets and other infested areas. Remove and cover food, cooking, and eating utensils before spraying cabinets. Do not restock shelves until surface dries completely. Some products are not suitable for use in residential kitchens or commercial food/feed preparation sites. Read the product label carefully
bifenthrin (Ortho)	Aerosol Spray, Liquid	
boric acid (Hot Shot)	Dust	
cyfluthrin (Bayer)	Liquid	
cypermethrin (Raid)	Aerosol	
imiprothin (Raid, Black Flag)	Aerosol Spray	Imoprothin is formulated with other pesticides in these products.

INSECT CONTROL

Table 5-20. Control of Household Pests–Products for Use by the General Public

Insecticide	Formulation	Precautions and Remarks
Cockroach (a) Indoors (continued)		
diatomaceous earth (PermaGuard)	Dust	Use diatomaceous earth in the same manner as boric acid powders. Some formulations contain pyrethrins and pyrethrum.
deltamethrin (Black Flag, Raid, Terro)	Aerosol Spray, Liquid	Place bait stations in infested areas; follow label instructions. Keep out of reach of children and pets. Sanitation is critical; before using baits, eliminate other food sources. Place bait stations in cabinets under sinks, behind stoves and refrigerators. Slow acting but gives long-lasting control. Force small amounts into all hidden nesting areas with dust applicator. Avoid overapplication and inhalation of dust. Some formulations may contain pyrethrins or pyrethrum. Do not contaminate food preparation or storage sites.
dinotefuran (Hot Shot)	Bait	
fipronil (Combat)	Bait, Bait Station	
hydramethylnon (Combat)	Bait	
hydroprene (Egg Stopper)	Bait Station	Hydroprene is an insect growth regulator and should be used with an adulticide.
lambda-cyhalothrin (Spectracide)	Liquid, Aerosol Spray	
lemongrass oil (Hot Shot)	Liquid, Aerosol Spray	
permethrin (Hot Shot, Bengal)	Aerosol Spray	
prallethrin (Hot Shot, Raid)	Aerosol Spray	
pyrethrins, pyrethrum	Aerosol Spray	
tetramethrin (Hot Shot)	Fogger	
chlorpyrifos (Hot Shot)	Bait	Apply products as directed on the label.
Cockroach (b) Outdoors		
bifenthrin (Ortho)	Liquid	Some species of cockroaches can live indoors and outdoors. Cockroaches that live outdoors tend to hide under mulch, ivy, and similar cover. Treat groundcover and along foundation walls, patios, and other areas where cockroaches are seen. Certain products cannot be used on or around edible plants. Read product labels for any limitations.
cyfluthrin (Bayer)	Liquid, Granule	
deltamethrin (Raid, Terro, Black Flag)	Liquid, Aerosol Spray	
diatomaceous earth (PermaGuard)	Dust	Apply products as directed on the label.
dinotefuran (Hot Shot)	Bait	
hydromethylnon (Combat)	Bait	
lambda-cyhalothrin (Spectracide)	Liquid, Granule	
Cricket (Indoors and in crawlspaces)		
allethrin (Ultrakill)	Aerosol	Crickets enter homes through basements and similar areas. Some formulations may be sprinkled along foundation. Read product label before using outdoors.
boric acid	Bait	
cyfluthrin (Bayer)	Liquid, Powder	
cypermethrin (Raid)	Aerosol Spray	Treat along foundation walls, patios, and other areas where crickets are seen.
deltamethrin (Raid, Black Flag)	Liquid	Apply products as directed on the label.
diatomaceous earth (PermaGuard)	Dust	Apply in a light 2- to 4-inch band around foundation. Do not use excessive amounts, and do not apply to foliage of ornamentals or to food crops.
imoprothrin (Raid, Black Flag)	Aerosol Spray	Imoprothin is formulated with other pesticides in these products.
lambda-cyhalothrin (Spectracide)	Liquid, Aerosol Spray	Apply products as directed on the label.
pyrethrins, pyrethrum (Hot Shot, Black Flag)	Aerosol Spray	
binfenthrin (Ortho)	Aerosol Spray	
prallethrin (Hot Shot, Raid)	Aerosol Spray	
Earwig (a) Indoors		
bifenthrin (Ortho)	Aerosol Spray, Liquid	
cyfluthrin (Bayer)	Liquid	
diatomaceous earth (PermaGuard)	Dust	
deltamethrin (Raid, Black Flag)	Liquid, Aerosol Spray	
imoprothrin (Raid, Black Flag)	Aerosol Spray	
lambda-cyhalothrin (Spectracide)	Liquid	
prallethrin (Hot Shot, Raid)	Aerosol Spray	
tetramethrin (Hot Shot)	Fogger	
Earwig (b) Outdoors		
bifenthrin (Ortho)	Granular, Liquid	Repeat treatments at 14-day intervals if necessary. Granular formulations are for outdoor use only and must be watered in or applied before rain.
cyfluthrin (Bayer)	Granular, Liquid	
cypermethrin (Raid)	Aerosol Spray	
diatomaceous earth (PermaGuard)	Dust	
gamma-cyhalothrin (Spectracide)	Liquid, Granule	
lambda-cyhalothrin (Spectracide)	Liquid	

Table 5-20. Control of Household Pests—Products for Use by the General Public

Insecticide	Formulation	Precautions and Remarks
Flea (a) Indoors		
allethrin (Ultrakill)	Aerosol	Only use products specifically labeled for flea control. Apply to carpets, crevices in flooring. Do not leave chemical residue on surface. Avoid accidental inhalation during application.
bifenthrin (Ortho)	Aerosol Spray, Liquid	
boric acid	Dust	Treat pet sleeping quarters and other localized areas, such as under cushions and furniture, as specified on label. Vacuum carpets and furniture before applying; dispose of contents properly. Sprays may be used for general area treatment. Also treat cracks, crevices, and similar areas only. Foggers are only effective when used in conjunction with sprays to other critical areas. Treat infested animals with properly labeled product for lasting control.
cyfluthrin (Bayer)	Liquid	
deltamethrin (Raid, Black Flag)	Liquid, Aerosol Spray	Apply as directed on the label.
lamda-cyhalothrin (Spectracide)	Liquid	
deltamethrin (Raid, Black Flag)	Aerosol Spray, Liquid	
d-phenothrin (Raid)	Aerosol	
pyrethrins, pyrethrum (Hot Shot, Black Flag)	Liquid, Fogger, Aerosol Spray	
tetramethrin (Hot Shot)	Fogger	
permethrin (Enforcer)	Liquid	
sumithrin (Enforcer)	Dust	
methoprene (Precor) pyriproxyfen	Aerosol Spray, Fogger, Liquid	Insect growth regulators that control immature fleas only. Usually formulated with an adulticide.
imiprothrin (Black Flag, Raid)	Aerosol Spray	
tetramethrin (Enforcer)	Aerosol Spray	
phenoxybenzyl (Enforcer)	Aerosol Spray	
Flea (b) Outdoors		
bifenthrin (Ortho)	Liquid	Concentrate on kennels and shaded areas where animals tend to rest of congregate. Apply liquid formulations with sufficient spray volume to saturate soil. Granular formulations must be watered in or applied before rain. Repeate as needed at 4- to 6-week intervals.
cyfluthrin (Bayer)	Liquid	
deltamethrin (Raid, Black Flag)	Liquid, Aerosol Spray	
diatomaceous earth (PermaGuard)	Dust	
gamma-cyhalothrin (Spectracide)	Liquid, Granule	Apply as directed on the label.
lambda-cyhalothrin (Spectracide, Enforcer)	Liquid	
Flies (a) Indoors		
dichlorvos (Vapona, Pest Strip, Ortho No-Pest Strip, Hot Shot)	Strip	Strips can only be used in unoccupied areas. Apply as a surface spray to areas or objects (such as garbage cans) infested with flies. Repeat treatments as may be necessary. See label before treating areas of vegetation.
lemongrass oil (Hot Shot)	Aerosol Spray	Sanitation in the area is essential for satisfactory control of flies.
pyrethrins, pyrethrum (various)	Aerosol Spray, Liquid	
d-phenothrin (Raid)	Aerosol	
tetramethrin (Hot Shot)	Fogger	
deltamethrin (Black Flag, Raid)	Liquid, Aerosol Spray	
Flies (b) Outdoors		
bifenthrin (Ortho)	Liquid	Apply as a surface spray to areas or objects (such as garbage cans) infested with flies. Repeat treatments may be necessary. See label before treating areas of vegetation.
cyfluthrin (Bayer)	Liquid	
deltamethrin (Black Flag, Raid, Spectracide)	Aerosol, Liquid	Sanitation in the area is essential for satisfactory control using any of these chemicals but particularly important with baits.
imidacloprid (Maxforce)	Bait	
lambda-cyhalothrin (Spectracide)	Liquid	Use as directed.
d-phenothrin (Raid)	Aerosol Spray	
prallethrin (Ultrakill)	Aerosol Spray	
Hornets, Mud Daubers, Wasps, Yellow Jackets (a) Indoors		
deltamethrin (Raid, Spectracide)	Liquid	
eugenol (Bioganic)	Aerosol Spray, Dust	
prallethrin (Raid, Hot Shot, Ultrakill, Spectracide)	Aerosol Spray	
d-phenothrin (Raid)	Aerosol Spray	
pyrethrins, pyrethrum	Aerosol Spray, Liquid	
tetramethrin (Hot Shot, Terro)	Fogger, Aerosol Spray	
cyfluthrin (Bayer)	Liquid	

Table 5-20. Control of Household Pests–Products for Use by the General Public

Insecticide	Formulation	Precautions and Remarks
Hornets, Mud Daubers, Wasps, Yellow Jackets (b) Nest and adjacent areas		
bifenthrin (Ortho)	Liquid	Apply to nest or opening after dark when insects have returned to nest. Re-treatment may be necessary. Most are packaged in pressurized containers that direct an insecticide stream of up to 10 feet. For yellowjackets and other soil-dwelling wasps, apply chemical to nests in soil.
carbaryl (Sevin)	Dust, Liquid	
cyfluthrin (Bayer)	Liquid	
deltamethrin (Black Flag, Raid)	Aerosol, Liquid	
diatomaceous earth (PermaGuard)	Dust	
eugenol (Bioganic)	Aerosol Spray, Dust	
lambda-cyhalothrin (Hot Shot)	Liquid	
phenothrin (Raid)	Aerosol Spray	
prallethrin (Ultrakill, Hot Shot, Raid, Spectracide)	Aerosol	
Lice: body, head, crab (on person)		
malathion (Ovide)	Liquid	Shampoo formulations. Thoroughly treat infested areas of body with lotion. Wash infested clothing with strong soap and very hot water. Dryclean woolens. Ovide requires a physician's prescription. **Insecticidal treatment of furniture, carpets, or other areas of the home is not needed.**
permethrin (Nix)	Liquid	
pyrethrins, pyrethrum (Rid)	Liquid	
Millipede (a) Indoors		
bifenthrin (Ortho)	Liquid	
cyfluthrin (Bayer)	Liquid	
diatomaceous earth (PermaGuard)	Dust	
imoprothrin (Raid, Black Flag)	Aerosol Spray, Liquid	
lambda-cyhalothrin (Spectracide)	Liquid	
mint oil (Victor)	Aerosol Spray	
deltamethrin (Black Flag, Raid)	Aerosol Spray	
prallethrin (Hot Shot, Raid)	Aerosol Spray	
pyrethrins, pyrethrum	Aerosol Spray	
Millipede (b) Outdoors		
bifenthrin (Ortho)	Granular, Liquid	Use as barrier treatment along foundation wall, door threshold, window ledges. Some sprays may damage vegetation under hot, humid conditions. Read label precautions. For lawn treatment, apply an insecticide band 10 to 15 feet wide. Apply liquid formulations with sufficient spray volume to saturate soil. Use granular formulations outdoors only; water in or apply before rain. Repeat as needed at 4- to 6-week intervals.
cyfluthrin (Bayer)	Granular, Liquid	
diatomaceous earth (PermaGuard)	Dust	
gamma-cyhalothrin (Spectracide)	Liquid, Granule	
Mosquitoes (a) Indoors		
cyfluthrin (Bayer)	Liquid	
deltamethrin (Black Flag, Raid)	Liquid	
lemongrass oil (Hot Shot)	Aerosol Spray	
tetramethrin (Hot Shot)	Fogger	
phenothrin (Raid)	Aerosol Spray	
pyrethrins, pyrethrum (Raid)	Aerosol Spray	
Mosquitoes (b) Outdoors (See also *Community Pest Control* Section)		
Bacillus thuringiensis (*Bti*)(Mosquito Dunks)	Solid	A biopesticide containing bacteria that kill mosquitoes and some biting flies. Place in small ponds, birdbaths, and ornamental pools (not swimming pools). Follow instructions for specifics of application.
bifenthrin (Ortho)	Liquid	Long-term control requires eliminating or cleaning mosquito breeding areas, such as discarded containers, ditches, and other artificial sources of standing water. Spraying nearby vegetation may eliminate some mosquito resting sites, but some formulation as may damage vegetation. Aerosols or foggers may be used for temporary relief when winds are insignificant. Use repellents on exposed body areas.
deltamethrin (Black Flag, Raid)	Aerosol Spray, Liquid	
gamma–cyhalothrin (Spectracide)	Liquid, Granule	
cyfluthrin (Bayer)	Liquid	
Pantry Pests (Pests in food storage areas)		
allethrin (Ultrakill)	Aerosol	Discard infested material. Overall treatment of storage shelves and pantry usually not necessary; treat cracks, crevices, moldings, and similar areas only. Remove and cover food, cooking, and eating utensils before spraying storage cabinets. Do not restock shelves until surfaces are dry. Cover treated shelves with shelf paper if desired. Some products are not suitable for use in residential kitchens or commercial food/feed preparation sites. Read the product label carefully. Imoprothin is formulated with other pesticides in these products.
cyfluthrin (Bayer)	Liquid	
deltamethrin (Raid, Black Flag)	Liquid	
diatomaceous earth (PermaGuard)	Dust	
imoprothrin (Black Flag, Raid)	Aerosol Spray	
lambda-cyhalothrin (Spectracide)	Liquid	
mint oil (Victor)	Aerosol Spray	
deltamethrin (Black Flag, Raid)	Aerosol Spray, Liquid	
pyrethrins, pyrethrum	Aerosol Spray	
bifenthrin (Ortho)	Aerosol Spray	

Table 5-20. Control of Household Pests–Products for Use by the General Public

Insecticide	Formulation	Precautions and Remarks
Silverfish		
bifenthrin (Ortho)	Liquid, Aerosol Spray	Apply to cracks and crevices, behind and underneath appliances. Spray along baseboards and other areas where silverfish are found.
cyfluthrin (Bayer)	Liquid	
deltamethrin (Raid, Black Flag)	Liquid	
diatomaceous earth (PermaGuard)	Dust	
imoprothrin (Raid, Hot Shot)	Aerosol Spray	Imoprothin is formulated with other pesticides in these products
lambda-cyhalothrin (Spectracide)	Liquid, Aerosol Spray	Follow label directions.
mint oil (Victor)	Aerosol Spray	
deltamethrin (Black Flag, Raid)	Aerosol, Liquid	
pyrethrins, pyrethrum	Aerosol Spray	
Sowbugs and Pillbugs (a) Indoors		
bifenthrin (Ortho)	Liquid	Clean up breeding and hiding places, and treat thoroughly. Outdoor barrier treatments along foundation and door thresholds are usually sufficient. Some products are not suitable for use in residential kitchens or commercial food/feed preparation sites. Read the product label carefully.
cyfluthrin (Bayer)	Liquid	
deltamethrin (Raid, Black Flag)	Liquid	
diatomaceous earth (PermaGuard)	Dust	Follow label directions.
lambda-cyhalothrin (Spectracide)	Liquid	
mint oil (Victor)	Aerosol Spray	
pyrethrins, pyrethrum	Aerosol Spray	
Sowbugs and Pillbugs (b) Outdoors		
bifenthrin (Ortho)	Granular, Liquid	
cyfluthrin (Bayer)	Granular, Liquid	
deltamethrin (Raid, Black Flag)	Liquid	
diatomaceous earth (PermaGuard)	Dust	
gamma-cyhalothrin (Spectracide)	Liquid, Granule	
lambda-cyhalothrin (Sepctracide)	Liquid, Aerosol Spray	
Spiders (a) Indoors		
bifenthrin (Ortho)	Liquid, Dust	Treat infested areas, along baseboards. Use foggers if rooms have been undisturbed for some time and spider populations are extensive. Some products are not suitable for use in residential kitchens or commercial food/feed preparation sites. Read the product label carefully.
cyfluthrin (Bayer)	Liquid	
cypermethrin (Raid)	Aerosol Spray	
diatomaceous earth (PermaGuard)	Dust	
imoprothrin (Raid Max, Black Flag)	Aerosol Spray	Imoprothin is formulated with other pesticides in these products.
lambda-cyhalothrin (Spectracide)	Liquid, Aerosol Spray	Follow label directions.
mint oil (Victor)	Aerosol Spray	
deltamethrin (Black Flag, Raid, Terro)	Aerosol Spray, Liquid	
pyrethrins, pyrethrum (Terro)	Aerosol Spray	
prallethrin (Hot Shot, Raid)	Aerosol Spray	
Spiders (b) Outdoors		
bifenthrin (Ortho)	Liquid, Granules	Apply as a barrier treatment along foundation. Spray corners of decks, eaves, porches and other areas where spiders tend to build webs. Webbing can be knocked down as an alternative. Exercise caution when spray in crawlspace. Avoid inhaling spray.
cyfluthrin (Bayer)	Liquid	
deltamethrin (Terro)	Liquid	
diatomaceous earth (PermaGuard)	Dust	Follow label directions.
lambda-cyhalothrin (Spectracide)	Liquid, Aerosol Spray	
Springtails (Indoors and outdoors)		
bifenthrin (Ortho)	Granular, Liquid	Apply as a barrier spray along foundation and entry points. Some products may be used indoors for temporary relief. Clean up moisture conditions that may attract insects indoors. Excess moisture may impede product effectiveness.
cyfluthrin (Bayer)	Liquid	
diatomaceous earth (PermaGuard)	Dust	
imoprothrin (Raid, Black Flag)	Aerosol Spray	Use indoors for temporary relief. Some products are not suitable for use in residential kitchens or commercial food/feed preparation sites. Read the product label carefully. Imoprothin is formulated with other pesticides in these products.
lambda-cyhalothrin (Spectracide)	Liquid	
mint oil (Victor)	Aerosol Spray	Follow label directions.
deltamethrin (Black Flag, Raid)	Aerosol	
pyrethrins, pyrethrum	Aerosol Spray	
gamma-cyhalothrin (Spectracide)	Liquid, Granules	

INSECT CONTROL

Table 5-20. Control of Household Pests–Products for Use by the General Public

Insecticide	Formulation	Precautions and Remarks
Stinging Caterpillars See *Trees and Woody Ornamentals* Section		
Stink Bugs (Indoors and outdoors)		
bifenthrin (Ortho)		
cyfluthrin (Bayer)		
cypermethrin (Raid)	Aerosol Spray	
deltamethrin (Raid)	Liquid	
gamma-cyhalothrin (Spectracide)		
Stored Food Pests See Pantry Pests.		
Ticks(Outdoors) See Brown Dog Tick and *Control of Insects on Pets* section		
Wasps, Yellow Jackets See Hornets, etc.		

Formulation Designations: Bait may be gel or granular; fogger is a total release aerosol; liquid for mixing with water or ready-to-use; powder for mixing with water.

Insect Control for Home Lawns

R. L. Brandenburg, Entomology Extension and Research

NOTE: Some products are for use only by professionals. Homeowner products are numerous and names change frequently, so it is not possible to list all homeowner products by brand names. When choosing a product to use at home, look at the label and use this table to compare the name of the active ingredients.

Table 5-21. Insect Control for Home Lawns

Pest Insecticide and Formulation	Amount per 1,000 Sq Ft	Precautions and Remarks
Ant (Also see Imported Fire Ant)		
carbaryl* (Sevin) 50 WP, 80 WSP and baits	See label	Treat mounds and surrounding area or apply broadcast.
clothianidin + bifenthrin (Aloft LC) G SC	 1.8 to 3.6 lb 0.27 to 0.54 fl oz	Toxic to fish and aquatic invertebrates. Do not apply near or allow runoff to surface waters or intertidal areas.
hydromethylnon* (Maxforce G) bait	See label	
pyrethroids* (Advanced Lawn, Bug-B-Gone, Deltaguard, Scimitar, Talstar, Tempo, Wisdom and others) Some ants are susceptible to fire ant products.	See label	Many pyrethroids are toxic to fish and aquatic invertebrates. Apply these products only as specified on the label.
Armyworm, Fall Armyworm, Cutworm		
azidirachtin* (Azatrol, Neemix, Turplex, etc.)	See label	
carbaryl* (Sevin) 50 WP, 80 WSP and baits	See label	Apply as a coarse spray in sufficient water for good coverage. Treat when first injury noted. Repeat as needed. Do not water into soil. Do not cut grass for 1 to 3 days after treatment.
chlorotraniliprole (Acelepryn) G SC	 1.15 to 2.3 lb 0.046 to 0.092 fl oz	Toxic to aquatic invertebrates, oysters and shrimp.
indoxacarb (Provaunt) WDG	0.046 to 0.092 oz	
pyrethroids* (Advanced Lawn, Bug-B-Gone, Deltaguard, Menace, Scimitar, Talstar, Tempo, Wisdom and others)	See label	Many pyrethroids are toxic to fish and aquatic invertebrates. Apply these products only as specified on the label.
spinosad A and D (Conserve) SC	0.25 to 1.25 fl oz	Rate varies with size and species.
thimethoxam + lambda-cyhalothrin (Tandem)	See label	Highly toxic to fish and aquatic invertebrates.
trichlorfon* (Dylox, Proxol) 80 SP	1.5 to 3 oz	
various entomogenous nematode and *B.t.* products	See label	
Bee and Wasp		
carbaryl* (Sevin) 50 WP	6 to 8 oz	Most of these are parasitic on soil pests, especially grubs; therefore they are beneficial. Sometimes there are so many bees and wasps burrowing in the soil that chemical treatments are necessary to prevent damage or reduce danger from stings. Spot spray ground nest openings. Bee, wasp, and hornet sprays in pressurized cans are also effective.
pyrethroids* (Advanced Lawn, Bug-B-Gone, Deltaguard, Scimitar, Talstar, Tempo, Wisdom and others)	See label	

INSECT CONTROL

Table 5-21. Insect Control for Home Lawns

Pest Insecticide and Formulation	Amount per 1,000 Sq Ft	Precautions and Remarks
Chinch Bug		
*Beauveria bassiana** (Naturalis-T)	See label	
carbaryl* (Sevin) 80 WSP	2.7 to 3.6 oz	
chlorotraniliprole (Acelepryn) G SC	1.15 to 2.3 lb 0.184 to 0.46 fl oz	Suppression only. Toxic to aquatic invertebrates, oysters and shrimp.
clothianidin + bifenthrin (Aloft LC) G SC	1.8 to 3.6 lb 0.27 to 0.54 fl oz	Toxic to fish and aquatic invertebrates. Do not apply near or allow runoff to surface or intertidal areas.
dinotefuran (Zylam 20 SG)	1.0 fl oz	For suppression, make application prior to hatching of first instar nymphs.
pyrethroids* (Advanced Lawn, Bug-B-Gone, Deltaguard, Menace, Scimitar, Talstar, Tempo, Wisdom and others)	See label	Many pyrethroids are toxic to fish and aquatic invertebrates. Apply these products only as specified on the label.
thimethoxam + lambda-cyhalothrin (Tandem)	See label	Apply when insects are first observed. Repeat applications may be necessary. Highly toxic to fish and aquatic invertebrates.
Grub, Green June Beetle (only) (Also see White Grub)		
carbaryl* (Sevin) 80 WSP	1.8 oz	Apply to the soil surface but do not water in.
Grub, White (Japanese beetle, Southern chafer, European chafer, billbug, green June beetle)		
carbaryl* (Sevin) 80 WSP	3.6 oz	
chlorotraniliprole (Acelepryn) G SC	1.15 to 2.3 lb 0.184 to 0.46 fl oz	Toxic to aquatic invertebrates, oysters and shrimp.
clothianidin (Arena) 0.25 G 50 WDG	1.84 to 3.67 lb 0.14 to 0.29 fl oz	Toxic to fish and aquatic invertebrates. Do not apply near or allow runoff to surface waters or intertidal areas.
clothianidin + bifenthrin (Aloft LC) G SC	1.8 to 3.6 lb 0.27 to 0.54 fl oz	Toxic to fish and aquatic invertebrates. Do not apply near or allow runoff to surface waters or intertidal areas.
dinotefuran (Zylam 20 SG)	1.0 fl oz	Make application prior to or during peak egg hatch.
imidacloprid (Advanced Lawn Grub Control, Merit, many others)	See label	
thiomethoxam (Meridian)	See label	Highly toxic to aquatic invertebrates.
Grub, White (Japanese beetle, Southern chafer, European chafer, billbug, green June beetle) (continued)		
thimethoxam + lambda-cyhalothrin (Tandem)	See label	Highly toxic to fish and aquatic invertebrates.
trichlorfon* (Proxol/Dylox) 80 SP	3.75 oz	
various entomogenous nematodes	See label	Must be Heterorhabditid species to be effective.
Imported Fire Ant		
acephate* (Ortho Fire Ant Killer and others)	1 to 2 tsp/ mound	Distribute uniformly over mound. For best results apply early in morning or late afternoon.
avermectin B1 (Ascend, Award II) 0.011% bait	See label	Apply as a mound treatment or broadcast bait.
carbaryl (Sevin)	See label	Use as a mound drench.
clothianidin + bifenthrin (Aloft LC SC)	2.3-3.6 lb	Toxic to fish and aquatic invertebrates. Do not apply near or allow runoff to surface waters or intertidal areas.
d-limolene (Orange Guard)	See label	Mound treatment. Acceptable to organic growers. May also be used around fruit and vegetable gardens.
fipronil 0.0143 G (Taurus G, Top Choice)	2 lb	Apply as a broadcast.
fipronil (Maxforce FC) bait	See label	Apply as a mound treatment or broadcast bait.
Fipronil + bifenthrin + lambda-cyahothrin (Taurus Trio G)	2 lb	Apply as a broadcast. Irrigate prior to treatment.
hydramethylnon* (Amdro Fire Ant Bait, Amdro Pro, Maxforce G)	See label	Follow label directions precisely. Use fresh bait. Repeat treatment usually required.
indoxacarb (Spectracide Fire Ant Once and Done) (Over 'n Out Fire Ant Killer Mound Treatment) (Advion)	See label	
Metaflumizone (Siesta) bait	See label	Mound or broadcast bait.
methoprene (Extinguish) bait	See label	Mound or broadcast. Follow label directions. Repeat treatments usually required.
methoprene + hydromethylnon (Extinguish Plus, Amdro Firestrike) bait	See label	Follow label directions precisely. Repeat treatments usually required. Use fresh bait. Found in broadcast or mound treatment packaging.
pyrethroids (Bayer Advanced, Menace, Ortho Fire Ant Killer, Mound Treatment, Talstar One, Tempo, Wisdom and others)	See label	Many pyrethroids are toxic to fish and aquatic invertebrates. Apply these products only as specified on the label.
pyriproxyfen (Distance) bait	See label	Mound or broadcast bait.
spinosad (Come and Get It Fire Ant Bait by Fertilome, Entrust, Payback, Green Light Fire Ant Control with Conserve, Green Light Fire Ant Killer with Spinosad Mound Drench) AUBURN UNIVERSITY HAS AN EXCELLENT PUBLICATION FOR HOMEOWNERS	See label	Acceptable to organic growers. Follow label directions precisely. Repeat treatments usually required. Use fresh bait. May also be used around fruit and vegetable gardens. http://www.aces.edu/pubs/docs/A/ANR-0175-A/ANR-0175-A.pdf

189

INSECT CONTROL

Table 5-21. Insect Control for Home Lawns

Pest Insecticide and Formulation	Amount per 1,000 Sq Ft	Precautions and Remarks
Mole Cricket		
carbaryl* baits	See label	
clothianidin + bifenthrin (Aloft LC) G SC	 1.8 to 3.6 lb 0.27 to 0.54 fl oz	Toxic to fish and aquatic invertebrates. Do not apply near or allow runoff to surface waters or intertidal areas. Application should be made during peak adult flight and egg lay.
dinotefuran (Zylam 20 SG)	1.0 fl oz	Make application prior to or during peak egg hatch.
Fipronil (Top Choice, Taurus G)	2 lb	Apply as a broadcast.
imidacloprid (Advanced Lawn Grub Control, Merit)	See label	
indoxacarb (Advion Insect Granules) bait	See label	
indoxacarb (Provaunt) WDG	0.275 oz	
pyrethroids* (Advanced Lawn, Bug-B-Gone, Deltaguard, Menace, Scimitar, Talstar, Tempo, Wisdom and others)	See label	Many pyrethroids are toxic to fish and aquatic invertebrates. Apply these products only as specified on the label.
thimethoxam + lambda-cyhalothrin (Tandem)	See label	Apply from first egg hatch to peak egg hatch. Highly toxic to fish and aquatic invertebrates.
Various entomogenous nematode products	See labels	Require irrigation.
Slug, Snail		
iron phosphate (Natria) bait		Apply in late afternoon.
measurol 2% B	1 lb	Apply in late afternoon.
metaldehyde	See label	Apply in late afternoon.
Sod Webworm (also Burrowing Sod Webworm)		
carbaryl* (Sevin) 80 WSP 50 WP	 3.6 oz 6.4 oz	Do not water in sprays. Use 6 gallons water plus the insecticide per 1,000 square feet. Treat in late afternoon. Do not cut grass for 1 to 3 days after treatment. Granules must be watered in.
dinotefuran (Zylam 20 SG)	1.0 fl oz	
pyrethroids* (Advanced Garden, Deltagard, Scimitar, Talstar, Tempo, Wisdom and others)	See label	Many pyrethroids are toxic to fish and aquatic invertebrates. Apply these products only as specified on the label.
spinosad A and D (Conserve) SC	0.25 to 1.25 fl oz	Rate varies with size and species.
thimethoxam + lambda-cyhalothrin (Tandem)	See label	Highly toxic to fish and aquatic invertebrates.
trichlorfon* (Dylox, Proxol) 80 SP	1.5 to 3 oz	Use sufficient water for good coverage.
various entomogenous nematode and *B.t.* products	See label	

* Several trade names available.Check label for active ingredient. Always follow label instructions.

VI — INSECT AND DISEASE CONTROL OF FRUITS

Apple Spray Program

J. F. Walgenbach, Entomology Research and Extension, and
S. M. Villani, Research and Extension Apple Pathologist

See *Integrated Orchard Management Guide for Commercial Apples in the Southeast* (AG-572) for more detailed information on apple disease and insect control. For a copy, contact Jim Walgenbach, 455 Research Drive, Mills River, NC 28759; jim_walgenbach@ncsu.edu. The guide is also available online at http://entomology.ces.ncsu.edu/apple-pest-management

Table 6-1. Apple Spray Program

Number and Time of Application	*Amount of Fungicide and Insecticide Per Acre*
Green-Tip Spray When buds show 0.25-inch new growth	**Fungicide: NOTE: Captan will cause injury if applied with or too close to oil applications.** **Apple Scab:** Apply dodine (Syllit FL) 1.5 to 3 pt + mancozeb 75 DF 3 lb *OR* dodine (Syllit FL) 1.5 to 3 pt + Captan 80 WDG 2.5 lb *OR* cyprodinil (Vangard 75WG) 5 oz *OR* cyprodinil (Vangard 75WG) 3 to 5 oz + Mancozeb 75DF 3 lb *OR* cyprodinil (Vangard 75WG) 3 to 5 oz + metiram (Polyram 80DF) 3 lb *OR* pyrimethanil (Scala SC) 7 to 10 oz *OR* pyrimethanil (Scala SC) 5 oz + Mancozeb 75DF 3 lb *OR* pyrimethanil (Scala SC) 5 oz + metiram (Polyram 80DF) 3 lb *OR* fluopyram/pyrimethanil (Luna Tranquility) 11.2 to 16 fl oz + mancozeb 75DF 3 lb *OR* penthiopyrad (Fontelis SC) 16 to 20 fl oz *OR* penthiopyrad (Fontelis SC) 16 to 20 fl oz + mancozeb 75 DF 3 lb *OR* Captan 80 WDG 5 lb *OR* metiram (Polyram 80DF) 3 to 6 lb *OR* mancozeb (Koverall) 3 to 6 lb. **Fire Blight:** Apply copper hydroxide/copper oxychloride (Badge SC) 3.5 to 7 pt (rate for silver to green tip only) *OR* copper hydroxide/copper oxychloride (Badge SC) 0.5 to 1.5 pt *OR* copper hydroxide/copper oxychloride (Badge X2, OMRI listed) 3.5 to 7 lb (rate for silver to green tip only) *OR* copper hydroxide/copper oxychloride (Badge X2, OMRI listed) 0.5 to 1.5 lb *OR* copper hydroxide (Kocide 3000) 3.5 to 7 lb (discontinue rate at ½" green tip) *OR* basic copper sulfate (Cuprofix Ultra 40 Disperss) 5 to 7.5 lb (apply rate between silver and green tip). **Phytophthora Rots:** Apply phosphorous acid (Prophyt) 2 to 4 pt/100 gal. **Insecticide:** Apply 3 gallons oil per 100 gallons water. For improved control of San Jose scale, add 1 quart chlorpyrifos (Lorsban 4EC) *OR* 4 ounces pyriproxyfen (Esteem 35 WP) *OR* 34.5 ounces buprofezin (Centaur 70WDG). This is an important spray for mite eggs and San Jose scale. If using mating disruption for codling moth and oriental fruit moth, dispensers (200 Isomate CM-OFM TT dispensers per acre or 1 CheckMate CM-OFM Puffer per acre) should be in place before bloom.
Half-Inch Green Spray One week after GREEN-TIP SPRAY	**Fungicide:** Use same as GREEN-TIP SPRAY. See warning about high application rates of copper. **Insecticide:** If an insecticide was not applied at green tip, use one of the products listed above; otherwise no insecticide is needed.
Tight Cluster Spray One week after HALF-INCH GREEN SPRAY	**Fungicide:** **Apple Scab:** Apply trifloxystrobin (Flint WG) 2.5 oz *OR* kresoxim-methyl (Sovran 50 WG) 3.2 to 6.4 oz *OR* fenbuconzole (Indar 2F) 6 to 8 fl oz *OR* fluopyram/pyrimethanil (Luna Tranquility) 11.2 to 16 fl oz *OR OR* fluopyram/trifloxystrobin (Luna Sensation) 4 to 5.8 fl oz *OR* fluxapyroxad/pyraclostrobin (Merivon) 4 to 5.5 fl oz *OR* penthiopyrad (Fontelis) 16 to 20 fl oz *OR* cyprodinil/difenoconazole (Inspire Super) 12 fl oz *OR* benzovindiflupyr (Aprovia) 5.5 to 7 fl oz *OR* captan (Captan 80WDG) 5 lb *Or* mancozeb (Koverall) 3 to 6 lb *OR* metiram (Polyram 80 DF) 3 to 6 lb. **Powdery Mildew:** Apply myclobutanil (Rally 40 WSP) 5 to 10 oz *OR* fenbuconazole (8 fl oz) *OR* triflumizole (Procure 480SC) 8 to 16 fl oz *OR* flutriafol (Topguard) 8 to 12 fl oz *OR* fluopyram/pyrimethanil (Luna Tranquility) 11.2 to 16 fl oz *OR OR* fluopyram/trifloxystrobin Luna Sensation (4 to 5.8 fl oz) *OR* fluxapyroxad/pyraclostrobin (Merivon) 4 to 5.5 fl oz *OR* penthiopyrad (Fontelis) 14 to 20 fl oz *OR* trifloxystrobin (Flint WG) 2 to 2.5 oz *OR* kresoxim-methyl (Sovran 50 WG) 4 to 6.4 oz *OR* benzovindiflupyr (Aprovia) 5.5 to 7 fl oz *OR* parrafinic oil (JMS Stylet Oil) 1 to 2 gal/100 gal *OR* sulfur (Microthiol Disperss) 10 to 20 lb. **Cedar Apple/Quince Rust:** Apply a DMI (FRAC GROUP 3) used for powdery mildew *OR* mancozeb (Koverall) 3 lb. **Black Rot/Frogeye Leaf Ppot:** captan (Captan 80 WDG) 2.5 to 5 lb *OR* thiophanate –methyl (Topsin 4.5 FL) 15 to 20 fl oz **Insecticide:** For rosy apple aphid, apply 5 ounces acetamiprid (Assail 30 SG). For rosy apple aphid and plant bugs, apply 4.4 ounces thiamethoxam (Actara 25 WP) *OR* 16 ounces fenpropathrin (Danitol 2.4 EC).
Pink Spray When blossom buds are pink, stems extended	**Fungicide:** Use same fungicides as TIGHT CLUSTER SPRAY. **Insecticide:** If an insecticide effective against rosy apple aphid and/or tarnished plant bug was not applied at TIGHT CLUSTER, apply one of the above materials.
Bloom Spray	**Fungicide:** Use same fungicide as TIGHT CLUSTER SPRAY. **Fire Blight Control:** Apply streptomycin (Firewall, Agrimycin) 24 oz *OR* kasugamycin (Kasumin 2L) 64 fl oz/100 gal *OR* oxytetracycline (Fireline) 12 oz/100 gal *OR* acibenzolar-S-methyl (Actigard 50WG) 1 to 2 oz + streptomycin (Firewall) 24 oz *OR* acibenzolar-S-methyl (Actigard 50WG) 1 to 2 oz + oxytetracycline (Fireline) 12 oz *OR Pantoea agglomerans* (Bloomtime Biological) 0.33 lb *OR* copper octanoate (Cueva) 2 qt *OR Bacillus amyloliquefaciens* (Double Nickel LC) 1 to 2 qt + copper octanoate (Cueva) 2 qt *OR Bacillus subtilis* (Serenade Optimum) 14 to 20 oz *OR* copper sulfate pentahydrate (MasterCop) 0.5 to 1.5 qt *OR* copper hydroxide/copper oxychloride (Badge SC) 0.5 to 1.5 pt *OR* copper hydroxide/copper oxychloride (Badge X2, OMRI listed) **Insecticide:** DO NOT USE an insecticide at BLOOM SPRAY.
Petal-Fall Spray When most petals have fallen	**Fungicide:** **Apple Scab, Powdery Mildew, Rusts:** Use same fungicide as TIGHT CLUSTER SPRAY. **Black and White Rot (Bot Rots):** Apply captan (Captan 80WDG) 5 lb *OR* trifloxystrobin (Flint 50WG) 2.5 oz *OR* kresoxim-methyl (Sovran 50 WG) 4 to 6.4 oz *OR* thiophanate-methyl (Topsin 4.5 FL) 15 to 20 fl oz *OR* fluazinam (Omega 500F) 13.8 fl oz. **Glomerella Leaf Spot and Bitter Rot:** Apply captan (Captan 80WDG) 5 lb *OR* phosphorous acid (ProPhyt) 4 pt + captan (Captan 80WDG) 3.75 lb *OR* phosphorous acid (ProPhyt) 3 to 4 pt + mancozeb (Koverall) 3 lb *OR* captan (Captan 80WDG) 2.5 lb + mancozeb (Koverall) 3 lb + phosphorous acid (ProPhyt) 3 to 4 pt *OR* captan (Captan 80WDG) 2.5 lb + mancozeb (Koverall) 3 lb *OR* fluxapyroxad/pyraclostrobin (Merivon) *OR* fluxapyroxad/pyraclostrobin (Merivon) + mancozeb (Koverall) 3 lb + phosphorous acid (ProPhyt) 4 pt *OR* boscalid/pyraclostrobin (Pristine) 14.5 to 18 oz *OR* trifloxystrobin (Flint 50WG) 3 oz + mancozeb (Koverall) 3 lb *OR* kresoxim-methyl (Sovran 50 WG) + mancozeb (Koverall) 3 lb *OR* fluazinam (Omega 500F) 13.8 fl oz. **Insecticide:** For plum curculio and Oriental fruit moth, apply 5 ounces indoxacarb (Avaunt 35WD) *OR* 3 pounds phosmet (Imidan 70 WP) or 4.5 ounces thiamethoxam (Actara 25 WP) *OR* 4 ounces clothianidin (Clutch 50 WDG). For preventive control of Euorpean red mite and leafminer, use 3 ounces abamectin (Agri-Mek 0.7SC) *PLUS* 0.25% horticultural spray oil (not a superior-type oil). If rosy apple aphid control is needed, apply 2.8 ounces imidacloprid (Admire 4.6SC).

Table 6-1. Apple Spray Program

Number and Time of Application	Amount of Fungicide and Insecticide Per Acre
First Cover Spray 8 to 10 days after PETAL-FALL SPRAY	**Fungicide:** **Powdery Mildew, Glomerella Leaf Spot, Black Rot, White Rot, Bitter Rot:** Refer to fungicides for PETAL FALL Application. Can also use Ziram 76DF (3 to 6 lb). **Flyspeck/Sooty Blotch:** Apply captan (Captan 80WDG) 2.5 to 5 lb *OR* trifloxystrobin (Flint 50WG) 1.5 to 2.5 oz *OR* kresoxim-methyl (Sovran 50 WG) 4 to 6.4 oz *OR* thiophanate methyl (Topsin 4.5FL) 15 to 20 fl oz *OR* ziram (Ziram 76DF) 6 lb *OR* cyprodinil/difenoconazole (Inspire Super) 12 fl oz *OR* fenbuconazole (Indar 2F) 6 to 8 fl oz *OR* benzovindiflupyr (Aprovia) 5.5 to 7 fl oz *OR* fluopyram/trifloxystrobin (Luna Sensation) 4 to 5.8 fl oz *OR* fluxapyroxad/pyraclostrobin (Merivon) 4 to 5.5 fl oz *OR* boscalid/pyraclostrobin (Pristine) 14.5 to 18.5 fl oz *OR* copper octanoate (Cueva) 2 qt *OR* copper octanoate (Cueva) 2 qt + *Bacillus amyloliquefaciens* (Double Nickel LC) 1 to 2 qt **Fire Blight (shoot blight/rat-tail bloom):** Apply prohexadione calcium (Apogee) 18 to 36 oz *OR* copper octanoate (Cueva) 2 qt *OR* copper octanoate (Cueva) 2 qt + *Bacillus amyloliquefaciens* (Double Nickel LC) 1 to 2 qt *OR* copper hydroxide/copper oxychloride (Badge X2, OMRI listed) 0.5 to 1.5 lb *OR* copper hydroxide/copper oxychloride (Badge SC) 0.5 to 1.5 pt **Insecticide:** For codling moth, apply 3 ounces chlorantraniliprole (Altacor 35 WDG) *OR* 5 ounces spinetoram (Delegate 25 WDG). If preventative control of European red mite is desired but was not applied at petal fall, apply 4 ounces clofentozine (Apollo SC) *OR* 4 ounces hexythiazox (Savey 50 DF) *OR* 3 ounces etoxazole (Zeal 72 WD).
Second Cover Spray 10 to 14 days after FIRST COVER SPRAY	**Fungicide:** Refer to relative effectiveness table for appropriate fungicides and to FIRST COVER SPRAYS (above) for summer disease control. **Insecticide:** Same as FIRST COVER SPRAY for control of codling moth.
Third Cover Spray 10 to 14 days after SECOND COVER SPRAY	**Fungicide:** Refer to relative effectiveness table for appropriate fungicides and to FIRST COVER SPRAYS (above) for summer disease control. **Insecticide:** For tufted apple bud moth, apply 12 ounces methoxyfenozide (Intrepid 2 F), 3 ounces chlorantraniliprole (Altacor 35 WDG) *OR* 5 ounces spinetoram (Delegate 25 WDG). On plantings susceptible to dogwood borer, apply 1 pound chlorpyrifos (Lorsban 4 EC) to trunk at base of tree using a handgun application anytime between third cover spray and mid July. In orchards with San Jose problems, apply 4 ounces pyriproxyfen (Esteem 35WP) or 34.5 ounces buprofezin (Centaur 70WDG).
Summer Cover Sprays 10- to 14-day intervals or as pest density and weather conditions dictate	**Fungicide:** Refer to relative effectiveness table for appropriate fungicides and to FIRST COVER SPRAYS (above) for summer disease control. **Insecticide:** Refer to relative effectiveness tables and AG-572 for appropriate insecticides and miticides for summer insect control. For second generation codling moth sprays (mid to late July), do not use the same insecticide used for first generation control (first and second cover sprays). Important sprays include codling moth in mid to late July, apple maggot in late July. For apple maggot, apply 2.8 ounces imidacloprid (Admire Pro) or 3 pounds phosmet (Imidan). In orchards not using mating disruption up to this point, apply 1.2 ounces/acre of Check OFM-F (sprayable pheromone for mating disruption) in mid to late July and again one month later for late-season oriental fruit moth control.

[1] Do not follow oil with captan or sulfur for 14 days.

Further Information

Producing Tree Fruit for Home Use. N.C. Cooperative Extension Service, AG-28; http://www.ces.ncsu.edu/depts/hort/hil/ag28.html.

Fire Blight of Apple and Pear. Plant Pathology Information Note 167 (http://www.ces.ncsu.edu/depts/pp/notes/oldnotes/fd3.htm).

A Grower's Guide to Apple Insects and Diseases in the Southeast. (http://ipm.ncsu.edu/apple/contents.html).

Copies of these publications are available from your county Cooperative Extension center.

Relative Effectiveness of Various Fungicides for Apple Disease Control

S. M. Villani, Research and Extension Apple Pathologist

(E = excellent; G=Good; F = Fair; P = Poor; NC = No control; ND = No data)

Table 6-2. Relative Effectiveness of Various Fungicides for Apple Disease Control

Fungicide and Rate of Usage Per Acre	FRAC Code	Days Between Last Spray and Harvest	Scab	Rusts	Brooks Spot	Black Rot/ White Rot	Glomerella/Bitter Rot	Sooty Blotch and Flyspeck	Powdery Mildew
benzovindiflupyr (Aprovia) 5.5 to 7 fl oz	7	30	E	F	ND	G-E	P-F	G-E	F-G
captan (Captan 80WDG) 5 lb	M4	0	G	P	G	G	G	F-G	NC
captan (Captan 80 WDG) 3.75 lb + potassium phosphite (ProPhyt 4.2 L) 2 to 3 qt	M4 33	0	F-G	P	G	G	G	G	NC
cyprodinil (Vangard 75 WG) 5 oz	9	0	F	NC	NC	NC	NC	NC	NC
difenoconazole + cyprodinil (Inspire Super) 12 fl oz	3 + 9	14	E	E	F	G	P	G	F
dodine (Syllit 3.4 FL) 1.5 to 3.0 pt + mancozeb (Koverall) 3 lb OR + captan (Captan 80WDG) 2.5 lb	M7 M3 M4	See label 77 (see label)	E E	P-F P	n/a n/a	n/a n/a	n/a n/a	n/a n/a	NC n/a
fenbuconazole (Indar 2F) 8 fl oz	3	14	E	E	F	P-F	P	F	G
fluazinam (Omega 500F) 13.8 fl oz	29	28	F	F	ND	F	F	F	P-F
fluopyram + trifloxystrobin (Luna Sensation) 4.0 to 5.8 fl oz	7 + 11	14	E	F	G-E	G	G	G-E	E
fluopyram + pyrimethanil (Luna Tranquility) 11.2 to 16 fl oz	7 + 9	72	E	F	ND	ND	ND	ND	G
fluxapyroxad + pyraclostrobin (Merivon) 4.4 to 5.5 fl oz	7 + 11	0	E	F	G-E	G-E	E	G-E	E
kresoxim-methyl (Sovran 50 WG) 4 to 6.4 oz [1]	11	30	E	P-F	G-E	G	F-G	G	G
mancozeb (Koverall) 3 lb	M3	77	F	F-G	F-G	F	G	P-F	P
metiram (Polyram 80 DF) 6 lb	M3	see label	G	F-G	F	P-F	F	P-F	P
myclobutanil (various brands) 5 to 10 oz [1]	3	14	G	E	P-F	P-F	P	P	E
penthiopyrad (Fontelis 1.67 SC) 14 to 20 fl oz	7	28	G-E	F	F	F-G	F-G	F-G	G
phosphorous acid (ProPhyt) 3 to 4 pt + mancozeb (Koverall) 3 lb OR + captan (Captan 80WDG) 3.75 lb	33 M3 M4	77 0	F-G F-G	F-G P	F-G G	F G	G G	P-F F-G	P P
phosphorous acid (ProPhyt) 3 to 4 pt + ziram (Ziram 76DF) 3 lb + captan (Captan 80WDG) 3.75 lb	33 M3 M4	14	G-E	F-G	G	G	G	F-G	P
phosphorous acid (ProPhyt) 4 pt + mancozeb (Koverall) 3 lb + fluxapyroxad + pyraclostrobin (Merivon) 5.5 fl oz	33 M3 7 + 11	77	E	F-G	G-E	G-E	E	E	G-E
phosphorous acid (ProPhyt) 4 pt + captan (Captan 80WDG) 2.5 lb + fluxapyroxad + pyraclostrobin (Merivon) 5.5 fl oz	33 M4 7 + 11	77	E	F	G-E	G-E	E	E	E
pyraclostrobin + boscalid (Pristine 38 WG) 14.4 to 18.4 oz	11 + 7	0	G-E	F	G-E	G-E	G-E	G-E	G
pyrimethanil (Scala 5 SC) 7 to 10 fl oz	9	72	F-G	NC	NC	NC	NC	NC	NC
sulfur (Microthiol Disperss) 10 to 20 lb	M2	see label	P-F	P-F	NC	NC	NC	P	G-E
thiophanate methyl (Topsin 4.5FL) 15 to 20 fl oz	1	1	—[3]	NC	G	G	P	G-E	P-F
trifloxystrobin (Flint 50WG) 1.5 to 3 oz	11	14	G	P-F	G-E	G	E	G	G-E
triflumizole (Procure 480SC) 8 to 16 fl oz [1]	3	14	F-G	E	G	P-F	P	P	G-E
ziram (Ziram 76 DF) 6 lb [2]	M3	14	F	G	G	F	G	G-E	P

[1] Use higher rate when the likelihood of disease is high.

[2] Combine Ziram with Topsin-M 70W at 8 to 12 ounces per acre to improve white rot, black rot, sooty blotch, and flyspeck control.

[3] Thiophanate methyl is not recommended for scab control in North Carolina because of scab resistance.

Relative Effectiveness of Various Insecticides for Apple Insect and Mite Control
J. F. Walgenbach, Entomology Research and Extension
(E – excellent; G – good; F – fair; P – poor; NC – no control or insufficient data)

Table 6-3A. Relative Effectiveness of Various Insecticides for Apple Insect and Mite Control

Insecticide, Brand Name, and Amount per Acre	IRAC MOA Group	Days Between Last Spray and Harvest	San Jose Scale	European Red Mite	Twospotted Spider Mite	Rosy Apple Aphid	Green Apple/Spirea Aphids	White Apple Leafhopper	Coddling Moth	Tufted Apple Bud Moth	Redbanded Leafroller	Oriental fruit moth
abamectin (Agri-Mek 0.7SC) 3.0 oz	6	28	NC	E	E	NC	NC	G	NC	NC	NC	NC
acetamiprid (Assail 30 SG) 5.0 oz	4A	7	P	NC	NC	E	E	E	G	P	P	G
B. thuringiensis (various brands) 1 lb	11	0	NC	NC	NC	NC	NC	NC	P	G	G	P
bifenazate (Acramite 50WS) 1 lb	UN	7	NC	E	E	NC	NC	NC	NC	NC	NC	NC
buprofezin (Centaur 70WDG) 34.5 oz	16	14	E	NC	NC	NC	NC	G	NC	NC	NC	NC
carbaryl (Sevin XLR) 4 pt	1A	1	P	NC	NC	NC	NC	E	G	P	G	G
chlorantraniliprole (Altacor 35WDG) 3 oz	28	14	NC	NC	NC	NC	NC	NC	E	E	E	E
chlorpyrifos (Lorsban 50W) 3 lb [1]	1B	>100	E	NC	NC	G	NC	NC	NC	NC	NC	NC
clofentezine (Apollo SC) 4 oz	10	45	NC	E	E	NC	NC	NC	NC	NC	NC	NC
clothianidin (Belay 2.13 SC) 6 oz	4A	7	ND	NC	NC	E	E	E	F	F	F	F
codling moth virus (CYD-X) 3 g	—	0	NC	NC	NC	NC	NC	NC	G	NC	NC	F
cyflumetofen (Nealta 1.67SC) 13.7 oz	25	7	NC	E	E	NC	NC	NC	NC	NC	NC	NC
diazinon (Diazinon 50WP) 4 lb	1B	21	G	P	P	F	F	F	F	P	F	F
esfenvalerate (Asana XL) 8 oz	3	21	NC	NC	NC	G	G	E	G	E	E	G
etoxazole (Zeal 72 WDG) 3 oz	10B	28	NC	G	G	NC	NC	NC	NC	NC	NC	NC
fenpropathrin (Danitol 2.4 EC) 16 oz	3	14	NC	G	G	E	G	E	G	E	E	G
fenpyroximate (Portal 0.4EC) 2 pt	21A	14	NC	E	E	NC	NC	G	NC	NC	NC	NC
flupyradifurone (Svanto 200SL) 10.5 oz	4D	14	G	NC	NC	E	E	E	NC	NC	NC	NC
gamma-cyhalothrin (Proaxis 0.5EC) 3 oz	3	14	NC	NC	NC	E	E	E	G	E	E	E
hexakis (Vendex 50W) 4 lb	12B	14	NC	G	G	NC	NC	NC	NC	NC	NC	NC
hexythiazox (Savey 50DF) 4 oz [1]	10A	28	NC	E	E	NC	NC	NC	NC	NC	NC	NC
imidacloprid (AdmirePro4) 2.8 oz	4A	7	NC	NC	NC	E	E	E	P	P	P	P
indoxacarb (Avaunt 30WDG) 5 oz	22A	28	NC	NC	NC	NC	NC	E	P	G	G	G
lambda-cyhalothrin (Karate 2.08CS) 2 oz	3	21	NC	NC	NC	E	E	E	G	E	E	G
methozyfenozide (Intrepid 2 F)16 oz	18	14	NC	NC	NC	NC	NC	NC	G	E	E	G
novaluron (Rimon 0.83 EC) 20 oz	15	14	NC	NC	NC	NC	NC	NC	G	E	E	E
oil, superior-type 3 gal/100 gal	—	NC	E	E	NC	P	NC	NC	P	P	P	P
oxamyl (Vydate 2L) 2 qt	1A	14	F	P	P	G	G	E	P	P	P	P
permethrin (Ambush 2E) 8 oz	3	>100	NC	NC	NC	G	G	E	G	E	E	G
phosmet (Imidan 70W) 3 lb	1B	7	P	NC	NC	NC	NC	NC	G	P	G	G
pyridaben (Nexter 75 WP) 4.4 oz	21A	25	NC	G	F	NC	P	G	NC	NC	NC	NC
pyriproxyfen (Esteem 35 WP) 5 oz	7C	35	E	NC	NC	P	P	NC	F	P	G	G
spinetoram (Delegate 25WDG) 5 oz	5	7	NC	NC	NC	NC	NC	NC	E	E	E	E
spirotetramat (Movento 2CS) 7.5 oz	23	7	G	NC	NC	E	E	G	NC	NC	NC	NC
thiomethoxam (Actara 25 WP) 4.5 oz	4A	35	G	NC	NC	E	E	E	P	P	P	P

Insects and Mites (Table continues on next page)

[1] Use prebloom only.

195

FRUITS, INSECT, DISEASE CONTROL

Table 6-3B. Relative Effectiveness of Various Insecticides for Apple Insect and Mite Control (continued)

Insecticide, Brand Name, and Amount per per Acre	IRAC MOA Group	Days Between Last Spray and Harvest	Spotted Tentiform Leafminer	Tarnished Plant Bug	Stink Bugs	Apple Maggot	Plum Curculio	Japanese Beetle	Woolly Apple Aphid	Stethorus punctum (Lady Beetle)	Amblyselus Fallacis (Mite)
abamectin (Agri-Mek 0.7SC) 3.0 oz	6	28	E	NC	NC	NC	NC	NC	NC	G	P
acetamiprid (Assail 30 SG) 5 oz	4A	7	G	E	G	G	F	E	P	E	G
B. thuringiensis (various brands) 1 lb	11	0	NC	NC	NC	NC	NC	NC	NC	NC	NC
bifenazate (Acramite 50WS) 1 lb	UN	7	NC	NC	NC	NC	NC	NC	NC	P	P
buprofezin (Centaur 70WDG) 34.5 oz	16	14	NC	NC	NC	NC	NC	NC	NC	P	P
carbaryl (Sevin XLR) 4 pt	1A	1	P	G	P	G	G	E	NC	E	E
Chlorantraniliprole (Altacore 35WDG) 3 oz	28	14	E	NC	NC	NC	P	NC	NC	P	P
Chlorpyrifos (Lorsban 4EC) 1 qt	1B	>100	NC	NC	P	NC	NC	NC	G	P	P
clofentezine (Apollo SC) 4 oz	10	45	NC	NC	NC	NC	NC	NC	NC	NC	P
clothianidin (Belay 2.13SC) 6 oz	4A	7	E	G	G	G	G	G	P	E	P
codling moth virus (CYD-X) 3 oz	—	0	NC	NC	NC	NC	NC	NC	NC	NC	NC
cyflumetofen (Nealta 1.67SC) 13.7 oz	25	7	NC	NC	NC	NC	NC	NC	NC	NC	P
diazinon (Diazinon 50WP) 4 lb	1B	21	P	G	G	G	G	G	E	G	G
esfenvalerate (Asana XL) 8 oz	3	21	G	G	P	G	G	G	NC	E	E
etoxazole (Zeal 72 WDG) 3 oz	10B	28	NC	NC	NC	NC	NC	NC	NC	P	P
fenpropathin (Danitol 2.4 EC) 16 oz	3	14	E	E	G	G	G	G	NC	E	E
flupyradifurone (Svanto 200SL) 10.5 oz	4D	14	NC	NC	NC	NC	NC	NC	G	P	P
fenpyroximate (Portal 0.4 EC) 2 pt	21A	14	NC	NC	NC	NC	NC	NC	NC	P	P
gamma-cyhalothrin(Proaxis 0.5EC) 3 oz	3	14	G	E	G	G	G	G	P	E	E
hexakis (Vendex 50W) 4 lb	12B	14	NC	NC	NC	NC	NC	NC	NC	NC	P
hexythiazox (Savey 50WP) 4 oz[1]	10A	28	NC	NC	NC	NC	NC	NC	NC	NC	P
imidacloprid (Admire Pro) 2.8 oz	4A	7	G	F	P	G	NC	G	P	E	P
indoxacarb (Avaunt 30WDG) 5 oz	22A	28	F	P	NC	P	E	G	NC	G	P
lambda-cyhalothrin (Karate 2.08CS) 2 oz	3	21	G	E	G	G	G	G	P	E	E
methoxyfenozide (Intrepid 2 F) 16 oz	18	14	G	NC	NC	NC	NC	NC	NC	P	P
novaluron (Rimon 0.83 EC) 20 oz	15	14	P	P	F	NC	NC	NC	NC	G	G
oil, superior-type 3 gal/100 gal water	—	NC	NC	NC	NC	NC	NC	NC	NC	NC	P
oxamyl (Vydate 2E) 2 qt	1A	14	G	G	P	NC	P	P	NC	P	P
permethrin (Ambush 2E) 8 oz	3	>100	G	E	G	G	G	NC	NC	E	E
phosmet (Imidan 70W) 3 lb	1B	7	P	G	P	E	G	E	NC	P	P
pyridaben (Nexter 75 WP) 4.4 oz	21A	25	NC	NC	NC	NC	NC	NC	NC	P	P
pyriproxyfen (Esteem 35 WP) 5 oz	7C	35	G	NC	NC	NC	NC	NC	P	F	P
spinetoram (Delegate 25WDG) 5 oz	5	7	E	NC	NC	P	P	NC	NC	G	P
spirotetramat (Movento 2CS) 7.5 oz	23	7	P	NC	P	NC	NC	NC	G	P	P
thiomethoxam (Actara 25 WP) 4.5 oz	4A	35	G	G	G	P	E	G	P	E	G

[1] Use prebloom only.

[2] Ratings for beneficial arthropods are based on toxicity to the organism; i.e., E implies excellent toxicity to the beneficial and will result in elimination of the natural enemy from an orchard. NC implies no control or toxicity and will result in preservation of the beneficial in an orchard.

Blueberry Management Program

H. J. Burrack, Entomology Extension and Research; and W. O. Cline, Plant Pathology Extension

The Insecticide Resistance Action Committee (IRAC) groups insecticides and the Fungicide Resistance Action Committee (FRAC) groups fungicides into mode of action (MOA) categories. These categories are listed following the pesticide and formulation names. To reduce the risk of resistance development, avoid successive applications of products with the same MOA. Organically acceptable insecticides (**OMRI** listed) are indicated in Precautions and Remarks.

Insecticides should only be applied if the pest of concern is present in economically damaging levels. If insect injury does not result in greater loss than the cost of treatment, treatment is not justified. Therefore, some degree of insect presence should be tolerated and insecticides should **not** be applied on a scheduled basis as may be appropriate for fungicides. Note that insecticides listed are acceptable for use on fruit to be marketed in the United States. If fruit is to be exported, check with purchasers to ensure that the materials you intend to use are acceptable for use on fruit in their target markets.

Fungicides are mainly protectants, and are usually applied prior to the appearance of disease symptoms, based on past history of the particular disease threat on a given cultivar, location and plant growth stage. Not all diseases are present on every farm. To avoid applying fungicides unnecessarily, learn to identify diseases by their symptoms, and keep records of those that occur on your farm.

Many pesticides have brand name and generic formulations. In general, information is provided for the most commonly used formulations of active ingredients available in multiple forumations. Carefully check the label of the product you plan to use in the event that it differs from those listed. The label is the law!

Table 6-4. Blueberry Management Program

Season Pest	Insecticide, Mode of Action Code, and Formulation	Amount of Formulation Per Acre	Restricted Entry Interval (REI) (hours)	Pre harvest Interval (PHI) (Days)	Precautions and Remarks
Dormant					
Scale insects	Oil superior-type, IRAC Unknown	1 to 3% vol/vol	4	0	Oil may be applied dormant or delayed dormant. Apply as needed for scale infestations. Reduce to 1% rate just before bloom. Do not apply oil when temperatures are expected to be higher than 65 degrees F or lower than 30 degrees F within 24 hours. Do not use within 14 days of lime-sulfur or Captan. Use 200 to 400 gallons water per acre. Some oils are **OMRI** listed; check labels.
Gall midge	**Blueberry gall midge** adults are tiny flies, and larvae are tiny white, carrot-shaped maggots that feed inside flower buds and leaf buds. Blueberry gall midge can be extremely injurious, especially to rabbiteye cultivars. Flies lay eggs in flower buds on warm winter days when bud scales initially begin to separate. **Gall midge sprays should be timed to protect the earliest flower buds which can realistically be expected to survive anticipated spring cold events.** Gall midge sprays also typically provide suppression of pre-bloom thrips population.				
	spinosad (IRAC 5) (Entrust SC)	4 to 6 fl oz	4	3	Entrust is **OMRI** listed. Do not apply more than 29 fl oz Entrust SC (0.45 lb active ingredient) per acre per year.
	Entrust (80W)	1.25 to 2 oz			Do not apply more than 9 oz Entrust 80W (0.45 lb active ingredient) per acre per year.
	spinetoram (IRAC 5) (Delegate)	3 to 6 oz	4	3	
	diazinon (IRAC 1) (Diazinon AG500)	1 pt per 100 gal water	5 days	7	Only one foliar application is allowed per year.
Delayed Dormant					
Exobasidium leaf and fruit spot	calcium polysulfide (FRAC M2) (Lime-Sulfur solution)	5 gal per acre in 50-70 gal of total spray volume	48	—	Apply at delayed dormant 1-2 weeks before leaf and/or flower buds begin to break. Exobasidium is not specifically on the label. However, when applied for Phomopsis, suppression of Exobasidium has been observed. **DANGER – calcium polysulfide products are caustic and can cause injury.** Calcium polysulfide products are also corrosive to metals and may permanently discolor or stain non-metal sprayer parts. Do not mix lime-sulfur solutions with acids or phosphate fertilizer products because deadly and potentially extremely flammable hydrogen sulfide gas may be emitted.
	calcium polysulfide (FRAC M2) (Sulforix)	1 gal per acre in sufficient water for coverage	48	—	Do not apply lime-sulfur or Sulforix within 14 days of an oil spray. Do not apply when air temperatures are above 85 degrees F. As a precaution, do not apply within 14 days of a Dormex spray.
Pre-Bloom Sprays - Green-tip on vegetative and flower buds					
Twig blight Mummy berry	fenbuconazole (FRAC 3) (Indar 75 WP) (Indar 2F)	2 oz 6 fl oz	12 12	30 30	
	pyraclostrobin + boscalid (FRAC 11+7) (Pristine 38 W)	18.5 to 23 oz	12	0	Do not make more than two sequential applications with any combination of strobilurin fungicides (Abound or Pristine) before alternation with a fungicide that has a different mode of action (Captan, Ziram, Switch). Do not make more than four applications of strobilurin fungicides per season. Do not tank mix Pristine with any other product.
	propiconazole (FRAC 3) (Orbit 3.6E, Tilt 3,6E, Bumper 41.8EC, Propimax EC)	6 fl oz	12	30	
	metconazole (FRAC 3) (Quash 50 WDG)	2.5 oz	12	7	May be applied by ground (min. 20 gpa) or air (min 10 gpa). Do not apply more than twice in a row, or more than 7.5 ounces per season, or more than three times per season.
	azoxystrobin + propiconazole (FRAC 3+11) (Quilt Xcel)	14 to 21 fl oz	12	30	Do not apply more than 82 fluid ounces per acre per season. Quilt Xcel may be applied by ground or air (minimum of 15 gpa).
Exobasidium leaf and fruit spot	The fungus Exobasidium causes green-to-pink spots on fruit that do not ripen normally, and spots on leaves that are light green above and white below. Affected berries are unsightly and not marketable. Fungicides applied for other diseases may provide some control. The disease is most severe in shaded areas with dense foliage and poor ventilation. For images of this disease, see: http://ncblueberryjournal.blogspot.com/2011/07/exobasidium-fruit-and-leaf-spot.html				

Table 6-4. Blueberry Management Program

Season Pest	Insecticide, Mode of Action Code, and Formulation	Amount of Formulation Per Acre	Restricted Entry Interval (REI) (hours)	Pre harvest Interval (PHI) (Days)	Precautions and Remarks
Pre-Bloom Sprays - Green-tip on vegetative and flower buds (continued)					
Thrips	Thrips rarely require treatment in southern high bush blueberries in North Carolina but can reach damaging levels in rabbiteye blueberries. Thrips present in densities greater than 2 per flower in open rabbiteye blooms may justify treatment. Begin sampling bloom clusters for thrips at Stage 3. Sample 2 to 3 times a week from Stage 3 up to bloom. A minimum of 10 flower clusters per acre should be observed and either placed in a closed plastic bag at room temperature, soaked in alcohol, or shaken onto a white sheet of paper.				
	Acetamiprid (IRAC 4A) (Assail 30SG)	4.5 to 5.3 oz	12	1	
	spinosad (IRAC 5) (Entrust SC)	4 to 6 fl oz	4	3	Entrust is **OMRI** listed. Do not apply more than 29 fl oz Entrust SC (0.45 lb active ingredient) per acre per year.
	Entrust (80W)	1.25 to 2 oz			Do not apply more than 9 oz Entrust 80W (0.45 lb active ingredient) per acre per year.
	spinetoram (IRAC 5) (Delegate)	3 to 6 oz	4	3	
Bloom Treatments - 10% to 20% bloom					
Pesticides can harm pollinating insects, so if pesticide applications are necessary during bloom, they should be made in the evening when bees are not foraging and to allow for the longest amount of dry time possible. See Table 5-1A. Relative Toxicity of Pesticides to Honey Bees for more information on specific active ingredients effects on bees.					
Twig blight	Same as Pre-Bloom Sprays				
Mummy berry	Same as Pre-Bloom Sprays				If mummy berry disease pressure is high this year or in previous years, apply fungicides every 7 to 10 days from budbreak through bloom. Foliar sprays using 25 to 50 gallons per acre are most effective.
Flower blight	Anticipate flower blight caused by the fungus *Botrytis cinerea* when excessive rain occurs during bloom, or following a freeze event that injures blossoms.				
	Fenhexamid FRAC 17 (Elevate 50 WDG)	1.5 lb	12	0	Elevate may not be applied by air.
	cyprodinil + fludioxonil (FRAC 9+12) (Switch 62.5 WG)	11 to 14 oz	12	0	Switch may not be applied by air.
	captan (FRAC M4) (Captan 50 WP)	4 lb	72	0	
	captan (FRAC M4) (Captec 4L)	2 qt	72	0	
	captan + fenhexamid (FRAC M4+ 17) (CaptEvate) 68 WG	3.5 to 4.7 lb	48	0	CaptEvate may not be applied by air. Do not use CaptEvate for more than two consecutive sprays.
Bloom Treatments - Full bloom					
Mummy berry	Same as PRE-BLOOM SPRAYS				Note that Indar should not be used alone at full bloom or between bloom and harvest. Tank mix with Captan, Captec, or Ziram.
Fruit rots	captan(FRAC M4) (Captan 50 WP)	4 lb	72	0	
	captan (FRAC M4) (Captec 4L)	2 qt	72	0	
	ziram (FRAC M3) (Ziram 76 DF)	3 lb	48	approx. 30	Ziram cannot be applied later than 3 weeks after full bloom.
	azoxystrobin (FRAC 11) (Abound 2.08 E)	6 to 15.5 fl oz	4	0	Do not make more than two sequential applications of any combination of strobilurin fungicides (Abound or Pristine) before alternating with a fungicide that has a different mode of action (Captan, Ziram, Switch). Do not make more than four applications per season.
	cyprodinil + fludioxonil (FRAC 9+12) (Switch 62.5 WG)	14 oz	12	0	
	pyraclostrobin + boscalid (FRAC 11+7) (Pristine38 W)	18.5 TO 23 oz	12	0	Do not tank mix Pristine with any other product (fungicide, insecticide, adjuvant, fertilizer, etc.)
	metconazole (FRAC 3) (Quash 50 WDG)	2.5 oz	12	7	May be applied by ground (min. 20 gpa) or air (min. 10 gpa). Do not apply more than twice in a row, or more than 7.5 ounces per season, or more than three times per season.
Flower blight	Same as Bloom Treatments (10% to 20% bloom)				
Petal Fall Treatments - Immediately after Bloom					
Fruit rots	Same as Bloom Treatments				Fruit rot treatments should be applied 7 to 10 days apart.
Thrips	Same as Pre-Bloom Treatments				
Plum curculio	Plum curculio is an infrequent pest of North Carolina blueberries. Petal fall treatments of the materials below will be effective against both plum curculio and fruitworms.				
	bifenthrin (IRAC 3) (Brigade) WSB	16 oz			Note that there are residue concerns for some IRAC Group 3A materials on fruit intended for export.
	chlorantraniliprole (IRAC 28) (Altacor)	3.0 to 4.5 oz	4	1	
	esfenvalerate (IRAC 3A) (Asana XL 0.66EC)	9.6 fl oz	12	14	Note that there are residue concerns for some IRAC Group 3A materials on fruit intended for export.
	zeta cypermethrin + bifenthrin (IRAC 3) (Hero)	4 to 10.3 fl oz	12	1	Hero is a premixed material and contains more than one active ingredient. Check labels carefully for the maximum amount of active ingredient than can be applied per acre per season. Note that there are residue concerns for some IRAC Group 3A materials on fruit intended for export.
	kaolin (IRAC unknown) (Surround) WP	25 to 50 lb	4	0	Surround acts like a barrier and masks fruit from pest recognition. Because of this barrier, fruit should be washed after harvest, and Surround may be most appropriate for processing fruit.

Table 6-4. Blueberry Management Program

Season Pest	Insecticide, Mode of Action Code, and Formulation	Amount of Formulation Per Acre	Restricted Entry Interval (REI) (hours)	Pre harvest Interval (PHI) (Days)	Precautions and Remarks
Petal Fall Treatments - Immediately after Bloom (continued)					
Cranberry fruitworm Cherry fruitworm	Fruitworm adults can be monitored with pheromone traps, and fruit should be observed for egg laying or evidence of tunneling. Treatments for fruitworms are most effective when timed to egg hatch, as larvae feed inside fruit. With the exception of Altacor, the materials listed below are not expected to have activity against plum curculio.				
	acetamiprid (IRAC 4A) (Assail) 30 SG	4.5 to 5.3 oz	12	1	
	carbaryl (IRAC 1A) (Sevin) XLR	1.5 to 2 qt	12	7	There are many carbaryl formulations.
	chlorantraniliprole (IRAC 28) (Altacor)	3.0 to 4.5 oz	4	1	Altacor is also also effective against plum curculio in blueberries.
	esfenvalerate, (IRAC 3A) (Asana XL) 0.66 EC	4.8 to 9.6 oz	12	14	Note that there are residue concerns for some IRAC Group 3A materials on fruit intended for export.
	indoxacarb (IRAC 22) (Avuant)	3.5 to 6.0 oz	12	7	
	methoxyfenozide (IRAC 18) (Intrepid) 2F	10 to 16 fl oz	4	7	
	novaluron (IRAC 15) (Rimon) 0.83 EC	20 to 30 fl oz	12	8	Rimon is not labeled for cherry fruitworm.
	pyriproxyfen (IRAC 7) (Knack)	16 fl oz	12	7	Knack is an insect growth regulator and application must be timed carefully to egg hatch.
	spinosad (IRAC 5) (Entrust SC)	4 to 6 fl oz	4	3	Entrust is **OMRI** listed. Do not apply more than 29 fl oz Entrust SC (0.45 lb active ingredient) per acre per year.
	Entrust (80W)	1.25 to 2 oz			Do not apply more than 9 oz Entrust 80W (0.45 lb active ingredient) per acre per year.
	spinetoram (IRAC 5) (Delegate) WG	3 to 5 oz	4	3	
	tebufenozide (IRAC 18) (Confirm) 2F	16 fl oz	4	14	
Leaf spots	fenbuconazole (FRAC 3) (Indar 75 WP) (Indar 2F)	2 oz 6 fl oz	12	30	Indar should not be used alone at full bloom or alone between bloom and harvest. Tank mix with Captan, Captec, or Ziram. Indar is usually limited to five applications per acre per year.
	pyraclostrobin + boscalid (FRAC 11+7) (Pristine 38 W)	18.5 to 23 oz	12	0	Do not make more than two sequential applications with any combination of strobilurin fungicides (Abound or Pristine) before alternation with a fungicide that has a different mode of action (Captan, Ziram, Switch). Do not make more than four applications of strobilurin fungicides per season. Do not tank mix Pristine with any other product (fungicide, insecticide, adjuvant, fertilizer, etc.)
	propiconazole (FRAC 3) (Orbit 3.6E, Tilt 3.6E, Banner 41.8 EC, Propimax EC)	6 fl oz	12	30	
	metconazole (FRAC 3) (Quash 50 WDG)	2.5 oz	12	7	May be applied by ground (minimum of 20 gpa) or air (minimum of 10 gpa). Do not apply more than twice in a row, or more than 7.5 ounces per season, or more than three times per season.
	azoxystrobin + propiconazole (FRAC 3+11) Quilt Xcel	14 to 21 fl oz	12	30	Do not apply more than 82 fluid ounces per acre per season. Quilt Xcel may be applied by ground or air (minimum of 15 gpa).
Fruit Ripening through Harvest					
Spotted wing drosophila	Spotted wing drosophila (SWD) females lay eggs in ripening and ripe soft skinned fruits, and larvae develop internally. Materials listed are likely to be effective against SWD based on current data. SWD treatments should begin when fruit start to ripen and continue weekly through the end of harvest. Rotate IRAC groups between successive sprays. Some management tools used for blueberry maggot are effective against SWD, and management of blueberry maggot and SWD should be integrated as much as feasible.				
	bifenthrin (IRAC 3A) (Brigade WSB)	5.3 to 16.0 oz	12	1	No more than 5 applications of Brigade can be made per season. Note that there are residue concerns for some IRAC Group 3A materials on fruit intended for export.
	cyantraniliprole (IRAC 28) (Exirel)	13.5 to 20.5 fl oz	12	3	Minimum number of days between treatments is five. Do not apply a total of more than 0.4 pound ai/A of CYAZYPYR® or cyantraniliprole containing products per year.
	fenpropathrin (IRAC 3A) (Danitol 2.4 EC)	10.33 to 16 fl oz	24	3	No more than 2 applications of Danitol can be made per season.
	malathion, (IRAC 1B) (Malathion) 8F	2.5 pt	12	1	There are several malathion formulations. No more than 2 applications of Malathion 8F can be made per year. No more than 5 pounds of malathion active ingredient from any source can be applied per acre per year. Use caution if this is the material of choice for multiple insect pests.
	(Malathion) ULV	10 fl oz	12	1	No more than 5 applications of Malathion ULV can be made per year. No more than 5 pounds of malathion active ingredient from any source can be applied per acre per year. Use caution if this is the material of choice for multiple insect pests.
	methomyl (IRAC 1A) (Lannate)	12 to 24 fl oz	48	3	No more than 4 applications of Lannate can be made per year.
	phosmet (IRAC 1B) (Imidan)	1.33 lb	24	3	No more than 5 applications of Imidan can be made per year.
	pyrthreins (IRAC 3) (Pyganic 1.4 EC)	16 to 64 fl oz	12	0	Pyganic is **OMRI** listed, but has limited residual activity and should not be used as the only SWD materials.

199

Table 6-4. Blueberry Management Program

Season Pest	Insecticide, Mode of Action Code, and Formulation	Amount of Formulation Per Acre	Restricted Entry Interval (REI) (hours)	Pre harvest Interval (PHI) (Days)	Precautions and Remarks
Fruit Ripening through Harvest (continued)					
Spotted wing drosophila (continued)	spinosad (IRAC 5) (Entrust SC)	4 to 6 fl oz	4	3	Entrust is **OMRI** listed. Do not apply more than 29 fl oz Entrust SC (0.45 lb active ingredient) per acre per year.
	Entrust (80W)	1.25 to 2 oz			Do not apply more than 9 oz Entrust 80W (0.45 lb active ingredient) per acre per year.
	spinetoram (IRAC 5) (Delegate)	3 to 6 oz	4	3	No more than 19.5 ounces of Delegate can be applied per acre per year.
	zeta cypermethrin (IRAC 3) (Mustang Max)	4 fl oz	12	1	No more than 6 applications of Mustang Max can be made per year. Note that there are residue concerns for some IRAC Group 3A materials on fruit intended for export.
	zeta cypermethrin + bifenthrin (IRAC 3) (Hero)	4 to 10.3 fl oz	12	1	No more than 46.35 fluid ounces of product can be applied oer acre per year. Note that there are residue concerns for some IRAC Group 3A materials on fruit intended for export.
Blueberry maggot	Blueberry maggot fliy activity typically begins in late May. Adults should be monitored with yellow sticky traps baited with ammonia food lures (ammonium acetate, ammonium carbonate, or ammonium bicarbonate). Check traps and change lures at least once per week. Treatments for blueberry maggot are not necessary unless adults have been observed in traps. Materials effective for SWD are also effective against blueberry maggot; additional treatments are not needed for blueberry maggot.				
	Acetamiprid (IRAC 4A) (Assail)	2.5 to 5.3 oz	12	1	Assail is effective against blueberry maggot, but should not be used alone for SWD management.
	esfenvalerate (IRAC 3A) (Asana XL) 0.66 EC	4.8 to 9.6 oz	12	14	Note that there are residue concerns for some IRAC Group 3A materials on fruit intended for export.
	fenpropathrin (IRAC 3A) (Danitol 2.4 EC)	10.33 to 16 fl oz	24	3	No more than 2 applications of Danitol can be made per season.
	imidacloprid (IRAC 4A) (Admire Pro)	2.1 to 2.8 fl oz (foliar)	12	3	Many formulations of imidacloprid are available. Admire Pro can be applied as either a soil or foliar treatment. Soil treatments are not recommended for blueberry maggot. Imidacloprid is not effective against SWD.
	malathion, (IRAC 1B) (Malathion) 8F	2.5 pt	12	1	There are several malathion formulations. No more than 2 applications of Malathion 8F can be made per year. No more than 5 pounds of malathion active ingredient from any source can be applied per acre per year. Use caution if this is the material of choice for multiple insect pests.
	(Malathion) ULV	10 fl oz	12	1	No more than 5 applications of Malathion ULV can be made per year. No more than 5 pounds of malathion active ingredient from any source can be applied per acre per year. Use caution if this is the material of choice for multiple insect pests.
	phosmet (IRAC 1B) (Imidan)	1.33 lb	24	3	No more than 5 applications of Imidan can be made per year.
	spinosad (IRAC 5) (Entrust SC)	4 to 6 fl oz	4	3	Entrust is **OMRI** listed. Do not apply more than 29 fl oz Entrust SC (0.45 lb active ingredient) per acre per year.
	Entrust (80W)	1.25 to 2 oz			Do not apply more than 9 oz Entrust 80W (0.45 lb active ingredient) per acre per year.
	spinetoram (IRAC 5) (Delegate)	3 to 6 oz	4	3	No more than 19.5 ounces of Delegate can be applied per acre per year.
	zeta cypermethrin (IRAC 3) (Mustang Max)	4 fl oz	12	1	No more than 6 applications of Mustang Max can be made per year. Note that there are residue concerns for some IRAC Group 3A materials on fruit intended for export.
	zeta cypermethrin + bifenthrin (IRAC 3) (Hero)	4 to 10.3 fl oz	12	1	No more than 46.35 fluid ounces of product can be applied oer acre per year. Note that there are residue concerns for some IRAC Group 3A materials on fruit intended for export.
Post Harvest					
Leaf Spots	Same as Petal Fall Treatments.				Leaf spot treatments should be applied every two weeks post harvest. Later leaf spot treatments may by be omitted if leaf spot incidence is low.
Blueberry bud mite	Only treat for blueberry bud mite if damage was a problem in the previous year. Many varieties are resistant to blueberry bud mite and do not typically require treatment.				
	Post harvest hedging, cultural control	NA	NA	NA	Hedging removes infested issue/
	oil superior-type IRAC unknown (many formulations)	2 gal	4	0	Bud mite treatments should be applied after harvest and again four weeks later.
Sharpnosed leafhoppers	Sharpnosed leafhopper vectors blueberry stunt disease. To reduce disease transmitting populations of sharpnosed leafhopper, treatments should be timed to their flight activity. Sharpnosed leafhoppers can be monitored with yellow sticky traps. If present, blueberry stunt infected plants should be removed from fields.				
	acetamiprid (IRAC 4A) (Assail)	2.5 to 5.3 oz	12	1	Allow 7 days between Assail treatments.
	esfenvalerate (IRAC 3A) (Asana XL) 0.66 EC	4.8 to 9.6 oz	12	14	Note that there are residue concerns for some IRAC Group 3A materials on fruit intended for export.
	imidacloprid (IRAC 4A) (Admire Pro)	2.1 to 2.8 fl oz (foliar)	12	3	Several formulations of imidacloprid are available. Admire Pro can be applied as either a soil or foliar treatment. Soil treatments are not recommended for sharpnosed leafhopper.
	thiamethoxam (IRAC 4A) (Actara)	3 to 4 oz	12	3	Allow 7 days between Actara treatments. Maximum 12 ounces per acre per season.

Table 6-4. Blueberry Management Program

Season Pest	Insecticide, Mode of Action Code, and Formulation	Amount of Formulation Per Acre	Restricted Entry Interval (REI) (hours)	Pre harvest Interval (PHI) (Days)	Precautions and Remarks
Post Harvest (continued)					
Japanese beetles	Japanese beetle feeding seldom requires treatment in North Carolina blueberries, and some pesticides applied for leafhoppers also control Japanese beetles.				
	esfenvalerate (IRAC 3A) (Asana XL) 0.66 EC	4.8 to 9.6 oz	12	14	Note that there are residue concerns for some IRAC Group 3A materials on fruit intended for export.
	imidacloprid (IRAC 4A) (Admire Pro)	2.1 to 2.8 fl oz (foliar)	12	3	Several formulations of imidacloprid are available. Admire Pro can be applied as either a soil or foliar treatment. Soil treatments are not recommended for Japanese beetle.
	phosmet (IRAC 1B) (Imidan)	1.33 lb	24	3	
Red humped and yellow necked caterpillars	Several species of caterpillars can feed on blueberries from late summer to early fall. These caterpillars can potentially defoliate bushes, but are often not widespread throughout the planting.				
	Hand removal	NA	NA	NA	Hand removal is often sufficient to control populations because they are typically clustered on single or a few bushes.
	Bacillus thuringiensis sub. *kurstaki* (Bt) (IRAC 11A) Dipel DF	0.5 to 1.0 lb	4	0	There are many Bt formulations. Dipel DF is **OMRI** listed.
	esfenvalerate (IRAC 3A) Asana 0.66EC	4.8 to 9.6 fl oz	12	14	Note that there are residue concerns for some IRAC Group 3A materials on fruit intended for export.
	tebufenozide (IRAC 18) (Confirm 2F)	16 fl oz	4	14	
Red imported fire ants	Fire ant baits should be applied when ants are actively foraging and take a few to several weeks to be fully effective. Fire ants can also be treated earlier in the season if present.				
	pyriproxyfen (IRAC 7C) (Esteem Ant Bait 0.5% B)	1.5 to 2 lb	12	1	Do not water for 24 hours after application.
	methoprene (IRAC 7C) (Extinguish Ant Bait 0.5% B)	1 to 1.5 lb	4	0	Extinguish can be applied as a mound treatment or broadcast. Extinguish is labeled for use on cropland, but Extinguish Plus is **NOT** labeled for use on cropland. Read labels carefully.
	spinosad (IRAC 5) (Entrust SC)	4 to 6 fl oz	4	3	Entrust is **OMRI** listed. Do not apply more than 29 fl oz Entrust SC (0.45 lb active ingredient) per acre per year.
	Entrust (80W)	1.25 to 2 oz			Do not apply more than 9 oz Entrust 80W (0.45 lb active ingredient) per acre per year.
Blueberry flea beetle	Blueberry flea beetles are an occasional pest in North Carolina blueberries and active primarily post harvest. Damage is typically not economically significant, but when new shoots are eaten in the fall, yield will be impacted.				
	acetamiprid (IRAC 4A) (Assail)	2.5 to 5.3 oz	12	1	Allow 7 days between Assail treatments.
	esfenvalerate (IRAC 3A) (Asana XL) 0.66 EC	4.8 to 9.6 oz	12	14	Note that there are residue concerns for some IRAC Group 3A materials on fruit intended for export.
	carbaryl (IRAC 1A) (Sevin 80S)	2.5 lb	12	7	
	imidacloprid (IRAC 4A) (Admire Pro)	2.1 to 2.8 fl oz (foliar)	12	3	Admire Pro can be applied as either a soil or foliar treatment. Soil treatments are not recommended for blueberry flea beetles.
	phosmet (IRAC 1B) (Imidan)	1.33 lb	24	3	No more than 5 applications of Imidan can be made per year.
	spinosad (IRAC 5) (Entrust SC)	4 to 6 fl oz	4	3	Entrust is **OMRI** listed. Do not apply more than 29 fl oz Entrust SC (0.45 lb active ingredient) per acre per year.
	Entrust (80W)	1.25 to 2 oz			Do not apply more than 9 oz Entrust 80W (0.45 lb active ingredient) per acre per year.
	spinetoram (IRAC 5) (Delegate)	3 to 6 oz	4	3	No more than 19.5 ounces of Delegate can be applied per acre per year.
	thiamethoxam (IRAC 4A) (Actara)	3 to 4 oz	12	3	Allow 7 days between Actara treatments. Maximum 12 oz per acre per season.
	zeta cypermethrin (IRAC 3) (Mustang Max)	4 fl oz	12	1	No more than 6 applications of Mustang Max can be made per year. Note that there are residue concerns for some IRAC Group 3A materials on fruit intended for export.
	zeta cypermethrin + bifenthrin (IRAC 3) (Hero)	4 to 10.3 fl oz	12	1	No more than 46.35 fluid ounces of product can be applied oer acre per year. Note that there are residue concerns for some IRAC Group 3A materials on fruit intended for export.

Further Information

Southeast Regional Blueberry Integrated Management Guide, www.smallfruits.org

NC Small Fruit and Specialty Crop IPM, www.ncsmallfruitsipm.blogspot.com

Caneberry Management Program

H. J. Burrack, Entomology Extension and Research, and W. O. Cline, Plant Pathology

The Insecticide Resistance Action Committee (IRAC) and Fungicide Resistance Action Committee (FRAC) group insecticides into mode of action categories. These categories are listed following the pesticide and formulation names. To reduce the risk of resistance development, avoid successive applications of insecticides with the same IRAC or FRAC designation for the same pest. Organically acceptable insecticides (**OMRI** listed) are indicated in Comments and Precautions.

Insecticides should only be applied if the pest of concern is present in economically damaging levels. If insect injury does not result in greater loss than the cost of treatment, treatment is not justified. Therefore, a degree of insect presence should be tolerated and insecticides should not necessarily be applied on a scheduled basis as may be appropriate for fungicides.

Pesticides should not be applied when bees are actively foraging. If necessary, apply insecticides and fungicides in the evening when bees are not active. Pay attention to pesticide label information regarding pollinator protection.

Many insecticide active ingredients are available in generic formulations. Generic products generally work similarly to their brand name counterparts, but formulation changes can impact efficacy and plant response. In general, information is provided for the most commonly used formulations of active ingredients available in multiple forumations. Carefully check the label of the product you plan to use in the event that it differs from those listed. The label is the law! Chemical names are subject to change; please check the active ingredient for all materials.

Table 6-5. Caneberry Management Program

SEASON Pest	Pesticide, Formulation, and IRAC	Rate of Formulation per Acre	Restricted Entry Interval (hours)	Preharvest Interval (PHI) (days)	Comments and Precautions
Late Winter or Early Spring When new growth is less than 0.5 inch long					
Anthracnose, Spur blight, Cane blight	liquid lime-sulfur, FRAC M2	6 to 12 gal	See label	See label	This is an important spray for anthracnose control. Make sure canes are thoroughly covered. Use a dilute spray solution.
	Copper-based products FRAC M1	See label	See label	See label	Many copper-based fungicides are available. May be used as a component in Bordeaux mixture, see footnote.
Raspberry crown borer	Removing infested plants is an important cultural control. In blocks with a history of raspberry crown borer, apply an insecticide either in late October to early November or early April (one application only) to provide a barrier for larvae boring into canes as they emerge from overwintering hibnacula. Follow label instructions for water volume and application methods.				
	bifenthrin, IRAC 3A (Brigade WSB)	16 oz	12	3	Do not exceed 0.2 lb bifenthrin (32 oz Brigade 10WSB) per acre per season.
	esfenvalerate, IRAC 3A (Asana XL)	9.6 fl oz	12	7	
	chlorantranilirpole, IRAC 28 (Altacor)	3-4.5 oz	4	3	
Just Before Blooms Open					
Anthracnose, Cane blight, Cane canker, Leaf spots, Spur blight	boscalid + pyraclostrobin (Pristine) 38 WDG, FRAC 7 + 11	18.5 to 23 oz	12	0	
	captan, FRAC M4 (Captan 50W) (Captan 80WDG) (Captec 4L)	4 lb 2.5 lb 2 qt	72	3	Do not apply more than 20 pounds of Captan 50W or 12.5 pounds Captan 80WDG per acre per season. Captan 50W is a 24C registration.
	pyraclostrobin (Cabrio 20EG), FRAC 11	14 oz	12	0	Do not make more than four applications collectively of the strobilurin fungicides (Abound, Qullt Xcel, Cabrio, Heritage, and Pristine) per season. Strobilurin (Group 11) fungicides will also control rusts and powdery mildew.
	azoxystrobin, FRAC 11 (Abound 2SC)	6.2 to 15.4 fl oz	4	0	Do not make more than four applications collectively of the strobilurin fungicides (Abound, Qullt Xcel, Cabrio, and Pristine) per season. Strobilurin (Group 11) fungicides will also control rusts and powdery mildew.
	azoxystrobin + propiconazole (Qullt Xcel), FRAC 11 + 3	14 to 21 fl oz	12	30	Do not make more than four applications collectively of the strobilurin fungicides (Abound, Qullt Xcel, Cabrio, and Pristine) per season. Strobilurin (Group 11) fungicides will also control rusts and powdery mildew.
Strawberry clipper weevil	Strawberry clipper weevil females lay their eggs in flower buds and clip the pedicle, causing the bud to wilt and drop off the plant. Some blackberry and raspberry varieties can compensate for bud injury, and strawberry clipper may not require treatment. **Do not apply insecticides when bees are foraging.**				
	bifenthrin, IRAC 3A (Brigade 10WSB)	8 to 16 oz	12	3	Do not exceed 0.2 lb bifenthrin (32 oz Brigade 10WSB) per acre per season.
	carbaryl, IRAC 1A (Sevin XLR)	1 to 2 qt	12	7	
	fenpropathrin, IRAC 3A (Danitol 2.4EC)	10.66 to 16 fl oz	24	3	Do not exceed 32 fluid ounces acre per season.
Gall midge	Gall midge larvae can feed on developing buds, and damage can appear similar to cold injury. Fields with a history of gall midge damage may require treatment, but this is rare. Confirm gall midge presence before considering a treatment targeting this pest alone.				
	bifenthrin, IRAC 3A (Brigade 2EC)	3.2 to 6.4 fl oz	12	3	Do not exceed 12.8 fluid ounces Brigade per acre per season.
	fenpropathrin, IRAC 3A (Danitol 2.4EC)	10.66 to 16 fl oz	24	3	Do not exceed 32 fluid ounces per acre per season.

Table 6-5. Caneberry Management Program

SEASON Pest	Pesticide, Formulation, and IRAC	Rate of Formulation per Acre	Restricted Entry Interval (hours)	Preharvest Interval (PHI) (days)	Comments and Precautions
Bloom and Petal Fall					
Pesticides may be hazardous to pollinators. When making any pesticide application during bloom, apply material in the evening when bees are not foraging to allow for as long a dry time as possible. See Table 5.1. Relative Toxicity of Pesticides to Bees for more information.					
Double blossom	Sprays during bloom are most important for control of double blossom. Begin sprays when first infected blossoms open and continue every 10 to 14 days through bloom. Rotate stobulrin (Group 11) fungicides with Switch to avoid resistance. It is important to protect primocanes as long as infected flowers continue to open.				
	azoxystrobin , FRAC 4 (Abound 2SC)	6.2 to 15.4 fl oz	4	0	
	boscalid + pyraclostrobin (Pristine 38WDG), FRAC 7 + 11	18.5 to 23 oz	12	0	Pristine will also control botrytis.
	cyprodinil + fludioxonil (Switch 62.5 WG), FRAC 9 + 12	11 to 14 oz	12	0	
	azoxystrobin + propiconazole (Quilt Xcel), FRAC 11 + 3	14 to 21 fl oz	12	30	
	Bordeaux mixture FRAC M1	See note at end of table	24	1	Crop injury may occur with Bordeaux mixture under slow drying conditions or in hot weather. Some injury often accompanies the use of copper fungicides; if injury is excessive, discontinue use.
Botrytis fruit rot	Apply at early bloom and repeat at full bloom. Rotate iprodione, Switch, and Elevate or CaptEvate to reduce the likelihood of resistance.				
	captan, FRAC M4 (Captan 50W) (Captan 80 WDG) (Captec 4L)	4 lb 2.5 lb 2 qt	72	3	
	fenhexamid (Elevate) 50 WDG, FRAC 17	1.5 lb	12	0	
	iprodione (several brands) 50 WG, FRAC 2 4F	1 to 2 lb 1 to 2 pt	24	0	
	cyprodinil + fludioxonil (Switch 62.5 WG), FRAC 9 + 12	11 to 14 oz	12	0	
	captan + fenhexamide (CaptEvate) 68WDG, FRAC M4 + 17	3.5 lb	48	30	CaptEvate is registered on raspberries only.
Cane canker, Cane blight, Spur blight	captan, FRAC M4 (Captan 50W) (Captan 80 WDG) (Captec 4L)	4 lb 2.5 lb 2 qt	72	3	
	pyraclostrobin (Cabrio 20EG), FRAC 11	14 oz	12	0	The strobilurin fungicides (Group 11) will also control rusts and powdery mildew.
	boscalid + pyraclostrobin (Pristine 38WDG), FRAC 7 + 11	18.5 to 23 oz	12	0	The strobilurin fungicides (Group 11) will also control rusts and powdery mildew.
	azoxystrobin, FRAC 11 (Abound 2SC)	6.2 to 15.4 fl oz	4	0	The strobilurin fungicides (Group 11) will also control rusts and powdery mildew.
	azoxystrobin + propiconazole (Quilt Xcel), FRAC 11 + 3	14 to 21 fl oz	12	30	The strobilurin fungicides (Group 11) will also control rusts and powdery mildew.
Rednecked cane borer	Scout canes during winter pruning. If 10% or greater of the primocanes per row, or more, of the primocanes than will be removed through pruning, have rednecked cane borer galls, control is justified. Treat after first bloom or when adults are observed.				
	bifenthrin, IRAC 3A (Brigade 10WSB)	8 to 16 oz	12	3	Do not exceed 0.2 lb bifenthrin (32 oz Brigade 10WSB) per acre per season.
Strawberry clipper weevil	See **JUST BEFORE BLOOMS OPEN**				
Post-Bloom					
Anthracnose, Leaf spots, Rusts, Powdery mildew	See **JUST BEFORE BLOOMS OPEN**				Applications for these diseases should be made every 14 days after petal fall until harvest
Double blossom	See **BLOOM AND PETAL FALL**				Additional treatments may be needed to protect primocanes if infected flowers continue to open.
Japanese beetles	Caneberries can tolerate some foliar feeding by Japanese beetles, but little work has been done to determine when foliar feeding impacts yield. Do not use Japanese beetle pheromone traps as they attract beetles from outside fields. Treatment for Japanese beetles is not recommended unless significant foliar loss occurs.				
	carbaryl, IRAC 1A (Sevin) 50 WP (Sevin) 4 XLR	2 lb 2 qt	12	7	
	fenpropathrin, IRAC 3A (Danitol) 2.4 EC	10.66 to 16 fl oz	24	3	Do not exceed 32 fluid ounces Danitol per acre per season.
	malathion, IRAC 1B (Malathion 8F)	2 pt	24	1	Make no more than 4 applications per year.
	zeta-cypermethrin (Mustang Max) IRAC 3	4 fl oz	12	1	Do not make more than 6 applications per season.
Leafrollers	Leafrolling caterpillars can feed on caneberry foliage. Foliage damage is typically not economically significant, but caterpillars can occasionally form webs on fruit. If caterpillars are impacting fruit, treatment may be justified.				
	Bacillus thuringiensis (Bt), IRAC 11A (Dipel DF)	0.5 to 1 lb	4	0	Dipel DF is **OMRI** listed.
	chlorantranilirpole, IRAC 28 (Altacor)	3-4.5 oz	4	3	

FRUITS, INSECT, DISEASE CONTROL

Table 6-5. Caneberry Management Program

SEASON Pest	Pesticide, Formulation, and IRAC	Rate of Formulation per Acre	Restricted Entry Interval (hours)	Preharvest Interval (PHI) (days)	Comments and Precautions
Post-Bloom (continued)					
Leafrollers (continued)	spinosad (IRAC 5) (Entrust SC)	4 to 6 fl oz	4	3	Entrust is **OMRI** listed. Do not apply more than 29 fl oz Entrust SC (0.45 lb active ingredient) per acre per year.
	Entrust (80W)	1.25 to 2 oz			Do not apply more than 9 oz Entrust 80W (0.45 lb active ingredient) per acre per year.
	spinetoram, IRAC 5 (Delegate)	3 to 6 oz	4	1	Do not exceed 19.5 ounces Delegate per acre per season.
Stink bugs, Plant bugs	Stink bug feeding does not typically damage berries, but they may be contamination pests during harvest. Plant bug and stink bugs may also feed on developing buds or shoots. Treatment is justified if insects are contaminating fruit.				
	esfenvalerate, IRAC 3A (Asana XL)	9.6 fl oz	12	7	Avoid applications when bees are foraging. Apply during evenings or early morning.
	fenpropathrin, IRAC 3A (Danitol 2.4EC)	10.66 to 16 fl oz	24	3	Do not exceed 32 fluid ounces per acre per season.
	thiamethoxam, IRAC 4A (Actara 25WDG)	3 oz	12	3	
Spider mites	There is no research-based treatment threshold for spider mites in caneberries, but treatment is recommended when a random sample of leaflets from the planting has an average of 10 motile mites. Leaflets should be examined with a minimum 10x hand lens to determine mite counts. Spider mites are more significant pests of raspberries than blackberries. Insecticides used against other pests may flare spider mite populations, particularly IRAC 1 and 3 materials. Observe plants for spider mites following treatment with these materials.				
	bifenazate, IRAC Unknown (Acramite 50WS)	1 lb	12	1	
	hexythiazox, IRAC 10A (Savey 50DG)	6 oz	12	3	Savey is primarily active against eggs and immature mites. Apply when populations are low.
	horticultural oils, IRAC Unknown (Saf-T-Side)	1 to 2% by volume	4	0	Summer oils are effective in moderating low mite populations pre-harvest. Use on a trial basis only until certain oil will not result in fruit finish problems. **Do not** use oils within 14 days of using any sulfur-containing material. **Do not** apply oils when temperatures will exceed 90 degrees F or dip below 50 degrees F.
	horticultural oils, IRAC Unknown (JMS Stylet Oil)	0.75 to 1.5% by volume	4	0	
	horticultural oils, IRAC Unknown (Organic JMS Stylet Oil)	0.75 to 1.5% by volume	4	0	Organic JMS Stylet Oil is **OMRI** listed.
Broad mites	Broad mites are emerging pests in blackberries and have been most problematic in primocane fruiting varieites. Significant populations have been observed in AR and NC. Broad mites cause leaf stunting and upward or downward cupping. Rule out other causes of leaf stunting (such as herbicide injury) and confirm broad mite presence before treating.				
	abamectin, IRAC 6 (Agri-Mek SC)				Agri-Mek SC has a Section 2(ee) label for use in caneberries, effective from July 6, 2016 to July 6, 2020 in AR, FL, IL, IN, NC, PA, and SC. Two Agri-Mek SC applications should be made 7 apart to be most effective.
Harvest					
Botrytis fruit rot	boscalid + paraclostrobin (Pristine 38 WDG), FRAC 7 + 11	18.5 to 23 oz	12	0	
	cyprodinil + fludioxonil (Switch 62.5 WG), FRAC 9 + 12	11 to 14 oz	12	0	
	fenhexamid (Elevate 50 WDG), FRAC 17	1.5 lb	12	0	
	iprodione (several brands) 50 WG, FRAC 2 4F	1 to 2 lb 1 to 2 pt	24	0	
	captan, FRAC M4 (Captan 50W) (Captan 80 WDG) (Captec 4L)	4 lb 2.5 lb 2 qt	72	3	Include captan in this spray if ripe rot is a problem. Pristine will also control ripe rot. Captan and CaptEvate have a 3-day PHI.
	Captan + fenhexamide (CaptEvate) 68WDG, FRAC M4 + 17	3.5 lb	48	3	CaptEvate is registered on raspberries only. Include captan in this spray if ripe rot is a problem. Pristine will also control ripe rot. Captan and CaptEvate have a 3-day PHI.
Flower thrips	Flower thrips can be a contamination pest at harvest. Fruit can be placed in a clear plastic bag before harvest and observed for flower thrips. There is no evidence at this time to suggest that flower thrips damage fruit or flowers and reduce yield.				
	acetamiprid IRAC 4A (Assail 30SG)	4.5 to 5.3 oz	12	1	
	spinosad (IRAC 5) (Entrust SC)	4 to 6 fl oz	4	3	Entrust is **OMRI** listed. Do not apply more than 29 fl oz Entrust SC (0.45 lb active ingredient) per acre per year.
	Entrust (80W)	1.25 to 2 oz			Do not apply more than 9 oz Entrust 80W (0.45 lb active ingredient) per acre per year.
	spinetoram, IRAC 5 (Delegate)	3 to 6 oz	4	1	Do not exceed 19.5 ounces Delegate per acre per season.
	zeta cypermethrin IRAC 3 (Mustang Max)	4.0 fl oz	12	1	

204

Table 6-5. Caneberry Management Program

SEASON Pest	Pesticide, Formulation, and IRAC	Rate of Formulation per Acre	Restricted Entry Interval (hours)	Preharvest Interval (PHI) (days)	Comments and Precautions
Harvest (continued)					
Spotted wing drosophila	Spotted wing drosophila (SWD) is a pest of soft skinned fruit. Female SWD lay eggs in ripe and ripening fruit, which can appear otherwise undamaged externally. Spotted wing drosophila are typically present in damaging densities during typical blackberry and raspberry harvest periods in North Carolina, so preventative treatment, beginning when fruit begin to change color, is recommended. Treatments should be applied at least every 7 days, and mode of action should be rotated between treatments. In addition to insecticide treatments, growers should also employ good cultural practices including: harvesting as frequently as possible, removing all ripe fruit from plants at each harvest, and storing all harvested from at temperatures below 41°F for as long as feasible before marketing.				
	bifenthrin, IRAC 3A (Brigade 10WSB)	8 to 16 oz	12	3	Do not exceed 0.2 lb bifenthrin (32 oz Brigade 10WSB) per acre per season.
	fenpropathrin, IRAC 3A (Danitol 2.4EC)	10.66 to 16 fl oz	24	3	Do not exceed 32 fluid ounces per acre per season.
	malathion, IRAC 1B (Malathion 8F)	2 pt	24	1	Make no more than 4 applications per year.
	spinosad (IRAC 5) (Entrust SC) Entrust (80W)	4 to 6 fl oz 1.25 to 2 oz	4	3	Entrust is **OMRI** listed. Do not apply more than 29 fl oz Entrust SC (0.45 lb active ingredient) per acre per year. Do not apply more than 9 oz Entrust 80W (0.45 lb active ingredient) per acre per year.
	spinetoram, IRAC 5 (Delegate)	3 to 6 oz	4	1	Do not exceed 19.5 ounces Delegate per acre per season.
	zeta cypermethrin IRAC 3 (Mustang)	4.3 fl oz	12	1	Do not exceed 25.8 fluid ounces Mustang per acre per season.
Japanese beetles and green June beetles	carbaryl, IRAC 1A (Sevin) 50 WP (Sevin) 4 XLR	 2 lb 2 qt	12	7	
	fenpropathrin, IRAC 3A (Danitol) 2.4 EC	10.66 to 16 fl oz	24	3	Do not exceed 32 fluid ounces Danitol per acre per season.
	zeta-cypermethrin (Mustang Max) IRAC 3	4 fl oz	12	1	Do not make more than 6 applications per season.
Just After Harvest and 14 Days Later					
Leaf spots	captan (Captan 50W) (Captan 80 WDG)	 4 lb 2.5 lb	72	3	
Japanese beetle	**See Post Bloom**				Japanses beetle treatments are necessary post harvest only if feeding is removing greater than 10% of foliage on primocanes.
Late October or Early November					
Raspberry crown borer	Raspberry crown borer treatments should be applied once per year, either in late fall or early spring. Applications during both time periods are not necessary.				
	bifenthrin, IRAC 3A (Brigade WSB) IRAC 3A	16 oz	12	3	
	chlorantranilirpole, IRAC 28 (Altacor)	3-4.5 oz	4	3	
	esfenvalerate, IRAC 3A (Asana XL)	9.6 fl oz	12	7	
Dormant					
Scale insects	Scale insects may be present on caneberries but are typically kept below economically damaging levels by parasitoids and predators. An open canopy minimizes scale populations. Examine plants after harvest and during pruning for scale, and if present in high numbers or resulting in sooty mold growth, consider a dormant season oil treatment.				
	horticultural oils IRAC Unknown (Saf T Side)	1 to 2% by volume	4	0	
	horticultural oils IRAC Unknown (JMS Stylet Oil)	0.75 to 1.5% by volume	4	0	
	horticultural oils IRAC Unknown (Organic JMS Stylet Oil)	0.75 to 1.5% by volume	4	0	Organic JMS Stylet Oil is **OMRI** listed.
Raspberry cane borer, Red neck cane borer	During winter pruning, examine canes for raspberry cane borer injury. Prune canes girdled by raspberry cane borer 2 to 3 cm below the lower girdle or gall. If evidence of boring is present below this cut, successive cuts should be made until no further injury is observed. Destroy or remove cuttings to prevent reinfestaion.				
Special Rust Sprays					
Cane and leaf rust, Orange rust	Begin applications in the spring just before orange rust pustules are formed on the lower leaf of brambles (use wild blackberries as indicators). Continue at 10- to 14-day intervals until the mean temperature remains above 77 degrees F. Infections can also occur in the late summer and fall.				
	boscalid + pyraclostrobin (Pristine 38 WG), FRAC 7 + 11	18.5 to 23 oz	12	0	Where orange rust has been a problem, alternate Rally and Cabrio or Pristine or azoxystrobin at 14-day intervals. For late leaf rust, begin when symptoms first appear, and continue on a 14-day interval.
	myclobutanil (several brands) 40 WSP, DF, WDG, FRAC 3	1.25 to 2.5 oz	24	1	
	pyraclostrobin (Cabrio 20EG), FRAC 11	14 oz	12	0	Where orange rust has been a problem, alternate Rally and Cabrio or Pristine or azoxystrobin at 14-day intervals.
	azoxystrobin, FRAC 11 (Abound 2SC)	6.2 to 15.4 fl oz	4	0	
	azoxystrobin + propiconazole (Quilt Xcel), FRAC 11 + 3	14 to 21 fl oz	12	30	

FRUITS, INSECT, DISEASE CONTROL

Table 6-5. Caneberry Management Program

SEASON Pest	Pesticide, Formulation, and IRAC	Rate of Formulation per Acre	Restricted Entry Interval (hours)	Preharvest Interval (PHI) (days)	Comments and Precautions
Special Treatments for Phytophthora Root Rot					
Phytophthora root rot	mefenoxam (Ridomil Gold SL), FRAC 4	—	48	45	Apply 0.25 pint per 1,000 linear feet of row in a 3 feet wide band in the spring and fall after harvest. Ridomil Gold is registered for raspberries only. 45-day phi.
	fosetyl Al (Aliette WSP), FRAC 33	5 lb	12	60	Begin when growth is 1 to 3 inches long and continue at 45- to 60-day intervals through the growing season. Registered for blackberries and raspberries. Maximum of four applications per year. 60-day phi.
	phosphite fungicides, FRAC 33	See label	See label	See label	Several phosphorus acid products are registered for control of Phytophthora root rot, including Prophyt and Agri-Fos. See label for recommendations.
Preplant Treatments for Nematodes					
Nematodes	1,3 dichloropene 37% + chloropicrin 57% (Pic-Clor 60 EC)	19.5 to 44.5 gal	5	NA	Preplant interval should be 4 to 8 weeks, or longer if dissipation is slow. See label for additional information.
	metam sodium (Vapam, Sectagon II, Busan 1020)	75 gal	See label	NA	Preplant interval is a minimum of 4 weeks.

More information available at http://rubus.ces.ncsu.edu.

Bordeaux mixture recipe is available at http://www.smallfruits.org/SmallFruitsRegGuide/Guides/2015/2015BrambleSpray%20Guide12_22_14.pdf

Bunch Grape Insect Management

H. J. Burrack, Entomology Extension and Research

With a few exceptions, which are noted, wine grapes should be treated for insects only damaging insect populations are present. Where treatment thresholds are known, these are provided. For many insect pests of wine grapes in the southeast, thresholds do not exist. Consult cooperative extension personnel for management recommendations if insects for which there are no thresholds are present.

The Insecticide Resistance Action Committee (IRAC) and the Fungicide Resistance Action Committee (FRAC) group insecticides and fungicides into mode of action categories. These categories are listed following the pesticide and formulation names. To reduce the risk of resistance development, avoid successive applications of insecticides or fungicides with the same IRAC or FRAC code for the same pest. Organically acceptable insecticides (OMRI listed) are indicated in Precautions and Remarks.

Some insecticide active ingredients are available in several formulations and under several trade names. For simplicity, the most common trade names and associated rates are listed. This is not intended to encourage the use of these products over generic versions.

PLEASE NOTE: This table does not contain information on wine grape disease management. For wine grape disease guidance, please refer to information provided online by Virginia Tech University at https://pubs.ext.vt.edu/456/456-017/Section-3_Grapes-1.pdf

Table 6-6. Bunch Grape Insect Management

When to Spray and Pest	Pesticide, Formulation, and IRAC, FRAC Code	Amount of Formulation to Use per Acre	Reentry Interval (REI) (hours)	Preharvest Interval (PHI) (days)	Precautions and Remarks
Bud Swell					
Grape flea beetle	Apply only if damaging numbers of adult beetles are present. If 4% or more of buds have been damaged by grape flea beetles, treatment is justified. Grape flea beetle adults emerge in early spring and feed on newly swollen buds and lay eggs. Larvae and adults from subsequent generations feed on leaves, but foliar feeding typically does not result in economically significant damage nor justify treatment.				
	carbaryl, IRAC 1A (Sevin XLR Plus)	1 to 2 qt	12	7	
	cyfluthrin IRAC 3 (Baythriod 2 EC)	2.4 to 3.2 fl oz	12	3	
	fenpropathrin, IRAC 3 (Danitol 2.4 EC)	8 fl oz	24	21	
	phosmet, IRAC 1B (Imidan 70WP)	2 lb	14	14	
Climbing cutworms	Scout for cutworm if damaged buds are observed. Look for cutworms at night. Cutworm treatment may be justified if greater than 4% of the buds examined are damaged and the variety does not have fruitful secondary buds. Spray in the evening if possible as cutworms are active at night.				
	Bacillus thuringiensis (Bt), IRAC 11	rates vary	4	0	Some Bt formulations are **OMRI** listed.
	carbaryl, IRAC 1A (Sevin XLR Plus)	1 to 2 qt	12	7	
	chlorantraniliprole IRAC 28 (Altacor)	2 to 4.5 oz	4	14	
	fenpropathrin, IRAC 3 (Danitol 2.4 EC)	10.66 to 21.33 fl oz	24	21	
	methoxyfenozide, IRAC 18 (Intrepid 2F)	12 to 16 fl oz	4	30	
	spinetoram IRAC 5 (Delegate 25 WG)	3 to 5 oz	4	7	
	spinosad, IRAC 5 (Entrust SC)	4 to 8 fl oz	4	7	Do not apply more than 29 fl oz Entrust SC (0.45 lb spinosad) per acre per season. Entrust is **OMRI** listed.
Mealybugs	Scout for mealybugs and European red mite under bark during dormant season. Use a minimum 10x hand lens to observe European red mite (ERM).				
	buprofezin, IRAC 16 (Applaud 70DF)	9 to 12 oz	12	30	
	clothianidin (IRAC 4A) Belay	6 fl oz (foliar) 6 to 12 fl oz (soil)	12	0 (foliar) 30 (soil)	Belay can be applied either to the soil or as a foliar spray. Soil applications are typically active for a longer period of time but must be made early in the year. Soil applications are more effective when made via drip irrigation.
	dinotefuran (IRAC 4A) (Venom 20SG)	0.44 to 0.66 lb (foliar) 1.13 to 1.32 lb (soil)	12	1 (foliar) 28 (soil)	Venom can be applied either to the soil or as a foliar spray. Soil applications are typically active for a longer period of time but must be made early in the year. Soil applications are more effective when made via drip irrigation.
	horticultural oils (Omni Supreme Spray Oil) (JMS Stylet Oil) (Organic JMS Stylet Oil)	0.5 to 1% by volume 1 to 2% by volume 1 to 2% by volume			If mealybugs or ERM were of economic concern during the previous season and present during dormant scouting, a dormant oil treatment may be justified during bud swell. **Do not** apply oil treatments in combination with sulfur or within 30 days of sulfur application. **Do not** apply oils when temperature will exceed 90 degrees F or dip below freezing. Organic JMS Stylet Oil is **OMRI** listed.
	imidacloprid (IRAC 4A) (Admire Pro)	1 to 1.4 fl oz (foliar) 7 to 14 fl oz (soil)	12	0 (foliar) 30 (soil)	Admire Pro can be applied either to the soil or as a foliar spray. Soil applications are typically active for a longer period of time but must be made early in the year. Soil applications are more effective when made via drip irrigation.

Table 6-6. Bunch Grape Insect Management

When to Spray and Pest	Pesticide, Formulation, and IRAC, FRAC Code	Amount of Formulation to Use per Acre	Reentry Interval (REI) (hours)	Preharvest Interval (PHI) (days)	Precautions and Remarks
At or Just Before Budburst					
Leafhopper/ sharpshooters (Pierce's Disease suppression)	Consider a soil application of a neonicotiniod (4A) insecticide for leafhoppers if plants symptomatic for Pierce's Disease have been observed in the vineyard or in nearby vineyards.				
	clothianidin (IRAC 4A) Belay	6 fl oz (foliar) 6 to 12 fl oz (soil)	12	0 (foliar) 30 (soil)	Belay can be applied either to the soil or as a foliar spray. Soil applications are typically active for a longer period of time but must be made early in the year. Soil applications are more effective when made via drip irrigation.
	cyfluthrin, IRAC 3 (Baythroid 2EC)	2.4 to 3.2 fl oz	12	3	
	dinotefuran (IRAC 4A) (Venom 20SG)	0.44 to 0.66 lb (foliar) 1.13 to 1.32 lb (soil)	12	1 (foliar) 28 (soil)	Venom can be applied either to the soil or as a foliar spray. Soil applications are typically active for a longer period of time but must be made early in the year. Soil applications are more effective when made via drip irrigation.
	imidacloprid (IRAC 4A) (Admire Pro)	1 to 1.4 fl oz (foliar) 7 to 14 fl oz (soil)	12	0 (foliar) 30 (soil)	Admire Pro can be applied either to the soil or as a foliar spray. Soil applications are typically active for a longer period of time but must be made early in the year. Soil applications are more effective when made via drip irrigation.
	acetamiprid, IRAC 4A (Assail 30SG)	2.5 oz	12	7	
Prebloom					
Flea beetle	See **Bud Swell** recommendations				
Grape berry moth	Grape berry moth is present in North Carolina, but it is not uniformly distributed in the state. If grape berry moth presence is suspected, observe flowers and fruit for injury and consider monitoring moth presence with pheromone baited traps.				
	chlorantraniliprole IRAC 28 (Altacor)	2 to 4.5 oz	4	14	chlorantraniliprole IRAC 28 (Altacor)
	fenpropathrin, IRAC 3 (Danitol 2.4 EC)	10.6 fl oz	24	21	
	methoxyfenozide, IRAC 18 (Intrepid 2F)	12 to 16 fl oz	4	30	Minimum application of Intrepid for airblast sprayers is 40 gallons per acre.
	indoxacarb (IRAC 22) (Avaunt 30DG)	5 to 6 oz	12	7	
	methomyl, IRAC 1A (Lannate SP) (Lannate LV)	0.5 to 1 lb 1.5 to 3 pt	7	14	
	pheromone (SPLAT-GBM)	1 kg	4	0	Apply SPLAT-GBM mating disruption when temperatures between 60-80 degrees F and no rain is expected within 1-2 hours. For high population densities, apply 1.0kg/A as 1,000 point sources of 1.0 g (1/4 tsp) throughout an acre. For low-moderate populations, apply 1.0 kg as 250 point sources of 2.5 g (1/2 tsp). See application information on label.
	spinosad, IRAC 5 (Entrust SC)	4 to 8 fl oz	4	7	Do not apply more than 29 fl oz Entrust SC (0.45 lb spinosad) per acre per season. Entrust is **OMRI** listed.
	spinetoram, IRAC 5 (Delegate)	3 to 5 oz	4	7	
Leafhoppers/ Sharpshooters (Pierce's Disease suppression) Initiation of foliar treatments should be based on trap captures.	**If foliar and soil applications of group 4A pesticides are part of a management plan for Pierce's disease (i.e., Admire Pro applied via drip and Venom foliar), at least one application of a different IRAC insecticide should occur as a rotation between these treatments.** Synthetic pyrethroid insecticides (Group 3) and organophosphates (Groups 1A and 1B) are broad spectrum insecticides and have the potential to flare spider mite populations. Observe spider mites before and after treatments to determine if these populations increase.				
	acetamiprid, IRAC 4A (Assail WSP)	2.5 oz	12	7	
	clothianidin (IRAC 4A) Belay	6 fl oz (foliar) 6 to 12 fl oz (soil)	12	0 (foliar) 30 (soil)	Belay can be applied either to the soil or as a foliar spray. Soil applications are typically active for a longer period of time but must be made early in the year. Soil applications are more effective when made via drip irrigation.
	cyfluthrin (IRAC 3) (Baythriod 2EC)	2.4 to 3.2 fl oz	12	3	
	dinotefuran (IRAC 4A) (Venom 20SG)	0.44 to 0.66 lb (foliar) 1.13 to 1.32 lb (soil)	12	1 (foliar) 28 (soil)	Venom can be applied either to the soil or as a foliar spray. Soil applications are typically active for a longer period of time but must be made early in the year. Soil applications are more effective when made via drip irrigation.
	fenpropathrin, IRAC 3 (Danitol 2.4 EC)	5.33 to 10.66 fl oz	24	21	
	imidacloprid (IRAC 4A) (Admire Pro)	1 to 1.4 fl oz (foliar) 7 to 14 fl oz (soil)	12	0 (foliar) 30 (soil)	Admire Pro can be applied either to the soil or as a foliar spray. Soil applications are typically active for a longer period of time but must be made early in the year. Soil applications are more effective when made via drip irrigation.
	carbaryl, IRAC 1A (Sevin 80S)	1.25 to 2.5 lb	12	7	**Foliar treatment**
	malathion, IRAC 1B (Malathion 57EC) (Malathion 5)	3 pt 3 pt	12	3	**Foliar treatment**

Table 6-6. Bunch Grape Insect Management

When to Spray and Pest	Pesticide, Formulation, and IRAC, FRAC Code	Amount of Formulation to Use per Acre	Reentry Interval (REI) (hours)	Preharvest Interval (PHI) (days)	Precautions and Remarks
Prebloom (continued)					
Grape phylloxera (foliar)	Grape phylloxera has root feeding and foliar feeding forms. Rootstocks used in grape propagation are resistant to root feeding forms and do not require treatment. Foliar phylloxera may be problematic in European-American hybrid varieties (i.e., Vidal, Seyval, etc.) and cause distinctive, wart-like galls on leaves. The mobile crawler stage of phylloxera is susceptible to insecticide treatment, but closed galls are not. Scouting for galls and crawlers should begin once leaves are expanded. If infested leaves are found in susceptible varieties, insecticide treatments should be timed to crawler emergence.				
	acetamiprid, IRAC 4A (Assail WSP)	2.5 oz	12	7	
	fenpropathrin, IRAC 3 (Danitol 2.4 EC)	5.33 to 10.66 fl oz	24	21	
	Imidacloprid, IRAC 4A (Admire Pro)	1 to 1.4 fl oz (foliar) 7 to 14 fl oz (soil)	12	0 (foliar) 30 (soil)	Admire Pro can be applied either to the soil or as a foliar spray. Soil applications are typically active for a longer period of time but must be made early in the year. Soil applications are more effective when made via drip irrigation.
	kaolin (Surround WP Crop Protectant)	25 lb	4		Surround is a barrier that reduces insect feeding. Harvest parameters may be altered and maturity may be delayed, especially in white wine varieties. Closely monitor harvest parameters to determine optimal time to harvest. Changes in harvest parameters can affect final taste. Wine grapes sprayed up to veraison will have minimal adherence to berries. Applications after veraison will adhere more on grape berries. Surround is **OMRI** listed.
	spirotetramat, IRAC 23 (Movento 2SC)	6 to 8 fl oz	24	7	Movento is also effective against root feeding forms.
Bloom					
Thrips	Thrips treatment may be justified if populations exceed an average of 10 thrips per cluster. To sample for thrips, beat blossom clusters over a white surface and count the number of thrips dislodged onto the surface. Count immediately after beating the blossom cluster. Sample at least 10 blossom clusters from different locations in the vineyard. During periods of heavy thrips pressure, a second application may be needed, but make it only if thrips numbers remain high. Wait at least 5 days before making a second application.				
	azadirachtin, IRAC Unknown (Aza-Direct)	1 to 2 pt	4		Aza-direct is **OMRI** listed. Data on thrips control are limited.
	dinotefuran, IRAC 4A (Venom)	3 to 5 oz	12	1	**Foliar treatment** Foliar applications of Group 4A insecticides should NOT be used following a long-acting soil application of any group 4A insecticide (i.e., Admire Pro, Venom, or Clutch).
	pyrethrins, IRAC 3 (Pyganic 1.4 EC) (Pyganic 5 EC)	16 to 64 fl oz 4.5 to 18 fl oz	12	0	Pyganic 1.4 EC and Pyganic 5 EC are **OMRI** listed. Data on thrips control are limited. Pyganic should be buffered to a pH between 5.5 and 7.
	spinosad, IRAC 5 (Entrust SC)	4 to 8 fl oz	4	7	Do not apply more than 29 fl oz Entrust SC (0.45 lb spinosad) per acre per season. Entrust is **OMRI** listed.
	spinetoram, IRAC 5, (Delegate)	3 to 5 oz	4	7	
Postbloom (immediately after bloom)					
Grape berry moth, grape flea beetle, and leafhoppers/ sharpshooters (Pierce's Disease suppression)	See **PREBLOOM** recommendatations.				
European red mite, twospotted spider mite	Sample for mites weekly using a minimum 10x hand lens. If greater than 50% of leaves observed have spider mites and no predatory mites are present, treatment is justified. Fast moving predatory mites can be distinguished from slower moving spider mites through direct observation. Rotate acaricides between IRACs to minimize selection for resistance.				
	abamectin, IRAC 6 (Agri-Mek 0.15EC) (many other formulations)	16 fl oz			
	bifenazate, IRAC Unknown (Acramite 50 WS)	1 lb	12	14	The reentry interval is 5 days for cane turning, tying, and girdling of table grapes. Minimum of 50 gallons per acre spray volume.
	dicofol, IRAC Unknown (Difocol 4EC) (Kelthane 50WSP)	2.5 pt 2.5 lb	12	7	Use Kelthane at 1 pound per acre on small vines. Do not make more than 2 applications per season.
	etoxazole, IRAC 10B (Zeal)	3 oz	12	28	Zeal is a growth regulator and kills eggs and young mites. It is most effective if applied when mite populations are low.
	fenbutatin-oxide, IRAC 12B (Vendex 50WP)	2.5 lb	48	28	Do not make more than 2 applications of Vendex per season.
	fenpyroximate, IRAC 21A (FujiMite 5EC)	2 pt	12	14	Do not apply more than 2 pints of FujiMite per acre per season. Apply in a minimum spray volume of 50 gallons per acre.
	horticultural oils, IRAC Unknown many materials, including (Saf T Side) (Glacial Spray Fluid)	1 to 2% by volume	4	0	Some oils are **OMRI** listed; check label. **Do not** use in combination with or immediately before or after spraying with fungicides such as Captan or any product containing sulfur. **Do not** use with carbaryl or dimethoate. **Do not** use with any product whose label recommends the use of no oils. **Do not** use in combination with NPK foliar fertilizer applications.

Table 6-6. Bunch Grape Insect Management

When to Spray and Pest	Pesticide, Formulation, and IRAC, FRAC Code	Amount of Formulation to Use per Acre	Reentry Interval (REI) (hours)	Preharvest Interval (PHI) (days)	Precautions and Remarks
Postbloom (immediately after bloom) (continued)					
European red mite, twospotted spider mite (continued)	pyridiben, IRAC 21 (Nexter 75WPSB) (Pyramite 60 WP)	5.2 oz 13.2 oz	12	7	The maximum amount of pyridiben allowed per acre per season is 26.4 ounces. Do not make more than two applications of pyridiben per season.
	spirodiclofen, IRAC 23 (Envidor 2SC)	18 fl oz	12	14	
Grape phylloxera (foliar form)	See **Prebloom** recommendations.				
1ˢᵗ Cover Spray (7 to 10 days after Postbloom Spray)					
Grape berry moth, leafhoppers/ sharpshooters (Pierce's Disease suppression)	See **Prebloom** recommendations				**If foliar and soil applications of group 4A pesticides are part of a management plan for Pierce's Disease (i.e., Admire Pro applied via drip and Venom foliar), at least one application of a different IRAC insecticide should occur as a rotation between these treatments.** Current information indicates that in areas where Pierce's Disease is a problem, controlling leafhoppers and sharpshooters through July reduces the risk of Pierce's Disease. See labels for preharvest intervals.
Japanese beetle, Green June beetles	Do not use Japanese beetle traps. Japanese beetle foliar feed only warrants treatment if it occurs on leaves below the top trellis wire. Green June beetles only require treatment if damageing otherwise sound fruit. Green June beetles are attracted to overripe, rotting fruit. Removal of damaged fruit will help reduce Green June beetle populations.				
	acetamiprid, IRAC 4A (Assail WSP)	1.1 oz	12	7	Foliar applications of Group 4A insecticides should NOT be used following a long-acting soil application of any group 4A insecticide (i.e., Admire Pro, Venom, or Clutch).
	azadirectin, IRAC Unknown (Aza-Direct)	1 to 2 pt oz	4	0	Aza-Direct is **OMRI** listed. Data on Japanese beetle control are limited.
	carbaryl, IRAC 1A (Sevin XLR Plus)	2 qt	12	7	Synthetic organophosphates (Groups 1A and 1B) are broad spectrum insecticides and have the potential to flare spider mite populations. Observe spider mites before and after treatments to determine if these populations increase.
	fenpropathrin, IRAC 3 (Danitol 2.4 EC)	10.6 to 21.3 fl oz	24	21	Do not apply Danitol within 21 days of harvest.
	indoxacarb, IRAC 22 (Avaunt 30DG)	3.5 to 6 oz	12	7	Avaunt is also very effective against caterpillar pests such as cutworms and grape berry moth.
	kaolin clay (Surround WP)	25 to 50 lb			Surround may delay fruit maturity, and therefore, anticipated harvest date. Fruit harvest characters should be carefully monitored if Surround is used to ensure timely harvest. Surround is **OMRI** listed.
	malathion, IRAC 1B (Malathion 8F)	1.88 pt	24	3	REI is 72 hours for girling and tying. Injury may occur to grapes for applications made after bloom. Check for phytoxicity in a small area before treating an entire field.
	phosmet, IRAC 1B (Imidan 70W)	1.33 to 2.125 lb	14	14	Rates are based on 200 gallons per acre spray volume.
Closing					
Japanese beetle, GreenJune beetle	Same as **1st Cover**				
Grape berry moth, Leafhopper/ sharpshooter (Pierce's Disease suppression)	Same as **Prebloom**				**Foliar applications of Group 4A insecticides should NOT be used following a long-acting soil application of any Group 4A insecticide (i.e., Admire Pro, Venom, or Clutch).** Current information indicates that in areas where Pierce's Disease is a problem, controlling leafhoppers and sharpshooters **through July** reduces the risk of Pierce's Disease. See labels for preharvest intervals.
2ⁿᵈ and Subsequent Cover Sprays (10- to 14-day intervals until the Preharvest Spray)					
Phylloxera, Japanese and June beetles	Same as **1st Cover**				Check labels for preharvest intervals.
Grape berry moth, Leafhopper/ sharpshooter (Pierce's Disease suppression)	Same as **Prebloom**				Current information indicates that in areas where Pierce's Disease is a problem, controlling leafhoppers and sharpshooters through July reduces the risk of Pierce's Disease. If Venom was applied as a soil treatment during prebloom, a second soil application is not permitted, but a foliar spray of Venom is permitted at this time. See label for further restrictions.
Mites	Same as **Postbloom**				
Grape rootworm, Southern grape rootworm	Grape rootworm larvae feed on roots. Adults are small, black weevils and make distinctive chain-like feeding markings on leaves. Foliar feeding does not results in yield reduction, but root feeding may reduce plant vigor over time. Treatments should be timed to adult activity, which typically peaks in June or July. Grape rootworms are sporadic pests in North Carolina and should not be treated preventatively.				
	carbaryl, IRAC 1A (Sevin XLR Plus)	2 qt	12 hours	7	

Table 6-6. Bunch Grape Insect Management

When to Spray and Pest	Pesticide, Formulation, and IRAC, FRAC Code	Amount of Formulation to Use per Acre	Reentry Interval (REI) (hours)	Preharvest Interval (PHI) (days)	Precautions and Remarks
Preharvest (10 to 14 days before harvest)					
Spotted wing drosophila	Spotted wing drosophila (SWD) is a recently detected invasive pest of ripe and ripening fruit. SWD do not necessarily cause significant damage in wine grapes. Growers are encouraged to monitor adult flies in vineyards and larvae in fruit and treat if present. Weekly or twice weekly treatments may be necessary to manage damaging populations. Insecticides effective against SWD may also have activity against leafhoppers.				
	beta-cyfulthrin, IRAC 3A (Baythroid XL 1EC)	1.6 to 3.2 fl oz	12	3	
	fenpropathrin, IRAC 3A (Danitol 2.4EC)	5.33 to 21.33 fl oz	24	12	
	imidacloprid + cyfluthrin, IRAC 4A + 3 (Leverage 2.4)	3 to 8 fl oz	12	3	
	malathion IRAC 1B (Malathion 5EC)	1.5 pt	12	3	
	spinosad, IRAC 5 (Entrust SC)	4 to 8 fl oz	4	7	Do not apply more than 29 fl oz Entrust SC (0.45 lb spinosad) per acre per season. Entrust is **OMRI** listed.
	spinetoram, IRAC 5 (Delegate)	3 to 5 oz	4	7	
	zeta-cypermethrin, IRAC 3 (Mustang)	2.15 to 4.3 fl oz	12	1	
Harvest					
Yellowjackets and bees	Check to make sure wasps are not nesting in vines. Spot treat or manually remove nests if present. Cover sprays for for wasps or bees is not recommended, because treatments with short PHI will not provide control, and only foraging worker wasps or bees will be killed, leaving the rest of the nest for reinfestation. Damaged fruit should be removed to reduce attraction for other bees and wasps.				
Spotted wing drosophila	See **Preharvest**				
Multicolored Asian lady beetle	MALB can be a contaminant pest at harvest. Sample at least 10 clusters per acre within a few days of harvest, place in a plastic bag for approximately 30 minutes and count beetles. Treatment thresholds vary by variety.				
	imidaclorpid, IRAC 4A (Admire Pro)	1 to 1.4 fl oz	12	0	Several concentrations of imidaclorpid (1.6F, 2F, 4F, and 4.6F) are available. Carefully read the label to determine the correct rate for target pests. Data on control with imiaclorpid are limited.
Postharvest (14- to 21-day intervals from harvest until first killing frost)					
Grape root borer	mating disruption (Isomate GRB)	100 dispensers	NA	NA	Dispensers should be placed prior to the beginning of grape root borer moth flight activity and be left in the vineyard until the end of flight activity. Moth flight timing varies between vineyards, but can be as early as July and last until October. Pheromone baited traps can help determine grape root borer populations and flight activity, but traps will not be effective if mating disruption is underway.
	Soil mounding, cultural control	NA			Use clean cultivation, mound soil (July 1 to Aug. 1) or at first moth emergence when using pheromone traps) or using tightly-sealed plastic mulch 3 feet from the base of vines. This practice will inhibit adult emergence from the soil when well timed. Mounded soil needs to be removed by Sept. 1.
	chloropyrifos (Lorsban Advanced) IRAC 1B	4.5 pt/100 gal water	24	35	Apply 2 quarts of mixture to soil at base of each vine. A single application should be sufficient, either pre or post harvest, depending upon grape root borer flight timing. Spray should not contact fruit or foliage. Application can be made with flood nozzles and low pressure (40 to 60 psi). The preharvest interval for Lorsban is 35 days.

For further information, see www.smallfruits.org.

Muscadine Grape Management Program

H. J. Burrack, Entomology Extension, and W. O. Cline, Plant Pathology Extension

Muscadine Disease Management Program

W. O. Cline, Plant Pathology Extension

For effective disease control, commercial growers should apply fungicides every two weeks from mid May through mid July, beginning prior to the onset of disease symptoms. Fungicides are generally not necessary in home plantings.

The Fungicide Resistance Action Committee groups fungicides into mode of action (MOA) categories. These categories are listed following the pesticide and formulation names. To reduce the risk of resistance development, avoid successive applications of products with the same MOA for the same pest. Organically acceptable insecticides (OMRI listed) are indicated in Precautions and Remarks.

Table 6-7A. Muscadine Disease Management Program

TIMING Pest(s)	Pesticide and Formulation	Amount of Formulation Per Acre	Restricted Entry Interval (REI)	Minimum Interval (Days) Between Application and Harvest; Preharvest Interval (PHI)	Precautions and Remarks
Shoots 6 to 10 inches					
Black rot, Bitter rot, Angular leaf spot, Powdery mildew	Azoxystrobin, FRAC 11 (Abound 2.08 SC)	11 to 15.4 fl oz	4 hrs	14	Do not make more than 2 sequential applications of strobilurin fungicides (Abound, Flint, or Pristine) before alternating with nonstrobilurin fungicides (Captan, Nova, Rally, or Topsin M).
	Myclobutanil, FRAC 3 (Nova, Rally 40 W)	3 to 5 oz	24 hrs	14	
	pyraclostrobin + boscalid, FRAC 11+7 (Pristine38 W)	8 to 12.5 oz*	12 hrs/5 days	14	The REI for Pristine is 12 hours for all crop uses except cane tying, cane turning or cane girdling. These operations, not normally performed on muscadines, require a five day (5d) re-entry interval. *Reccomended rates; higher rates up to 23 ounces per acre can be used when disease pressure is high.
	thiophanate-methyl, FRAC 1 (Topsin M70 WSB)	1 to 1.5 lb	48 hrs	7	
	Trifloxystrobin, FRAC 11 (Flint 50 WG)	2 oz	12 hrs	14	Do not apply Flint fungicide to Concord grapes or injury may occur.
	EBDCs, FRAC M3 (Manzate Prostick, Penncozeb 75 DF, Dithane M45)	1.5 to 4 lb	24 hrs	66	Cannot be used within 66 days of harvest.
Powdery mildew only	wettable sulfur, FRAC M2 (Microthiol, other brands) 80 to 92% S	2 to 5 lb	24 hrs	1	Must be applied every 7 to 10 days. Dilute sulfur in 100 gallons of water per acre. Sulfur corrodes sprayers and trellis wires.
Bloom					
Black rot, Bitter rot, Angular leaf spot, Powdery mildew	Same as **Shoots 6 to 10 inches** recommendations				
Fruit rots, Sooty blotch	Fry is susceptible to sooty blotch.				
	Azoxystrobin, FRAC 11 (Abound 2.08 SC)	11 to 15.4 fl oz	4 hrs	14	
	Captan, FRAC M4 (Captan 50 WP) (Captec 4L)	2 to 4 lb	48 hrs	2 (re-entry)	Do not make more than 2 sequential applications of strobilurin fungicides (Abound, Flint, or Pristine) before alternating with nonstrobilurin fungicides (Captan, Nova, Rally, or Topsin M).
	Captan, FRAC M4 (Captec 4L)	2 qt	48 hrs	2 (re-entry)	
	pyraclostrobin + boscalid, FRAC 11+7 (Pristine 38 W)	8 to 12.5 oz	12 hrs/5 days	14	The REI for Pristine is 12 hours for all crop uses except cane tying, cane turning or cane girdling. These operations, not normally performed on muscadines, require a five day (5d) re-entry interval.
	Trifloxystrobin, FRAC 11 (Flint 50 WG)	2 oz	12 hrs	14	Do not apply Flint fungicide to Concord grapes or injury may occur.
Every 2 weeks until harvest					
Same as sprays for BLOOM					Tank mix Topsin M or Nova, Rally with Captan or Captec, OR alternate Topsin M or Nova, Rally with Abound, Flint, or Pristine.

FRUITS, INSECT, DISEASE CONTROL

Relative Effectiveness of Various Fungicides for Muscadine Grape Disease Control

Table 6-7B. Relative Effectiveness of Various Fungicides for Muscadine Grape Disease Control

(— = ineffective or injurious; +++++ = very effective or very safe)

Fungicide	Angular Leafspot	Bitter Rot	Powdery Mildew	Ripe Rot	Macrophoma Rot	Black Rot	Plant Safety
azoxystrobin (Abound)	++++	+++	+++	++++	++++	++++	+++++
captan (Captan, Captec)	+++	+	++	+++	+++	+++	+++++
myclobutanil (Nova, Rally) 40 W	++	++	++++	–	+	++++	+++++
pyraclostrobin + boscalid (Pristine)	++++	+++	+++	++++	++++	++++	+++++
EBDCs (Manzate, Penncozeb, Dithane, others)	+++	+++	++	+++	+++	+++	++++
sulfur (various)	–	–	++++	–	–	–	+++
thiophanate-methyl (Topsin M)	+++	++	+++	–	+	+++	+++++
trifloxystrobin (Flint)	+++	+++	++++	++++	++++	++++	+++++

Muscadine Insect Management

H. J. Burrack, Entomology Extension

Insect management differs from disease management because some insect injury can be tolerated before economic damage occurs. Apply insecticides only if potentially damaging populations are present. Sampling techniques and tools are described when available.The Insecticide Resistance Action Committee (IRAC) groups insecticides into mode of action (MOA) categories. These categories are listed following the pesticide and formulation names. To reduce the risk of resistance development, avoid successive applications of products with the same MOA for the same pest. Organically acceptable insecticides (OMRI listed) are indicated in Precautions and Remarks.

Many insecticide active ingredients come in several formulations, or generic versions of the same formulation. In general, information is provided for the most commonly used formulations of active ingredients available in multiple forumations. Carefully check the label of the product you plan to use in the event that it differs from those listed. The label is the law!

Table 6-7C. Muscadine Insect Management

Pest	Pesticide and Formulation	Amount of Formulation Per Acre	Restricted Entry Interval (REI)	Minimum Interval (Days) Between Application and Harvest; Preharvest interval (PHI)	Precautions and Remarks
Aphids	Aphids are not common pests in North Carolina muscadines and are typically only problematic in spring on new growth. Aphid populations in late summer do not typically justify treatment. Treatment is only justified when sooty mold is present or new growth is deformed.				
	acetamiprid, IRAC 4A (Assail 30 SG)	2.5 oz	12 hours		
	imidacloprid, IRAC 4A (Admire Pro)	1 to 1.4 fl oz	12	0	There are many formulations of imidacloprid. Read labels carefully for rate information.
	fenpropathrin, IRAC 3 (Danitol 2.4) EC	10.66 to 21.33 fl oz	24 hours	21	Do not exceed 2.66 pints of Danitol per acre per season. Make no more than 2 applications of Danitol per season.
Climbing Cutworms	Scout for cutworm if damaged buds are observed. Look for cutworms at night. Cutworm treatment may be justified if greater than 4% of the buds examined are damaged and the variety does not have fruitful secondary buds. Spray in the evening if possible as cutworms are active at night. Only treat if cutworms are present.				
	Bacillus thuringiensis (Bt), IRAC 11 (many formulations)	rates vary	4 hours	0	Many Bt formulations are OMRI listed.
	carbaryl, IRAC 1A (Sevin XLR Plus)	1 to 2 qt	12 hours	7	
	fenpropathrin, IRAC 3 (Danitol 2.4 EC)	10.66 to 21.33 fl oz	24 hours	21	Do not exceed 2.66 pints of Danitol per acre per season. Make no more than 2 applications of Danitol per season.
	methoxyfenozide, IRAC 18 (Intrepid 2F)	12 to 16 fl oz	4 hours	30	Minimum application for airblast sprayers of 40 gpa.
	rynaxypyr, IRAC 28 (Altacor)	3 to 4.5 fl oz	4 hours	14	Use between 100-200 gallons per acre total spray volume.
	spinosad , IRAC 5 (Entrust 80 WP) (Entrust 2SC)	1.25 to 2.5 oz 4 to 8 fl oz	4 hours	7	Entrust is **OMRI** listed.
	spinetoram, IRAC 5 (Delegate)	3 to 5 oz	4 hours	7 days	Do not exceed 5 applications of Delegate per year or 19.5 ounces per acre per crop year.
Grape Berry Moth	Grape berry moth is present in NC, but it is not uniformly distributed in the state. If grape berry moth presence is suspected, observe flowers and fruit for injury and consider monitoring moth presence with pheromone baited traps.				
	bifenthrin, IRAC 3 (Brigade 2EC)	3.2 to 6.4 fl oz	12 hours	30	

FRUITS, INSECT, DISEASE CONTROL

Table 6-7C. Muscadine Insect Management

Pest	Pesticide and Formulation	Amount of Formulation Per Acre	Restricted Entry Interval (REI)	Minimum Interval (Days) Between Application and Harvest; Preharvest interval (PHI)	Precautions and Remarks
Grape Berry Moth (continued)	bifenthrin + imidacloprid, IRAC 3 + 4A (Brigadier)	3.8 to 6.4 fl oz	12 hours	30	
	clothianidin, IRAC 4A (Clutch 50 WDG)	3 oz	12 hours	0	
	indoxacarb IRAC 22 (Avaunt)	5 to 6 oz	12 hours	7	
	methoxyfenozide, IRAC 18 (Intrepid 2F)	12 to 16 fl oz	4 hours	30	
	spinosad , IRAC 5 (Entrust 80 WP) (Entrust 2SC)	1.25 to 2.5 oz 4 to 8 fl oz	4 hours	7	Entrust is **OMRI** listed.
	phosmet, IRAC 1B (Imidan 70 WP)	1.33 to 2.125 lb	14 days	14	Do not apply more than 6.5 pounds Imidan per acre per year.
	methomyl, IRAC 1A (Lannate SP)	0.5 to 1 lb	7 days	1, fresh market 14, wine	
	fenpropathrin, IRAC 3 (Danitol 2.4 EC)	10.66 to 21.33 fl oz	24 hours	21	Do not exceed 2.66 pints of Danitol per acre per season. Make no more than 2 applications of Danitol per season.
	pyriproxyfen IRAC 7C (Esteem 0.83EC)	16 fl oz	12 hours	21	
	rynaxypyr, IRAC 28 (Altacor)	3 to 4.5 fl oz	4 hours	14	Use between 100-200 gallons per acre total spray volume.
Grape Flea Beetle	Grape flea beetle larvae feed on developing buds during bud swell. If greater than 4% of buds observed are damaged by grape flea beetles, treatment may be justified. Apply only if damaging numbers of adult beetles are present.				
	bifenthrin + imidacloprid, IRAC 3 + 4A (Brigadier)	3.8 to 6.4 fl oz	12 hours	30	
	carbaryl, IRAC 1A (Sevin XLR Plus)	2 qt	12 hours	7	
	fenpropathrin, IRAC 3 (Danitol 2.4 EC)	10.66 to 21.33 fl oz	24 hours	21	Do not exceed 2.66 pints of Danitol per acre per season. Make no more than 2 applications of Danitol per season.
	phosmet, IRAC 1B (Imidan 70 WP)	1.33 to 2.125 lb	14 days	14	Do not apply more than 6.5 pounds Imidan per acre per year.
Grape rootworm, Southern grape rootworm	Grape rootworm larvae feed on roots. Adults are small, black weevils that make distinctive chain-like feeding markings on leaves. Foliar feeding does not result in yield reduction, but root feeding may reduce plant vigor over time. Treatments should be timed to adult activity, which typically peaks in June or July. Grape rootworms are sporadic pests in North Carolina and should not be treated preventatively.				
	carbaryl, IRAC 1A (Sevin XLR Plus)	2 qt	12 hours	7	
Leafhoppers, Sharpshooters	Leafhoppers are important vectors of Pierce's Disease in *Vinifera* grapes, but Pierce's Disease is not a common problem of muscadine grapes. Var. Carlos has been observed with Pierce's Disease symptoms, but the disease does not appear to persist in plants over-winter. Therefore, leafhoppers should not be preventatively treated in muscadines.				
	Large leafhopper populations may result in leaf stippling and yellowing, and populations of this size may result in economic damage and justify treatment.				
	abamectin IRAC 6 (Agri-Mek 0.15EC, many formulations)	8 to 16 fl oz	12 hours	28	
	acetamiprid, IRAC 4A (Assail 30 SG)	2.5 oz	12 hours		
	bifenthrin, IRAC 3 (Brigade 2EC)	3.2 to 6.4 fl oz	12 hours	30	
	bifenthrin + imidacloprid, IRAC 3 + 4A (Brigadier)	3.8 to 6.4 fl oz	12 hours	30	
	dinotefuran, IRAC 4A (Venom)	1 to 3 oz	12 hours	28	Venom may be applied as a foliar spray at 1 to 3 ounces or to the soil at 5 to 6 ounces. See label for details.
	clothianidin IRAC 4A (Clutch 50 WDG)	6 oz	12 hours	30	Clutch is applied to the soil either via drip or trickle irrigation.
	fenpropathrin, IRAC 3 (Danitol 2.4 EC)	5.3 to 10.6 fl oz	24 hours	21	Do not exceed 2.66 pints of Danitol per acre per season. Make no more than 2 applications of Danitol per season.
	imidacloprid, IRAC 4A (Admire Pro)	1 to 1.4 fl oz (foliar) 7 to 14 oz (soil)	12 hours	0 30	Admire Pro is applied to as a foliar spray and as a soil treatment See label for application details.
	malathion, IRAC 1B (57 EC or Malathion 5)	3 pt	12 hours	3	Malathion may cause injury to berries if applied after bloom. Rates are based on 200 gpa spray volumes.
	phosmet, IRAC 1B (Imidan 70 W)	1.33 to 2.125 lb	14 days	14	Do not apply more than 6.5 pounds Imidan per acre per year.
	thiamethoxam IRAC 4A (Actara)	1.5 to 3.4 oz	12 hours	5	
Japanese Beetle, Green June Beetle	Foliar feeding by Japanese beetles in established vineyards does not justify treatment unless it occurs on leaves below the top trellis wire. Fruit feeding by Japanese beetles is rare. Green June beetles are attracted to damaged, decomposing fruit but may also feed on undamaged fruit once in the vineyard. Removal of damaged fruit that will not be harvested will minimize Green June beetle populations and should be conducted before considering pesticide application.				
	acetamiprid, IRAC 4A (Assail 30 SG)	2.5 oz			Southeastern data for Assail on Japanese beetles are limited.
	bifenthrin, IRAC 3 (Brigade 2EC)	3.2 to 6.4 fl oz	12 hours	30	
	bifenthrin + imidacloprid, IRAC 3 + 4A (Brigadier)	3.8 to 6.4 fl oz	12 hours	30	
	carbaryl, IRAC 1A (Sevin XLR Plus)	2 qt	12 hours	7	
	clothianidin, IRAC 4A (Clutch 50 WDG)	3 oz	12 hours	0	

Table 6-7C. Muscadine Insect Management

Pest	Pesticide and Formulation	Amount of Formulation Per Acre	Restricted Entry Interval (REI)	Minimum Interval (Days) Between Application and Harvest; Preharvest interval (PHI)	Precautions and Remarks
Japanese Beetle, Green June Beetle (continued)	fenpropathrin, IRAC 3 (Danitol 2.4 EC)	10.66 to 21.33 fl oz	24 hours	21	Do not exceed 2.66 pints of Danitol per acre per season. Make no more than 2 applications of Danitol per season.
	imidacloprid, IRAC 4A (Admire Pro)	1 to 1.4 fl oz (foliar)	12 hours	0	Admire Pro is applied to as a foliar spray and as a soil treatment. Soil treatments are not recommended for Japanese and green June beetles.
	indoxacarb IRAC 22 (Avaunt)	5 to 6 oz	12 hours	7	
	Kaolin IRAC NA (Surround)	25 to 50 lb	4 hours	0	Surround creates a barrier on plants and reduces insect attraction and feeding. In order to be effective, leaf and/or fruit surfaces should be coated thoroughly. Material should be reapplied at least 7 days apart to maintain coverage. **Not recommended** for use in fresh market grapes after bloom as white residue can remain until harvest.
	phosmet, IRAC 1B (Imidan 70 W)	1.5 to 3.4 oz	12 hours	5	Do not apply more than 6.5 pounds Imidan per acre per year.
Spider Mites	Sample for mites using a minimum 10x hand lens. There is no clearly defined threshold for mites in muscadine grapes. Treatment for *Vinifera* grapes is recommended when greater than 50% of leaves are infested. Fast moving predatory mites can be distinguished from slower moving spider mites through direct observation. Some insecticides, such as carbaryl, can flare mite populations, and care should be used with these materials when mites are present. Rotate acaricides between MOAs to minimize selection for resistance.				
	abamectin, IRAC 6 (Agri-Mek 0.15EC) (many other formulations)	16 fl oz	12 hours	28	Abamectin is a restricted use product. Do not reapply within 21 days of initial application. Abamectin is an EC (emulsifiable concentrate), which can cause phytoxicity in some crops. Check for possible plant injury before treating an entire field.
	bifenazate, IRAC Unknown (Acramite 50 WS)	1 lb	12 hours	14	The reentry interval is 5 days for can turning, tying, and girdling. Apply in a minimum spray volume of 50 gallons per acre.
	etoxazole, IRAC 10B (Zeal)	3 oz	12 hours	28	Zeal is a growth regulator and kills eggs and young mites. It is most effective if applied when mite populations are low.
	fenpyroximate, IRAC 21A (Portal 5EC)	2 pt	12 hours	14	Do not apply more than 2 pints of Portal per acre per season. Apply in a minimum spray volume of 50 gallons per acre.
	pyridiben IRAC 21 (Nextor 75 WSB)	8.8 to 10.67 oz	12 hours	7	The maximum amount of pyridiben allowed per acre per season is 26.4 ounces. Do not make more than two applications of pyridiben per season.
Stink bugs	Stink bugs are not a common pest of muscadine grapes, and there is not evidence that they directly damage fruit. If present at harvest, they may contaminate fruit, and under this scenario, they may justify treatment.				
	fenpropathrin, IRAC 3 (Danitol 2.4 EC)	10.66 to 21.33 fl oz	24 hours	21	Do not exceed 2.66 pints of Danitol per acre per season. Make no more than 2 applications of Danitol per season. Fenpropathrin (and other Group 3 materials
	phosmet, IRAC 1B (Imidan 70 W)	1.33 to 2.125 lb	14 days	14	Do not apply more than 6.5 pounds Imidan per acre per year.
Grape Root Borer	mating disruption (Isomate GRB)	100 dispensers	NA	NA	Dispensers should be placed prior to the beginning of grape root borer moth flight activity and be left in the vineyard until the end of flight activity. Moth flight timing varies between vineyards, but can be as early as July and last until October. Pheromone baited traps can help determine grape root borer populations and flight activity, but traps will not be effective if mating disruption is underway.
	Cultivation or soil mounding	NA	NA	NA	Use clean cultivation, mound soil (July 1 or at first moth emergence when using pheromone traps) or using tightly-sealed plastic mulch 3 feet from the base of vines. This practice will inhibit adult emergence from the soil when well timed. Mounded soil needs to be removed by Sept. 1.
	chlorpyrifos, IRAC 1A (Lorsban Advanced)	4.5 pt per 100 gal	24 hours	35	Apply 2 quarts of mixture to soil at the base of each vine. Make a single application 35 days before harvest. Spray should not contact fruit or foliage. Application can be made with flood nozzles and low pressure (40 to 60 psi).
Red Imported Fire Ant	Bait treatments can effectively manage fire ants, but they typically take 2 to 4 weeks to reach full efficacy. Baits must be applied when ants are actively foraging. Test for foraging by placing food near the nest. Check for ant activity after 30 minutes.				
	pyriproxyfen IRAC 7C (Esteem Ant Bait)	1.5 to 2 lb	12 hours	1	Do not exceed 0.22 pound of active ingredient per season.
	methoprene MOA 7A (Extinguish Professional Fire Ant Bait)	1 to 1.5 lb	4 hours	0	

Further Information

Muscadine Grape Diseases and Their Control, Plant Pathology Information Note 145, https://www.ces.ncsu.edu/depts/pp/notes/Fruit/fdin012/fdin012.htm

Southeast Regional Muscadine Integrated Management Guide, www.smallfruits.org

Peach and Nectarine Spray Guide
D. F. Ritchie, Plant Pathology Extension, and J. Walgenbach, Entomology Extension

Although many pesticides are registered for disease and insect control on peaches, the following spray program lists the ones that have performed well under North Carolina conditions. The rates of pesticides recommended should give control when pest pressure is moderate to severe, assuming they are applied correctly. Where the rate is given as a range, the lower rate can be used when pest pressure is low; the higher rate should be used when pest pressure is great. Thus, the following spray program is intended to be only a guide since pest and orchard conditions can vary from orchard to orchard and year to year.

The rates given are based on the use of rate per acre; 75 to 125 gallons of water per acre provides optimal spray coverage for pest/disease control in most orchards.

Note: For imported fire ant, treat active mounds off season with directed bait formulations like Clinch, Esteem, Extinguish, and Logic. Insect growth regulators will give complete control after 30 days. Always follow label directions for best results.

Table 6-8. Peach and Nectarine Spray Guide

When to Spray	Pest	Pesticide	REI (hrs)	PHI (days)	Formulation Per Acre	Remarks
Dormant Before buds swell in late winter	Leaf curl	**Fungicide:** chlorothalonil (Bravo Weather Stik) 6 F OR	12	NA	4 pt	Other chemicals registered for leaf curl include copper-containing compounds (consult labels). Copper provides adequate leaf-curl control when at least 4 pounds/acre is applied before bud-swell. To control white peach scale, two dormant oil sprays 2 weeks apart are necessary. Oil will NOT control leaf curl.
		ziram (Ziram) 76 DF	48	14	5 lb	
	Scale insect, mite	**Insecticide:** oil, superior-type +	4	NA	4 gal	Addition of an insecticide with oil will improve control of scales. Where white peach scale is a problem, two applications of oil at 14- to 21-day intervals may improve control. Insecticides should be added with the second application. Application of Centaur, Esteem or Movento (see Shuck Split to Shuck Fall) may be delayed to after bloom. NOTE: Heavy reliance on pyrethroid insecticides may flare scale populations, particulary San Jose scale.
		buprofezin (Centaur) 70WDG OR	12	14	25 oz	
		chlorpyrifos (Lorsban) 4E OR	96	NA	2 pt	
		pyriproxyfen (Esteem) 35WP	12	14	4 to 5 oz	
Bloom	Brown rot, blossom blight	**Fungicide:** captan (Captan, Captec) 50 WP, 4L OR	24-96	0	5 lb, 2.5 qt	Fungicide sprays at full pink to early bloom and again at full bloom may reduce blossom blight, but another spray may be needed if bloom extends beyond 2 weeks. Demethylation inhibiting (DMI) fungicides (Elite, Indar, Nova, Orbit, Quash) are effective against blossom blight but are prone to resistance problems if used regularly. Resistance to any one of the DMI fungicides results in cross-resistance to the others. **It is recommended that DMI fungicides be saved for preharvest sprays and that they not be used in bloom and cover sprays.** Do not use more than one application of thiophanate-methyl or if resistant strains are present.
		chlorothalonil (Bravo Weather Stik, Echo 720) 6 F OR	12	NA	3.125 pt	
		thiophanate-methyl (Topsin M, T-Methyl) +	12	1	1.0 lb +	
		captan (Captan, Captec) 50WP, 4L	12	0	4 lb, 2 qt	
		OR cyprodinil (Vangard) 75 WG	12	0	5 oz	Vangard is another alternative to DMI fungicides during bloom.
		Insectcide: None				
Petal-Fall After petals are off but before fruit are showing	Scab, brown rot	**Fungicide:** captan (Captan, Captec) 50 WP, 4L OR	12	0	5 lb, 2.5 qt	Including a fungicide at petal fall may enhance scab control.
		sulfur	24	0	9 lb actual sulfur	
	Plum curculio, catfacing insect, Oriental fruit moth	**Insecticide:** Pyrethroid (see list in Table 6-9) OR	12-24	3-14	—	For the list of pyrethroids registered see IRAC MOA Group 3A in Table 6-9.
		phosmet (Imidan) 50 WP OR	24	14	3 lb	Imidan has a 4-day re-entry interval, 14 days for general public (i.e., pick-your-own customers).
		thiamethoxam (Actara) 25 WDG OR	12	14	5 oz	Do not apply more than 11 ounces of Actara per acre per season.
		Indoxacarb (Avaunt) 30WG	12	14	5 oz	Avaunt is not recommended for catfacing insects (plant bugs and stink bugs), plum curculio and oriental fruit moth only.
Shuck Split to Shuck Fall After fruit are showing, but before 75% of the fruit have shucks off	Scab, brown rot	**Fungicide:** captan (Captan, Captec) 50 WP, 4 L OR	12	0	5 lb, 2.5 qt	Very critical period for start of scab control. Tank-mix of thiophanate-methyl 0.75 pound a.i./acre (Topsin M, T-Methyl) with captan or sulfur first two sprays enhances scab control. **Chlorothalonil cannot be used later than shuck split.**
		chlorothalonil (Bravo Weather Stik, Echo 720) 6 F OR	12	NA	4 pt	
		sulfur	24	0	9 lb actual sulfur	
	Plum curculio, catfacing insect	**Insectcide:** Same as in Petal-Fall				When cool weather delays shuckoff, a second application of insecticide in 7 to 10 days may be necessary to control catfacing insects.

Table 6-8. Peach and Nectarine Spray Guide

When to Spray	Pest	Pesticide	REI (hrs)	PHI (days)	Formulation Per Acre	Remarks
Cover Sprays Begin 7 to 10 days after shuck fall, continue 10 to 14 days, stopping at least 2 weeks before harvest	Scab, brown rot	**Fungicide:** Same as SHUCK SPLIT				First through third cover sprays are very important for scab control on peach. Chlorothalonil used at shuck split can give 3 weeks of scab control. NOTE: **Chlorothalonil cannot be used after the shuck split spray.**
	Plum curculio, Stink bugs, Oriental fruit moth	**Insecticide:** Pyrethroid (see list in Table 6-9) OR	12-24	3-14	—	Do NOT apply esfenvalerate or permethrin within 2 weeks of harvest.
		indoxacarb (Avaunt) 30 WG OR	12	14	6 fl oz	Do NOT make more than four applications of Avaunt per season.
		methomyl (Lannate) 2.4 L OR	96	4	3 pt	
		phosmet (Imidan) 50 WP OR	24	14	3 lb	Do NOT apply phosmet within 3 weeks of harvest.
		spinetoram (Delegate) 25 WDG OR	12	7	5 fl oz	Delegate and Altacor are primarily for oriental fruit moth control.
		chlorantraniliprole (Altacor) 35 WDG OR	4	10	3 oz	
		Cyantraniliprole (Exirel) 0.83SE	12	3	17 oz	
	San Jose Scale	buprofezin (Centaur) 70WDG OR	12	14	25 oz	In most orchards treated with oil and a scale-active insecticide before bloom, additional control measures are usually not necessary. However, in orchards with a history of San Jose scale, targeting first generation crawlers at the second or third cover spray will help to minimize the potential for damage.
		pyriproxyfen (Esteem) 35WP OR	12	14	4 to 5 oz	
		spirotetramat (Movento) 2F	24	7	6 oz	
Preharvest Begin 2 to 3 weeks before harvest; apply fungicides at 7- to 10-day intervals. In periods of high disease pressure, closer spray intervals may be necessary.	Brown rot	**Fungicide:** azoxystrobin (Abound) 2.08 F OR	4	0	12 to 15 fl oz	NOTE: Check product label for any preharvest interval (PHI) or reentry interval (REI) times and other restrictions.
		azoxystrobin + difenoconazole (Quadris Top) 2.71SC OR	12	0	12 to 14 fl oz	azoxystrobin (PHI = 0 day, RE = 4 hr)
		difenoconazole + cyrodinil (Inspire Super) 2.82EW OR	12	2	16 to 20 fl oz	azoxystrobin+difenoconazole(PHI =0 day,REI=12 hr) difenoconazole + cyrodinil (PHI=2 days, REI=12 hr)
		fenbuconazole (Indar) 75 WSP OR	12	0	2 oz pouch	fenbuconazole (PHI = 0 day, REI = 12 hr)
		trifloxystrobin+fluropyram (Luna Sensation) 4.2SC OR	12	1	5.0 to 7.6 fl oz	trifloxystrobin + fluropyram (PHI = 1 day, REI = 12 hr) metconazole (PHI =14 days, REI = 12 hr)
		metconazole (Quash) 50 WDG OR	12	14	3.5 to 4.0 oz	penthiopyrad (PHI=0 day, RE=12 hr)
		penthiopyrad (Fontelis) 1.67SC OR	12	0	14-20 fl oz	propiconazole (PHI = 0 day, REI = 24 hr) pyraclostrobin + boscalid (PHI= 0 day, REI= 12 hr)
		propiconazole (Bumper, Orbit, PropiMax) 3.6 EC OR	24	0	4 fl oz	pyraclostrobin+fluxapyroxad (PHI=0 day, REI=12 hr) tebuconazole (PHI = 0 day, REI = 12 hr)
		pyraclostrobin + boscalid (Pristine) 38 WG OR	12	0	10.5 to 14.5 oz	
		pyraclostrobin+fluxapyroxad (Merivon) 500 SC OR	12	0	4 to 6.7 fl oz	Preharvest use of propiconazole, azoxystrobin + difenoconazole, and difenoconazole + cyrodinil, is limited to 2 applications.
		tebuconazole (Elite, Orius, Tebuzol) 45 DF OR	12	0	4 to 8 oz	Do not make more than 2 sequential applications of these fungicides before alternating with a fungicide having a different mode of action.
		tebuconazole + trifloxystrobin (Adament) 50 WG	24	1	4 to 6 oz	
	June beetle, Japanese beetle	**Insecticide:** carbaryl (Sevin) 80 WSP OR	12	3	2.5 lb	Do NOT apply Sevin less than 3 days before harvest.
		imidacloprid (Admire Pro) 4.6SC	12	0	1.4 fl oz	Admire has a 0-day preharvest interval, but a 12-hour re-entry interval.
Borer Spray	Peachtree borer	**Insecticide:** chlorpyrifos (Lorsban) 4 EC OR	96	NA	1 qt/100 gal	Tree trunks and limbs should be sprayed to drip, after harvest or after August 1, whichever comes last. Best control results when applied the week of Sept. 1.
		pyrethroid (see Table 6-9)	12-24	3-14		
Special Spray	Spider mite	**Miticide:**				When mite populations increase to large numbers they may cause severe injury. Examine outer leaves for mites and mottled appearance, especially during hot, dry periods.
		Preventive Spray: hexythiazox (Savey) 50 DF OR	12	28	3 to 6 oz	Apollo and savey are ovicides and should be applied early in the season. Apollo and Savey have the same mode of action, so if resistance develops to one compound, populations will be resistant to both products. Do not apply more than once per year, and preferably use once every other year.
		clofentezine (Apollo) SC	12	21	4 to 8 oz	
		Curative Spray: bifenazate (Acramite) 50 WP	12	3	1 lb	Do not apply within 3 days of harvest.
		abamectin (Agri-Mek) 0.7SC	12	21	2.25 to 4.25 fl oz	Include 0.25% horticultural oil (not dormant oil) or a nonionic surfactant with Agri-Mek. 21 day PHI.

Relative Effectiveness and Safety of Various Insecticides for Peach Insects

J. F. Walgenbach, Entomology Extension

(— = ineffective or does not apply; +++++ = very effective)

Table 6-9. Relative Effectiveness and Safety of Various Insecticides for Peach Insects

IRAC† MOA Group	Insecticide Formulation and Rate per 100 Gallons Water	Days Between Last Spray and Harvest	Plum Curculio	Oriental Fruit Moth	Peachtree Borer	Catfacing insects (stink bugs)	Scales (White Peach, San Jose)	Beetles (June, Japanese)	Safety*
1A	methomyl (Lannate 2.4 L) 1 pt	4	+++	++++	+	+++	+	++	Danger, Restricted
	carbaryl (Sevin 80 SP) 1.25 lb	3	+	++	+	+	++	+++++	Caution
1B	phosmet (Imidan 50 WP) 1.5 lb	14	+++	++++	+	+	+	+++	Warning
	chlorpyrifos (Lorsban 4.0 EC) 3 qt	Prebloom and postharvest only	—	—	+++++	—	+++++	—	Danger, Restricted
3A	beta-cyfluthrin (Baythroid 1EC) 2.4 oz	7	++++	+++++	++	++++	—	+++	Warning, Restricted
	Cyfluthrin (Tombstone 2EC) 2.4 oz	7	++++	+++++	++	+++++	—		Danger, Restricted
	esfenvalerate (Asana 0.66 EC) 5.8 oz	14	++++	+++++	++	++	—	++++	Warning, Restricted
	fenpropathin (Danitol 2.4 EC) 16 oz	3	++++	+++++	++	+++++	—	+++++	Warning, Restricted
	gamma-cyhalothrin (Proaxis 0.5EC) 3.8 oz	14	++++	+++++	++	+++++	—	+++++	Caution, Restricted
	lambda-cyhalothrin (Karate 2.08CS) 1.9 oz	14	++++	+++++	++	+++++	—	+++++	Warning, Restricted
	permethrin (Pounce 2.0 EC, 25 WP) 6 oz	7	++++	+++++	++	++++	—	++++	Warning, restricted
4A	acetamiprid (Assail 30 SG) 7 oz	7	++	++++	+	++	+++	++++	Caution
	Chlothianidin (Belay SL) 6 oz	21	++++	+	+	++++	—	+++	Caution
	dinotefuran (Scorpion 35SL) 5.25 oz (Venom 70SG) 4 oz	3	++++	+	+	++++	+	+++	Caution
	imidacloprid (Provado 1.6F)3 oz	0	+	—	—	+	—	+++	Caution
	thiamethoxam (Actara 25WDG) 2.5 oz	14	+++++	+++	—	++++	+++	++++	Caution
16	buprofezin (Centaur 70WSB) 17 oz	14	—	—	—	—	+++++	—	Caution
28	chlorantraniliprole (Altacor 35WDG) 2.5 oz	10	—	+++++	++	—	—	—	Caution
	cyantraniliprole (Exirel 0.83SE) 17 oz	3	+++	+++++	++	—	—	+++	Caution
22	indoxacarb (Avaunt 30 DG) 5 oz	14	+++++	++++	++	—	—	+++	Caution
7C	pyriproxyfen (Esteem 35 WP) 5 oz	14	—	++	—	—	+++++	—	Caution
5	spinetoram (Delegate 25WDG) 2.5 oz	7	+	+++++	—	—	—	+	Caution
23	Spirotetramat (Movento 2SC) 8 oz	7	—	—	—	—	++++	—	Caution
—	oil superior — 2 gal	—	—	—	—	—	+++++	—	Caution

† Insecticide Resistance Action Committee (IRAC) mode of action (MOA) group.

* Relative Toxicity (Safety):

    Danger = most toxic to man

    Caution = least toxic to man

    Restricted = restricted use compound; may be applied only by licensed pesticide operators

Relative Effectiveness of Chemicals for Disease Control on Peaches and Nectarines

D. F. Ritchie, Plant Pathology Extension

(E = excellent; G = good; F = Fair; P = poor; NC = no control; NA = not applicable; ND = no data)

Table 6-10. Relative Effectiveness of Chemicals for Disease Control on Peaches and Nectarines

Fungicide/Bactericide and Rate per Acre per 100 gallons	Mode of Action Code	Days Between Last Spray and Harvest	Reentry Interval (REI)* (hours)	Leaf Curl	Blossom Blight	Brown Rot	Peach Scab	Bacterial Spot
azoxystrobin (Abound) 2.08 F — 12 fl oz	11	0	4	NA	F-G	G	G	NC
azoxystrobin+difenoconazole (Quadris Top)2.71SC — 14 fl oz	11, 3	0	12	NA	G	G-E	G-E	NC
captan (Captan, Captec) 50 WP, 4L — 5lb, 2.5 qt	M4	0	24 to 96	NA	F	F-G	G	NC
chlorothalonil (Bravo Weather Stik, Echo 720) 6 F — 4 pt	M5	NA	12**	G	F-G	NA	G	NC
copper 3000; Cuprofix ULTRA 40D) — 4 to 8 lb ***	M1	at least 21	12 to 24	F-G	NA	NA	NA	G
cyprodinil (Vangard) 75 WG — 5 oz	9	NA	12	NA	G	NA	NA	NC
difenoconazole + cyrodinil (Inspire Super) 2.82EW — 20 fl oz	3, 9	2	12	NA	G	G-E	G-E	NC
fenbuconazole (Indar) 75 WSP — 2 oz	3	0	12	NA	G	G	P-F	NC
fenhexamid (Elevate) 50 WDG — 1.5 lb	17	0	12	NA	ND	F	P-F	NC
trifloxystrobin + fluropyram (Luna Sensation) 4.2SC 6.0 fl oz	11, 7	1	12	NA	G	G-E	G	NC
iprodione (Rovral) — 1.5 lb, 1.5 pt ****	2	NA	12	NA	G	NA	NC	NC
metconazole (Quash) 50WDG	3	14	12	NA	G	G	F	NC
myclobutanil (Rally)40 WP — 4 oz	3	0	24	NA	G	F	NC-P	NC
oxytetracycline (FireLine, Mycoshield) 17 WP — 0.75 lb	41	21	12	NA	NA	NA	NA	F-G
penthiopyrad (Fontelis) 1.67SC 20 fl oz	7	0	12	NA	ND	F-G	P-F	NC
propiconazole (Bumper, Orbit, PropiMax) 3.6 EC — 4 fl oz	3	0	24	NA	G	G	NC-P	NC
pyraclostrobin + boscalid (Pristine) 38 WG — 10.5 to 14.5 oz	11, 7	0	12	NA	G	G-E	F	NC
pyraclostrobin+fluxapyroxad (Merivon) 500SC — 6.7 fl oz	11, 7	0	12	NA	G	E	G	NC
pyrimethanil (Scala SC) 60 SC — 1 pt	9	2	12	NA	ND	P-F	ND	NC
Sulfur — 9 lb actual sulfur	M2	0	24	NA	P-F	P	F-G	NC
tebuconazole (Elite, Orius, Tebuzol) 45 DF, WP — 4 oz	3	0	12	NA	G	G	P	NC
tebuconazole + trifloxystrobin (Adament) 50WG	3, 11	1	24	NA	G	G	P-F	NC
thiophanate-methyl (Topsin M) 70 WP, 4.5FL — 1 lb, 1.5 pt + captan (Captan) 50 WP — 4 lb	1, M4	1	12 24 to 96	NA	G	F-G	G	NC
ziram (Ziram) 76 DF — 5 lb	M3	14	48	G	P	P	P-F	P

* REI = reentry interval. Hours between last spray and reentry without using personal protective equipment. This time interval can vary depending on product formulation, always consult label of product being used.

** Consult chlorothalonil label for REI precautions related to risk of eye injury.

*** Rate of copper stated is for dormant spray. Rates must be sequentially greatly reduced to lessen foliar injury when used during the growing season.

**** Rovral is not registered for use after bloom.

Mode of Action Codes – fungicides having the same code have a similar mode of action and thus are not appropriate mixing or alternating partners for use in resistance management.

Nematode Control on Peaches

D. F. Ritchie, Plant Pathology Extension

Preplant Soil Fumigation — In light, sandy soil where root-knot and ring nematodes are present, preplant soil fumigation is imperative. If the nematode assay indicates the presence of root-knot or ring nematodes, it may be advantageous to fumigate the entire orchard site in **October to mid-November** before planting the trees in late winter to early spring. If the nematode assay does not indicate the presence of root-knot or ring nematodes, an 8- to 10-foot strip to be used for the tree row may be fumigated.

Table 6-11. Preplant Soil Fumigation

Materials	Rate/treated acre*
1,3 dichloropropene (Telone II) OR	27 to 36 gallons
metam-sodium (Vapam, Sectagon II, Busan 1020) tarped	75 to 100 gallons

* Rate will vary depending on soil type. Follow manufacturer's directions for rate and application procedures.

Postplant Treatment (Bearing and Nonbearing Trees) — NO MATERIALS REGISTERED for postplant use.

Further Information

Southeastern Peach Growers' Handbook (http://www.ent.uga.edu/peach/peachhbk/toc.htm)

2017 Southeastern Peach, Nectarine and Plum Pest Management and Culture Guide. University of Georgia Bulletin 1171 (http://www.ent.uga.edu/peach/PeachGuide.pdf) updated annually

219

Strawberry Disease Control

F. J. Louws, Plant Pathology Extension (with major input from Guido Schnabel and Chuck Johnson)

For more information and details, see the *Southeast Regional Strawberry Integrated Management Guide*, which is online at http://www.smallfruits.org/SmallFruitsRegGuide/.

Pre-Planting Disease Control

Table 6-12A. Pre-Planting Disease Control

Management Options	Amount of Formulation per Acre	Effectiveness (+) or Importance (*)	Comments
Anthracnose Angular leaf spot Phytophthora crown rot Fusarium wilt (not reported in Eastern U.S.) Viruses	Disease-free plants	***** +++++	Use of certified plants or plants produced in a similarly stringent program is the most important method to prevent these diseases.
Nematodes	Sample soil	***	Sample soils for nematode analysis through local state services to determine which fumigant or IPM management plan may be required.
Nematodes and soilborne pathogens (Pythium, Phytophthora, Fusarium, Rhizoctonia)	Crop rotation and cover crop selection	*** +++	Selected summer cover crops and rotating fields to other crops for 2 to 3 years can suppress nematode populations and reduce black root rot and other disease problems.
Weeds Root and crown rot disorders Nematodes (Black root rot; Phytophthora crown rot)	Pre-plant fumigation and laying down plastic mulch	+++++	See fumigation table below. Consult with custom applicators and/or extension agents for product and rate recommendations.

Pre-Planting and Early Post-planting: Nematode Management

Table 6-12B. Pre-planting and Early Post-planting: Nematode Management

Management Options	Amount of Formulation per Acre	Effectiveness (+) or Importance (*)	REI	PHI	Comments
Nimitz or Fluensulfone 480EC	3.5 - 7 pt per treated acre	See comments	0 hr	0 days	Nimitz is a "traditional contact nematicide." It has not been extensively tested on strawberry in the Southeast and Mid-Atlantic states, but research on other crops in these areas and on strawberry elsewhere suggests moderate to good activity — not quite as effective as soil fumigant standards — against most major plant-parasitic nematode species. Apply via drip or incorporated spray at least 7 days before planting; only 1 application per year. Soil temperature must be 60 degrees F or above. Soil incorporation in the top 6-8 inches is critical. Irrigating (0.5-1 inches) 2-5 days after application is recommended.
Majestene (heat-killed *Burkholderia* spp. strain A396)	4 – 8 qt (8 – 16 pt or 1 - 2 gal)	See comments	4 hr	0 days	Majestene is a biological nematicide approved for organic strawberry production. It has not been extensively field-tested on strawberry in the Southeast and Mid-Atlantic states, but research to date suggests useful activity against root-knot, lesion, sting, stunt, ring, and reniform nematodes. Can be applied as a pre-plant incorporated, in-furrow or banded spray as long as spray volume is sufficient to thoroughly soak the root zone. However, Majestene can also be drip-applied prior to planting, at planting or shortly thereafter, and again later in the season. Higher rates are likely more effective, and repeated applications also increase the extent and duration of nematode control. If nematode populations are high, another product may also be necessary for control.

Fumigants

New labels require extensive risk mitigation measures including fumigant management plans (FMPs), buffer restrictions, worker protection safety standards and other measures. Details are on the labels and see http://www2.epa.gov/soil-fumigants. Some fumigants are registered on multiple crops but with crop or soil-type specific rates; others are registered for specific crops and/or in certain states only. Follow all labels carefully.

Relative Efficacy of Currently Registered Fumigants or Fumigant Combinations for Managing Soilborne Nematodes, Diseases, and Weeds in Plasticulture Strawberries

Table 12C. Relative Efficacy of Currently Registered Fumigants or Fumigant Combinations for Managing Soilborne Nematodes, Diseases, and Weeds in Plasticulture Strawberries[1]

Product	Rate per Treated Acre[2]		Nematodes	Disease	Nutsedge	Weeds: Annual
	Volume (gal)	Weight (lb)				
Telone II (1,3-D)	15-27	153-275	+++++	+	-	-
Telone EC[3]	9-24[4]	91-242[4]	+++++	+	-	-
Telone C17 (1,3-D + chloropicrin)	32.4-42	343-445	+++++	+++	+	+
Telone C35 (1,3-D + chloropicrin)	39–50	437-560	+++++	+++++	+	++
InLine (1,3-D + chloropicrin)[3]	29–57.6 (See Label)	325-645 (See Label)	+++++	+++++	+	+++
Pic-Clor 60 (chloropicrin + 1,3-D)	48.6	588	+++++	+++++	+	+++
Pic-Clor 60 EC[3]	42.6	503	+++++	+++++	+	+++
Metam potassium[5]	30-62	318-657	++	+++	+	++++
Metam sodium[5] (MS)	37.5-75	379-758	++	+++	+	++++
Chloropicrin + MS[5]	19.5–31.5 + 37.5-75	275-444 + 379-758	++	+++++	++	++++
Chloropicrin	48.6	150-350	+	+++++	—	—
Tri-Pic 100EC[3]	8-24	100-300	+	+++++	—	—
Paladin (dimethyl disulphide)[6]	35.0–51.3	310-455	++++	++++	++++	+++[5]
Paladin EC[3,6]	37.0-54.2	326-479	++++	++++	++++	+++[5]
Dominus (allyl isothiocyanate)[7]	25-40[4]	212-340[4]	++	+++	+	+++

[1] Fumigants with lower efficacy against weeds may require a complementary herbicide or hand-weeding program, although use of virtually impermeable film (VIF) or totally impermeable film (TIF) may increase weed control, particularly with Telone C35 or Paladin. Refer to the Herbicide Recommendation section of this guide for directions pertaining to herbicide applications. Telone can persist more than 21 days under cool or wet soil conditions.

[2] Rates can sometimes be reduced if products are applied with VIF or TIF.

[3] Product is formulated for application through drip lines under a plastic mulch; efficacy is dependent on good distribution of the product in the bed profile.

[4] Labelled rates are per *broadcast-equivalent* acre, NOT per treated acre.

[5] Metam potassium can be Metam KLR, K-Pam, Sectagon K54 or other registered formulations, and should be used in soils with high sodium content. Metam sodium can be Vapam, Sectagon 42, Metam CLR or other registered formulations.

[6] Paladin should be applied with 21% chloropicrin and VIF or TIF to enhance disease control, and has low efficacy on certain small seeded broadleaf weeds and grasses. Paladin may not be registered in all States.

[7] Dominus is registered but there is limited experience with the product through University or independent trials in our region; growers may want to consider this on an experimental basis. Planting interval is 10 days. The active ingredient allyl isothiocyanate is similar to the active ingredient in metam sodium products (methyl isothiocyanate) and is likely to behave in a similar manner with a similar pest control profile.

Planting and Early Post-Planting: Disease Control

General Pesticide Information

FRAC/IRAC/HRAC codes — these acronyms refer to industry-sponsored committees addressing resistance to crop protection materials; Fungicide Resistance Action Committee (FRAC), Insecticides Resistance Action Committee (IRAC) and Herbicide Resistance Action Committee (HRAC). Pesticides affect their target pest in a variety of ways, and the way a pesticide kills the target organism is called the *mode of action* **(MOA)**. Although pesticides have different names and may have different active ingredients, they may have the same MOA. Over time, pests can become resistant to a pesticide, and typically this resistance applies to all pesticides with the same MOA. When rotating pesticides, it is important to select pesticides with different MOAs. The FRAC/IRAC/HRAC have grouped crop protection materials into groups with shared MOAs and given them numerical designations, which appear on pesticide labels. The code **UN** means the MOA is unknown. *When selecting pesticides, avoid successive applications of materials in the same MOA group to minimize potential resistance development.* More information about this topic can be found at www.frac.info, www.irac-online.org, andwww.HRACglobal.com.

Organic Materials Review Institute (OMRI; www.omri.org) listed materials are acceptable for production systems certified as organic. Organically acceptable materials (**OMRI** listed) are in the comments section.

Generics: Many pesticide active ingredients are available in generic formulations. For brevity, these formulations are not generally listed. Trade names are listed to aid in identifying products and not intended to promote the use of these products or to discourage use of generic products. Generic products generally work similarly to their brand name counterparts, but formulation changes can impact efficacy and plant response. As with any new chemical, read and follow all label instructions. Chemical names are subject to change; please check the active ingredient for all materials.

Pesticide Environmental Stewardship website is found at http://pesticidestewardship.org/Pages/default.aspx. Information on proper pesticide use and handling, calibration of equipment, reading pesticide labels, disposal, handling spills, and other topics is presented.

Pre-plant dips: Several products are registered for plant dips to manage pathogens or to protect plants just prior to field setting, but only a limited amount of research has been done with plant dips. In general, these treatments are not recommended except under specific

Chapter VI — 2017 N.C. Agricultural Chemicals Manual

circumstances; for example, if a disease has been diagnosed to be on the transplants. Products not labeled for dip treatments should not be used for dips, since poor plant performance has been observed in research trials.

Abound or Azaka — Mix 5 to 8 fl oz/100 gal of water. Dip plants for 2 to 5 minutes. Transplant treated plants as quickly as possible. This treatment has been developed for bare root transplants with a known problem of anthracnose. The dip is a whole plant dip, and some growers do not re-use the water for fear of spreading bacterial angular leaf spot and other diseases. It is reasonable to expect these fungicides to have some Rhizoctonia suppressive activity, but there are no research results to demonstrate a benefit. For managing Rhizoctonia, a root dip should suffice, rather than dipping whole plants. Rhizoctonia (and the black root rot problem) builds up over time, and it is doubtful that a root dip would offer much benefit for season long control. Growers must ensure root dip waste is properly disposed.

Switch — Switch offers options for treating plants known to be infected with Colletotrichum species and has shown good efficacy in reducing losses due to the crown rot pathogen in bare root transplants (Colletotrichum gloeosporioides). Use 5 to 8 fl oz/100 gal water. Wash transplants to remove excess soil prior to dipping. Completely immerse planting stock in dip solution. Dip or expose plants for a minimum of 2 to 5 minutes. Do not reuse solution. Growers must ensure proper disposal of root dip waste. Plant treated plants as quickly as possible. Delayed planting could cause plant stunting.

Phosphites — Dip plants in 2.5 lb/100 gal (Aliette), 2 pints/100 gal (ProPhyt), or 2.5 pints/100 gal (Phostrol) for 15 to 30 minutes and then plant within 24 hours after treatment. This treatment should help to suppress Pythium and Phytophthora problems.

Little data are available for other plant dip products, including **Oxidate**, and it is doubtful that they offer management of root diseases. In most cases, root pathogens are internal to the tissue and are not controlled by surface disinfectants.

Fungicide Resistance Management Recommendations

Botrytis cinerea historically has a high potential to develop resistance, and recent data suggest a high percentage of strains are resistant to several important fungicides. Therefore, it is important to give these recommendations serious consideration:

1. Limit the number of times fungicides of the same class (same FRAC code) are applied in 1 year.

2. Tank-mix a broad spectrum fungicide such as captan or Thiram with Topsin-M (a benzimidazole fungicide) as Topsin-M does no longer have Botrytis activity due to resistance, but is helpful for several early season foliar diseases if present.

3. Resistance profiles vary from farm-to-farm. Sample gray mold populations for their resistance profile through Clemson University (schnabe@clemson.edu)

It is currently suggested that the strobilurin (now called QoI; FRAC code 11) fungicides (Abound, Azaka, Cabrio, Pristine, Merivon, and Quadris Top) be saved for use in controlling anthracnose diseases when there is a high potential for disease pressure. Captan or Thiram should help suppress anthracnose when utilized in Botrytis or other disease control applications, but the QoI materials are currently the most efficacious materials for control of anthracnose. Some of these QoI materials may have activity against multiple pathogens other than the anthracnose pathogens, but unless anthracnose occurs in conjunction with these other diseases of concern, it is suggested that the QoIs not be used. With only 4-5 total applications of the QoI fungicides per crop, it is imperative that they be utilized effectively. Also, resistance management is extremely important with the QoIs; make sure to follow all resistance management guidelines. Recently, we have documented reduced activity with azoxystrobin (Abound, Azaka) with certain strains of the anthracnose fruit rot (AFR) pathogen. Cabrio, Merivon, or Pristine have offered better control of AFR in recent research efforts.

Powdery mildew — Monitor the field for the first signs of powdery mildew (leaf distortion and discoloration). Mildew in the fall does not appear to cause significant damage and may not reappear in the spring. *Therefore, most growers will not need to spray for powdery mildew.* However, fields have been observed in the fall with severe foliar disease incidence, and plant productivity may then be hampered, justifying control measures. Likewise, if powdery mildew pressure occurs in the spring and affects the fruit, the fruit will have a dull appearance and be unmarketable unless managed well. High tunnels favor powdery mildew. Certain fungicides such as the QoI materials and Protocol are registered for powdery mildew, but are not recommended due to resistance selection.

Anthracnose — Most plantings are rarely at risk for anthracnose. Thus, anthracnose fungicides may not be needed. In most cases, contaminated plant sources are identified before or soon after planting. Know your plant source. If present, anthracnose on plants can cause petiole lesions (black sunken areas) stunting and plant death. Fall fungicide applications will be required for *Colletotrichum* only if plant source problems are identified, usually appearing as symptomatic plants or assayed for quiescent infections. Research results show that QoIs are more effective against the fruit rot pathogen ('*acutatum*') compared to the crown rot pathogen ('*gloeosporioides*'). Captan, Topsin M or Switch are as effective as the QoIs for controlling the crown rot pathogen. In general, it is most effective to save the QoI (Group 11) chemistry for spring applications and protect the fruit if anthracnose ('*acutatum*') is known to be present. Failure in management of some '*acutatum*' populations has been observed with Abound or similar azoxystrobin products (see above).

Table 6-13A. Planting and Early Post-Planting: Disease Management

Management Options	Amount of Formulation per Acre	Effectiveness (+) or Importance (*)	REI	PHI	Comments (FRAC/IRAC Code)
Red stele; *Phytophthora* crown/root rots					
mefenoxam (Ridomil Gold SL)	1 pt	++++	12 hr	0 days	Apply in sufficient water in drip applications to move the fungicide into the root zone. Use proportionately less Ridomil Gold for band treatments. **Do not exceed 3 pts/year.** **FRAC–4**
mefenoxam (Ultra Flourish)	2 pt	++++	12 hr	0 days	Apply in sufficient water to move the fungicide into the root zone. Use proportionately less mefenoxam for band treatments. **Do not exceed 6 pts/year. FRAC–4**
metalaxyl (MetaStar and generics)	2 qt/treated A	++++	48 hr	0 days	Apply in sufficient water to move the fungicide into the root zone. **Do not exceed 6 qt/treated A/year. FRAC–4**
phosphites, e.g. Aliette ProPhyt, Phostrol	Various rates; see label	++	12 hr	0 days	Phosphite-based chemicals are not as effective as Ridomil Gold. Consider phosphites if the pathogen is known to be resistant to mefenoxam or if root systems are poor AND foliage is healthy for chemical uptake. **FRAC–33**
***Rhizoctonia* sp.(seedling root; basal stem rot)**					
Abound FL, Azaka	0.40 to 0.80 fl oz/1,000 row feet	++	4 hr	0 days	This is a drip irrigation application method. Can be considered especially for plug plants with poor root systems or plants placed into non-fumigated beds or beds with excess water in heavy soils. **FRAC–11**
Powdery mildew only					

PM is not a common problem at this time of year; PM may come on transplants but usually does not persist nor does it present an economic problem in open fields. There is more risk of PM in high tunnels.

Management Options	Amount of Formulation per Acre	Effectiveness (+) or Importance (*)	REI	PHI	Comments (FRAC/IRAC Code)
Procure 50WS Procure 480SC	4 to 8 oz 4 to 8 fl oz	+++++	12 hr	1 day	Check label for prohibited rotational crops. Do not plant leafy or fruiting vegetables within 30 days after application. Do not plant bulb or root vegetables within 60 days after application. Do not plant cotton, small cereal grains and all other crops not registered within one year application. **FRAC–3**
Rally 40WSP	2.5 to 5 oz	+++++	24 hr	0 days	Rally is registered for control of leaf spot, leaf blight, and powdery mildew. Do not apply more than 30 oz per year. **FRAC–3**
Sulfur (multiple formulations)	5 to 10 lb	+++	24 hr	1 day	Spray as needed. Avoid using in middle of a hot sunny day that may cause leaf burning. See label. **FRAC–M2**
Quintec	4 to 6 fl oz	+++++	24 hr	1 day	Do not use more than 4 times per crop and no more than 2 times in a row. Rotate with other mildewcides. For food and feed crops not registered on the label, the re-plant interval is 30 days after application. See label. **FRAC–13**
Protocol	1.33 pt	+++	24 hr	1 day	Premix of two active ingredients, thiophanate-methyl (**FRAC–1**) and propiconazole (**FRAC–3**). No more than 2 sequential applications should be made before alternating with fungicides that have a different mode of action.
Anthracnose ('acutatum')					
Pristine WG	18.5 to 23 oz	+++++	12 hr	0 days	Premix of two active ingredients, pyraclostrobin (**FRAC–11**) and boscalid (**FRAC–7**). See resistance management notes above.
Merivon	5.5 to 8 fl oz	+++++	12 hr	0 days	Premix of two active ingredients, pyraclostrobin (**FRAC–11**) and fluxapyroxad (**FRAC–7**). See resistance management notes on page 20.
Cabrio 20EG	12 to 14 oz	+++++	24 hr	0 days	Active ingredient, Pyraclostrobin (**FRAC-11**)
Abound FL, Azaka	6.2 to 15.4 fl oz	+++++	4 hr	0 days	Failure in management of some 'acutatum' populations has been observed with Abound and similar products. **FRAC–11**
Tilt and multiple generics	4 fl oz	+++	12 hr	0 days	No more than 2 sequential applications should be made before alternating with fungicides that have a different mode of action. **FRAC–3**
Quadris Top	12 to 14 fl oz	+++	12 hr	0 days	Premix of two active ingredients, azoxystrobin (**FRAC–11**) and difenoconazole (**FRAC–3**). No more than 2 sequential applications should be made before alternating with fungicides that have a different mode of action.
Protocol	1.33 pt	+++	24 hr	1 day	Premix of two active ingredients, thiophanate-methyl (**FRAC–1**) and propiconazole (**FRAC–3**). No more than 2 sequential applications should be made before alternating with fungicides that have a different mode of action.
Anthracnose ('gloeosporioides' crown rot)					
Captan 50W	3 to 6 lb (50W)	++	24 hr	1 day	In plantings known to be infected with the anthracnose crown rot pathogen, consider applying captan plus Topsin-M at 10- to 14-day intervals, for a total of 2 to 3 applications in the fall. **FRAC–M4**
Captan 80WDG	1.9-3.8 lb (80W)				
Captec 4L	2.5 qt	++	24 hr	1 day	**FRAC–M4**
Topsin-M 70WP	1 lb	++	12 hr	1 day	See note (page 20) on resistance management. **FRAC–1**
Quadris Top	12 to 14 fl oz	+++	12 hr	0 days	Same as above. **FRAC–3, FRAC–11**

New Leaf Growth to Pre-Bloom: Disease Management

Table 6-13B. New Leaf Growth to Pre-bloom: Disease Management

Management Options	Amount of Formulation per Acre	Effectiveness (+) or Importance (*)	REI	PHI	Comments (FRAC/IRAC Code)
Botrytis crown rot					
Botrytis crown rot may occur during warm winter periods after early bloom is frost killed and colonized by *Botrytis*. The pathogen typically grows down the flower stem (peduncle) and colonizes the upper crown tissue, causing death of the leaf petioles, particularly if plants are large or planted densely.					
Rovral 4F and generics (iprodione)	1.5 to 2 pt	++++	24 hr	NA	Do not apply after first fruiting flower, and do not make more than 1 application of Rovral per season. Crown rot control during the early winter and prior to bloom may be the most effective use of the one Rovral application allowed in strawberries. **FRAC–2**
Switch 62.5 WG	11 to 14 oz	++++	12 hr	0 days	See resistance management information on page 8. **FRAC–12, FRAC–9**
Captan 50W	3 to 6 lb (50W)	++	24 hr	1 day	See notes below. **FRAC–M4**
Captan 80WDG	1.9 - 3.8 lb (80WDG)				
Botrytis					
Remove dead and dying leaves just before bloom		** +++			Symptomatic leaf removal is effective but may not be economical if fungicides are heavily used for Botrytis management. If anthracnose fruit rot is present, hand-pruning plants may create more anthracnose disease problems. Do not use QoI products- these should be saved for use as fruit develop and to avoid selection of resistant populations.

Leaf spots, Leaf blights and Powdery Mildew generally do not become economically important diseases in the fall or early spring. Thus, fungicides are generally not required for these problems. Thresholds have not been established, so the need for fungicides should be determined on a farm-by-farm basis depending on the disease pressure present. Phomopsis and leaf spot may be associated with plant sources; therefore, disease incidence can vary from year to year. Warm wet weather favors disease progress. See previous notes on powdery mildew under "Planting and Early Post-planting: Disease Management" (page 8). In the spring, monitor fields closely observing the underside of strawberry leaves to determine if powdery mildew is present.

Management Options	Amount of Formulation per Acre	Effectiveness (+) or Importance (*)	REI	PHI	Comments (FRAC/IRAC Code)
Phomopsis leaf blight					
Captan 50W Captan 80WDG	3 to 6 lb (50W) 1.9 - 3.8 lb (80WDG)	++ ++	24 hr 24 hr	1 day 1 day	When foliar symptoms appear, make 1 or 2 captan applications plus Topsin-M at a 10- to 14-day interval for better control than captan products alone would provide. Do not apply more than 24 lb captan active ingredient per acre per year. **FRAC–M4**
Captec 4L	2.5 qt	++	24 hr	1 day	
Topsin-M 70WP	1 lb	++	12 hr	1 day	See note above on page 8 resistance management. **FRAC–1**
Rally 40WSP	2.5 to 5 oz	++++	24 hr	0 days	Rally is registered for control of leaf spot, leaf blight, and powdery mildew. Do not apply more than 30 oz per acre. **FRAC 3**
Common leaf spot, leaf scorch, leaf blight (e.g. Mycosphaerella, Phomopsis, Gnomonia)					
Captan 50W or Captan 80 WDG plus Topsin-M 70WP	1 lb (50W); 1.6 lb (80WDG) 1 lb	+++	24 hr 12 hr	1 day 1 day	When foliar symptoms appear, make 1 or 2 captan applications plus Topsin-M at a 10- to 14-day interval for better control than captan products alone would provide. Do not apply more than 24 lb captan active ingredient per acre per year. Do not tank mix captan products with highly alkaline pesticides, such as Bordeaux mixture. See resistance management notes on page 8. **FRAC–M4, FRAC–1**
Captan 50W (alone) Captan 80 WDG (alone)	1.5 lb (50W) 2.4 lb (80WDG)	++	24 hr	1 day	**FRAC–M4**
Thiram 24/7 Thiram Granuflo	2.6 qt (24/7) 4.4. lb	++	24 hr 24 hr	1 day 3 days	**FRAC–M3**
Rally 40WSP	2.5 to 5 oz	++++	24 hr	0 days	Rally is registered for control of leaf spot, leaf blight, and powdery mildew. Do not apply more than 30 oz per year. **FRAC–3**
Powdery mildew only					
Procure 480SC	4 to 8 fl oz	+++++	12 hr	1 day	Check label for prohibited rotational crops. Do not plant leafy or fruiting vegetables within 30 days after application. Do not plant bulb or root vegetables within 60 days after application. Do not plant cotton, small cereal grains and all other crops not registered within one year application. **FRAC–3**
Rally 40WSP	2.5 to 5 oz	+++++	24 hr	0 days	Rally is registered for control of leaf spot, leaf blight, and powdery mildew. Do not apply more than 30 oz per year. **FRAC–3**
Quintec	4 to 6 fl oz	+++++	24 hr	1 day	Do not use more than 4 times per crop and no more than 2 times in a row. Rotate with other mildewcides. Rotation to non-registered crops less than 30 days after application is prohibited. **FRAC–3**
Torino	3.4 oz	?	4 hr	0 days	Do not make more than two (2) applications per year. Do not apply more than once every 14 days. **FRAC–U6**
Tilt and other generics	4 fl oz	+++	12 hr	0 days	No more than 2 sequential applications should be made before alternating with fungicides that have a different mode of action. **FRAC–3**

Table 6-13B. New Leaf Growth to Pre-bloom: Disease Management

Management Options	Amount of Formulation per Acre	Effectiveness (+) or Importance (*)	REI	PHI	Comments (FRAC/IRAC Code)
Angular (bacterial) leaf spot (*Xanthomonas fragariae*)					
Basic copper sulfate (various formulations)	2 to 3 lb/100 gal	+	48 hr	0 hr	Angular (bacterial) leaf spot can be a serious problem during cool, wet conditions. These compounds provide some control unless conditions highly favor disease. Repeat applications at 7 to 10 day intervals. Discontinue when phytotoxicity appears, usually after 4 to 5 applications.
copper hydroxide (various formulations)	0.35 to 0.58 a.i. (various formulations)	+	24 hr	0 days	
copper salts of fatty and rosin acids (various formulations)	3 – 4 pts (various formulations)	+	12 hr	0 days	NOTE: All copper sulfate, copper hydroxide and other copper products labeled for strawberry can be used, but check label for the proper rate because different products will contain different percents of active ingredient. **FRAC–M1. FRAC–M1.**
cuprous oxide (various formulations)	1.05 – 4.2 lbs a.i. (various formulations)		12 hr	0 days	
Actigard 50WG	0.5 to 0.75 oz./a	+		0 days	Labeled for suppression; Do not apply to stressed plants. **DO NOT EXCEED MAXIMUM RATE.** Actigard is a plant activator and has no direct activity on the bacteria. See supplemental label for details. **FRAC–21**
Red stele; Phytophthora crown/root rots					
mefenoxam (Ridomil Gold SL and other formulations)	1 pt	++++	12 hr	0 days	Strawberry plants initiate considerable root growth in the early spring. Time control applications in problem fields when new growth begins in the spring. Apply in sufficient water to move the fungicide into the root zone. Use proportionately less fungicide for band treatments (e.g., for drip applications). **FRAC–4**
Ultra Flourish	2 pt				
metalaxyl (MetaStar and generics)	2 qt/treated A	++++	48 hr	0 days	
phosphites (e.g., Aliette, ProPhyt, Phostrol)	Various rates; see label	++	12 hr	0 days	The phosphite-based chemicals are not as effective as Ridomil Gold. Consider phosphites if the pathogen is known to be resistant to mefenoxam or if strawberry plants have poor root systems but sufficient foliage for chemical uptake. **FRAC–33**

Early Bloom (10%) and into Harvest: Disease Management

F. J. Louws, Plant Pathology Extension

The primary diseases of concern at early bloom and into harvest are **Botrytis fruit rot** and **anthracnose fruit rot (AFR)**. Most growers rarely experience anthracnose problems and may not need an anthracnose management program. Several **key principles** should be kept in mind:

1. Abound, Azaka, Cabrio, Merivon, Pristine, Luna Sensation belong to the same family of chemicals (QoI; Group 11 chemistry). Pyraclostrobin (Cabrio, Merivon, and Pristine) has offered better control of AFR in recent research efforts. No more than 2 applications of a Group 11 fungicide should be made per season for resistance management. Strategic timing is necessary. Pristine, Luna Sensation and Merivon also have a second chemical that has good broad spectrum activity against a number of diseases, especially those caused by Botrytis.

2. Captan, Thiram, and Switch offer a broad spectrum of disease control. Switch has not performed well against AFR in NC research.

3. Polyoxin D (PhD; OSO 5%SC; Tavano 5%SC) is as effective as captan for Botrytis at high label rates and can help reduce the number the number of captan sprays. Consider substituting Polyoxin D up to 2 times for captan or thiram. Polyoxin D has low activity against AFR.

4. Elevate should not be used more than twice per season due to resistance concerns. It is effective against Botrytis but no other fungal pathogens.

5. High risk fungicides of the same chemical class should not be applied in consecutive applications.

6. CaptEvate is a premix of captan and Elevate which has good broad-spectrum activity.

7. Bloom sprays are the most important for managing Botrytis, because 90% of fruit infection occurs through the flower at bloom. Recent research suggests bloom sprays are also critical for anthracnose fruit rot control.

8. Fruit rot diseases develop rapidly during wet periods or in poorly ventilated locations. Control is easier when initiated before the problem develops. Spray coverage is important and dependent on nozzle condition, tractor speed, pressure, and plant density. Spray coverage can be checked with water sensitive cards.

9. Botrytis has acquired resistance to several fungicides. Tests can be secured through Clemson University to help determine farm-specific recommendations. In the absence of such tests, growers should rely primarily on captan for gray mold control. For instructions on sampling contact Dr. Schnabel at schnabe@clemson.edu.

For growers who have a received a resistance profile report, follow the recommendations in that report.

For growers who do not have a report and who adopt a conservative (low risk) fungicide program, apply sprays every 7 to 10 days according to ONE of the following suggested schedules.

SCHEDULE 1. For cases when there is no risk of anthracnose and growers need to focus on gray mold control (most fields):
Rotate two or more of the following: polyoxin D; thiram; captan; CaptEvate; Switch; captan + Fontelis, But only thiram or captan can be applied more than twice per season due to resistance management.

Options: For a reduced fungicide program, initiate applications at FIRST bloom as above but apply subsequent sprays before predicted wet weather that favors Botrytis; end applications about 26 to 30 days before expected final harvests. Increase the time between spray applications when dry weather persists. Research trials have documented that 4 sprays during bloom are sufficient to offer season-long Botrytis fruit rot control. Also, consult available forecasting models (see above).

SCHEDULE 2. For cases where anthracnose fruit rot risk is high and gray mold control is also needed:
Application 1: At FIRST bloom apply captan or thiram tank mixed with Cabrio.
Application 2: Apply ONE of these alternatives: CaptEvate OR captan.
Application 3: Same as Application 1.
Application 4 and weekly: Rotate two or more of the following: captan; captan+ Cabrio.

In other words, there should be continuous coverage with Captan, a FRAC 11 or FRAC 3 products, or the combination. Follow **key principle 1** above. During periods of cool wet weather and during bloom, incorporate Switch for better Botrytis control. Pristine, Merivon, or Cabrio show the best efficacy under high anthracnose (AFR) pressure in research studies.

Table 6-14. Early Bloom (10%) and into Harvest: Disease Management

Management Options	Amount of Formulation per Acre	Effectiveness (+) or Importance (*)	REI	PHI	Comments (FRAC/IRAC Code)
Botrytis gray mold					
Captan 50W Captan 80WDG	3 to 6 lb (50W) or 1.9-3.8 lb (80W)	+++	24 hr	1 day	See suggested schedule above on page 20. Do not apply more than 24 lb of captan active ingredient per acre per year. **FRAC–M4**
Captec 4L	2.5 qt	+++	24 hr	1 day	
Switch 62.5WG	11 to 14 oz	+++++	12 hr	0 days	Do not apply more than twice per season due to resistance management. See resistance management notes on page 8 and 20. **FRAC–12, FRAC–9**
Ph-D WDG	6.2 oz	+++	4 hr	0 days	Do not apply more than twice per season due to resistance management. **FRAC–19**
OSO 5%SC Tavano 5%SC	6.5 – 13 fl oz	+++	4 hr	0 days	Do not apply more than twice per season due to resistance management. **FRAC–19**
Thiram 75 WDG Thiram 24/7	4.4 lb (WDG) 2.6 qt (24/7)	+++	24 hr	3 days	Make 3 to 5 applications at 10-day intervals. Thiram is a broad spectrum fungicide similar to captan. **FRAC–M3**
Elevate 50WDG	1.5 lb	+++++	4 hr	0 days	Do not apply more than twice per season due to resistance management. Under light pressure, 1.0 lb plus captan may be used (see label). **FRAC–17**
Fontelis	16 to 24 fl oz	+++++	12 hr	0 days	Do not apply FRAC-7 products more than twice per season due to resistance management. Some matted row cultivars may show phytotoxicity (see label). **FRAC-7**
CaptEvate 68 WDG	3.5 to 5.25 lb	+++++	24 hr	0 days	CaptEvate is a combination product of captan plus Elevate. Do not make more than 2 consecutive applications before switching to a fungicide with a different mode of action. Do not apply more than 21.0 lb/acre/season. With plastic mulch, do not apply within 16 feet of naturally vegetated or aquatic areas. **FRAC–M4, FRAC–17**
Scala	18 fl oz 9 fl oz	+++	12 hr	1 day	Use lower rate only in a tank mix with another fungicide active against gray mold (e.g. captan or Thiram). **FRAC–12**
Luna Tranquility	16-27 fl oz	+++++	12 hr	0 day	Do not use any FRAC 11 or 7 products more than twice per season for resistance management. **FRAC–11, FRAC–7**
Luna Sensation	6-7.6 fl oz	+++++	12 hr	0 day	Do not use any FRAC 11 or 7 products more than twice per season for resistance management. **FRAC–11, FRAC–7**
Fracture	24.4 – 36.6 fl oz	+	4 hr	1 day	Active ingredient is a protein extract of sweet white Lupin seeds. Some efficacy can be expected at the highest rate.
Botrytis and Anthracnose (acutatum)					
Products in this section are labeled for both botrytis and anthracnose.					
Pristine WG	18.5 to 23 oz	+++++	12 hr	0 days	Do not apply more than 2 applications per acre per crop year. See page 20. **FRAC–11, FRAC–7**
Luna Sensation	6-7.6 fl oz	+++++	12 hr	0 day	Do not use any FRAC 11 or 7 products more than twice per season for resistance management. **FRAC–11, FRAC–7**
Merivon	8 to 11 fl oz	+++++	12 hr	0 days	Do not apply more than 2 applications per acre per crop year. See page 20. **FRAC–11, FRAC–7**
Captan 50W Captan 80 WDG	3 to 6 lb (50W) 1.9-3.8 lb (80WDG)	+++	24 hr	1 day	For better control and resistance management, use captan applications plus Topsin-M (see label). See suggested schedule above. Do not apply more than 24 lb of captan active ingredient per acre per year. **FRAC–M4**

Table 6-14. Early Bloom (10%) and into Harvest: Disease Management

Management Options	Amount of Formulation per Acre	Effectiveness (+) or Importance (*)	REI	PHI	Comments (FRAC/IRAC Code)
Anthracnose (acutatum)					
Abound 2.08 F Azaka	6.2 to 15.4 fl oz	+++ (failure found in some fields)	4 hr	4 hr	See notes on page 20 to manage risk of developing fungicide resistance. In recent research, Abound and similar products has performed less well than Cabrio/Pristine. **FRAC–11**
Luna Sensation	6-7.6 fl oz	+++++	12 hr	0 day	Do not use any FRAC 11 or 7 products more than twice per season for resistance management. **FRAC–11, FRAC–7**
Merivon	5.5 to 8 fl oz	++++	12 hr	0 days	See notes on page 20 to manage risk of developing fungicide resistance. **FRAC–11, FRAC–7**
Pristine WG	18.5 to 23 oz	++++	12 hr	0 days	See notes on page 20 to manage risk of developing fungicide resistance. **FRAC–11, FRAC–7**
Cabrio EG	12 to 14 oz	++++	12 hr	0 days	See notes on page 20 to manage risk of developing fungicide resistance. **FRAC–11**
Tilt and multiple generics	4 fl oz	++?	12 hr	0 days	Registered for Anthracnose Fruit Rot only. No more than 2 sequential applications should be made before alternating with fungicides that have a different mode of action. Not registered for Anthracnose crown rot control. **FRAC–3**
Quadris Top	12 to 14 fl oz	+++	12 hr	0 days	Premix of two active ingredients, azoxystrobin (**FRAC–11**) and difenoconazole (**FRAC–3**). No more than 2 applications should be made per season for resistance management.
Protocol	1.33 pt	+++	24 hr	1 day	Premix of two active ingredients, thiophanate-methyl (**FRAC–1**) and propiconazole (**FRAC–3**). No more than 2 applications should be made per season for resistance management.
Anthracnose ('gloeosporioides' crown rot)					
Captan 50W	3 to 6 lb (50W)	++	24 hr	1 day	In plantings known to be infected with the anthracnose crown rot pathogen, consider applying captan plus Topsin-M at 10- to 14-day intervals, for a total of 2 to 3 applications in the fall. **FRAC–M4**
Captan 80WDG	1.9-3.8 lb (80W)				
Captec 4L	2.5 qt	++	24 hr	1 day	**FRAC–M4**
Topsin-M 70WP	1 lb	++	12 hr	1 day	See note (page 20) on resistance management. **FRAC–1**
Quadris Top	12 to 14 fl oz	+++	12 hr	0 days	Same as above. **FRAC–3, FRAC–11**
Powdery mildew (only)					
Procure 50WS Procure 480SC	4 to 8 oz 4 to 8 fl oz	+++++	12 hr	1 day	Check label for prohibited rotational crops. Do not plant leafy or fruiting vegetables within 30 days after application. Do not plant bulb or root vegetables within 60 days after application. Do not plant cotton, small cereal grains and all other crops not registered within one year application. **FRAC–3**
Rally 40WSP	2.5 to 5 oz	+++++	24 hr	0 days	Rally is registered for control of leaf spot, leaf blight, and powdery mildew. Do not apply more than 30 oz per year. **FRAC–3**
Quintec	4 to 6 fl oz	+++++	24 hr	1 day	Do not use more than 4 times per crop and no more than 2 times in a row. Rotate with other mildewcides. Rotation to all other crops within 1 year after application, unless Quintec is registered for use on those crops, is prohibited. **FRAC–13**
Torino	3.4 oz	?	4 hr	0 days	Do not make more than two (2) applications per year. Do not apply more than once every 14 days. **FRAC–U6**
Powdery mildew and Anthracnose (acutatum)					
Abound 2.08 F Azaka	6.2 to 15.4 fl oz	++++	4 hr	4 hr	See notes on page 20 to manage risk of developing fungicide resistance. **FRAC–11**
Pristine WG	18.5 to 23 oz	++++	12 hr	0 days	See notes on page 20 to manage risk of developing fungicide resistance. **FRAC –11, FRAC–7**
Luna Sensation	6-7.6 fl oz	+++++	12 hr	0 day	Do not use any FRAC 11 or 7 products more than twice per season for resistance management. **FRAC–11, FRAC–7**
Cabrio EG	12 to 14 oz	++++	12 hr	0 days	See notes on page 20 to manage risk of developing fungicide resistance. DO NOT EXCEED 1.5 QT/YEAR. **FRAC–11**
Tilt and multiple generics	4 fl oz	+++	12 hr	0 days	Registered for Anthracnose Fruit Rot only. No more than 2 sequential applications should be made before alternating with fungicides that have a different mode of action. Not registered for Anthracnose crown rot control. **FRAC–3**
Quadris Top	12 to 14 fl oz	+++	12 hr	0 days	Premix of two active ingredients, azoxystrobin (**FRAC–11**) and difenoconazole (**FRAC–3**). No more than 2 applications should be made per season for resistance management.

Relative Effectiveness of Various Chemicals for Strawberry Disease Control

F. J. Louws, Plant Pathology Extension

(— = ineffective; +++ = very effective; ? = efficacy unknown)

Table 6-15. Relative Effectiveness of Various Chemicals for Strawberry Disease Control

Pesticide	Anthracnose (crown rot)	Anthracnose (fruit rot)	Gray mold	Powdery mildew	Common leaf spot	Leaf blight and fruit rot	Leather rot	Mucor fruit rot	Rhizopus rot	Angular leaf spot	Phytophthora crown rot	Red stele root rot
Strobilurins: azoxystrobin (Abound; Azaka)	++	+	+	+	+	—	+++	—	—	—	—	—
pyraclostrobin (Cabrio)	++	+++	+	+	+	—	+++	—	—	—	—	—
pyraclostrobin + boscalid (Pristine) pyraclostrobin + fluxapyroxad (Merivon) trifloxystrobin + fluopyram (Luna Sensation)	++	+++	+++^{R}	+	+++	+++	—	?	?	—	—	—
acibenzolar-S-methyl (Actigard)	—	—	—	—	—	—	—	—	—	+	—	—
azoxystrobin + difenoconazole (Quadris Top)	++	++	+	++	++	?	+	—	—	—	—	—
azoxystrobin + propiconazole (QuiltXcel)	+++	++	—	++	?	?	—	—	—	—	—	—
fluopyram + pyrimethanil (Luna Tranquility)	—	—	+++^{R}	++	—	—	—	—	—	—	—	—
BLAD (Fracture)	?	?	+	+	?	?	—	?	?	—	—	—
captan (Captan and generics)	++	++	++	—	++	+	+	+	+	—	—	—
copper	—	—	—	—	+^{P}	—	+^{P}	—	—	+^{P}	—	—
cyprodinil + fludioxinil (Switch)	++	+	+++^{R}	?	+?	+?	—	?	?	?	—	—
fenhexamide (Elevate)	—	—	+++^{R}	—	—	—	—	—	—	—	—	—
fenhexamide + captan (CaptEvate)	+	++	+++	—	++	+	+	+	+	—	—	—
fosetyl-Al (Aliette)	—	—	—	—	—	—	++	—	—	—	++	++
iprodione (Rovral and generics)	—	—	+++^{R}	—	++	—	—	X	—	—	—	—
mefenoxam (Ridomil) or similar products	—	—	—	—	—	—	—	+++^{R}	—	—	+++	+++
metalaxyl (MetaStar) or similar products	—	—	—	—	—	—	—	+++^{R}	—	—	+++	+++
myclobutanil (Rally)	—	—	—	+++^{R}	++^{R}	++^{R}	—	—	—	—	—	—
penthiopyrad (Fontelis)	—	—	+++^{R}	++^{R}	—	—	—	—	—	—	—	—
phosphites (ProPhyt; Phostrol and others)	—	—	—	—	—	—	++	—	—	—	++	++
Polyoxin D (Ph-D; OSO; Tavano)	?	?	++	—	?	?	?	?	?	?	?	?
propiconazole (Tilt; other generics)	+	++	-	++^{R}	++?	?	-	-	-	-	-	-
pyrimethanil (Scala)	—	—	++^{R}	—	—	—	—	—	—	—	—	—
quinoxyfen (Quintec)	—	—	—	+++	—	—	—	—	—	—	—	—
sulfur	—	—	—	++^{P}	—	—	—	—	—	—	—	—
thiophanate-methyl (Topsin M)	++^{R}	—	++^{R}	+^{R}	++	++	—	X	—	—	—	—
thiophanate-methyl + propiconazole (Protocol)	++^{R}	++	++^{R}	++^{R}	++	++	—	X	—	—	—	—
thiram (Thiram)	++	++	++	—	+	+	+	+	+	—	—	—
triflumizole (Procure)	—	—	?	+++^{R}	?	?	—	—	—	—	—	—

^{R} = Not effective if pathogen is resistant to the fungicide.

^{P} = Phytotoxicity could occur.

X = Chemical use increases problem

Strawberry Insect Management

H. J. Burrack, Entomology Extension

Examine strawberry plants for insects and mites prior to and following transplant. Consider treating if damaging populations of early season pests (cutworms and spider mites) are present. Initiate a weekly insect sampling program in early spring, prior to flowering. Base insecticide treatments on comparison of field counts to treatment thresholds, when available.

Insecticide Resistance Action Committee (IRAC) mode of action (MOA) groupings are listed following insecticide names. Materials in the same IRAC grouping have the same mode of action. When selecting insecticides, avoid successive applications of materials in the same IRAC group to minimize potential resistance development. Organically acceptable materials (**OMRI** listed) are noted under Precautions and Remarks.

Many insecticide active ingredients are available in generic formulations. Generic products generally work similarly to their brand name counterparts, but formulation changes can impact efficacy and plant response. In general, information is provided for the most commonly used formulations of active ingredients available in multiple forumations. Carefully check the label of the product you plan to use in the event that it differs from those listed. The label is the law!

Table 6-16. Strawberry Insect Control

Season Pest	Insecticide, Formulation, and IRAC Group	Amount of Formulation per Acre	Reentry Interval (hours)	Pre harvest interval (days)	Precautions and Remarks
Post Transplant					
Cyclamen mite	Cyclamen mites are rare in North Carolina strawberries and are typically introduced on infested plants. Insepect plants closely upon receipt and post transplant.				
	fenpyroximate IRAC 21 (Portal)	2 pt	12	1	
	imidacloprid, IRAC 4A (Admire Pro)	10.3 to 14 fl oz (soil)	12	14	Admire Pro can be applied to the soil in transplant water or through drip irrigation or as a foliar treatment. Foliar treatments are not recommended for use against cyclamen mites. Apply in transplant water or through irrigation. Do not apply when bees are foraging or within 10 days of bloom.
Crickets	Cricket feeding on foliage rarely requires treatment, but crickets may occationally damage fruit when grown in high tunnels during the winter and early spring.				
	carbaryl, IRAC 1A (Sevin 4 XLR)	1 to 2 qt	12	7	Many formulations of carbaryl are available.
	fenproparthrin, IRAC 3A (Danitol 2.4 EC)	10.67 to 21.33 fl oz	24	2	Do not make more than 2 total applications. Apply in at least 100 gallons of water per acre.
	malathion, IRAC 1B (Malathion) 57 EC	1.5 to 3 pt	12	3	
Cutworm	Small cutworms feed on leaves before damaging crowns. If cutworms are suspected but caterpillars are not observed, check plants in evening because larvae are nocturnal.				
	carbaryl, IRAC 1A (Sevin 4 XLR)	1 to 2 qt	12	7	Many formulations of carbaryl are available. Foliar applications for carbaryl can flare spider mites. Apply late in afternoon when plants clipped at the base are first noticed.
	fenproparthrin, IRAC 3A (Danitol)	10.67 to 21.33 fl oz	24	2	Do not make more than 2 total applications.
	methoxyfenozide, IRAC 18 (Intrepid)	6 to 12 fl oz	4	3	
	spinosad, IRAC 5 (Entrust SC)	4 to 8 fl oz	4	1	Rotate to a different class of insect control products after 2 successive applications of spinosad. Do not apply more than 18 fl oz Entrust SC (0.28 lb spinosad) per acre per crop. Entrust is **OMRI** listed.
	malathion, IRAC 1B (Malation 8 Flowable)	1.5 to 2 pt	12	3	Malathion 8 Flowable can be applied via drip lines, allowing treatment under plastic.
	Bacillus thuringiensis (Bt), IRAC 11B2 (Dipel DF)	0.5 to 1.0 lb	4	0	Dipel DF is **OMRI** listed.
Slugs and snails	metaldehyde (Deadline Bullets)	0.4 to 1.6 lb	NA	See label	Repeated applications may be necessary.
	iron phosphate (Sluggo)	10 to 44 lb	NA	1	Repeated applications may be necessary. Soil should be moist with no standing water when product is applied. Sluggo is **OMRI** listed.
Preharvest					
Red imported fire ants	Baits treatments will control entire mounts, but treatments take between twoand four weeks to be fully effective. Treat active mounds off season or before picking begins with directed bait formulations. Ensure that ants are actively foraging before applying baits. If mounds develop during harvest, drench treatments may reducue activity temporarily. Consult your cooperative extension agent for mountd drench recommendations.				
	pyriproxyfen, IRAC 7C (Esteem Ant Bait 0.5% B)	1.5 to 2 lb	12	1	Do not water for 24 hours after application.
	methoprene, IRAC 7C (Extinguish Ant Bait 0.5% B)	1 to 1.5 lb	4	0	Extinguish can be applied as a mound treatment or broadcast. Extinguish is labeled for use on cropland, but Extinguish Plus is **not** labeled for use on cropland. Read labels carefully.
Aphids	Aphids are typically infrequent pests in strawberries. If aphids are present preflowing in numbers greater than 10 per newlty expanded leaf, they should be managed before bloom. Harvest period populations are often controlled by natural enemies. Aphids typically only warrant preventative treatment, via soil applied insecticides, in nursery production to prevent virus transmission.				
	imidacloprid, IRAC 4A (Admire Pro, soil) (Admire Pro, foliar)	10.5 to 14 fl oz / 1.3 fl oz	12 / 12	14 / 7	Admire Pro can be used as a soil or foliar treatment. Soil applications should be made through the irrigation system. **Do not** make imidacloprid applications when bees are foraging or within 10 days of bloom.
	insecticidal soap (M-pede)	2.5 fl oz	12	0	Rate is per 100 gallons of water. Test for phytotoxicity effects on a limited area before widespread use. **OMRI** listed.
	malathion, IRAC 1B (Malathion) 57 EC	1.5 pt	12	3	Only 4 applications can be made per year.
	thiamethoxam, IRAC 4A (Actara)	1.5 to 3 oz	12	3	Do not apply material immediately prior to bud opening, during bloom, or when bees are foraging.

Table 6-16. Strawberry Insect Control

Season Pest	Insecticide, Formulation, and IRAC Group	Amount of Formulation per Acre	Reentry Interval (hours)	Pre harvest interval (days)	Precautions and Remarks
Preharvest (continued)					
Strawberry weevil (clipper)	Preventative treatments for strawberry clipper are not recommended. **Materials effective against strawberry clipper are also toxic to bees.** Following pollinator protection language on pesticide labels carefully.				
	bifenthrin, IRAC 3A (Brigade WSB)	6.4 to 32 oz	12	0	Do not apply when bees are foraging. Do not apply more than 80 ounces of product per acre per year.
	carbaryl, IRAC 1A (Sevin 4 XLR)	1 to 2 qt	12	1	Do not apply when bees are foraging, but Sevin XLR is relatively less bee toxic compared to other carbaryl formulations when dry.
	fenpropathrin, IRAC 3A (Danitol 2.4EC)	16 to 21.66 fl oz	24	2	Do not apply when bees are foraging.
Two-spotted spider mite	Coverage is important for spider mite management. Materials should generally be used at the high label rate, in high volumes of water (200 gallons per acre recommended), and applied using high pressure or electrostatic equipment. Mites should be treated if they exceed 5 per leaflet prior to harvest.				
	abamectin, IRAC 6 (Agri-Mek 0.15 EC)	16 oz	12	3	Make 2 applications 7 to 10 days apart when mites first appear. Do not exceed 64 fluid ounces per acre in a growing season. Do not apply in less than 100 gallons of water per acre. Do not repeat treatment within 21days of second application. Do not use in strawberry nurseries.
	acequinocyl, IRAC 20B (Kanemite 15 SC)	31 oz	12	1	Allow 21 days between treatments. Do not make more than 2 applications per season.
	bifenazate, IRAC Unknown (Acarmite 50WP)	1 lb	12	1	Use only 2 applications per year. Use in a minimum of 100 gallons per acre.
	etoxazole, IRAC 10B (Zeal)	3 oz	12	1	Zeal is an ovicide/larvicide and should be applied early in the mite life cycle.
	fenpyroximate IRAC 21 (Portal)	2 pt	12	1	
	hexythiazox, IRAC 10A (Savey 50 WP)	7 oz	12	3	One application per season. Will control eggs and suppress small mites. Do not use in nurseries.
	mineral/petroleum oils (Organic JMS Stylet Oil)	0.75% by volume	4	4 hours	There are numerous oils registered in strawberries. Oils are effective only if very good coverage is achieved. Oils should not be applied 48 hours or less before freezing temperature, at temperatures over 90 degrees F, or to water-stressed plants. Because oils lack the residual activity of conventional acaricides, they may need to be applied repeatedly to control mites. Organic JMS Stylet Oil is **OMRI** listed.
	predatory mites (*Phytoseiulus persimilis* and others)	30,000 to 60,000	NA	NA	Release 2 to 3 mites per plant when mite populations are low. Predatory mite releases must be initiated at or before Twospotted spider mites reach threshold levels (2 to 5 mites per leaflet), and spider mite populations must be followed closely after predatory mite releases. Consult commercial insectaries for predatory mite release rate and species recommendations. Other predatory mite species may also provide good control of twospotted spider mites in NC strawberries.
	rosemary & peppermint oils (Ecotec) (Ecotrol)	32 to 64% by volume	0	0	Because oils lack the residual activity of conventional acaricides, they may need to be applied repeatedly for control. Ecotec and Ecotrol are **OMRI** listed.
	spiromesifen, IRAC 23 (Oberon 2SC)	16 fl oz	12	3	Do not apply more than 48 fluid ounces or make more than three applications per season.
	sucrose octanoate (SucraShield)	0.8 to 1.0% v/v	48	0	Data on SucraShield for spider mite management are limited. Apply in a volume of 100 to 200 gallons per acre. SucraShield is **OMRI** listed.
Whitefly	spiromesifen, IRAC 23 (Oberon 2SC)	12 to 16 fl oz	12	3	Do not apply more than 48 fluid ounces or make more than three applications per season.
	thiamethoxam, IRAC 4A (Actara)	1.5 to 3 oz	12	3	**Do not** apply material immediately prior to bud opening, during bloom, or when bees are foraging.
Harvest					
Sap beetle	**Materials effective against sap beetles are toxic to bees.** Follow pollinator protection language on labels carefully.				
	Bait buckets and fruit removal	NA	NA	NA	Cultural control is the most effective form of sap beetle management. Sap beetles are attracted to the odor of overripe fruit. Thorough picking will reduce sap beetle populations and can eliminate the need for treatment. Culls should be disposed of offsite or buried. Bucket traps baited with rotting fruit or bread dough placed outside the field will attract sap beetles and can be used to determine when populations are present or to lure insects from field. Buckets should be checked and emptied at least weekly. Baits should be disposed of offsite or buried.
	novaluron IRAC 15 (Rimon 0.83 EC)	12 fl oz	12	1	Rimon is an insect growth regulator and is effective at reducing populations of immature sap beetles.
Spotted wing drosophila	Female spotted wing drosophila (SWD) lay eggs in ripening and ripe soft skinned fruits. SWD injury in spring bearing strawberries has been inconsistent in previous years, but summer and fall fruiting strawberries are at high risk of infestation. If SWD are active during strawberry harvest, treatments should be applied weekly and reapplied in the event of rain. **Many materials effective against SWD are toxic to bees.** Follow pollinator protection language on labels carefully. Apply SWD treatments in the evening or night, when bees are not actively foraging.				
	bifenthrin, IRAC 3A (Brigade WSB)	6.4 to 32 oz	12	0	There are many bifenthrin formulations. Do not apply when bees are foraging. Do not apply more than 80 ounces of product per acre per year. Brigade is effective against adult sap beetles.
	fenpropathrin, IRAC 3A (Danitol 2.4 EC)	16 to 21.33 fl oz	24	2	
	malathion, IRAC 1B (Malathion 57 EC)	1.5 to 3 pt	12	3	

Table 6-16. Strawberry Insect Control

Season Pest	Insecticide, Formulation, and IRAC Group	Amount of Formulation per Acre	Reentry Interval (hours)	Pre harvest interval (days)	Precautions and Remarks
Harvest (continued)					
Spotted wing drosophila (continued)	spinosad, IRAC 5 (Entrust SC)	4 to 8 fl oz	4	1	Rotate to a different class of insect control products after 2 successive applications of spinosad. Do not apply more than 18 fl oz Entrust SC (0.28 lb spinosad) per acre per crop. Entrust is **OMRI** listed.
	spinetoram, IRAC 5 (Radiant SC)	6 to 10 fl oz	4	1	
Corn earworm, European corn borer	Corn earworm and European corn borer larvae can feed on strawberry fruit. This damage is most common in warm years. Watch for eggs on strawberry fruit near the stem end. Adult moths can be monitored using pheromone traps conditions are appropriate for infestation. **Many materials effective against caterpillar pests are toxic to bees.** Follow pollinator protection language on labels carefully.				
	chlorantraniliprole IRAC 28 (Coragen)	3.5 to 5 fl oz	4	1	
	novaluron IRAC 15 (Rimon 0.83 EC)	9 to 12 fl oz	12	1	Rimon treatments must be timed to egg hatch.
	Bacillus thuringiensis (Bt), IRAC 11B2 (Dipel DF)	0.5 to 1.0 lb	4	0	Dipel DF is **OMRI** listed.
Tarnished plant bugs or Lygus bugs	Lygus bugs are typically only present in North Carolina strawberries at the end of the spring season, although they may be more problematic in day neutral, ever-bearing, or other strawberry season extension systems. Lygus bug injury results in malformed fruit and can resemble poor pollination. Lygus injury can be distinguished from poor pollination based on seed size. The seeds of Lygus damaged fruit are all the same size, while poor pollination results in varied seed sizes. **Many materials effective against lygus bugs are toxic to bees.** Follow pollinator protection language on labels carefully.				
	novaluron IRAC 15 (Rimon 0.83 EC)	9 to 12 fl oz	12	1	
	bifenthrin, IRAC 3A (Brigade WSB)	6.4 to 32 oz	12	0	There are many bifenthrin formulations. Do not apply when bees are foraging. Do not apply more than 80 ounces of product per acre per year. Brigade is effective against adult sap beetles.
	fenpropathrin, IRAC 3A (Danitol 2.4 EC)	16 to 21.33 fl oz	24	2	**Do not** apply when bees are foraging.
Flower thrips	Treatment is only necessary when thrips injury is present on berries. Thrips injury, which resembles bronzing on the stem end of berries, will typically not be present until the end of the season, if at all. **Materials effective against thrips are toxic to bees.** Follow pollinator protection language on labels carefully.				
	spinosad, IRAC 5 (Entrust SC)	4 to 8 fl oz	4	1	Rotate to a different class of insect control products after 2 successive applications of spinosad. Do not apply more than 18 fl oz Entrust SC (0.28 lb spinosad) per acre per crop. Entrust is **OMRI** listed.
	spinetoram, IRAC 5 (Radiant SC)	6 to 10 fl oz	4	1	
Spittlebug	Spittlebugs are occasional pests in strawberries and should only be treated if directly damaging fruit.				
	fenpropathrin, IRAC (Danitol 2.4 EC)	10.67 oz	24	2	Do not apply when bees are foraging. Do not make more than two applications.
	malathion (several products) 57 EC	1.5 pt	12	3	Do not apply when bees are foraging.

Further Information

Southeast Regional Strawberry Integrated Management Guide, http://www.smallfruits.org

Strawberry Growers Information Portal, http://strawberries.ces.ncsu.edu

VII — CHEMICAL WEED CONTROL

Chemical Weed Control in Field Corn

W. J. Everman, Crop Science Department

NOTES: A mode of action code has been added to the Herbicide and Formulation column of this table. Use MOA codes for herbicide resistance management. See Table 7-10, Herbicide Resistance Management, for details.

Control of witchweed is part of the State/Federal Quarantine Program. Contact the N.C. Department of Agriculture, Plant Industry Division, at 1-800-206-9333.

Table 7-1A. Chemical Weed Control in Field Corn

Herbicide, Mode of Action Code[1] and Formulation	Amount of Formulation Per Acre	Pounds Active Ingredient Per Acre	Precautions and Remarks
No-Till Burndown, Emerged annual weeds, top-kill and suppression of perennials			
glyphosate, MOA 9 (numerous brands and formulations)	See label	0.38 to 1.13 (lb a.e.)	Glyphosate is available as an isopropylamine salt and a potassium salt. Glyphosate formulations and application rates should be compared on the basis of pounds of glyphosate acid equivalent (a.e.) per gallon and per acre, respectively. The rate in the preceeding column is expressed as a.e. See TABLE 7-10 for glyphosate rate conversions. Apply before crop emerges. Glyphosate rate depends on weed species and weed size; see labels for suggested rates. See comments on labels concerning nitrogen as the carrier. Weed control may be decreased when nitrogen or other liquid fertilizers are used as carriers. Apply in 10 to 20 gallons of water per acre using flat fan nozzles. For residual grass and broadleaf weed control, glyphosate can be tank mixed with most preemergence corn herbicides and herbicide combinations. See the section on Corn—Preemergence. Refer to specific product labels for application rates, weeds controlled, application directions, and precautions. Adjuvant recommendations vary according to the glyphosate product used. See label of brand used for specific recommendations. May tank mix glyphosate with Harmony SG at 0.45 to 0.9 ounce per acre to improve control of curly dock, Carolina geranium, henbit, and wild garlic. Tank mix can be applied anytime prior to corn emergence. See Harmony SG label for details. May tank mix Resolve with glyphosate for improved control of Italian ryegrass and henbit. Glyphosate and the above glyphosate tank mixes will not control field pansy. A tank mix of Gramoxone plus atrazine should be used where field pansy is present. Glyphosate-resistant horseweed (marestail) is now common in eastern North Carolina counties. A tank mix of glyphosate or Gramoxone plus either 1.5 to 2 pints of 2,4-D or 0.5 pint of Clarity is suggested. Apply these tank mixes 7 to 14 days ahead of planting. If horseweed is present at planting time, a tank mix of Gramoxone plus atrazine is suggested.
paraquat, MOA 22 (Gramoxone Inteon) 2 SL (Firestorm) 3 SL (Parazone) 3 SL	2 to 4 pt 1.33 to 2.67 pt 1.33 to 2.67 pt	0.5 to 1 (lb a.e.)	Apply before, during, or after planting but before crop emerges using clean water or clear fertilizer solution as the carrier. Apply in a minimum of 10 GPA (20 to 40 preferred) using flat fan nozzles. Add either a nonionic surfactant at 1 pint per 100 gallons or a crop oil concentrate at 1 gallon per 100 gallons. Use 0.5 to 0.64 pound a.e. on weeds 1 to 3 inches, 0.75 pound a.e. on weeds 3 to 6 inches, and 1 pound a.e. on weeds 6 inches or taller. Use 0.5 pound a.e. for rye cover crop or 0.75 pound a.e. for wheat cover crop. Rainfast within 30 minutes. For residual grass and broadleaf weed control, paraquat can be tank mixed with most preemergence corn herbicides and herbicide combinations. See the section on Corn—Preemergence. Refer to specific product labels for application rates, weeds controlled, application directions, and precautions. Better and more consistent burndown will be achieved with mixtures of paraquat plus atrazine than with paraquat alone.
glufosinate-ammonium, MOA 10 (Liberty 280 SL)	29 to 36 fl oz	0.53 to 0.66	Liberty can be applied prior to emergence of transgenic or non-transgenic hybrids to control emerged weeds. See label for adjuvant use. If applied preplant, no additional applications can be made following crop emergence. Day time temperatures should be above 75 degrees F for control by Liberty. Thorough spray coverage is necessary: a minimum of 20 GPA with flat fan nozzles is suggested.
carfentrazone-ethyl, MOA 14 (Aim EC)	1.0 to 2.0 oz	0.016 to 0.031	Apply alone or with other herbicides or fertilizers as a burndown to control or suppress weeds. For optimum performance apply to actively growing weeds. Coverage is essential for good control. Broad-spectrum control of annual and perennial weeds requires a tank mix with herbicides such as glyphosate, Liberty, Gramoxone, 2,4-D or dicamba. Add either a nonionic surfactant at 2 pints per 100 gallons, or crop oil concentrate at 1 to 2 gallons per 100 gallons, or methylated seed oil.
pyraflufen-ethyl, MOA 14 (ET) 1 SL	0.5 to 2.0 fl oz	0.003 to 0.015 (lb a.i.)	ET can be use for suppression of small emerged summer annual and winter weeds. See label for adjuvant and spray volume recommendations. Research with ET is limited in North Carolina.
No-Till Burndown, Glyphosate-resistant horseweed (marestail)			
flumioxazin, MOA 14 (Valor SX) S1 WDG + paraquat, MOA 22 (Gramoxone Inteon) 2.5 SL or glyphosate, MOA 9 (numerous brands and formulations)	2.0 oz 2 to 4 pt See label	0.063 0.5 to 1.0 (lb a.e.) 0.38 to 1.13 (lb a.e.)	Corn may be planted 7 days after application if a minimum of 25% of the soil surface is covered with residue and a minimum of ¼ inch of rainfall has occurred between application and planting, otherwise corn must be planted 14 to 30 days after application. Can be applied with other herbicides, including Clarity, 2,4-D and atrazine. Apply with non-ionic surfactant at 1 quart100 gallons. Carefully follow label directons for sprayer cleaning after each day's use.
glyphostate, MOA 9 (numerous brands and formulations) + atrazine, MOA 5 (numerous brands) + 2,4-D, MOA 4 (numerous brands)	See labels	0.75 to 1.13 (lb a.e.) + 1 to 2 + 0.75 to 1	See comments for glyphosate alone. Apply mixtures containing 2,4-D at least 7 to 14 days ahead of corn planting.
glyphostate, MOA 9 (numerous brands and formulations) + atrazine, MOA 5 (numerous brands) + dicamba, MOA 4 (numerous brands)	See labels	0.75 to 1.13 (lb a.e.) + 1 to 2 + 0.25	See comments for glyphosate alone. Mixtures containing dicamba may be applied to medium- to fine-textured soils before or during planting. Do not apply to coarse-textured soils with less than 2.5% organic matter. Avoid contact of the herbicide with the seed by planting corn at least 1.5 inches deep and ensuring the furrow is closed.

CHEMICAL WEED CONTROL

Table 7-1A. Chemical Weed Control in Field Corn

Herbicide, Mode of Action Code[1] and Formulation	Amount of Formulation Per Acre	Pounds Active Ingredient Per Acre	Precautions and Remarks
No-Till Burndown, Glyphosate-resistant horseweed (marestail) (continued)			
glufosinate-ammonium, MOA 10 (Liberty 280 SL) + atrazine, MOA 5 (numerous brands)	29 fl oz + 1 to 2 qt	0.53 + 1 to 2	Daytime temperatures should be above 75 degrees F for control by Liberty. Thorough spray coverage is necessary; a minimum of 20 GPA with flat-fan nozzles is suggested.
No-Till Burndown or Preemergence			
saflufenacil, MOA 14 (Sharpen)	2.0 to 3.5 fl oz	0.045 to 0.078	Sharpen can be used burndown for control of glyphosate-resistant horseweed. See label for adjuvant selection (burndown) and application rate based on soil texture and organic matter content. A tank mix of glyphosate, paraquat, or glufosinate should be used in burndown applications to provide control of grass and additional broadleaf weed species. Do not apply where an organophosphate or carbamate insecticide has been applied.
dimethenamid-p, MOA 15 + saflufenacil, MOA 14 (Verdict)	10 to 16 oz	0.44 to 0.65 (lb a.i.)	Verdict can be used to control a range of grass and broadleaf weeds, including glyphosate-resistant horseweed. See label for adjuvant selection (burndown) and application rate based on soil texture and organic matter content. Corn injury from Verdict can occur when organphosphate or carbamate insecticides are applied to corn.
imazethapyr, MOA 2 + saflufenacil, MOA 14 (Optill)	2.0 oz	0.085 (lb a.l.)	**Use only on clearfield hybrids.** See Optill label for application with other herbicides. Apply prior to corn emergence to prevent injury. Do not apply if an organophosphate or carbamate insecticide has been used.
rimsulfuron, MOA 2 + thifensulfuron-methyl, MOA 2 (Leadoff) 33.4 WDG	1.5 to 2.7 oz	0.031 to 0.056	Leadoff can be applied from fall up to planting to control a range of grass and broadleaf weeds. Including a non-selective herbicide to provide additional control of emerged weeds is recommended. See label for recommended adjuvants. The addition of atrazine will provide added residual and burndown activity.
thifensulfuron, MOA 2 + tribenuron, MOA 2 (FirstShot SG with TotalSol) 50 WDG	0.5 to 0.8 oz	0.008 to 0.013 + 0.008 to 0.013	FirstShot can be applied from fall up to 14 days before corn planting to control a range of broadleaf weeds. Including a non-selective herbicide to provide additional control of emerged weeds is recommended. See label for recommended adjuvants. The addition of atrazine will provide added residual and burndown activity.
oxyfluorfen, MOA 14 (Goal 2XL) 2L (GoalTender) 4L	1 to 2 pt 0.5 to 1 pt	0.25 to 0.5	Goal 2XL can be applied at 30 days or more prior to corn planting to control common winter and summer annual weed species. After application but prior to corn planting, at least three significant rainfalls (0.25 inch or greater) must occur, otherwise Goal 2XL must be incorporated into the soil to a depth of 2.5 inches to avoid crop injury. A tank mix combination with glyphosate, paraquat, glufosinate, 2,4-D, or dicamba is recommended. See label for tankmix recommendations. Do not use corn plants from a treated field for green chop, ensilage, forage, or fodder.
Preemergence, Annual grasses; control of suppression of yellow nutsedge			
acetochlor, MOA 15 (Degree) 3.8 FME (Harness) 7 EC (Surpass EC) 6.4 EC (TopNotch) 3.2 FME	2.25 to 5 pt 1.25 to 2.75 pt 1.5 to 3 pt 4 to 6 pt	1.1 to 2.4 1.1 to 2.4 1.2 to 2.4 1.6 to 2.4	Controls most annual grasses, pigweed, and nightshade. Does not adequately control Texas panicum, seedling johnsongrass, and shattercane. Controls yellow nutsedge when incorporated; suppresses yellow nutsedge if applied preemergence. Do not apply to sands with less than 3% organic matter, loamy sands with less than 2% organic matter, or sandy loams with less than 1% organic matter if groundwater depth is 30 feet or less. Read label and adjust rates for soil texture and organic matter. Degree can be used at rates of 4.5 to 6.25 pints on soils with 6% to 10% organic matter and at 6.25 pints on soils with greater than 10% organic matter. Harness can be used at rates of 2.5 to 3.4 pints on soils with 6% to 10% organic matter and 3.4 pints on soils with greater than 10% organic matter. Surpass can be used at rates up to 3.75 pints on soils with greater than 7% organic matter. These herbicides can be shallowly incorporated; see labels for details. See labels for rotational crops. May be tank mixed with atrazine or simazine for broadleaf weed control. Each herbicide contains the safener dichlormid. May be applied to emerged corn up to 11 inches tall. Does not control emerged weeds. See labels for tank mix options to control emerged weeds.
alachlor, MOA 15 (Micro-Tech) 4 FME	2 to 4 qt	2 to 4	Controls most annual grasses and pigweed. At higher rates, controls nightshade. Does not adequately control Texas panicum, seedling johnsongrass, and shattercane. Generally less effective on yellow nutsedge than dimethenamid or metolachlor. Read labels and adjust rates for soil texture and organic matter. May be shallowly incorporated; see labels for details. May be mixed with atrazine or simazine for broadleaf weed control. May be applied to emerged corn up to 5 inches tall. Does not control emerged weeds. See labels for tank mix options to control emerged weeds.
dimethenamid-P, MOA 15 (Outlook) 6.0 EC	12 to 21 fl oz	0.56 to 1.0	Use 12 to 18 fluid ounces on soils with less than 3% organic matter and 14 to 21 fluid ounces on soils with greater than 3% organic matter. Controls most annual grasses and pigweed. At higher rates, controls nightshade and yellow nutsedge. Better yellow nutsedge control if incorporated; see label for incorporation details. Does not adequately control Texas panicum, seedling johnsongrass, and shattercane. Read label and adjust rates for soil texture and organic matter. May be mixed with atrazine or simazine for broadleaf weed control. Do not apply to sandy soil with less than 3% organic matter where depth to groundwater is 30 ft or less. May be applied to emerged corn up to 12 inches tall. Does not control emerged weeds. See labels for tank mix options to control emerged weeds.
metolachlor, MOA 15 (Me-Too-Lachlor II) 7.8 EC (Parallel) 7.8 EC	1 to 2 pt	0.98 to 1.95	See comments for s-metolachlor products. Products containing s-metolachlor are more active on weeds per unit of formulated product than those containing metolachlor. In general, it takes 1.5 pints of a metolachlor product to get the activity one would get from 1 pint of an s-metolachlor product.
S-metolachlor, MOA 15 (Brawl II) 7.64 EC (Cinch) 7.64 EC (Dual II Magnum) 7.64 EC (Medal II) 7.64 EC	1 to 2 pt	0.96 to 1.91	Controls most annual grasses and pigweed. At higher rates, controls nightshade and yellow nutsedge. Better yellow nutsedge control if incorporated; see label for incorporation details. Does not adequately control Texas panicum, seedling johnsongrass, and shattercane. Read labels and adjust rates for soil texture and organic matter. May be mixed with atrazine or simazine for broadleaf weed control. May be applied to emerged corn up to 40 inches tall. Direct if corn is taller than 5 inches Does not control emerged weeds. See labels for tank mix options to control emerged weeds.
pyroxasulfone, MOA 15 (Zidua) 85 WG	1.5 to 4 oz	0.0796 to 0.213	Use 1.5 to 2.75 ounces per acre on coarse soils, 2 to 3 ounces on medium soils, and up to 4 ounces per acre on fine textured soils. Controls most annual grasses, pigweed, and nightshade. Provides suppression of Texas panicum, seedling johnsongrass, and shattercane. May be mixed with atrazine or Sharpen for broadleaf weed control. May be applied to emerged corn up to the V4 stage. Does not control emerged weeds. See label for tank mix options to control emerged weeds.
pyroxasulfone, MOA 15 + fluthiacet-methyl, MOA 14 (Anthem) 2.15	5 to 13 oz		Use 5 to 8 ounces per acre on coarse soils, 6.5 to 10 ounces on medium soils, and 9 to 13 ounces on fine textured soils. Control most annual grasses, pigweed, and nightshade. Provides suppression of Texas panicum, seedling johnsongrass, and shattercane. May be applied to emerged corn through V4 stage. Will provide control of some small emerged broadleaf weeds (<2"). See label for tank mix options.

CHEMICAL WEED CONTROL

Table 7-1A. Chemical Weed Control in Field Corn

Herbicide, Mode of Action Code[1] and Formulation	Amount of Formulation Per Acre	Pounds Active Ingredient Per Acre	Precautions and Remarks
Preemergence, Annual broadleaf weeds			
atrazine, MOA 5 (AAtrex 4L) 4 F (AAtrex Nine-O) 90 WDG	1 to 2 qt 1.1 to 2.2 lb	1 to 2	Do not exceed 1.6 pounds a.i. on highly erodible soils (as defined by the NRCS) with less than 30% plant residue cover. Do not exceed 2 pounds a.i. on any soil. See labels for comments on rotational crops. May be applied preplant incorporated; see labels for details. May be tank mixed with preemergence grass control herbicides. Generic brands of atrazine are available, including products containing 5 pounds per gallon. See label for details on set-back requirements from streams and lakes.
isoxaflutole, MOA 27 (Balance FLEXX) 2 L	3 to 6 fl oz	0.047 to 0.09	Controls most broadleaf weeds, some annual grasses. Do not exceed 3 fluid ounces on coarse soils with 1.5% organic matter or less. May be tank mixed with preemergence grass control herbicides. Addition of atrazine will help extend residual control of broadleaf weeds such as morningglory species. May be applied postemergence up to 2-leaf collar corn. See labels for comments on rotational crops. May be applied preplant incorporated; see labels for details. See label for pH, groundwater, and soil texture recommendations.
mesotrione, MOA 27 (Callisto) 4 F	6 to 7.7 oz	0.19 to 0.24	Controls pigweed, lambsquarters, jimsonweed, common ragweed, smartweed, velvetleaf, and nightshade. Does not control sicklepod or prickly sida. Not adequately effective on cocklebur or morningglory. Callisto is generally more effective when applied postemergence. No rotational restrictions for small grains or for other crops planted the following spring. Can mix with various preemergence grass control herbicides or with atrazine or atrazine-containing products. See precautions on label concerning use of Counter and Lorsban.
Preemergence, Most annual grasses and broadleaf weeds			
acetochlor, MOA 15 + atrazine, MOA 5 (Degree Xtra) 4.04 FME	2.9 to 3.7 qt	2 to 2.5 + 1 to 1.25	Controls most broadleaf weeds and annual grasses. Does not adequately control Texas panicum, seedling johnsongrass, or shattercane. Do not apply to sands with less than 3% organic matter, loamy sands with less than 2% organic matter, or sandy loams with less than 1% organic matter if groundwater depth is 30 ft or less. Read labels and adjust rates for soil texture and organic matter. See labels for comments on rotational crops. May be incorporated; see labels for details. See labels for details on set-back requirements from streams and lakes. Do not exceed 1.6 pounds a.i. atrazine on highly erodible soils (as defined by NRCS) with less than 30% plant residue cover. Degree Xtra contains 2.7 pounds acetochlor and 1.34 pounds atrazine per gallon. Harness Xtra and Keystone NXT contain 3.1 pounds acetochlor and 2.5 pounds atrazine per gallon. FulTime contains 2.4 pounds acetochlor and 1.6 pounds atrazine per gallon. These products and certain tank mixes may also be applied early postemergence; see labels for details.
acetochlor, MOA 15 + atrazine, MOA 5 (FulTime) 4 F	2.5 to 5 qt	1.5 to 3 + 1 to 2	
acetochlor, MOA 15 + atrazine, MOA 5 (Harness Xtra) 5.6 F	1.4 to 3 qt	1.1 to 2.3 + 0.9 to 1.9	
acetochlor, MOA 15 + atrazine, MOA 5 (Keystone NXT) 5.6 F	1.4 to 3 qt	1.1 to 2.3 + 0.9 to 1.9	
alachlor, MOA 15 + atrazine, MOA 5 (Bullet) 4 FME (Lariat) 4 F	2.5 to 4.5 qt	1.56 to 2.81 + 0.94 to 1.69	Controls most broadleaf weeds and annual grasses. Does not adequately control Texas panicum, seedling johnsongrass, or shattercane. Read label and adjust rates for soil texture and organic matter. Do not exceed 4.25 quarts on highly erodible soils (as defined by the NRCS) with less than 30% plant residue cover. See labels for comments on rotational crops. May be applied preplant incorporated; see labels for details. See label for details on set-back requirements from streams and lakes. Bullet and Lariat contains 2.5 pounds alachlor and 1.5 pounds atrazine per gallon. May be applied early postemergence; see labels for details.
atrazine, MOA 5 (AAtrex 4L) 4 F (AAtrex Nine-O) 90 WDG + simazine, MOA 5 (Princep 4L) 4 F (Princep Caliber 90) 90 WDG	2 to 3 pt 1.1 to 1.6 lb + 1 to 1.44 qt 1.1 to 1.6 lb	1 to 1.5 1 to 1.44 + 1 to 1.44	Controls most annual broadleaf weeds plus crabgrass, goosegrass, fall panicum, and foxtails. Does not control Texas panicum, broadleaf signalgrass, seedling johnsongrass, or shattercane. Can be incorporated; see labels for details. Can use a 1:2 ratio of atrazine to simazine on more severe annual grass problems. If using 1:2 ratio, atrazine rates are 0.66 to 0.96 pound a.i. and simazine rates are 1.34 to 1.92 pounds a.i. Read label and adjust rates to soil texture. See labels for rotational restrictions and other precautions. Generic brands of simazine and atrazine are available, including atrazine products containing 5 pounds per gallon. See atrazine label for details on set-back requirements from streams and lakes.
dimethenamid-P, MOA 15 + atrazine, MOA 5 (Guardsman Max) 5 F	2.4 to 4.6 pt	0.5 to 1 + 1 to 1.9	Controls most broadleaf weeds and annual grasses. Does not adequately control Texas panicum, seedling johnsongrass, or shattercane. Adjust rate for soil texture and organic matter according to label. Do not exceed 3.8 pints on highly erodible soils (as defined by the NRCS) with less than 30% plant residue cover. See label for comments on rotational crops. May be applied preplant incorporated; see label for details. See label for set-back requirements from streams and lakes. Guardsman Max contains 1.7 pounds of dimethenamid-P and 3.3 pounds of atrazine per gallon. May be applied early postemergence; see label for details.
metolachlor, MOA 15 + atrazine, MOA 5 (Parallel Plus) 5.5 F	1.4 to 2.83 qt	0.95 to 1.9 + 1 to 2	See below comments for s-metolachlor plus atrazine. Products containing s-metalachlor are more active on weeds per unit of formulated product than those containing metolachlor.
S-metolachlor, MOA 15 + atrazine, MOA 5 (Bicep II Magnum) 5.5 F (Brawl II ATZ) 5.5 F (Cinch ATZ) 5.5 F (Medal II AT) 5.5 F	1.3 to 2.6 qt	0.78 to 1.56 + 1 to 2	Controls most broadleaf weeds and annual grasses. Does not adequately control Texas panicum, seedling johnsongrass, or shattercane. Rates for coarse-textured soils may not give sufficient control of heavy fall panicum, broadleaf signalgrass, and other grassy weeds. Cultivation and/or an additional herbicide application may be needed. Read label and adjust rates for soil texture and organic matter. Do not exceed 2.1 quarts on highly erodible soils (as defined by the NRCS) with less than 30% plant residue cover. See labels for comments on rotational crops. May be applied preplant incorporated; see labels for details. See label for details on set-back requirements from streams and lakes. These products contain 2.4 pounds of S-metolachlor and 3.1 pounds atrazine per gallon. May be applied early postemergence; see label for details.
S-metolachlor, MOA 15 + atrazine, MOA 5 + mesotrione, MOA 27 (Lexar) 3.7 L	3 to 3.5 qt	1.3 to 1.5 + 1.3 to 1.5 + 0.17 to 0.20	Controls most broadleaf weeds and annual grasses. Does not adequately control Texas panicum, seedling johnsongrass, or shattercane. May not adequately control cocklebur, morningglory, or sicklepod. Use 3 quarts on soils with less than 3% organic matter; use 3.5 quarts on soils with greater than 3% organic matter. Not recommended on soils with greater than 10% organic matter. See label for setback requirements from streams and lakes. May be applied postemergence to corn up to 12 inches tall; see label for tank mixes to control grasses when applying postemergence. Lexar contains 1.74 pounds S-metolachlor, 1.74 pounds atrazine, and 0.224 pound mesotrione per gallon.
isoxaflutole, MOA 27 + thiencarbazone-methyl, MOA 2 (Corvus)	3.33 to 5.6 fl oz	0.068 to 0.115	Controls most broadleaf weeds and annual grasses. Do not exceed 3.33 fluid ounces on coarse soils with 2.0% organic matter or less. The addition of atrazine will help extend residual control of broadleaf weeds such as morningglory species. May be applied postemergence up to 2-leaf collar corn. See label for pH, groundwater, and soil texture specific recommendations.

Table 7-1A. Chemical Weed Control in Field Corn

Herbicide, Mode of Action Code[1] and Formulation	Amount of Formulation Per Acre	Pounds Active Ingredient Per Acre	Precautions and Remarks
Preemergence, Most annual grasses and broadleaf weeds (continued)			
pyroxasulfone, MOA 15 + fluthiacet-methyl, MOA 14 + atrazine, MOA 5 (Anthem ATZ)	1.75 to 4 pt	0.985 to 2.253	Controls most broadleaf weeds and annual grasses. Does not adequately control Texas panicum, seedling johnsongrasss, or shattercane. Do not exceed 4.0 pt on highly erodible soils (as defined by the NRCS) with less than 30% plant residue cover. See labels for comments on rotational crops. May be applied preplant on medium and fine textured soils; see label for details. See label for details on set-back requirements from streams and lakes. Use 1.75 to 2.0 pints on coarse soils, 2 to 2.75 pints on medium soils, and up to 4 pints on fine soils preemergence. May be applied postemergence up to 4-leaf corn; see label for details.
S-metolachlor, MOA 15 + atrazine, MOA 5 + mesotrione, MOA 27 + Bicyclopyrone, MOA 27 (Accuron) 3.44 L	2.5 to 3.0 qt	1.34 to 1.61 + 0.625 to 0.75 + 0.15 to 0.18 + 0.038 to 0.045	Controls most broadleaf weeds and annual grasses. Does not adequately control Texas panicum, seedling johnsongrass, or shattercane. May need a tank-mix for control of heavy infestations of broadleaf signalgrass. Use 2.5 quarts on soils with less than 3% organic matter; use 3.0 quarts on soils with greater than 3% organic matter. Not recommended on soils with greater than 10% organic matter. See label for setback requirements from streams and lakes. May be applied postemergence to corn up to 12 inches tall; see label for tank mixes to control grasses when applying postemergence. Accuron contains 2.14 pounds S-metolachlor, 1.0 pound atrazine, 0.24 pound mesotrione, and 0.06 pounds bicyclopyrone per gallon.
acetochlor, MOA 15 + flumetsulam, MOA 2 + clopyralid, MOA 4 (SureStart) 4.25 L	1.5 to 3.0 qt	1.41 to 2.81 + 0.045 to 0.09 + 0.143 to 0.285	Controls most broadleaf weeds and annual grasses. Does not adequately control Texas panicum, seedling johnsongrass, or shattercane. Do not apply to sands with less than 3% organic matter, loamy sands with less than 2% organic matter, or sandy loams with less than 1% organic matter if groundwater depth is 30 feet or less. Read label and adjust rates for soil texture and organic matter. Tobacco, cotton, and peanuts cannot be planted for 18 to 26 months after application of SureStart, see label for comments on rotational crops. May be incorporated; see labels for details. This product and certain tank mixes may also be applied early postemergence; see label for details.
acetochlor, MOA 15 + mesotrione, MOA 27 + clopyralid, MOA 4 (Resicore) 3.29 L	2.25 to 3.0 qt	1.58 to 2.1 + 0.169 to 0.225 + 0.107 to 0.143	Controls most broadleaf weeds and annual grasses. Does not adequately control Texas panicum, seedling johnsongrass, or shattercane. Do not apply to sands with less than 3% organic matter, loamy sands with less than 2% organic matter, or sandy loams with less than 1% organic matter if groundwater depth is 30 ft or less. Read label and adjust rates for soil texture and organic matter. Tobacco, cotton, and peanuts cannot be planted for 18 months after application of Resicore, see label for comments on rotational crops. May be incorporated; see labels for details. This product and certain tank mixes may also be applied early postemergence; see label for details.
pendimethalin, MOA 3 (Prowl) 3.3 EC (Prowl H₂O) 3.8 L + atrazine, MOA 5 (AAtrex 4L) 4 F (AAtrex Nine-O) 90 WDG	1.8 to 3.6 pt 2 to 4 pt + 1 to 2 qt 1.1 to 2.2 lb	0.75 to 1.5 0.95 to 1.9 + 1 to 2	Annual grass control is more variable and sometimes less acceptable with this combination than with alternatives. Suggested for use only on fields with light annual grass pressure. Do not apply to excessively wet soils. Read label and adjust rates to soil texture and organic matter. Do not exceed 1.6 pounds a.i. atrazine on highly erodible soils (as defined by the NRCS) with less than 30% plant residue cover. See labels for comments on rotational crops. Do NOT incorporate. See atrazine label for details on set-back requirements from streams and lakes. Generic brands of atrazine (including products containing 5 pounds/gallon) and pendimethalin (such as Acumen, Helena Pendimethalin, Pendant, Pendimax, and Stealth, all containing 3.3 pounds/gallon) are available.
glufosinate-ammonium, MOA 10 (Liberty 280 SL)	29 to 36 fl oz	0.53 to 0.66 (lb a.i.)	Liberty can be applied prior to emergence of transgenic or nontransgenic hybrids. See comments under **No-Till Burndown** for more details.
Early Postemergence, Small annual broadleaf and grass weeds			
acetochlor, MOA 15 + atrazine, MOA 5 (Degree Xtra) 4.04 FME	2.9 to 3.7 qt	2 to 2.5 + 1 to 1.25	Apply as a very early postemergence application to weeds no larger than two leaves and before corn exceeds 11 inches. Adjust rates for soil types as specified on labels. See remarks concerning soil type limitations, set-back requirements from streams and lakes, and rotational crops under Field Corn — Preemergence. May be tank mixed with several other herbicides to control emerged weeds. If an atrazine-containing herbicide was applied earlier, the total amount of atrazine per acre per season should not exceed 2.5 pounds a.i.
(FulTime) 4L	2.5 to 5 qt	1.5 to 3 + 1 to 2	
(Harness Xtra) 5.6 L	1.4 to 3 qt	1.1 to 2.3 + 0.9 to 1.9	
(Keystone NXT) 5.6 L	1.4 to 3 qt	1.1 to 2.3 + 0.9 to 1.9	
alachlor, MOA 15 + atrazine, MOA 5 (Bullet) 4 FME	2.5 to 4.5 qt	1.6 to 2.8 + 0.9 to1.7	Apply as a very early postemergence application to weeds no larger than two leaves and before corn exceeds 5 inches See remarks for Bullet under Field Corn—Preemergence. If an atrazine-containing herbicide was applied earlier, the total amount of atrazine per acre per season should not exceed 2.5 pounds a.i. See labels for details on set-back requirements from streams and lakes.
atrazine, MOA 5 (AAtrex 4L) 4 F (AAtrex Nine-O) 90 WDG	2 qt 2.2 lb	2	Atrazine can be sprayed overtop of corn as an early postemergence treatment. Must be applied before weeds are over 1.5 inches tall to be effective and before corn exceeds 12 inches tall. Not effective during drought. Add 1 quart per acre of crop oil concentrate. If an earlier application was made, the total atrazine applied may not exceed 2.5 pounds a.i. per acre per year. See label for details on set-back requirements from streams and lakes. May be tank mixed with preemergence grass control herbicides. When tank mixing, check respective labels for application rates, weeds controlled, specific application directions, and precautions. Generic brands are available, including products containing 5 pounds per gallon.
dimethenamid-P, MOA 15 + atrazine, MOA 5 (Guardsman Max) 5F	2.4 to 4.6 pt	0.5 to 1 + 1 to 1.9	May be applied to corn up to 12 inches tall. Application must be made before weeds are greater than 1.5 inches tall or in a tank mix with products that will control emerged weeds. If an atrazine-containing herbicide was applied earlier, the total amount of atrazine per acre per season should not exceed 2.5 pounds a.i. See label for details on set-back requirements from streams and lakes.
metolachlor, MOA 15 + atrazine, MOA 5 (Parallel Plus) 5.5 F	1.4 to 2.83 qt	0.95 to 1.9 + 1 to 2	See below comments for s-metolachlor plus atrazine products. Products containing s-metolachlor are more active on weeds per unit of formulated product than those containing metolachlor.
S-metolachlor, MOA 15 + atrazine, MOA 5 (Bicep II Magnum) 5.5 L (Brawl II ATZ) 5.5 F (Cinch ATZ) 5.5 L (Medal II AT) 5.5 F	1.6 to 2.6 qt	0.96 to 1.56 + 1.24 to 2	Apply as a very early postemergence application to weeds no larger than two leaves and before corn exceeds 5 inches See remarks for these products under Field Corn—Preemergence. If an atrazine-containing herbicide was applied earlier, the total amount of atrazine per acre per season should not exceed 2.5 pounds a.i. See label for details on set-back requirements from streams and lakes.

CHEMICAL WEED CONTROL

Table 7-1A. Chemical Weed Control in Field Corn

Herbicide, Mode of Action Code[1] and Formulation	Amount of Formulation Per Acre	Pounds Active Ingredient Per Acre	Precautions and Remarks
Early Postemergence, Small annual broadleaf and grass weeds (continued)			
S-metolachlor, MOA 15 + atrazine, MOA 5 + mesotrione, MOA 27 (Lexar) 3.7 L	3 to 3.5 qt	1.3 to 1.5 + 1.3 to 1.5 + 0.17 to 0.20	May be applied to corn up to 12 inches tall. Add nonionic surfactant according to label directions. Do not apply to Counter-treated corn. Application to corn treated with other organophosphate insecticides may cause injury.
topramezone, MOA 27 + dimethenamid-P, MOA 15 (Armezon PRO) 5.35 EC	14 to 20	0.01 to 0.016 + 0.57 to 0.82	May be applied to corn from emergence up to the 8-leaf stage or 30-inches-tall corn. Armezon PRO rates vary by soil texture and organic matter, refer to label for recommendations. Add methylated seed oil or crop oil concentrate at 0.5 to 1.0 gallon per 100 gallons of water or nonionic surfactant at 0.25 to 0.5 gallon per 100 gallons of water. Oil-type adjuvants are not recommended when tank-mixing with atrazine. Add nitrogen fertilizer at 1.25 to 2.5 gallons per 100 gallons of water or 8.5 to 17 pounds of ammonium sulfate per 100 gallons.
S-metolachlor, MOA 15 + atrazine, MOA 5 + mesotrione, MOA 27 + Bicyclopyrone, MOA 27 (Accuron) 3.44 L	2.5 to 3.0 qt	1.34 to 1.61 + 0.625 to 0.75 + 0.15 to 0.18 + 0.038 to 0.045	May be applied to corn up to 12 inches tall. Add nonionic surfactant according to label directions. Do not apply to Counter-treated corn. Application to corn treated with other organophosphate insecticides may cause injury.
Postemergence, Annual broadleaf weeds			
bentazon, MOA 6 (Basagran) 4 SL	1.5 to 2 pt	0.75 to 1	Apply overtop or directed. Drop nozzles suggested after corn is 8 inches tall to ensure better weed coverage. Controls many broadleaf weeds such as cocklebur, jimsonweed, smartweed, velvetleaf, prickly sida, spurred anoda, and spreading dayflower. See label for weeds controlled and recommended weed size for treatment. Add crop oil concentrate at 2 pints per acre. May be tank mixed with atrazine.
acetochlor, MOA 15 + flumetsulam, MOA 2 + clopyralid, MOA 4 (SureStart) 4.25 L	1.5 to 3.0 qt		Apply as a very early postemergence application to broadleaf weeds no larger 2 inches and before corn exceeds 11 inches. Established or germinated grass weeds present at application will not be controlled. May be tank mixed with several other herbicides to control emerged weeds. Adjust rates for soil types as specified on label. See remarks concerning soil type limitations and rotational crops under Field Corn — Preemergence. Tobacco, cotton, and peanuts cannot be planted for 18 to 26 months after application of SureStart, see label for comments on rotational crops.
acetochlor, MOA 15 + mesotrione, MOA 27 + clopyralid, MOA 4 (Resicore) 3.29 L	2.25 to 3.0 qt	1.58 to 2.1 + 0.169 to 0.225 + 0.107 to 0.143	Apply as a very early postemergence application to broadleaf weeds no larger 3 inches and before corn exceeds 11 inches. Control of emerged grass weeds will not be consistent. May be tank mixed with several other herbicides to control emerged weeds. Adjust rates for soil types as specified on label. See remarks concerning soil type limitations and rotational crops under Field Corn — Preemergence. Tobacco, cotton, and peanuts cannot be planted for 18 months after application of Resicore, see label for comments on rotational crops.
bromoxynil, MOA 6 (Buctril) 2 EC (Buctril) 4 EC	1 to 1.5 pt 0.5 to 0.75 pt	0.25 to 0.38	Can be applied overtop of corn from four-leaf stage up to tasseling. Drop nozzles suggested after corn is 8 inches tall for better coverage on weeds. Controls most broadleaf weeds if treated when small. See label for weeds controlled and recommended weed size for treatment. Does not control sicklepod, prickly sida, spurred anoda, or croton. Marginally effective on pigweed and morningglory unless treated very timely. Crop oil or surfactant not necessary when applying Buctril alone. Can tank mix with Accent, atrazine, Banvel, Clarity, or 2,4-D for broader spectrum control. Primary advantage over 2,4-D or Clarity is safety when sensitive crops are nearby. Will cause some burn on corn foliage.
carfentrazone, MOA 14 (Aim EC) 2 L	0.5 fl oz	0.008	Controls velvetleaf, morningglory, redroot pigweed, lambsquarters, and nightshade. See label for weed size to treat. Apply before corn exceeds V8 stage (8 leaves with collars). Add nonionic surfactant at 1 quart per 100 gallons. May be mixed with 2,4-D amine, Accent, atrazine, Banvel, Callisto, Clarity or Distinct. May be mixed with Lightning for Clearfield corn, Liberty for Liberty Link corn, or glyphosate for Roundup Ready corn.
dicamba, dimethylamine salt, MOA 4 (Banvel) 4 SL (Diablo) 4 SL (Dicamba DMA Salt) 4 SL (Rifle) 4 SL (Sterling) 4 SL	0.5 pt	0.25	Apply overtop of corn from spike stage until 8 inches tall. On corn 8 to 36 inches tall, Banvel or Clarity can be applied at 0.5 pint using drop nozzles. Carefully follow all precautions on label concerning drift to sensitive crops. Dicamba is more effective than 2,4-D on smartweed, sicklepod, nightshade, burcucumber, and pokeweed.
dicamba, diglycolamine salt, MOA 4 (Clarity) 4 SL	0.5 pt	0.25	
dicamba, sodium salt, MOA 4 + diflufenzopyr, sodium salt (Distinct) 76.4 WDG	4 oz	0.125 + 0.05	Apply to corn 4 to 36 inches tall. Drop nozzles suggested on corn taller than 10 inches Drop nozzles must be used on corn taller than 24 inches Rate can be increased to 6 ounces on corn shorter than 10 inches Add nonionic surfactant at 1 quart per 100 gallons spray solution plus either 5 quarts 30% UAN or 5 pounds ammonium sulfate per 100 gallons. Do not add crop oil. Control of annual weeds similar to that by Banvel or Clarity. Distinct may be somewhat more effective on perennial broadleaf weeds. Carefully follow all precautions on label concerning drift to sensitive crops.
dicamba, sodium salt, MOA 4 + diflufenzopyr, sodium salt + safener (Status) 61.1 WDG	5 to 10 oz	0.14 to 0.28 + 0.053 to 0.106	Apply to corn 4 to 36 inches tall. Drop nozzles suggested on corn taller than 24 inches Add nonionic surfactant at 1 quart per 100 gallons spray solution plus either 5 quarts 30% UAN or 5 pounds ammonium sulfate per 100 gallons. Do not add crop oil. Potential for crop injury from Status is much less than from dicamba products without safener.
flumiclorac pentyl ester, MOA 14 (Resource) 0.86 EC	4 to 8 fl oz	0.027 to 0.054	Can be applied overtop of corn from the two-leaf through the ten-leaf stage at 4 to 6 fluid ounces per acre. At 4 to 6 fluid ounces, Resource controls velvetleaf and small lambsquarters, ragweed, smooth pigweed, and Palmer amaranth. When applying overtop, add nonionic surfactant at 1 quart/100 gallons spray solution. Resource can be directed at 4 to 8 fluid ounces per acre. At 8 fluid ounces, Resource controls velvetleaf and small cocklebur, lambsquarters, ragweed, jimsonweed, Palmer amaranth, redroot and smooth pigweed, and prickly sida. See label for recommended weed sizes for treatment. When directing, add 2 pints per acre of crop oil concentrate. For broader spectrum control, Resource may be tank mixed with atrazine, Accent, Banvel, Buctril, Clarity, or 2,4-D. May be mixed with glyphosate on Roundup Ready corn, with Liberty on Liberty Link corn, and with Lightning on Clearfield corn.

Table 7-1A. Chemical Weed Control in Field Corn

Herbicide, Mode of Action Code[1] and Formulation	Amount of Formulation Per Acre	Pounds Active Ingredient Per Acre	Precautions and Remarks
Postemergence, Annual broadleaf weeds (continued)			
mesotrione, MOA 27 (Callisto) 4 F	3 fl oz	0.094	Can be applied overtop or with drop nozzles until corn is 30 inches tall or has eight leaves. Add crop oil concentrate at 1 gallon per 100 gallons spray solution. Do not use methylated seed oil or adjuvant blends containing methylated seed oil. Controls most broadleaf weeds. Partial control of common ragweed and morningglory. Does not control sicklepod or prickly sida. Can tank mix with atrazine, Accent Q, or Steadfast Q. See precautions on labels of these products. Can be mixed with Liberty on Liberty Link corn or glyphosate on Roundup Ready corn. No rotational restrictions for small grains or other crops planted the following spring. Rainfast in 1 hour. See precautions on label concerning use of Counter and Lorsban.
thifensulfuron methyl, MOA 2 (Harmony SG) 50 WDG	0.125 oz	0.0039	Apply only to corn with 2 to 6 leaves (1 to 4 collars) but not larger than four collars or 12 in tall. Add either nonionic surfactant at 1 quart per 100 gallons or crop oil concentrate at 1 gallon per 100 gallons. Also, add a nitrogen-containing fertilizer according to label directions. Controls lambsquarters, pigweeds, smartweed, and velvetleaf; see label for recommended weed size for treatment. Harmony SG will also control curly dock and burcucumber. See label for comments concerning injury when used in conjunction with insecticides. May be mixed with atrazine.
2,4-D amine, MOA 4 (various brands) 3.8 SL	0.5 to 1 pt	0.24 to 0.48	Use 0.5 pint overtop when corn is 4 to 5 inches tall and weeds are small. Increase rate to 1 pint as corn reaches 8 inches Use drop nozzles and direct spray toward base of corn if over 8 inches tall. Do not cultivate for about 10 days after spraying as corn may be brittle. Reduce rate of 2,4-D if extremely hot or soil is wet. For better control of sicklepod and horsenettle, add a nonionic surfactant to 1 pint of 2,4-D and direct spray. Not adequately effective on smartweed, nightshade, burcucumber, or pokeweed. Use extreme caution to avoid drift to sensitive crops such as cotton and tobacco. **Use of ester formulations or acid + ester formulations (such as Weedone 638) or 2,4-D is not suggested if sensitive crops, especially cotton or tobacco, are located within 1 mile of the corn.**
rimsulfuron, MOA 2 + mesotrione, MOA 27 (Realm Q) 38.75% WDG	4 oz	0.014 + 0.0031	Apply overtop or with drop nozzles to corn up to 20 inches or 7 leaf collars. Controls redroot and smooth cocklebur, common ragweed, pigweed, and velvetleaf. Partial control of morningglory. Also controls 1- to 2-inches fall panicum. Provides short-term residual control of lambsquarters, nightshade, redroot pigweed, and pigweed. Does not control sicklepod or prickly sida. Add 1 quart nonionic surfactant per 100 gallons and 2 quarts/acre UAN. May tank mix with other postemergence corn herbicides (except Basgran), including glyphosate on Roundup Ready corn and Liberty on Liberty Link corn. See label statement concerning sensitive hybrids. See rotational restrictions on label.
rimsulfuron, MOA 2 + thifensulfuron, MOA 2 (Resolve Q) 22.4% WDG	1.25 oz	0.014 + 0.0031	Apply overtop or with drop nozzles to corn up to 20 inches or 7 leaf collars. Controls redroot and smooth pigweed and velvetleaf. Suppresses cocklebur, smartweed, lambsquarters, common ragweed, and morningglory. Also controls 1- to 2-inches tall panicum. Provides short-term residual control of lambsquarters, nightshade, redroot pigweed, and smooth pigweed. Add 1 quart nonionic surfactant per 100 gallons and 2 quart/acre UAN. May tank mix with other postemergence corn herbicides (except Basgran), including glyphosate on Roundup Ready corn and Liberty on Liberty Link corn. See label statement concerning sensitive hybrids. See rotational restrictions on label.
tembotrione, MOA 27 (Laudis) 3.5 L	3 fl oz	0.082	Can be applied overtop or with drop nozzles to corn from emergence up to V8 stage. Add methylated seed oil at 1 gallon per 100 gallons of spray solution. Also add 1.5 quarts/A UAN. Controls most broadleaf weeds. Does not control sicklepod or prickly sida and only suppresses morningglory. Controls or suppresses some grasses. See label for weeds controlled and recommended size for treatment. Can tank mix with atrazine, Accent Q, Buctril, or Steadfast Q. Can tank mix with Liberty on Liberty Link corn or glyphosate on Roundup Ready corn. Rain-free in 1 hour. See label for rotational restrictions.
tembotrione, MOA 27 + thiencarbazone-methyl, MOA 2 (Capreno) 2.88 + 0.57 SC	3 fl oz	0.003 + 0.015 (lb a.i.)	Apply postemergence over the top prior to V6 stage. Use drop nozzles for better coverage of later stages of corn. Apply with crop oil concentrate at 1 gallon/100 gallons and UAN at 1.5 quarts/A. See comments under Laudis for spectrum of control and possible tank mixture. See label for precautions when organophosphate insecticide is used in corn.
topramezone, MOA 27 (Impact) 2.8 L (Armezon) 2.8 L	0.75 fl oz	0.016	Can be applied overtop or with drop nozzles to corn from emergence until 45 days prior to harvest. Add crop oil concentrate or methylated seed oil at 1 gallon per 100 gallons spray solution. Also add 1.25 to 2.5 gallons UAN per 100 gallons spray solution. See label for adjuvant recommendations in tank mixes. Controls most broadleaf weeds. Does not control sicklepod and only suppresses morningglory. Controls or suppresses some grasses. See label for weeds controlled and recommended size for treatment. Can tank mix with glyphosate on Roundup Ready corn, Liberty on Liberty Link corn, or Lightning on Clearfield corn. Rain-free in 1 hour. See label for rotational restrictions.
pyraflufen-ethyl, MOA 14 (ET)	0.5 to 2.0 fl oz	0.0008 to 0.0003 (lb a.i.)	ET can be used for limited suppression of small broadleaf weeds up to V4 stage. Do not apply with crop oil concentrate. Some leaf speckling can ocur but is transient. See label for adjuvant and spray volume recommendations. Research with ET is limited in North Carolina.
Postemergence, Annual grasses, broadleaf weeds, and johnsongrass			
nicosulfuron, MOA 2 + rimsulfuron, MOA 2 (Steadfast Q) 37.7 WDG	1.5 oz	0.024 + 0.011	Apply to corn up to 20 inches tall with 6 or fewer leaf collars. Add 1 gallon crop oil concentrate per 100 gallons or 1 quart nonionic surfactant per 100 gallons. Also add nitrogen fertilizer according to label directions. Controls johnsongrass and most annual grasses and broadleaf weeds. May not adequately control crabgrass, goosegrass, Palmer amaranth, and sicklepod. See label for weeds controlled and weed size to treat. Do not apply to corn treated with Counter 15G; see label for precautions concerning other organophosphate insecticides. See label for comments on susceptible hybrids. Can tank mix with atrazine, Callisto, Clarity, or Distinct. See label precautions for tank mixes. Steadfast Q contains 25% nicosulfuron and 12.5% rimsulfuron.
nicosulfuron, MOA 2 + mesotrione, MOA 27 (Revulin Q) 51.2 WDG	3.4 to 4.0 oz	0.031 to 0.036 + 0.078 to 0.092	Apply to corn up to 20 inches tall with 6 or fewer leaf collars. Add 1 gallon crop oil concentrate per 100 gallons or 1 quart nonionic surfactant per 100 gallons. Also add nitrogen fertilizer according to label directions. Controls johnsongrass and most annual grass and broadleaf weeds. May not adequately control goosegrass, sicklepod, and Palmer amaranth greater than 3 inches tall. See label for weeds controlled and weed size to treat. Do not apply to corn treated with Counter 15G; see label for precautions concerning tankmixing with foliar applied organophosphate insecticides. Can tank mix with atrazine, glyphosate, or glufosinate. See label precautions for tank mixes. Revulin Q contains 14.4% nicosulfuron and 36.8% mesotrione.
Postemergence, Annual broadleaf weeds and some annual grasses: Clearfield hybrids Only			
imazethapyr, MOA 2 + imazapyr, MOA 2 (Lightning) 70 WDG	1.28 oz	0.042 + 0.014	**USE ONLY ON CLEARFIELD HYBRIDS.** Make only one application per year. Can be applied anytime up to 45 days prior to harvest. Use of drop nozzles will give better coverage in larger corn. Controls most annual broadleaf weeds and certain annual grasses; see label for weeds controlled and recommended growth stage for application. Not adequately effective on ragweed. Season-long sicklepod control may require a pre-emergence application of atrazine or a lay-by application of an appropriate herbicide. Suppresses yellow and purple nutsedge. Add either 1 quart per 100 gallons nonionic surfactant or 1 gallon per 100 gallons of crop oil concentrate or methylated seed oil. Also add 1 to 2 quarts per acre of UAN or 2.5 pounds per acre ammonium sulfate. May be tank mixed with most other corn herbicides; see labels for details. See Lightning label for rotational restrictions.

CHEMICAL WEED CONTROL

CHEMICAL WEED CONTROL

Table 7-1A. Chemical Weed Control in Field Corn

Herbicide, Mode of Action Code[1] and Formulation	Amount of Formulation Per Acre	Pounds Active Ingredient Per Acre	Precautions and Remarks
Postemergence, Annual broadleaf weeds and some annual grasses: **Clearfield hybrids Only** (continued)			
imazethapyr, MOA 2 + saflufenacil, MOA 14 (Optill)	2.0 oz	0.085 (lb a.i.)	USE ONLY ON CLEARFIELD HYBRIDS. See Optill label for application with other herbicides. Apply prior to corn emergence to prevent injury. Do not apply if an organophosphate or carbamaate insecticide has been used.
Postemergence, Annual grasses and annual broadleaf weeds: **Liberty Link hybrids Only**			
glufosinate-ammonium, MOA 10 (Liberty 280) 2.34 SL	22 fl oz	0.40	**Use only on Liberty Link hybrids.** May be applied overtop from corn emergence until corn is in V7 stage (7 leaves with collar) or 24 inches tall, whichever comes first. Apply with drop nozzles to corn 24 to 36 inches tall. Rate depends on weed species and weed size; see label for details. Controls most annual grass and broadleaf weeds, but only marginally effective on goosegrass. Not effective on dayflower. Timing of application is critical for pigweed control. Use of drop nozzles in corn over 8 inches tall may improve spray coverage. May make two applications per year, but do not exceed 44 fluid ounces per season. Add 3 pounds per acre of ammonium sulfate. Do not add surfactant or crop oil. May be tank mixed with most postemergence corn herbicides; see respective labels for details. Tank mixes of Liberty plus atrazine have been most effective.
Postemergence, Annual grasses and broadleaf weeds, johnsongrass, and suppression of perennial broadleaf weeds: **Roundup Ready Hybrids Only**			
glyphosate, MOA 9 (numerous brands and formulations)	See label	0.75 to 1.125 (lb a.e.)	**Apply only to Roundup Ready (glyphosate-resistant) hybrids.** Glyphosate is available as an isopropylamine salt and a potassium salt. Glyphosate formulations and application rates should be compared on the basis of pounds of glyphosate acid equivalent (a.e.) per gallon and per acre, respectively. See Table 7-10 for glyphosate rate conversions. Glyphosate controls most annual weeds; exceptions include dayflower, Florida pusley, and hemp sesbania. Timely application is critical for morningglory control. Glyphosate also controls johnsongrass and suppresses other perennial weeds. See label of brand you use for recommended rates and sizes of weeds to treat. Adjuvant recommendations vary according to glyphosate product. See label of brand used for specific recommendations. Apply overtop from corn emergence through the V8 stage (8 leaves with collars) or until corn reaches 30 inches, whichever comes first. Drop nozzles suggested on corn 24 to 30 inches Apply only with drop nozzles when corn is 30 to 48 inches Make multiple applications at least 10 days apart. Do not exceed a total of 2.25 pounds a.e. per acre in crop. For resistance managment, do not rely entirely on glyphosate. Herbicides with other modes of action should be included in the program. Such herbicides can be preemergence, mixed with glyphosate, or lay-by. See comments on resistance management in Table 7-10. Any registered soil-applied herbicide or lay-by herbicide can be used on Roundup Ready corn. Aim, atrazine, Bullet, Callisto, Clarity, Degree, Degree Xtra, Distinct, Harness, Harness Xtra, Impact, Laudis, Micro-tech, Resolve, Resolve Q, Resource, Status or 2,4-D can be mixed with glyphosate applied postemergence. When using a tank mix, follow all directions and precautions on the respective labels, especially corn stage for application.
glyphosate, MOA 9 + s-metolachlor, MOA 15 + mesotrione, MOA 27 (Halex GT) 4.39 L	3.6 to 4 pt	0.94 to 1.05 + 0.94 to 1.05 + 0.094 to 0.105	**Apply only to hybrids designated Roundup Ready or Agrisure GT.** Apply from corn emergence up to 30 inches or 8 leaf collars. Add nonionic surfactant at 1 quart/100 gallons spray solution. Also add AMS according to label directions. Do not substitute UAN for AMS. May tank mix with atrazine. See precautions on label when using Halex GT in conjunction with insecticides.
Postemergence, Yellow nutsedge			
bentazon, MOA 6 (Basagran) 4 SL	1.5 to 2 pt	0.75 to 1	Apply overtop of corn or directed when yellow nutsedge is 6 to 8 inches tall. If needed, make a second application 7 to 10 days later. Add 2 pints per acre of crop oil concentrate.
Postemergence, Yellow and purple nutsedge			
halosulfuron (Sandea) 75 WDG	0.67 to 1.33 oz	0.03 to 0.06	Apply overtop or with drop nozzles to corn from spike stage until layby. Add nonionic surfactant at 1 quart/100 gallons spray solution.
Postemergence, Yellow and purple nutsedge: **Clearfield hybrids Only**			
imazethapyr, MOA 2 + imazapyr, MOA 2 (Lightning) 70 WDG	1.28 oz	0.042 + 0.014	**Use only on Clearfield hybrids.** Apply when nutsedge is 1 to 3 inches tall. Add surfactant and nitrogen-containing fertilizer as specified on the label. See comments on Lightning in section on broadleaf weed control. Label claims supression only.
Postemergence, Yellow and purple nutsedge; **Roundup Ready hybrids Only**			
glyphosate, MOA 9 (numerous brands and formulations)	See label	0.75 to 1.125 (lb a.e.)	**Apply only to Roundup Ready hybrids.** Two applications of glyphosate may be necessary for nutsedge control. See previous comments under Roundup Ready corn.
Postemergence, Annual grasses			
foramsulfuron, MOA 2 (Option) 35 WDG	1.5 oz	0.033	Apply overtop to corn in V1 to V6 stage. Can be applied with drop nozzles to corn through V8 stage. Add a methylated seed oil at 1.5 pints per acre and either 1.5 to 2 quart per acre of 30% UAN or 1.5 to 3 pounds per acre of ammonium sulfate. Do not cultivate for 7 days before or after application. Controls annual grasses; see comments on label for crabgrass and broadleaf signalgrass. Also controls small broadleaf weeds, such as burcucumber, cocklebur, pigweed, lambsquarters, common ragweed, velvetleaf, and nightshade. May be applied twice per season. May be tank mixed with atrazine, Callisto, or Distinct. Do not apply to corn treated with Counter, Dyfonate, or Thimet. See comments on label concerning sensitive hybrids.
nicosulfuron, MOA 2 (Accent) 75 WDG (Accent Q) 54.5 WDG	0.67 oz 0.9 oz	0.031	Can be applied overtop or with drop nozzles to corn up to 20 inches tall. If corn is 20 to 36 inches tall, apply only with drop nozzles and avoid spraying into the corn whorl. Do not apply if corn is greater than 36 inches Add either a crop oil concentrate at 1 gallon per 100 gallons or a nonionic surfactant at 1 quart per 100 gallons spray solution. See label concerning additional adjuvants. Do not cultivate for 10 days before application. Controls ryegrass, small broadleaf signalgrass, foxtails, fall panicum, Texas panicum, barnyardgrass, shattercane, and seedling johnsongrass. May not adequately control crabgrass and goosegrass. Also controls small burcucumber, jimsonweed, morningglory, pigweed, and smartweed. Can be applied twice, but do not exceed 1.33 ounces per acre per year. Reduced rates of 1/3 to 1/2 ounce may be applied under certain conditions; see label for details. May be tank mixed with atrazine, Callisto, Clarity, or Distinct for improved broadleaf control. See label for comments concerning injury when used in conjunction with insecticides.

Table 7-1A. Chemical Weed Control in Field Corn

Herbicide, Mode of Action Code[1] and Formulation	Amount of Formulation Per Acre	Pounds Active Ingredient Per Acre	Precautions and Remarks
Postemergence, Johnsongrass and shattercane			
formasulfuron, MOA 2 (Option) 35WDG	1.5 oz	0.033	Apply before seedling and rhizome johnsongrass exceed 16 inches tall. See comments for Option under Annual Grasses.
nicosulfuron, MOA 2 (Accent) 75 WDG (Accent Q) 54.5 WDG	0.67 oz 0.9 oz	0.031	Apply when seedling johnsongrass is 4 to 12 inches tall, rhizome johnsongrass is 8 to 18 inches tall, or shattercane is 4 to 12 inches tall. See other comments for Accent under Annual Grasses.
Postemergence, Johnsongrass: **Roundup Ready hybrids Only**			
glyphosate, MOA 9 (numerous brands and formulations)	See label	0.75 (lb a.e.)	**Apply only to Roundup Ready hybrids.** See previous comments under Roundup Ready corn.
Postemergence, Bermudagrass: **Roundup Ready hybrids Only**			
glyphosate, MOA 9 (numerous brands and formulations)	See label	0.75 to 1.125 (lb a.e.)	**Apply only to Roundup Ready hybrids.** See previous comments under Roundup Ready corn. Two applications are usually required for adequate control.
Lay-by, Annual broadleaf weeds; control or suppression of perennial broadleaf weeds			
2,4-D amine, MOA 4 (various brands) 3.8 SL	0.5 to 1 pt	0.24 to 0.48	Apply with drop nozzles. Do not apply to corn in the tassel to dough stage. May add 1 quart of nonionic surfactant per 100 gallons spray solution. Surfactant may increase control of sicklepod and perennial weeds. Corn hybrids vary in sensitivity; check with seed dealer for sensitivity of hybrid used. Use extreme caution to avoid drift to sensitive crops, such as cotton and tobacco. **Use of ester formulations or acid + ester formulations (such as Weedone 638) or 2,4-D is not suggested if sensitive crops, especially cotton or tobacco, are located within 1 mile of the corn.** Liquid nitrogen may be used as the carrier. When using 2,4-D amine, mix 1 pint of herbicide in at least 2 quarts of water, and add this mixture to the spray tank with considerable agitation until thoroughly mixed. Do not allow nitrogen-herbicide mixture to stand in the sprayer.
dicamba, dimethylamine salt, MOA 4 (Banvel) 4 SL	0.5 pt	0.25	Apply as directed spray using water as the carrier to corn up to 36 inches tall. Do not apply within 15 days of tassel emergence. Add nonionic surfactant at 1 pint per 100 gallons for Clarity or 2 pints per 100 gallons for Banvel, Distinct, or Status. See comments on labels concerning addition of UAN or AMS. Follow precautions on labels concerning drift to sensitive crops.
dicamba, digycolamine salt, MOA 4 (Clarity) 4 SL	0.5 pt	0.25	
dicamba, sodium salt, MOA 4 + diflufenzopyr, sodium salt (Distinct) 61.1 WDG	4 oz	0.125 + 0.053	
dicamba, sodium salt, MOA 4 + diflufenzopyr, sodium salt + safener (Status) 61.1 WDG	5 to 10 oz	0.14 to 0.28 + 0.053 to 0.107	
ametryn, MOA 5 (Evik) 80 WDG	2 lb	1.6	Apply as a directed spray after corn is at least 15 inches tall. Do not apply Evik within 3 weeks of tasseling. Add nonionic surfactant at 2 quart per 100 gallons spray solution. Evik and Linex may be applied using liquid nitrogen as the carrier. Add surfactant when using nitrogen as the carrier.
linuron, MOA 7 (Linex) 4 L	1.5 pt	0.75	Note that current labeled rates of Linex have been reduced from previous years.
Preharvest, Annual grasses and johnsongrass			
sodium chlorate (Defol 750) 7.5 L	3.2 qt	6	Apply on warm, sunny day at least 14 days before anticipated harvest. Apply by ground or air after corn reaches hard dough or dent stage. Add surfactant or crop oil according to label directions. Thorough spray coverage essential.
Preharvest, Broadleaf weeds			
2, 4-D, amine, MOA 4 (various brands)	1 to 2 pt	0.48 to 0.95	Suppresses perennial broadleaf weeds and controls many annual broadleaf weeds. Apply after hard dough or dent stage by ground or air. Avoid drift to sensitive crops.
carfentrazone, MOA 14 (Aim) 2 EC	1.9 fl oz	0.03	Desiccates morningglory, cocklebur, and pigweed. Apply 3 or more days ahead of harvest. Add 1 gallons crop oil concentrate per 100 gallons spray solution. Thorough coverage is critical; use minimum of 20 GPA for ground application. May be applied by air. For dense morningglory infestations, two applications at 1 ounce/acre may be more effective.
Preharvest, Annual grasses and broadleaf weeds			
paraquat, MOA 22 (Gramoxone Inteon) 2 SL	1.2 to 2 pt	0.3 to 0.5	Apply after black layer has formed and at least 7 days prior to harvest. Add nonionic surfactant at 1 quart per 100 gallons spray solution. Generic brands of paraquat containing 3 pounds active ingredient per gallon are available These products would be applied at 0.8 to 1.3 pints per acre.
Preharvest, Annual grasses, johnsongrass, and broadleaf weeds			
glyphosate, MOA 9 (numerous brands and formulations)	See label	0.75 (lb a.e.)	Apply after kernel fill is complete (black layer formed) and grain moisture is 35% or less. Apply at least 7 days prior to harvest. Maximum rate for aerial application varies by product; see label of brand used. Avoid drift to other crops and desirable vegetation.
Postharvest, Horsenettle and other perennial and annual broadleaf weeds			
2,4-D amine, MOA 4 (various brands) 3.8 SL + dicamba, MOA 4 (Banvel) 4 SL (Clarity) 4 SL	2 pt + 1 to 2 pt 1 to 2 pt	0.95 + 0.5 to 1	This is an effective way to reduce perennial broadleaf weeds in succeeding crops. Follow label precautions on dicamba label concerning drift to sensitive crops. Delay small grain seeding at least 20 days.
Postharvest, Bermudagrass, other annual and perennial weeds			
glyphosate, MOA 9 (numerous brands and formulations)	See label	1.5 (lb a.e.)	This is an effective way to reduce perennial weeds in succeeding crops. Apply at least 10 to 14 days before killing frost. Rate can be increased up to 3.75 pounds a.e. Include adjuvant according to the label for the brand used. Dicamba may be mixed with glyphosate.

CHEMICAL WEED CONTROL

CHEMICAL WEED CONTROL

Weed Response to Preemergence Herbicides — Corn

W. J. Everman, Crop Science Department

Ratings based upon average to good soil and weather conditions for herbicide performance and upon proper application rate, technique, and timing.

Table 7-1B. Weed Response to Preemergence Herbicides in Corn

Species	Atrazine	Atrazine + Simazine	Bicep II Magnum, Brawl II, Cinch ATZ, or Medal II AT	Bullet or Lariat	Callisto	Corvus	Dual II Magnum, Brawl II, Cinch, or Medal II	Outlook	Guardsman Max	Degree, Harness, Surpass, or TopNotch	Degree Xtra, Fultime, or Harness Xtra	Micro-Tech	Lexar	Accuron	Prowl H2O	Zidua
Grasses:																
Bermudagrass	N	N	N	N	N	N	N	N	N	N	N	N	N	N	N	N
Broadleaf signalgrass	P	P	G	FG	P	—	G	FG	FG	G	G	FG	G	G	P	GE
Crabgrass	G[1]	G[2]	E	E	F	E	E	E	E	E	E	E	E	E	F	E
Fall panicum	N	FG	E	E	PN	GE	E	E	E	E	E	E	E	E	PF	E
Foxtails	F	FG	E	E	PN	E	E	E	E	E	E	E	E	E	PF	E
Goosegrass	F	FG	E	E	PN	GE	E	E	E	E	E	E	E	E	PF	E
Johnsongrass																
Seedling	N	N	PF	PF	N	—	PF	PF	PF	PF	PF	PF	F	F	PF	PF
Rhizome	N	N	N	N	N	—	N	N	N	N	N	N	N	N	N	N
Shattercane	N	N	P	P	N	—	P	P	P	P	P	P	PF	PF	PF	P
Texas panicum	N	N	PF	PF	N	G	PF	PF	PF	PF	PF	PF	PF	PF	PF	F
Sedges:																
Nutsedge																
Yellow	N	N	F	PF	P	F	FG[3]	F	F	PF	PF	PF	FG	FG	N	F
Purple	N	N	N	N	—	—	N	N	N	N	N	N	N	N	N	N
Broadleaf Weeds:																
Balloonvine	G	GE	G	G	—		N	N	G	N	G	N	G	--	N	N
Burcucumber[4]	F	FG	F	F	—	—	N	N	F	N	F	N	F	--	N	N
Cocklebur	G	GE	G	G	PF	G	N	N	G	N	G	N	G	E	N	N
Eastern black nightshade	E	E	E	E	E	E	F	F	E	F	E	FG	E	E	N	F
Florida beggarweed	G	GE	G	G	—	—	F	F	G	F	G	F	G	G	N	—
Florida pusley	E	E	E	E	—	GE	GE	GE	E	GE	E	GE	E	E	G	—
Hemp sesbania	F	F	F	F	—	—	N	N	F	N	F	N	F	G	N	—
Jimsonweed	E	E	E	E	G	G	N	N	E	N	E	N	E	E	N	F
Lambsquarters	E	E	E	E	E	E	F	FG	E	F	E	F	E	E	G	FG
Morningglory	G	G	G	G	FG	G	N	N	G	N	G	N	G	G	N	N
Pigweed	E	E	E	E	E	E	G	GE	E	GE	E	GE	E	E	FG	E
Prickly sida	E	E	E	E	—	G	P	P	E	P	E	P	E	E	N	P
Ragweed																
Common	E	E	E	E	F	G	PF	F	E	PF	E	PF	E	E	N	F
Giant	FG	G	G	G	—	—	N	N	G	N	G	N	FG	G	N	N
Sicklepod	G	GE	G	G	P	G	N	N	G	N	G	N	G	G	N	N
Smartweed	G	GE	G	G	GE	—	N	N	G	N	G	N	G	G	N	F
Tropic croton	G	GE	G	G	PN	G	N	N	G	N	G	N	G	G	N	—
Velvetleaf	G	G	G	G	E	G	N	N	G	N	G	N	G	G	N	F

[1] No control of smooth crabgrass.

[2] Poor to fair on smooth crabgrass.

[3] Dual is normally good on yellow nutsedge when incorporated

[4] Multiple flushes of germination; one application of any herbicide will seldom be adequate.

Key:

E = excellent control, 90% or better

G = good control, 80% to 90%

F = fair control, 50% to 80%

P = poor control, 25% to 50%

N = no control, less than 25%

Weed Response to Postemergence Herbicides — Corn
W. J. Everman, Crop Science Department

Ratings based upon average to good soil and weather conditions for herbicide performance and upon proper application rate, technique, and timing.

Table 7-1C. Weed Response to Postemergence Herbicides — Corn

Species	Accent	Aim[1]	Atrazine[1]	Banvel, Clarity, Distinct, or Status	Basagran	Buctril	Callisto	Capreno	2,4-D	Evik[2]	Glyphosate[5]	Harmony SG	Liberty[3]	Lightning[4]	Linex[2]	Resource	Revulin Q	Steadfast Q
Grasses:																		
Bermudagrass	N	N	N	N	N	N	N	N	N	N	F[7]	N	NP	N	N	N	N	N
Broadleaf signalgrass	GE	N	F	N	N	N	P	GE	N	E	E	N	GE	G	GE	N	G	G
Crabgrass	PF	N	FG[6]	N	N	N	F	GE	N	GE	E	N	G	PF	GE	N	PF	PF
Fall panicum	G	N	P	N	N	N	P	E	N	GE	E	N	E	PF	GE	N	G	G
Foxtails	G	N	G	N	N	N	P	E	N	E	E	N	E	G	E	N	G	G
Goosegrass	P	N	G	N	N	N	P	E	N	GE	E	N	PF	P	G	N	P	P
Johnsongrass																		
Seedling	E	N	P	N	N	N	P	E	N	GE	E	N	GE	GE	G	N	E	E
Rhizome	GE	N	N	N	N	N	N	N	N	P	E	N	F[11]	G[8]	NP	N	GE	G
Shattercane	E	N	P	N	N	N	—	E	N	G	E	N	—	G	FG	N	E	E
Texas panicum	G	N	NP	N	N	N	—	GE	N	G	E	N	—	PF	G	N	G	G
Sedges:																		
Nutsedge																		
Yellow	P	N	PF	N	G[12]	N	F	F	N	F	FG[7]	N	P	F	F	N	P	P
Purple	N	N	N	N	N	N	F	F	N	PF	G	N	P	FG	P	N	N	N
Broadleaf Weeds:																		
Balloonvine	—	—	G	G	P	—	—	—	G	—	—	—	—	—	—	P	—	—
Burcucumber[9]	F	N	FG	F	P	F	—	—	P	F	E	PF	G	P	F	FG	F	F
Cocklebur	F	N	E	E	E	E	E	GE	E	E	E	FG	E	E	E	G	E	F
Eastern black nightshade	N	G	GE	E	P	G	G	E	F	G	FG	P	G	GE	PF	FG	G	P
Florida beggarweed	—	—	G	G	N	E	—	—	FG	E	G	—	E	—	E	P	—	—
Florida pusley	N	FG	G	G	PN	E	—	GE	G	E	P	—	FG	FG	G	—	—	—
Hemp sesbania	F	—	FG	E	P	G	—	—	E	F	P	—	—	—	—	P	—	—
Jimsonweed	FG	N	E	E	E	E	E	E	E	E	E	F	G	E	E	G	E	E
Lambsquarters	P	G	E	E	FG	E	E	GE	E	E	E	E	E	G	E	G	E	F
Morningglory	F	G	G	E	P	G	GE	GE	E	E	FG[10]	FG	E	G	E	FG	GE	G
Pigweed	G	G	E	E	N	F	E	E	E	E	E	E	FG	E	E	G	E	G
Prickly sida	P	—	GE	G	G	F	P	G	G	GE	G	P	GE	G	GE	—	P	P
Ragweed																		
Common	P	P	GE	E	G	E	FG	GE	E	E	E	F	E	PF	E	G	FG	P
Giant	P	N	F	GE	GE	E	—	G	E	G	G	P	G	P	G	P	—	P
Sicklepod	F	—	G	GE	N	N	P	G	G	G	E	P	E	F[13]	GE	N	P	F
Smartweed	G	—	G	E	E	GE	G	E	F	G	G	E	E	GE	GE	P	G	G
Tropic croton	—	—	G	GE	F	FG	—	G	G	G	G	P	—	P	G	P	—	—
Velvetleaf	F	E	G	G	G	G	G	E	G	G	E	G	G	GE	G	E	G	F

[1] Assumes addition of crop oil concentrate.

[2] Apply directed only.

[3] Apply only to Liberty Link hybrids.

[4] Apply only to Clearfield corn hybrids.

[5] Apply only to Roundup Ready (glyphosate-resistant) hybrids. See comments on resistance management in TABLE 7-10.

[6] No control of smooth crabgrass.

[7] Control is good with two applications of glyphosate.

[8] Follow-up treatment with Accent may be needed for acceptable control.

[9] Multiple flushes of germination; one application of any herbicide will seldom be adequate.

[10] With good application timing and a follow-up application as needed, morningglory control can be good.

[11] Liberty applied twice is usually good on johnsongrass.

[12] Two applications may be needed for good control.

[13] Sicklepod control by Lightning can be erratic. For more consistent control, mix atrazine, Banvel, Clarity, Distinct, Marksman, or 2,4-D with Lightning.

Key:
E = excellent control, 90% or better
G = good control, 80% to 90%
F = fair control, 50% to 80%
P = poor control, 25% to 50%
N = no control, less than 25%

CHEMICAL WEED CONTROL

Chemical Weed Control in Cotton

A. C. York, William Neal Reynolds Professor Emeritus, Crop and Soil Sciences Department

Table 7-2A. Brand Names and Formulations of Active Ingredients Listed in Table 7-2B.

Active ingredient(s)	Brand name(s)[1]	Formulation[2]	Mode of Action
acetochlor	Warrant	3 CS	15
acetochlor + fomesafen, premix	Warrant Ultra	3.45 CS (2.82 + 0.63)	15 + 14
carfentrazone	Aim EC	2 EC	14
clethodim	Intensity One, Select Max, TAPOUT	0.97 EC	1
	Arrow 2EC, Agri Star Clethodim 2EC, Avatar S2, Dakota, Grassout Max, Intensity, Section 2 EC, Select 2EC, Shadow, Tide Clethodim 2EC, Volunteer, Willowood Clethodim 2EC	2 EC	1
	Shadow 3 EC, Section Three	3 EC	1
2,4-D	Numerous brand names, formulations	EC or S	4
dicamba, diglycolamine salt	Clarity, Clash, Detonate, Dicamba HD, Dicash DGA-4, Sterling Blue, Strut	4[3] S	4
dicamba acid + 2,4-D ester	Burnmaster	4.07 EC (1 + 3.077)	4 + 4
	Spitfire	3.57 EC (0.5 + 3.07)	4 + 4
dimethenamid-P	Outlook	6 EC	15
diuron	Direx 4L, Diuron 4L, Parrot 4L, Sekor 4L, SuperDi 4L	4 F	7
	Diuron 80, Diuron 80 DF, Diuron 80 WDG, Karmex DF, Parrot DF	80 WDG	7
diuron + linuron, premix	Layby Pro	4 F (2 + 2)	7 + 7
fluazifop p-butyl	Fusilade DX	2 EC	1
flumiclorac pentyl ester	Resource	0.86 EC	14
flumioxazin	Outflank, Panther, Rowel, Valor SX, Warfox	51 WDG	14
	Panther SC	4 F	14
flumioxazin + pyroxasulfone, premix	Fierce	76 WDG (33.5 + 42.5)	14 + 15
fluometuron	Cotoran 4L	4 F	7
fomesafen	Battle Star, Foma 1.88, Rumble, Willowood Fomesafen 1.88	1.88 S	14
	Dawn, Foma 2.0, Fomesafen 2 SL, Reflex, Ringside, Shafen, Top Gun, Willowood Fomesafen 2 SL	2 S	14
	Sinister	2.87 S	14
fomesafen + glyphosate, premix	Flexstar GT3.5	2.82 S (0.56 + 2.26[3])	14 + 9
glufosinate-ammonium	Agri Star Surmise, Cheetah, Forfeit 280, Interline, Liberty 280 SL, Refer 280 SL, Willowood Glufosinate 280SL	2.34 S	10
glyphosate	See Table 7-9.	S	9
glyphosate + S-metolachlor, premix	Sequence	5.25 EC (2.25[3] + 3)	9 + 15
lactofen	Cobra	2 EC	14
MSMA	MSMA 6 Plus	6 S	17
	MSMA 6.6	6.6 S	17
paraquat	Gramoxone SL	2 S	22
	Bonedry, Devour, Firestorm, Helmquat 3SL, Paraquat Concentrate, Para-Shot 3.0, Parazone 3SL, Quik-Quat, Willowood Paraquat 3SL	3 S	22
pendimethalin	Acumen, Framework, Helena Pendimethalin, Pendipro 3.3 EC, Prowl 3.3 EC, Stealth	3.3 EC	3
	Prowl H2O, Satellite Hydrocap	3.8 AS	3
prometryn	Caparol 4L, Cotton Pro	4 F	5
prometryn + trifloxysulfuron, premix	Suprend	80 WDG (79.3 + 0.7)	5 + 2
pyraflufen ethyl	ET Herbicide/Defoliant	0.208 EC	14
pyrithiobac sodium	Dupont Staple LX, Pyrimax 3.2 L, Pysonex	3.2 S	2
pyroxasulfone	Zidua	85 WDG	15
quizalofop p-ethyl	Assure II, Se-Cure EC, Targa	0.88 EC	1
rimsulfuron + thifensulfuron, premix	Leadoff	33.4 WDG (16.7 + 16.7)	2 + 2
	Crusher	50 WDG (25 + 25)	2 + 2
S-metolachlor	Brawl, Charger Basic, Dual Magnum, Medal	7.62 EC	15
	Brawl II, Dual II Magnum, Medal II	7.64 EC	15
S-metolachlor + fomesafen, premix	Prefix	5.29 EC	14 + 15
saflufenacil	Sharpen	2.85 SC	14
sethoxydim	Nufarm Sethoxydim SPC, Poast Plus	1 EC	1
	Poast	1.5 EC	1
thifensulfuron + tribenuron, premix	Audit 1:1, Edition Broadspec, Firstshot, Rapport Broadspec	50 WDG (25 + 25)	2 + 2
	Harmony Extra SG	50 WDG (33.33 + 16.67)	2 + 2
	Nimble, Treaty Extra, T-Square, Volta Extra	75 WDG (50 + 25)	2 + 2
trifloxysulfuron	Envoke	75 WDG	2
trifluralin	Treflan HFP, Trifluralin 4 EC, Trifluralin HF, Triflurex HFP, Trust	4 EC	3

[1] Brands registered for sale and use in cotton in North Carolina in 2016 (http://www.kellysolutions.com/nc/searchbychem.asp); labels available at www.cdms.net.

[2] AS, aqueous suspension; CS, capsule suspension; EC, emulsifiable concentrate; F, flowable; S, solution; SC, suspension concentrate; WDG, water dispersible granule.

[3] Glyphosate rate expressed as acid equivalent.

NOTE: A mode of action code (MOA) has been added to the Herbicide and Formulation column of the following table. Use MOA codes for herbicide resistance management. See Table 7-10, Herbicide Resistance Management, for details.

Table 7-2B. Chemical Weed Control in Cotton

Herbicide, Mode of Action Code and Formulation	Amount of Formulation Per Acre	Pounds Active Ingredient Per Acre	Precautions and Remarks
Early Preplant Burndown, Burndown of emerged annual weeds in no-till, strip-till, or stale seedbed systems			
glyphosate; MOA 9	See label	0.56 to 1.13 (lb a.e.)	Apply any time prior to planting to control emerged weeds. See labels for weeds controlled, application rates for specific weeds, and application directions and precautions. Does not adequately control cutleaf eveningprimrose, field pansy, or Carolina geranium, and may not adequately control wild radish. Glyphosate is available in several formulations. Glyphosate formulations and application rates should be compared on the basis of pounds of glyphosate acid equivalent (a.e.) per gallon and per acre, respectively. Rates in the preceeding column are expressed as a.e. See Table 7-9 for glyphosate rate conversions. Adjuvant recommendations vary according to the glyphosate product used. See label of brand used for specific recommendations. Cover crops: Wheat < 12 inches: 0.56 lb a.e. Wheat > 12 inches: 0.75 lb a.e. Rye < 18 inches: 0.56 lb a.e. Rye > 18 inches: 0.75 lb a.e. **See comments under EARLY PREPLANT BURNDOWN—Glyphosate-resistant horseweed.**
glyphosate; MOA 9 + Aim (2 EC); MOA 14	See label + 0.5 to 1 fl oz	0.56 to 1.13 (lb a.e.) + 0.008 to 0.016	See comments for glyphosate alone. Aim contains carfentrazone; See Table 7-2A. Aim added to glyphosate will increase speed of control and may improve control of some species, but overall long-term control is generally not improved. This tank mix will not control cutleaf eveningprimrose, wild radish, or glyphosate-resistant horseweed. There is no waiting period between application and cotton planting.
glyphosate; MOA 9 + Burnmaster (4.07 EC) or Spitfire (3.57 EC); MOA 4	See label + 1.5 to 2 pt or 1.5 to 2 pt	0.56 to 1.13 (lb a.e.) + 0.76 to 1.02 or 0.67 to 0.89	See comments for glyphosate alone. Burnmaster contains 1.0 lb a.e./gal of dicamba acid plus 3.07 lb a.e./gal of 2,4-D ester. Spitfire contains 0.5 lb a.e./gal of dicamba acid plus 3.07 lb a.e./gal of 2,4-D ester. Following application of either Burnmaster or Spitfire and a mimimum of 1 inch rainfall, a waiting period of at least 30 days is required before planting. At the 2 pt rate, Burnmaster or Spitfire will control glyphosate-resistant horseweed.
glyphosate; MOA 9 + Clarity (4 S); MOA 4	See label + 8 fl oz	0.56 to 1.13 (lb a.e.) + 0.25	See comments for glyphosate alone. Clarity is formulated as the diglycolamine salt of dicamba. This salt is preferred rather than the dimethylamine salt. Other brands containing dicamba diglycolamine salt are listed in Table 7-2A. Following application of dicamba and a minimum of 1 inch rainfall, a waiting period of at least 21 days is required before planting. Dicamba controls or suppresses several annual broadleaf weeds, and it suppresses Carolina geranium and curly dock. See Table 7-2E for weed response. Dicamba is somewhat less effective on cutleaf eveningprimrose than 2,4-D. This tank mixture will control glyphosate-resistant horseweed.
glyphosate; MOA 9 + 2,4-D; MOA 4	See label + See label	0.56 to 1.13 (lb a.e.) + 0.24 to 0.95 (lb a.e.)	See comments for glyphosate alone. Most, but not all, brands of 2,4-D may be applied at least 30 days ahead of cotton planting. 2,4-D is typically applied at 0.48 lb a.e (1 pt/acre of 3.8 lb/gal formulation). See Table 7-2E for weed response. Excellent control of cutleaf eveningprimrose can be obtained with 2,4-D at 0.18 to 0.24 lb a.e. Glyphosate plus 2,4-D is not effective on Carolina geranium. At higher rates (0.95 lb a.e.; 2 pt/acre of 3.8 lb/gal formulation), this tank mix will control small glyphosate-resistant horseweed. Amine and ester formulations of 2,4-D mixed with glyphosate are similarly effective. An amine formulation is preferred if sensitive vegetation is nearby. **See comments under EARLY PREPLANT BURNDOWN—Glyphosate-resistant horseweed.**
glyphosate; MOA 9 + ET (0.208 EC); MOA 14	See label + 0.5 to 2 fl oz	0.56 to 1.13 (lb a.e.) + 0.0008 to 0.0032	See comments for glyphosate alone. ET contains pyraflufen ethyl. ET added to glyphosate will increase speed of control and may improve control of some species, but overall long-term control is generally not improved. This tank mix will not control cutleaf eveningprimrose, wild radish, or glyphosate-resistant horseweed. There is no waiting period between application and cotton planting.
glyphosate; MOA 9 + Firstshot (50 WDG); MOA 2 + 2	See label + 0.8 oz	0.56 to 1.13 (lb a.e.) + 0.025	Firstshot is a 1:1 ratio premix of thifensulfuron plus tribenuron. Other brands of this premix are listed in Table 7-2A. This treatment should be applied at least 14 days prior to planting. Compared to glyphosate alone, the tank mix is more effective on Carolina geranium, curly dock, henbit, swinecress, Virginia pepperweed, wild mustard, and wild radish. Add nonionic surfactant according to the Firstshot label. This tank mix is not effective on cutleaf eveningprimrose or glyphosate-resistant horseweed.
glyphosate; MOA 9 + Harmony Extra (50 WDG); MOA 2 + 2	See label + 0.75 oz	0.56 to 1.13 (lb a.e.) + 0.023	Harmony Extra is a 2:1 ratio premix of thifensulfuron plus tribenuron formulated as 50 WDG. This treatment should be applied at least 14 days prior to planting. Compared to glyphosate alone, the tank mix is more effective on Carolina geranium, curly dock, henbit, swinecress, Virginia pepperweed, wild mustard, and wild radish. See Table 7-2E for weed response. Add nonionic surfactant according to the Harmony Extra label. This tank mix is not effective on cutleaf eveningprimrose or glyphosate-resistant horseweed. Other brands of thifensulfuron plus tribenuron 2:1 ratio, formulated as 75 WDG, include Nimble, Treaty Extra, T-Square, Volta Extra. The equivalent rate of these brands is 0.5 oz/acre.
glyphosate; MOA 9 + Leadoff (33.4 WDG); MOA 2 + 2	See label + 1.5 oz	0.56 to 1.13 (lb a.e.) + 0.031	Leadoff is a 1:1 ratio premix of rimsulfuron plus thifensulfuron. Can be applied from late fall to 30 days prior to planting. Controls emerged winter annual weeds plus provides residual control of later emerging winter weeds. See Leadoff label for adjuvant recommendations. 2,4-D can also be included in the mixture. Leadoff does not substitute for a flumioxazin (Valor, others) application. The best use of Leadoff is a late fall or winter application (December to early March) followed by another burndown application containing flumioxazin 2 to 4 weeks ahead of planting. Crusher 50 WDG also contains a 1:1 ratio of rimsulfuron plus thifensulfuron. The equivalent rate of Crusher is 1 oz/acre.
glyphosate; MOA 9 + Resource (0.86 EC); MOA 14	See label + 2 to 4 fl oz	0.56 to 1.13 (lb a.e.) + 0.013 to 0.027	See comments for glyphosate alone. Resource contains flumiclorac pentyl ester. Resource added to glyphosate will increase speed of control and may improve control of some species, but overall long-term control is generally not improved. This tank mix will not control cutleaf eveningprimrose, wild radish, or glyphosate-resistant horseweed. There is no waiting period between application and cotton planting.

Table 7-2B. Chemical Weed Control in Cotton

Herbicide, Mode of Action Code and Formulation	Amount of Formulation Per Acre	Pounds Active Ingredient Per Acre	Precautions and Remarks
Early Preplant Burndown, Burndown of emerged annual weeds in no-till, strip-till, or stale seedbed systems (continued)			
glyphosate; MOA 9 + Valor SX (51 WDG); MOA 14	See label + 1 to 2 oz	0.56 to 1.13 (lb a.e.) 0.031 to 0.063	See comments for glyphosate alone. Valor SX contains flumioxazin. Other brands of flumioxazin are listed in Table 7-2A. In no-till or stale seedbed system, a minimum of 14 days must pass and a 1-inch rainfall must occur between flumioxazin application and cotton planting when flumioxazin is applied at 1 oz/acre; 21 days must pass when applied at 1.5 to 2 oz/acre. If a strip-till operation occurs between flumioxazin application and cotton planting, the waiting interval can be reduced to 14 days for 2 oz flumioxazin. However, strip-tilling after flumioxazin application will reduce or eliminate weed control in the tilled strip. Compared to glyphosate alone, the tank mix will improve control of cutleaf eveningprimrose and wild radish. However, this tank mix is less effective than glyphosate plus 2,4-D on primrose and wild radish. Dicamba or 2,4-D may be added to this mixture. Regardless of glyphosate product used, a nonionic surfactant at 1 qt/100 gal. is recommended on flumioxazin labels. Applied at 1 oz/acre, flumioxazin will give 2 to 4 weeks residual control of lambsquarters, pigweed, prickly sida, spurge, and Florida pusley. At 2 oz/acre, flumioxazin will give 6 to 8 weeks residual control of these species. Application to cover crop or dense stand of winter weeds may reduce residual control. This tank mixture will not control glyphosate-resistant horseweed. **See comments under EARLY PREPLANT BURNDOWN—Glyphosate-resistant horseweed.** **Carefully follow label directions for cleaning out the sprayer after each day's use.**
paraquat; MOA 22 2 lb/gal formulations 3 lb/gal formulations	 2.6 to 4 pt 1.7 to 2.7 pt	0.65 to 1	Apply any time prior to planting to control emerged weeds. Add nonionic surfactant at 1 pt per 100 gal or crop oil concentrate at 1 gal per 100 gal. Follow directions and precautions on label. Usually not adequately effective on cutleaf eveningprimrose, horseweed, or larger wild mustard or wild radish. Apply 0.65 lb a.i. for wheat and 0.5 lb a.i. for rye cover crops. Best control of small grain cover crops will be achieved if paraquat is applied at the boot stage or later. See Table 7-2A for brands.
paraquat; MOA 22 2 lb/gal formulations 3 lb/gal formulations + diuron, MOA 7 4 F formulations 80 WDG formulations	 2.6 to 4 pt 1.7 to 2.7 pt + 1 to 2 pt 0.63 to 1.25 lb	0.65 to 1 + 0.5 to 1	See comments for paraquat alone. See diuron label for use rates on various soils. Apply 15 to 45 days ahead of planting. If Cotoran is applied preemergence, reduce Cotoran rate to account for residual activity of diuron. When mixed with crop oil concentrate and applied in April, this combination has given good control of common weeds, including cutleaf eveningprimrose. See Table 7-2A for brands of paraquat and diuron.
Early Preplant Burndown, Glyphosate-resistant horseweed			
glyphosate; MOA 9 + 2,4-D; MOA 4 + Valor SX (51 WDG); MOA 14	See label + See label + 2 oz	0.56 to 1.13 (lb a.e.) + 0.95 (lb a.e.) + 0.063	Glyphosate-resistant horseweed is very common in eastern North Carolina and is beginning to become a problem in the Piedmont. See previous comments concerning waiting intervals between application of 2,4-D, dicamba, and flumioxazin and planting. The 2,4-D or dicamba is needed in the mixture to control emerged resistant horseweed, and the flumioxazin will control horseweed germinating after this application. Dicamba may be somewhat more effective than 2,4-D on horseweed. Better results will be obtained if treatment is applied prior to first of April.
glyphosate; MOA 9 + Clarity (4 SL); MOA 4 + Valor SX (51 WDG); MOA 14	See label + 8 fl oz + 1 to 2 oz	0.56 to 1.13 (lb a.e.) + 0.25 + 0.063	Valor SX contains flumioxazin; Clarity contains dicamba diglycolamine salt; Liberty contains glufosinate. See Table 7-2A for other brands containing these active ingredients. Sharpen contains saflufenacil. Do not mix flumioxazin with Sharpen. See Sharpen label for specifics on adjuvant selection. **After applying Sharpen, wait to plant cotton until at least 42 days and an accumulation of 1 inch of rainfall has occurred.** Do not apply Sharpen to soils classified as sand with less than 1.5% organic matter.
glyphosate; MOA 9 + Sharpen (2.85 F); MOA 14	See label + 1.0 fl oz	0.56 to 1.13 (lb a.e.) + 0.022	Glufosinate (Liberty, others) is recommended only for fields where growers have failed to control glyphosate-resistant horseweed and it is too late to use 2,4-D or dicamba. Best results with glufosinate will be obtained if sprayed when daytime temperatures exceed 75 degrees. If greater than 29 oz applied preplant, the seasonal total applied cannot exceed 72 fl oz.
Liberty (2.34 S); MOA 10	29 to 43 fl oz	0.53 to 0.79	
At Planting Burndown, Burndown of cover crops and weeds at planting			
glyphosate; MOA 9	See label	0.56 to 1.13 (lb a.e.)	See Table 7-9 for glyphosate rate conversions. If an early burndown treatment was applied (see COTTON—Early Preplant Burndown), apply glyphosate or paraquat in combination with desired residual herbicides at planting. Glyphosate or paraquat may be tank mixed with registered preemergence herbicides and applied after planting but before cotton emergence. See suggested rates and precautions on labels of tank-mix partners. If an early burndown treatment was not used, apply glyphosate or paraquat 7 to 21 days ahead of planting. If weeds are emerged at planting, make a second application in combination with desired residual herbicides. See comments on residual herbicides under COTTON—Preemergence. Glyphosate and paraquat rates depend upon weed species and size; see labels for recommended rates. Add nonionic surfactant at 1 pt per 100 gal or crop oil concentrate at 1 gal per 100 gal spray solution to paraquat. Need for adjuvants with glyphosate depends upon brand used; see specific labels for details. Cover crops: Wheat < 12 inches: glyphosate, 0.56 lb a.e. or paraquat, 0.65 lb a.i. Wheat > 12 inches: glyphosate 0.75 lb a.e. or paraquat 0.65 lb a.i. Rye < 18 inches: glyphosate, 0.56 lb a.e. or paraquat 0.5 lb a.i. Rye > 18 inches: glyphosate, 0.75 lb a.e. or paraquat 0.5 lb a.i.
paraquat; MOA 22 2 lb/gal formulations 3 lb/gal formulations	 2.6 to 4 pt 1.7 to 2.7 pt	0.65 to 1	
Liberty (2.34 S); MOA 10	29 to 43 fl oz	0.53 to 0.79	See comments under EARLY PREPLANT BURNDOWN for more details. Suggested only if glyphosate-resistant horseweed is a problem. Liberty contains glufosinate. Other brands of glufosinate are listed in Table 7-2A.
Preplant Incorporated, Annual grasses and certain small-seeded broadleaf weeds			
Prowl 3.3 EC; MOA 3	1.2 to 3.6 pt	0.5 to 1.5	Prowl 3.3 EC and Prowl H2O contain pendimethalin; Treflan contains trifluralin; see Table 7-2A for other brands of pendimethalin and trifluralin. Consult labels for application rates and for time, method, and depth of incorporation. Deep incorporation, especially on sandy soils, may cause stunting and delayed crop development. Incorporation of trifluralin can be delayed 24 hr; pendimethalin incorporation can be delayed 7 days. Immediate incorporation is suggested.
Prowl H2O (3.8 AS); MOA 3	2 to 4 pt	0.95 to 1.9	
Treflan (4EC); MOA 3	1 to 2 pt	0.5 to 1	
Preemergence, Annual grasses and pigweed			
Warrant (3 CS); MOA 15	3 pt	1.125	Warrant contains acetochlor in an encapsulated formulation. It can be applied in combination with another preemergence herbicide such as diuron, fluometuron (Cotoran), or fomesafen (Reflex, others).

246

Table 7-2B. Chemical Weed Control in Cotton

Herbicide, Mode of Action Code and Formulation	Amount of Formulation Per Acre	Pounds Active Ingredient Per Acre	Precautions and Remarks
Preemergence, Annual grasses, pigweed, and lambsquarters			
Prowl 3.3 EC; MOA 3	2.4 to 3.6 pt	1.0 to 1.5	See labels for rates on specific soils. May be mixed with diuron, fluometuron (Cotoran), fomesafen (Reflex, others), or pyrithiobac (Staple LX, others).
Prowl H2O (3.8 AS); MOA 3	2.1 to 4 pt	1.0 to 1.9	
Preemergence, Annual grasses and various broadleaf weeds			
Warrant Ultra (3.45 CS); MOA 15 + 14	3 pt	1.29	Warrant Ultra is a premix formulation containing 2.82 lb/gal of acetochlor plus 0.63 lb/gal fomesafen. Use preemergence only on coarse-textured soils.
Preemergence, Annual broadleaf weeds			
Cotoran (4 F); MOA 7	1 to 2 qt	1 to 2	Use lower end of rate range on lighter soils. May be tank mixed with pendimethalin (Prowl, others), fomesafen (Reflex, others), pyrithiobac (Staple LX, others), or Warrant.
diuron, MOA 7 4 F formulations 80 WDG formulations	1 to 2 pt 0.63 to 1.25 lb	0.5 to 1	See labels for rates on specific soils. May be mixed with pendimethalin (Prowl, others), fomesafen (Reflex, others), pyrithiobac (Staple LX, others), or Warrant. See rotational restrictions and maximum seasonal use rates on label. See Table 7-2A for brands of diuron.
Reflex (2 S); MOA 14	1 pt	0.25	Reflex contains fomesafen. Other brands containing 1.88 or 2 lb/gal of fomesafen are listed in Table 7-2A. Suggested primarily for control of Palmer amaranth. Label restricts preemergence application only to coarse-textured soils. May be tank mixed with diuron, Cotoran, pendimethalin (Prowl, others), pyrithiobac (Staple LX, others), or Warrant. See labels for specific comments on tank mixing.
Sinister (2.87 S); MOA 14	0.7 pt	0.25	Sinister contains fomesafen. Suggested primarily for control of Palmer amaranth. Label restricts preemergence application only to coarse-textured soils. May be tank mixed with diuron, Cotoran, pendimethalin (Prowl, others), pyrithiobac (Staple LX, others), or Warrant. See labels for specific comments on tank mixing.
Staple LX (3.2 SL); MOA 2	1.7 to 2.1 fl oz	0.0425 to 0.053	Staple contains pyrithiobac. Other brands of pyrithiobac are listed in Table 7-2A. Do not apply pyrithiobac preemergence on soils with less than 0.5% organic matter. May tank mix with diuron, Cotoran, pendimethalin (Prowl, others), or fomesafen (Reflex, others). Palmer amaranth biotypes resistant to pyrithiobac are very common in North Carolina.
Post-emergence Overtop, Annual broadleaf weeds, any cultivar			
Envoke (75 WDG); MOA 2	0.1 oz	0.0047	Envoke contains trifloxysulfuron. May be applied overtop cotton after it has a minimum of five true leaves up to 60 days prior to harvest. On larger cotton, directed application is preferred for better coverage of weeds. Add nonionic surfactant at 0.25% by volume (1 qt per 100 gal). May make two applications, but do not exceed 0.0188 lb a.i./acre per year of trifloxysulfuron from the combined use of all trifloxysulfuron-containing products (Envoke and Suprend). Do not mix with other pesticides when applying overtop of cotton. See label for rotational restrictions and weeds controlled. Controls most broadleaf weeds with timely application; common exceptions include prickly sida, jimsonweed, copperleaf, and spurred anoda. Reduced growth of cotton, due to shortened internodes, is sometimes observed. Shortened internodes are more likely on smaller cotton. Envoke may also be applied overtop at 0.15 oz/acre if needed for larger weeds.
			Pyrithiobac (Staple, others) and Envoke are ALS inhibitors. Biotypes of Palmer amaranth, common ragweed, and cocklebur resistant to ALS inhibitors have been found in North Carolina; ALS-resistant Palmer amaranth is very common. To aid in resistance management, it is suggested that an ALS inhibitor be applied only once per year.
Staple LX (3.2 S); MOA 2	2.6 to 3.8 fl oz	0.065 to 0.095	Staple contains pyrithiobac. Other brands of pyrithiobac are listed in Table 7-2A. May be applied overtop of cotton from cotyledonary stage up to 60 days prior to harvest. Avoid application during or shortly after cool weather. Add nonionic surfactant at 0.25% by volume (1 qt per 100 gal). Do not add crop oil. May make two applications per year, not exceeding a total of 5.1 fl oz. May be tank mixed with most insecticides, but do not tank mix with any product containing malathion. Tank mixing with post-emergence grass control herbicides is discouraged. See label for rotational restrictions and weeds controlled. Timing of application is very important for most weeds. Apply before susceptible broadleaf weeds exceed 4 inches tall. Does not control tall morningglory, lambsquarters, or common ragweed. Only suppresses sicklepod. Pyrithiobac and Envoke are ALS inhibitors. Biotypes of Palmer amaranth, common ragweed, and cocklebur resistant to ALS inhibitors have been found in North Carolina; ALS-resistant Palmer amaranth is very common. To aid in resistance management, it is suggested that an ALS inhibitor be applied only once a year.
Envoke (75 WDG); MOA 2 + Staple LX (3.2 S); MOA 2	0.1 oz + 1.3 to 1.9 fl oz	0.0047 + 0.033 to 0.048	See comments for Envoke and Staple applied alone. Compared with Envoke alone, tank mix is more effective on eclipta, jimsonweed, and spurred anoda. Compared with Staple alone, tank mix is more effective on ragweed, lambsquarters, tall morningglory, and sicklepod. Add nonionic surfactant at 0.25% by volume.
Post-emergence Overtop, Annual grasses, any cultivar			
Assure II (0.88 EC); MOA 1	7 to 8 fl oz	0.05 to 0.06	Assure II contains quizalofop p-ethyl. Others brands are listed in Table 7-2A. Apply to actively growing grass not under drought stress. See label for maximum weed size to treat and suggested rate. Apply in 10 to 40 gpa. Add either crop oil concentrate at 1% by volume (1 gal per 100 gal) or nonionic surfactant at 0.25% by volume (1 qt per 100 gal). A second application may be made if needed. May use 5 oz per acre for seedling johnsongrass or shattercane.
Fusilade DX (2 EC); MOA 1	8 to 12 fl oz	0.125 to 0.188	Apply to actively growing grass not under drought stress. Suggested application rate varies by species and weed size; see label. Apply in 5 to 40 gpa at 40 to 60 psi. Add either crop oil concentrate at 1% by volume (1 gal per 100 gal) or nonionic surfactant at 0.25% by volume (1 qt per 100 gal). Second application may be made if necessary. May use 6 oz per acre for seedling johnsongrass or shattercane.
Poast (1.5 EC) or Poast Plus (1.0 EC) MOA 1	16 fl oz or 24 lf oz	0.19	Apply to actively growing grass not under drought stress. Consult label for maximum grass size to treat. Apply in 5 to 20 gpa at 40 to 60 psi. Add 2 pt per acre of crop oil concentrate. A second application may be made if necessary. Consult label for special rates for early treatment or rescue treatment. Poast and Poast Plus contain sethoxydim. See Table 7-2A for other brands of sethoxydim.
Select (2 EC) or Select Max (0.97 EC) MOA 1	6 to 8 fl oz or 9 to 16 fl oz	0.094 to 0.125 or 0.068 to 0.121	Contains clethodim; see Table 7-2A for other brands. Apply to actively growing grass not under drought stress. See labels for maximum weed size to treat and suggested rate. Apply in 10 to 40 GPA. Add adjuvant according to label. A second application may be made if needed.
Post-emergence Overtop, Bermudagrass and Johnsongrass, any cultivar			
Assure II (0.88 EC); MOA 1	10 fl oz	0.07	Assure II contains quizalofop p-ethyl. Others brands are listed in Table 7-2A. Apply to actively growing bermudagrass when runners are up to 6 inches or johnsongrass 10 to 24 inches tall. A second application of 7 fl oz per acre may be applied if needed when bermudagrass regrowth is up to 6 inches or johnsongrass regrowth is 6 to 10 inches. Add either a crop oil concentrate at 1% by volume (1 gal per 100 gal) or a nonionic surfactant at 0.25% by volume (1 qt per 100 gal).

CHEMICAL WEED CONTROL

CHEMICAL WEED CONTROL

Table 7-2B. Chemical Weed Control in Cotton

Herbicide, Mode of Action Code and Formulation	Amount of Formulation Per Acre	Pounds Active Ingredient Per Acre	Precautions and Remarks
Post-emergence Overtop, Bermudagrass and Johnsongrass, any cultivar (continued)			
Fusilade DX (2 EC); MOA 1	12 fl oz	0.19	Apply to actively growing bermudagrass when runners are 4 to 8 inches long or johnsongrass 8 to 18 inches tall and before boot stage. If regrowth occurs, make a second application of 8 fl oz when bermudagrass runners are 4 to 8 inches or johnsongrass regrowth is 6 to 12 inches. Add either crop oil concentrate at 1 qt per acre or nonionic surfactant at 0.25% by volume (1 qt per 100 gal).
Poast (1.5 EC) or Poast Plus (1.0 EC) MOA 1	24 fl oz or 36 fl oz	0.28	Apply to actively growing bermudagrass before runners exceed 6 inches or to johnsongrass up to 25 inches tall. If regrowth occurs or new plants emerge, make a second application before bermudagrass runners exceed 4 inches or johnsongrass regrowth exceeds 12 inches. See labels for second application rates. Add 2 pt of crop oil concentrate per acre. Poast and Poast Plus contain sethoxydim. See Table 7-2A for other brands of sethoxydim.
Select (2 EC) or Select Max (0.97 EC) MOA 1	8 to 16 fl oz or 12 to 32 fl oz	0.125 to 0.25 or 0.091 to 0.24	Apply to actively growing bermudagrass when runners are up to 6 inches or johnsongrass is 12 to 24 inches tall. A second application may be applied if needed when bermudagrass regrowth is up to 6 inches or jonhsongrass regrowth is 6 to 18 inches. See directions for second application rates. Add adjuvant according to label. Use the higher rate under heavy grass pressure or on larger grass. Contains clethodim; see Table 7-2A for other brands.
Post-emergence Overtop, Annual broadleaf weeds and most annual grasses, LibertyLink, Glytol LibertyLink, or XtendFlex Cultivars only			
Liberty (2.34 S); MOA 10	29 to 43 fl oz	0.53 to 0.79	Contains glufosinate; see Table 7-2A for other brands. Apply overtop or directed from cotton emergence until the early bloom stage. Good spray coverage is critical. Use flat-fan nozzles and a minimum of 15 GPA. Better coverage may be obtained on larger cotton with a semi-directed application. An adjuvant is not necessary. Application time of day is important. Two hours of sunshine before a morning application is suggested. Do not apply later than 1 hour before sunset. Multiple applications are allowed. Liberty at 22 to 29 fl oz can be applied three times, with a seasonal maximum of 87 fl oz. If applied at rates greater than 29 oz, only two applications are allowed and the total rate per season cannot exceed 72 fl oz. Liberty controls most annual grass and broadleaf weeds, although timing of application on pigweed (including Palmer amaranth) and grasses (especially goosegrass) is critical. Preemergence herbicides are encouraged to help in control of pigweed and grasses. Liberty is generally more effective on broadleaf weeds than grasses. Broadleaf weeds should be 2 to 3 inches tall and grasses 1 to 2 inches tall. Postemergence grass control herbicides, such as clethodim, fluazifop, quizalofop, and sethoxydim, should not be mixed with glufosinate. Applications of postemergence grass herbicides and glufosinate should be separated by at least 5 days.
Liberty (2.34 S); MOA 10 + Dual Magnum, 7.62 EC; MOA 15	29 to 43 fl oz + 1 to 1.33 pt	0.53 to 0.79 + 0.95 to 1.27	See comments for Liberty applied alone. Dual contains S-metolachlor; see Table 7-2A for other brands. S-metolachlor will not control emerged weeds, but it can provide residual control of susceptible species such as annual grasses and pigweed species. This treatment may cause foliar burn on the crop. Burn may be enhanced if applied to cotton with dew, under extremely high temperatures, or when mixed with insecticides or adjuvants. Several products containing metolachlor (not S-metolachlor) are available. Metolachlor products are less effective per unit of formulated product than those with S-metolachlor. In general, it takes 1.5 pt of a metolachlor product to give the activity one gets from 1 pt of S-metolachlor.
Liberty (2.34 L); MOA 10 + Outlook (6 EC); MOA 15	29 to 43 fl oz + 12 to 16 fl oz	0.53 to 0.79 + 0.56 to 0.75	See comments for Liberty applied alone. Outlook will not control emerged weeds, but it can provide residual control of susceptible species such as annual grasses and pigweed species. This treatment may cause foliar burn on the crop. Burn may be enhanced if applied to cotton with dew, under extremely high temperatures, or when mixed with insecticides or adjuvants.
Liberty (2.34 L); MOA 10 + Staple LX (3.2 S); MOA 2	29 to 43 fl oz + 1.3 to 3.8 fl oz	0.53 to 0.79 + 0.033 to 0.095	See comments for Liberty applied alone. Staple contains pyrithiobac. See Table 7-2A for other brands of glufosinate and pyrithiobac. See directions on pyrithiobac label concerning adjuvant usage. Staple will improve control of non-ALS-resistant Palmer amaranth plus provide residual control.
Liberty (2.34 L); MOA 10 + Warrant (3 CS); MOA 15	29 to 43 fl oz + 3 pt	0.53 to 0.79 + 1.125	See comments for Liberty applied alone. Warrant will not control emerged weeds, but it can provide residual control of susceptible species such as annual grasses and pigweed species. This treatment may cause foliar burn on the crop. Burn may be enhanced if applied to cotton with dew, under extremely high temperatures, or when mixed with insecticides or adjuvants.
Post-emergence Overtop, Annual broadleaf weeds and most annual grasses, Phytogen Widestrike Cultivars			
Liberty (2.34 S); MOA 10	29 fl oz	0.53	Liberty contains glufosinate; see Table 7-2A for other brands of glufosinate. Phytogen cultivars with the Widestrike trait can be treated with glufosinate. Tolerance to glufosinate in these cultivars is not complete, and varying levels of crop injury may be observed. Greater injury can be expected when glufosinate is mixed with insecticides or other herbicides. **Growers assume the liability** of crop injury when cotton with the Widestrike trait is treated with glufosinate. It is suggested that the rate not exceed 29 oz per application with a maximum of two applications per year. It is also suggested that glufosinate not be applied beyond the 8-leaf stage of cotton and that AMS not be included in the application. See above comments for use of glufosinate on LibertyLink cultivars, including the statement on application time of day.
Post-emergence Overtop, Annual grasses, broadleaf weeds, perennial grasses, and nutsedge; suppression of perennial broadleaf weeds, Roundup Ready Flex, Glytol LibertyLink, or XtendFlex Cultivars only			
glyphosate; MOA 9	See labels	0.56 to 1.13 (lb a.e.)	Glyphosate formulations and application rates should be compared on the basis of pounds of glyphosate acid equivalent (a.e.) per gallon and per acre, respectively. Rates in the preceeding column are expressed as a.e. See TABLE 7-9 for glyphosate rate conversions. Glyphosate controls most annual weeds; exceptions include dayflower, dove weed, Florida pusley, and hemp sesbania. Timely application is critical for morningglory control. Multiple applications are needed for nutsedge and bermudagrass. See label of brand you use for recommended rates and sizes of weeds to treat. Adjuvant recommendations vary according to glyphosate product. See label of brand used for specific recommendations. Glyphosate-resistant Palmer amaranth is common in North Carolina, and glyphosate-resistant common ragweed is present in several counties. Continued heavy reliance on herbicide programs based predominantely on glyphosate will enhance selection for resistant biotypes. Other chemistry, including preemergence herbicides, tank mixes with glyphosate, and layby herbicides in addition to glyphosate, is recommended as part of a resistance-management strategy. See section on Herbicide Resistance and TABLE 7-10A.
glyphosate; MOA 9 + Dual Magnum (7.62 EC); MOA 15	See labels + 1 to 1.33 pt	0.56 to 1.13 (lb a.e.) + 0.95 to 1.27	See comments for glyphosate applied alone. Dual Magnum contains S-metolachlor; see Table 7-2A for other brands. S-metolachlor will not control emerged weeds, but it can provide residual control of susceptible species such as annual grasses and pigweed species. This treatment may cause foliar burn on the crop. Burn may be enhanced if applied to cotton with dew, under extremely high temperatures, or when mixed with insecticide or adjuvant. Several products containing metolachlor (not S-metolachlor) are available. Metolachlor products are less effective per unit of formulated product than those with S-metolachlor. In general, it takes 1.5 pt of a metolachlor product to give the activity one gets from 1 pt of S-metolachlor.

Table 7-2B. Chemical Weed Control in Cotton

Herbicide, Mode of Action Code and Formulation	Amount of Formulation Per Acre	Pounds Active Ingredient Per Acre	Precautions and Remarks
Post-emergence Overtop, Annual grasses, broadleaf weeds, perennial grasses, and nutsedge; suppression of perennial broadleaf weeds, Roundup Ready Flex, Glytol LibertyLink, or XtendFlex Cultivars only (continued)			
glyphosate; MOA 9 + Envoke (75 WDG); MOA 2	See labels + 0.1 oz	0.56 to 1.13 (lb a.e.) + 0.0047	See comments for glyphosate and Envoke applied alone. See Envoke label and glyphosate label for suggestions on adjuvant usage. Tank mix can be applied from 5-leaf cotton stage until 60 days prior to harvest. For better crop safety, however, cotton should have at least 7 to 8 leaves at time of treatment.
glyphosate; MOA 9 + Outlook (6 EC); MOA 15	See labels + 12 to 16 fl oz	0.56 to 1.13 (lb a.e.) + 0.56 to 0.75	See comments for glyphosate applied alone. Can apply from first true leaf to mid-bloom stage. Outlook will not control emerged weeds, but it will provide residual control of susceptible species such as annual grasses and pigweed species. Optimum timing is 2- to 3-leaf cotton, and before weeds emerge. Make only one application per year. Outlook plus glyphosate may cause some foliar burn on cotton; mixing with insecticides or adjuvants may increase burn. Suggested rates are 12 oz on coarse soils, 14 oz on medium soils, and 16 oz on fine soils.
glyphosate; MOA 9 + Staple LX (3.2 S;) MOA 2	See labels + 1.3 to 3.8 fl oz	0.56 to 1.13 (lb a.e.) + 0.033 to 0.095	See comments for glyphosate applied alone. Staple contains pyrithiobac; see Table 7-2A for other brands. Can apply overtop from cotyledonary stage cotton until 60 days prior to harvest. See directions on pyrithiobac labels concerning adjuvant usage. Palmer amaranth resistant to both pyrithiobac and glyphosate is widespread in North Carolina.
glyphosate; MOA 9 + Warrant (3 CS); MOA 15	See labels + 3 pt	0.56 to 1.13 (lb a.e.) + 1.125	See comments for glyphosate applied alone. Apply after cotton is completely emerged but before first bloom. Warrant will not control emerged weeds, but it will provide residual control of susceptible species such as annual grasses and pigweed species. Optimum timing is 2- to 3-leaf cotton, and before weeds emerge. A second application can be made if directed to the soil surface. Warrant plus glyphosate may cause some foliar burn on cotton; mixing with insecticides or adjuvants may increase burn. Warrant rate can be increased to 4 pt on some soils; see label for details.
Sequence (5.25 L); MOA 9 + 15	2.5 pt	0.70 (lb a.e.) + 0.94	Sequence is a premix containing 2.25 lb a.e./gal glyphosate and 3 lb/gal S-metolachlor. See comments for glyphosate alone and glyphosate + S-metolachlor. Apply to cotton in cotyledonary stage up to 10-leaf stage, but not to cotton taller than 12 inches.
Post-emergence Overtop, Volunteer Roundup Ready corn, any cultivar			
Assure II (0.88 EC); MOA 1	5 to 8 fl oz	0.034 to 0.055	Contains quizalofop; see Table 7-2A for other brands. See above comments comments for quizalofop. See quizalofop label for application rates on various sizes of corn and for adjuvant recommendations. Can mix with glyphosate.
Fusilade DX (2 EC); MOA 1	4 to 6 fl oz	0.063 to 0.094	See above comments for Fusilade. See label for application rates on various sizes of corn and for adjuvant recommendations. Can mix with glyphosate.
Poast (1.5 EC) or Poast Plus (1.0 EC) MOA 1	16 fl oz or 24 fl oz	0.189	Contains sethoxydim; see Table 7-2A for other brands. See above comments comments for sethoxydim. See sethoxydim label for application rates on various sizes of corn and for adjuvant recommendations. Can mix with glyphosate.
Select (2 EC) or Select Max (0.97 EC) MOA 1	4 to 8 fl oz or 6 to 12 fl oz	0.063 to 0.125 or 0.045 to 0.106	Contains clethodim; see Table 7-2A for other brands. See clethodim labels for application rates on various sizes of corn and for adjuvant recommendations. Can mix with glyphosate.
Post-emergence Overtop, Volunteer Roundup Ready soybean, any cultivar			
Envoke (75 WDG); MOA 2	0.1 oz	0.0047	See above comments for Envoke. Cotton should have at least five leaves, and the soybean should have no more than four to five trifoliate leaves. Not adequately effective on soybean with the STS trait.
Post-emergence Directed, Cocklebur, small annual grasses, and nutsedge, any cultivar			
MSMA; MOA 17 6 lb/gal formulation 6.6 lb/gal formulation	2.67 pt 2.5 pt	2	Do not apply overtop at these rates. MSMA can be directed alone or mixed with other postemergence broadleaf herbicides on cotton at least 3 inches tall up to first bloom. Do not apply MSMA after first bloom. Adequate control of nutsedge usually requires two applications. Follow label directions for use of adjuvants.
Post-emergence Directed, Annual broadleaf weeds, small annual grasses, and nutsedge, any cultivar			
Caparol (4 F); MOA 5 + MSMA; MOA 17 6 lb/gal formulation 6.6 lb/gal formulation	1.3 to 2.8 pt + 2.67 pt 2.5 pt	0.65 to 1.4 + 2	Caparol contains prometryn; see Table 7-2A for other brands. Apply 1.3 pt Caparol as directed spray only to cotton at least 6 inches tall. Increase to higher rate for the soil type after cotton is at least 12 inches tall. See label for rates on various soil types. Add 2 qt nonionic surfactant per 100 gal spray solution. Do not apply after first bloom. Aim at 1 fl oz or Cobra at 6 to 12.5 fl oz may be added to improve control of larger morningglory. Cotton should be at least 16 inches tall when when applying Aim. Do not allow Aim to contact green stem tissue.
Cobra (2 EC); MOA 14 + MSMA; MOA 17 6 lb/gal formulation 6.6 lb/gal formulation	12.5 oz + 2.67 pt 2.5 pt	0.2 + 2	Apply as directed spray or with hooded sprayer. Cotton should be at least 6 to 8 inches tall, preferably larger. See Cobra label for weeds controlled, directions on weed size and application rates, and use of surfactant or crop oil. Do not apply MSMA after first bloom.
Cobra (2 EC); MOA 14 + diruron, 4 F; MOA 7 + MSMA; MOA 17 6 lb/gal formulation 6.6 lb/gal formulation	6 to 12.5 fl oz + 0.8 to 1.2 pt + 2.67 pt 2.5 pt	0.094 to 0.2 + 0.4 to 0.6 + 2	See Table 7-2A for brands of diuron. Apply as directed spray or with hooded sprayer. Cotton should be at least 12 inches tall. See Cobra label for weeds controlled and directions on weed size and application rates. Add 1 qt per acre of crop oil concentrate. See rotational restrictions on diuron label. Do not apply MSMA after first bloom.
Cotoran (4 L); MOA 7 + MSMA; MOA 17 6 lb/gal formulation 6.6 lb/gal formulation	1 to 2 qt + 2.67 pt 2.5 pt	1 to 2 + 2	Apply as a directed spray only to cotton at least 3 inches tall up to first bloom. Follow label directions for weed size and addition of surfactant. See Cotoran label for maximum application rates per season and rotational restrictions. S-metolachlor (Dual Magnum, others) may be added.
diuron (4 F) or diuron (80 WDG); MOA 7 + MSMA; MOA 17 6 lb/gal formulation 6.6 lb/gal formulation	1.6 to 2.4 pt or 1 to 1.5 lb + 2.67 pt 2.5 pt	0.8 to 1.2 + 2	See Table 7-2A for brands of diuron. Apply as directed spray only to cotton at least 12 inches tall. Adjust rate according to soil type. See application precautions on label. Add nonionic surfactant at 1 to 2 quarts per 100 gal spray solution or crop oil concentrate at 1 gal per 100 gal spray solution. See label for rotational restrictions. Do not apply MSMA after first bloom. Aim at 1 fl oz or Cobra at 6 to 8 fl oz per acre may be added to improve control of larger morningglory. Cotton should be at least 16 inches tall when applying Aim. Do not allow Aim to contact green stem tissue.

Table 7-2B. Chemical Weed Control in Cotton

Herbicide, Mode of Action Code and Formulation	Amount of Formulation Per Acre	Pounds Active Ingredient Per Acre	Precautions and Remarks
Post-emergence Directed, Annual broadleaf weeds, small annual grasses, and nutsedge, any cultivar (continued)			
Fierce (76 WDG); MOA 14 + 15 + MSMA; MOA17 6.0 lb/gal formulation 6.6 lb/gal formulation	3 oz + 2.67 pt 2.5 pt	0.143 + 2	Friece is a premix containing 33.5% flumioxazin plus 42.5% pyroxasulfone. Can be applied with hooded sprayer after cotton is at least 6 inches tall. Do not allow spray solution to contact cotton. Can be applied with layby applicator after cotton is at least 16 inches tall, but do not contact more than the lower 2 inches of cotton stalk. Add non-ionic surfactant according to label. Do not use crop oil, methylated seed oil, or organo-silicant adjuvants. Controls emerged broadleaf weeds normally controlled by flumioxazin plus MSMA and gives residual control of annual grasses, Palmer amaranth and other pigweed species, nightshade, eclipta, Florida pusley, and most annual grass species.
Layby Pro (4 L); MOA 7 + MSMA; MOA 17 6 lb/gal formulation 6.6 lb/gal formultion	2 pt + 2.67 pt 2.5 pt	1.0 + 2	Layby Pro contains 2 lb/gal diuron plus 2 lb per gal linuron. Apply as directed spray only to cotton at least 15 inches tall. See application precautions on label. Add crop oil concentrate at 1 gal per 100 gal spray solution. See label for rotational restrictions. Do not apply MSMA after first bloom. Aim at 1 fl oz per acre may be added to improve control of larger morningglory. Do not allow Aim to contact green stem tissue.
Suprend (80 WDG); MOA 5 + 2 + MSMA; MOA 17 6 lb/gal formulation 6.6 lb/gal formulation	1 to 1.25 lb + 2.67 pt 2.5 pt	0.8 to 1 + 2	Suprend is a premix product containing 79.3% prometryn plus 0.7% trifloxysulfuron. Apply as directed spray to cotton at least 6 inches tall, preferably taller. Add nonionic surfactant at 1 qt per 100 gal spray solution. See rotational restrictions on label. Do not apply MSMA after first bloom. Do not exceed 0.0188 lb a.i./acre per year of trifloxysulfuron from the combined use of all trifloxysulfuron-containing products (Envoke and Suprend).
Valor SX (51 WDG); MOA 14 + MSMA; MOA17 6.0 lb/gal formulation 6.6 lb/gal formulation	2 oz + 2.67 pt 2.5 pt	0.064 + 2	Valor contains flumioxazin; see Table 7-2A for other brands. Apply as directed spray only to cotton at least 16 inches tall. Direct the spray to the lower 2 inches of the cotton stem. Do not allow spray solution to contact green portion of stem. Add nonionic surfactant at 1 qt per 100 gal spray solution. Do not use crop oil concentrate, methylated seed oil, organo-silicone adjuvants, or any adjuvant product containing any of these. Do not apply MSMA after first bloom. No rotational restrictions of concern in North Carolina. May be applied under a hood on cotton at least 6 inches tall. Do not allow spray solution to contact cotton.
Post-emergence Directed, Annual grasses and broadleaf weeds, nutsedge, and suppression of perennial weeds; Roundup Ready Flex, Glytol LibertyLink, or XtendFlex Cultivars only			
Glyphosate; MOA 9	See labels	0.75 to 1.13 (lb a.e.)	Glyphosate alone can be directed up to 7 days prior to harvest. When using glyphosate alone, contact with the cotton is not of concern; the primary reason to direct is to obtain better coverage of weeds under the crop canopy. Use of other herbicides, in addition to glyphosate, is recommended to aid in resistance management. See the section on Herbicide Resistance Management and TABLE 7-10A. When tank mixing, follow directions on label of tank mix partner concerning cotton size for application, application directions (including allowable contact with cotton plant), and rotational restrictions. Glyphosate-resistant Palmer amaranth is widespread in North Carolina, and glyphosate-resistant common ragweed is present in several counties.
Glyphosate; MOA 9 + Aim (2 EC); MOA 14	See labels + 1 to 1.5 fl oz	0.75 to 1.13 (lb a.e.) + 0.016 to 0.024	Cotton should be at least 16 inches tall. Extreme care should be exercised in application; see directions and precautions on Aim label. Contact on green stem tissue will lead to severe injury. Add crop oil concentrate according to the Aim label. See comments on Aim label concerning sprayer clean-out. Compared to glyphosate alone, this combination controls larger morningglories. See above comments for glyphosate alone.
glyphosate; MOA 9 + Caparol (4 F); MOA 5	See labels + 0.5 to 1	0.75 to 1.13 (lb a.e.) + 1 to 2 pt	Caparol contains prometryn; see Table 7-2A for other brands. Direct to cotton at least 6 to 8 inches tall. Use 1 to 1.3 pt prometryn on cotton 6 to 12 inches tall; rate can be increased to 2 pt on cotton at least 12 inches tall. Add surfactant according to the label of the brand of glyphosate used. See precautions and rotational restrictions on prometryn label. Compared to glyphosate alone, this combination will improve control of larger morningglory and may provide residual control of small-seeded broadleaf weeds, such as pigweed. This mixture may give less control of larger grasses than glyphosate alone under drier conditions.
glyphosate MOA 9 + diuron (4 F) or diuron (80 WDG); MOA 7	See labels + 1 to 1.5 pt or 0.63 to 0.94 lb	0.75 to 1.13 (lb a.e.) + 0.5 to 0.75	See Table 7-2A for brands of diuron. Use 1 pt of diuron 4F or 0.63 lb diuron 80WDG on cotton 8 to 12 inches tall. Increase rate to 1.5 pt or 0.94 lb on cotton greater than 12 inches. See comments for glyphosate applied alone. Add surfactant according to the label of the glyphosate brand used. Compared to glyphosate alone, this combination controls larger morningglories and provides residual control of small-seeded broadleaf weeds, such as pigweed. This tank mix may give less control of larger grasses than glyphosate alone under dry conditions. See diuron label for rotational restrictions.
glyphosate MOA 9 + Dual Magnum (7.62 EC); MOA 9	See labels + 1 to 1.33 pt	0.75 to 1.13 (lb a.e.) + 0.95 to 1.27	Dual Magnum contains s-metolachlor; see Table 7-2A for other brands of s-metolachlor. Can be applied to cotton 3 inches tall through layby. See comments for glyphosate applied alone. S-metolachlor does not improve control of emerged weeds, but it can give residual control of annual grasses, pigweed species, and spreading dayflower plus suppression of yellow nutsedge. Do not apply to sand or loamy sand soils.
glyphosate MOA 9 + Envoke (75 DF); MOA 2	See labels + 0.1 to 0.2 oz	0.75 to 1.13 (lb a.e.) + 0.0047 to 0.0094	Direct to cotton from 6 inches tall through layby. Add nonionic surfactant according to the Envoke label. Compared to glyphosate alone, the combination is more effective on nutsedge and morningglory and provides residual control of susceptible broadleaf weeds. See comments above for glyphosate alone.
glyphosate; MOA 9 + Fierce (76 WDG); MOA 14 + 15	See label + 3 oz	0.75 to 1.13 (lb a.e.) + 0.143	Fierce is a premix containing 33.5% flumioxazin plus 42.5% pyroxasulfone. See comments for glyphosate applied alone and Fierce applied alone. Can be applied with hooded sprayer after cotton is at least 6 inches tall. Do not allow spray solution to contact cotton. Can be applied with layby applicator after cotton is at least 16 inches tall, but do not contact more than the lower 2 inches of cotton stalk. Add non-ionic surfactant according to premix label. Do not use crop oil, methylated seed oil, or organo-silicant adjuvants.
glyphosate; MOA 9 + Outlook (6 EC); MOA 15	See labels + 12 to 16 fl oz	0.75 to 1.13 (lb a.e.) + 0.56 to 0.75	Can be directed to cotton up to mid-bloom. See comments for glyphosate applied alone. Outlook does not improve control of emerged weeds, but it can give residual control of annual grasses and pigweed species. Suggested rates are 12 oz on coarse soils, 14 oz on medium soils, and 16 oz on fine soils.
glyphosate; MOA 9 + Prefix (5.29 EC); MOA 14 + 15	See labels + 2 to 2.33 pt	0.75 to 1.13 (lb a.e.) + 1.32 to 1.54	Prefix contains fomesafen plus S-metolachlor. Use as a layby application to cotton with a minimum of 4 inches of bark on the stem. Do not use Prefix at layby if fomesafen (Reflex, others) was used preemergence. See Prefix label for suggestions on adjuvant.
glyphosate; MOA 9 + Reflex (2 L); MOA 14	See label + 1 to 1.5 pt	0.75 to 1.13 (lb a.e.) + 0.25 to 0.375	Reflex contains fomesafen; see Table 7-2A for other brands. Use as a layby application to cotton with a minimum of 4 inches of bark on the stem. Add surfactant or crop oil according to the fomesafen label. May include prometryn, diuron, S-metolachlor, Envoke, Layby Pro, or Suprend in the mixture. Do not use fomesafen at layby if fomesafen was used preemergence.

CHEMICAL WEED CONTROL

250

Table 7-2B. Chemical Weed Control in Cotton

Herbicide, Mode of Action Code and Formulation	Amount of Formulation Per Acre	Pounds Active Ingredient Per Acre	Precautions and Remarks
Post-emergence Directed, Annual grasses and broadleaf weeds, nutsedge, and suppression of perennial weeds; Roundup Ready Flex, Glytol LibertyLink, or XtendFlex Cultivars only (continued)			
glyphosate MOA 9 + Suprend (80 WDG); MOA 5 + 2	See labels + 1 to 1.25 lb	0.75 to 1.13 (lb a.e.) + 0.8 to 1.0	Suprend contains prometryn plus trifloxysulfuron. Direct to cotton at least 6 to 8 inches tall. Add surfactant according to label of glyphosate brand used. See precautions and rotational restrictions on Suprend label. Compared to glyphosate alone, this combination will improve control of larger morningglory and nutsedge, and may provide residual control of small-seeded broadleaf weeds, such as pigweed. This mixture may give less control of larger grasses than glyphosate alone under drier conditions.
glyphosate; MOA 9 + Valor SX (51 WDG); MOA 14	See label + 1 to 2 oz	0.75 to 1.13 (lb a.e.) + 0.031 to 0.063	Valor contains flumioxazin; see Table 7-2A for other brands. Cotton should be at least 16 inches tall. Direct the spray to the lower 1 to 2 inches of the cotton stem; minimize cotton contact as much as possible. Do not allow spray solution to contact green portion of stem. See comments above for glyphosate alone. Add nonionic surfactant at 1 qt per 100 gal spray solution. DO NOT use crop oil concentrate, methylated seed oil, organo-silicone adjuvants, or any adjuvant product containing any of these. No rotational restrictions of concern in North Carolina. Compared with glyphosate alone, the combination will give better control of larger morningglories plus residual control of susceptible broadleaf weeds. May be applied under a hood on cotton at least 6 inches tall. Do not allow spray solution to contact cotton.
glyphosate; MOA 9 + Warrant (3.0 CS); MOA 9	See labels + 3 pt	0.75 to 1.13 (lb a.e.) + 1.125	Can be directed to cotton up to first bloom. See comments for glyphosate applied alone. Warrant does not improve control of emerged weeds, but it can give residual control of annual grasses and pigweed species.
glyphosate; MOA 9 + Warrant Ultra (3.45 CS); MOA 9 + 14	See labels +	0.75 to 1.13 (lb a.e.) +	Warrant Ultra contains acetochlor plus fomesafen. Use as a layby application to cotton with a minimum of 4 inches of bark on the stem. Avoid contact with cotton foliage. Do not use Warrant Ultra at layby if Warrant Ultra or fomesafen (Reflex, others) was used preemergence. Do not exceed a maximum of 3 lb active acetochlor per year from all applications.
glyphosate; MOA 9 + Zidua (85 WDG); MOA 15	See labels + 0.75 to 2.1 oz	0.75 to 1.13 (lb a.e.) + 0.040 to 0.112	Labeled rate of Zidua is 0.75 to 1.5 oz on coarse- or medium-textured soils and 1.5 to 2.1 oz on fine-textured soils. Do not use on soils with greater than 10% organic matter. Direct to minimize contact with cotton foliage when cotton is from 5-leaf stage to beginning of bloom stage. Do not apply overtop. Zidua does not control emerged weeds but gives residual control of annual grasses and pigweed species.
Sequence (5.25 L); MOA 9 + 15	2.5 pt	0.70 (lb a.e.) + 0.94	Sequence contains 2.25 lb a.e./gal glyphosate plus 3 lb/gal S-metolachlor. Direct to cotton up to 12 inches tall. Do not add adjuvants or tank mix with other products. Compared with glyphosate alone, Sequence will give residual control of annual grasses, pigweed species, and spreading dayflower plus suppression of nutsedge. See comments above for glyphosate alone.
Post-Emergence (hooded sprayers), Annual grasses, broadleaf weeds, and sedges			
glyphosate; MOA 9	See labels	0.75 (lb a.e.)	On varieties not resistant to glyphosate, hoods must be kept close to the ground so that no spray solution contacts the crop. Speed should not exceed 5 mph. Use 5 to 10 gpa and maximum pressure of 25 psi. Do not use liquid nitrogen as the carrier. Other herbicides as discussed in the section on directed application may be mixed with glyphosate to improve burndown and to provide residual control.
paraquat; MOA 22 2.0 lb/gal formulations 3.0 lb/gal formulations	 1.2 to 2.4 pt 0.8 to 1.6 pt	0.3 to 0.6	See Table 7-2A for brands of paraquat. Hoods should be kept as close to the ground as possible. Do NOT allow the spray solution to contact cotton plants. Apply in a minimum of 10 gpa at maximum pressure of 25 psi. Do not exceed 5 mph. It is suggested that cotton be at least 6 inches tall. Add nonionic surfactant or crop oil concentrate according to the paraquat label. Control will generally be much better if diuron, Cotoran, or prometryn is mixed with paraquat. Diuron, Cotoran, or prometryn may also provide residual control.
Harvest Aid, Annual grasses and broadleaf weeds			
glyphosate; MOA 9	See labels	0.75 to 1.5 (lb a.e.)	Apply to any cultivar after at least 60% of the bolls are open. May be tank mixed with some defoliants; see labels for details. Include nonionic surfactant according to the label of the glyphosate brand used. Can be applied to Roundup Ready Flex, Xtend Flex, or GlyTol LIbertyLink cotton up to 7 days before harvest.
paraquat; MOA 22 2.0 lb/gal formulations 3.0 lb/gal formulations	 1 to 2 pt 0.67 to 1.33 pt	0.25 to 0.5	See Table 7-2 for paraquat brands. Defoliate cotton as normal. After at least 75 to 80% of the bolls are open, the remaining bolls expected to be harvested are mature, and most of the cotton leaves have dropped, apply paraquat in a minimum of 20 gal per acre and add 1 pt nonionic surfactant per 100 gal. Wait 5 days before picking, then pick as soon as possible.

CHEMICAL WEED CONTROL

251

Weed Response to Preplant Incorporated, Preemergence, and Postemergence Overtop Cotton Herbicides

A. C. York, William Neal Reynolds Professor Emeritus, Crop and Soil Sciences Department

Ratings based upon average to good soil and weather conditions for herbicide performance and upon proper application rate, technique, and timing.

Table 7-2C. Weed Response to Preplant Incorporated, Preemergence, and Postemergence Overtop Herbicides in Cotton

Species	PPI Prowl or Treflan	PRE Warrant	diuron	Cotoran	Reflex	Prowl	Staple	POST Select	Fusilade	Assure II	Poast, Poast Plus	Liberty	Liberty + Staple	glyphosate	glyphosate + Staple	glyphosate + Envoke	Staple	Envoke
Bermudagrass	N	N	N	N	N	N	N	G[1]	G[1]	G[1]	F[1]	N	N	F[1]	F[1]	F[1]	N	N
Broadleaf signalgrass	G	G	P	P	FG	F	P	E	GE	G	E	G	G	E	E	E	N	N
Crabgrass	E	E	FG	FG	FG	G	P	GE	G	G	GE	G	G	E	E	E	N	P
Crowfootgrass	E	E	FG	FG	ND	G	P	G	F	G	FG	G	G	E	E	E	N	N
Fall panicum	G	E	P	F	ND	F	PF	E	GE	GE	E	G	G	E	E	E	N	NP
Foxtails	E	E	ND	FG	ND	G	P	E	E	E	E	G	G	E	E	E	N	NP
Goosegrass	E	E	F	F	ND	G	PF	GE	G	G	GE	P	P	E	E	E	N	NP
Johnsongrass Seedling	E	F	P	P	ND	G	FG	E	E	E	E	G	G	E	E	P	P	
Rhizome	P	N	N	N	N	N	N	GE	GE	E	G	F[2]	F[2]	GE	GE	GE	NP	P
Texas panicum	G	P	P	P	F	F	N	E	G	G	E	G	G	E	E	E	N	NP
Nutsedge Purple	N	N	N	N	ND	N	F	N	N	N	N	PF	F	FG[1]	FG[1]	GE	PF	FG
Yellow	N	PF	N	N	GE	N	F	N	N	N	N	PF	F	F[1]	FG[1]	GE	PF	G
Cocklebur	N	N	F	FG	G	N	NP	N	N	N	N	E	E	E	E	E	G	GE
Common purslane	E	G	E	E	G	G	G	N	N	N	N	FG	FG	FG	G	ND	F	ND
Common ragweed glyphosate-susceptible	N	P	G	E	G	N	NP	N	N	N	N	E	E	E	E	E	P	G
glyphosate-resistant	N	P	G	E	G	N	NP	N	N	N	N	E	E	N	P	G	P	G
Cowpea	N	N	P	P	ND	N	FG	N	N	N	N	G	E	GE	E	E	G	G
Crotalaria	N	N	G	G	ND	N	ND	N	N	N	N	ND	G	G	G	G	G	ND
Eclipta	P	FG	G	G	GE	P	ND	N	N	N	N	G	E	E	E	E	G	PF
Florida beggarweed	P	P	G	GE	P	N	G	N	N	N	N	G	E	E	E	E	G	GE
Florida pusley	E	E	PF	FG	P	FG	F	N	N	N	N	F	F	PF	PF	PF	NP	P
Hemp sesbania	N	N	P	P	P	N	P	N	N	N	N	ND	ND	PF	GE	ND	GE	ND
Jimsonweed	N	N	G	G	ND	N	FG	N	N	N	N	E	E	E	E	E	E	N
Lambsquarters	GE	P	E	E	E	G	G	N	N	N	N	E	E	G	G	E	N	G
Morningglory tall	P	P	F	G	PF	P	P	N	N	N	N	E	E	FG	FG	E	P	G
other species	P	P	F	G	PF	P	F	N	N	N	N	E	E	FG	GE	E	G	G
Palmer amaranth glyphosate-susceptible	FG	G	G	PF	E	PF	G	N	N	N	N	G	G	E	E	E	F	PF
glyphosate-resistant	FG	G	G	PF	E	PF	G	N	N	N	N	G	G	N	F	PF	F	PF
ALS-resistant	FG	G	G	PF	E	PF	N	N	N	N	N	G	G	E	E	E	N	N
glyphosate- & ALS-resistant	FG	G	G	PF	E	PF	N	N	N	N	N	G	G	N	N	N	N	N
Pigweed, redroot	E	GE	GE	GE	E	FG	E	N	N	N	N	G	GE	E	E	E	G	FG
Prickly sida	N	P	F	G	ND	N	G	N	N	N	N	FG	FG	FG	FG	FG	F	N
Sicklepod	N	N	F	G	P	N	PF	N	N	N	N	E	E	E	E	E	PF	E
Smartweed	N	N	G	G	ND	N	G	N	N	N	N	GE	E	G	E	E	G	G
Spurge	N	F	F	PF	ND	N	G	N	N	N	N	FG	G	G	G	G	FG	ND
Spurred anoda	N	N	F	F	ND	N	E	N	N	N	N	P	G	E	E	E	G	P
Tropic croton	N	N	FG	FG	FG	N	FG	N	N	N	N	G	G	E	E	E	P	PF
Velvetleaf	N	N	PF	F	ND	N	E	N	N	N	N	F	G	E	E	E	G	G
Volunteer peanuts	N	N	P	PF	P	N	P	N	N	N	N	GE	GE	F	F	FG	P	PF

[1] Two applications may be required for acceptable control.
[2] Acceptable johnsongrass control can be obtained with two applications of Liberty.

Key:

E = excellent control, 90% or better
G = good control, 80% to 90%
F = fair control, 50% to 80%
ND = data not available

P = poor control, 25% to 50%
N = no control, less than 25%
ND = data not available

Weed Response to Postemergence-Directed Cotton Herbicides

A. C. YORK, William Neal Reynolds Professor Emeritus, Crop and Soil Sciences Department

Ratings based upon average to good soil and weather conditions for herbicide performance and upon proper application rate, technique, and timing.

Table 7-2d. Weed Response to Postemergence-Directed Herbicides in Cotton

Species	diuron+ MSMA	Valor or Fierce + MSMA	Cotoran + MSMA	Cobra + MSMA	Caparol + MSMA	Layby Pro+ MSMA	Suprend+ MSMA	glyphosate + Aim	glyphosate + diuron	glyphosate + Valor or Fierce	glyphosate + Caparol	glyphosate + Suprend	paraquat + diuron[1]
Bermudagrass	N	N	N	N	N	N	N	F^2	F^2	F^2	F^2	F^2	P
Broadleaf signalgrass	F	F	F	F	F	G	F	E	GE	E	GE	GE	GE
Crabgrass	F	F	F	PF	FG	G	FG	E	GE	E	GE	GE	G
Crowfootgrass	F	F	F	PF	FG	FG	FG	E	GE	E	GE	GE	G
Fall panicum	F	F	F	PF	FG	FG	FG	E	GE	E	GE	GE	G
Foxtails	F	F	F	PF	FG	FG	FG	E	GE	E	GE	GE	G
Goosegrass	F	F	F	PF	FG	FG	FG	E	GE	E	GE	GE	G
Johnsongrass													
Seedling	F	F	F	PF	FG	FG	FG	E	GE	E	GE	GE	G
Rhizome	P	P	P	P	P	P	P	GE	G	GE	G	G	P
Texas panicum	F	PF	P	P	F	F	F	E	GE	E	GE	GE	G
Nutsedge													
Purple	F^3	FG^3	F^3	F^3	F^3	F^3	E	FG^2	FG^2	FG^2	FG^2	GE	PF
Yellow	G	G	FG^3	FG^3	FG^3	G	E	F^2	F^2	F^2	F^2	E	PF
Cocklebur	E	E	E	E	E	E	E	E	E	E	E	GE	E
Common purslane	G	G	FG	G	FG	G	ND	FG	GE	GE	GE	E	E
Common ragweed													
glyphosate-susceptible	E	GE	GE	E	E	E	E	E	E	E	E	E	G
glyphosate-resistant	E	GE	GE	E	E	E	E	PF	PF	P	P	G	G
Cowpea	G	G	G	FG	G	G	E	GE	GE	E	GE	E	E
Crotalaria	G	ND	G	G	G	G	E	G	G	ND	G	G	ND
Eclipta	E	E	G	E	G	E	E	E	E	E	E	E	G
Florida beggarweed	E	E	E	E	E	E	E	E	E	E	E	E	E
Florida pusley	F	FG	F	F	F	F	F	G	G	GE	G	G	P
Hemp sesbania	PF	ND	PF	F	PF	PF	ND	GE	ND	ND	ND	ND	FG
Jimsonweed	G	E	GE	GE	G	G	G	E	E	E	E	E	E
Lambsquarters	G	FG	G	F	G	G	GE	GE	GE	GE	GE	E	G
Morningglory	GE	E	G	E	G	GE	E	E	GE	E	GE	E	G
Palmer amaranth													
glyphosate-susceptible	G	G	FG	G	F	G	GE	E	E	E	E	E	GE
glyphosate-resistant	G	G	FG	G	F	G	GE	PF	F	PF	PF	F	GE
ALS-resistant	G	G	FG	G	F	G	F	E	E	E	E	E	GE
glyphosate- & ALS-resistant	G	G	FG	G	F	G	F	PF	F	PF	PF	P	GE
Pigweed, redroot	GE	GE	G	G	G	GE	GE	E	E	E	E	E	GE
Prickly sida	GE	GE	FG	GE	GE	GE	GE	FG	G	GE	G	G	FG
Sicklepod	GE	GE	G	PF	GE	GE	E	E	E	E	E	E	E
Smartweed	F	G	G	F	F	F	ND	GE	G	G	G	E	GE
Spurge	G	G	PF	G	G	G	ND	GE	GE	G	GE	E	ND
Spurred anoda	F	G	FG	F	F	F	ND	E	E	E	E	E	G
Tropic croton	G	E	G	E	G	G	GE	E	E	E	E	E	G
Velvetleaf	F	G	F	F	F	G	F	E	E	E	E	E	ND
Volunteer peanuts	G	FG	FG	PF	FG	G	G	FG	G	FG	FG	FG	F

[1] Apply only with hooded spray; avoid contact with cotton.
[2] Good control usually obtained with two applications of glyphosate.
[3] Good control of nutsedge can usually be obtained with two applications of MSMA.

Key:

E = excellent control, 90% or better
G = good control, 80% to 90%
F = fair control, 50% to 80%
P = poor control, 25% to 50%
N = no control, less than 25%
ND = data not available

CHEMICAL WEED CONTROL

Weed Response to Burndown Cotton Herbicides
A. C. YORK, William Neal Reynolds Professor Emeritus, Crop and Soil Sciences Department

Ratings based upon average to good soil and weather conditions for herbicide performance and upon proper application rate, technique, and timing.

Table 7-2E. Weed Response to Burndown Herbicides in Cotton

Species	2,4-D[1]	glyphosate	glyphosate + Clarity[2]	glyphosate + 2,4-D[1]	glyphosate + Valor + Clarity[2]	glyphosate + Valor + 2,4-D[1,3]	glyphosate + Leadoff[4]	glyphosate + Sharpen[5]	glyphosate + Firstshot or Harmony Extra[6]	paraquat	paraquat + diuron[7]
Annual bluegrass	N	E	E	E	E	E	E	E	E	GE	E
Little barley	N	E	E	E	E	E	E	E	E	G	E
Buttercups	G	E	E	E	E	E	E	E	E	E	E
Carolina geranium	PF	PF	G	F	G	F	G	ND	GE	GE	E
Chickweed	P	E	E	E	E	E	E	E	E	E	E
Cudweed	NP	E	E	E	E	E	E	E	E	FG	G
Curly dock	F	F	GE	FG	F	FG	F	ND	E	NP	P
Cutleaf eveningprimrose	E	PF	G	E	FG	E	PF	ND	F	F[8]	G[8]
Field pansy	P	F	F	F	F	F	ND	ND	F	G	GE
Henbit	PF	G	E	E	E	E	E	ND	E	E	E
Horseweed (marestail)											
glyphosate-susceptible	GE[9]	GE	E	E[9]	E	E	GE	E	GE	PF	G
glyphosate-resistant	GE[9]	N	E	E[9]	N	GE[9]	P	E	P	PF	G
Prickly lettuce	G	E	E	E	E	E	E	E	E	P	PF
Ryegrass	N	G	G	G	G	G	GE	G	G	F	FG
Speedwell	PF	E	E	E	E	E	E	E	E	E	E
Swinecress	F	FG	FG	G	ND	E	E	E	E	E	E
Vetch	E	F	E	E	FG	E	ND	ND	G	PF	F
Virginia pepperweed	GE	G	GE	E	GE	E	G	ND	G	G	G
Wheat or rye cover crop	N	E	E	E	E	E	E	E	E	G[11]	GE[11]
Wild mustard, wild radish	FG[10]	FG	G	E	G	E	FG	ND	GE	FG	G

[1] Apply 2,4-D at least 30 days ahead of planting.

[2] Following application of Clarity (dicamba) and a minimum of 1 inch of rainfall, a minimum 21-day waiting period is required before planting.

[3] A minimum of 14 days must pass and 1 inch of ranfall must occur between application of Valor (flumioxazin) at 1 oz and planting. Delay planting 21 days after application of 1.5 to 2 oz. The waiting interval can be reduced to 14 days if a strip tillage operation will occur after application and before planting.

[4] Apply Leadoff from late fall to 30 days prior to planting. See comments in Table 7-2B. May include 2,4-D or dicamba.

[5] A minimum of 42 days must pass and 1 inch of rainfall must occur between Sharpen application and cotton planting.

[6] Delay cotton planting 14 days after Firstshot or Harmony Extra application.

[7] Diuron should be applied 15 to 45 days ahead of planting.

[8] This level of control requires that the primrose be blooming when treated.

[9] This level of control requires 2 pt of 2,4-D.

[10] Wild radish and wild mustard control by 2,4-D is good if application is made before plants begin flowering. Use 1 pt per acre of 2,4-D to control these species.

[11] Wheat or rye must have visible seedheads for this level of control.

Key:

E = excellent control, 90% or better
G = good control, 80% to 90%
F = fair control, 50% to 80%
P = poor control, 25% to 50%
N = no control, less than 25%
ND = data not available

Chemical Weed Control in Peanuts

D. L. Jordan, Crop Science Department

Control of witchweed is part of the State/Federal Quarantine Program. Contact the N.C. Department of Agriculture, Plant Industry Division, at 1-800-206-9333. Note that the active ingredient and examples of tradenames are listed for each herbicide. This approach is used for the simplicity of presentation but does not imply that other formulations of the same active ingredient are not equally effective. See Table 7-3D for products other than those listed in 7-3A, 7-3B, and 7-3C that are equally effective.

Table 7-3A. Chemical Weed Control in Peanuts

Herbicide and Formulation	Pounds Active Ingredient Per Acre	Precautions and Remarks
Preplant Incorporated, Annual grasses and small-seeded broadleaf weeds		
alachlor, MOA 15 (Intrro 4 EC)	2 to 3 (2 to 3 qt)	Incorporate no deeper than 2 inches; see label for specific instructions. Unless shallowly incorporated, Intrro is more consistently effective when applied preemergence. Weak on Texas panicum. Do not apply more than 3 qt of Intrro per acre per season. **Before using Intrro, check with buyers to determine if there are marketing restrictions on Intrro-treated peanuts.**
acetochlor, MOA 15 (Warrant 3 ME)	0.94 to 1.5 (1.25 to 2 qt)	Apply and incorporate in top 2 inches of soil. Do not apply more than 4 qt of Warrant per acre per year.
ethalfluralin, MOA 3 (Sonalan 3 EC)	0.56 to 0.75 (1.5 to 2 pt)	Controls common annual grasses including Texas panicum. Use 3 pt Prowl or 2 pt ethalfluralin for control of broadleaf signalgrass, Texas panicum, and fall panicum. Incorporate 3 inches deep for Texas panicum; otherwise, incorporate 2 to 3 inches deep. See labels for maximum waiting period between application and incorporation. Immediate incorporation is best. Dual Magnum, Outlook, or Warrant may be tank mixed with Prowl or Sonalan to suppress yellow nutsedge.
pendimethalin, MOA 3 (Prowl H2O 3.8 EC) (Prowl 3.3 EC)	0.71 to 1.43 (1.5 to 3 pt) (1.7 to 3.5 pt)	
Peplant Incorporated, Annual grasses, small-seeded broadleaf weeds, and nutsedge		
dimethenamid, MOA 15 (Outlook 6.0 L)	0.75 to 1 (16 to 21 fl oz)	Apply and incorporate in top 2 inches of soil within 14 days of planting. Use high rate of Dual Magnum, Dual, or Outlook for yellow nutsedge and broadleaf signalgrass. Not effective on purple nutsedge. Weak on Texas panicum. May be tank mixed with Prowl or Sonalan.
metolachlor, MOA 15 (Dual Magnum 7.62 EC) (Dual 8 EC)	0.95 to 1.27 (1 to 1.33 pt) (1.5 to 2 pt)	
Preplant Incorporated, Broadleaf weeds and suppression of nutsedge		
diclosulam, MOA 2 (Strongarm 84 WDG)	0.024 (0.45 oz)	Effective on common cocklebur, morningglory, common ragweed, eclipta, and common lambsquarters. Suppresses yellow and purple nutsedge. Does not control sicklepod. More effective when applied in combination with Dual, Outlook, Warrant, Prowl, or Sonalan. See label for rotation restrictions, especially corn and grain sorghum. Growers are cautioned that Strongarm can occasionally injure cotton the following year on soils with a shallow hardpan (less than 10 inches) and/or loam soils. Cotton grown under early season stress resulting from conditions such as excessively cool, wet, dry, or crusted soils may be particularly susceptible to carryover of Strongarm. The rotation interval between applying Strongarm to peanut and then planting cotton is 18 months in Camden, Currituck, Pasquotank, and Perquimans counties. Some weed species have developed resistance to Strongarm including common rageweed and Palmer amaranth.
Preplant Incorporated, Annual grasses, broadleaf weeds, and suppression of nutsedge		
diclosulam, MOA 2 Strongarm + pendimethalin, MOA 3 (Prowl H2O 3.8 EC) (Prowl 3.3 EC) or ethalfluralin, MOA 3 (Sonalan 3 EC) or metolachlor, MOA 15 (Dual Magnum 7.62 EC) (Dual 8 EC) or dimethenamid (Outlook 6.0 L) or acetochlor (Warrant 3 ME)	0.024 (0.45 oz) + 0.71 to 1.43 (1.5 to 3 pt) (1.7 to 3.5 pt) or 0.56 to 0.75 (1.5 to 2 pt) or 0.95 to 1.27 (1 to 1.33 pt) (1.5 to 2 pt) or 0.75 to 1 (16 to 21 fl oz) or 0.95 to 1.5 (1.24 to 2 qt)	Effective on annual grasses, common cocklebur, common ragweed, eclipta, morningglory, and common lambsquarters. Supresses purple and yellow nutsedge. Does not control sicklepod. See Strongarm label for rotation restrictions.
PPI followed by PRE, Annual grasses, broadleaf weeds, and supression of nutsedge		
pendimethalin, MOA 3 (Prowl H2O 3.8 EC) (Prowl 3.3 EC) or ethalfluralin, MOA 3 (Sonalan 3 EC) or metolachlor, MOA 15 (Dual Magnum 7.62 EC) (Dual 8 EC) or dimethenamid, MOA 15 (Outlook 6.0L) or acetochlor, MOA 15 (Warrant 3 ME) followed by diclosulam, MOA 2 (Strongarm 84 WDG) or flumioxazin, MOA 14 (Valor SX 51 WDG)	0.71 to 1.43 (1.5 to 3 pt) (1.7 to 3.5 pt) or 0.56 to 0.75 (1.5 to 2 pt) or 0.95 to 1.27 (1 to 1.33 pt) (1.5 to 2 pt) or 0.75 to 1 (16 to 21 oz) or 0.95 to 1.5 (1.24 to 2 qt) 0.024 0.45 oz or 0.063 (2 oz)	Controls most broadleaf weeds. Will not control sicklepod and is marginal on certain large-seeded broadleaf weeds. **Do not incorporate** Valor SX. Valor SX should be applied to the soil surface immediately after planting. Significant injury can occur if flumioxazin is incorporated or applied 3 or more days after planting. Significant injury from Valor SX has been noted in some years even when applied according to label recommendations. However, injury is generally transient and does not affect yield. See previous comments about cotton response to Strongarm applied the previous year on some soils. Up to 3 oz per acre of Valor SX can be applied to peanut but injury potential increases. **See product label for sprayer cleanup before other uses.**

Table 7-3A. Chemical Weed Control in Peanuts

Herbicide and Formulation	Pounds Active Ingredient Per Acre	Precautions and Remarks
Split application (PPI + POST), Most broadleaf weeds and nutsedge		
imazethapyr, MOA 2 (Pursuit 2 AS)	0.031 + 0.031 (2 + 2 oz)	Effective on most common broadleaf weeds and yellow and purple nutsedge. Does not control eclipta, lambsquarters, ragweed, or croton. Pursuit will usually control seedling johnsongrass and foxtails. For control of other annual grasses, Pursuit may be tank mixed with Dual Magnum, Dual, Outlook, Prowl H2O, Prowl, or Sonalan and incorporated. See label for incorporation directions and rotational restrictions. Some weed species have developed resistance to Pursuit. Research in N.C. has generally shown more effective control of a broader spectrum of weeds with split applications of half of the Pursuit applied preplant incorporated followed by the other half applied early postemergence.
Preemergence, Annual grasses and small-seeded broadleaf weeds		
alachlor, MOA 15 (Intrro 4 EC)	2 to 3 (2 to 3 qt)	Apply as soon after planting as possible. All four herbicides are weak on Texas panicum. **Before using Inntro, check with buyers to determine if there are marketing restrictions on Intrro-treated peanuts.**
dimethenamid, MOA 15 (Outlook 6.0 L)	0.75 to 1 (16 to 21 fl oz)	
metolachlor, MOA 15 (Dual Magnum 7.62 EC) (Dual 8 EC)	0.95 to 1.27 (1 to 1.33 pt) (1.5 to 2 pt)	
acetochlor (Warrant 3 ME)	0.95 to 1.5 (1.25 to 2 qt)	
Preemergence, Broadleaf weeds		
flumioxazin, MOA 14 (Valor SX 51 WDG)	0.063 2 oz	Apply within 2 days after planting. Significant injury can occur if Valor SX is incorporated or applied 3 or more days after seeding. Controls carpetweed, common lambsquarters, Florida pusley, nightshade, prickly sida, and spotted spurge. Does not control sicklepod, yellow and purple nutsedge, or annual grasses. Morningglory control is marginal where Valor SX is applied at 2 oz per acre. Significant injury from Valor SX has been noted in some years even when applied according to label. However, injury is generally transient and does not affect yield. Injury may occur if excessive and forceful rainfall occurs when peanut is emerging. Peanut recovers from injury by midseason in most instances. Up to 3 oz per acre of Valor SX can be applied to peanut, but injury potential increases. **See product label for comments on sprayer cleanup before other uses.**
Preemergence, Annual grasses, broadleaf weeds, and suppression of nutsedge		
flumioxazin, MOA 14 (Valor SX 51 WDG) + metolachlor, MOA 15 (Dual Magnum 7.62 EC) (Dual 8 EC) or dimethenamid, MOA 15 (Outlook 6.0L) or acetlochlor, MOA 15 (Warrant 3 ME)	0.063 (2 oz) + 0.95 to 1.27 (1 to 1.33 pt) 1.5 to 2 pt or 0.75 to 1 (16 to 21 fl oz) or 0.94 to 1.5 (1.25 to 2 qt)	Apply within 2 days after planting. Significant injury can occur if applied 3 or more days after planting. The combination of Valor SX and Dual, Dual Magnum, Warrant, or Outlook does not control sicklepod but will control annual grasses (except Texas panicum) and will suppress yellow nutsedge. Valor SX and Warrant will not suppress yellow nutsedge. Significant injury from Valor SX has been noted in some years even when applied according to label recommendations. However, injury is generally transient and does not affect yield. Injury may occur if excessive and forceful rainfall occurs when peanut is emerging. Peanut recovers from injury by midseason in most instances. Up to 3 oz per acre of Valor SX can be applied to peanut but injury potential increases. **See product label for comments on sprayer cleanup before other uses.**
diclosulam, MOA 2 (Strongarm 84 WDG)	0.024 (0.45 oz)	Effective on common cocklebur, morningglory, common ragweed, eclipta, and common lambsquarters. Suppresses yellow and purple nutsedge. Does not control sicklepod. More effective when applied in combination with Dual, Dual Magnum, Outlook, Prowl, Sonalan, or Warrant. See label for rotation restrictions, especially corn and grain sorghum. See previous comments on possible cotton injury from Strongarm applied the previous year on some soils.
sulfentrazone, MOA 14 + carfentrazone, MOA 14 (Spartan Charge (0.35 + 3.15 F)	0.07 to 0.12 (3 to 5 fl oz)	Do not apply Spartan Charge after peanuts crack soil. Application immediately after planting is advised. See label for specific rates based on soil texture and organic matter content. See product label for comments on applicaton with other herbicides. Rotation restriction for planting cotton following Spartan Charge at recommended rates for peanut is 12 months.
diclosulam, MOA 2 (Strongarm 84 WDG) + metolachlor, MOA 15 (Dual Magnum 7.62 EC) (Dual 8 EC) or dimethenamid, MOA 15 (Outlook 6.0 L) or acetolchlor, MOA 15 (Warrant 3 ME)	0.024 (0.45 oz) + 0.95 to 1.27 (1 to 1.33 pt) 1.5 to 2 pt or 0.75 to 1 (16 to 21 oz) or 0.94 to 1.5 (1.25 to 2 qt)	Effective on annual grasses, common cocklebur, common ragweed, eclipta, morningglory, and common lambsquarters. Supresses purple and yellow nutsedge. Does not control sicklepod. See label for rotation restrictions. Some weed species have developed resistance to Strongarm. See previous comments on carryover potential to cotton on some soils and restrictions on planting corn or grain sorghum after use in peanut.
Preemergence, Most annual broadleaf weeds and nutsedge		
imazethapyr, MOA 2 (Pursuit 2 AS)	0.063 (4 fl oz)	Effective on most common broadleaf weeds and yellow and purple nutsedge. Does not control ragweed, eclipta, lambsquarters, or croton. Pursuit may be tank mixed with Dual, Dual Magnum, Warrant, ot Outlook for annual grass control. See label for **rotational restrictions.** Some weed species have developed resistance to Pursuit. Research in N.C. has generally shown more effective control of a broader spectrum of weeds with split applications of half of the Pursuit applied preplant incorporated followed by the other half applied early postemergence.
Cracking stage, Emerged annual grasses and broadleaf weeds		
paraquat, MOA 22 (Gramoxone 2.5 SL) (Parazone 3 SL)	0.13 (8 oz) (5.4 oz)	Apply at ground cracking to control small emerged annual grasses and broadleaf weeds. May be tank mixed with Dual, Dual Magnum, Outlook, or Warrant for residual control. Tank mix may increase injury to emerged peanuts. Add 1 pint nonionic surfactant per 100 gallons spray solution. Follow safety precautions on label. Applying Basagran at 0.5 pt per acre will reduce injury.
Cracking stage and Postemergence, Additional residual control of annual grasses and certain small-seeded broadleaf weeds		
alachlor, MOA 15 (Intrro 4 EC)	2 to 3 (2-3 qt)	Use as a supplement to preplant or preemergence herbicides to provide additional residual control of annual grasses and certain small-seeded broadleaf weeds such as pigweed and eclipta. This treatment will not control emerged grasses or broadleaf weeds. See product labels for recommended tank mixtures with contact and systemic herbicides with foliar activity on weeds.
dimethenamid, MOA 15 (Outlook 6.0L)	0.75 to 1 (16 to 21 oz)	
metolachlor, MOA 15 (Dual Magnum 7.62 EC) (Dual 8 EC)	0.95 1 pt 1.5 pt	

Table 7-3A. Chemical Weed Control in Peanuts

Herbicide and Formulation	Pounds Active Ingredient Per Acre	Precautions and Remarks
Cracking stage and Postemergence, Additional residual control of annual grasses and certain small-seeded broadleaf weeds (continued)		
acetochlor, MOA 15 (Warrant 3 ME)	0.95 to 1.5 (1.25 to 2 qt)	
Cracking stage, Most annual broadleaf weeds and nutsedge		
imazethapyr, MOA 2 (Pursuit 2 AS)	0.063 (4 oz)	Effective on most common broadleaf weeds and yellow and purple nutsedge. Does not control ragweed, eclipta, lambsquarters, or croton. If weeds are emerged, add surfactant or crop oil according to label directions. See label for rotational restrictions. Pursuit may be tank mixed with paraquat. Some weed species have developed resistance to Pursuit.
Cracking stage, Some emerged broadleaf weeds and suppression of eclipta and yellow nutsedge		
diclosulam, MOA 2 (Strongarm 84 WDG)	0.024 (0.45 oz)	Strongarm can be applied through the cracking stage. Add 1 quart nonionic surfactant per 100 gallons. The spectrum of weeds controlled is much narrower when applied to emerged weeds. Strongarm will not control emerged common lambsquarters or pigweeds but will control common ragweed and morningglories and will suppress yellow nutsedge and eclipta. See product labels for information on mixing Strongarm with other herbicides. Some weed species have developed resistance to Strongarm. See product label for carryover potential to cotton, corn, and grain sorghum. Strongarm suppresses emerged marestail and dogfennel more effectively than other postemergene broadleaf herbicides when applied to small weeds.
Postemergence, Annual broadleaf weeds		
acifluorfen, MOA 14 (Ultra Blazer 2 L)	0.25 to 0.38 (1 to 1.5 pt)	Apply when weeds are small and actively growing. Use minimum of 20 GPA and high pressure (40 to 60 psi). See label for species controlled, maximum weed size to treat, and addition of surfactant. Do not apply more than 2 pints per acre per season. May make sequential applications of 0.25 pound followed by 0.25 pound per acre. Allow at least 15 days between sequential applications.
acifluorfen, MOA 14 (Ultra Blazer 2 L) + 2,4-DB, MOA 4 (Butyrac 200 2 L)	0.25 to 0.38 (1 to 1.5 pt) + 0.25 (16 fl oz)	Addition of 2,4-DB to Ultra Blazer improves control of certain weeds when weed size exceeds that specified on the Ultra Blazer label. See label suggestions on use of surfactant or crop oil. Apply when peanuts are at least 2 weeks old and before pod filling begins.
bentazon, MOA 6 (Basagran 4 L)	0.75 to 1 (1.5 to 2 pt)	Apply when weeds are small and actively growing. Use minimum of 20 GPA and high pressure (40 to 60 psi). See label for addition of oil concentrate, species controlled, and maximum weed size to treat. Basagran may also be applied at 1 pint per acre for control of cocklebur, jimsonweed, and smartweed 4 inches or less. Do not apply more than 4 pints of bentazon per acre per season.
bentazon, MOA 6 (Basagran 4 L) + acifluorfen, MOA 14 (Ultra Blazer 2 L)	0.5 to 1 (1 to 2 pt) + 0.25 to 0.38 (1 to 1.5 pt)	See above comments for Ultra Blazer and Basagran. See labels for weeds controlled, maximum weed size to treat, and use of adjuvants. Can be applied as a tank mixture or as Storm 4L.
bentazon, MOA 6 + acifluorfen, MOA 14 (Storm 4L)	0.5 + 0.25 (1.5 pt)	These rates of bentazon and acifluorfen (Ultra Blazer and Basagran) may not provide consistent control of lambsquarters, prickly sida, spurred anoda, and morningglory.
bentazon, MOA 6 (Basagran 4 L) + acifluorfen, MOA 14 (Ultra Blazer 2 L) + 2,4-DB, MOA 4 (Butyrac 200 2 L)	0.5 (1 pt) + 0.25 (1 pt) + 0.125 to 0.25 (8 to 16 fl oz)	Adding 2,4-DB will improve control of larger morningglory, cocklebur, common ragweed, pigweed, jimsonweed, and citron. Add surfactant or crop oil according to label directions. Apply when peanuts are at least 2 weeks old. Do not apply after pod filling begins. See comments for Ultra Blazer and Basagran alone.
bentazon, MOA 6 (Basagran 4 L) + 2,4-DB, MOA 4 (Butyrac 200 2 L)	0.75 to 1 1.5 to 2 pt) + 0.125 (8 fl oz)	Addition of 2,4-DB to Basagran improves control of morningglories. See above comments for Basagran. Add surfactant or crop oil according to label directions. Do not make more than two applications per year. Apply when peanuts are at least 2 weeks old and not within 45 days of harvest.
imazapic, MOA 2 (Cadre 2 AS) (Impose 2 AS)	0.063 (4 fl oz)	Controls most broadleaf weeds except ragweed, croton, lambsquarters, and eclipta. Apply before weeds exceed 2 to 4 inches; see label for specific weed sizes to treat. Add nonionic surfactant at 1 quart per 100 gallons or crop oil concentrate at 1 quart per acre. A soil-applied grass control herbicide should be used. However, Cadre will usually control escaped broadleaf signalgrass, large crabgrass, fall panicum, and Texas panicum but not goosegrass. Cadre can be mixed with Cobra, Ultra Blazer, and 2,4-DB. See label for rotational restrictions. Some weed species have developed resistance to Cadre.
imazethapyr, MOA 2 (Pursuit 2 L)	0.063 (4 fl oz)	Effective on most common broadleaf weeds and yellow and purple nutsedge. Does not control eclipta, lambsquarters, ragweed, or croton. Apply when weeds are 3 inches tall or less. Add surfactant or crop oil according to label directions. See label for rotational restrictions. Pursuit rmay be tank mixed with Basagran, Ultra Blazer, Gramoxone, and 2,4-DB. Some weed species have developed resistance to Pursuit.
2,4-DB, MOA 4 (Buryrac 200 2 L)	0.2 to 0.25 (12 to 16 fl oz)	Effective on cocklebur and morningglory; pitted morningglory may be only partially controlled. Best results achieved when applied to small weeds. May use two applications per year. Do not apply within 45 days before harvest.
lactofen, MOA 14 (Cobra 2 EC)	0.2 (12.5 fl oz)	Apply after peanuts have at least six true leaves. Apply to actively growing peanut. Controls most annual broadleaf weeds. See label for species controlled and maximum weed size to treat. Add nonionic surfactant at 1 quart per 100 gallons or crop oil concentrate or methylated seed oil at 1 to 2 pints per acre. See label on when to use various adjuvants. Allow at least 14 days between applications. Can be tank mixed with Basagran, Pursuit, Cadre, 2,4-DB, and/or Select.
lactofen, MOA 14 (Cobra 2 EC) + bentazon, MOA 6 (Basagran 4 L)	0.2 (12.5 fl oz) + 0.75 to 1 (1.5 to 2 pt)	See above comments for Basagran and Lactofen alone. See labels for weeds controlled, maximum weed size to treat, and use of adjuvants.

CHEMICAL WEED CONTROL

Table 7-3A. Chemical Weed Control in Peanuts

Herbicide and Formulation	Pounds Active Ingredient Per Acre	Precautions and Remarks
Postemergence, Annual broadleaf weeds (continued)		
lactofen, MOA 14 (Cobra 2 EC) + bentazon, MOA 6 (Basagran 4 L) + 2,4-DB, MOA 4 (Butyrac 200 2 L)	0.2 (12.5 fl oz) + 0.75 to 1 (1.5 to 2 pt) + 0.125 to 0.25 (8-16 fl oz)	Adding 2,4-DB will improve control of larger morningglory, cocklebur, common ragweed, jimsonweed, and citron. See above comments for bentazon, lactofen, and 2,4-DB. See labels for weeds controlled, maximum weed size to treat, and use of adjuvants.
lactofen, MOA 14 (Cobra 2 EC) + imazapic, MOA 2 (Cadre 2 AS) (Impose 2 AS)	0.2 (12.5 fl oz) + 0.063 (4 fl oz)	See above comments for imazapic and lactofen. See labels for weeds controlled, maximum weed size to treat, and use of adjuvants. Some weed species have developed resistance to Cadre.
lactofen, MOA 14 (Cobra 2 EC) + imazethapyr, MOA 2 (Pursuit 2 AS)	0.2 (12.5 fl oz) + 0.063 (4 fl oz)	See above comments for imazethapyr and lactofen. See labels for weeds controlled, maximum weed size to treat, and use of adjuvants. Some weed species have developed resistance to Pursuit.
Postemergence, Annual grasses and broadleaf weeds		
paraquat, MOA 22 (Gramoxone 2 SL) (Parazone 3 SL)	0.13 (8 fl oz) (5.4 fl oz)	See label for weeds controlled and maximum weed size to treat; best results if weeds 1 inches or less. A postemergence application may be made following an at-crack application. Do not make more than two applications per season, do not apply later than 28 days after ground cracking, and do not apply if peanuts are under stress or have significant injury from thrips feeding. Gramoxone is more effective when applied within 2 weeks after peanut emergence. Add 1 pint of nonionic surfactant per 100 gallons of spray solution. Will cause foliar burn on peanuts, but peanuts recover, and yield is not affected. Follow all safety precautions on label.
paraquat, MOA 22 (Gramoxone 2 SL) (Parazone 3 SL) + bentazon, MOA 6 (Basagran 4 L)	0.13 (8 oz) (5.4 oz) + 0.25 to 0.75 (0.5 to 1.5 pt)	See previous comments for paraquat alone. Adding Basagran improves control of common ragweed, prickly sida, smartweed, lambsquarters, and cocklebur and reduces injury to peanuts from paraquat. May be applied any time from ground cracking up to 28 days after ground cracking. Add 1 pint of nonionic surfactant per 100 gallons of spray solution.
paraquat, MOA 22 (Gramoxone 2 SL) (Parazone 3 SL) + bentazon, MOA 6 + acifluorfen, MOA 14 (Storm 4 L)	0.13 (8 fl oz) (5.4 fl oz) + 0.5 + 0.25 1 pt	See previous comments for paraquat alone. Storm improves control of common ragweed, smartweed, lambsquarters, common cocklebur, tropic croton, and spurred anoda. May be applied anytime from ground cracking up to 28 days after ground cracking. Add 0.5 pint of nonionic surfactant per 100 gallons of spray solution. The mixture of Gramoxone SL and Storm is more injurious than these herbicides applied alone.
Postemergence, Florida beggarweed		
chlorimuron, MOA 2 (Classic 0.25 DF)	0.008 (0.5 oz)	Use only for control of Florida beggarweed. Apply from 60 days after crop emergence to within 45 days of harvest. Application to peanuts less than 60 days old will result in crop injury and yield reduction. Apply before Florida beggarweed has begun to bloom and before it has reached 10 inches tall. Larger beggarweed may only be suppressed. Add 1 quart of nonionic surfactant per 100 gallons spray solution; do not add crop oil. May be tank mixed with 2,4-DB; see label for rates and precautions. **Recommended as a salvage treatment only.**
Postemergence, Yellow nutsedge		
bentazon, MOA 6 (Basagran 4 L)	0.75 to 1 (1.5 to 2 pt)	Apply when nutsedge is 6 to 8 inches tall. A repeat application 7 to 10 days later may be needed. Adding crop oil concentrate at 1 quart per acre will increase control. Do not apply more than 2 pints of Basagran per season. Not effective on purple nutsedge.
Postemergence, Yellow and purple nutsedge		
imazapic, MOA 2 (Cadre 2 AS) (Impose 2 AS)	0.063 (4 fl oz)	Apply postemergence when nutsedge is 4 inches or less. Add nonionic surfactant at 1 quart per 100 gallons or crop oil concentrate at 1 quart per acre. See label for rotational restrictions.
imazethapyr, MOA 2 (Pursuit 2 AS)	0.063 (4 fl oz)	Apply before nutsedge is larger than 3 inches tall. Add surfactant at 1 quart per 100 gallons or crop oil concentrate at 1 quart per acre. Do not mix with Basagran for nutsedge control. See label for rotational restrictions. A split application with half of the Pursuit applied preplant incorporated and half applied early postemergence may be more effective than applying all of the Pursuit at one time.
Postemergence, Annual grasses		
clethodim, MOA 1 (Select Max 0.97 EC) (Select 2 EC)	0.094 to 0.125 (9 to 16 fl oz) (6 to 8 fl oz)	Apply Select and Poast to actively growing grass not under drought stress. Consult labels for maximum grass size to treat. Apply in 5 to 20 GPA at 40 to 60 psi. Do not cultivate within 7 days before or after application. Add 2 pints crop oil to Poast. See label for adjuvant use with Select or Select Max. Some broadleaf/sedge herbicides and fungicides can reduce the efficacy of Select and Poast when applied in tank mixtures. See product labels for specific instructions concerning compatibility with other chemicals. See 2017 Peanut Information AG-331 for specific pesticides that reduce control by these herbicides.
sethoxydim, MOA 1 (Poast 1 EC) (Poast Plus 1.5 EC)	0.19 (1.5 pt) (1 pt)	
Postemergence, Bermudagrass		
clethodim, MOA 1 (Select Max 0.97 EC) (Select 2 EC)	0.125 to 0.25 (12 to 32 fl oz) (8 to 16 fl oz)	Apply to actively growing bermudagrass before runners exceed 6 inches In most cases, a second application will be needed. Make second application if regrowth occurs. See comments under annual grasses for adjuvant selection and tank mixing for these herbicides.
sethoxydim, MOA 1 (Poast 1 EC) (Poast Plus 1.5 EC)	0.28 (2.25 pt) (1.5 pt)	

Table 7-3A. Chemical Weed Control in Peanuts

Herbicide and Formulation	Pounds Active Ingredient Per Acre	Precautions and Remarks
Postemergence, Rhizome johnsongrass		
clethodim, MOA 1 (Select Max 0.97 EC) (Select 2 EC)	0.125 to 0.25 (12 to 32 fl oz) (8 to 16 fl oz)	Apply to actively growing johnsongrass before it exceeds 25 inches tall. Add 2 pints per acre of crop oil concentrate. A second application of the same rates can be made if needed before new plants or regrowth exceeds 12 inches.
sethoxydim, MOA 1 (Poast 1 EC) (Poast Plus 1.5 EC)	0.28 (2.25 pt) (1.5 pt)	
Postemergence, Suppression of large Palmer amaranth and other pigweed species that are resistant to the ALS inhibiting herbicides imazapic, chlorimuron, imazethapyr, and dicolsulam		
2,4-DB, MOA 4 (Buryrc 200 2 SL) + lactofen, MOA 14 (Cobra 2 EC) or acifluorfen, MOA 14 (Ultra Blazer 2 L)	0.25 (16 fl oz) + 0.20 (12.5 fl oz) or 0.38 (1.5 pt)	Suppresses and does not completely control Palmer amaranth and other pigweed species that exceed 8 inches. Suppression of weeds exceeding 12 inches will be less than suppression of smaller weeds. Do not expect suppression to exceed 60%. Applying 2,4-DB 3 to 4 days prior to Ultra Blazer or Cobra may be more effective than tank mixtures of 2,4-DB with Ultra Blazer or Cobra. Cobra is generally more effective on larger Palmer amaranth and other pigweed species than Ultra Blazer. Apply crop oil concentrate at 1 gallon per 100 gallons water with acifluorfen or lactofen. Do not apply adjuvant with 2,4-DB alone. See product labels for comments on spray volume and effects on peanut especially during pod set and pod fill. Higher spray volumes are more effective by increasing spray coverage of the contact herbicides Ultra Blazer and Cobra.
2,4-DB, MOA 4 (Butyrac 200 2 SL) then lactofen, MOA 14 (Cobra 2 EC) or acifluorfen, MOA 14 (Ultra Blazer 2 L)	0.25 (16 fl oz) then 0.20 (12.5 fl oz) or 0.38 (1.5 pt)	Two applications of 2,4-DB spaced 10 to 14 days apart will suppress Palmer amaranth and other pigweed species. Although suppression by 2,4-DB is lower than sequential or tank mix application of 2,4-DB and acifluorfen or lactofen within two weeks after application, suppression by sequential applications of 2,4-DB 4 to 5 weeks after initial application is only slightly lower than suppression by sequential or tank mix applilcation of 2,4-DB and Ultra Blazer or Cobra. For more information on managing herbicide-resistant weeds in peanut, see AG-331, *2017 Peanut Information.*
2,4-DB, MOA 4 (Butyrac 200 2 L) then 2,4-DB, MOA 4 (Butyrac 200 2 L)	0.25 (16 oz) then 0.25 (16 oz)	
paraquat, MOA 22 (Gramoxone SL)	See comments	Apply in a roller/wiper implement. Best control achieved when at least 60% coverage of weed foliage occurs. Do not allow paraquat to contact peanut foliage. Mix 1 part Gramoxone SL (other formulations may not be labeled) with 1 to 1.5 parts water to prepare 40 to 50% solution. Add nonionic surfactant at 1 quart per100 gallons. Adjust equipment to apply up to 2 pints per acre of the herbicide-water mixture.
Postemergence, Late-season residual control of annual grasses and certain small-seeded weeds		
dimethenamid, MOA 15 (Outlook 6.0 L)	0.75 to 1 (16 to 21 fl oz)	Will not control emerged grasses or weeds; apply following a cultivation or appropriate postemergence herbicide if emerged grasses or broadleaf weeds are present. Benefit likely only on very sandy fields heavily infested with annual grasses that receive above normal rainfall during the first 4 to 5 weeks of the growing season. Lay-by of Dual Magnum, Outlook, or Warrant may also be of value in fields with a history of eclipta problems; the application must be made before eclipta emerges. Rates are on a broadcast basis; apply in an 18-inch band to row middles. See labels for preharvest intervals.
metolachlor, MOA 15 (Dual Magnum 7.62 EC)	0.64 to 0.84 (0.67 to 0.88 pt)	
acetochlor, MOA 15 (Warrant 3 ME)	0.95-1.5 (1.25 to 2 qt)	

Weed Response to Preplant Incorporated, Preemergence, and At-Cracking Herbicides in Peanuts

D. L. Jordan, Crop Science Department

Ratings based upon average to good soil and weather conditions for herbicide performance and upon proper application rate, technique, and timing.

Table 7-3B. Weed Response to Preplant Incorporated, Preemergence, and At-Cracking Herbicides in Peanuts

Herbicides Key: PPI = Preplant Incorporated; PRE = Preemergence; AC= At-Cracking; POST = Postemergence

	Prowl or Sonalan PPI	Prowl or Sonalan + Dual Magnum or Dual PPI	Prowl or Sonalan + Outlook PPI	Dual Magnum or Dual PPI	Warrant PPI	Outlook PPI	Strongarm PPI or PRE	Prowl or Sonalan + Strongarm PPI	Dual Magnum, Dual or Outlook PPI or PRE	Pursuit PPI + POST	Dual Magnum or Dual PRE	Intrro PRE	Warrant PRE	Outlook PRE	Valor SX PRE	Prowl or Sonalan PPI + Valor SX PRE	Dual Magnum, Dual, Outlook or Warrant + Valor SX PRE	Dual Magnum or Dual AC[1]	Intrro AC[1]	Outlook AC[1]	Gramoxone SL AC	Strongarm AC[2]	Gramoxone + Strongarm AC[2]
Bermudagrass	N	N	N	N	N	N	N	N	N	N	N	N	N	N	N	N	N	N	N	N	P	N	P
Black nightshade	N	F	F	F	F	F	N	N	F	G	F	FG	FG	F	E	E	E	F	FG	F	PF	N	G
Broadleaf signalgrass	G	E	E	G	FG	FG	P	G	G	G	G	FG	FG	FG	P	G	FG	G	FG	FG	E	N	GE
Carpetweed	G	G	G	FG	FG	FG	G	G	G	FG	FG	FG	FG	G	—	G	G	FG	FG	G	FG	—	G
Cocklebur	N	N	N	N	N	N	G	G	G	GE	N	N	N	N	PF	PF	PF	N	N	N	E	E	E
Common ragweed	N	P	PF	PF	PF	F	G	G	GE	P	PF	PF	PF	F	FG	G	GE	PF	PF	F	F	E	E
Crabgrass	E	E	E	E	E	E	P	E	E	F	E	E	E	E	PF	E	E	E	E	E	G	N	G
Crowfootgrass	E	E	E	E	E	E	—	—	—	—	E	E	E	E	PF	G	G	E	E	E	E	N	GE
Dayflower	P	GE	—	GE	—	—	G	G	GE	—	GE	—	—	—	F	F	GE	GE	—	—	—	—	G
Eclipta	N	G	G	G	FG	G	GE	GE	GE	P	FG	FG	FG	FG	G	G	GE	FG	FG	FG	FG	NP	FG
Fall panicum	G	E	E	E	E	E	P	E	E	PF	E	E	E	E	PF	FG	GE	E	E	E	E	N	GE
Florida beggarweed	N	PF	PF	F	F	F	F	F	F	P	F	F	F	F	G	GE	E	F	F	F	E	FG	G
Foxtails	E	E	E	E	E	E	P	E	E	G	E	E	E	E	PF	E	E	E	E	E	E	N	GE
Goosegrass	E	E	E	E	E	E	P	E	E	PF	E	E	E	E	PF	GE	E	E	E	E	E	N	GE
Jimsonweed	N	N	N	N	N	N	GE	GE	GE	G	N	N	N	N	N	G	GE	N	N	N	E	—	E
Johnsongrass, Seedling	G	G	G	PF	PF	PF	N	G	PF	GE	PF	PF	PF	PF	N	FG	PF	PF	PF	PF	E	N	GE
Johnsongrass, Rhizome	P	PF	PF	N	N	N	P	N	FG	N	N	N	N	N	N	N	N	N	N	N	P	N	P
Lambsquarters	G	NG	G	F	F	FG	FG	GE	GE	FG	F	F	F	FG	GE	GE	GE	F	F	FG	F	N	G
Morningglory	P	P	P	N	N	N	G	G	G	G	N	N	N	N	FG	G	G	N	N	N	F	GE	E
Nutsedge, Yellow	N	G	FG	G	N	FG	FG	FG	G	FG	FG	P	N	F	P	PF	FG	FG	N	F	PF	PF	G
Nutsedge, Purple	N	N	N	N	N	N	FG	FG	FG	FG	N	N	—	N	P	P	P	N	N	N	PF	NP	PF
Pigweed	G	E	E	G	G	G	G	E	E	E	G	GE	G	GE	E	E	E	G	GE	GE	E	NP	E
Prickly sida	N	P	P	P	P	P	FG	FG	FG	G	P	P	P	P	FG	G	G	P	P	P	F	—	G
Purslane	G	GE	GE	G	FG	G	—	G	G	—	G	G	FG	G	G	GE	GE	GE	P	P	—	—	—
Sicklepod	N	NP	NP	NP	NP	NP	P	P	P	P	NP	PF	NP	NP	P	PF	PF	NP	PF	NP	G	N	G
Smartweed	N	N	N	N	N	N	G	G	G	G	N	N	N	N	—	—	—	N	N	N	G	—	E
Spurge spp.	P	F	F	PF	P	PF	—	—	—	P	F	P	P	F	G	G	G	N	N	N	F[2]	—	F[2]
Spurred anoda	N	N	N	N	N	N	FG	FG	FG	G	N	N	N	N	F	FG	FG	N	N	N	P	—	G
Texas panicum	G	G	G	PF	PF	PF	P	G	PF	PF	PF	PF	PF	PF	G	E	PF	PF	PF	PF	E	N	GE
Tropic croton	N	N	N	N	N	N	PF	PF	PF	P	N	N	N	N	—	—	—	N	N	N	F	—	F
Velvetleaf	N	N	N	N	N	N	GE	GE	GE	FG	N	N	N	N	F	FG	FG	N	N	N	F	—	FG

[1] Residual control only.
[2] Assumes weeds are 1- to 2-inches tall or smaller.

Key:
E = excellent control, 90% or better
F = fair control, 50% to 80%
G = good control, 80% to 90%
P = poor control, 25% to 50%
N = no control, less than 25%

CHEMICAL WEED CONTROL

Weed Response to Postemergence Herbicides in Peanuts

D. L. Jordan, Crop Science Department

Ratings based upon average to good soil and weather conditions for herbicide performance and upon proper application rate, technique, and timing.

Table 7-3C. Weed Response to Postemergence Herbicides — Peanuts

Herbicides Key: PPI = Preplant Incorporated; PRE = Preemergence; AC = At-Cracking; POST = Postemergence

Species	Butyrac 200	Gramoxone [1]	Gramoxone + Basagran	Gramoxone + Storm	Basagran	Basagran + buryac 200	Ultra Blazer	Ultra Blazer + Butyrac 200	Ultra Blazer + Basagran [2]	Storm	Storm + Butyrac 200	Pursuit + Butyac 200	Cadre or Impose	Cobra	Cobra + Ultra Blazer	Cobra + Basagran + Butyrac 200	Cobra + Cadre or Impose	Cobra + Pursuit	Poast or Poast Plus	Select or Select Max
Bermudagrass	N	P	P	P	N	N	N	N	P	N	N	N	N	N	N	N	N	N	FG	G
Black nightshade	N	PF	PF	G	P	P	G[1]	G[1]	G[1]	G[1]	G[1]	G	G	G[1]	G[1]	G[1]	G	G	N	N
Broadleaf signalgrass	N	GE	E	GE	N	N	NP	NP	P	NP	NP	G	G	N	N	N	G	G	E	E
Carpetweed	P	FG	FG	G	P	P	GE	E	E	E	G	G	FG	FG	G	G	G	G	N	N
Cocklebur	E	G	E	E	E	G	E	G	E	E	E	E	E	E	G	G	E	E	N	N
Common ragweed	PF	F	G	E	G[4]	G[4]	E[1]	E[1]	E[1]	E[1]	E[1]	P	PF	E	E	E	E	E	N	N
Crabgrass	N	G	G	G	N	N	N	N	N	N	N	FG	FG	N	N	N	FG	FG	GE	GE
Crowfootgrass	N	GE	G	GE	N	N	P	P	P	P	P	G	N	N	N	N	G	P	F	G
Dayflower	—	G	G	FG	G	G	—	—	G	FG	FG	—	G	—	G	G	G	—	N	N
Eclipta	P	F	F	FG	FG	FG	G	G	G	FG	FG	P	F	G	G	G	G	G	N	N
Fall panicum	N	GE	G	GE	N	N	PF	PF	P	PF	PF	PF	G	N	N	N	G	PF	E	E
Florida beggarweed	P	G	GE	G	N	P	PF	F	F	P	P	P	F	F	F	F	F	F	N	N
Foxtails	N	GE	G	GE	N	N	PF	PF	P	PF	PF	G	G	N	N	N	G	G	E	E
Goosegrass	N	GE	G	GE	N	N	N	N	N	N	N	N	F	N	N	N	F	N	GE	GE
Jimsonweed	P	G	E	E	E	E	E	E	E	E	E	G	E	E	E	E	E	E	N	N
Johnsongrass, Seedling	N	GE	GE	GE	N	N	P	P	P	P	P	GE	E	N	N	N	E	GE	E	E
Johnsongrass, Rhizome	N	P	P	P	N	N	N	N	N	N	N	F	FG	N	N	N	FG	F	G	GE
Lambsquarters	PF	F	G	G	FG	G[4]	G	G	GE	G	G	P	PF	P	FG	G	PF	P	N	N
Morningglory, Pitted	FG	F	FG	E	P	G	E	E	E	E	E	G	GE	G	G	G	GE	G	N	N
Morningglory, Others	E	F	FG	E	P	E	GE	E	E	GE	E	E	G	G	G	G	E	G	N	N
Nutsedge, Yellow	N	PF	FG	G	G[3]	G	N	N	G	F	F	F	G	N	G[3]	G[3]	G	F	N	N
Nutsedge, Purple	N	PF	PF	PF	NP	P	N	N	P	N	N	FG	G	N	P	P	G	FG	N	N
Pigweed	PF	G	G	E	N	P	E	E	E	E	E	E	E	E	E	E	E	E	N	N
Prickly sida	F	F	G	G	G	G	N	F	G	FG	G	P	G	G	G	G	G	G	N	N
Purslane	FG	—	G	G	G	G	E	E	E	GE	GE	FG	—	E	E	E	E	E	N	N
Sicklepod	G[3]	G	G	G	N	G[6]	NP	G[6]	NP	NP	G[6]	G[6]	E	P	P	G[6]	E	F	N	N
Smartweed	PF	G	E	E	E	E	GE	E	E	E	E	G	F	F	E	E	F	G	N	N
Spurge spp.	P	F[1]	F[1]	F[1]	P	P	F[1]	F[1]	F[1]	PF[1]	PF[1]	PF[1]	—	F[1]	F[1]	F[1]	F[1]	—	N	N
Spurred anoda	P	P	FG	G	G	GE	P	P	G	F	F	F	G	F	G	GE	G	F	N	N
Texas panicum	N	GE	G	GE	N	N	NP	NP	NP	NP	NP	NP	G	N	N	N	G	NP	E	E
Tropic croton	PF	F	F	G	F	F	G	G	G	G	G	P	P	G	G	G	G	G	N	N
Velvetleaf	P	F	G	FG	G	G	PF	PF	FG	FG	FG	FG	G	G	G	G	G	G	N	N

[1] Assumes weeds are 1 to 2 inches tall or smaller.
[2] Assumes optimum rates and ratios of Basagran and Blazer; see labels.
[3] Two applications, 10 to 14 days apart.
[4] Assumes optimum conditions and addition of crop oil concentrate.
[5] Ratings assume weeds in one- to two-leaf stage.
[6] Assumes follow-up treatment with 2,4-DB.

Key:
E = excellent control, 90% or better
G = good control, 80% to 90%
F = fair control, 50% to 80%
P = poor control, 25% to 50%
N = no control, less than 25%

CHEMICAL WEED CONTROL

Brand Names and Formulations of Active Ingredients

Table 7-3D. Brand names and formulations of active ingredients listed in Table 7-3A.

Active ingredient(s)	Brand name(s)[a]	Formulation[b]	Mode of Action
Acetochlor	Warrant	3 ME	15
Acifluorfen	Ultra Blazer		22
Bentazon	Basagran		
Carfentrazone	Aim EC	2 EC	14
Clethodim	Intensity One, Select Max, Shadow Ultra, TAPOUT	0.97 EC	1
	Arrow 2EC, Agri Star Clethodim 2EC, AmTide Clethodim 2 EC, Avatar, Avatar S2, Clethodim, Dakota, Grassout Max, Intensity, PS Clethodim, Section 2 EC, Select, Shadow, Tide Clethodim, Volunteer, Willowood Clethodim	2 EC	1
	Shadow 3 EC, Section Three	3 EC	1
Dimethenamid-P	Outlook	6 EC	15
Flumioxazin	Outflank, Panther, Rowel, Valor SX, Warfox	51 WDG	14
Lactofen	Cobra	2 EC	14
Paraquat	Gramoxone SL, Gramoxone Inteon	2 S	22
	Bonedry, Devour, Firestorm, Helmquat, Paraquat Concentrate, Para-Shot, Parazone, Willowood Paraquat	3 S	22
Pendimethalin	Acumen, Framework, Helena Pendimethalin, Pendipro 3.3 EC, Prowl 3.3 EC, Stealth	3.3 EC	3
	Prowl H_2O, Satellite Hydrocap	3.8 AS	3
pyraflufen ethyl	ET Herbicide/Defoliant	0.208 EC	14
S-metolachlor	Brawl, Dual Magnum, Medal	7.62 EC	15
	Brawl II, Cinch, Dual II Magnum, Medal II	7.64 EC	15
Metolachlor	Parallel PCS, Stalwart	8 EC	15
sulfentrazone + carfentrazone	Spartan Charge		14
sethoxydim	Nufarm Sethoxydim SPC, Poast Plus	1 EC	1
	Poast	1.5 EC	1
2,4-DB	Albaugh Butyrac 175 Broadleaf Herbicide, 2,4-DB 175 Herbicide	1.75 SC	4
	Albaugh Butyrac 200 Broadleaf Herbicide, 2,4-DB 200 Herbicide, 2,4-DB 200	2 SC	4

[a] Brands listed were registered for sale and use in peanut in North Carolina in 2015 according to http://www.kellysolutions.com/nc/searchbychem.asp
[b] AS, aqueous suspension; CS, capsule suspension; EC, emulsifiable concentrate; F, flowable; ME, microencapsulated; S, solution; SC, suspension concentrate; WDG, water dispersible granule.

Chemical Weed Control in Sorghum

W. J. Everman, Crop Science Department

NOTE: A mode of action code has been added to the Herbicide and Formulation column of this table. Use MOA codes for herbicide resistance management. See Table 7-10, Herbicide Resistance Management, for details.

Table 7-4. Chemical Weed Control in Sorghum

Herbicide, Mode of Action Code[1], and Formulation	Amount of Formulation Per Acre	Pounds Active Ingredient Per Acre	Precautions and Remarks
Grain Sorghum No-Till Burndown, Emerged annual broadleaf and grass weeds, suppression or control of perennials			
glyphosate, MOA 9 (numerous brands and formulations)	See label	0.56 to 1.13 (lb a.e.)	Glyphosate is available as an isopropylamine salt and a potassium salt. Glyphosate formulations and application rates should be compared on the basis of pounds of glyphosate acid equivalent (a.e.) per gallon and per acre, respectively. Rate in the preceding column is expressed as a.e. See Table 7-10 for glyphosate rate conversions. Apply before crop emerges. Glyphosate rate depends upon weed species and weed size; see labels for suggested rates. Higher rates can be applied for perennial weeds; see labels for details. See comments on labels concerning nitrogen as the carrier. Apply in 10 to 20 gallons of water per acre using flat fan nozzles. For residual grass and broadleaf weed control, glyphosate products may be tank mixed with most preemergence herbicides. See the section on Grain Sorghum—Preemergence. Refer to specific product labels for application rates, weeds controlled, application directions, and precautions. Adjuvant recommendations vary according to the glyphosate product used. See label of brand used for specific recommendations.
Grain Sorghum No-Till Burndown, Emerged annual broadleaf and grass weeds, top-kill of perennials			
paraquat, MOA 22 (Gramoxone Inteon) 2 SL	2 to 4 pt	0.5 to 1	Apply before, during, or after planting but before crop emerges using clean water or clear fertilizer solution as the carrier. Apply in a minimum of 10 GPA (20 to 40 preferred) using flat fan nozzles. Add either a nonionic surfactant at 1 pint per 100 gallons or crop oil concentrate at 1 gallon per 100 gallons. Use 0.5 to 0.64 pound a.i. on weeds 1 to 3 inches, 0.75 pound a.i. on weeds 3 to 6 inches, and 1 pound a.i. on weeds 6 inches or larger. Use 0.5 pound a.i. for rye cover crop or 0.75 pound a.i. for wheat cover crop. Rainfast within 30 minutes. For residual grass and broadleaf weed control, paraquat can be tank mixed with most preemergence sorghum herbicides and herbicide combinations. See the section on Grain Sorghum—Preemergence, Conventionally Planted. Refer to specific product labels for application rates, weeds controlled, application directions, and precautions. Better control of emerged weeds will be obtained with tank mixtures of Gramoxone plus an atrazine-containing product. Generic brands of paraquat containing 3 pounds active per gallon may be applied at 1.3 to 2.7 pints.
Grain Sorghum No-Till Burndown or Premergence			
saflufenacil, MOA 14 (Sharpen)	1.0 to 2.0 fl oz	0.027 to 0.054 (lb a.i.)	Sharpen can be applied to control glyphosate-resistant marestail prior to grain sorghum emergence. See label for application with other herbicides and specifics on adjuvant selection. To avoid injury potential wth burndown or preemergence applications, consult local seed company for possible injury of grain sorghum hybrids or varieties.
thifensulfuron, MOA 2 + tribenuron, MOA 2 (FirstShot SG with TotalSol) 50 WDG	0.5 to 0.8 oz	0.008 to 0.013 + 0.008 to 0.013	FirstShot should be applied 14 days prior to sorghum planting. Tank mix with glyphosate or paraquat for broad spectrum burndown. See label for specific rates and soil type restrictions. See label for adjuvant recommendations.
Grain Sorghum Preemergence, Annual broadleaf weeds and certain annual grasses			
atrazine, MOA 5 (AAtrex) 4 F (AAtrex Nine-O) 90 WDG	1 to 2 qt 1.1 to 2.2 lb	1 to 2	Controls most broadleaf weeds and large crabgrass, crowfootgrass, foxtails, goosegrass, and sandbur. Does not control broadleaf signalgrass, fall panicum, Texas panicum, seedling johnsongrass, or shatter-cane. Do not use on sand, loamy sand, or sandy loam soils. Do not use on medium- or fine-textured soils with less than 1% organic matter. On highly erodible soils (defined by NRCS) with less than 30% plant residue cover, do not exceed 1.6 pounds active ingredient. See labels for details on set-back requirements from streams and lakes. See labels for comments on rotational crops. For improved grass control, atrazine may be tank mixed with s-metolachlor, alachlor, or dimethenamid if the seed have been properly treated with a safener. See comments for s-metolachlor, alachlor, or dimethenamid applied preemergence. Generic brands of atrazine are available.
Grain Sorghum Preemergence, Annual grasses and small-seeded broadleaf weeds			
alachlor, MOA 15 (Intro) 4 EC (Micro-Tech) 4 FME	1.5 to 2.5 qt	1.5 to 2.5	Controls most annual grasses and pigweed. Does not control seedling johnsongrass, shattercane, or Texas panicum. Use only on grain sorghum planted with seed properly treated with a safener containing the active ingredient flurazole. Rate depends upon soil texture; see label for details. May be tank mixed with atrazine for broadleaf weed control. See comments for atrazine applied preemergence.
dimethenamid-P, MOA 15 (Outlook) 6.0 EC	12 to 21 fl oz	0.56 to 0.98	Use 12 to 18 fluid ounces on soils with less than 3% organic matter or 14 to 21 fluid ounces on soils with greater than 3% organic matter. Controls most annual grasses and pigweed. Does not control seedling johnsongrass, shattercane, or Texas panicum. Use only on grain sorghum planted with seed properly treated with Concep or Screen protectant. Rate depends upon soil texture and organic matter; see label for details. May be tank mixed with atrazine for broadleaf weed control. See comments for atrazine applied preemergence.
metolachlor, MOA 15 (Me-Too-Lachlor) 8 EC (Parallel) 7.8 EC (Parallel PCS) 8 EC (Stalwart) 8 EC	1 to 1.67 pt	1 to 1.67	See comments for s-metolachlor products. Products containing s-metolachlor are more active on weeds per unit of formulated product than those containing metolachlor. In general, it takes 1.5 pints of a metolachlor product to get the activity one would get from 1 pint of an s-metolachlor product.
S-metolachlor, MOA 15 (Brawl) 7.62 EC (Brawl II) 7.64 EC (Cinch) 7.64 EC (Dual Magnum) 7.62 EC (Dual II Magnum) 7.64 EC (Medal) 7.62 EC (Medal II) 7.64 EC	1 to 1.67 pt	0.95 to 1.6	Controls most annual grasses and pigweed. Does not control seedling johnsongrass, shattercane, or Texas panicum. Use only on grain sorghum planted with seed properly treated with Concep or Screen protectant. Rate depends upon soil texture and organic matter; see label for details. May be tank mixed with atrazine for broadleaf weed control. See comments for atrazine applied preemergence.

CHEMICAL WEED CONTROL

Table 7-4. Chemical Weed Control in Sorghum

Herbicide, Mode of Action Code[1], and Formulation	Amount of Formulation Per Acre	Pounds Active Ingredient Per Acre	Precautions and Remarks
Grain Sorghum Preemergence, Annual grasses and broadleaf weeds			
alachlor, MOA 15 + atrazine, MOA 5 (Bullet) 4 FME (Lariat) 4 F	2.5 to 4 qt	1.56 to 2.5 + 0.94 to 1.5	Controls most annual grasses and broadleaf weeds. Does not control seedling johnsongrass, shattercane, or Texas panicum. Use only on grain sorghum planted with seed properly treated with Concep or Screen protectant. Rate depends upon soil texture and organic matter; see label for details. See label for comments on rotational crops and details on set-back requirements from streams and lakes.
dimethenamid-P, MOA 15 + atrazine, MOA 5 (Guardsman Max) 5F	2.5 to 4.6 pt	0.5 to 1 + 1 to 1.9	Controls most annual grasses and broadleaf weeds. Does not control seedling johnsongrass, shattercane, or Texas panicum. Use only with Concep-treated seed. Apply only to medium- or fine-textured soils. Rate depends on soil texture and organic matter; see label for details. See label for comments on rotational crops and set-back requirements from streams and lakes. May be applied postemergence to sorghum up to 12 inches tall.
S-metolachlor, MOA 15 + atrazine, MOA 5 (Bicep II Magnum) 5.5 F (Brawl II ATZ) 5.5 F (Cinch ATZ) 5.5 F (Medal II AT) 5.5 F	1.6 to 2.1 qt	0.96 to 1.26 + 1.24 to 1.63	Controls most annual grasses and broadleaf weeds. Does not control seedling johnsongrass, shattercane, or Texas panicum. Use only with Concep-treated seed. Apply only to medium- and fine-textured soils with at least 1% organic matter. See label for comments on rotational crops and details on set-back requirements from streams and lakes. May be applied postemergence to sorghum up to 12 inches tall.
dimethenamid-P, MOA 15 + saflufenacil, MOA 14 (Verdict) + dimethenamid-P, MOA 15 (Outlook) 6.0 EC	10 to 18 fl oz + 4 to 12 fl oz	0.045 to 0.08 + 0.4 to 0.7 + 0.19 to 0.56	Tank mix with Outlook at 4 to 10 fluid ounces on coarse soils or 6 to 12 fluid ounces on medium to fine soils to ensure grass control. The tank mix controls most annual grasses and broadleaf weeds. Does not control seedling johnsongrass, shattercane or Texas panicum. Use only with Concep-treated seed. Apply only where organic matter is greater than 1.5%. Rate depends on soil texture; see label for details. See label for comments on rotational crops. DO NOT apply to emerged sorghum, or severe injury may occur.
Grain Sorghum Postemergence, Annual grass and small broadleaf weeds			
quinclorac, MOA 4, 26 (Facet L) 1.5 L	22 to 32 fl oz	0.26 to 0.375	Apply from preemergence up to 12 inches tall. Grass and broadleaf weeds must be under 2 inches tall. Controls small barnyardgrass, broadleaf signalgrass, large crabgrass, and foxtail species. See label for list of broadleaf weeds controlled. Add 1 quart per acre of crop oil concentrate or 1 to 2 pints per acre of methylated seed oil. May be tankmixed with atrazine, 2,4-D, dicamba, Peak, or Buctril. See label for rotation restrictions.
Grain Sorghum Postemergence, Annual broadleaf weeds			
atrazine, MOA 5 (AAtrex) 4 F (AAtrex Nine-O) 90 WDG	1.2 qt 1.3 lb	1.2	Apply after sorghum reaches the three-leaf stage but before it exceeds 12 inches tall. Do not use on sand or loamy sand soil. Broadleaf weeds must be 4 inches tall or less. See label for list of weeds controlled. Add 1 quart per acre of crop oil concentrate. If a postemergence application is made following an at-planting application, do not exceed a total of 2.5 pounds active ingredient per acre per season. See label for details on set-back requirements from streams and lakes. Generic brands of atrazine are available.
bentazon, MOA 6 (Basagran) 4 SL	1.5 to 2 pt	0.75 to 1	Apply overtop or directed any time prior to heading. See label for weeds controlled and recommended weed size for treatment. Adding crop oil concentrate at 1 to 2 pt per acre will improve control. Do not apply more than 2 pints Basagran per acre per season. Basagran also controls or suppresses yellow nutsedge. May be tank mixed with atrazine. When tank mixing, see respective labels for application rates, directions, and precautions.
bromoxynil, MOA 6 (Buctril) 2 EC (Buctril 4 EC) 4 EC	1.5 pt 0.75 pt	0.375	Can apply overtop of sorghum from the four-leaf stage until the preboot stage. Use of drop nozzles is suggested after sorghum is 6 to 8 inches tall to ensure better weed coverage. An adjuvant is not needed. Controls cocklebur, morningglory, lambsquarters, ragweed, jimsonweed, smartweed, velvetleaf, and very small pigweed. See label for recommended weed size for treatment. Do not apply when sorghum foliage is wet. May be tank mixed with atrazine, Banvel, Clarity, or 2,4-D. When tank mixing, see respective labels for application rates, directions, and precautions.
bromoxynil, MOA 6 + pyrasulfotole MOA 27 (Huskie) 2.06 EC	13 to 16 oz	0.21 to 0.26	Appy overtop between 3 leaf stage up to 30 inches and/or flag leaf emergence, whichever comes first. Best control occurs when weeds are 4 inches tall or less. Controls cocklebur, morningglory, lambsquarters, ragweed, jimsonweed, smartweed, velvetleaf, and pigweed. Transitory leaf burn will occur after Huskie application to grain sorghum. Stunting and yellowing can also occour, however these symptoms generally dissipate within 21 days and do not affect yield. May be tank mixed with atrazine, 2,4-D or dicamba as needed for additional broadleaf control. May also be tankmixed with Bicep II Magnum, Dual II Magnum, Guardsman Max, Outlook, Starane, and Warrant for additional control. When tank mixing, see respective labels for application rates, directions, and precautions.
carfentrazone, MOA 14 (Aim) 2 EC	0.5 fl oz	0.008	Apply from sorghum emergence through six-leaf stage. Add nonionic surfactant according to label directions. Controls small lambsquarters, morningglory, pigweed. Aim at rates up to 1 fluid ounce can be applied with drop nozzles.
dicamba, MOA 4 (Banvel) 4 S L (Clarity) 4 SL	0.5 pt 0.5 pt	0.25	Apply from spike stage until sorghum is 8 inches tall. May be tank mixed with atrazine or Buctril. When tank mixing, see respective labels for application rates, directions, and precautions. **Carefully follow all precautions on labels to avoid drift to sensitive crops.**
dicamba, MOA 4 + atrazine, MOA 5 (Marksman) 3.2 F	2 pt	0.28 + 0.53	Controls most broadleaf weeds. Apply when sorghum has two to five leaves (about 2 to 8 inches tall). Do not add surfactant or crop oil. Do not apply in vicinity of dicamba-sensitive crops. See label for details on set-back requirements from streams and lakes.
prosulfuron, MOA 2 (Peak) 57 WDG	0.75 to 1 oz	0.027 to 0.036	Controls pigweed, lambsquarters, cocklebur, morningglory, jimsonweed, ragweed, smartweed, sicklepod, and velvetleaf. Apply to sorghum 5 to 30 inches tall. Use drop nozzles if sorghum is over 20 inches. Add nonionic surfactant at 1 quart per 100 gallons or crop oil concentrate at 1 quart per acre. See label for rotational restrictions. May tank mix with atrazine, Banvel, Buctril, Marksman, or 2,4-D. See labels for details. See Peak label for rotational restrictions.
2,4-D amine formulation, MOA 4 (various brands) 3.8 SL	0.5 pt	0.24	Can apply overtop of sorghum 6 to 15 inches tall. Wait until secondary roots are well established. Sorghum is less tolerant of 2,4-D than is corn. Use drop nozzles as soon as possible and certainly after sorghum is 8 inches tall. Note that 2,4-D rates listed here are less than rates on most labels. Less than label-recommended rates are suggested to avoid injury to the crop. Do not apply during boot, flowering, or early dough stages. May be applied in nitrogen solution at lay-by. When mixing 2,4-D amine in nitrogen solution, add 1 pint of 2,4-D amine to 4 pints of water and mix. Then add this mixture to the nitrogen solution in the spray tank with considerable agitation until thoroughly mixed. Do not allow mixture to stand in sprayer. **Use extreme caution to avoid drift to sensitive crops such as cotton and tobacco. Ester formulations of 2,4-D may be applied to sorghum. However, use of ester formulations of 2,4-D or acid/ester mixtures, such as Weedone 638, is not suggested if sensitive crops are located within 1 mile of the sorghum.**

Table 7-4. Chemical Weed Control in Sorghum

Herbicide, Mode of Action Code[1], and Formulation	Amount of Formulation Per Acre	Pounds Active Ingredient Per Acre	Precautions and Remarks
Grain Sorghum Postemergence-Directed, Annual grass and broadleaf weeds			
linuron, MOA 7 (Linex) 4 L	1 to 2 pt	0.5 to 1	Apply as directed spray in 25 to 40 gallons per acre of water. Add 1 pint of nonionic surfactant per 25 gallons of spray mixture. For application with precision directed equipment, apply 0.5 pound active per acre when sorghum is 12 inches tall and weeds are up to 2 inches tall. Apply 0.5 to 1 pound active per acre when sorghum is 15 inches tall and weeds are 2 to 4 inches tall.
Forage Sorghum Preemergence, Annual broadleaf weeds and certain annual grasses			
atrazine, MOA 5 (AAtrex) 4 F (AAtrex Nine-O) 90 WDG	1 to 2 qt 1.1 to 2.2 lb	1 to 2	Controls most broadleaf weeds and large crabgrass, crowfootgrass, foxtails, goosegrass, and sandbur. Does not control broadleaf signalgrass, fall panicum, Texas panicum, seedling johnsongrass, or shatter-cane. Do not use on sand, loamy sand, or sandy loam soils. Do not use on medium- or fine-textured soils with less than 1% organic matter. Do not exceed 1.6 pounds active ingredient per acre on highly erodible soils (as defined by the NRCS) with less than 30% plant residue cover. See labels for comments on rotational crops. See label for details on set-back requirements from streams and lakes. For improved grass control, atrazine may be tank mixed with Cinch, Dual Magnum, or Dual II Magnum; see comments for Cinch, Dual Magnum, or Dual II Magnum applied preemergence. Generic brands of atrazine are available.
metolachlor, MOA 15 (Me-Too-Lachlor) 8 EC (Parallel) 7.8 EC (Parallel PCS) 8 EC (Stalwart) 8 EC	1 to 1.67 pt	1 to 1.67	See comments for s-metolachlor products. Products containing s-metolachlor are more active on weeds per unit of formulated product than those containing metolachlor. In general, it takes 1.5 pints of a metolachlor product to get the activity one would get from 1 pint of an s-metolachlor product.
S-metolachlor, MOA 15 (Brawl) 7.62 EC (Brawl II) 7.64 EC (Cinch) 7.64 EC (Dual Magnum) 7.62 EC (Dual II Magnum) 7.64 EC (Medal) 7.62 EC (Medal II) 7.62 EC	1 to 1.67 pt	0.95 to 1.6	Controls most annual grasses and pigweed. Does not control seedling johnsongrass, shattercane, or Texas panicum. Use only on sorghum planted with seed properly treated with Concep or Screen protectant. Rate depends upon soil texture and organic matter; see label for details. May be tank mixed with atrazine for broadleaf weed control. See comments for atrazine applied preemergence.
S-metolachlor, MOA 15 + atrazine, MOA 5 (Bicep II Magnum) 5.5 F (Brawl II ATZ) 5.5 F (Cinch ATZ) 5.5 F (Medal II AT) 5.5 F	1.6 to 2.1 qt	0.96 to 1.26 + 1.24 to 1.63	Controls most annual grasses and broadleaf weeds. Does not control seedling johnsongrass, shattercane, or Texas panicum. Use only with Concep- or Screen-treated seed. Apply only to medium- and fine-textured soils with at least 1% organic matter. See label for comments on rotational crops and details on set-back requirements from streams and lakes.
Forage Sorghum Postemergence, Annual broadleaf weeds			
atrazine, MOA 5 (AAtrex) 4 F (AAtrex Nine-O) 90 WDG	1.2 qt 1.3 lb	1.2	Apply after sorghum reaches the three-leaf stage but before it exceeds 12 inches tall. Do not use on sand or loamy sand soil. Broadleaf weeds must be 4 inches tall or less. See label for list of weeds controlled. Add 1 quart per acre of crop oil concentrate. If a postemergence application is made following an at-planting application, do not exceed a total of 2.5 pounds active ingredient per acre per season. Do not graze or feed forage from treated areas for 21 days following application. See label for details on set-back requirements from streams and lakes. Generic brands of atrazine are available.
bentazon, MOA 6 (Basagran) 4 SL	1.5 to 2 pt	0.75 to 1	Apply overtop or directed any time prior to heading. See label for weeds controlled and recommended weed size for treatment. Adding crop oil concentrate at 1 to 2 pints per acre will improve control. Do not apply more than 2 pints per acre per season. Basagran also controls or suppresses yellow nutsedge. May be tank mixed with atrazine. When tank mixing, see respective labels for application rates and directions and precautions. Do not graze treated fields for at least 12 days following Basagran application.
bromoxynil, MOA 6 (Buctril) 2 EC (Buctril 4 EC) 4 EC	1.5 pt 0.75 pt	0.375	Can apply overtop of sorghum from the four-leaf stage until the preboot stage. See label for weeds controlled and recommended weed size for treatment. Do not apply when sorghum foliage is wet. May be tank mixed with atrazine, Banvel, Clarity, or 2,4-D. When tank mixing, see respective labels for application directions, precautions, and weeds controlled. Do not cut for feed or fodder or graze within 30 days of application.
carfentrazone, MOA 14 (Aim) 2 EC	0.5 fl oz	0.008	Apply from sorghum emergence through six-leaf stage. Add nonionic surfactant according to label directons. Controls small lambsquarters, morningglory, and pigweed.
dicamba, MOA 4 (Banvel) 4 SL (Clarity) 4 SL	0.5 pt 0.5 pt	0.25	Apply from spike stage until sorghum is 8 inches tall. May be tank mixed with atrazine or Buctril. When tank mixing, see respective labels for application rates and directions and precautions. Do not cut for silage prior to mature grain stage. Do not remove animals from treated areas for slaughter prior to 30 days after application. For lactating dairy animals, wait 7 days before grazing or 37 days before harvest for hay. There is no waiting period between treatment and grazing for non-lactating animals. **Carefully follow all precautions on labels to avoid drift to sensitive crops.**
dicamba, MOA 4 + atrazine, MOA 5 (Marksman) 3.2 F	2 pt	0.28 + 0.53	Controls most broadleaf weeds. Apply when sorghum has two to five leaves (is about 2 to 8 inches tall). Do not add surfactant or crop oil. Do not apply in the vicinity of dicamba-sensitive crops. See label for set-back requirements from streams and lakes.
2,4-D amine formulation, MOA 4 (various brands) 3.8 SL	0.5 pt	0.24	Can apply overtop of sorghum 6 to 15 inches tall. Wait until secondary roots are well established. Sorghum is less tolerant of 2,4-D than is corn. Note that 2,4-D rates listed here are less than rates on most labels. Less than label-recommended rates are suggested to avoid injury to the crop. Do not apply during boot, flowering, or early dough stages. Do not forage or feed sorghum fodder for 7 days following application. **Use extreme caution to avoid drift to sensitive crops such as cotton and tobacco. Ester formulations of 2,4-D may be applied to sorghum. However, use of ester formulations of 2,4-D or acid/ester mixes, such as Weedone 638, is not suggested if sensitive crops, especially cotton and tobacco, are located within 1 mile of the sorghum.**

[1] Mode of Action (MOA) code developed by the Weed Science Society of America. See Table 7-10, Herbicide Resistance Management, for details.

CHEMICAL WEED CONTROL

Chemical Weed Control in Soybeans

W. J. Everman, Crop Science Department

NOTES: A mode of action code has been added to the Herbicide and Formulation column of this table. Use MOA codes for herbicide resistance management. See Table 7-10, Herbicide Resistance Management, for details.

Control of witchweed is part of the State/Federal Quarantine Program. Contact the N.C. Department of Agriculture, Plant Industry Division, at 1-800-206-9333.

Table 7-5A. Chemical Weed Control in Soybeans

Herbicide, Mode of Action Code[1], and Formulation	Amount of Formulation Per Acre	Pounds Active Ingredient Per Acre	Precautions and Remarks
Preplant (foliar application), Conventional or Reduced Tillage, Control or suppression of emerged weeds to reduce tillage operations			
glyphosate, MOA 9 (numerous brands and) formulations)	See label	0.38 to 1.13 (lb a.e.)	Glyphosate is available as an isopropylamine salt and a potassium salt. Glyphosate formulations and application rates should be compared on the basis of pounds of glyphosate acid equivalent (a.e.) per gallon and per acre, respectively. Rate in the preceding column is expressed as a.e. See Table 7-10 for glyphosate rate conversions. Recommended rates depend upon weed species and size; see labels for details. Higher rates can be used for specific situations. Delay tillage at least 3 days after application. Adjuvant recommendations vary by glyphosate brand; follow directions on label of brand used. May add 0.75 to 1 pint of 2,4-D for improved control of specific broadleaf weeds. Delay planting at least 7 days after application of ester formulations of 2,4-D or 15 days after application of amine formulations. Use only a brand of 2,4-D with the preplant application included on the label. Follow all precautions on the 2,4-D label. Use of an ester formulation of 2,4-D is discouraged within 1 mile of cotton.
Preplant Incorporated, Annual grasses			
ethalfluralin, MOA 3 (Sonalan) 3 EC	1.5 to 3 pt	0.56 to 1.12	Controls common annual grasses plus pigweed and lambsquarters. Incorporate in top 2 to 3 inches of seedbed within 2 days of application; immediate incorporation suggested. For broadleaf weed control, Sonalan may be tank mixed with most broadleaf herbicides. When tank mixing, see respective labels for application rates, weeds controlled, specific application directions, and precautions.
metolachlor, MOA 15 (Me-Too-Lachlor) 8 EC (Parallel PCS) 8 EC (Parrlay) 8 EC (Stalwart) 8 EC	1 to 2 pt	1 to 2	See comments for s-metolachlor products. Products containing s-metolachlor are more active on weeds per unit of formulated product than those containing metolachlor. In general, it takes 1.5 pints of metolachlor product to get the activity one would get from 1 pint of s-metolachlor product.
pendimethalin, MOA 3 (Prowl) 3.3 EC (Prowl H₂O) 3.8 L	1.2 to 3.6 pt 1.5 to 3 pt	0.5 to 1.5 0.7 to 1.4	Controls common annual grasses plus pigweed and lambsquarters. Incorporate in top 2 to 3 inches of seedbed within 7 days of application; immediate incorporation suggested. For broadleaf weed control, pendimethalin may be tank mixed with most broadleaf herbicides. When tank mixing, see respective labels for application rates, weeds controlled, specific application directions, and precautions. Generic brands are available.
S-metolachlor, MOA 15 (Brawl) 7.62 EC (Cinch) 7.64 EC (Dual Magnum) 7.62 EC (Dual II Magnum) 7.64 EC (Medal) 7.62 EC (Medal II) 7.64 EC	1 to 2 pt	0.95 to 1.91	Controls annual grasses and pigweed. At higher rates, controls nightshade and yellow nutsedge. Better yellow nutsedge control if incorporated; see labels for incorporation details. Except for yellow nutsedge, preemergence application preferred. Does not adequately control Texas panicum, seedling johnsongrass, and shattercane. Read labels and adjust rates for soil texture and organic matter. These herbicides may be applied at rates up to 2.5 pints on soils with 6% to 20% organic matter. For broadleaf weed control, S-metolachlor may be tank mixed with most broadleaf herbicides; do not mix with Valor. When tank mixing, see respective labels for application rates, weeds controlled, specific application directions, and precautions.
trifluralin, MOA 3 (Treflan) 4 EC (Treflan HFP) 4 EC	1 to 2 pt	0.5 to 1	Controls common annual grasses plus pigweed and lambsquarters. Incorporate in top 2 to 3 inches of seedbed within 8 hr of application; immediate incorporation suggested. For broadleaf weed control, Treflan may be tank mixed with most broadleaf herbicides. When tank mixing, see respective labels for application rates, weeds controlled, specific application directions, and precautions. Generic brands are available.
Preplant Incorporated, Annual broadleaf weeds			
imazaquin, MOA 2 (Scepter) 70 WDG	2.8 oz	0.123	Controls most broadleaf weeds; a follow-up post-emergence herbicide application often needed for adequate sicklepod control. Follow all precautions on the label, including rotational restrictions. For annual grass control, Scepter may be tank mixed with alachlor, pendimethalin, S-metolachlor, or trifluralin. When tank mixing, see respective labels for application rates, weeds controlled, specific application directions, and precautions.
metribuzin, MOA 5 (Sencor) 75 WDG	0.33 to 0.67 lb	0.25 to 0.5	Controls many broadleaf weeds. Will not adequately control cocklebur or morningglory. Acceptable control of sicklepod may require a follow-up postemergence herbicide application. Activity of metribuzin is highly dependent upon soil texture and organic matter. Follow label directions for application rates, soil type restrictions, etc. Do not use on sand with less than 1% organic matter. Do not use on loamy sand or sandy loam soils with less than 0.5% organic matter. Some varieties are particularly sensitive to metribuzin; see labels for details. Soybeans may be injured when metribuzin is applied to soil treated with organophosphate insecticides and/or nematicides; see precautions on label. For annual grass control, Sencor may be tank mixed with alachlor, pendimethalin, S-metolachlor, or trifluralin. When tank mixing see respective labels for application rates, weeds controlled, specific application directions, and precautions.
dimethenamid-P, MOA 15 (Outlook) 6.0 EC	14 to 21 fl oz	0.66 to 0.98	Incorporate 2 inches deep. Not effective on purple nutsedge. Follow label carefully for use rates on various soil types. Do not apply to sandy soils if organic matter is less than 3% and depth to groundwater is 30 feet or less.
metolachlor, MOA 15 (Me-Too-Lachlor) 8 EC (Parallel PCS) 8 EC (Parrlay) 8 EC (Stalwart) 8 EC	1 to 2 pt	1 to 2	See comments for s-metolachlor products. Products containing s-metolachlor are more active on weeds per unit of formulated product than those containing metolachlor. In general, it takes 1.5 pints of a metolachlor product to get the activity one would get from 1 pint of an s-metolachlor product.
S-metolachlor, MOA 15 (Brawl) 7.62 EC (Cinch) 7.64 EC (Dual Magnum) 7.62 EC (Dual II Magnum) 7.64 EC (Medal) 7.62 EC (Medal II) 7.64 EC	1.33 to 2 pt	1.27 to 1.91	Incorporate 2 inches deep. Not effective on purple nutsedge.

266

Table 7-5A. Chemical Weed Control in Soybeans

Herbicide, Mode of Action Code[1], and Formulation	Amount of Formulation Per Acre	Pounds Active Ingredient Per Acre	Precautions and Remarks
Burndown, No-Till Planting, Emerged grass and broadleaf weeds			
glyphosate, MOA 9 (numerous brands and formulations)	See label	0.56 to 1.13 (lb a.e.)	Glyphosate is available as an isopropylamine salt and a potassium salt. Glyphosate formulations and application rates should be compared on the basis of pounds of glyphosate acid equivalent (a.e.) per gallon and per acre, respectively. The rate in the preceding column is expressed as a.e. See TABLE 7-10 for glyphosate rate conversions. Apply before crop emergence. Rate depends upon weed species and size; see labels for details. Higher rates (up to 3.75 pounds acid equivalent) may be used for perennial weeds. Adjuvant recommendations vary by glyphosate brand. See label of brand used for specific recommendations. For residual grass and broadleaf weed control, glyphsate may be tank mixed with most preemergence soybean herbicides. Refer to the label of the tank mix partner for application rates, directions, limitations, weeds controlled, and precautions.
glyphosate, MOA 9 + fomesafen, MOA 14 (Flexstar GT) 3.29 L	3 to 4.5 pt	1 to 1.55 (lb a.e.) + 0.25 to 0.37	Apply before crop emergence. See label for adjuvant suggestions. May mix with 2,4-D or dicamba for improved burndown of specific weeds. See waiting intervals between application and planting on labels for 2,4-D or dicamba. Do not exceed 4.5 pints per acre of Fllexstar GT per year. Also, do not exceed 0.375 pound a.i. of fomesafen per year from all sources.
paraquat, MOA 22 (Gramoxone Inteon) 2 SL	2 to 4 pt	0.5 to 1	Apply before crop emergence. Use 2 pints on weeds 1 to 3 inches, 3 pints on weeds 3 to 6 inches, and 4 pints on weeds 6 inches or taller. Use 2 pints for rye cover crop and 2.5 to 3 pints on wheat cover crops. Add crop oil concentrate or nonionic surfactant according to label directions. Generic brands of paraquat containing 3 pounds active per gallon are available. Apply these products at two-thirds of the rates mentioned here. Residual herbicides for grass and broadleaf weed control may be tank mixed with Gramoxone. Control of cutleaf eveningprimrose, wild radish, and most broadleaf weeds will be increased by adding 2,4-D at 0.75 to 1 pint. Delay planting at least 7 days after application of ester formulations of 2,4-D or 15 days after application of amine formulations of 2,4-D. Use of ester formulations is discouraged if sensitive crops, especially cotton and tobacco, are located within 1 mile.
Burndown, No-Till Planting, Cutleaf eveningprimrose, wild radish, and vetch, plus other weeds controlled by glyphosate			
glyphosate, MOA 9 (numerous brands and formulations) + 2,4-D, MOA 4 (numerous brands and formulations) 3.8 SL	See label + 0.5 to 1 pt	0.56 to 1.13 (lb a.e.) + 0.24 to 0.48	See comments for glyphosate alone. Apply ester formulations of 2,4-D at least 7 days ahead of planting. Apply amine formulations of 2,4-D at least 15 days ahead of planting. Plant soybeans at least 1 inch deep. See comments on 2,4-D labels concerning use on coarse-textured soils with less than 1% organic matter. For 2,4-D formulations other than 3.8 pounds per gallon, adjust rate accordingly. Use 0.5 pint 2,4-D for primrose; use 1 pint for other weeds. Use of ester formulations is discouraged if senstive crops, particularly cotton or tobacco, are located within 1 mile.
pyraflufen-ethyl, MOA 14 (ET) 1 SL	0.5 to 2.0 fl oz	0.003 to 0.015 (lb a.i.)	ET can be used for limited suppression of small emerged summer annual and winter weeds. See label for adjuvant and spray volume recommedations Research with ET is limited in North Carolina.
Burndown, No-Till Planting, Glyphosate-resistant horseweed plus other weeds			
glyphosate, MOA 9 (numerous brands and formulations) + 2,4-D, MOA 4 (numerous brands and formulations) 3.8 SL + flumioxazin, MOA 14 (Valor SX) 51 WDG	See label + 1.5 to 2 pt + 2 to 3 oz	0.56 to 1.13 (lb a.e) + 0.71 to 0.95 + 0.064 to 0.096	Glyphosate-resistant horseweed (marestail) is relatively common in eastern North Carolina, and continued spread is anticipated. See comments for glyphosate alone. Acceptable control of glyphosate-resistant horseweed requires both a residual herbicide (Valor SX, Valor XLT, Envive, Leadoff, or Trivence) and either 2,4-D or Clarity. An alternative approach would be application of glyphosate plus either 2,4-D or Clarity preplant followed by Gramoxone plus a residual herbicide at planting. Do NOT till or otherwise disturb the soil surface following application of Valor SX, Valor XLT, Envive, Leadoff, or Trivence.
glyphosate, MOA 9 (numerous brands and formulations) + 2,4-D, MOA 4 (numerous brands and formulations) 3.8 SL + flumioxazin, MOA 14, + chlorimuron, MOA 2 (Valor XLT) 40.3 WDG	See label + 1.5 to 2 pt + 3 to 5 oz	0.56 to 1.13 (lb a.e) + 0.71 to 0.95 + 0.056 to 0.094 + 0.019 to 0.032	2,4-D rates suggested for horseweed should be applied at least 30 days ahead of planting. For 2,4-D formulations other than 3.8 pounds per gallon, adjust rate accordingly. Use of ester formulations of 2,4-D is discouraged if sensitive crops, especially cotton and tobacco, are located within 1 mile. Following application of Clarity and accumulation of at least 1 inch rainfall, delay soybean planting at least 14 days. Follow precautions on Clarity label concerning drift to sensitive crops. Leadoff must be applied at least 30 days ahead of planting. Trivence may be applied any time from spring, up to 3 days after planting. See label for recommended adjuvants. Rates are dependant upon soil texture and organic matter, see label for specific rate recommendations. Afforia may be applied at 2.5 to 3.75 oz 7 days prior to planting soybean, the rate must be reduced to 2.5 oz if applying less than 7 days prior to planting. Rates are dependant upon soil texture and organic matter, see label for specific rate recommendations. See label for crop rotation restrictions. Horseweed cannot be controlled with a burndown prior to planting double-crop soybeans because the combine cuts off the horseweed, leaving little to no foliage to spray. If horseweed is present in wheat, apply 0.75 to 0.9 ounce of either Harmony SG or Harmony Extra with TotalSol plus 3 ounces of Clarity in February or early March.
glyphosate, MOA 9 (numerous brands and formulations) + 2,4-D, MOA 4 (numerous brands and formulations) 3.8 SL + flumioxazin, MOA 14, + pyroxasulfone, MOA 15 (Fierce) 76 WDG	See label + 1.5 to 2 pt + 3 to 3.75 oz	0.56 to 1.13 (lb a.e) + 0.71 to 0.95 + 0.06 to 0.079 + 0.0796 to 0.0996	
glyphosate, MOA 9 (numerous brands and formulations) + 2,4-D, MOA 4 (numerous brands and formulations) 3.8 SL + flumioxazin, MOA 14 + chlorimuron, MOA 2 + thifensulfuron, MOA 2 (Envive) 41.3 WDG	See label + 1.5 to 2 pt + 2.5 to 4 oz	0.56 to 1.13 (lb a.e) + 0.71 to 0.95 + 0.046 to 0.074 + 0.0214 to 0.023 + 0.0045 to 0.007	

Table 7-5A. Chemical Weed Control in Soybeans

CHEMICAL WEED CONTROL

Herbicide, Mode of Action Code[1], and Formulation	Amount of Formulation Per Acre	Pounds Active Ingredient Per Acre	Precautions and Remarks
Burndown, No-Till Planting, Glyphosate-resistant horseweed plus other weeds (continued)			
glyphosate, MOA 9 (numerous brands and formulations)	See label	0.56 to 1.13 (lb a.e)	
+	+	+	
2,4-D, MOA 4 (numerous brands and formulations) 3.8 SL	1.5 to 2 pt	0.71 to 0.95 +	
+			
rimsulfuron, MOA 2	1.5	0.0155 +	
+			
thifensulfuron, MOA 2 (Leadoff) 33.4 WDG		0.0155	
glyphosate, MOA 9 (numerous brands and formulations)	See label	0.56 to 1.13 (lb a.e)	
+	+	+	
2,4-D, MOA 4 (numerous brands and formulations) 3.8 SL	1.5 to 2 pt	0.71 to 0.95 +	
+			
flumioxazin, MOA 14		0.048 to 0.070 +	
+			
chlorimuron, MOA 2		0.0146 to 0.021 +	
+			
metribuzin, MOA 5 (Trivence) 61.3 DG	6 to 8.7 oz	0.167 to 0.243	
glyphosate, MOA 9 (numerous brands and formulations)	See label	0.56 to 1.13 (lb a.e)	
+	+	+	
2,4-D, MOA 4 (numerous brands and formulations) 3.8 SL	1.5 to 2 pt	0.71 to 0.95 +	
+	+		
flumioxazin, MOA 14		0.071 to 0.096 +	
+			
thifensulfuron, MOA 2	2.5 to 3.75 oz	0.008 to 0.012 +	
+			
tribenuron, MOA 2 (Afforia) 50.8 DG		0.008 to 0.012	
saflufenacil, MOA 14 (Sharpen) 3.42 SL	1.0 fl oz	0.027 (lb a.i.)	Sharpen can be applied to control glyphosate-resistant marestail. Applying Sharpen with other herbicides will broaden the spectrum of control. See label for specific information on adjuvant selection. **Interval between application and soybean planting for Sharpen varies by soil texture and organic matter content. See Sharpen label for specific information.**
glufosinate, MOA 10 (Liberty) 2.34 SL	29 to 36 fl oz	0.53 to 0.66	Liberty 280 SL can be applied prior to emergence of any transgenic or conventional soybean variety to control emerged weeds. See label for adjuvant use. In crop applications to **Liberty-Link soybeans** can be made at 22-29 fluid ounces following a burndown application with a maximum seasonal use of 65 ounces per acre. Thorough spray coverage is essential. Apply in minimum of 15 GPA; dense weed canopies require 20 to 40 GPA. Poor performance is likely if daytime temperatures are less than 75 degrees F or if weeds are drought stressed.
oxyfluorfen, MOA 14 (Goal 2XL) 2L (GoalTender) 4L	1 to 2 pt 0.5 to 1 pt	0.25 to 0.5	Goal 2XL can be applied at 7 days or more prior to soybean planting to control common winter and summer annual weed speciesA tank mix combination with glyphosate, paraquat, glufosinate, 2,4-D or dicamba is recommended. See label for tankmix recommendations.
Burndown, No-Till Planting, Curly dock, vetch, and Carolina geranium plus other weeds controlled by glyphosate			
glyphosate, MOA 9 (numerous brands and formulations)	See label	0.56 to 1.13 (lb a.e.)	See comments for glyphosate alone. Harmony SG can be applied anytime prior to soybean planting. Soybean planting should be delayed at least 14 days after application of Harmony Extra. Soybean may be planted 1 day after 0.5 ounce of FirstShot is applied, however planting should be delayed at least 7 days if higher rates are applied.
+	+	+	
thifensulfuron, MOA 2 (Harmony SG) 50 WDG	0.75 oz	0.023	
glyphosate, MOA 9 (numerous brands and formulations)	See label	0.56 to 1.13 (lb a.e.)	
+	+	+	
thifensulfuron, MOA 2		0.016 +	
+			
tribenuron, MOA 2 (Harmony Extra SG with TotalSol) 50 WDG	0.75 oz	0.008	
glyphosate, MOA 9 (numerous brands and formulations)	See label	0.56 to 1.13 (lb a.e.)	
+	+	+	
thifensulfuron, MOA 2		0.008 to 0.013 +	
+			
tribenuron, MOA 2 (FirstShot SG with TotalSol) 50 WDG	0.5 to 0.8 oz	0.008 to 0.013	

Table 7-5A. Chemical Weed Control in Soybeans

Herbicide, Mode of Action Code[1], and Formulation	Amount of Formulation Per Acre	Pounds Active Ingredient Per Acre	Precautions and Remarks
Burndown, No-Till Planting, Italian ryegrass, wheat, barley, and rye			
glyphosate, MOA 9 (numerous brands and formulations) + clethodim, MOA 1 (Select MAX) 0.97 EC	See label + 9 to 16 fl oz	0.56 to 1.13 (lb a.e.) + 0.067 to 0.12 (lb a.i.)	Apply to weeds 2 to 6 inches tall. See label for instructions on adjuvant used depending o glyphosate formulation.
Preemergence, No-Till or Conventional, Any Cultivar, Annual grasses			
alachlor, MOA 15 (Intrro) 4 EC (Micro-Tech) 4 FME	2 to 3 qt	2 to 3	Controls annual grasses except Texas panicum, shattercane, and seedling johnsongrass. Also controls pigweed and nightshade. For broadleaf weed control, alachlor may be tank mixed with most broadleaf herbicides; do not mix with Valor SX, Valor XLT, or Envive unless applied 14 or more days ahead of planting. When tank mixing, see respective labels for application rates, weeds controlled, specific application directions, and precautions. May also be shallowly incorporated; see labels for details. Generic brands of alachlor are available.
clomazone, MOA 13 (Command 3 ME) 3 FME	1.3 to 3.3 pt	0.5 to 1.25	Controls most annual grasses; shattercane and Texas panicum are only suppressed. Also controls a number of broadleaf weeds. See label for weeds controlled. Read the label carefully and follow all precautions on label pertaining to off-site movement, buffer zones, drift control agents, and rotational restrictions. For broader spectrum control, Command 3 ME may be tank mixed with most broadleaf herbicides. When tank mixing, see respective labels for application rates, weeds controlled, specific application directions, and precautions.
dimethenamid-P, MOA 15 (Outlook) 6.0 EC	12 to 21 fl oz	0.56 to 0.98	Use 12 to 18 fluid ounces on soils with less than 3% organic matter or 14 to 21 fluid ounces on soils with greater than 3% organic matter. Controls annual grasses except seedling johnsongrass, Texas panicum, and shattercane. Also controls pigweed and nightshade. See label for application directions and rates for various soils. May also be shallowly incorporated; see label for details. For broadleaf weed control, Outlook may be tank mixed with most broadleaf herbicides; do not mix with Valor SX, Valor XLT, or Envive unless applied 14 or more days ahead of planting. Do not apply to sandy soils if organic matter is less than 3% and depth to groundwater is 30 feet or less.
metolachlor, MOA 15 (Me-Too-Lachlor) 8 EC (Parallel PCS) 8 EC (Parrlay) 8 EC (Stalwart) 8 EC	1 to 2 pt	1 to 2	See comments for s-metolachlor products. Products containing s-metolachlor are more active on weeds per unit of formulated product than those containing metolachlor. In general, it takes 1.5 pints of a metolachlor product to get the activity one would get from 1 pint of s-metolachlor product.
S-metolachlor, MOA 15 (Brawl) 7.62 EC (Cinch) 7.64 EC (Dual Magnum) 7.62 EC (Dual II Magnum) 7.64 EC (Medal) 7.62 EC (Medal II) 7.64 EC	1 to 2 pt	0.95 to 1.91	Controls annual grasses except Texas panicum, shattercane, and seedling johnsongrass. Also controls pigweed and nightshade on mineral soils. May also be shallowly incorporated; see label for details. For broadleaf weed control, S-metolachlor may be tank mixed with most broadleaf herbicides; do not mix with Valor SX or Valor XLT unless applied 14 or more days ahead of planting. When tank mixing, see respective labels for application rates, weeds controlled, specific application directions, and precautions.
pendimethalin, MOA 3 (Prowl) 3.3 EC (Prowl H2O) 3.8 L	1.2 to 3 pt 1.5 to 2.5 pt	0.5 to 1.2 0.7 to 1.2	Preemergence application of pendimethalin suggested only where annual grass pressure is expected to be light. Pendimethalin generally performs better when incorporated. For broadleaf weed control, pendimethalin may be tank mixed with most broadleaf herbicides. When tank mixing, see respective labels for application rates, weeds controlled, specific application directions, and precautions. Generic brands are available.
pyroxasulfone, MOA 15 (Zidua) 85 WG	1.5 to 3.5 oz	0.0796 to 0.186	Use 1.5 to 2.1 ounces per acre on coarse soils, 2 to 3 ounces on medium soils, and up to 3.5 ounces per acre on fine textured soils. Controls most annual grasses, pigweed, and nightshade. Provides suppression of Texas panicum, seedling johnsongrass, and shattercane. May be tank mixed with most broadleaf herbicides; when tank mixing, see respective labels for application rates, weeds controlled, specific application directions, and precautions. May be applied to emerged soybean at the first trifoliate leaf stage to third trifoliate leaf stage. Do not apply from emergence through unifoliate stage or unacceptable injury may occur. Does not control emerged weeds. See label for tank mix options to control emerged weeds.
pyroxasulfone, MOA 15 + fluthiacet-methyl, MOA 14 (Anthem) 2.15 L	6 to 11 oz	0.101 to 0.185	Use 6 to 6.5 ounces per acre on coarse soils, 6.5 to 9.5 ounces on medium soils, and up to 11 ounces per acre on fine textured soils. Controls most annual grasses, pigweed, and nightshade. Provides suppression of Texas panicum, seedling johnsongrass, and shattercane. May be tank mixed with most broadleaf herbicides; when tank mixing, see respective labels for application rates, weeds controlled, specific application directions, and precautions. May be applied to emerged soybean up to third trifoliate leaf stage. May provide some control of emerged broadleaf weeds (less than 2 inches). See label for tank mix options to control emerged weeds.
Preemergence, No-Till or Conventional, Any Cultivar, Annual broadleaf weeds			
clomazone, MOA 13 (Command 3 ME) 3 FME	1.3 to 3.3 pt	0.5 to 1.25	Command controls selected broadleaf weeds such as balloonvine, velvetleaf, spurred anoda, prickly sida, croton, Pennsylvania smartweed, common ragweed, lambsquarters, and jimsonweed. It also controls most annual grasses. Command does not control pigweed, morningglory, sicklepod, nightshade, and ladysthumb. See label for specific weeds controlled and rates for specific weeds. Read label carefully and follow all precautions on label pertaining to off-site movement, buffer zones, drift control agents, and rotational restrictions. For broader spectrum control, Command may be tank mixed with a number of soil-applied herbicides; see label for details. When tank mixing, see respective labels for application rates, weeds controlled, specific application directions, and precautions.
flumetsulam, MOA 2 (Python) 80 WDG	0.8 to 1.33 oz	0.04 to 0.067	Controls most broadleaf weeds; control of ragweed, cocklebur, and morningglory can be variable. Rates of 1.25 to 1.33 ounces suggested for sicklepod. Acceptable control of sicklepod may require a follow-up postemergence herbicide application. May be mixed with registered soil-applied grass control herbicides. See label for weeds controlled, **rotational restrictions,** and restrictions on soil type and organic matter.

CHEMICAL WEED CONTROL

269

Table 7-5A. Chemical Weed Control in Soybeans

Herbicide, Mode of Action Code[1], and Formulation	Amount of Formulation Per Acre	Pounds Active Ingredient Per Acre	Precautions and Remarks
Preemergence, No-Till or Conventional, Any Cultivar, Annual broadleaf weeds (continued)			
flumioxazin, MOA 14 (Valor SX) 51 WDG (Rowel) 51 WDG	2 to 3 oz	0.063 to 0.094	Rate depends on weed species and soil texture; follow label directions when selecting rate. May be tank mixed with Prowl or Command for annual grass control. Valor SX, Valor XLT, Envive, Surveil or Gangster should not be mixed with alachlor, metolachlor, S-metolachlor, or dimethenamid-P and applied preemergence. Combinations of Valor SX, Valor XLT, Envive, Surveil or Gangster plus alachlor, metolachlor, S-metolachlor, or dimethenamid-P can be applied 14 or more days ahead of planting. Do not apply Valor XLT within 14 days before or after application of organophosphate insecticide or any variety that is not DuPont BOLT, STS or STS/RR due to injury potential.
flumioxazin, MOA 14 + chlorimuron, MOA 2 (Valor XLT) 40.3 WDG (Rowel FX) 40.3 WDG	3 to 5 oz	0.056 to 0.094 + 0.019 to 0.032	
flumioxazin, MOA 14 + chlorimuron, MOA 2 + thifensulfuron, MOA 2 (Envive) 41.3 WDG	2.5 to 4 oz	0.046 to 0.074 + 0.0214 to 0.023 + 0.0045 to 0.007	
flumioxazin, MOA 14 + cloransulam, MOA 2 (Surveil) 48 WG	2.8 to 4.2 oz	0.063 to 0.095 + 0.021 to 0.032	
flumioxazin, MOA 14 + cloransulam, MOA 2 (Gangster, co-pack of Gangster V [51% flumioxazin] and Gangster FR [84% cloransulam])	1.5 to 3 oz + 0.3 to 0.6 oz	0.047 to 0.094 + 0.016 to 0.032	
imazaquin, MOA 2 (Scepter) 70 WDG	2.8 oz	0.123	Controls most broadleaf weeds if adequate rainfall received for activation. A follow-up postemergence herbicide application often needed for adequate sicklepod control. Follow all precautions on the label, including rotational restrictions. For annual grass control, Scepter may be tank mixed with alachlor, Command, dimethenamid-P, pendimethalin, or S-metolachlor. When tank mixing, see respective labels for application rates, weeds controlled, specific application directions, and precautions.
linuron, MOA 7 (Linex) 4 L	0.66 to 3 pt	0.33 to 1.5	Rate depends greatly on soil texture and organic matter content; follow label directions carefully when selecting rates. Do not use on sand or loamy sand soils or any soil with less than 0.5% organic matter. Linuron controls pigweed, lambsquarters, and common ragweed. For annual grass control, linuron may be tank mixed with alachlor, Command, dimethenamid-P, pendimethalin, or S-metolachlor. When tank mixing, see respective labels for application rates, weeds controlled, specific application directions, and precautions. Generic brands may be available.
metribuzin, MOA 5 (Sencor) 75 DF	0.33 to 0.67 pt	0.25 to 0.5	Rate depends greatly on soil texture and organic matter content; follow label directions carefully when selecting rates. Do not use Sencor on sand soils with less than 1% organic matter or on any soil with less than 0.5% organic matter. Some varieties are particularly sensitive to metribuzin; see labels for details. Soybeans may be injured when metribuzin is applied to soil treated with organophosphate insecticides and/or nematicides. Does not adequately control cocklebur or morningglory. Adequate sicklepod control may require a follow-up postemergence herbicide application. For annual grass control, metribuzin may be tank mixed with alachlor, Command, dimethenamid-P, pendimethalin, or S-metolachlor. When tank mixing, see respective labels for application rates, weeds controlled, specific application directions, and precautions.
sulfentrazone, MOA 14 + metribuzin, MOA 5 (Authority MTZ) 45 WDG	12 to 20 oz	0.135 to 0.225 + 0.20 to 0.34	Rate depends upon soil texture and organic matter; see label for application rates. Controls most broadleaf weeds, including Palmer amaranth, morningglory, and cocklebur; sicklepod suppressed. See statement on label concerning sensitive varieties. See label for rotational restrictions.
Preemergence, No-Till or Conventional, Any Cultivar, Annual grasses and broadleaf weeds — packaged herbicide mixtures			
s-metolachlor, MOA 15 + fomesafen, MOA 14 (Prefix) 5.29 L	2 to 3 pt	1.09 to 1.63 + 0.24 to 0.36	Controls most annual grasses (except Texas panicum, seedling johnsongrass, shattercane) and broadleaf weeds, such as pigweed species (including Palmer amaranth), lambsquarters, common ragweed, Florida pusley, smartweed, and nightshade. Does not control sicklepod, and it only suppresses morningglory, cocklebur, and prickly sida.
metribuzin, MOA 5 + S-metolachlor, MOA 15 (Boundary) 7.8 L	1 to 2.5 pt	0.19 to 0.47 + 0.79 to 1.96	See comments for metribuzin applied preemergence. Rate depends on soil texture and organic matter content; follow label directions carefully when selecting rate. Do not use on coarse-textured soils with less than 0.5% organic matter. Controls weeds normally controlled by Sencor and Dual Magnum. May be mixed with Command, FirstRate, Prowl, Python, or Scepter. Followup postemergence herbicide needed in most cases.
imazethapyr, MOA 2 + saflufenacil, MOA 14 (OpTILL)	2 oz	0.022 + 0.63	Optill is labeled for application up to soybean emergence. Application to emerged soybean can result in significant injury. See lable for adjuvant selection.
imazethapyr, MOA 2 + saflufenacil, MOA 14 + pyroxasulfone, MOA 15 (Zidua PRO) 4.09 SC	4.5 oz + + 6 fl oz	0.0169 to 0.0225 + 0.0468 to 0.062 + 0.08 to 0.107	Zidua PRO is labeled for burndown application up to preemergence. Application to emerged soybean can result in significant injury. Delay soybean planting 30 days after application on coarse soils where organic matter is ≤ 2.0%. See lable for adjuvant selection. See label for tank mixing information.
sulfentrazone, MOA 14 + s-metolachlor. MOA 15 (Broadaxe) 7 EC (Broadaxe XC) 7 EC	19 to 38.7 fl oz	0.104 to 0.212 + 0.94 to 1.90	Controls most annual grasses (except Texas panicum, seedling johnsongrass, shattercane) and broadleaf weeds, such as pigweed species (including Palmer amaranth), lambsquarters, morningglory, prickly sida, smartweed, and nightshade. Also provides control of yellow nutsedge. Does not control sicklepod or cocklebur.
acetochlor, MOA 15 + fomesafen, MOA 14 (Warrant Ultra) 3.45 L	48 to 70 fl oz	1.06 to 1.54 + 0.236 to 0.345	May be applied preplant, at-planting, preemergence or postemergence. May only be applied ONCE per growing season. Controls most annual grasses (except Texas panicum, seedling johnsongrass, shattercane) and broadleaf weeds, such as pigweed species (including Palmer amaranth), lambsquarters, common ragweed, Florida pusley, smartweed, and nightshade. Does not control sicklepod, and it only suppresses morningglory, cocklebur, and prickly sida. See label for rate and timing information.

Table 7-5A. Chemical Weed Control in Soybeans

Herbicide, Mode of Action Code[1], and Formulation	Amount of Formulation Per Acre	Pounds Active Ingredient Per Acre	Precautions and Remarks
Preemergence, No-Till or Conventional, Any Cultivar, Annual grasses and broadleaf weeds — packaged herbicide mixtures			
sulfentrazone, MOA 14 + cloransulam-methyl, MOA 2 (Sonic) 78 WG	6.45 to 8 oz	0.25 to 0.31 + 0.032 to 0.04	May be applied preplant incorporated, preplant, or preemergence within 3 days after planting. Apply 6.45 oz where organic matter is less than 3%. May be applied alone or in tank mix combination with other registered herbicides. Controls several annual grasses (except barnyardgrass, Texas panicum, seedling johnsongrass, shattercane) and broadleaf weeds such as pigweed species (including Palmer amaranth), lambsquarters, horseweed, cocklebur, morningglory, smartweed, and nightshade. Also provides control of purple and yellow nutsedge. Does not control sicklepod.
Preemergence, No-Till or Conventional, Any Cultivar, Nutsedge			
dimethenamid-P, MOA 15 (Outlook) 6.0 EC	14 to 21 fl oz	0.66 to 1	Dimethenamid and metolachlor control or suppress only yellow nutsedge. These herbicides are more effective on yellow nutsedge when incorporated. However, these herbicides applied preemergence may provide adequate control of lighter infestations of yellow nutsedge. Neither product controls purple nutsedge. Follow labels carefully for use rates on various soil types. Do not apply Outlook to sand soils if organic matter is less than 3% and depth to groundwater is 30 feet or less. Generic brands containing metolachlor, not S-metolachlor, are available. See previous comments concerning these products. Prefix, which contains s-metolachlor, will suppress or control yellow nutsedge.
S-metolachlor, MOA 15 (Brawl) 7.62 EC (Cinch) 7.64 EC (Dual Magnum) 7.62 EC (Dual II Magnum) 7.64 EC (Medal) 7.62 EC (Medal II) 7.64 EC	1.33 to 2 pt	1.27 to 1.91	
sulfentrazone, MOA 14 + metribuzin, MOA 5 (Authority MTZ) 45 WDG	12 to 20 oz	0.135 to 0.225 + 0.20 to 0.34	See comments under Annual Grasses and Broadleaf Weeds. Controls yellow and purple nutsedge.
sulfentrazone, MOA 14 + s-metolachlor. MOA 15 (Broadaxe) 7 EC (Broadaxe XC) 7 EC	19 to 38.7 fl oz	0.104 to 0.212 + 0.94 to 1.90	See comments under Annual Grasses and Broadleaf Weeds. Controls yellow and purple nutsedge.
Postemergence Overtop; Roundup Ready Cultivars, Annual grasses and broadleaf weeds plus suppression of perennial weeds—**Roundup Ready cultivars Only**			
glyphosate, MOA 9 (numerous brands and formulations)	See label	0.75 to 1.5 (lb a.e)	**Apply only to Roundup Ready Cultivars. See comments on resistance management in TABLE 7-10.** A preemergence herbicide is highly recommended to control weeds not controlled by glyphosate (such as Florida pusley), to reduce early season weed competition, to broaden the window of application for glyphosate, and to aid in resistance management. Any registered soil-applied herbicide can be used on Roundup Ready soybeans. Glyphosate controls most annual weeds; exceptions include dayflower, hemp sesbania, and Florida pusley. Timely application required for morningglory control. Can be applied from cracking stage throughout flowering. Multiple applications can be made, but do not exceed 2.2 pounds a.e. per acre per year during this period. Total glyphosate use (preplant, in-crop, and preharvest) should not exceed 6 pounds a.e. per acre per year. Glyphosate is available as an isopropylamine salt and a potassium salt. Glyphosate formulations and application rates should be compared on the basis of pounds of glyphosate acid equivalent (a.e.) per gallon and per acre, respectively. Rate in the preceding column is expressed as a.e. See Table 7-10 for glyphosate rate conversions. Adjuvant recommendations vary by glyphosate product: see label of brand used for details. Rate depends upon weed species and size; see labels for details. Timely application is encouraged. The first application should be made 18 to 20 days after planting. Repeat applications can be made if needed. The following products can be mixed with at least some of glyphosate brands. Refer to label of tank mix partner or glyphosate product used for timing of application, weed sizes, and use of adjuvants. **Ultra Blazer** (1 pint): improves control of hemp sesbania, black nightshade, and larger morningglory. Minor antagonism sometimes noted on grasses and pigweed. Use 1.5 pints to control glyphosate-resistant Palmer amaranth up to 4 inches **Classic** (0.25 to 0.33 oz): improves control of hemp sesbania, spreading dayflower, and larger morningglory. **FirstRate** (0.2 to 0.3 fluid ounce): improves control of spreading dayflower, dove weed, and larger morningglory. **Harmony SG** (0.125 ounces/acre): improves control of lambsquarters and velvetleaf. Controls glyphosate-resistant Palmer amaranth unless it is also ALS resistant. Apply before Palmer amaranth exceeds 8 inches Apply after first trifoliate has fully expanded. See label for use of ammonium sulfate. Some soybean injury can be expected. **Reflex, Flexstar** (6 to 12 fluid ounces): improves control of hemp sesbania, black nightshade, and larger morningglory. Antagonism sometimes noted on grasses and pigweed. Must be applied at 16 fluid ounces to control glyphosate-resistant Palmer amaranth 4 inches tall, or at 24 fluid ounces if 6 inches tall. **Resource** (2 to 4 fluid ounces): improves control of larger morningglory. Minor antagonism sometimes observed on pigweed. Apply at 6 to 8 ounces for glyphosate-resistant Palmer amaranth up to 4 inches **Synchrony XP** (0.375 ounce): improves control of lambsquarters, morningglory, and velvetleaf. Controls glyphosate-resistant Palmer amaranth unless it is also ALS resistant. Apply before Palmer amaranth exceeds 4 inches **Storm** (0.75 to 1.5 pints): improves control of hemp sesbania, black nightshade, and larger morningglory. Minor antagonism sometimes noted on grasses and pigweed.
glyphosate, MOA 9 + fomesafen, MOA 14 (Flexstar GT) 3.29 L	3 to 4.5 pt	1 to 1.5 (lb a.e.) + 0.25 to 0.37	Controls grasses and most annual broadleaf weeds. See label for suggested rates according to weed size. Also see label for adjuvant recommendations. Apply with flat fan nozzles. Do not exceed 4.5 pints per acre of Flexstar GT per year. Also do not exceed 0.375 pound a.i. of fomesafen per year from all sources.
glyphosate isopropylamine salt, MOA 9 + imazethapyr, MOA 2 (Extreme) 2.17 SL	3 pt	0.56 + 0.064	**Apply only to Roundup Ready cultivars.** Apply before soybean bloom and make only one application per year. Add nonionic surfactant at 1 pint per 100 gallons spray solution plus 2.5 pounds of ammonium sulfate or 1 to 2 quarts of UAN. See label for application directions, precautions, and rotational restrictions. May be more effective on yellow nutsedge and morningglory than glyphosate alone.
glyphosate, MOA 9 + acetachlor, MOA 15 (Warrant) 3.0 ME	See label 1.5 qt	0.56 to 0.75 (lb a.e.) + 1.1	**Apply only to Roundup Ready cultivars.** Apply overtop soybean with glyphosate at V2-V3 soybean for best results. Apply prior to R3. Warrant can be directed at V5-V6. Warrant provides residual control only.

CHEMICAL WEED CONTROL

Table 7-5A. Chemical Weed Control in Soybeans

Herbicide, Mode of Action Code[1], and Formulation	Amount of Formulation Per Acre	Pounds Active Ingredient Per Acre	Precautions and Remarks
Postemergence Overtop; Roundup Ready Cultivars, Volunteer Roundup Ready corn in Roundup Ready soybeans			
glyphosate, MOA 9 (numerous brands and formulations) + clethodim, MOA 1 (Select) 2 EC (Select Max) 0.97 EC	See label + 4 to 8 fl oz 6 to 12 fl oz	0.56 to 0.75 (lb a.e.) + 0.063 to 0.125 0.045 to 0.106	See comments for glyphosate alone. For corn up to 12 inches tall, apply 4 to 6 ounces of Select or 6 ounces of Select Max. For corn up to 24 inches tall, apply 6 to 8 ounces of Select or 9 ounces of Select Max. For corn up to 36 inches, apply 12 ounces Select Max. Add 2.5 pounds per acre ammonium sulfate or equivalent. If brand of glyphosate used does not contain surfactant, add nonionic surfactant at 0.25 to 0.5% by volume. If applying Select or Select Max alone, see labels for adjuvant recommendations.
glyphosate, MOA 9 (numerous brands and formulations) + fluazifop-p-butyl, MOA 1 (Fusilade DX) 2 EC	See label + 4 to 6 fl oz	0.56 to 0.75 (lb a.e.) + 0.063 to 0.094	See comments for glyphosate alone. Apply 4 ounces Fusilade for corn less than 12 inches Increase rate to 6 ounces for corn up to 24 inches Add any adjuvants suggested on the label of the glyphosate product used. Additionally, add 0.25% by volume of crop oil concentrate. If applying Fusilade alone, see label for adjuvant recommendations.
glyphosate, MOA 9 (numerous brands and formulations) + quizalofop-p-ethyl, MOA 1 (Assure II) 0.88 EC	See label + 5 to 8 fl oz	0.56 to 0.75 (lb a.e.) + 0.034 to 0.055	See comments for glyphosate alone. Apply Assure at 4 ounces to corn up to 12 inches, 5 ounces to corn up to 18 inches, and 8 ounces to corn up to 30 inches If the brand of glyphosate contains adjuvant, add 0.125% nonionic surfactant by volume. If the brand of glyphosate does not contain adjuvant, add surfactant according to the glyphosate label. If applying Assure alone, see label for adjuvant recommendations.
Postemergence Overtop; Liberty Link Cultivars, Annual grasses and broadleaf weeds			
glufosinate-ammonium, MOA 10 (Liberty SL)	29 to 36 fl oz	0.53 to 0.66 (lb a.i.)	**Apply only to Liberty Link cultivars.** Can be applied as single or sequential applications prior to V4 stage of soybean. Do not apply more than 36 ounces as a single application. If applied as a burndown prior to planting, Liberty SL can be applied in season to soybean with a maximum seasonal use of 65 ounces/acre. See product label for possible tank mixtures with other herbicides.
Postemergence Overtop; Any Cultivar, Annual grasses			
clethodim, MOA 1 (Select) 2 EC (Select Max) 0.97 EC	6 to 8 fl oz 9 to 16 fl oz	0.094 to 0.125 0.068 to 0.121	Apply to actively growing grasses not under drought stress. See label for specific rates and weed size to treat. Add crop oil concentrate at 1 quart per acre to Select. To Select Max, add nonionic surfactant at 0.25% volume, crop oil concentrate at 1% by volume, or methylated seed oil at 1% by volume. Do not cultivate for 7 days before or after application. Generic brands are available.
fluazifop-p-butyl, MOA 1 (Fusilade DX) 2 EC	6 to 12 fl oz	0.094 to 0.188	Apply to actively growing grass not under drought stress. Suggested application rate varies by species; see label for application directions, rates, maximum weed sizes to treat, etc. Add either 1% crop oil concentrate (1 gallon per 100 gallons) or 0.25% nonionic surfactant (1 quart per 100 gallons). Do not cultivate for 7 days before or after application.
quizalofop p-ethyl, MOA 1 (Assure II) 0.88 EC	5 to 8 fl oz	0.034 to 0.055	Apply to actively growing grass not under drought stress. Suggested application rate varies by species; see label for application directions, rates, maximum weed sizes to treat, etc. Add 1% (1 gallon per 100 gallons) crop oil concentrate or 0.25% (1 quart per 100 gallons) nonionic surfactant. Do not cultivate for 7 days before or after application. Generic brands are available.
sethoxydim, MOA 1 (Poast) 1.5 EC (Poast Plus) 1 EC	16 fl oz 24 fl oz	0.19	Apply to actively growing grass not under drought stress. Consult label for maximum grass size to treat, application directions, etc. Add 2 pints per acre of crop oil concentrate. Do not cultivate for 7 days before or after application.
Postemergence Overtop; Any Cultivar, Annual broadleaf weeds			
acifluorfen, MOA 14 (Ultra Blazer) 2 SL	0.5 to 1.5 pt	0.13 to 0.38	See label for weeds controlled, recommended rates for specific weeds, and maximum weed size to treat. Label recommends nonionic surfactant at 1 to 2 pints per 100 gallons spray solution. For broader spectrum control, acifluorfen may be tank mixed with Basagran, Classic, FirstRate, Pursuit, Raptor, Resource, Scepter, Synchrony, or 2,4-DB. When tank mixing, see respective labels for application rates, weeds controlled, maximum weed size to treat, specific application directions, and precautions.
acifluorfen, MOA 14 + bentazon, MOA 6 (Storm) 4 SL	1.5 pt	0.25 + 0.5	See label for weeds controlled and maximum weed size to treat. Add 1 pint per acre of crop oil concentrate or nonionic surfactant at 1 quart per 100 gallons. For broader spectrum control, Storm may be tank mixed with Basagran, Classic, FirstRate, Pursuit, Raptor, Resource, or Scepter. See respective labels for application rates, weeds controlled, maximum weed size to treat, specific application directions, and precautions.
bentazon, MOA 6 (Basagran) 4 SL	1 to 2 pt	0.5 to 1	See label for weeds controlled, recommended rates for specific weeds, and maximum weed size to treat. Add 1.25% by volume (not to exceed 2 pints per acre) of crop oil concentrate when treating for lambsquarters, common ragweed, or hemp sesbania. If velvetleaf is primary target, add 0.5 to 1.0 gallon per acre of liquid nitrogen instead of crop oil. For broader spectrum control, Basagran may be tank mixed with Classic, Cobra, FirstRate, Flexstar, Pursuit, Raptor, Reflex, Resource, Scepter, Storm, Ultra Blazer, or 2,4-DB. When tank mixing, see respective labels for application rates, weeds controlled, maximum weed size to treat, specific application directions, and precautions.
cloransulam-methyl, MOA 2 (FirstRate) 84 WDG	0.3 oz	0.016	Controls cocklebur, jimsonweed, morningglory, ragweed, smartweed, velvetleaf, spreading dayflower, dove weed, and small horseweed. See label for recommended weed size to treat. FirstRate will usually control sicklepod in the cotyledonary to first leaf stage; larger sicklepod will not be controlled. Add either nonionic surfactant at 1 to 2 pints per 100 gallons or crop oil concentrate at 1.2 gallons per 100 gallons. If velvetleaf is the target weed, also add 2.5 gallons 30% UAN per 100 gallons. FirstRate can be applied twice per season. For broader spectrum control, FirstRate may be tank mixed with Basagran, Classic, Cobra, Flexstar, Pursuit, Raptor, Reflex, Resource, Storm, Synchrony, or Ultra Blazer.
chlorimuron ethyl, MOA 2 (Classic) 25 WDG	0.5 to 0.75 oz	0.008 to 0.012	See label for weeds controlled, recommended rates for specific weeds, maximum weed size to treat, rotational restrictions, and sprayer cleanup. Add 0.25% by volume (1 quart per 100 gallons) of nonionic surfactant. Under hot, dry conditions, 1% crop oil concentrate may be used instead of surfactant; crop oil increases potential for injury. See label for specific adjuvant recommendations when treating velvetleaf. For broader spectrum control, Classic may be tank mixed with Basagran, Ultra Blazer, Cobra, FirstRate, Harmony GT, Reflex, Resource, Storm, or Ultra Blazer. When tank mixing, see respective labels for application rates, weeds controlled, maximum weed size to treat, specific application directions, and precautions.
chlorimuron ethyl, MOA 2 + thifensulfuron methyl, MOA 2 (Synchrony XP) 28.4 WDG	0.375 to 1.125 oz	0.005 to 0.015 + 0.0016 to 0.0049	**For non-BOLT, STS cultivars, use only 0.375 ounce rate.** Rate can be increased to 1.125 ounces on BOLT, STS cultivars. See label for weeds controlled, maximum weed size to treat, and rotational restrictions. Add crop oil concentrate at 1% by volume except when tank mixing with a product whose label precludes use of crop oil concentrate; in that case, use nonionic surfactant at 0.25% by volume. Under dry conditions, adding 2 quarts per acre of UAN may enhance control. Synchrony may be tank mixed with Cobra, FirstRate, Flexstar, Harmony GT, Reflex, Resource, or Ultra Blazer.

Table 7-5A. Chemical Weed Control in Soybeans

Herbicide, Mode of Action Code[1], and Formulation	Amount of Formulation Per Acre	Pounds Active Ingredient Per Acre	Precautions and Remarks
Postemergence Overtop; Any Cultivar, Annual broadleaf weeds (continued)			
flumiclorac pentyl ester, MOA 14 (Resource) 0.86 EC	4 to 8 fl oz	0.027 to 0.054	Suggested for use where velvetleaf is a problem. Excellent control of velvetleaf. Also controls small lambsquarters, pigweed species, prickly sida, and common ragweed. See label for weeds controlled and recommended weed size for treatment. Add 1 quart per acre of crop oil concentrate. Resource may be tank mixed with Basagran, Classic, Cobra, FirstRate, Flexstar, Harmony GT, Pursuit, Raptor, Reflex, Scepter, Storm, Synchrony, or Ultra Blazer.
fomesafen, MOA 14 (Flexstar) 1.88 SL (Reflex) 2 SL (Dawn) 2 SL (Rythm) 1.88 SL	1 to 1.5 pt	0.25 to 0.38 0.24 to 0.35 0.24 to 0.35 0.25 to 0.38	See labels for weeds controlled, recommended rates for specific weeds, maximum weed size to treat, and rotational restrictions. Add 1% crop oil concentrate by volume (4 quarts per 100 gallons) or 0.25% nonionic surfactant (1 quart per 100 gallons). For broader spectrum control, Reflex or Flexstar may be tank mixed with Basagran, Classic, FirstRate, Harmony GT, Pursuit, Raptor, Resource, Scepter, Synchrony, or 2,4-DB. When tank mixing, see respective labels for application rates, weeds controlled, maximum weed size to treat, specific application directions, and precautions. Flexstar is somewhat more active than Reflex and can be more effective on lambsquarters, prickly sida, spurred anoda, and velvetleaf. Foliar burn on the crop may also be greater with Flexstar under conditions of high moisture and high temperatures. See label for tax mix partners with Dawn and Rythm.
s-metolachlor, MOA 15 + fomesafen, MOA 14 (Prefix) 5.29 L	2 to 2.33 pt	1.09 to 2.27 + 0.24 to 0.28	Apply from cracking to third trifoliate of soybean. Add 0.25% by volume of nonionic surfactant; do not use crop oil. Do not exceed 3 pints per acre of Prefix per year, and do not exceed 0.375 pound a.i. of fomesafen from all sources combined.
fomesafen, MOA 14 + fluthiacet methyl, MOA 14 (Marvel) 3 L	5 to 7.25 oz	0.117 to 0.17	Apply from preplant through full flowering (prior to R3). Add 0.25% by volume of nonionic surfactant, 0.5% by volume crop oil concentrate, or 0.5% by volume methylated seed oil. Crop oil or methylated seed oil are recommended under dry conditions. Do not exceed 0.375 pound a.i. of fomesafen from all sources combined. See label for tankmix partners.
imazamox, MOA 2 (Raptor) 1S L	5 fl oz	0.04	Controls many common broadleaf weeds. Foxtails, fall panicum, broadleaf signalgrass, seedling johnsongrass, and shattercane usually adequately controlled. Does not control sicklepod. May not adequately control ragweed, prickly sida, or Palmer amaranth. May tank mix with Basagran, FirstRate, Flexstar, Reflex, Resource, Storm, or Ultra Blazer for improved control of ragweed and Palmer amaranth; tank mixes may reduce grass control. Suppresses yellow and purple nutsedge. Add either crop oil concentrate at 2 pints per acre or nonionic surfactant at 1 quart per 100 gallons. See label concerning addition of nitrogen-containing fertilizer.
imazaquin, MOA 2 (Scepter) 70 WDG	1.4 to 2.8 oz	0.063 to 0.125	See label for weeds controlled, recommended rates for specific weeds, maximum weed size to treat, and rotational restrictions. Primarily for control of cocklebur and pigweed. Add 0.25% by volume (1 quart per 100 gallons) nonionic surfactant. Alternatively, a crop oil concentrate can be used at the rate recommended on crop oil label. For broader spectrum control, Scepter may be tank mixed with Basagran, Cobra, Flexstar, Reflex, Resource, Storm, or Ultra Blazer. When tank mixing, see respective labels for application rates, weeds controlled, maximum weed size to treat, specific application directions, and precautions.
imazethapyr, MOA 2 (Pursuit) 70 WDG	1.44 oz	0.063	See label for weeds controlled, recommended rates for specific weeds, maximum weed size to treat, and rotational restrictions. Also suppresses johnsongrass, broadleaf signalgrass, and foxtails. Add 0.25% by volume (1 quart per 100 gallons) nonionic surfactant or 1.5 to 2 gallons per acre of crop oil concentrate. For broader spectrum control, Pursuit may be tank mixed with Basagran, Cobra, FirstRate, Flexstar, Harmony GT, Reflex, Resource, Storm, or Ultra Blazer. When tank mixing, see respective labels for application rates, weeds controlled, maximum weed size to treat, specific application directions, and precautions.
lactofen, MOA 14 (Cobra) 2 EC	6 to 12.5 fl oz	0.094 to 0.2	See label for weeds controlled, recommended rates, weed size to treat, and recommended adjuvants. At higher rates, Cobra usually causes excessive foliar burn on soybeans. Lower rates tank mixed with other herbicides may be of some value in specific situations. Cobra may be tank mixed with Basagran, Classic, FirstRate, Pursuit, Resource, Scepter, Synchrony, or 2,4-DB. See labels for weeds controlled and specific use directions.
thifensulfuron methyl, MOA 2 (Harmony SG) 50 WDG	0.125 oz	0.004	See label for weeds controlled, maximum weed size to treat, and sprayer cleanup. Add 0.125% to 0.25% by volume (1 to 2 pints per 100 gallons) of nonionic surfactant when applying Harmony SG alone or 0.125% in tank mixes. Under dry or cool conditions, a crop oil concentrate may be used; see label for details. In addition to surfactant or crop oil, Harmony SG label specifies use of an ammonium nitrogen fertilizer. This is usually of value only when treating for velvetleaf. For broader spectrum control, Harmony SG may be tank mixed with Classic, Flexstar, Pursuit, Reflex, Resource, or Synchrony. When tank mixing, see respective labels for application rates, weeds controlled, maximum weed size to treat, specific application directions, and precautions.
pyraflufen-ethyl, MOA 14 (ET) 1 SL	0.5 to 2.0 fl oz	0.003 to 0.015 (lb a.i.)	ET can be applied from emergence to V6 soybean to suppress small broadleaf weeds. Some leaf speckling can occur but is transient. Do no apply crop oil concentrate. See label for adjuvant and spray volume recommendations. Research with ET is limited in North Carolina.

Table 7-5A. Chemical Weed Control in Soybeans

Herbicide, Mode of Action Code[1], and Formulation	Amount of Formulation Per Acre	Pounds Active Ingredient Per Acre	Precautions and Remarks
Postemergence Overtop; Any Cultivar, Annual grasses and broadleaf weeds—tank mixtures			
quizalofop p-ethyl, MOA 1 (Assure II) + Basagran, Classic, Cobra, FirstRate, Flexstar, Harmony SG, Pursuit, Raptor, Reflex, Scepter, Storm, Synchrony STS, or Ultra Blazer (See TABLE 7-10 for MOAs)	See labels	See labels	The listed two-way tank mixes are covered on one or more of the respective labels. Consult the labels of the products to be used for specific application rates, directions, precautions, and adjuvant usage. Formulations and active ingredients of the various products can be found elsewhere in this publication. A number of three-way tank mixes (not listed here) also are registered.
fenoxaprop-ethyl, MOA 1 (Fusilade DX) + Basagran, Classic, Cobra, Flexstar, Pursuit, Raptor, Reflex, Scepter, Storm, Synchrony STS, or Ultra Blazer (See TABLE 7-10 for MOAs)			While mixing postemergence grass and broadleaf herbicides is convenient and saves time and trips across the field, best results often are obtained when the grass and broadleaf herbicides are applied separately. Antagonism of the grass herbicide (reduced grass control) often occurs when the grass herbicide is mixed with a broadleaf herbicide. Antagonism is more likely to occur under marginal spraying conditions, such as large grasses and dry weather. Some of the broadleaf herbicides also are more antagonistic than others. The antagonism may be partially or completely overcome by increasing the rate of the grass herbicide; labels for some of the grass herbicides suggest increased rates when tank mixing. The adjuvants needed for good activity of the grass herbicide also may enhance crop injury from the broadleaf herbicide; follow label directions carefully for use of adjuvants. Tank mixing should be considered only when the optimum timing for application of the grass and broadleaf herbicides coincides. Tank mixes generally should not be used when treating for rhizome johnsongrass or bermudagrass.
sethoxydim, MOA 1 (Poast or Poast Plus) + Basagran, Classic, Cobra, FirstRate, Flexstar, Pursuit, Reflex, Resource, Scepter, Storm, Synchrony STS, or Ultra Blazer (See TABLE 7-10 for MOAs)			If sequential applications are made, the recommended waiting interval between application of the grass and broadleaf herbicides varies depending upon the herbicides used and the order in which they are applied. See the labels for specific recommendations. However, the following are general guidelines: 1) If the grass herbicide is applied first, the broadleaf herbicide can be applied 24 hr later; 2) if Basagran or Resource is applied first, the grass herbicide can be applied 24 hr later; 3) if Classic, FirstRate, Harmony SG, or Synchrony STS is applied first, wait at least 3 days before applying the grass herbicide;
clethodim, MOA 1 (Select or Select Max) + Basagran, Classic, Cobra, FirstRate, Flexstar, Pursuit, Raptor, Reflex, Resource, Storm, Synchrony STS, or Ultra Blazer (See TABLE 7-10 for MOAs)			4) if Pursuit, Raptor, or Scepter is applied first, wait at least 5 days before applying the grass herbicide; and 5) if Cobra, Flexstar, Reflex, Storm, or Ultra Blazer is applied first, delay application of the grass herbicide until the grass resumes active growth with development of new leaves.
Postemergence Overtop; Any Cultivar, Annual broadleaf weeds, Rhizome johnsongrass			
clethodim, MOA 1 (Select) 2 EC (Select Max) 0.97 EC	8 to 16 fl oz 12 to 32 fl oz	0.125 to 0.25 0.091 to 0.24	Apply when johnsongrass is 12 to 24 inches tall. If needed, make second application of 6 to 8 oz of Select or 9 to 24 fluid ounces of Select Max when regrowth is 6 to 18 inches Add crop oil concentrate at 1 quart per acre to Select. To Select Max, add nonionic surfactant at 0.25% by volume, crop oil concentrate at 1% by volume, or methylated seed oil at 1% by volume. Generic brands available.
fluazifop p-butyl, MOA 1 (Fusilade DX) 2 EC	12 fl oz	0.19	Apply when johnsongrass is 8 to 18 inches tall and before boot stage. Add either a nonionic surfactant at 0.25% by volume (1 quart per 100 gallons) or a crop oil concentrate at 1% by volume (1 gallons per 100 gallons). If needed, make second application of 8 fluid ounces when regrowth is 6 to 12 inches
quizalofop p-ethyl, MOA 1 (Assure II) 0.88 EC	10 fl oz	0.07	Apply when johnsongrass is 10 to 24 inches tall. If needed, make second application of 7 fluid ounces per acre when regrowth is 6 to 10 inches Add either crop oil concentrate at 1% (1 gallon per 100 gallons) or nonionic surfactant at 0.25% (1 quart per 100 gallons). Generic brands available.
sethoxydim, MOA 1 (Poast) 1.5 EC (Poast Plus) 1 EC	24 fl oz 36 fl oz	0.28	Apply to actively growing johnsongrass 20 to 25 inches tall. Add 2 pints per acre of crop oil concentrate. A second application of 16 ounces of Poast or 24 ounces of Poast Plus may be made when regrowth is 12 inches.
Postemergence Overtop; Any Cultivar, Bermudagrass			
clethodim, MOA 1 (Select) 2 EC (Select Max) 0.97 EC	8 to 16 fl oz 12 to 32 fl oz	0.125 to 0.25 0.091 to 0.24	Apply before bermudagrass runners exceed 6 inches If needed, make second application of 8 to 16 ounces of Select or 12 to 32 fluid ounces of Select Max when regrowth is less than 6 inches Add crop oil concentrate at 1 quart per acre to Select. To Select Max, add nonionic surfactant at 0.25% by volume, crop oil concentrate at 1% by volume, or methylated seed oil at 1% by volume. Generic brands available.
fluazifop p-butyl, MOA 1 (Fusilade DX) 2 EC	12 fl oz	0.19	Apply when bermudagrass runners are 4 to 8 inches If regrowth occurs, apply 8 fluid ounces when regrowth is 4 to 8 inches Add crop oil concentrate at 1% by volume (1 gallon per 100 gallons) or nonionic surfactant at 0.25% by volume (1 quart per 100 gallons).
quizalofop p-ethyl, MOA 1 (Assure II) 0.88 EC	10 fl oz	0.07	Apply when bermudagrass is 3 inches tall or has up to 6 inches runners. If regrowth occurs, make second application of 7 fluid ounces/acre when runners are 6 inches Add either crop oil concentrate at 1% (1 gallon per 100 gallons) or nonionic surfactant at 0.25% (1 quart per 100 gallons). Generic brands available.
sethoxydim, MOA 1 (Poast) 1.5 EC (Poast Plus) 1 EC	24 fl oz 36 fl oz	0.28	Apply to actively growing bermudagrass before runners exceed 6 inches Add 2 pints per acre of crop oil concentrate. A second application of 16 ounces of Poast or 24 ounces Poast Plus may be made when regrowth is 4 inches.
Postemergence Overtop; Any Cultivar, Nutsedge			
bentazon, MOA 6 (Basagran) 4 SL	1.5 to 2 pt	0.75 to 1	For yellow nutsedge only; Basagran does not control purple nutsedge. Apply when yellow nutsedge is 6 to 8 inches tall. Add 2 pints per acre of crop oil concentrate. If needed, make second application of same rate 7 to 10 days later.
chlorimuron ethyl, MOA 2 (Classic) 25 WDG	0.5 to 0.75 oz	0.008 to 0.012	Controls yellow nutsedge; suppresses purple nutsedge. Apply when yellow nutsedge is 2 to 4 inches tall. Add surfactant according to label directions.
chlorimuron ethyl, MOA 2 + thifensulfuron methyl, MOA 2 (Synchrony STS SP) 42 WDG	0.5 oz	0.01 + 0.003	Controls yellow nutsedge; suppresses purple nutsedge. Apply when yellow nutsedge is 2 to 3 inches tall. Add crop oil concentrate according to label directions. **Apply only to BOLT, STS soybean.**
imazethapyr, MOA 2 (Pursuit) 70 WDG	1.44 oz	0.063	Apply when nutsedge is 1 to 3 inches tall. Add surfactant or crop oil according to label directions. Pursuit is more effective on purple nutsedge than on yellow nutsedge.

Table 7-5A. Chemical Weed Control in Soybeans

Herbicide, Mode of Action Code[1], and Formulation	Amount of Formulation Per Acre	Pounds Active Ingredient Per Acre	Precautions and Remarks
Late Postemergence Overtop, Salvage Treatment; Any Cultivar, Cocklebur and morningglory			
2,4-DB, MOA 4 (various brands) 2 SL 1.75 SL 75 WP	 1 pt 1.1 pt 0.33 lb	0.25	Spray overtop soybeans from 1 week before bloom up to midbloom. This treatment may be used when needed as an aid to control cocklebur and morningglory and as a supplement to but not a replacement for early postemergence treatments. Salvage treatment only—substantial crop injury may occur. Do not add surfactant or crop oil. Not suggested for use on soybeans to be saved for seed.
Postemergence Directed; Any Cultivar, Small grasses and broadleaf weeds			
metribuzin, MOA 5 (Sencor) 75 WDG	0.33 to 0.67 lb	0.25 to 0.5	Apply only as a directed spray to soybeans at least 8 inches tall. Do not spray higher than 2 to 3 inches on the soybean stem. Add surfactant according to label directions. Do not use if soil has been wet for 2 to 3 days. Do not use on sandy soils or any soil having less than 0.5% organic matter. Some varieties of soybeans are sensitive to metribuzin; see label for details.
Postemergence Directed; Any Cultivar, Cocklebur and morningglory			
2,4-DB, MOA 4 (various brands) 2 SL 1.75 SL 75 WP	 13 fl oz 15 fl oz 4.3 oz	0.2	Soybeans must be at least 8 inches tall. Contact no more than the lower third of soybean plant. Follow other precautions on label.
Postemergence with Wiper Applicators; Any Cultivar, Certain weeds taller than crop, especially grasses			
glyphosate, MOA 9 (numerous brands and formulations)	Not applicable; see label	Not applicable; see label	Apply glyphosate above crop with wiper-type applicator. Follow label directions carefully. Do not let glyphosate contact crop plants. Johnsongrass and tall annual grasses such as fall panicum are very susceptible; broadleaf weeds are less susceptible. Use only as supplement to a good early season weed management program.
Postemergence with Hooded Sprayer; Any Cultivar, Annual and perennial grasses and broadleaf weeds			
glyphosate, MOA 9 (numerous brands and formulations)	See label	0.38 to 1.13 (lb a.e.)	Glyphosate is available as an isopropylamine salt and a potassium salt. Glyphosate formulations and application rates should be compared on the basis of pounds of glyphosate acid equivalent (a.e.) per gallon and per acre, respectively. Rate in the preceding column is expressed as a.e. See TABLE 7-10 for glyphosate rate conversions. Glyphosate rate depends upon weed species and size; see labels for specific rates. Higher rates can be used for perennial weeds; see labels for details. Keep hoods as close to the ground as possible. Contact of spray with foliage of non Roundup Ready soybeans will cause severe injury.
Harvest Aid; Any Cultivar, Annual and perennial weeds			
glyphosate, MOA 9 (numerous brands and formulations)	See label	0.75 to 2.25 (lb a.e.)	Glyphosate is available as an isopropylamine salt and a potassium salt. Glyphosate formulations and application rates should be compared on the basis of pounds of glyphosate acid equivalent (a.e.) per gallon and per acre, respectively. Rate in the preceding column is expressed as a.e. See TABLE 7-10 for glyphosate rate conversions. See labels for weeds controlled, maximum weed size to treat, and specific application rates for various species. Apply after pods have set and lost all green color. Apply at least 7 days before harvest. Can be applied by ground or air. Do not apply to soybeans grown for seed.
sodium chlorate (Defol 5) 5 L	4.8 qt	6	Apply when soybeans are mature and ready for harvest. Apply 7 to 10 days before anticipated harvest date. Apply in a minimum of 20 gallons of water per acre. Will control most grass and broadleaf weed species and desiccate soybean prior to harvest. Do not graze or harvest treated soybean for forage.
Harvest Aid; Any Cultivar, Annual grasses and broadleaf weeds			
paraquat, MOA 22 (Gramoxone Inteon) 2 SL	8 to 16 fl oz	0.13 to 0.25	Apply when pods are fully developed and at least one-half of leaves have dropped and leaves left on plants are turning yellow. Can be applied by ground or air. Generic brands containing 3 pounds active per gallon are available. This product would be used at 5.3 to 10.7 fluid ounces.
Harvest Aid; Any Cultivar, Annual broadleaf weeds			
carfentrazone, MOA 14 (Aim) 2 EC	1.5 fl oz	0.023	Desiccates morningglory, pigweed, and cocklebur. Apply 3 or more days ahead of harvest. Add 1 gallon crop oil concentrate per 100 gallons spray solution. Thorough coverage is essential; use a minimum of 20 GPA by ground equipment. May be applied by air. May tank mix with Gramoxone. Tank mixes with Gramoxone should be applied 15 days ahead of harvest.
saflufenacil, MOA 14 (Sharpen) 2.85 SC	1.0 to 2.0 fl oz	0.022 to 0.044	Apply when soybeans have reached physiological maturity, no green color on pods and seeds. Apply 3 or more days ahead of harvest, allow up to 10 days for optimum desiccation. See label for specific directions on indeterminant and determinant varieties. Add 1 gallon methylated seed oil per 100 gallons spray solution plus ammonium-based adjuvant system (AMS or UAN) for optimum desiccation. Do not apply to soybeans grown for seed.

[1] Mode of Action (MOA) code developed by the Weed Science Society of America. See Table 7-10, Herbicide Resistance Management, for details.

CHEMICAL WEED CONTROL

Weed Response to Preplant Incorporated and Preemergence Herbicides in Soybeans

W. J. Everman, Crop Science Department

Ratings are based upon average to good soil and weather conditions for herbicide performance and upon proper application rate, technique, and timing.

Table 7-5B. Weed Response to Preplant Incorporated and Peremergence Herbicides in Soybeans

Herbicides
Key: PPI = Preplant Incorporated; PRE = Preemergence

Species	Prowl or Treflan PPI	Sonalan PPI	Authority MTZ PRE	Command PRE	Dual Magnum, Dual II Magnum PRE	Outlook PRE	Intrro or Micro-Tech PRE	Reflex PRE	Sencor PRE	Linex PRE	Prefix PRE	Prowl PRE	Python PRE	Scepter PRE	Valor SX PRE	Valor XLT PRE	Envive PRE
Bermudagrass	N	N	N	PF	N	N	N	N	N	N	N	N	N	N	N	N	N
Broadleaf signalgrass	G	G	F	E	G	FG	FG	FG	PF	P	G	P	N	PF	N	N	N
Crabgrass	E	E	F	E	E	E	E	FG	F	FG	E	F	P	NP	N	N	N
Fall panicum	G	G	P	E	E	E	E	—	NP	F	E	PF	N	NP	N	N	N
Foxtails	E	E	F	E	E	E	E	—	NP	FG	E	F	P	FG	N	N	N
Goosegrass	E	E	FG	E	E	E	E	—	F	FG	E	PF	P	NP	N	N	N
Johnsongrass, Seedling	G	G	—	G	PF	PF	PF	—	PF	NP	PF	PF	N	FG	N	N	N
Johnsongrass, Rhizome	P	P	N	N	N	N	N	—	N	N	N	N	N	N	N	N	N
Shattercane	G	G	—	F	P	P	P	—	N	N	P	PF	N	F	N	N	N
Texas panicum	G	G	PF	F	PF	PF	PF	F	N	PF	F	PF	N	NP	N	N	N
Nutsedge, Yellow	N	N	E	N	FG[3]	F	P	GE	N	N	GE	N	N	PF	N	N	N
Nutsedge, Purple	N	N	E	N	N	N	N	—	N	N	—	N	N	NP	N	N	N
Balloonvine	N	N	—	G	N	N	N	—	G	F	—	N	P	F	—	—	—
Eastern black nightshade	N	F	—	P	F	F	FG	—	N	NP	—	P	PF	PF	E	E	E
Burcucumber [1]	N	N	—	NP	N	N	N	—	P	N	—	N	P	PF	—	—	—
Cocklebur	N	N	G	F	N	N	N	G	PF	N	G	N	G	E	P	FG	FG
Cowpea	N	N	—	N	N	N	N	—	PF	N	—	N	NP	N	—	—	—
Crotalaria	N	N	—	—	N	N	N	—	NP	N	—	N	—	—	—	—	—
Florida beggarweed	N	N	G	FG	F	F	F	P	G	F	F	N	F	F	E	E	E
Florida pusley	E	E	—	P	G	G	G	G	G	G	E	G	G	GE	GE	GE	GE
Hemp sesbania	N	N	GE	N	N	N	PF	P	GE	F	P	N	N	N	G	G	G
Jimsonweed	N	N	E	G	N	N	N	—	G	F	—	N	G	FG	E	E	E
Lambsquarters	G	G	E	G	F	FG	F	E	E	GE	E	G	E	G	E	E	E
Morningglory	P	P	E	P[2]	N	N	N	PF	G	P	PF	N	F	G	G	G	G
Palmer amaranth	G	G	G	N	FG	G	GE	E	G	PF	E	PF	G [4]	GE [4]	E	E	E
Pigweed, Redroot and Smooth	G	G	E	N	G	GE	GE	E	E	E	E	FG	E	E	E	E	E
Prickly sida	N	N	GE	E	P	P	P	—	G	F	—	N	E	G	E	E	E
Ragweed, Common	N	N	GE	G	PF	F	PF	G	G	G	G	N	FG	G	G	G	G
Ragweed, Giant	N	N	—	PF	N	N	N	G	P	P	G	N	PF	G	F	F	F
Sicklepod	N	N	G	NP	NP	PF	NP	PF	G	NP	P	N	G	FG	PF	F	F
Smartweed	N	N	E	E	N	N	N	—	G	F	—	N	G	G	F	—	—
Spurred anoda	N	N	G	E	N	N	N	—	G	P	—	N	E	P	E	E	E
Tropic croton	N	N	—	E	N	N	N	FG	FG	PF	FG	N	—	NP	E	E	E
Velvetleaf	N	N	GE	E	N	N	N	—	G	P	—	N	GE	F	F	G	G

[1] Multiple flushes of germination; one application of any herbicide will seldom be adequate.

[2] Fair on pitted morningglory.

[3] Good on yellow nutsedge when incorporated.

[4] Palmer amaranth resistant to ALS inhibitors is common in NC. This ALS-inhibiting herbicide will perform poorly on resistant biotypes.

E = excellent control, 90% or better
G = good control, 80% to 90%
F = fair control, 50% to 80%
P = poor control, 25% to 50%
N = no control, less than 25%

CHEMICAL WEED CONTROL

Weed Response to Postemergence Herbicides in Soybeans

W. J. Everman, Crop Science Department

Ratings based upon average to good soil and weather conditions for herbicide performance and upon proper application rate, technique, and timing.

Table 7-5C. Weed Response to Postemergence Herbicides in Soybeans

Species	Assure II	Fusilade	Poast	Select, Select Max	Basagran	Classic	Cobra	FirstRate	Flexstar	Flexstar GT	Glyphosate[1]	Harmony SG	Pursuit	Raptor	Reflex	Resource	Scepter	Storm	Synchrony STS[2]	Ultra Blazer	Liberty
Bermudagrass	G	G	FG	G	N	N	N	N	N	—	G[3]	N	N	N	N	N	N	N	N	N	N
Broadleaf signalgrass	GE	GE	E	E	N	N	N	N	N	E	E	N	G	G	N	N	N	NP	N	NP	G
Crabgrass	G	G	GE	GE	N	N	N	N	N	E	E	N	PF	PF	N	N	N	N	N	N	FG
Fall panicum	E	E	E	E	N	N	P	N	N	E	E	N	F	G	N	N	N	P	N	P	G
Foxtails	E	E	E	E	N	N	P	N	N	E	E	N	G	G	N	N	N	P	N	P	G
Goosegrass	GE	GE	GE	GE	N	N	N	N	N	E	E	N	NP	NP	N	N	N	N	N	N	P
Johnsongrass, Seedling	E	E	E	E	N	NP	P	N	N	E	E	N	GE	GE	N	N	N	P	NP	P	G
Johnsongrass, Rhizome	E	GE	G	GE	N	N	N	N	N	E	E	N	G[5]	G[5]	N	N	N	N	N	N	F
Shattercane	E	E	E	E	N	N	P	N	N	E	E	N	G	G	N	N	N	P	N	P	—
Texas panicum	G	G	E	E	N	N	N	N	N	E	E	N	PF	PF	N	N	N	N	N	N	G
Nutsedge, Purple	N	N	N	N	NP	PF	N	PF	N	G	G	N	G	N	N	N	N	N	PF	N	P
Nutsedge, Yellow	N	N	N	N	G[3]	G	N	PF	F	—	FG[4]	N	FG	FG	F	N	P	F	G	N	P
Balloonvine	N	N	N	N	P	FG	GE	P	G	E	G	—	P	—	G	P	P	G	FG	GE	—
Eastern black nightshade	N	N	N	N	P	F	G	N	G	E	FG	N	G	G	G	P	P	G	F	G	—
Burcucumber[6]	N	N	N	N	P	G	G	F	FG	G	F	G	PF	PF	FG	F	P	F	G	FG	—
Cocklebur	N	N	N	N	E	E	GE	E	E	E	E	FG	E	E	E	G	E	E	E	G	E
Cowpea	N	N	N	N	N	GE	F	P	PF	E	E	—	N	—	P	—	N	P	GE	PF	G
Crotalaria	N	N	N	N	P	G	G	—	G	E	G	—	N	—	G	—	N	E	G	E	—
Florida beggarweed	N	N	N	N	N	E	FG	FG	P	G	G	—	N	—	P	P	NP	P	E	PF	G
Hemp sesbania	N	N	N	N	P	E	G	PF	E	G	PF	—	N	—	E	P	N	GE	E	E	—
Jimsonweed	N	N	N	N	E	E	GE	E	E	E	E	F	GE	E	E	G	NP	E	E	E	E
Lambsquarters	N	N	N	N	FG	N	P	PN	F	E	E	E	PF	G	PF	G	P	G	E	G	E
Morningglory	N	N	N	N	P	G	G	E	GE	E	FG[8]	FG	FG	FG	GE	FG	P	GE	G	GE	E
Palmer amaranth	N	N	N	N	N	F[11]	G	P	G	E[10]	E[10]	GE[11]	GE[11]	—	G	FG	G[11]	G	E[11]	G	FG
Pigweed, Redroot or Smooth	N	N	N	N	N	G	E	P	GE	E	E	E	E	E	GE	G	E	E	E	E	G
Prickly sida	N	N	N	N	G	N	G	P	F	G	G	N	P	G	NP	N	PF	FG	P	N	G
Ragweed, Common	N	N	N	N	G[9]	G	E	E	GE	E	E	F	PF	F	GE	G	F	E	G	E	E
Ragweed, Giant	N	N	N	N	GE	FG	G	GE	E	E	G	P	F	F	E	P	P	GE	FG	GE	E
Sicklepod	N	N	N	N	N	G	NP	F[7]	P	E	E	P	N	N	P	N	FG	NP	G	NP	E
Smartweed	N	N	N	N	E	E	F	E	G	E	G	E	GE	G	G	P	FG	E	E	GE	GE
Spurred anoda	N	N	N	N	G	F	F	F	F	E	E	N	F	F	P	P	NP	F	F	P	P
Tropic croton	N	N	N	N	F	NP	G	P	G	E	E	N	N	N	G	P	N	G	NP	G	G
Velvetleaf	N	N	N	N	G	F	G	G	F	E	E	G	F	E	P	E	NP	FG	G	PF	—

[1] Apply to Roundup Ready (glyphosate-resistant) cultivars only.

[2] Apply only to BOLT, STS cultivars.

[3] Assumes two applications.

[4] Yellow nutsedge control is good with two applications of glyphosate.

[5] Follow-up treatment with a postemergence grass herbicide may be necessary.

[6] Multiple flushes of germination; one application of any herbicide will seldom be adequate.

[7] FirstRate is good on sicklepod if applied at cotyledonary to first leaf stage.

[8] With good timing and a follow-up application as needed, morningglory control can be good.

[9] Assumes addition of crop oil concentrate.

[10] Palmer amaranth resistant to glyphosate is common in NC. Glyphosate will perform poorly on resistant biotypes.

[11] Palmer amaranth resistant to ALS-inhibiting herbicides is common in NC. ALS-inhibiting herbicides will perform poorly on resistant biotypes.

Key:
E = excellent control, 90% or better
G = good control, 80% to 90%
F = fair control, 50% to 80%
P = poor control, 25% to 50%
N = no control, less than 25%

CHEMICAL WEED CONTROL

Chemical Weed Control in Sunflowers

D. L. Jordan, Crop Science Department

NOTE: A mode of action code has been added to the Herbicide and Formulation column of this table. Use MOA codes for herbicide resistance management. See Table 7-10, Herbicide Resistance Management, for details.

Table 7-6. Chemical Weed Control in Sunflowers

Herbicide, Mode of Action Code, and Formulation	Amount of Formulation Per Acre	Pounds Active Ingredient Per Acre	Precautions and Remarks
Preplant Foliar, Burndown of weeds and cover crops before planting			
glyphosate, MOA 9 (numerous brands and formulations)	See label	0.75 (lb a.e.)	Apply before or after planting but before sunflowers emerge. See labels for suggested weed sizes to treat and for application rates. Add nonionic surfactant or crop oil concentrate according to Gramoxone label. The need for an adjuvant with glyphosate depends upon the brand used; see the label of the brand used.
paraquat, MOA 22 (Gramoxone Inteon) 2 SL	2 to 4 pt	0.5 to 1	
EPTC, MOA 8 (Eptam) 7 EC	2.5 to 3.5 pt	2.2 to 3	Controls annual grasses and small-seeded broadleaf weeds. Incorporate immediately after application according to label directions.
Preplant Incorporated, Annual grasses and small-seeded broadleaf weeds			
ethalfluralin, MOA 3 (Sonalan HFP) 3 EC	1.5 to 3 pt	0.56 to 1.13	Controls common annual grasses plus pigweed and lambsquarters. Incorporate into top 2 to 3 inches of seedbed. See label for application rate based upon soil texture. Generic brands of pendimethalin and trifluralin are available.
pendimethalin, MOA 3 (Prowl) 3.3 EC (Prowl H₂0) 3.8 L	1.2 to 3.6 pt 1.5 to 3.0 pt	0.5 to 1.5 0.71 to 1.43	
trifluralin, MOA 3 (Treflan HFP) 4 EC	1 to 2 pt	0.5 to 1	
Preplant Incorporated, Annual grasses, pigweed, and yellow nutsedge			
S-metolachlor, MOA 15 (Brawl) 7.62 EC (Dual Magnum) 7.62 EC	1 to 2 pt	0.95 to 1.91	Controls common annual grasses except Texas panicum, seedling johnsongrass, and shattercane. Also controls pigweed. At higher rates, controls yellow nutsedge. Incorporate into top 2 to 3 inches of seedbed. See label for application rate based upon soil texture.
Preemergence, Annual grasses and small-seeded broadleaf weeds			
pendimethalin, MOA 3 (Prowl H₂0) 3.8 L (Prowl) 3.3 EC	1.5 to 3 pt 1.2 to 3.6 pt	0.71 to 1.43 0.5 to 1.5	See above comments for pendimethalin applied preplant incorporated. Pendimethaln is more consistently effective when incorporated.
S-metolachlor, MOA 15 (Brawl) 7.62 EC (Dual Magnum) 7.62 EC	1 to 2 pt	0.95 to 1.91	See above comments for Dual Magnum applied preplant incorporated. Dual Magnum is more consistently effective on yellow nutsedge when incorporated.
Preemergence, Annual broadleaf weeds and nutsedge			
sulfentrazone, MOA 14 (Spartan) 4 F	3 to 8 oz	0.094 to 0.25	Controls nutsedge and most common annual broadleaf weeds. Only fair control of cocklebur, and no control of ragweed or sicklepod. May tank mix with other registered preemergence herbicides for annual grass control. Adjust Spartan and Spartan Charge application rate according to soil texture and organic matter as specified on label. May also be shallowly incorporated in the top 2 inches of seed bed. Do not plant corn, sweet potatoes, or cotton for 10, 12, and 12 months, respectively, after Spartan or Spartan Charge application. See label for rotational restrictions on other crops.
sulfentrazone, MOA 14 + carfentrazone, MOA 14 (Spartan Charge) 0.35 + 3.15 F	3.75 to 10.2 fl oz		
Postemergence, Annual grasses			
clethodim, MOA 1 (Select) 2 EC (Select Max) 0.97 EC	6 to 8 fl oz 9 to 16 fl oz	0.094 to 0.125 0.068 to 0.121	Apply to actively growing grasses not under drought stress. See label for grass size, application rates, and directions. Add 2 pints of crop oil concentrate to Select or Poast. To Select Max, add one of the following: nonionic surfactant at 0.25% by volume; crop oil concentrate at 1.0% by volume; or methylated seed oil at 1% by volume. Other formulation sof clethodim are available.
sethoxydim, MOA 1 (Poast) 1.5 EC	16 fl oz	0.19	
Postemergence, Bermudagrass			
clethodim, MOA 1 (Select) 2 EC (Select Max) 0.97 EC	6 to 8 fl oz 12 to 32 fl oz	0.094 to 0.125 0.091 to 0.24	Apply before bermudagrass runners exceed 6 inches If needed, make second application of 8 to 16 fluid ounces of Select or 12 to 32 fluid ounces of Select Max when regrowth is less than 6 inches Add crop oil concentrate to Select at 1 quart per acre. To Select Max, add nonionic surfactant at 0.25% by voume, crop oil concentrate at 1.0% by voume, or methylated seed oil at 1% by volume. Other formulation sof clethodim are available.
sethoxydim, MOA 1 (Poast) 1.5 EC	24 fl oz	0.28	Apply before bermudagrass runners exceed 6 inches If needed, make second application of 16 fluid ounces per acre when regrowth is less than 4 inches Add crop oil concentrate at 1 quart per acre.
clethodim, MOA 1 (Select) 2 EC (Select Max) 0.97 EC	8 to 16 fl oz 12 to 32 fl oz	0.125 to 0.25 0.091 to 0.24	Apply when johnsongrass is 12 to 24 inches tall. If needed, make second application of Select at 6 to 8 fluid ounces or 9 to 24 fluid ounces of Select Max when regrowth is 6 to 18 inches Add crop oil concentrate to Select at 1 quart per acre. To Select Max, add nonionic surfactant at 0.25% by voume, crop oil concentrate at 1.0% by voume, or methylated seed oil at 1% by volume. Other formulation sof clethodim are available.
sethoxydim, MOA 1 (Poast) 1.5 EC	24 fl ounces	0.28	Apply when johnsongrass is 20 to 25 inches tall. If needed, make second application of 16 fluid ounces per acre when regrowth is 12 inches Add crop oil concentrate at 1 quart per acre.
Postemergence, Annual grasses and broadleaf weeds: Clearfield Cultivars Only			
imazamox, MOA 2 (Beyond) 1 L	4 fl oz	0.031	**APPLY ONLY TO CLEARFIELD CULTIVARS.** Beyond will severely injure or kill non-Clearfield cultivars. Apply to sunflower in the 2- to 8-leaf stage when broadleaf weeds are 3 inches or less. Add crop oil concentrate or nonionic surfactant plus nitrogen according to label directions. Controls many common broadleaf weeds plus some annual grasses. See label for weeds controlled.
Postemergence with Hooded Sprayer, Annual broadleaf weeds, annual and perennial grasses			
glyphosate, MOA 9 (numerous brands and formulations)	See label	0.75 (lb a.e.)	Glyphosate is available as an isopropylamine salt and a potassium salt. Glyphosate formulations and application rates should be compared on the basis of pounds of glyphosate acid equivalent (a.e.) per gallon and per acre, respectively. Rate in the preceding column is expressed as a.e. Apply to row middles using hooded or shielded sprayer that allows no contact of sunflowers by spray solution.

Chemical Weed Control in Tobacco

L. R. Fisher, M. C. Vann, M. D. Inman, R. W. Seagroves, and D. S. Whitley, Crop and Soil Sciences Department

NOTE: A mode of action code has been added to the Herbicide and Formulation column of this table. Use MOA codes for herbicide resistance management. See Table 7-10, Herbicide Resistance Management, for details.

Table 7-7A. Chemical Weed Control in Tobacco

Herbicide, Mode of Action Code,[1] and Formulation	Amount of Formulation Per Acre	Pounds Active Ingredient Per Acre	Precautions and Remarks
Flue-Cured Field (before transplanting), Most annual grasses and some broadleaf weeds plus nutsedge suppression			
pebulate, MOA 8 (Tillam) 6 EC	2.7 qt	4	Apply to soil surface before bedding and immediately incorporate according to label instructions. Transplant as soon as possible. Early season stunting may occur under unfavorable growing conditions. Does not control cocklebur, morningglory, ragweed, or perennial weeds. Cultivate tobacco at least twice. See label for tank mixes with other pesticides.
Flue-Cured Field (before transplanting), Some annual grasses and some broadleaf weeds			
Napropamide, MOA 17 (Devrinol) 2 XT (Devrinol) 50 DF	2 to 4 qt 2 to 4 lb (broadcast, see label for band application)	1 to 2	Lower rates usually adequate for most soils. Apply to soil surface and incorporate according to label instructions. Some early season stunting may occur under unfavorable growing conditions. Does not control cocklebur, morningglory, or perennial weeds. Gives some suppression of ragweed. NOTE: Do not seed crops not specified on label for 12 months after application.
Flue-Cured Field (before transplanting), Most annual grasses and some broadleaf weeds			
pendimethalin, MOA 3 (Prowl) 3.3 EC (Prowl) H$_2$O	2.4 to 3.0 pt 2.0 to 2.5 pt	1.0 to 1.25 0.95 to 1.19	Can be applied up to 60 days before transplanting. Apply before bedding and incorporate into soil according to label instructions. Some early season stunting may occur under unfavorable growing conditions. Does not control cocklebur, morningglory, ragweed, or perennial weeds.
Flue-Cured Field (before transplanting), Annual grasses and some broadleaf weeds			
clomazone, MOA 13 (Command) 3 ME	2 to 2.67 pt	0.75 to 1 lb	Excellent annual grass control plus control of certain broadleaf weeds, such as prickly sida, jimsonweed, tropic croton, smartweed, and common ragweed. Partial control of cocklebur; does not control pigweed, sicklepod, or morningglory. Some whitening of lower leaves may occur but plants should recover. Do not plant small grains or alfalfa in the fall or following spring after Command application. Apply no more than once per season.
Flue-Cured Field (before transplanting), Broadleaf weeds, nutsedges, and some grasses			
sulfentrazone, MOA 14 (Spartan) 4F	4.5 to 12 fl oz	0.14 to 0.38 lb	Excellent control of pigweed, morningglories, and nutsedges. Application rate is based on soil type and organic matter. See Spartan 4 F or Spartan Charge label for rate determination and application methods. Early season stunting may occur especially when incorporated. Rainfall or irrigation needed within 7 to 10 days of application for maximum weed control, particularly when surface applied. Observe rotational crop guidelines on label.
sulfentrazone + carfentrazone, 0.35 + 3.15 MOA 14 (Spartan Charge)	5.7 to 15.2 fl oz	0.16 to 0.41 lb	
Flue-Cured Field (after transplanting), Most annual grasses and some broadleaf weeds			
Napropamide, MOA 17 (Devrinol) 2 XT (Devrinol) 50 DF	2 to 4 qt 2 to 4 lb (broadcast, see label for band application)	1 to 2 lb	Apply overtop immediately after transplanting tobacco. See remarks for Devrinol under "Before Transplanting." NOTE: Do not seed crops not specified on label for 12 months after application. Small grain seeded for cover crop in fall may be stunted. Do not use small grain for food or feed.
Flue-Cured Field (after transplanting)			
clomazone, MOA 13 (Command) 3 FME	2 to 2.67 pt	0.75 to 1.0 lb	Excellent annual grass control plus control of certain broadleaf weeds, such as prickly sida, jimsonweed, tropic croton, smartweed, and common ragweed. Partial control of cocklebur; does not control pigweed, sicklepod, or morningglory. Make a single broadcast application in a minimum of 20 gallons of water. Apply no more than once per season. Apply over the top of tobacco plants immediately or up to 7 days after transplanting but prior to emergence of weeds. Some whitening of lower leaves may occur, but plants should recover. Do not plant small grains or alfalfa in the fall or following spring after Command application.
Flue-Cured Field (after transplanting), Postemergence control of annual grasses			
sethoxydim, MOA 1 (Poast) 1.5 EC	1 to 1.5 pt	0.19 to 0.28	Apply to actively growing grass not under drought stress. Apply in 5 to 20 gallons of spray at 40 to 60 psi. Add 2 pints of crop oil concentrate per acre. Do not apply within 42 days of harvest. Do not apply more than 4 pints per acre per season. Complete coverage of grass required for control.
Flue-Cured Field (after transplanting), Postemergence control of some broadleaf weeds			
Carfentrazone MOA 14 (Aim) 2 EC	0.8 to 1.5 oz	0.0125 to 0.024 lb	Apply using SHIELDED SPRAYER or HOODED SPRAYER to emerged, actively growing weeds PRIOR TO LAYBY. Do not apply when conditions favor drift. MUST PREVENT CONTACT OF SPRAY SOLUTION WITH TOBACCO PLANT. See Label for further instruction.
Flue-Cured Lay-by, Most annual grasses and some broadleaf weeds			
Napropamide, MOA 17 (Devrinol), 2 XT (Devrinol), 50 DF	2 to 4 qt 2 to 4 lb (band, see label for band application)	1 to 2	Apply in a band to row middles immediately after last cultivation. Lower rates usually adequate for most tobacco soils. Incorporate lightly or sprinkler irrigate, if no rainfall within 3 days after application. Do not apply more than a total of 4 lb of Devrinol per acre in a season. See remarks for Devrinol under "Before Transplanting" and "After Transplanting."
pendimethalin, MOA 3 (Prowl) 3.3 EC (Prowl) H$_2$O	1.8 to 2.4 pt 1.5 to 2.0 pt	0.75 to 1.0 0.71 to 0.95	Apply to row middles immediately after last cultivation. Avoid contact with tobacco leaves. Use higher rate on medium- or fine-textured soils where grass infestation is heavy or if no herbicide was used previously. Rainfall or irrigation is needed within 7 days. Does not control emerged weeds.
Flue-Cured after first harvest, Postemergence control of some broadleaf weeds			
Carfentrazone MOA 14 (Aim) 2 EC	0.8 to 1.5 oz	0.0125 to 0.024 lb	Apply AFTER FIRST HARVEST for control of actively growing, emerged weeds. Position nozzles 3 to 4 inches above the soil and directed underneath the crop canopy. Do not apply when conditions favor drift. MUST PREVENT CONTACT OF SPRAY SOLUTION WITH TOBACCO PLANT. See label for further instruction.

CHEMICAL WEED CONTROL

CHEMICAL WEED CONTROL

Table 7-7A. Chemical Weed Control in Tobacco

Herbicide, Mode of Action Code,[1] and Formulation	Amount of Formulation Per Acre	Pounds Active Ingredient Per Acre	Precautions and Remarks
Burley Field (before transplanting), Most annual grasses and some broadleaf weeds plus nutsedge suppression			
pebulate, MOA 8 (Tillam) 6 EC	2.7 qt	4	See remarks for Tillam under Flue-Cured. Tank mix suppresses hairy galinsoga under ideal conditions (ample rainfall after application).
Burley Field (before transplanting), Some annual grasses and broadleaf weeds			
napropamide, MOA 17 (Devrinol), 2 XT (Devrinol), 50 DF	2 to 4 qt 2 to 4 lb (broadcast, see label for band application)	1 to 2	See remarks for Devrinol under Flue-Cured. Suppresses hairy galinsoga under ideal conditions (ample rainfall after application).
Burley Field (before transplanting), Most annual grasses and some broadleaf weeds			
pendimethalin, MOA 3 (Prowl) 3.3 EC (Prowl) H₂O	2.4 to 3.0 pt 2.0 to 2.5 pt	1.0 to 1.25 0.95 to 1.19	See remarks for Prowl under Flue-Cured. The higher labelled rates may be needed for soils where burley tobacco is grown in N.C. Add hairy galinsoga to weeds not controlled.
Burley Field (before transplanting), Annual grasses and some broadleaf weeds			
clomazone, MOA 13 (Command 3 ME) 3 FME	2 to 2.67 pt	0.75 to 1 lb	Excellent annual grass control plus control of certain broadleaf weeds, such as prickly sida, jimsonweed, tropic croton, smartweed, and common ragweed. Partial control of cocklebur; does not control pigweed, sicklepod, or morningglory. Some whitening of lower leaves may occur but plants should recover. Do not plant small grains or alfalfa in the fall or following spring after Command application. Apply no more than once per season.
Burley Field (before transplanting), Broadleaf weeds, nutsedges, and some grasses			
sulfentrazone, MOA 14 (Spartan) 4F sulfentrazone + carfentrazone, 0.35 + 3.15 MOA 14 (Spartan Charge)	4.5 to 12 fl oz 5.7 to 15.2 fl oz	0.14 to 0.38 lb 0.16 to 0.41 lb	Excellent control of pigweed, morningglories, and nutsedges. Application rate is based on soil type and organic matter. See Spartan 4 F or Spartan Charge label for rate determination and application methods. Early season stunting may occur especially when incorporated. Rainfall or irrigation needed within 7 to 10 days of application for maximum weed control, particularly when surface applied. Observe rotational crop guidelines on label.
Burley Field (after transplanting), Most annual grasses and some broadleaf weeds			
napropamide, MOA 17 (Devrinol), 2 XT (Devrinol) 50 DF	2 to 4 qt 2 to 4 lb (broadcast, see label for band application)	1 to 2 lb	Apply overtop immediately after transplanting tobacco. See remarks for Devrinol under "After Transplanting" in flue-cured section. Suppresses hairy galinsoga under ideal conditions (ample rainfall after application).
Burley Field (after transplanting), Annual grasses and some broadleaf weeds			
clomazone, MOA 13 (Command) 3 ME	2 to 2.67 pt	0.75 to 1 lb	Excellent annual grass control plus control of certain broadleaf weeds, such as prickly sida, jimsonweed, tropic croton, smartweed, and common ragweed. Partial control of cocklebur; does not control pigweed, sicklepod, or morningglory. Make a single broadcast application in a minimum of 20 gallons of water. Apply no more than once per season. Apply over the top of tobacco plants immediately, or up to 7 days after transplanting, but prior to emergence of weeds. Some whitening of lower leaves may occur, but plants should recover. Do not plant small grains or alfalfa in the fall or following spring after applying Command.
Burley Field (after transplanting), Postemergence control of annual grasses			
sethoxydim, MOA 1 (Poast) 1.5 EC	1 to 1.5 pt	0.19 to 0.28	Apply to actively growing grass not under drought stress. Apply in 5 to 20 gallons of spray at 40 to 60 psi. Add 2 pints of crop oil concentrate per acre. Do not apply within 42 days of harvest. Do not apply more than 4 pints per acre per season. Complete coverage of grass required for control.
Burley Field (after transplanting), Postemergence control of some broadleaf weeds			
Carfentrazone MOA 14 (Aim) 2 EC	0.8 to 1.5 oz	0.0125 to 0.024 lb	Apply using SHIELDED SPRAYER or HOODED SPRAYER to emerged, actively growing weeds PRIOR TO LAYBY. Do not apply when conditions favor drift. MUST PREVENT CONTACT OF SPRAY SOLUTION WITH TOBACCO PLANT. See Label for further instruction.
Burley Lay-by, Most annual grasses and some broadleaf weeds			
napropamide, MOA 17 (Devrinol), 2 XT (Devrinol) 50 DF	2 to 4 qt 2 to 4 lb (band, see label for band application)	1 to 2	Apply in a band to row middles immediately after last cultivation. Lower rates usually adequate for most tobacco soils. Incorporate lightly or sprinkler irrigate if no rainfall within 3 days after application. Do not apply more than a total of 4 pounds of Devrinol 50 WP per acre in a season. See remarks for Devrinol under "After Transplanting" in flue-cured section. Suppresses hairy galinsoga under ideal conditions (ample rainfall after application).
pendimethalin, MOA 3 (Prowl) 3.3 EC (Prowl) H₂O	1.8 to 2.4 pt 1.5 to 2.0 pt	0.75 to 1.0 0.71 to 0.95	Apply to row middles immediately after last cultivation. Avoid contact with tobacco leaves. Use higher rate on medium- or fine-textured soils where grass infestation is heavy or if no herbicide was used previously. Rainfall or irrigation is needed within 7 days. Does not control emerged weeds.

[1] Mode of Action (MOA) code developed by the Weed Science Society of America. See Table 7-10, Herbicide Resistance Management, for details.

280

Weed Response to Herbicides in Tobacco

L. R. Fisher, M. C. Vann, M. D. Inman, R. W. Seagroves, and D. S. Whitley, Crop and Soil Sciences Department

Ratings based upon average to good soil and weather conditions for herbicide performance and upon proper application rate, technique, and timing.

Table 7-7B. Weed Response to Herbicides in Tobacco

Species	Command	Devrinol	Poast	Prowl	Spartan	Tillam	Aim
Barnyardgrass	E	GE	E	GE	F	GE	N
Bermudagrass	PF	P	FG	P	P	P	N
Broadleaf signalgrass	E	G	E	G	F	P	N
Crabgrass	E	E	GE	E	F	E	N
Crowfootgrass	E	E	FG	E	F	E	N
Fall panicum	E	G	E	GE	—	G	N
Foxtails	E	E	E	E	F	E	N
Goosegrass	E	E	GE	E	F	G	N
Sandbur	G	—	FG	G	—	G	P
Seedling johnsongrass	G	F	E	G	—	G	N
Texas panicum	G	—	E	G	F	P	N
Nutsedge	P	P	N	P	E	FG	N
Cocklebur	F	P	N	P	FG	P	G
Common purslane	FG	E	N	P	G	G	G
Ragweed Common	G	F	N	P	P	P	N
Giant	PF	PF	N	P	—	P	N
Hairy galinsoga	G	PF	N	P	G	P	P
Jimsonweed	G	P	N	P	—	P	G
Lambsquarters	G	G	N	G	E	G	G
Morningglory	P	P	N	P	E	P	E
Pigweed	P	G	N	G	E	G	E
Prickly sida	E	P	N	P	G	P	P
Sicklepod	P	P	N	P	P	P	P
Smartweed	G	P	N	P	E	P	G

KEY

E = excellent control, 90% or better
G = good control, 80% to 90%
F = fair control, 50% to 80%
P = poor control, 25% to 50%
N = no control, less than 25%

CHEMICAL WEED CONTROL

Chemical Weed Control in Wheat, Barley, Oats, Rye, and Triticale
W. J. Everman, Crop Science Department

NOTE: A mode of action code has been added to the Herbicide and Formulation column of this table. Use MOA codes for herbicide resistance management. See Table 7-10, Herbicide Resistance Management, for details.

Table 7-8A. Chemical Weed Control in Wheat, Barley, Oats, Rye, and Triticale

Herbicide, Mode of Action Code[1], and Formulation	Amount of Formulation per Acre	Pounds Active Ingredient per Acre	Precautions and Remarks
Wheat Preplant No-Till, Emerged annual broadleaf and grass weeds, volunteer corn, top-kill of perennials			
paraquat, MOA 22 (Gramoxone Inteon) 2 SL (Gramoxone Max) 3 SL	2 to 4 pt	0.5 to 1	Rate depends upon weed size; see label. Apply before crop emerges. Add nonionic surfactant at 1 pint per 100 gallons spray solution or crop oil concentrate at 1 gallon per 100 gallons spray solution. See application directions on label. May be tank mixed with Hoelon. Generic formulations of paraquat containing 3 pounds active per gallon are available. Apply these products at 1.3 to 2.7 pints.
Wheat Preplant No-Till, Emerged annual broadleaf and grass weeds, control or suppression of perennial weeds			
glyphosate, MOA 9 (numerous brands and formulations)	See label	0.38 to 1.13 (lb a.e.)	Glyphosate is available as an isopropylamine salt and a potassium salt. Glyphosate formulations and application rates should be compared on the basis of pounds of glyphosate acid equivalent (a.e.) per gallon and per acre, respectively. Rate in the preceding column is expressed as a.e. See Table 7-10 for glyphosate rate conversions. Rate depends upon weed species and size; see label. Apply before crop emergence. Adjuvant recommendations vary by glyphosate brand; see label of brand used for details. Select or Select Max may be mixed with glyphosate to control volunteer Roundup Ready corn. For corn up to 12 inches tall, apply 4 to 6 fluid ounces of Select or 6 fluid ounces of Select Max. For corn up to 24 inches increase Select rate to 6 to 8 fluid ounces or Select Max rate to 9 fluid ounces. Select or Select Max must be applied at least 30 days ahead of wheat planting. Valor SX at 1 to 2 ounces per acre will suppress ryegrass and bluegrass and controls several broadleaf weeds. Do not till after application and apply at least 30 days prior to planting.
Wheat Preplant No-Till or Preemergence, Broadleaf weeds			
saflufenacil, MOA 14 (Sharpen) 3.42	1.0 to 2.0 fl oz	0.027 to 0.054	See label for broadleaf weeds controlled. Sharpen does not control grasses. Apply with amonium sulfate (1 to 2 gallons/100 gallons) and methylated seed oil (1 gallons/100 gallons). Do not apply if wheat has germinated. See label for tank mixtures.
flumioxazin, MOA 14 (Valor SX) 51 WDG + paraquat, MOA 22 (Gramoxone Inteon) 2.5 SL OR glyphosate, MOA 9 (numerous brands and formulations)	2.0 oz 2 to 4 pt See label	0,063 0.5 to 1 0.38 to 1.13 (lb a.e.)	Use only on no-till or minimum till fields where stubble from the previous crop has not been incorporated. Valor SX must be applied at least 30 days ahead of planting wheat. Residual control of broadleaf weeds and Italian ryegrass. Apply with nonionic surfactant at 1 quart/100 gallons. Can be applied with nitrogen carriers. Do no perform tillage after application. Carefully follow label directions for sprayer cleaning after each day's use.
thifensulfuron, MOA 2 + tribenuron, MOA 2 (FirstShot SG with TotalSol) 50 WDG	0.5 to 0.8 oz	0.008 to 0.013 + 0.008 to 0.013	FirstShot does not control grasses. May be applied up to planting to control emerged broadleaf weeds. Add nonionic surfactant at 2 to 4 pints per 100 gallons spray solution or crop oil concentrate or methylated seed oil at 1 gallon per 100 gallons spray solution. In addition, add nitrogen fertilizer at a rate of 2 quarts per acre or ammonium sulfate at 2 pounds per acre. See label for tank mixtures.
Wheat Preplant No-Till or Preemergence, Emerged annual broadleaf weeds			
pyraflufen-ethyl, MOA 14 (ET) 1 SL	0.5 to 2.0 fl oz	0.003 to 0.015 (lb a.i.)	ET can be used for limited suppression of small emerged summer annual and winter weeds. See label for adjuvant and spray volume recommendations. Research with ET is limited in North Carolina.
Wheat Preemergence, Italian ryegrass and annual broadleaf weeds			
chlorsulfuron, MOA 2 + metsulfuron methyl, MOA 2 (Finesse) 75 WDG	0.5 oz	0.0195 + 0.0039	Ryegrass control is variable; expect only suppression. May stunt wheat on sandy soils. Suggested primarily for fields with Hoelon-resistant ryegrass. Also controls most annual broadleaf weeds. Do not use where a later application of Osprey or PowerFlex is anticipated. Plant only STS soybeans following wheat harvest. May cause severe injury in non-STS soybeans.
pyroxasulfone, MOA 15 + carfentrazone, MOA 14 (Anthem Flex) 4 SC	2.0 to 4.5 fl oz	0.058 to 0.13 + 0.004 to 0.009	Apply to wheat as a preemergence treatment from 2.0 to 4.5 fluid ounces depending on soil type. Read label and adjust rates for soil texture. Under extended periods of dry weather, adequate weed control may not be achieved, 0.5 inch of rainfall may be necessary for activation and optimal weed control. Do not apply to broadcast seeded wheat. Do not apply preemergence if 0.25 inch or more rain is expected within 48 hours of application. Plant wheat a minimum of 1 inch deep, but not over 1.5 inches deep.
Wheat Spike Stage, Italian ryegrass			
flufenacet, MOA 15 + metribuzin, MOA 5 (Axiom) 68 WDG	4 to 10 oz	0.136 to 0.034 + 0.34 to 0.085	Apply to wheat in the spike stage. Preemergence application can cause severe injury on coarse-textured soils. Application rate depends on soil type; see label. In general, North Carolina research has shown best results with 6 to 7 ounces on coarse soils and 8 to 9 ounces on medium and heavy soils. If rainfall is received timely, Axiom controls ryegrass well. It also controls chickweed, henbit, and wild radish.
pyroxasulfone, MOA15 (Zidua) 85 WG	0.7 to 2.0 oz/A	0.037 to 0.106	Apply to wheat when 80% of germinated wheat seeds have a shoot at least 1/2-inch long until wheat spiking. May be applied as a broadcast spray to wheat at spiking up to the fourth-tiller. Will not control emerged weeds. Sequential applicatioin may be applied, but do not exceed a total of 2.25 ounces/acre in a season.
pyroxasulfone, MOA 15 + carfentrazone, MOA 14 (Anthem Flex) 4 SC	2.5 to 4.5 fl oz	0.073 to 0.13 + 0.005 to 0.009	Apply to wheat when 80% of germinated wheat seeds have a shoot at least 1/2-inch long until wheat spiking. May be applied as a broadcast spray to wheat at spiking up to the fourth-tiller. Early postemergence broadcast applications will provide suppression of emerged chickweed and henbit. Do not apply more than a maximum cumulative amount of 4.55 fluid ounces per acre in one cropping season.
flumioxazin, MOA 14 + pyroxasulfone, MOA 15 (Fierce) 76 WDG	1.5 oz	0.07	Apply to wheat when 95% of wheat is in the spike to 2-leaf stage of growth for control of ryegrass and wild radish. Wheat seed must be planted between 1 and 1.5 inches deep or injury may occur. Do not apply to fields where wheat seed has been broadcast and shallow incorporated. Do not tank mix with any adjuvant, fertilizer, or pest control product or severe injury will occur. Rainfall of at least 0.5 inch within 10 days after application is necessary for activation. Avoid applications to heavy sand and low organic matter areas or excessive injury may occur following heavy rainfall.

Table 7-8A. Chemical Weed Control in Wheat, Barley, Oats, Rye, and Triticale

Herbicide, Mode of Action Code[1], and Formulation	Amount of Formulation per Acre	Pounds Active Ingredient per Acre	Precautions and Remarks
Wheat Postemergence, Italian ryegrass			
diclofop-methyl, MOA 1 (Hoelon) 3 EC	1.33 to 2.67 pt	0.5 to 1	Apply when ryegrass is in the one- to five-leaf stage. See label for specific rates depending upon weed size. Application to smaller ryegrass is more effective. Make only one application per season. Do not tank mix with broadleaf herbicides or use liquid nitrogen as the carrier. Apply before the first wheat node (joint) develops. May add 1 to 2 pints per acre of crop oil concentrate under dry conditions or when ryegrass is large. In most cases, crop oil is not necessary. See precautions on label concerning temperatures. Biotypes of ryegrass resistant to Hoelon are common within the state. Hoelon should not be applied to fields where resistance is suspected.
mesosulfuron, MOA 2 (Osprey) 4.5 WDG	4.75 oz	0.013	Apply when ryegrass is in one-leaf to two-tiller stage. Add adjuvant as directed on label. In North Carolina, nonionic surfactant at 1 to 2 quarts per 100 gallons spray solution plus 1 to 2 quarts of 30% liquid nitrogen per acre is preferred. See label for broadleaf weeds controlled. For additional control, Osprey may be mixed with Harmony Extra. Do not tank mix with 2,4-D or dicamba. Do not apply using liquid nitrogen as the carrier. Do not topdress wheat within 14 days of Osprey application. In fields with Hoelon-susceptible ryegrass, it is recommended that Osprey or PowerFlex and Hoelon or Axial be used on an alternating basis (i.e., rotated) as part of a resistance management strategy. See comments for Axial. Ryegrass with multiple resistance to Axial, PowerFlex, Hoelon, Osprey, and PowerFlex occurs in North Carolina.
pinoxaden, MOA 1 (Axial XL) 0.42 EC	16.4 fl oz	0.054	Apply to wheat with two or more leaves when ryegrass has one to five leaves on the main stem. More effective when applied to smaller ryegrass. May be tank mixed with Harmony Extra. When mixing, add Harmony Extra first, then Axial. No adjuvants are necessary. May be applied in water-nitrogen mixtures containing up to 50% liquid nitrogen by volume. Add water to tank, then add Axial. Mix thoroughly, and then add the nitrogen. Axial and Hoelon have the same mode of action. Ryegrass resistant to Hoelon may be cross-resistant to Axial, although in some cases Axial will control Hoelon-resistant ryegrass. If Hoelon resistance is expected, trial use of Axial on limited acreage is suggested to determine if the biotype is susceptible to Axial.
pyroxsulam, MOA 2 (PowerFlex) 7.5 WDG	3.5 oz	0.0164	Can be applied to wheat from the 3-leaf stage until jointing. Apply after the majority of the ryegrass has emerged but before it exceeds the 2-tiller stage. Add nonionic surfactant at 1 to 2 quarts/100 gallons spray solution. See label for broadleaf weeds controlled. For additional control, PowerFlex may be mixed with Harmony Extra. Do not mix with dicamba or with amine formulations of 2,4-D or MCPA. Can be applied in water-nitrogen mixtures containing up to 50% liquid nitrogen by volume, or a maximum of 30 pounds/acre. If applying in liquid nitrogen, reduce surfactant rate to 1 pint/100 gallons. Rainfast in 4 hours. Do not apply to wet foliage. Current labeling specifies a 5-month rotation for soybeans. Limited research in North Carolina has shown no problems with soybeans double-cropped behind PowerFlex-treated wheat. Corn, cotton, or peanuts can be planted 9 months after application. See comments under mesosulfuron (Osprey) and pinoxaden (Axial) concerning resistance management.
Wheat Postemergence, Cheat			
pyroxsulam, MOA 2 (PowerFlex) 7.5 WDG	3.5 oz	0.0164	See comments for PowerFlex under Italian Ryegrass.
Wheat Postemergence, Annual bluegrass			
mesosulfuron, MOA 2 (Osprey) 4.5 WDG	4.75 oz	0.013	See comments for Osprey under Italian Ryegrass. Apply to bluegrass from the 1-leaf to 2-tiller stage. Application when plants are about the size of a quarter coin has worked well.
Wheat Postemergence, Wild garlic, curly dock, and most winter annual broadleaf weeds except cornflower and vetch			
thifensulfuron-methyl, MOA 2 + tribenuron-methyl, MOA 2 (Harmony Extra SG with TotalSol) 50 WDG	0.45 to 0.9 oz	0.0094 to 0.0188 + 0.0047 to 0.0094	Apply after the two-leaf stage of wheat but before flag leaf is visible. Use 0.45 to 0.6 ounce for most winter annual weeds. Use 0.75 to 0.9 ounce for wild garlic and wild radish. Wild garlic should be less than 12 inches tall and should have 2 to 4 inches of new growth. Control is enhanced when application is made during warm temperatures (50 degrees F or more) to actively growing garlic plants. Add 1 quart of nonionic surfactant per 100 gallons of spray solution. Liquid nitrogen may be used as the carrier. May tank mix Harmony Extra with 0.125 to 0.375 pound active ingredient of 2,4-D for improved control of wild radish. Follow mixing instructions on the label when using nitrogen as the carrier or when mixing with 2,4-D. Reduce surfactant rate according to label instructions when using nitrogen as the carrier or when mixing with 2,4-D. Do not tank mix with Hoelon. May be tank mixed with Axial or Osprey.
Wheat Postemergence, Most winter annual broadleaf weeds except chickweed, henbit, and knawel			
2,4-D amine, MOA 4 (various brands) 3.8 SL	1 pt	0.48	Apply after wheat is fully tillered (usually 4 to 8 inches tall; stages 4 and 5 on Feekes scale) but before jointing. Spraying wheat too young or after jointing can cause deformed heads, reduced yields, and uneven ripening. Better results are obtained when day-time temperatures are above 50 degrees F. Increase the rate of 2,4-D by 50% to control corn cockle. Liquid nitrogen may be used as the carrier for 2,4-D. Ester formulations can be added directly to the nitrogen. If using amine formulation, premix in water (1 part 2,4-D to 4 parts water) and add mixture to nitrogen with strong agitation. Amine formulations give less burn than ester formulations in nitrogen.
2,4-D ester, MOA 4 (various brands) 3.8 SL	1 pt	0.48	
2,4-D ester, MOA 4 (various brands) 5.7 SL	0.67 pt	0.48	
2,4-D acid/ester, MOA 4 (Weedone 638) 2.8 SL	1 pt	0.35	
Wheat Postemergence, Most winter annual broadleaf weeds			
dicamba, MOA 4 (Banvel) 4 SL (Clarity) 4 SL	0.25 pt	0.125	Apply after wheat is fully tillered but before jointing. Better results will be obtained if applied when daytime temperatures are above 50 degrees F. Liquid nitrogen may be used as the carrier.
dicamba, MOA 4 (Banvel) 4 SL (Clarity) 4 SL + 2,4-D amine, MOA 4 (various brands) 3.8 SL OR 2,4-D ester, MOA 4 (various brands) 3.8 SL	0.25 pt + 0.5 to 0.75 pt or 0.5 to 0.75 pt	0.125 + 0.24 to 0.36 or 0.24 to 0.36	Apply after wheat is fully tillered (usually 4 to 8 inches tall; stages 4 and 5 on Feekes scale) but before jointing. Compared to dicamba alone, tank mixture is more effective on buttercup, cornflower, field pennycress, Virginia pepperweed, shepherdspurse, wild mustard, and wild radish. Use this tank mix only if both herbicides are necessary for weed control. Tank mix may injure wheat.
halauxifen-methly, MOA 2 + florasulam, MOA 4 (Quelex) 20WG	0.75 oz	0.009	Apply from 2-leaf to flag leaf stage when weeds are actively growing in the 2 to 4 leaf stage. May be tank mixed with other herbicides labeled for use on wheat. Do not apply within 60 days of crop harvest. Add 1.6 to 4 pt of nonionic surfactant per 100 gallons of spray solution, or a crop oil concentrate or methylated seed oil at 4 to 8 pt per 100 gallons. Liquid nitrogen may be used as the carrier, see label for guidelines.
pyraflufen-methyl, MOA 14 (ET) 1 SL	0.5 to 2.0 fl oz	0.003 to 0.015 (lb a.i.)	ET can be used to suppress small annual winter weeds. Although application is registered to flag leaf appearance, coverage of small weeds is necessary but difficult when wheat is tall. See label for adjuvant and spray volume and carrier recommendations. Research with ET is limited in North Carolina.

CHEMICAL WEED CONTROL

Table 7-8A. Chemical Weed Control in Wheat, Barley, Oats, Rye, and Triticale

Herbicide, Mode of Action Code[1], and Formulation	Amount of Formulation per Acre	Pounds Active Ingredient per Acre	Precautions and Remarks
Wheat Preharvest, Annual broadleaf and grass weeds, suppression of perennial weeds			
glyphosate, MOA 9 (numerous brands and formulations)	See label	0.75 (lb a.e.)	Apply after hard dough stage of grain (30% or less grain moisture) and at least 7 days before harvest. Do not apply to wheat grown for seed. Glyphosate is available as an isopropylamine salt and a potassium salt. Glyphosate formulations and application rates should be compared on the basis of pounds of glyphosate acid equivalent (a.e.) per gallonslon and per acre, respectively. Rate in the preceding column is expressed as a.e. See Table 7-10 for glyphosate rate conversions.
Wheat Preharvest, Annual broadleaf weeds			
2,4-D amine, MOA 4 (various brands) 3.8 SL	1 to 2 pt	0.48 to 0.95	Apply when grain is in hard dough stage or later. Do not allow drift to sensitive crops (be especially careful with ester formulations). Amine formulations strongly encouraged if sensitive crops are nearby, especially cotton and tobacco.
2,4-D ester, MOA 4 (various brands) 3.8 SL	1 to 2 pt	0.48 to 0.95	
2,4-D ester, MOA 4 (various brands) 5.7 SL	0.67 to 1.3 pt	0.48 to 0.95	
2,4-D acid/ester, MOA 4 (Weedone 638) 2.8 SL	1 to 2 pt	0.35 to 0.7	
Barley Preplant No-Till, Emerged annual broadleaf and grass weeds, volunteer corn, top-kill of perennials			
paraquat, MOA 22 (Gramoxone Inteon) 2 SL	2 to 4 pt	0.5 to 1	Rate depends upon weed size; see label. Apply before crop emerges. Add nonionic surfactant at 1 pint per 100 gallons spray solution or crop oil concentrate at 1 gallon per 100 gallons spray solution. See application directions on label. Generic formulations of paraquat containing 3 pounds active per gallon are available. Apply these products at 1.3 to 2.7 pints.
Barley Preplant No-Till, Emerged annual broadleaf and grass weeds, control or suppression of perennial weeds			
glyphosate, MOA 9 (numerous brands and formulations)	See label	0.56 to 1.13 (lb a.e.)	Glyphosate is available as an isopropylamine salt and a potassium salt. Glyphosate formulations and application rates should be compared on the basis of pounds of glyphosate acid equivalent (a.e.) per gallon and per acre, respectively. Rate in the preceding column is expressed as a.e. See Table 7-10 for glyphosate rate conversions. Rate depends upon weed species and size; see labels. Apply before crop emergence. Adjuvant recommendations vary by glyphosate brand; see label of brand used for details. Select or Select Max may be mixed with glyphosate to control volunteer Roundup Ready corn. For corn up to 12 inches tall, apply 4 to 6 fluid ounces of Select or 6 fluid ounces of Select Max. For corn up to 24 inches increase Select rate to 6 to 8 fluid ounces or Select Max rate to 9 fluid ounces. Select or Select Max must be applied at least 30 days ahead of barley planting.
thifensulfuron, MOA 2 + tribenuron, MOA 2 (FirstShot SG with TotalSol) 50 WDG	0.5 to 0.8 oz	0.008 to 0.013 + 0.008 to 0.013	FirstShot does not control grasses. May be applied up to planting to control emerged broadleaf weeds. Add nonionic surfactant at 2 to 4 pints per 100 gallons spray solution or crop oil concentrate or methylated seed oil at 1 gallon per 100 gallons spray solution. In addition, add nitrogen fertilizer at a rate of 2 quarts per acre or ammonium sulfate at 2 pounds per acre. See label for tank mixtures.
Barley Postemergence, Italian ryegrass			
diclofop-methyl, MOA 1 (Hoelon) 3 EC	1.33 to 2.67 pt	0.5 to 1	Apply when ryegrass is in the one- to four-leaf stage and after tiller initiation but prior to jointing of barley. Make only one application per year. Do not tank mix with broadleaf herbicides or use liquid nitrogen as the carrier. Do not add crop oil. Apply only to the following varieties of barley: Anson, Boone, Henry, Milton, Molly Bloom, Mulligan, Nomini, Pennco, Starling, Sussex, and Wysor. Cold (lower than 40 degrees F) and/or prolonged wet conditions increase barley sensitivity to Hoelon. Biotypes of ryegrass resistant to Hoelon are becoming more common within the state. Hoelon should not be applied to fields where resistance is suspected.
pinoxaden, MOA 1 (Axial XL) 0.42 EC	16.4 fl oz	0.054	Apply to barley with two or more leaves when ryegrass has one to five leaves on the main stem. More effective when applied to smaller ryegrass. May be tank mixed with Harmony Extra. When mixing, add Harmony Extra first, then Axial. May be applied in water/nitrogen mixtures containing up to 50% liquid nitrogen by volume. Add water to tank, then add Axial. Mix thoroughly, and add the nitrogen. Axial and Hoelon have the same mode of action. Ryegrass resistant to Hoelon may be cross-resistant to Axial, although in some cases Axial will control Hoelon-resistant ryegrass. If Hoelon resistance is expected, trial use of Axial on limited acreage is suggested to determine if the ryegrass biotype is susceptible to Axial.
Barley Postemergence, Wild garlic, curly dock, and most winter annual broadleaf weeds except cornflower and vetch			
thifensulfuron-methyl, MOA 2 + tribenuron-methyl, MOA 2 (Harmony Extra SG with TotalSol) 50 WDG	0.45 to 0.9 oz	0.0094 to 0.0188 + 0.0047 to 0.0094	Apply after the two-leaf stage of barley but before flag leaf is visible. Use 0.45 to 0.6 ounce for most winter annual weeds. Use 0.75 to 0.9 ounce for wild garlic and wild radish. Wild garlic should be less than 12 inches tall and should have 2 to 4 inches of new growth. Control is enhanced when application is made during warm temperatures (50 degrees F or more) to actively growing garlic plants. Add 1 quart of nonionic surfactant per 100 gallons of spray solution. Liquid nitrogen may be used as the carrier. May tank mix Harmony Extra with 0.125 to 0.375 pound a.i. of 2,4-D for improved control of wild radish. Follow mixing instructions on the label when using nitrogen as the carrier or when mixing with 2,4-D. Reduce surfactant rate according to label instructions when using nitrogen as the carrier or when mixing with 2,4-D. Do not tank mix with Hoelon.
Barley Postemergence, Most winter annual broadleaf weeds except chickweed, henbit, and knawel			
2,4-D amine, MOA 4 (various brands) 3.8 SL	1 pt	0.48	Apply after barley is fully tillered but before jointing. Spraying barley too young or after jointing can cause deformed heads, reduced yields, and uneven ripening. Better results are obtained when day-time temperatures are above 50 degrees F. Increase the rate of 2,4-D by 50% to control corn cockle. Liquid nitrogen may be used as the carrier for 2,4-D. Ester formulations can be added directly to the nitrogen. If using amine formulation, premix in water (1 part 2,4-D to 4 parts water) and add mixture to nitrogen with strong agitation. Amine formulations give less burn than ester formulations in nitrogen.
2,4-D ester, MOA 4 (various brands) 3.8 SL	1 pt	0.48	
2,4-D ester, MOA 4 (various brands) 5.7 SL	0.67 pt	0.48	
2,4-D acid/ester, MOA 4 (Weedone 638) 2.8 SL	1 pt	0.35	
Barley Postemergence, Most winter annual broadleaf weeds			
dicamba, MOA 4 (Banvel) 4 SL (Clarity) 4 SL	0.25 pt	0.125	Apply before jointing stage of growth. Risk of crop injury is least if applied after winter dormancy and before grain begins to joint. Better results will be obtained if applied when daytime temperatures are above 50 degrees F. Liquid nitrogen may be used as the carrier.
dicamba, MOA 4 (Banvel) 4 SL (Clarity) 4 SL + 2,4-D amine, MOA 4 (various brands) 3.8 SL OR 2,4-D ester, MOA 4	0.25 pt 0.25 pt + 0.5 pt or	0.125 + 0.24 or	Apply after barley is fully tillered but before jointing. Compared to dicamba alone, tank mixture is more effective on buttercup, cornflower, field pennycress, Virginia pepper-weed, shepherdspurse, and wild mustard. Use this tank mix only if both herbicides are necessary for weed control. Tank mix may injure barley.

Table 7-8A. Chemical Weed Control in Wheat, Barley, Oats, Rye, and Triticale

Herbicide, Mode of Action Code[1], and Formulation	Amount of Formulation per Acre	Pounds Active Ingredient per Acre	Precautions and Remarks
(various brands) 3.8 SL	0.5 pt	0.24	
Barley Postemergence, Most winter annual broadleaf weeds (continued)			
halauxifen-methly, MOA 2 + florasulam, MOA 4 (Quelex) 20WG	0.75 oz	0.009	Apply from 2-leaf to flag leaf stage when weeds are actively growing in the 2 to 4 leaf stage. May be tank mixed with other herbicides labeled for use on barley. Do not apply within 60 days of crop harvest. Add 1.6 to 4 pt of nonionic surfactant per 100 gallons of spray solution, or a crop oil concentrate or methylated seed oil at 4 to 8 pt per 100 gallons. Liquid nitrogen may be used as the carrier, see label for guidelines.
Barley Preharvest, Annual broadleaf weeds			
2,4-D amine, MOA 4 (various brands) 3.8 SL	1 to 2 pt	0.48 to 0.95	Apply when grain is in hard dough stage or later. Do not allow drift to sensitive crops (be especially careful with ester formulations). Amine formulations strongly encouraged if sensitive crops are nearby, especially cotton or tobacco.
2,4-D ester, MOA 4 (various brands) 3.8 SL	1 to 2 pt	0.48 to 0.95	
2,4-D ester, MOA 4 (various brands) 5.7 SL	0.67 to 1.3 pt	0.48 to 0.95	
2,4-D acid/ester, MOA 4 (Weedone 638) 2.8 SL	1 to 2 pt	0.35 to 0.7	
Barley Preharvest, Annual broadleaf and grass weeds, suppression of perennial weeds			
glyphosate, MOA 9 (numerous brands and formulations)	See label	0.75 (lb a.e.)	Apply after hard dough stage of grain (and 20% or less moisture) and at least 7 days before harvest. Do not apply to barley grown for seed. Glyphosate is available as an isopropylamine salt and a potassium salt. Compare glyphosate formulations and application rates on the basis of glyphosate acid equivalent (a.e.) per gallon and per acre, respectively. Rate in the preceding column is expressed as a.e. See Table 7-10 for glyphosate rate conversions.
Oats Preplant No-Till, Emerged annual broadleaf and grass weeds, volunteer corn, control or suppression of perennial weeds			
glyphosate, MOA 9 (numerous brands and formulations)	See label	0.56 to 1.13 (lb a.e.)	Glyphosate is available as an isopropylamine salt and a potassium salt. Glyphosate formulations and application rates should be compared on the basis of pounds of glyphosate acid equivalent (a.e.) per gallon and per acre, respectively. Rate in the preceding column is expressed as a.e. See Table 7-10 for glyphosate rate conversions. Rate depends upon weed species and size; see labels. Apply before crop emergence. Adjuvant recommendations vary by glyphosate brand; see label of brand used for details. Select or Select Max may be mixed with glyphosate to control volunteer Roundup Ready corn. For corn up to 12 inches tall, apply 4 to 6 fluid ounces of Select or 6 fluid ounces of Select Max. For corn up to 24 inches, increase Select rate to 6 to 8 fluid ounces or Select Max rate to 9 fluid ounces. Select or Select Max must be applied at least 30 days ahead of planting oats.
Oats Postemergence, Wild garlic, curly dock, and most winter annual broadleaf weeds except cornflower and vetch			
thifensulfuron-methyl, MOA 2 + tribenuron-methyl, MOA 2 (Harmony Extra SG with TotalSol) 50 WDG	0.45 to 0.6 oz	0.0094 to 0.0125 + 0.0047 to 0.0063	Apply after the two-leaf stage of oats but before flag leaf is visible. Wild garlic should be less than 12 inches tall and should have 2 to 4 inches of new growth. Control is enhanced when application is made during warm temperatures (50 degrees F or more) to actively growing garlic plants. Add 1 quart of nonionic surfactant per 100 gallons of spray solution. Liquid nitrogen may be used as the carrier. May tank mix Harmony Extra with 0.125 to 0.375 pound a.i. of 2,4-D for improved control of wild radish. Follow mixing instructions on the label when using nitrogen as the carrier or when mixing with 2,4-D. Reduce surfactant rate according to label instructions when using nitrogen as the carrier or when mixing with 2,4-D. Oats are more sensitive to 2,4-D than wheat.
Oats Postemergence, Most winter annual broadleaf weeds except chickweed, henbit, and knawel			
2,4-D amine, MOA 4 (various brands) 3.8 SL	1 pt	0.48	Apply after oats are fully tillered but before jointing. Spraying oats too young or after jointing can cause deformed heads, reduced yields, and uneven ripening. Also, oats are less tolerant of 2,4-D than wheat. Better results are obtained when daytime temperatures are above 50 degrees F. Liquid nitrogen may be used as the carrier for 2,4-D. Premix in water (1 part 2,4-D to 4 parts water) and add mixture to nitrogen with strong agitation.
Oats Postemergence, Most winter annual broadleaf weeds			
dicamba, MOA 4 (Banvel) 4 SL (Clarity) 4 SL	0.25 pt 0.25 pt	0.125	Apply before jointing stage of growth. Risk of crop injury is least if applied after winter dormancy and before grain begins to joint. Better results will be obtained if applied when daytime temperatures are above 50 degreees F. Liquid nitrogen may be used as the carrier.
Oats Preharvest, Annual broadleaf weeds			
2,4-D amine, MOA 4 (various brands) 3.8 SL	1 to 2 pt	0.48 to 0.95	Apply when grain is in hard dough stage or later. Do not allow drift to sensitive crops, especially cotton and tobacco. Amine formulations strongly encouraged if sensitive crops are nearby, especially cotton or tobacco.
2,4-D ester, MOA 4 (various brands) 3.8 SL	1 to 2 pt		
Rye Preplant No-Till, Emerged annual broadleaf and grass weeds, control or suppression of perennial weeds			
glyphosate, MOA 9 (numerous brands and formulations)	See label	0.56 to 1.13 (lb a.e.)	Glyphosate is available as an isopropylamine salt and a potassium salt. Glyphosate formulations and application rates should be compared on the basis of pounds of glyphosate acid equivalent (a.e.) per gallon and per acre, respectively. Rate in the preceding column is expressed as a.e. See Table 7-10 for glyphosate rate conversions. Rate depends upon weed species and size; see labels. Apply before crop emergence. Adjuvant recommendations vary by glyphosate brand; see label of brand used for details. Select or Select Max may be mixed with glyphosate to control volunteer Roundup Ready corn. For corn up to 12 inches tall, apply 4 to 6 fluid ounces of Select or 6 fluid ounces of Select Max. For corn up to 24 inches, increase Select rate to 6 to 8 fluid ounces or Select Max rate to 9 fluid ounces. Select or Select Max must be applied at least 30 days ahead of planting rye.
Rye Postemergence, Most winter annual broadleaf weeds except chickweed, henbit, and knawel			
2,4-D amine, MOA 4 (various brands) 3.8 SL	1 pt	0.48	Apply after rye is fully tillered but before jointing. Spraying rye too young or after jointing can cause deformed heads, reduced yields, and uneven ripening. Better results are obtained when daytime temperatures are above 50°F. Increase the rate of 2,4-D by 50% to control corn cockle. Liquid nitrogen may be used as the carrier for 2,4-D. Ester formulations can be added directly to the nitrogen. If using amine formulation, premix in water (1 part 2,4-D to 4 parts water) and add mixture to nitrogen with strong agitation. Amine formulations give less burn than ester formulations in nitrogen.
2,4-D ester, MOA 4 (various brands) 3.8 SL	1 pt		
2,4-D ester, MOA 4 (various brands) 5.7 SL	0.67 pt		
Rye Preharvest, Annual broadleaf weeds			
2,4-D ester, MOA 4 (various brands) 3.8 SL	1 to 2 pt	0.48 to 0.95	Apply when grain is in hard dough stage or later. Do not allow drift to sensitive crops, especially cotton and tobacco. Amine formulations are strongly encouraged if sensitive crops are nearby, especially cotton or

CHEMICAL WEED CONTROL

285

Table 7-8A. Chemical Weed Control in Wheat, Barley, Oats, Rye, and Triticale

Herbicide, Mode of Action Code[1], and Formulation	Amount of Formulation per Acre	Pounds Active Ingredient per Acre	Precautions and Remarks
2,4-D ester, MOA 4 (various brands) 5.7 SL	0.67 to 1.3 pt		tobacco.
Triticale Preplant No-Till, Emerged annual broadleaf and grass weeds, control or suppression of perennial weeds			
glyphosate, MOA 9 (numerous brands and formulations)	See label	0.56 to 1.13 (lb a.e.)	Glyphosate is available as an isopropylamine salt and a potassium salt. Glyphosate formulations and application rates should be compared on the basis of pounds of glyphosate acid equivalent (a.e.) per gallon and per acre, respectively. Rate in the preceding column is expressed as a.e. See Table 7-10 for glyphosate rate conversions. Rate depends upon weed species and size; see labels. Apply before crop emergence. Adjuvant recommendations vary by glyphosate brand; see label of brand used for details. Select or Select Max may be mixed with glyphosate to control volunteer Roundup Ready corn. For corn up to 12 inches tall, apply 4 to 6 fluid ounces of Select or 6 fluid ounces of Select Max. For corn up to 24 inches, increase Select rate to 6 to 8 fluid ounces or Select Max rate to 9 fluid ounces. Select or Select Max must be applied at least 30 days ahead of planting triticale.
thifensulfuron, MOA 2 + tribenuron, MOA 2 (FirstShot SG with TotalSol) 50 WDG	0.5 to 0.8 oz	0.008 to 0.013 + 0.008 to 0.013	FirstShot does not control grasses. May be applied up to planting to control emerged broadleaf weeds. Add nonionic surfactant at 2 to 4 pints per 100 gallons spray solution or crop oil concentrate or methylated seed oil at 1 gallon per 100 gallons spray solution. In addition, add nitrogen fertilizer at a rate of 2 quarts per acre or ammonium sulfate at 2 pounds per acre. See label for tank mixtures.
Triticale Preemergence, Italian ryegrass and annual broadleaf weeds			
chlorsulfuron, MOA 2 + metsulfuron methyl, MOA 2 (Finesse) 75 WDG	0.5 oz	0.0195 + 0.0039	See comments under Wheat, Preemergence.
Triticale Postemergence, Wild garlic and annual broadleaf weeds			
thifensulfuron-methyl, MOA 2 + tribenuron-methyl, MOA 2 (Harmony Extra SG with TotalSol) 50 WDG	0.45 to 0.9 oz	0.0094 to 0.0188 + 0.0047 to 0.0094	Apply after two-leaf stage of triticale but before flag leaf is visible. See comments for Harmony Extra under Wheat-Postemergence.
Triticale Postemergence, Annual broadleaf weeds			
2,4-D, MOA 4 (Amine 4 2,4-D) 3.8 SL (Weedar 64) 3.8 L	1 pt	0.48	See comments for 2,4-D under Wheat- Postemergence.
dicamba, MOA 4 (Banvel) 4 SL (Clarity) 4 SL	0.25 oz	0.125	See comments for dicamba under Wheat- Postemergence.
halauxifen-methly, MOA 2 + florasulam, MOA 4 (Quelex) 20WG	0.75 oz	0.009	See comments for dicamba under Wheat- Postemergence.

[1] Mode of Action (MOA) code developed by the Weed Science Society of America. See Table 7-10, Herbicide Resistance Management, for details.

CHEMICAL WEED CONTROL

Weed Response to Herbicides in Small Grains

W. J. Everman, Crop Science Department

Ratings based upon average to good soil and weather conditions for herbicide performance and upon proper application rate, technique, and timing.

Table 7-8B. Weed Response to Herbicides in Small Grains

Species	Herbicide									
	Axial	Axiom	Banvel/Clarity	Finesse[2]	2,4-D	Harmony Extra	Hoelon	Osprey	PowerFlex[7]	Zidua
Annual bluegrass	N	G[4]	N	N	N	N	N	G	-	E[4]
Annual ryegrass	GE[6]	G[4]	N	F	N	N	E[1]	E[5]	E[5]	E[4]
Buttercup	N	-	F	G	G	G	N	-	-	-
Chickweed, common	N	G	G	G	P	G	N	FG	-	F[4]
Cornflower	N	-	FG	F	G	P	N	P	-	-
Curly dock	N	N	F	-	P	E	N	P	-	N
Cutleaf eveningprimrose	N	-	G	-	E	G	N	P	-	-
Field pennycress	N	-	F	G	G	G	N	-	-	-
Henbit	N	GE[4]	F	G	P	G	N	G	-	F[4]
Knawel	N	-	G	-	P	G	N	-	-	-
Shepherd's-purse	N	-	FG	G	GE	E	N	-	-	-
Swinecress	N	-	G	-	G	E	N	E	-	-
Vetch	N	-	E	-	G	P	N	N	-	-
Virginia pepperweed	N	-	F	-	E	G	N	-	-	-
Wild garlic	N	N	F	P	F	E	N	P	-	-
Wild mustard	N	G[4]	F	G	GE	G	N	E[3]	GE	P[4]
Wild radish	N	G[4]	F	G	GE	G	N	E[3]	GE	P[4]

[1] A biotype of ryegrass resistant to Hoelon is common in North Carolina, especially in the piedmont. If resistance is suspected, do not apply Hoelon.

[2] Applied preemergence.

[3] Rating assumes mustard or radish is 1 to 2 inches Osprey is less effective on larger plants.

[4] Assumes adequate rainfall for activation prior to weed emergence.

[5] A biotype of ryegrass resistant to Osprey and PowerFlex has been found in the southern piedmont.

[6] May not control Hoelon-resistant ryegrass. See comments under "Wheat-Postemergence."

[7] Inadequate research has been conducted in North Carolina to determine response of most broadleaf weeds to PowerFlex. The label claims control of a number of broadleaf species, including Carolina geranium, chickweed, hairy bittercress, field pennycress, sheperdspurse, buttercup, Virginia pepperweed, and vetch.

Key:
E = excellent control, 90% or better
G = good control, 80% to 90%
F = fair control, 50% to 80%
P = poor control, 25% to 50%
N = no control, less than 25%

CHEMICAL WEED CONTROL

Glyphosate Formulations

W. J. Everman, Crop Science Department

Table 7-9. Glyphosate Formulations

Brand Names[1]	Formulation	Active Ingredient (lb formulated salt/gal)	Acid Equivalent	Equivalent Rates	
				lb a.e./acre	fl oz product/acre
Abundit Extra AgSaver Glyphosate 41% Plus Alecto 41-S Alecto 41HL Alligare Glyphosate 4 Plus Buccaneer Buccaneer Plus Credit 41 Credit 41 Extra Crop$mart Glyphosate 41% Extra Envy Four Power Plus Glyfine Plus Glyfos Glyfos X-TRA Glyphogan Plus Gly Star Original Gly Star Gold Gly Star Plus Helosate Plus Advanced Honcho Honcho Plus Imitator Plus Mad Dog Mad Dog Plus Meychem 41% Makaze Makaze Yield Pro	Isopropylamine salt	4	3	0.375 0.56 0.75 1.125	16 24 32 48
Buccaneer 5 Extra Credit Plus	Isopropylamine salt	5.0	3.75	0.375 0.56 0.75 1.125	12.8 19.1 25.6 38.4
Buccaneer 5 Extra Glyfine 5 Plus	Isopropylamine salt	5.4	4	0.375 0.56 0.75 1.125	12 18 24 36
Roundup POWERMAX Roundup POWERMAX II Roundup WEATHERMAX	Potassium salt	5.5	4.5	0.375 0.56 0.75 1.125	10.7 15.9 21.3 32
Touchdown HiTech	Potassium salt	6	5	0.375 0.56 0.75 1.125	9.6 14.3 19.2 28.8
Touchdown Total Traxion	Potassium salt	5	4.17	0.375 0.56 0.75 1.125	11.5 17.2 23 34.5
Showdown	Isopropylamine salt + Monoammonium salt	3.64 + 0.33	3	0.375 0.56 0.75 1.125	16 24 32 48
Nufarm Credit Nufarm Credit Extra	Ammonium salt + Potassium salt	1.8 + 1.6	3	0.375 0.56 0.75 1.125	16 24 32 48
Credit Xtreme	Isopropylamine salt + Potassium salt	3.33 + 2.5	4.5	0.375 0.56 0.75 1.125	10.7 15.9 21.3 32

[1] Other brands may be available. See comments on resistance management in Table 7-10. Warning: Some generic brands of glyphosate are of inferior quality. They may be somewhat less effective on weeds, and they may cause injury to Roundup Ready crops, especially cotton.

CHEMICAL WEED CONTROL

Herbicide Resistance Management

W. J. Everman and F. H. Yelverton, Crop Science Department

Herbicide resistance is becoming a serious problem in North Carolina and across the country. Herbicide resistance is not a new phenomenon in North Carolina. Goosegrass resistant to dinitroaniline herbicides was first reported in North Carolina in the 1970s. Since then, smooth pigweed and common lambsquarters resistant to triazines, cocklebur resistant to organoarsenicals and ALS inhibitors, Palmer amaranth resistant to ALS inhibitors, Italian ryegrass resistant to ACCase inhibitors and ALS inhibitors, and common ragweed resistant to ALS inhibitors have been observed. Of greatest concern is weed resistance to glyphosate. Horseweed (2003), Palmer amaranth (2005), common ragweed (2006), and Italian ryegrass (2009). resistant to glyphosate have been found. Resistance of common lambsquarters to glyphosate is suspected. Weed resistance to glyphosate is a highly significant concern in light of the extensive use of glyphosate for burndown in conservation tillage systems and in Roundup Ready cotton, soybeans, and corn.

Crop rotation, along with appropriate herbicide rotation, should be employed to the extent possible. Additionally, cultivation, where feasible, can be very helpful in herbicide resistance management. However, the most important component of a resistance management strategy is rotation of herbicide modes of action and use of multiple herbicide modes of action within each crop.

Mode of action describes the process whereby a herbicide kills susceptible plants. Table 7-10 lists the mode of action, along with the chemical family, of all the herbicides likely to be used on agronomic and horticultural crops in North Carolina. Each herbicide mode of action is assigned a numerical code for ease of use. Wherever possible, at least two modes of action should be used within each crop. This can be accomplished by preemergence herbicide applications followed by postemergence applications and by tank mixtures of herbicides with two or more modes of action. Also, within a rotation, one should try to avoid dependance on herbicides with the same mode of action in all crops in the rotation. For example, in a corn and soybean rotation, it is best not to use an ALS inhibitor (#2) in each crop. Alternatively, if an ALS inhibitor is used in each crop, herbicides with other modes of action should also be included. Similarly, it would be best to not rely exclusively on glyphosate in both crops.

In Roundup Ready corn and soybeans, it is recommended that glyphosate plus herbicides with at least one other mode of action be used. In Roundup Ready cotton, it is recommended that at least two modes of actions, in addition to glyphosate, be used.

Table 7-10A. Herbicide Modes of Action

Brand Names	Active Ingredient(s)	Chemical Family	Mode of Action[1]
AAtrex	atrazine	triazine	5
Accent Q	nicosulfuron	sulfonylurea	2
Acumen	pendimethalin	dinitroalinine	3
Aim	carfentrazone	triazolinone	14
Alachlor	alachlor	chloroacetamide	15
Alanap	naptalam	phthalamate simicarbazone	19
Armezon	topramazone	triketone	27
Arrow	clethodim	cyclohexanedione	1
Atrazine	atrazine	triazine	5
Assure II	quizalofop	aryloxyphenoxy-propionate	1
Authority Assist	sulfentrazone + imazethapyr	triazolinone + imidazdinone	14 + 2
Authority First	sulfentrazone + cloransulam	triazolinone + triazolopyrimidine	14 + 2
Authority MTZ	sulfentrazone + metribuzin	triazolinone + triazinone	14 + 5
Axial	pinoxaden	phenylpyrazoline	1
Axiom	flufenacet + metribuzin	oxyacetamide + triazinone	15 + 5
Balance FLEXX	Isoxaflutole	Isoxazole	27
Banvel	dicamba	benzoic acid	4
Banvel-K + Atrazine	dicamba + atrazine	benzoic acid + triazine	4 + 5
Basagran	bentazon	benzothiadiazinone	6
Beyond	imazamox	imidazolinone	2
Bicep II Magnum	s-metolachlor + atrazine	chloroacetamide + triazine	15 + 5
Blazer	acifluorfen	diphenylether	14
Boundary	s-metolachlor + metribuzin	chloroacetamide + triazinone	15 + 5
Brawl, Brawl II	s-metolachlor	chloroacetamide	15
Brawl II ATZ	s-metolachlor + atrazine	chloroacetamide + triazine	15 + 5
Breakfree ATZ	s-metolachlor + atrazine	chloroacetamide + triazine	15 + 5
Break-Up	pronamide	benzamide	3
Buctril	bromoxynil	nitrile	6
Bullet	alachlor + atrazine	chloroacetamide + triazine	15 + 5
Butoxone	2,4-DB	phenoxy-carboxylic acid	4
Butyrac	2,4-DB	phenoxy-carboxylic acid	4
Cadre	imazapic	imidazolinone	2
Callisto	mesotrione	triketone	27
Camix	mesotrione + s-metolachlor	triketone + chloroacetamide	27 + 15
Canopy	metribuzin + chlorimuron	triazine + sulfonylurea	5 + 2

Table 7-10A. Herbicide Modes of Action

Brand Names	Active Ingredient(s)	Chemical Family	Mode of Action[1]
Canopy EX	chlorimuron + tribenuron	sulfonylurea + sulfonylurea	2 + 2
Canopy XL	sulfentrazone + chlorimuron	diphenylether + sulfonylurea	14 + 2
Caparol	prometryn	triazine	5
Capreno	tembotrione + thiencarbazone	benzoyl pyrazole + triazolone	27 + 2
Celebrity, Celebrity Plus	nicosulfuron + dicamba	sulfonylurea + benzoic acid	2 + 4
Charger Basic	s-metolachlor	chloroacetamide	15
Cinch	s-metolachlor	chloroacetamide	15
Cinch ATZ	s-metolachlor + atrazine	chloroacetamide + triazine	15 + 5
Clarity	dicamba	benzoic acid	4
Classic	chlorimuron	sulfonylurea	2
Clethodim	clethodim	cyclohexanedione	1
Clopyr AG	clopyralid	pyridine carboxylic acid	4
Cobra	lactofen	diphenylether	14
Command	clomazone	isoxazolidinone	13
Confidence	acetochlor	chloroacetamide	15
Confidence Xtra	acetochlor + atrazine	chloroacetamide + triazine	15 + 5
Corvus	Isoxaflutole + thiencarbazone-methyl	isoxazole + triazolone	27 + 2
Cotoran	fluometuron	urea	7
Cotton-Pro	prometryne	triazine	5
Curbit	ethalfluralin	dinitroaniline	3
Dacthal	DCPA	benzoic acid	3
Dawn	fomesafen	diphenyl ether	14
Define	flufenacet	oxyacetamide	15
Degree	acetochlor	chloroacetamide	15
Degree Xtra	acetochlor + atrazine	chloroacetamide + triazine	15 + 5
Devrinol	napropamide	acetamide	15
Diablo	dicamba	benzoic acid	4
Dicamba DMA Salt	dicamba	benzoic acid	4
Direx	diuron	urea	7
Distinct	dicamba + diflufenzopyr	benzoic acid + semicarbazone	4 + 19
Diuron	diuron	urea	7
Double Team	acetochlor + atrazine	chloroacetamide + triazine	15 + 5
DSMA, numerous brands	DSMA	organoarsenical	17
Dual II, Dual II Magnum	s-metolachlor	chloroacetamide	15
Envive	flumioxazim + chlorimuron + thifensulfuron	n-phenylphthalimide + triazolopyrimidine + sulfonylurea	14 + 2 + 2
Envoke	trifloxysulfuron	sulfonylurea	2
Eptam	EPTC	thiocarbamate	8
Equip	formasulfuron + iodosulfuron	sulfonylurea	2 + 2
Eradicane	EPTC	thiocarbamate	8
Establish	dimethenamid	chloroacetamide	15
Establish ATZ	dimethenamid + atrazine	chloroacetamide + triazine	15 + 5
ET	pyraflufen ethyl	phenylpyrazole	14
Evik	ametryne	triazine	5
Expert	glyphosate + s-metolachlor + atrazine	glycine + chloroacetamide + triazine	9 + 15 + 5
Express	tribenuron	sulfonylurea	2
Extreme	glyphosate + imazethapyr	glycine + imidazolinone	9 + 2
Fierce	flumioxazin + pyroxasulfone	n phenylphthalimide + chloroacetamide	14 + 15
Finesse	chlorsulfuron + metsulfuron	sulfonylurea + sulfonylurea	2 + 2
Firestorm	paraquat	bipyridylium	22
Firstrate	cloransulam	triazolopyrimidine	2
Flexstar	fomesafen	diphenylether	14
Fultime	acetochlor + atrazine	chloroacetamide + triazine	15 + 5
Fusilade DX	fluazifop	aryloxyphenoxy-propionate	1
Fusion	fluazifop + fenoxaprop	aryloxyphenoxy-propionate + aryloxyphenoxy-propionate	1 + 1
Galigan	oxyfluorfen	diphenylether	14

Table 7-10A. Herbicide Modes of Action

Brand Names	Active Ingredient(s)	Chemical Family	Mode of Action[1]
Gangster	flumioxazin + cloransulam	n-phenylphthalimide + triazolopyrimidine	14 + 2
Guardsman Max	dimethenamide + atrazine	chloroacetamide + triazine	15 + 5
Glyphosate (numerous brands)	glyphosate	glycine	9
Goal	oxyflurofen	diphenylether	14
Goal Tender	oxyfluorfen	diphenylether	14
Gramoxone (Inteon)	paraquat	bipyridylium	22
Halex GT	s-metolachlor + glyphosate + mesotrione	chloroacetamide + glycine + triketone	15 + 9 + 27
Harmony Extra	thifensulfuron + tribenuron	sulfonylurea + sulfonylurea	2 + 2
Harmony GT, Harmony SG	thifensulfuron	sulfonylurea	2
Harness	acetochlor	chloroacetamide	15
Harness Xtra	acetochlor + atrazine	chloroacetamide + triazine	15 + 5
Hoelon	diclofop	aryloxyphenoxy-propionate	1
Ignite, Ignite 280	glufosinate	phosphinic acid	10
Impact	topramazone	triketone	27
Intrro	alachlor	chloroacetamide	15
Karmex	diuron	urea	7
Kerb	pronamide	benzamide	3
Keystone NXT	acetochlor + atrazine	chloroacetamide + triazine	15 + 5
Lariat	alachlor + atrazine	chloroacetamide + triazine	15 + 5
Laudis	tembotrione	benzoyl pyrazole	27
Layby Pro	diuron + linuron	urea + urea	7 + 7
Leadoff	rimsulfuron + thifensulfuron	sulfonylurea	2 + 2
Lexar	mesotrione + s-metolachlor + atrazine	triketone + chloroacetamide + triazine	27+15+5
Liberty	glufosinate	phosphinic acid	10
Liberty ATZ	glufosinate + atrazine	phosphinic acid + triazine	10 + 5
Lightning	imazethapyr + imazapyr	imidazolinone + imidazolinone	2 + 2
Linex	linuron	urea	7
Lorox	linuron	urea	7
Lumax	mesotrione + s-metolachlor + atrazine	triketone + chloroacetamide + atrazine	27+15+5
Marksman	dicamba + atrazine	benzoic acid + triazine	4 + 5
Matrix	rimsulfuron	sulfonylurea	2
Medal, Medal II	s-metolachlor	chloroacetamide	15
Medal II AT	s-metolachlor + atrazine	chloroacetamide + triazine	15 + 5
Me-Too-Lachlor	metolachlor	chloroacetamide	15
Metribuzin	metribuzin	triazinone	5
Metri DF	metribuzin	triazinone	5
Micro-Tech	alachlor	chloroacetamide	15
Moxy	bromoxynil	nitrile	6
MSMA (numerous brands)	MSMA	organoarsenical	17
OpTILL	imazethapyr + saflufenacil	imidazolinone + pyrimidinedione	2 + 14
Option	foramsulfuron	sulfonylurea	2
Osprey	mesosulfuron	sulfonylurea	2
Outlook	dimethenamid	chloroacetamide	15
OxiFlo	oxyfluorfen	diphenylether	14
Parallel, Parallel PCS	metolachlor	chloroacetamide	15
Parallel Plus	metolachlor + atrazine	chloroacetamide + atrazine	15 + 5
Parazone	paraquat	bipyridylium	22
Parrlay	metolachlor	chloroacetamide	15
Peak	prosulfuron	sulfonylurea	2
Pendant	pendimethalin	dinitroaniline	3
Pendimax	pendimethalin	dinitroaniline	3
Permit	halosulfuron	sulfonylurea	2
Poast, Poast Plus	sethoxydim	cyclohexanedione	1
PowerFlex	pyrosulam	triazolopyrimidine	2
Prefar	bensulide	phosphorodithioate	8

CHEMICAL WEED CONTROL

Table 7-10A. Herbicide Modes of Action

Brand Names	Active Ingredient(s)	Chemical Family	Mode of Action[1]
Prefix	s-metolachlor + fomesafen	chloroacetamide + diphenylether	15 + 14
Princep	simazine	triazine	5
Prometryn	prometryn	triazine	5
Prowl, Prowl H$_2$O	pendimethalin	dinitroaniline	3
Pursuit	imazethapyr	imidazolinone	2
Pyramin	pyrazon	pyridazinone	6
Python	flumetsulam	triazolopyrimidine	2
Raptor	imazamox	imidazolinone	2
Realm Q	rimsulfuron + mesotrione	sulfonylurea + triketone	2 + 27
Reflex	fomesafen	diphenylether	14
Resicore	acetochlor + mesotrione + clopyralid	chloroacetamide + triketone + pyridine carboxylic acid	15 + 27 + 4
Resolve	rimsulfuron	sulfonylurea	2
Resolve Q	rimsulfuron + thifensulfuron	sulfonylurea + sulfonylurea	2 + 2
Resource	flumiclorac-pentyl	n-phenylphthalimide	14
Ro-Neet	cycloate	thiocarbamate	8
Roundup (and other brands)	glyphosate	glycine	9
Rythm	fomesafen	diphenyl ether	14
Sandea	halosulfuron	sulfonylurea	2
Scepter	imazaquin	imidazolinone	2
Select, Select Max	clethodim	cyclohexanedione	1
Sencor	metribuzin	triazinone	5
Sequence	glyphosate + s-metolachlor	glycine + chloroacetamide	9 + 15
Sharpen	saflufenacil	pyrimidinedione	14
Simazine	simazine	triazine	5
Sim-Trol	simazine	triazine	5
Sinbar	terbacil	uracil	5
Sonalan	ethalfluralin	dinitroaniline	3
Sonic	sulfentrazone + cloransulam	triazolinone + triazolopyrimidine	14 + 2
Spartan	sulfentrazone	triazolinone	14
Spartan Charge	sulfentrazone + carfentrazone	triazolinone + triazolinone	14 + 14
Spin-Aid	phenmedipham	phenyl-carbamate	6
Squadron	imazaquin + pendimethalin	imidazolinone + dinitroaniline	2 + 3
Stalwart, Stalwart C	metolachlor	chloroacetamide	15
Stalwart Xtra	metolachlor + atrazine	chloroacetamide + triazine	15 + 5
Staple	pyrithiobac	pyrimidinyl(thio)benzoate	2
Status	dicamba + diflufenzopyr	benzoic acid + semicarbazone	4 + 19
Steadfast Q	nicosulfuron + rimsulfuron	sulfonylurea + sulfonylurea	2 + 2
Steadfast ATZ	nicosulfuron + rimsulfuron + atrazine	sulfonylurea + sulfonylurea + triazine	2 + 2 + 5
Stealth	pendimethalin	dinitroaniline	3
Sterling	dicamba	benzoic acid	4
Stinger	clopyralid	pyridine carboxylic acid	4
Storm	acifluorfen + bentazon	diphenylether + benzothiadiazinone	14 + 6
Strategy	ethalfluralin + clomazone	dinitroaniline + isoxazolidinone	3 + 13
Strongarm	diclosulam	triazolopyrimidine	2
Suprend	prometryn + trifloxysulfuron	triazine + sulfonylurea	5 + 2
Sutan+	butylate	thiocarbamate	8
SureStart	acetochlor + flumetsulam + clopyralid	chloroacetamide + triazolopyrimidine + pyridine carboxylic acid	15 + 2 + 4
Surpass	acetochlor	chloroacetamide	15
Synchrony XP	chlorimuron + thifensulfuron	sulfonylurea + sulfonylurea	2 + 2
Targa	quizalofop	aryloxyphenoxy-propionate	1
Tillam	pebulate	thiocarbamate	8
TopNotch	acetochlor	chloroacetamide	15
Treflan	trifluralin	dinitroaniline	3
Triangle	metolachlor + atrazine	chloroacetamide + triazine	15 + 5
Trifluralin	trifluralin	dinitroaniline	3

Table 7-10A. Herbicide Modes of Action

Brand Names	Active Ingredient(s)	Chemical Family	Mode of Action[1]
Trigger	clethodim	cyclohexanedione	1
Trilin	trifluralin	dinitroaniline	3
Trizmet II	atrazine + metolachlor	triazine + chloroacetamide	5 + 15
Trust	trifluralin	dinitroaniline	3
Ultra Blazer	acifluorfen	diphenylether	14
Valor SX	flumioxazin	n-phenylphthalimide	14
Valor XLT	flumioxazin + chlorimuron	n-phenylphthalimide + sulfonylurea	14 + 2
Vision	dicamba	benzoic acid	4
Volley	acetochlor	chloroacetamide	15
Volley ATZ	acetochlor + atrazine	chloroacetamide + triazine	15 + 5
Volunteer	clethodim	cyclohexanedione	1
Warrant	acetachlor	chloroacetamide	15
Weedmaster	2,4-D + dicamba	phenoxy-carboxylic acid + benzoic acid	4 + 4
Yukon	halosulfuron + dicamba	sulfonylurea + benzoic acid	2 + 4
Zidua	pyroxasulfone	chloroacetamide	14
2,4-D (numerous brands)	2,4-D	phenoxy-carboxylic acid	4
2,4-DB (numerous brands)	2,4-DB	phenoxy-carboxylic acid	4

Mode of Action Code Key:
1   ACCase inhibition
2   ALS inhibition
3   Microtubule assembly inhibition
4   Synthetic auxin
5   Photosystem II inhibition, different binding behavior than groups 6 and 7
6   Photosystem II inhibition, different binding behavior than groups 5 and 7
7   Photosystem II inhibition, different binding behavior than groups 5 and 6
8   Inhibition of lipid synthesis - not ACCase inhibition
9   EPSP synthase inhibition
10 Glutamine synthase inhibition
12 Inhibition of carotenoid biosynthesis at PDS
13 Inhibition of carotenoid biosynthesis, unknown target
14 PPO inhibition
15 Inhibition of very long-chain fatty acids
17 Unknown mode of action
19 Auxin transport inhibition
22 Photosystem I electron diversion
27 Inhibition of HPPD

CHEMICAL WEED CONTROL

Herbicide Modes of Action for Hay Crops, Pastures, Lawns and Turf

Table 7-10B. Herbicide Modes of Action for Hay Crops, Pastures, and Lawns and Turf

Brands listed were registered for sale and use in North Carolina in 2015 according to http://www.kellysolutions.com/nc/searchbychem.asp. Active ingredients enclosed within parentheses are prepackaged herbicides.

Active Ingredient(s)	Brand Names	Chemical Family	Mode of Action[1]
(2, 4-D + mecoprop + dichlorprop)	Spoiler, Triamine	phenoxycarboxylic acid + phenoxyalkanoic acid + chlorinated phenoxy	4 + 4 + 4
2,4-D amine	(4 SL): various trade names	phenoxy-carboxylic acid	4
(2,4-D + aminopyralid)	GrazonNext HL	phenoxy-carboxylic acid + pyridinecarboxylic acid	4 + 4
(2,4-D ester + carfentrazone-ethyl)	Rage D-Tech	phenoxy-carboxylic acid + triazolinone	4 + 14
(2,4-D + clopyralid + dicamba)	Millennium Ultra 2	phenoxycarboxylic acid + pyridinecarboxylic acid + benzoic acid	4 + 4 + 4
(2,4-D amine + cloypyralid)	Curtail	phenoxy-carboxylic acid + pyridinecarboxylic acid	4 + 4
(2,4-D + dicamba)	Brash, Range Star, Rifle-D, Weedmaster	phenoxy-carboxylic acid + benzoic acid	4 + 4
(2,4-D + fluroxypyr + dicamba)	E-2, Escalade 2	phenoxy + pyridinecarboxylic acid + benzoic acid	4 + 4 + 4

Table 7-10B. Herbicide Modes of Action for Hay Crops, Pastures, and Lawns and Turf

Brands listed were registered for sale and use in North Carolina in 2015 according to http://www.kellysolutions.com/nc/searchbychem.asp. Active ingredients enclosed within parentheses are prepackaged herbicides.

Active Ingredient(s)	Brand Names	Chemical Family	Mode of Action[1]
(2,4-D + glyphosate)	Campaign	phenoxy-carboxylic acid + glycine	4 + 9
(2,4-D + mecoprop + dicamba)	3-D, Three-Way Selective Herbicide, Triplet SF, various Trimec formulations	phenoxycarboxylic acid + phenoxyalkanoic acid + benzoic acid	4 + 4 + 4
(2,4-D + picloram)	Grazon P+D, Troooper P+D	phenoxy-carboxylic acid + pyridinecarboxylic acid	4 + 4
(2,4-D + triclopyr)	Candor, Crossbow	phenoxy-carboxylic acid + pyridinecarboxylic acid	4 + 4
2,4-DB	(1.75 EC): 2,4-DB 175 (2 EC): 2,4-DB 200, Butyrac 200	phenoxy-carboxylic acid	4
amicarbazone	Xonerate	triazolinone	5
aminopyralid	Milestone	pyradinecarboxylic acid	4
(aminopyralid + metsulfuron methyl)	Chaparral	pyradinecarboxylic acid + sulfonylurea	4 + 2
asulam	Asulox	carbamate	18
atrazine	(4 L): AAtrex 4 L, Atrazine 4 L (90 DF, 90 WDG): Aatrex Nine-O, Atrazine 90 WDG, Atrazine 90 DF	triazine	5
benefin	Lebanon Balan 2.5 G, The Andersons Crabgrass Preventer with 2.5 Balan	dinitroaniline	3
(benefin + trifluralin)	Fertilizer with Team Pro 0.86%	dinitroaniline + dinitroaniline	3 + 3
bensulide	(4 EC): Bensumec 4 LF (8.5 G): Weedgrass Preventer (12.5 G): Pre-San Granular	organophosphorus	8
(bensulide + oxadiazon)	Goosegrass / Crabgrass Control	organophosphorus + oxadiazole	8 + 14
bentazon	Basagran Sedge Control, Basagran T/O, Bentazon 4, Lescogran	benzothiadiazole	6
bispyribac-sodium	Velocity	pyrimidinyloxybenzoic acid	2
bromoxynil	(2 EC): Broclean, Maestro 2 EC (4 EC): Buctril 4 EC	nitrile	6
carfentrazone-ethyl	Aim EC, QuickSilver T&O	triazinone	14
(carfentrazone + 2,4-D ester + mecoprop + dicamba)	(2.2 EC): Speed Zone (0.81 EC): Speed Zone Southern	triazinone + phenoxycarboxylic acid + phenoxyalkanoic acid + benzoic acid	14 + 4 + 4 + 4
(carfentrazone + MCPA + mecoprop + dicamba)	Power Zone	triazinone + phenoxy + phenoxyalkanoic acid + benzoic acid	14 + 4 + 4 + 4
(carfentrazone + quinclorac)	SquareOne	triazinone + quinoline carboxylic acid	14 + (27 + 4)
chlorsulfuron	Alligare Chlorsulfuron 75, Telar XP	sulfonylurea	2
clethodim	(2 EC): Arrow 2 EC, Avatar S2, Cleanse 2 EC, Clethodim 2 LC, Clethodim 2 E, Dakota, Section 2 EC, Shadow, Tide Clethodim 2 EC, Select 2 EC, Volunteer, Willowood Clethodim 2 EC (0.97 EC): Envoy Plus, TapOut	cyclohexanedione	1
cloypyralid	Lontrel T&O	pyridinecarboxylic acid	4
dicamba / diglycolamine	Banvel, Clarity, Clash, Detonate, Rifle, Strut, Topeka, Vanquish	benzoic acid	4
diclofop-methyl	Illoxan	aryloxyphenoxy propionate	1
(diflufenzopyr-sodium + dicamba)	Overdrive	semicarbazone + benzoic acid	19 + 4
dimethenamid	Tower	chloroacetamide	15
(dimethenamid + pendimethalin)	Freehand	chloroacetamide + dinitroaniline	15 + 3
diquat	Diquash, Diquat., Diquat SPC, Harvester, Priceto Diquat 2 L, Reward LS, Tribune	bipyridilium	22
diruon	Direx 4 L, Diuron 4 L	phenylurea	7
dithiopyr	(2 EW, 2 L): Armortech CGC 2 L, Dimension 2 EW, Dithiopyr 2 L (40 WP): Armortech CGC 40, Dimension Ultra, Dithiopyr 40 WSP	pyridine	4
EPTC	Eptam 7-E	thiocarbamate	8
ethofumesate	(1.5 EC): Progress (4 EC): Phoenix Thrasher, PoaConstrictor, Prograss SC	benzofuranes	8
fenoxaprop	Acclaim Extra	aryloxyphenoxy propionate	1
flazasulfuron	Katana	sulfonylurea	2
florasulam	Defendor	triazolopyrimidine	2
fluazifop	Fusilade II	aryloxyphenoxy propionate	1
flumioxazin	Sureguard	N-phenylphthalimide	14
foramsulfuron	Revolver	sulfonylurea	2
glufosinate	Finale	organophosphorus	10
glyphosate – Roundup formulations	(4 SL, 5 SL, 5.4 SL, 5.5 SL): various trade names	glycine	9

Table 7-10B. Herbicide Modes of Action for Hay Crops, Pastures, and Lawns and Turf

Brands listed were registered for sale and use in North Carolina in 2015 according to http://www.kellysolutions.com/nc/searchbychem.asp. Active ingredients enclosed within parentheses are prepackaged herbicides.

Active Ingredient(s)	Brand Names	Chemical Family	Mode of Action[1]
glyphosate – Touchdown formulations	(3 SL): Touchdown Pro (4.17 SL): Touchdown Total (5 SL): Touchdown HiTech		
(glyphosate + diquat)	Razor Burn, Roundup QuikPRO	glycine + bipyridilium	9 + 22
halosulfuron	HiYield Nutsedge Control, Nutgrass Killer II, Profine 75, Sandea	sulfonylurea	2
imazapic	Imazapic 2 SL, Impose, Panoramic, Plateau	imadazolinone	2
imazaquin	Image 70 DG	imidazolinone	2
imazethapyr	Pursuit, Slay, Thunder	Imidazolinone	2
imazosulfuron	Celero	sulfonylurea	2
indaziflam	(20 WSP): Specticle 20 WSP (0.0224 G): Specticle G (0.622F): Specticle Flo	benzamide	21
(indaziflam + diquat bromide + glyphosate)	Specticle Total	benzamide + bipyridilium + glycine	21 + 22 + 9
isoxaben	Gallery, Isoxaben	benzamide	21
(MCPA + mecoprop + dicamba)	Ortho Weed B Gon Pro Southern, Tri-Power	mcpa + phenoxyalkanoic acid + benzoic acid	4 + 4 + 4
(MCPA amine + fluroxypyr ester + dicamba)	Change Up	phenoxy + pyridinecarboxylic acid + benzoic acid	4 + 4 + 4
(MCPA amine + fluroxypyr ester + triclopyr amine)	Battleship III	phenoxy + pyridinecarboxylic acid + pyridinecarboxylic acid	4 + 4 + 4
(MCPA amine + triclopyr amine + dicamba)	Horsepower, Lesco Eliminate	phenoxy + pyridinecarboxylic acid + benzoic acid	4 + 4 + 4
(MCPA ester + triclopyr ester + dicamba)	Cool Power, Lesco Three-Way Ester II, Monterey Spurge Power	phenoxy + pyridinecarboxylic acid + benzoic acid	4 + 4 + 4
mecoprop	MCPP-p4 Amine, Mecomec 2.5, Mecomec 4	phenoxyalkanoic acid	4
mesotrione	Tenacity	benzoylcyclohexanedione	27
metolachlor	Pennant Magnum	chloroacetamide	15
metribuzin	(75 DF): Dimetric DF, Glory, Metribuzin 75, Tricor DF 4 L, 4 F): Glory 4L, Tricor 4 F	triazinone	5
metsulfuron methyl	Accurate, Amtide MSM 60 DF, Amtide MSM Turf, Manor, MSM 60 DF, MSM Turf, Ortho Weed B Gon Pro St. Augustine, Plotter, Purestand, Rometsol	sulfonylurea	2
(metsulfuron + 2,4-D + dicamba)	Cimarron Max	sulfonylurea + phenoxy-carboxylic acid + benzoic acid	2 + 4 + 4
(metsulfuron + chlorsulfuron)	Chisom, Cimarron Plus	sulfonylurea + sulfonylurea	2 + 2
(metsulfuron methyl + rimsulfuron)	Negate	sulfonylurea + sulfonylurea	2 + 2
monosodium methylarsonate	(6 SL): Target 6 Plus, Drexel MSMA 6 Plus (6.6 SL): Target 6.6 Plus, Drexel MSMA 6.6 Plus	organic arsenical	17
napropamide	Devrinol 50 DF	acetamide	15
(nicosulfuron + metsulfuron methyl)	Pastora	sulfonylurea + sulfonylurea	2 + 2
oryzalin	(4 AS, 4 L): Monterey Weed Impede 4 AS, Oryzalin 4 AS, Phoenix Harrier 4 L, Surflan AS (85WDG): Surflan WDG	dinitroaniline	3
oxadiazon	(2 G): Oxadiazon 2 G, Ronstar G (50 WP): Oxadiazon 50 WSB, Ronstar 50 WSB (3.17 SC): Oxadiazon SC, Phoenix Starfighter L, Ronstar Flo	oxadiazole	14
(oxadiazon + prodiamine)	Pro-mate Ronstar + Barricade 1.2 G, Regalstar II	oxadiazole + dinitroaniline	14 + 3
paraquat	(2 SL): Cyclone SL 2.0, Gramoxone Inteon, Gramoxone SL (3 SL): Bonedry, Firestorm, Helmquat 3 SL, Parazone 3 SL	bipyridilium	22
pendimethalin	(3.8 CS): HydroCap, Pendulum AquaCap, Pre-M AquaCap, Prowl H2O, Satellite HydroCap (3.3 EC): Drexel Pin-Dee 3.3 T&O (2 G): Pendulum 2 G (0.86 G): fertilizers – Pendimethalin, Pre-M, Propendi, Pro-mate Pendi (1.29 G): Step 1 Crabgrass Preventer, Turf Builder with Halts	dinitroaniline	3
penoxsulam	(0.31 L): Sapphire (0.03 G): Harrels Fert. with Penoxsulam, lesco LockUp, Pennington Seed LockUp (014 G): Harrels Fert. with Penoxsulam	triazolopyrimidine	2
(picloram + fluroxypyr)	Surmount	pyradinecarboxylic acid + pyradinyloxyacetic acid	4 + 4
prodiamine	(65 WDG): Armortech Kade, Barricade, Calvacade, Halts Pro, Phoenix Knighthawk, ProClipse, Prodiamine, Quali-Pro Prodiamine, RegalKade, Resolute, Stonewall (4 FL): Barricade, Evade, Resolute (0.5 G): RegalKade, Signature Crabgrass Preventer, Turf Pride	dinitroaniline	3
pronamide	(50 WP): Kerb 50-W (3.3 SC): Kerb SC T&O, Willowood Pronamide 3.3SC	benzamide	3
pyraflufen ethyl	Octane 2% SC	phenylpyrazole	14

CHEMICAL WEED CONTROL

Table 7-10B. Herbicide Modes of Action for Hay Crops, Pastures, and Lawns and Turf

Brands listed were registered for sale and use in North Carolina in 2015 according to http://www.kellysolutions.com/nc/searchbychem.asp. Active ingredients enclosed within parentheses are prepackaged herbicides.

Active Ingredient(s)	Brand Names	Chemical Family	Mode of Action[1]
quinclorac	(75 DF): Armortech Quinclorac Pro, Quinclorac, Quinclorac SPC, Quinstar (1.5 SL): Drive XLR8, Quinclorac 1.5 L	quinoline carboxylic acid	(27 + 4)
(quinclorac + mecoprop + dicamba)	Onetime	quinoline carboxylic acid + phenoxyalkanoic acid + benzoic acid	(27 + 4) + 4 + 4
(quinclorac + sulfentrazone + 2,4-D amine + dicamba)	Q4 Plus	quinoline carboxylic acid + triazinone + phenoxy-carboxylic acid + benzoic acid	(27 + 4) + 14 + 4 + 4
rimsulfuron	Rimsulfuron	sulfonylurea	2
sethoxydim	(1.5 EC): Poast (1 EC): Poast Plus, Segment, Sethoxydim SPC	cyclohexanedione	1
siduron	Tupersan	phenylurea	7
simazine	(4 L): Princep Liquid, Simazine, Sim-Trol (90 DF, 90 WDG): Simazine, Sim-Trol, Princep Caliber 90	triazine	5
sulfentrazone	Dismiss Turf	triazinone	14
(sulfentrazone + 2,4-D + mecoprop + dicamba)	Surge	triazinone + phenoxycarboxylic acid + phenoxyalkanoic acid + benzoic acid	14 + 4 + 4 + 4
(sulfentrazone + imazethapyr)	Dismiss South	triazinone + imidazolinone	14 + 2
(sulfentrazone + metsulfuron-methyl)	Blindside	triazinone + sulfonylurea	14 + 2
(sulfentrazone + prodiamine)	Echelon	triazinone + dinitroaniline	14 + 3
(sulfentrazone + quinclorac)	Solitare	triazinone + quinoline carboxylic acid	14 + (27 + 4)
sulfosulfuron	Certainty, Outrider	sulfonylurea	2
tebuthiuron	(20 P): Spike 20 P, Alligare 20 P (80 DF, 80 WG): Spike 80 DF, Alligare 80 WG	thiadiazolyurea	7
terbacil	Sinbar 80 WDG	uracil	5
(thiencarbazone + foramsulfuron + halosulfuron)	Tribute Total	triazolinone + sulfonylurea + sulfonylurea	14 + 2 + 2
(thiencarbazone + iodosulfuron + dicamba)	Celsius WG	triazolinone + sulfonylurea + benzoic acid	14 + 2 + 4
(thiencarbazone + iodosulfuron + foramsulfuron)	Derigo	triazolinone + sulfonylurea + sulfonylurea	14 + 2 + 2
topramezone	Pylex	benzoylpyrazole	27
triclopyr	Monterey Turflon Ester, Remedy Ultra, Turflon Ester Ultra	pyradinecarboxylic acid	4
(triclopyr + cloypyralid)	Confront, 2-D	pyradinecarboxylic acid + pyradinecarboxylic acid	4 + 4
(triclopyr + fluroxypyr)	PastureGard HL	pyradinecarboxylic acid + pyradinyloxyacetic acid	4 + 4
(triclopyr + sulfentrazone + 2,4-D + dicamba)	Tzone SE	pyridinecarboxylic acid + triazinone + phenoxy-carboxylic acid + benzoic acid	4 + 14 + 4 + 4
trifloxysulfuron	Monument	sulfonylurea	2
trifluralin	(10 G): Treflan TR-10, Trifluralin 10 G (4 EC): Treflan HFP, Trifluralin 4 EC	dinitroaniline	3

Chemical Weed Control in Clary Sage

R. B. Batts, Horticultural Science Department

Table 7-11. Chemical Weed Control in Clary Sage

Herbicide, Mode of Action Code[1], and Formulation	Amount of Formulation Per Acre	Pounds Active Ingredient Per Acre	Precautions and Remarks
Preplant Injection, Weeds, diseases, and nematodes.			
metam sodium (Vapam HL) 42% (Sectagon-42) 42%	37.5 to 75 gal 15 to 75 gal	15.7 to 31.5 6.3 to 31.5	Rates are dependent on soil types and weeds known to be present. Apply when soil moisture is at field capacity. Read label thoroughly for regulatory, safety and application instructions. Interval to planting is usually 14 to 21 days, but can be as long as 30 days in certain environments.
metam potassium (K-PAM HL, Sectagon-K54) 54%	30-60 gal	16.2 to 32.4	
Preplant and preemergence, Annual and perennial grass and broadleaf weeds. Stale bed application.			
Flumioxazin, MOA 14 (Valor SX) 51 WDG	1 to 3 oz	0.03 to 0.095	Apply preplant, 30 days or more before planting. Application rate varies based on preplant interval. Treated area must receive a minimum of 0.5 inches of rainfall/irrigation begrween application and planting. Sequential preplant applications are prohibitied. Some soil texture restrictions apply. See label for details.
glyphosate, MOA 9 (various brands) 4 SL (various brands) 5 SL (Roundup Weather Max) 5.5 L	1 to 3 pt 0.8 to 2.4 pt 11 to 32 oz	0.5 to 1.5	Apply to emerged weeds before crop emergence. Perennial weeds may require higher rates of glyphosate. Consult the manufacturer's label for rates for specific weeds. Certain glyphosate formulations require the addition of surfactant. Adding nonionic surfactant to glyphosate formulation formulated with nonionic surfactant may result in reduced weed control.
d-limonene, MOA unclassified (AvengerAG) 55%	14 to 20 %	—	Apply broadcast as 1:6 ratio (14%) of Greenmatch:water in a minimum of 60 gallons per acre. For spot application on difficult to control weeds, use 1:4 ratio (20%). For best results, apply to dry foliage with temperatures above 50 degrees F. Do not exceed 8.5 gallons Greenmatch per acre per application.
pelargonic acid, MOA unclassified (Scythe) 4.2 EC	3 to 10%	—	Label recommends application in 75 to 200 gallons of solution per acre when applied alone. When tank mixing with other herbicides, delivery rates can be reduced to 10 to75 gallons per acre. See label for recommended ratio to apply on different types of weeds and other instructions.
Postemergence, residual weed control.			
Flumioxazin, MOA 14 (Valor SX) 51 WDG	1 to 2	0.03 to 0.06	Apply to crop that is a minimum of 4 inches in diameter. Do not apply after crop begins to enter reproductive stage. Do not add adjuvant, fertilizer or other pest control products. Do not use on clary sage grown for food or feed purposes.
Postemergence, Cutleaf eveningprimrose and certain other broadleaf weeds.			
linuron, MOA 7 (Linex) 4L	1 to 1.5 pt	0.5 to 0.75	Crop must be a minimum of 4 inches in diameter for application. Do not use on clary sage grown for food or feed purposes. Do not use on sands or loamy sands or on soils with less than 1% organic matter. Temporary yellowing or stunting of crop may occur.
Postemergence, Henbit and other winter annual broadleaf weeds.			
oxyfluorfen, MOA 14 (Goal 2XL) 2EC (Galigan, Oxystar) 2E GoalTender) 4E Galigan H$_2$O) 4EC	0.5 to 1 pt 0.5 to 1 pt 0.25 to 0.5 pt 0.25 to 0.5 pt	0.12 to 0. 25	Apply to 2-4 leaf henbit. Additional applications may be needed for subsequent henbit emergence. Do not apply more than 6 pints of Goal 2XL or Galigan per acre per year. Do not apply more than 3 pints of GoalTender or Galigan H$_2$O per acre per year. Clary sage may exhibit phytotoxicity on leaf margins after application, but recovery should occur quickly.
Postemergence, Emerged weeds.			
Paraquat MOA 22 (Parazone) 3SL	3 to 32 oz	0.07 to 0.75	Application rate varies based on size and physiological status of the crop. See product label for specific rates and timings. Do not exceed 1.125 pounds paraquat from this and any other registered paraquat product per growing season. Do not use clary sage for food or feed.
Paraquat MOA 22 (Parazone) 3SL	See comments	See comments	**Rope wick/carpet roller applicator.** Mix 1 part Parazone with 3-4 parts water to prepare a 25-33% solution. Add nonionic surfactant at 0.25% v/v, or 1 quart per 100 gallons. Do not allow mixture to contact crop. Do not apply Parazone in this manner within 30 days of harvest.
pelargonic acid, MOA unclassified (Scythe) 4.2 EC	3 to 10%	—	**Hooded/shielded application to row middles. Do not allow crop contact.** Label recommends application in 75 to 200 gallons of solution per acre when applied alone. When tank mixing with other herbicides, delivery rates can be reduced to 10 to 75 gallons per acre. See label for recommended ratio to apply on different types of weeds and other instructions.
Postemergence, Emerged annual and perennial grasses.			
clethodim, MOA 1 (Arrow, Clethodim, others) 2EC	6 to 8 oz	0.09 to 0.125	Arrow and Clethodim require addition of crop oil concentrate to spray mixture. See label for precautions regarding crop oil concentrates. Do not apply within 14 days of harvest.
(Select Max) 1EC	9 to 32 oz	0.07 to 0.25	Select Max allows the use of a nonionic surfactant, methylated seed oil, or crop oil concentrate in the mixture. Label suggests different rate ranges for annual and perennial grasses. See label for details. Do not apply within 14 days of harvest.
(Intensity One, Tapout) 1EC	9 to 16 oz	0.07 to 0.125	Label recommends the inclusion of nonionic surfactant at 0.25% v/v. Do not apply more than 64 fluid ounces per acre per crop. Do not apply within 14 days of harvest.

[1] Mode of Action (MOA) code developed by the Weed Science Society of America. See Table 7-10, Herbicide Resistance Management, for details.

CHEMICAL WEED CONTROL

Chemical Weed Control in Small Fruit Crops

W. E. Mitchem and K. M. Jennings, Horticultural Science Department

NOTE: A mode of action code has been added to the Herbicide and Formulation column of this table. Use MOA codes for herbicide resistance management. See Table 7-10, Herbicide Modes of Action, for details.

Table 7-12A. Chemical Weed Control in Small Fruit Crops

Timing/Targeted Weeds	Herbicide, Mode of Action Code[1] and Formulation	Amount of Formulation Per Acre	Pounds Active Ingredient Per Acre	Precautions and Remarks
Blackberries				
PREPLANT Annual and perennial weeds	glyphosate, MOA 9 (various brands) 4 SL (various brands) 5 SL (Roundup WeatherMax) 5.5 L	1 to 3 pt 0.8 to 2.4 pt 11 to 32 oz	0.5 to 1.5	Apply to emerged weeds at least 30 days before crop transplanting. Perennial weeds may require higher rates of glyphosate. Consult the manufacturer's label for rates for specific weeds. See label for further instructions.
PREEMERGENCE Annual grass and small seeded broadleaf weeds	napropamide, MOA 15 (Devrinol) 50 WDG	8 lb	4	Apply to soil surface free of weeds and plant residue. Rainfall or overhead irrigation is needed within 1 to 2 days of application for effective herbicide performance. In new plantings allow soil to settle after transplanting before applying Devrinol.
	oryzalin, MOA 3 (Oryzalin or Surflan) 4 AS	2 to 6	2 to 6	Oryzalin may be tank mixed with paraquat for non-selective POST weed control. Total use rate cannot exceed 12 quarts/acre per year. Sequential applications may be used along as there is 2.5 months between applications. Tank mix with simazine for broad spectrum PRE weed control.
PREEMERGENCE and POSTEMERGENCE Annual broadleaf weeds	mesotrione, MOA 27 (Callisto) 4L	3 to 6 oz	0.094 to 0.185	May be applied as split applications of 3 oz per acre followed by 3 oz per acre. If two applications are made, there must be at least 14 days between applications. Do not apply more than 6 ounces per acre per year. Do not apply after the onset of bloom stage or illegal residues may occur. The addition of crop oil concentrate at a rate of 1% v/v (1 gallon per 100 gallons of spray solution) is recommened for POST weed control.
PREEMERGENCE Annual broadleaf and grass weeds. Nutsedge suppression	norflurazon, MOA 12 (Solicam) 80 DF	1.25 to 5	1 to 4	Apply to dormant caneberries established at least 1 year. Tank mix with paraquat for control of emerged weeds. The addition of simaziine will expand spectrum of PRE control. Solicam has a 60 day PHI. In areas prone to soil movement injury to ground cover planted in row middles can occur.
PREEMERGENCE and POSTEMERGENCE Annual broadleaf and grass weeds	terbacil, MOA 5 (Sinbar) 80 WDG	1 to 2	0.8 to 1.6	Apply as a directed spray in early fall, winter, or spring before fruit set. Do not allow contact with desirable foliage. Sinbar has a 70 day PHI. Do not use on loamy sand and sandy soils or on soils having less tna 1% organic matter. Do not use on eroded areas where subsoil is exposed. Tank mix with paraquat for non-selective POST weed control. Sinbar has POST emergence activity on some weeds like horseweed.
PREEMERGENCE Broadleaf and some grass weeds	simazine, MOA 5 (Princep or Simazine) 4L (Princep or Simazine) 90 DF	1 to 4 qt 1.1 to 4.4 lb	1 to 4	Use half rate on caneberries established less than 6 months. Rate is soil texture dependent therefore higher rates cannot be used sandy loam, loamy sand, or sand soils. See label for soil texture precautions. Apply as a single application in spring or as a split application of 2 quarts/acre in spring followed by 2 quarts/acre in the fall. Do not apply when fruit is present or illegal residues may occur. Tank mix with oryzalin or norflurazon for expanded residual control of annual grasses like crabgrass and goosegrass. Simazine may be applied in combination with paraquat for non-selective POST weed control.
PREEMERGENCE Annual and perennial broadleaf weeds as well as grass weeds	dichlobenil, MOA 20 (Casoron) 4G (Casoron) CS	100 lb 1.4 to 2.8 gal	4 2 to 4	Granular formulation should be applied in January or February. Rainfall or snow is needed for activation. Warm temperatures increase herbicide loss due to volatilization. The liquid formulation should be used during warmer temperatures (up to 70 degrees F) as it is less likely to volatilize. As with the granular formulation, it needs to be activated with rainfall. The liquid formulation may be tank mixed with other herbicides like paraquat.
PREEMERGENCE + POSTEMERGENCE Controls annual broadleaf and some grass weeds	Sulfentrazone + Carfentrazone, MOA 14 (Zeus Prime XC)	8 to 15 fl. oz	0.22 to 0.41	Use in blackberry plantings that have been establish at least 2 years. When applied in a band treating less than 50% of the vineyard floor, Zeus Prime may be applied twice per year so long as use rate does not exceed 15 fluid ounces per acre on a broadcast basis. Rainfall is needed within two weeks of application to ensure herbicide activation. Applications made after bud break must be made using hooded application equipment. There must be at least 60 days between sequential applications. Zeus Prime has a 3 day PHI. Tank mix with non-selective POST herbicides for broad spectrum control of emerged weeds. The addition of oryzalin is necessary for expanded residual control of annual grass weeds. **Zeus will control yellow nutsedge.**
	rimsulfuron, MOA 2 (Solida) 25 WG (Matrix)	4 oz	0.063	Rimsulfuron has POST and PRE activity on broadleaf and some grass weeds. For broad spectrum residual control of annual grass weeds, tank mix rimsulfuron with oryzalin or Sinbar. For nonselective POST weed control, tank mix rimsulfuron with paraquat. Do not treat blackberry plantings established less than 1 year. Rainfall for herbicide activation is necessary within 2 to 3 weeks of application. Do not apply within 21 days of harvest. The pH of spray solution should be in the range of 4 to 8. Rimsulfuron may be applied as a sequential application so long as total use rate does not exceed 4 ounces per acre per year and application is made in band to less than 50% of orchard floor. Allow at least 30 days between sequential applications. To reduce the risk of primocane injury, apply prior to primocane emergence or wait until primocanes are 3 feet tall. If primocanes are emerged at the time of application, chlorosis and stunting is likely but in most instances those primocanes recover.
POSTEMERGENCE Annual broadleaf weeds like pigweed, morningglory, lambsquarter, purslane, nightshade, tropical spiderwort, and smartweed	carfentrazone, MOA 14 (Aim) 2 EC	0.8 to 2 oz	0.013 to 0.03	Do not allow sray solution to contact desirable vegetation, foliage, flowers, or fruit. Every precaution should be taken to avoid herbicide injury related to herbicide drift. Use rate should not exceed 25 ounces/acre per year and there must be at least a 14 day interval between applications. The addition of a non-ionic surfactant at 0.25% v/v (1 quart per 100 gallons of spray solution) or crop oil concentrate at 1 to 2% v/v (1 to 2 gallons per 100 gallons of spray solution is necessary for optimum herbicide performance. Aim has a 15 day PHI. Aim may be used to suppress primocane emergence. See label for instructions and rate information relative to primocane suppression.

Table 7-12A. Chemical Weed Control in Small Fruit Crops

Timing/Targeted Weeds	Herbicide, Mode of Action Code[1] and Formulation	Amount of Formulation Per Acre	Pounds Active Ingredient Per Acre	Precautions and Remarks
Blackberries (continued)				
POSTEMERGENCE Non-selective broadleaf and grass weed control	paraquat, MOA 22 (Firestorm, Parazone, Paraquat Concentrate) 3 SL (Gramoxone) 2 SL	1.3 to 2.7 pt 2 to 4 pt	0.5 to 1.0	Do not allow herbicide to directly contact desirable foliage or green canes. Young plants must be shielded or severe crop injury or death will result. The addition of a non-ionic surfactant at 0.25% v/v (1 quart/100 gallons of spray solution) is necessary for adequate control. Paraquat can be tank mixed with PRE herbicides. DO NOT make more than 5 applications per year.
POSTEMERGENCE Annual and perennial grass weeds	clethodim, MOA 1 (Select Max or Intensity One) 1 EC (Select, Arrow, Volunteer, Clethodim) 2 EC	9 to 16 oz 6 to 8 oz	0.07 to 0.125	Low rates are for annual grass weeds. Use higher rates and sequential applications for perennial grass (bermudagrass or johnsongrass) control. Select Max and Intensity One need to be applied in combination with a non-ionic surfactant. Select and other generic formulations of clethodim require the addition of either a non-ionic surfactant or a crop oil concentrate. See label for specific information related to spray adjuvants. **The Select Max formulation is labeled fo bearing caneberries and has a 7 day PHI. ALL other clethodim formulations are registered for use in non-bearing caneberries ONLY.**
	fluazifop, MOA 1 (Fusilade DX) 2 EC	12 to 24 oz	0.19 to 0.38	Sequential applications will be necessary for perennialgrass control. The addition of a non-ionic surfactant (1 qt/100 gal of spray solution) or crop oil concentrate (1 gallon/100 gallons of spray solution) is necessary. **Use in non-bearing caneberry plantings only.**
	sethoxydim, MOA 1 (Poast) 2 EC	1 to 2.5 pt	0.25 to 0.63	Sequential application will be necessary for perennial grass control. The addition of a non-ionic surfactant (1 qt/100 gal of spray solution) or crop oil concentrate (1 gal/100 gal of spray solution) is necessary for optimum herbicide performance. Poast has a 50 day PHI.
Blueberries				
PREPLANT Annual and perennial weeds	glyphosate, MOA 9 (various brands) 4 SL (various brands) 5 SL (Roundup WeatherMax) 5.5 L	1 to 3 pt 0.8 to 2.4 pt 11 to 32 oz	0.5 to 1.5	Apply to emerged weeds at least 30 days before crop transplanting. Perennial weeds may require higher rates of glyphosate. Consult the manufacturer's label for rates for specific weeds. See label for further instruction.
PREEMERGENCE Annual weeds and some perennial (goldenrod) weeds	hexazinone, MOA 5 (Velpar) 2 SL (Velpar) 75 DF	0.5 to 1 gal 1.3 to 2.6 lb	1 to 2	Apply as a directed spray to soil and weeds before blueberry leaf emergence but at least 90 days before harvest. Use lower rates on poorly drained or sandy soils. Bushes must be established for at least 3 years.
PREEMERGENCE Annual weeds (crabgrass, chickweed) and some perennial (dogfennel) weeds	dichlobenil, MOA 20 (Casoron) 4 G (Casoron CS) 1.4 L	100 to 150 lb 1.4 to 2.8 gal	4 to 6 2 to 3.92	Apply in the early winter, no later than mid-February, to plants that have been established 1 year or longer (Casoron CS formulation). Casoron 4G may be used in blueberry planted at least 4 weeks earlier.
PREEMERGENCE Annual broadleaf weeds (Maryland meadowbeauty, pigweed spp., morningglory spp.) and some annual grasses (large crabgrass)	flumioxazin, MOA 14 (Chateau) 51 WDG	6 to 12 oz	0.19 to 0.375	Do not apply to blueberries established less than 2 years unless they are protected from spray contact by nonporous wrap, grow tubes, or waxed containers. Do not apply more than 12 ounces per acre during a 12-month period. If a sequential application is applied it must occur no earlier than 60 days after the first application. Do not apply more than 6 ounces per application to bushes less than 3 years old on soils having a sand plus gravel content greater than 80%. Apply at the base of the bush. Chateau should be tank-mixed with a registered burndown herbicide to control emerged weeds. Residual weed control will be reduced if vegetation prevents Chateau from reaching soil surface. PHI = 7 days.
PREEMERGENCE Directed to the soil surface Annual grass and broadleaf weeds, yellow nutsedge suppression	S-metolachlor, MOA 15 (Dual Magnum) 7.62 EC	0.67 to 1.33 qt	0.64 to 1.27	This is a Section 24C special local needs label. Growers must obtain label prior to making Dual Magnum applications. Growers must obtain label at www.farmassist.com. Avoid contact with crop foliage, or crop injury may occur. See label for rate range based on soil type. Blueberry plants less than one year may be more sensitive.
PREEMERGENCE Annual broadleaf weeds (morningglory, chickweed) and some annual grasses (large crabgrass)	diuron, MOA 7 (Karmex) 80 DF (Direx) 4 L	1.5 to 2 lb 1.2 to 1.6 qt	1.2 to 1.6	Use only in fields that have been established for at least 1 year. Apply as a band treatment at the base of bushes. The addition of a surfactant will kill many small emerged weeds. May be applied in the spring and again in the fall after harvest.
PREEMERGENCE Annual broadleaf weeds (pigweed, common purslane) and some annual grasses (crabgrass spp., fall panicum)	simazine, MOA 5 (Princep Caliber 90) 90 WDG (Princep) 4 L	2.2 to 4.4 lb 2 to 4 qt	2 to 4	Apply half the maximum annual application in the spring before buds break and weeds emerge, and half after harvest. Do not apply more than 1 pound a.i. simazine on newly planted blueberries. Apply in minimum of 40 gallons of water per acre.
PREEMERGENCE Directed Underneath Bushes, Broadleaf weeds and some annual grasses	rimsulfuron, MOA 2 (Matrix) 25 WG (Solida) 25 WG	4 oz	0.063	Application after bud break may cause temporary chlorosis and/or stunting of leaves contacted by the spray solution. Rimsulfuron has POST and PRE activity on broadleaf and some grass weeds. For broad spectrum residual control, tank mix rimsulfuron with oryzalin. For nonselective POST weed control, tank mix rimsulfuron with glyphosate, paraquat, or Rely. Do not treat blueberries established less than 1 year. The best residual control is obtained when at least 0.5 inches of rain or overhead irrigation comes within the first week after application. The pH of spray solution should be in the range of 4 to 8. Do not use on soils classified as a Sand. Rimsulfuron may be applied as a sequential application so long as total use rate does not exceed 4 ounces per acre per year and application is made in band to less than 50% of floor. Do not apply within 21 days of first harvest.

CHEMICAL WEED CONTROL

Table 7-12A. Chemical Weed Control in Small Fruit Crops

Timing/Targeted Weeds	Herbicide, Mode of Action Code[1] and Formulation	Amount of Formulation Per Acre	Pounds Active Ingredient Per Acre	Precautions and Remarks
Blueberries (continued)				
PREEMERGENCE Annual broadleaf weeds (chickweed, red sorrel from seed) and some annual and perennial grasses	pronamide, MOA 3 (Kerb) 3.3 SC	2.5 to 5 pt	1 to 2	Apply in the fall or winter. May be applied to newly planted bushes as long as roots are well established. Do not apply more than 2 lb ai or make more than one application per year.
PREEMERGENCE Most annual broadleaf and grass weeds plus many perennials	terbacil, MOA 5 (Sinbar) 80 WDG	0.5 to 2 lb	0.4 to 1.6	Apply as directed spray in early spring or in fall after harvest. May be applied before weeds emerge or shortly after they emerge. Use only in plantings established 1 year or longer. Do not use on sandy soils with less than 3% organic matter. This herbicide can be very active but injurious on blueberries. See label for further information.
PREEMERGENCE Annual broadleaf weeds	mesotrione, MOA 27 (Callisto) 4L	3 to 6 oz	0.094 to 0.185	May be applied as a split application of 3 ounces per acre followed by 3 ounces per acre no less than 14 days apart. Do not apply more than 6 oz per acre per year. Do not apply after the onset of bloom stage or illegal residues may occur.
PREEMERGENCE Broadleaf weeds including corn spurry, common cocklebur, dayflower, horseweed, smartweed, wild mustard, wild radish, and pigweed	halosulfuron, MOA 2 (Sandea) 75 DG	0.5 to 1 oz	0.024 to 0.047	Do not apply to plants established less than 1 year. Apply as a directed treatment to avoid contact with the crop. Occasional injury may occur. For nutsedge control, apply Sandea postemergence to the nutsedge (see Sandea listed below for postemergence control). PHI = 14 days
PREEMERGENCE Annual grasses and small seeded broadleaf weeds	napropamide, MOA 15 (Devrinol, Devrinol DF-XT) 50 DF (Devrinol) 10 G (Devrinol 2-XT)	8 lb 40 lb 8 qt	4	Apply to weed-free soil surface. Enough irrigation or rainfall to wet the soil to a depth of 4 inches is necessary within 24 hours of application. Apply as a directed spray to the base of the blueberry plant. May be used on first-year plantings. If using Devrinol DF-XT or Devrinol 2-XT, time between application and irrigation is extended to 48 hours.
PREEMERGENCE Will suppress yellow and purple nutsedge	norflurazon, MOA 12 (Solicam) 78.6 WDG	2.5 to 5 lb	2 to 4	Apply as a directed spray from fall to early spring when the crop is dormant and before weeds emerge. Make only one application per year. Blueberries must be established 6 months prior to Solicam use. Application of Solicam may result in temporary bleaching or chlorosis of the leaves. Preharvest interval is 60 days.
PREEMERGENCE Annual broadleaf (morningglory, pigweed spp., common purslane) and annual grasses (crabgrass spp., barnyardgrass)	oryzalin, MOA 3 (Oryzalin or Surflan) 4 AS	2 to 4 qt	2 to 4	This treatment may be used on first year plants.
PREEMERGENCE Broadleaf, grasses and sedge weeds	Sulfentrazone, MOA 14 (Zeus XC)	12 oz	0.375	Apply to crop that has been growing for at least 3 years and is in good condition. Aviod direct or indirect spray contact to foliage and green bark. Apply as a uniform broadcast soil application in a minimum of 10 gallons of water per acre. For best control, Zeus XC should be applied when there are no weeds present. If weeds are present, tank mix with a postemergence herbicide to eliminate emerged weeds. Do not apply more than 0.375 lb ai per acre per 12-month period. Appropriate soil moisture at time of application or at least ½ inch of rainfall or irrigation within two weeks after application is required to achieve desired level of weed control. Preharvest interval is 3 days.
PREEMERGENCE Broadleaf weeds	Isoxaben, MOA 21 (Trellis)	1.33 lb	1.0	APPLY ONLY ON ESTABLISHED NON-BEARING PLANTS. Apply as a uniform broadcast soil application in a minimum of 10 gallons of water per acre. For best control, Trellis should be applied when there are no weeds present. Appropriate soil moisture at time of application or at least ½ inch of rainfall or irrigation within 21 days after application is required to achive desired level of weed control.
PREEMERGENCE Most annual grasses and some broadleaf weeds	Pendimethalin, MOA 3 (Prowl H₂O)	2 to 6.3 qt	1.9 to 5.96	Do not exceed a total of 6.3 qt per acre per year in blueberries. Do not apply over the top of bushes with leaves, shoots, or buds. Contact by the spray mixture may cause injury. PHI = 30 days.
POSTEMERGENCE NON-SELECTIVE Most annual broadleaves and grasses and a few perennials	glufosinate, MOA 10 (Reckon 280 SL) 2.34 SL	48 to 82 oz	0.88 to 1.5	Apply as a directed spray to emerged weeds in a minimum of 20 gallons water per acre. Do not allow spray to contact desirable foliage or green bark. Preharvest interval is 14 days. The addition of a spray grade ammonium sulfate will enhance activity on difficult to control weeds. The use of additional surfactants or crop oil is not needed and/or may increase potential for crop injury.
	glyphosate, MOA 9 (various brands) 4 SL (various brands) 5 SL (Roundup WeatherMax) 5.5 L	1 to 3 pt 0.8 to 2.4 pt 11 to 32 oz	0.5 to 1.5	DO NOT SPRAY GREEN CANES, BARK, OR FOLIAGE. Apply as a directed shielded spray to base of established plants. Do not apply within 14 days of harvest. Wiper applications may also be used. Perennial weeds may require higher rates of glyphosate. Certain glyphosate formulations require the addition of a surfactant. See label for specific rates for herbicide and surfactant.
POSTEMERGENCE NON-SELECTIVE Contact kill of all green foliage	paraquat, MOA 22 (Firestorm, Parazone) 3 SL (Gramoxone Inteon, Gramoxone SL, Gramoxone SL 2.0) 2 SL	1.3 to 2.7 pt 2 to 4 pt	0.56 to 1	Apply as a directed spray to weeds before new canes emerge. Avoid paraquat contact with new canes, as injury will occur. Use of paraquat in rabbiteye blueberry can increase incidence of stem blight if herbicide contacts green stems. Rabbiteye producers should consider using other non-selective herbicides.
POSTEMERGENCE	pelargonic acid, MOA 27 (Scythe) 4.2 SL	3 to 10% soln.		Apply as a directed or shielded spray.
POSTEMERGENCE Broadleaf weeds up to 4 inches tall or 3 inches in diameter	carfentrazone-ethyl, MOA 14 Aim 2EC	1 to 2 oz	0.015 to 0.031	Apply as a hooded spray with application equipment designed to prevent spray deposit on green stems, leaf tissue, flowers, or fruit. Use in established fields only; do not use on newly set plants. May be used alone or tank-mixed with other herbicides. May be a good option for sodded middles as it does not control grasses. Add crop oil concentrate at 1% by volume (1 gal/100 gal of spray solution) or a nonionic surfactant at 0.25% by volume (1 qt/100 gal of spray solution).

Table 7-12A. Chemical Weed Control in Small Fruit Crops

Timing/Targeted Weeds	Herbicide, Mode of Action Code[1] and Formulation	Amount of Formulation Per Acre	Pounds Active Ingredient Per Acre	Precautions and Remarks
Blueberries (continued)				
POSTEMERGENCE Yellow nutsedge and some broadleaf weeds	bentazon, MOA 6 (Basagran) 4SL	1.5 to 2 pt	0.75 to 1	NONBEARING ONLY. For yellow nutsedge control, two applications may be needed. Apply when plants are 6 to 8 inches tall. If needed, make a second application at the same rate 7 to 10 days later. Add crop oil concentrate to the spray soultion at a rate of 2 pints in 20 to 50 gallons of water per acre.
POSTEMERGENCE Yellow and purple nutsedge, pigweed, common ragweed, wild radish, wild mustard velvetleaf, smartweed, common cocklebur, dayflower, rice flatsedge	halosulfuron, MOA 2 (Sandea) 75 DG	0.75 to 1 oz	0.036 to 0.047	Do not apply to plants established less than 1 year. Apply as a directed treatment to avoid contact with the crop. Occasional injury may occur. See label for further instructions regarding nutsedge control. PHI = 14 days.
POSTEMERGENCE Annual and perennial grasses	clethodim, MOA 1 (Arrow, Select, and others) 2 EC (Select Max and others) 1 EC	6 to 8 oz 9 to 16 oz	0.09 to 0.125 0.07 to 0.125	Select formulation is for use on nonbearing crop only (within one year of harvest). Select Max formulation may be applied as a directed spray to nonbearing and bearing crop. Select Max formulation requires the use of a nonionic surfactant rather than crop oil concentrate. PHI for Select Max is 14 days.
	fluazifop, MOA 1 (Fusilade DX) 2 EC	16 to 24 oz	0.25 to 0.38	USE ON NONBEARING CROP ONLY. Postemergence grass control. Check label for specific rates and timings. Do not apply within 1 year of the first harvest. Use of a crop oil or surfactant will be necessary. Sequential applications are necessary for adequate control of perennial grasses.
	sethoxydim, MOA 1 (Poast) 1.5 EC	1.5 to 2.5 pt	0.3 to 0.5	Check label for specific rates and timings. Use a crop oil at a rate of 1 qt per acre. May be used on bearing blueberries. PHI = 30 days.
Grapes				
PREEMERGENCE Directed Underneath Vines, Annuals and many perennials	dichlobenil, MOA 20 (Casoron) 4 G (Casoron CS) 1.4 CS	100 to 150 lb 1.4 to 2.8 gal	4 to 6 2 to 4	Do not apply Casoron 4G within 4 weeks of transplanting. Apply in January and February. High rate is necessary for perennial weed control. Casoron CS may only be used in vineyards established at least 1 year.
PREEMERGENCE Directed Underneath Vines, Annual broadleaf and grass weeds	simazine, MOA 5 (Princep Simazine) 90 WDG (Princep Simazine) 4 L	2.2 to 4.4 lb 2 to 4 qt	2 to 4	Apply before germination of annual weeds. Do not apply in vineyards less than 3 years old. Tank mix with glyphosate, paraquat, or glufosinate for POST weed control. Tank mixing simazine with oryzalin or Prowl H2O will improve residual control of annual grasses and certain broadleaf weeds.
	flumioxazin, MOA 14 (Chateau) 51 WDG (Tuscany) 51 WDG	6 to 12 oz	0.19 to 0.375	Apply as a directed spray using hooded or shielded application equipment. The trunks of grape vines established less than 2 years must be shielded from contact with spray solution using grow tubes. Do not apply after flowering unless using hooded or shielded application equipment and applicator can insure spray material does not contact fruit or desirable foliage. DO NOT tank mix with glyphosate when applying Chateau after bud break due to increased injury potential. . Do not apply more than 6 ounces per acre per application to vines less than 3 years old on soils having a sand plus gravel content greater than 80%. DO NOT apply sequential applications closer than 30 days apart. Total use rate cannot exceed 24 ounces/acre per year. Chateau has a 60 day PHI.
	diuron, MOA 7 (Diuron, Karmex DS, Karmex XP)	2 to 3 lb	1.6 to 2.4	Apply before germination of annual weeds. Vineyards must be at least 3 years old. Higher rate may be used on soils with greater than 2% organic matter and high clay content. Do not use on sandy loam or coarser soils. Tank mix with glyphosate, paraquat, or glufosinate for POST weed control. Applications in vineyards having less than 2% organic matter may cause injury if heavy rainfall occurs soon after application. This risk is assumed by user.
PREEMERGENCE Directed Underneath Vines, Annual grasses and small-seeded broadleaf weeds	oryzalin, MOA 3 (Oryzalin, Surflan) 4 AS	2 to 4 qt	2 to 4	Apply once soil has settled after transplanting. Multiple applications per year are permitted; see label for details. Apply in combination with Trellis in newly planted vineyards for improved control of broadleaf weeds. Sequential applications may be used so long as total use rate does not exceed 12 quarts per acre per year. Allow 2.5 months between applications. In established planting oryzaln may be tank mixed with simazine for broad spectrum residual control of annual weeds. Apply in combination with paraquat, glyphosate, or glufosinate for non-selective POST weed control.
	pendimethalin, MOA (Prowl) H2O 4E	2 to 6.3 qt	2 to 6	In newly planted grapes allow soil to settle after transplanting before applying Prowl. Use only during dormancy (prior to bud swell) when applying around newly planted and 1 year old vines. In bearing vineyards apply any time after harvest, during winter dormancy, and spring. Use rate cannot exceed 6.3 quarts/acre per year. Prowl H_2O has a 90-day PHI. Tank mix with simazine or rimsulfuron for expanded residual control of broadleaf weeds. Prowl may be applied incombination with paraquat, glyphosate, or glufosinate for non-selective POST weed control.
	norflurazon, MOA 12 (Solicam) 80 WDG	1.25 to 5 lb	1 to 4	Vines must be established at least 2 years in the field before application. DO NOT use on nursery stock. Loss of pigment in leaf veins will occur in coarse-textured soils when applied within 3 months after budbreak. Rate is soil-texture dependent. 60-day PHI.
PREEMERGENCE Directed Underneath Vines, Annual grasses and broadleaf weeds	oryzalin, MOA 3 (Oryzalin, Surflan) 4 AS + simazine, MOA 5 (Princep) 90 WDG (Princep) 4 L	2 to 4 qt + 2.2 to 4.4 lb 2 to 4 qt	2 to 4 + 2 to 4	Tank mix for use before weed emergence. See comments for oryzalin and simazine.

CHEMICAL WEED CONTROL

301

Table 7-12A. Chemical Weed Control in Small Fruit Crops

Timing/Targeted Weeds	Herbicide, Mode of Action Code[1] and Formulation	Amount of Formulation Per Acre	Pounds Active Ingredient Per Acre	Precautions and Remarks
Grapes (continued)				
PREEMERGENCE + POSTEMERGENCE Directed Underneath Vines for control of broadleaf weeds and some annual grasses	Sulfentrazone + Carfentrazone (Zeus Prime XC)	8 to 15 fl. oz	0.22 to 0.41	Use in vineyards that have been establish at least 2 years. When applied in a band treating less than 50% of the vineyard floor, Zeus Prime may be applied twice per year so long as use rate does not exceed 15 fluid ounces per acre on a broadcast basis. Rainfall is needed within two weeks of application to ensure herbicide activation. Applications made after bud break must be made using hooded application equipment. There must be at least 60 days between sequential applications. Zeus Prime has a 3 day PHI. Tank mix with non-selective POST herbicides for broad spectrum control of emerged weeds. The addition of oryzalin is necessary for expanded residual control of annual grass weeds. **Zeus will control yellow nutsedge.**
PREEMERGENCE Directed Underneath Vines, Broadleaf weeds and some annual grasses	rimsulfuron, MOA 2 (Matrix) 25 WG (Pruvin) 25 WG (Solida) 25 WG	4 oz	0.063	Rimsulfuron has POST and PRE activity on broadleaf and some grass weeds. For broad spectrum residual control, tank mix rimsulfuron with oryzalin, Prowl H$_2$O, or diuron. For nonselective POST weed control, tank mix rimsulfuron with glyphosate, paraquat, or Rely. Do not treat vineyards established less than 1 year. Rainfall for herbicide activation is necessary within 2 to 3 weeks of application. Do not apply within 14 days of harvest. The pH of spray solution should be in the range of 4 to 8. Rimsulfuron may be applied as a sequential application so long as total use rate does not exceed 4 ounces/acre per year and application is made in band to less than 50% of vineyard floor.
POSTEMERGENCE Directed Underneath Vines, Non-selective weed control	paraquat, MOA 22 (Firestorm) 3 SL (Paraquat Concentrate) 3 SL (Parazone) 3 SL (Gramoxone Inteon) 2 SL	1.7 to 2.7 pt 2.5 to 4 pt	0.66 to 1	Apply in 20 gallons per acre spray mix when grass and weeds are 1 to 6 inches high and succulent for best results. Direct spray with low pressure to avoid contact with foliage or bark less than 1 year old. Add a nonionic surfactant at a rate of 32 ounces per 100 gallons of spray solution. May be used for sucker suppression. See label for details.
	glyphosate, MOA 9 various brands and formulations)	See label	1 to 2	DO NOT SPRAY GREEN BARK OR FOLIAGE. Apply preplant or as a directed spray to base of established vines. Do not treat within 14 days of harvest. Wiper applications may also be used. Perennial weeds may require higher rates of glyphosate. See label for specific rates. Do not apply in late summer or fall. Some formulations may require the addition of a surfactant.
	glufosinate, MOA 10 (Reckon, Rely, Lifeline) 2.34 SL	48 to 82 oz	0.88 to 1.5	Apply as a directed spray to emerged weeds in a minimum of 20 gallons water per acre with a minimum of 30 psi spray pressure when weeds are 1 to 6 inches high. For spot application, use 1.7 ounces per gal of water and spray to wet but not runoff; however, spot spray solution should not contact vines or injury can occur. Do not allow spray to contact desirable foliage or green bark. Do not apply within 14 days of harvest. See label for specific rates. Do not make more than 3 applications per year. The addition of a spray grade ammonium sulfate will enhance Rely 280 activity on difficult to control weeds. The use of additional surfactants or crop oil is not needed and/or may increase potential for crop injury.
POSTEMERGENCE, Directed Underneath Vines, Annual broadleaf weeds	carfentrazone-ethyl, MOA 14 (Aim) 2 EC	0.5 to 1.6 oz	0.008 to 0.025	Apply as a directed spray or as a hooded spray. DO NOT allow spray solution to contact green tissue, leaves, flowers, or fruit. Aim may be used alone or tank mixed with other herbicides; see label for tank mixing instructions. Aim controls cocklebur, pigweed, nightshade, velvetleaf, carpetweed, and spreading dayflower. Do not apply within 3 days of harvest. Apply in minimum spray volume of 20 GPA. Apply in combination with crop oil concentrate at 1% v/v (1 gal/100 gal of spray solution) or a nonionic surfactant at 0.25% v/v (1 qt/100 gal of spray solution). Do not use on newly transplanted vines.
POSTEMERGENCE, Directed Underneath Vines, Annual and perennial grasses	clethodim, MOA 1 (Arrow, Select, others) 2 EC (Select Max and others) 1 EC	6 to 8 oz 9 to 16 oz	0.07 to 0.125	USE ON NONBEARING CROP ONLY. Postemergence grass control. Very effective in controlling bluegrass. Do not apply within 1 year of harvest. See label for all other instructions. Sequential applications necessary for adequate control of perennial grass weeds. Always apply 80% active ingredient nonionic surfactant at a rate of 0.25% volume per volume (1 pt/50 gal of spray).
	fluazifop, MOA 1 (Fusilade DX) 2 EC	16 to 24 oz	0.25 to 0.38	Sequential applications will be necessary for perennial grass control. Check label for rates and timings for specific weeds. Do not apply within 50 days of harvest. Use of a crop oil or surfactant will be necessary..
	sethoxydim, MOA 1 (Poast) 1.5 EC	1.5 to 2.5 pt	0.3 to 0.5	Postemergence grass control. Check label for rates and timings for specific grasses. Use a crop oil at a rate of 1 quart per acre. Do not apply within 50 days of harvest. Sequential applications necessary for adequate control of perennial grass weeds.
Strawberries (matted row)				
PREEMERGENCE Most annual grasses and small-seeded broadleaf weeds	DCPA, MOA 3 (Dacthal) W-75 (Dacthal) 6 F	8 to 12 lb 8 to 12 pt	6 to 9	Apply over the top of newly planted transplants or in fall or early spring for preemergence weed control. Do not apply after first bloom through harvest.
	napropamide, MOA 15 (Devrinol, Devrinol DF-XT) 50 DF (Devrinol) 2 EC	4 to 8 lb 8 qt	2 to 4	Apply to established plants before weed emergence anytime, except the interval between bloom and harvest. Apply to established plantings in fall to early spring prior to bloom. See label for notes on irrigation requirement.
PREEMERGENCE Most annual broadleaf weeds and grass weeds	terbacil, MOA 5 (Sinbar) 80 WDG	2 to 6 oz	0.1 to 0.3	For planting year: apply 2 to 3 ounces of Sinbar per acre after transplanting but before new runner plants start to root. If strawberry transplants are allowed to develop new foliage prior to Sinbar application, apply 0.5 to 1 inch irrigation or rainfall immediately after application. For control of winter weeds: apply 2 to 6 ounces Sinbar per acre in late summer or early fall. If crop is not dormant, the application must be followed immediately by 0.5 to 1 inch irrigation or rainfall. To extend weed control through harvest of the following year, apply 2 to 4 ounces Sinbar per acre just prior to mulching in the late fall. For harvest years: after postharvest renovation and before new growth begins in midsummer, apply 4 to 6 ounces of Sinbar per acre. To extend weed control through harvest of the following year, apply 4 to 6 ounces of Sinbar per acre just prior to mulching in late fall. Do not apply within 110 days of harvest. See label for more information.

Table 7-12A. Chemical Weed Control in Small Fruit Crops

Timing/Targeted Weeds	Herbicide, Mode of Action Code[1] and Formulation	Amount of Formulation Per Acre	Pounds Active Ingredient Per Acre	Precautions and Remarks
Strawberries (matted row) (continued)				
PREEMERGENCE Henbit, chickweed, cutleaf evening primrose, wild radish	flumioxazin, MOA 14 (Chateau) 51 SW	3 oz	0.096	**Crop row.** Apply to **dormant** strawberries for the preemergence control of weeds. Crop oil concentrate at 1% v/v or nonionic surfactant at 0.25% v/v may be added to help control small emerged broadleaf weeds. **Row middle.** Use a hooded or shielded applicator. DO NOT apply over strawberries. Apply prior to weed emergence. Crop spotting may occur if adjuvant is used. DO NOT APPLY AFTER FRUIT SET.
PREEMERGENCE Yellow nutsedge, purple nutsedge, corn spurry, yellow woodsorrel, henbit, chickweed	sulfentrazone, MOA 14 (Spartan) 4F	4 to 8 oz		Please refer to label for soil type restrictions.
POSTEMERGENCE Broadleaf weeds including vetch, clover, dock, sowthistle, and thistle	clopyralid, MOA 4 (Stinger) 3 EC	Spring: 0.67 pt Post harvest 0.33 to 0.67 pt	0.125 to 0.25	Apply postemergence in spring or postharvest to control emerged broadleaf weeds in established strawberries. Do not tankmix with surfactant or other pesticides. PHI = 30 days.
POSTEMERGENCE Annual broadleaf weeds	2,4-D amine, 4, MOA 1 (Amine 4 2,4-D Weed Killer) 4 SL	2 to 3 pt	1 to 1.5	Apply to well-established strawberries after harvest and before runners form or when crop is dormant. Not more than two treatments per year. Do not apply during bud, flower, or fruit stage. Timing is very critical to avoid damage. Do not apply unless possible injury to crop is acceptable.
	acifluorfen, MOA 14 (UltraBlazer) 2 L	0.5 to 1.5 pt	0.125 to 0.375	**Crop row.** Apply after last harvest or following bed renovationA second application can be made in late fall or early spring when plants are dormant. **Row middle.** May be applied up to 1.5 pints/acre. PHI = 120 days.
POSTEMERGENCE NON-SELECTIVE Contact kill of all green foliage	paraquat, MOA 22 (Firestorm) 3 SL (Parazone) 3 SL (Gramoxone Inteon) 2 L	1.3 pt 2 pt	0.5 to 1	For control of emerged broadleaf and grass weeds, use shields and direct spray between the rows to prevent contact with strawberry foliage. Use a nonionic surfactant at a rate of 16 to 32 ounces per 100 gal spray mix or 1 gallon approved crop oil concentrate per 100 gal spray mix. Do not apply within 21 days of harvest.
POSTEMERGENCE Annual and perennial grasses only	clethodim, MOA 1 (Arrow, Clethodim, Intensity, Select) 2 EC (Intensity One, Select Max) 1 EC	6 to 8 oz 9 to 16 oz	0.094 to 0.125 0.07 to 0.125	Apply postemergence for control of emerged grasses in strawberries. With Arrow and Select, add 1 gallon crop oil concentrate per 100 gallons spray mix. With Select Max, add 0.25% nonionic surfactant, 1 quart per 100 gallons spray mix. Very effective in controlling annual bluegrass. Apply to actively growing grasses not under drought stress. Do not apply within 4 days of harvest.
	fluazifop, MOA 1 (Fusilade DX) 2 EC	16 to 24 oz	0.25 to 0.38	USE ON NONBEARING CROP ONLY. Postemergence grass control. Check label for rates and timings for specific weeds. Do not apply within 1 year of the first harvest. Use of a crop oil or surfactant will be necessary.
	sethoxydim, MOA 1 (Poast) 1.5 EC	1 to 1.5 pt	0.2 to 0.3	Apply to emerged grasses. Consult manufacturer's label for specific rates and best times to treat. Add 1 quart of crop oil concentrate per acre. Do not apply on days that are unusually hot and humid. Do not apply within 7 days of harvest.
Strawberries (plasticulture) Preplant				
PREEMERGENCE Yellow and purple nutsedge, broadleaf and grass weeds	EPTC, MOA 8 (Eptam) 7E	3.5 to 7 pt	3 to 6	No research has been conducted in NC. For best control of nutsedge, soil must have enough moisture for tuber sprouting. Allow 10 to 14 days for nutsedge tuber sprouting to occur, then lightly till to destroy shoots and dry the soil surface. Apply and incorporate Eptam 7E to prevent volatilization, immediately incorporate into soil to a depth of approximately 2 to 4 inches. If possible, use a leveling device behind the incorporating equipment to leave soil surface as smooth as possible. Field traffic, excessive rainfall, or irrigation and other soil disturbances will reduce the level of nutsedge suppression. To avoid injury to following crops, irrigating at least 30 days prior to planting is recommended. Do not plant crops not on Eptam 7E label for 45 days after application.
PREEMERGENCE Broadleaf weeds	acifluorfen, MOA 14 (Ultra Blazer) 2 L	0.5 to 1.5 pt	0.125 to 0.375	**Crop row.** Make one banded application before laying plastic mulch and after final land preparation, and prior to transplanting the crop. For best results, avoid soil disturbance during laying of plastic and planting of crop. **Row middles between plastic mulch rows.** Apply as a direct-shielded application to strawberry row middles between mulched beds. DO NOT ALLOW ULTRA BLAZER TO CONTACT STRAWBERRY PLANTS. Limited research has been conducted with Ultra Blazer in North Carolina
PREEMERGENCE Annual grasses and broadleaf weeds	napropamide, MOA 15 (Devrinol, Devrinol 2-XT) 2 EC (Devrinol, Devrinol DF-XT) 50 DF	8 qt 8 lb	4	Devrinol applied to bed before laying the plastic has potential to injure strawberry plants. For plantbed treatment, preplant incorporate to weed-free soil before laying plastic mulch. Soil should be well worked yet moist enough to permit a thorough incorporation to a depth of 2 inches incorporated within 24 hours of application before laying of plastic mulch. If weed pressure is from small-seeded annuals, apply Devrinol to the suface of the bed immediately in front of the laying of the plasic mulch. If soil is dry, water or sprinkler irrigate with sufficent water to wet to a depth of 2 to 4 inches before covering with plastic mulch. Lay the plastic mulch over the treated soil on the same day as the Devrinol application.
PREEMERGENCE Broadleaf weeds including Carolina geranium and cutleaf eveningprimrose, and a few annual grasses	oxyfluorfen, MOA 14 (Goal) 2 XL	up to 2 pt	up to 0.5	Apply to the soil surface of pre-formed beds at least 30 days prior to transplanting crop for control of many broadleaf weeds that will emerge from hole near crop. While incorporation is not necessary, it may result in less crop injury. Soil disturbance after application will reduce weed control. Plastic mulch can be applied any time after applying Goal, but best results are likely if it is applied soon after Goal.
PREEMERGENCE Annual broadleaf weeds including cutleaf evening primrose, henbit, chickweed, horseweed, wild radish and some annual grasses	flumioxazin, MOA 14 (Chateau SW) 51 WDG	3 oz	0.096	**Crop row.** Apply a minimum of 30 days prior to transplanting and prior to plastic mulch being laid. **Row middles between plastic mulch rows.** Apply only to row middles. DO NOT APPLY over top of strawberries. Apply prior to weed emergence and prior to fruit set. Crop spotting may occur if an adjuvant is added. Application after fruit set may result in spotting of fruit and should be avoided. Do not allow spray drift to come in contact with fruit or foliage.

CHEMICAL WEED CONTROL

CHEMICAL WEED CONTROL

Table 7-12A. Chemical Weed Control in Small Fruit Crops

Timing/Targeted Weeds	Herbicide, Mode of Action Code[1] and Formulation	Amount of Formulation Per Acre	Pounds Active Ingredient Per Acre	Precautions and Remarks
Strawberries (plasticulture) Preplant (continued)				
PREEMERGENCE Yellow nutsedge, purple nutsedge, corn spurry, yellow woodsorrel, henbit, chickweed	sulfentrazone, MOA 14 (Spartan) 4F	4 to 8 oz	0.125 to 0.25	Please refer to label for soil type restrictions.
POSTEMERGENCE Annual and perennial grasses only	sethoxydim, MOA 1 (Poast) 1.5 EC	1 to 2.5 pt	0.2 to 0.5	Apply as a postemergence application to kill emerged grasses. Most effective on actively growing grasses. See label for specific rates and best times to treat. Add 1 quart per acre of crop oil concentrate to spray solution. Very effective control of ryegrass but will not control sedges. Also, effective on volunteer small grains (wheat, etc.). Do not tank mix with other pesticides. PHI = 7 days.
	clethodim, MOA 1 (Arrow, Clethodim, Intensity, Select) 2 EC	6 to 8 oz	0.094 to 0.125	Apply as a postemergence application to kill emerged grasses. Use high rate and sequential applications for perennial grasses (bermudagrass or johnsongrass). With Arrow, Clethodim, Intensity, and Select, add 1 gallon crop oil concentrate per 100 gallons spray mix. With Intensity One and Select Max, add 0.25% nonionic surfactant, 1 quart per 100 gallons spray mix. Very effective in controlling annual bluegrass. Apply to actively growing grasses not under drought stress. PHI = 4 days.
	(Intensity One, Select Max) 1 EC	9 to 16 oz	0.07 to 0.125	
POSTEMERGENCE Broadleaf weeds including vetch, clover, dock, sowthistle, and thistle	clopyralid, MOA 4 (Stinger) 3 EC	0.33 to 0.5 pt	0.125 to 0.187	**Crop row.** Apply postemergence over crop for postemergence control. Do not use with other pesticides or surfactants. Do not apply within 30 days of harvest. DO NOT compost treated vegetation if compost will be used on sensitive plants
Strawberries (plasticulture) Row Middles				
PREEMERGENCE Annual grasses and small-seeded broadleaf weeds	napropamide, MOA 15 (Devrinol, Devrinol DF-XT) 50 DF	8 lb	4	Apply as a banded preemergence treatment to the middles between plastic before weed emergence. Tank mixture with paraquat will provide pre- and postemergence weed control. Rainfall or irrigation within 24 hours after Devrinol application is needed for optimum control. Effective on volunteer small grains (wheat, etc.) if applied before emergence.
	(Devrinol, Devrinol 2-XT) 2EC	8 qt		
PREEMERGENCE Annual grasses and small-seeded broadleaf weeds	Pendimethalin, MOA 3 (Prowl H$_2$O) 3.8 EC	1.5 pt	0.72	Avoid contact with strawberry plant. See label for more information. PHI = 35 days.
POSTEMERGENCE Annual broadleaf weeds	carfentrazone-ethyl, MOA 14 (Aim) 1.9 EW or 2 EC	up to 2 oz	up to 0.031	Apply post directed using hooded sprayers for control of emerged weeds. If crop is contacted, burning of contacted area will occur. Most effective on weeds less than 4 inches tall or rosettes less than 3 inches across. Use a crop oil concentrate at up to 1 gallon per 100 gallons solution or a nonionic surfactant at 2 pint per 100 gallons of spray solution. Coverage is essential for good weed control. Does not control grass weeds. Can be tank mixed with other registered herbicides. PHI = 0 days.
POSTEMERGENCE Broadleaf weeds including vetch, clover, dock, sowthistle, thistle	clopyralid, MOA 4 (Stinger) 3 EC	0.33 to 0.67 pt	0.125 to 0.25	Do not tank mix with other pesticides. Do not include an adjuvant. PHI = 30 days.
POSTEMERGENCE NON-SELECTIVE Contact kill of green foliage	paraquat, MOA 22 (Firestorm, Parazone) 3 SL (Gramoxone Inteon, Gramoxone SL, Gramoxone SL 2.0) 2 SL	1.3 pt 2 pt	0.5	Apply as a banded treatment using shields to the middles between plastic to kill emerged weeds. To avoid injury, do not allow spray to contact strawberry plants. Add a nonionic surfactant at a rate of 16 to 32 ounces per 100 gallons or 1 gallon approved crop oil concentrate per 100 gallons spray solution. Do not apply more than three times per season. PHI = 21 days.
POSTEMERGENCE NON-SELECTIVE Most emerged weeds	glyphosate, MOA 9 (Roundup Weather Max) 5.5L	11 to 22 oz	0.5 to 0.94	Apply as a hooded spray in row middles or shielded spray in row middles or wiper applications in row middles or post harvest. To prevent severe injury to crop, do not let herbicide contact foliage, green shoots or stems, exposed roots, or fruit of crop. PHI = 14 days.
POSTEMERGENCE	pelargonic acid, MOA 27 (Scythe) 4.2 SL	3 to 10% soln.		Apply with hooded or shielded sprayer for weed control in row middles. See label for directions.
POSTEMERGENCE Annual and perennial grasses only	sethoxydim, MOA 1 (Poast) 1.5 EC	1 to 2.5 pt	0.2 to 0.5	Apply as a postemergence application to kill emerged grasses. Most effective on actively growing grasses. See label for specific rates and best times to treat. Add 1 quart per acre of crop oil concentrate to spray solution. Very effective control of ryegrass but will not control sedges. Also, effective on volunteer small grains (wheat, etc.). PHI = 7 days.
	clethodim, MOA 1 (Arrow, Clethodim, Intensity, Select) 2 EC	6 to 8 oz	0.094 to 0.125	Apply as a postemergence application to kill emerged grasses. With Arrow, Clethodim, Intensity, and Select, add 1 gallon crop oil concentrate per 100 gallons spray mix. With Intensity One and Select Max, add 0.25% nonionic surfactant, 1 quart per 100 gallons spray mix. Very effective in controlling annual bluegrass. Apply to actively growing grasses not under drought stress. PHI = 4 days.
	(Intensity One, Select Max) 1 EC	9 to 16 oz	0.07 to 0.125	

[1] Mode of action (MOA) code developed by the Weed Science Society of America. See TABLE 7-10, Herbicide Modes of Action, for details.

Chemical Weed Control in Tree Fruit Crops

W. E. Mitchem, Horticultural Science Department

NOTE: A mode of action code has been added to the Herbicide and Formulation column of this table. Use MOA codes for herbicide resistance management. See Table 7-10, Herbicide Resistance Management, for details.

Table 7-12B. Chemical Weed Control in Fruit Crops—Tree Fruits

Weed	Herbicide, Mode of Action,* and Formulation	Amount of Formulation Per Acre	Pounds Active Ingredient Per Acre	Precautions and Remarks
Apples, Preemergence Directed Underneath Tree				
Annual and perennial grass and broadleaf weeds	dichlobenil, MOA 20 (Casoron) 4 G (Casoron) 1.4 CS	100 to 150 lb 1.4 to 2.8 gal	4 to 6 2 to 3.92	For best results apply Casoron 4G in January or February. In order to prevent loss to volatilization, Casoron CS should be applied when temperatures are less than 70 degrees F. Casoron CS may be tank mixed with glyphosate or other herbicides registered for use in apples. Casoron 4G can be used in newly planted orchards once trees have been planted for 4 weeks. Casoron CS should only be used in established orchards 1 year after transplanting.
Annual grasses and broadleaf weeds	diuron, MOA 7 (Diuron, Karmex DF) 80 WDG (Direx, Diuron) 4L	2 to 4 lb 1.6 to 3.2 qt	1.6 to 3.2	Apply in spring (March thru May) to trees established in the orchard for at least 1 year. Best results occur if rainfall occurs within 2 weeks of application. DO NOT treat varieties grafted on full-dwarf rootstocks. When using sequential applications allow at least 90 days between applications and total use rate cannot exceed 4 pounds/acre per year.
	flumioxazin, MOA 14 (Chateau) 51 SW (Tuscany) 51 WG	6 to 12 oz	0.19 to 0.38	Chateau is for newly planted and established orchards. Shield the trunks of trees established less than 1 year from contact with spray solution. Tank mix with glyphosate, glufosinate, or paraquat for POST weed control. After budbreak, only tank mix with glufosinate or paraquat. Do not apply more than 6 ounces per acre to trees planted less than 3 years in soil having a sand plus gravel content more than 80%. Sequential applications are very effective. Do not apply within 60 days of harvest. Chateau may only be applied after final harvest and no later than pink flower bud in bearing orchards. Do not use more than 24 ounces of Chateau in a 12-month period. When applying after bud break in non-bearing orchards use hooded application equipment.
	indaziflam, MOA 29 (Alion) 1.67 SC	3.5 to 6.5 oz	0.046 to 0.085	Use in orchards established 3 years or more. See label for details pertaining to replants in established orchards. Allow at least 90 days between applications. Use rate cannot exceed 3.5 fluid ounces per acre per application on soils having less than 1% organic matter. On soils with an organic matter content from 1 to 3%, no more than 5 fluid ounces/acre can be applied in a single application and the total use rate for the year cannot exceed 8.5 fluid ounces/acre. In order to apply more than 5 fluid ounces/acre in a single application soil organic matter must be >3%. Do not use on soils that have a 20% or greater gravel content. Do not use in orchards with open channels or cracks in soil. Alion has a 14-day PHI. Tank mix glyphosate, glufosinate, or paraquat for nonselective POST weed control.
Annual grasses and small seeded broadleaf weeds	norflurazon, MOA 12 (Solicam) 80 WDG	2.5 to 5 lb	2 to 4	Can be tank mixed with Karmex, Goal, paraquat, Prowl, glyphosate, Princep, rimsulfuron, or oryzalin. Do not apply to newly transplanted trees until ground has settled. Rate is soil texture dependent. See label for details. PHI is 60 days. Multiple applications can be made per season so long as cumulative rate does not exceed maximum use rate for soil texture and crop.
	oryzalin, MOA 3 (Oryzalin or Surflan) 4 AS	2 to 6 qt	2 to 6	Allow soil to settle around newly transplanted trees before application. Rainfall (1/2 to 1") with 14 days of application is necessary for optimum herbicide performance. Oryzalin may be tank mixed with Goal, paraquat, glyphosate, Solicam, and simazine. May be applied sequentially. See label for details. Sequential applications may be used so long as total use rate does not exceed 12 quarts per acre per year. Allow 2.5 months between applications.
	pendimethalin, MOA 3 (Prowl H$_2$O) 4 AS	2 to 4.2 qt	2 to 4	Most effective when adequate rainfall or irrigation is received within 7 days after application. Do not apply to newly transplanted trees until ground has settled. Rainfall within 7 to 10 days of application is necessary for herbicide activation. Tank mix with paraquat for POST weed control. 60-day preharvest interval (PHI). May be applied as a sequential application as long as rate does not exceed 4.2 quarts/acre per year. Allow 30 days between applications
Broadleaf weeds and some annual grasses	rimsulfuron, MOA 2 (Matrix) 25 WG (Solida) 25 WG (Pruvin) 25 WG	4 oz	0.063	For broad spectrum residual control, tank mix with diuron, Sinbar, oryzalin or Prowl H$_2$O. For nonselective POST control, apply in combination with glyphosate or paraquat. Rimsulfuron does have POST activity on certain broadleaf weeds (see label for list). Rimsulfuron will control emerged horseweed less than 3 inches tall when applied in combination with a non-ionic surfactant and a spray grade ammonium sulfate (2 pound/acre). **DO NOT** treat orchards established less than 1 year. Rainfall within 2 to 3 weeks of application is necessary for herbicide activation. Spray solutions having a pH of less than 4.0 or greater than 8.0 will result in herbicide degradation. Rimsulfuron has a 7-day PHI for apples. Rimsulfuron may be applied as a sequential application so long as total use rate does not exceed 4 ounces/acre per year and application is made in a band on less than 50% of orchard floor. Allow at least 30 days between applications.
Annual broadleaf and grass weeds	simazine, MOA 5 (Princep, Simazine) 4 L 90 WDG	2 to 4 qt 2.2 to 4.4 lb	2 to 4	Apply preemergence to trees that have been established 1 year or more. Apply with glyphosate, paraquat, or glufosinate for postemergence weed control. PHI for simazine is 150 days. Tank mixing simazine with oryzalin, Solicam, or Prowl H$_2$O will expand residual control of annual grasses and certain broadleaf weeds.
Annual broadleaf weeds and yellow nutsedge	sulfentrazon + carfentrazone-ethyl, MOA 14 (Zeus Prime XC) 3.5 XC	7.7 to 15.1 oz	0.21 to 0.41	Use only in orchards established 3 years of longer. Avoid contact with green tissue or bark. Sequential applications of Zeus Prime can be applied whent directed as a banded application (50% band or less of orchard floor) so long as total use rate does not exceed 15.1 fl. oz/A on a broadcast basis with a year and the second application is not applied within 60 days of the initial application. Zeus Prime has a 14 day PHI. For optimum residual control of annual grass weeds tank mix with oryzalin, pendimethalin, or norflurazon. Tank mix glyphosate, glufosinate, or paraquat for non-selective POST weed control.

CHEMICAL WEED CONTROL

Table 7-12B. Chemical Weed Control in Fruit Crops—Tree Fruits

Weed	Herbicide, Mode of Action,* and Formulation	Amount of Formulation Per Acre	Pounds Active Ingredient Per Acre	Precautions and Remarks
Apples, Preemergence Directed Underneath Tree (continued)				
Broadleaf and grass weeds	terbacil, MOA 5 (Sinbar) 80 WDG	2 to 4 lb	1.6 to 3.2	Use only on trees that have been established 3 years or more. Tank mixing Sinbar with Karmex allows Sinbar to be used in orchards established 1 year or longer. Rate varies with soil organic matter. See label for details. Apply no more than 3 lb unless soil organic matter is greater than 2%. Do not use on sand or loamy sand soils. Do not use on soils having less than 1% organic matter. Sinbar has a 60-day PHI. Tank mix with glyphosate, paraquat or glufosinate for non-selective postemergence weed control.
Most annual broadleaf weeds and grass weeds in **NEWLY PLANTED NON-BEARING ORCHARDS**	terbacil, MOA 5 (Sinbar) 80 WDG	0.5 to 1.0 lb	0.4 to 0.8	Apply once adequate rainfall has occurred allowing the soil to settle after transplanting. Apply no more than 1 pound per acre per year. For best results, apply 0.5 pound in spring followed by another 0.5 pound when control from initial application fails. Do not use on soils coarser than sandy loams of soils with less than 1% organic matter. Tank mix with paraquat for non-selective postemergence weed control.
Apples, Preemergence Tank Mixes				
Many annuals and perennial grass and broadleaf weeds	diuron, MOA 7 (Diuron or Karmex DF or XP) 80 WDG (Direx) 4 L + terbacil, MOA 5 (Sinbar) 80 WDG	1 to 2 lb 1.6 to 3.2 qt + 1 to 2 lb	0.8 to 1.6 0.5 to 1 + 0.8 to 1.6	DO NOT treat varieties grafted on full-dwarf rootstocks. Use only on trees established in orchard for 1 year. See labels for details. See labels for soil texture and organic matter restrictions.
Annual grasses and broadleaf weeds	norflurazon, MOA 12 (Solicam) 80 WDG + simazine, MOA 5 (Princep, Simazine) 4 L 90 WDG	2.5 to 5 lb + 2 to 4 qt 2.2 to 4.4 lb	2 to 4 + 2 to 4	See labels for details. Apply in combination with paraquat, glyphosate, or glufosinate for postemergence control.
	oryzalin, MOA 3 (Oryzalin, Surflan) 4 AS + simazine, MOA 5 (Princep, Simazine) 4 L 90 WDG	2 to 4 qt + 2 to 4 qt 2.2 to 4.4 lb	2 to 4 + 2 to 4	See labels for details. Apply in combination with paraquat, glyphosate, or glufosinate for postemergence control.
	rimsulfuron, MOA 2 (Matrix) 25 WG (Solida) 25 WG (Pruvin) 25 WG + terbacil, MOA 5 (Sinbar) 80 WDG	2 oz + 1 to 2 lb	0.031 + 0.8 to 1.6	See each product label for use precautions. Tank mix with glyphosate, glufosinate, or paraquat for non-selective POST weed control. See label for soil texture and organic matter restrictions.
	norflurazon, MOA 12 (Solicam) 80 WDG + diuron, MOA 7 (Diuron or Karmex DF or XP) 80 WDG	2.5 to 5 lb + 2 to 4 lb	2 to 4 + 1.6 to 3.2	Use only on trees established 1 year or more. Do not treat varieties grafted on full-dwarf rootstocks. See label for details. Apply in combination with glyphosate, paraquat, or glufosinate for postemergence weed control.
	rimsulfuron, MOA 2 (Matrix) 25 WG (Solida) 25 WG (Pruvin) 25 WG + oryzalin, MOA 3 (Oryzalin, Surflan) 4 AS	4 oz + 2 to 4 qt	0.0625 + 2 to 4	Tank mix with glyphosate, glufosinate, or paraquat for non-selective POST weed control.
	rimsulfuron, MOA 2 (Matrix) 25 WG (Solida) 25 WG (Pruvin) 25 WG + diuron, MOA 7 (Diuron or Karmex DF or XP) 80 WDG	4 oz + 2 to 4 lb	0.0625 + 1.6 to 3.2	Tank mix with glyphosate, glufosinate, or paraquat for non-selective POST weed control.
Annual broadleaf and grass weeds as well as yellow nutsedge	sulfentrazone + carfentrazone-ethyl, MOA 14 (Zeus Prime) 3.5 XC + oryzalin, MOA 3 (Oryzalin, Suflan) 4 AX	7.7 to 15.1 oz	0.21 to 0.41	
Apples, Postemergence, Directed Underneath Tree				
Broadleaf weeds including morningglory, pigweed, dayflower, lambsquarters, and prickly lettuce	carfentrazone-ethyl, MOA 14 (Aim) 2 EC	0.5 to 2 oz	0.008 to 0.032	Apply alone or tank mixed with other herbicides. Apply in a minimum spray volume of 20 gpa. Applications can be made with boom equipment, hooded sprayers, or shielded sprayers. Do not allow Aim to contact green bark, desirable foliage, flowers, or fruit. Contact with fruit or foliage will result in spotting and leaf necrosis. The trunks of trees less than 2 years old must be protected from Aim. Do not apply within 3 days of harvest. Sequential applications may be used so long as there is 14 days between applications and total use rate for year does not exceed 7.9 ounces/acre. Best results are obtained when applied to weeds in the 2- to 3-leaf stage. Apply in combination with a nonioinic surfactant (1 quart/100 gallons of spray solution) or crop oil concentrate (1 gallon/100 gallons of spray solution).

Table 7-12B. Chemical Weed Control in Fruit Crops—Tree Fruits

Weed	Herbicide, Mode of Action,* and Formulation	Amount of Formulation Per Acre	Pounds Active Ingredient Per Acre	Precautions and Remarks
Apples, Postemergence, Directed Underneath Tree (continued)				
Broadleaf weeds including perennials like blackberry, horsenettle, poison ivy, Virginia creeper, and white clover.	fluroxypyr, MOA 4 (Starane Ultra) 2.8	0.7 to 1.4 pt	0.35 to 0.70	DO NOT apply during bloom or to trees less than 4 years old. Make only one application per year. Starane may be tank mixed with other herbicides registered for use on apples. Starane has a 14-day PHI. **This use is not listed on the Starane label because it is on a supplemental label.**
Broadleaf weeds including horseweed, morningglory, pigweed, ragweed, smartweed and purslane	saiflufenacil, MOA 14 (Treevix) 70 WG	1 oz	0.044	Do not apply more than 3 ounces/acre per year. Allow at least 21 days between applications. Treevix has a 0 day PHI. Treevix may be tank mixed with glyphosate, glufosinate, Poast, and oxyfluorfen. Treevix provides excellent control of horseweed, purslane, morningglory species, ragweed, and smartweed. The addition of methylated seed oil at 1% v/v (1 gallon per 100 gallons of spray solution) plus ammonium sulfate at 8.5 to 17 lbs/100 gallons of spray solution is required for optimum herbicide performance.
Most annual broadleaf and grass weeds plus many perennials	glufosinate, MOA 10 (Lifeline, Reckon, Rely) 2.34 L	48 to 82 oz	0.88 to 1.5	DO NOT SPRAY GREEN BARK OR FOLIAGE. Glufosinate should not be used on trees within 1 year of transplanting. Apply in a minimum of 20 gallons of water per acre as a directed spray under trees. Repeat applications may be necessary for control of perennial weeds. Glufosinate can be tank mixed with diuron, Sinbar, Solicam, oryzalin, Devrinol, Goal, rimsulfuron, and simazine. Glufosinate has a 14 day PHI. DO NOT allow spot spray applications to directly contact tree or suckers. The addition of ammonium sulfate will enhance glufosinate activity on difficult to control species; however, the addition of surfactants and crop oil will increase risk of crop injury.
	glyphosate, MOA 9 (various brands and formulations)	See label	1 to 2	DO NOT SPRAY GREEN BARK OR FOLIAGE. Trees are more susceptible to injury from midsummer until dormant. Repeat applications may be necessary for control of perennial weeds. Can be tank mixed with Goal, Karmex, simazine, Solicam, and oryzalin. Check label for specifics. Generic glyphosate formulations may require the addition of a surfactant at 0.5% by volume (2 quart per 100 gallons of spray solution). See label for spray additive information and for detailed restriction information.
Yellow and purple nutsedge, horsenettle, and pokeweed	halosulfuron, MOA 2 (Sandea) 75 WDG	0.5 to 1 oz	0.023 to 0.047	Apply halosulfuron to actively growing weeds. Do not apply to apple trees established less than 1 year. Do not apply more than 2 ounces/acre per 12-month period. Avoid herbicide contact with tree foliage. Addition of a nonionic surfactant is necessary for optimum herbicide performance. Sequential applications may be more effective on yellow nutsedge than one application. When using sequential applications use at least 0.75 ounces/acre per application. Halosulfuron may be tank mixed with glyphosate for broad spectrum POST weed control.
Broadleaf and some small annual grass weeds	paraquat, MOA 22 (Gramoxone) SL 2 SL (Firestorm) 3 SL (Paraquat Concentrate) 3 SL (Parazone) 3 SL	2.5 to 4 pt 1.7 to 2.7 pt	0.6 to 1	Apply when grass and weeds are 1 to 6 inches high and succulent for best results. Direct spray with low pressure to avoid contact with tree foliage or bark less than 1 year old. Young trees must be shielded to prevent spray contact with bark. Add surfactant at 0.25% by volume (2 pints per 100 gallons). Paraquat may be tank mixed with Goal, Karmex, simazine, Sinbar, Solicam, and oryzalin. Paraquat is a restricted use pesticide.
Broadleaf weeds	2,4-D amine, MOA 4 (various manufacturers and brands) 3.8 SL	1 to 3 pt	0.95 to 1.4	Apply any time during the growing season to actively growing broadleaf weeds except during apple bloom. Trees must be at least 1 year old. Do not apply more than two applications per crop cycle (75-day interval between applications) or within 14 days of harvest. Some formulations limit rate to 2 pints per acre. See label for details.
Broadleaf weeds including clover, horseweed, thistle, dandelion and mugwort	clopyralid, MOA 4 (Stinger) 3EC	0.33 to 0.66 pt	0.125 to 0.25	Use on trees established that have been established in the orchard at least 1 year or longer. No more than two applications of Stinger can be applied per crop year and total use rate within that time frame cannot exceed 2/3 pt per acre. DO NOT apply within 30 days of harvest. DO NOT apply during bloom and avoid contact with foliage, fruit, and tree trunk.
Grasses	clethodim, MOA 1 (Arrow, Clethodim, Intensity, or Select) 2 EC (Select Max or Intensity One) 1 EC	6 to 8 oz 12 to 16 oz	0.094 to 0.125	Apply to actively growing grasses not under stress. See label for rate and optimum grass size to treat. Multiple applications may be necessary to control perennial grass weeds. Use in **NONBEARING TREES ONLY**. When using the 2 EC formulations, add crop oil concentrate at 1% by volume (1 gallon per 100 gallons of spray solution). When using the 1 EC formulations, add nonionic surfactant at 0.25% by volume (1 qt per 100 gallons).
	fluazifop, MOA 1 (Fusilade DX) 2 EC	12 to 24 oz	0.125 to 0.38	**NONBEARING TREES ONLY.** Apply to actively growing grasses not under stress. See label for rate and optimum grass size to treat. Multiple applications may be necessary to control perennial grass weeds. Add crop oil at 1% by volume (1 gallon per 100 gallons).
	sethoxydim, MOA 1 (Poast) 1.5 EC	1.0 to 2.5 pt	0.19 to 0.47	Apply to actively growing grasses not under stress. See label for rate and optimum grass size to treat. Multiple applications may be necessary to control perennial grass weeds. Add Dash adjuvant at 1 pint per acre or crop oil concentrate at 1 quart per acre. Do not apply within 14 days of harvest. Do not apply more than 7.5 pints per acre per year.
Apples, Ground Cover Suppression				
Suppression of fescue, orchardgrass, and bluegrass	glyphosate, MOA 9 (various brands and formulations)	Rates and application time vary for each grass species. See label for details.		Mow one time in spring. Apply 3 to 4 days after mowing. **Caution:** This treatment will normally discolor the grass. **DO NOT** apply after seedhead emergence. See label for details.
Peaches, Preemergence Directed UnderneathTree				
Annual grasses and some broadleaf weeds	diuron, MOA 7 (Direx , Diuron) 4 L (Diuron, Karmex DF) 80 DF	1.6 to 2.2 qt 2 to 2.75 lb	1.6 to 2.2	Apply in spring to trees at least 3 years old. Rate is soil texture dependent. May be tank mixed with Sinbar, Solicam, glyphosate, or paraquat. Karmex DF, Karmex XP, and Direx 4L have a 20-day PHI. Other formulations of diuron have a 90-day PHI.

Table 7-12B. Chemical Weed Control in Fruit Crops—Tree Fruits

Weed	Herbicide, Mode of Action,* and Formulation	Amount of Formulation Per Acre	Pounds Active Ingredient Per Acre	Precautions and Remarks
Peaches, Preemergence Directed UnderneathTree (continued)				
Annual grasses and some broadleaf weeds (continued)	flumioxazin, MOA 14 (Chateau) 51 SW (Tuscany) 51 WG	6 to 12 oz	0.19 to 0.38	Chateau is for newly planted and established orchards. Shield or protect trees established less than 1 year from contact with spray solution. Tank mix with paraquat for non-selective POST weed control. Do not apply more than 6 ounces per acre to trees planted less than 3 years in soil having a sand plus gravel content more than 80%. Sequential applications are very effective. Due to the potential for crop injury, Chateau should not be applied in bearing orchards after budbreak until after final harvest. Do not apply within 60 days of harvest. Do not use more than 24 ounces per acre per year. In non-bearing orchards Chateau may be applied after bud break however application equipment should be hooded.
	indaziflam, MOA 29 (Alion) 1.67 SC	3.5 to 6.5 oz	0.046 to 0.085	Use in orchards established 3 years or longer. See label for details regarding the management of replants in established orchards. Use rate cannot exceed 3.5 fluid ounces/acre per application on soils having less than 1% organic matter. On soils with an organic matter content from 1 to 3%, no more than 5 fluid ounces/acre can be applied in a single application and the total use rate for the year cannot exceed 8.5 fluid ounces/acre. In order to apply more than 5 fluid ounces/acre in a single application soil organic matter must be >3%. Do not use on soils with 20% or more gravel content. Allow at least 90 days between applications. Research has shown Alion applied in the fall followed by a late spring application will provide summer long control of annual broadleaf and grass weeds. Do not treat soil around trees with cracks or channels, or with depressions. Tank mix Alion with glyphosate or paraquat for nonselective POST weed control. Alion has a 14-day PHI.
	norflurazon, MOA 12 (Solicam) 80 WDG	2.5 to 5 lb	2 to 4	Can be tank mixed with Karmex, Goal, glyphosate, paraquat, Prowl, rimsulfuron, simazine, Sinbar, or oryzalin. Rate is soil texture dependent. See label for details. Do not apply within 6 months of transplanting. PHI is 60 days. Multiple applications can be made per season so long as total use rate does not exceed maximum use rate for soil texture and crop.
	oryzalin, MOA 3 (Oryzalin or Surflan) 4 AS	2 to 6 qt	2 to 6	Allow soil to settle around newly transplanted trees before application. Oryzalin may be tank mixed with glyphosate, paraquat, simazine, and Solicam. Sequential applications permitted as long as there is 2.5 months between the applications. See label for details. In newly planted orchards may be tank mixed with Gallery for broad spectrum preemergence control. Sequential applications may be used if total use rate does not exceed 12 quarts per acre per year.
	pendimethalin, MOA 3 (Prowl H₂O) 4 AS	2 to 4 qt	2 to 4	Most effective when adequate rainfall or irrigation is received within 7 days after application. Do not apply to newly transplanted trees until ground has settled around roots. Apply with paraquat to control emerged weeds. Prowl H₂O has a 60-day preharvest interval (PHI). Pendimethalin may be applied as sequential applications so long as total amount used does not exceed 4.2 qt/A per year. Allow at least 30 days between applications.
Broadleaf and some grass weeds	rimsulfuron, MOA 2 (Matrix) 25 WG (Solida) 25 WG (Pruvin) 25 WG	4 oz	0.063	For broad spectrum PRE control, tank mix with diuron, Sinbar, oryzalin or Prowl H₂O. For nonselective POST control, apply with glyphosate or paraquat. Rimsulfuron does have POST activity on certain broadleaf weeds (see lable for list). Rimsulfuron will control emerged horseweed less than 3 inches tall when applied in combination with a non-ionic surfactant and a spray grade ammonium sulfate (2 pounds/acre). Do NOT treat orchards established less than 1 year. Rainfall within 2 to 3 weeks of application is necessary for herbicide activation. Spray solutions having a pH lower than 4.0 or higher than 8.0 will result in herbicide degradation. Rimsulfuron has a 14-day PHI for stone fruit, and sequential applications can be made so long as total use rate does not exceed 4 ounces/acre per year and application is made in a band on less than 50% of orchard floor. Allow at least 30 days between applications.
Broadleaf and grass weed control for NEWLY PLANTED NON-BEARING ORCHARDS	terbacil, MOA 5 (Sinbar) 80 WDG	0.5 to 1.0 lb	0.4 to 0.8	Apply once soil has settled after transplanting. Apply no more than 1 lb per acre per year. For best results, apply 0.5 lb in the spring followed by another 0.5 lb when control from initial application fails. Do not apply on soils coarser than sandy loam. Do not use on soils having less than 1% organic matter.
Annual broadleaf and grass weeds	simazine, MOA 5 (Princep, Simazine) 4 L 90 WDG	1.6 to 4 qt 1.8 to 4.4 lb	1.6 to 4	Apply in early spring before weed emergence. Use only on trees established 1 year or more. Do not use on sand or loamy sand soils. Tank mixing simazine with oryzalin, Prowl H₂O, or Solicam will improve residual control of annual grasses and certain broadleaf weeds.
Annual broadleaf and grass weeds plus many perennial grasses	terbacil, MOA 5 (Sinbar) 80 WDG	2 to 4 lb	1.6 to 3.2	Use on trees established 3 years, however when tank mixed with Karmex XP Sinbar may be applied in orchards established 1 year or longer. Sinbar may only be used on soils with at least 1% organic matter. Unless soil organic matter is greater than 2% do not exceed 3 pounds/acre. Do not use on sand or loamy sand soils. Sinbar is an excellent choice for tank mixing with diuron or rimsulfuron for extended broad spectrum residual control of those products. Sinbar has a 60-day PHI.
Peaches, Preemergence Tank Mixes				
Many annual and perennial grasses and broadleaf weeds	diuron, MOA 7 (Diuron or Karmex DF or XP) 80 WDG + terbacil, MOA 5 (Sinbar) 80 WDG	1 to 2 lb + 1 to 2 lb	0.8 to 1.6 + 0.8 to 1.6	Use only under trees established in the orchard for at least 1 year. Apply to soils having at least 1% organic matter. See label for details.
	oryzalin, MOA 3 (Oryzalin or Surflan) 4 AS + simazine, MOA 5 (Princep, Simazine) 4 L 90 WDG	2 to 4 qt + 1.6 to 4 qt 1.75 to 4.4 lb	2 to 4 + 1.6 to 4	Tank mix for use before weed emergence. Tree must be established at least 1 year.

Table 7-12B. Chemical Weed Control in Fruit Crops—Tree Fruits

Weed	Herbicide, Mode of Action,* and Formulation	Amount of Formulation Per Acre	Pounds Active Ingredient Per Acre	Precautions and Remarks
Peaches, Preemergence Tank Mixes (continued)				
Many annual and perennial grasses and broadleaf weeds (continued)	norflurazon, MOA 12 (Solicam) 80 WDG + simazine, MOA 5 (Princep, Simazine) 4 L 90 WDG	2.5 to 5 lb + 2 to 4 qt 2.2 to 4.4 lb	2 to 4 + 2 to 4	See labels for details.
	rimsulfuron, MOA 2 (Matrix) 25 WG (Solida) 25 WG (Pruvin) 25 WG + terbacil, MOA 5 (Sinbar) 80 WDG	2 oz + 1 to 2 lb	0.063 + 0.8 to 1.6	See labels for use precautions and details.
	rimsulfuron, MOA 2 (Matrix) 25 WG (Solida) 25 WG (Pruvin) 25 WG + diuron, MOA 7 (Diuron or Karmex DF or XP) 80 WDG	4 oz + 2 to 4 lb	0.0625 + 1.6 to 3.2	Tank mix with glyphosate, glufosinate, or paraquat for non-selective POST weed control.
	norflurazon, MOA 12 (Solicam) 80 WDG + diuron, MOA 7 (Diuron or Karmex DF) 80 WDG	2.5 to 5 lb + 2 to 4 lb	2 to 4 + 1.6 to 3.2	See labels for details. Trees must be established at least 3 years.
	rimsulfuron, MOA 2 (Matrix) 25 WG (Solida) 25 WG (Pruvin) 25 WG + oryzalin, MOA 3 (Oryzalin, Surflan) 4 AS	4 oz + 2 to 4 qt	0.063 + 2 to 4	Tank mix with glyphosate or paraquat for non-selective POST weed control.
Peaches, Postemergence, Directed Underneath Tree				
Broadleaf weeds including morningglory, pigweed, lambsquarters, cocklebur, smartweed, and dayflower	carfentrazone-ethyl, MOA 14 (Aim) 2 EC	0.5 to 1.6 oz	0.008 to 0.025	Apply alone or tank mixed with other herbicides. Apply in a minimum spray volume of 20 gpa. Applications can be made with boom equipment, hooded sprayers, or shielded sprayers. Do not allow Aim to contact green bark, desirable foliage, flowers, or fruit of the crop. Contact with fruit or foliage will result in spotting and leaf necrosis. The trunks of trees established less than 2 years must be protected. Do not apply within 3 days of harvest. Best results are obtained when applied to weeds in the 2- to 3-leaf stage. Sequential applications may be used so long as there is atleast 14 days between applications and total use rate for year does not exceed 7.9 ounces/acre per year. Apply in combination with a nonioinic surfactant (1 quartt per 100 gallons of spray solution) or crop oil concentrate (1 gallon per 100 gallons of spray solution).
Kill all green foliage on contact	paraquat, MOA 22 (Gramoxone SL) 2 SL (Firestorm) 3 SL (Paraquat Concentrate) 3 SL (Parazone) 3 SL	2.5 to 4 pt 1.7 to 2.7 pt	0.66 to 1	Apply when grass and weeds are 1 to 6 inches high and succulent for best results. Direct spray with low pressure to avoid contact with tree foliage or bark. Add surfactant at 0.25% by volume (2 pints per 100 gallons) for best results. Paraquat may be tank mixed with Goal, Karmex, simazine, Sinbar, Solicam, and oryzalin. Paraquat is a restricted use pesticide. Newly planted trees can be severely injured by paraquat, so use a shield or wrap to protect the tree from spray. Do not make more than three applications per year. Paraquat has a 14-day PHI. Paraquat has a 14-day PHI for peach and 28-day PHI for nectarine.
Non-selective weed control	glyphosate, MOA 9 (various brands and formulations)	See label	1	Do not apply in orchards established less than 2 years. Applications must be made with shielded sprayer. Low hanging limbs and suckers must be removed at least 10 days prior to application. DO NOT use glyphosate 90 days past bloom. DO NOT allow glyphoste to contact foliage or bark; EXTREME care must be taken to prevent injury. See label for details. Some glyphosate formulations may require the addition of a surfactant.
Most annual broadleaf and grass weeds plus many perennials	glufosinate, MOA 10 (Lifeline) 2.34 L (Reckon 280 SL) 2.34 L (Rely 280) 2.34 L	48 to 82 oz	0.88 to 1.5	**DO NOT SPRAY GREEN BARK, UNCALLUSED BARK OR DESIRABLE FOLIAGE UNLESS TREES ARE PROTECTED.** Glufosinate should not be used on trees within 1 year of transplanting. Apply in a minimum of 20 gallons of water per acre as a directed spray under trees. Repeat applications may be necessary for control of perennial weeds. Glufosinate can be tank mixed with diuron, Sinbar, Solicam, oryzalin, Devrinol, Goal, rimsulfuron, and simazine. Glufosinate has a 14 day PHI. DO NOT apply more than 164 fl oz per acre within a 12 month period. There must be at least 28 days between applications. Glufosinate formulations contain surfactant therefore **additional nonionic surfactants or crop oils are not necessary and may increase potential for injury.**
Grasses	clethodim, MOA 1 (Arrow, Clethodim, Intensity, or Select) 2 EC (Select Max or Intensity One) 1 EC	6 to 8 oz 12 to 16 oz	0.094 to 0.125	Apply to actively growing grasses not under stress. See label for rate and optimum grass size to treat. Multiple applications may be necessary to control perennial grass weeds. When using 2 EC formulation chemicals, add crop oil concentrate at 1% by volume (1 gallon per 100 gallons). When using 1 EC formulations, use a nonionic surfactant at 0.25% by volume rather than crop oil. **Select Max has a 14-day PHI for peach. Unless otherwise stated on label, all other clethodim products are for non-bearing orchards ONLY.**
	fluazifop, MOA 1 (Fusilade DX) 2 EC	8 to 24 oz	0.125 to 0.38	Apply to actively growing grasses not under stress. See label for rate and optimum grass size to treat. Multiple applications may be necessary to control perennial grass weeds. Add crop oil at 1% by volume (1 gallon per 100 gallons). Do not apply within 14 days or harvest. Do not apply more than 72 fl ounces per acre per year.

Table 7-12B. Chemical Weed Control in Fruit Crops—Tree Fruits

Weed	Herbicide, Mode of Action,* and Formulation	Amount of Formulation Per Acre	Pounds Active Ingredient Per Acre	Precautions and Remarks
Peaches, Postemergence, Directed Underneath Tree (continued)				
Grasses (continued)	sethoxydim, MOA 1 (Poast) 1.5 EC	1.0 to 2.5 pt	0.19 to 0.47	Apply to annual grasses up to 12 inches tall. For perennial grasses apply early in the growth cycle at the high use rate. Multiple applications may be necessary for perennial grass weeds. Add Dash adjuvant at 1 pint per acre or crop oil concentrate at 1 quart per acre. Do not apply within 25 days of harvest. Do not apply more than 5 pints per acre per year.
Broadleaf weeds	2,4-D amine, MOA 4 (Weedar 64) (various brands) 3.8 SL	1 to 3 pt	0.95 to 1.4	Do not apply within 40 days of harvest. Do not apply more than twice a year and allow 75 days between applications. Trees must be at least 1 year old. Use when trees are dormant. Some formulations limit rate to 2 pints per acre. See label for details.
Broadleaf weeds including clover, horseweed, dock, thistle and mugwort	clopyralid, MOA 4 (Stinger) 3 EC	0.33 to 0.66 pt	0.125 to 0.25	Multiple applications may be used as long as total amount does not exceed maximum rate. Use at least 10 GPA of spray solution. Stinger may be tank mixed with preemergence herbicides. Do not apply within 30 days of harvest. Do not apply more than twice.
Pecans, Preemergence, Directed UnderneathTree				
Broadleaf weeds and annual grasses	diuron, MOA 7 (Diruon or Karmex DF) 80 WDG (Direx) 4 L	2 to 4 lb 1.6 to 3.2 qt	1.6 to 3.2	Do not apply to trees less than 3 years old. Rate is soil texture dependent. Do not use on soils with less than 0.5% organic matter.
Annual grass and some broadleaf weeds	norflurazon, MOA 12 (Solicam) 80 WDG	2.5 to 5 lb	2 to 4	Do not apply when nuts are on the ground. Rate is soil texture dependent. See label for details. Do not apply within 6 months of planting. PHI is 60 days. Multiple applications can be made per season so long as total use rate does not exceed maximum use rate for soil texture and crop.
	oryzalin, MOA 3 (Oryzalin or Surflan) 4 AS	2 to 6 qt	2 to 4	Allow soil to settle around newly transplanted trees before application. Oryzalin may be tank mixed with Goal, glyphosate, paraquat, simazine, and Solicam. Sequential application permitted. See label for details. Sequential applications may be used so long as total use rate does not exceed 12 qt per acre per year and there is a minimum of 2.5 months between applications.
	pendimethalin, MOA 3 (Prowl H$_2$O) 4 AS	2 to 6 qt	2 to 6	Most effective when adequate rainfall or irrigation is received within 7 days after application. Do not apply to newly transplanted trees until ground has settled around roots. Apply with paraquat to control emerged weeds. Prowl H2O has a 60 day PHI. Pendimethalin may be applied in sequential applications so long as total use rate does not exceed maximum of application rate on the label and there is at least 30 days between applications.
Broadleaf weeds	isoxaben, MOA 21 (Trellis) 75 DF	0.66 to 1.33 lb	0.5 to 1.0	Trellis may applied to newly planted pecan orchards once soil has been settled by rainfall. Rainfall is needed with 21 days of application in order to activate Trellis. Failure to get activation within 21 days will reduce herbicide performance. DO NOT apply within 60 days of harvest. Trellis may be tank mixed with oryzalin for expanded residual control of annual grass weeds. The addition of glyphosate, paraquat or Rely will be necessary for non-selective POST weed control.
Annual broadleaf weeds and grass weeds	simazine, MOA 5 (Princep, Simazine) 4 L 90 WDG	2 to 4 qt 2.2 to 4.4 lb	2 to 4	Apply preemergence to weeds under trees that have been established 2 years or more. Do not apply when nuts are on the ground. Do not use on sand or loamy sand soils. Tank mixing simazine with oryzalin, Solicam, or Prowl H2O will improve residual control of annual grasses and certain broadleaf weeds.
	flumioxazin, MOA 14 (Chateau) 51 SW (Tuscany) 51 WDG	6 to 12 oz	0.19 to 0.38	Chateau may be applied in newly planted and established orchards. Trees established less than 1 year must be shielded from contact with spray solution to prevent injury. Do not apply more than 6 ounces per acre to trees planted less than 3 years in soil having a sand plus gravel content more than 80%. Sequential applications are very effective; however, allow 60 days between applications. Do not apply after bud break through final harvest unless using shielded application equipment. When applying Chateau after bud break DO NOT tank mix with glyphosate or 2,4-D amine. When tank mixed with glyphosate and/or 2,4-D amine there is the potential for drift injury. Chateau has a 60-day PHI. Use rate cannot exceed 24 ounces per acre in a 12-month period.
	indaziflam, MOA 29 (Alion) 1.67 SC	3.5 to 6.5 oz	0.045 to 0.085	Use in orchards established 3 years or longer. See label for details regarding the management of replants in established orchards. Do not use on soils having a 20% or greater gravel content. Use rate cannot exceed 3.5 fluid ounces per acre per application on soils having less than 1% organic matter. On soils with an organic matter content from 1 to 3%, no more than 5 fluid ounces/acre can be applied in a single application and the total use rate for the year cannot exceed 8.5 fluid ounces/acre. In order to apply more than 5 fluid ounces/acre in a single application soil organic matter must be >3%. Allow at least 90 days between applications. Do not treat soil around trees with cracks or channels, or with depressions. Tank mix Alion with glyphosate, glufosinate, or paraquat for nonselective POST weed control. Alion has a 14-day PHI.
	rimsulfuron, MOA 2 (Matrix) 25 WG (Solida) 25 WG (Pruvin) 25 WG	4 oz	0.063	For broad spectrum PRE control, tank mix with, diuron, oryzalin or Prowl H$_2$O. For nonselective POST control, apply in combination with glyphosate or paraquat. Rimsulfuron does have POST activity on certain broadleaf weeds (see lable for list). Rimsulfuron will control emerged horseweed less than 3 inches tall when applied in combination with a non-ionic surfactant and a spray grade ammonium sulfate (2 pounds per acre). Do NOT treat orchards established less than 1 year. Rainfall within 2 to 3 weeks of application is necessary for herbicide activation. Spray solutions having a pH lower than 4.0 or higher than 8.0 will result in herbicide degradation. Rimsulfuron has a 14-day PHI for Pecan. Rimsulfuron may be applied as a sequential application so long as total use rate does not exceed 4 ounces/acre per year and application is made in a band on less than 50% of orchard floor. Allow at least 30 days between applications.

Table 7-12B. Chemical Weed Control in Fruit Crops—Tree Fruits

Weed	Herbicide, Mode of Action,* and Formulation	Amount of Formulation Per Acre	Pounds Active Ingredient Per Acre	Precautions and Remarks
Pecans, Preemergence, Tank Mix Options				
Annual broadleaf weeds and grass weeds	diuron, MOA 7 (Diuron or Karmex DF or XP) 80 WDG + norflurazon, MOA 12 (Solicam) 80 WDG	2 to 4 lb + 2.5 to 5 lb	1.6 to 3.2 + 2 to 4	Trees must be established in the orchard for 3 years.
	norflurazon, MOA 12 (Solicam) 80 F + simazine, MOA 5 (Princep, Simazine) 4 L 90 WDG	2.5 to 5 lb + 2 to 4 qt 2.2 to 4.4 lb	2 to 4 + 2 to 4	Trees must be established for at least 2 years. See labels for details.
	oryzalin, MOA 3 (Oryzalin, Surflan) 4 AS + simazine, MOA 5 (Princep, Simazine) 4 L 90 WDG	2 to 4 qt + 2 to 4 qt 2.2 to 4.4 lb	2 to 4 + 2 to 4	See label for details.
Pecans, Postemergence, Directed Underneath Tree				
Broadleaf weeds including morningglory, pigweed, lambsquarters, cocklebur, smartweed, and dayflower	carfentrazone-ethyl, MOA 14 (Aim) 2 EC	0.5 to 2 oz	0.008 to 0.031	Apply alone or tank mixed with other herbicides. Apply in a minimum spray volume of 20 gpa. Applications can be made with boom equipment, hooded sprayers, or shielded sprayers. Do not allow Aim to contact green bark, desirable foliage, flowers, or fruit of the crop. Contact with fruit or foliage will result in spotting and leaf necrosis. The trunks of trees less than 2 years old must be protected from direct contact with Aim. Do not apply within 3 days of harvest. Sequential applications may be used so long as total use rate does not exceed 7.9 ounces/acre per year and there is 14 days between applications. Best results are obtained when applied to weeds in the 2- to 3-leaf stage. Apply in combination with a nonioinic surfactant (1 quart per 100 gallons of spray solution) or crop oil concentrate (1 gallon per 100 gallons of spray solution).
Most annual broadleaf and grass weeds plus many perennials	glufosinate, MOA 10 (Lifeline, Reckon, and Rely) 2.34L	48 to 82 oz	0.88 to 1.5	DO NOT SPRAY GREEN BARK OR FOLIAGE. Glufosinate should not be used on trees within 1 year of transplanting. Do not make more than 3 applications per year. Apply in a minimum of 20 gallons of water per acre as a directed spray under trees. Repeat applications may be necessary for control of perennial weeds. Glufosinate can be tank mixed with diuron, Solicam, Surflan, Devrinol, Goal, rimsulfuron, and simazine. Do not apply within 14 days of harvest. The addition of ammonium sulfate will enhance glufosinate activity on difficult to control species, however the addition of non-ionic surfactants or crop oil will increase the risk of crop injury.
	glyphosate, MOA 9 (various brands and formulations)	See label	1 to 2	DO NOT SPRAY GREEN BARK OR FOLIAGE. Repeat applications may be necessary for control of perennial weeds. Tank mix with Goal, Karmex, simazine, Solicam, and Surflan. Check label for details. Generic glyphosate formulations may require the addition of surfactant at 0.5% by volume (2 qt per 100 gal). See label to determine if surfactant is needed for the formulation you use.
Yellow and purple nutsedge, horsenettle, pokeweed and other broadleaf weeds.	halosulfuron, MOA 2 (Sandea) 75 WDG	0.66 to 1.33 oz	0.032 to 0.063	Use on trees established in orchard at least 12 months. Avoid contacting bark or foliage or severe injury or death may occur. The addition of 0.25% surfactant (1 quart per 100 gallons of spray solution) will be necessary for adequate control. Do not make more than two applications per year. Use no more than 1 ounces per acre on soils classified as sand, loamy sand, or sandy loam. Sandea has a 1-day PHI. User assumes risk when treating trees recovering from certain stress conditions. Sandea may be tank mixed with glyphosate to control weeds other than nutsedge.
Annual broadleaf and grass weeds	paraquat, MOA 22 (Firestorm) 3 SL (Paraquat Concentrate) 3 SL (Parazone) 3 SL (Gramoxone Inteon) 2 SL	1.75 to 2.7 pt 2.5 to 4.0 pt	0.66 to 1	Apply when grass and weeds are 1 to 6 inches high and succulent for best results. Direct spray with low pressure to avoid contact with tree foliage or bark less than 1 year old. Add surfactant at 0.25% by volume (2 pints per 100 gallon) or 1% crop oil concentrate (1 gallon per 100 gallons) for best results. Paraquat may be tank mixed with Goal, Karmex, simazine, Solicam, and oryzalin. Paraquat is a restricted use pesticide.
Broadleaf weeds	2,4-D amine, MOA 4 (various brands) 3.8 SL	2 to 3 pt	0.95 to 1.4	Apply anytime during the growing season to actively growing broadleaf weeds except during bloom. Do not apply more than two applications per year. Allow at least 30 days between sequential applications. Do not use within 60 days of harvest. Do not apply to trees less than 1 year old. Some formulations may limit use rate to 2 pints per acre. Refer to product label for details.
Grasses	clethodim, MOA 1 (Arrow, Intensity, or Select) 2 EC (Select Max or Intensity One) 1 EC	6 to 8 oz 12 to 16 oz	0.094 to 0.125	**NONBEARING TREES ONLY.** Apply to actively growing grasses not under stress. See label for rate and optimum grass size to treat. Multiple applications may be necessary to control perennial grass weeds. For 2 EC formulation chemicals, add crop oil concentrate at 1% by volume (1 gallon per 100 gallons). For 1 EC formulation chemicals, a nonionic surfactant at 0.25% by volume may be used rather than crop oil.
	fluazifop, MOA 1 (Fusilade DX) 2 EC	8 to 24 oz	0.125 to 0.38	Postemergence grass control. Annuals up to 12 inches tall and 6 to 10 inches new growth on perennials. Multiple applications may be necessary to control perennial grass weeds. Add crop oil at 1% by volume (1 gallon per 100 gallons). Limited to 72 ounces per year. Do not apply within 30 days of harvest.
	sethoxydim, MOA 1 (Poast) 1.5 EC	1.0 to 2.5 pt	0.19 to 0.47	Apply to annual grasses up to 12 inches tall. For perennial grasses, apply early in the growth cycle at the high use rate. Multiple applications may be necessary to control perennial grass weeds. Add Dash adjuvant at 1 pint per acre or crop oil at 1 quart per acre. Do not apply within 15 days of harvest. Do not apply more than 10 pints per year.

CHEMICAL WEED CONTROL

311

Table 7-12B. Chemical Weed Control in Fruit Crops—Tree Fruits

Weed	Herbicide, Mode of Action,* and Formulation	Amount of Formulation Per Acre	Pounds Active Ingredient Per Acre	Precautions and Remarks
Pecans, Ground Cover Suppression				
Groundcover suppression in row middles	glyphosate, MOA 9 (various brands) 4 SL (various brands) 5 SL	Rate and application times different for grass species. See label.		See label directions specific for each grass species. **DO NOT** apply after seedhead emergence. See label for details.
	(Roundup WeatherMax) 5.5 SL	See label	See label	

* Mode of action (MOA) code developed by the Weed Science Society of America. See Table 7-10, Herbicide Resistance Management, for details.

Chemical Weed Control in Hay Crops and Pastures
F. H. Yelverton and T. W. Gannon, Crop Science Department

Note: A mode of action code has been added to the Herbicide and Formulation column of this table. Use MOA codes for herbicide resistance management. See Table 7-10B, Herbicide Modes of Action for Hay Crops, Lawns and Turf for details concerning active ingredients, brand names, chemical families, and modes of action.

Table 7-13. Chemical Weed Control in Hay Crops and Pastures

Weed	Herbicide, Mode of Action, and Formulation	Amount of Formulation per Acre	Pounds Active Ingredient per Acre	Precautions and Remarks
Alfalfa, Birdsfoot Trefoil, Clovers, Lespedeza Preplant				
Certain annual grass weeds, broadleaf weeds, and nutsedge species	EPTC, MOA 8 (7 EC)	3.5 pt	3	Use on clay and clay loam soils of piedmont. Incorporate into soil immediately after application. See label for directions. Temporary crop stunting may occur if conditions for germination and growth are not optimum. Do not use if grain or grass crop is to be planted with the legume.
Allfalfa, Preplant or Preemergence				
Various grass and broadleaf weeds	paraquat, MOA 22 (2 SL) (3 SL)	2.5 to 4 pt 1.67 to 2.67 pt	0.625 to 1	Apply prior to crop emergence. Add nonionic surfactant at 1 to 2 pints per 100 gallons.
Alfalfa, Established Preemergence				
Crabgrass, foxtails, and other annual grasses	trifluralin, MOA 3 (10 G) (4 EC)	20 lb 2 qt	2	A single rainfall of 0.5 inch or more after application is required to activate trifluralin. Apply 2 quarts trifluralin HFP if chemigation or water incorporated.
Alfalfa, Seedling				
Seedling broadleaf weeds, such as burcucumber, cocklebur, jimsonweed, lambsquarters, velvetleaf, Virginia pepperweed, shepherd's-purse, wild radish, and species of morningglory, mustard, nightshade, pigweed, ragweed, and smartweed	bromoxynil, MOA 6 (2 EC) (4 EC)	1 to 1.5 pt 0.5 to 0.75 pt	0.25 to 0.375	Apply in fall or spring to seedling alfalfa with a minimum of 4 trifoliate leaves and to weeds not greater than 4 leaf stage, 2 inches in height, or 1 inch in diameter, whichever comes first. Unacceptable crop injury can occur 3 days after application if temperatures exceed 70 degrees F. For chemigation only, apply to alfalfa with 2 trifoliate leaves at 2 pints/A to most susceptible weeds not greater than 8 leaf stage, 4 inches in height, or 2 inches in diameter, whichever comes first. Unacceptable crop injury can occur 3 days after application if temperatures exceed 85 degrees F. Bromoxynil can be tank mixed with 2,4-DB 200 or imazethapyr 2 AS. Do not apply in warm humid conditions or to alfalfa under any kind of stress. Do not add surfactant unless specified. Do not cut for feed or graze spring-treated alfalfa until 30 days after treatment. Wait until spring, or 60 days after treatment, for winter-treated alfalfa.
Alfalfa, Birdsfoot TrefoilL, Clovers, Seedling				
Certain broadleaf weeds such as cocklebur, lambsquarters, morningglory, pigweed, ragweed, smartweed, curly dock, shepherdspurse, and wild mustard	2,4-DB, MOA 4 (2 EC)	2 to 6 pt	0.5 to 1.5	Apply postemergence when weeds are less than 3 in. tall and legume has at least 2 to 4 trifoliate leaves. Do not graze or feed seedling legume crops to livestock within 60 days after application.
Alfalfa, Clover, Birdsfoot Trefoil, Crown Vetch, Established and Seedling				
Ryegrass species, annual bluegrass, perennial bluegrass, orchardgrass, chickweed, and volunteer grain	pronamide, MOA 3 (3.3 SC) (50-W)	1.25 to 5 pt 1 to 4 lb	0.516 to 2.06 0.5 to 2	Use preemergence or postemergence to the weeds only on established legume plantings or on new plantings after the legume has reached the trifoliate leaf stage or beyond. Controls henbit, shepherdspurse, wild radish and wild mustard with preemergence applications. Apply from Oct. 15 to Jan. 15. Optimum herbicidal activity occurs when applications are made under cool temperatures (55 degrees F or less) and followed by rainfall or irrigation. Do not graze or harvest for forage or dehydration within 120 days of treatment.
Alfalfa, Established and Seedling				
Lambsquarters, pigweed, ragweed, morningglory, and smartweed	2,4-DB, MOA 4 (2 EC)	2 to 6 pt	0.5 to 1.5	Apply postemergence when weeds are less than 3 inches tall. Do not graze established alfalfa or cut for hay within 30 days after application.

Table 7-13. Chemical Weed Control in Hay Crops and Pastures

Weed	Herbicide, Mode of Action, and Formulation	Amount of Formulation per Acre	Pounds Active Ingredient per Acre	Precautions and Remarks
Alfalfa, Established and Seedling (continued)				
Crabgrass, foxtails, seedling johnsongrass and certain broadleaf weeds such as chickweed, cocklebur, henbit, jimsonweed, morningglory, wild mustard, nightshade, pepperweed, pigweed, ragweed, smartweed, spurge, and Russian thistle	imazethapyr, MOA 2 (2 AS)	3 to 6 fl oz	0.048 to 0.095	Apply postemergence when seedling alfalfa or clover is in the second trifoliate stage or larger. Can also be applied postemergence to established alfalfa or clover in the fall, in the spring to dormant or semi-dormant alfalfa or clover, or between cuttings. Application should be made before significant alfalfa or clover growth or regrowth to allow herbicide to reach target weeds. Use 80% active nonionic surfactant at 1 quart per 100 gallons of water or a crop oil concentrate at 1 quart per acre. Weeds should be 1 to 3 inches in height. Imazethapyr will reduce growth of perennial grasses (fescue, etc.) that are present in the stand. See label for weeds controlled and other precautions.
Annual bluegrass, barnyardgrass, crabgrass, crowfootgrass, foxtail species, goosegrass, Italian ryegrass, seedling johnsongrass, fall panicum, Texas panicum, sandbur, signalgrass, and certain broadleaf weeds such as palmer amaranth, common chickweed, henbit, lambsquarters, pigweed species, and smartweed	pendimethalin, MOA 3 (3.8 CS)	1.1 to 4.2 qt	1.045 to 4	Use on alfalfa grown for forage, hay, or seed. Apply 1.1 to 4.2 quarts/acre prior to weed emergence in fall after last cut, during winter dormancy, in the spring, or between cuttings before alfalfa reaches 6 inches when grown for forage or hay. Apply same rates for alfalfa grown for seed production when dormant or before alfalfa exceeds 10 inches after first or second cut. Use drop nozzles to minimize foliar contact. Apply 1.1 to 2.1 pints/acre to seedling alfalfa in second trifoliate stage before 6 inches of growth. Do not harvest alfalfa forage or hay less than 28 days after applying 2.1 quarts/acre or less than 50 days after applying more than 2.1 quarts/acre. Do not harvest alfalfa seed less than 90 days after application. Some stunting and chlorosis of alfalfa may occur after postemergence applications.
Annual and perennial grasses	clethodim, MOA 1 (2 EC)	8 to 16 fl oz	0.125 to 0.250	Apply postemergence for annual grasses in seedling alfalfa at 6 to 8 fluid ounces per acre or in established alfalfa at 8 ounces per acre. Apply postemergence for bermudagrass and rhizome johnsongrass at 8 to 16 fluid ounces per acre. Add a crop oil concentrate at 1 quart per acre. Can be applied at any stage of alfalfa growth. Apply to actively growing grasses not under drought stress. Be sure grasses have leaves present for contact by the spray solution. Do not apply within 15 days of grazing, feeding, or harvesting (cutting) alfalfa for forage or hay. Select may be tank mixed with 2,4-DB or imazethapyr. When tank mixing, see respective labels for application rates, weeds controlled, maximum weed size to treat, specific application directions, and precautions.
	sethoxydim, MOA 1 (1.5 EC) (1 EC)	1 to 2.5 pt 1.5 to 3.5 pt	0.19 to 0.47	Apply postemergence for annual grasses at 0.19 pound a.i. (the lower rate) per acre and for bermudagrass and johnsongrass at 0.47 pound a.i. (the higher rate) per acre. Add 2 pints crop oil concentrate per acre. Use 10 to 20 gallons of spray solution per acre. Can be applied at any stage of alfalfa growth. Do not apply to weedy grasses or alfalfa under stress. Be sure grasses have leaves present for contact by the spray solution. Do not apply within 7 days of grazing, feeding, or cutting for (undried) forage, or within 14 days of cutting alfalfa for (dry) hay.
Alfalfa, Established Dormant				
Winter annual weeds, such as chickweed, henbit, bittercress, pepperweed, shepherds purse, yellow rocket, and ryegrass	metribuzin, MOA 5 (75 DF) (4 L, 4 F)	0.67 lb 0.5 to 2 pt	0.5 0.25 to 1	Good results have been obtained in N.C. when the herbicide was applied from Nov. 20 to Dec. 20. Do not graze or harvest within 28 days after application. In alfalfa-grass mixtures, it will provide partial reduction of forage grass stands.
	paraquat, MOA 22 (2 SL) (3 SL)	0.75 to 2 pt 0.7 to 1.3 pt	0.1875 to 0.5 0.263 to 0.488	Apply up to 1.25 pints per acre on fall-seeded newly established stands less than 1 year old. Apply up to 2 pints per acre on established stands. Tank mix with metribuzin will improve vegetation control. Apply late fall to winter after last fall cutting and before first spring cutting. Alfalfa must be dormant to avoid injury. There is a 60-day grazing or preharvest interval. Add 1 to 2 pints per acre nonionic surfactant per 100 gallons.
	terbacil, MOA 5 (80 WDG)	0.5 to 1.5 lb	0.4 to 1.2	Apply to established alfalfa stands at least 1 year old before or shortly after weed growth begins. Weeds have been controlled with an application from mid-November through February. Do not use terbacil on alfalfa-grass mixtures on sand or loamy sand soils or on soils containing less than 1% organic matter.
Alfalfa, Between Cuttings (even first year alfalfa)				
Various grass and broadleaf weeds	paraquat, MOA 22 (2 SL) (3 SL)	1 pt 0.7 pt	0.25 0.263	Apply immediately after hay or silage removal but no more than 5 days after a cutting. Apply up to 2 times for first year alfalfa and 3 times for established alfalfa. There is a 30 day grazing or preharvest interval. Add 1 to 2 pints per acre nonionic surfactant per 100 gallons.
Lespedeza, Preplant				
Certain annual grass and broadleaf weeds	EPTC, MOA 8 (7 EC)	3.5 pt	3	See remarks under alfalfa.
No-Till Alfalfa or No-Till Pasture Reseeding				
Complete kill of existing sod	glyphosate, MOA 9 (4 SL) (5.5 SL)	1 to 5 qt 2.75 pt	1 to 5 1.89	Broadcast spray 10 to 14 days before planting. Provides better control of perennial weeds. Check label for rate according to weeds present.
	paraquat, MOA 22 (2 SL) (3 SL)	2 to 4 pt 1.67 to 2.67 pt	0.5 to 1 0.625 to 1	Broadcast spray in 20 to 30 gallons of water per acre. Make 2 applications if needed. If spraying following hay harvest, allow enough regrowth to provide leaf area to absorb the herbicide. Add 1 pint of a nonionic surfactant per 100 gallons of water.
No-Till Pasture Reseeding with Grasses or Clover				
Suppression of existing sod and undesirable emerged broadleaf and grass weeds to permit pasture reseeding	paraquat, MOA 22 (2 SL) (3 SL)	1 to 2 pt 0.7 to 1.3 pt	0.25 to 0.5 0.263 to 0.488	Rates are per sprayed acre. Usually band sprayed for planting clover into existing grass sod. Vegetation should not exceed 3 inches in height at time of treatment. Add 1 to 2 pints of a nonionic surfactant per 100 gallons of water. Spray bermudagrass or bahiagrass sod in late summer to early fall.
Pastures, Clover, Velvetbean, Lespedeza, Lupine, Sainfoin, Trefoil, Vetch, Crown Vetch, Milk Vetch				
Various grass and broadleaf weeds	paraquat, MOA 22 (2 SL) (3 SL)	0.75 to 2 pt 0.7 to 1.3 pt	0.1875 to 0.5 0.263 to 0.488	Apply up to 1.2 pints per acre to fall-seeded newly established stands less than 1 year old. Apply up to 2 pints per acre to established stands. Apply when dormant in late fall or winter after last cutting and before first spring cutting. Do not apply if regrowth is >2 inches. Make only 1 application per season. There is a 60 day grazing or preharvest interval. Add 1 to 2 pints per acre nonionic surfactant per 100 gallons.

CHEMICAL WEED CONTROL

CHEMICAL WEED CONTROL

Table 7-13. Chemical Weed Control in Hay Crops and Pastures

Weed	Herbicide, Mode of Action, and Formulation	Amount of Formulation per Acre	Pounds Active Ingredient per Acre	Precautions and Remarks
Pastures, Ladino Clover, Orchardgrass, Fescue, and other grasses				
Curly dock, ragweed, bitterweed, pigweed, dandelion, and other broadleaf weeds	2,4-D amine, MOA 4 (4 SL)	1 to 2 pt	0.5 to 1	Spray when weeds are 4 to 8 inches tall and before heading. Clover may be stunted and growth retarded 3 to 6 weeks. Use lower rate in warm, wet weather. For wild garlic, apply late February or early March. Repeat for 2 years. **Do not graze dairy animals on treated areas within 7 days after application.** Remove meat animals from treated areas for 3 days before slaughter. Withdrawal is not necessary if more than 2 weeks have elapsed since treatment. Do not cut treated grass for hay within 30 days after application.
Wild garlic	2,4-D amine, MOA 4 (4 SL)	1 qt	1	
Perennial Grasses, Rangeland, Permanent Grass Pastures				
Many annual and perennial grass and broadleaf weeds and nutsedge species (goosegrass not controlled)	imazapic, MOA 2 (2 SL)	4 to 12 fl oz	0.0625 to 0.1875	Apply to common and coastal bermudagrass varieties. Jiggs bermudagrass is more sensitive than other types. Expect 30 to 45 days of bermudagrass suppression. Do not apply 1) to drought stressed bermudagrass, 2) during spring transition, 3) to newly aerated fields for 30 days, 4) to newly sprigged or seeded bermudagrass, or 5) to World Feeder bermudagrass varieties. To speed bermudagrass recovery, apply with nitrogen fertilizer and do not add a spray adjuvant. If spray carrier is water, add a nonionic surfactant at 0.25% by volume or a methylated seed oil at 1.5 to 2 pints per acre. Imazapic also controls winter weeds when applied to dormant bermudagrass, and can be mixed with glyphosate at this time. There is a 7-day hay restriction.
Johnsongrass, kyllinga species, purple and yellow nutsedge	sulfosulfuron, MOA 2 (75 WG)	1.33 oz	0.0625	Apply to established bermudagrass and bahiagrass pastures. A second application can be made 40 days after initial application if needed, but do not exceed 2.66 ounces per acre per year. Apply a nonionic surfactant at 0.25% v/v. There are no grazing restrictions. Do not harvest for hay within 14 days of application. Johnsongrass is best controlled if sulfosulfuron is applied at 18 to 24 inches and up to heading stage.
Amaranth and pigweed species, ladysthumb, wild mustard, wild radish, maypop passionflower, pokeweed, ragweed species, hemp sesbania, velvetleaf, purple and yellow nutsedge	halosulfuron, MOA 2 (75 DF)	0.67 to 1.33 oz	0.0314 to 0.0628	Listed weeds are controlled postemergence if small and actively growing at application. Check label for weeds that are controlled preemergence. The following adjuvants can be used with halosulfuron; nonionic surfactant applied at 0.25 to 0.5% v/v, crop oil concentrate or methylated seed oil at 1% v/v, high quality granular spray grade ammonium sulfate at 2 to 4 pounds per acre or liquid ammonium sulfate at equivalent nitrogen rate of 2 to 4 lb per acre. If nutsedge populations resprout or reemerge, a second spot application not to exceed 0.75 ounce per treated acre in these areas can be made. There are no grazing or slaughter restrictions. There is a 37 day pre-harvest interval.
Emerged annual, biennial, and perennial broadleaf weeds and certain woody species	[picloram + 2,4-D], MOA 4 + 4 (0.54 + 2 lb/gal SL)	1 to 8 pt	0.3175 to 2.54	Due to crop sensitivity, [picloram + 2,4-D] should not be used in cotton- or tobacco-growing regions of the state. Do not graze lactating dairy animals for 7 days after application. Do not harvest grass for hay for 30 days after application. Meat animals must be withdrawn from treated forage at least 3 days before slaughter. There are no other grazing restrictions for non-lactating dairy animals or other livestock. Newly seeded grasses may be injured. Check label for livestock transfer restrictions due to possible urine and manure contamination with picloram. Check label for all other restrictions.
	[picloram + fluroxypyr], MOA 4 + 4 (1.19 + 0.96 lb/gal EC)	1.5 to 6 pt	0.40 to 1.60	Due to crop sensitivity, [picloram + fluroxypyr] should not be used in cotton- or tobacco-growing regions of the state. Do not allow lactating dairy animals to graze or consume harvested forage within 14 days after application. There are no grazing restrictions for nonlactating dairy animals or other livestock. Do not harvest hay within 7 days after application. Withdraw meat animals from treated forage at least 3 days before slaughter. Newly seeded or sprigged grasses may be injured. Check label for livestock transfer restrictions due to possible urine and manure contamination with picloram. Check label for all other restrictions.
Mustard, radish, cocklebur, vetch, and other susceptible broadleaf weeds	2,4-D amine, MOA 4 (4 SL)	1 to 2 pt	0.5 to 1	Do not spray in seedling stages or just before heading. Apply after the perennial grass seedlings have reached the 2- to 4-leaf stage. For wild garlic apply February or March. Repeat for 2 years. **Do not graze dairy animals on treated areas within 7 days after application.** Remove meat animals from treated areas for 3 days before slaughter. Withdrawal is not necessary if more than 2 weeks have elapsed since treatment. Do not cut treated grass for hay within 30 days after application.
Wild garlic	2,4-D amine, MOA 4 (4 SL)	3 qt	3	
Many broadleaf weeds including certain ones resistant to 2,4-D	dicamba, MOA 4 (4 SL)	0.5 to 2 pt	0.25 to 1	Rate dependent on weed species and size. See label for specific rates and precautions concerning grazing.
Many broadleaf weeds including dogfennel, thistles, and horsenettle	dicamba (4 SL) + 2,4-D amine (4 SL) (a tank mix) MOA 4 + 4	0.5 pt + 1.5 pt	0.25 + 0.75	The tank-mix combination will control a greater number of broadleaf weeds than either herbicide alone. Observe each label for restrictions on grazing and cutting for hay. For 1 pint of dicamba or 2 pints of [2,4-D amine + dicamba], do not graze lactating dairy animals for 7 days or harvest hay for 37 days. No grazing restrictions for other livestock; however, meat animals must be removed 30 days before slaughter.
	[dicamba + 2,4-D amine], MOA 4 + 4 (a premix) (1 + 2.87 lb/gal SL)	1 to 2 pt	0.13 to 0.25 + 0.36 to 0.72	
	[2,4-D + triclopyr], MOA 4 + 4 (2 + 1 lb/gal EC)	1 to 6 qt	0.5 to 3 + 0.25 to 1.5	Rate depends on weeds to be controlled. Woody plant control requires 6 quarts or more. Consult label for specific rates. Withdraw livestock from treated forage at least 3 days before slaughter the year of treatment. Do not graze lactating dairy animals on treated areas for 14 days following treatment. Do not harvest grass for hay from treated areas for 1 year following treatment for lactating dairy animals. Wait 7 days for other livestock.
	metsulfuron methyl, MOA 2 (60 WDG)	0.1 to 1 oz	0.0038 to 0.038	Bermudagrass, bluegrass, orchardgrass, bromegrass, and timothy are tolerant. Metsulfuron methyl may cause stunting and seedhead suppression of tall fescue. Therefore, do not exceed 0.4 ounce product per acre. Pensacola bahiagrass controlled at 0.3 ounce product per acre in established bermudagrass. Also controls wild garlic. Alfalfa, clover, and ryegrass are highly sensitive. Use a nonionic surfactant of at least 60% active ingredient with a hydrophilic/lipophilic balance greater than 12 at 1 to 2 pints per 100 gallons of spray solution. Use a COC of 80% quality or MSO with 15% surfactant emulsifiers at 1 gallon per 100 gallons spray solution. Metsulfuron methyl has no grazing restrictions.

Table 7-13. Chemical Weed Control in Hay Crops and Pastures

Weed	Herbicide, Mode of Action, and Formulation	Amount of Formulation per Acre	Pounds Active Ingredient per Acre	Precautions and Remarks
Perennial Grasses, Rangeland, Permanent Grass Pastures (continued)				
	[metsulfuron methyl + chlorsulfuron], MOA 2 + 2 (48% + 15% WDG)	0.125 to 1.25 oz	0.005 to 0.05	Tolerant to native grasses, such as bluestems, blue grama, and buffalograss, as well as bermudagrass, bluegrass, orchardgrass, bromegrass (but not Matua), and fescue. To minimize fescue injury, do not exceed 0.5 ounce per acre and use a nonionic surfactant unless liquid nitrogen is the carrier. Apply to bermudagrass 2 months after establishment and fescue 24 months after establishment. Generally, treat actively growing weeds less than 4 inches tall or 4 inches in diameter. However, this product also provides preemergence control. Unless otherwise recommended, use a nonionic surfactant at 0.25% v/v or crop oil concentrate at 1% v/v. There are no grazing or hay harvest restrictions.
Many annual and perennial broadleaf weeds and woody brush	[metsulfuron methyl + dicamba + 2,4-D amine], MOA 2 + 4 + 4 (a 2-part product) [60%+(1+2.87) lb/gal]	0.25 to 1 oz + 1 to 4 pt	0.009375 to 0.038 + 0.48 to 1.94	Observe same precautions as metsulfuron methyl except for the following changes or additions: Tall fescue: do not exceed 0.25 ounce per acre Part A + 1 pint per acre Part B. Nonlactating meat animals: remove 30 days prior to slaughter. Lactating dairy animals: 7-day grazing and 37-day hay restriction.
Henbit, common chickweed, mustards, buttercup, Carolina geranium, pigweed species, common lambsquarters, and other susceptible broadleaf weeds	chlorsulfuron, MOA 2 (75 DF)	0.25 to 1 oz	0.0117 to 0.047	Treat perennial weeds in bud to bloom stage or fall rosette stage. There are no grazing or hay restrictions with rates up to 1.33 ounces per acre. For bermudagrass and orchardgrass, apply a maximum of 1 ounce per acre. Apply up to 0.5 ounce per acre on tall fescue. Spot treat with 1.33 ounces per acre if grass injury can be tolerated. Use a high quality spray adjuvant for improved postemergence control, but do not use LI-700 or other acidifying spray adjuvants.
Buttercup species, cocklebur, henbit, horsenettle, horseweed, ragweed, thistles, and other susceptible broadleaf weeds	aminopyralid, MOA 4 (2 SL)	3 to 7 fl oz	0.04688 to 0.10938	Due to crop sensitivity, use extreme caution around sensitive crops, including but not limited to alfalfa, cotton, potatoes, soybeans, tobacco, and other broadleaf or vegetable crops, fruit trees, or ornamental plants. Do not use aminopyralid-treated plant residues, including hay or straw from treated areas, or manure from animals that have grazed treated areas in compost or mulch that will be in contact with susceptible broadleaf plants. Hay treated within the preceding 18 months can only be used on the farm or ranch where the product was applied. There are no other restrictions on grazing or hay harvest following aminopyralid applications. Check product label for list of precautions.
	[aminopyralid + 2,4-D amine], MOA 4 + 4 (0.41 + 3.33 lb/gal L)	1.2 to 2.1 pt	0.56 to 0.975	Do not use on areas where loss of desirable broadleaf forage plants (legumes) cannot be tolerated. Do not use hay, straw, or manure from farm animals that have grazed forage or eaten hay harvested from treated areas within previous 3 days in compost or mulch that will be in contact with susceptible broadleaf plants. Wait 7 days after application to harvest forage for hay. Wait 30 days to make second application. Do not transfer grazing animals from treated areas to sensitive broadleaf crop areas without allowing for 3 days grazing on nontreated areas. Hay treated within the preceding 18 months can only be used on the farm or ranch where the product was applied.
	[aminopyralid + metsulfuron methyl], MOA 4 + 2 (71.58 WG)	1 to 3.3 oz	0.0447 to 0.1476	[Aminopyralid + metsulfuron methyl] is effective on Pensacola bahiagrass. At higher rates, [aminopyralid + metsulfuron methyl] may stunt tall fescue, cause yellowing, or cause seedhead suppression. Follow label precautions to minimize these symptoms. Include 1% crop oil concentrate, 0.25% non-ionic surfactant, 0.5% methylated seed oil, or 2 quarts/acre urea ammonium nitrate. Can spot spray less than 50% of an acre with up to 6.6 ounces/acre. There are no grazing or hay harvest restrictions. Do not use on grasses grown for seed. Do not overseed 4 months after treatment. Aminopyralid Precautions: Do not transfer grazing animals from treated areas to sensitive broadleaf crop areas without allowing for 3 days grazing on nontreated areas. Do not use hay, straw, or manure from animals that have grazed forage or eaten hay from treated areas within previous 3 days in compost or mulch that will contact susceptible broadleaf plants. Do not spread manure on land used for growing susceptible broadleaf crops if animals have consumed treated hay within 3 previous days. Conduct a field bioassay before planting a broadleaf crop. Hay treated within the preceding 18 months can only be used on the farm or ranch where the product was applied.
Catchweed bedstraw, common lambsquarters, mustard spp., amaranthus spp., velvetleaf, bittercress, shepherds-purse, annual sowthistle, corn spurry, Russian thistle, redstem filaree	carfentrazone ethyl, MOA 14 (2 EC)	0.5 to 2 fl oz	0.0078125 to 0.03125	Use in grasses grown for forage, fodder, hay, seed, and sod. There are no grazing or hay restrictions. Add a nonionic surfactant at 0.25% v/v or crop oil concentrate or methylated seed oil at 1% v/v with or without a high quality sprayable liquid N fertilizer at 2 to 4% v/v or ammonium sulfate at 2 to 4 pounds per acre. Apply to weeds up to 4 inches tall. In overseeded pastures, Aim can be applied to barley, millet, oats, rye, teosinte, triticale, and wheat from prior to planting up to joint stage. Tank mix 2,4-d amine or ester for extended broadleaf weed control but don't harvest for forage within 7 days of application.
Catchweed bedstraw, common lambsquarters, mustard spp., nightshade spp., amaranthus spp., velvetleaf, bittercress, shepherds-purse, annual sowthistle, corn spurry, Russian thistle, redstem filaree, and many other weeds susceptible to 2,4-D	[carfentrazone ethyl + 2,4-D ester], MOA 14 + 4 (0.13 + 5.92 lb/gal EC)	0.5 to 2 pt	0.378125 to 1.5125	Use in grasses grown for forage, fodder, hay, seed, and sod. Restrictions after application include: 3-day slaughter; 7-day dairy grazing; and 30-day hay harvest. Add a nonionic surfactant at 0.25% v/v or crop oil concentrate or methylated seed oil at 1.5 to 2% v/v with or without a high quality sprayable liquid N fertilizer at 2 to 4% v/v or ammonium sulfate at 2 to 4 pounds per acre. Apply to weeds up to 6 inches tall. In overseeded pastures, 8 to 16 ounces of [carfentrazone ethyl + 2,4-D ester] can be applied to barley, oats, rye, and wheat from 3 tiller stage up to joint stage. Use only a nonionic surfactant with or without a fertilizer solution as described above in small grains. Do not graze dairy or meat animals being finished for slaughter for 14 days following application. Do not feed treated straw to livestock.
Buttercup species, cocklebur, dogfennel, henbit, horsenettle, horseweed, select amaranth and pigweed species, other susceptible broadleaf weeds.	[diflufenzopyr sodium + dicamba], MOA 19 + 4 (21.3% + 55% DF)	4 to 8 oz	0.05 + 0.125 to 0.1 + 0.25	Use caution around sensitive broadleaf crops, including but not limited to alfalfa, clover, and lespedeza. There are no hay harvesting or grazing restrictions for pasture and rangelands treated with [diflufenzopyr sodium + dicamba]; however, there is a 30-day crop rotation or planting restriction. [Diflufenzopyr + dicamba] is for use on established grasses; do not apply to newly seeded grasses or to small grains grown for pasture.

315

Table 7-13. Chemical Weed Control in Hay Crops and Pastures

Weed	Herbicide, Mode of Action, and Formulation	Amount of Formulation per Acre	Pounds Active Ingredient per Acre	Precautions and Remarks
Perennial Grasses, Rangeland, Permanent Grass Pastures (continued)				
Many broadleaf weeds including most legumes, cocklebur, curly dock, horseweed, jimsonweed, lambsquarters, prickly lettuce, mustards, nightshades, redroot pigweed, plantains, wild radish, sicklepod, sowthistles, and thistles	[2,4-D amine + cloypyralid], MOA 4 + 4 (2 +0.38 lb/gal)	2 to 4 qt	1.19 to 2.38	Apply to established grass pastures. Allow 7 days grazing on nontreated pasture before livestock transfer to sensitive broadleaf crop areas. Do not use plant residues or manure from treated areas for composting or mulching near sensitive plants. Avoid movement of treated soil and minimize spray drift when possible. There is a 14-day grazing restriction for lactating cattle, a 30-day restriction for haying, and a 7-day restriction for slaughter. Addition of surfactant is usually not necessary.
Many woody plants, such as poplar, sumac, sassafras, wax myrtle, oaks, red maple, locust, eastern persimmon, and broadleaf weeds such as poison oak, poison ivy, blackberry, clover, curly dock, multiflora rose, lespedeza, mustard, plantain, and vetch	triclopyr, MOA 4 (4 EC)	1 to 3 pt	0.5 to 1.5	Apply to established grass pastures. Rate depends on weed species to be controlled. See label for rates for specific weeds and dilution rates for woody plant control. For lactating dairy animals do not graze until next growing season. There are no grazing restrictions for other livestock. Portions of grazed area may be treated if comprised of no more than 10 percent of the total grazable area. There is a 3 day slaughter and 14 day hay harvest restriction. Addition of 2,4-D at 1 pint per acre enhances weed spectrum. A surfactant or oil-based carrier is recommended, but liquid nitrogen carrier may be used as well.
	[triclopyr + fluroxypyr], MOA 4 + 4 (3 + 1 lb/gal EC)	0.75 to 1.5 pt	0.38 to 0.75	Apply to established grass pastures. In general, apply 0.75 to 1 pint per acre for annual broadleaf weeds, 1 to 1.5 pints per acre for biennial and perennial broadleaf weeds, and 1 to 4 pints per acre for woody plants. Do not graze or harvest green forage for lactating dairy animals during the same growing season. There are 14-day haying and 3-day slaughter restrictions. A nonionic surfactant at 1 to 2 quarts per 100 gallons of spray solution may improve control of drought-stressed weeds.
Multiflora rose	dicamba, MOA 4 (4 SL)	1 to 2 gal	4 to 8	Mix 1 gallon of dicamba herbicide with 99 gallons of water. Add a nonionic surfactant at the rate of 2 quarts per 100 gallons of spray solution to improve wetting. Apply 100 to 200 gallons of spray solution per acre. Completely wet foliage and stems, allowing spray solution to run down the stem. Apply at full vegetative stage before bloom. For spot treatment, mix 3 tbsp of dicamba per 1 gallon of water, or directly apply concentrate to root area with a spotgun applicator. Use 1 fluid ounce of dicamba per 10 feet of canopy diameter and apply before bud-break in spring. Do not graze dairy animals for 60 days after treatment. There is no waiting period between treatment and grazing for animals other than dairy animals.
	glyphosate, MOA 9 (4 SL, 5.5 SL)	1% solution		Apply 1% solution of glyphosate in water with a handgun. Spray foliage completely. Apply after full bloom until August 1. Do not graze livestock for 10 days following treatment.
	metsulfuron methyl, MOA 2 (60 WDG)	0.5 oz	0.019	Broadcast application: treat in spring when multiflora rose is fully leafed but less than 3 ft tall. Spot application: treat spring through summer at 1 ounce per 100 gallons of spray solution. Add surfactant. Metsulfuron methyl has no grazing restrictions.
	tebuthiuron, MOA 7 (20 P) (80 DF)	20 lb 2.5	4 2	Apply in the spring just before growth begins or during periods of vigorous growth, late spring or early summer (March 15 to June 1). Best to apply to individual clumps on basis of area (sq ft) covered. Spread uniformly over the plant roots. Check label carefully for restrictions and precautions on use. Do not contaminate water used for irrigation.
	[2,4-D + triclopyr], MOA 4+4 (2 + 1 lb/gal EC)	1.5 gal	4.5	Mix 1 to 1.5 gallons of [2,4-D + triclopyr] with water to make 100 gallons of total spray solution. Spray to give thorough coverage of foliage, wetting leaves and stems to the drip point. The best time for treatment is during early to mid-flowering stage. See restrictions for [2,4-D + triclopyr] on grazing and hay harvest listed previously.
Bermudagrass, Newly Sprigged				
Annual and perennial grass and broadleaf weeds	diuron, MOA 7 (4 L)	0.8 to 2.4 qt	0.8 to 2.4	Apply after planting, before bermudagrass or weed emergence. Apply 0.4 to 0.8 quart per acre if annual weeds are up to 4 inches tall with an NIS at 2 quarts per 100 gallons of water. If bermudagrass has emerged at treatment, expect temporary burn. Plant sprigs 2 inches deep to reduce crop injury potential. Do not graze or feed livestock within 70 days of application.
Bermudagrass, Dormant				
Emerged winter annual broadleaf and grass weeds	metsulfuron methyl, MOA 2 (60 WDG)	0.1 to 0.3 oz	0.0038 to 0.011	Controls many winter annual broadleaf weeds.
	paraquat, MOA 22 (2 SL) (3 SL)	1 to 2 pt 0.7 to 1.3 pt	0.25 to 0.5 0.263 to 0.488	Apply in February or March in dormant bermudagrass and Coastal bermudagrass pastures. Add 1 pint of a nonionic surfactant per 100 gallons of water. Do not mow for hay until 40 days after treatment.
Barnyardgrass, crabgrass, crowfootgrass, foxtail species, goosegrass, seedling johnsongrass, fall panicum, Texas panicum, sandbur, signalgrass, and certain broadlead weeds such as palmer amaranth, lambsquarters, pigweed species, and smartweed	pendimethalin, MOA 3 (3.8 CS)	1.1 to 4.2 qt	1.045 to 4	Apply to established bermudagrass pasture and hay fields in winter dormancy. Apply as a single full rate or in two split applications with a half rate at the onset of winter dormancy and another half rate prior to spring greenup. Do not harvest for forage or graze until 45 days after treatment. Do not harvest for hay until 60 days after harvest. Observe a plant back interval of 270 days after treatment. Use of pendimethalin on rangeland is prohibited.

Table 7-13. Chemical Weed Control in Hay Crops and Pastures

Weed	Herbicide, Mode of Action, and Formulation	Amount of Formulation per Acre	Pounds Active Ingredient per Acre	Precautions and Remarks
Bermudagrass, Coastal				
Various annual grass and broadleaf weeds	glyphosate MOA 9 (4 SL) (5.5 SL)	1 pt 0.727 pt	0.5	May be applied to coastal bermudagrass prior to spring growth or immediately after first cutting. Cannot be applied prior to spring growth and immediately after first cutting in the same year. Remove domestic livestock from area before making applications. When applying prior to spring growth, apply in late winter or early spring but before new coastal bermudagrass growth begins in spring. Applications to new growth can damage bermudagrass. Wait 60 days after making this application before grazing or harvesting the treated area. When applied after first cutting, apply after the first bermudagrass cutting when the bermudagrass has not yet begun to grow. Applications made after regrowth can damage the bermudagrass. Wait 28 days after making this application before grazing or harvesting the treated area.
Suppresses large crabgrass, goosegrass. Controls barnyardgrass, broadleaf signalgrass, foxtail species, johnsongrass up to 18 inches, panicum species, Italian ryegrass, sandbur, volunteer cereals, Pensacola bahiagrass, wild garlic, many broadleaf weeds such as bitter sneezeweed, buttercup, geranium, chickweed, curly dock, dandelion, dog fennel, henbit, horseweed, jimsonweed, lambs-quarters, morningglory, pigweed, plantain, smartweed, wild mustard.	[nicosulfuron + metsulfuron methyl], MOA 2 + 2 (56.2% + 15% WG)	1 to 1.5 oz	0.0445 to 0.0668	There are no grazing or hay restrictions. [Nicosulfuron + metsulfuron methyl] provides pre and post broadleaf weed control and only post grass weed control. Apply to established bermudagrass. Temporary crop injury may occur if treated on new growth more than 2 inches or after 7 days following harvest. Do not exceed 2.5 ounce/acre per year if sequential applications are needed; 0.25% non-ionic surfactant is preferred but can apply with 1% crop oil concentrate or 2 quarts/acre urea ammonium nitrate. Check label for acceptable tank mix partners. [Nicosulfuron + metsulfuron methyl] will control crabgrass and sandbur up to 2 inches and goosegrass up to 2 tillers. FIFRA Section 2 (ee) allows glyphosate to be tankmixed with [Nicosulfuron + metsulfuron methyl] for improved control or suppression of crabgrass, sandbur, foxtail, rescuegrass, Japanese brome, little barley, and ryegrass. Apply 2.5 to 4.1 ounce a.i. per acre glyphosate. Expect temporary yellowing or stunting of bermudagrass. Add nonionic surfactant at 0.25% v/v.
Hybrid Bermudagrass, (Coastal, Tifton 44)				
Young annual broadleaf weeds	2,4-D amine, MOA 4 (4 SL)	1 qt	1	Apply after sprigging. Gives little preemergence weed control. Later applications may be needed to control broadleaf weeds.
Sorghum-Sudan Hybrids, Preemergence				
Annual broadleaf and grass weeds	atrazine, MOA 5 (4 L) (90 DF, 90 WDG)	3.2 to 4 pt 1.8 to 2.2 lb	1.6 to 2	Use only on silt-loam, clay loam, and clay soils with more than 1% organic matter. Use lower rates on soils 1% to 1.5% organic matter and higher rates on soils having more than 1.5% organic matter. On highly erodible soils (as defined by SCS), if conservation tillage is practiced, leaving at least 30% of soil covered with plant residues at planting, apply a maximum of 2 pounds active per acre as broadcast spray. If soil coverage with plant residue is less than 30% at planting, a maximum of 1.6 pounds active per acre may be applied. On soils not highly erodible, apply 2 pounds active per acre as a broadcast spray.

Chemical Weed Control in Lawns and Turf

F. H. Yelverton and T. W. Gannon, Crop Science Department

Note: A mode of action code has been added to the Herbicide and Formulation column of this table. Use MOA codes for herbicide resistance management. See Table 7-10B, Herbicide Modes of Action for Hay Crops, Lawns and Turf for details concerning active ingredients, brand names, chemical families and modes of action.

Several of the preemergence herbicides are available on fertilizer carriers for homeowner application.

Table 7-14. Chemical Weed Control in Lawns and Turf

Herbicide and Formulation	Amount of Formulation Per 1,000 sq ft	Amount of Formulation per Acre	Pounds Active Ingredient per Acre	Precautions and Remarks
Preemergence Control, Smooth and Large Crabgrass, Goosegrass, Foxtails, other annual grasses				
benefin, MOA 3 (2.5 G)	2.75 lb	120 lb	3	Safe to apply to all established turfgrass except bentgrass. Do not apply in the spring to lawns seeded the previous fall or to golf course greens. Do not use on newly sprigged turfgrasses.
[benefin + trifluralin], MOA 3 + 3 (0.86 G)	8 lb	349 lb	3	Use on lawns and golf course fairways of bahiagrass, bentgrass, bermudagrass, centipedegrass, fescue, perennial ryegrass, St. Augustinegrass, and zoysiagrass.
bensulide, MOA 8 (4 EC) (8.5 G, 12.5 G)		Varies, several concentrations available	10	May be applied to all established turfgrass and dichondra, residential lawns, and golf course greens and tees. Limit 2 applications per year to greens and tees. Do not use on newly sprigged turfgrasses. Not effective for goosegrass control.
[bensulide + oxadiazon], MOA 8 + 14 (6.56 G)	2.6 lb	116 lb	6 + 1.5	Controls crabgrass and goosegrass. Use on established bermudagrass, zoysiagrass, tall fescue, bentgrass, perennial bluegrass, or perennial ryegrass fairways and tees. Use also on bermudagrass and bentgrass greens.
dithiopyr, MOA 4 (2 EW, 2 L) (40 WP)	0.75 fl oz 0.46 oz	1 qt 20 oz	0.5	May be applied to most all cool-season and warm-season turfgrasses except colonial bentgrass. See label for injury precautions regarding certain varieties. Also controls pre-tillered crabgrass. Split applications recommended in southern and coastal regions of the state (0.25 pound a.i. at 8-week intervals). Timely irrigation or rainfall is critical for activation.

CHEMICAL WEED CONTROL

Table 7-14. Chemical Weed Control in Lawns and Turf

Herbicide and Formulation	Amount of Formulation Per 1,000 sq ft	Amount of Formulation per Acre	Pounds Active Ingredient per Acre	Precautions and Remarks
Preemergence Control, Smooth and Large Crabgrass, Goosegrass, Foxtails, other annual grasses (continued)				
indaziflam, MOA 21 (20 WSP)	0.057 to 0.115 oz	2.5 to 5 oz	0.03125 to 0.0625	Use only on established turf (1 year after seeding) such as bermudagrass, zoysiagrass, centipedegrass, St. Augustinegrass, seashore paspalum, and bahiagrass. Labeled for commercial and residential lawns, golf courses (roughs, tees, fairways), sod farms, athletic fields, parks and cemeteries. Use a minimum of 2.5 ounces per acre for crabgrass, annual bluegrass and broadleaf weed control and a minimum of 3.75 ounces per acre for goosegrass, annual sedge and kyllinga species control. Apply up to 2.5 ounces per acre on centipedegrass and St. Augustinegrass due to tolerance concerns. For all other tolerant turfgrasses, do not exceed 5 ounces per acre in a single application or 7.1 ounces per acre within a calendar year. There is an 8 month overseeding restriction following a 2.5 ounces per acre application. Can sprig 2 months following application, or if sprigged first, wait 4 months before spraying. Can sod 4 months following application, or if sodded first wait 2 months after rooting before spraying.
(0.622F)	0.69 to 0.23 fl oz	3 to 10 fl oz	0.01458 to 0.0486	Use up to 6 fluid ounces per acre on common bermudagrass, centipedegrass and St. Augustinegrass and 10 fluid ounces per acre on hybrid bermudagrass, zoysiagrass and bahiagrass established 16 months in areas such as golf course roughs and fairways, residential and commercial turf, sod farms, athletic fields, parks and cemeteries. 10 fluid ounces per acre needed for annual sedge and kyllinga species control. Don't exceed 18.5 fluid ounces per acre per year. Do not vertical mow 1 month before or after application. Irrigate within 2 days of treatment for maximum benefit. Check label for split or multiple application rates and timings. Delay overseeding 10 months if 4.5 to 6 fluid ounces used and 12 months if 6 to 9 fluid ounces used. For sod production, only apply to bermudagrass, zoysiagrass or bahiagrass. Apply if 80% ribbon coverage and before 4 months prior to harvest. Wait 6 month after treatment if sodding bare ground. Apply to actively growing sod established for 3 months.
(0.0224 G)	2.9 to 4.6 lb	125 to 200 lb	0.028 to 0.045	Use on same warm season turf species established at least 16 months and sites as above. Do not exceed 400 pounds product per year. Allow a 15 feet buffer from cool season turf areas. Do not apply upslope from cool season turf.
metolachlor, MOA 15 (7.62 EC)	0.96 fl oz	2.6 pt	2.48	Apply to established bermudagrass, centipedegrass, St. Augustinegrass, bahiagrass, and zoysiagrass. Can apply up to 4.2 pints per acre per year to same area used for commercial sod production.
napropamide, MOA 15 (50 DF)	1.5 to 2.2 oz	4 to 6 lb	2 to 3	Use in established bahiagrass, bermudagrass, centipedegrass, St. Augustinegrass, and tall fescue.
oryzalin, MOA 3 (4 AS, 4 L)	1.5 fl oz	2 qt	2	Use on established bahiagrass, centipedegrass, tall fescue, St. Augustinegrass, zoysiagrass, and bermudagrass except greens and tees. A total of 3 quart per acre may be used if application is split by applying 1.5 quarts per acre followed by 1.5 quarts per acre 8 to 10 weeks later. Follow label directions. Do not apply in the spring or summer to tall fescue reseeded the previous fall.
oryzalin, MOA 3 (85 WDG)	0.64 to 0.88 oz	1.75 to 2.4 lb	1.4875 to 2.04	Observe same turf tolerances and tall fescue precautions as above. Successful preemergence activity should occur if activated by 0.5 inch of water within 21 days of application. Apply 2.4 pounds per acre as a single application or 1.75 pounds per acre in sequential applications spaced 12 weeks apart.
oxadiazon, MOA 14 (2 G)	2.3 to 4.6 lb	100 to 200 lb	2 to 4	Use in established perennial bluegrass, perennial ryegrass, bentgrass, bermudagrass, tall fescue, zoysiagrass, and St. Augustinegrass. Red fescue is not tolerant. Do not apply to dichondra, centipedegrass, putting greens or tees, or to newly seeded areas. Do not apply to bentgrass mowed at less than 3/8 inch. Do not apply to wet turf. Rainfall or irrigation after application will improve weed control activity. May be applied when sprigging bermudagrass and zoysiagrass. Do not apply to home lawns.
oxadiazon, MOA 14 (50 WP)	1.5 to 2.2 oz	4 to 6 lb	2 to 3	Use in dormant, established bermudagrass, St. Augustinegrass, and zoysiagrass in fairways and parks. Should be applied at least 2 to 3 weeks before greenup of turf. May be applied when sprigging bermudagrass and zoysiagrass. Do not use on home lawns.
oxadiazon, MOA 14 (3.17 SC)	1.85 to 2.8 fl oz	2.52 to 3.81 qt	2 to 3	Use in dormant, established bermudagrass, St. Augustinegrass, and zoysiagrass in fairways and parks. May apply 2 lb a.i. per acre when sprigging bermudagrass. Apply at least 2 to 3 weeks before greenup of turf. Do not use on home lawns.
[oxadiazon + prodiamine], MOA 14 + 3 (1.2 G)	4.5 lb	200 lb	2 + 0.4	Use on turf, golf courses (excluding putting greens) of established bermudagrass, zoysiagrass, St. Augustinegrass, ryegrass, centipedegrass, bentgrass, bluegrass, and tall fescue. Contains 38% N. Apply to dry foliage.
pendimethalin, MOA 3 (2 G) (0.86 G) (1.29 G)	1.72 to 3.44 lb 2.67 to 5.34 lb 2.67 lb	75 to 150 lb 116 to 232 lb 116 lb	1.5 to 3 1 to 2 1.5	Use on established bahiagrass, bermudagrass, centipedegrass, fine fescue, Kentucky bluegrass, perennial ryegrass, St. Augustinegrass, tall fescue, and zoysiagrass. Do not use on winter-overseeded grasses. Wait 4 months after treatment to seed or sod. Do not apply to newly seeded turf until after the 4th mowing. Do not apply to newly sprigged turf until 5 months establishment.
pendimethalin, MOA 3 (3.8 CS)	1.15 to 2.3 fl oz	3.1 to 6.3 pt	1.5 to 3	Use on noncropland as well as established nonresidential and residential turf areas mowed at least 4 times consisting of bahiagrass, bermudagrass, buffalograss, centipedegrass, St. Augustinegrass, zoysiagrass, Kentucky bluegrass, perennial ryegrass, bentgrass, established *Poa annua* (0.5 inch height or taller), fine fescue, and tall fescue. Do not use on bentgrass or *Poa annua* greens and tees. If lower rate is applied initially, repeat in 6-8 weeks for extended control. Do not reseed or overseed into treated turfgrass for 3 months, or sprig turfgrass for 5 months following application. Do not exceed 4.2 pints per acre on residential and sod farm turfgrass.

Table 7-14. Chemical Weed Control in Lawns and Turf

Herbicide and Formulation	Amount of Formulation Per 1,000 sq ft	Amount of Formulation per Acre	Pounds Active Ingredient per Acre	Precautions and Remarks
Preemergence Control, Smooth and Large Crabgrass, Goosegrass, Foxtails, other annual grasses (continued)				
[pendimethalin + dimethemamid], MOA 3 + 15 (1.75 G)	2.3 to 4.6 lb	100 to 200	1.75 to 3.5	Use on residential, commercial, recreational, sod farm and golf course turf, excluding greens. Tolerant turf species include bermudagrass, centipedegrass, St. Augustinegrass, seashore paspalum and zoysiagrass. For extended control, make sequential applications within 5 to 8 weeks not to exceed 400 pounds per acre. Irrigate within 24 hours of application for optimum control. Following application, wait 3 months to overseed, reseed or sprig. If sprigged first, wait 2 months for root establishment to treat. On new sod, mow at least twice before application. On new seedlings, mow at least 4 times before application. Wait 2 weeks after aerification or verticutting before applying.
prodiamine, MOA 3 (65 WG) (4 FL)	0.185 to 0.83 oz 0.23 to 1.1 fl oz	0.5 to 2.3 lb 0.625 to 3 pt	0.325 to 1.5 0.3125 to 1.5	May be used on established bahiagrass, bermudagrass, centipedegrass, St. Augustinegrass, zoysia, tall fescue, creeping red fescue, perennial bluegrass and ryegrass, and creeping bentgrass. Do not apply to greens. May apply when sprigging or plugging bermudagrass, up to 0.8 pound product per acre.
prodiamine, MOA 3 (0.5 G)	1.5 to 6.9 lb	64 to 300 lb	0.32 to 1.5	See precautions for prodiamine 65 WG and 4 FL above except may be used on established turf only. Do not apply more than 150 pounds per acre per application. Do not make more than two applications per calendar year. Wait at least 60 days after initial application before making a second application. Prodiamine is coated on a 32-3-12 dry fertilizer carrier.
siduron, MOA 7 (50 WP)	7.3 oz	20 lb	10	Use only on bluegrass, fescue, perennial ryegrass, and certain bentgrasses (check label). Can be used at the rate of 8 pounds of formulation when seeding bentgrass, bluegrass, fescue, and ryegrass. Can be used in newly sprigged or established zoysia. Do not use on bermudagrass, carpetgrass, centipedegrass.
Preemergence Control, Goosegrass				
dimethenamid, MOA 15 (6 L)	0.48 to 0.73 fl oz	21 to 32 fl oz	1 to 1.5	Use on residential, commercial, recreational, sod farm and golf course turf, excluding greens. Apply 21 ounces to established bentgrass, bluegrass species, fescue species and perennial ryegrass maintained at 0.5 inch cut, but expect yellowing and stand reduction. Apply 32 ounces to bahiagrass, bermudagrass species, centipedegrass, St. Augustinegrass, seashore paspalum and zoysiagrass. For extended control, make sequential applications within 5 to 8 weeks at 32 fluid ounces per acre rate. Irrigate within 24 hours of application for optimum control. Following application, wait 6 weeks to overseed or reseed, wait 2 months to sprig, wait 2 mowings for new sod, and wait 4 mowings for newly seeded turf.
Preemergence Control, Annual Bluegrass (Poa annua)				
[benefin + trifluralin], MOA 3 + 3 (0.86 G)	4 to 8 lb	174 to 349 lb	1.5 to 3	Apply during late summer before Poa annua germinates. Do not apply to turf areas that are to be overseeded.
bensulide, MOA 8 (4 EC) (8.5 G, 12.5 G)		several concentrations available	12.5	See section on preemergence control of crabgrass and goosegrass or product labels for turfgrass tolerance, precautions and remarks for the listed preemergence annual bluegrass herbicides.
dithiopyr, MOA 4 (2 EW, 2 L) (40 WP)	0.75 fl oz 0.46 oz	1 qt 20 oz	0.5	Timely irrigation or rainfall is critical for activation.
indaziflam, MOA 21 (20 WSP)	0.057 to 0.115 oz	2.5 to 5 oz	0.031 to 0.063	
(0.622 F)	0.138 to 0.23 fl oz	6 to 10 fl oz	0.029 to 0.049	
(0.0224 G)	2.9 to 4.6 lb	125 to 200 lb	0.028 to 0.045	
metolachlor, MOA 15 (7.62 EC)	0.48 to 0.96 fl oz	1.3 to 2.6 pt	1.24 to 2.48	
napropamide, MOA 15 (50 DF)	1.5 to 2.25 oz	4 to 6 lb	2 to 3	
oryzalin, MOA 3 (4 AS)	1.1 fl oz	1.5 qt	1.5	
(85 WDG)	0.64 to 0.88 oz	1.75 to 2.4 lb	1.4875 to 2.04	Apply full rate unless potentially thin turfgrass cover is a problem caused by dense poa infestation.
oxadiazon, MOA 14 (2 G)	2.3 to 4.6 lb	100 to 200 lb	2 to 4	
pendimethalin, MOA 3 (2 G) (0.86 G) (1.29 G) (3.8 CS)	1.72 to 3.44 lb 2.67 to 5.34 lb 2.67 lb 1.15 to 1.55 fl oz	75 to 150 lb 116 to 232 lb 116 lb 3.1 to 4.2 pt	1.5 to 3 1 to 2 1.5 1.5 to 2	
(pendimethalin + dimethemamid), MOA 3 + 15 (1.75 G)	2.3 to 4.6 lb	100 to 200	1.75 to 3.5	
prodiamine, MOA 3 (65 WG) (4 FL)	0.185 to 0.83 oz 0.23 to 1.1 fl oz	0.5 to 2.3 lb 0.625 to 3 pt	0.325 to 1.5 0.3125 to 1.5	
pronamide, MOA 3 (3.3 SC)	0.46 to 1.29 fl oz	1.25 to 3.5 pt	0.5 to 1.5	Not for home use. Can be applied from Sept. 15 to Feb. 1 for preemergence or postemergence annual bluegrass control in bermudagrass, zoysiagrass, centipedegrass, and St. Augustinegrass grown for sod, nonresidential or industrial sites, golf course turf, and stadium or professional athletic fields. 1.25 to 2.5 pints per acre provides preemergence to pre tiller stage control. 2 to 2.5 pints per acre provides postemergence control from early tiller to early seedhead stage. 2.5 to 3.5 pints per acre for postemergence control at seedhead stage. Henbit and chickweed species controlled at preemergence timings. Can be used for removal of overseeded grasses; do not overseed if it is desired to maintain a stand. Do not overseed treated area within 90 days of treatment. Injury symptoms from postemergence applications can take up to 5 weeks to develop.

CHEMICAL WEED CONTROL

Table 7-14. Chemical Weed Control in Lawns and Turf

Herbicide and Formulation	Amount of Formulation Per 1,000 sq ft	Amount of Formulation per Acre	Pounds Active Ingredient per Acre	Precautions and Remarks
Preemergence Control, Annual Bluegrass in Overseeded Bermudagrass				
benefin, MOA 3 (2.5 G)	2.75 lb	120 lb	3	Apply in late summer before *Poa annua* germinates. Perennial ryegrass can be overseeded 6 weeks after benefin is applied.
dithiopyr, MOA 4 (2 EW, 2 L) (40 WP)	0.75 fl oz 0.46 oz	1 qt 20 oz	0.5	Apply in late summer before *Poa annua* germinates. Perennial ryegrass can be overseeded 6 to 8 weeks after application. Apply only on well-established bermudagrass. Do not reapply in fall or winter after overseeding unless injury can be tolerated.
prodiamine, MOA 3 (65 WG)	0.213 to 0.367 oz	0.58 to 1 lb	0.37 to 0.65	Use on golf courses (excluding putting greens) when overseeding with perennial ryegrass at a minimum seeding rate of 350 pounds per acre. Apply 8 to 10 weeks before overseeding and expect 70% or greater control. For best potential control, use higher rate and shorter time interval before overseeding. However, this could increase ryegrass seedling mortality or temporarily reduce root growth.
Preemergence and Postemergence Control, Annual Bluegrass				
ethofumesate, MOA 8 (1.5 EC)	2 fl oz	2.67 qt	1	For control of annual bluegrass in dormant bermudagrass overseeded with perennial ryegrass or in established perennial ryegrass turf. Rates are per application. The first application should be 30 to 45 days after overseeding with perennial ryegrass. The second application should be 21 to 28 days later. Do not apply ethofumesate to overseeded bermudagrass after Jan. 1 in N.C.
ethofumesate, MOA 8 (4 SC)	0.55 to 1.47 fl oz	1.5 to 4 pt	0.75 to 2	Must be professionally applied to residential and nonresidential turf including golf courses and sod farms. May be applied to established perennial ryegrass, Kentucky bluegrass, creeping bentgrass, tall fescue, St. Augustinegrass, and dormant bermudagrass. Do not apply to putting greens. Delay application at least 8 weeks after a pgr application. Fall annual bluegrass control best during period of maximum germination. Spring applications most effective following fall applications. For overseeded bermudagrass, apply 1 to 2 weeks after perennial ryegrass emergence and repeat at 21- to 28-day intervals. Do not apply to bermudagrass 4 weeks prior to breaking winter dormancy.
Postemergence Control and Seedhead Suppression, Annual Bluegrass in Overseeded Bermudagrass Fairways, Tees				
bispyribac-sodium, MOA 2 (17.6 SG)	0.046 to 0.138 oz	2 to 6 oz	0.021875 to 0.065625	Do not apply to putting greens, ryegrass mowed to less than 0.375 inch, or non-overseeded bermudagrass. Apply between Feb. 1 and March 15. Make first application when annual bluegrass begins flowering. If actively flowering, use the low rate and re-treat in 28 to 35 days. If not actively flowering, use the low rate and retreat in 14 to 21 days with the low rate. Do not apply if air temperature is less than 50 degrees F within 3 days after application. Check label for further special instructions.
amicarbazone, MOA 5 (70 WG)	0.023 to 0.23 oz	1 to 10 oz	0.044 to 0.44	Also tolerant to 6 month established turfgrasses such as bahiagrass, centipedegrass, seashore paspalum, St. Augustinegrass, zoysiagrass, bentgrass, Kentucky bluegrass, perennial ryegrass, fine and tall fescue. Labeled for use on golf course, sod farm, residential, commercial, athletic field and roadside turf. Bentgrass tees: 1 ounce/acre at 7 day intervals for 4 applications. Bentgrass roughs and fairways: 2 to 3 ounces/acre for 14 to 21 day intervals for 2 applications. Cool season turf: 2 to 4 ounces/acre for 14 to 21 day intervals for 2 applications. Warm season turf: 3 to 10 ounces/acre for 14 to 21 day intervals for 2 applications not to exceed 10 ounces/acre per year. Allow 4 weeks before cutting or lifting sod. Allow 1 week before overseeding winter grasses.
Postemergence Control, Annual Bluegrass, Overseeded Perennial Ryegrass, Tall Fescue, *Poa trivialis*				
flazasulfuron, MOA 2 (25 DG)	0.011 to 0.069 oz	0.5 to 3 oz	0.0078 to 0.0469	For use on well established bermudagrass, zoysiagrass, centipedegrass, and seashore paspalum grown on turf including golf courses (including fairways, roughs, greens (bermudagrass and seashore paspalum only), tees, collars and approaches), industrial parks, tank-sod- and seed farms, cemeteries, athletic field and commercial lawns. Residential turf applications are limited to spot applications. Apply a maximum of 1.5 ounces per acre on fully green centipedegrass and seashore paspalum. 3 ounces per acre needed for annual bluegrass control and best if applied in spring. 0.5 to 1.5 ounces per acre will control perennial and Italian ryegrass. For clumpy ryegrass, use 1.5 to 3 ounces per acre. 1.5 ounces per acre needed for tall fescue control. 2.25 to 3 ounces per acre needed for *poa trivialis* control. Include a nonionic surfactant at 0.25% by volume.
foramsulfuron, MOA 2 (0.19 SC)	0.2 to 0.6 fl oz	8.8 to 26.2 fl oz	0.013 to 0.039	For use on bermudagrass and zoysiagrass grown on home lawns, golf courses and sod farms. Do not use on warm season turfgrass collars surrounding bentgrass greens. May be applied up to 1 week prior to overseeding. Do not apply within 2 weeks of bermudagrass sprigging. Apply in 25 to 60 gallons water per acre. Rainfast after 2 hours. Surfactant not required.
rimsulfuron, MOA 2 (25 DF)	0.011 to 0.092 oz	0.5 to 4 oz	0.0078 to 0.0625	May be applied to bermudagrass, zoysiagrass and centipedegrass on professionally managed sports facilities at professional and collegiate levels, golf courses, sod farms, roadsides, industrial and commercial lawns. For annual bluegrass control, apply November through December and again February through March if needed at 2 ounces per acre. May be applied 10 to 14 days prior to overseeding. For overseeded removal, apply 2 ounces per acre 3 to 4 weeks before desired removal date, and repeat 3 weeks later if needed. For weed control along roadsides, apply 4 ounces per acre if single application only. A nonionic surfactant at 0.25% by volume or an oil adjuvant such as crop oil concentrate and modified seed oil at 1% by volume are required. Do not apply to cool-season turfgrasses, residential lawns or newly sprigged/sodded bermudagrass.

320

Table 7-14. Chemical Weed Control in Lawns and Turf

Herbicide and Formulation	Amount of Formulation Per 1,000 sq ft	Amount of Formulation per Acre	Pounds Active Ingredient per Acre	Precautions and Remarks
Postemergence Control, Annual Bluegrass, Overseeded Perennial Ryegrass, Tall Fescue, *Poa trivialis* (continued)				
[metsulfuron + rimsulfuron], MOA 2 + 2 (37 WG)	0.0344352 oz	1.5 oz	0.0346875	Use on well established bermudagrass and zoysiagrass grown on nonresidential turf including golf courses, sod farms, industrial and commercial lawns, and professionally managed college and professional sports fields. Overseeding can occur 2 months after application. Include a nonionic surfactant at 0.25% by volume.
sulfosulfuron, MOA 2 (75 DG)	0.017 to 0.046 oz	0.75 to 2 oz	0.035 to 0.09375	May be applied to certain ornamental native grasses and also bermudagrass species, zoysiagrass, centipedegrass, St. Augustinegrass, and kikuyugrass grown on sod farms, golf courses (excluding greens), commercial and residential turf that is highly managed, and other noncrop areas. Use 1.5 to 2 ounces per acre for fall annual bluegrass control 7 to 10 days before overseeding. Use 0.75 to 1.25 ounces per acre for fall or winter control in nonoverseeded bermudagrass, and reapply if needed but not before 21 days after initial application. For tall fescue control, two applications may be required at 4- to 10-week intervals. Perennial ryegrass control not as complete as with foramsulfuron, rimsulfuron, or trifloxysulfuron. Use a nonionic surfactant at 0.25% by volume. Do not exceed 2.66 ounces per acre per year.
trifloxysulfuron, MOA 2 (75 WG)	0.0023 to 0.0129 oz	0.1 to 0.56 oz	0.0047 to 0.0263	May be applied to residential bermudagrass and zoysiagrass and also on golf courses, sod farms, and other nonresidential turf areas. A nonionic surfactant at 0.25 to 0.5% by volume is recommended. Temporary discoloration may occur if used with MSO or COC. May be applied 3 weeks prior to overseeding. Use rates of 0.1 to 0.3 ounces per acre to remove overseeded perennial ryegrass and *Poa trivialis* to aid bermudagrass spring transition. Labeled turf species can be seeded or sprigged into treated areas 4 weeks after application.
Preemergence and Postemergence Control, Annual Bluegrass and certain winter annual broadleaf weeds				
atrazine, MOA 5 (4 L) (90 DF,90 WG)	0.75 to 1.5 fl oz 0.025 to 0.05 lb	1 to 2 qt 1.1 to 2.2 lb	1 to 2	Use on centipedegrass, St. Augustinegrass, and dormant bermudagrass. Apply Nov. 15 to Dec. 31. Follow label directions.
simazine, MOA 5 (90 WDG, 90 DF) (4 L)	0.4 to 0.8 oz 0.75 to 1.5 fl oz	1.1 to 2.2 lb 1 to 2 qt	1 to 2	Use on bermudagrass, centipedegrass, St. Augustinegrass, and zoysiagrass. See label for instructions on newly sprigged turfgrass or on hybrid bermudagrass. Apply Nov. 15 to Dec. 15. Follow label directions.
Preemergence Control, Certain Broadleaf Weeds				
isoxaben, MOA 21 (75 DF, 75 WG)	0.25 to 0.5 oz	0.66 to 1.33 lb	0.5 to 1	All established turfgrasses are tolerant. However, do not apply to putting greens or turfgrass grown for seed. Check label for specific weeds controlled.
pendimethalin, MOA 3 (3.8 CS)	1.15 to 1.55 fl oz	3.1 to 4.2 pt	1.5 to 2	See section on preemergence control of crabgrass or product label for turfgrass tolerance. Provides preemergence control of summer broadleaf weeds, such as prostrate spurge, prostrate knotweed, and purslane species, as well as winter broadleaf weeds, such as yellow woodsorrel, hop clover, cudweed species, common chickweed, lawn burweed, henbit, and corn speedwell when applied before expected germination.
Preemergence and Postemergence Control Crabgrass, Goosegrass, Other Annual Grasses, Broadleaf Weeds, Sedges				
mesotrione, MOA 27 (4 SC)	0.092 to 0.183 fl oz	4 to 8 fl oz	0.125 to 0.25	Use on residential turf, golf courses (not greens) and sod farms for pre- and postemergence weed control. Tolerant turfgrasses include St. Augustinegrass, centipedegrass, tall fescue, fine fescue, Kentucky bluegrass, and perennial ryegrass. Add a nonionic surfactant and repeat application after 2 to 3 weeks for improved postemergence control. Tank mix with prodiamine 65 WG for extended preemergence grassy weed control. Can be applied at seeding to all tolerant grasses except fine fescue. After turf germination, wait 4 weeks or until turf has been mowed twice before making a postemergence application. Also controls henbit, chickweed, dandelion, white clover, Florida betony, Florida pusley, ground ivy, oxalis, wild violet, creeping bentgrass, and yellow nutsedge.
[sulfentrazone + prodiamine], MOA 14 + 3 (4 SC)	0.184 to 0.826 fl oz	0.5 to 2.25 pt	0.25 to 1.125	For use in residential and institutional lawns, athletic fields, golf course fairways and roughs, roadsides, utility right-of-ways, railways, and industrial areas. Apply to turf following a second mowing if a good root system has been established. Apply up to 12 fluid ounces per acre to bentgrass at 0.5 inch or higher, fine fescue, and perennial ryegrass. Apply 18 to 24 fluid ounces per acre to perennial bluegrass, tall fescue, and all warm season grasses except St. Augustinegrass (do not apply) and bermudagrass (apply 18 to 36 fluid ounces per acre). For sod production, apply 6 months after establishment, and do not harvest within 3 months. Do not apply with adjuvants or surfactants. [Sulfentrazone + prodiamine should not be applied to cool-season turf with N-containing fertilizers unless some short-term discoloration is tolerable.
Postemergence Control, Crabgrass, Goosegrass				
fenoxaprop, MOA 1 (0.57 EC)	0.3 to 0.9 fl oz	0.8 to 2.4 pt	0.057 to 0.174	Use only on perennial ryegrass, fine fescue, tall fescue, Kentucky bluegrass, and zoysiagrass. Reduced vigor or discoloration can occur. Rate depends upon leaf number or tillers of grass weeds and turf tolerance. Check label. A second application may be applied after 14 days.
	0.08 fl oz	3.5 fl oz	0.016	Apply only to established Penncross bentgrass maintained at a minimum cutting height of at least 0.25 inch. Bentgrass should be established for one growing season. Do not apply to greens. Applications should be made at a minimum of 21-day intervals, beginning in the spring when grassy weeds first emerge and are not larger than two-leaf. Repeat applications throughout the summer as new infestations of one- to two-leaf grassy weeds occur. See label for restrictions.
metribuzin, MOA 5 (75 DF)	0.12 to 0.24 oz	0.33 to 0.67 lb	0.25 to 0.5	Recommended for application by commercial applicators only on established bermudagrass turf (such as parks, athletic fields, golf course fairways, cemeteries, and sod farms) that has a mowing height of 0.5 inch or greater. Apply when turf is vigorously growing and not under stress. Repeat if necessary in 7 to 10 days. Do not make more than two applications per season. Do not apply to greens, tees, or aprons.

CHEMICAL WEED CONTROL

CHEMICAL WEED CONTROL

Table 7-14. Chemical Weed Control in Lawns and Turf

Herbicide and Formulation	Amount of Formulation Per 1,000 sq ft	Amount of Formulation per Acre	Pounds Active Ingredient per Acre	Precautions and Remarks
Postemergence Control, Crabgrass, Goosegrass (continued)				
sethoxydim, MOA 1 (1 EC)	0.8 to 1.38 fl oz	2.25 to 3.75 pt	0.28 to 0.47	Use in seedling and established centipedegrass and fine fescues. Apply 2.25 pint to grasses up to 6 inches and 3.75 pints to grasses up to 12 inches if turf is tolerant. Does not control yellow and purple nutsedge, annual bluegrass or broadleaf weeds. Apply no sooner than 3 weeks after spring greenup of centipedegrass. Apply before crabgrass becomes extensively tillered. Delay all treatments until newly planted centipedegrass has 3 inches of new stolon growth. Do not mow within 7 days before or after application. Two applications 3 weeks apart will suppress bahiagrass. Additives or adjuvants not required.
Postemergence Control, Smooth and Large Crabgrass, Barnyardgrass, White and Hop Clover, Common Dandelion, Dollarweed, Foxtails				
quinclorac, MOA (27 + 4) (75 DF) (1.5 SL)	0.367 oz 1.45 fl oz	1 lb 2 qt	0.75	For use in residential and nonresidential turf that is established or newly seeded, overseeded, or sprigged. Refer to label for specific varieties. Apply to common and hybrid bermudagrass, Kentucky bluegrass, annual bluegrass, buffalograss, tall fescue, annual and perennial ryegrass, creeping bentgrass, and zoysiagrass. Can also be applied to fine fescue but must be in a blend. Some discoloration of hybrid bermudagrass, creeping bentgrass or fine fescue may occur. Do not apply to bahiagrass, centipedegrass, St. Augustinegrass, or dichondra. Do not use on golf course greens or collars. The addition of methylated seed oil (1.5 pints per acre or 0.55 ounces per 1,000 square feet) or a crop oil concentrate (2 pint per acre or 0.73 ounces per 1,000 square feet) is required for control. Application to weeds under stress will result in poor control. Irrigation 24 hours prior to application is recommended if drought conditions exist. Some ornamental plants are sensitive to quinclorac. See label for further precautions.
Postemergence Control, Smooth and Large Crabgrass, Barnyardgrass, Foxtails, and many broadleaf weeds				
[quinclorac + sulfentrazone + 2,4-D amine + dicamba], MOA (27 + 4) + 14 + 4 + 4 (1.79 L)	1.8 to 3 fl oz	5 to 8 pt	1.12 to 1.79	For use in fully dormant bermudagrass as well as actively growing bermudagrass after spring greenup but use only 5 to 7 pints per acre. Also labeled in fully dormant zoysiagrass s well as cool-season turf including annual bluegrass and ryegrass, perennial bluegrass and ryegrass, and fescue species. Do not apply to bahiagrass, bentgrass (creeping, Seaside, Colonial), centipedegrass, St. Augustinegrass, carpetgrass, and golf course greens, tees, and collars. May be applied to home lawns. Apply to seedling grasses after second or third mowing, or 28 days after emergence. Wait 3 to 4 weeks after sodding, sprigging, or plugging operations to apply. Wait 4 weeks after application to seed.
[quinclorac + mecoprop + dicamba], MOA (27 + 4) + 4 + 4 (2.45 SL)	0.5 to 1.45 fl oz	0.68 to 2 qt	0.4165 to 1.225	For use in warm- and cool-season residential and non-residential turf, including but not limited to commercial property, parks, roadsides, schools, athletic fields, cemeteries, and golf courses. May be applied to species of bermudagrass, bluegrass, fescue, and ryegrass as well as creeping bentgrass, seashore paspalum, and zoysiagrass. Use with methylated seed oil at 1.5 pints per acre. Allow 28 days of seedling or sprig growth before application. If treating first, allow 28 days before seeding or sprigging. Do not apply to golf course collars or greens or to turf grown for sod. Use low rate in 2 split applications when treating creeping bentgrass.
[carfentrazone + quinclorac], MOA 14 + (27 + 4) (75 WG)	0.184 to 0.413 oz	8 to 18 oz	0.35 to 0.79	Can use up to 12 ounces per acre 7 days after emergence from seed or sod installment on bluegrass and fescue species and perennial ryegrass; 18 ounces per acre can be used 7 days after seed, sod or sprig operations on bermudagrass species, centipedegrass and seashore paspalum. Wait 14 days after emergence for zoysiagrass. May apply to residential, commercial, and institutional lawns, athletic fields, sod farms, and golf course fairways and roughs. Adjuvants not required but may help on mature weeds.
[sulfentrazone + quinclorac], MOA 14 + (27 + 4) (75 WG)	0.367 to 0.735 oz	1 to 2 lb	0.75 to 1.5	Use up to 21 ounces per acre on well-established tall fescue, Kentucky bluegrass and perennial ryegrass; up to 32 ounces per acre on well-established bermudagrass, centipedegrass, zoysiagrass and seashore paspalum. May be applied to residential, commercial, and institutional lawns, athletic fields, sod farms, and golf course fairways and roughs. After treatment, wait at least 1 month before reseeding, overseeding (use slit seeder for best results), or sprigging. Wait at least 3 months for sod establishment and do not spray within 3 months of harvest. Controls goosegrass in the 1 to 4 leaf stage. Yellow nutsedge and kyllinga species are also controlled. Do not apply with a spray adjuvant.
[fenoxaprop + fluroxypyr + dicamba], MOA 1 + 4 + 4 (0.75 EC)	1.3 to 1.5 fl oz	3.5 to 4 pt	0.33 to 0.375	Tolerant turfgrass species include zoysiagrass, Kentucky bluegrass, perennial ryegrass, fine and tall fescue. May be applied to golf courses excluding greens and tees, athletic fields, commercial and residential turf. Sod farm use is not permitted. Best grass weed control will be achieved when treated from 1 leaf to 4 tiller stage. Do not apply more than 15 pints per acre per year. Do not reapply within 14 days of an application. Surfactant not required. Spot treat using 0.6 to 1 fluid ounces per 1 gallon water.
Postemergence Control, Large Crabgrass, Carpetgrass, Bull Paspalum, Bahiagrass, Foxtails, and many broadleaf weeds, including Chamberbitter, Corn Speedwell, Dichondra, Dollarweed, Doveweed, Florida Betony, Florida Pusley, Lespedeza, Oxalis, Spurge, Virginia Buttonweed, Kyllinga				
[thiencarbazone-methyl + iodosulfuron-methyl + dicamba], MOA 14 + 2 + 4 (68 WG)	0.057 to 0.113 oz	2.5 to 4.9 oz	0.106 to 0.208	For use by licensed applicators in residential and commercial lawns, golf courses (excluding greens), sports fields, parks, recreational areas, roadsides, school grounds, and sod farms. Provides up to 60 days residual control. Use on bermudagrass, zoysiagrass, centipedegrass, and St Augustinegrass. Apply maximum 7.4 ounces per acre per season. Safe to use at high temperatures. Ryegrass can be overseeded 2 weeks after application. Apply 30 days prior to seeding bermudagrass or zoysiagrass. Wait 2 weeks after bermudagrass seedling emergence or sprigging operation before applying. For zoysiagrass, wait 3 weeks after seedling emergence before applying. A nonionic surfactant or methylated seed oil at 0.25% v/v is required for optimum control.

Table 7-14. Chemical Weed Control in Lawns and Turf

Herbicide and Formulation	Amount of Formulation Per 1,000 sq ft	Amount of Formulation per Acre	Pounds Active Ingredient per Acre	Precautions and Remarks
Postemergence Control or Suppression of Summer Weeds Such as Crabgrass Species, Goosegrass, Dallisgrass, Virginia Buttonweed, Doveweed, Florida Pusley, Nutsedge and Kyllinga Species; Winter Weeds Such as *poa annua, poa trivialis*, Tall Fescue, Henbit, Corn Speedwell, and Species of Ryegrass, Chickweed, and Clover				
[thiencarbazone-methyl + foramsulfuron + halosulfuron], MOA 14 + 2 + 2 (60.5 WG)	0.0735 oz	3.2 oz	0.121	Apply to well-established residential and commercial bermudagrass and zoysiagrass (Emerald, Meyer, Zeon) lawns, golf courses (excluding greens), athletic fields, sod farms, roadsides, parks, cemeteries and recreational areas. Do not exceed 3.2 ounces per acre per application or 6.4 ounces per acre yearly. Use 0.25 to 0.5% by volume nonionic surfactant or 0.5 to 1% by volume methylated seed oil. After application, wait 12 weeks to overseed ryegrass or bermudagrass. Wait 1 month after bermudagrass seedling emergence and 2 weeks after sprigging or sodding bermudagrass before treating. Temporary stunting and yellowing may last up to 2 weeks but turf will recover. Crabgrass and goosegrass are controlled up to 2 tiller stage.
Postemergence Control, Goosegrass				
diclofop-methyl, MOA 1 (3 EC)	0.75 to 1 fl oz	32 to 43 fl oz	0.75 to 1	Apply in established bermudagrass. Rate depends on number of goosegrass leaves from one to four leaves. Check label for specific rates.
foramsulfuron, MOA 2 (0.19 SC)	0.39 fl oz	17 fl oz	0.025	For use on bermudagrass and zoysiagrass grown on home lawns, golf courses and sod farms. See precautions listed under annual bluegrass section. For goosegrass control, apply 17 fl ounces per acre on plants up to 2 tillers followed by 17 fluid ounces per acre 2 weeks later.
sulfentrazone, MOA 14 (4 SC)	0.275 fl oz	0.75 pt	0.375	May be applied to home lawns. For use on creeping bentgrass, tall and fine fescue, perennial ryegrass, Kentucky bluegrass, and all warm-season turf species except St. Augustinegrass. See precautions listed under purple and yellow nutsedge section. For goosegrass control, apply 0.75 pint per acre on plants up to 2 tillers.
Postemergence Control, Bahiagrass, Crabgrass, Dallisgrass, Goosegrass, Nutsedge, Annual Sedges, Sandbur				
MSMA, MOA 17 (6 SL, 6.6 SL)		several concentrations	1.82 to 4.5	MSMA is only registered for golf course, sod farm, and highway right-of-way use. Bermudagrass, bluegrass and zoysiagrass are tolerant. Injury may result on bentgrass, fescue and also St. Augustinegrass grown for commercial sod production only. Do not use on carpetgrass or centipedegrass. MSMA restrictions: For existing golf courses, spot treat (100 square feet per spot) not to exceed 25% of total acreage. For new courses, make 1 broadcast application per year. For sod farms, make 1 to 2 broadcast applications per year and maintain 25 feet buffer around permanent water bodies. For highway rights of way, make 2 broadcast applications and maintain 100 feet buffer around permanent water bodies.
Postemergence Control, Crabgrass, Goosegrass, Sandbur, Dallisgrass				
MSMA, MOA 17 (6 SL, 6.6 SL) + metribuzin, MOA 5 (75 DF)		several concentrations + 0.17 to 0.33 lb	1.5 to 2 + 0.125 to 0.25	See remarks for MSMA and metribuzin. The combination improves goosegrass control. Should be applied to bermudagrass only.
Postemergence Control, Crabgrass, Goosegrass, Sandbur				
asulam, MOA 18 (3.34 SL)	1.8 fl oz	5 pt	2	Use only on St. Augustinegrass and Tifway 419 turf. On golf courses, use only on fairways and roughs.
Postemergence Control, Crabgrass and Foxtail Species, Goosegrass, Broadleaf Signalgrass, Japanese Stilgrass **Postemergence Suppression, Creeping Bentgrass, Common Bermudagrass, Dallisgrass, Nimblewill**				
topramezone, MOA 27 (2.8 L)	0.023 to 0.034 fl oz	1 to 1.5 fl oz	0.021875 to 0.0328125	Labeled for broadcast treatment use in residential and athletic field turf, as well as in nonresidential turf sites including sod farms, golf courses (excluding greens and collars), parks, roadsides, cemeteries, and commercial properties. Tolerant turf species include Kentucky bluegrass, tall and fine fescue, perennial ryegrass, and centipedegrass at seeding and then anytime beyond 28 days after seeding. Add crop oil concentrate or methylated seed oil for enhanced control at 0.5 to 1% by volume. Don't apply greater than 2 fluid ounces per acre per application or 4 fluid ounces per acre per year. Bleaching intensity of susceptible weeds reduced and broadleaf weed spectrum increased if tankmixed with quinclorac, [quinclorac + mecoprop + dicamba] or triclopyr. For suppression of above-listed weeds, add triclopyr at 1 pound ae per acre and make either 2 or 3 applications at 3 to 4 week intervals depending on topramezone rate. Creeping bentgrass is marginally tolerant to topramezone at 0.25 fluid ounces per acre. Test on a small area before large-scale use. Sequential applications may be required to achieve desired level of weed control.
Postemergence Control, Yellow Nutsedge, Annual Sedge				
bentazon, MOA 6 (4 SL)	0.75 to 1.5 fl oz	1 to 2 qt	1 to 2	For control of yellow nutsedge in established bluegrass, fescues, bentgrass, ryegrass, bermudagrass, bahiagrass, St. Augustinegrass, centipedegrass, and zoysiagrass. Apply to yellow nutsedge when actively growing under good soil moisture conditions. Additional applications may be made at intervals of 10 to 14 days until nutsedge is controlled.
Postemergence Control, Purple and Yellow Nutsedge, Kyllinga Species				
flazasulfuron, MOA 2 (25 DG)	0.034 to 0.069 oz	1.5 to 3 oz	0.023 to 0.0469	For use on well established bermudagrass, zoysiagrass, centipedegrass and seashore paspalum grown on nonresidential turf including golf course fairways, roughs and tees, and industrial parks, tank-sod- and seed farms, cemeteries, athletic field and commercial lawns. Apply a maximum of 1.5 ounces per acre on fully green centipedegrass and seashore paspalum. 3 ounces per acre needed for perennial nutsedge and some annual sedge species control. Repeat applications in 2 to 6 weeks when nutsedge or sedge growth is evident. 1.5 to 2.25 ounces per acre will control kyllinga species. Maintain a 25 feet nontreated border beside susceptible turf species. Can overseed in 2 weeks if applied up to 1.5 ounces per acre. Wait 4 weeks if applied more than 1.5 ounces per acre. Include a nonionic surfactant at 0.25% by volume.

CHEMICAL WEED CONTROL

Table 7-14. Chemical Weed Control in Lawns and Turf

Herbicide and Formulation	Amount of Formulation Per 1,000 sq ft	Amount of Formulation per Acre	Pounds Active Ingredient per Acre	Precautions and Remarks
Postemergence Control, Purple and Yellow Nutsedge, Kyllinga Species (continued)				
imazaquin, MOA 2 (70 DG)	0.128 to 0.256 oz	0.357 to 0.714 lb	0.25 to 0.5	Use on bermudagrass, centipedegrass, St. Augustinegrass, and zoysiagrass. Do not apply during spring greenup. Temporary yellowing may occur. Add a nonionic surfactant at 2 pt per 100 gal of spray solution. Addition of MSMA at 1.5 lb active per acre will improve sedge control in MSMA tolerant turfgrasses.
imazosulfuron, MOA 2 (75 WG)	0.184 to 0.322 oz	8 to 14 oz	0.38 to 0.66	May be applied to established (two mowings) residential and commercial bermudagrass, zoysiagrass, centipedegrass, St. Augustinegrass, creeping bentgrass, Kentucky bluegrass, perennial ryegrass, tall fescue, and fine fescue. Do not apply to putting greens. Reapply 3 weeks after initial application when using the 8 ounces per acre rate. Reapply as needed 3 weeks after initial application when using rates above 8 ounces per acre. Wait 4 weeks to seed or sod after application. Use an 80% active nonionic surfactant at 0.25% by volume. For spot treatment, add 0.25 to 0.33 oz in 1 to 2 gallons of water per 1000 square feet. Add 2 teaspoons nonionic surfactant per gallon.
halosulfuron, MOA 2 (75 WDG)	0.9 g	0.67 to 1.33 oz	0.031 to 0.062	May be applied to established residential and commercial bermudagrass, bahiagrass, zoysiagrass, centipedegrass, St. Augustinegrass, creeping bentgrass, Kentucky bluegrass, perennial ryegrass, tall fescue, and fine fescue. Apply broadcast when sedges have reached the 3- to 8-leaf stage. Use lower rate for light infestations and higher rate for heavy infestations. A second treatment will usually be required 6 to 10 weeks after the initial treatment. Use an 80% active nonionic surfactant at 2 quarts per 100 gallons of spray solution (0.5% by volume). Do not exceed 1 to 2 pints of surfactant per acre. Do not apply to putting greens. Halosulfuron only suppresses green kyllinga.
MSMA, MOA 17 (6 SL, 6.6 SL)		several concentrations	2 to 3	See remarks for MSMA above. Will require at least 2 applications 7 to 10 days apart.
sulfosulfuron, MOA 2 (75 DG)	0.017 to 0.029 oz	0.75 to 1.25 oz	0.035 to 0.059	May be applied to certain ornamental native grasses and also bermudagrass species, zoysiagrass, centipedegrass, St. Augustinegrass, and kikuyugrass grown on sod farms, golf courses (excluding greens), commercial and residential turf that is highly managed, and other noncrop areas. Use 0.75 to 1.25 ounces per acre, and repeat in 4 to 10 weeks if needed. Use a nonionic surfactant at 0.25% by volume.
trifloxysulfuron, MOA 2 (75 WG)	0.0023 to 0.0129 oz	0.1 to 0.56 oz	0.0047 to 0.0263	May be applied to residential bermudagrass and zoysiagrass and on golf courses, sod farms, and other nonresidential turf areas. A nonionic surfactant at 0.25 to 0.5% by volume is recommended. Temporary discoloration may occur if used with MSO or COC. Use rates of 0.33 to 0.56 ounces per acre for sedge and kyllinga species control. Labeled turf species can be seeded or sprigged into treated areas 4 weeks after application. Repeat application may be needed in 4 to 6 weeks.
Postemergence Control, Purple and Yellow Nutsedge, Kyllinga Species, and various broadleaf weeds				
sulfentrazone, MOA 14 (4 SC)	0.092 to 0.275 fl oz	0.25 to 0.75 pt	0.125 to 0.375	May be applied to home lawns. For use on creeping bentgrass, tall and fine fescue, perennial ryegrass, Kentucky bluegrass, and all warm-season turf species except St. Augustinegrass. Wait 3 months to seed, overseed, or sprig unless overseeding bermudagrass with perennial ryegrass, which only requires a 4- to 6-week waiting period after application. Apply to seedling grasses after second mowing and to new sod 6 months after establishment.
[sulfentrazone + imazethapyr], MOA 14 + 2 (4 SC)	0.22 to 0.33 fl oz	9.5 to 14.4 fl oz	0.29 to 0.45	May be applied to home lawns, athletic fields, sod farms, golf course fairways and roughs, and various non-crop sites. For use on bahiagrass, bermudagrass, centipedegrass, and zoysiagrass. Do not apply to soils classified as sand with less than 1% organic matter. Do not reseed, overseed, or sprig within 1 month of application. Expect slight perennial ryegrass injury if overseeded 2 to 4 weeks after application. Allow 3 month sod establishment before treatment.
[sulfentrazone + metsulfuron], MOA 14 + 2 (66 WG)	0.075 to 0.23 oz	3.25 to 10 oz	0.134 to 0.413	May be applied to established residential, commercial and institutional lawns, athletic fields, sod farms, and golf course fairways and roughs. Use up to 6.5 ounces per acre on Kentucky bluegrass and tall fescue and 10 ounces per acre on bermudagrass, centipedegrass, St. Augustinegrass and zoysiagrass. Do not reseed, overseed, or sprig within 1 month of application. Expect slight perennial ryegrass injury if overseeded 6 to 8 weeks after application. Allow 3 months sod establishment before treatment. No adjuvant needed.
Postemergence Control, Bahiagrass, Crabgrass, Yellow and Purple Nutsedge, Annual Sedge, Kyllinga Species				
imazapic, MOA 2 (2 AS)	0.092 to 0.184 fl oz	4 to 8 fl oz	0.063 to 0.125	For use on unimproved centipedegrass after complete greenup only. Not for use in home lawns. Do not use on other turfgrass species. A repeat application may be needed on tough to control perennial weeds such as bahiagrass. The highest labeled rate may discolor centipedegrass by causing a red color.
Postemergence Control, Dandelion, Carpetweed, Carolina Cranesbill, Curly Dock, Plantain, Dichondra, Shepherds-Purse, Yellow Rocket				
2,4-D amine, MOA 4 (4 SL)	3 to 4 tsp	1.5 to 2 pt	0.75 to 1	Cut rate one-half for bentgrass, carpetgrass, centipedegrass, and St. Augustinegrass. Spray when weeds are young and actively growing. To reduce danger of injury to flowers and ornamentals by spray drift, use low pressure and do not spray on windy days.
Postemergence Control, Common Chickweed, Mouseear Chickweed, Creeping Charlie or Ground Ivy, Dandelion, Lespedeza, Black Medic, Spotted Spurge, Hop or White Clover				
mecoprop, MOA 4 (1.9 L) (1.16 L) (1.74 L)	1 to 1.5 fl oz 1.5 to 2.25 fl oz 0.75 to 1.5 fl oz	2.7 to 4 pt 4 to 6 pt 2 to 4 pt	0.64 to 0.95 0.58 to 0.87 0.43 to 0.87	Observe same precaution as for 2,4-D. May be used on bentgrass, carpetgrass, centipedegrass, St. Augustinegrass, and other turf grasses.

Table 7-14. Chemical Weed Control in Lawns and Turf

Herbicide and Formulation	Amount of Formulation Per 1,000 sq ft	Amount of Formulation per Acre	Pounds Active Ingredient per Acre	Precautions and Remarks
Postemergence Control, Chickweed, White Clover, Dandelion, Curly Dock, Hawkweed, Henbit, Knotweed, Red Sorrel, Knawel, Spurweed, Spotted Spurge, Wild Strawberry, Yarrow				
dicamba, MOA 4 (4 SL)	1 to 2 tsp	0.5 to 1 pt	0.25 to 0.5	Apply as foliar spray to growing weeds. Prevent injury to ornamentals. Avoid rooting zone of shallow-rooted trees and shrubs.
diglycolamine, MOA 4 (4 SL)	1 to 4.5 tsp	0.5 to 2 pt	0.25 to 1	Do not exceed 1 pint per acre on bentgrass, carpetgrass, buffalograss, and St. Augustinegrass. Apply to newly seeded grasses after the second mowing. Do not exceed 0.25 pint per acre on extended sensitve plant roots on sandy soils and 0.5 pint per acre on clay soils.
Postemergence Control, All Weeds Listed Under 2,4-D Amine, MCPP, Dicamba, and Diglycolamine Sections				
[2,4-D amine + MCPP + dicamba], MOA 4 + 4 + 4 (various formulations)	See individual label	See individual label	See individual label	Check individual labels for specific rates, instructions and precautions. Generally, 1) apply to grass seedlings after second mowing; 2) apply to sodded, sprigged, or plugged turf 3 to 4 weeks after operations; and 3) wait 3 to 4 weeks after application to seed. Many products labeled for tall fescue, perennial ryegrass, perennial bluegrass, bermudagrass, and St. Augustinegrass. Some products labeled for bentgrass putting greens, bahiagrass, zoysiagrass, and centipedegrass. Some products labled for home use when applied by a commercial applicator.
[2,4-D amine + MCPP + dichlorprop], MOA 4+ 4 + 4 (4.11 L) (2.48 L)	0.62 to 1.47 fl oz 0.64 to 1.47 fl oz	\ 1.7 to 4 pt 1.75 to 4 pt	0.873 to 2.06 0.543 to 1.24	
[MCPA + MCPP + dicamba], MOA 4+4+4 (4 L)	0.7 to 1.5 fl oz	2.5 to 4.1 pt	1.25 to 2.05	
Postemergence Control, Curly Dock, Broadleaf Dock, Galinsoga, Nightshade, Clover (Red, Hop, White, Sweet), Goldenrod, Musk Thistle, Speedwells, Common Vetch, Hairy Buttercup, Broadleaf Plantain				
clopyralid, MOA 4 (3 EC)	0.1 to 0.5 fl oz	0.25 to 1.33 pt	0.09 to 0.5	Do not apply to home lawns. May be used on bentgrass, Kentucky bluegrass, creeping, red, chewings, sheep and tall fescue, perennial ryegrass, bermudagrass, bahiagrass, buffalograss, centipedegrass, zoysiagrass, and St. Augustinegrass. Do not apply to putting greens and tees. Should be applied in a minimum of 20 gallons of water per acre. Surfactants are not necessary. Do not apply to exposed roots of certain trees and shrubs (legumes such as acacia, locust, mimosa, redbud, or mesquite) or *Tilia* spp. Do not use treated clippings for mulching and compost during the growing season of application.
All Weeds Listed Under 2,4-D Amine, Clopyralid, Dicamba, and Diglycolamine Sections				
[2,4-D amine + clopyralid + dicamba], MOA 4 + 4 + 4 (3.56 L)	0.55 to 1.1 fl oz	1.5 to 3 pt	0.67 to 1.34	Do not apply to home lawns. Use on perennial bluegrass, ryegrass, and fescue species, bentgrass (excluding greens and tees), bermudagrass, zoysiagrass, and bahiagrass. Do not apply to seedling grasses until well established. Wait 3 to 4 weeks after application to seed.
Postemergence Control, Virginia Buttonweed, Chickweed Species, White Clover, Dandelion, Henbit, Ground Ivy, Prostrate Knotweed, Matchweed, Black Medic, Plantain Species, Common Woodsorrel				
[2,4-D amine + fluroxypyr + dicamba], MOA 4 + 4 + 4 (4 SL)	0.36 to 1.1 fl oz	1 to 3 pt	0.5 to 1.5	Use on perennial bluegrass and ryegrass, tall fescue, creeping bentgrass (excluding greens and tees), bermudagrass species, bahiagrass, zoysiagrass, and St. Augustinegrass in residential, industrial, and institutional lawns, parks, cemeteries, athletic fields, golf courses, and sod farms. Use on St. Augustinegrass sod farms only. Apply 1 to 2 pints per acre on creeping bentgrass and 1.5 to 1.8 pints per acre on warm season turf grown for sod. Apply 2 to 3 pints per acre to all other turf areas. For non-turf areas, rate can be increased to 2 to 5 pinst per acre. Application can be made to grass seedlings after second mowing and to newly sodded, sprigged, or plugged grasses 3 to 4 weeks after operations.
[MCPA amine + fluroxypyr ester + dicamba], MOA 4 + 4 + 4 (4.8 SL)	0.73 to 1.1 fl oz	2 to 3 pt	1.2 to 1.8	Same turf tolerances and uses as [2,4-D amine + fluroxypyr + dicamba] in addition to centipedegrass. Only spot treat St. Augustinegrass when temperature exceeds 80 degrees F. Do not apply more than two applications per year totaling 3 pints per acre. For non-turf areas, rate can be increased to 2 to 5 pints per acre. Application can be made to grass seedlings after second mowing and to newly sodded, sprigged, or plugged grasses 3 to 4 weeks after operations. Sod farm rates include 1.25 pints per acre for creeping bentgrass, 2 to 3 pints per acre for all other cool season grasses listed on label and 1.5 to 1.8 pints per acre for all warm season grasses listed on label.
Postemergence Control, Winter and Summer Annual Broadleaf Weeds				
bentazon + atrazine, MOA 6 + 5 Create by tank mixing			0.5 to 0.75 + 0.5 to 0.75	Apply to bermudagrass, centipedegrass, St. Augustinegrass, and zoysiagrass. Check individual labels for weeds controlled and weed size for proper application.
Postemergence Control, Black Medic, White, Hop Clover, Buckhorn Plantain, Common Chickweed, Mouseear Chickweed, Henbit, Spurweed (Lawn Burweed), Broadleaf Plantain, Dandelion, False Dandelion, Lespedeza, Prostrate Spurge, Wild Violet				
[triclopyr + cloypyralid], MOA 4 + 4 (3 SL)	0.37 to 0.74 fl oz	1 to 2 pt	0.28 to 0.56 + 0.09 to 0.19	Do not apply to home lawns. May be used on centipedegrass, bermudagrass, zoysiagrass, tall fescue, creeping red fescue, chewing fescue, Kentucky bluegrass, perennial ryegrass. Repeat treatment may be necessary for prostrate spurge and wild violet. Quali-Pro formulation: maintain 0.5 inch height for warm season turf. Do not apply to bermudagrass sod farms. Wait 3 weeks to reseed. Do not use grass clippings for compost or mulch.
[MCPA ester + triclopyr ester + dicamba], MOA 4 + 4 + 4 (3.6 EC)	0.91 to 1.29 fl oz	2.5 to 3.5 pt	1.125 to 1.575	May be applied to home lawns by a commercial applicator. Not for use on turf grown for resale or other commercial use as sod or seed production. Use on perennial bluegrass, ryegrass, fescue species, bentgrass (excluding greens and tees), bermudagrass, zoysiagrass, and bahiagrass. Do not apply to seedling grasses until well established. Wait 3 to 4 weeks after application to seed.

CHEMICAL WEED CONTROL

CHEMICAL WEED CONTROL

Table 7-14. Chemical Weed Control in Lawns and Turf

Herbicide and Formulation	Amount of Formulation Per 1,000 sq ft	Amount of Formulation per Acre	Pounds Active Ingredient per Acre	Precautions and Remarks
Postemergence Control, Black Medic, White, Hop Clover, Buckhorn Plantain, Common Chickweed, Mouseear Chickweed, Henbit, Spurweed (Lawn Burweed), Broadleaf Plantain, Dandelion, False Dandelion, Lespedeza, Prostrate Spurge, Wild Violet (continued)				
[MCPA amine + triclopyr amine + dicamba], MOA 4 + 4 + 4 (4.56 L)	0.73 to 1.1 fl oz	2 to 3 pt	1.14 to 1.71	
[MCPA amine + fluroxypyr ester + triclopyr amine], MOA 4+4+4 (3.41 L)	0.37 to 1.47 fl oz	1 to 4 pt	0.42625 to 1.705	Apply by a commercial applicator to residential, industrial, and institutional lawns, sod farms, parks, cemeteries, athletic fields, roadsides, and golf courses excluding greens and tees. May apply to bentgrass, Kentucky bluegrass, perennial ryegrass, fescue species, bahiagrass, bermudagrass, centipedegrass, and zoysiagrass. Do not spray on warm season turf less than 0.5 inch and do not exceed 3 pints per acre. Generally apply 3 to 4 pints per acre except on fairway bentgrass, which can only tolerate 2 pints per acre. Wait 3 to 4 weeks after application to reseed. Check label for spray adjuvant recommendation.
Postemergence Control, Plantain, Chickweed, Dandelion, Purslane, and Thistle Species, Ground Ivy, Lawn Burweed, Henbit, Corn Speedwell, Spotted Spurge				
carfentrazone-ethyl, MOA 14 (1.9 EW)	0.0126 to 0.048 fl oz	0.55 to 2.1 fl oz	0.008 to 0.031	May be applied to bahiagrass, bermudagrass, buffalograss, centipedegrass, St. Augustinegrass, zoysiagrass, Kentucky bluegrass, tall fescue, fine fescue, perennial ryegrass, and bentgrass. To expand the weed spectrum and extend control of the weeds listed here and on the label, carfentrazone-ethyl can be tank mixed with the entire range of phenoxy products—amines, esters, and other salts—and is also compatible with dicamba, atrazine, glyphosate, glufosinate, clopyralid, triclopyr, and MSMA. When applied alone, add 0.12 to 0.25% nonionic surfactant.
Postemergence Control, White Clover, Dandelion, Ground Ivy, Spurges, Plantains, Chickweeds, Henbit, Lawn Burweed, Woodsorrels, Dollarweed, Poison Ivy, Poison Oak, Corn Speedwell, Wild Strawberry, Wild Violet, Virginia Pepperweed, Shepherd's Purse				
[carfentrazone + 2,4-D ester + MCPP + dicamba], MOA 14 + 4 + 4 + 4 (2.2 EC)	0.75 to 1.8 fl oz	2 to 5 pt	0.55 to 1.375	May be used on annual and perennial bluegrass, annual and perennial ryegrass, tall and fine fescue, creeping and colonial bentgrass, common and hybrid bermudagrass, and zoysiagrass. For use in ornamental turf, golf courses, lawns, sod farms, cemeteries, and parks. Optimum results when applied when temperatures are between 45 and 75 degrees F but may be applied up to 90 degrees F. Lower rates may be used in cooler weather. Rainfast within 3 hr and may reseed after 2 week. May apply 3 to 4 wk after sodding, sprigging, or plugging. Also may be used on bahiagrass, buffalograss, St. Augustinegrass, centipedegrass, seashore paspalum, and kikuyugrass. May reseed after 1 week.
[carfentrazone + 2,4-D ester + MCPP + dicamba], MOA 14 + 4 + 4 + 4 (0.81 EC)	0.55 to 2.2 fl oz	1.5 to 6 pt	0.1519 to 0.6075	
[carfentrazone + MCPA ester + MCPP + dicamba], MOA 14 + 4 + 4 + 4 (2.91 EC)	0.75 to 2.2 fl oz	2 to 6 pt	0.7275 to 2.1825	Same precautions and turf uses as [carfentrazone + 2,4-D ester + MCPP + dicamba] 2.2 EC except cannot be applied to creeping and colonial bentgrass.
penoxsulam + sulfentrazone + 2,4-D + dicamba	1.0 to 2.2 fl oz	2.7 to 6 pt	0.271 to 0.602	May be used on established Kentucky bluegrass, annual bluegrass, perennial ryegrass, annual ryegrass, tall fescue, common and hybrid bermudagrass, zoysiagrass, centipedegrass, and seashore paspalum. Do not apply more than 2.7 pints per acre on fescue or perennial ryegrass unless turf injury can be tolerated. Do not apply more than 3.5 pints per acre on St. Augustinegrass unless injury, discoloration, stunting, and thinning can be tolerated.
[sulfentrazone + 2,4-D amine + MCPP + dicamba], MOA 14 + 4 + 4 + 4 (2.18 SL)	0.92 to 1.84 fl oz	2.5 to 5 pt	0.68 to 1.36	Apply 2.5 to 3.25 pints per acre on warm season turf including bermudagrass species, zoysiagrass, bahiagrass, and buffalograss. Apply 3.25 to 4 pints per acre on cool season turf including species of bluegrass, ryegrass, fescue, and bentgrass (excluding greens and tees). 4 to 5 pints per acre needed to control corn speedwell and wild violet. Turf areas include residential, ornamental, institutional, and sod farms. Apply to grass seedlings after second mowing. Apply to sodded, sprigged, or plugged areas 3 to 4 weeks after operations. Treated areas may be reseeded 3 weeks after application.
[triclopyr ester + sulfentrazone + 2,4-D ester + dicamba], MOA 4 + 14 + 4 + 4 (2.51 EC)	0.75 to 1.5 fl oz	2 to 4 pt	0.628 to 1.26	Apply 2 to 2.25 pints per acre on fully dormant bermudagrass, zoysiagrass, and bahiagrass. Apply 3.25 to 4 pints per acre on annual and perennial bluegrass and ryegrass, and tall, red, and fine fescue. Rainfast within 3 hours. Approved turf areas include residential, ornamental, institutional, noncropland, and sod farms. Apply to grass seedlings after the second or third mowing. Apply to sodded, sprigged, or plugged areas 3 to 4 weeks after operations. Treated areas may be reseeded 3 weeks after application.
Postemergence Control, Chickweed, Clover, Plantain and Dandelion Species, Florida Betony, Dollarweed, Groud Ivy, Lespedeza, and Yellow Woodsorrel				
florasulam, MOA 2 (0.42 SC)	0.09 fl oz	4 fl oz	0.013125	Can be used on all established major warm and cool season turfgrass species in residential lawns, golf courses (excluding putting greens), sports fields, sod farms and commercial turf areas. Controls Carolina geranium, species of chickweed, clover and dandelion, vetch, dollarweed and common groundsel. Do not exceed 3 applications or 12 fluid ounces per acre per year. Apply to newly seeded or sprigged turf after third mowing or when tillering and secondary root development has occurred. Wait 4 weeks to reseed. When used alone, add a nonionic surfactant at 0.2% by volume.
penoxsulam, MOA 2 (0.014 G) (0.03 G)	3.4 to 10.3 lb 1.7 to 4.6 lb	150 to 450 lb 75 to 200 lb	0.02 to 0.06	May be applied to residential and commercial lawns, golf courses (excluding greens and tees), parks, athletic fields, and sod farms. Use on turf that has been mowed at least 3 times or sprigs that have developed secondary root oyotomo. Apply up to 75 pounds per acre of 0.03 G or 150 pounds per acre of 0.014 G to perennial ryegrass and tall fescue. Apply up to 150 pounds per acre of 0.03 G or 300 pounds per acre of 0.014 G to bentgrass, Kentucky bluegrass, and fine fescue. Apply up to 200 pounds per acre of 0.03 G or 450 pounds per acre of 0.014 G to bermudagrass, centipedegreass, zoysiagrass, and St. Augustinegrass. Do not apply to dormant centipedegrass. Reapply at 4 weeks if needed but do not exceed 300 pounds per acre of 0.03 G or 650 pounds per acre of 0.014 G per season. After treatment, wait 3 to 4 weeks to reseed.
(0.31 L)	0.092 to 0.55 fl oz	0.25 to 1.5 pt	0.01 to 0.058	Same statement as above concerning turf uses and reseeding intervals. Bermudagrass and kikuyugrass are the only warm season grasses labeled for use. Apply up to 1 pint per acre on bentgrass, 1.5 pint per acre on bermudagrass and kikuyugrass and 2 pints per acre on tall fescue and perennial ryegrass. Do not apply greater than 2.3 pints per acre per year. Surfactant not required.

Table 7-14. Chemical Weed Control in Lawns and Turf

Herbicide and Formulation	Amount of Formulation Per 1,000 sq ft	Amount of Formulation per Acre	Pounds Active Ingredient per Acre	Precautions and Remarks
Carpetweed, Chickweed, Dandelion, Curly Dock, Cutleaf Eveningprimrose, Henbit, Knotweed, Common Mallow, Poison Ivy, and Annual Sowthistle				
pyraflufen ethyl, MOA 14 (0.177 SC)	0.016 to 0.092 fl oz	0.7 to 4 fl oz	0.000938 to 0.0055	Used in established sod farm and ornamental turf by commercial applicators and professional landscapers only. Turf can be newly seeded, sodded, or sprigged as long as it is established and not under stress. Tolerant turfgrasses include bermudagrass, centipedegrass, St. Augustinegrass, zoysiagrass, tall fescue, perennial ryegrass, perennial bluegrass, and creeping bentgrass (not greens or tees). Apply 1 to 4 fluid ounces alone to 3- to 6-inch tall weeds. For larger weeds and broader spectrum control, apply 0.75 to 1.5 fluid ounces and tank mix with 2,4-D, mecoprop, dicamba, chloroprop, MCPA, triclopyr, or fluroxypyr.
Postemergence Control, Bahiagrass, Perennial Ryegrass, Wild Garlic, Spurweed, Henbit, Miscellaneous Other Broadleaf Weeds				
metsulfuron, MOA 2 (60 WDG)	0.003 to 0.02 oz	0.125 to 1 oz	0.005 to 0.038	May be applied to established bermudagrass, zoysiagrass (Meyer or Emerald), St. Augustinegrass, Kentucky bluegrass or fine fescue. Do not apply to turf less than 1 year old. Do not exceed 0.5 ounces per acre on centipedegrass, fine fescue, or Kentucky bluegrass. See label for a complete list of weeds controlled. The addition of 0.25% nonionic surfactant will enhance control. May be used for removal of perennial ryegrass from overseeded warm-season turf species. For bahiagrass control, use 0.25 to 0.75 ounces per acre after spring greenup but before seedhead development. A repeat treatment may be necessary in 4 to 6 weeks.
metsulfuron (Patriot) 60 WDG	0.007 to 0.046 oz	0.33 to 2 oz	0.012 to 0.075	Apply to unimproved industrial turf only. Use maximum of 0.5 ounce per acre for fescue and bluegrass and 2 ounces per acre for bermudagrass.
[metsulfuron + rimsulfuron], MOA 2 + 2 (37 WG)	0.0344352 oz	1.5 oz	0.0346875	See comments under postemergence annual bluegrass control. For bahiagrass control, a repeat treatment may be necessary 4 to 6 weeks after initial application.
Postemergence Control, Wild Garlic, Wild Onion				
imazaquin, MOA 2 (70 DG)	0.128 to 0.256 oz	0.357 to 0.714 lb	0.25 to 0.5	Use on bermudagrass, centipedegrass, St. Augustinegrass, and zoysiagrass. Do not apply during spring greenup. Temporary yellowing may occur. Add a nonionic surfactant at 2 pints per 100 gallons of spray solution.
2,4-D amine, MOA 4 (4 SL)	2.2 fl oz	3 qt	3	Apply in fall when garlic is young and actively growing. Add a wetting agent to keep spray from bouncing off garlic leaves. Repeat treatment for 2 years. Avoid spray drift which can injure susceptible plants. Use on bluegrass, fescue, bermudagrass, or zoysia. For more susceptible grasses, uses spot treatment below.
	Spot treatment			One tbsp of 1% 2,4-D solution per garlic clump or use pressurized applicator. Apply December to April. Use as spot treatment for widely scattered clumps in small areas. Avoid excessive spraying as turfgrass injury may result.
Postemergence Control of Various Grass and Broadleaf Weeds in Unimproved Turf and Other Noncrop Areas				
glyphosate, MOA 9 (5.5 SL) (5 SL) (4 SL)	0.14 to 1.01 0.12 to 0.87 fl oz 0.75 to 2.94 fl oz	0.375 to 2.75 pt 0.3125 to 2.375 pt 1 to 4 qt	0.26 to 1.89 0.2 to 1.48 0.5 to 4	Check specific labels for correct rates. Apply to dormant or actively growing well established bermudagrass and bahiagrass. Bahiagrass growth will be suppressed if treated after spring greenup and before seedhead formation. Treat winter annual weeds when less than 6 inches tall. Higher rates are needed for more mature plants. Apply in 10 to 40 gallons of water per acre and use an NIS at 2 quarts per 100 gallons of spray solution.
[glyphosate + 2,4-D amine], MOA 9 + 4 (1.2 + 1.9 lb/gal SL)	0.55 to 1.47 fl oz	1.5 to 4 pt	0.58 to 1.55	Apply in 15 to 30 gallons of water per acre. May be applied to highly maintained dormant bermudagrass at 2 to 4 pt per acre. In low maintenance bermudagrass, sulfometuron can be added at 0.25 to 1 ounce per acre when dormant or actively growing. Apply 2 to 4 pints per acre on dormant bahiagrass and 1.5 to 2 pints per acre on actively growing bahiagrass. Tank mix with sulfometuron if needed. Check label for sulfometuron rates. Tall fescue applications can be made in the spring or summer at 2 to 3 pints per acre with or without sulfometuron. Spray tall fescue at 4 to 6 inches tall and before seedhead emergence to minimize injury.
sulfosulfuron, MOA 2 (75 WG)	0.017 to 0.046 oz	0.75 to 2 oz	0.035 to 0.094	May be used in well established dormant and actively growing bermudagrass and bahiagrass. Wait 30 days to re-treat if needed; do not exceed 2.66 ounces per acre per year. If treating weeds postemergence, use an NIS at 2 quarts per 100 gallons spray solution unless tank mixed with glyphosate. Sulfosulfuron can be tank mixed with [glyphosate + 2,4-D amine], metsulfuron, sulfometuron, and chlorsulfuron, but check label for proper turf species and timing. Expect temporary injury or discoloration with tank mix partners. For well established tall fescue, do not exceed 1 ounce per acre per year, and do not tank mix. Effective on johnsongrass.
[thiencarbazone-methyl + iodosulfuron-methyl + foramsulfuron], MOA 14 + 2 + 2 (36.4 WDG)	0.069 to 0.138 oz	3 to 6 oz	0.068 to 0.137	May be applied to unimproved bermudagrass, zoysiagrass, centipedegrass and bare ground sites on private, public and military land for control of many annual and perennial broadleaf and grass weeds. Check label for complete weed listing, rate needed and recommended adjuvant. Repeat application in 4 to 6 weeks if weed regrowth occurs not to exceed 6 ounces product per year. Spot treatment (spray-to-wet) rate is 3 to 6 ounces product per 25- to 100-gallon solution. Nonionic surfactant is generally recommended at 0.25 to 0.5%. Use 0.5 to 1% methylated seed oil for difficult to control broadleaf weeds and perennial grasses. Increased control may be achieved with 1.5 to 3 pounds per acre ammonium sulfate in high humidity climates or 1.5 to 2 quarts per acre urea ammonium nitrate in low humidity climates. Do not use an organosilicone surfactant.

CHEMICAL WEED CONTROL

CHEMICAL WEED CONTROL

Table 7-14. Chemical Weed Control in Lawns and Turf

Herbicide and Formulation	Amount of Formulation Per 1,000 sq ft	Amount of Formulation per Acre	Pounds Active Ingredient per Acre	Precautions and Remarks
Postemergence Control in Dormant Warm Season Turf Annual Bluegrass, Various Other Winter Annual Weeds				
diquat, MOA 22 (2 SL)	0.4 to 0.75 fl oz	1 to 2 pt	0.25 to 0.5	Apply in 20 to 100 gallons spray mix as a broadcast application. Add 1 to 2 pints of a nonionic surfactant per 100 gallons of solution. Bermudagrass must be dormant. More than one application may be needed.
flumioxazin, MOA 14 (51 WG)	0.1837 to 0.2755 oz	8 to 12 oz	0.255 to 0.3825	Use on completely dormant bermudagrass turf including residential and commercial lawns, golf courses (excluding greens), sod farms, roadsides, athletic fields, parks and schools. Add 0.25% by volume nonionic surfactant for postemergence applications. Provides preemergence control of annual grasses such as crabgrass, goosegrass, foxtail species, barnyardgrass and annual bluegrass. Does control annual bluegrass postemergence along with many common winter annual broadleaf weeds such as chickweed species, henbit, Carolina geranium and hairy bittercress. Allow a 15 feet buffer zone when applying upslope from bentgrass greens or bermudagrass greens overseeded with *Poa trivialis*. To limit potential lateral movement, do not apply to saturated soil.
glyphosate - Roundup, MOA 9 (4 SL) (5 SL) (5.5 SL)	0.37 fl oz 0.29 fl oz 0.27 fl oz	1 pt 0.8 pt 0.73 pt	0.5	Check specific labels for correct rates. Apply in 5 to 40 gallons water per acre with 0.5% by volume of a nonionic surfactant. Application to actively growing annual bluegrass must be made before initiation of bermudagrass greenup in the spring.
glyphosate - Touchdown, MOA 9 (3 LC) (4.17 LC) (5 LC)	0.18 to 1.47 fl oz 0.13 to 1.06 fl oz 0.11 to 0.88 fl oz	0.5 to 4 pt 0.36 to 2.88 pt 0.3 to 2.4 pt	0.1875 to 1.5	Apply to dormant bermudagrass and bahiagrass before spring greenup. Apply in 10 to 40 gallons water per acre. Will control winter annual weeds up to 6 inches tall and 4- to 6-leaf tall fescue. Use a 75% active ingredient nonionic surfactant at 0.25% by volume or dry ammonium sulfate at 0.5% by weight.
[glyphosate + diquat], MOA 9 + 22 (76 WG) (4.21 SL)	0.11 to 0.37 fl oz 0.18 to 0.62 fl oz	5 to 16 oz 8 to 27 fl oz	0.24 to 0.76 0.26 to 0.89	Apply to dormant bermudagrass and bahiagrass not grown for research, sale, or other commercial uses, such as sod, seed production. Apply in 10 to 80 gallons water per acre. Rates greater than 9 ounces per acre of 76 WG product or 15 fluid ounces per acre of 4.21 SL product may cause injury or delay greenup in highly maintained areas. Controls tall fescue.
metribuzin, MOA 5 (75 WDF)	0.25 oz	0.67 lb	0.5	For application by commercial applicators to dormant bermudagrass turf. Broadcast spray before greenup of turf. Do not apply to greens, tees, or aprons. Controls common chickweed, corn speedwell, henbit, parsley-piert, and spurweed.
Suppression/Control, Bermudagrass				
fenoxaprop, MOA 1 (0.57 EC)	0.46 fl oz	1.25 pt	0.089	Use on Kentucky bluegrass, perennial ryegrass, fine and tall fescue, and zoysiagrass. Apply June 1, July 1, Aug. 1, Sept. 1, repeat for 2 years. Can be tankmixed with 1 pt per acre triclopyr following the same schedule as above. Apply June 1 and Aug. 1 for 2 years if tank mixed with 1 quart per acre triclopyr. Zoysia may show discoloration but should recover in 10 to 14 days following tankmix applications.
fluazifop, MOA 1 (2 EC)	0.05 to 0.14 fl oz	2 to 6 oz	0.03 to 0.09	Use on tall fescue or zoysia. For fescue, apply 5 to 6 ounces per acre during warm weather in early spring when bermudagrass is breaking dormancy; repeat in fall when bermudagrass is preparing for dormancy. For zoysia, apply 4 ounces per acre on June 1, Aug. 1; repeat for 2 years. Can tank-mix with 1 quart per acre triclopyr following schedule above. Zoysia or tall fescue may show slight discoloration but should recover in 10 to 14 days. Add a nonionic surfactant at 0.25% v/v. Apply in a minimum of 30 gallons of water per acre.
siduron, MOA 7 (50 WP)	0.5 to 1 lb	21.78 to 43.56 lb	10.88 to 21.78	Apply as 8- to 12-inch band treatment with a single nozzle sprayer along putting green perimeter to suppress bermudagrass stolon encroachment. Initiate in March or April, and continue subsequent applications at 4- to 5-week intervals.
triclopyr, MOA 4 (4 EC)	0.73 fl oz	1 qt	1.0	Use on perennial bluegrass, perennial ryegrass, tall fescue or ornamental turf including sod farms and golf courses. Do not apply to zoysia unless injury can be tolerated. Apply June 1, July 1, Aug. 1, Sept. 1, repeat for 2 years. Can be tank-mixed with fenoxaprop or fluazifop at rates, timings listed above. New low-odor formulation uses methylated seed oil solvents instead of petroleum distillates.
Postemergence Control Bermudagrass				
clethodim, MOA 1 (0.97 EC)	0.4 to 0.8 fl oz	17 to 34 fl oz	0.125 to 0.25	For use on sod farms only. Do not apply to centipedegrass being grown for seed. Do not apply until 3 weeks after full greenup of centipedegrass in spring. Do not mow for 1 week before and after application. The addition of a nonionic surfactant at 0.25 % solution (1 pint per 50 gallons water) or a crop oil concentrate at 1% solution (2 quart per 50 gallons water) is necessary for control. A repeat application usually 3 to 4 weeks after the first application will be required for bermudagrass control. Use higher rates for more established bermudagrass. Do not apply more than 68 ounces of clethodim per acre per year. Some discoloration of centipedegrass will occur at higher rates.
Preplant Control or Lawn Renovation — Emerged Annual and Perennial Grass and Broadleaf Weeds				
glyphosate - Roundup, MOA 9 (4 SL) (5 SL) (5.5 SL)	0.75 to 3 fl oz 0.54 to 2.17 fl oz 0.54 to 2.14 fl oz	1 to 4 qt 0.8 to 3.2 qt 0.73 to 2.91 qt	1 to 4	Where existing vegetation is growing in a field or unmowed situation, apply to actively growing weeds at the stages according to label. Where existing vegetation is growing under mowed turfgrass management, apply after omitting at least one regular mowing to allow sufficient growth for good interception of the spray. Tillage or renovation techniques such as vertical mowing, coring, or slicing should be delayed for 7 days after application. Desirable turfgrass may be established following treatment.
glyphosate - Touchdown, MOA 9 (3 LC) (4.17 LC) (5 LC)	0.18 to 1.47 fl oz 0.13 to 1.06 fl oz 0.11 to 0.88 fl oz	0.5 to 4 pt 0.36 to 2.88 pt 0.3 to 2.4 pt	0.1875 to 1.5	Same remarks as glyphosate, above. In addition, use a 75% active ingredient nonionic surfactant at 0.25% by volume or dry ammonium sulfate at 0.5% by weight.

Table 7-14. Chemical Weed Control in Lawns and Turf

Herbicide and Formulation	Amount of Formulation Per 1,000 sq ft	Amount of Formulation per Acre	Pounds Active Ingredient per Acre	Precautions and Remarks
Preplant Control or Lawn Renovation — Emerged Annual and Perennial Grass and Broadleaf Weeds (continued)				
[glyphosate + diquat], MOA 9 + 22 (76 WG) (4.21 SL)	1.65 to 4.5 oz 2.75 to 5.5 fl oz	4.5 to 12.25 lb 3.75 to 7.5 qt	3.4 to 9.3 3.95 to 7.89	Generally use the 75 WG product at 4.5 pounds per acre on annuals, 9 pounds per acre on perennials, and 12.25 pounds per acre on dusty or stressed plants, dense stands, or difficult-to-control perennials. Generally use the 4.21 SL product at 3.75 quarts per acre on annuals and 7.5 quarts per acre on perennials. Do not use on turf grown for research, for sale, or for commercial uses, such as sod or seed production. Do not use if renovating bermudagrass or kikuyugrass sods. Delay tillage for 7 days after application.
[indaziflam + diquat dibromide + glyphosate], MOA 21 + 22 + 9 (1.958 SL)	1 pt	5.44 gal	10.66	For nonselective preemergence and postemergence control in noncrop areas. Reapply 4 months after initial application if needed not to exceed 1 quart per 1000 square feet per year. Apply 1 pint in 1 gallon of water to cover 1000 square feet. Do not seed for 12 months after application.
Trimming and Edging and Control of Emerged Weeds				
diquat, MOA 22 (2 SL)	0.4 to 0.75 fl oz	1 to 2 pt	0.25 to 0.5	Add nonionic surfactant at 0.25 ounce per gallon of water. Water volumes above 15 gal per acre should be used. For spot sprays, use 0.3 to 0.75 fluid ounce per gallon.
glufosinate, MOA 10 (1 SL)	2.2 to 4.4 fl oz	3 to 6 qt	0.75 to 1.5	Rate depends on weed to be controlled and stage of growth. Consult label. For spot or directed spray use 1.5 to 4 fluid ounces per gallon of water.
glyphosate + diquat, MOA 9 + 22 (76 WG) (4.21 SL)	1.65 to 4.5 oz 2.75 to 5.5 fl oz	4.5 to 12.25 lb 3.75 to 7.5 qt	3.4 to 9.3 3.95 to 7.89	May be used in general noncrop areas. Do not use on plants grown for sale or other commercial uses, such as seed production. See rate comments in lawn renovation section. For spray to wet treatments, apply the 76 WG product at 1.2 ounces per gal of water for annuals and 1.5 ounces per gal of water for perennials. Apply the 4.21 SL product at 2 fluid ounces per gallon of water for annuals and 2.5 fluid ounces per gal water for perennials. For directed spot treatment of perennials using hand-held low volume equipment, apply 4 to 8 ounces per gallon of water.

Chemical Weed Control in Ornamentals

J. C. Neal, Horticultural Science Department

Table 7-15. Chemical Weed Control in Ornamentals

More detailed information about herbicides labeled for use in nursery crops and landscape plantings is available on the Weed Management in Nurseries, Landscapes and Christmas Trees web portal (https://weeds.ces.ncsu.edu).

Weed	Herbicide and Formulation	Amount Formulation Per Acre	Pounds Active Ingredient Per Acre	Precautions and Remarks
Preplant to all Ornamentals				
Most annuals and perennials	dazomet MOA 27 (Basamid granular)	218 to 525 lb	216 to 520	Preplant soil fumigant. Dazomet is a restricted-use pesticide. Follow label directions and restrictions. Incorporate to 8 inches deep. Drench with water or cover with plastic for best results. Planting must be delayed until fumigant has dissipated from the soil. This may take a month or more.
	diquat dibromide MOA 22 (Reward) 2 L	1 to 2 qt	0.5 to 1	Non-selective, contact-type, postemergence control of seedling weeds. A nonionic surfactant should be added to the spray solution. Apply for full coverage and thorough weed contact. Retreatment will be necessary for established weeds. May be used in landscapes, nurseries and greenhouses.
	glufosinate MOA 10 (Finale) 1 L	3 to 6 qt	0.75 to 1.5	Non-selective postemergence control of weeds. Thorough coverage is essential. Apply in a minimum of 20 gallons of water per acre. No residual control. Repeat applications may be necessary for control of perennial weeds.
	glyphosate MOA 9 (Roundup Pro, Touchdown Pro, and many others)	1 to 5 qt	1 to 5	Non-selective, systemic herbicide. Apply to emerged weeds prior to planting ornamentals.
	paraquat MOA 22 (Gramoxone Extra) 2.5 L	2 to 3 pt	0.6 to 0.9	Non-selective, contact herbicide. Apply when grass and weeds are 1 to 6 inches high and succulent for best results. Direct spray with low pressure to avoid contact with foliage or bark of crop less than 1 year old. Add nonionic surfactant, 0.25% by volume (2 pints per 100 gallons of water). Not for use in landscapes or greenhouses.
Post-plant Preemergence Weed Control				
Annual grasses and broadleaf weeds (preemergence) See label for susceptible species	benefin + oryzalin MOA 3 (XL) 1 + 1 G	200 to 300 lb	2 to 3 + 2 to 3	Apply preemergence to weeds. May be applied in spring and fall to ornamental plants.
	dichlobenil MOA 20 (Casoron) 4 G	100 to 150 lb	4 to 6	Do not use on fir, hemlock, *Ilex crenata, I. rotunda,* or *I. vomitoria.* Do not use more than 6 pounds per acre on azalea, rhododendron, boxwood, holly, euonymus, forsythia, leucothoe, ivy, lilac, heather, or any plantings less than 1 year old. Do not use in seedbeds, cutting, or transplant beds. Do not apply until 4 weeks after transplanting any plants. Apply in winter. http://content.ces.ncsu.edu/casoron-dichlobenil
	dithiopyr MOA 4 (Dimension) 2 EW 40 WP	2 pt 20 oz	0.5	Preemergence control of annual grasses and some small seeded broadleaf weeds in turf, landscape plantings, and nurseries. Use as a directed application around ornamental plants unless specified otherwise on the product label. See label for tolerant species and restrictions. http://content.ces.ncsu.edu/dimension-dithiopyr

329

Table 7-15. Chemical Weed Control in Ornamentals

More detailed information about herbicides labeled for use in nursery crops and landscape plantings is available on the Weed Management in Nurseries, Landscapes and Christmas Trees web portal (https://weeds.ces.ncsu.edu).

Weed	Herbicide and Formulation	Amount Formulation Per Acre	Pounds Active Ingredient Per Acre	Precautions and Remarks
Post-plant Preemergence Weed Control (continued)				
Annual grasses and broadleaf weeds (preemergence) See label for susceptible species (continued)	dimethenamid-p MOA 15 (Tower) 6 EC	21 to 36 oz	1 to 1.5	Preemergence control of annual sedges, annual grasses and many annual broadleaf weeds in woody landscape plantings, field and container nurseries. Suppression of yellow nutsege. Avoid foliar treatments over the top of early spring growth flushes as injury to ornamental plants can occur. http://content.ces.ncsu.edu/tower-dimethenamid-p
	dimethenamid-p + pendimethalin MOA 15 + 3 (Freehand) 1.75 (0.75 + 1) G	100 to 200 lb	1.75 to 3.5	Preemergence control of annual grasses and many broadleaf weeds from seed as well as suppression of yellow nutsedge in container and field nurseries, and woody landscape plantings. http://content.ces.ncsu.edu/freehand-dimethenamid-p-pendimethalin
	flumioxazin MOA 14 (Broadstar) G	150	0.375	Preemergence control of most annual broadleaf and annual grasses in container and field-grown woody nursery crops. Not for use in landscape plantings. See label for species and precautions. Do not apply to wet foliage or newly potted liners. http://content.ces.ncsu.edu/broadstar-flumioxazin
	flumioxazin MOA 14 (Sureguard) DG	0.75 lb	0.375	Preemergence control of most annual broadleaf and annual grasses, and early postemergence control of seedling broadleaf weeds in field and container-grown woody nursery crops and certain landscape plantings. See label for species and precautions. http://content.ces.ncsu.edu/sureguard-flumioxazin
	imazaquin MOA 2 (Image) 1.5 LC	1 to 1.3 qt	0.4 to 0.5	Apply as a directed spray away from rooting zone. Labeled for over-the-top sprays on a few species.
	Indaziflam MOA 21 (Marengo) 0.0224% G	100 to 200 lb	0.022 to 0.044	Preemergence control of many annual weeds in container or field grown nursery crops. Not labeled for use in landscape plantings. See label for specific species. http://content.ces.ncsu.edu/marengo-indaziflam
	(Marengo) 0.622 SC	7.5 to 15.5 fl oz	0.036 to 0.075	Preemergence weed control in field grown nursery crops. Use as a directed spray. Also for use as a ground treatment in container nurseries and greenhouses (and similar covered structures). When treating inside covered houses, structures should have no crop present at the time of treatment. See label for full details.
	Indaziflam MOA 21 (Specticle) 0.0224% G	100 to 200 lb	0.022 to 0.044	Preemergence control of many annual weeds in established woody landscape plantings. Do not use in areas planted to or to be planted to bedding plants. Labeled for use on a limited number of herbaceous perennials. The GR formulation is less injurious to ornamental plants than the SC formulation. The SC formulation should be applied as a directed spray, avoiding contact with the foliage of ornamental plants. Check labels for details.
	(Specticle) 0.622 SC	6 to 12 fl oz	0.029 to 0.058	
	isoxaben MOA 21 (Gallery) 75 DF (Gallery) 4.16 SC	0.66 to 1.33 lb 16 to 31 fl oz	0.5 to 1	Use as preemergence control of broadleaf weeds in many field and container grown ornamentals, turf, and landscape plantings. Generally used in combination with another herbicide for broader spectrum weed control. http://content.ces.ncsu.edu/gallery-isoxaben
	isoxaben + prodiamine MOA 21 + 3 (Gemini) 3.7 SC (1.5 + 2.2)	43.5 to 87 fl oz	1.25 (0.5 + 0.75) to 2.5 (1 + 1.5)	Preemergence control of weeds in field- and container-grown ornamentals and landscape plantings. See label for specific species.
	isoxaben + trifluralin MOA 21 + 3 (Snapshot TG) 2.5 G	100 to 200 lb	2.5 to 5	Preemergence control of weeds in field- and container-grown ornamentals and landscape plantings. See label for specific species. https://content.ces.ncsu.edu/snapshot-tg-isoxaben-trifluralin
	S-metolachlor MOA 15 (Pennant Magnum) 7.62 EC	1.3 to 2.6 pt	1.2 to 2.5	Preemergence control of annual grasses, annual sedges, and some annual broadleaf weeds including dovenweed, as well as suppression of yellow nutsedge. Apply to soil surface immediately after planting. Avoid foliar treatments over the top of early spring growth flushes as injury to ornamental plants can occur. https://content.ces.ncsu.edu/pennant-magnum-s-metolachlor
	napropamide MOA 15 (Devrinol) 50 DF	8 to 12 lb	4 to 6	Apply preemergence to weeds and as a directed spray in ornamentals. Can be used in field or container nurseries or landscape plantings. If broadcast over top of ornamentals, irrigate soon after application to reduce risk of foliar injury. https://content.ces.ncsu.edu/devrinol-napropamide
	oryzalin MOA 3 (Surflan) 4 AS	2 to 4 qt	2 to 4	Preemergence to weeds in field or container nurseries or landscape plantings. Apply only to established plantings. Do not use in seedbeds or transplant beds. Not recommended for use on soils containing more than 3% organic matter. Use higher rate for longer term control. Do not apply on hemlock. http://content.ces.ncsu.edu/surflan-oryzalin
	oxadiazon MOA 14 (Ronstar) 2 G	100 to 200 lb	2 to 4	Apply preemergence to weeds. Can be used on container- and field-grown ornamentals. Repeat applications are labeled for some species. Injury has been observed on ajuga, liriope, mondo, and fig. Granules may burn tender foliage of several species if irrigation is not used to wash them off. Caution: Plants that trap granules in leaf axil can be injured.
	oxyfluorfen MOA 14 (Goal) 2 XL (Galigan) 2E	5 to 10 pt	1 to 2	Preemergence and postemergence control of many broadleaf and grass weeds in conifers and dormant deciduous trees. Do not apply when conifers have young tender growth. Lower rates are used in conifer seedbeds and for postemergence treatments. https://content.ces.ncsu.edu/goal-goaltender-oxyfluorfen
	oxyfluorfen + oryzalin MOA 14 + 3 (Rout) 3 (2+1) G	100 lb	3 (2 + 1)	Apply preemergence to weeds. Can be used on container and field-grown ornamentals. Repeat applications are labeled. Injury is to be expected to herbaceous plants or to plants with leaf orientation that might trap granules. Check label for genera of plants on which it can be used.
	oxyfluorfen + oxadiazon MOA 14 (Regal OO) 3 (2 + 1) G	100 lb	3 (2 + 1)	Apply preemergence to weeds. May be used on container- or field-grown woody ornamentals, including liner production. Injury is to be expected to herbaceous plants or to plants with leaf orientation that might trap granules. Check label for genera of plants on which it can be used. https://content.ces.ncsu.edu/regal-o-o-oxyfluorfen-oxadiazon
	oxyfluorfen + pendimethalin MOA 14 + 3 (Ornamental Herbicide 2) 3 (2+1) G	100 lb	3 (2 + 1)	Apply preemergence to weeds. Can be used on container- and field-grown ornamentals. Repeat applications are labeled. Injury is to be expected to herbaceous plants or to plants with leaf orientation that might trap granules. Check label for genera of plants on which it can be used. https://content.ces.ncsu.edu/regal-o-o-oxyfluorfen-oxadiazon

Table 7-15. Chemical Weed Control in Ornamentals

More detailed information about herbicides labeled for use in nursery crops and landscape plantings is available on the Weed Management in Nurseries, Landscapes and Christmas Trees web portal (https://weeds.ces.ncsu.edu).

Weed	Herbicide and Formulation	Amount Formulation Per Acre	Pounds Active Ingredient Per Acre	Precautions and Remarks
Post-plant Preemergence Weed Control (continued)				
Annual grasses and broadleaf weeds (preemergence) See label for susceptible species (continued)	oxyfluorfen + prodiamine MOA 14 + 3 (Biathlon) 2.75 G	100 lb	2.75 (2 + 0.75)	Apply preemergence to weeds. Can be used on container- and field-grown ornamentals. Repeat applications are labeled. Injury is to be expected to herbaceous plants or to plants with leaf orientation that might trap granules. Check label for genera of plants on which it can be used. https://content.ces.ncsu.edu/biathlon-oxyfluorfen-prodiamine
	oxyfluorfen + trifluralin MOA 14 + 3 (HGH 75) 5 (2 + 3) G	100 lb	5 (2 + 3)	Preemergence control of annual grasses and broadleaf weeds in container- and field-grown woody ornamentals and landscape beds. Injury is to be expected to herbaceous ornamentals or plants with leaf orientation that might trap granules. https://content.ces.ncsu.edu/hgh-75-oxyfluorfen-trifluralin
	pendimethalin MOA 3 (Corral, Pendulum) several formulations	See label	2 to 4	Preemergence control of annual grasses and some broadleaf weeds in turf, landscape plantings, container and field-grown nursery crops, and Christmas trees. Pendulum Aqua Cap is labeled only for turf and landscape uses. See labels for details. https://content.ces.ncsu.edu/pendulum-aquacap-corral-pendimethalin
	prodiamine MOA 3 (Barricade) 65 WG, 4 FL (Regalkade) 0.5 G	1 to 1.15 lb 21 to 48 oz 150 lb	0.65 to 0.75	Apply preemergence to weeds. Labeled for use in turf, landscape plantings, and nurseries. See label for tolerant species and restrictions. https://content.ces.ncsu.edu/barricade-prodiamine-regalkade-g-prodiamine
	pronamide MOA 3 (Kerb) 50 WP	2 to 4 lb	1 to 2	Pre and postemergence control of cool-season grasses and some annual broadleaf weeds from seed. Apply in late winter just before rain or snowfall. Not recommended for soils that are high in muck or peat. Check label for use restrictions.
	simazine MOA 5 (Princep) 4 L	2 to 3 qt	2 to 3	Apply preemergence to weeds in field nurseries and Christmas trees. Injury has occurred on azaleas, Japanese holly, euonymus, lilac, privet, pittosporum, mock orange, hemlock, boxwood, and several other broadleaf species. High rates will injure Fraser fir. https://content.ces.ncsu.edu/princep-simazine-simazine
	trifluralin MOA 3 (Preen) 1.47G (Treflan) 5 G	136 to 272 lb 80 lb	2 to 4 4	Preemergence to weeds. Irrigate after application. May injure some azalea cultivars.
Post-Plant, Postemergence Selective Grass Control				
Annual and perennial grasses (postemergence) See label for tolerant species	clethodim MOA 1 (Envoy and others)	8 to 34 fl oz	0.06 to 0.25	Postemergent grass control. Annuals 2 to 6 inches tall, perennials at 4 to 12 inches new growth. Add nonionic surfactant at 0.25% v/v (2 pints per 100 gallons) to final spray. https://content.ces.ncsu.edu/envoy-plus-clethodim
	fenoxaprop-P MOA 1 (Acclaim Extra) .57EC	13 to 39 oz	0.06 to 0.17	Apply to emerged grass using at least 40 gpa. Can be used overtop of many flowers and woody ornamentals. Check label. Injury has been observed on Bar Harbor juniper, philodendron, Salvia, Podocarpus, and Pittosporum when sprayed with this product.
	fluazifop-P MOA 1 (Fusilade II) 2 EC	2 to 3 pt	0.25 to 0.4	Postemergence grass control. Annuals not over 2 to 8 inches tall, perennials at 4 to 12 inches new growth. Consult label for tolerant species. Use nonionic surfactant and no oil. https://content.ces.ncsu.edu/fusilade-ii-fluazifop-p-butyl
	sethoxydim MOA 1 (Segment, Sethoxydim) 1 EC	36 to 60 oz	0.3 to 0.5	Postemergence grass control. Annuals up to 12 inches tall and 6 to 10 inches new growth on perennials. https://content.ces.ncsu.edu/segment-sethoxydim
Post-Plant, Postemergence Weed Control				
Annual grasses and broadleaf weeds (postemergence)	asulam MOA 18 (Asulox) 3.34 L	77 to 128 oz	2 to 7	Apply postemergence to weeds in many conifers.
	bentazon MOA 6 (Basagran TO) 4 L	1.5 to 2 pt	0.75 to 1	Postemergent directed to many established ornamentals for yellow nutsedge and seedling broadleaf weed control.
	clopyralid MOA 4 (Lontrel) 3 L	4 to 11 oz	0.09 to 0.25	Postemergence control of legume and many aster weeds. Can be used as a directed spray around on several field-grown woody ornamentals. Can be applied overtop of actively growing conifers transplanted 1 year or more. Apply when weeds are young and actively growing. https://content.ces.ncsu.edu/lontrel-clopyralid
	dichlobenil MOA 20 (Casoron) 4 G	100 to 150 lb	4 to 6	Pre and Postemergence control of many annual and perennial weeds. Do not use on fir, hemlock, Ilex crenata, I. rotunda, or I. vomitoria. Do not use more than 6 pounds per acre on azalea, rhododendron, boxwood, holly, euonymus, forsythia, leucothoe, ivy, lilac, heather, or any plantings less than 1 year old. Do not use in seedbeds, cutting, or transplant beds. Do not apply until 4 weeks after transplanting any plants. Winter applications are best. https://content.ces.ncsu.edu/casoron-dichlobenil
	diquat dibromide (Reward) 2 L	1 to 2 qt	0.5 to 1	Non-selective, contact-type, postemergence control of seedling weeds. A nonionic surfactant should be added to the spray solution. Apply for full coverage and thorough weed contact. Retreatment will be necessary for established weeds. May be used in landscapes, nurseries and greenhouses.
	flumioxazin (Sureguard) DG	0.75 lb	0.375	Early-postemergence and residual control of many seedling annual broadleaf weeds in field and container-grown woody nursery crops and certain landscape plantings. See label for species and precautions. https://content.ces.ncsu.edu/sureguard-flumioxazin
	glufosinate (Finale) 1 L	3 to 6 qt	0.75 to 1.5	Non-selective postemergence control of weeds. Thorough coverage is essential. Apply in a minimum of 20 gallons of water per acre. No residual control. Repeat applications may be necessary for control of perennial weeds.
	glyphosate (Many formulations are available including Roundup-Pro Max, Touchdown Pro, and others)	1 to 5 qt	1 to 5	Non-selective postemergence, systemic control of weeds. DO NOT SPRAY GREEN BARK OR FOLIAGE of crop. Exercise extreme caution in applications near small plants. Use of a shielded sprayer can increase crop safety. Apply in 20 to 30 gallons of water per acre as a directed spray under shrubs or trees. No residual control. Repeat applications may be necessary for control of perennial weeds. https://content.ces.ncsu.edu/glyphosate-1
	imazaquin (Image) 1.5 LC	1 to 1.3 qt	0.4 to 0.5	Apply as a directed spray away from rooting zone. May injure landscape plants through foliar or root absorption. Read the label carefully before use.
	oxyfluorfen (Goal) 2 EC	1 to 2 pt	0.25 to 0.5	Pre- and postemergence control. Apply 1 to 2 pints of Goal 2 EC per acre as a postemergence application on some conifers. Add 0.25% (v/v) nonionic surfactant.

Table 7-15. Chemical Weed Control in Ornamentals

More detailed information about herbicides labeled for use in nursery crops and landscape plantings is available on the Weed Management in Nurseries, Landscapes and Christmas Trees web portal (https://weeds.ces.ncsu.edu).

Weed	Herbicide and Formulation	Amount Formulation Per Acre	Pounds Active Ingredient Per Acre	Precautions and Remarks
Post-Plant, Postemergence Weed Control (continued)				
Annual grasses and broadleaf weeds (postemergence) (continued)	paraquat (Gramoxone Extra) 2.8 L	2 to 3 pt	0.6 to 0.9	Non-selective postemergence control of weeds. Apply when grass and weeds are 1 to 6 inches high and succulent for best results. Direct spray with low pressure to avoid contact with foliage or bark of crop less than 1 year old. Add wetting agent to make 0.25% (2 pints per 100 gallons) by volume of spray for best results. Not for use in landscapes.
	pelargonic acid (Scythe)	3 to 10% by volume	na	Non-selective, contact-type control of seedling broadleaf and grass weeds. Use as a directed spray avoiding contact with foliage and stems of desirable plants. Thorough spray coverage is required. Use the lower concentration for small, succulent seedling weeds. Higher concentrations are needed for larger weeds. Repeated applications are generally required. May be used in landscapes, nurseries and greenhouses.
Sedges (postemergence)	bentazon (Basagran T/O) 4 L	1.5 to 2 pt	0.75 to 1	Postemergent directed spray. For best results add 1 quart per acre crop oil concentrate. For yellow nutsedge and annual sedges; does not control purple nutsedge. May be used in nurseries or landscape plantings.
	halosulfuron (Sledgehammer) 75 DF	0.67 to 1.33 oz	0.031 to 0.062	Early postemergence control of yellow and purple nutsedge. Use only as a directed spray around established woody plants. Add 0.25% nonionic surfactant. For use in landscape plantings only. Not labeled for use in nurseries. https://content.ces.ncsu.edu/sedgehammer-halosulfuron
	imazaquin (Image) 1.5 LC	1 to 1.3 pt	0.4 to 0.5	Pre- or early postemergence control of purple nutsedge. Use as a directed spray. Labeled as an over the top spray on a few species. Add a nonionic surfactant. For use in landscape plantings only. Read label carefully before use.
	sulfentrazon (Dismiss) 4L	6 oz	0.188	Postemergence suppressionl of yellow and purple nutsedge, as well as several broadleaf weeds. Apply as a directed spray avoiding contact with foliage of desirable plants. Reapply when regrowth of treated weeds is observed. Do not exceed 12 ounces per acre per year. May be used in landscape plantings, as well as field and container nurseries

CHEMICAL WEED CONTROL

332

Chemical Weed Control in Vegetable Crops

K. M. Jennings, D. W. Monks, and W. E. Mitchem, Horticultural Science

NOTE: A mode of action code (MOA) has been added to the Herbicide and Formulation column in this table. Use MOA codes for herbicide resistance.

Table 7-16. Chemical Weed Control in Vegetable Crops

Weed	Herbicide, Mode of Action Code* and Formulation	Amount of Formulation Per Acre	Pounds Active Ingredient Per Acre	Precautions and Remarks
Asparagus (seeded and new crown plantings), Preemergence				
Contact kill of all green foliage, stale bed application	paraquat, MOA 22 (Firestorm, Parazone) 3 SL (Gramoxone SL) 2 SL	1.7 to 2.7 pt 2.5 to 4 pt	0.6 to 1	Apply to emerged weeds before crop emergence as a broadcast or band treatment over a preformed row. Row should be formed several days ahead of planting and treating to allow maximum weed emergence. Use a nonionic surfactant at a rate of 16 to 32 ounces per 100 gallons spray mix or 1 gallon approved crop oil concentrate per 100 gallons spray mix. No more than 3 applications per year.
Annual and perennial grass and broadleaf weeds, stale bed application	glyphosate, MOA 9 (numerous brands and formulations)	See labels	See labels	Apply to emerged weeds before crop emergence. Row should be formed several days ahead of planting and treating to allow maximum weed emergence. Perennial weeds may require higher rates. The need for an adjuvant depends on brand used.
Annual grasses and small-seeded broadleaf weeds	linuron, MOA 7 (Lorox DF) 50 WDG	1 to 2 lb	0.5 to 1	**Preemergence application.** Plant seed 0.5 inch deep in coarse soils. Apply to soil surface. See label for further instruction. **Postemergence application.** Apply when ferns are 6 to 18 inches tall. Make one or two applications, but do not exceed 2 pounds active ingredient total per acre. Do not use fertilizer and surfactant or crop oil, as injury will occur. Use the lower rate on coarse soils. Not recommended on sand or loamy sand soils. Do not apply within 1 day of harvest.
Annual grasses and certain broadleaf weeds	pendimethalin, MOA 3 (Prowl H₂O) 3.8 AS	8.2 pt	3.9	Newly planted crown asparagus only. Do not apply to newly seeded asparagus. Newly planted crowns must be covered with at least 2 to 4 inches of soil prior to application. **Do not apply Prowl H20 at more than 2.4 pints per acre in sandy soils. See label for more information.**
Asparagus (seeded and new crown plantings), Postemergence				
Annual and perennial grasses	clethodim, MOA 1 (Intensity One, Select Max) 1 EC (Arrow) 2 EC	9 to 16 oz 6 to 8 oz	0.07 to 0.125 0.094 to 0.125	Apply to emerged grasses. Consult the manufacturer's label for best times to treat specific grasses. For Select Max, add 2 pints nonionic surfactant per 100 gallons spray mixture. With sethoxydim, add 1 quart crop oil concentrate per acre. Adding crop oil to Poast may increase the likelihood of crop injury at high air temperatures. With fluazifop, add 1 quart of nonionic surfactant or 1 gallon crop oil concentrate per 100 gallons of spray mix. PHI is 1 day.
	fluazifop, MOA 1 (Fusilade DX) 2 EC	6 to 16 oz	0.1 to 0.25	
	sethoxydim, MOA 1 (Poast) 1.5 EC	1.5 to 2.5 pt	0.3 to 0.5	
Asparagus (established - at least 2 years old) Preemergence				
Annual grasses and small-seeded broadleaf weeds	linuron, MOA 7 (Lorox DF) 50 WDG	1 to 2 lb	0.5 to 1	Apply before spear emergence or immediately after a cutting. Do not use a surfactant or fertilizer solution in spray mixture. Use the lower rates on coarse soils. Not recommended for sand or loamy sand soils. Repeat applications may be made but do not exceed 4 pounds per acre per year.
	napropamide, MOA 15 (Devrinol DF-XT) 50 DF (Devrinol 2-XT) 2 EC	8 lb 2 gal		Apply to the soil surface in spring before weed and spear emergence. Do not exceed 8 pounds per acre per year. See XT labels for information regarding delay in irrigation event.
	trifluralin, MOA 3 (Trilin, Treflan HFP, Treflan) 4 EC	1 to 4 pt	0.5 to 2	In winter or early spring, apply to dormant asparagus after ferns are removed but before spear emergence, or apply after harvest in late spring or early summer. In a calendar year, the maximum rate is 2 pints per acre for coarse soils, 3 pints on medium soils and 4 pints on fine soils. See label for further restrictions on rates for soil types.
Annual grasses and certain broadleaf weeds	pendimethalin, MOA 3 (Prowl H₂O) 3.8 AS	8.2 pt	3.9	Apply at least 14 days prior to the first spear harvest or after seasonal harvest complete. Do not apply over the top of emerged spears as severe injury may occure. **Do not apply Prowl H20 at more than 2.4 pints per acre in sandy soils.**
Annual broadleaf and grass weeds	diuron, MOA 7 (Karmex) 80 DF (Direx) 4 L	1 to 4 lb 0.8 to 3.2 qt	0.8 to 3.2	Apply in spring before spear emergence but no *earlier* than 4 weeks before spear emergence. A second application may be made immediately after last spear. **For the majority of N.C. plantings, a 1 to 2 pounds per acre dosage of 80 DF or 0.8 to 1.6 quarts rate of Direx should be used.** Diuron also controls small emerged weeds but less effectively.
	flumioxazin, MOA 14 (Chateau) 51 SW	6 oz	0.188	Apply only to dormant asparagus no sooner than 14 days before spears emerge or after the last harvest. Do not apply more than 6 ounces per acre during a single growing season. Provides residual weed control. Can be tank mixed with paraquat for control of emerged weeds. Apply in a minimum of 15 gallons spray mix per acre. Add a nonionic surfactant at 1 quart per 100 gallons of spray mix. A spray-grade nitrogen source (either ammonium sulfate at 2 to 2.5 pounds per acre or 28 to 32 percent nitrogen solutions at 1 to 2 quarts per acre) may be added to increase herbicidal activity.
	metribuzin, MOA 5 (Metribuzin) 75 WDG (TriCor DF) 75 WDG	1.3 to 2.67 lb 1.3 to 2.67 lb	1 to 2	Make a single application to small emerged weeds and the soil surface in early spring before spear emergence or following last cutting. Do not apply within 14 days of harvest or after spear emergence. For the majority of N.C. plantings, the low rate should be used. Do not make postharvest applications until after the last harvest of spears. Tricor DF may be used in sprinkler irrigation. A split application can be used. See label for rates.
	terbacil, MOA 5 (Sinbar) 80 WDG	see label for rate	see label for rate	Apply in spring before weed emergence and spear emergence or immediately after last clean-cut harvest. Use the lower rate on sandy soils and the higher rate on silty or clay soils. Do not use on soils containing less than 1% organic matter nor on gravelly soils or eroded areas where subsoil or roots are exposed. Do not harvest within 5 days after application. See label about rotation restrictions.
Grasses and broadleaf weeds	norflurazon, MOA 12 (Solicam) 80 DF	2.5 to 5 lb	2 to 4	Either rainfall or irrigation necessary for herbicide activation with in 4 weeks after application. Salicom DF rates depend on type of soil. Solicam may be tank mixed with other herbicide registered for use in asparagus. See label for rates and tank mix information. PHI 14 days

CHEMICAL WEED CONTROL

Table 7-16. Chemical Weed Control in Vegetable Crops

Weed	Herbicide, Mode of Action Code* and Formulation	Amount of Formulation Per Acre	Pounds Active Ingredient Per Acre	Precautions and Remarks
Asparagus (established - at least 2 years old) Preemergence (continued)				
Annual broadleaf weeds	mesotrione, MOA 27 (Callisto) 4 L	3 to 7.7 fl oz	0.093 to 0.25	**Preemergence application: apply as a spring application prior to spear emergence, after final harvest, or both.** For optimum preemergence weed control callisto must be applied after fern mowing, disking or other tillage operations but before to spear emergence. **Directed or semidirected application:** Apply after final harvest with care to minimize contact with any standing asparagus spears to avoid crop injury. Do not make more than two applications per year or apply more than 7.7 ounces per acre per year.
Asparagus (established - at least 2 years old), Postemergence				
Broadleaf weeds including trumpetcreeper	2,4-D, MOA 4 (Amine 4 and various other brands) 3.8 SL	1.5 to 2 qt	1.5 to 2	Apply in spring before spear emergence or immediately following a clean cutting. Make no more than two applications during the harvest season and these should be spaced at least 1 month apart. Postharvest sprays should be directed under ferns, avoiding contact with ferns, stems, or emerging spears. Add a nonionic surfactant at a rate of 1 quart per 100 gallons spray mix. **Do not apply if sensitive crops are planted nearby or if conditions favor drift. PHI 3 days**
Broadleaf weeds including trumpetcreeper, annual sowthistle, black mustard, nettleleaf goosefoot, and wild radish	dicamba, diglycolamine salt, MOA 4 (Clarity) 4 L	8 to 16 oz	0.25 to 0.5	Apply to emerged and actively growing weeds in 40 to 60 gallons of diluted spray per treated acre immediately after cutting in the field but at least 24 hours before the next cutting. If spray contacts emerged spears, twisting of spears may occur. Discard twisted spears. See label for more information. **Follow precautions on label concerning drift to sensitive crops. PHI = 1 day**
Contact kill of emerged annual weeds, suppression of emerged perennial weeds, and contact kill of volunteer ferns	paraquat, MOA 22 (Firestorm, Parazone) 3 SL (Gramoxone SL) 2 SL	1.7 to 2.7 pt 2.5 to 4 pt	0.6 to 1	Apply to control emerged weeds (including volunteer ferns). Apply in a minimum of 20 gallons spray mix per acre to control weeds before spears emerge or after last harvest. Do not apply within 6 days of harvest. Use a nonionic surfactant at a rate of 1 quart per 100 gallons spray mix or 1 gallon approved crop oil concentrate per 100 gallons spray mix.
Volunteer ferns (seedling) and certain broadleaf weeds	linuron, MOA 7 (Lorox DF) 50 WDG	2 lb	1	Apply before cutting season or immediately after. Do not apply within 1 day of harvest. Lorox will also control emerged annual broadleaf weeds that are up to 3 inches in height. Lorox can also be applied as a directed spray to the base of the ferns.
Annual and perennial grass and broadleaf weeds Established volunteer ferns	glyphosate, MOA 9 (numerous brands and formulations)	See labels	See labels	Apply to emerged weeds up to 1 week before spear emergence or immediately after last cutting has removed all above-ground parts or as a directed spray to mature fern. Avoid contact with the stem to reduce risk of injury. Perennial weeds may require higher rates of glyphosate. For spot treatment, apply immediately after cutting but prior to emergence of new spears. Certain glyphosate formulations may require the addition of a surfactant. Adding nonionic surfactant to glyphosate formulated with nonionic surfactant may result in reduced weed control.
Yellow and purple nutsedge, non-ALS resistant pigweed, cocklebur, ragweed, wild radish	halosulfuron, MOA 2 (Profine 75) 75 DF (Sandea) 75 DF	0.5 to 1.5 oz	0.024 to 0.072	**Postemergence and post-transplant.** Apply before or during harvesting season. **Do not use nonionic surfactant or crop oil or unacceptable crop injury may occur.** Without the addition of a nonionic surfactant, postemergence weed control may be reduced. Do not exceed 1 ounce per acre per year. Do not harvest within 24 hours of application. **Postharvest.** Apply after final harvest with drop nozzles to limit contact with crop. Contact with the fern may result in temporary yellowing. Add a nonionic surfactant at 1 quart per 100 gallons of spray mixture. Under heavy nutsedge pressure, split applications will be more effective; see label for details. Do not exceed 1 ounce per acre per year.
Annual and perennial grasses	clethodim, MOA 1 (Intenstiy One, Select Max) 1 EC (Arrow) 2 EC	9 to 16 oz 6 to 8 oz	0.07 to 0.125 0.094 to 0.125	For Select Max, add 2 pints nonionic surfactant per 100 gallons spray mixture. DO NOT USE CLETHODIM WITHIN 1 DAY OF HARVEST.
	fluazifop, MOA 1 (Fusilade DX) 2 EC	6 to 16 oz	0.1 to 0.25	Apply to emerged grasses. Consult the manufacturer's label for best times to treat specific grasses. With sethoxydim, add 1 qt crop oil concentrate per acre. With fluazifop, add 1 quart nonionic surfactant or 1 gallon crop oil concentrate per 100 gallons of spray mix. Adding crop oil to Poast may increase the likelihood of crop injury at high air temperature. DO NOT USE FLUAZIFOP OR SETHOXYDIM WITHIN 1 DAY OF HARVEST.
	sethoxydim, MOA 1 (Poast) 1.5 EC	1.5 to 2.5 pt	0.3 to 0.5	
Beans, Preplant and Preemergence				
Contact kill of all green foliage, stale bed application	paraquat, MOA 22 (Firestorm, Parazone) 3 SL (Gramoxone SL) 2 SL	1.3 to 2.7 pt 2 to 4 pt	0.5 to 1	**Lima or snap beans only.** Apply in a minimum of 10 gallons spray mix per acre to emerged weeds before crop emergence as a broadcast or band treatment over a preformed row. Use sufficient water to give thorough coverage. Row should be formed several days ahead of planting and treating to allow maximum weed emergence. Use a nonionic surfactant at a rate of 16 to 32 ounces per 100 gallons spray mix or 1 gallon approved crop oil concentrate per 100 gallons spray mix.
Most broadleaf weeds less than 4 inches tall or rosettes less than 3 inches in diameter; does not control grasses	carfentrazone-ethyl, MOA 14 (Aim) 1.9 EW or 2 EC	up to 2 oz	up to 0.031	Legume vegetable group (Group 6) such as but not limited to edamame, kidney bean, lima bean, pinto bean, snap bean, soybean, and wax bean. Apply prior to or no later than one day after planting. Use a nonionic surfactant or crop oil with Aim. See label for rate. Coverage is essential for good weed control. Can be tank mixed with other registered burndown herbicides.
Annual and perennial grass and broadleaf weeds, stale bed application	glyphosate, MOA 9 (numerous brands and formulations)	See labels	See labels	**Various beans are covered.** Apply to emerged weeds before crop emergence. Perennial weeds may require higher rates of glyphosate. Consult the manufacturer's label for rates for specific weeds. Certain glyphosate formulations may require the addition of a surfactant. Adding nonionic surfactant to glyphosate formulated with nonionic surfactant may result in reduced weed control. See label for details.
Annual grasses and small-seeded broadleaf weeds	ethalfluralin, MOA 3 (Sonalan HFP) 3 EC	1.5 to 3 pt	0.6 to 1.1	**Dry beans only.** See label for specific bean. Apply preplant and incorporate into the soil 2 to 3 inches deep using a rototiller or tandem disk. If groundcherry or nightshade is a problem, the rate range can be increased to 3 to 4.5 pints per acre. For broader spectrum control, Sonalan may be tank mixed with Eptam or Dual. Read the combination product labels for directions, cautions, and limitations before use.
	dimethenamid, MOA 15 (Outlook) 6.0 EC	12 to 18 oz	0.55 to 0.85	**Dry beans only.** See label for specific bean. Apply preplant incorporated, preemergence to the soil surface after planting, or early postemergence (first to third trifoliate stage). Dry beans may be harvested 70 or more days after Outlook application. For soils having 3% or greater organic matter, see label for rate. See label for further instructions, including those for tank mixtures.

Table 7-16. Chemical Weed Control in Vegetable Crops

Weed	Herbicide, Mode of Action Code* and Formulation	Amount of Formulation Per Acre	Pounds Active Ingredient Per Acre	Precautions and Remarks
Beans, Preplant and Preemergence (continued)				
Annual grasses and small-seeded broadleaf weeds (continued)	trifluralin, MOA 3 (Treflan, Trifluralin, Trifluralin HF, other brands) 4 EC	1 to 1.5 pt	0.5 to 0.75	**Dry, lima, or snap beans only.** See label for specific bean. Apply preplant and incorporate into the soil 2 to 3 inches deep within 8 hr. Incorporate with a power-driven rototiller or by cross disking.
	pendimethalin, MOA 3 (Prowl H₂0) 3.8 AS	1.5 to 3 pt	0.75 to 1.5	**Edible beans: dry, lima, or snap beans, and certain others.** See label for specific bean. Apply preplant and incorporate into the soil 2 to 3 inches using a power-driven rototiller or by cross disking. **Do not apply after seeding.**
	S-metolachlor, MOA 15 (Brawl, Dual Magnum, Medal) 7.62 EC (Brawl II, Dual II Magnum, Medal II) 7.64 EC	1 to 2 pt	0.95 to 1.91	**Dry, lima, or snap beans, and certain others.** See label for specific bean, and specific rate based on soil texture. Apply preplant incorporated or preemergence to the soil surface after planting.
Annual grasses and broadleaf weeds	clomazone, MOA 13 (Command) 3ME	0.4 to 0.67 pt	0.15 to 0.25	**Snap beans (succulent) only.** Apply to the soil surface immediately after seeding. Offers weak control of pigweed. See label for further instructions. Do not apply within 45 days of harvest. Limited research has been done on this product in this crop in North Carolina.
Yellow and purple nutsedge, grasses and some small-seeded broadleaf weeds	EPTC, MOA 8 (Eptam) 7 EC	2.25 to 3.5 pt	2 to 3	**Dry or snap beans only.** See label for specific bean. Apply preplant and incorporate immediately to a depth of 3 inches or may be applied at lay-by as a directed application before bean pods start to form to control late season weeds. See label for instructions on incorporation. May be tank mixed with Prowl. Do not use on black-eyed beans, lima beans, or other flat-podded beans except Romano.
Many broadleaf weeds	fomesafen, MOA 14 (Reflex 2 EC)	1 to 1.5 pt	0.25 to 0.375	**Dry bean and snap beans only.** Apply preplant surface or preemergence. Total use per year cannot exceed 1.5 pints per acre. See label for further instructions and precautions.
Yellow and purple nutsedge, common cocklebur, and other broadleaf weeds	halosulfuron-methyl, MOA 2 (Profine 75) 75 DG (Sandea) 75 DG	0.5 to 0.75 oz	0.024 to 0.036	**Dry beans and succulent snap beans including lima beans only.** Apply after seeding but prior to cracking. Do not apply more than 0.67 ounce product per acre to dry bean. Data are lacking on runner-type snap beans. See label for other instructions.
Broadleaf weeds including morningglory, pigweed, smartweed, and purslane	imazethapyr, MOA 2 (Pursuit) 2 EC	1.5 oz	0.023	**Dry beans and lima beans only.** See label for specific bean. Apply preemergence or preplant incorporated. Pursuit should be applied with a registered preemergence grass herbicide. **Snap beans only.** Apply preemergence or preplant incorporated. For preplant incorporated application, apply within 1 week of planting. May be used with a registered grass herbicide. Reduced crop growth, quality, yield, and/or delayed crop maturation may result. Do not apply within 30 days of harvest of snapbeans.
Beans, Postemergence				
Annual broadleaf weeds and yellow nutsedge	bentazon, MOA 6 (Basagran) 4 SL	1 to 2 pt	0.5 to 1	**Dry, lima, or snap beans only.** Apply overtop of beans and weeds when beans have one to two expanded trifoliate leaves. Two applications spaced 7 to 10 days apart may be made for nutsedge control. Do not apply more than 2 quarts per season or within 30 days of harvest. Use of crop oil as an adjuvant will improve weed control but will likely increase crop injury. See label regarding crop oil concentrate use. Do not within 30 days of harvest.
Many broadleaf weeds	fomesafen, MOA 14 (Reflex 2 EC)	0.75 to 1 pt	0.188 to 0.25	**Dry or snap beans only.** See label for specific bean. Apply postemergence to dry beans or snap beans that have at least one expanded trifoliate leaf. Include a nonionic surfactant at 1 quart per 100 gallons spray mixture. Total use per year cannot exceed 1.5 pints per acre. Do not apply within 45 days of dry bean harvest or 30 days of snap bean harvest. **Postemergence application of fomesafen can cause significant injury to the crop.** See label for further information.
Most broadleaf weeds less than 4 inches tall or rosettes less than 3 inches in diameter. Does not control grasses	carfentrazone-ethyl, MOA 14 (Aim) 1.9 EW or 2 EC	up to 2 oz	up to 0.031	**Edible beans: edamame, kidney bean, lima bean, pinto bean, snap bean and wax bean only.** Apply post-directed using hooded sprayers for control of emerged weeds. If crop is contacted, burning of contacted area will occur. Use a nonionic surfactant or crop oil with Aim. See label for rate. Coverage is essential for good weed control. Can be tank mixed with other registered herbicides.
Yellow and purple nutsedge	EPTC, MOA 8 (Eptam) 7 EC	3.5 pt	3	**Green or dry beans only.** See label for specific bean. Do not use on lima bean or pea. Apply and incorporate at last cultivation as a directed spray to soil at the base of crop plants before pods start to form.
Yellow and purple nutsedge, common cocklebur, and other broadleaf weeds	halosulfuron-methyl, MOA 2 (Profine 75) 75 DG (Sandea) 75 DG	0.5 to 0.67 oz	0.024 to 0.031	**Succulent snap beans, including lima beans.** Apply after crop has reached 2-to 3-trifoliate leaf stage but prior to flowering. Postemergence application may cause significant but temporary stunting and may delay crop maturation. Use directed spray to limit crop injury. Do not apply within 30 days of harvest. See label for further precautions. Data lacking on runner-type snap beans.
Annual broadleaf weeds, including morningglory, pigweed, smartweed, and purslane	imazethapyr, MOA 2 (Pursuit) 2.EC	1.5 to 3 oz	0.023 to 0.047	**Dry beans and snap beans only.** See label for specific bean. Use only 1.5 ounces EC formulation on snap bean and up to 3 ounces on dry beans. Apply postemergence to 1- to 3-inches weeds (one to four leaves) when beans have at least one fully expanded trifoliate leaf. Add nonionic surfactant at 2 pints per 100 gallons of spray mixture with all postemergence applications. For snap beans, allow at least 30 days between application and harvest. For dry bean, do not apply within 60 days of harvest. See label for instructions on use.
Most emerged weeds	glyphosate, MOA 9 (Roundup PowerMax) 5.5 L (Roundup WeatherMax) 5.5 L	11 to 22 oz	0.5 to 0.94	**Row middles only.** See label for specific bean. Apply as a hooded spray in row middles, as shielded spray in row middles, as wiper applications in row middles, or postharvest. Spot treatment is allowed in some bean crops. To avoid severe injury to crop, do not allow herbicide to contact foliage, green shoots, stems, exposed, roots, or fruit of crop. Do not apply within 14 days of harvest.
Annual and perennial grasses	sethoxydim, MOA 1 (Poast) 1.5 EC	1 to 1.5 pt	0.2 to 0.3	**Dry or succulent beans only.** See label for specific bean. For succulent beans, products with quizalofop are limited to snap beans. Apply to emerged grasses. Consult manufacturer's label for specific rates and best times to treat. With sethoxydim, add 1 quart of crop oil concentrate per acre. With quizalofop, add 1 gal oil concentrate or 1 quart nonionic surfactant per 100 gallons spray. Adding crop oil to Poast may increase the likelihood of crop injury at high air temperatures. Do not apply on days that are unusually hot and humid. Do not apply within 15 days and 30 days of harvest for succulent and dry beans, respectively.
	quizalofop p-ethyl, MOA 1 (Assure II) 0.88 EC (Targa) 0.88 EC	6 to 12 oz	0.04 to 0.08	

335

Table 7-16. Chemical Weed Control in Vegetable Crops

Weed	Herbicide, Mode of Action Code* and Formulation	Amount of Formulation Per Acre	Pounds Active Ingredient Per Acre	Precautions and Remarks
Beans, Postemergence (continued)				
Annual and perennial grasses (continued)	clethodim, MOA 1 (Arrow, Clethodim, Intensity, Select) 2 EC	6 to 16 oz	0.094 to 0.25	**Dry or succulent beans.** See label for specific bean. Select is registered for dry beans only. Apply postemergence for control of emerged grasses. See label for specific rate for crop. For Arrow, Clethodim, or Select, add a crop oil concentrate at 1 quart per acre. For Select Max or Intensity One, add 2 pints nonionic surfactant per 100 gallons spray mixture. Adding crop oil may increase the likelihood of crop injury at high air temperatures. Very effective in controlling annual bluegrass. Apply to actively growing grasses not under drought stress. See label for minimum time from application to harvest.
	(Intensity One, Select Max) 1 EC	9 to 16 oz	0.07 to 0.125	
Beets (Garden or Table), Preplant				
Most emerged weeds except for resistant pigweed, primrose, or spiderwort	glyphosate, MOA 9 (numerous brands and formulations)	See labels	See labels	**Garden beets only.** Apply to emerged weeds before seeding or after seeding but before crop emergence. Perennial weeds may require higher rates. Certain glyphosate formulations may require the addition of a surfactant. Adding nonionic surfactant to glyphosate formulated with nonionic surfactant may result in reduced weed control.
Emerged broadleaf weeds	pyraflufen, MOA 14 (ET Herbicide) 0.208 EC	0.5 to 2 fl oz	0.0008 to 0.003	**Garden beets only.** Apply as a preplant burndown treatment in a minimum of 10 gallons per acre. **Addition of a crop oil concentrate at 1 to 2% is recommended for optimum weed control. See label for additional information.**
Emerged broadleaf and grass weeds	pelargonic, (Scythe) 4.2 EC	3 to 10% v/v		Apply as a preplant burndown treatment or prior to crop emergence from seed.
Beets (Garden or Table), Preemergence				
Annual grasses (crabgrass spp., foxtail spp., barnyardgrass, annual ryegrass, annual bluegrass) and broadleaf weeds (Lamium spp., lambsquarters, common purslane, redroot pigweed, shepherdspurse)	cyclohexylethylthio-carbamate, MOA 3 (Ro-Neet) 6E	0.5 to 0.67 gal	3 to 4	Use on mineral soils only. Use higher dosage rate on heavier soils. Read label for further instructions.
Beets (Garden or Table), Postemergence				
Broadleaf weeds including sowthistle clover, cocklebur, jimsonweed, and ragweed	clopyralid, MOA 4 (Solix 3, Stinger) 3EC	0.25 to 0.5 pt	0.093 to 0.187	Apply to beets having 2 to 8 leaves when weeds are small and actively growing. Will control most legumes. Do not apply within 30 days of harvest. Do not apply more than 0.5 pint per acre per year. See label for information regarding rotational restrictions. The PHI is 30 days.
Broadleaf weeds including wild mustard, common lambsquarters, common chickweed, purslane suppression	phenmedipham, MOA 6 (Spin-Aid) 1.3 EC	1.5 to 3 pt	0.25 to 0.5	Red garden beets only. Apply to red garden beets in the 2 to 6 leaf stage. Rate is dependent on crop stage. See label for specific rate. Best control occurs when applied to seeds in cotyledon to 2 leaf stage. Minor crop stunting may be observed for approximately 10 days. Do not add spray adjuvant. Do not apply within 60 days of harvest.
Broadleaf weeds including wild mustard, sheperd's and velvetleaf	triflusulfuron methyl, MOA 2 (Upbeet) 50 DF	0.5 oz	0.0156	**Garden beets.** Apply when beets are at the 2 to 4 leaf stage. Additional applications may be made at the 4 to 6, and 6 to 8 leaf stages. See label for information on adjuvants. Total amount of Upbeet must not exceed 1.5 ounces per acre per growing season. Do not apply within 30 days of harvest.
Annual and perennial grasses	sethoxydim, MOA 1 (Poast) 1.5 EC	1 to 1.5 pt	0.2 to 0.3	Apply to emerged grasses. Consult manufacturer's label for specific rates and best times to treat. Add 1 quart of crop oil concentrate per acre. Adding crop oil to Poast may increase the likelihood of crop injury at high air temperatures. Do not apply Poast on days that are unusually hot and humid. Do not apply within 60 days of harvest.
	clethodim, MOA 1 (Arrow, Clethodim, Intensity, Select) 2 EC	6 to 8 oz	0.094 to 0.125	Apply postemergence for annual grasses at 6 to 8 ounces per acre or bermudagrass and johnsongrass at 8 ounces per acre. For Arrow, Clethodim, or Select, add a crop oil concentrate at 1% per spray volume. For Select Max, add 2 pints nonionic surfactant per 100 gallons spray mixture. Adding crop oil may increase the likelihood of crop injury at high air temperatures. Very effective in controlling annual bluegrass. Apply to actively growing grasses not under drought stress. Do not apply within 30 days of harvest.
	(Select Max, Intensity One) 1 EC	9 to 16 oz	0.07 to 0.125	
Beets (Garden or Table), Row Middles Only				
Most emerged weeds except resistant pigweed	glyphosate, MOA 9 (numerous brands and formulations)	See labels	See labels	Apply as a hooded spray in row middles, as shielded spray in row middles, as wiper applications in row middles, or postharvest. To avoid severe injury to crop, do not allow herbicide to contact foliage, green shoots, stems, exposed, roots, or fruit of crop. The need for an adjuvant depends on brand used. Do not apply within 14 days of harvest.
Annual broadleaf weeds including morningglory, spiderwort, and very small pigweed	carfentrazone-ethyl, MOA 14 (Aim) 1.9 EW or 2 EC	up to 2 oz	up to 0.031	Apply post-directed using hooded sprayers for control of emerged weeds. If crop is contacted, burning of contacted area will occur. Use a crop oil concentrate or a nonionic surfactant with Aim. See label for directions. Coverage is essential for good weed control. Can be tank mixed with other registered herbicides.
Broccoli – See Cole Crops				
Cabbage – See Cole Crops				
Cantaloupes (Muskmelons), Preplant and Preemergence				
Suppression or control of most annual grasses and broadleaf weeds, full rate required for nutsedge control	metam sodium (Vapam HL) 42%	37.5 to 75 gal	15.7 to 31.5	Rates are dependent on soil type and weeds present. Apply when soil moisture is at field capacity (100 to 125%). Apply through soil injection using a rotary tiller or inject with knives no more than 4 inches apart; follow immediately with a roller to smooth and compact the soil surface or with mulch. May apply through drip irrigation prior to planting a second crop on mulch. Plant back interval is often 14 to 21 days and can be 30 days in some environments. See label for all restrictions and additional information.
Most broadleaf weeds less than 4 inches tall or rosettes less than 3 inches in diameter; does not control grasses	carfentrazone-ethyl, MOA 14 (Aim) 1.9 EW or 2 EC	up to 2 oz	up to 0.031	**Transplant crop: apply no later than one day before transplanting crop.** **Seeded crop: apply no later than 7 days before seeding crop.** Use a crop oil at up to 1 gallon per 100 gallons of spray solution or a nonionic surfactant at 2 pints per 100 gallons of spray solution. Coverage is essential for good weed control. Can be tank mixed with other registered burndown herbicides.

Table 7-16. Chemical Weed Control in Vegetable Crops

Weed	Herbicide, Mode of Action Code* and Formulation	Amount of Formulation Per Acre	Pounds Active Ingredient Per Acre	Precautions and Remarks
Cantaloupes (Muskmelons), Preplant and Preemergence (continued)				
Contact kill of all green foliage, stale bed application	paraquat, MOA 22 (Firestorm, Parazone) 3 SL (Gramoxone SL) 2 SL	1.3 to 2.7 pt 2 to 4 pt	0.5 to 1	Apply in a minimum of 10 gallons spray mix per acre to emerged weeds before crop emerges or before transplanting as a broadcast or band treatment over a preformed row. Seedbeds or plantbeds should be formed as far ahead of treatment as possible to allow maximum weed emergence. Use a nonionic surfactant at a rate of 16 to 32 ounces per 100 gallons spray mix or 1 gallon approved crop oil concentrate per 100 gallons spray mix.
Annual and perennial grass and broadleaf weeds, stale bed application	glyphosate, MOA 9 (numerous brands and formulations	See labels	See labels	Apply to emerged weeds at least 3 days before seeding or transplanting. Perennial weeds may require higher rates of glyphosate. Consult manufacturer's label for rates for specific weeds. When applying Roundup before transplanting crops into plastic mulch, carefully remove residues of this product from the plastic prior to transplanting. To prevent crop injury, residues can be removed by 0.5 inch rainfall or by applying water via a sprinkler system. Certain glyphosate formulations may require the addition of a surfactant. Adding nonionic surfactant to glyphosate formulated with nonionic surfactant may result in reduced weed control.
Emerged broadleaf weeds	pyraflufen, MOA 14 (ET Herbicide) 0.208 EC	0.5 to 2 fl oz	0.0008 to 0.003	Apply as a preplant burndown treatment in a minimum of 10 gallons per acre. **Addition of a crop oil concentrate at 1 to 2% is recommended for optimum weed control. See label for additional information.**
Emerged broadleaf and grass weeds	pelargonic, (Scythe) 4.2 EC	3 to 10% v/v		Apply before crop emergence and control emerged weeds. There is no residual activity. May be tank mixed with soil residual compounds. See lable for instruction. May also be used as a banded spray between row middles. Use a shielded sprayer directed to the row middles to reduce drift to the crop.
Annual grasses and small-seeded broadleaf weeds	bensulide, MOA 8 (Prefar) 4 EC	5 to 6 qt	5 to 6	Registered for cucurbit vegetable group (Crop grouping 9). Apply preplant and incorporate into the soil 1 to 2 inches (1 inch incorporation is optimum) with a rototiller or tandem disk, or apply preemergence after seeding and follow with irrigation. Check replant restrictions for small grains and other crops on label.
Annual grasses and broadleaf weeds; weak on pigweed and morningglory	clomazone, MOA 13 (Command) 3 ME	0.4 to 0.67 pt	0.15 to 0.25	Apply immediately after seeding or just prior to transplanting with transplanted crop. Roots of transplants must be below the chemical barrier when planting. See label for further instruction.
Annual grasses and some small-seeded broadleaf weeds	ethalfluralin, MOA 3 (Curbit) 3 EC	3 to 4.5 pt	1.1 to 1.7	Apply postplant to seeded crop prior to crop emergence or as a banded spray between rows after crop emergence or transplanting. See label for timing. Shallow cultivation, irrigation, or rainfall within 5 days needed for good weed control. Do not use under mulches, row covers, or hot caps. Under conditions of unusually cold or wet soil and air temperatures, crop stunting and injury may occur. Crop injury can occur if seeding depth is too shallow.
Annual grasses and broadleaf weeds	ethalfluralin, MOA 3 + clomazone, MOA 13 (Strategy) 2.1 L	2 to 6 pt	0.4 to 1.2 + 0.125 to 0.375	Apply to the soil surface immediately after seeding crop for preemergence control of weeds. **Do not apply prior to planting crop. Do not soil incorporate.** May also be used as a **banded** treatment **between** rows after crop emergence or transplanting. Do not apply over or under plastic mulch. See label for application rate according to soil type and crop restriction instructions.
Yellow and purple nutsedge and broadleaf weeds	halosulfuron-methyl, MOA 2 (Profine 75) 75 DG (Sandea) 75 DG	0.5 to 0.75 oz	0.024 to 0.036	Apply after seeding or prior to transplanting crop. For transplanted crop, do not transplant until 7 days after application. Rate can be increased to 1 ounce of product per acre to middles between rows. Do not apply within 57 days of harvest.
Cantaloupes (Muskmelons), Postemergence				
Annual grasses and some small-seeded broadleaf weeds	DCPA, MOA 3 (Dacthal) W-75 (Dacthal) 6 F	8 to 10 lb 6 to 10 pt	4.5 to 7.5	**Not labeled for transplanted crop.** To improve preemergence control of late emerging weeds. Apply only when crop has 4 to 5 true leaves, is well-established, and growing conditions are favorable. Will not control emerged weeds. Incorporation not recommended.
	trifluralin, MOA 3 (Treflan HFP, Trifluralin, Trifluralin HF) 4EC	1 to 2 pt	0.5 to 0.75	**Apply as a directed spray to soil between rows after crop emergence when crop plants have reached** three to four true leaf stage of growth. Avoid contacting foliage as slight crop injury may occur. Set incorporation equipment to move treated soil around base of crop plants. Do not apply within 30 days of harvest. Will not control emerged weeds.
Yellow and purple nutsedge and broadleaf weeds including cocklebur, galinsoga, smartweed, ragweed, wild radish, and pigweed	halosulfuron-methyl, MOA 2 (Profine 75) 75 DG (Sandea) 75 DG	0.5 to 0.75 oz	0.024 to 0.036	Apply postemergence only after the crop has reached 3 to 5 true leaves but before first female flowers appear. Do not apply sooner than 14 days after transplanting. Use nonionic surfactant at 1 quart per 100 gallons of spray solution with all postemergence applications. Avoid over-the-top applications during late summer when temperature and humidity are high. Do not apply within 57 days of harvest.
Most broadleaf weeds less than 4 inches tall or rosettes less than 3 inches in diameter; does not control grasses	carfentrazone-ethyl, MOA 14 (Aim) 1.9 EW or 2 EC	up to 2 oz	up to 0.031	Apply post-directed using hooded sprayers for control of emerged weeds. If crop is contacted, burning of contacted area will occur. Use a crop oil concentrate or a nonionic surfactant with Aim. See label for directions. Coverage is essential for good weed control. Can be tank mixed with other registered herbicides. PHI = 0 days.
Most emerged weeds	glyphosate, MOA 9 (numerous brands and formulations)	See labels	See labels	Apply as a hooded spray in row middles, as shielded spray in row middles, as wiper applications in row middles, or postharvest. To avoid severe injury to crop, do not allow herbicide to contact foliage, green shoots, stems, exposed, roots, or fruit of crop. Do not apply within 14 days of harvest.
Annual and perennial grasses only	sethoxydim, MOA 1 (Poast) 1.5 EC	1 to 1.5 pt	0.2 to 0.3	Apply to emerged grasses. Consult manufacturer's label for specific rates and best times to treat. Add 1 quart of crop oil concentrate per acre. Adding crop oil to Poast may increase the likelihood of crop injury at high air temperatures. Do not apply Poast on days that are unusually hot and humid. Do not apply within 3 days of harvest.
	clethodim, MOA 1 (Arrow, Clethodim, Intensity, Select) 2 EC (Intensity One, Select Max) 1 EC	6 to 8 oz 9 to 16 oz	0.094 to 0.125 0.07 to 0.125	Apply postemergence for control of grass in cantaloupes (muskmelons). For Arrow, Clethodim, or Select, add 1 gallon crop oil concentrate per 100 gallons spray mix. For Select Max, add 2 pints nonionic surfactant per 100 gallons spray mixture. Adding crop oil may increase the likelihood of crop injury at high air temperatures. Very effective in controlling annual bluegrass. Apply to actively growing grasses not under drought stress. Do not apply within 14 days of harvest.

CHEMICAL WEED CONTROL

Table 7-16. Chemical Weed Control in Vegetable Crops

Weed	Herbicide, Mode of Action Code* and Formulation	Amount of Formulation Per Acre	Pounds Active Ingredient Per Acre	Precautions and Remarks
Carrots, Preplant				
Contact kill of all green foliage, stale bed application	paraquat, MOA 22 (Firestorm, Parazone) 3 SL (Gramoxone SL) 2 SL	1.3 to 2.7 pt 2 to 4 pt	0.5 to 1	Apply to emerged weeds before crop emergence as a broadcast or band treatment over a preformed row. Use sufficient water to give thorough coverage. Row should be formed several days ahead of planting and treating to allow maximum weed emergence. Use a nonionic surfactant at a rate of 16 to 32 ounces per 100 gallons spray mix or 1 gallon approved crop oil concentrate per 100 gallons spray mix.
Most emerged weeds except resistant pigweed	glyphosate, MOA 9 (numerous brands and formulations)	See labels	See labels	Apply to emerged weeds before seeding or crop emergence. Perennial weeds may require higher rates. Certain glyphosate formulations require the addition of surfactant. Adding nonionic surfactant to glyphosate formulated with nonionic surfactant may result in reduced weed control.
Emerged broadleaf and grass weeds	pelargonic, (Scythe) 4.2 EC	3 to 10% v/v		Apply as a preplant burndown or prior to emergence of plants from seed. There is no residual activity. May be tank mixed with soil residual compounds. See label for instruction. May also be used as a banded spray between row middles. Use a shielded sprayer directed to the row middles to reduce drift to the crop.
Carrots, Preplant incorporated (PPI) or Preemergence (PRE)				
Annual grasses and small-seeded broadleaf weeds	trifluralin, MOA 3 (Treflan, Trifluralin) 4 EC	1 to 2 pt	0.5 to 1	Apply preplant and incorporate into the soil 2 to 3 inches within 8 hr. Use lower rate on coarse soils with less than 2% organic matter.
Broadleaf and grass weeds	pendimethalin, MOA 3 (Prowl H$_2$O) 3.8 AS	2 pt	0.95	Apply postplant within 2 days after planting but prior to crop emergence. See label for instruction on layby treatment. PHI interval is 60 days.
Broadleaf and grass weeds	prometryn, MOA 5 (Caparol) 4L	2 to 4 pt	1 to 2	Apply as preemergence and or postemergence over the top to carrot. Make Post application through the six leaf stage of carrot. See label for application rate and crop rotation restrictions. PHI is 30 days.
Carrots, Postemergence				
Annual grasses and broadleaf weeds	linuron, MOA 7 (Lorox DF) 50 WDG	1.5 to 3 lb	0.75 to 1.5	Apply as a broadcast spray after carrots are at least 3 inches tall. If applied earlier crop injury may occur. Avoid spraying after three or more cloudy days. Repeat applications may be made, but do not exceed 4 pounds of Lorox DF per acre per season. Do not use a surfactant or crop oil. Carrot varieties vary in their resistance, therefore determine tolerance to Lorox DF before adoption as a field practice to prevent potential crop injury. See label for further directions. PHI = 14 days.
Annual broadleaf weeds and some grasses	metribuzin, MOA 5 (Dimetric, Metribuzin, TriCor DF) 75 WDG (Metri, TriCor 4F) 4 F	0.33 lb 0.5 pt	0.25 0.25	Apply overtop when weeds are less than 1 inch tall and carrots have 5 to 6 true leaves. A second application may be made after a time interval of at least 3 weeks. Do not apply unless 3 sunny days precede application. Do not apply within 3 days of other pesticide applications. Preharvest interval is 60 days.
Annual and perennial grasses	clethodim, MOA 1 (Arrow, Clethodim, Intensity, Select) 2 EC (Select Max) 1 EC	6 to 8 oz 9 to 16 oz	0.094 to 0.125 0.07 to 0.125	Apply to actively growing grasses not under drought stress. With Arrow, Clethodim, or Select, add 1 gallon crop oil concentrate per 100 gallons spray mix. With Select Max, add 2 pints nonionic surfactant per 100 gallons of spray mixture. Adding crop oil may increase the likelihood of crop injury at high air temperatures. Do not mix with other pesticides. Very effective in controlling annual bluegrass. Do not apply within 30 days of harvest.
	fluazifop, MOA 1 (Fusilade DX) 2 EC	6 to 16 oz	0.1 to 0.25	Apply to actively growing grasses not under drought stress. Up to 48 ounces of Fusilade DX may be applied per year. See label for rates for specific weeds. Add 1 gallon crop oil concentrate or 1 quart nonionic surfactant per 100 gallons spray mix. Do not mix with other pesticides. Do not apply within 45 days of harvest.
	sethoxydim, MOA 1 (Poast) 1.5 EC	1 to 1.5 pt	0.2 to 0.3	Apply to actively growing grasses not under drought stress. Consult manufacturer's label for specific rate and best times to treat. Add 1 quart of crop oil concentrate per acre. Adding crop oil may increase the likelihood of crop injury at high air temperatures. Do not apply on days that are unusually hot and humid. Do not apply with other pesticides. Do not apply within 30 days of harvest.
Carrots, Row Middles				
Most broadleaf weeds less than 4 inches tall or rosettes less than 3 inches in diameter; does not control grasses	carfentrazone-ethyl, MOA 14 (Aim) 1.9 EW or 2 EC	up to 2 ounces	up to 0.031	Apply as a hooded spray in row middles for control of emerged weeds. If crop is contacted, burning of contacted area will occur. Use a crop oil concentrate or a nonionic surfactant with Aim. See label for directions. Coverage is essential for good weed control. Can be tank mixed with other registered herbicides.
Most emerged weeds except resistant pigweed	glyphosate, MOA 9 (numerous brands and formulations)	See labels	See labels	Apply as a hooded spray in row middles, as shielded spray in row middles, as wiper applications in row middles, or postharvest. To avoid severe injury to crop, do not allow herbicide to contact foliage, green shoots or stems, exposed roots, or fruit of crop. Do not apply within 14 days of harvest.
Cauliflower – See Cole Crops				
Celery, Preplant				
Annual and perennial grass and broadleaf weeds, stale bed application	glyphosate, MOA 9 (numerous brands and formulations)	See labels	See labels	Apply to emerged weeds before crop emergence. Perennial weeds may require higher rates. Consult the manufacturer's label for rates for specific weeds. Certain glyphosate formulations require the addition of surfactant. Adding nonionic surfactant to glyphosate formulated with nonionic surfactant may result in reduced weed control.
Cutleaf evening primrose, carolina geranium, henbit, and a few grasses	oxyfluorfen, MOA 14 (Goaltender) 4 F (Goal 2 XL) 2 EC	up to 1 pt up to 2 pt	up to 0.5	**Transplants only.** Apply to soil surface of pre-formed beds at least 30 days prior to transplanting. No research has been conducted in North Carolina, therefore, try on a limited number of acres first.
Emerged broadleaf and grass weeds	pelargonic, (Scythe) 4.2 EC	3 to 10% v/v		Apply as a preplant burndown. There is no residual activity. May be tank mixed with soil residual compounds. See lable for instruction. May also be used as a banded spray between row middles. Use a shielded sprayer directed to the row middles to reduce drift to the crop.

Table 7-16. Chemical Weed Control in Vegetable Crops

Weed	Herbicide, Mode of Action Code* and Formulation	Amount of Formulation Per Acre	Pounds Active Ingredient Per Acre	Precautions and Remarks
Celery, Preplant incorporate (PPI) or Preemergence (PRE)				
Annual grasses and small-seeded broadleaf weeds	trifluralin, MOA 3 (Treflan, Treflan HFP, Trifluralin) 4 EC	1 to 2 pt	0.5 to 1 lb	Apply incorporated to direct seeded or transplant celery before planting, at planting, or immediately after planting. Incorporate within 8 hours of application. Use lower rate on coarse soils with less than 2% organic matter.
	Bensulide (Prefar) 4-E	5 to 6 qt	5 to 6	**Transplants only.** Apply PRE after planting, irrigate immediately. See label for rotation restrictions.
Celery, Postemergence				
Annual broadleaf and grass weeds	linuron, MOA 7 (Lorox DF) 50 WDG	1.5 to 3 lb	0.75 to 1.5	Apply after celery is transplanted and established but before celery is 8 inches tall. Grasses should be less than 2 inches in height, and broadleaf weeds should be less than 6 inches tall. Do not tank mix with other products including surfactant or crop oil. Avoid spraying after 3 or more cloudy days or when temperature exceeds 85 F. Not recommended for sands or loamy sand soil. Preharvest interval is 45 days.
Annual and perennial grasses only	clethodim, MOA 1 (Arrow, Clethodim, Intensity, Select) 2 EC\n\n(Select Max) 1 EC	6 to 8 oz\n\n9 to 16 oz	0.094 to 0.125\n\n0.07 to 0.125	Apply to actively growing grasses not under drought stress. With Arrow, Clethodim, or Select, add 1 gallon crop oil concentrate per 100 gallons spray mix. With Select Max, add 2 pints of nonionic surfactant per 100 gallons spray mixture. Very effective in controlling annual bluegrass. Apply to actively growing grasses not under drought stress. Adding crop oil may increase the likelihood of crop injury at high air temperature. Do not apply within 30 days of harvest.
	sethoxydim, MOA 1 (Poast) 1.5 EC	1 to 1.5 pt	0.2 to 0.3	Apply to actively growing grasses not under drought stress. Consult label for specific rates and best times to treat. Add 1 quart of crop oil concentrate per acre. Adding crop oil to Poast may increase the likelihood of crop injury at high air temperatures. Do not apply Poast on unusually hot and humid days. Do not apply within 30 days of harvest.
Most broadleaf weeds less than 4 inches tall or rosettes less than 3 inches in diameter; does not control grasses	carfentrazone-ethyl, MOA 14 (Aim) 1.9 EW or 2 EC	up to 2 oz	up to 0.031	Apply post-directed using hooded sprayers for control of emerged weeds. If crop is contacted, burning of contacted area will occur. Use a crop oil concentrate or a nonionic surfactant with Aim. See label for directions. Coverage is essential for good weed control. Can be tank mixed with other registered herbicides.
Most emerged weeds	glyphosate, MOA 9 (numerous brands and formulations)	See labels	See labels	**Row middles only.** Apply as a hooded spray in row middles, as shielded spray in row middles, as wiper applications in row middles, or postharvest. To avoid severe injury to crop, do not allow herbicide to contact foliage, green shoots, stems, exposed, roots, or fruit of crop. Do not apply within 14 days of harvest.
Cole Crops: Broccoli, Cabbage, Cauliflower — Preplant and Preemergence				
Contact kill of all green foliage, stale bed application	paraquat, MOA 22 (Firestorm, Parazone) 3 SL\n\n(Gramoxone SL) 2 SL	1.3 to 2.7 pt\n\n2 to 4 pt	0.5 to 1	Apply in minimum of 10 gallons spray mix per acre to emerged weeds before crop emergence or transplanting as a broadcast or band treatment over a preformed row. Use sufficient water for thorough coverage. Row should be formed several days ahead of planting and treated to allow maximum weed emergence. Use nonionic surfactant at rate of 16 to 32 ounces per 100 gallons spray mix or 1 gallon approved crop oil concentrate per 100 gallons spray mix.
Most broadleaf weeds less than 4 inches tall or rosettes less than 3 inches in diameter; does not control grasses	carfentrazone-ethyl, MOA 14 (Aim) 2 EC	up to 2 oz	up to 0.031	Apply no later than one day before transplanting or seven days before seeding. See label for rate for crop oil or nonionic surfactant. Coverage is essential for good weed control. See label for more information.
Annual and perennial grass and broadleaf weeds, stale bed application	glyphosate, MOA 9 (numerous brands and formulations)	See labels	See labels	Apply to emerged weeds before crop emergence or before transplanting. Perennial weeds may require higher rates of glyphosate. Consult the manufacturer's label for rates for specific weeds. When applying Roundup before transplanting crops into plastic mulch, care must be taken to remove residues of this product from the plastic prior to transplanting. To prevent crop injury, residues can be removed by 0.5 inch natural rainfall or by applying water via a sprinkler system. Certain glyphosate formulations may require the addition of a surfactant. Adding nonionic surfactant to glyphosate formulated with nonionic surfactant may result in reduced weed control.
Emerged broadleaf and grass weeds	pelargonic, (Scythe) 4.2 EC	3 to 10% v/v		Also labeled for collards, kale, mustard/turnip greens. Apply as a preplant burndown or prior to emergence of plants from seed. There is no residual activity. May be tank mixed with soil residual compounds. See label for instruction. May also be used as a banded spray between row middles. Use a shielded sprayer directed to the row middles to reduce drift to the crop.
Annual grasses and small-seeded broadleaf weeds	bensulide, MOA 8 (Prefar) 4 EC	5 to 6 qt	5 to 6	Also labeled for Chinese broccoli, broccoli raab, Chinese cabbage (bok choy, Napa), Chinese mustard cabbage (gai choy), and kohlrabi. Apply preplant or preemergence after planting. With preemergence application, irrigate immediately after application. See label for more directions.
	trifluralin, MOA 3 (Treflan HFP, Trifluralin, Trifluralin HF) 4 EC	1 to 1.5 pt	0.5 to 0.75	Also labeled for brussels sprouts. Apply and incorporate prior to transplanting. **Caution:** If soil conditions are cool and wet, reduced stands and stunting may occur. Direct seeded cole crops exhibit marginal tolerance to higher than recommended rates.
	DCPA, MOA 3 (Dacthal) W-75 (Dacthal) 6 F	8 to 10 lb\n8 to 10 pt	6 to 7.5	Also labeled for brussels sprouts and all other Brassica (cole) leafy vegetables in this crop group. Apply immediately after seeding or transplanting. May also be incorporated.
Annual grasses and broadleaf weeds; weak on pigweed	clomazone, MOA 13 (Command) 3ME	0.67 pt\n\n0.67 to 1.3 pt	0.25\n\n0.25 to 0.50	**Direct seeded cabbage only.** Apply to the soil surface immediately after seeding. Offers weak control of pigweed. See label for further instructions. Limited research has been done on this product in this crop in North Carolina.\n\n**Transplanted cabbage only.** Apply broadcast to the soil prior to transplanting cabbage. See label for further instructions. Limited research has been conducted with this product on this crop in North Carolina.
Hairy galinsoga, common lambsquarters, redroot pigweed, and Palmer amaranth	sulfentrazone, MOA 14 (Spartan) 4 F	2.25 to 4.5 oz	0.07 to 0.14	**Cabbage only (Transplanted Processing only).** May be applied 60 days prior to planting up to planting time. Application rate depend on soil type.

Table 7-16. Chemical Weed Control in Vegetable Crops

Weed	Herbicide, Mode of Action Code* and Formulation	Amount of Formulation Per Acre	Pounds Active Ingredient Per Acre	Precautions and Remarks
Cole Crops: Broccoli, Cabbage, Cauliflower — Preplant and Preemergence (continued)				
Annual grasses and small-seeded broadleaf weeds, including galinsoga, common ragweed, and smartweed	napropamide, MOA 15 (Devrinol DF) 50 DF (Devrinol DF-XT) 50 DF (Devrinol 2-XT) 2 EC	4 lb 4 lb 4 qt	2 2 2	Includes brussels sprouts. Apply to weed-free soil just after seeding or transplanting as a surface application. Light cultivations, rainfall, or irrigation will be necessary within 24 hours to activate this chemical.
Many broadleaf weeds, including galinsoga, common ragweed, and smartweed	oxyfluorfen, MOA 14 (Goal 2 XL, Galigan) 2 EC (GoalTender) 4 E	1 to 2 pt 0.5 to 1 pt	0.25 to 0.5	**Transplants only.** Surface apply before transplanting. Do not incorporate or knock the bed off after application. *Do not spray over the top of transplants.* Oxyfluorfen is weak on grasses. Expect to see some temporary crop injury.
Cole Crops: Broccoli, Cabbage, Cauliflower — Postemergence				
Most broadleaf weeds less than 4 inches tall or rosettes less than 3 inches in diameter; does not control grasses	carfentrazone-ethyl, MOA 14 (Aim) 1.9 EW or 2 EC	up to 2 oz	up to 0.031	Apply post-directed using hooded sprayers for control of emerged weeds. If crop is contacted, burning of contacted area will occur. Use crop oil concentrate at up to 1 gallon per 100 gallons solution or a nonionic surfactant at 2 pints per 100 gallons of spray solution. Coverage is essential for good weed control. Can be tank mixed with other registered herbicides.
Broadleaf weeds including sowthistle, clover, cocklebur, jimsonweed, and ragweed	clopyralid, MOA 4 (Stinger) 3 EC	0.25 to 0.5 pt	0.09 to 0.187	Labeled for broccoli, cabbage, cauliflower, broccoli raab, brussels sprouts, cavalo broccolo, Chinese cabbage (bok choy), Chinese broccoli, Chinese mustard, and Chinese cabbage (Napa). Apply to crop when weeds are small and actively growing. Will control most legumes. Do not apply within 30 days of harvest.
Most emerged weeds	glyphosate, MOA 9 (numerous brands and formulations)	See labels	See labels	**Row middles only.** Apply as a hooded spray in row middles, as shielded spray in row middles, as wiper applications in row middles, or postharvest. To avoid severe injury to crop, do not allow herbicide to contact foliage, green shoots, stems, exposed, roots, or fruit of crop. Do not apply within 14 days of harvest.
Annual and perennial grasses only	clethodim, MOA 1 (Arrow, Clethodim, Intensity, Select) 2 EC (Select Max) 1 EC	6 to 8 oz 9 to 16 oz	0.094 to 0.125 0.07 to 0.125	Apply to emerged grasses. Consult manufacturer's label for specific rates and best times to treat. For sethoxydim, add 1 quart of crop oil concentrate per acre. For Arrow, Clethodim, or Select, add crop oil concentrate at 1 gallon per 100 gallons of spray solution. For Select Max, add 2 pints nonionic surfactant per 100 gallons of spray mixture. Adding crop oil to Poast or Select may increase the likelihood of crop injury at high air temperature. Do not apply Poast or Select plus crop oil on days that are unusually hot and humid. Do not apply within 30 days of harvest.
	sethoxydim, MOA 1 (Poast) 1.5 EC	1 to 1.5 pt	0.2 to 0.3	
Corn (sweet), Preplant Burndown				
Most broadleaf weeds less than 4 inches tall or rosettes less than 3 inches in diameter; does not control grasses	carfentrazone-ethyl, MOA 14 (Aim) 1.9 EW or 2 EC	up to 2 oz	up to 0.031	Apply prior to planting or within 24 hours after planting. Use a crop oil concentrate or a nonionic surfactant with Aim. For optimum performance, make applications to actively growing weeds up to 4 inches high or rosettes less than 3 inches across. Coverage is essential for good weed control. Optimum braod spectrum control of annual and perennial weeds requires a tank mix with burndown herbicides such as glyphosate, paraquat or 2,4-D. Must be applied prior to the preharvest interval of 14 leaf collars. See label for directions.
Contact kill of all green foliage, stale bed and minimum tillage application	paraquat, MOA 22 (Firestorm, Parazone) 3 SL (Gramoxone SL) 2 SL	1.5 to 2.7 pt 2.4 to 4 pt	0.6 to 1	Apply in a minimum of 20 gallons spray mix per acre to emerged weeds before crop emergence as a broadcast or band treatment over a preformed row. Seedbeds should be formed several days ahead of planting and treating to allow maximum weed emergence. Plant with a minimum of soil movement for best results. Use a nonionic surfactant at a rate of 16 to 32 ounces per 100 gallons spray mix or 1 gallon approved crop oil concentrate per 100 gallons spray mix. May be tank mixed with preemergence sweet corn herbicides and herbicide combinations. See section on Corn (sweet), Preemergence. Check label for directions and specific rates.
Annual and perennial grass and broadleaf weeds, stale bed application	glyphosate, MOA 9 (numerous brands and formulations)	See labels	See labels	Apply to emerged weeds before crop emergence. Do not feed crop residue to livestock for 8 weeks following treatment. Perennial weeds may require higher rates of glyphosate. Consult manufacturer's label for rates for specific weeds. Check label for directions. Certain glyphosate formulations require addition of surfactant. Adding nonionic surfactant to glyphosate formulated with nonionic surfactant may result in reduced weed control. Glyphosate-resistant horseweed (marestail) is now common in eastern North Carolina counties. If horseweed is present at planting time, a tank mixture of paraquat and atrazine is suggested.
Broadleaf weeds	2,4-D amine 4, MOA 4 (various brands)	1 to 2 pt	0.5 to 1	May be tank mixed with glyphosate for broad spectrum weed control including glyphosate-resistant horseweed (marestail). See label for planting restrictions if applied prior to planting.
Corn (sweet), Preemergence				
Most annual grass weeds, including fall panicum, broadleaf signalgrass, and small-seeded broadleaf weeds	alachlor, MOA 15 (Micro-Tech) 4 FME	2 to 4 qt	2 to 4	Apply to soil surface immediately after planting. Higher rates will improve control of ragweed and lambsquarter. May be tank mixed with atrazine, glyphosate, or simazine. Various other brands are available. Check label for directions.
	dimethenamid, MOA 15 (Outlook) 6.0 EC	12 to 21 oz	0.56 to 1.0	Apply to soil surface immediately after planting. May be tank mixed with atrazine, glyphosate, or paraquat. See label for other herbicides that may be tank mixed to broaden weed control spectrum.
	metolachlor, MOA 15 (Me-Too-Lachlo rII) 7.8 EC (Parallel) 7.8 EC	1 to 2 pt	0.98 to 1.98	See comments for *s*-metolachlor products. Products containing s-metolachlor are more active on weeds per unit of formulated product than those containing metolachlor. See label for all instructions.
	S-metolachlor, MOA 15 (Brawl II, Dual II Magnum, Medal II) 7.64 EC	1 to 2 pt	0.95 to 1.91	Apply to soil surface immediately after planting. May be tank mixed with atrazine, glyphosate, or simazine. Check label for directions. Rate is soil-texture and organic-matter dependent. See label for details.
	pyroxasulfone, MOA 15 (Zidua) 85 WG	1.5 to 4 oz	0.0796 to 0.213	Rate ranges based on soil texture. See label for specific rate relating to your fields. Sweet corn seed must be planted a minimum of 1-inch deep. Provides suppression of Texas panicum, seedlinhg johnsongrass, and shattercane. Do not harvest sweet corn ears for human consumption less than 37 days after application of this herbicide. See label regarding tank mixtures for broader spectrum control and/or control of emerged weeds. Research with Zidua in sweet corn is limited in North Carolina.

Table 7-16. Chemical Weed Control in Vegetable Crops

Weed	Herbicide, Mode of Action Code* and Formulation	Amount of Formulation Per Acre	Pounds Active Ingredient Per Acre	Precautions and Remarks
Corn (sweet), Preemergence (continued)				
Most annual broadleaf and grass weeds	atrazine, MOA 5 (various brands) 4 F (various brands) 90 WDG	1 to 2 qt 1.1 to 2.2 lb	1 to 2	Apply to soil surface immediately after planting. Shallow cultivations will improve control. Check label for restrictions on rotational crops. See label for reduced rate if soil coverage with plant residue is less than 30% at planting. Does not control fall panicum or smooth crabgrass. May be tank mixed with metolachlor, alachlor, glyphosate, paraquat, bentazon, or simazine. Check label for directions.
	alachlor, MOA 15 + atrazine, MOA 5 (Bullet or Lariat) 4 F	2.5 to 4.5 qt	1.56 to 2.81 + 0.94 to 1.69	Apply to soil surface immediately after planting. Soil texture and organic matter influence application rate. See label for further instruction.
	dimethenamid, MOA 15 + atrazine, MOA 5 (Guardsman Max) 5 F	2.4 to 4.6 pt	0.51 to 1 + 1 to 1.9	Apply to soil surface immediately after planting. Does not control Texas panicum, seedling johnsongrass, or shattercane adequately. Adjust rate for soil texture and organic matter according to label. See label for reduced rate if soil coverage with plant residue is less than 30% at planting. See labels for comments on rotational crops. See label for additional instructionjs.
	S-metolachlor, MOA 15 + atrazine, MOA 5 (Bicep II Magnum) 5.5 F	1.3 to 2.6 qt	0.78 to 1.56 + 1 to 2	Apply to soil surface immediately after planting. Does not adequately control Texas panicum, seedling johnsongrass, or shattercane. May not adequately control cocklebur, morningglory, or sicklepod. Cultivation or other herbicides may be needed. See label for rates based on soil texture and organic matter and for information on setback requirements from streams and lakes. See label for reduced rate if soil coverage with plant residue is less than 30% at planting and for comments on rotational crops.
Grass and broadleaf weeds	pendimethalin, MOA 3 (Prowl H2O) 3.8 AS	2 to 4 pt	1 to 2	Apply preemergence before crop germinates, or postemergence until sweet corn is 20 to 24 inches tall or in the V8 growth stage. Do not apply in reduced, minimum or no-till sweet corn. See label for use rates accoding to organic matter. See label for additional information. See label for tank mix options.
Broadleaf and grass weeds	Simazine, MOA 5 (Princep) 4L	2 qt	2	Apply preemergence before weeds and crop emerge. See label for tank mix options. PHI is 45 days.
Corn (sweet), Postemergence				
Most annual broadleaf and grass weeds	atrazine, MOA 5 (various brands) 4 L (various brands) 90 WDG	2 qt 2.2 lb	2	Apply overtop before weeds exceed 1.5 inches in height. See label for additional information in controlling larger weeds. See label for amount of oil concentrate to add to spray mix. See label on setback requirements from streams and lakes.
Annual grasses and broadleaf weeds	dimethenamid, MOA 15 (Outlook) 6.0 EC + atrazine, MOA 5 (AAtrex) 4 F or 90 WDG	8 to 21 oz + See label for rate	0.375 to 1 + See label for rate	Apply overtop corn before crop reaches 12 inches tall and before weeds exceed the two-leaf stage. Larger weeds will not be controlled. Good residual control of annual grass and broadleaf weeds. Do not apply within 50 days of sweet corn ear harvest. Do not apply to corn 12 inches or taller. Also available as the commercial products Guardsman or LeadOff.
	S-metolachlor, MOA 15 (Dual II Magnum) 7.64 EC + atrazine, MOA 5 (AAtrex) 4 F (AAtrex) 90 WDG	1 to 1.67 pt + 1 to 2 qt 1.3 to 2.2 lb	0.95 to 1.58 + 1 to 2	Apply overtop corn (5 inches or less) before weeds exceed the two-leaf stage. Larger weeds will not be controlled. Do not apply within 30 days of sweet corn ear harvest. Good residual control of annual grass and broadleaf weeds. Also available as Bicep II or Bicep II Magnum.
Cocklebur, common ragweed, jimsonweed, Pennsylvania smartweed, velvetleaf, yellow nutsedge, and morningglory	bentazon, MOA 6 (Basagran) 4 SL	1 to 2 pt	0.5 to 1	Apply early postemergence overtop when weeds are small and corn has one to five leaves. See label for rates according to weed size and special directions for annual morningglory and yellow nutsedge control. Use a crop oil at a rate of 1 quart per acre.
Many broadleaf weeds	mesotrione, MOA 27 (Callisto) 4 EC	3 oz	0.094	Apply overtop corn 30 inches or less or 8 leaves or less to control emerged broadleaf weeds. Use nonionic surfactant at 2 pints per 100 gallons of spray solution. DO NOT add VAN or AMS adjuvants when making post application in sweetcorn or severe injury will occur. Most effective on small weeds, however, if weeds are greater than 5 inches or for improved control of certain weeds, certain atrazine formulations may be mixed with this herbicide. See label for further information. Do not apply within 45 days of harvest.
Annual broadleaf weeds and some grasses	tembotrione, MOA 27 (Laudis) 3.5 L	3 fl oz	0.082	Can be applied overtop or with drop nozzles to sweet corn from emergence up to V7 stage. Controls most broadleaf weeds.Does not control sicklepod or prickly sida and only suppresses morningglory.Controls or suppresses some grasses. See label for weeds controlled and recommended size for treatment. Herbicide sensitivity in all hybrids and inbreds of sweet corn has not been tested. See label for information on adjuvant use. May be tankmixed with atrazine to increase weed spectrum and consistency of control. If tankmixed with atrazine do not apply if corn is 12 inches tall or greater. See label for further restrictions and instructions.
	Topramezone, MOA 27 (Impact) 2.8 L	0.75 fl oz	0.016	Can be applied overtop or with drop nozzles to sweet corn from emergence until 45 days prior to harvest. Does not control sicklepod and only suppresses morningglory. Controls or suppresses some grasses. See label for weeds controlled and recommended size for treatment. This product has not been tested on all inbred line for tolerance. See label for information on adjuvant use. See label for further restrictions and instructions. Do not apply within 45 days of sweet corn ear harvest.
Velvetleaf, spreading dayflower, morningglory species, and redroot pigweed. Will not control grasses	fluthiacet-methyl, MOA 14 (Cadet) 0.91 L	0.6 to 0.9 oz	0.0042 to 0.06	**Processing sweet corn only.** Apply to small weeds, generally about 2 inches tall. Will control large velvetleaf up to 36 inches. See label for information on adjuvant use. See label for further restrictions and instructions.
Annual broadleaf weeds	fluthiacet-methyl, MOA 14 + mesotrione, MOA 27 (Solstice)	2.5 to 3.15 fl oz	0.004 to 0.0053 0.074 to 0.0931	Apply up to the V8 growth stage (or 30 inches tall). See label for crop rotation restrictions. Do not include nitrogen based adjuvants (UAN or AMS) when making postemergence application or severe injury will occur. Use nonionic surfactant at 1 qt per 100 gallons of spray. Do not apply within 40 days of sweet corn ear harvest. See label for further instructions.

CHEMICAL WEED CONTROL

Table 7-16. Chemical Weed Control in Vegetable Crops

Weed	Herbicide, Mode of Action Code* and Formulation	Amount of Formulation Per Acre	Pounds Active Ingredient Per Acre	Precautions and Remarks
Corn (sweet), Postemergence (continued)				
Velvetleaf, pigweed, nightshade, morningglory, common lambsquarters	carfentrazone-ethyl, MOA 14 (Aim) 2.0 EC	0.5 oz	0.008	Apply postemergence to actively growing weeds less than 4 inches high (rosettes less than 3 inches across) up to the 14-leaf collar stage of corn. Rates above 0.5 oz will aid in controlling larger weeds and certain weeds (see label for specific rate). Directed sprays will lessen the chance of crop injury and allow later application. Coverage of weeds is essential for control. Use nonionic surfactant (2 pints per 100 gallons of spray) with all applications. Under dry conditions, the use of crop oil concentrate may improve weed control. Mix with atrazine to improve control of many broadleaf weeds. Limited information is available concerning the use of this product in sweet corn. Do not apply more than 2 ounces per acre per season. Do not apply within 3 days of sweet corn ear harvest.
Broadleaf weeds including sowthistle, clover, cocklebur, jimsonweed, ragweed, Jerusalem artichoke, and thistle	clopyralid, MOA 4 (Stinger) 3 EC	0.25 to 0.67 pt	0.095 to 0.25	**Processing sweet corn only.** Apply to sweet corn when weeds are small (less than 5-leaf stage) and actively growing. Addition of surfactants, crop oils, or other adjuvants is not usually necessary when using Stinger. Use of adjuvants may reduce selectivity to the crop. Do not apply to sweet corn over 18 inches tall. Will control most legumes. Do not apply within 30 days of harvest.
Cocklebur, passionflower (maypop), pigweed, pokeweed, ragweed, smartweed (Pennsylvania), velvetleaf	halosulfuron-methyl, MOA 2 (Profine 75, Sandea) 75 WDG	0.67 oz	0.032	Apply over the top or with drop nozzles to sweet corn from spike to lay-by for control of emerged weeds. Add nonionic surfactant at 1 to 2 quarts per 100 gallons of spray solution. See label for all instructions and restrictions. Do not apply within 30 days of harvest.
Cocklebur, pigweed, lambsquarters, morningglory, sicklepod, and many other annual broadleaf weeds	2,4-D amine, MOA 4 (various brands) 3.8 SL	0.5 to 1 pt	0.24 to 0.48	Use 0.5 pt of 2,4-D overtop when corn is 4 to 5 inches tall and weeds are small. Increase rate to 1 pt as corn reaches 8 inches. Use drop nozzles and direct spray toward base if corn is over 8 inches tall. Do not cultivate for about 10 days after spraying as corn may be brittle. Reduce rate of 2,4-D if extremely hot and soil is wet. For better sicklepod and horsenettle control, add a nonionic surfactant when using a directed spray at a rate of 1 quart per 100 gallons spray solution. Do not apply within 45 days of sweet corn harvest.
Annual grasses and broadleaf weeds	paraquat, MOA 22 (Firestorm, Parazone) 3 SL (Gramoxone SL) 2 SL	0.7 to 1.3 pt 1 to 2 pt	0.25 to 0.5	DO NOT SPRAY OVERTOP OF CORN OR SEVERE INJURY WILL OCCUR. Make a postdirected application in a minimum of 20 gallons spray mix per acre to emerged weeds when the smallest corn is **at least 10 inches tall.** Use nonionic surfactant at a rate of 16 to 32 ounces per 100 gallons spray mix or 1 gallon approved crop oil concentrate per 100 gallons spray mix. Use of a hooded or shielded sprayer will reduce crop injury.
Certain grasses, including barnyardgrass, foxtails, Texas panicum, and johnsongrass; and broadleaf weeds, including burcucumber, jimsonweed, pigweed, pokeweed, and smartweeds	nicosulfuron, MOA 2 (Accent) 75 WDG	0.67 oz	0.031	Apply to sweet corn up to 12 inches tall or up to and including 5 leaf collars. For corn 12 to 18 inches tall, apply only with drop nozzles. Sweet corn hybrids vary in their sensitivity to Accent. Do not apply to Merit sweet corn variety. Contact company representative for information on other local hybrids that have been evaluated with Accent. Accent may be applied to corn previously treated with Fortress, Aztec, or Force, or non-organophosphate soil insecticides regardless of soil type. See label for more information on use of soil insecticides with Accent. Label prohibits application of Accent to corn previously treated with Counter insecticide, and also indicates that applying Accent to corn previously treated with Counter 20 CR, Lorsban, or Thimet may result in unacceptable crop injury, especially on soils with less than 4% organic matter. See label for information on use of adjuvants.
Cucumbers, Preplant and Preemergence				
Suppression or control of most annual grasses and broadleaf weeds, full rate required for nutsedge control	metam sodium (Vapam HL) 42%	37.5 to 75 gal	15.7 to 31.5	Rates are dependent on soil type and weeds present. Apply when soil moisture is at field capacity (100 to 125%). Apply through soil injection using a rotary tiller or inject with knives no more than 4 inches apart; follow immediately with a roller to smooth and compact the soil surface or with mulch. May apply through drip irrigation prior to planting a second crop on mulch. Plant back interval is often 14 to 21 days and can be 30 days in some environments. See label for all restrictions and additional information.
Emerged broadleaf and grass weeds	pelargonic, (Scythe) 4.2 EC	3 to 10% v/v		Apply before crop emergence and control emerged weeds. There is no residual activity. May be tank mixed with soil residual compounds. See label for further instructions. May also be used as a banded spray between row middles. Use a shielded sprayer directed to the row middles to reduce drift to the crop.
Most broadleaf weeds less than 4 inches tall or rosettes less than 3 inches in diameter, does not control grasses.	carfentrazone-ethyl, MOA 14 (Aim) 1.9 EW or 2 EC	up to 2 oz	up to 0.031	**Aim 1.9 EW is registered for application in transplant production systems only. Aim 2 EC is registered in seeded and transplant production systems.** Apply no later than one day before transplanting or no later than 7 days before seeding crop. See label for information about application timing. Use a crop oil at up to 1 gallon per 100 gallons of spray solution or a nonionic surfactant at 2 pints per 100 gallons of spray solution. Coverage is essential for good weed control. Can be tank mixed with other registered burndown herbicides.
Contact kill of all green foliage, stale bed application	paraquat, MOA 22 (Firestorm, Parazone) 3 SL (Gramoxone SL) 2 SL	1.3 to 2.7 pt 2 to 4 pt	0.5 to 1	Apply in a minimum of 10 gallons spray mix per acre to emerged weeds before crop emergence as a broadcast or band treatment over a preformed row. Use sufficient water to give thorough coverage. Row should be formed several days ahead of planting and treating to allow maximum weed emergence. Use a nonionic surfactant at a rate of 16 to 32 ounces per 100 gallons spray mix or 1 gallon approved crop oil concentrate per 100 gallons spray mix.
Annual and perennial grass and broadleaf weeds, stale bed application	glyphosate, MOA 9 (numerous brands and formulations)	See labels	See labels	Apply to emerged weeds at least 3 days before seeding or transplanting. Perennial weeds may require higher rates of glyphosate. Consult the manufacturer's label for rates for specific weeds. When applying Roundup before transplanting crops into plastic mulch, care must be taken to remove residues of this product from the plastic prior to transplanting. To prevent crop injury, residues can be removed by 0.5 inch natural rainfall or by applying water via a sprinkler system. Certain glyphosate formulations require the addition of surfactant. Adding nonionic surfactant to glyphosate formulated with nonionic surfactant may result in reduced weed control.
Annual grasses and small-seeded broadleaf weeds	bensulide, MOA 8 (Prefar) 4 EC	5 to 6 qt	5 to 6	Registered for cucurbit vegetable group (Crop grouping 9). Apply preplant and incorporate into the soil 1 to 2 inches (1 inch incorporation is optimum) with a rototiller or tandem disk, or apply to the soil surface after seeding and follow with irrigation within 36 hours of application. Check replant restrictions for small grains on label.
Annual grasses and some small-seeded broadleaf weeds; weak on pigweed	clomazone, MOA 13 (Command) 3 ME	0.4 to 1 pt	0.15 to 0.375	Apply immediately after seeding. See label for further information.

Table 7-16. Chemical Weed Control in Vegetable Crops

Weed	Herbicide, Mode of Action Code* and Formulation	Amount of Formulation Per Acre	Pounds Active Ingredient Per Acre	Precautions and Remarks
Cucumbers, Preplant and Preemergence (continued)				
Annual grasses and some small-seeded broadleaf weeds	ethalfluralin, MOA 3 (Curbit) 3 EC	3 to 4.5 pt	1.1 to 1.7	Apply postplant to seeded crop prior to crop emergence or as a banded spray between rows after crop emergence or transplanting. See label for timing. Shallow cultivation, irrigation, or rainfall within 5 days is needed for good weed control. Do not use under mulches, row covers, or hot caps. Under conditions of unusually cold or wet soil and air temperatures, crop stunting or injury may occur. Crop injury can occur if seeding depth is too shallow.
Annual grasses and broadleaf weeds	ethalfluralin, MOA 3 + clomazone, MOA 13 (Strategy) 2.1 L	2 to 6 pt	0.4 to 1.2 + 0.125 to 0.375	Apply to the soil surface immediately after crop seeding for preemergence control of weeds. **Do not apply prior to planting crop. Do not soil incorporate.** May also be used as a **banded** treatment **between** rows after crop emergence or transplanting. Do not apply over or under plastic mulch.
Yellow and purple nutsedge and broadleaf weeds	halosulfuron-methyl, MOA 2 (Profine 75) 75 DG (Sandea) 75 DG	0.5 to 0.75 oz	0.024 to 0.036	Apply after seeding but before soil cracking or prior to transplanting crop. For transplanting, do not transplant until 7 days after application. For seeded or transplanting cucumbers in plasticulture, do not plant within 7 days of Sandea application. Rate can be increased to 1 ounce of product per acre to middles between rows.
Cucumbers, Postemergence				
Annual grasses and small-seeded broadleaf weeds	trifluralin, MOA 3 (Treflan HFP, Trifluraline, Trifluralin HF) 4EC	1 to 2 pt	0.5 to 0.75	**Will not control emerged weeds. Row middles only.** To improve preemergence control of late emerging weeds. **Apply as a directed spray to soil between rows after crop emergence when crop plants have reached** three to four true leaf stage of growth. Avoid contacting crop foliage as slight crop injury may occur. Set incorporation equipment to move treated soil around base of crop plants. Do not apply within 30 days of harvest.
Yellow and purple nutsedge and broadleaf weeds, including cocklebur, galinsoga, smartweed, ragweed, wild radish and pigweed	halosulfuron-methyl, MOA 2 (Profine 75, Sandea) 75 DG	0.5 to 0.75 oz	0.024 to 0.036	Apply postemergence only after the crop has reached 3 to 5 true leaves but before first female flowers appear. Do not apply sooner than 14 days after transplanting. Use nonionic surfactant at 1 quart per 100 gallons of spray solution with all postemergnce applications. Do not apply within 14 days of harvest.
Most broadleaf weeds less than 4 inches tall or rosettes less than 3 inches in diameter; does not control grasses	carfentrazone-ethyl, MOA 14 (Aim) 1.9 EW or 2 EC	up to 2 oz	up to 0.031	Apply post-directed using hooded sprayers for control of emerged weeds. If crop is contacted, burning of contacted area will occur. Use crop oil concentrate at up to 1 gallon per 100 gallons solution or a nonionic surfactant at 2 pints per 100 gallons of spray solution. Coverage is essential for good weed control. Can be tank mixed with other registered herbicides.
Most emerged weeds	glyphosate, MOA 9 (numerous brands and formulations)	See labels	See labels	**Row middles only.** Apply as a hooded spray in row middles, as shielded spray in row middles, as wiper applications in row middles, or postharvest. To avoid severe injury to crop, do not allow herbicide to contact foliage, green shoots, stems, exposed, roots, or fruit of crop. Do not apply within 14 days of harvest.
Annual and perennial grasses only	sethoxydim, MOA 1 (Poast) 1.5 EC	1 to 1.5 pt	0.2 to 0.3	Apply to emerged grasses. Consult manufacturer's label for specific rates and best times to treat. Add 1 quart of crop oil concentrate per acre. Adding crop oil to Poast may increase the likelihood of crop injury at high air temperatures. Do not apply Poast on days that are unusually hot and humid. Do not apply within 14 days of harvest.
	clethodim, MOA 1 (Arrow, Clethodim, Intensity, Select) 2 EC	6 to 8 oz	0.094 to 0.125	Control of emerged grasses. For Arrow, Clethodim, and Select, add 1 gallon crop oil concentrate per 100 gallons spray mix. For Select Max and Intensity One, add 2 pints nonionic surfactant per 100 gallons spray mixture. Adding crop oil may increase the likelihood of crop injury at high air temperatures. Very effective in controlling annual bluegrass. Apply to actively growing grasses not under drought stress. Do not apply within 14 days of harvest.
	(Select Max, Intensity One) 1 EC	9 to 16 oz	0.07 to 0.125	
Eggplant, Preplant				
Suppression or control of most annual grasses and broadleaf weeds, full rate required for nutsedge control	metam sodium (Vapam HL) 42%	37.5 to 75 gal	15.7 to 31.5	Rates are dependent on soil type and weeds present. Apply when soil moisture is at field capacity (100 to 125%). Apply through soil injection using a rotary tiller or inject with knives no more than 4 inches apart; follow immediately with a roller to smooth and compact the soil surface or with mulch. May apply through drip irrigation prior to planting a second crop on mulch. Plant back interval is often 14 to 21 days and can be 30 days in some environments. See label for all restrictions and additional information. Chloropicrin (150 pounds per acre broadcast) will also be needed when laying first crop mulch to control nutsedge.
Contact kill of all green foliage, stale bed application	paraquat, MOA 22 (Firestorm) 3 SL (Parazone) 3 SL (Gramoxone SL) 2 SL	1.3 to 2.7 pt 2 to 4 pt	0.5 to 1	Apply in a minimum of 10 gallons spray mix per acre to emerged weeds before transplanting as a broadcast or band treatment over a preformed row. Use sufficient water to give thorough coverage. Row should be formed several days ahead of planting and treating to allow maximum weed emergence. Use a nonionic surfactant at a rate of 16 to 32 ounces per 100 gallons spray mix or 1 gallon approved crop oil concentrate per 100 gallons spray mix.
Most broadleaf weeds less than 4 inches tall or rosettes less than 3 inches in diameter; does not control grasses	carfentrazone-ethyl, MOA 14 (Aim) 1.9 EW or 2 EC	up to 2 oz	up to 0.031	**Aim 1.9 EW is registered for application in transplant production systems only. Aim 2 EC is registered in seeded and transplant production systems.** Apply no later than one day before transplanting crop (Aim 1.9 EW or Aim 2 EC) or no later than 7 days before seeding crop (Aim 2 EC only). See label for information about application timing. Use a crop oil at up to 1 gallon per 100 gallons of spray solution or a nonionic surfactant at 2 pints per 100 gallons of spray solution. Coverage is essential for good weed control. Can be tank mixed with other registered burndown herbicides.
Annual and perennial grass and broadleaf weeds, stale bed application	glyphosate, MOA 9 (numerous brands and formulations	See labels	See labels	Apply to emerged weeds at least 3 days before seeding or transplanting. Perennial weeds may require higher rates of glyphosate. Consult the manufacturer's label for rates for specific weeds. When applying Roundup before transplanting crops into plastic mulch, care must be taken to remove residues of this product from the plastic prior to transplanting. To prevent crop injury, residues can be removed by 0.5 inch natural rainfall or by applying water via a sprinkler system. Certain glyphosate formulations require the addition of surfactant. Adding nonionic surfactant to glyphosate formulated with nonionic surfactant may result in reduced weed control.

Table 7-16. Chemical Weed Control in Vegetable Crops

Weed	Herbicide, Mode of Action Code* and Formulation	Amount of Formulation Per Acre	Pounds Active Ingredient Per Acre	Precautions and Remarks
Eggplant, Preemergence				
Annual grasses and small-seeded broadleaf weeds	bensulide, MOA 8 (Prefar) 4 EC	5 to 6 qt	5 to 6	Apply preplant incorporated (1 inch incorporation is optimum) or preemergence after planting. With preemergence application, irrigate immediately after application. See label for more directions.
Annual grasses and some broadleaf weeds	napropamide, MOA 15 (Devrinol, Devrinol DF-XT) 50 DF (Devrinol 2-XT) 2 EC	2 to 4 lb 2 to 4 qt	1 to 2	**Transplanted eggplant only.** Apply preplant and incorporate into soil 1 to 2 inches using a rototiller or tandem disk. Shallow cultivations or irrigation will improve control. See label for small grains replanting restrictions. May also be applied in the **row middles** between plastic covered beds. See label for more information. See XT labels for information regarding delay in irrigation event.
Eggplant, Postemergence				
Annual grasses and small-seeded broadleaf weeds	DCPA, MOA 3 (Dacthal) W-75 (Dacthal) 6 F	6 to 10 lb 6 to 10 pt	4.5 to 7.5	Application confined to a period of 4 to 6 weeks after transplanting. To improve preemergence control of late emerging weeds. Apply to weed-free soil over the top of transplants.
Annual and perennial grasses only	sethoxydim, MOA 1 (Poast) 1.5 EC	1 to 1.5 pt	0.2 to 0.3	Apply to emerged grasses. Consult manufacturer's label for specific rates and best times to treat. Add 1 quart of crop oil concentrate per acre. Adding crop oil to Poast may increase the likelihood of crop injury at high air temperatures. Do not apply Poast on days that are unusually hot and humid. Do not apply within 20 days of harvest.
	clethodim, MOA 1 (Arrow, Clethodim, Intensity, Select) 2 EC	6 to 8 oz	0.094 to 0.125	Apply postemergence for for control of grasses. With Arrow, Clethodim, or Select, add 1 gallon crop oil concentrate per 100 gallons spray mix. With Select Max, add 2 pints of nonionic surfactant per 100 gallons spray mixture. Adding crop oil may increase the likelihood of crop injury at high air temperature. Very effective in controlling annual bluegrass. Apply to actively growing grasses not under drought stress. Do not apply within 20 days of harvest.
	(Select Max, Intensity One) 1 EC	9 to 16 oz	0.07 to 0.125	
Eggplant, Row Middles				
Most broadleaf weeds less than 4 inch tall or rosettes less than 3 inches in diameter; does not control grasses	carfentrazone-ethyl, MOA 14 (Aim) 1.9 EW or 2 EC	up to 2 oz	up to 0.031	Apply post-directed using hooded sprayers for control of emerged weeds. If crop is contacted, burning of contacted area will occur. Use crop oil concentrate at up to 1 gallon per 100 gallons solution or a nonionic surfactant at 2 pints per 100 gallons of spray solution. Coverage is essential for good weed control. Can be tank mixed with other registered herbicides. PHI 0 days.
Most emerged weeds	glyphosate, MOA 9 ((numerous brands and formulations)	See labels	See labels	Apply as a hooded spray in row middles, as shielded spray in row middles, as wiper applications in row middles, or postharvest. To avoid severe injury to crop, do not allow herbicide to contact foliage, green shoots or stems, exposed roots, or fruit of crop. Do not apply within 14 days of harvest.
Yellow and purple nutsedge and broadleaf weeds	halosulfuron-methyl, MOA 2 (Profine 75) 75 DG (Sandea) 75 DG	0.5 to 1 oz	0.024 to 0.048	Apply between rows as a postemergence spray. Do not allow spray to contact crop or plastic mulch. Early season application will give postemergence and preemergence control. For postemergence applications, use nonionic surfactant at 1 quart per 100 gallons of spray solution. Preharvest interval is 30 days.
Contact kill of all green foliage	paraquat, MOA 22 (Firestorm, Parazone) 3SL	1.3 pt	0.5	Apply in 10 gallons spray mix as a shielded spray to emerged weeds between rows of eggplant. Use a nonionic surfactant at a rate of 16 to 32 ounces per 100 gallons spray mix or 1 gallon approved crop oil concentrate per 100 gallons spray mix. Do not allow spray to contact crop or injury will result.
	(Gramoxone SL) 2 SL	2 pt		
Garlic, Preplant or Preemergence				
Annual and perennial grass and broadleaf weeds	glyphosate, MOA 9 (numerous brands and formulations)	See labels	See labels	Stale bed application. Apply to emerged weeds at least 3 days before planting. Perennial weeds may require higher rates of glyphosate. Consult the manufacturer's label for rates for specific weeds. Certain glyphosate formulations require the addition of surfactant. Adding nonionic surfactant to glyphosate formulated with nonionic surfactant may result in reduced weed control.
	paraquat, MOA 22 (Firestorm, Parazone) 3 SL	1.7 to 2.7 pt	0.6 to 1	Apply in a minimum of 20 gallons spray mix per acre to emerged weeds before crop emergence as a broadcast or band treatment over a preformed row. Row should be formed several days ahead of planting and treating to allow maximum weed emergence. Use a nonionic surfactant at a rate of 16 to 32 ounces per 100 gallons spray mix or 1 gallon approved crop oil concentrate per 100 gallons spray mix. Do not apply within 60 days of harvest.
	(Gramoxone SL) 2 SL	2.5 to 4 pt		
Most broadleaf weeds less than 4 inches tall or rosettes less than 3 inches in diameter; does not control grasses	carfentrazone-ethyl, MOA 14 (Aim) 2.0 EC	Up to 2oz	0.031	Apply no later than 30 days before planting. See label for specific Aim rate relating to weed species and proper adjuvant and rate. Coverage is essential for good weed control. Can be tank mixed with other registered burndown herbicides.
Emerged broadleaf weeds	pyraflufen, MOA 14 (ET Herbicide) 0.208 EC	0.5 to 2 fl oz	0.0008 to 0.003	Apply as a preplant burndown treatment in a minimum of 10 gallon solution per acre. See label for information on use of adjuvant.
Emerged broadleaf and grass weeds	pelargonic, (Scythe) 4.2 EC	3 to 10% v/v		Apply as a preplant burndown treatment or use in row middles using shielded sprayer.
Annual grasses and small-seeded broadleaf weeds	bensulide, MOA 8 (Prefar) 4 EC	5 to 6 qt	5 to 6	Apply preplant incorporated (1 inch incorporation is optimum) or preemergence after planting. With preemergence application, irrigate immediately after application. See label for more directions.
Annual broadleaf weeds	oxyfluorfen, MOA 14 (Galigan, Goal 2XL) 2 E	1 to 2 pt	0.25 to 0.5	**Transplanted garlic only.** For use on a fallow bed. Garlic may be planted immediately following application of 1 pt of product. For rates above 1 pt do not plant within 30 days. PHI 60 days.

Table 7-16. Chemical Weed Control in Vegetable Crops

Weed	Herbicide, Mode of Action Code* and Formulation	Amount of Formulation Per Acre	Pounds Active Ingredient Per Acre	Precautions and Remarks
Garlic, Preemergence				
Annual grasses and small-seeded broadleaf weeds	dimethenamid-P, MOA 15 (Outlook) 6 EC	12 to 21 oz	0.6 to 1	For preemergence weed control. Apply after crop has reached 2 true leaves until a minimum of 30 days before harvest. If applications are made to transplanted crop, DO NOT apply until transplants are in the ground and soil has settled around transplants with several days to recover.
	flumioxazin, MOA 14 (Chateau) 51 SW	6 oz	0.188	For preemergence weed control. Apply prior to garlic and weed emergence. Application should be made within 3 days after planting garlic.
	pendimethalin, MOA 3 (Prowl) 3.3 EC (Prowl H₂O) 3.8 AS	1.2 to 3.6 pt 1.5 to 3 pt	0.5 to 1.5 0.75 to 1.5	For preemergence weed control. Apply preemergence after planting but prior to weed and crop emergence, or postemergence to garlic in the 1 to 5 true leaf stage. Prowl can be applied sequentially by applying preemergence followed by a postemergence application. Preharvest interval is 45 days.
Garlic, Postemergence				
Most annual broadleaf weeds	oxyfluorfen, MOA 14 (Galigan) 2 E (Goal 2 XL) 2 EC (GoalTender) 4 E	0.5 pt 0.5 pt 0.25 pt	0.12	**Transplanted dry bulb only.** May be used as a postemergence spray to both the weeds and crop after the garlic has at least two fully developed true leaves. Some injury to garlic may result. Injury will be more severe if the chemical is applied during cool, wet weather. Weeds should be in the 2- to 4-leaf stage for best results. Preharvest interval is 60 days.
Annual and perennial grasses only	clethodim, MOA 1 (Arrow, Clethodim, Intensity, Select) 2 EC (Select Max, Intensity One) 1 EC	6 to 16 oz 9 to 32 oz	0.09 to 0.25 0.07 to 0.25	Apply to emerged grasses. Consult manufacturer's label for specific rates and best times to treat. With Arrow, Clethodim, or Select, add 1 gallon crop oil concentrate per 100 gallons spray mix. With Select Max, add 2 pints of nonionic surfactant per 100 gallons spray mixture. Adding crop oil may increase the likelihood of crop injury at high air temperatures. Do not apply Arrow, Clethodim, or Select on unusually hot and humid days. Do not apply within 45 days of harvest. Very effective in controlling annual bluegrass.
	fluazifop, MOA 1 (Fusilade DX) 2 EC	6 to 16 oz	0.1 to 0.25	Apply to emerged grasses. Consult label for specific rates and best times to treat. Add 1 gallon crop oil concentrate or 1 quart nonionic surfactant per 100 gallons spray mix. Do not apply on unusually hot and humid days. Do not apply within 45 days of harvest.
	sethoxydim, MOA 1 (Poast) 1.5 EC	1 pt	0.2	Apply to emerged grasses. Consult manufacturer's label for specific rates and best times to treat. Add 1 quart of crop oil concentrate per acre. Adding crop oil to Poast may increase the likelihood of crop injury at high air temperatures. Do not apply Poast on days that are unusually hot and humid. Do not apply within 30 days of harvest.
Garlic, Row Middles				
Most emerged weeds	glyphosate, MOA 9 (numerous brands and formulations)	See labels	See labels	**Row middles only.** Apply as a hooded spray in row middles, as shielded spray in row middles, as wiper applications in row middles, or post harvest. To avoid severe injury to crop, do not allow herbicide to contact foliage, green shoots, stems, exposed, roots, or fruit of crop. Do not apply within 14 days of harvest.
Most broadleaf weeds less than 4 inches tall or rosettes less than 3 inches in diameter; does not control grasses	carfentrazone-ethyl, MOA 14 (Aim) 1.9 EW or 2 EC	up to 2 oz	up to 0.031	Apply post-directed using hooded sprayers for control of emerged weeds. If crop is contacted, burning of contacted area will occur. Use a nonionic surfactant or crop oil with Aim. See label for rate. Coverage is essential for good weed control. Can be tank mixed with other registered herbicides.
Greens (Collard, kale, mustard, and turnip greens or roots), Preplant				
Emerged broadleaf and grass weeds	pelargonic, (Scythe) 4.2 EC	3 to 10% v/v		Apply as preplant burndown emerged weeds. See lable for instruction. May also be used as a banded spray between row middles. Use a shielded sprayer directed to the row middles to reduce drift to the crop.
Contact kill of all green foliage, stale bed application	paraquat, MOA 22 (Firestorm, Parazone) 3 SL (Gramoxone SL) 2 SL	1.3 to 2.7 pt 2 to 4 pt	0.5 to 1	**Collard and turnip only.** Apply in a minimum of 10 gallons spray mix per acre to emerged weeds before crop emergence or transplanting as a broadcast or band treatment over a preformed row. Use sufficient water to give thorough coverage. Row should be formed several days ahead of planting and treating to allow maximum weed emergence. Use a nonionic surfactant at a rate of 16 to 32 ounces per 100 gallons spray mix or 1 gallon approved crop oil concentrate per 100 gallons spray mix.
Annual and perennial grass and broadleaf weeds, stale bed application	glyphosate, MOA 9 (numerous brands and formulations)	See labels	See labels	Apply to emerged weeds before crop emergence. Do not feed crop residue to livestock for 8 weeks following treatment. Perennial weeds may require higher rates of glyphosate. Consult the manufacturer's label for rates for specific weeds. Certain glyphosate formulations require the addition of surfactant. Adding nonionic surfactant to glyphosate formulated with nonionic surfactant may result in reduced weed control.
Annual grasses and small-seeded broadleaf weeds	trifluralin, MOA 3 (Treflan) 4 EC	1 to 1.5 pt	0.5 to 0.75	Greens: collard, kale, mustard and turnip (fresh or processing). Apply preplant and incorporate into the soil 2 to 3 inches within 8 hr using a rototiller or tandem disk. Do not use if turnip roots are to be consumed.
	bensulide, MOA 8 (Prefar) 4 EC	5 to 6 qt	5 to 6	Brassica (cole) leafy vegetable group. Not labeled for turnip. Apply preplant or preemergence after planting. With preemergence application, irrigate immediately after application. See label for more directions.
	DCPA, MOA 3 (Dacthal) W-75 (Dacthal) 6 F	6 to 10 lb 6 to 10 pt	4.5 to 7.5	Also labeled for broccoli raab (raab, raab salad), and hanover salad. Apply immediately after seeding. May also be incorporated.
Greens, Postemergence				
Broadleaf weeds including sowthistle clover, cocklebur, jimsonweed, and ragweed	clopyralid, MOA 4 (Stinger) 3 EC	0.25 to 0.5 pt	0.11 to 0.187	**Kale, collards, mustard, turnip, mizuna, mustard spinach, and rape. See label to determine if other Brassica (Cole) leafy vegetables are registered.** Apply to crop when weeds are small and actively growing. Will control most legumes. For kale, collards, mustard, and turnip (roots), do not apply within 30 days of harvest. For turnip tops, do not apply within 15 days of harvest. Mustard green injury observed in some research trials.
Annual and perennial grasses only	clethodim, MOA 1 (Arrow, Clethodim, Intensity, Select) 2 EC (Select Max) 1 EC	6 to 8 oz 9 to 16 oz	0.094 to 0.125 0.07 to 0.125	Apply postemergence for grass control. With Arrow, Clethodim, or Select, add 1 gallon crop oil concentrate per 100 gallons spray mix. With Select Max, add 2 pints nonionic surfactant per 100 gallons spray mixture. Adding crop oil may increase the likelihood of crop injury at high air temperatures. Very effective in controlling annual bluegrass. Apply to actively growing grasses not under drought stress. Do not apply within 14 days of harvest of green crops. Do not apply within 30 days of harvest of turnips grown for roots.

CHEMICAL WEED CONTROL

345

CHEMICAL WEED CONTROL

Table 7-16. Chemical Weed Control in Vegetable Crops

Weed	Herbicide, Mode of Action Code* and Formulation	Amount of Formulation Per Acre	Pounds Active Ingredient Per Acre	Precautions and Remarks
Greens, Postemergence (continued)				
Annual and perennial grasses only (continued)	sethoxydim, MOA 1 (Poast) 1.5 EC	1 to 1.5 pt	0.2 to 0.3	ALSO LABELED FOR RAPE GREENS. Do not apply within 14 days of harvest of turnip and 30 days of harvest of other greens. Apply to emerged grasses. Consult manufacturer's label for specific rates and best times to treat. Add 1 quart crop oil concentrate per acre. Adding crop oil to Poast may increase the likelihood of crop injury at high air temperatures. Do not apply on unusually hot and humid days.
Greens, Row middles				
Most broadleaf weeds less than 4 inches tall or rosettes less than 3 inches in diameter; does not control grasses	carfentrazone-ethyl, MOA 14 (Aim) 1.9 EW or 2 EC	up to 2 oz	up to 0.031	Apply post-directed using hooded sprayers for control of emerged weeds. If crop is contacted, burning of contacted area will occur. Use a nonionic surfactant or crop oil with Aim. See label for rate. Coverage is essential for good weed control. Can be tank mixed with other registered herbicides.
Most emerged weeds	glyphosate, MOA 9 (numerous brands and formulations)	See labels	See labels	**Not labeled for turnip greens.** Apply as a hooded spray in row middles, as shielded spray in row middles, as wiper applications in row middles, or postharvest. To avoid severe injury to crop, do not allow herbicide to contact foliage, green shoots or stems, exposed roots, or fruit of crop. Do not apply within 14 days of harvest.
Lettuce, Preplant				
Suppression or control of most annual grasses and broadleaf weeds, full rate required for nutsedge control	metam sodium (Vapam HL) 42%	37.5 to 75 gal	15.7 to 31.5	Rates are dependent on soil type and weeds present. Apply when soil moisture is at field capacity (100 to 125%). Apply through soil injection using a rotary tiller or inject with knives no more than 4 inches apart; follow immediately with a roller to smooth and compact the soil surface or with mulch. May apply through drip irrigation prior to planting a second crop on mulch. Plant back interval is often 14 to 21 days and can be 30 days in some environments. See label for all restrictions and additional information.
Contact kill of all green foliage, stale bed application	paraquat, MOA 22 (Firestorm, Parazone) 3 SL	1.3 to 2.7 pt	0.5 to 1	Apply in a minimum of 10 gallons spray mix per acre to emerged weeds before crop emerges as a broadcast or band treatment over a preformed row. Row should be formed several days ahead of planting and treating to allow maximum weed emergence. Use a nonionic surfactant at a rate of 16 to 32 ounces per 100 gallons spray solution or 1 gallon approved crop oil concentrate per 100 gallons spray mix.
	(Gramoxone SL) 2 SL	2 to 4 pt		
Annual and perennial grass and broadleaf weeds, stale bed application	glyphosate, MOA 9 (numerous brands and formulations)	See labels	See labels	Apply to emerged weeds before crop emergence. Do not feed crop residue to livestock for 8 weeks following treatment. Perennial weeds may require higher rates of glyphosate. Consult the manufacturer's label for rates for specific weeds. Certain glyphosate formulations require the addition of surfactant. Adding nonionic surfactant to glyphosate formulated with nonionic surfactant may result in reduced weed control.
Lettuce, Preplant or preemergence				
Annual grasses and small-seeded broadleaf weeds	benefin, MOA 3 (Balan) 60 WDG	2 to 2.5 lb	1.2 to 1.5	Apply preplant and incorporate 2 to 3 inches deep with a rototiller or tandem disk before seeding or transplanting.
	bensulide, MOA 8 (Prefar) 4 EC	5 to 6 qt	5 to 6	Apply preplant incorporated (1 inch incorporation is optimum) or preemergence after planting. With preemergence application, irrigate immediately after application. See label for more directions.
Most annual grasses and broadleaf weeds	pronamide, MOA 3 (Kerb) 50 WP	2 to 4 lb	1 to 2 lb	**Do not apply to leaf lettuce.** Also labeled in endive, escarole or radicchio greens. Can be used preplant or preemergence. Application can also be made postemergence to head lettuce but should be made before weed germination if possible or before weeds are beyond the two-leaf stage. Moisture is necessary to activate. Do not apply within 55 days of harvest. Make only one application per crop. Consult label for planting restrictions for rotational crops.
Lettuce, Postemergence				
Annual and perennial grasses only	sethoxydim, MOA 1 (Poast) 1.5 EC	1 to 1.5 pt	0.2 to 0.3	Apply to emerged grasses. Arrow, Clethodim, and Select are only registered for leaf lettuce. Consult manufacturer's label for specific rates and best times to treat. For sethoxydim, add 1 quart of crop oil concentrate per acre. Use of Poast or clethodim with crop oil may increase the likelihood of crop injury at high air temperatures. For Arrow, Clethodim, or Select, add 1 gallon crop oil concentrate per 100 gallons spray solution. With Select Max, add 2 pints nonionic surfactant per 100 gallons spray mixture. Do not apply on unusually hot and humid days. Do not apply sethoxydim within 30 days of harvest on head lettuce or within 15 days of harvest on leaf lettuce. For clethodim, do not apply within 14 days of harvest.
	clethodim, MOA 1 (Arrow, Clethodim, Intensity, Select) 2 EC	6 to 8 oz	0.09 to 0.125	
	(Select Max, Intensity One) 1 EC	9 to 16 oz	0.07 to 0.125	
Most emerged weeds	glyphosate, MOA 9 (numerous brands and formulations)	See labels	See labels	**Row middles only.** Apply as a hooded spray in row middles, as shielded spray in row middles, as wiper applications in row middles, or postharvest. To avoid severe injury to crop, do not allow herbicide to contact foliage, green shoots, stems, exposed, roots, or fruit of crop. Do not apply within 14 days of harvest.
Most broadleaf weeds less than 4 inches tall or rosettes less than 3 inches in diameter; does not control grasses	carfentrazone-ethyl, MOA 14 (Aim) 1.9 EW or 2 EC	up to 2 oz	up to 0.031	Apply post-directed using hooded sprayers for control of emerged weeds. If crop is contacted, burning of contacted area will occur. Use a nonionic surfactant or crop oil with Aim. See label for rate. Coverage is essential for good weed control. Can be tank mixed with other registered herbicides.
Okra, Preplant and Preemergence				
Annual and perennial grass and broadleaf weeds, stale bed application	glyphosate, MOA 9 (numerous brands and formulations)	See labels	See labels	Apply to emerged weeds before crop emergence. Perennial weeds may require higher rates. Consult the manufacturer's label for rates for specific weeds. Certain glyphosate formulations require the addition of surfactant. Adding nonionic surfactant to glyphosate formulated with nonionic surfactant may result in reduced weed control.
Most broadleaf weeds less than 4 inches tall or rosettes less than 3 inches in diameter; does not control grasses	carfentrazone-ethyl, MOA 14 (Aim) 1.9 EW or 2 EC	up to 2 oz	up to 0.031	Apply no later than 1 day before transplanting crop. Use a nonionic surfactant or crop oil with Aim. See label for rate. Coverage is essential for good weed control. Can be tank mixed with other registered burndown herbicides. PHI 0 days.
Annual grasses and small-seeded broadleaf weeds	trifluralin, MOA 3 (Treflan, Treflan HFP, Trifluralin, Trilin) 4 EC	1 to 2 pt	0.5 to 1	Apply preplant and incorporate into the soil 2 to 3 inches within 8 hr using a rototiller or tandem disk.

Table 7-16. Chemical Weed Control in Vegetable Crops

Weed	Herbicide, Mode of Action Code* and Formulation	Amount of Formulation Per Acre	Pounds Active Ingredient Per Acre	Precautions and Remarks
Okra, Preplant and Preemergence (continued)				
Broadleaf and grass weeds	prometryn, MOA 5 (Caparol) 4L	1.5 to 3 pt	1 to 2	Apply preemergence and or post-directed to okra. Make a single preemergence application of Caparol at 3 pints per acre after planting before crop emergence or 1 preemegence application at 1.5 pt followed by one post-directed application at 1.5 pints per acre (okra 7-9 leaf stage). Do not exceed 3 pints per acre of Caparol per season. See label for crop rotation restrictions. PHI is 14 days.
Annual broadleaf weeds including pigweed spp.	mesotrione, MOA 27 (Callisto) 4 L	6 oz	0.19	May be applied as row middle or hooded POST-directed application but not both. For preemergence row middle application, apply as a banded application to the row middles prior to weed emergence. Leave 1 foot of untreated area over the okra row or 6 inches on each side of the planted row. Do not apply Callisto directly over the planted row or severe injury may occur. Injury risk is greatest on coarse textured soils (sand, sandy loam or loamy sand).
Okra, Postemergence				
Annual and perennial grasses only	sethoxydim, MOA 1 (Poast) 1.5 EC	1 to 1.5 pt	0.3 lb	Apply to actively growing grasses not under drought stress. Do not apply on days that are unusually hot and humid. Do not apply within 14 days of harvest.
Most broadleaf weeds less than 4 inches tall or rosettes less than 3 inches in diameter; does not control grasses	carfentrazone-ethyl, MOA 14 (Aim) 1.9 EW or 2 EC	up to 2 oz	up to 0.031	Apply post-directed using hooded sprayers for control of emerged weeds. If crop is contacted, burning of contacted area will occur. Use a nonionic surfactant or crop oil with Aim. See label for rate. Coverage is essential for good weed control. Can be tank mixed with other registered herbicides. PHI 0 days.
Most emerged weeds	glyphosate, MOA 9 (numerous brands and formulations)	See labels	See labels	**Row middles only.** Apply as a hooded spray in row middles, as shielded spray in row middles, as wiper applications in row middles, or postharvest. To avoid severe injury to crop, do not allow herbicide to contact foliage, green shoots or stems, exposed roots, or fruit of crop. Do not apply within 14 days of harvest.
Annual broadleaf weeds including pigweed 3 inches or less	mesotrione, MOA 27 (Callisto) 4 L	3 oz	0.093	May be applied as a row middle or hooded POST-directed application but not both. For postemergence hooded application, okra must be at least 3 inches tall. Minimize amount of Callisto that contacts okra foliage or crop injury will occur. PHI 28 days.
Yellow and purple nutsedge and broadleaf weeds	halosulfuron-methyl, MOA 2 (Sandea) 75 DG	0.5 to 1 oz	0.024 to 0.048	Apply to row middles as a postemergence shielded or hooded spray to avoid contact of herbicide with planted crop. In plasticulture, do not allow spray to contact plastic. Do not apply more than 2 ounces per acre per 12-month period. PHI 30 days.
Onions, Preplant and Preemergence				
Suppression or control of most annual grasses and broadleaf weeds, full rate required for nutsedge control	metam sodium (Vapam HL) 42%	37.5 to 75 gal	15.7 to 31.5	**Dry bulb and green onion.** Rates are dependent on soil type and weeds present. Apply when soil moisture is at field capacity (100 to 125%). Apply through soil injection using a rotary tiller or inject with knives no more than 4 inches apart; follow immediately with a roller to smooth and compact the soil surface or with mulch. May apply through drip irrigation prior to planting a second crop on mulch. Plant back interval is often 14 to 21 days and can be 30 days in some environments. See label for all restrictions and additional information.
Contact kill of all green foliage, stale bed application	paraquat, MOA 22 (Firestorm, Parazone) 3 SL (Gramoxone SL 2.0) 2 SL	1.7 to 2.7 pt 2.5 to 4 pt	0.65 to 1	**Seeded onion only.** Apply in a minimum of 10 gallons spray mix per acre to emerged weeds before crop emergence or transplanting as a broadcast or band treatment over a preformed row. Row should be formed several days ahead of planting and treating to allow maximum weed emergence. Use a nonionic surfactant at a rate of 16 to 32 ounces per 100 gallons spray mix or 1 gallon approved crop oil concentrate per 100 gallons spray mix. PHI 60 days.
Annual and perennial grass and broadleaf weeds	glyphosate, MOA 9 (numerous brands and formulations)	See labels	See labels	Apply to emerged weeds before crop emergence. Perennial weeds may require higher rates of glyphosate. Consult the manufacturer's label for rates for specific weeds. Use on direct seeded onions only. Certain glyphosate formulations require the addition of surfactant. Adding nonionic surfactant to glyphosate formulated with nonionic surfactant may result in reduced weed control.
Most broadleaf weeds less than 4 inches tall or rosettes less than 3 inches in diameter; does not control grasses	carfentrazone-ethyl, MOA 14 (Aim) 2.0 EC	up to 2 oz	0.031	Apply no later than 30 days before planting. See label for specific Aim rate relating to weed species and proper adjuvant and rate. Coverage is essential for good weed control. Can be tank mixed with other registered burndown herbicides.
Annual grasses and small-seeded broadleaf weeds	bensulide, MOA 8 (Prefar) 4 E	5 to 6 qt	5 to 6	**Dry bulb only.** Apply preplant incorporated (1 inch incorporation is optimum) or preemergence after planting. With preemergence application, irrigate immediately after application. See label for more directions and for rotation restrictions.
	DCPA, MOA 3 (Dacthal) W-75 (Dacthal) 6 F	8 to 10 lb 8 to 10 pt	6 to 7.5	**Dry bulb and green.** Apply immediately after seeding or transplanting and/or at layby. See label for timing layby treatments.
Annual broadleaf weeds	oxyfluorfen, MOA 14 (Galigan) 2 E (Goal 2 XL) 2 EC (GoalTender) 4 E	1 to 2 pt 1 to 2 pt 1 pt	0.25 to 0.5 0.25 to 0.5 0.5	**Transplanted dry bulb only.** Apply as a single application immediately (within 2 days) after transplanting for preemergence control of weeds. Injury can occur if applications are made during cool, wet weather or prior to the full development of the true leaves. See label for rates and instructions for use. Do not apply within 45 days of harvest.
Most annual grasses and some broadleaf weeds	pendimethalin, MOA 3 (Prowl) 3.3 EC (Prowl) 3.8 AS	See label	See label	**Dry bulb and green onion (chives, leeks, spring onions, scallions, Japanese bunching onions, green shallots, and green eschalots). Prowl 3.3 EC is not registered for green onion.** For preemergence weed control. Apply when onions have two to nine true leaves (dry bulb) and two to three leaves (green onion) but prior to weed emergence. For green onion the soil must be a muck soil or be a mineral soil with at least 3% organic matter. See label for additional information on rate depending on soil type. PHI for dry bulb onion is 45 days. PHI for green onion is 30 days.
	dimethenamid-P, MOA 15 (Outlook) 6 EC	12 to 21 oz	0.6 to 1	**Dry bulb and green onion (leeks, spring onions or scallions, Japanese bunching onions, green shallots or eschalots).** For preemergence weed control. Apply after crop has reached 2 true leaves until a minimum of 30 days before harvest. If applications are made to transplanted crop, DO NOT apply until transplants are in the ground and soil has settled around transplants with several days to recover. See label for further details.

347

Table 7-16. Chemical Weed Control in Vegetable Crops

Weed	Herbicide, Mode of Action Code* and Formulation	Amount of Formulation Per Acre	Pounds Active Ingredient Per Acre	Precautions and Remarks
Onions, Postemergence				
Most annual broadleaf weeds	oxyfluorfen, MOA 14 (Galigan) 2 E (Goal 2 XL) 2 EC (GoalTender) 4 E	0.5 pt 0.5 pt 0.25 pt	0.12	**Dry bulb only.** May be used as a postemergence spray to both the weeds and crop after the onions have at least two fully developed true leaves. Some injury to onions may result. Injury will be more severe if the chemical is applied during cool, wet weather. Weeds should be in the two- to four-leaf stage for best results. Do not make more than four applications per year. Do not apply within 45 days of harvest.
Common lambsquarters, common chickweed, common purslane, black nightshade, ladysthumb, Pennsylvania smartweed, redroot pigweed, and some annual grasses	ethofumesate, MOA 8 (Nortron) 4 SC	16 to 32 oz	0.5 to 1	Apply at planting or just after planting prior to weed emergence. Can be used postemergence at 16 oz per acre. See label for more information. Rainfall of at least 0.5 inch is needed for activation.
Most broadleaf weeds less than 4 inches tall or rosettes less than 3 inches in diameter; does not control grasses	carfentrazone-ethyl, MOA 14 (Aim) 1.9 EW or 2 EC	up to 2 oz	up to 0.031	Apply post-directed using hooded sprayers for control of emerged weeds. If crop is contacted, burning of contacted area will occur. Use a nonionic surfactant or crop oil with Aim. See label for rate. Coverage is essential for good weed control. Can be tank mixed with other registered herbicides.
Most emerged weeds	glyphosate, MOA 9 (numerous brands and formulations)	See labels	See labels	**Row middles only.** Apply as a hooded spray in row middles, as shielded spray in row middles, as wiper applications in row middles, or postharvest. To avoid severe injury to crop, do not allow herbicide to contact foliage, green shoots, stems, exposed, roots, or fruit of crop. Do not apply within 14 days of harvest.
Annual and perennial grasses only	fluazifop, MOA 1 (Fusilade DX) 2 EC	6 to 16 oz	0.1 to 0.25	**Dry bulb only.** Apply to emerged grasses. Consult manufacturer's label for specific rates and best times to treat. Add 1 gallon crop oil concentrate or 1 quart nonionic surfactant per 100 gallons spray mix. Do not apply on days that are unusually hot and humid. Do not apply within 45 days of harvest.
	sethoxydim, MOA 1 (Poast) 1.5 EC	1 to 1.5 pt	0.2 to 0.3	**Dry bulb and green.** Apply to emerged grasses. Consult manufacturer's label for specific rates and best times to treat. Add 1 quart of crop oil concentrate per acre. Adding crop oil to Poast may increase the likelihood of crop injury at high air temperatures. Do not apply Poast on days that are unusually hot and humid. Do not apply within 30 days of harvest.
Annual and perennial grasses only (continued)	clethodim, MOA 1 (Arrow, Clethodim, Intensity, Select) 2 EC (Intensity One, Select Max) 1 EC	6 to16 oz 9 to 32 oz	0.09 to 0.25 0.07 to 0.25	**Dry bulb only.** Apply to emerged grasses. Consult label for specific rates and best times to treat. With Arrow, Clethodim, or Select, add 1 gallon crop oil concentrate per 100 gallons spray mix. With Select Max or Intensity One, add 2 pints nonionic surfactant per 100 gallons spray mixture. Adding crop oil may increase the likelihood of crop injury at high air temperatures. Do not apply Selecton unusually hot and humid days. Do not apply within 45 days of dry bulb onion harvest. Intensity One may be applied to dry bulb onion or green onions (Leeks, scallions or spring onions, Japanese bunching onion, shallots or eschalots). Do not exceed 16 ounces of Intensity One per acre on green onions. Do not apply Intensity One herbicide within 14 days of green onion harvest.
Peas, Green, Preplant and Preemergence				
Contact kill of all green foliage, stale bed application	paraquat, MOA 22 (Firestorm,Parazone) 3 SL (Gramoxone SL) 2 SL	1.3 to 2.7 pt 2 to 4 pt	0.5 to 1	Apply in minimum of 10 gallons spray mix per acre to emerged weeds before crop emergence as broadcast or band treatment over preformed row. Use sufficient water for thorough coverage. Row should be formed several days before planting and treating to allow maximum weed emergence. Use nonionic surfactant at a rate of 16 to 32 ounces per 100 gallons spray mix or 1 gallon approved crop oil concentrate per 100 gallons spray mix.
Most broadleaf weeds less than 4 inches tall or rosettes less than 3 inches in diameter; does not control grasses	carfentrazone-ethyl, MOA 14 (Aim) 1.9 EW or 2 EC	up to 2 oz	up to 0.031	Apply prior to planting or emergence of crop. Use a nonionic surfactant or crop oil with Aim. See label for rate. Coverage is essential for good weed control. Can be tank mixed with other registered burndown herbicides.
Annual and perennial grass and broadleaf weeds	glyphosate, MOA 9 (numerous brands and formulations)	See labels	See labels	Apply to emerged weeds before crop emergence. Do not feed crop residue to livestock for 8 weeks following treatment. Perennial weeds may require higher rates of glyphosate. Consult the manufacturer's label for rates for specific weeds. Certain glyphosate formulations require the addition of a surfactant. Adding nonionic surfactant to glyphosate formulated with nonionic surfactant may result in reduced weed control.
Annual grasses and small-seeded broadleaf weeds	pendimethalin, MOA 3 (Prowl H₂O) 3.8 AS	1.5 to 3 pt	0.75 to 1.5	**Southern peas (cowpeas) and snap beans only.** Apply preplant and incorporate into the soil 2 to 3 inches using a power driven rototiller or by cross disking. **Do not apply after seeding. Do not apply when air temperature is below 45 degrees F.**
Broadleaf weeds	saflufenacil, MOA 14 (Sharpen) 3.42 SL	1 oz	0.027	**Dry field pea, edible pea (sugar snap, English pea, garden pea, green pea, marrowfat pea) and chickpea only:** Apply as a preplant/ preemergence burndown of small actively growing broadleaf weeds. Can also be used preplant incorporated or preemergence in edible pea. See label for directions. Do not apply more than 2 fluid ounces per acre per season.
Annual grasses and small-seeded broadleaf weeds	trifluralin, MOA 3 (Treflan, Trifluralin, Trifluralin HF, other brands) 4 EC	1 to 1.5 pt	0.5 to 0.75	**English peas only.** Apply preplant and incorporate to a depth of 2 to 3 inches within 8 hr with a rototiller or tandem disk.
Annual grasses and broadleaf weeds; weak on pigweed	clomazone, MOA 13 (Command) 3ME	1.3 pt	0.5	Apply to the soil surface immediately after seeding. See label for further instruction. Limited research has been done on this product in this crop in North Carolina.
Annual grasses, small-seeded broadleaf weeds, and suppression of yellow nutsedge	S-metolachlor, MOA 15 (Brawl, Dual Magnum, Medal) 7.62 EC (Brawl II, Dual II Magnum, Medal II) 7.64 EC	1 to 2 pt	0.95 to 1.91	Apply to soil surface immediately after seeding. Shallow cultivations will improve control. See label for specific rate.
Annual broadleaf weeds including morningglory, pigweed, smartweed, and purslane	imazethapyr, MOA 2 (Pursuit) 2 EC	up to 3 oz	Up to 0.047	**English peas only.** Apply preplant incorporated or to soil surface immediately after planting. See label for more details.

Table 7-16. Chemical Weed Control in Vegetable Crops

Weed	Herbicide, Mode of Action Code* and Formulation	Amount of Formulation Per Acre	Pounds Active Ingredient Per Acre	Precautions and Remarks
Peas, Green, Postemergence				
Annual broadleaf weeds and yellow nutsedge	bentazon, MOA 6 (Basagran) 4 SL	1 to 2 pt	0.5 to 1	Apply overtop of peas when weeds are small and peas have at least three pairs of leaves (four nodes). **Do not add crop oil concentrate to spray mix.** Do not apply within 10 days of harvest. Do not apply when peas are in bloom.
Most broadleaf weeds less than 4 inches tall or rosettes less than 3 inches in diameter; does not control grasses	carfentrazone-ethyl, MOA 14 (Aim) 1.9 EW or 2 EC	up to 2 oz	up to 0.031	Apply post-directed using hooded sprayers for control of emerged weeds. If crop is contacted, burning of contacted area will occur. Use a nonionic surfactant or crop oil with Aim. See Label for rate. Coverage is essential for good weed control. Can be tank mixed with other registered herbicides.
Most emerged weeds	glyphosate, MOA 9 (numerous brands and formulations)	See labels	See labels 0.5 to 0.94	**Row middles only.** Apply as a hooded spray in row middles, as shielded spray in row middles, as wiper applications in row middles, or post harvest. To avoid severe injury to crop, do not allow herbicide to contact foliage, green shoots or stems, exposed roots, or fruit of crop. Do not apply within 14 days of harvest.
Annual and perennial grasses	sethoxydim, MOA 1 (Poast) 1.5 EC	1 to 1.5 pt	0.2 to 0.3	Apply to emerged grasses. Consult manufacturer's label for specific rates and best times to treat. With sethoxydim, add 1 quart of crop oil concentrate per acre. Adding crop oil to Poast or Assure II may increase the likelihood of crop injury at high air temperatures. With quizalofop, add 1 gallon oil concentrate or 1 quart nonionic surfactant per 100 gallons spray. Do not apply Poast or Assure II on days that are unusually hot and humid. Do not apply sethoxydim within 15 days or Assure II or Targa within 30 days of harvest.
	quizalofop p-ethyl, MOA 1 (Assure II) 0.88 EC (Targa) 0.88 EC	6 to 12 oz	0.04 to 0.08	
Annual broadleaf weeds including morningglory, pigweed, smartweed, and purslane	imazethapyr, MOA 2 (Pursuit) 2 EC	up to 3 oz	Up to 0.047	**See label for pea type.** Apply postemergence to 1- to 3-inch weeds (one to four leaves) when peas are at least 3 inches high but prior to five nodes and before flowering. Add nonionic surfactant at 2 pints per 100 gallons of spray mix. See label for crop rotation restrictions. PHI 30 days.
Broadleaf and grass weeds	imazamox, MOA 2 (Raptor) 1 SL	4 fl oz	0.31	**Dry peas only:** Apply postemergence before bloom stage but after dry peas have at least 3 pairs of leaves. See label for further information.
Peas, Southern (cowpeas, blackeyed peas), Preplant or Preemergence				
Contact kill of all green foliage, stale bed application	paraquat, MOA 22 (Firestorm, Parazone) 3 SL	1.3 to 2.7 pt	0.5 to 1	Apply in a minimum of 20 gallons spray solution to emerged weeds before crop emergence as a broadcast or band treatment over a preformed row. Use sufficient water to give thorough coverage. Row should be formed several days ahead of planting and treating to allow maximum weed emergence. Use a nonionic surfactant at a rate of 16 to 32 ounces per 100 gallons spray mix or 1 gallon approved crop oil concentrate per 100 gal spray mix.
	(Gramoxone SL) 2 SL	2 to 4 pt		
Most broadleaf weeds less than 4 inches tall or rosettes less than 3 inches in diameter; does not control grasses	carfentrazone-ethyl, MOA 14 (Aim) 1.9 EW or 2 EC	Up to 2 oz	Up to 0.031	Apply prior to planting or emergence of crop. Use a nonionic surfactant or crop oil with Aim. See label for rate. Coverage is essential for good weed control. Can be tank mixed with other registered burndown herbicides.
Annual and perennial grass and broadleaf weeds, stale bed application.	glyphosate, MOA 9 (numerous brands and formulations)	See labels	See labels	Apply to emerged weeds before crop emergence. Do not feed crop residue to livestock for 8 weeks following treatment. Perennial weeds may require higher rates of glyphosate. Consult the manufacturer's label for rates for specific weeds. Certain glyphosate formulations require the addition of a surfactant. Adding nonionic surfactant to glyphosate formulated with nonionic surfactant may result in reduced weed control.
Annual grasses and small-seeded broadleaf weeds	pendimethalin, MOA 3 (Prowl H₂O) 3.8 AS	1.5 to 3 pt	0.75 to 1.5	**Not labeled for blackeyed peas.** Apply preplant and incorporate into the soil 2 to 3 inches using a power driven rototiller or by cross disking. **Do not apply after seeding.**
	trifluralin, MOA 3 (Treflan HFP, Trifluralin, Trifluralin HF) 4 EC	1 to 2 pt	0.5 to 1	Apply preplant and incorporate into the soil 2 to 3 inches deep within 8 hr with a rototiller or tandem disk.
Annual grasses and broadleaf weeds	clomazone, MOA 13 (Command) 3ME	0.4 to 0.67 pt	0.15 to 0.25	**Succulent southern peas only.** Apply to the soil surface immediately after seeding. Offers weak control of pigweed. See label for further instruction. Limited research has been done on this product in this crop in North Carolina.
Annual grasses, small-seeded broadleaf weeds, and suppression of yellow nutsedge	S-metolachlor, MOA 15 (Brawl, Dual Magnum, Medal) 7.62 EC (Brawl II, Dual II Magnum, Medal II) 7.64 EC	1 to 2 pt	0.95 to 1.91	Apply to soil surface immediately after planting. Shallow cultivations will improve control. May also be soil incorporated before planting.
Annual grasses and broadleaf weeds including morningglory, pigweed, smartweed, and purslane	imazethapyr, MOA 2 (Pursuit) 2 EC	Up to 4 oz	Up to 0.063	Apply preemergence or preplant incorporated. See label for rate for specific pea species.
Peas, Southern, Postemergence				
Annual broadleaf weeds and yellow nutsedge	bentazon, MOA 6 (Basagran) 4 SL	1 to 2 pt	0.5 to 1	Apply overtop of peas when weeds are small and peas have at least three pairs of leaves (four nodes). **Do not add crop oil concentrate to spray mix.** See label for weeds controlled with Basagran. Do not apply within 30 days of harvest. Do not apply when peas are in bloom.
Annual broadleaf weeds including morningglory, pigweed, smartweed, and purslane	imazethapyr, MOA 2 (Pursuit) 2 EC	Up to 4 oz	Up to 0.063	**Southern peas and certain dry peas.** Apply postemergence to 1- to 3-inch weeds (one to four leaves) when peas are at least 3 inches in height but prior to five nodes and flowering. Add nonionic surfactant at 2 pints per 100 gallons of spray mixture with all postemergence applications. Do not apply within 30 days of harvest. See label for rate for specific pea species.
Most broadleaf weeds less than 4 inches tall or rosettes less than 3 inches in diameter; does not control grasses	carfentrazone-ethyl, MOA 14 (Aim) 1.9 EW or 2 EC	up to 2 oz	up to 0.031	Apply post-directed using hooded sprayers for control of emerged weeds. If crop is contacted, burning of contacted area will occur. Use a nonionic surfactant or crop oil with Aim. See label for rate. Coverage is essential for good weed control. Can be tank mixed with other registered herbicides.

CHEMICAL WEED CONTROL

Table 7-16. Chemical Weed Control in Vegetable Crops

Weed	Herbicide, Mode of Action Code* and Formulation	Amount of Formulation Per Acre	Pounds Active Ingredient Per Acre	Precautions and Remarks
Peas, Southern, Postemergence (continued)				
Most emerged weeds	glyphosate, MOA 9 (numerous brands and formulations)	See labels	See labels	**Row middles only.** Apply as a hooded spray in row middles, as shielded spray in row middles, as wiper applications in row middles, or post harvest. To avoid severe injury to crop, do not allow herbicide to contact foliage, green shoots, stems, exposed, roots, or fruit of crop. Do not apply within 14 days of harvest.
Annual and perennial grasses	quizalofop p-ethyl, MOA 1 (Assure II, Targa) 0.88 EC	6 to 12 oz	0.04 to 0.08	Apply to emerged grasses. Consult manufacturer's label for specific rates and best times to treat. With sethoxydim, add 1 quart of crop oil concentrate per acre. With quizalofop, add 1 gallon oil concentrate or 1 quart nonionic surfactant per 100 gallons spray. Adding crop oil to Assure II or Poast may increase the likelihood of crop injury at high air temperatures. Do not apply Assure II or Poast on days that are unusually hot and humid. With sethoxydim, do not apply within 15 days and 30 days of harvest for succulent and dry peas, respectively. With quizalofop, do not apply within 60 days of harvest of dry Southern peas, or within 30 days of harvest of succulent Southern peas.
	sethoxydim, MOA 1 (Poast) 1.5 EC	1 to 1.5 pt	0.2 to 0.3	
	clethodim, MOA 1 (Intensity One, Select Max) 1 EC	9 to 16 oz	0.07 to 0.125 lb	For Select Max or Intensity One, add 2 pints nonionic surfactant per 100 gallons spray mixture. Apply before bloom. Do not make more than one application per acre per season. Do not apply clethodim within 21 days of harvest.
Peppers, Preplant				
Suppression or control of most annual grasses and broadleaf weeds, full rate required for nutsedge control	metam sodium (Vapam HL) 42%	37.5 to 75 gal	15.7 to 31.5	Rates are dependent on soil type and weeds present. Apply when soil moisture is at field capacity (100 to 125%). Apply through soil injection using a rotary tiller or inject with knives no more than 4 inches apart; follow immediately with a roller to smooth and compact the soil surface or with mulch. May apply through drip irrigation prior to planting a second crop on mulch however adhere to label guidelines on crop plant back interval. Plant back interval is often 14 to 21 days and can be 30 days in some environments. See label for all restrictions and additional information. Chloropicrin (150lb/A broadcast) will also be needed when laying first crop mulch to control nutsedge.
Contact kill of all green foliage, stale bed application	paraquat, MOA 22 (Firestorm, Parazone) 3 SL	1.3 to 2.7 pt	0.5 to 1	Apply in a minimum of 10 gallons of spray mix per acre to emerged weeds before transplanting as a broadcast or band treatment over a preformed row. Row should be formed several days ahead of planting and treating to allow maximum weed emergence. Use a nonionic surfactant at a rate of 16 to 32 ounces per 100 gallons spray mix or 1 gallon approved crop oil concentrate per 100 gallons spray mix.
	(Gramoxone SL) 2 SL	2 to 4 pt		
Most broadleaf weeds less than 4 inches tall or rosettes less than 3 inches in diameter; does not control grasses	carfentrazone-ethyl, MOA 14 (Aim) 1.9 EW or 2 EC	up to 2 oz	up to 0.031	**Transplanted crop. Apply no later than 1 day before transplanting crop.** **Seeded crop. Apply no later than 7 days before planting seeded crop.** Use a nonionic surfactant or crop oil. See label for rate. Coverage of weed is essential for good weed control. Can be tank mixed with other registered burndown herbicides.
Annual and perennial grass and broadleaf weeds, stale bed application	glyphosate, MOA 9 (numerous brands and formulations)	See labels	See labels	Apply to emerged weeds at least 3 days before seeding or transplanting. When applying Roundup before transplanting crops into plastic mulch, care must be taken to remove residues of this product from the plastic prior to transplanting. To prevent crop injury, residues can be removed by 0.5 inch natural rainfall or by applying water via a sprinkler system. Perennial weeds may require higher rates of glyphosate. Consult the manufacturer's label for specific weeds. Certain glyphosate formulations require the addition of a surfactant. Adding nonionic surfactant to glyphosate formulated with nonionic surfactant may result in reduced weed control.
Broadleaf weeds including Carolina geranium and cutleaf eveningprimrose and a few annual grasses	oxyfluorfen, MOA 14 (Goal) 2XL (GoalTender) 4 F	up to 2 pt / up to 1 pt	0.5 lb	**Plasticulture only.** Apply to soil surface of pre-formed beds at least 30 days prior to transplanting crop. While incorporation is not necessary, it may result in less crop injury. Plastic mulch can be applied any time after application but best results are likely if applied soon after application.
Palmer amaranth, redroot pigweed, smooth pigweed, Galinsoga sp., black nightshade, Eastern black nightshade, common purslane, partial control of yellow nutsedge	fomesafen, MOA 14 (Reflex) 2 EC	1 to 1.5 pt	0.25 to 0.375	This is a Section 24(c) special local needs label for transplanted pepper in North Carolina. Growers must obtain the label at Farmassist.com prior to making an application of Reflex. **See label for further instructions.** **Plasticulture In-row Application for Transplanted Pepper.** Apply after final bed formation and the drip tape is laid but prior to laying plastic mulch. Avoid soil disturbance after application. Unless restricted by other products such as fumigants, pepper may be transplanted immediately following the application of Reflex and the application of the mulch. **Bareground for Transplanted Pepper.** Apply pretransplant up to 7 days prior to transplanting pepper. Weed control will be reduced if soil is disturbed after application. During the transplanting operation make sure the soil in the transplant hole settles flush or above the surrounding soil surface. Avoid cultural practices that may concentrate Reflex-treated soil around the transplant root ball. An overhead irrigation or rainfall event between Reflex herbicide application and transplanting will ensure herbicide activation and will likely reduce the potential for crop injury due to splashing. **Plasticulture Row Middle Application.** Apply to row middles with a hooded or shielded sprayer. Avoid drift of herbicide on mulch. If drift occurs, 0.5 inch of rain or irrigation must occur prior to transplanting. **Carryover is a large concern; see label for more information.**
Annual grasses and small-seeded broadleaf weeds	clomazone, MOA 13 (Command) 3 ME	0.67 to 2.67 pt	0.25 to 1	**Not labeled for banana pepper.** Apply preplant before transplanting. Weak on pigweed. See label for instructions on use.
	napropamide, MOA 15 (Devrinol, Devrinol DF-XT) 50 DF (Devrinol, Devrinol 2-XT) 2 EC	2 to 4 lb / 2 to 4 qt	1 to 2	**Bare ground:** Apply preplant and incorporate into the soil 1 to 2 inches as soon as possible with a rototiller or tandem disk. Can be used on direct-seeded or transplanted peppers. See label for instructions on use. **Plasticulture:** Apply to a weed-free soil before laying plastic mulch. Soil should be well worked yet moist enough to permit a thorough incorporation to a depth of 2 inches. Mechanically incorporate or irrigate within 24 hours after application. If weed pressure is from small seeded annuals, apply to the surface of the bed immediately in front of the laying of plastic mulch. If soil is dry, water or sprinkle irrigate with sufficient water to wet to a depth of 2 to 4 inches before covering with plastic mulch. **Between rows:** Apply to a weed free soil surface between the rows (bareground or plastic mulch). Mechanically incorporate or irrigate Devrinol into the soil to a depth of 1 to 2 inches within 24 hours of application. See XT labels for information regarding delay in irrigation event.

Table 7-16. Chemical Weed Control in Vegetable Crops

Weed	Herbicide, Mode of Action Code* and Formulation	Amount of Formulation Per Acre	Pounds Active Ingredient Per Acre	Precautions and Remarks
Peppers, Preplant (continued)				
Annual grasses and small-seeded broadleaf weeds (continued)	pendimethalin, MOA 3 (Prowl H$_2$O) 3.8	1 to 3 pt	0.5 to 1.5	May be applied in chili pepper, cooking pepper, pimento, Jalapeno, and sweet pepper. Do not apply more than 3 pints per acre per season. See label for specific use rate for your soil type. Avoid direct contact with pepper foliage or stems. Do not apply within 70 days of harvest. See label for further instructions and precautions. **Between rows.** Can be applied as a post-directed spray on the soil at the base of the plant beneath plants and between rows. **In-row.** May be applied as a broadcast preplant incorporated surface application prior to transplanting peppers. Limited research has been conducted in NC.
	trifluralin, MOA 3 (Treflan, Treflan HFP, Trifluralin HF) 4 EC	1 to 2 pt	0.5 to 1	Apply pretransplant, and incorporate to a depth of 2 to 3 inches within 8 hr with a rototiller or tandem disk.
Peppers, Preplant and Preemergence				
Annual grasses and small seeded broadleaf weeds	bensulide, MOA 8 (Prefar) 4 EC	5 to 6 qt	5 to 6	Apply preplant incorporated (1 inch incorporation is optimum) or preemergence. With preemergence application, irrigate immediately after application. See label for directions.
Peppers, Preplant or Postemergence				
Annual grass and broadleaf weeds, yellow nutsedge suppression	S-metolachlor, MOA 15 (Dual Magnum) 7.62 EC	8 to 12 oz	0.47 to 0.7	**Bell pepper transplants only.** This is a Section 24© special local needs label. Growers must obtain label prior to making Dual Magnum applications. Growers must obtain label at www.farmassist.com Option 1: Apply 8 to 12 ounces to the soil surface of pre-formed beds prior to laying plastic. Insure the plastic laying process does not incorporate or disturb the treated bed. Option 2: Apply 12 ounces overtop of bell pepper between 1 and 3 weeks after planting. Does not control emerged weeds. Limited data are available for NC. Do not apply more than 12 ounces per acre as it is likely that injury will occur including decreased crop vigor. Read label for further instructions.
Peppers, Postemergence				
Annual and perennial grasses only	sethoxydim, MOA 1 (Poast) 1.5 EC	1 to 1.5 pt	0.2 to 0.3	Apply to emerged grasses. Consult manufacturer's label for specific rates and best times to treat. Add 1 quart of crop oil concentrate per acre. Adding crop oil to Poast may increase the likelihood of crop injury at high air temperatures. Do not apply Poast on days that are unusually hot and humid. Do not apply within 7 days of harvest.
	clethodim, MOA 1 (Arrow, Clethodim, Intensity, Select) 2 EC (Select Max, Intensity One) 1 EC	6 to 8 oz 9 to 16 oz	0.094 to 0.125 0.07 to 0.125	Apply postemergence to control grasses. With Arrow, Clethodim, or Select, add 1 gallon crop oil concentrate per 100 gallons spray mix. With Select Max, add 2 pints nonionic surfactant per 100 gallons spray mixture. Adding crop oil may increase likelihood of crop injury at high air temperatures. Very effective in controlling annual bluegrass. Apply to actively growing grasses not under drought stress. Do not apply within 20 days of harvest.
Broadleaf, grass (suppression only), and yellow nutsedge	imazosulfuron, MOA 2 (League) 0.5 DF	4 to 6.4 oz	0.19 to 0.3	**Pepper (Bell and non-bell).** Apply to pepper plants that are well established and at least 10 inches tall. Apply directed to the base of the plants stem, no higher than 2 inches from the soil surface and do not contact fruit. Consult label for approved surfactants and crop rotation restrictions. PHI 21 days.
Peppers, Row Middles				
Most broadleaf weeds less than 4 inches tall or rosettes less than 3 inches in diameter; does not control grasses	carfentrazone-ethyl, MOA 14 (Aim) 1.9 EW or 2 EC	up to 2 oz	up to 0.031	Apply post-directed using hooded sprayer for control of emerged weeds. If crop is contacted, burning of contacted area will occur. Use a nonionic surfactant or crop oil with Aim. See label for rate. Coverage is essential for good weed control. Can be tank mixed with other registered herbicides. PHI 0 days.
Most emerged weeds	glyphosate, MOA 9 (numerous brands and formulations)	See labels	See labels	Apply as a hooded spray in row middles, as shielded spray in row middles, as wiper applications in row middles, or postharvest. To avoid severe injury to crop, do not allow herbicide to contact foliage, green shoots, stems, exposed, roots, or fruit of crop. Do not apply within 14 days of harvest.
Yellow and purple nutsedge and broadleaf weeds	halosulfuron-methyl, MOA 2 (Profine 75, Sandea) 75 DG	0.5 to 1 oz	0.024 to 0.048 lb	Apply to row middles as a postemergence spray. In plasticulture, do not allow spray to contact plastic. Early season application will give postemergence and preemergence control. Do not apply within 30 days of harvest. For postemergence applications, use nonionic surfactant at 1 quart per 100 gallons of spray solution.
Contact kill of all green foliage	paraquat, MOA 22 (Firestorm, Parazone) 3 SL (Gramoxone SL) 2 SL	1.3 pt 2 pt	0.5	Apply in a minimum of 20 gallons spray mix per acre as a shielded spray to emerged weeds between rows of peppers. Use a nonionic surfactant at a rate of 16 ounces per 100 gallons spray mix. Do not apply more than three times per season.
Potatoes, Irish, Preplant and Preemergence				
Contact kill of all green foliage, stale bed application	paraquat, MOA 22 (Firestorm, Parazone) 3 SL (Gramoxone SL) 2 SL	0.7 to 1.3 pt 1 to 2 pt	0.25 to 0.5	Apply in a minimum of 20 gallons spray mix per acre to emerged weeds up to ground cracking before crop emergence. May be used instead of the drag-off operation to kill emerged weeds before the application of preemergence herbicides. Use a nonionic surfactant at a rate of 16 to 32 ounces per 100 gallons spray mix or 1 gallon approved crop oil concentrate per 100 gallons spray mix.
Most broadleaf weeds less than 4 inches tall or rosettes less than 3 inches in diameter; does not control grasses	carfentrazone-ethyl, MOA 14 (Aim) 1.9 EW or 2 EC	up to 2 oz	up to 0.031	Apply prior to planting, or within 1 day after planting crop. Use a nonionic surfactant or crop oil with Aim. See label for rate. Coverage of weed is essential for good weed control. Can be tank mixed with other registered burndown herbicides.
Annual and perennial grass and broadleaf weeds, stale bed application	glyphosate, MOA 9 (numerous brands and formulations)	See labels	See labels	Apply to emerged weeds before crop emergence. Do not feed crop residue to livestock for 8 weeks following treatment. Perennial weeds may require higher rates of glyphosate. Consult the manufacturer's label for rates for specific weeds. Certain glyphosate formulations require the addition of a surfactant. Adding nonionic surfactant to glyphosate formulated with nonionic surfactant may result in reduced weed control.

CHEMICAL WEED CONTROL

Table 7-16. Chemical Weed Control in Vegetable Crops

Weed	Herbicide, Mode of Action Code* and Formulation	Amount of Formulation Per Acre	Pounds Active Ingredient Per Acre	Precautions and Remarks
Potatoes, Irish, Preplant and Preemergence (continued)				
Annual grasses and small-seeded broadleaf weeds	pendimethalin, MOA 3 (Prowl) 3.3 EC (Prowl H₂O) 3.8 AS	1.8 to 3.6 pt 1.5 to 3 pt	0.75 to 1.5 0.75 to 1.5	Apply just after planting or drag-off to weed-free soil before crop emerges or from emergence until crop reaches 6 inches tall.
Annual grasses and small-seeded broadleaf weeds, plus yellow nutsedge suppression	S-metolachlor, MOA 15 (Brawl, Dual Magnum, Medal) 7.62 EC (Brawl II, Dual II Magnum, Medal II) 7.64 EC	1 to 2 pt	0.95 to1.91	Apply just after planting or drag-off to weed-free soil before crop emerges. Dual Magnum can also be applied at lay-by for control of late season weeds. Do not apply within 60 days after the at-planting to drag-off application, or within 40 days after a lay-by application. See label for further instruction.
	dimethenamid-P, MOA 15 (Outlook) 6 EC	12 to 21 oz	0.6 to 1	Apply just after planting or drag-off to weed-free soil before crop emerges. See label for further instruction. PHI 40 days.
Annual grasses, most broadleaf weeds, plus yellow and purple nutsedge suppression	EPTC, MOA 8 (Eptam) 7 EC	3.5 pt	3	Apply preplant and incorporate into the soil 2 to 3 inches with a rototiller or tandem disk. The variety "Superior" has been shown to be sensitive to Eptam. See label for specific methods of incorporation. For late season preemergence nutsedge control, apply and incorporate as a directed spray to the soil on both sides of the crop row. See label for more detail.
Most annual broadleaf weeds and some annual grasses	flumioxazin, MOA 14 (Chateau) 51 SW	1.5 oz	0.047	Apply immediately after hilling. A minimum of 2 inches of soil must cover the vegetative portion of the potato plant at the time of application of Chateau. DO NOT apply to emerged potatoes. DO NOT incorporate Chateau or weed control will be reduced. Can be tank mixed with burndown herbicides if weeds present at application. See label for further instructions.
	linuron, MOA 7 (Lorox DF) 50 WDG (Linex) 4L	1.5 to 3 lb 1.5 to 3 pt	0.75 to 1.5	Apply just after planting or drag-off or hilling but before crop emerges. If emerged weeds are present, add 1 pint surfactant for each 25 gallons spray mixture. Weeds may be up to 3 inches tall at time of application.
	metribuzin, MOA 5 (TriCor DF, Dimetric DF, and other trade names) 75 WDG	0.33 to 1.33 lb	0.23 to 1	Apply just after planting or drag-off but before crop emerges. Weeds may be emerged at time of application. On sand soils or sensitive varieties, do not exceed 0.67 pound per acre. See label for list of sensitive varieties.
	rimsulfuron, MOA 2 (Matrix, Pruvin) 25 WDG	1 to 1.5 oz	0.016 to 0.023	Apply after drag-off or hilling but before potatoes and weeds emerge. If emerged weeds are present, add surfactant. See label for specific rate. Can be tank mixed with Eptam, Prowl, Sencor, Lorox, or Dual Magnum. See label for further instructions.
Broadleaf, grass and nutsedge weeds	fomesafen, MOA 14 (Reflex) 2 EC	1 pt	0.25	Applt as a preemergence after planting but prior to potato emergence. Do not apply as s pre-plant incorporated application or emerged potato plant severe crop injury may occur. Do not exceed Reflex at 1 pint per acre per season. PHI 70 days.
Potatoes, Irish, Postemergence				
Most annual broadleaf weeds and some annual grasses	metribuzin, MOA 5 (TriCor DF, Dimetric DF and other trade names) 75 WDG	0.33 to 0.67 lb	0.25 to 0.5	Do not use on early maturing smooth-skinned white or red-skinned varieties. Apply only if there have been at least three successive days of sunny weather before application. Treat before weeds are 1 inch tall. Treatment may cause some chlorosis or minor necrosis. Do not apply within 60 days of harvest.
	rimsulfuron, MOA 2 (Matrix, Pruvin) 25 WDG	1 to 1.5 oz	0.016 to 0.023	Apply to young actively growing weeds after crop emergence.More effective on small weeds. Add nonionic surfactant at 1 to 2 pints per 100 gallons water. Can be tank mixed with Eptam or Sencor or some foliar fungicides. See label for further instructions. PHI 60 days.
Most broadleaf weeds less than 4 inches tall or rosettes less than 3 inches in diameter; does not control grasses	carfentrazone-ethyl, MOA 14 (Aim) 1.9 EW or 2 EC	up to 2 oz	up to 0.031	Apply post-directed using hooded sprayers for control of emerged weeds. If crop is contacted, burning of contacted area will occur. Use a nonionic surfactant or crop oil with Aim. See label for rate. Coverage is essential for good weed control. Can be tank mixed with other registered herbicides. Do not apply within 7 days of harvest.
Most emerged weeds	glyphosate, MOA 9 ((numerous brands and formulations)	See labels	See labels	**Row middles only.** Apply as a hooded spray in row middles, as shielded spray in row middles, as wiper applications in row middles, or postharvest. To avoid severe injury to crop, do not allow herbicide to contact foliage, green shoots, stems, exposed, roots, or fruit of crop. Do not apply within 14 days of harvest.
Annual and perennial grasses only	clethodim, MOA 1 (Arrow, Clethodim, Intensity. Select 2 EC (Intensity One, Select Max) 1 EC	6 to 8 oz 9 to 32 oz	0.094 to 0.125 0.07 to 0.25	Apply postemergence for control of grasses. With Arrow, Clethodim, Intensity or Select, add 1 quart crop of oil concentrate per acre. With Intensity One orSelect Max, nonionic surfactant at 2 pints per 100 gallons spray mixture can be used instead of crop oil concentrate. Adding crop oil may increase the likelihood of crop injury at high air temperatures. Very effective in controlling annual bluegrass. Apply to actively growing grasses not under drought stress. Do not apply within 30 days of harvest.
	sethoxydim, MOA 1 (Poast) 1.5 EC	1 to 1.5 pt	0.2 to 0.3	Apply to emerged grasses. Consult manufacturer's label for specific rates and best times to treat. Add 1 quart of crop oil concentrate per acre. Adding crop oil to Poast may increase the likelihood of crop injury at high air temperatures. Do not apply on days that are unusually hot and humid. Do not apply within 30 days of harvest.
Pumpkins, Preplant and Preemergence				
Contact kill of all green foliage, stale bed application	paraquat, MOA 22 (Firestorm, Parazone) 3 SL (Gramoxone SL) 2 SL	1.3 to 2.7 pt 2 to 4 pt	0.5 to 1	Apply in a minimum of 20 gallons spray mix per acre to emerged weeds before crop emergence or transplanting as a band or broadcast treatment over a preformed row. Use sufficient water to give thorough coverage. Row should be formed several days ahead of planting or treating to allow maximum weed emergence. Use a nonionic surfactant at a rate of 16 to 32 ounces per 100 gallons spray solution or 1 gallon approved crop oil concentrate per 100 gallons spray mix.
Most broadleaf weeds less than 4 inches tall or rosettes less than 3 inches in diameter; does not control grasses	carfentrazone-ethyl, MOA 14 (Aim) 1.9 EW or 2 EC	up to 2 oz	up to 0.031	**Not registered for use on seeded crop.** Apply prior to transplanting crop.Use a nonionic surfactant or crop oil with Aim. See label for rate. Coverage is essential for good weed control. Can be tank mixed with other registered burndown herbicides.

Table 7-16. Chemical Weed Control in Vegetable Crops

Weed	Herbicide, Mode of Action Code* and Formulation	Amount of Formulation Per Acre	Pounds Active Ingredient Per Acre	Precautions and Remarks
Pumpkins, Preplant and Preemergence (continued)				
Annual and perennial grass and broadleaf weeds, stale bed application	glyphosate, MOA 9 (numerous brands and formulations)	See labels	See labels	Apply to emerged weeds at least 3 days before seeding or transplanting. Perennial weeds may require higher rates of glyphosate. Consult the manufacturer's label for rates for specific weeds. Certain glyphosate formulations require the addition of a surfactant. Adding nonionic surfactant to glyphosate formulated with nonionic surfactant may result in reduced weed control.
Annual grasses and some small-seeded broadleaf weeds	bensulide, MOA 8 (Prefar) 4 EC	5 to 6 qt	5 to 6	Registered for cucurbit vegetable group (Crop grouping 9). Apply preplant and incorporate into the soil 1 to 2 inches (1 inch incorporation is optimum) with a rototiller or tandem disk, or apply to the soil surface after seeding and follow with irrigation. Check replant restrictions for small grains on label. See label for use rate if Prefar 4 EC is used.
	ethalfluralin, MOA 3 (Curbit) 3 EC	3 to 4.5 pt	1.1 to 1.7	Apply postplant to seeded crop prior to crop emergence, or as a banded spray between rows after crop emergence or transplanting. See label for timing. Shallow cultivation, irrigation, or rainfall within 5 days is needed for good weed control. Do not use under mulches, row covers, or hot caps. Under conditions of unusually cold or wet soil and air temperatures, crop stunting or injury may occur. Crop injury can occur if seeding depth is too shallow.
	ethalfluralin, MOA 3 + clomazone, MOA 13 (Strategy) 2.1 L	2 to 6 pt	0.4 to 1.2 + 0.125 to 0.375	Apply to the soil surface immediately after crop seeding for preemergence control of weeds. **Do not apply prior to planting the crop. Do not soil incorporate**. May also be used as a **banded** treatment **between** rows after crop emergence or transplanting.
Yellow and purple nutsedge suppression, non-ALS resistant pigweed, wild radish, and ragweed	halosulfuron-methyl, MOA 2 (Profine 75) 75 DG (Sandea) 75 DG	0.5 to 0.75 oz	0.024 to 0.036 lb	**Direct-seeded pumpkin or winter squash:** Apply after seeding but prior to soil cracking. **Transplanted pumpkin and winter squash:** Apply 7 days prior to transplanting. See label for specific rate. See label for crop rotational restrictions and other information. **Post Transplant in pumpkin and winter squash:** Can be applied as an over-the-top application, a directed spray application, or with crop shields. Apply to transplants that are established, actively growing and in the 3 to 5 true leaf stages or no sooner than 14 days after transplanting unless local conditions demonstrate safety at an earlier interval, but before first female flowers appear. **Row middle/furrow applications in direct-seeded and transplant pumpkin and winter squash:** Apply between rows of direct-seeded or transplanted crop while avoiding contact of the herbicide with the planted crop. If plastic is used on the planted row, adjust equipment to keep the application off the plastic. Reduce rate and spray volume in proportion to area actually sprayed. Rate can be increased to 1 oz per acrea if needed for row middle/furrow applications.
Annual broadleaf, grass and yellow nutsedge	S-metolachlor, MOA 15 (Brawl™, Dual Magnum)	1 to 1.33 pt	0.95 to 1.26	Apply as interrow or interhill application. Leave a 1 foot untreated area over the seeded row (6 in. on either side of the row). **Application made as a broadcast spray over the planted row or hill or directly to crop foliage will increase the risk of injury to the crop.** Apply before weeds emerge. See label for further instructions.
Pumpkins, Postemergence				
Yellow and purple nutsedge suppression, non-ALS resistant pigweed, wild radish, and ragweed	halosulfuron-methyl, MOA 2 (Profine 75) 75 DG (Sandea) 75 DG	0.5 to 0.75 oz	0.024 to 0.036 lb	**Direct-seeded pumpkin and winter squash:** Apply after crop has reached the 2 to 5 true leaf stage, preferably 4 to 5 true leaves, but before first female flowers appear. Do not apply within 30 days of harvest. **Post Transplant in pumpkin and winter squash:** Can be applied as an over-the-top application, a directed spray application, or with crop shields. Apply to transplants that are established, actively growing and in the 3 to 5 true leaf stages or no sooner than 14 days after transplanting unless local conditions demonstrate safety at an earlier interval, but before first female flowers appear. Do not apply within 30 days of harvest. **Row middle/furrow applications in direct-seeded and transplant pumpkin or winter squash:** Apply between rows of direct-seeded or transplanted crop while avoiding contact of the herbicide with the planted crop. If plastic is used on the planted row, adjust equipment to keep the application off the plastic. Reduce rate and spray volume in proportion to area actually sprayed. Rate can be increased to 1 oz per acrea if needed for row middle/furrow applications. Do not apply within 30 days of harvest.
Annual and perennial grasses only	clethodim, MOA 1 (Arrow, Clethodim, Intensity, Select) 2 EC (Intensity One, Select Max) 1 EC (Select Max) 1 EC	6 to 8 oz 9 to 16 oz	0.094 to 0.125 0.07 to 0.125	Apply postemergence for control of grasses. With Arrow, Clethodim, or Select, add 1 gallon crop oil concentrate per 100 gallons spray mix. With Select Max or Intensity One, add 2 pints of nonionic surfactant per 100 gallons spray mixture. Adding crop oil concentrate may increase the likelihood of crop injury at high air temperatures. Very effective in controlling annual bluegrass. Apply to actively growing grasses not under drought stress. Do not apply within 14 days of harvest.
	sethoxydim, MOA 1 (Poast) 1.5 EC	1 to 1.5 pt	0.2 to 0.3	Apply to emerged grasses. Consult manufacturer's label for specific rates and best times to treat. Add 1 quart of crop oil concentrate per acre. Crop oil may increase the likelihood of crop injury at high temperatures. Do not apply Poast on days that are unusually hot and humid. Do not apply within 14 days of harvest.
Pumpkins, Row Middles				
Annual grasses and some small-seeded broadleaf weeds	trifluralin, MOA 3 (Treflan) 4 EC (Treflan HFP) 4 EC	1 to 1.5 pt	0.5 to 0.75	**Row middles.** To improve preemergence control of late emerging weeds. Apply after emergence when crop plants have reached the three to four true leaf stage of growth. Apply as a directed spray to soil between the rows. Avoid contacting foliage as slight crop injury may occur. Set incorporation equipment to move treated soil around base of crop plants. Do not apply within 30 days of harvest.
Most broadleaf weeds less than 4 inches tall or rosettes less than 3 inches in diameter; does not control grasses	carfentrazone-ethyl, MOA 14 (Aim) 1.9 EW or 2 EC	up to 2 oz	up to 0.031	Apply post-directed using hooded sprayers for control of emerged weeds. If crop is contacted, burning of contacted area will occur. Use a nonionic surfactant or crop oil with Aim. See label for rate. Coverage is essential for good weed control. Can be tank mixed with other registered herbicides. PHI 0 days.

CHEMICAL WEED CONTROL

353

Table 7-16. Chemical Weed Control in Vegetable Crops

Weed	Herbicide, Mode of Action Code* and Formulation	Amount of Formulation Per Acre	Pounds Active Ingredient Per Acre	Precautions and Remarks
Pumpkins, Row Middles (continued)				
Most emerged weeds	glyphosate, MOA 9 (numerous brands and formulations)	See labels	See labels	**Row middles.** Apply as a hooded spray in row middles, as shielded spray in row middles, as wiper applications in row middles, or post harvest. To avoid severe injury to crop, do not allow herbicide to contact foliage, green shoots, stems, exposed, roots, or fruit of crop. Do not apply within 14 days of harvest.
Yellow and purple nutsedge and broadleaf weeds	halosulfuron-methyl, MOA 2 (Sandea, Profine 75) 75 DG	0.5 to 1 oz	0.024 to 0.048 lb	**Row middles.** Apply to row middles as a postemergence spray. In plasticulture, do not allow spray to contact plastic. Early season application will give postemergence and preemergence control. Do not apply within 30 days of harvest. For postemergence applications, use nonionic surfactant at 1 quart per 100 gallons of spray solution.
Radish, Preplant				
Annual and perennial grass and broadleaf weeds	glyphosate, MOA 9 (numerous brands and formulations)	See labels	See labels	Apply to emerged weeds before planting. Perennial weeds may require higher rates of glyphosate. Consult the manufacturer's label for rates for specific weeds. Certain glyphosate formulations may require addition of a surfactant. Adding nonionic surfactant to glyphosate formulated with nonionic surfactant may result in reduced weed control.
Annual grasses and broadleaf weeds	trifluralin, MOA 3 (Treflan, Treflan HFP, Trifluralin, Trifluralin HF) 4 EC	1 to 1.5 pt	0.5 to 0.75	Apply preplant and incorporate immediately after application for preemergence weed control. Low rate should be used on coarse-textured soil.
Radish, Postemergence				
Most broadleaf weeds less than 4 inches tall or rosettes less than 3 inches in diameter; does not control grasses	carfentrazone-ethyl, MOA 14 (Aim) 1.9 EW or 2 EC	up to 2 oz	up to 0.031	Apply post-directed using hooded sprayers for control of emerged weeds. If crop is contacted, burning of contacted area will occur. Use a nonionic surfactant or crop oil with Aim. See label for rate. Coverage is essential for good weed control. Can be tank mixed with other registered herbicides.
Annual and perennial grasses	clethodim, MOA 1 (Arrow, Clethodim, Intensity, Select) 2 EC	6 to 8 oz	0.94 to 0.125	Apply postemergence to emerged grasses. See label for rates for specific grasses. With Arrow, Clethodim, or Select, add crop oil concentrate at 1 gallon per 100 gallons of spray solution. With Select Max, add nonionic surfactant at 2 pints per 100 gallons spray mixture. Do not spray within 15 days of harvest.
	(Select Max, Intensity One) 1 EC	9 to 16 oz	0.07 to 0.125	
Spinach, Preemergence				
Annual and perennial grass and broadleaf weeds, stale bed application	glyphosate, MOA 9 ((numerous brands and formulations)	See labels	See labels	Apply to emerged weeds before crop emergence. Do not feed residue to livestock for 8 weeks. Perennial weeds may require higher rates of glyphosate. Consult the manufacturer's label for rates for specific weeds. Certain glyphosate formulations require the addition of a surfactant. Adding nonionic surfactant to glyphosate formulated with nonionic surfactant may result in reduced weed control.
Annual grasses (crabgrass spp., foxtail spp., barnyardgrass, annual ryegrass, annual bluegrass) and broadleaf weeds (Lamium spp., lambsquarters, common purslane, redroot pigweed, shepherdspurse)	cyclohexylethylthio-carbamate, MOA 3 (Ro-Neet) 6E	2 qt	3	Use on sandy mineral soils only. Read label for further instructions.
Spinach, Postemergence				
Broadleaf weeds including sowthistle clover, cocklebur, jimsonweed, and ragweed	clopyralid, MOA 4 (Stinger) 3 EC	0.17 to 0.33 pt	0.0625 to 0.125 lb	Apply to spinach in the 2- to 5-leaf stage when weeds are small and actively growing. Will control most legumes. See label for more precautions. Do not apply within 21 days of harvest.
Broadleaf weeds	phenmedipham, MOA 6 (Spin-aid) 1.3 EC	3 to 6 pt	0.5 to 1	**For processing spinach only.** Do not use when expected high temperatures will be above 75 degrees F. For best results, spray when weeds are in the two-leaf stage. Use the 6 pints rate only on well-established crops that are not under stress. Do not apply within 21 days of harvest. Spinach plants must have more than six true leaves.
Most broadleaf weeds less than 4 inches tall or rosettes less than 3 inches in diameter; does not control grasses	carfentrazone-ethyl, MOA 14 (Aim) 1.9 EW or 2 EC	up to 2 oz	up to 0.031	Apply post-directed using hooded sprayers for control of emerged weeds. If crop is contacted, burning of contacted area will occur. Use a nonionic surfactant or crop oil with Aim. See label for rate. Coverage is essential for good weed control. Can be tank mixed with other registered herbicides.
Most emerged weeds	glyphosate, MOA 9 (numerous brands and formulations)	See labels	See labels	**Row middles only.** Apply as a hooded spray in row middles, as shielded spray in row middles, as wiper applications in row middles, or post harvest. To avoid severe injury to crop, do not allow herbicide to contact foliage, green shoots, stems, exposed, roots, or fruit of crop. Do not apply within 14 days of harvest.
Annual and perennial grasses only	sethoxydim, MOA 1 (Poast) 1.5 EC	1 to 1.5 pt	0.2 to 0.3	Apply to emerged grasses. Consult manufacturer's label for specific rates and best times to treat. For sethoxydim, add 1 quart of crop oil concentrate per acre. For Arrow, Clethodim, or Select, add 1 gallon of crop oil concentrate per 100 gallons spray solution. For Select Max, add nonionic surfactant at 2 pints per 100 gallons of spray mixture. Adding crop oil to Poast or Select may increase the likelihood of crop injury at high air temperatures. Do not apply Poast, Arrow, Clethodim, or Select on days that are unusually hot and humid. Do not apply sethoxydim within 15 days of harvest or clethodim within 14 days of harvest.
	clethodim, MOA 1 (Arrow, Clethodim, Intensity, Select) 2 EC	6 to 8 oz	0.094 to 0.125	
	(Intensity One, Select Max) 1 EC	9 to 16 oz	0.07 to 0.125	

Table 7-16. Chemical Weed Control in Vegetable Crops

Weed	Herbicide, Mode of Action Code* and Formulation	Amount of Formulation Per Acre	Pounds Active Ingredient Per Acre	Precautions and Remarks
Squash, Preplant and Preemergence				
Suppression or control of most annual grasses and broadleaf weeds, full rate required for nutsedge control	metam sodium (Vapam HL) 42%	37.5 to 75 gal	15.7 to 31.5	Rates are dependent on soil type and weeds present. Apply when soil moisture is at field capacity (100 to 125%). Apply through soil injection using a rotary tiller or inject with knives no more than 4 inches apart; follow immediately with a roller to smooth and compact the soil surface or with mulch. May apply through drip irrigation prior to planting a second crop on mulch. Plant back interval is often 14 to 21 days and can be 30 days in some environments. See label for all restrictions and additional information.
Emerged broadleaf and grass weeds	pelargonic, (Scythe) 4.2 EC	3 to 10% v/v		Apply before crop emergence and control emerged weeds. There is no residual activity. May be tank mixed with soil residual compounds. See lable for instruction. May also be used as a banded spray between row middles. Use a shielded sprayer directed to the row middles to reduce drift to the crop.
Contact kill of all green foliage, stale bed application	paraquat, MOA 22 (Firestorm, Parazone) 3 SL (Gramoxone SL) 2 SL	1.3 to 2.7 pt 2 to 4 pt	0.5 to 1	Apply in a minimum of 10 gallons spray mix per acre to emerged weeds before transplanting or crop emergence as a band or broadcast treatment over a preformed row. Use sufficient water to give thorough coverage. Row should be formed several days ahead of planting or treating to allow maximum weed emergence. Use a nonionic surfactant at a rate of 16 to 32 ounces per 100 gallons spray mix or 1 gallon approved crop oil concentrate per 100 gallons spray mix.
Most broadleaf weeds less than 4 inches tall or rosettes less than 3 inches in diameter; does not control grasses	carfentrazone-ethyl, MOA 14 (Aim) 1.9 EW or 2 EC	up to 2 oz	up to 0.031	Not registered for seeded crop. Apply prior to transplanting crop. Use a nonionic surfactant or crop oil with Aim. See label for rate. Coverage is essential for good weed control. Can be tank mixed with other registered burndown herbicides.
Annual and perennial grass and broadleaf weeds, stale bed application	glyphosate, MOA 9 (numerous brands and formulations)	See labels	See labels	Apply to emerged weeds at least 3 days before seeding or transplanting. When applying Roundup before transplanting crops into plastic mulch, care must be taken to remove residues of this product from the plastic prior to transplanting. To prevent crop injury, residues can be removed by 0.5 inch natural rainfall or by applying water via a sprinkler system. Perennial weeds may require higher rates of glyphosate. Consult the manufacturer's label for rates for specific weeds. Certain glyphosate formulations require the addition of a surfactant. Adding nonionic surfactant to glyphosate formulated with nonionic surfactant may result in reduced weed control.
Annual grasses and small-seeded broadleaf weeds	bensulide, MOA 8 (Prefar) 4 EC	5 to 6 qt	5 to 6	Registered for cucurbit vegetable group (Crop grouping 9). Apply preplant and incorporate into the soil 1 to 2 inches (1 inch incorporation is optimum) with a rototiller or tandem disk, or apply to the soil surface after seeding and follow by irrigation. Check replant restrictions for small grains on label.
	ethalfluralin, MOA 3 (Curbit) 3 EC	1.5 to 2 pt	0.56 to .75	**For squash grown on bare ground only.** Apply to the soil surface immediately after seeding. Seed must be covered with soil to prevent crop injury. For coarse-textured soils, use lowest rate of rate range. Shallow cultivation, irrigation, or rainfall within 5 days is needed for good weed control. If weather is unusually cold or soil wet and cold, crop stunting or injury may occur. Crop injury can also occur if seeding depth is too shallow. See label for further precautions and instruction.
		3 to 4.5 pt	1.1 to 1.7	For squash grown on plastic only. Apply to soil surface between the rows of black plastic immediately after seeding or transplanting. **Do not use under mulches, row covers, or hot caps.** Do not apply prior to planting or over plastic. See label for further instruction.
Annual grasses and broadleaf weeds	ethalfluralin, MOA 3 + clomazone, 13 (Strategy) 2.1 L	2 to 6 pt	0.4 to 1.2 + 0.125 to 0.375	Apply to the soil surface immediately after crop seeding for preemergence control of weeds. **Do not apply prior to planting crop. Do not soil incorporate.** May also be used as a **banded** treatment **between** rows after crop emergence or transplanting.
Suppression of annual grasses and broadleaf weeds; weak on pigweed and morningglory	clomazone, MOA 13 (Command) 3 ME	0.67 to 1.3 pt	0.25 to 0.48	Apply immediately after seeding or prior to transplanting. Seeds and roots of transplants must be below the chemical barriers when planting. Command should only be applied between rows when squash is grown on plastic. Some cultivars may be sensitive to Command (see label). Use lower rates on coarse soils. Higher rates can be used on winter squashes. See label about rotation restrictions.
Yellow and purple nutsedge and broadleaf weeds	halosulfuron-methyl, MOA 2 (Sandea, Profine 75) 75 DG	0.5 to 1 oz	0.024 to 0.048	**Row middles only.** Apply to row middles as preemergence spray. In plasticulture, do not allow spray to contact plastic. Early season application will give postemergence and preemergence control. Do not apply within 30 days of harvest. For postemergence applications, use nonionic surfactant at 1 quart per 100 gallons of spray solution. **WINTER SQUASH: Direct seeded:** Apply 0.5 to 0.75 ounce per acre Sandea as preemergence after planting, but before soil cracking **OR** postemergence after crop has reached 2 to 5 true leaf stage, but before first female flowers appear. **Transplanted:** Apply 0.5 to 0.75 ounce per acre Sandea prior to transplant. Transplanting should not be made sooner than 7 days after application. May be applied after crop emergence over the top of the crop when plants reach the four to five true leaf stage but before first female flowers appear. See label for further instructions.
Squash, Postemergence				
Annual grasses and small-seeded broadleaf weeds	trifluralin, MOA 3 (Treflan) 4 EC (Treflan HFP) 4 EC	1 to 1.5 pt	0.5 to 0.75	**Row middles only.** To improve preemergence control of late emerging weeds. **Apply as a directed spray to soil between rows after crop emergence when crop plants have reached** three to four true leaf stage of growth. Avoid contacting foliage as slight crop injury may occur. Set incorporation equipment to move treated soil around base of crop plants. Do not apply within 30 days of harvest. Will not control emerged weeds.
Most broadleaf weeds less than 4 inches tall or rosettes less than 3 inches in diameter; does not control grasses	carfentrazone-ethyl, MOA 14 (Aim) 1.9 EW or 2 EC	up to 2 oz	up to 0.031	Apply post-directed using hooded sprayers for control of emerged weeds. If crop is contacted, burning of contacted area will occur. Use a nonionic surfactant or crop oil with Aim. See label for rate. Coverage is essential for good weed control. Can be tank mixed with other registered herbicides. PHI 0 days.
Most emerged weeds	glyphosate, MOA 9 (numerous brands and formulations)	See labels	See labels	**Row middles only.** Apply as a hooded spray in row middles, as shielded spray in row middles, as wiper applications in row middles, or post harvest. To avoid severe injury to crop, do not allow herbicide to contact foliage, green shoots, stems, exposed, roots, or fruit of crop. Do not apply within 14 days of harvest.

Table 7-16. Chemical Weed Control in Vegetable Crops

Weed	Herbicide, Mode of Action Code* and Formulation	Amount of Formulation Per Acre	Pounds Active Ingredient Per Acre	Precautions and Remarks
Squash, Postemergence (continued)				
Yellow and purple nutsedge and broadleaf weeds	halosulfuron-methyl, MOA 2 (Sandea, Profine 75) 75 DG	0.5 to 1 oz	0.024 to 0.048 lb	**Row middles only.** Apply to row middles as postemergence spray. In plasticulture, do not allow spray to contact plastic. Early season application will give postemergence and preemergence control. Do not apply within 30 days of harvest. For postemergence applications, use nonionic surfactant at 1 quart per 100 gallons of spray solution.
Annual and perennial grasses only	clethodim, MOA 1 (Arrow, Clethodim, Intensity, Select) 2 EC (Intensity One, Select Max) 1 EC	6 to 8 oz 9 to 16 oz	0.094 to 0.125 0.07 to 0.125	Apply postemergence for control of grasses. With Arrow, Clethodim, or Select, add 1 gallon crop oil concentrate per 100 gallons spray mix. With Select Max or Intensity One, add 2 pints nonionic surfactant per 100 gallons spray mixture. Adding crop oil may increase likelihood of crop injury at high air temperatures. Very effective control of annual bluegrass. Apply to actively growing grasses not under drought stress. Do not apply within 14 days of harvest.
	sethoxydim, MOA 1 (Poast)1.5 EC	1 to 1.5 pt	0.2 to 0.3	Apply to emerged grasses. Consult manufacturer's label for specific rates and best times to treat. Add 1 quart of crop oil concentrate per acre. Adding crop oil to Poast may increase the likelihood of crop injury at high air temperatures. Do not apply Poast on days that are unusually hot and humid. Do not apply within 14 days of harvest.
Sweetpotato, Preplant				
Annual and perennial grass and broadleaf weeds, stale seed bed application	glyphosate, MOA 9 (numerous brands and formulations)	See labels	See labels	Apply to emerged weeds before transplanting. Perennial weeds may require higher glyphosate rates. Consult label for rates for specific weeds. Certain glyphosate formulations may require the addition of a surfactant. Adding nonionic surfactant to glyphosate formulated with nonionic surfactant may result in reduced weed control.
Suppression or control of most annual grasses and broadleaf weeds, full rate required for nutsedge control	metam sodium (Vapam HL) 42%	37.5 to 75 gal	15.7 to 31.5	Rates are dependent on soil type and weeds present. Plant back interval is often 14 to 21 days and can be 30 days in some environments. See label for all restrictions and additional information.
Annual broadleaf weeds including Palmer amaranth and other pigweeds, smartweed, morningglory, wild mustard, wild radish, common purslane, common lambsquarters	flumioxazin, MOA 14 (Valor SX) 51 WDG	3 oz	0.094	Apply 2 to 5 days prior to transplanting crop for control of many annual broadleaf weeds and annual sedges. Movement of soil during transplanting should not occur or reduced weed control may result. Do not use on greenhouse-grown transplants. Do not apply postemergence or serious crop injury will occur. Do not use on transplants harvested more than 2 days prior to transplanting. Do not use on transplant propagation beds. See label for further instructions.
Sweetpotato, Preemergence				
Annual grass and broadleaf weeds, Palmer amaranth, yellow nutsedge suppression	S-metolachlor, MOA 15 (Dual Magnum) 7.62 EC	0.75 pt	1.0	This is a Section 24C special local needs label. Growers must check farmassist.com website to make sure Dual Magnum is registered for use in their state. Obtain label from www.farmassist.com prior to making Dual Magnum applications. Apply over top of sweetpotatoes after transplanting but prior to weed emergence. Do not apply preplant. Do not incorporate after application. Injury potential is greatest when applied to sands or loamy sands especially if a heavy rainfall event occurs following application. See label for further information.
Annual grasses such as large crabgrass and broadleaf weeds including velvetleaf, purslane, prickly sida	clomazone, MOA 13 (Command) 3 ME	up to 2 pt	up to 0.75	Apply preplant or after transplanting prior to weed emergence for preemergence control. Weak on pigweed. The label allows up to 4 pt per acre. See label for other instructions and precautions.
Annual grasses including large crabgrass and broadleaf weeds including purslane, Florida pusley, common lambsquarters	DCPA, MOA 3 (Dacthal) W-75 (Dacthal) 6 F	8 to 10 lb 8 to 10 pt	6 to 7.5	Apply to the soil surface immediately after transplanting. May also be applied at layby for preemergence weed control late in the growing season. Do not apply in plant beds or crop injury will occur.
Annual grasses including crabgrass, foxtail, goosegrass, fall panicum and broadleaf weeds including pigweed, Florida pusley, purslane	napropamide, MOA 15 (Devrinol, Devrinol DF-XT) 50 DF (Devrinol, Devrinol 2-XT) 2 EC	2 to 4 lb 2 to 4 qt	1 to 2	**Plant Beds.** Apply to the soil surface after sweetpotato roots are covered with soil but prior to soil cracking and sweetpotato plant emergence. Does not control emerged weeds. Check label for more information. **Production Fields.** Apply to the soil surface immediately after transplanting. If rainfall does not occur within 24 hr, shallow incorporate or irrigate with sufficient water to wet the soil to a depth of 2 to 4 inches Check label for more information. See XT labels for information regarding delay in irrigation event.
Sweetpotato, Postemergence				
Annual and perennial grasses only	clethodim, MOA 1 (Arrow, Clethodim, Intensity, Select) 2 EC (Select Max, Intensity One) 1 EC	6 to 16 oz 9 to 32 oz	0.094 to 0.25 0.07 to 0.25	Apply to actively growing grasses not under drought stress. For Arrow, Clethodim, or Select, add 1 gallon crop oil concentrate per 100 gallons spray mix. For Select Max, add 2 pints nonionic surfactant per 100 gallons spray mixture. Adding crop oil may increase the likelihood of crop injury at high air temperatures. Very effective in controlling annual bluegrass. Do not apply within 30 days of harvest.
	fluazifop, MOA 1 (Fusilade DX) 2 EC	6 to 16 oz	0.1 to 0.25	Apply to actively growing grasses not under drought stress. Consult manufacturer's label for specific rates and best times to treat. Add 1 gallon crop oil concentrate or 1 quart nonionic surfactant per 100 gallons spray mix. Do not apply Fusilade on days that are unusually hot and humid. Do not apply within 14 days of harvest.
	sethoxydim, MOA 1 (Poast) 1.5 EC	1 to 1.5 pt	0.2 to 0.3	Apply to actively growing grasses not under drought stress. Add 1 quart of crop oil concentrate per acre. Adding crop oil to Poast may increase the likelihood of crop injury at high air temperatures. Do not apply Poast on days that are unusually hot and humid. Do not apply within 30 days of harvest.

Table 7-16. Chemical Weed Control in Vegetable Crops

Weed	Herbicide, Mode of Action Code* and Formulation	Amount of Formulation Per Acre	Pounds Active Ingredient Per Acre	Precautions and Remarks
Sweetpotato, Row Middles				
Most broadleaf weeds less than 4 inches tall or rosettes less than 3 inches in diameter; does not control grasses	carfentrazone-ethyl, MOA 14 (Aim) 1.9 EW or 2 EC	up to 2 oz	up to 0.031	Apply post-directed using hooded sprayers for control of emerged weeds. If crop is contacted, burning of contacted area will occur. Use a nonionic surfactant or crop oil with Aim. See label for rate. Coverage is essential for good weed control. Can be tank mixed with other registered herbicides.
Most emerged weeds	glyphosate, MOA 9 ((numerous brands and formulations)	See labels	See labels	Apply as a hooded spray in row middles, as shielded spray in row middles, as wiper applications in row middles, or postharvest. To avoid severe injury to crop, do not allow herbicide to contact foliage, green shoots, stems, exposed, roots, or fruit of crop. May cause cracking of sweetpotato storage roots if spray solution comes in contact with sweetpotato foliage. Do not apply within 14 days of harvest.
Tomato, Preplant				
Suppression or control of most annual grasses and broadleaf weeds, full rate required for nutsedge control	metam sodium (Vapam HL) 42%	37.5 to 75 gal	15.7 to 31.5	Rates are dependent on soil type and weeds present. Apply when soil moisture is at field capacity (100 to 125%). Apply through soil injection using a rotary tiller or inject with knives no more than 4 inches apart; follow immediately with a roller to smooth and compact the soil surface or with mulch. May apply through drip irrigation prior to planting a second crop on mulch; however, adhere to label guidelines on crop plant back interval. Plant back interval is often 14 to 21 days and can be 30 days in some environments. See label for all restrictions and additional information. Chloropicrin (150 pounds per acre broadcast) will also be needed when laying first crop mulch to control nutsedge.
Most broadleaf weeds less than 4 inches tall or rosettes less than 3 inches in diameter; does not control grasses	carfentrazone-ethyl, MOA 14 (Aim) 1.9 EW or 2 EC	up to 2 oz	up to 0.031	**Transplanted crop: Apply no later than 1 day before transplanting.** **Seeded crop (Aim 2EC only): Apply no later than 7 days before planting seeded crop.** Use a nonionic surfactant or crop oil with Aim. See label for rate. Coverage is essential for good weed control. Can be tank mixed with other registered burndown herbicides.
Contact kill of all green foliage, stale bed application	paraquat, MOA 22 (Firestorm, Parazone) 3 SL	1.3 to 2.7 pt	0.5 to 1	Apply to emerged weeds in a minimum of 20 gallons spray mix per acre before crop emergence as a broadcast or band treatment over a preformed row. Row should be formed several days ahead of planting and treating to allow maximum weed emergence. Use a nonionic surfactant at a rate of 16 to 32 ounces per 100 gallons spray mix or 1 gallon approved crop oil concentrate per 100 gallons spray mix.
	(Gramoxone SL) 2 SL	2 to 4 pt		
Broadleaf weeds including Carolina geranium and cutleaf eveningprimrose and a few annual grasses	oxyfluorfen, MOA 14 (Goal) 2 XL	up to 2 pt	0.5 lb	**Plasticulture only.** Apply to soil surface of pre-formed beds at least 30 days prior to transplanting crop. While incorporation is not necessary, it may result in less crop injury. Plastic mulch can be applied any time after application but best results are likely if applied soon after application.
Annual grasses and small-seeded broadleaf weeds including common lambsquarters, pigweed, carpetweed, and common purslane	napropamide, MOA 15 (Devrinol, Devrinol DF-XT) 50 DF (Devrinol, Devrinol 2-XT) 2 EC	2 to 4 lb 2 to 4 qt	1 to 2	**Bare ground:** Apply preplant and incorporate into the soil 1 to 2 inches as soon as possible with a rototiller or tandem disk. Can be used on direct-seeded or transplanted tomatoes. See label for instructions on use. **Plasticulture:** Apply to a weed-free soil before laying plastic mulch. Soil should be well worked yet moist enough to permit a thorough incorporation to a depth of 2 inches. Mechanically incorporate or irrigate within 24 hours after application. If weed pressure is from small seeded annuals, apply to the surface of the bed immediately in front of the laying of plastic mulch. If soil is dry, water or sprinkle irrigate with sufficient water to wet to a depth of 2 to 4 inches before covering with plastic mulch. **Between rows:** Apply to a weed free soil surface between the rows (bareground or plastic mulch). Mechanically incorporate or irrigate Devrinol into the soil to a depth of 1 to 2 inches within 24 hours of application. See XT labels for information regarding delay in irrigation event.
	pendimethalin, MOA 3 (Prowl H$_2$O) 3.8	1 to 3 pt	0.5 to 1.5	**Plasticulture In-row:** May be applied as a preplant surface application or a preplant incorporated application prior to transplanting tomato. Limited research has been conducted in NC. **Bareground In-row:** May be applied as a broadcast preplant surface application or preplant incorporated application prior to transplanting tomato. **Post-directed spray:** May be applied as a post-directed spray on the soil at the base of the plant, beneath plants, and between rows. Avoid direct contact with tomato foliage or stems. Do not apply over the top of tomato. Do not apply within 21 days of harvest. Do not apply more than 3 pt per acre per season. See label for specific use rate for your soil type. Emerged weeds will not be controlled. See label for further instructions and precautions.
	trifluralin, MOA 3 (Treflan HFP, Trifluralin, Trifluralin HF, various other trade names),) 4 EC	1 pt	0.5	**Transplant tomato.** Apply pretransplant and incorporate into the soil 2 to 3 inches within 8 hr using a rototiller or tandem disk. Can be applied post-plant as a directed spray to soil between the rows and beneath plants and then incorporated.
Yellow and purple nutsedge and broadleaf weeds including pigweed, wild radish, common ragweed, suppression of purslane	halosulfuron-methyl, MOA 2 (Profine, Sandea) 75 DG	0.5 to 1 oz	0.024 to 0.048	For pretransplant application under plastic mulch, apply to pre-formed bed just prior to plastic mulch application and delay transplanting at least 7 days. Can be applied for pretransplant application in bareground tomato. Early season application will give postemergence and preemergence control. The 1 ounce rate is for preemergence and postemergence control in row middles only. For postemergence applications, use nonionic surfactant at 1 quart per 100 gallons of spray solution. Do not apply within 30 days of harvest.
Yellow nutsedge, annual grasses, and broadleaf weeds including pigweed, Palmer amaranth, Florida pusley, Hairy galinsoga, Eastern black nightshade, and carpetweed	S-metolachlor, MOA 15 (Brawl, Dual Magnum) 7.62 EC	1 to 2 pt	0.95 to 1.50	Apply preplant or postdirected to transplants after the first settling rain or irrigation. In plasticulture, apply preformed beds just prior to applying plastic mulch. Lower rates of rate range for S-metolachlor are safest to tomato. May also be used to treat row middles in bedded tomato. Minimize contact with crop. Do not apply within 90 days of harvest. Also registered for use in row middles, and in seeded crop. See label for further instructions.

Table 7-16. Chemical Weed Control in Vegetable Crops

Weed	Herbicide, Mode of Action Code* and Formulation	Amount of Formulation Per Acre	Pounds Active Ingredient Per Acre	Precautions and Remarks
Tomato, Preplant (continued)				
Palmer amaranth, redroot pigweed, smooth pigweed, Galinsoga sp., black nightshade, Eastern black nightshade, common purslane, partial control of yellow nutsedge	fomesafen, MOA 14 (Reflex) 2 EC	1 to 1.5 pt	0.25 to 0.375	This is a Section 24(c) special local needs label for transplanted tomato in North Carolina. Growers must obtain the label at Farmassist.com prior to making an application of Reflex. **See label for further instructions.** **Plasticulture In-row Application for Transplanted Tomato.** Apply after final bed formation and the drip tape is laid but prior to laying plastic mulch. Avoid soil disturbance after application. Unless restricted by other products such as fumigants, tomato may be transplanted immediately following the application of Reflex and the application of the mulch. **Bareground for Transplanted Tomato.** Apply pretransplant up to 7 days prior to transplanting tomato. Weed control will be reduced if soil is disturbed after application. During the transplanting operation, make sure the soil in the transplant hole settles flush or above the surrounding soil surface. Avoid cultural practices that may concentrate Reflex-treated soil around the transplant root ball. An overhead irrigation or rainfall event between Reflex herbicide application and transplanting will ensure herbicide activation and will likely reduce the potential for crop injury due to splashing. **Plasticulture Row Middle Application.** Apply to row middles with a hooded or shielded sprayer. Avoid drift of herbicide on mulch. If drift occurs, 0.5 inch of rain or irrigation must occur prior to transplanting. **Carryover is a large concern; see label for more information.**
Annual grasses and broadleaf weeds including jimsonweed, common ragweed, smartweed, and velvetleaf	metribuzin, MOA 5 (TriCor DF, Metribuzin) 75 WDG	0.33 to 0.67 lb	0.25 to 0.5	Apply to soil surface and incorporate 2 to 4 inches deep before transplanting. See label for instructions.
Broadleaf weeds including Carolina geranium and cutleaf eveningprimrose and a few annual grasses	oxyfluorfen, MOA 14 (Goal) 2XL (GoalTender) 4 F	up to 2 pt up to 1 pt	0.5 lb	**Plasticulture (fallow beds) only.** Apply to soil surface of pre-formed beds at least 30 days prior to transplanting crop. While incorporation is not necessary, it may result in less crop injury. Plastic mulch can be applied any time after application but best results are likely if applied soon after application.
Broadleaf, grass (suppression),yellow nutsedge (PRE or POST), purple nutsedge (POST only)	imazosulfuron, MOA 2 (League) 0.5 DF	4 to 6.4 oz	0.19 to 0.3	Apply to planting beds before plastic is laid. Tomato may be transplanted 1 day after application. Postemergence application may be made on direct-seeded plants that are well established (4 to 5 leaf stage) and in transplanted tomato 3 to 5 days after transplanting. 10 inches tall, **if a pretransplant application was not made.** Consult label for approved surfactants and crop rotation restrictions. PHI 21 days.
Tomato, Postemergence				
Annual grasses and small-seeded broadleaf weeds	DCPA, MOA 3 (Dacthal) W-75 (Dacthal) 6 F	6 to 10 lb 6 to 10 pt	4.5 to 7.5	Apply over the top of transplants only between 4 to 6 weeks after transplanting to improve preemergence control of late emerging weeds. Will not control emerged weeds.
Yellow and purple nutsedge and broadleaf weeds	halosulfuron-methyl, MOA 2 (Sandea, Profine 75) 75 DG	0.5 to 1 oz	0.024 to 0.048 lb	Apply no sooner than 14 days after transplanting For postemergence applications, use nonionic surfactant at 1 quart per 100 gallons of spray solution. Some weeds, such as nutsedge, may require two applications of Sandea; if a second application is needed, spot-treat only weed-infested areas. Do not apply within 30 days of harvest. See label for further instructions.
Annual grasses and broadleaf weeds, including cocklebur, common ragweed, smartweed, velvetleaf, jimsonweed, yellow nutsedge, and morningglory	metribuzin, MOA 5 (TriCor DF, Metribuzin) 75 WDG	0.33 to 1.33 lb	0.25 to 1	Use either as a broadcast or directed spray but do not exceed 0.5 pound a.i. with a broadcast spray. Do not apply within 7 days of harvest. Do not exceed 1 pound a.i. per year. Do not apply as a broadcast spray unless 3 sunny days precede application.
Most broadleaf weeds including wild radish, common purslane, redroot and smooth pigweed	rimsulfuron, MOA 2 (Matrix) 25 WDG (Pruvin) 25 WDG	1 to 2 oz	0.25 to 0.5 oz	Apply in tomatoes after the crop has at least two true leaves and weeds are small (1 inch or less) and actively growing. Add nonionic surfactant at 1 quart per 100 gallons of spray solution. Do not apply within 45 days of tomato harvest. See label for further instruction.
Yellow nutsedge, morningglory, common cocklebur, common lambsquarters, and other broadleaf weeds	trifloxysulfuron-sodium, MOA 2 (Envoke) 75 DG	0.1 to 0.2 oz	0.0047 to 0.0094	Apply post-directed to tomato grown on plastic for control of nutsedge and certain broadleaf weeds. Crop should be transplanted at least 14 days prior to application. The application should be made prior to fruit set and at least 45 days prior to harvest. Use nonionic surfactant at 1 quart per 100 gallons spray solution with all applications.
Annual and perennial grasses only	clethodim, MOA 1 (Arrow, Clethodim, Intensity, Select) 2 EC (Select Max, Intensity One) 1 EC	6 to 16 fl oz 9 to 32 oz	0.094 to 0.25 0.07 to 0.25	Apply to actively growing grasses not suffering from drought stress. With Arrow, Clethodim, or Select, add a crop oil concentrate at 1% by volume (1 gallon per 100 gallons spray mix). With Select Max, add 2 pints of nonionic surfactant per 100 gallons spray mixture. Adding crop oil may increase the likelihood of crop injury at high air temperatures. Do not apply on unusually hot and humid days. Very effective in controlling annual bluegrass. Do not apply within 20 days of harvest.
	sethoxydim, MOA 1 (Poast) 1.5 EC	1 to 1.5 pt	0.2 to 0.3	Apply to actively growing grasses not under drought stress. Add 1 quart of crop oil concentrate per acre. Adding crop oil to Poast may increase the likelihood of crop injury at high air temperatures. Do not apply Poast on days that are unusually hot and humid. Do not apply within 20 days of harvest.
Tomato, Row Middles				
Yellow nutsedge, morningglory, common cocklebur, common lambsquarters, and other broadleaf weeds	trifloxysulfuron-sodium, MOA 2 (Envoke) 75 DG	0.1 to 0.2 oz	0.0047 to 0.0094	Crop should be transplanted at least 14 days prior to application. Use nonionic surfactant at 1 quart per 100 gallons spray solution with all applications. The application should be made prior to fruit set and at least 45 days prior to harvest. See label for information on registered tank mixes. Tank mixtures with Select or Poast may reduce grass control. See label for more information.
Yellow and purple nutsedge and broadleaf weeds	halosulfuron-methyl, MOA 2 (Profine, Sandea) 75 DG	0.5 to 1 oz	0.024 to 0.048 lb	For postemergence applications, use nonionic surfactant at 1 quart per 100 gallons of spray solution. Some weeds, such as nutsedge, may require two applications of Sandea; if a second application is needed, spot-treat only weed-infested areas. Do not apply within 30 days of harvest. See label for further instructions.

CHEMICAL WEED CONTROL

Table 7-16. Chemical Weed Control in Vegetable Crops

Weed	Herbicide, Mode of Action Code* and Formulation	Amount of Formulation Per Acre	Pounds Active Ingredient Per Acre	Precautions and Remarks
Tomato, Row Middles (continued)				
Most broadleaf weeds less than 4 inches tall or rosettes less than 3 inches in diameter; does not control grasses	carfentrazone-ethyl, MOA 14 (Aim) 1.9 EW or 2 EC	up to 2 oz	up to 0.031	Apply post-directed using hooded sprayers for control of emerged weeds. If crop is contacted, burning of contacted area will occur. .Use a nonionic surfactant or crop oil with Aim. See label for rate. Coverage is essential for good weed control. Can be tank mixed with other registered herbicides. PHI 0 days.
Annual grasses and small-seeded broadleaf weeds	napropamide, MOA 15 (Devrinol, Devrinol DF-XT) 50 DF (Devrinol, Devrinol 2-XT) 2 EC	2 to 4 lb 2 to 4 qt	1 to 2	**Plasticulture:** Apply to a weed-free soil surface. Apply within 24 hours of rainfall, or mechanically incorporate or irrigate into the soil to a depth of 1 to 2 inches
	pendimethalin, MOA 3 (Prowl H₂O) 3.8	1 to 3 pt	0.5 to 1.5	Post-directed spray on the soil at the base of the plant, beneath plants and between rows. Avoid direct contact with tomato foliage or stems. Do not apply more than 3 pints per acre per season. See label for specific use rate for your soil type. Emerged weeds will not be controlled. Avoid direct contact with tomato foliage or stems. Do not apply within 21 days of harvest. See label for further instructions and precautions.
Contact kill of all green foliage	paraquat, MOA 22 (Firestorm, Parazone) 3 SL (Gramoxone SL) 2 SL	1.3 pt 2 pt	0.5 to 1	Apply for control of emerged weeds between rows of tomatoes. Do not allow spray to contract crop or injury will occur. Do not make more than 3 applications per season. Do not apply within 30 days of harvest.
Watermelon, Preplant				
Suppression or control of most annual grasses and broadleaf weeds, full rate required for nutsedge control	metam sodium (Vapam HL) 42%	37.5 to 75 gal	15.7 to 31.5	Rates are dependent on soil type and weeds present. Apply when soil moisture is at field capacity (100 to 125%). Apply through soil injection using a rotary tiller or inject with knives no more than 4 inches apart; follow immediately with a roller to smooth and compact the soil surface or with mulch. May apply through drip irrigation prior to planting a second crop on mulch. Plant back interval is often 14 to 21 days and can be 30 days in some environments. See label for all restrictions and additional information.
Contact kill of all green foliage, stale bed application	paraquat, MOA 22 (Firestorm, Parazone) 3 SL (Gramoxone SL) 2 SL	1.3 to 2.7 pt 2 to 4 pt	0.5 to 1	Apply in a minimum of 10 gallons spray mix per acre to emerged weeds before crop emergence or transplanting as a broadcast or band treatment over a preformed row. Row should be formed several days ahead of planting and treating to allow maximum weed emergence. Plant with a minimum of soil movement for best results. Use a nonionic surfactant at a rate of 16 to 32 ounces per 100 gallons spray mix or 1 gallon approved crop oil concentrate per 100 gallons spray mix.
Most broadleaf weeds less than 4 inches tall or rosettes less than 3 inches in diameter; does not control grasses	carfentrazone-ethyl, MOA 14 (Aim) 1.9 EW or 2 EC	up to 2 oz	up to 0.031	**Transplants only.** Apply prior to transplanting of crop. Use a nonionic surfactant or crop oil with Aim. See label for rate. Coverage is essential for good weed control. Can be tank mixed with other registered burndown herbicides.
Annual and perennial grass and broadleaf weeds, stale bed application	glyphosate, MOA 9 (numerous brands and formulations)	See labels	See labels	Apply to emerged weeds at least 3 days before seeding or transplanting. When applying Roundup before transplanting crops into plastic mulch, care must be taken to remove residues of this product from the plastic prior to transplanting. To prevent crop injury, residues can be removed by 0.5 inch natural rainfall or by applying water via a sprinkler system. Perennial weeds may require higher rates of glyphosate. Consult the manufacturer's label for rates for specific weeds. Certain glyphosate formulations require the addition of a surfactant. Adding nonionic surfactant to glyphosate formulated with nonionic surfactant may result in reduced weed control.
Annual grasses	bensulide, MOA 8 (Prefar) 4 E	5 to 6 qt	5 to 6	Registered for cucurbit vegetable group (Crop grouping 9). Apply preplant and incorporate into the soil 1 to 2 inches (1 inch incorporation is optimum) with a rototiller or tandem disk, or apply to the soil surface after seeding and follow with irrigation. Check replant restrictions for small grains on label.
Watermelon, Preemergence				
Annual grasses and broadleaf weeds	clomazone, MOA 13 (Command) 3 ME	0.4 to 0.67 pt	0.15 to 0.25	Apply immediately after seeding, or just prior to transplanting. Roots of transplants must be below the chemical barrier when planting. Offers weak control of pigweed. See label for instructions.
Annual grasses and some small-seeded broadleaf weeds	ethalfluralin, MOA 3 (Curbit) 3 EC	3 to 4.5 pt	1.1 to 1.7	Apply postplant to seeded crop prior to crop emergence, or as a banded spray between rows after crop emergence or transplanting. See label for timing. Shallow cultivation, irrigation, or rainfall within 5 days is needed for good weed control. Do not use under mulches, row covers, or hot caps. Under conditions of unusually cold or wet soil and air temperatures, crop stunting or injury may occur. Crop injury can occur if seeding depth is too shallow.
Annual grasses and broadleaf weeds	ethalfluralin, MOA 3 + clomazone, MOA 13 (Strategy) 2.1 L	2 to 6 pt	0.4 to 1.2 + 0.125 to 0.375	Apply to soil surface immediately after crop seeding for preemergence control of weeds. **Do not apply prior to planting. Do not incorporate. Do not apply under mulch.** May also be used as a banded treatment between rows after crop emergence or transplanting.
Broadleaf weeds	terbacil, MOA 5 (Sinbar) 80 WP	2 to 4 oz	0.1 to 0.2	Apply after seeding but before crop emerges, or prior to transplanting crop. With plasticulture, Sinbar may be applied preemergence under plastic mulch or to row middles. May be applied over plastic mulch prior to transplanting, or prior to punching holes into the plastic mulch for transplanting. Sinbar must be washed off the surface of the plastic mulch with a minimum of 0.5 inch of rainfall or irritation prior to punching transplant holes or transplanting watermelon. Do not apply within 70 days of harvest. See label for further instructions.
Yellow and purple nutsedge suppression, pigweed and ragweed control	halosulfuron-methyl, MOA 2 (Profine 75, Sandea) 75 DG	0.5 to 0.75 oz 0.5 to 1 oz	0.024 to 0.036 0.024 to 0.048	**Bareground.** Apply after seeding but before cracking or prior to transplanting crop. **Plasticulture.** Application may be made to preformed beds prior to laying plastic. If application is made prior to planting, wait 7 days after application to seed or transplant. Stunting may occur but should be short lived with no negative effects on yield or maturity in favorable growing conditions. **See label for information on rotation and other restrictions.** **Row middles only.** Apply to row middles as a preemergence spray. In plasticulture, do not allow spray to contact plastic. Early season application will give postemergence and preemergence control. Do not apply within 57 days of harvest. For postemergence applications, use nonionic surfactant at 1 quart per 100 gallons of spray solution.

CHEMICAL WEED CONTROL

359

CHEMICAL WEED CONTROL

Table 7-16. Chemical Weed Control in Vegetable Crops

Weed	Herbicide, Mode of Action Code* and Formulation	Amount of Formulation Per Acre	Pounds Active Ingredient Per Acre	Precautions and Remarks
Watermelon, Postemergence				
Annual grasses and some small seeded broadleaf weeds	DCPA, MOA 3 (Dacthal) W-75 (Dacthal) 6 F	8 to 10 lb 8 to 10 pt	6 to 7.5	**Not labeled for transplanted crop.** To improve preemergence control of late emerging weeds, apply only when crop has 4 to 5 true leaves, is well-established, and growing conditions are favorable. Will not control emerged weeds. Incorporation not recommended.
Annual and perennial grasses only	clethodim, MOA 1 (Arrow, Clethodim, Intensity, Select) 2 EC (Intensity One, Select Max) 1 EC	6 to 8 oz 9 to 16 oz	0.094 to 0.125 0.07 to 0.125	Apply postemergence for control of grasses. With Arrow, Clethodim, or Select, add 1 gallon crop oil concentrate per 100 gallons spray mix. With Select Max, add 2 pints nonionic surfactant per 100 gallons spray mixture. Adding crop oil may increase the likelihood of crop injury at high air temperatures. Very effective in controlling annual bluegrass. Apply to actively growing grasses not under drought stress. Do not apply within 14 days of harvest.
	sethoxydim, MOA 1 (Poast) 1.5 EC	1 to 1.5 pt	0.2 to 0.3	Apply to emerged grasses. Consult manufacturer's label for specific rates and best times to treat. Add 1 quart of crop oil concentrate per acre. Adding crop oil to Poast may increase the likelihood of crop injury at high air temperatures. Do not apply Poast on days that are unusually hot and humid. Do not apply within 14 days of harvest.
Watermelon, Row Middles				
Annual grasses and some small-seeded broadleaf weeds	trifluralin, MOA 3 (Treflan HFP, Trifluralin, Trifluralin HF) 4 EC	1 to 2 pt	0.5 to 0.75	To improve preemergence control of late emerging weeds. Apply after emergence when crop plants have reached the three to four true leaf stage of growth. Apply as a directed spray to soil between the rows. Avoid contacting foliage as slight crop injury may occur. Set incorporation equipment to move treated soil around base of crop plants. Do not apply within 60 days of harvest. Will not control emerged weeds.
Broadleaf weeds	terbacil, MOA 5 (Sinbar) 80 WP	2 to 4 oz	0.1 to 0.2	With plasticulture, Sinbar may be applied to row middles. Do not apply within 70 days of harvest. See label for further instructions.
Most broadleaf weeds less than 4 inches tall or rosettes less than 3 inches in diameter; does not control grasses	carfentrazone-ethyl, MOA 14 (Aim) 1.9 EW or 2 EC	up to 2 oz	up to 0.031	Apply post-directed using hooded sprayers for control of emerged weeds. If crop is contacted, burning of contacted area will occur. Use a nonionic surfactant or crop oil with Aim. See label for rate. Coverage is essential for good weed control. Can be tank mixed with other registered herbicides.
Most emerged weeds	glyphosate, MOA 9 (numerous brands and formulations)	See labels	See labels	Apply as a hooded spray in row middles, as shielded spray in row middles, as wiper applications in row middles, or postharvest. To avoid severe injury to crop, do not allow herbicide to contact foliage, green shoots, stems, exposed, roots, or fruit of crop. Do not apply within 14 days of harvest.
Yellow and purple nutsedge and broadleaf weeds	halosulfuron-methyl, MOA 2 (Profine 75, Sandea) 75 DG	0.5 to 1 oz	0.024 to 0.048	Apply to row middles as a postemergence spray. In plasticulture, do not allow spray to contact plastic. Early season application will give postemergence and preemergence control. For postemergence applications, use nonionic surfactant at 1 quart per 100 gallons of spray solution. Do not apply within 57 days of harvest.

* Mode of action (MOA) code developed by the Weed Science Society of America.

Chemical Weed Control in Forest Stands

Compiled by C. Lambert and M. Megalos, NCSU Extension Forestry

With professional assistance by:
Travis Rogers, Dow AgroSciences
Robert York, BASF Corporation
Michael Fleming, Nufarm Americas, Inc

Herbicides can be a safe, effective alternative to mechanical or manual forest weed control methods if used according to label directions. To avoid drift, volatilization, or direct damage to areas adjacent to sprayed forests, it is essential to follow label recommendations related to:

1. **Adjuvants**—such as drift control agents or surfactants which increase uptake.

2. **Spray Equipment**—specifically, type of aircraft, spray boom, pressure (psi), or other spray equipment restrictions.

3. **Weather Conditions**—such as temperature, relative humidity, wind speed, and wind direction.

4. **Buffer Strips**—width and location relative to dwellings, other occupied structures, streams, lakes, ditches, gardens, and boundaries.

> **READ THE LABEL AND <u>ALWAYS</u> USE MANFACTURERS RECOMMENDATIONS.**
> **THEY WILL CHANGE MORE OFTEN THAN RECOMMENDATONS BELOW!**
> **These recommendations may not apply to Christmas trees and woody ornamentals.**

Additional recommendations for minor forestry uses such as hardwood plantations and specific species may be found in Chapter 17, "Weed Control in Field- and Container-Grown Ornamentals," in the *Nursery Crops Production Manual.* * Consult a county Extension agent, professional forester, or weed control specialist before spraying forest stands.

Contact the Department of Horticultural Science, Box 7609, NC State University, Raleigh, NC 27695-7609. The manual sells for $35. Individual chapter prices vary.

CHEMICAL WEED CONTROL

Table 7-17. Pine Release

Type Application / Plants Controlled	Herbicide and Formulation	Amount of Formulation Per Acre	Uses, Remarks, and Precautions
Foliar Spray			
Many broadleaf weeds, grasses, and woody species (black locust, cherry, oaks, persimmon, maple, sassafras, sumac, sweetgum, yellow-poplar)	Glyphosate, MOA 9 (Rodeo or Aquaneat) 5.4L	0.75 qt Rate may vary	Use at end of first growing season for first-year plantings after planted pine seedlings buds have hardened off. May be tank mixed with Oust XP and/or Arsenal or Arsenal AC or Chopper Gen2 for many pines. See labels for mix details.
	OR Other brands	1.125 to 1.875 qt	Use late summer or early fall after conifers have hardened off (second and subsequent growing seasons).
		12 to 18 oz	Herbaceous release (early season) for loblolly and longleaf pines. May be tank mixed with Oust XP and/or Arsenal or Arsenal AC for many species. See labels for mix details.
Woody plants (oaks, sweetgum, elm, and sumac)	Hexazinone, MOA 5 (Velpar L) 2EC	4 to 6 pt (first-year plantings) 4 to 8 pt (established trees)	Use spring to early summer. All rates depend on soil type. Use lower rates on coarser soils. See label.
	OR (Velpar DF) 75%	1.33 to 1.8 lb (first-year plantings) 1.33 to 2.66 (established trees)	
Many grasses, broadleaf weeds, vines, brambles, woody brush, and trees	Imazapyr, MOA 2 Arsenal AC (53.1%) or Arsenal 2NS (27.8%) or Chopper Gen2 (26.7%) or Polaris AC Complete)	12 to 20 oz or 24 to 40 oz or 24 to 80 oz or 48 to 80 oz for loblolly 12 to 20 oz for loblolly	See label for rates for specific species of pine. May be tank mixed with Rodeo (only loblolly after late summer hardening off), Oust XP, and, for loblolly only, Oust Extra.
Many grasses, broadleaf weeds, vines, brambles, woody brush, and trees	Imazapyr, MOA 2 (63.2%) or Arsenal AC (53.1%) or Arsenal 2NS (27.8%) or Chopper Gen2 (26.7%) + Metsulfuron methyl, MOA 2 (9.5%) (Lineage Clearstand)	9 to 16 oz	Labeled for lobolly and slash pines. Can use up to 19 oz for mid rotation release of pines.
Herbaceous plants (crabgrass, dog fennel, fescue, willoweed [fireweed], goldenrod, horseweed, Kentucky bluegrass, yellow nutsedge, panicums [broadleaf, fall, narrow], pokeweed, ragweed, white snakeroot, and yellow sweetclover)	Sulfometuron methyl, MOA 2 (Oust XP 75% or Spyder)	2 to 8 oz	Rates depend on soil type and species. Labeled for loblolly, slash, longleaf, Virginia, and white pines. May be tank mixed with Velpar and/or Arsenal for some species. See labels. Do not exceed 8 oz per acre per year.
Many annual grasses and broadleaf weeds	Hexazinone, MOA 5 63.2% + Sulfometuron methyl, MOA 2 11.8% (Oustar)	10 to 19 oz (first-year control) + 12 to 24 oz (after first year)	Use spring to early summer. All rates depend on soil type. For loblolly, slash, and longleaf pine only. See labels.

Table 7-17. Pine Release

Type Application / Plants Controlled	Herbicide and Formulation	Amount of Formulation Per Acre	Uses, Remarks, and Precautions
Foliar Spray (continued)			
Certain hardwoods, weeds, and grasses	Sulfometuron methyl, MOA 2 (15%) + Metsulfuron methyl, MOA 2 (56.25%) (Oust Extra or Spyder Extra)	2.66 to 4 oz	Loblolly and slash only. May be tank mixed with Arsenal AC for loblolly only. Use lower rates on coarser soils.
Herbaceous weeds	Imazapyr, MOA 2, (Arsenal 2NS (27.8%) or Arsenal AC (53.1%) or Chopper Gen2 (26.7%) or Polaris AC Complete)	8 to 20 oz 4 to 10 oz 8 to 20 oz 4 to 10 oz	May be tank mixed with Oust see label for rates for specific species of pine.
Broadleaf herbaceous plants such as bedstraw, hemp dogbane, sericea lespedeza, cocklebur, coffeeweed, common and giant ragweed, dog fennel, grape, marestail, morning glory, goldenrod, blackberry, ironweed, wild carrot, common mullein, mustard, vetch, and prickly pear cactus.	Fluroxypyr MOA 4, (Vista XRT) OR Comet 1.5lb/g fluroxypyr)	6 to 23 fl oz .67 to 2.3 pts	Do not apply as an over-the-top broadcast treatment during active terminal growth (from initiation of budbreak/growth flush until seasonal terminal growth has hardened off and overwintering buds have formed.) Directed spray applications may be made to pines during periods of active growth, but care should be taken to avoid spray contact with actively growing foliage.
Broadleaf herbaceous plants such as ragweed, coffeeweed, clover, marestail, knapweed, partridge pea, sicklepod, smartweed, thistle, and morning glory	Clopyralid MOA 2, (Transline or Clean Slate)	8 to 21 fl oz	Labeled for use on first year pine seedlings and older trees. Will not control henbit or chickweed.
Herbaceous weeds (ragweed, marestail, morning glory, sicklepod, pigweed, and blackberry.) See label for complete list of susceptible broadleaf weeds	Aminopyralid (Milestone)	5 to 7 fl oz	Labeled for use for release of longleaf pine only. Do not apply over the top of other desirable pine species. May be tank mixed with approved products for use over longleaf. Applications can be performed over the top of longleaf from first year after planting to 7 feet in height. Newly planted seedlings should be at least three months old before application and not under stress. Read and follow label recommendations. Do not add surfactant/adjuvant.
Preemergent/Foliar			
Most annual grasses and many broadleaf weeds	Pendimethalin, MOA 3 (Pendulum) 2 G OR (Pendulum) 3.3 EC OR Pendulum AquaCap (38.7%)	100 to 200 lb 2.4 to 4.8 qt 2.1 to 4.2 qt	Apply at time of planting or to established trees. Planting slit must be closed to avoid root contact. Labeled for loblolly, white and Virginia pines and a number of hardwood species. See label for details. See label for details.
Spot-gun			
Woody plants (cherry, blackgum, dogwood, elm, hawthorn, hickory, oaks, maple, sweetgum, and sumac)	Hexazinone, MOA 5 (Velpar L) 2EC OR (Velpar DF) 75%	3 to 10 qt 1.3 to 5.3 lb	Apply in calibrated spots on grid pattern. Rate per acre and grid pattern depend on soil texture and species composition. Apply with exact delivery handgun. Apply late winter to early summer. Do not apply spots within 36 inches or directly upslope from seedlings. Poor results may occur if site is burned 3 to 6 months before treatment or on stump sprouts less than 1 year old.

Forest Site Preparation, Stand Conversion, Timber Stand Improvement

Compiled by C. Lambert and M. Megalos, NCSU Extension Forestry

Table 7-18. Forest Site Preparation, Stand Conversion, Timber Stand Improvement

Type Application	Plants Controlled	Herbicide and Formulation	Amount of Formulation Per Acre	Uses, Remarks, and Precautions
Foliar Spray	For control of many broadleaf weeds, annual and perennial grasses, brush, vines and brambles	Imazapyr, MOA 2 (63.2%) + metsulfuron methy, MOA 2 (9.5%) (Lineage Clearstand) OR Tank Mix Polaris AC Complete + Patriot OR Imazapyr, MOA 2 Arsenal AC (53.1%) or Arsenal 2NS (27.8%) or Chopper Gen2 (26.7%) + metsulfuron methy, MOA 2 (9.5%)	8 to 25 oz for loblolly and slash pine 10 to 24 oz + 1 to 2 oz 20 to 40 oz or 40 to 80 oz or 32 to 64 oz See label	Can tank mix with KRENITE-S at 4 to 6 quarts for control of natural pines. For loblolly and slash pine. See label for details.

Table 7-18. Forest Site Preparation, Stand Conversion, Timber Stand Improvement

Type Application	Plants Controlled	Herbicide and Formulation	Amount of Formulation Per Acre	Uses, Remarks, and Precautions
Foliar Spray	For control of natural pines	Ammonium salt of fosamine, MOA 17 (41.2%) (Krenite-S)	4 to 6 qts	Can be tank mixed with most other site prep products. Need highest label rate for control of Virginia pine.
		Detail	2 oz/A	See label for further instructions. Detail should be used in combination with 4 qt/A of glyphosate
Foliar Spray	Control of woody brush including but not limited to loblolly pine, oak species, legumes, maple, cherry, poplar, blackberry, etc.	Glyphosate, MOA 9 (Accord XRT II) 5.07 lb/gal + Aminopyralid, MOA 4 (Milestone) + Imazapyr, MOA 2 Arsenal AC (53.1%) or Arsenal 2NS (27.8%) or Chopper Gen2 (26.7%) or Polaris AC Complete)	6 to 7 qts + 7 fl oz + 12 to 40 oz or 24 to 80 oz or 24 to 64 oz or 12 to 40 oz	Recommended for use from June – July 31. Offers broad spectrum hardwood and natural pine control. See label for details.
Foliar Spray	Control of woody brush including but not limited to southern pines, oak species, legumes, maple, cherry, blackberry, poplar,gallberry, wax myrtle, yaupon, bay spp.etc.	Glyphosate, MOA 9 (Accord XRT II) 5.07 lb/gal + Aminopyralid, MOA 4 (Milestone) + Triclopyr ester , MOA 4(Forestry Garlon XRT) + Imazapyr, MOA 2 (Arsenal 2NS (27.8%) or Arsenal AC (53.1%) or Chopper Gen2 (26.7%) or Polaris AC Complete)	5 qts + 7 fl oz + 21 fl oz + 16 to 40 oz or 24 to 80 oz or 24 to 64 oz or 16 to 40 oz	Recommended for use from Aug. 1 to Oct. 15th. Offers broad spectrum hardwood and natural pine control. See label for details.
Foliar Spray	Control of woody brush including but not limited to southern and Virginia pine, oak spp., cherry, maple, poplar, legumes, wax myrtle, blackberry, etc.	Glyphosate (Accord XRT II 5.07 lb/gal or Aquaneat) + Saflufenacil MOA14 (Detail) + Triclopyr ester, MOA 4 (Forestry Garlon XRT) + Imazapyr, MOA 2 Arsenal AC (53.1%) or Arsenal 2NS (27.8%) or Chopper Gen2 (26.7%) or Polaris AC Complete)	6 qts + 2 fl oz + 21 fl oz + 12 to 40 oz or 24 to 80 oz or 24 to 64 oz or 12 to 40 oz	Recommended for use from June 1 to Sept. 15. Recommended when VA pine are part of the target species. See label for details.
Foliar Spray	Woody brush, trees, vines, grasses, and broadleaf weeds (Alder, berries [blackberry, dewberry, raspberry], elderberry, honeysuckle,, maples [red, sugar], oaks [red, Northern pine, white], multi flora rose, poison ivy, poison oak, trumpet creeper, willow)	Glyphosate, MOA 9 (Accord XRT II) 5.07lb/gal OR Razor Pro OR Other brands	2% solution as a high volume spray using a spray to wet basis Rates may vary	Apply over actively growing plants when foliage has fully developed. See labels for appropriate rates for weeds and woody brush. See labels.
Foliar Spray	Kudzu, wisteria	Aminopyralid, MOA 4 (Milestone)	7 fl oz	0.5 ounce/gallon of water for spot treatments (assuming 20 gpa). See label for full details. Do not apply over the top of susceptible tree species.
	Kudzu	Clopyralid, MOA 4 (Transline) Clean Slate	21 fl oz	Application June to October to actively growing plants
Directed Foliar Spray	Woody vines and brush	Aminopyralid MOA 4 + Triclopyr Amine, MOA 4 (Capstone)	5 to 7%	Excellent control of woody vines such as kudzu, wisteria, blackberry, honeysuckle, and small seedling brush. For example, cherry, maple, southern pine, oak spp., and legumes, etc. Does not hurt most grass species.
	Over 50 species of woody brush, broadleaf perennials, etc.	2,4-D, 2,4-DP (Patron 170)	1.5% to 4.25%	Selective to most grasses control of over 50 broadleaf perennial and woody species including poison ivy, kudzu, blackberry, dewberry,honeysuckle, locust, and many more.
Cut Stumps	Sweetgum, poplar, oak, sycamore. Suppression of dogwood, blackgum, hickory, and red maple	Glyphosate, MOA 9 (Rodeo or Aquaneat) 5.4L	Spray or brush 50 to 100% solution to freshly cut stumps	Best used mid growing season. Avoid spring sapflow. Treat stumps immediately after cutting. Treat entire stump surface for small trees and just cambium for large trees.
		OR Other brands	Rates may vary	See labels.
Cut Stump and Basal treatment	Works on more than 90 woody brush and vine species. See label for a complete list	Triclopyr ester, MOA 4 (Garlon 4 Ultra) or Relegate or Triclopyr ester (Pathfinder II)	20 to 25% Garlon 4 Ultra + 75 to 80% basal oil or use Pathfinder II as a ready-to-use	Treat the lower 12-15 inches of the target stem down to the ground line making sure to treat completely around the stem. Treat to wet not to the point of runoff. See label for full details. May be used on smooth bark species up to 6 inches in basal diameter and up to 4 inches in basal diameter for rough bark species. Can be performed all year except when the bark is wet and anytime after cutting. .

CHEMICAL WEED CONTROL

Table 7-18. Forest Site Preparation, Stand Conversion, Timber Stand Improvement

Type Application	Plants Controlled	Herbicide and Formulation	Amount of Formulation Per Acre	Uses, Remarks, and Precautions
Foliar Spray	Annual bluegrass, ash, aspens (bigtooth, trembling), asters, balsam poplar, barnyardgrass, bentgrass, birch, common groundsel, common ragweed, elksedge, elm, false dandelion, fleabane, flowering dogwood, foxtail, hawthorn, hazel, hickory, oaks, oxeye daisy, Pennsylvania smartweed, pinegrass, red maple, sourwood sweetgum, velvetgrass, wild carrot, wild cherry, and willows. Treatment provides partial control of Canada thistle (suppression only), catsear, crabgrass, curly dock, dandelion, willowweed (fireweed) fescue, goldenrod, heath aster, honeysuckle, horseweed, orchardgrass, and perennial grasses (quackgrass and ryegrass)	Hexazinone, MOA 5 (Velpar L) 2 EC OR (Velpar DF) (75%)	4 to 10 qt 2 1/3 to 6 2/3 lb	Labeled for longleaf pine, shortleaf pine, slash pine, Virginia pine, loblolly pine, Scotch pine, and white spruce. All rates depend on soil type. Use lower rates on coarser soils. Read and follow label directions.
Spotgun	Many woody plants (black cherry, blackgum dogwood, elm, hawthorn, hickory, oaks, maple, sweetgum, and sumac)	Hexazinone, MOA 5 (Velpar L) 2 EC OR (Velpar DF) (75%)	4 to 10 qt Mix 2 2/3 lb/gal of water. Apply suspension at 3 to 10 qt/acre, using grid pattern. See label for grid spacing.	Use an exact-delivery handgun applicator and calibrate for precise delivery of the undiluted product. Selection of the rate per acre and "grid" pattern will depend on soil texture and woody plant species composition. Use the higher rates on fine-textured soils and when the major component of the hardwoods is a difficult-to-control species such as blackgum, dogwood, hickory, or red maple. Use the lower rates on medium-to-coarse textured soils where elm, cherry, oak, and sweetgum are dominant. Results may be unsatisfactory where stump sprouts of less than one year's growth predominate. Labeled for loblolly, slash, shortleaf, and longleaf pines. Suspensions of DF require intermittent agitation.
Broadcast Granular	Many woody plants (American elder, balsam poplar, birch, black cherry, blackgum, hickory, boxelder, brambles [blackberry, dewberry, and raspberry], cherry, chokeberry, cottonwood, dogwood Eastern red cedar, elm, green ash, hawthorn, hornbeam, mulberry, multiflora rose, Norway maple, red maple, oaks [black, blackjack, bluejack, post, southern red, Turkey, water, and white oak], Russian olive, sumac, sweetgum, white ash, wild plum, willow, sourwood, and pine)	Hexazinone, MOA 5 (Velpar ULW) 75% granule (no longer commercially available)	2.5 to 6.3 lb	May be applied broadcast or to individual plant stems. Rainfall is required to dissolve the granules and move the herbicide into the root zone. Apply in spring when hardwood leaves are 50% developed but no later than June 15. Results are best on coarse-to-medium textured soils. Not recommended for poorly drained or marsh sites. Labeled for loblolly, longleaf, shortleaf, slash, and Virginia pine.
Foliar Spray	Many grasses, broadleaf annual, and perennial weeds, vines, brambles, woody brush, and trees	Imazapyr, MOA 2 Arsenal AC (53.1%) or Arsenal (27.8%) or Chopper Gen2 (26.7%) or Polaris AC Complete)	12 to 40 oz or 24 to 80 oz or 24 to 64 oz or 12 to 40 oz Add 0.5% v/v nonionic surfactant to mix.	Broadcast spray during growing season to prepare site for many conifers. May be tank mixed with Accord XRT II, Oust, Oust Extra (loblolly only), and others. See labels. Not labeled for hardwood site preparation.
Tree Injection, Frill, or Hack 'N' Squirt	Brush species listed above	Imazapyr, MOA 2 Arsenal AC (53.1%) or Arsenal 2NS (27.8%) or Chopper Gen2 (26.7%) or Polaris AC Complete) OR Aminopyralid, MOA 4 + triclopyr amine, MOA 4 (Capstone) OR Triclopyr choline, MOA 4 (Vastlan)	25% solution in water Use undiluted and apply 1 ml of undiluted product into each cut 40% solution mixed with water	Use hatchet or tree injector and space sloping cuts evenly around the trunk. Apply 1 ml/3 in. DBH. Best used mid growing season but dormant applications may be used. Avoid spring sapflow. Nearby desirable hardwoods may be injured or killed. See label for details Use hatchet or machete and space cuts no more than 2 inches apart and at a 45 degree angle. Make cuts around the tree trunk at a convenient height so that the cuts overlap slightly and make a continuous circle around the trunk. Avoid treatment during periods of heavy sap flow such as the spring. Use hatchet or machete and space cuts no more than 2 inches apart and at a 45 degree angle. Make cuts around the tree trunk at a convenient height so that the cuts overlap slightly and make a continuous circle around the trunk. Avoid treatment during periods of heavy sap flow such as the spring.
Tree Injection Frill, or Hack 'N Squirt	Sweetgum, poplar, oak, sycamore. Suppression of dogwood, blackgum, hickory, and red maple	Glyphosate, MOA 9 (Rodeo) 5.4 lb/gal or Aquaneat OR Imazapyr, MOA 2 Arsenal AC (53.1%) or Arsenal 2NS (27.8%) or Chopper Gen2 (26.7%) OR Other brands	1 ml/2 to 3 in. of trunk diameter, or use continuous frill and a diluted mix. 1 mL of solution in each cut site Rates for other brands may vary	Use hatchet or tree injector and space sloping cuts evenly around the trunk. Best used mid growing season. Avoid spring sapflow. See label for details See labels.

Table 7-18. Forest Site Preparation, Stand Conversion, Timber Stand Improvement

Type Application	Plants Controlled	Herbicide and Formulation	Amount of Formulation Per Acre	Uses, Remarks, and Precautions
Cut Stumps	Brush species listed above	Imazapyr, MOA 2 Arsenal AC (53.1%) or Arsenal 2NS (27.8%) or Chopper Gen2 (26.7%) or Polaris AC Complete)	25% solution in water 6 oz/gal water on freshly cut stump surface.	Arsenal AC Dilute solution mix 4-6 ounces in 1 gallon of water. Concentrate solution mix 1 quart with 1 pint of water. See label for details. Arsenal (27.8%) Dilute solution mix 8-12 ounces in 1 gallon of water. Concentrate solution mix 2 quarts with 1 pint of water. See label for details. Chopper Gen2 Dilute solution mix 8-12 ounces in 1 gallon of water. Concentrate solution mix 2 quarts with 1 pint of water. See label for details. Best used during growing season; avoid spring sapflow. Treat stumps promptly after cutting. Nearby desirable hardwoods may be killed or injured.
		OR Aminopyralid, MOA 4 + triclopyr amine, MOA 4 (Capstone)	Use 100% solution.	Best used immediately after cut. Avoid spring sap flow. Spray or brush cambium layer immediately after cut.
		OR Triclopyr ester, MOA 4 (Garlon 4 Ultra 4lb/gal or Relegate, contains no petrolium distilates)	Use 20 to 25% solution mixed with 75 to 80% basal oil.	Can be used all year except when bark is wet. Treat the cambium layer and root collar down to the ground line. Treat to wet and not to the point of run-off. Can be used up to several weeks after cutting.
Foliar Spray	Many species of brush and broadleaf weeds, including: **BRUSH:** Ash, aspen, birches, brambles, cedar, cherry, dogwood, elms, gooseberry, honeylocust, multiflora rose, oak, poplar, shortleaf pine, sumac, wild plum, willows, sycamore, spruce, blackberry, black locust, buckbrush, cottonwood, honeysuckle. **WEEDS:** bedstraw, burdock, chicory, dock, kudzu, morning glory, poison ivy, poison oak, thistles, trumpet vine	Imazapyr, MOA 2 (Chopper GEN2 or Polaris SP) + Glyphosate, MOA 9 (Accord XRT II)	32 to 64 oz + 2 qts	Apply during growing season to prepare site for many conifers. A seed oil mix that is 12 to 50% of volume is recommended. Use 5 to 40 gallons mix/acre for good coverage. Not labeled for hardwood site preparation. Tank mixing Accord XRT II with Chopper treatments will enhance brown-out of target vegetation thus enabling landowner to more effectively burn.
	Most grasses, broadleaf annual and perennial weeds, vines, brambles, hardwoods and pines	Imazapyr, MOA 2 Arsenal AC (53.1%) or Arsenal 2NS (27.8%) or Chopper Gen2 (26.7%) or Polaris SP) + Glyphosate, MOA 9 (Accord XRT II)) + Aminopyralid, MOA 4 (Milestone)	16 to 40 oz or 24 to 80 oz or 24 to 64 oz or 32 to 64 oz + 6 to 7.5 qts + 7 fl oz	Effective broad-spectrum site preparation for hardwoods and natural pine.
Basal Bark	Controls over 90 woody plants and vines	Triclopyr, MOA 4 (Pathfinder II) OR Triclopyr, MOA 4 (Garlon 4 Ultra or Relegate)	100% solution ready to use formulation of Garlon 4 Ultra 20-25% solution mixed with 75-80% basal oil	Spray or wet lower 12 to 15 inches of stems. Best used on rough bark species no more than 4 inches diameter and smooth bark species no more than 6 inches diameter. Perform year round except when bark is wet. Spray to wet but not to the point of run-off.
Foliar Spray	Many woody species and broadleaf weeds, grasses	Glufosinate-ammonium, MOA 10 (Derringer F) 1 lb a.i./gal	2 to 6 qt	May be used to prepare site for planting hardwoods and conifers. Use higher rate for woody or heavy dense brush. See label for specific details. May be tank mixed with Arsenal AC and Chopper for conifer site preparation.
Foliar Spray	Alder, birch, blackberry, black cherry, blackjack oak, black locust, currant, fir, gooseberry, hemlock, honeysuckle, oaks, pine, poison ivy, poison oak, poplar, red elm, red maple, serviceberry, spruce, sycamore, tulip poplar, willow, winged elm	Dicamba, MOA 4 + 2,4-D, MOA 4 (Veteran 720) 2.9 lb a.i./gal	1 gal	Apply 1 gallon in 20 to 100 gallons water per acre depending upon density of brush. Do not apply more than 1 gallon per acre per year.
Foliar Spray	Many brush species (alder, ash, basswood, beech, birch, blackberry, black gum, cedar, cherry, cottonwood, dogwood, elm, grape, hawthorn, hemlock, hickory, honeysuckle, hornbeam, poison ivy, kudzu, locust, maple, oaks, persimmon, pines, poplar, sassafras, sumac, sweetgum, sycamore, willow) Controls many woody plants including yaupon, gallberry, wax myrtle, Baccharis spp., and many bay species.	Triclopyr (choline), MOA 4 (Vastlan) 4 lb a.i./gal OR Triclopyr (ester), MOA 4 (Forestry Garlon XRT) 6.3 lb/gal	0.5 to 1.5 gal 21 fl. oz. to 1 gal	Ground applications: Mix product in water (20 to 100 gallons; enough for good coverage) + 0.5% v/v nonionic surfactant. May also be tank mixed with other products. See label for details. Use higher rates for woody plant control.
Tree Injection, Frill, Hack 'N' Squirt	Same as above	Triclopyr (choline), MOA 4 (Vastlan) 4 lb a.i./gal	Mix 40:60 ratio with water to make a 40% solution	Inject 0.5 ml undiluted or 1 ml diluted/inch tree diameter. Space cuts evenly. Best used during growing season.
Cut Stump	Same as above	Triclopyr (choline), MOA 4 (Vastlan) 4 lb a.i./gal	Use 40% solution mixed with water	Spray or paint fresh cut stumps immediately after cut. Best used during the growing season but avoid period of spring sap flow.
Cut Surface Treatment	Same as above	Dicamba, MOA 4 (Banvel CST) 1 S	See label	Inject 1 ml undiluted at 1- to 2-inch intervals. Spray or paint frill or girdle OR spray or paint cut stumps with undiluted herbicide

Table 7-18. Forest Site Preparation, Stand Conversion, Timber Stand Improvement

CHEMICAL WEED CONTROL

Type Application	Plants Controlled	Herbicide and Formulation	Amount of Formulation Per Acre	Uses, Remarks, and Precautions
Basal Bark or Dormant Stem Treatment	Same as above	Dicamba, MOA 4 + 2,4-D(ester), MOA 4 (Banvel 520) 2.9 lb total a.i./gal	1 to 5 gal	For hydraulic sprayers use 1 to 3 gallons Banvel 520 per 100 gallons water or oil to water emulsion per acre. For backpack sprayers mix 8 to 16 gallons Banvel per 100 gallons water or oil to water emulsion and apply 30 gallons mix per acre. **BASAL BARK**—Apply year around to lower 1.5 to 2 feet of stem, allowing runoff. **DORMANT STEM**—Apply anytime brush is dormant by thoroughly wetting stems to point of runoff.
Foliar Spray	**WOODY**: Alder, ash, beech, birch, blackberry, blackgum, cherry, cottonwood, dogwood, elderberry, elm, hawthorn, hickory, hornbeam, locust, maples, mulberry, oaks, persimmon, pine, poison oak, poplar, sassafras, sumac, sweetbay, magnolia, sweetgum, sycamore, tulip poplar, willow, winged elm **ANNUAL AND PERENNIAL BROADLEAF WEEDS**: Bindweed, burdock, Canada thistle, chicory, curly dock, dandelion, field bindweed, lambsquarter, plantain, ragweed, smartweed, tansy ragwort, vetch, wild lettuce	Triclopyr-(choline), MOA 4 (Vastlan) 4 lb/gal	0.3 to 1.5 gal	**Aerial Application**—Mix 0.75-1 gallon herbicide with 0.25 to 1 pint agricultural surfactant. For broader spectrum control, tank mix with other suitable herbicides. Apply in 10 to 30 gal water per acre. **Ground High-Volume Application**—Mix 0.5 to 1 gallon herbicide in water to make 100 gallons solution + 8-16 oz adjuvant. For broader spectrum control, tank mix with other suitable herbicides. **Ground Low-Volume Directed Spot Application**—Mix 0.5-2% solution (0.64 – 2.5 fl oz herbicide/gallon of water) with 1 % (1.28 fl oz surfactant/gal of water)
Tree Injection, Frill, Hack 'N' Squirt	See woody plants above	Aminopyralid, MOA 4 + triclopyr amine, MOA 4 (Capstone) OR Picloram, MOA 4 + 2,4-D, MOA 4 (Pathway) 1.25 lb a.i./gal	Apply as ready to use mixtures (100% solution)	Inject Capstone undiluted making cuts no more than 2 inches apart. Use 1 ml/inch of Pathway undiluted. Space injections evenly. Best used during the summer or fall.
Frill or Girdle	See woody plants above	Same as above	Use undiluted	Make a single hack girdle or "frill" of overlapping cuts through the bark completely around the tree. Spray or paint the cuts with undiluted product using enough volume to wet the treated areas.
Cut Stump	See woody plants above	Aminopyralid, MOA 4 + triclopyr amine, MOA 4 (Capstone) OR Picloram, MOA 4 + 2,4-D, MOA 4 (Pathway) 1.25 lb a.i./gal	Apply as ready to use mixtures (100% solution)	Spray or paint cambium of freshly cut stumps with undiluted herbicide solution.
Foliar Spray	Same as above	Triclopyr (ester), MOA 4 (Forestry Garlon XRT) 6.3 lb a.i./gal OR 4 lb/gal triclopyr (Relegate)	21 fl oz to 1 gal 32 oz to 1.5 gal	Use higher rates to control woody species. Apply in enough water for good coverage, usually 20 to 40 gallons per acre. A nonionic surfactant (0.5 to 1.0%) is recommended. Tank mixes may provide additional benefits. Will not control most native grasses. Complete coverage of target vegetation is necessary for best results.
Basal Bark	Same as above	Same as above OR Triclopyr, MOA 4 (Pathfinder II) 0.75 lb a.i./gal	13% Forestry Garlon XRT + 87% basal oil Apply as a ready to use formulation (100% solution)	Spray 13% solution of Forestry Garlon XRT in oil penetrant (diesel or vegetable oil) to lower 12 to 18 inches of tree trunk to point of runoff. Most effective on trees less than 6 inches in diameter with smooth or thin bark. Apply to dry bark in winter or summer. Avoid spring sap flow
Foliar	Many annual and biennial broadleaf weeds and woody brush	Picloram, MOA 4 (Tordon K) 2 lb a.i./gal	0.25 to 4 qt	Restricted use chemical. See label for specific rates for given situations, tank mixes, and volumes per acre.
Foliar	Many annual and biennial broadleaf and woody brush species	Diglycolamine salt of dicamba, MOA 4 (Vanquish) 4 lb a.i./gal	0.5 to 4 pt	Use lower rates for annual and biennial weeds and heavy rates for woody brush. Many tank mixes are available. Apply in 15 or more gallons of water/acre with a nonionic surfactant (0.5 to 1% v/v).
Tree Injection, Frill, Hack 'N' Squirt	Many species of hardwood trees and brush	Same as above	See label	Apply undiluted product using injector or hatchet with one cut per inch tree diameter. Or apply diluted mixture (1:1-3 parts water) to overlapping cuts (frill).
Cut Stump	Same as above	Same as above	See label	Dilute 1 to 1 with water. Spray or paint the freshly cut stump cambium.
Dormant Stem	Many species of hardwood trees and brush. Controls woody vines and conifers	Triclopyr ester (Garlon 4 Ultra) 4 lb/gal	2-3 gallons of Garlon 4 Ultra + 3 gal of Crop oil per 100 gallons of water	Spray upper and lower stems of dormant brush to wet. See label for more details.
Cut Surface and Basal Spray	Many species of hardwood trees and brush	Triclopyr ester, MOA 4 + 2,4-D, MOA 4 (Crossbow or Candor)	See label	Read label carefully as rates and dilutions vary widely depending on tree species to control and type of application.

Aquatic Weed Control

R. J. Richardson, Crop Science Department, and K. D. Getsinger, U.S. Army Engineer Research and Development Center, Vicksburg, MS, and Adjunct Professor, Crop Science Department, NC State University

Several options, including hand removal, cultural, mechanical, biological, and chemical control techniques are available for the management of aquatic weeds. The applicator should choose the most efficacious, environmentally acceptable and cost-effective alternative that is available for a particular weed problem. The particular management strategy to use in a given situation will depend on the intended use of the body of water, fish, and wildlife populations that may be impacted, type of environment in which the weed problem occurs, and the particular weed species of concern. Before selecting your management strategy, **be sure to have the weed(s) of concern identified by a qualified individual**.

Assistance in weed identification is available from the Cooperative Extension center in your county. Additional information on management techniques also may be obtained from the county Extension center; ask for AG-437, *Weed Management in Small Ponds*; AG-438, *Weed Control in Irrigation Water Supplies;* and AG-449, *Hydrilla, A Rapidly Spreading Aquatic Weed in North Carolina*. Information on pond construction, stocking, and general pond management may be found in AG-424, *Pond Management Guide*. Additional information may be found on the Aquatic Weed Management Web site:

http://www.weedscience.ncsu.edu/aquaticweeds.

For the purpose of description and management, aquatic weeds may be grouped either on the basis of their botanical relationships or on the basis of their growth habits. Most plants in each group are managed similarly, with some exceptions.

Table 7-19A. Aquatic Weed Groups — Grouping of Aquatic Plants on the Basis of Botanical Relationships

Category and Description	Examples
Algae — These plants may be either microscopic or visible to the naked eye, exist as single cells or occur in clusters or filaments containing many cells, and may be either free floating (planktonic) or attached to the soil, rocks, or vegetation. Filamentous algae may be unbranched, slightly or highly branched, or net-like. Some planktonic algae are mobile. Certain types of algae (macroalgae) may be large, very coarse, and resemble submersed vascular plants. Most algae (except macroalgae) usually require magnification to be identified accurately. Algae do not contain vascular (water conducting) tissues, consequently all chemicals used for algae control have only contact activity. Algae reproduce by cell division, fragmentation, and sexually by spores.	**Filamentous Algae** Bluegreens or Cyanobacteria Giant *Lyngbya* Green algae *Oedogonium* *Hydrodictyon* (water net) *Spirogyra* *Pithophora* **Planktonic Algae** Bluegreens or Cyanobacteria *Lyngbya* *Anabaena* *Oscillatoria* *Microcystis* Euglenoids (*Euglena*) **Macroalgae** Muskgrass (*Chara*) Stonewort (*Nitella*)
Mosses — These plants are visible to the naked eye and resemble delicate, leafy submersed plants. The mosses lack vascular tissues or roots, but usually are attached to the soil. Mosses reproduce sexually by spore production.	*Fontinalis* *Sphagnum* (peat moss)
Ferns — These plants are visible to the naked eye, either free floating or rooted to the bottom, occasionally forming loosely consolidated floating mats. Ferns have vascular tissues and reproduce by vegetative propagation and sexually by spores.	Giant salvinia (*Salvinia molesta*) Mosquito fern (*Azolla* spp.) Water clover (*Marsilea quadrifolia*) Water spangles (*Salvinia minima*)
Vascular flowering plants — These plants may be rooted or unrooted, free floating, submersed, floating-leaved, or emergent. Most reproduce vegetatively by means of rhizomes, stolons, and various other vegetative perennating structures including turions and tubers. Most also produce flowers and may set seeds. This group has a vascular system that shows varying degrees of development from rudimentary in the case of the duckweeds and submersed species to very complex and highly developed in emergent plants and includes annual and perennial herbaceous forms and several woody species.	Bald cypress (*Taxodium distichum*) Bladderwort (*Utricularia* spp.) Bulrushes (*Scirpus* spp.) Cattail (*Typha* spp.) Duckweed (*Lemna* spp. and *Spirodela* spp.) Hydrilla (*Hydrilla verticillata*) Naiads (*Najas* spp.) Pondweeds (*Potamogeton* spp.) Rushes (*Juncus* spp.) Spikerushes (*Eleocharis* spp.) Waterhyacinth (*Eichhornia crassipes*) Watermilfoils (*Myriophyllum* spp.)
Submersed plants — Plants in this group grow beneath the surface of the water and may be rooted to the bottom or free floating, with or without roots. Flowers usually are produced above the surface of the water and occasionally may be supported by specialized floatation structures. Some species will produce emergent floral spikes that extend several inches above the surface of the water and are covered with bracts that resemble leaves. Submersed plants usually have poorly developed vascular systems and very limited structural tissue and depend on the water's buoyancy for support. Filamentous algae and macroalgae also could be considered submersed plants.	American elodea (*Elodea canadensis* and *E. nuttallii*) Bladderwort (*Utricularia* spp.) Brazilian elodea (*Egeria densa*) Brittle naiad (*Najas minor*) Coontail (*Ceratophyllum demersum*) Creeping rush (*Juncus repens*) Eurasian watermilfoil (*Myriophyllum spicatum*) Fanwort (*Cabomba caroliniana*) Hydrilla (*Hydrilla verticillata*) Parrotfeather (*Myriophyllum aquaticum*) Pondweeds (*Potamogeton* spp.) Proliferating spikerush (*Eleocharis baldwinii*) Southern naiad (*Najas guadalupensis*) Variable-leaf milfoil (*Myriophyllum heterophyllum*) Widgeongrass (*Ruppia maritima*) Wild celery (*Vallisneria americana*)

CHEMICAL WEED CONTROL

CHEMICAL WEED CONTROL

Table 7-19B. Aquatic Weed Groups — Grouping of Aquatic Plants on the Basis of Growth Habit

NOTE: Some species have growth habits that overlap and may be listed more than once.

Category and Description	Examples
Submersed plants — Plants in this group grow beneath the surface of the water and may be rooted to the bottom or free floating, with or without roots. Flowers usually are produced above the surface of the water and occasionally may be supported by specialized floatation structures. Some species will produce emergent floral spikes that extend several inches above the surface of the water and are covered with bracts that resemble leaves. Submersed plants usually have poorly developed vascular systems and very limited structural tissue and depend on the water's buoyancy for support. Filamentous algae and macroalgae also could be considered submersed plants.	American elodea (*Elodea canadensis* and *E. nuttallii*) Bladderwort (*Utricularia* spp.) Brazilian elodea (*Egeria densa*) Brittle naiad (*Najas minor*) Coontail (*Ceratophyllum demersum*) Creeping rush (*Juncus repens*) Eurasian watermilfoil (*Myriophyllum spicatum*) Fanwort (*Cabomba caroliniana*) Hydrilla (*Hydrilla verticillata*) Parrotfeather (*Myriophyllum aquaticum*) Pondweeds (*Potamogeton* spp.) Proliferating spikerush (*Eleocharis baldwinii*) Southern naiad (*Najas guadalupensis*) Variable-leaf milfoil (*Myriophyllum heterophyllum*) Widgeongrass (*Ruppia maritima*) Wild celery (*Vallisneria americana*)
Free-floating plants — Plants in this group float on the surface of the water and may lie flat on the water or be raised well above the surface. These plants, with the exception of the duckweeds, watermeal, and mosquito ferns, have well-developed vascular systems and substantial supportive tissues. Most form true roots. Flowers extend above the surface of the water in the flowering plants.	Duckweeds (*Lemna* spp. and *Spirodela* spp.) Floating heart (*Nymphoides aquatica*) Frogbit (*Limnobium spongia*) Giant salvinia (*Salvinia molesta*) Mosquito fern (*Azolla caroliniana*) Waterhyacinth (*Eichhornia crassipes*) Waterlettuce (*Pistia stratiotes*) Watermeal (*Wolffia* spp.)
Floating leaf plants — These plants are rooted in the bottom and have their leaves attached to long, tough stems that extend to the surface from depths up to 6 ft or more. The leaves float directly on the surface of the water. Mature leaves of some species may push well above the surface into an emergent position. Most of these plants have extensive root and rhizome systems and well-developed vascular systems and supportive tissues. Flowers float just above the surface or are extended well above the surface on a tough stem. A few nonvascular representatives.	American lotus (*Nelumbo lutea*) Fragrant waterlily (*Nymphaea odorata*) Illinois pondweed (*Potamogeton illinoiensis*) Spatterdock (*Nuphar luteum*) Water clover (*Marsilea quadrifolia*) Watershield (*Brasenia schreberi*)
Emergent plants — These plants grow rooted in the bottom with their leaves and green stems extending well above the surface of the water. A few species also may form floating mats. All have extensive root and rhizome systems and well-developed vascular systems and supportive tissues. Reproduction occurs vegetatively by rhizomes and stolons; floating mat-forming species also reproduce readily by stem fragmentation. Most flower prolifically and form many seeds.	**Broadleaf Species** Arrow arum (*Peltandra virginica*) Arrowhead (*Sagittaria* spp.) Asian spiderwort (*Murdannia keisak*) Frogbit (*Limnobium spongia*) Lizard's tail (*Saururus cernuus*) Pickerelweed (*Pondederia cordata*) Smartweeds (*Polygonum* spp.) **Mat-forming Broadleaf Species** Alligatorweed (*Alternanthera philoxeroides*) Creeping waterprimrose (*Ludwigia hexapetala*) Water pennywort (*Hydrocotyle* spp.) Water willow (*Justicia americana*) **Sedges, Rushes, Spikerushes, and Grasses** Bulrush (*Scirpus* spp.) Cattail (*Typha* spp.) Common reed (*Phragmites australis*) Flat sedge (*Carex* spp.) Foursquare (*Eleocharis quadrangulata*) Maidencane (*Panicum hemitomon*) Rushes (*Juncus* spp.) Sedge (*Cyperus* spp.) Soft rush (*Juncus effusus*) Softstem bulrush (*Scirpus validus*) Southern wildrice (*Zizaniopsis miliacea*) Spikerushes (*Eleocharis* spp.) Threesquare bulrush (*Scirpus americanus*) Torpedograss (*Panicum repens*) Water paspalum (*Paspalum repens*) Woolgrass (*Scirpus cyperinus*) **Other Common Species** Bur-reed (*Sparganium americanum*) Scouring rush (*Equisetum hymale*)
Woody plants — These are obligate, aquatic species of trees usually growing totally flooded or in saturated soils, but occasionally occur in upland areas (usually planted there). Some form systems of "knees" to provide aeration for the root systems. They are deciduous, dropping leaves in the autumn, and are rarely if ever vegetative during winter months.	Bald cypress (*Taxodium distichum*) Pond cypress (*Taxodium ascendens*) Tupelo (*Nyssa aquatica*)

Biological Control of Aquatic Weeds with Triploid Grass Carp

While the triploid, sterile grass carp is a cost-effective control method, it is best suited for use in small ponds, where submersed aquatic plants are not required for fish and wildlife habitat. Grass carp are effective on most submersed weeds. They generally are less effective on algae and weeds in the floating and emergent groups. Refer to the chart below for information on the relative effectiveness of grass carp for different weeds.

Grass carp are normally stocked at 15 fish per acre in small ponds. In larger ponds, they are usually stocked at 15 to 20 fish per vegetated acre. Large fish (minimum of 8 to 10 inches long) should be stocked to prevent loss due to predation by large bass and wading birds. If the surface of the pond is completely covered with vegetation, some limited herbicide application or mechanical removal of weeds from a portion of the pond will be necessary before stocking to allow oxygen to reach the underlying water. Grass carp may be stocked at any time of the growing season, but best results are usually obtained by a late summer or fall stocking.

No permit is required to purchase up to 150 triploid grass carp for stocking a private pond. At a stocking rate of 15 fish per acre of water, 150 triploid grass carp are adequate to control vegetation in a 10-acre pond. A permit from the Wildlife Resources Commission is required for larger stockings. Grass carp may be purchased from a licensed distributor. For a list of North Carolina vendors, see http://www.ncagr.com/aquacult/grasscarp.htm. Permits, a list of certified distributors, and additional information on stocking of triploid grass carp may be obtained from the Wildlife Resources Commission, Chief of Inland Fisheries, 1721 Mail Service Center, Raleigh, NC 27699-1721, or call at (919) 707-0220.

Table 7-20. Biological Control of Aquatic Weeds with Triploid Grass Carp

Weed	Relative Effectiveness	Comments
Algae Filamentous (green and bluegreen) and planktonic	Poor	High stocking rates (60 to 75 or more fish per acre) with small fish (4 to 6 inches size) are required to achieve temporary control; control usually decreases as fish grow larger and are unable to feed on the algae.
Macroalgae Chara and Nitella	Good to Excellent	Chara usually is beneficial to fish and wildlife.
Floating and Floating-Leaved Weeds Duckweeds, watermeal	Poor	Small fish at very high stocking rates (see filamentous algae above) may give control; larger fish at normal stocking rates usually are not effective.
Water ferns (Azolla and Salvinia)	Fair to Poor	
Alligatorweed, water lilies, water primrose, lotus, watershield, spadderdock, waterhyacinth	Poor	Grass carp may feed lightly on weeds in this group, but control is usually unacceptable.
Emergent and Marginal Weeds Cattails, rushes, common reed, bulrushes, pickerelweed, pennywort, arrowhead	Poor	Grass carp may feed lightly on weeds in this group, but control is usually unacceptable.
Submersed Weeds	Good to Excellent	Most rooted and free-floating submersed weeds in ponds are readily controlled with triploid grass carp; control may be poorer on the watermilfoils, particularly Eurasian waterfoil.

Chemical Control of Aquatic Plants

Table 7-21. Chemical Control of Aquatic Plants

Herbicide, Formulation, and Mode of Action Code	Amount of Formulation	Active Ingredient Rate or Concentration	Precautions and Remarks[2]
Algae, blue-green			
copper sulfate (various)	See label	0.5 to 1 ppm	Apply crystals or powder at early stage of growth by any method to give rapid and uniform dispersion. For best results, apply on a clear day. Do not apply to muddy water. **Warning: Copper is toxic to fish.** Formulated copper products have a greater margin of safety to fish.
sodium carbonate peroxyhydrate (various)	See label	0.3 to 1.7 ppm	Apply with 8 to 10 hours of daylight remaining. Do not reapply within 48 hours.
Algae, filamentous and planktonic			
copper complex (various)	0.6 gal/acre ft	0.2 ppm	Dilute with water in ratio of at least 9-to-1 and apply uniformly. For best results, apply on a clear day and break up floating mats of filamentous algae before treatment. **Warning: Copper is toxic to fish.**
copper sulfate (various)	See label	0.5 to 1 ppm	Same as under Algae, blue-green. For best results break up floating mats of filamentous algae before treatment. **Warning: Copper is toxic to fish.** Formulated copper products have a greater margin of safety to fish.
diquat (Reward) 2 lb/gal MOA 22	See label	0.18 to 0.37 ppm	For certain filamentous algae—*Pithophora* spp. and *Spirogyra* spp. Check label for application instructions. For best results, break up floating mats before treatment.
Algae, macro, chara, nitella			
copper complex (Cutrine-Plus Granular) 3.7 G (Cutrine-Plus) 0.9 lb/gal (K-Tea) 0.8 lb/gal	60 lb/surface acre 1.2 gal/acre ft 1.7 to 3.4 gal/acre ft	2.2 lb/acre 0.4 ppm 0.5 to 1.0 ppm	Distribute granular formulation evenly over infested area when plants are young. If chara is in water less than 3 ft deep or growth is near the surface, the liquid formulation may be used. Dilute with water in ratio of at least 9-to-1 and apply uniformly. **Warning: Copper is toxic to fish.**
Algae, Pithophora and cladophoraa			
flumioxazin (Clipper) 51% MOA 14	6 to 12 oz/A	3 to 6 ai/A or 100 to 400 ppb	Early morning applications may be more effective. If vegetation is dense, treat in sections to avoid reducing dissolved oxgen. Water pH greater than 7.5 will reduce effectiveness.

Table 7-21. Chemical Control of Aquatic Plants

Herbicide, Formulation, and Mode of Action Code	Amount of Formulation	Active Ingredient Rate or Concentration	Precautions and Remarks[2]
Floating Weeds (except watermeal)			
2,4-D amine (various) MOA 4	See label	2 to 4 lb/acre	Thorough wetting of foliage is essential. Apply in 100 to 400 gallons of water per acre. Use low pressure, large nozzle, and spray thickener. **For use on nonirrigation ditchbanks only. Do not apply to water.**
bispyribac (Tradewind) 80% MOA 2	1 to 2 oz/A	0.8 to 1.6 oz ai/A	Controls duckweed, mosquito fern, salvinia, water hyacinth, water lettuce, and water pennywort. Apply with at least 30 gpa water volume. Include appropriate adjuvant.
carafentrazone (Stingray) 1.9 lb/gal, MOA 14	3.4 to 13.5 fl oz/acre	0.05 to 0.2 lb/acre	Controls water lettuce, waterhyacinth, salvia, duckweed, mosquito fern, and water spinach. Rates vary according to target species. Methylated seed oil or nonionic surfactant recommended.
diquat (Reward) 2 lb/gal MOA 22	0.5 to 0.75 gal/ surface acre	1 to 1.5 lb/acre	Weeds controlled: pennywort, salvinia, waterhyacinth, waterlettuce. Apply in a spray volume of 150 to 200 gallons of water per acre plus 1 pint nonionic surfactant.
	1 gal/surface acre	2 lb/acre	For duckweed control, apply in a spray volume of 50 to 150 gallons of water per acre. Take care to cover all plants on water and damp marginal areas. Will require retreatment. A nonionic surfactant at 0.5% by volume will be needed.
glyphosate (various) MOA 9	See label	See label	For control of waterlillies, spadderdock, and lotus, apply as foliar spray on a calm day when there is little to no wave action. Vegetation must be on or above the surface for treatment to be effective. A nonionic surfactant (e.g. Cide-Kick or Induce) is required with Rodeo. If applying from a boat, take care not to create waves that may wash the herbicide off floating leaves. Will not control small floating plants, such as azolla, duckweed, or watermeal.
imazamox (Clearcast) 1 lb/gal, MOA 2	32 to 64 fl oz/acre	0.25 to 0.5 lb ai/acre 50 to 150 ppb	See label for specific weeds controlled. A nonionic surfactant or methylated seed oil is recommended for foliar applications. Spot treatments may be made with up to 5% solution by volume.
imazapyr (Habitat) MOA 2	1 to 4 pt/acre	0.25 to 1.5 lb/acre	Rates vary according to target species. Retreatment of some plants may be required. A nonionic surfactant is recommended. Will not control small floating plants, such as azolla, duckweed, or watermeal.
penoxsulam (Galleon) 2 lb/gal, MOA 2	2 to 5.6 fl oz/acre	0.03 to 0.09 lb/acre 5 to 150 ppb	A nonionic surfactant is recommended for foliar applications.
triclopyr (Renovate 3) MOA 4	0.5 to 2 gal/acre	1.5 to 6 lb/acre	Rates vary according to target species. Addition of approved nonionic surfactant is recommended.
topramezone (Oasis) 29.7%	up to 16 fl oz/acre	up to 0.35 lb/acre	Use of an aquatic safe surfactant is required for all foliar applications. Check label for specific irrigation restrictions.
Floating Weeds (watermeal and others)			
flumioxazin (Clipper) 51% MOA 14	6 to 12 oz/acre	3 to 6 ai/A or 100 to 400 ppb	Early morning applications may be more effective. If vegetation is dense, treat in sections to avoid reducing dissolved oxgen. Water pH greater than 7.5 will reduce effectiveness. A follow-up application may be needed for watermeal control.
fluridone (Sonar) 4 AS MOA 12	Ponds: 0.16 to 1.5 qt/acre	0.16 to 1 lb/acre 10 to 90 ppb	Use the maximum labelled rate for the average depth of your pond. Do not apply when there is substantial outflow from the pond. Do not apply as a spot treatment. See label for specific weeds controlled. For watermeal, use 45 to 90 ppb. Other floating species may be controlled with lower rates. Do not use treated water for irrigation for 7 to 30 days. See label for irrigation precautions. **Warning: 30 days may be insufficient restriction if pond water will be used to irrigate very sensitive crops, such as tobacco, tomatoes, or peppers.**
Emergent, Marginal, and Ditchbank Weeds			
2,4-D amine (various) MOA 4	See label	2 to 4 lb/acre	Thorough wetting of foliage is essential. Apply in 100 to 400 gallons of water per acre. Use low pressure, large nozzle and spray thickener. **For use on nonirrigation ditchbanks only. Do not apply to water.**
2,4-D granular (Navigate) 20 G (2,4-D Gran 20) 20 G MOA 4	150 to 200 lb/surface acre	30 to 40 lb/acre	Weeds controlled: arrowhead, bulrush, creeping waterprimrose, pickerelweed, smartweed, spadderdock, waterchestnut, waterlily, watershield. Rate depends upon species and depth of water. Check label. Apply early, when weeds are actively growing, with a rotary seeder. Spadderdock may require retreatment.
bispyribac (Tradewind) 80% MOA 2	1 to 2 oz/A	0.8 to 1.6 oz ai/A	Controls alligatorweed and parrotfeather. Apply with at least 30 gpa water volume. Include appropriate adjuvant.
carafentrazone (Stingray) 1.9 lb/gal, MOA 14	6.7 to 13.5 fl oz/acre	0.2 lb/acre	Suppresses alligatorweed and waterprimrose.
diquat (Reward) 2 lb/gal (Weedtrine) 0.4 lb/gal MOA 22	1 gal/surface acre	2 lb/acre	For control of cattails in ponds or lakes. For top kill, apply in 100 gal of water per acre with 0.25% to 0.5% nonionic surfactant. Apply before flowering for best results. Retreat as needed.
flumioxazin (Clipper) 51% MOA 14	6 to 12 oz/A	3 to 6 ai/A or 100 to 400 ppb	Early morning applications may be more effective. If vegetation is dense, treat in sections to avoid reducing dissolved oxgen. Ensure adequate coverage of dense vegetation or a follow-up application may be necessary.
glyphosate (various) MOA 9	See label	See label	Rates vary according to target species. Retreatment of alligatorweed is necessary. Aquatic-approved nonionic surfactant (e.g. Cide-Kick or Induce) required with Rodeo. Note: The use of very hard water or water containing high concentrations of iron to prepare spray solutions may result in reduced efficacy of glyphosate.
imazamox (Clearcast) 1 lb/gal MOA 2	32 to 64 fl oz/acre	0.25 to 0.5 lb ai/acre 50 to 500 ppb	See label for specific weeds controlled. A nonionic surfactant or methylated seed oil is recommended for foliar applications. Spot treatments may be made with up to 5% solution by volume. Rates vary according to target species. Retreatment of some plants may be required. A nonionic surfactant is recommended.
imazapyr (Habitat) MOA 2	1 to 6 pt/acre	0.25 to 1.5 lb/acre	Rates vary according to target species. Retreatment of some plants may be required. A nonionic surfactant is recommended.
penoxsulam (Galleon) 2 lb/gal, MOA 2	2 to 5.6 fl oz/acre	0.03 to 0.09 lb/acre 5 to 500 ppb	See label for specific weeds controlled and application details.
triclopyr (Renovate 3) MOA 4	0.5 to 2 gal/ acre	1.5 to 6 lb/acre	Rates vary according to target species. Addition of an approved nonionic surfactant is recommended.
topramezone (Oasis) 29.7%	up to 16 fl oz/acre	up to 0.35 lb/acre	Use of an aquatic safe surfactant is required for all foliar applications. Check label for specific irrigation restrictions.

Table 7-21. Chemical Control of Aquatic Plants

Herbicide, Formulation, and Mode of Action Code	Amount of Formulation	Active Ingredient Rate or Concentration	Precautions and Remarks[2]
Submersed Weeds[3]			
2,4-D granular (Navigate) 20 G, MOA 4	100 to 200 lb/ surface acre	20 to 40 lb/acre	Rate depends upon weed to be controlled and depth of water. Check labels for species and rates. Apply uniformly with a rotary seeder.
bispyribac (Tradewind) 80% MOA 2	See label	10 to 45 ppb	Controls hydrilla, sago pondweed, and Eurasian watermilfoil. Do not apply in areas of high water flow or water diffusion. Refer to label for specific details on application rate based on water volume.
carfentrazone (Stingray) 1.9 lb/gal MOA 4	0.286 to 5.75 gal/acre	200 ppb	Controls Eurasian watermilfoil. Apply in spring or early summer as a subsurface application or with an appropriate adjuvant to ensure sinking and mixing of the spray mix.
diquat (Reward) 2 lb/gal MOA 22	1 to 2 gal/ surface acre	2 to 4 lb/acre	Weeds controlled: bladderwort, coontail, elodea, naiads, pond weeds. Apply early in season by pouring directly into water in strips 40 feet apart. Later in season, as weeds reach surface, pour in strips 20 feet apart or inject a dilute solution. Not effective in muddy water.
endothall (Aquathol K) 4.2 lb/gal (Aquathol Super K) 63 G	0.3 to 2.6 gal/acre ft 2.2 to 17.6 lb/acre ft	0.5 to 5 ppm	Weeds controlled: bass weed, bur reed, coontail, hydrilla (Aquathol K only), pondweeds, watermilfoil, water star grass. Rate depends upon weed species and type of treatment. Spot or marginal treatments require higher rates. Aquathol Granular is especially useful for spot or marginal treatments.
flumioxazin (Clipper) 51% MOA 14	See label	100 to 400 ppb	Early morning applications may be more effective. If vegetation is dense, treat in sections to avoid reducing dissolved oxygen. Water pH greater than 7.5 will reduce effectiveness.
fluridone (Sonar) AS MOA 12	Ponds: 0.16 to 1 qt/acre Lakes: 0.2 to 4 qt/acre Canals: 2 qt/acre	0.16 to 1 lb/acre 0.2 to 4 lb/acre 2 lb/acre	Do not use water for irrigation for 7 to 30 days. See label for specific irrigation precautions. Application to canals should be made only if water flow can be restricted. **Warning: 30 days may be insufficient restriction if applied to small ponds and pond water will be used to irrigate very sensitive crops, such as tobacco, tomatoes, or peppers.**
(Sonar SRP) MOA 12	Ponds: 3.2 to 30 lb/acre Lakes: 4 to 80 lb/acre Canals: 40 lb/acre Rivers: 40 lb/acre	0.16 to1.5 lb/acre 0.2 to 4 lb/acre 2 lb/acre 2 lb/acre	
imazamox (Clearcast) 1 lb/gal, MOA 2	See label	50 to 500 ppb	Rates vary according to target species and depth to be treated. See label for specific weeds controlled and application details.
penoxsulam (Galleon) 2 lb/gal, MOA 2	See label	5 to 150 ppb	Rates vary according to target species and depth to be treated. See label for specific weeds controlled and application details.
triclopyr (Renovate 3 or OTF), MOA 4	See label	1.5 to 6 lb/acre 0.5 to 2.5 ppm	Rates vary according to target species and depth to be treated. See label for specific weeds controlled and application details.
topramezone (Oasis) 29.7%	up to 16 fl oz/acre	up to 0.35 lb/acre	Use of an aquatic safe surfactant is required for all foliar applications. Check label for specific irrigation restrictions.

[1] Mode of Action (MOA) code developed by the Weed Science Society of America. Cooper compounds, endothall. and sodium carbonate peroxyhydrate have not been assigned codes.

[2] Also see comments for specific herbicides under "Table 7-25. Labeled Sites and Restrictions."

[3] Grass carp give cost-effective control on the majority of the weeds in this group and should be given consideration *before* using herbicides. See text at beginning of this section under *Biological Control of Aquatic Weeds with Triploid Grass Carp*. A permit is required to purchase more than 150 grass carp or for stocking in impoundments larger than 10 acres. Grass carp usually are **not effective** on filamentous algae, duckweed, watermeal, or any of the plants in the floating or emergent groups.

CHEMICAL WEED CONTROL

Table 7-22. Waiting Period (in Days) Before Using Water After Application of Herbicides for Aquatic Weed Control

Herbicide	Irrigation[1]	Fish Consumption	Watering Livestock	Swimming
2,4-D (various formulations and manufacturers)	Water use restrictions vary by formulation and manufacturer. In general, if water is used for irrigating sensitive crops, 2,4-D should not be used. Turfgrasses are generally tolerant to low concentrations of 2,4-D. Also, many 2,4-D formulations are NOT labelled for aquatic use. Read the label before purchasing and/or use.			
Bispyribac (Tradewind)	Do not irrigate until concentrations are < 1 ppb	No restrictions	Do not water livestock until concentrations are ≤ 1 ppb	No restrictions
carfentrazone (Stingray)	1 to 14[2]	No restrictions	0 to 1	No restrictions
copper (Copper sulfate pentahydrate, including Bluestone and EarthTec; and complexed copper formulations, including Algae-Pro, Captain, Clearigate, Cutrine-Plus, Cutrine-Plus Granular, K-Tea, Komeen, etc.)	No restrictions	No restrictions	No restrictions	No restrictions
diquat (Reward)	3 to 5[3]	No restrictions	1	No restrictions
endothall (Aquathol K) (Aquathol Super K) (Hydrothol 191) (Hydrothol 191 granular)	No restrictions for many situations. See label for specific restrictions	No restrictions	7 to 25	No restrictions
Flumioxazin (Clipper)	0 to 5[3]	No restrictions	No restrictions	No restrictions
fluridone (Sonar 4AS, Sonar SRP)	7 to 30[3]	No restrictions	No restrictions	No restrictions
Glyphosate (AquaMaster, Aqua Neat, Rodeo, Touchdown Pro)	No restrictions	No restrictions	No restrictions	No restrictions
imazamox (Clearcast)	0+[3]	No restrictions	No restrictions	No restrictions
Imazapyr (Habitat)	120	No restrictions	No restrictions	No restrictions
penoxsulam (Galleon)	Do not irrigate food crops until residues ≤ 1 ppb	No restrictions	No restrictions	No restrictions
sodium carbonate peroxyhydrate (GreenClean Pro, Pak 27)	No restrictions	No restrictions	No restrictions	No restrictions
topramezone (Oasis)	See label for specific irrigation restrictions	No restrictions	No restrictions	No restrictions
triclopyr (Renovate 3, Renovate OTF)	120 0 to established grass	No restrictions	Next growing season for lactating dairy animals	No restrictions

[1] Irrigation restrictions may be removed for specific products if a laboratory assay of treated water meets a standard as stated on the product label.

[2] Do not use treated water for irrigation in commercial nurseries or greenhouses.

[3] Refer to product label for specific restrictions.

CHEMICAL WEED CONTROL

Table 7-23. Effectiveness of Herbicides and Triploid Grass Carp for Control of Weeds Commonly Found in N.C. Ponds

Weeds	2,4-D	bispyribac	carfentrazone	copper compounds	diquat	diquat +copper	endothall Aquathol	endothall Hydrothol	flumioxazin	fluridone	glyphosate	imazamox	imazapyr	peroxide compounds	penoxsulam	triclopyr	triploid grass carp
Algae																	
Planktonic	NR	ID	NR	G	P	G	NR	P	ID	NR	NR	NR	NR	G	NR	NR	NR
Filamentous	NR	ID	NR	G	E	E	NR	E	G	NR	NR	NR	NR	ID	NR	NR	P
Chara / Nitella	NR	ID	ID	G	G	E	NR	G	P	NR	NR	NR	NR	ID	NR	NR	E
Floating Plants																	
Azolla (mosquito fern)	NR	G	F	F	E	E	NR	NR	ID	E	NR	ID	NR	NR	G	NR	P
Duckweed	P	G	G	P	G	G	NR	NR	E	E	NR	NR	NR	NR	G	P	P
Frogbit	F	ID	ID	NR	E	E	NR	NR	G	NR	P	E	E	NR	ID	G	P
Salvinia, common	NR	G	G	P	E	E	NR	NR	G	E	G	E	ID	NR	ID	NR	P
Salvinia, giant	NR	G	G	P	E	E	F	NR	F	E	G	P	G	NR	E	NR	P
Waterhyacinth	E	G	G	NR	G	G	NR	NR	P	F	G	E	G	NR	E	E	P
Watermeal	NR	NR	NR	NR	P	P	NR	NR	G	G	NR	NR	NR	NR	P	NR	P
Water lettuce	NR	G	G	NR	G	G	G	G	E	NR	E	G	E	NR	E	NR	P
Emersed Plants																	
Alligatorweed	P	G	F	NR	NR	NR	NR	NR	F	F	G	G	G	NR	G	G	P
American lotus	G	ID	NR	NR	NR	NR	NR	NR	ID	G	E	F	G	NR	ID	G	P
Cattail	F	ID	NR	NR	F	F	NR	NR	P	G	E	G-E	E	NR	ID	F	P
Creeping waterprimrose	E	ID	F	NR	NR	NR	NR	NR	ID	F	E	F	E	NR	G	E	P
Floating hearts	P	ID	NR	NR	F	F	E	E	ID	F	G	G	G	NR	F	P	P
Fragrant waterlily	G	ID	NR	NR	NR	NR	NR	NR	ID	G	E	G	E	NR	ID	G	P
Grass species	NR	ID	NR	NR	F	F	NR	NR	NR	F	E	F	E	NR	ID	NR	P
Parrotfeather	E	G	F	NR	NR	NR	NR	NR	F	NR	F	G	E	NR	G	E	NR
Phragmites (Common reed)	NR	ID	NR	NR	NR	NR	NR	NR	P	NR	G	F-G	E	NR	NR	F	P
Pickeralweed	G	ID	NR	NR	NR	NR	NR	NR	ID	NR	F	E	E	NR	ID	G	P
Rush	NR	ID	NR	NR	NR	NR	NR	NR	ID	NR	G	ID	G	NR	ID	F	P
Spatterdock	G	ID	NR	NR	NR	NR	NR	NR	ID	G	E	G	E	NR	ID	F	P
Smartweeds	F	ID	NR	NR	F	F	NR	NR	ID	F	G	G	G	NR	F	G	P
Waterpennywort	G	G	NR	NR	F	F	NR	NR	G	G	E	E	E	NR	F	G	P
Watershield	E	ID	NR	NR	F	F	NR	NR	ID	F	E	G	G	NR	ID	E	P
Submersed Plants																	
Bladderwort	P	ID	ID	NR	F	F	P	P	ID	E	NR	F-G	NR	NR	ID	P	E
Cabomba	NR	ID	ID	NR	F	F	F	F	G	F	NR	F	NR	NR	ID	NR	F
Coontail	G	ID	ID	NR	E	E	E	E	G	E	NR	NR	NR	NR	ID	G	E
Egeria (Brazilian elodea)	NR	ID	ID	F	E	E	P	P	ID	E	NR	ID	NR	NR	G	NR	E
Eurasian watermilfoil	E	G	G	NR	G	G	E	NR	G	E	NR	F	NR	NR	G	E	P
Hydrilla, monoecious	NR	G	ID	F	G	E	E	E	G	E	NR	F	NR	NR	G	NR	E
Naiad, brittle	NR	ID	ID	G	E	E	E	E	G	E	NR	ID	NR	NR	F	NR	E
Naiad, Southern	NR	ID	ID	G	P	G	P	P	G	G	NR	ID	NR	NR	F	NR	E
Parrotfeather	E	G	ID	NR	G	G	E	E	G	E	NR	F	NR	NR	G	E	F
Pondweed species	NR	G	ID	NR	E	E	E	E	G	E	NR	G	NR	NR	G	NR	E
Proliferating spikerush	NR	ID	ID	NR	NR	NR	NR	NR	P	F	NR	F	NR	NR	F	NR	E
Variable leaf milfoil	E	ID	G	NR	E	E	E	E	E	G	NR	NR	NR	NR	NR	E	P

Key: NR = Not Recommended; P = Poor; G=Good ; ID = Insufficient Data; F = Fair; E = Excellent

CHEMICAL WEED CONTROL

Pond Dyes

Pond dyes may be used to prevent the growth of filamentous algae and submersed macrophyte vegetation. Pond dyes are not herbicides and do not directly kill aquatic plants. They function by blocking light penetration to the bottom of the pond. As a result, these products are most effective when applied very early in the growing season.

The use of a pond dye in aquacultural ponds usually is not recommended, as they tend to inhibit phytoplankton productivity that is needed to produce oxygen and provide food for zooplankton, which are the major food of fry and the smaller juvenile fishes. Application rates usually are about one part per million or 1 gallon per acre for a pond averaging 4 feet deep (i.e., 1 gallon per 4 acre-feet of water) for algae and most submersed weeds. For hydrilla, the rate needs to be doubled, due to its ability to grow at very low light levels. Several of the available pond dyes are registered by the EPA for aquatic weed control. Pond dyes *should not be applied to drinking water supplies or to streams or any body of water where there is any substantial outflow.*

Table 7-24. Pond Dyes

Examples of Pond Dyes	EPA Registered
Admiral Liquid	Yes
Aquashade	Yes

Table 7-25. Labeled Sites and Restrictions

Herbicide and Formulation	Labeled Sites	Restrictions
2, 4-D amine (Weedar 64) 3.8 lb a.i./gal Other formulations	potable water reservoirs, farm and fish ponds, lakes, golf course water hazards, fish hatcheries	Delay the use of treated waters for irrigation and domestic purposes for 3 weeks after application unless an assay indicates that chemical water concentrates are below the minimum amount as specified on the product label. Do not treat irrigation ditches where water will be used for overhead irrigation of susceptible crops. Refer to specific product label for restrictions.
2,4-D granular (Navigate) 20 G	ponds and lakes	Do not apply to water used for irrigation, agricultural sprays, watering dairy animals, or domestic water supplies.
Bispyribac (Tradewind) 80%	Bayous, canals, fresh water ponds, lakes, marshes, and reservoirs	Do not irrigate until water concentrations are less than 1 ppb. Do not treat water used for crawfish production.
copper-complex (Cutrine-Plus) 0.9 lb/gal (Cutrine-Plus) 3.7 G (K-Tea) 0.8 lb/gal **copper sulfate**	potable water reservoirs, farm and fish ponds, lakes, golf course water hazards, fish hatcheries	No restrictions on use of treated water. Check tolerance of crop to copper applied in irrigation water. Trout are very susceptible to copper. Toxicity to other fish increases with decreasing hardness of water.
carfentrazone (Stingray) 1.9 lb/gal	ponds, lakes, reservoirs, marshes, wetlands, drainage ditches, canals, streams, rivers, etc.	Irrigation: Do not use treated water in commercial nurseries or greenhouses. Field crops may be irrigated after 1 day if less than 20% of surface area was treated, or after 14 days if treatment was 20% or more of surface area or until an assay indicates that chemical water concentrates are below a minimum amount as specified on the product label. Treated water may be used for turf irrigation with no restriction if less than 20% of the total water body was treated. A 14-day restriction applies for larger area treatments. Do not apply within 0.25 miles an active potable water intake (upstream only in flowing waters), or turn intake off for at least 24 hours as specified on product label. Do not drink or water livestock for 1 day if 20% or more of total surface area was treated. Applicators must be licensed or certified by the state.
diquat (Reward) 2 lb/gal	lakes, still ponds, ditches, laterals, waterways	Apply only to still water and/or public waters. Do not apply to turbid waters. Do not use treated water for irrigation of food crops, preparation of agricultural sprays, or for drinking for 5 days after application. Turf and nonfood crops may be irrigated 3 days after treatment. Do not use water for livestock for one day after treatment. Water use restrictions may be removed if an approved assay is conducted and water concentration is less than the maximum contaminant level as specified on product label.
dyes (Admiral Liquid) (Aquashade)	ponds and lakes with little to no outflow	No not apply to water bodies not under direct control of user. Do not apply to water that will be used for human consumption.
endothall (Aquathol K) 4.23 lb/gal (Aquathol Super K granular) 63% (Hydrothol 191) 2 lb a.i./gal (Hydrothol granular) 11.2%	drainage canals, lakes, ponds	Restrictions up to 25 days may apply to waters used for domestic uses, irrigation, or watering livestock. Setback distance of at least 600 feet from functioning potable water intakes may also apply. Refer to specific product label for current restrictions on domestic use, irrigation, livestock use, and setback distance. Hydrothol formulations may kill fish when rates exceed 0.3 ppm. Check label for drinking water restrictions. Fish may be killed by rates exceeding 0.3 ppm. Irrigation and animal consumption restrictions of 7 to 25 days, depending on rate.
Flumioxazin (Clipper) 51%	Bayous, canals, fresh water ponds, lakes, marshes, and reservoirs	Do not irrigate from treated water for at least 5 days. Do not treat water used for crawfish production.
fluridone (Sonar 4 AS or SRP)	lakes, ponds, canals	Treated ponds may not be used for irrigaton for 7 to 30 days. See label for irrigation precautions. [1]
glyphosate (AquaMaster) 5.4 lb a.i./gal (AquaNeat) 5.4 lb a.i./gal (Rodeo) 5.4 lb a.i./gal (Touchdown Pro) 3 lb a.e./gal Other formulations	all bodies of fresh water and all types of aquatic sites	Do not apply within 0.5 mile of an active potable water intake (upstream only in flowing waters), or turn intake off for at least 48 hours as specified on product label. Refer to specific product label for restrictions.
imazamox (Clearcast)	in and around aquatic and noncropland sites	Irrigation: Do not apply to water to be used for irrigation of greenhouse or nursery plants. Do not irrigate from still or quiescent water bodies within 24 hours of application. Do not irrigate if concentrations exceed 50 ppb.
imazapyr (Habitat)	in and around standing and flowing waters, including estuarine and marine sites	Irrigation: Do not use treated water for 120 days following application or until an assay indicates that chemical water concentrations are below a minimum amount as specified on the product label. Do not apply within 0.5 mile of an active potable water intake (upstream only in flowing waters), or turn intake off for at least 48 hours as specified on product label. Do not apply to irrigation ditches or canals within 1 mile of an active irrigation water intake unless the irrigation restrictions can be observed. Applicators must be licensed or certified by the state.
penoxsulam (Galleon)	in and around quiescent water bodies and exposed sediments of de-watered areas	Do not apply to flowing water. Irrigation: Do not apply to water to be used for irrigation of greenhouse or nursery plants. Do not irrigate established food crops, other than rice, if concentrations exceed 1 ppb. Do not irrigate established rice if concentrations in treated water exceed 30 ppb. No restrictions on use of treated water for turf irrigation, if concentrations are less than 30 ppb. Consult SePRO for other situations/commodities.

Table 7-25. Labeled Sites and Restrictions

Herbicide and Formulation	Labeled Sites	Restrictions
sodium carbonate peroxyhydrate (GreenClean Pro) (PAK 27)	ponds, lakes, lagoons, canals, ditches, etc.	Do not apply to treated, finished drinking water reservoirs.
triclopyr (Renovate 3) 3 lb/gal (Renovate OTF) 10 G	quiescent and slow-moving waters; non-irrigation canals	Irrigation: Do not use treated water for 120 days following application or until treated water has a non-detectable triclopyr level by an assay as specified on the product label. No restriction on irrigation of established grass. Applications around potable water intakes must observe minimum setback distances and/or minimum water concentrations as specified on the product label. Do not apply directly to or allow to come in direct contact with grapes, tobacco, vegetable crops, flowers, and other desirable broadleaf plants. Do not apply to esturarine or marine sites; do not apply directly to un-impounded rivers or streams; and do not apply to irrigation ditches or canals. Do not allow lactating dairy animals to graze treated areas until the next growing season after application unless spot-treatment was applied to less than 10% of total grazable area. Animals for slaughter must be removed from the treated area for at least 3 days.

[1] Water use restrictions for irrigation vary with formulation. See label for precautions. A 30-day restriction may be insufficient if applied to small ponds intended for irrigation of very sensitive crops, such as tobacco, tomatoes, or peppers.

Chemical Control of Specific Weeds

F. H. Yelverton and T. W. Gannon, Crop and Soil Sciences; D. W. Monks, Horticultural Science

Table 7-26. Chemical Control of Specific Weeds

Herbicide and Formulation	Amount of Formulation	Time of Application	Precautions and Remarks
Artichoke, Betony or Florida Betony			
dichlobenil (Casoron) 4 G (Gordon's Barrier) 4 G (Proteam Zapzit) 4 G	100 to 250 lb/acre	Any time. Fall or winter best.	*Do not use on cropland.* See Precautions and Remarks under **Weed Control in Woody Ornamentals.** In order to comply with state noxious weed regulations, contact your local plant protection specialist or NCDA&CS weed specialist at 919-707-3749.
mesotrione (Tenacity) 4 SC	4 to 8 fl oz/acre	Apply to young, actively growing weeds.	Use a nonionic surfactant at 0.25% v/v. Reapply in 3 weeks. Controls Florida betony.
Bamboo (Cane)			
glyphosate (Accord) 4 SL (Roundup PRO) 4 SL and various other trade names	2% solution sprayed to wet	Midsummer to early fall while actively growing.	For large canes; first cut canes and allow to regrow to 4 to 6 feet in height. Avoid drift to desirable vegetation. Use Accord for forestry and utility rights-of-way sites.
	1 part herbicide + 2 parts water, applied with sponge	Whenever new shoots are in the "husk" stage (before leaves open) and are 12 feet to 24 inches in height.	Wear rubber gloves. Wipe entire shoot with a sponge dampened with the herbicide. Sponge should not be dripping wet. Do not allow contact with desirable vegetation; avoid dripping onto grass.
Bermudagrass			
glyphosate + fluazifop (Roundup PRO) 4 SL + (Fusilade DX) 2 EC	2 to 0.75 qt/acre	See Precautions and Remarks	Apply when bermudagrass is actively growing. Repeat applications when bermudagrass regrows. Wait 30 days after last application to seed, sprig, or sod new bermudagrass.
Dazomet (Basamid)	218 to 525 lb/acre	See Precautions and Remarks	Restricted Use Pesticide. May be used to control many annual and perennial weed species including bermudagrass. Apply to properly prepared soil and incorporate physically or with water. Consult and follow label directions.
Berries (*Rubus* spp.)			
glyphosate (Accord) 4 SL (Roundup PRO) 4 SL and various other trade names	3 to 4 qt/acre as broadcast spray or 1% to 1.5% solution with handheld equipment	After bloom stage	Use higher rate for plants that have reached the woody stage of growth. Best results are obtained when sprayed in late summer after berries are formed. Use Accord for forestry and utility rights-of-way sites. Roundup PRO should be used for industrial and other noncropland areas. Use Roundup ULTRA on agricultural areas.
triclopyr amine (Garlon 3 A) 3 SL	1 gal/100 gal water	After leaves fully expand in spring and before leaf color change in fall	See comments under kudzu.
Florida Betony			
atrazine (AAtrex 4L) 4L	1 to 2 qt/A	Fall and/or winter	May be used on centipedegrass, St. Augustinegrass and dormant bermudagrass. Do not apply after Dec. 31 on bermudagrass unless delay in greenup is acceptable.
triclopyr + clopyralid (Confront) 3 SL	1 to 2 pt/A	Fall and/or winter	Do not apply to home lawns. May be used on centipedegrass, bermudagrass, and zoysiagrass. Do not apply to bermudagrass during transition. Repeat applications may be required.
MCPA amine + triclopyr amine + dicamba (Horsepower) 4.56 L	2 to 3 pt/A	Fall and/or winter	May be applied to home lawns by a commercial applicator. Not for use on turf grown for resale or other commercial uses such as sod or seed production. May be used on bermudagrass, zoysiagrass and bahiagrasss.
MCPA ester + triclopyr ester + dicamba (Coolpower) 3.6 L	2.5 to 3.5 pt/A	Fall and/or winter	May be applied to home lawns by a commercial applicator. Not for use on turf grown for resale or other commercial uses such as sod or seed production. May be used on bermudagrass, zoysiagrass and bahiagrasss.

Table 7-26. Chemical Control of Specific Weeds

Herbicide and Formulation	Amount of Formulation	Time of Application	Precautions and Remarks
Honeysuckle			
dicamba (Banvel) 4 SL	1 gal/100 gal water	When actively growing, prior to bloom.	Using hand-held equipment, spray to wet leaves. Add a nonionic surfactant at the rate of 2 quarts per 100 gallons of finished spray solution to improve wetting. Keep spray off desired plants. Do not spray in rooting zone of desired plants.
2,4-D amine (various) 4 SL	1.5 qt/50 gal water		
glyphosate (Accord) 4 SL (Roundup PRO) 4 SL and various other trade names	3 to 4 qt/acre as broadcast spray or 1% to 1.5% solution with handheld equipment	When plants are actively growing at or beyond the bloom stage of growth.	Use the higher rate for plants that have reached the woody stage of growth. Ensure thorough spray coverage with hand-held equipment. Use Accord for forestry and utility rights-of-way sites. Roundup PRO should be used for industrial and other noncropland areas. Use Roundup ULTRA on agricultural areas.
Kudzu			
aminopyralid (Milestone) 2 SL	7 oz/A	Apply to young, actively growing plants	May be used in permanent grass pastures, rangeland, noncrop areas, nonirrigation ditch banks and other natural areas. Due to crop sensitivity, use extreme caution around sensitive crops including but not limited to alfalfa, cotton, potatoes, soybeans, tobacco and other broadleaf or vegetable crops, fruit trees, or ornamental plants. Do not use aminopyralid-treated plant residues, including hay or straw from treated areas, or manure from animals that have grazed treated areas in compost or mulch that will be in contact with susceptible broadleaf plants. There are no restrictions on grazing or hay harvest following aminopyralid applications. Check and follow label directions for completed list and precautions.
clopyralid (Transline) 3 SL	1 to 3 qt/100 gal	During active growth.	Spray to wet leaves. Do not apply more than 1.337 pints per acre.
dicamba (Banvel) 4 SL	1 gal/100 gal water	When actively growing, before bloom.	Using hand-held equipment, spray to wet leaves. Keep spray off desired plants. Do not spray in rooting zones of desired plants. Add a nonionic surfactant at the rate of 2 quarts per 100 gallons of finished spray solution to improve wetting.
(Vanquish) 4 SL	0.5 gal/100 gal water	Dormant season; just prior to budbreak (early to mid-March).	Do not spray in rooting zones of desired plants. Regrowth should be sprayed in mid- to late summer with glyphosate or clopyralid.
fosamine (Krenite S) 4 SL	1.5 to 3 gal/acre	August through September.	Spray to wet leaves thoroughly. Good coverage is necessary.
glyphosate (Accord) 4 SL (Roundup PRO) 4 SL and various other trade names	4 qt/acre as broadcast spray or 2% solution with handheld equipment	When actively growing at or beyond bloom stage of growth.	Repeat applications are necessary to maintain control. Ensure thorough spray coverage with hand-held equipment. Apply before frost in the fall. Use Accord for forestry and utility rights-of-way sites. Roundup PRO should be used for industrial and other noncropland areas. Use Roundup ULTRA on agricultural areas.
metsulfuron methyl (Escort) 60 WDG	3 to 4 oz/acre	When actively growing.	Add 1 quart nonionic surfactant per 100 gallons of water plus a drift control agent. Do not apply by air.
sulfometuron methyl (Oust) 75 WDG	12 oz/acre per 100 gal water	When actively growing.	Add 1 quart nonionic surfactant per 100 gallons of water plus a drift control agent. Do not apply by air.
triclopyr amine (Garlon 3 A) 3 SL	2 gal/100 gal water	During mid-season when plants are actively growing.	Spray to wet leaves thoroughly. Do not allow to drift; product is very toxic to tobacco, soybeans, many other broadleaf crops, trees, and ornamentals. Most grasses very resistant.
Mugwort (*Artemisia vulgaris*)			
dichlobenil (Casoron) 4 G (Gordon's Barrier) 4 G (Proteam Zapzit) 4 G	100 to 250 lb/acre	Any time. Fall or winter best.	*Do not use on cropland.* See Precautions and Remarks under **Weed Control in Woody Ornamentals.**
EPTC (Eptam 7-E) 7 EC	6.75 pt/acre	Spring or fall.	Plow area before treatment. Incorporate chemical into soil immediately after application. Rototilling is the preferred method, but deep cross-disking is satisfactory. Treatment most effective when soil is moist but not wet. Under normal conditions the herbicide will be dissipated in 8 to 12 weeks. Tilling the soil several times before planting will help dissipate chemical from soil.
Multiflora Rose (See publication AG-536 for more details)			
2,4-D + triclopyr (Crossbow) 3 EC	1 to 1.5 gal in 100 gal water (handgun application) or a 1% to 1.5% solution for smaller amounts	At fall vegetative stage prior to full bloom	Spray to wet all leaves and green stems to drip point. Use low spraying pressure to prevent drift. For best results, apply when plants are actively growing during the early to mid-flowering stage.
	Undiluted herbicide		Small plants may be controlled by a thinline basal application of undiluted herbicide across all stems at a height where the stems are less than 0.5 inch in diameter. Apply approximately 20 ml of undiluted product per bush. Treat when bark is dry and rain is not forecast. For bushes with more than 3 or 4 stems, coverage of each stem may be difficult; basal bark or dormant stem applications may be more effective (see these sections under **Chemical Control of Woody Vegetation**).
			Warning: *Restrictions on grazing or harvesting of green forage:* Do not graze lactating dairy animals or harvest green forage for 14 days following treatment with 2 gallons per acre or less; with treatment rates greater than 2 gallons per acre, do not graze or harvest green forage until the following growing season. For other livestock, no grazing restrictions apply at rates under 2 gallons per acre. Above 2 gallons per acre, do not graze or harvest green forage from treated areas for 14 days after treatment. *Restrictions on haying (harvesting of dried forage):* For lactating dairy animals, do ot harvest hay until the next growing season. For other livestock, do not harvest hay for 7 days after treatment at rates under 2 gallons per acre. Above 2 gallons per acre, do not harvest hay for 14 days after treatment. *Slaughter restrictions:* Withdraw livestock from grazing treated grass or treated hay at least 3 days before slaughter. This restriction applies to grazing during the season following treatment or hay harvested during the season following treatment.
dicamba (Banvel) 4 SL	1 gal/100 gal water	At full vegetative stage, prior to bloom.	Can be used in pastures and noncropland. A maximum of 200 gallons of spray solution can be used per acre. Spray with a handgun and completely wet foliage and stems, allowing spray solution to run down the stems. Add a nonionic surfactant at the rate of 2 quarts/100 gallons of finished spray solution to improve wetting. Do not graze dairy animals for 60 days. There is no waiting period between treatment and grazing beef cattle or other livestock. Do not spray desired plants or in rooting zone of desired plants. Follow-up treatments may be necessary in subsequent years.

Table 7-26. Chemical Control of Specific Weeds

Herbicide and Formulation	Amount of Formulation	Time of Application	Precautions and Remarks
Multiflora Rose (See publication AG-536 for more details) (continued)			
fosamine (Krenite S) 4 SL	1.5 to 3 gal/ 100 gal water	Apply to foliage during the 2-month period before fall leaf coloration.	Thoroughly and uniformly cover plants without drenching. Use 1 to 2 quarts per 100 gallons of a penetrating type, oil-based surfactant to improve activity. Use in noncropland, fence lines, etc.
gyphosate (Accord) 4 SL (Roundup PRO) 4 SL and various other trade names	1 gal/100 gal water	Apply to foliage after full bloom until August 1.	Use handgun and thoroughly cover bush. May be used in noncropland and pasture. Do not graze livestock for 10 days following treatment. Use Accord for forestry and utility rights-of-way sites. Roundup PRO should be used for industrial and other noncropland areas. Use Roundup ULTRA on agricultural areas.
metsulfuron methyl (Escort) 60 WDG	0.5 to 1 oz/acre	Early spring after bushes are fully leaved out.	Escort is not labelled for pastures; Cimarron may be used to control roses in pastures. For effective broadcast treatments, rose bushes should not be taller than 3 feet. For spot treatment, apply as foliar spray to runoff, and do not exceed 75 gallons total spray per acre. Use 1 pint to 1 quart surfactant per 100 gallons spray.
metsulfuron methyl (Cimmaron) 60 WDG	0.3 oz/acre for broadcast; 1 oz per 100 gal water for spot treatment		
tebuthiuron (Spike) 20 P	20 lb/acre	Apply after ground thaws in spring before or soon after leaf flush. May require 2 yr for kill of large canes.	Broadcast over root zone. Do not apply near desirable trees or shrubs. Ground may be bare for 3 to 5 years where applied.
(Spike) 80 WP	5 lb/acre		Apply in water with a backpack sprayer as a band at the base of bushes or lace overtop of bushes. Same precautions as above for Spike 20 P.
Nutsedge			
2,4-D amine (various brands) 4 SL	1 qt/acre	Early in growing season following a thorough disking. Repeat at 3-week intervals for 3 treatments.	Corn crop can be produced while using 2,4-D. Apply preemergence rate. Follow 3 to 4 weeks later with rate suggested for corn and repeat. Rate suggested can be used following tobacco harvest.
dichlobenil (Casoron)	See section on **Weed Control in Woody Ornamentals**. Do not use on row crop land.		
glyphosate (Accord) 4 SL (Roundup PRO) 4 SL and various other trade names	3 qt/acre	See remarks.	Apply when plants are in flower or when new nutlets can be found at rhizome tips. Tillage two weeks after application will improve control. Repeat treatment will be required for long-term control. Use Accord for forestry and utility rights-of-way sites. Roundup PRO should be used for industrial and other noncropland areas. Use Roundup ULTRA on agricultural areas.
bentazon (Basagran) 4 SL	Follow label directions	Follow label directions	See weed control for specific field, turf, or vegetable crop for nutsedge suppression. Bentazon and S-metolachlor control yellow nutsedge only.
chlorimuron (Classic) 25 WDG	Follow label directions	Follow label directions	See weed control for specific field, turf, or vegetable crop for nutsedge suppression.
EPTC (Eptam 7-E) 7 EC	Follow label directions	Follow label directions	See weed control for specific field, turf, or vegetable crop for nutsedge suppression.
halosulfuron (Permit, Sandea, Sedgehammer) 75 DF	Follow label directions	Follow label directions	See weed control for specific field, turf, or vegetable crop for nutsedge suppression.
imazaquin (Image) 70 DG	Follow label directions	Follow label directions	See weed control for specific field, turf, or vegetable crop for nutsedge suppression.
imazethapyr (Pursuit) 2 AS	Follow label directions	Follow label directions	See weed control for specific field, turf, or vegetable crop for nutsedge suppression.
S-metolachlor (Dual Magnum, Dual II Magnum)	Follow label directions	Follow label directions.	See weed control for specific field, turf, or vegetable crop for nutsedge suppression. Bentazon and S-metolachlor control yellow nutsedge only.
sulfosulfuron (Certainty) 75 WG	1.25 oz/acre	Apply in May for June.	Add a nonionic surfactant at 0.25% v/v. Make a second application in 6 to 10 weeks, if needed. Certainty and Monument control yellow nutsedge, purple nutsedge, and kyllinga species.
trifloxysulfuron (Monument) 75 WG	0.45 to 0.56 oz/acre	Apply in May for June.	Add a nonionic surfactant at 0.25% v/v. Make a second application in 6 to 10 weeks, if needed. Certainty and Monument control yellow nutsedge, purple nutsedge, and kyllinga species.
Poison Ivy and Poison Oak			
2,4-D amine (various brands) 4 SL	2 qt/100 gal water	Apply in late spring or early summer when the plants are growing rapidly.	Apply only to plant material to be killed. Apply as a wetting spray. Avoid drift. Repeat in 6 to 8 weeks if needed. Use Accord for forestry and utility rights-of-way sites. Roundup PRO should be used for industrial and other noncropland areas. Use Roundup ULTRA on agricultural areas.
dicamba (Banvel) 4 SL	1 gal/100 gal water	At full vegetative stage, before bloom.	See comments for honeysuckle.
glyphosate (Accord) 4 SL (Roundup PRO) 4 SL and various other trade names	4 to 5 qt/acre as a broadcast spray or 2% solution with handheld equipment	After leaves fully expand in the spring and before leaf color changes in the fall.	
triclopyr amine (Garlon 3 A) 3 SL	1 gal/100 gal water	After leaves fully expand in spring and before leaf color changes in fall.	See comments for Kudzu.
Tree of Heaven (*Ailanthus altissima*)			
metsulfuron methyl (Escort) 60 WDG	1 oz/100 gal water for high-volume treatment 1 oz/20 gal water for low-volume treatment	Mid-summer	For right-of-way use only. Escort is not labeled for pastures.

Chemical Control of Woody Plants

R. J. Richardson, F. H. Yelverton, and T. W. Gannon, Crop and Soil Sciences Department

CHEMICAL WEED CONTROL

Table 7-27. Chemical Control of Woody Plants

Herbicide and Formulation	Amount of Formulation	Precautions and Remarks
Foliage Treatment, most woody species: ash, red maple, and persimmon generally resistant. Rhododendron resistant		
2,4-D amine (various brands) 4 SL, MOA 4	2 gallons in 100 gallons water	Use amine formulations to reduce vapor drift hazard. Use low spraying pressure to prevent spray drift. Wet foliage and stems thoroughly. Most effective results obtained by spraying within 6 weeks after plants have reached full-leaf stage. This treatment used primarily on trees or brush less than 6 feet tall. Only certain brands of 2,4-D can be used on ditch banks or near other bodies of water; check labels.
2,4-D low volatile ester or oil-soluble amine (various brands and concentrations)	varies	Use as invert emulsion to reduce drift hazards See remarks for 2,4-D amine.
2,4-D + triclopyr (Crossbow) EC	1 to 1.5 gallons in 100 gallons water (handgun application)	Spray to wet all leaves and green stems to drip point. Use low spraying pressure to prevent drift. For best results, apply when plants are actively growing after full leaf in spring to early summer. This treatment is used primarily on trees and brush less than 6 feet tall.
2.0 + 1.0 pound/gallon, MOA 4	1.5 to 4 gallons in water to deliver 15 to 30 gallons total spray/acre	For application via boom or other broadcast spray equipment. For aerial application (helicopter only), use Nalcotrol to prevent drift. See label for specific information. **Warning:** *Restrictions on grazing or harvesting of green forage:* Do not graze lactating dairy animals or harvest green forage for 14 days following treatment with 2 gallons per acre or less; with treatment rates greater than 2 gallons per acre, do not graze or harvest green forage until the following growing season. For other livestock, no grazing restrictions apply at rates under 2 gallons per acre. Above 2 gallons per acre, do not graze or harvest green forage from treated areas for 14 days after treatment. *Restrictions on haying (harvesting of dried forage):* For lactating dairy animals, do not harvest hay until the next growing season. For other livestock, do not harvest hay for 7 days after treatment at rates under 2 gallons per acre. Above 2 gallons per acre, do not harvest hay for 14 days after treatment. *Slaughter restrictions:* Withdraw livestock from grazing treated grass or treated hay at lest 3 days before slaughter. This restriction applies to grazing during the season following treatment or hay harvested during the season following treatment.
fosamine (Krenite S) 4 SL	1.5 to 3 gallons in 100 gallons water	Apply to foliage during the 2-month period prior to fall leaf coloration. Thoroughly and uniformly cover plants without drenching. Add surfactant WK at the rate of 1 quart per 100 gallons of spray. Surfactant WK is not needed with Krenite S. Rate and gallonage depend on plant size and species to be controlled. Check label. Use in noncropland, fence lines, etc.
dicamba (Banvel) 4 SL	1 gallon in 100 gallons	Apply when leaves are fully developed. Spray with a handgun to completely wet foliage, and allow spray to run down the stem. Add a nonionic surfactant at the rate of 2 quarts per 100 gallons of finished spray solution to improve wetting. Retreatment may be required, but do not exceed 2 gallons per treated acre during one growing season. Keep spray off desired plants. Do not spray in rooting zone of desired plants.
triclopyr (Garlon 3A) 3 SL (Garlon) 4.4 EC	2 to 3 gallons in 100 gallons water 1 to 3 gallons in 100 gallons water	Spray to thoroughly wet leaves, stems, and root collars. Can be mixed with other woody plant herbicides. See label. Avoid drift.
Foliage Treatment, woody brush and trees		
2,4-D amine (DMA 4 IVM) 3.8 SL, MOA 4	2 to 8 pints/acre	Apply when weeds are small and actively growing before bud stage. Bienennial and perennial species are best controlled in seedling to rosette stage before flower stalks appear.
dicamba (Vanquish) 4 SL, MOA 4	0.5 to 4 pints in 25 to 200 gallons water	For low volume applications, apply 3 to 5% v/v rate. Check product label for tank mix partners for woody brush and vines.
glyphosate (Accord Concentrate) 5.4 SL, MOA 9	5 to 8% solution	If brush has been mowed or trees cut, wait until regrowth reaches recommended stage before treating. Apply as a low volume directed spray on at least 50% of the targeted foliage using a lateral zigzag motion from top to bottom. Spray to wet, not runoff. Add NIS at 2 quarts per 100 gallons of spray solution.
metsulfuron methyl (Escort XP) 60 DF, MOA 2	0.33 to 4 ounces/acre in 10 to 50 gallons water	For industrial, noncrop sites on young, actively growing weeds and brush. High volume ground application: mix 0.5 to 3 ounces per 100 gallons spray solution, and apply at 100 to 400 gallons per acre. Low volume and ultra-low volume ground applications: mix 4 to 8 ounces per 100 gallons spray solution, and apply at 10 to 50 gallons per acre.
triclopyr (Remedy) 4 EC, MOA 4	2 pints in 10 gallons water/acre	Treat after rapid growth period in spring when leaf tissue is fully expanded and terminal growth has slowed. During drought or for hard-to-control weeds, add 2 to 3 quarts of 2,4-D low volatile ester to spray solution.
triclopyr + fluroxypyr (PastureGard) 2 EC 1.5 + 0.5 pounds/gallon, MOA 4	3 to 8 pints/acre	Broadcast applications: treat in late spring through summer when leaves are fully expanded and terminal growth has slowed. If brush has been mowed, allow 9 to 12 months of regrowth before treating. NIS or liquid fertilizer at 1 to 2 quarts per 100 gallons of spray solution may improve control. High volume foliar treatment of individual plants: apply 1 to 2 gallons of PastureGard plus 1 quart NIS per 100 gallons of spray solution.
Foliage Treatment, black locust, honey locust, mimosa, redbud, and wisteria		
aminopyralid (Milestone VM) 2 SL, MOA 4	4 to 7 fluid ounces/Acre	Treat when weeds are actively growing. Include a non-ionic surfactant. Avoid mowing for 14 days after application.
Foliage Treatment, numerous woody species		
aminopyralid + triclopyr (Milestone VM Plus) 1.1 SL, MOA 4	6 to 9 pints/acre	Treat when weeds are actively growing. Include a non-ionic surfactant.
Foliage Treatment, most vegetation		
imazapyr (various) 2 SL, MOA 2	0.5 to 5% v/v 0.6 to 6.4 fluid ounces/gallon	Most effective with 1% methylated seed oil.
Basal Stem Treatment, most woody species; black locust resistant		
2,4-D low volatile ester (various brands) 4 SL, MOA 4	2 gallons in 100 gallons high quality mineral oil	Spray lower 12 inches of stem or trunk and let some solution run into ground. May be used any time of year, but is much more effective during dormant season. One growing season required before plants die completely. This treatment used primarily on plants less than 6 inches in diameter. Root suckering species may be resistant. *Both dormant stem and basal treatments useful to farmers and landowners because during winter there is less hazard to crops and more labor probably available.* Do not use around the home or ditch banks.
triclopyr (Garlon) 4.4 EC, MOA 4	1 to 3 gallons in 100 gallons high quality mineral oil	

Table 7-27. Chemical Control of Woody Plants

Herbicide and Formulation	Amount of Formulation	Precautions and Remarks
Basal Stem Treatment, most woody species; black locust resistant (continued)		
2,4-D + triclopyr (Crossbow) EC 2.0 + 1.0 pound/gallon, MOA 4	4 gallons in high quality mineral oil to make 100 gallons spray	Spray basal portions of trees or brush to a height of 15 to 20 inches from the ground. Thoroughly wet all basal bark areas, including crown and ground sprouts and ground area at base of stems or trunk. For trees larger than 6 to 8 inches diameter, use stump treatment. Winter and early spring treatments give best results. See warning for livestock and haying usage for Crossbow listed above under "Most Woody Species."
Basal Stem Treatment, most woody species		
imazapyr (Stalker) 2 SL, MOA 2	8 to 12 fluid ounces in 1 gallon high quality mineral oil	Treat lower 18 inches of stem. May be used on stems up to 4 inches DBH. Do not apply to point of dripping or puddling.
Basal Stem Treatment, woody brush and trees		
2,4-D amine (DMA 4 IVM) 3.8 SL, MOA 4	8 qt in 100 gal water or 2.6 fl oz in 1 gal water	Thoroughly wet the base and root collar of all stems until the spray accumulates around the root collar at the ground line. Wetting the stems will aid in control.
triclopyr (Remedy) 4 EC, MOA 4	2 gallons in 98 gallons high quality mineral oil	Spray basal 15 to 20 inches of plant to point of runoff at soil surface.
triclopyr + fluroxypyr (PastureGard) 2 EC 1.5 + 0.5 pounds/gallon, MOA 4	50% PastureGard + 50% high quality mineral oil	Apply at any time to stems less than 6 inches in diameter except when snow or water prevents spraying to ground line. Use solid cone or flat fan nozzles at low pressure. Spray to wet but not runoff.
Dormant Stem Treatment, most woody species		
2,4-D + triclopyr (Crossbow) EC 2.0 + 1.0 lb/gal, MOA 4	1 to 4 gallons in high quality mineral oil to make 100 gallons spray	Thoroughly wet upper and lower stems, including root collar and any ground sprouts. Treat when brush is dormant and the bark is dry, but not when snow or water prevents spraying to ground line. Best results occur with late-winter to early spring applications. Brush over 8 feet in height is difficult to control with this method. See warning for livestock and haying usage for Crossbow listed above under "Most Woody Species."
Dormant Stem Treatment, woody brush and trees		
triclopyr (Remedy) 4 EC, MOA 4	3 to 6 quarts in high quality mineral oil to make 100 gallons spray	Treat any time brush is dormant and most foliage has dropped. Use 20 to 40 psi with knapsack or power spraying equipment. Do not apply if snow or water prevents spraying to ground line. Wet stems to point of runoff and ground below the plant for root suckering species, such as sumac, sassafras, or locust.
Stump Treatment To Prevent Regrowth, most woody species		
2,4-D low volatile ester (various brands) 4 SL, MOA 4	3 gallons in 100 gallons high quality mineral oil	Soak freshly cut stumps with spray solution to prevent sprouting, or use AMS crystals on stump. Hasten decay of stump by covering with layers of soil and a nitrogen fertilizer. Keep moist.
2,4-D + triclopyr (Crossbow) EC 2.0 + 1.0 lb/gal, MOA 4	4 gallons in high quality mineral oil to make 100 gallons spray	Cut down trees and treat stumps, including the freshly cut surface, bark, crown, and ground sprouts. Winter and early spring treatments (before growth begins) give best results.
dicamba (Banvel) 4 SL, MOA 4	16.5 gal in 100 gal water	Spray or paint freshly cut surface with the solution. Area adjacent to bark should be thoroughly wet.
Stump Treatment To Prevent Regrowth, Woody brush and trees		
2,4-D amine (DMA 4 IVM) 3.8 SL, MOA 4	8 qt in 100 gal water or 2.6 fl oz in 1 gal water	Apply as soon as possible after cutting trees. Thoroughly soak entire stump including cut surface, bark, and exposed roots.
dicamba (Vanquish) 4 SL, MOA 4	1 gal in 1 to 3 gal water	NIS or oil may be added to enhance control. Make application within 30 minutes of cutting. Area adjacent to the bark should be thoroughly wet.
triclopyr (Remedy) 4 EC, MOA 4	20 to 30 gallons in high quality mineral oil to make 100 gallons spray	Treat with a backpack or knapsack sprayer using low pressure and a solid cone or flat fan nozzle. Spray stump sides and outer portion of cut surface but not to point of runoff. Apply anytime except when snow or water prevent spraying to ground line.
Stump Treatment To Prevent Regrowth, Woody brush and trees (continued)		
triclopyr + fluroxypyr (PastureGard) 2 EC 1.5 + 0.5 lb/gal, MOA 4	50% PastureGard + 50% high quality mineral oil	Apply to freshly cut stumps using solid cone or flat fan nozzles at low pressure. Wet stump sides, root collar, and outer portion of cut surface but not to point of runoff. Apply anytime except when snow or water prevent spraying to ground line.
Stump Treatment To Prevent Regrowth, woody species, such as alder, dogwood, hickory, maple, oak, poplar, sweet gum, sycamore, and willow		
glyphosate (Accord Concentrate) 5.4 SL, MOA 9	50 to 100% solution	Treat freshly cut stumps or resprouts. Apply to freshly cut stumps immediately after cutting or reduced performance may occur.
Stump Treatment, numerous wood species		
aminopyralid + triclopyr (Milestone VM Plus) 1.1 SL, MOA 4	apply undiluted	Apply as soon as possible after cutting stems.
Stump Treatment, most woody species		
imazapyr (Stalker) 2 SL, MOA 2	8 to 16 ounces in 1 gallon high quality mineral oil	Apply as soon as possible after cutting stems.
Soil Treatment Beneath Woody Plants, most woody species		
hexazinone (Velpar L) 2 SL, MOA 5	2 to 4 gallons in 100 gallons water	Apply as a coarse spray, using a handgun applicator. Direct spray beneath plants to be controlled. Apply during the period between late winter and early summer. Do not apply in vicinity of desirable plants.
bromacil (Hyvar X-L) 2 SL, MOA 5	varies	Apply as a coarse spray, using a handgun applicator. Use at least 200 gallons of spray per acre. Direct spray beneath plants to be controlled just before or during the period of active growth. Do not apply in vicinity of desirable plants. Rates depend on species to be controlled. Check label.
tebuthiuron (Spike) 20 P, MOA 7	5 to 30 pounds/acre	Rates depend on species to be controlled. Check label for specific rates. Apply when ground is not frozen. Do not apply to the root zone of desirable trees or shrubs or where runoff can carry the herbicide to desired plants.

CHEMICAL WEED CONTROL

Total Vegetation Control in Noncropland

R. J. Richardson, F. H. Yelverton, L. S. Warren Jr., and T. W. Gannon, Crop and Soil Sciences Department

Controlling all weeds for extended periods is expensive; it is practical only where complete vegetation control is desirable and soil erosion is not an important factor. Such areas are around signposts and buildings, along highways and railroads, under guardrails and fences, and in parking lots. Do not use any of the treatments where adjacent trees, ornamentals, or crops might be affected. Roots of nearby desirable plants, especially trees and shrubs, may grow into an area that has been treated and be killed.

Effective rates for control vary with the weed species, degree of infestation, soil type, and environmental conditions. The lower rates are generally applied to annuals, biennials, shallow-rooted perennials, and seedling perennials, whereas the higher rates are applied to established, deep-rooted, and other hard-to-kill perennials. For specific details, read and follow directions on the label.

Table 7-28. Total Vegetation Control in Noncropland

Herbicide and Formulation	Amount of Formulation	Precautions and Remarks
bromacil (Hyvar X) 80 WP (Hyvar X-L) 2 SL	3 to 30 pounds/acre 1.5 to 12 gallons/acre	For handgun sprayer, use at least 200 gallons of water per acre. For treating small areas, use a hand sprayer or sprinkling can. For retreatment apply 2 to 6 pounds of active material per acre when annual weeds and grasses reappear on sites where weed growth has been controlled. Rates depend on weeds to be controlled. Check label.
bromacil + diuron (Krovar I) 80 WDG	4 to 30 pounds/acre	Rates depend on weeds to be controlled. Check label. For retreatment use 4 to 6 pounds per acre.
diuron (Karmex) 80 WDG	5 to 15 pounds/acre	For most annual weeds.
	20 to 60 pounds/acre	For perennial weeds. Addition of paraquat at the rate of 0.5 pound per acre plus nonionic surfactant will provide quick kill of existing vegetation and allow lower rates of diuron to be used.
imazapic + glyphosate (Journey) 0.75 + 1.5 lb/gal AS	0.33 to 1 quart/acre	For broadcast or spot treatment of annual and perennial grass and broadleaf weeds and vine species. May be applied preemergence or postemergence but postemergence is preferred. Depending on weed species, use a 5 to 20% active MSO at 1 to 2 pints per acre or a NIS at least 60% active with HLB ratio between 12 and 17 at 0.25% v/v. Nitrogen-based liquid fertilizers may be added but not substituted for spray adjuvants. For extended residual control, tank mix with Arsenal, Endurance, Escort, Karmex, Krova, Oust, Pendulum, Roundup Pro, Sahara, Tordon, Vanquish, or 2,4-D. Spot treat with 0.8 to 17 ounces per gallon + 1% v/v MSO.
glyphosate (Roundup PRO) 4 SL (Roundup ULTRA) 4 SL	4 quarts/100 gallons water or 2 to 3 ounces/gallon water/1,000 square feet	For industrial and nonagricultural uses in hand-held, high-volume equipment, mix 4 quarts of Roundup PRO in 100 gallons of water and spray to wet. For farmstead weed control, use 1% to 2% solution of Roundup ULTRA. Use a 2% solution for perennial weeds.
imazapyr (Arsenal) 2 SL	0.25 to 0.5 gallons/acre	See label instructions.
imazapic + glyphosate (Journey) 0.75 + 1.5 lb/gal AS	0.33 to 1 quart/acre	For broadcast or spot treatment of annual and perennial grass and broadleaf weeds and vine species. May be applied preemergence or postemergence but postemergence is preferred. Depending on weed species, use a 5 to 20% active MSO at 1 to 2 pints per acre or a NIS at least 60% active with HLB ration between 12 and 17 at 0.25% v/v. Nitrogen-based liquid fertilizers may be added but not substituted for spray adjuvants. For extended residual control, tank mix with Arsenal, Endurance, Escort, Karmex, Krovan, Oust, Pendulum, Roundup Pro, Sahara, Tordon, Vanquish, or 2,4-D. Spot treat with 0.8 to 17 ounces per gallon + 1% v/v MSO.
oryzalin + glyphosate (Surflan AS Specialty) 4 AS + (Roundup PRO) 4 SL	4 to 6 quart/acre + 4 to 6 quart/acre	For safer total vegetation control in vicinity of desirable vegetation. Rates depend upon size and weeds to be controlled. See Surflan label. Apply in 100 gallons water per acre.
prometon (Pramitol 25 E) 2 EC	5 to 30 gallons/acre	For annuals, use 5 to 7.5 gallons per acre. For perennials, use 20 to 30 gallons per acre. Apply in 50 to 100 gallons of water. For faster knockdown of established weeds and grasses, apply in 100 to 200 gallons oil. For maintenance application in following seasons, reduce rate in half.
(Pramitol) 5 PS	0.5 to 2 pounds/100 square feet	Pellets containing 5% prometon, 0.5% simazine, 40% sodium chlorate, and 50% sodium metaborate. For maintenance use 1 pound per 100 square feet.
sulfometuron methyl + chlorsulfuron (Landmark MP) 50 + 25 DG (Landmark II MP) 56.25 + 18.75 DG	4.5 to 9 ounces/acre 2.66 to 10 ounces/acre	Controls many annual and perennial grass and broadleaf weeds on terrestrial noncrop sites, including public, private, and military lands. Do not apply to recreation areas or paved surfaces. Can be applied to areas where temporary surface water has collected. Treat weeds preemergence or early postemergence when actively germinating or growing.

VIII — PLANT GROWTH REGULATORS

PLANT GROWTH REGULATORS

Growth Regulators for Cotton

K. L. Edmisten and G. D. Collins, Crop Science Department

Table 8-1. Growth Regulators for Cotton

Chemical and Formulation	Amount of Formulation Per Acre	Remarks
To suppress excessive vegetative growth		
mepiquat chloride (various brands) 0.35 lb/gal	0.5 to 1 pt	Base rate and timing on field-conditions. Most consistent results occur when applied to cotton that is at least 24 inches tall at early bloom (5 to 6 white blooms per 25 feet of row). A follow-up application may be warranted 10 to 14 days later if excessive growth continues.
mepiquat pentaborate (Pentia)	0.82 lb/gal 4 to 24 fl oz	When early season conditions promote excessive prebloom growth, mepiquat chloride may be applied at the rate of 0.125 to 0.25 pint beginning at match head square (first square 0.125 to 0.25 inch in diameter). Repeat applications can be made when renewed growth occurs. Follow label directions. Mepiquat chloride will consistently suppress vegetative growth. Other benefits may include easier scouting, better insecticide coverage, less boll rot, earlier maturity, easier defoliation, and more efficient use of ground sprayers and low drum pickers. Yield response is inconsistent.
cyclanilid 1.84 lb/gal + mepiquat chloride 0.736 lb/gal (Stance)	2 to 3 fl oz	Use the 2 fluid ounces rate if applied at match-head square and the 2.5 to 3 ounces rate on applications following match-head square.
To stimulate boll opening and enhance defoliation		
ethephon (various brands) 6 lb/gal	0.66 to 1.33 qt	Use higher rate during cool weather. Prep boll opener will accelerate boll opening and enhance the activity of defoliants. It will not stimulate boll maturity. Micronaire may be reduced if Prep is applied to cotton that is less than 60 percent open.

For more information, see Extension publication AG-417, *Cotton Information,* which is available at your county Cooperative Extension center.

Guide for use of Defoliants on Cotton

K. L. Edmisten and G. D. Collins, Crop Science Department

Apply defoliants when at least 60%, preferably 70 to 75%, of the bolls are open and the remaining bolls expected to be harvested are mature. A boll is mature enough for defoliation when it is too hard to be squeezed between thumb and fingers, when it is too hard to be sliced with a sharp knife, and when the seed coats turn light brown. Apply defoliants in a volume of 15 to 25 GPA by ground or at least 5 GPA by air. Consult the label for recommended rates when tank mixed with other harvest aids and for maximum season-long rates.

Table 8-2. Guide for use of Defoliants on Cotton

Chemical Name	Brand Name and Formulation	Amount of Formulation Per Acre	Remarks
carfentrazone	Aim (2 lb/gal)	1 to 1.5 oz	Aim, like other herbicidal defoliants, can cause desiccation; however, the desiccation appears to be more transitory than that achieved with some of the other herbicidal defoliants. Aim appears to desiccate mature morningglories very well, and it does not seem like the addition of ethephon-type products is needed to improve morningglory desiccaton. The label recommends a minimum of 10 gallons per acre for ground applications and the use of oil concentrate at 1% by volume (1 gallon per 100 gallons of spray solution). Do not use crop oil concentrate with Aim in mixtures with CottonQuik or FirstPick. Aim can be tank mixed with other defoliant products if boll opening or regrowth control is desired. Lower rates may be needed in defoliation mixtures.
ethephon + cyclanilide	Finish (6 lb/gal) (various brands)	1.33 to 2 pt	Use higher rates in cool weather. Finish is defoliant and boll opener. Finish also provides regrowth control. Terminal regrowth control is stronger than basal regrowth control. Finish will provide acceptable regrowth control in many situations. In situations where extended regrowth control is needed (in the 20- to 28-day range), Roundup (in conventional cotton only) or Dropp would provide more acceptable regrowth control. Finish performance may benefit from the addition of a low rate of a standard defoliant in situations where cotton is actively growing with juvenile growth. Use compatibility agent when mixing with Def 6 or Folex.
glyphosate	Roundup (various brands)	24 to 32 oz	Based on limited trials in N.C. concerning varieties that do not convey tolerance to glyphosate, Roundup provides effective regrowth control. Roundup provides very little defoliation and should be used in combination with a defoliant. Experience in N.C. is limited to tank mixes with 1.5 pints of Def or Folex per acre. Although Roundup prevents regrowth, it will not defoliate juvenile growth that may occur prior to defoliation. Roundup also provides some weed control when defoliating weedy cotton. Roundup will not provide regrowth control on Roundup Ready cotton.
mono (N,Nimethylalkyamine) salt of endothall	Accelerate (0.52 lb/gal)	1.0 to 1.5 pt	Accelerate may be added to Def or Folex at 1.5 pints per acre to speed leaf drop by approximately 25% during the first few days of defoliant activity. The rate of leaf drop after 7 to 10 days has generally not been improved with Accelerate. Always add Accelerate to organic phosphates (Def or Folex) previously tank mixed with water.
pyraflufen-ethyl	ET (0.2 lb/gal)	1.5 to 2.5 oz	Like all herbicidal defoliants, ET can cause desiccation, although the desiccation tends to be transitory compared to some herbicidal defoliants. ET should help with desiccation of morningglory and other weeds listed on the herbicide label. ET can be mixed with other harvest-aid materials to provide boll opening or regrowth control. Consult the label for additive recommendations. In general, the use of a crop oil concentrate at 0.5% by volume is recommended. Do not use crop oil concentrate with ET in mixtures with CottonQuik or FirstPick. Use lower rates in warm weather.

Table 8-2. Guide for use of Defoliants on Cotton

Chemical Name	Brand Name and Formulation	Amount of Formulation Per Acre	Remarks
flumiclorac-pentyl	Resource (0.86 lb/gal)	4 to 6 fl oz	Resource is an herbicidal defoliant that, like all herbicidal defoliants, can cause dessication, although the dessication tends to be more transitory than with some herbicidal defoliants. Under ideal defoliation conditions (warm, sunny days), add a NIS at 1 quart per 100 gallons of spray solution. Under dry or cool weater, use a methylated seed oil (MSO) or aganosilicone spray solution. Apply in a minimum of 10 gallons per acre for ground applications and a minimum of 5 gallons per acre for aerial applications. Do not use flood jet or air induction nozzles. Resource can be tank mixed with othe products if boll opening or regrowth control is desired. Resource only needs a 1 hour rain-free period. Preharvest interval (PHI) is 7 days.
fluthiacetmethyl	Blizzard (0.91 lb/gal)	0.5 to 0.66 fl oz	Blizzard is a PPO-inhibitor herbicidal-type defoliant. Experience with Blizzard has been limited in North Carolina. It can be tank mixed with ethephon-based products. Similar to other PPO-inhibitor defoliants, Blizzard should be very useful in desiccating juvenile foliage and as a second application prior to harvest. Add a crop oil concentrate or surfactant to tank mixes containing Blizzard according to label directions.
N-phenyl-N,1,2,3-thiadiazol-5-ylurea	Dropp (50% WP) Free Fall (50% WP)	0.2 to 0.4 lb	Dropp is very effective at regrowth suppression. However, the activity of Dropp is reduced to a greater extent by cool weather than the activity of Def 6, Folex, or Harvade. Dropp requires the longest rain-free period of any of the defoliants—24 hours. Follow label directions and safety precautions.
thidiazuron	Dropp SC Freefall SC (4 lb/gal) (various brands)	1.6 to 6.4 oz	Dropp Ultra is a mixture of Dropp and Diuron. Dropp Ultra may cause desiccation under very hot conditions, especially with high humidity. The addition of other defoliants is not recommended except for Prep or generic ethephon products. The addition of crop oil concentrate may improve performance in cooler weather. The addition of silicon surfactants is not recommended with Dropp Ultra.
Thidiazuron + diuron	Ginstar (1 lb/gal of thidiazuron plus 0.5 lb/gal of diuron) (various brands)	0.4 to 1 pt	Ginstar is a mixture of Dropp and Diuron with an emulsifier that makes it more likely to desiccate than Dropp Ultra. Do not use Ginstar if temperatures are over 90 degrees F. The addition of other defoliants is not recommended except for Prep or generic ethephon products. Ginstar rates should not exceed 10 ounces per acre in N.C. except under very adverse conditions.
urea sulfate and ethephon	CottonQuik or FirstPick	1.75 to 3.5 qt	Use higher rates in cool weather. The 2 quarts rate is recommended for most situations. Use a low rate of a standard defoliant with FirstPick unless the cotton is well cut-out with no juvenile growth.
tribufos	Def 6 or Folex (6 lb/gal)	1.33 to 1.5 pt	Use lower range of rates if crop is well matured as indicated by yellow- or red-tinged foliage or when temperatures are warm. Use higher recommended rates when plants are still green and actively growing or in dry weather or cool weather; needs a 1-hour rain-free period. Follow label directions and safety precautions.
2,3-dihydro-5,6-dimethyl 1,4 dithiin 1,1,4,4-tetraoxide	Harvade (5 lb/gal)	0.5 pt	The addition of 1 pint of crop oil concentrate per acre is necessary for acceptable results. Rainfall within 6 hours may reduce activity. Results similar to those from Def 6 or Folex. Follow label directions and safety precautions.

Harvest Aids, Preharvest Dessicants, and Postharvest Dessicants

K. M. Jennings, Horticultural Science Department

All crops. Maximum coverage of plants is essential for adequate chemical desiccation. Dense vine growth and/or heavy weed growth necessitates use of maximum permitted rates and high spray gallonage per acre. Observe instructions for use of wetting agents.

Potato desiccation. Vine destruction in North Carolina has usually been accomplished by mechanical methods immediately before digging. Several chemicals are also available for preharvest destruction. All of these chemicals must be used in advance of digging. Vine killing with chemicals often aids skin set, depending upon maturity of the crop. Inhibition of heat sprouts has also been observed with chemical vine killing. Browning of the vascular ring of potato tubers sometimes occurs after using vine kills.

Table 8-3. Harvest Aids and Preharvest Desiccants

Crop Chemical and Formulation	Amount of Formulation per Acre	Pounds Active Ingredient per Acre	Precautions and Remarks
Chili Pepper			
Sodium chlorate (Defol 750) 7.5 L (Defol 5) 5 L	1.6 to 4 qt 4 to 10 qt	3 to 7.5	**Processing only.** Apply 10 days before harvest. Consult with processor before applying. See label for further instructions.
Dry Bean, Southern Pea, Guar Bean[1]			
paraquat (Firestorm) 3 SL (Gramoxone SL) 2 SL (Gramoxone SL 2.0) 2 SL	0.8 to 1.3 pt 1.2 to 2 pt 1.2 to 2.0 pt	0.3 to 0.5	Apply paraquat for weed and bean desiccation. Apply when the crop is mature and **at least 80%** of the pods are yellowing and mostly ripe with no more than 40% (bush-type beans) or 30% (vine-type beans) of the leaves still green. Add nonionic surfactant at 1 quart per 100 gallons spray mix. Do not harvest or graze for at least 7 days after application. Make ground application in at least 20 gallons of water per acre. See label for further instructions.
sodium chlorate (Defol 750) 7.5 L (Defol 5) 5 L	3.2 qt 4.8 qt	6	Apply approximately 7 to 10 days before harvest. Do not graze livestock on treated fields or feed treated fodder or forage to livestock. See label for further instructions.

PLANT GROWTH REGULATORS

Table 8-3. Harvest Aids and Preharvest Desiccants

Crop Chemical and Formulation	Amount of Formulation per Acre	Pounds Active Ingredient per Acre	Precautions and Remarks
Potato			
diquat (Reglone) 2 L	1 to 2 pt	0.25 to 0.5	**Desiccation of potato vines.** Apply at least 7 days before harvest in 20 gallons of water per acre with ground equipment or 5 gallons of water per acre with aerial equipment. If vine growth is very dense, make a second application 5 days after the first application. Do not apply to drought-stressed potatoes.
glufosinate (Rely 280) 2.34 SL	1.3 pt	0.38	**Desiccation of potato vines.** Use sufficient water for thorough coverage of potato vines. For best results, apply at the beginning of the natural senescence of potato vines. See label for further instructions.
sodium chlorate (Defol 750) 7.5 L (Defol 5) 5 L	3.2 qt 4.8 qt	6	Apply 10 days before harvest. DO NOT apply under conditions of extreme heat during the middle of the day.
pelargonic acid (Scythe) 4.2 L	7 to 10% solution		Preharvest interval is 24 hours. See label for further instructions.
pyraflufen (ET) 0.208 L (ETX) 0.335 L	2.75 to 5.5 oz 1.7 to 3.4 oz	0.045 to 0.009	Apply in early stage of crop senescence. Two applications may be necessary for complete desiccation. Preharvest interval is 7 days. See label for further information.
Sunflower			
paraquat (Firestorm) 3 SL (Gramoxone SL) 2 SL (Gramoxone SL 2.0) 2 SL	0.8 to 1.3 pt 1.2 to 2 pt 1.2 to 2 pt	0.3 to 0.5	**Preharvest dessication.** Apply when sunflower seeds reach maturity (when seed moisture is 35% or lower). For many varieties, this corresponds to the time when the back of the heads are yellow and bracts are turning brown. Do not graze treated areas or feed treated forage to livestock. Do not apply within 7 days of harvest.
sodium chlorate (Defol 750) 7.5 L (Defol 5) 5 L	3.2 qt 4.8 qt	6	Apply 7 or more days before harvest of mature sunflower. Do not graze livestock on treated fields or feed treated forage to livestock within 14 days of application. See label for further instructions.
Sweetpotato			
pelargonic acid (Scythe) 4.2 L	7 to 10% solution		Preharvest interval is 24 hours. See label for further instructions.
Tomato			
paraquat (Firestorm) 3 SL (Gramoxone SL) 2 L (Gramoxone SL 2.0) 2 SL	1.6 to 2.5 pt 2.4 to 3.75 pt 2.4 to 3.75 pt	0.6 to 0.94	**Postharvest desiccation of crop and weeds.** Add a nonionic surfactant at 1 pint per 100 gallons spray solution. Apply in a minimum of 40 gallons spray solution. See label for further information.

[1] Sweet lupin, white sweet lupin, white lupin, grain lupin, adzuki beans, asparagus beans, black beans, broad beans, field beans, garbanzo beans, kidney beans, lablab beans lima beans, moth beans, mung beans, navy beans, pinto beans, rice beans, tepary beans, urd beans, guar beans, blackeyed peas, chickpeas, cowpeas, crowder peas, southern peas, and catjang.

Growth-Regulating Chemicals for Apples

M. L. Parker and J. D. Obermiller, Horticultural Science Department

Table 8-4. Growth-Regulating Chemicals for Apples

Purpose Material	Amount of Formulation per 100 Gallons	Minimum Interval (Days) Between Application and Harvest	Precautions and Remarks
To improve shape and increase fruit weight of responsive cultivars of apples, such as Delicious, Gala, or Ginger Gold			
Perlan or Promalin	1 to 2 pts	—	Apply between king bloom opening and full-bloom stage as a fine mist. Can be applied as a single spray at 1 or 2 pints per 100 gallons or as two split applications at 1 pint per 100 gallons. Do not exceed the maximum indicated rate of 2 pints per acre for the combined sprays. Do not apply Promalin later than full bloom because only late blossoms will be affected and early blooms may be chemically thinned. Do not apply Promalin when air temperatures are lower than 40 degrees F or higher than 90 degrees F. For optimum results, have the water pH near neutral (pH 7) and always below 8.5.
To increase lateral bud break and shoot growth and improve branch angle on nonbearing and nursery trees			
Promalin and nonionic surfactant or Perlan	0.25 to 1 pt Promalin per 5 gallons; use surfactant at 0.2 to 0.3% (v/v) or 1.25 fl. oz. per 5 gallons	—	Apply to previous season's leader growth with thorough coverage when new terminal growth is 1 to 3 inches long or nursery trees have reached the height where branching is required. Apply using a hand held sprayer and ensure thorough wetting of the foliage and bark. May also be applied as a latex paint mixture with a brush or sponge at 0.2 to 0.33 pints per pint of latex paint. Do not apply the Promalin-latex paint mixture after bud break.
To increase branching of nursery stock and young (nonbearing) trees, to improve branch angles, stimulate bud break and improve tree structure			
MaxCel	128 fl. oz./40 gal of water	—	Make the first of 3-4 applications to new growth at the height where branching is desired and apply at every 10-12 inches of new growth.
To promote lateral branching on current season's terminal growth			
Promalin without surfactant or MaxCel	4 fl. oz. per 5 gallons (125 ppm) 3.2 fl. oz. per gallon (500 ppm)		Apply to every 8 to 10 inches of new terminal growth in conjunction with removal of at least half of each immature terminal leaf. Do not damage the growing point (Summer Knipping). Apply 2 to 3 times with a single nozzle directed to the shoot tip, beginning when the leader has reached the height at which lateral branches are required. Apply at 7 to 10 day intervals.

PLANT GROWTH REGULATORS

Table 8-4. Growth-Regulating Chemicals for Apples

Purpose Material	Amount of Formulation per 100 Gallons	Minimum Interval (Days) Between Application and Harvest	Precautions and Remarks
To increase fruit set after frost			
Promalin or Perlan	16 to 32 fl. oz.	—	Apply within 24 hours after a frost event when the majority of the crop is between early bloom and full bloom. Apply in 75 to 150 gallons of water per acre. Do not apply to frozen foliage, blossoms or developing fruit. Allow trees to completely thaw prior to application. Do not use a surfactant.
To suppress russet formation on Golden Delicious and other russet sensitive varieties			
Novagib 10L or ProVide 10SG	20 to 33 fl. oz. Novagib 10L per acre in 100 gallons of water 2.1 to 3.5 oz. of ProVide 10SG per acre in 100 gallons of water	—	Recommended application interval is at petal fall (PF), then 10, 20, and 30 days after petal fall. There is an 80 fluid ounces per acre limit per season for Novagib 10L. Apply as a fine mist. **Do not** apply until runoff. Make 2 to 4 applications. Always make the first ProVide application at the beginning of petal fall, and continue at 7 to 10 day intervals. Direct 85% of the spray volume to the upper two-thirds of the tree.
To reduce the occurrence of Stayman fruit cracking			
Novagib 10L or ProVide 10SG	32 to 64 fl. oz. of Novagib 10L 1.8 to 3.5 oz. of ProVide 10SG		Apply in 3 to 6 consecutive applications at 14 to 21 day intervals beginning 2 to 3 weeks before fruit cracking is expected to occur. Apply at 50% of Tree Row Volume. **Do not** treat for cracking suppression apples that have received ProVide 10SG applications to suppress russet during the same growing season. Always apply when conditions favor slow drying.
To thin Golden Delicious fruit on trees where a heavy fruit set has occurred and to aid in encouraging annual bearing			
Fruitone L or PoMaxa alone or with a wetting agent, which will double the effective activity	2 to 6 fl. oz. (5 to 15 ppm) surfactant, if used, is 1 pt/100 gal	2	Until experience is gained in your orchard, use all concentrations of all thinning chemicals only on a trial basis on a few trees. Optimum thinning with Fruitone L or PoMaxa is achieved when fruit diameter is 5 to 10 mm. Under optimum growing conditions, the best size for thinning usually occurs between 15 to 21 days after full bloom. Applying NAA too early will result in sticking the fruit on rather than removing it. Applications are most effective at temperatures from 70 degrees F to 75 degrees F. Do not exceed 161 fluid ounces of Fruitone L or PoMaxa per acre per season.
To thin heavily set non-spur Red Delicious and Rome and to aid in encouraging annual bearing			
carbaryl Carbaryl 50WP or Carbaryl 4L/Sevin XLR	1 to 2 lb 1 to 2 pt	— 3	Apply at 8 to 12 mm average fruit diameter for optimum thinning. Applying carbaryl to Golden Delicious may cause russet.
To thin heavily set spur strains of Red Delicious and to aid in return bloom			
carbaryl Carbaryl 50WP or Carbaryl 4L/Sevin XLR + naphthalene-acetic acid (Fruitone L or PoMaxa)	2 lb (1 lb a.i.) 2 pt (1 lb a.i.) + 2 fl. oz. Fruitone L or PoMaxa (5 ppm)	— 3 2	Apply at 8 to 10 mm average fruit diameter. Application of Fruitone L or PoMaxa at larger fruit size to Red Delicious may cause "nubbins" and half-grown fruit to stick on the tree through to harvest. Apply as a dilute spray to runoff.
carbaryl Carbaryl 50WP or Carbaryl 4L/Sevin XLR + ethephon (Ethrel)	2 lb (1 lb a.i.) 2 pt (1 lb a.i.) + 1.5 pt	3	Apply at 12 to 15 mm average fruit diameter. Apply only as a dilute spray to runoff (for use only on spur strains of Delicious). Applying Ethrel when daily maximum temperatures exceed 80 degrees F may result in excessive thinning. Do not apply Ethrel if daily maximum temperature exceeds 90 degrees F.
To defruit trees too young or too small to begin bearing or where fruit load needs to be eliminated (all varieties)			
naphthalene-acetic acid (Fruitone L or PoMaxa) + carbaryl Carbaryl 50WP or Carbaryl 4L/Sevin XLR + ethephon (Ethrel)	2 to 4 fl. oz. (5-10 ppm) + 2 lb (1 lb a.i.) 2 pt (1 lb a.i.) + 2 pt	2 3	Apply dilute at 8 to 10 mm average fruit diameter. Caution: This will temporarily suppress vegetative growth.
Post-bloom thinning agent for cultivars such as Gala, Fuji, Spur Red Delicious			
N-(phenylmethyl)-1H-purine-6-amine Exilis Plus or MaxCel + Carbaryl 50 WP or Carbaryl 4L/Sevin XLR	46-122 fl. oz. 48 to 128 fl. oz. 2 lb (1 lb a.i.) 2 pt (1 lb a.i.)	86 86 3	Make 1 to 2 applications of 75 to 200 ppm when king fruit are 5 to 10 mm in diameter. Do not apply more than 296 fluid ounces of Exilis Plus per acre per season. Do not apply Exilis Plus if the temperature is below 60 degrees F. Make 1 to 2 applications of 75-200 ppm when king fruit are 5 to 15 mm in diameter. Applications will be most effective when the maximum temperature is above 65 degrees F on the day of application, and the following 2-3 days. Do not apply more than 308 fluid ounces of MaxCel per acre per season.
To thin Stayman, Rome, McIntosh, Jonathan, or Gala			
carbaryl Carbaryl 50WP or Carbaryl 4L/Sevin XLR + naphthalene-acetic acid or carbaryl Carbaryl 50 WP or Carbaryl 4L/Sevin XLR + naphthalene-acetic acid (Fruitone L or PoMaxa)	2 lb (1 lb. a.i) 2 pt (1 lb a.i.) + 2 to 4 fl. oz. Fruitone L or PoMaxa (5 to 10 ppm) or 1 to 2 lb (0.5 to 1 lb a.i.) 1 to 2 pt (0.5 to 1 lb a.i.) + 1 to 2 fl. oz. Fruitone L or PoMaxa (2.5 to 5 ppm)	3 2 3 2	Apply as a dilute spray when fruit are 9 to 11 mm average fruit diameter.

PLANT GROWTH REGULATORS

385

Table 8-4. Growth-Regulating Chemicals for Apples

Purpose Material	Amount of Formulation per 100 Gallons	Minimum Interval (Days) Between Application and Harvest	Precautions and Remarks
To increase return bloom for the following season, especially on heavily cropped trees			
naphthalene-acetic acid (Fruitone L or PoMaxa)	2 to 8 fl. oz. Fruitone L or PoMaxa (5 to 20 ppm) or 1.0 to 4.0 fl. oz. Fruitone L or PoMaxa (2.5 to 10 ppm) + surfactant	2	After the chemical fruit thinning activity window is past (typically 6 to 8 weeks after petal fall) use biweekly applications of naphthalene-acetic acid (NAA) in the next 3 to 4 cover sprays. (Typically two applications in June and two applications in July for Southeastern U.S. apple-growing areas). These NAA applications may be tank mixed with routine pesticide cover sprays. Even when used at low rates, reduced fruit quality such as early ripening or water core or leaf drop can result on sensitive varieties such as Early McIntosh or other early summer varieties.
To reduce vegetative growth and to reduce later season tree canopy volume and density for improving pesticide efficiency			
prohexadione-calcium (Apogee)	3 to 6 oz. biweekly or 8 to 12 oz. monthly	45	Apply as a sequential application every other week beginning when new shoots are 1 inch long using 3 to 6 ounces in 100 gallons per acre or apply as a sequential monthly application beginning when new shoots are 1 inch long using 8 to 12 ounces in 100 gallons. Do not tank mix Apogee with calcium nutrient sprays; but Apogee can be tank mixed in pesticide cover sprays. Do not apply more than 99 ounces of Apogee per acre per season. Use an acidifier if the pH of the spray water is greater than 7.0.
To aid in preventing preharvest fruit drop			
naphthalene-acetic acid (Fruitone L or PoMaxa)	8 to 32 fl. oz. per acre	2	Apply at anticipated start of fruit drop. One application will normally prevent fruit drop for 7 to 10 days. If necessary, repeat applications can be made at weekly intervals. Apply only when temperature is 70 degrees F or higher for maximum effectiveness. Use sufficient water to ensure adequate coverage of fruit.
To aid in delaying the onset of preharvest fruit drop before loosening begins			
naphthalene-acetic acid (Fruitone L or PoMaxa)	2 fl. oz. (5 ppm) applied as weekly sprays beginning 4 weeks before anticipated harvest	2	Preloading with the low rate of NAA is a more effective drop control program than waiting to use a higher rate as the fruit begins to loosen. Preloading may also increase return bloom of Golden Delicious and Red Delicious.
To delay preharvest fruit drop and fruit maturity to allow time for added fruit size increase			
aminoethoxyvinylglycine or AVG (ReTain)	Apply one 333-g pouch (50 g a.i.) per acre	7	Apply 28 days before anticipated harvest in most years. This is the most effective stop-drop control program; however, fruit maturity and harvest date will be significantly delayed. Do not apply ReTain to plants under stress. Avoid ReTain application during the heat of the day. Maintain solution pH between 6 and 8. Do not apply ReTain if rain is expected within 8 hours of application. Delay ReTain application until 2 to 3 weeks before normal harvest in hot years. For optimal response, use ReTain with a 100% organosilicone surfactant at a final surfactant concentration of 0.05 to 0.1% (v/v) in the spray tank. To prevent possible spotting, use the 0.05% (v/v) concentration when high temperature (in excess of 86 degrees F) weather conditions prevail or are anticipated. On sensitive varieties such as Gala, the amount of a.i. per acre can be reduced to 25 g. Very good drop control and maintenance of fruit firmness on the tree is achieved by combining ReTain with 10-20 ppm NAA (4-8 fluid ounces Fruitone L or PoMaxa per 100 gallons) applied 2 weeks before harvest. Higher application rates of NAA may cause fruit softening at temperatures above 85 degrees F.
To control suckers from the ground around the trunk of apple trees			
naphthalene-acetic acid (NAA), ethyl ester (Tre-Hold sprout inhibitor A-112)	Apply a 1 % (v/v) solution	—	Apply a 1 % (v/v) solution during the dormant season prior to the green tip stage or during the summer pruning season when new sucker growth is 4 to12 inches long. Apply as a low-pressure, large droplet, directed spray with hand-held equipment. A thorough application, giving complete wetting and coverage, is necessary for good results. Do not allow spray to drift onto tree foliage or fruiting spurs. For best results, cut off woody sucker growth at ground level during the dormant season. Do not apply during the period from bloom to 4 weeks after bloom.
To control water sprout regrowth around pruning cuts			
naphthalene-acetic acid (NAA), ethyl ester (Tre-Hold sprout inhibitor A-112)	Use 0.5 % (v/v) on newly planted trees Use 1.0 % (v/v) on established plantings	—	Apply with a cloth or brush as a localized application to the pruning cut and surrounding bark any time after pruning and before growth starts in the spring. Do not spray Tre-Hold up in the trees. Do not allow Tre-Hold to contact buds or fruiting spurs. Tre-Hold use in the tree is not recommended when green growth is present. One to 4 pints of light-colored latex (water based) paint may be added per gallon to mark completed application. Thorough application, giving complete wetting and coverage, is necessary for good results.
To aid in controlling scald on stored apples			
diphenylamine (DPA)	1,000 ppm; amount per 100 gallons depends on the a.i. in the formulated product. Consult the product label.	—	Apply as a dip or spray to harvested fruit. For maximum effect, uncooled fruit should be sprayed or dipped as soon after harvest as possible. The longer treatment is delayed after harvest, the less its effectiveness. Fruit should be thoroughly covered; however, fruit should not be dipped for longer than 30 seconds to prevent excess residue. Best control is obtained when both the fruit and the solution are at room temperature.
To maintain apple flesh firmness, reduce scald, and maintain fruit acidity.			
1-methylcyclopropene (1-MCP) (SmartFresh or SmartTabs)	Sufficient to achieve a final concentration of 1 ppm of the a.i.	—	SmartFresh/SmartTabs is a postharvest treatment that is introduced into the atmosphere of an airtight facility or container in which the fruit is held for 24 hours. Fruit must be treated within 3 to 5 days of harvest. The amount of product used depends upon the volume (cubic feet) of the treatment container. After treatment, the fruit is held in regular cold storage.

PLANT GROWTH REGULATORS

Growth Regulators for Floricultural Crops in Greenhouses

B. E. Whipker, Horticultural Science Department

This table lists labeled rates of plant growth regulators (PGR) for greenhouse crops. These rates are guidelines based on labeled recommendations, research at North Carolina State University, and recommendations by suppliers. Read the label for a complete listing of precautions. The degree of control can vary by a number of factors, including: plant type, cultivar, stage of development, fertilization program, growing temperatures, and crop spacing. When using a PGR for the first time, it is good to test the rate on a few plants prior to spraying the entire crop. Keep accurate records and adjust rates for your location. Additional information about plant growth regulators is available at http://www.pgrinfo.com.

General Recommendations: Plug culture and flat culture have different recommended rates. The rates in this table include recommendations for both plug (lower rates) and flat culture (higher rates). Apply ALL foliar sprays of plant growth regulators using 0.5 gallons per 100 square feet of bench area.

Table 8-5. Growth Regulators for Floricultural Crops in Greenhouses

Crop	Purpose	Chemical	Rate	Precautions and Remarks
ABUTILON	To control plant growth	Citadel/Cycocel	750 to 1,500 ppm spray	
		Dazide/B-Nine	2,500 ppm spray	Rate for use on plugs.
		Piccolo/Piccolo 10 XC /Bonzi/Paczol	5 ppm spray	Can be applied once plant fills the pot, 2 to 3 weeks after transplanting.
	To increase branching	Florel/Collate	250 to 500 ppm spray	Applied 2 weeks after transplanting. Follow with a pinch if needed.
ACHILLEA	To control plant growth	Dazide/B-Nine	2,500 ppm spray	One or 2 sprays may be needed to keep plants more compact.
		Piccolo/Piccolo 10 XC /Bonzi/Paczol/Downsize	0.5 to 1 ppm drench	Apply to moderately moist substrate.
ACHMELLA OLERAEA	To control plant growth	Piccolo/Piccolo 10 XC /Bonzi/Paczol	15 ppm spray	Apply 2 weeks after transplant. Repeat a week later or a week after pinch if needed.
AGASTACHE	To control plant growth	Citadel+Dazide/Cycocel+B-Nine	3,000 ppm + 1,500 ppm spray	Rates for compact genetics needing slight growth control.
AGERATUM	To control plant growth	Abide/A-Rest	7 to 26 ppm spray	
		Dazide/B-Nine	2,500 to 5,000 ppm spray	One or 2 sprays may be needed to keep plants more compact.
		Piccolo/Piccolo 10 XC /Bonzi/Paczol	15 to 45 ppm spray	High rates of Piccolo 10 XC may delay flowering. Late applications and overdosing may cause slow growth on transplantation. This can be avoided by using multiple applications of 25% to 50% of the specified rate and monitoring plant growth.
		Citadel/Chlormequat E-Pro /Cycocel	800 to 1,500 ppm spray	
		Concise/Sumagic	2 to 30 ppm spray	Cultivar response rates vary. Use lower rates to hold plants.
		Topflor	20 to 60 ppm spray	Based on NC State University trials. Adjust rates for other locations.
AGERATUM, Plugs	To control plant growth	Piccolo/Piccolo 10 XC /Bonzi/Paczol	5 to 10 ppm spray	Timing of application should normally begin at the 1 to 2 true leaf stage.
ALCEA ROSEA	To control plant growth	Piccolo/Piccolo 10 XC /Bonzi/Paczol	30 to 50 ppm spray	
		Piccolo/Piccolo 10 XC /Bonzi/Paczol/ Downsize	0.12 to 0.24 mg a.i. (1 to 2 ppm) drench for a 6-in. pot; apply 4 fl. oz./6-in. pot	
ALTERNANTHERA (Joseph's coat)	To control plant growth	Abide/A-Rest	25 to 132 ppm spray	
			0.25 to 0.5 mg a.i. (2 to 4 ppm) for a 6-in. pot (1 to 2 fl. oz./gal of drench solution: apply 4 fl. oz./6-in. pot)	Drench volumes and mg a.i. vary with pot size.
		Citadel/Chlormequat E-Pro /Cycocel	Spray	Apply only if needed. Not recommended on some cultivars due to potential phytotoxicity.
		Dazide/B-Nine	5,000 ppm spray	
		Florel/Collate	500 ppm spray	To keep plants more compact. Based on Texas A&M University trials.
		Piccolo/Piccolo 10 XC /Bonzi/Paczol/Downsize	30 to 45 ppm spray	Rate for *Alternanthera dentata*.
			4 ppm drench	To keep plants more compact. Apply to moderately moist substrate
ALYSSUM	To control plant growth	Piccolo/Piccolo 10 XC /Bonzi/Paczol	40 to 60 ppm spray	
		Concise/Sumagic	5 to 25 ppm spray	
		Dazide/B-Nine	2,500 ppm spray	
ALYSSUM, Plugs	To control plant growth	Piccolo/Piccolo 10 XC /Bonzi/Paczol	10 to 20 ppm spray	Timing of application should normally begin at the 1 to 2 true leaf stage.

PLANT GROWTH REGULATORS

Table 8-5. Growth Regulators for Floricultural Crops in Greenhouses

Crop	Purpose	Chemical	Rate	Precautions and Remarks
AMARYLLIS	To control plant growth	Piccolo/Piccolo 10 XC /Bonzi/Paczol	23.66 mg a.i. (200 ppm) drench for a 6-in. pot (6.4 fl. oz./gal. of drench solution; apply 4 fl. oz./6-in. pot)	Drench volumes and mg a.i. vary with pot size.
			100 ppm bulb soak	
ANAGALLIS	To control plant growth	Piccolo/Piccolo 10 XC /Bonzi/Paczol/Downsize	0.5 ppm drench	To keep plants more compact. Apply to moderately moist substrate.
ANEMONE	To control plant growth	Piccolo/Piccolo 10 XC /Bonzi/Paczol/Downsize	2 ppm drench	Rates for 'Mona Lisa' series. Apply about 6 weeks after transplant when the foliage has covered the pot and the first visible flower bud is showing. Rates up to 4 ppm can be used after conducting your own trial. Apply one week earlier during warm weather if needed.
ANGELONIA	To control plant growth	Citadel + Dazide/Cycocel + B-Nine	1,500 to 3,000 ppm Dazide/B-Nine + 750 to 1,000 ppm Citadel/Cycocel applied as a tank-mix spray	At planting, soft pinch to promote lateral shoot development.
		Citadel/Cycocel	1,500 ppm spray	
		Concise/Sumagic	10 to 20 ppm spray	Based on NC State University trials.
		Dazide/B-Nine	3,000 ppm spray	
		Florel/Collate	Spray	Not recommended.
		Topflor	45 to 60 ppm spray	Based on NC State University trials.
AQUILEGIA	To control plant growth	Dazide/B-Nine	3,000 to 5,000 ppm spray	
ARGYRANTHEMUM	To control plant growth	Citadel/Cycocel	750 to 1,500 ppm spray	
		Citadel+Dazide/Cycocel+B-Nine	750 to 1,000 ppm + 1,000 to 2000 ppm spray	Rates for compact genetics needing slight growth control.
		Concise/Sumagic	3 to 40 ppm spray	Based on NC State University trails conducted during late spring. Trial rates of 3 to 5 ppm for compact genetics.
		Piccolo/Piccolo 10 XC /Bonzi/Paczol	5 to 10 ppm spray	Rates for compact genetics needing slight growth control.
			1 to 5 ppm drench	Rates for compact genetics needing slight growth control.
		Dazide/B-Nine	1,500 to 2,500 ppm spray	
		Topflor	50 to 75 ppm spray	Based on NC State University trials conducted during late spring. A slightly phytotoxicity occurred with rates greater than 40 ppm, but damage was quickly hidden by new leaf growth.
	To induce basal branching	Collate/Florel	500 ppm spray	Apply one week after establishment.
ASCLEPIAS	To control plant growth	Piccolo/Piccolo 10 XC /Bonzi/Paczol	30 to 60 ppm spray	
ASTER NOVI-BELGII (Perennial)	To control plant growth	Concise/Sumagic	80 to 160 ppm spray	
		Dazide/B-Nine	1,500 to 5,000 ppm spray	
		Piccolo/Piccolo 10 XC /Bonzi/Paczol	160 ppm spray	Use lower rates of 5 to 10 ppm later in the season.
			12 to 16 ppm drench	
ASTER, Bedding Plant (*Callistephus chinensis*)	To control plant growth	Abide/A-Rest	7 to 26 ppm spray	
		Dazide/B-Nine	2,500 to 5,000 ppm spray	
ASTER, Cut (*Callistephus chinensis*)	To promote stem elongation and break dormancy	Florgib/ProGibb T&O	50 to 100 ppm spray	Make one to three applications during the early vegetative period at 2- to 3-week intervals. Apply when plants are 2 to 6 inches tall.
ASTERISCUS MARITIMUS (Compact Gold Coin)	To control plant growth	Dazide/B-Nine	750 to 1,500 ppm spray	
		Citadel/Cycocel	800 to 1,500 ppm spray	
		None	None	Plants grown with good light and optimal growing conditions generally do not need PGRs.
ASTILBE	To control plant growth	Concise/Sumagic	25 ppm drench	Apply just prior to flower stem elongation.
		Dazide/B-Nine	5,000 ppm spray	1 or 2 sprays can be used to keep plants more compact, begin once flower stalks show color. 1 to 2 week delay in flowering possible.
		Piccolo/Piccolo 10 XC /Bonzi/Paczol	30 ppm drench	Apply just prior to flower stem elongation.
AZALEA	To control plant growth	Abide/A-Rest	26 ppm spray	
		Concise	5 to 15 ppm spray	Apply as a uniform spray at a volume of 1.5 qt. per 100 sq. ft. of bench area approximately 4 to 6 weeks after the final pinch. Shorter-growing cultivars (Gloria, Solitaire): use 10 ppm. If a second application is required 2 to 3 weeks later, use 5 to 10 ppm. Taller-growing cultivars (Prize): use 10 ppm. If a second application is required 2 to 3 weeks later, use 10 to 15 ppm.

PLANT GROWTH REGULATORS

Table 8-5. Growth Regulators for Floricultural Crops in Greenhouses

Crop	Purpose	Chemical	Rate	Precautions and Remarks
AZALEA (continued)	To promote flower initiation	Dazide/B-Nine	1,500 to 2,500 ppm spray	Apply solution when new growth from final pinch is 1 to 2 in. long.
		Citadel/Chlormequat E-Pro /Cycocel	1,000 to 4,000 ppm spray	Optimum rates are generally between 1,000 and 2,000 ppm. Two to six multiple sprays may be needed. Make first application when new growth is approximately 2 in. long.
	To prevent flower bud initiation during vegetative growth	GibGro	130 to 850 ppm spray	Apply two to three sprays at 2- to 3-week intervals.
		Florgib/ProGibb T&O	100 to 750 ppm spray	Apply a first application beginning 2 to 3 weeks after pinching. Weekly applications can continue for 1 to 2 additional weeks, for a maximum of three total applications.
	For partial or full substitution of cold treatment	GibGro	265 to 1,055 ppm spray	Spray timing, concentration and number of applications vary with cultivar, as well as intended degree of cold substitution. Consult label for exact recommendations. Not labeled for California.
		Florgib/ProGibb T&O	250 to 500 ppm spray	Spray timing, concentration and number of applications vary with cultivar, as well as intended degree of cold substitution. Consult label for exact recommendations.
	To promote lateral shoot growth on vegetative plants	Off-Shoot-O	Use a 3 to 5% solution in greenhouses; use a 5 to 7% solution outdoors. Apply as a foliar spray.	Efficacy is related to relative humidity and temperature. Spray a few plants to check activity prior to treating the entire crop; effect should be visible in about 1 hr. Be certain chemical covers shoot tip. Ineffective if microscopic flower buds are present.
	To increase lateral branching	Augeo	3,125 to 6,250 ppm spray	
		Florel/Collate	2,500 to 5,000 ppm spray	
	To control plant growth, reduce bypass shoot elongation and promote flower bud initiation	Piccolo/Piccolo 10 XC /Bonzi/Paczol	100 to 200 ppm spray	To control plant growth and promote flower bud initiation, apply after final shaping when new growth is 1.5 to 2 in. long. To reduce bypass shoot development, apply after bud set when bypass shoots are barely visible, or about 5 to 7 weeks prior to cooling.
		Piccolo/Piccolo 10 XC /Bonzi/Paczol/ Downsize	0.59 to 1.77 mg a.i. (5 to 15 ppm) drench for a 6-in. pot; apply 4 fl. oz./6-in. pot)	Drench volumes mg a.i. vary with pot size.
	To control plant growth	Concise/Sumagic	10 to 15 ppm spray	Apply at 1.5 qt per 100 sq. ft. of bench area.
BACOPA (SUTERA)	To control plant growth	Dazide/B-Nine	750 to 1,500 ppm spray	At planting, soft pinch to promote lateral shoot development. Initially try with lower rate.
		Piccolo	4 to 8 ppm liner root soak	Irrigation of the liners occured within 24 hours prior to application, which results in a moderately dry substrate (the stage the plants would be watered but not wilted). Soak for a minimum of 30 to 60 seconds. Transplant after 3-hour waiting period. Rates based on Michigan State University trials.
		Piccolo/Piccolo 10 XC /Bonzi/Paczol	1 to 2 ppm drench	
		Florel/Collate	150 to 200 ppm spray	Early spray will increase branching and reduce early flowering.
	To increase lateral branching	Florel/Collate	150 to 200 ppm spray	
BEDDING PLANTS (Not specifically listed in this table)	To control plant growth	Abide/A-Rest	6 to 66 ppm spray; use 15 ppm spray as a base rate and adjust as needed	
			0.06 to 0.12 mg a.i. drench for a 4-in. pot; apply 2 fl. oz./4-in. pot)	Drench volumes and mg a.i. vary with pot size.
		Citadel + Dazide/Cycocel + B-Nine	800 to 5,000 ppm + 1,000 to 1,500 ppm Cycocel applied as a tank-mix spray	Use the highest rate of Cycocel that doesn't cause excessive leaf yellowing, and then adjust the B-Nine/Dazide rate up and down within the labeled range to attain the desired level of height control.
		Piccolo/Bonzi/Paczol	5 to 90 ppm spray. Use 30 ppm spray as a base rate and adjust as needed.	Conduct trials on a small number of plants, adjusting the rates as needed for desired final plant height and duration of height control. Not recommended for use on fibrous begonia or vinca.
		Piccolo/Bonzi/Paczol/ Downsize	0.118 mg a.i. drench for a 6-in. pot; apply 4 fl. oz./6-in. pot)	Drench applications are recommended only for bedding plants in 6-in. or larger containers. Not recommended for use on fibrous begonia or vinca.
		Citadel/Cycocel	800 to 1,500 ppm spray	Conduct trials on a small number of plants, adjusting the rates as needed for desired final plant height and duration of height control.

PLANT GROWTH REGULATORS

Table 8-5. Growth Regulators for Floricultural Crops in Greenhouses

Crop	Purpose	Chemical	Rate	Precautions and Remarks
BEDDING PLANTS (continued)	To control plant growth	Concise/Sumagic	1 to 50 ppm spray	Conduct trials on a small number of plants, adjusting the rates as needed for desired final plant height and duration of height control. Apply spray as elongation begins (plant height about 2 to 4 in.).
			0.1 to 2 ppm drench	
		Piccolo 10 XC	15 to 30 ppm spray	General starting point for conducting trials for plants not specifically on the label. Use lowest rate in the Northern Belt Region and the upper rate in the Sunbelt Region.
			1 ppm drench	General starting point for conducting trials for plants not specifically on the label.
	To promote plant growth and overcome over-application of gibberellin-inhibiting PGRs	Florgib/ProGibb T&O	1 to 25 ppm spray	Conduct trials on a small number of plants initially using 1 ppm unless previous experience warrants higher use rates. Following assessment of plant response, and if desired results were not evident, reapplication or an increase in rate may be warranted. Consult the label for additional precautions.
		Fascination/Fresco	1 to 5 ppm spray	Conduct trials on a small number of plants initially using 1 ppm unless previous experience warrants higher use rates. Following assessment of plant response, and if desired results were not evident, reapplication or an increase in rate may be warranted. The most common rates for use are 3 to 5 ppm. SEE LABEL FOR ADDITIONAL PRECAUTIONS BEFORE USE.
	To induce lateral or basal branching	Configure	50 to 500 ppm spray	The supplemental label allows legal use on greenhouse-grown plants not specifically listed on the original label. See label for trialing suggestions and precautions.
BEDDING PLANT PLUGS (Not specifically listed in this table)	To control plant growth	Abide/A-Rest	3 to 35 ppm spray	
			Drench plug flats with a 0.5 to 1 ppm solution	For uniform application, use a subirrigation delivery system. Plug trays should not be excessively dry prior to the subirrigation treatment. Plants should develop one to two true leaves prior to first application.
		Dazide/B-Nine	1,500 to 2,500 ppm spray	Conduct trials on a small number of plants, adjusting the rate as needed for desired final plant height and duration of height control. Can be used at the beginning of the true first leaf stage through the finishing stage.
		Citadel + Dazide/Cycocel + B-Nine	800 to 5,000 ppm Dazide/B-Nine + 1,000 to 1,500 ppm Citadel/Cycocel applied as a tank-mix spray	Use the highest rate of Citadel/Cycocel that doesn't cause excessive leaf yellowing, and then adjust the B-Nine/Dazide rate up and down within the labeled range to attain desired level of height control.
	To control plant growth	Piccolo/Piccolo 10 XC /Bonzi/Paczol	1 to 20 ppm spray. Use 5 ppm spray as a base rate and adjust as needed.	Conduct trials on a small number of plants, adjusting the rate as needed for desired final plant height and duration of height control. Plants should develop one to two true leaves prior to first application.
		Cycocel	400 to 1,500 ppm spray	Conduct trials on a small number of plants, adjusting the rates as needed for desired final plant height and duration of height control.
		Concise/Sumagic	0.5 to 10 ppm spray	Conduct trials on a small number of plants, adjusting the rates as needed for desired final plant height and duration of height control. Plugs can be especially sensitive to Concise/Sumagic.
BEGONIA, Hiemalis (Elatior)	To control plant growth	Citadel/Cycocel	500 to 1,000 ppm spray	Applied 1 week after short days begin in summer or when short days begin in winter. Late applications can result insufficient flower stalk elongation.
	To increase lateral branching	Augeo	781 to 1,562 ppm spray	
BEGONIA, Seed (Wax)	To control plant growth	Abide/A-Rest	3 to 15 ppm spray	Use lower half of rate range for plugs and upper range for finishing plants.
		Dazide/B-Nine	2,500 to 5,000 ppm spray	
		Florel/Collate	500 ppm spray	Apply to increase lateral branching, prevent flower initiation and development, and inhibit internode elongation.
		Concise/Sumagic	Sprays	Not registered for use. Can result in excessive control.
		Piccolo/Piccolo 10 XC /Bonzi/Paczol	Sprays	Not registered for use. Can result in excessive control.
		Topflor	Sprays	Not registered for use. Can result in excessive control.
		Citadel/Cycocel	500 ppm spray	
		Citadel + Dazide/Cycocel + B-Nine	1,000 to 1,250 ppm Dazide/B-Nine + 800 to 1,250 ppm Citadel/Cycocel applied as a tank-mix spray	
BEGONIA, Tuberous	To control plant growth	Citadel/Cycocel	250 to 500 ppm spray	Rate can be used on Stage 4 plugs or beginning 2 weeks after transplanting.
		Citadel/Cycocel	1,000 ppm spray	Rate for actively growing plants.
		Dazide/B-Nine	2,500 ppm spray	Rate for actively growing plants.

Table 8-5. Growth Regulators for Floricultural Crops in Greenhouses

Crop	Purpose	Chemical	Rate	Precautions and Remarks
BEGONIA, Vegetative	To control plant growth	Citadel/Cycocel	750 to 1,000 ppm spray	
BEGONIA, Vegetative (Dragon Wing)	To control plant growth	Piccolo/Piccolo 10 XC /Bonzi/Paczol	3 to 5 ppm spray	For 4-in. pots, apply a weekly 3 ppm spray starting 2 weeks after transplanting for 3 weeks. For 6-in. pots, use 5 ppm starting 2 weeks after transplant. A second and third application may be useful.
BELLIS	To control plant growth	Dazide/B-Nine	2,500 ppm spray	If needed.
		Concise/Sumagic	5 ppm spray	If needed.
BIDENS	To control plant growth	Dazide/B-Nine	1,500 to 2,500 ppm spray	At planting, soft pinch to promote lateral shoot development.
		Concise/Sumagic	1 to 5 ppm spray	Rates for genetics needing slight growth control.
			0.25 ppm drench	Rates for genetics needing slight growth control.
	To increase lateral branching	Florel/Collate	300 to 500 ppm spray	
BOUGAINVILLEA	To control plant growth	Abide/A-Rest	50 ppm drench	
		Piccolo/Piccolo 10 XC /Bonzi/Paczol	25 to 100 ppm drench	
	To increase lateral branching	Augeo	400 to 1,600 ppm spray	Cultivar response rates vary. Conduct your own trials to determine suitability and appropriate timing.
			1,600 ppm drench	Cultivar response rates vary. Conduct your own trials to determine suitability and appropriate timing.
BRACHYSCOME	To control plant growth	Florel/Collate	500 to 1,000 ppm spray	To keep plants more compact. Based on Texas A&M University trials.
		Dazide/B-Nine	2,500 to 5,000 ppm spray	
BRACTEANTHA, BRACTEATA	To control plant growth	Dazide/B-Nine	2,500 ppm spray	
		Piccolo/Piccolo 10 XC /Bonzi/Paczol	20 to 30 ppm spray	
			1 ppm drench	
		Concise/Sumagic	10 to 20 ppm spray	
	To increase lateral branching	Florel/Collate	300 to 500 ppm	
BROMELIAD	To promote flower initiation	Florel/Collate	2,471 ppm spray	Cultivar response rates vary. Conduct your own trials to determine suitability and appropriate timing.
BROWALLIA	To control plant growth	Dazide/B-Nine	2,500 to 5,000 ppm spray	
BULB CROPS (Not specifically listed in this table)	To control plant growth	Abide/A-Rest	25 to 50 ppm spray	
			0.25 mg a.i. (2 ppm) drench for a 6-in. pot; apply 4 fl. oz./ 6-in. pot)	Drench volumes and mg a.i. vary with pot size.
		Piccolo/Piccolo 10 XC /Bonzi/Paczol	100 ppm spray	Conduct trials on a small number of plants, adjusting the rate as needed for desired final plant height and duration of height control.
			1.183 mg a.i. (10 ppm) drench for a 6-in. pot; apply 4 fl. oz./ 6-in. pot)	Drench volumes and mg a.i. vary with pot size.
			20 ppm bulb soak	Soak for 15 min. Conduct trials on a small number of bulbs, adjusting the rate and soaking period (up to 1 hour) as needed for desired final plant height.
		Concise/Sumagic	2.5 to 20 ppm spray	Conduct trials on a small number of plants, adjusting the rate as needed for desired final plant height and length of height control.
			1 to 3 ppm drench	Drench volumes and mg a.i. vary with pot size. Application should be made when newly emerged shoots are 1 to 2 in. tall.
			1 to 10 ppm bulb soak	Soak for 1 to 5 min. Conduct trials on a small number of bulbs, adjusting the rate and soaking period as needed for desired final plant height.
	To promote plant growth and overcome over-application of gibberellin- inhibiting PGRs.	Fascination	1 to 25 ppm spray	Conduct trials on a small number of plants initially using 1 ppm, unless previous experience warrants higher use rates. Following assessment of plant response, and if desired results were not evident, reapplication or an increase in rate may be warranted. The most common rates for use are 3 to 5 ppm. SEE LABEL FOR ADDITIONAL PRECAUTIONS BEFORE USE.
CALADIUM	To control plant growth	Dazide/B-Nine	2,500 to 5,000 ppm spray	
		Piccolo/Bonzi/Paczol/ Downsize	100 to 200 ppm spray (3.2 to 6.4 fl oz/gal)	Make first spray application when plants are 2 to 4 in. tall.
			0.24 to 1.77 mg a.i. (5 to 15 ppm) drench for a 6-in. pot; apply 4 fl. oz./6-in. pot)	Make first application when plants are 1 to 2 in. tall. Drench volumes and mg a.i. vary with pot size.

PLANT GROWTH REGULATORS

Table 8-5. Growth Regulators for Floricultural Crops in Greenhouses

Crop	Purpose	Chemical	Rate	Precautions and Remarks
CALADIUM (continued)	To control plant growth	Piccolo/Bonzi/Paczol	60 ppm tuber soak	Soak tubers for 30 min. prior to planting.
		Piccolo 10 XC	100 to 200 ppm spray	Spray applications of Piccolo 10 XC are the least desirable method for controlling bulb plant height and must be applied sequentially to maximize uniformity of the crop. Begin spray applications when plants reach a height of 2 to 4 in.
			2 to 16 ppm drench	Drench volume varies with pot size. Begin drench applications when plants reach a height of 1 to 2 in.
		Topflor	0.5 to 2 mg a.i. drench for a 6-in. pot	Based on NC State University trials. Adjust rates for other locations. Use lower rates for less vigorous cultivars.
CALCEOLARIA	To control plant growth	Citadel/Cycocel	400 to 800 ppm spray	Use to control internode length. Apply 400 ppm when flower buds 1-in. in diameter. Repeat 2 weeks later if needed.
		Dazide/B-Nine	1,000 to 1,500 ppm spray	Use to control internode length.
CALENDULA	To control plant growth	Dazide/B-Nine	2,500 to 5,000 ppm spray	Can be used when the visible flower bud is pea sized. Rates of 3,500 ppm be used 4 to 5 weeks after germination (when 3 to 4 mature leaves formed).
			2,500 to 5,000 ppm spray	Plugs: Use 2,500 ppm with Stage 1 and 5,000 ppm with Stages 2 or 3.
		Concise/Sumagic	1 ppm spray	Plugs: Use at Stages 2 or 3.
		Piccolo/Piccolo 10 XC /Bonzi/Paczol	4 ppm spray	Plugs: Use at Stages 2 or 3.
CALIBRACHOA	To control plant growth	Dazide/B-Nine	2,500 to 5,000 ppm spray	At planting, soft pinch to promote lateral shoot development. Multiple applications may be required.
		Citadel + Dazide/Cycocel + B-Nine	2,500 ppm Dazide + 500 to 1,500 ppm Citadel applied as a tank-mix spray	
		Concise/Sumagic	10 to 25 ppm spray	Try lower rate initially. Apply 2 weeks after transplanting.
		Piccolo/Piccolo 10 XC /Bonzi/Paczol	3 to 50 ppm spray	Use rates of 3 to 5 ppm for compact genetics needing slight growth control.
		Piccolo/Piccolo 10 XC /Bonzi/Paczol/Downsize	3 to 8 ppm drench	Rates for compact genetics needing slight growth control. Begin with 1 to 2 ppm to determine suitable rates.
		Florel/Collate	300 to 500 ppm spray	Early spray will increase branching and reduce early flowering.
		Topflor	5 to 10 ppm spray	
CALLA LILY (*Zantedeschia*)	To control plant growth	Piccolo/Piccolo 10 XC /Bonzi/Paczol/Downsize	0.59 to 1.77 mg a.i. (5 to 15 ppm) drench for a 6-in. pot; apply 4 fl. oz./6-in. pot	Make first application when plants are 1 to 2 in. tall. Drench volumes and mg a.i. vary with pot size.
		Piccolo/Bonzi/Paczol	20 ppm rhizome/tuber soak	Soak the rhizomes/tubers for 15 min. prior to planting.
		Concise/Sumagic	1 to 2 mg a.i. drench (8.45 to 16.9 ppm); apply 4 fl. oz./6-in. pot)	Optimal rate based on NC State University trials. Adjust rate for plant vigor. Drench volumes and mg a.i. vary with pot size.
		Topflor	1 to 2.25 mg a.i drench for a 6-in. pot	Based on NC State University trials. Adjust rates for other locations.
	To promote flowering	Florgib/ProGibb T&O	500 ppm rhizome/tuber soak	Soak the rhizomes or tubers for 10 min. prior to planting. See label for details.
CAMPANULA	To control plant growth	Dazide/B-Nine	2,500 to 5,000 ppm spray	Use at visible bud.
		Topflor	10 to 30 ppm spray	Use at visible bud.
CANNA LILY	To control plant growth	Topflor	50 to 80 ppm spray	
CELOSIA	To control plant growth	Abide/A-Rest	7 to 26 ppm spray	
		Dazide/B-Nine	2,500 to 5,000 ppm spray	
		Piccolo/Piccolo 10 XC /Bonzi/Paczol	15 to 45 ppm spray	
		Citadel/Chlormequat E-Pro /Cycocel	800 to 1,500 ppm spray	
		Concise/Sumagic	10 to 20 ppm spray	
		Topflor	10 to 40 ppm spray	Based on NC State University trials. Adjust rates for other locations.
CELOSIA, Plugs	To control plant growth	Piccolo/Piccolo 10 XC /Bonzi/Paczol	5 to 10 ppm spray	Timing of application should normally begin at the 1 to 2 true leaf stage.
CENTAUREA	To control plant growth	Abide/A-Rest	10 to 15 ppm spray	
		Dazide/B-Nine	2,500 to 5,000 ppm spray	
CENTRADENIA HYBRID	To control plant growth	None	None	Plants grown with good light and optimal growing conditions generally do not need PGRs.

PLANT GROWTH REGULATORS

Table 8-5. Growth Regulators for Floricultural Crops in Greenhouses

Crop	Purpose	Chemical	Rate	Precautions and Remarks
CHRISTMAS CACTUS (*Schlumbergera spp.*)	To increase branching under vegetative conditions	Configure	100 ppm spray	After planting when new vegetative growth begins, uniformly apply 1 to 2 quarts of finished spray solution to 100 sq. ft. of area.
	To increase the number of flower buds under reproductive conditions	Configure	100 to 200 ppm spray	Apply as a uniform foliar spray after the start of short days following leveling, or when flower buds become visible. See the label for specific guidelines based on lighted or natural-season growth plants.
CHRYSANTHEMUM, Cut	To reduce "neck" stretching	Dazide/B-Nine	2,500 ppm spray	Spray upper foliage 5 weeks after start of short-day treatment.
	To elongate peduncles of pompom-type mums	Florgib/ProGibb T&O	25 to 60 ppm spray	Use a single application 4 to 5 weeks after initiation of short days. Direct spray solution towards the flower buds. See label for precautions.
CHRYSANTHEMUM, Perennial	To control plant growth	Piccolo/Piccolo 10 XC /Bonzi/Paczol	50 to 200 ppm spray	
		Piccolo/Piccolo 10 XC /Bonzi/Paczol/Downsize	0.12 to 0.48 mg a.i. (1 to 4 ppm) drench for a 6-in. pot; apply 4 fl. oz./6-in. pot	
CHRYSANTHEMUM, Potted	To control plant growth	Abide/A-Rest	25 to 50 ppm spray	
			0.25 to 0.5 mg a.i. drench for a 6-in. pot; apply 4 fl. oz./6-in. pot	Apply when plants are 2 to 6 in. in height (about 2 weeks after pinch). Drench rates and application volumes vary with pot size.
		Dazide/B-Nine	1,000 ppm preplant foliar dip	Rooted cuttings can be dipped in solution to thoroughly wet leaves and stems and then potted. Allow foliage to dry before watering in. For unrooted cuttings, dip stems in solution, remove to flat, cover to prevent dehydration and hold overnight under cool conditions. Stick the next day.
			2,500 to 5,000 ppm spray	Spray when new growth from pinch is 1 to 2 in. long. Some varieties may require another application 3 weeks later.
		Piccolo/Piccolo 10 XC /Bonzi/Paczol	50 to 200 ppm spray	Applications should begin when axillary shoots are 2 to 3 in. long. Sprays can be applied earlier to vigorous cultivars if additional control is desired. Sequential applications of lower rates generally provide more uniformly shaped plants than single-spray applications. Uniform application of both sprays and drenches is critical for uniform crop development.
		Piccolo/Piccolo 10 XC /Bonzi/Paczol/Downsize	0.118 to 0.473 mg a.i. (1 to 4 ppm) drench for a 6-in. pot; apply 4 fl. oz./6-in. pot	Drench volumes and mg a.i. vary with pot size. Begin when the axillary shoots are to 2 to 3 in. long. Uniform application is required.
		Concise	5 to 10 ppm dip treatment on cuttings	Apply when the lateral shoots are 1.5 to 2.0 in. tall (about 7 to 14 days after pinching). Test for cultivar sensitivity. Multiple applications of the lower label rate may elicit a more satisfactory response and/or increasing the spray volume from 2 qts/100 sq. ft. to 3 qts/100 sq. ft. For Florida only: use a foliar spray concentration between 5 to 10 ppm (1.3 to 2.56 fl. oz./gal). For medium to tall cultivars, increase the spray volume to 3 qts/100 sq. ft.
			2.5 to 10 ppm spray	Apply as a dip treatment on unrooted cuttings followed by a foliar spray in the low rate range. On rooted cuttings, use a solution of 2.5 ppm or less, followed by a foliar spray in the low rate range.
		Concise/Sumagic	2.5 to 10 ppm spray	
		Topflor	7.5 to 25 ppm spray	Based on NC State University trials. Adjust rates for other locations. Use lower rates for less vigorous cultivars.
CHRYSANTHEMUM, Garden	To control plant growth	Concise	5 to 10 ppm dip treatment on cuttings	Apply when the lateral shoots are 1.5 to 2.0 in. tall (about 7 to 14 days after pinching). Test for cultivar sensitivity. Multiple applications of the lower label rate may elicit a more satisfactory response and/or increasing the spray volume from 2 qts/100 sq. ft. to 3 qts/100 sq. ft. For Florida only: use a foliar spray concentration between 5 to 10 ppm (1.3 to 2.56 fl. oz./gal). For medium to tall cultivars, increase the spray volume to 3 qts/100 sq. ft.
		Concise/Sumagic	2.5 to 10 ppm spray	
	To increase lateral branching	Florel/Collate	500 ppm spray	Florel and Collate applications will provide some growth retardant effects and delay flowering. Read the label for restrictions on timing of applications.
CHRYSOCEPHALUM APICULATUM	To control plant growth	Dazide/B-Nine	2,500 ppm spray	Plants pinched and grown with good light and optimal growing conditions generally do not need PGRs.
CLARKIA (*Godetia*)	To control plant growth	Concise/Sumagic	15 to 25 ppm drench	Trial rates for cultivar response. Rates based on older cultivars.
		Dazide/B-Nine	3,000 ppm foliar spray	Trial rates for cultivar response. Rates based on older cultivars.
		Piccolo/Piccolo 10 XC /Bonzi/Paczol	20 to 30 ppm drench	Trial rates for cultivar response. Rates based on older cultivars.

PLANT GROWTH REGULATORS

Table 8-5. Growth Regulators for Floricultural Crops in Greenhouses

Crop	Purpose	Chemical	Rate	Precautions and Remarks
CLEMATIS	To control plant growth	Abide/A-Rest	25 to 132 ppm spray	
			0.25 to 0.5 mg a.i. (2 to 4 ppm) drench for a 6-in. pot; apply 4 fl. oz./6-in. pot	Drench volumes and mg a.i. vary with pot size.
CLEOME	To control plant growth	Abide/A-Rest	7 to 26 ppm spray	
		Citadel/Chlormequat E-Pro /Cycocel	800 to 1,500 ppm spray	
		Dazide/B-Nine	4,000 to 5,000 ppm spray	Multiple applications may be required, make them at 7 to 10 day intervals.
		Piccolo/Piccolo 10 XC /Bonzi/Paczol	20 to 30 ppm spray	Multiple applications may be required, make them at 7 to 10 day intervals.
CLERODENDRUM	To control plant growth	Abide/A-Rest	50 ppm spray	
			0.9 mg a.i. drench	
		Piccolo/Piccolo 10 XC /Bonzi/Paczol	100 ppm drench	
			0.5 mg a.i. drench	
	To increase lateral branching	Augeo	1,042 to 2,083 ppm spray	
COLEUS PLUGS, Seed	To control plant growth	Piccolo/Piccolo 10 XC /Bonzi/Paczol	5 to 10 ppm spray	Timing of application should normally begin at the 1 to 2 true leaf stage.
COLEUS, Seed	To control plant growth	Dazide/B-Nine	2,500 to 5,000 ppm spray	
		Piccolo/Piccolo 10 XC /Bonzi/Paczol	15 to 30 ppm spray	
		Citadel/Chlormequat E-Pro /Cycocel	400 to 3,000 ppm spray	
		Concise/Sumagic	10 to 20 ppm spray	
		Topflor	20 to 40 ppm spray	Based on NC State University trials. Adjust rates for other locations.
COLEUS, Vegetative	To control plant growth	Citadel + Dazide/Cycocel + B-Nine	2,500 to 4,000 ppm + 1,000 to 1,500 ppm Cycocel applied as a tank-mix spray	Scheduling the crop to avoid excessive stretch is the most effective means of controlling growth.
		Piccolo/Piccolo 10 XC /Bonzi/Paczol	5 to 30 ppm spray	
			1 to 2 ppm drench	
		Citadel/Chlormequat E-Pro /Cycocel	800 to 1,500 ppm spray	
		Concise/Sumagic	5 to 20 ppm spray	Use rates of 5 to 10 ppm for compact genetics needing slight growth control.
		Collate/Florel	500 ppm spray	
COLUMBINE	To control plant growth	Abide/A-Rest	65 to 132 ppm spray	
			0.25 to 0.5 mg a.i. (2 to 4 ppm) drench for a 6-in. pot; apply 4 fl. oz./6-in. pot	Drench volumes and mg a.i. vary with pot size.
CONEFLOWER (Echinacea spp.)	To control plant growth	Concise/Sumagic	30 to 40 ppm spray	
	To increase branching	Configure	300 to 900 ppm spray	Apply after plant establishment and resumption of growth (i.e., approximately 2 weeks after potting). Apply in a uniform spray volume of 2 qts/100 sq. ft. of area. Application timing and rate may vary with cultivar.
CONSOLIDA (Larkspur)	To control plant growth	Abide/A-Rest	35 to 132 ppm spray	
			0.25 to 0.5 mg a.i. drench for a 6-in. pot (1 to 2 fl. oz./gal of drench solution; apply 4 fl. oz./6-in. pot)	Drench volumes and mg a.i. vary with pot size.
		Concise/Sumagic	5 ppm drench	
		Dazide/B-Nine	2,500 to 5,000 ppm spray	
		Piccolo/Piccolo 10 XC /Bonzi/Paczol	30 to 60 ppm spray	
CONSOLIDA (Larkspur), Cut	To promote growth and stem elongation	Florgib/ProGibb T&O	50 to 100 ppm spray	Apply when plants are 4 to 8 in. tall. Apply at 2- to 3-week intervals. See label for precautions.
COREOPSIS	To control plant growth	Concise/Sumagic	2 to 4 ppm spray	Rates for compact genetics needing slight growth control.
		Piccolo/Piccolo 10 XC /Bonzi/Paczol	3 to 100 ppm spray	Use rates of 3 to 6 ppm for compact genetics needing slight growth control.
		Piccolo/Piccolo 10 XC /Bonzi/Paczol/Downsize	0.59 to 1.18 mg a.i. (5 to 10 ppm) drench for a 6-in. pot; apply 4 fl. oz./6-in. pot	Rates for vigorous genetics needing moderate growth control.
		Topflor	2 to 4 ppm spray	Rates for compact genetics needing slight growth control.

PLANT GROWTH REGULATORS

394

Table 8-5. Growth Regulators for Floricultural Crops in Greenhouses

Crop	Purpose	Chemical	Rate	Precautions and Remarks
CORNFLOWER (Centaurea)	To control plant growth	Abide/A-Rest	7 to 26 ppm spray	
		Dazide/B-Nine	2,500 to 5,000 ppm spray	
COSMOS	To control plant growth	Dazide/B-Nine	2,500 to 5,000 ppm spray	
CROSSANDRA	To control plant growth	Dazide/B-Nine	2,500 to 5,000 ppm spray	Apply after pinch when new growth is 2 in. long.
		Piccolo/Piccolo 10 XC /Bonzi/Paczol	50 ppm spray	Apply 2 weeks after pinch.
CUPHEA	To control plant growth	Dazide/B-Nine	1,500 to 2,500 ppm spray	PGRs not required on compact cultivars.
		Piccolo/Piccolo 10 XC /Bonzi/Paczol	1 to 5 ppm spray	Initially, test on a few plants to determine rate for optimum control. Cuphea is sensitive to excessive rates.
		Piccolo/Piccolo 10 XC /Bonzi/Paczol/Downsize	0.25 to 2 ppm drench	Use rates of 0.25 to 0.5 ppm for compact genetics needing slight growth control. Use 2 ppm for vigorous cultivars grown in the south.
DAFFODIL	To control plant growth	Piccolo/Piccolo 10 XC /Bonzi/Paczol/Downsize	2.37 to 4.73 mg a.i. (20 to 40 ppm) drench for a 6-in. pot; apply 4 fl. oz./6-in. pot	Make first application when plants are 1 to 2 in. tall. Drench volumes and mg a.i. vary with pot size.
		Piccolo/Piccolo 10 XC /Bonzi/Paczol	80 ppm bulb soak	Soak bulbs for 1 hr. prior to planting. Ten minute soaks of 400 ppm provided excellent results in NC State University trials.
		Florel/Collate	2,000 ppm spray	Controls plant height and stem topple. Apply when shoots are 3 to 4 in. tall. See label for cultivar differences in rates.
DAHLIA, Bedding Plant	To control plant growth	Abide/A-Rest	7 to 26 ppm spray	
		Dazide/B-Nine	2,500 to 5,000 ppm spray	
		Citadel + Dazide/Cycocel + B-Nine	2,500 to 4,000 ppm + 1,000 to 1,500 ppm Cycocel applied as a tank-mix spray	
		Piccolo/Piccolo 10 XC /Bonzi/Paczol	15 to 45 ppm spray	
		Citadel/Chlormequat E-Pro /Cycocel	800 to 1,500 ppm spray	
		Concise/Sumagic	10 to 20 ppm spray	
DAHLIA PLUGS, Bedding Plant	To control plant growth	Piccolo/Piccolo 10 XC /Bonzi/Paczol	5 to 10 ppm spray	Timing of application should normally begin at the 1 to 2 true leaf stage.
DAHLIA, Tuberous	To control plant growth	Abide/A-Rest	0.25 to 0.5 mg a.i. (2 to 4 ppm) drench for a 6-in. pot; apply 4 fl. oz./6-in. pot	Drench volumes and mg a.i. vary with pot size.
		Piccolo/Piccolo 10 XC /Bonzi/Paczol/Downsize	1.18 to 4.73 mg a.i. (10 to 40 ppm) drench for a 6-in. pot; apply 4 fl. oz./6-in. pot	
		Piccolo/Piccolo 10 XC /Bonzi/Paczol	Greater than 40 ppm tuber soak	Soak tubers for 20 min. prior to planting.
		Concise/Sumagic	0.25 to 0.5 mg a.i. drench (2.1 to 4.2 ppm); apply 4 fl. oz./6-in. pot	Optimal rate based on NC State University trials. Adjust rate for plant vigor. Drench volumes and mg a.i. vary with pot size.
		Topflor	0.25 to 2 mg a.i. (2.1 to 16.9 ppm) drench for a 6-in. pot	Based on NC State University trials. Adjust rates for other locations. Use lower rates for less vigorous cultivars.
DELPHINIUM	To control plant growth	Abide/A-Rest	35 to 132 ppm spray	
			0.25 to 0.5 mg a.i. drench for a 6-in. pot (1 to 2 fl. oz./gal of drench solution; apply 4 fl. oz./6-in. pot)	Drench volumes and mg a.i. vary with pot size.
		Concise/Sumagic	5 ppm drench	
		Dazide/B-Nine	2,500 to 5,000 ppm spray	
		Piccolo/Piccolo 10 XC /Bonzi/Paczol	30 to 60 ppm spray	
DELPHINIUM, Cut	To promote plant growth and stem elongation	Florgib/ProGibb T&O	50 to 100 ppm spray	Apply when plants are 4 to 8 in. tall. More than one application is possible at 2- to 3-week intervals. See label for precautions.
DIANTHUS, Bedding Plant	To control plant growth	Abide/A-Rest	7 to 26 ppm spray	
		Dazide/B-Nine	2,500 to 5,000 ppm spray	
		Piccolo/Piccolo 10 XC /Bonzi/Paczol	5 to 60 ppm spray	Cultivar response rates vary. Conduct your own trials to determine suitability and appropriate timing. Some series recommend the use of 5 to 8 ppm sprays.
		Citadel/Chlormequat E-Pro /Cycocel	800 to 1,500 ppm spray	
		Concise/Sumagic	3 to 5 ppm spray	
DIANTHUS PLUGS, Bedding plant	To control plant growth	Piccolo/Piccolo 10 XC /Bonzi/Paczol	10 to 20 ppm spray	Timing of application should normally begin at the 1 to 2 true leaf stage.

PLANT GROWTH REGULATORS

395

Table 8-5. Growth Regulators for Floricultural Crops in Greenhouses

Crop	Purpose	Chemical	Rate	Precautions and Remarks
DIANTHUS, Cut	To promote plant growth and stem elongation	Florgib/ProGibb T&O	50 to 100 ppm spray	Apply when plants are 4 to 8 in. tall. More than one application is possible at 2- to 3-week intervals. See label for precautions.
DIANTHUS, Pot	To control plant growth	Concise/Sumagic	15 ppm spray	
		Piccolo/Piccolo 10 XC /Bonzi/Paczol	15 ppm spray	
DIASCIA Hybrid	To control plant growth	Dazide/B-Nine	1,250 to 5,000 ppm	At planting, soft pinch to promote lateral shoot development. Use higher rates on vigorous cultivars.
		Concise/Sumagic	5 to 15 ppm spray	Use lower rates to ensure taller flower spikes.
		Florel/Collate	200 to 500 ppm spray	Use 2 weeks after pinch.
		Piccolo/Piccolo 10 XC /Bonzi/Paczol	30 ppm spray	
			1 to 2 ppm drench	
DIASCIA, Seed	To control plant growth	Abide/A-Rest	20 ppm spray	Start application 7 to 10 days after transplant. Repeat 7 days later.
		Concise/Sumagic	5 to 10 ppm spray	To hold plants under warm conditions. Use caution, plants very responsive.
		Dazide/B-Nine	3,000 to 5,000 ppm spray	Start application 7 to 10 days after transplant.
		Piccolo/Piccolo 10 XC /Bonzi/Paczol	10 to 20 ppm spray	To hold plants under warm conditions. Use caution, plants very responsive.
***DICENTRA SPECTABILIS* (Bleeding Heart)**	To control plant growth	Abide/A-Rest	65 to 132 ppm spray	
			0.25 to 0.5 mg a.i. drench for a 6-in. pot; apply 4 fl. oz./6-in. pot)	Drench volumes and mg a.i. vary with pot size.
		Dazide/B-Nine	2,500 to 5,000 ppm spray	Apply as new sprouts emerge from the pot. Repeat if needed due to non-uniform emergence.
DICHONDRA ARGENTEA	To control plant growth	Citadel+Dazide/Cycocel+B-Nine	1,000 ppm + 5,000 ppm spray	Also increases branching and improves silver color.
		Dazide/B-Nine	5,000 ppm spray	Also increases branching and improves silver color. Apply 2 weeks after transplanting.
DIGITALIS	To control plant growth	Piccolo/Piccolo 10 XC /Bonzi/Paczol	80 to 160 ppm spray	
		Piccolo/Piccolo 10 XC /Bonzi/Paczol/Downsize	0.24 to 0.48 mg a.i. (2 to 4 ppm) drench for a 6-in. pot; apply 4 fl. oz./6-in. pot	
DOROTHEANTHUS BELLIDIFORMIS	To control plant growth	None	None	Plants pinched and grown with good light and optimal growing conditions generally do not need PGRs.
DRACAENA	To control plant growth	Abide/A-Rest	25 to 132 ppm spray	
			0.25 to 0.5 mg a.i. (2 to 4 ppm) drench for a 6-in. pot; apply 4 fl. oz./6-in. pot	Drench volumes and mg a.i. vary with pot size.
DUSTY MILLER (*Senecio cineraria*)	To control plant growth	Dazide/B-Nine	2,500 to 5,000 ppm spray	
		Concise/Sumagic	30 ppm spray	
EASTER LILY (See Lily, Easter)				
ECHEVERIA spp	To induce offsets and induce flower development	Configure	100 to 400 ppm spray	Based on NC State University trials when applied 2 weeks after potting. A slight increase in offsets occured along with the induction of flowering.
EGGPLANT	To control plant growth	Sumagic	2 to 10 ppm spray	See label for application suggestions and precautions. Make initial foliar applications when 2 to 4 true leaves are present. Apply uniformly as a foliar spray using 2 qt/100 sq. ft. Sequential applications at lower recommended rates will generally provide more growth control than a single high rate application. First-time users should apply the lowest recommended rate in order to determine optimal rate for individual cultivars under local environmental conditions. If additional growth control is required, a sequential spray application at the lowest recommended rate should be made 7 to 14 days after the initial application. If multiple applications are made to the transplants, the total amount of Sumagic applied may not exceed that from a single application of a 10 ppm spray. The final application may not occur later than 14 days after the 2 to 4 true leaf stage.
ERYSIMUM	To control plant growth	None	None	Plants grown with good light and optimal growing conditions generally do not need PGRs.
EUPATORIUM	To control plant growth	Piccolo/Piccolo 10 XC /Bonzi/Paczol	>240 ppm spray	
		Piccolo/Piccolo 10 XC /Bonzi/Paczol/Downsize	0.96 to 1.18 mg a.i. (8 to 10 ppm) drench for a 6-in. pot; apply 4 fl. oz./6-in. pot	

PLANT GROWTH REGULATORS

Table 8-5. Growth Regulators for Floricultural Crops in Greenhouses

Crop	Purpose	Chemical	Rate	Precautions and Remarks
EUPHORBIA HYPERICIFOLIA HYBRID	To control plant growth	Dazide/B-Nine	2,500 ppm spray	Plant growth slow early on. Apply PGRs if control is needed.
		Citadel+Dazide/Cycocel+B-Nine	750 ppm + 2,500 ppm spray	
		Florel/Collate	Spray	Not recommended.
		Piccolo/Piccolo 10 XC /Bonzi/Paczol	0.5 to 2 ppm drench	Can be applied 3 to 4 weeks before finish, using the lower rate in the North and higher rate in the South.
EVOLVULUS	To control plant growth	None	None	Plants grown with good light and optimal growing conditions generally do not need PGRs.
EXACUM	To control plant growth	Dazide/B-Nine	2,500 to 5,000 ppm spray	
		Piccolo/Piccolo 10 XC /Bonzi/Paczol	75 ppm spray	
			0.25 to 0.75 mg a.i. drench for a 6-in. pot	
		Topflor	25 to 50 ppm spray	Based on NC State University trials. Adjust rates for other locations.
			0.01 to 0.03 mg a.i. (0.08 to 0.25 ppm) drench for a 6-in. pot	Based on NC State University trials. Adjust rates for other locations. Exacum is very responsive to Topflor drenches, so start trials with lower rates.
FATSHEDERA	To control plant growth	Abide/A-Rest	65 to 132 ppm spray	
			0.25 to 0.5 mg a.i. drench for a 6-in. pot; apply 4 fl. oz./6-in. pot	Drench volumes and mg a.i. vary with pot size.
FELICIA	To control plant growth	Citadel+Dazide/Cycocel+B-Nine	1,000 to 1,500 ppm + 2,500 to 4,000 ppm spray	Pinch plant as needed to improve shape.
		Citadel/Cycocel	1,500 ppm spray	Applied to pinched plants.
FLOWERING/ FOLIAGE PLANTS, Herbaceous Species (Not specifically listed in this table)	To control plant growth	Abide/A-Rest	20 to 50 ppm spray	Recommended starting rate for an Abide/A-Rest spray on a new herbaceous flowering or foliage species is 33 ppm (16 fl. oz./gal).
			0.125 to 0.25 mg a.i. (1 to 2 ppm) drench for a 6-in. pot; apply 4 fl. oz./6-in. pot	Drench volumes and mg a.i. vary with pot size.
		Piccolo/Piccolo 10 XC /Bonzi/Paczol	30 ppm spray	Conduct trials on a small number of plants, adjust-ing the rate as needed for desired final plant height and length of height control.
		Piccolo/Piccolo 10 XC /Bonzi/Paczol/Downsize	0.118 mg a.i. (1 ppm) drench for a 6-in. pot; apply 4 fl. oz./ 6-in. pot	Drench volumes and mg a.i. vary with pot size. Conduct trials on a small number of plants.
		Citadel/Cycocel	800 to 3,000 ppm spray	Optimum rate depends on species, desired amount of height control and environmental conditions. The suggested initial rate for small-scale trials is 1,250 ppm. Example: herbaceous species known to respond to Cycocel are—Achimenes, Aster, Astilbe, Begonia (hiemalis), Begonia (tuberous), Calceolaria, Carnation, Chrysanthemum, Columbine, Easter lily, *Gynura aurantiaca*, Ivy, Kalanchoe, *Lilium spp.*, Morning glory, Pachystachys, *Pilea spp.*, Pentas, *Salvia spp.*, Schefflera, *Sedum spp.* and Sunflower.
			2,000 to 4,000 ppm drench	Drench volumes vary with pot size. See label for recommended volumes. Herbaceous species known to respond to Cycocel are listed above.
		Concise/Sumagic	5 to 40 ppm spray	Conduct trials on a small number of plants, adjusting the rate as needed for desired final plant height and length of height control.
			0.1 to 1 ppm drench	Drench volumes and mg a.i. vary with pot size.
	To promote plant growth and overcome over-applications of gibberellin-inhibiting PGRs	Florgib/ProGibb T&O	1 to 25 ppm spray	Conduct trials on a small number of plants initially using 1 ppm, unless previous experience warrants higher use rates. Following assessment of plant response, and if desired results are not evident, reapplication or an increase in rates may be warranted. Consult the label for additional precautions.
		Fascination	1 to 25 ppm spray	Conduct trials on a small number of plants initially using 1 ppm, unless previous experience warrants higher use rates. Following assessment of plant response, and if desired results were not evident, reapplication or an increase in rate may be warranted. The most common rates for use are 3 to 5 ppm. SEE LABEL FOR ADDITIONAL PRECAUTIONS BEFORE USE.
	To induce lateral or basal branching	Configure	50 to 500 ppm spray	The supplemental label allows legal use on greenhouse grown plants not specifically listed on the original label. See label for trialing suggestions and precautions.
FLOWERING/ FOLIAGE PLANTS, Woody Species (Not specificallylisted in this table)	To control plant growth	Abide/A-Rest	50 ppm spray	
			0.25 mg a.i. (2 ppm) drench for a 6-in. pot; apply 4 fl. oz./ 6-in. pot	Drench volumes and mg a.i. vary with pot size.

PLANT GROWTH REGULATORS

Table 8-5. Growth Regulators for Floricultural Crops in Greenhouses

Crop	Purpose	Chemical	Rate	Precautions and Remarks
FLOWERING/ FOLIAGE PLANTS, Woody Species (Not specifically listed in this table) (continued)	To control plant growth	Dazide/B-Nine	2,500 to 7,500 ppm spray	Two or more applications may be necessary if new growth begins to stretch or for enhanced coloration.
		Piccolo/Piccolo 10 XC /Bonzi/Paczol	50 ppm spray	Conduct trials on a small number of plants, adjust-ing the rate as needed for desired final plant height and length of height control.
		Piccolo/Piccolo 10 XC /Bonzi/Paczol/Downsize	0.237 mg a.i. drench for a 6-in. pot; apply 4 fl. oz./6-in. pot	Drench volumes and mg a.i. vary with pot size.
		Citadel/Cycocel	800 to 3,000 ppm spray	Optimum rate depends on species, desired amount of height control and environmental conditions. The suggested initial rate for small-scale trials is 1,250 ppm. Example: woody species known to respond to Cycocel are—*Barleria cristata*, Bougainvillea, Camellia, Gardenia, Fuchsia, Hollies, Hydrangea, Lantana, *Pseuderanthemum lactifolia*, Rhododendron and Roses (potted).
			2,000 to 4,000 ppm drench	Drench volumes vary with pot size. See label for recommended volumes. Woody species known to respond to Cycocel are listed above.
		Concise/Sumagic	20 to 50 ppm spray	Conduct trials on a small number of plants, adjusting the rate as needed for desired final plant height and length of height control.
			0.5 to 2 ppm drench	Drench volumes and mg a.i. vary with pot size.
FREESIA	To control plant growth	Abide/A-Rest	100 to 200 ppm corm soak	Soak corms in the solution for 1 hr. before planting. Cultivar response varies, so conduct your own trials.
		Piccolo/Piccolo 10 XC /Bonzi/Paczol/Downsize	0.22 to 0.48 mg a.i. (2 to 4 ppm) drench for a 6-in. pot; apply 4 fl. oz./6-in. pot	To increase lateral branching.
		Piccolo/Piccolo 10 XC /Bonzi/Paczol	50 to 200 ppm corm soak	Soak corms in the solution for 1 hr. before planting. Cultivar response varies, so conduct your own trials.
FUCHSIA	To control plant growth	Abide/A-Rest	25 to 75 ppm spray	May also increase flowering.
		Dazide/B-Nine	1,250 to 2,500 ppm spray	
		Piccolo/Piccolo 10 XC /Bonzi/Paczol	5 to 10 ppm spray	Make applications prior to visible bud to avoid delay.
		Concise/Sumagic	2 to 5 ppm spray	Make applications prior to visible bud to avoid delay.
	To increase lateral branching	Augeo	781 to 2,343 ppm spray	
		Florel/Collate	500 ppm spray	Florel and Collate applications will provide some growth retardant effects and delay flowering. Read the label for restrictions on timing of applications.
	To promote stem elongation for topiary	Florgib/ProGibb T&O	200 to 400 ppm spray	For use on upright growing cultivars used for topiary. Weekly sprays can be used, maximum 3 applications.
GARDENIA	To control plant growth	Abide/A-Rest	50 ppm spray	
			0.25 mg a.i. (2 ppm) drench for a 6-in. pot; apply 4 fl. oz./ 6-in. pot	Drench volumes and mg a.i. vary with pot size.
		Dazide/B-Nine	5,000 ppm spray	Spray when plants are at two-thirds final market size.
		Piccolo/Piccolo 10 XC /Bonzi/Paczol	12 ppm drench	Flower delay possible. Apply prior to floral initiation (short days) or 6 weeks after pinching.
		Topflor	100 to 200 ppm spray	Apply prior to floral initiation (short days) or 6 weeks after pinching.
	To increase lateral branching	Augeo	2,343 to 4,687 ppm spray	
GAURA	To control plant growth	Dazide/B-Nine	3,000 to 4,000 ppm spray	
		Piccolo/Piccolo 10 XC /Bonzi/Paczol	30 to 50 ppm spray	
		Piccolo/Piccolo 10 XC /Bonzi/Paczol/Downsize	3.54 mg a.i. (30 ppm) drench for a 6-in. pot; apply 4 fl. oz./ 6-in. pot	
		Concise/Sumagic	10 to 30 ppm spray	
GAZANIA	To control plant growth	Citadel/Chlormequat E-Pro /Cyclocel	1,500 ppm spray	Make applications prior to visible bud to avoid delay.
		Dazide/B-Nine	2,500 ppm spray	Make applications prior to visible bud to avoid delay.
GERANIUM	To control plant growth	Abide/A-Rest	26 to 66 ppm spray	
		Piccolo/Bonzi/Paczol	5 to 30 ppm spray	Apply to zonal geraniums when new growth is 1.5 to 2 in. long. Apply to seed geraniums approximately 2 to 4 weeks after transplanting.
		Concise	3 to 8 ppm spray	Use lower rates for less vigorous plants and higher rates for more vigorous growing plants. Flower delay on some cultivars can occur when using rates >6 ppm.

PLANT GROWTH REGULATORS

Table 8-5. Growth Regulators for Floricultural Crops in Greenhouses

Crop	Purpose	Chemical	Rate	Precautions and Remarks
GERANIUM (continued)	To control plant growth	Citadel/Chlormequat E-Pro /Cycocel	800 to 1,500 ppm spray	Make first application 2 to 4 weeks after planting plugs or rooted cuttings (after stems have started elongating). Multiple applications may be needed.
		Piccolo 10 XC	10 to 30 ppm spray	See Piccolo remarks for GERANIUM. Early applications may require lower rates to avoid overdosing. Piccolo 10 XC will reduce late stretch when applied as the flower stems begins to elongate.
		Concise/Sumagic	3 to 6 ppm spray for cutting geraniums and 2 to 4 ppm spray for seed geraniums	
		Topflor	15 to 25 ppm spray	Apply to zonal geraniums when new growth is 1.5 to 2 in. long.
	To promote earlier flowering in seed geraniums	Citadel/Chlormequat E-Pro /Cycocel	1,500 ppm spray	Make two applications at 35 and 42 days after seeding. Treated plants should flower earlier and be more compact and more well-branched than untreated plants.
		Florgib/ProGibb	5 to 15 ppm spray	Make a single foliar application when first flower bud set is noted. Spray the entire plant until runoff. See label for precautions.
	To increase flower number and size in cutting geranium	Florgib/ProGibb T&O	1 to 5 ppm spray	Make a single foliar application when first flower bud set is noted. Spray the entire plant until runoff. See label for precautions.
	To increase lateral branching	Florel/Collate	300 to 500 ppm spray	Labeled for zonal and ivy geraniums. Use the lower concentration for ivy geraniums. Florel and Collate will also provide some growth retardant effect and delay flowering. Read the label for restrictions on timing of applications.
GERANIUM, IVY	To control plant growth	Citadel/Chlormequat E-Pro /Cyclocel	750 to 1,500 ppm spray	
	To increase branching	Augeo	1,562 ppm spray	Labeled for ivy geraniums only.
		Florel/Collate	200 to 300 ppm spray	
GERANIUM, Seed	To promote earlier flowering	Citadel	1,500 ppm spray	See label. Make two spray applications at 35 and 42 days after seeding. Plants flower quicker, are compact and have increased lateral breaks.
	To control plant growth	Concise	2 to 4 ppm spray	Apply when plant height is approximately 4 in. tall.
GERBERA DAISY	To control plant growth	Abide/A-Rest	25 to 132 ppm spray	Do not apply when flower stems are visible.
			0.25 to 0.5 mg a.i. drench for a 6-in. pot; apply 4 fl. oz./6-in. pot	Drench volumes and mg a.i. vary with pot size. Do not apply when flower stems are visible.
		Dazide/B-Nine	1,200 to 5,000 ppm spray	Do not apply when flower stems are visible. Apply lower rate at 10 to 14 interval if needed.
GLADIOLUS	To control plant growth	Abide/A-Rest	1.5 mg drench per 0.5 gal. pot	For container grown plants.
		Piccolo/Piccolo 10 XC /Bonzi/Paczol	2.5 to 5.0 mg drench per 0.5 gal. pot	For container grown plants.
GLOXINIA (Sinningia speciosa)	To control peduncle length	Dazide/B-Nine	1,250 ppm spray	PGRs may not be required on compact cultivars. Make first application when the leaves reach the side of the pot. A repeat application can be made 7 to 10 days later if needed. Flower streaking can develop if PGR applied to when buds showing color. Phytotoxicity may occur at rates >1,250 ppm.
		Piccolo/Piccolo 10 XC /Bonzi/Paczol	30 ppm spray	Can be applied when buds grow above the foliage.
			4 to 8 ppm drenches	For elongation control late in the season (10 weeks after transplant).
GOMPHRENA	To control plant growth	Dazide/B-Nine	2,500 to 5,000 ppm spray	
		Citadel/Chlormequat E-Pro /Cycocel	800 to 1,5,00 ppm spray	
GOODENIA	To control plant growth	None	None	Plants grown with good light and optimal growing conditions generally do not need PGRs.
GRAPE IVY	To increase lateral branching	Augeo	781 to 1,562 ppm spray	
GROUNDCHERRY	To control plant growth	Sumagic	2 to 10 ppm spray	See precautions listed with EGGPLANT.
GYPSOPHILA	To accelerate plant growth, increase stem and flower number and increase flower uniformity	Florgib/ProGibb T&O	150 to 500 ppm spray	Make 3 to 4 foliar applications after 4 weeks of new growth has occurred after pinching. Use 2-week intervals between sprays. See label for precautions.
HELENIUM AMARUM	To control plant growth	Dazide/B-Nine	5,000 ppm spray	Apply after plant established (2 weeks after transplant).
HELICHRYSUM PETIOLARE/ H. ITALICUM (Licorice plant)	To control plant growth	Piccolo/Piccolo 10 XC /Bonzi/Paczol	1 ppm drench	Plants grown with good light and optimal growing conditions generally do not need PGRs.
	To increase lateral branching	Florel/Collate	300 to 500 ppm spray	Make first application after 2 weeks. Repeat in 2 weeks if needed (with larger pots).

PLANT GROWTH REGULATORS

Table 8-5. Growth Regulators for Floricultural Crops in Greenhouses

Crop	Purpose	Chemical	Rate	Precautions and Remarks
HELICONIA	To control plant growth	Piccolo/Piccolo 10 XC /Bonzi/Paczol	15 to 30 ppm spray	Apply when axillary shoots are 4 to 6-in. high after removal of primary shoot (2 to 3 months after planting). Cultivar variation possible, so conduct your own trials to determine optimal rates.
			0.375 mg a.i. drench / 6-in. pot	Apply when axillary shoots are 4 to 6-in. high after removal of primary shoot (2 to 3 months after planting). Cultivar variation possible, so conduct your own trials to determine optimal rates.
HELIOTROPIUM ARBORESCENS	To control plant growth	Citadel/Chlormequat E-Pro /Cycocel	500 ppm spray	Rate for compact genetics needing slight growth control.
		Citadel+Dazide/Cycocel+B-Nine	750 to 1,000 ppm + 1,500 to 3,000 ppm spray	Rate for compact genetics needing slight growth control.
HIBISCUS MOSCHEUTOS	To control plant growth	Citadel/Cycocel	1000 ppm foliar spray	Multiple applications may be required.
		Concise/Sumagic	15 ppm foliar spray	
HIBISCUS ROSA-SINENSIS	To control plant growth	Dazide/B-Nine	2,500 to 5,000 ppm spray	
		Piccolo/Piccolo 10 XC /Bonzi/Paczol	5 to 150 ppm spray	Application should be made when laterals are 1 to 4 in. long. Single applications control lateral growth for 3 to 6 weeks.
		Concise	10 ppm spray	Apply within 7 days after pruning. Make additional applications as necessary to obtain desired results. Florida only: use a foliar spray concentration between 5 to 10 ppm and apply a uniform spray volume of 3 qts/100 sq. ft.
		Citadel/Chlormequat E-Pro /Cycocel	200 to 600 ppm spray	Multiple applications starting prior to first pinch are recommended. See label for additional precautions. Avoid applications after flower buds are visible.
		Concise/Sumagic	0.025 to 0.2 mg a.i. drench per pot	
HOLLY	To control plant growth	Abide/A-Rest	50 ppm spray	
			0.25 mg a.i. (2 ppm) drench for a 6-in. pot; apply 4 fl. oz./6-in. pot	Drench volumes and mg a.i. vary with pot size.
HOLLYHOCK	To control plant growth	Piccolo/Bonzi/Paczol	30 to 50 ppm spray	
		Concise/Sumagic	5 to 40 ppm spray	
HOSTA	To promote lateral growth on finished plants	Configure	1,000 to 3,000 ppm spray	Apply in a uniform spray volume. Application is most effective when plants are fully established prior to application (i.e. at least 3 to 4 weeks after potting), when there is evidence of surface root development but before flower initiation.
	To increase production of offsets for propagation	Configure	1,000 to 3,000 ppm spray	Apply in a uniform spray volume to fully established, actively growing stock plants. Repeat the application at 30-day intervals during the growing season. Offsets may be harvested at any time. Treatment effects may vary by Hosta cultivar and may respond differently to a given rate. Multiple applications at 30-day intervals using lower rates may be more effective than a single application at a higher rate. Conduct trials on a small number of plants under actual use conditions to establish the proper use rates and timings.
HYACINTH	To reduce stem topple	Florel/Collate	1,000 ppm spray	To reduce stem topple at time of full flower, apply foliar spray before florets have opened.
	To control plant growth	Piccolo/Bonzi/Paczol	100 ppm bulb soak	Ten minute soaks provided excellent results in NC State University trials. Cultivar response varied.
		Concise/Sumagic	20 to 40 ppm bulb soak	Two to ten minute preplant soaks provided excellent results in NC State University trials. Cultivar response varied.
		Topflor	0.5 to 1 mg a.i. (4.2 to 8.45 ppm) drench for a 6-in. pot	Based on NC State University trials. Adjust rates for other locations.
			10 to 25 ppm bulb soak	Two to ten minute preplant soaks provided excellent results in NC State University trials. Cultivar response varied.
HYBRID LILY (See Lily, Hybrid)				
HYDRANGEA	To control plant growth	Abide/A-Rest	50 ppm spray	
			0.25 mg a.i. (2 ppm) drench for a 6-in. pot; apply 4 fl. oz./ 6-in. pot	Drench volumes and mg a.i. vary with pot size.
		Dazide/B-Nine	1,250 to 7,500 ppm spray	Use lower rate in spring when 4 to 5 pairs of leaves are visible and new growth is starting to unfold, but not later than 4 weeks after initiation of forcing. Use higher rate for summer when regrowth after pinching is 1 to 2 in. long.
		Topflor	100 to 200 ppm spray	
HYPOESTES	To control plant growth	Chlormequat E-Pro	800 to 1,500 ppm spray	Initially apply after second set of leaves have developed. If needed reapply 2 weeks later.
		Citadel/Cycocel	400 to 1,500 ppm spray	Initially apply after second set of leaves have developed. If needed reapply 2 weeks later.
		Dazide/B-Nine	1,000 ppm spray	Initially apply after second set of leaves have developed. If needed reapply 2 weeks later.

PLANT GROWTH REGULATORS

Table 8-5. Growth Regulators for Floricultural Crops in Greenhouses

Crop	Purpose	Chemical	Rate	Precautions and Remarks
IMPATIENS, Seed	To control plant growth	Abide/A-Rest	10 to 44 ppm spray	
		Piccolo/Piccolo 10 XC /Bonzi/Paczol	10 to 45 ppm spray	
		Concise/Sumagic	5 to 10 ppm spray	
		Topflor	20 to 60 ppm spray	Based on NC State University trials. Adjust rates for other locations.
	To increase branching	Florel/**Collate**	100 to 300 ppm spray	Use if better branching needed.
IMPATIENS PLUGS, Seed	To control plant growth	Piccolo/Piccolo 10 XC /Bonzi/Paczol	0.5 to 10 ppm spray (0.015 to 0.32 fl. oz./gal)	Timing of application should normally begin at the 1 to 2 true leaf stage.
IMPATIENS, Vegetative	To control plant growth	Piccolo/Bonzi/Paczol	2 to 15 ppm spray	Cultivars' response to PGRs varies, so test a few plants to determine rate for optimum control.
			0.5 to 1 ppm drench	Drench volumes and mg a.i. vary with pot size. See label for recommended volumes.
		Florel/Collate	100 to 300 ppm spray	Will improve branching.
IMPATIENS, Seashell-type	To control plant growth	Piccolo/Bonzi/Paczol	5 to 8 ppm spray	Apply when plants have reached 75% of finished height. Don't apply to plants under stress. Recommendations based on Michigan trials.
IOCHROMA	To control plant growth	Dazide/B-Nine	5,000 ppm spray	
		Piccolo/Piccolo 10 XC /Bonzi/Paczol	2 ppm spray	
IPOMOEA	To control plant growth	Concise/Sumagic	10 to 25 ppm spray	Not needed if optimal scheduling is used. If needed, apply when plants have reached 75% of finished growth. Recommendations based on NC State University trials.
		Dazide/B-Nine	2,500 ppm spray	Apply as needed.
		Florel/Collate	500 to 1000 ppm spray	Will improve branching and control growth.
		Piccolo/Piccolo 10 XC /Bonzi/Paczol	8 ppm drench	Applied to plugs prior to transplanting
IRESINE HYBRID	To control plant growth	Citadel+Dazide/Cycocel+B-Nine	1,000 to 1,500 ppm + 2,500 to 4,000 ppm spray	
		Piccolo/Piccolo 10 XC /Bonzi/Paczol	5 to 10 ppm spray	
		Piccolo/Piccolo10XC/Bonzi/Paczol/ Downsize	1 to 3 ppm drench	
JACOBINIA (Pink)	To control plant growth	Piccolo/Piccolo 10 XC /Bonzi/Paczol	5 to 10 ppm spray	
		Piccolo/Piccolo 10 XC /Bonzi/Paczol/Downsize	0.06 to 0.12 mg a.i. (0.5 to 1 ppm) drench for a 6-in. pot; apply 4 fl. oz./6-in. pot	
JERUSALEM CHERRY (*Solanum pseudocapsicum*)	To control plant growth	Citadel/Chlormequat E-Pro	800 to 1,500 ppm spray	
		Citadel/Cycocel	400 to 1,500 ppm spray	
	To promote stem elongation for topiary	Florgib/ProGibb T&O	250 ppm spray	For plants grown in 6-in. pots and with 4 to 6-in. of growth, apply 2 foliar sprays 10 days apart to promote stem elongation for topiary plants. Stake plants to support stem.
KALANCHOE	To control plant growth	Abide/A-Rest	50 ppm spray	Apply with axillary growth begins and repeat 20 to 30 days after short days begin. Trial to determine optimal rates and timing for your location.
		Dazide/B-Nine	2,500 to 5,000 ppm spray	Rates and timing vary with the season and cultivar. Applications typically begin 2 weeks after pinching. Apply sprays every 7 days in the summer, 10 to 15 days in the spring and fall, and 14 to 21 days in the winter. Trial to determine optimal rates and timing for your location.
		Piccolo/Piccolo 10 XC /Bonzi/Paczol	2 to 4 ppm spray	Trial to determine optimal rates and timing for your location.
	To increase lateral branching	Augeo	1,042 to 2,343 ppm spray	
	To control peduncle length	Dazide/B-Nine	1,200 to 5,000 ppm spray	Phytotoxicity possible if B-Nine/Dazide accumulates in cupped areas of certain cupped-leafed varieties.
LACHENALIA sp.	To control plant growth	Concise/Sumagic	20 ppm corm soaks	Rates based on trials at Cornell University.
		Piccolo/Piccolo 10 XC /Bonzi/Paczol	100 to 200 ppm spray	Rates based on trials at Cornell University.
			1 to 2 mg a.i./pot drench	Rates based on trials at Cornell University.
LAMIUM	To control plant growth	Concise/Sumagic	5 ppm spray	
		Piccolo/Piccolo 10 XC /Bonzi/Paczol	30 ppm spray	
			1 ppm drench	
	To increase lateral branching	Collate/Florel	500 ppm spray	Improves branching and produces compact growth.

PLANT GROWTH REGULATORS

401

Table 8-5. Growth Regulators for Floricultural Crops in Greenhouses

Crop	Purpose	Chemical	Rate	Precautions and Remarks
LANTANA	To control plant growth	Citadel + Dazide/Cycocel + B-Nine	2,500 to 5,000 ppm + 1,000 to 1,500 ppm Cycocel applied as a tank-mix spray	Cultivar response varies.
		Piccolo/Bonzi/Paczol	20 to 40 ppm spray	
		Concise/Sumagic	10 to 20 ppm spray	
	To increase lateral branching	Augeo	781 to 1,562 ppm spray	
		Florel/Collate	500 ppm spray	Florel and Collate applications will provide some growth retardant effects and delay flowering. Read the label for restrictions on timing of applications.
LAURENTIA AXILLARIS	To control plant growth	Abide/A-Rest	2 to 4 ppm spray	
		Dazide/B-Nine	2,500 ppm spray	
		Piccolo/Piccolo 10 XC /Bonzi/Paczol	1 to 2 ppm drench	
LIATRIS	To control plant growth	Abide/A-Rest	25 to 132 ppm spray	
			0.25 to 0.5 mg a.i. (2 to 4 ppm) drench for a 6-in. pot; apply 4 fl. oz./6-in. pot)	Drench volumes and mg a.i. vary with pot size.
		Dazide/B-Nine	2,500 to 5,000 ppm spray	
LILY, Easter	To control plant growth	Abide/A-Rest	30 to 132 ppm spray. Use 50 ppm spray as a base rate and adjust as needed.	Apply when newly developing shoots are 2 to 3 in. long; a second application when shoots average 6 in. long may be needed.
			0.25 to 0.5 mg a.i. (2 to 4 ppm) drench for a 6-in. pot; apply 4 fl. oz./6-in. pot	Single drench should be applied when shoots average 3 to 5 in. long. Drench volumes and mg a.i. vary with pot size.
		Concise	3 to 15 ppm spray	Apply when shoots average 3 in. tall. It is best to make only one foliar application per crop.
			0.03 to 0.06 mg a.i. (0.23 to 0.5 ppm) drench for a 6-in. pot; apply 4 fl. oz./6-in. pot	Apply when shoots average 3 in. tall. Use lower rates on cultivars such as Nellie White and higher rates for Ace. For Florida only: use a solution concentration of between 0.05 to 0.12 mg a.i. (0.4 to 1.0 ppm) drench for a 6-in. pot (0.11 to 0.26 fl. oz./gal of drench solution, apply 4 fl. oz./6-in. pot).
		Concise/Sumagic	3 to 15 ppm spray	Apply when shoots average 3 in. tall.
			0.03 to 0.06 mg a.i. (0.25 to 0.5 ppm) drench for a 6-in. pot; apply 4 fl. oz./6-in. pot	Drench volumes and mg a.i. vary with pot size.
	To prevent leaf yellowing	Fresco/Fascination	5 to 10 ppm spray	Apply early season (7 to 10 days PRIOR to visible bud stage) and mid-season (7 to 10 days AFTER visible bud stage). Apply spray only to lower leaves to minimize stem elongation. See label.
	To prevent leaf yellowing and prolong flowering	Fresco/Fascination	100 ppm spray	Apply late season (when first bud reaches at least 3 in. in length) and no more than 14 days prior to placement in a cooler or shipping. Apply to foliar and flower buds. See label.
LILY, Hybrid	To control plant growth	Piccolo/Piccolo 10 XC /Bonzi/Paczol	200 to 500 ppm spray	Make first spray application when plants are 2 to 4 in. tall.
		Piccolo/Bonzi/Paczol	5 to 30 ppm bulb soak	Soak bulbs in the solution for 15 min. prior to planting.
		Piccolo/Piccolo 10 XC /Bonzi/Paczol/Downsize	0.25 to 0.5 mg a.i. (4 to 30 ppm) drench for a 6-in. pot; apply 4 fl. oz./6-in. pot	Single drench should be applied when shoots average 3 to 5 in. long. Drench volumes and mg a.i. vary with pot size and cultivar.
		Concise	2.5 to 20 ppm spray	Conduct a trial to determine optimal rates for each cultivar and adjust the rate as needed. Spray when shoots average 3 in. tall. If a second application is needed or a split application is made, it should be applied when the shoots average 6 in. tall. Usually two applications of foliar sprays at a lower rate are more effective than one application at a higher rate. Avoid applications after visible bud stage.
			1 to 3 ppm drench	Drench volume varies with pot size. Applications should be made when newly emerged shoots are 1 to 2 in. tall.
			1 to 10 ppm bulb soak	Treatment soak time should range from 1 to 5 minutes. Soak time will vary depending on bulb size, cultivar, and final desired height. Lower rates may require longer soak times (5 to 10 minutes) than higher rates (1 minute).
		Concise/Sumagic	3 to 15 ppm spray	Apply when shoots average 3 in. tall.
			0.03 to 0.06 mg a.i. (0.25 to 0.5 ppm) drench for a 6-in. pot; apply 4 fl. oz./6-in. pot	Drench volumes and mg a.i. vary with pot size.
		Topflor	0.25 to 0.5 mg a.i. (2.1 to 4.2 ppm) drench for a 6-in. pot	Based on NC State University trials. Adjust rates for other locations and plant response.
	To prevent leaf yellowing	Fresco/Fascination	5 to 10 ppm spray	Apply early season (7 to 10 days PRIOR to visible bud stage) and mid-season (7 to 10 days AFTER visible bud stage). Apply spray only to lower leaves to minimize stem elongation. See label.

PLANT GROWTH REGULATORS

Table 8-5. Growth Regulators for Floricultural Crops in Greenhouses

Crop	Purpose	Chemical	Rate	Precautions and Remarks
LILY, Hybrid (continued)	To prevent leaf yellowing and prolong flowering	Fresco/Fascination	100 ppm spray	Apply late season (when first bud reaches at least 3 in. in length) and no more than 14 days prior to placement in a cooler or shipping. Apply to foliar and flower buds. See label.
LILY, Oriental	To control plant growth	Piccolo/Bonzi/Paczol	100 to 200 ppm bulb soak	Ten minute preplant soaks provided excellent results in NC State University trials. Cultivar response varied.
		Concise	2.5 to 10 ppm spray	See Concise label comments for Hybrid lilies.
			1 to 10 ppm bulb soak	See Concise label comments for Hybrid lilies.
		Concise/Sumagic	1 to 10 ppm bulb soak	See Concise label comments for Hybrid lilies. Ten minute preplant soaks of 5 ppm provided excellent results in NC State University trials. Cultivar response varied.
		Piccolo 10 XC	200 to 500 ppm spray	Begin spray applications when plants reach a height of 2 to 4 inches.
			4 to 30 ppm drench	Drench volume varies with pot size. Begin drench applications when plants reach a height of 1 to 2 inches.
		Topflor	0.5 mg a.i. drench (4.2 ppm); apply 4 fl. oz./6-in. pot	Optimal rate based on NC State University trials. Adjust rate for plant vigor. Drench volumes and mg a.i. vary with pot size.
			25 ppm bulb soak	Ten minute preplant soaks provided excellent results in NC State University trials. Cultivar response varied.
	To prevent leaf yellowing	Fresco/Fascination	100 ppm spray	Apply early season (7 to 10 days PRIOR or AFTER visible bud stage). Apply spray only to lower leaves to minimize stem elongation. See label.
	To prevent leaf yellowing and prolong flowering	Fresco/Fascination	100 ppm spray	Apply late season (no more than 14 days prior to placement in a cooler or shipping). Apply to foliar and flower buds. See label.
LINARIA HYBRIDA (Baby snapdragon)	To control plant growth	Dazide + Citadel/B-Nine + Cycocel	2,500 ppm Dazide/B-Nine + 300 to 500 ppm Citadel/Cycocel applied as a tank-mix spray	Controlled plant growth, but did not strengthen stems as well as paclobutrazol sprays.
		Piccolo/Piccolo 10 XC /Bonzi/Paczol	10 to 30 ppm spray	Use 10 ppm 1 week after transplant. Make a second application of 20 to 30 ppm once the secondary shoots are 2 inches long. Stengthened stems and improved flower coloration.
LINER DIPS	To control plant growth	Piccolo	0.5 to 8 ppm preplant liner dip	See label: for detailed recommendations for chemical application techniques, adjusting rates for northern or southern locations, and the specific rates for achieving the desired level of activity.
LIPSTICK VINE	To increase lateral branching	Augeo	521 to 1,042 ppm spray	
LISIANTHUS (Eustoma)	To control plant growth	Abide/A-Rest	0.5 mg a.i. drench	Cultivar response varies.
		Concise/Sumagic	5 to 10 ppm spray	Cultivar response varies.
		Dazide/B-Nine	2,500 to 5,000 ppm spray	Cultivar response varies.
		Piccolo/Piccolo 10 XC /Bonzi/Paczol	4 to 16 ppm drench	Cultivar response varies.
LOBELIA	To control plant growth	Dazide/B-Nine	1,500 to 2,500 ppm spray	
		Concise/Sumagic	1 to 10 ppm spray	
		Piccolo/Piccolo 10 XC /Bonzi/Paczol	4 ppm spray	
			1 ppm drench	Can be used 3 to 5 weeks before sale to control stretch.
LOBULARIA	To control plant growth	Piccolo	4 to 8 ppm liner root soak	Irrigation of the liners occured within 24 hours prior to application, which results in a moderately dry substrate (the stage the plants would be watered but not wilted). Soak for a minimum of 30 to 60 seconds. Transplant after 3-hour waiting period. Rate based on North Carolina State University trials with Snow Princess.
			75 to 100 ppm spray	Sprays less effective than preplant liner soaks or substrate drenches. Rate based on North Carolina State University trials with Snow Princess.
			2 to 4 ppm drench	Drench volume varies with pot size. Rate based on North Carolina State University trials with Snow Princess.
		Concise	0.5 to 1 ppm liner root soak	Irrigation of the liners occured within 24 hours prior to application, which results in a moderately dry substrate (the stage the plants would be watered but not wilted). Soak for a minimum of 30 to 60 seconds. Transplant after 3-hour waiting period. Rate based on North Carolina State University trials with Snow Princess.
			20 to 25 ppm spray	Sprays less effective than preplant liner soaks or substrate drenches. Rate based on North Carolina State University trials with Snow Princess.
			1 to 2 ppm drench	Drench volume varies with pot size. Rate based on North Carolina State University trials with Snow Princess.
		Topflor	10 ppm spray	

PLANT GROWTH REGULATORS

Table 8-5. Growth Regulators for Floricultural Crops in Greenhouses

Crop	Purpose	Chemical	Rate	Precautions and Remarks
MANDEVILLA SANDERI (Dipladenia)	To control plant growth	None	None	Plants grown with good light and optimal growing conditions generally do not need PGRs.
MARIGOLD	To control plant growth	Abide/A-Rest	13 to 33 ppm spray	
		Dazide/B-Nine	2,500 to 5,000 ppm spray	
		Piccolo/Piccolo 10 XC /Bonzi/Paczol	15 to 60 ppm spray	Use 15 to 30 ppm for French type and 30 to 60 ppm for African type (apply at an early stage of plant growth for African type with good stem coverage, especially for vigorous varieties).
		Citadel/Chlormequat E-Pro /Cycocel	800 to 1,500 ppm spray	
		Concise/Sumagic	10 to 20 ppm spray	
		Topflor	20 to 60 ppm spray	Based on NC State University trials. Adjust rates for other locations.
MARIGOLD, Plugs	To control plant growth	Piccolo/Piccolo 10 XC /Bonzi/Paczol	5 to 20 ppm spray	Timing of application should normally begin at the 1 to 2 true leaf stage. Use 5 to 10 ppm for French types and 10 to 20 ppm for African types.
MATTHIOLA, Bedding Plant (Stock)	To control plant growth	Dazide + Citadel/B-Nine + Cycocel	800 to 5,000 ppm Dazide/B-Nine + 1,000 to 1,500 ppm Citadel/Cycocel applied as a tank-mix spray	
MATTHIOLA, Cut (Stock)	To promote growth and stem elongation	Florgib/ProGibb T&O	50 to 100 ppm spray	Apply when plants are 4 to 8 in. tall. Apply at 2- to 3- week intervals. See label for precautions.
MELAMPODIUM	To control plant growth	Dazide/B-Nine	2,500 ppm spray	Use when plants reach 75% of marketable size to tone.
MIMULUS	To control plant growth	Dazide/B-Nine	2,500 ppm spray	Use if needed. Delay in flowering possible with multiple applications.
MONARDA	To control plant growth	Piccolo/Piccolo 10 XC /Bonzi/Paczol	60 to 160 ppm spray	
		Piccolo/Piccolo 10 XC /Bonzi/Paczol/Downsize	>0.48 mg a.i. (>4 ppm) drench for a 6-in. pot; apply 4 fl. oz./6-in. pot	
		Concise/Sumagic	15 to 30 ppm spray	
MONSTERA	To control plant growth	Abide/A-Rest	25 to 132 ppm spray	
			0.25 to 0.5 mg a.i. (2 to 4 ppm) drench for a 6-in. pot; apply 4 fl. oz./6-in. pot	Drench volumes and mg a.i. vary with pot size.
MONTBRETIA	To control plant growth	Piccolo/Bonzi/Paczol	20 to 30 ppm corm soak	Soak corms in the solution for 15 min. prior to planting.
NARSISSUS	To control plant growth	Florel/Collate	500 to 2,000 ppm spray	For types requiring a vernalization period (*Narcissus* hybrids), apply when new leaves reach 3 to 4 inches of height. For paperwhite narcissus (*Narcissus tazetta*), apply 2,000 ppm when the new leaves are 3 to 4 inches tall. Cultivar response varies, so conduct your own trial to determine suitable concentrations. Results based on Cornell University trials.
NASTURTIUM	To control plant growth	Citadel/Chlormequat E-Pro /Cycocel	800 to 1,500 ppm spray	Use only on non-food plants.
NEMESIA	To control plant growth	Dazide/B-Nine	2,500 to 5,000 ppm spray	Use on compact varieties to tone and hold crop.
		Piccolo/Bonzi/Paczol	10 to 20 ppm spray	Based on NC State University trials.
		Florel/Collate	250 to 500 ppm spray	Make final application 4 to 6 weeks before sale.
		Concise/Sumagic	3 to 30 ppm spray	In NC State University trials, 5 ppm worked well on Vanilla Sachet.
		Topflor	2.5 to 5 ppm spray	Recommendation based on NC State University trials with Vanilla Sachet.
NEPTHYTIS, Green and Green Gold	To control plant growth	Abide/A-Rest	25 to 132 ppm spray	
			0.25 to 0.5 mg a.i. (2 to 4 ppm) drench for a 6-in. pot; apply 4 fl. oz./6-in. pot	Drench volumes and mg a.i. vary with pot size.
NEW GUINEA IMPATIENS	To control plant growth	Piccolo/Piccolo 10 XC /Bonzi/Paczol	0.25 to 15 ppm spray	Apply 2 to 4 weeks after transplanting. Cultivars' response to PGRs varies greatly. Test a few plants to determine rate for optimal control.
		Piccolo/Bonzi/Paczol	0.25 to 2 ppm drench	Drench volumes vary with pot size. See label for recommendations. Cultivars response to PGRs varies greatly. Test a few plants to determine rate for optimal control.
		Florel/Collate	100 to 300 ppm spray	To increase lateral branching and reduce premature flowering, don't apply within 8 weeks of desired flower date.
		Topflor	5 to 15 ppm spray	Apply 2 to 4 weeks after transplanting. Cultivars' response to PGRs varies greatly. Test a few plants to determine rate for optimal control.

PLANT GROWTH REGULATORS

Table 8-5. Growth Regulators for Floricultural Crops in Greenhouses

Crop	Purpose	Chemical	Rate	Precautions and Remarks
NEW GUINEA IMPATIENS, Plugs	To control plant growth	Piccolo 10 XC	0.25 to 5 ppm spray	Timing of application should normally begin at the 1 to 2 true leaf stage.
NICOTIANA	To control plant growth	Dazide/B-Nine	2,500 to 5,000 ppm spray	Higher initial rates can be used after the plant becomes established. Use lower rate with multiple applications at 3 week interval.
NOLANA PARADOXA	To control plant growth	Florel/Collate	500 ppm spray	To keep plants more compact. Based on Texas A&M University trials.
OENOTHERA	To control plant growth	Concise/Sumagic	5 to 10 ppm spray	Apply if needed.
ORNAMENTAL CABBAGE and KALE (Non-food)	To control plant growth	Dazide/B-Nine	2,500 to 5,000 ppm spray	Use the higher rates for more vigorous types/cultivars. Multiple applications may be needed. Recommendation based on North Carolina conditions.
		Concise/Sumagic	2.5 to 8 ppm spray	Use higher rates for more vigorous cultivars. Cultivar response can vary. Recommendation based on North Carolina conditions.
ORNAMENTAL PEPPERS (*Capsicum*) (Non-food)	To control plant growth	Piccolo/Bonzi/Paczol	20 ppm foliar spray	Recommendation based on North Carolina conditions for a moderately vigorous cultivar.
		Concise/Sumagic	5 to 15 ppm spray	
ORNAMENTAL VEGETABLES (Non-food)	To control plant growth	Dazide/B-Nine	2,500 to 5,000 ppm spray	Use the higher rates for more vigorous types/cultivars like kale Red Bor. Multiple applications may be needed. Recommendation based on North Carolina conditions.
		Concise/Sumagic	10 to 25 ppm spray	Use higher rates for more vigorous cultivars. Recommendation based on North Carolina conditions.
ORNITHOGALUM	To increase stem length	Florgib/ProGibb T&O	100 ppm dip	Soak the bulbs for 20 minutes prior to potting.
OSTEOSPERMUM	To control plant growth	Citadel/Cycocel	750 to 1,500 ppm spray	Two applications may be required. Two applications of 1,500 ppm (with the first applied at the start and the second at the end of the vernalization period) provided excellent results in NC State University trials.
			1,500 to 3,000 ppm drench	Drench volumes vary with pot size. See label for recommended volumes.
		Concise/Sumagic	8 ppm spray	Recommendation based on European trials on a cultivar with prostrate growth. Rates less than 24 ppm were not effective in NC State University trials.
			0.25 to 2 ppm drench; apply 3 fl. oz./5-in pot	One application of 1 to 2 ppm (at the start of vernalization) or two applications of 1 ppm (at the start of vernalization) and 0.5 ppm (at the end of the vernalization period) provided excellent results in NC State University trials for 4.5-in. production.
		Dazide/B-Nine	2,500 to 4,000 ppm spray	Can be applied 3 or 4 times (weekly) after pinch.
		Dazide + Citadel/B-Nine + Cycocel	1,500 to 3,000 ppm Dazide/B-Nine + 1,000 to 1,500 ppm Citadel/Cycocel applied as a tank-mix spray	Multiple sprays required. Stop applications after visible bud to avoid flower delay and smaller flowers. Not effective in NC State University trials.
		Piccolo	4 to 8 ppm liner root soak	Irrigation of the liners occured within 24 hours prior to application, which results in a moderately dry substrate (the stage the plants would be watered but not wilted). Soak for a minimum of 30 to 60 seconds. Transplant after 3-hour waiting period. Rate based on Michigan State University trials.
		Piccolo/Bonzi/Paczol	27 to 54 ppm drench (8 to 16 mg a.i.) during production	Drench volumes vary with pot size. See label for recommended volumes. (based on NC State University trials)
			2 to 3 ppm drench (0.236 to 0.35 mg a.i.) for holding plants	
		Piccolo/Piccolo 10 XC /Bonzi/Paczol	15 to 30 ppm spray	
		Topflor	20 to 60 ppm spray	
			1 to 2 ppm drench; apply 3 fl. oz./5-in pot	One application of 1 to 2 ppm (at the start of vernalization) or two applications of 1 ppm (at the start of vernalization) and 0.5 ppm (at the end of the vernalization period) provided excellent results in NC State University trials for 4.5-in. production.
OTACANTHUS	To control plant growth	Dazide/B-Nine	2,500 ppm spray	Make first application when new growth appears after pinching. A second application may be used if a second pinch is planned.
OTOMERIA	To control plant growth	Dazide/B-Nine	1,700 ppm spray	Apply 1 to 3 times if needed to tone the plant.
OXALLIS	To control plant growth	Abide/A-Rest	33 ppm spray	To limit petiole stretch.
		Concise/Sumagic	0.1 mg a.i. / 4.5-in. pot drench	
		Piccolo/Piccolo 10 XC /Bonzi/Paczol	1 to 4 ppm sprays	Rates for *O. regnellii*.
			1 to 10 ppm preplant dip	Dip for 5 minutes. Rates for *O. regnellii*.

PLANT GROWTH REGULATORS

Table 8-5. Growth Regulators for Floricultural Crops in Greenhouses

Crop	Purpose	Chemical	Rate	Precautions and Remarks
PANSY	To control plant growth	Abide/A-Rest	3 to 15 ppm spray	
		Piccolo/Piccolo 10 XC /Bonzi/Paczol	5 to 15 ppm spray	Apply when plants are 2 in. in diameter. Use higher rates for higher temperatures and more vigorous cultivars. Late applications may delay flowering.
		Concise/Sumagic	1 to 6 ppm spray	Apply when plants are 3 to 4 in. tall. Use higher rates for higher temperatures and more vigorous cultivars. Late applications may delay flowering.
		Topflor	2.5 to 7.5 ppm spray	Based on NC State University trials. Adjust rates for other locations. Pansies are very responsive to Topflor, so start trials with lower rates.
PANSY PLUGS	To control plant growth	Piccolo/Piccolo 10 XC /Bonzi/Paczol	1 to 5 ppm spray	Timing of application should normally begin at the 1 to 2 true leaf stage. Pansies are sensitive as plugs, so determine optimal rates.
PENNISETUM GLAUCUM	To control plant growth	Collate/Florel	500 ppm spray	Apply first appliation 4 weeks after sowing or 1 week after transplant. If needed a second application can be made 10 to 14 days later. Promotes side shoot production more than providing height control.
		Piccolo/Piccolo 10 XC /Bonzi/Paczol	6 to 8 ppm drench	For direct sown seed, apply palcobutrazol 4 weeks after sowing. A second application possible 10 days later if needed.
			3 to 5 ppm drench	For plugs, apply 1 week after transplant.
***PENNISETUM SETACEUM* 'Rubrum'**	To control plant growth	Concise/Sumagic	5 ppm spray	First application can be made 21 days after transplanting. Repeat If need 14 days later.
PENSTEMON HARTWEGII	To control plant growth	Citadel+Dazide/Cycocel+B-Nine	1,000 ppm + 2,500 ppm spray	Rates for moderately vigorous cultivars. Up to 2 sprays may be needed.
		Concise/Sumagic	5 to 10 ppm spray	Rates for moderately vigorous cultivars. Up to 2 sprays may be needed.
		Dazide/B-Nine	2,500 ppm spray	Rates for moderately vigorous cultivars. Up to 2 sprays may be needed.
		Florel/Collate	Spray	Not recommended because of flower delay.
PENTAS	To control plant growth	Abide/A-Rest	2 to 4 ppm spray	
		Citadel/Cycocel	1,000 to 1,500 ppm spray	
		Dazide/B-Nine	2,500 to 5,000 ppm spray	
		Piccolo/Piccolo 10 XC /Bonzi/Paczol	2 to 3 ppm spray	
PEPINO	To control plant growth	Sumagic	2 to 10 ppm spray	See precautions listed with EGGPLANT.
PEPPER	To control plant growth	Sumagic	2 to 10 ppm spray	See precautions listed with EGGPLANT.
PERENNIALS (Not specifically listed in this table)	To induce lateral or basal branching	Configure	50 to 500 ppm spray	The supplemental label allows legal use on greenhouse grown plants not specifically listed on the original label. See label for trialing suggestions and precautions.
PERICALLIS (Cineraria)	To control plant growth	Dazide/B-Nine	2,000 ppm spray	Apply every 14 days if needed.
PERILLA	To control plant growth	Concise/Sumagic	3 to 5 ppm spray	Apply if needed.
		Dazide/B-Nine	2,000 to 4,000 ppm spray	Apply 1 to 3 times as needed.
		Dazide + Citadel/B-Nine + Cycocel	2,500 to 4,000 ppm + 1,000 to 1,500 ppm Citadel/Cycocel applied as a tank-mix spray	
		Piccolo/Bonzi/Paczol	10 to 20 ppm spray	
PETUNIA, Seed	To control plant growth	Abide/A-Rest	10 to 26 ppm spray	
		Dazide/B-Nine	2,500 to 5,000 ppm spray	
		Piccolo/Piccolo 10 XC /Bonzi/Paczol	15 to 45 ppm spray	
		Concise/Sumagic	25 to 50 ppm spray	
		Topflor	20 to 60 ppm spray	Based on NC State University trials. Adjust rates for other locations.
PETUNIA PLUGS, Seed	To control plant growth	Piccolo/Piccolo 10 XC /Bonzi/Paczol	5 to 10 ppm spray	Timing of application should normally begin at the 1 to 2 true leaf stage.
PETUNIA, Vegetative	To control plant growth	Abide/A-Rest	10 to 26 ppm spray	Multiple applications may be required.
		Dazide/B-Nine	2,500 to 5,000 ppm spray	
		Dazide/B-Nine + Bonzi/Piccolo/Paczol	2,500 ppm Dazide spray + 40 ppm Bonzi/Piccolo/Paczol applied as a tank-mix spray	Recommendation based on NC State University trials.
		Dazide/B-Nine + Topflor	2,500 ppm Dazide spray + 15 to 30 ppm Topflor applied as a tank-mix spray	Recommendation based on NC State University trials.

PLANT GROWTH REGULATORS

Table 8-5. Growth Regulators for Floricultural Crops in Greenhouses

Crop	Purpose	Chemical	Rate	Precautions and Remarks
PETUNIA, Vegetative (continued)	To control plant growth	Piccolo/Bonzi/Paczol	5 to 45 ppm spray	An application at 2 to 4 ppm can be made 1 to 2 weeks after transplanting, followed by a 20 to 30 ppm spray 2 to 3 weeks later. Cultivars' responses to PGRs vary. Test a few plants to determine rate for optimal control. Finished plants can be maintained and have prolonged shelf life when 5 to 10 ppm sprays are applied on full-grown, mature plants. Recommendations based on Michigan conditions.
		Concise/Sumagic	20 to 50 ppm spray	20 ppm worked well in NC State University trials.
		Piccolo	12 ppm liner root soak	Irrigation of the liners occured within 24 hours prior to application, which results in a moderately dry substrate (the stage the plants would be watered but not wilted). Soak for a minimum of 30 to 60 seconds. Transplant after 3-hour waiting period. Rate based on Michigan State University trials with petunia multiflora prostrate Wave Purple.
		Topflor	15 to 60 ppm spray	Recommendation based on NC State University trials.
	To increase lateral branching	Florel/Collate	300 to 500 ppm spray	
PHALAENOPSIS Orchids	To increase flower number and earlier flowering	Configure	200 to 400 ppm spray	Apply Configure 1 week after the start of forcing (cooling). Cultivar response varies. Some cultivars are sensitive to Configure and distorted flower stalks may form, so conduct your own trials to determine suitability. Recommendation based on Michigan State University trials.
	To control inflorescence length	Concise/Sumagic	100 to 200 ppm spray	Apply when the flower spike length is 1 inch (3 cm).
		Piccolo/Piccolo 10 XC /Bonzi/Paczol	250 ppm spray	Apply when the flower spike length is 1 inch (3 cm).
PHILODENDRON	To control plant growth/vine control	Abide/A-Rest	25 to 132 ppm spray	
			0.25 to 0.5 mg a.i. (2 to 4 ppm) drench for a 6-in. pot; apply 4 fl. oz./6-in. pot	Drench volumes and mg a.i. vary with pot size.
		Citadel/Cycocel	3,000 ppm spray	
		Dazide/B-Nine	2,500 to 7,500 ppm spray	
PHLOX DRUMMONDII	To control plant growth	Dazide/B-Nine	2,500 to 5,000 ppm spray	
PHLOX MACULATA, (Hybrid)	To control plant growth	Concise/Sumagic	5 to 10 ppm spray	
		Dazide/B-Nine	2,500 to 5,000 ppm spray	
		Topflor	10 to 15 ppm spray	
PILEA	To control plant growth	Abide/A-Rest	25 to 132 ppm spray	
			0.25 to 0.5 mg a.i. (2 to 4 ppm) drench for a 6-in. pot; apply 4 fl. oz./6-in. pot	Drench volumes and mg a.i. vary with pot size.
PLATYCODON	To control plant growth	Abide/A-Rest	100 ppm spray	PGRs usually not required.
		Dazide/B-Nine	1,500 to 5,000 ppm spray	PGRs usually not required. High rates have been reported to cause edge burn.
PLECTRANTHUS	To control plant growth	Dazide + Citadel/B-Nine + Cycocel	1,500 to 2,500 ppm + 750 to 1,000 ppm Citadel/Cycocel applied as a tank-mix spray	Cultivars' responses to PGRs vary. Test a few plants to determine rate for optimal control. See label.
		Piccolo/Bonzi/Paczol	5 to 20 ppm spray	Cultivars' responses to PGRs vary.
PLUMBAGO AURICULATA	To control plant growth	Collate/Florel	1,000 ppm spray	Pinching plants help improve the overall form. In additon, to further enhance secondary shoots, apply PGR 1 week before pinch.
POINSETTIA	To control plant growth	Abide/A-Rest	0.06 to 0.25 mg a.i. (2 to 4 ppm) drench for a 6-in. pot; apply 4 fl. oz./6-in. pot	Drench volume and mg a.i. vary with pot size. Start with lower rates.
		Dazide/B-Nine	2,000 to 3,000 ppm spray	Not effective in NC State University studies.
		Dazide + Citadel/B-Nine + Cycocel	800 to 2,500 ppm + 1,000 to 1,500 ppm Citadel/Cycocel applied as a tank-mix spray	Use the higher rates of this tank-mix spray on stock plants and for finishing crops in very warm regions. Outside of very warm areas, use the lower rates. Late applications can delay flowering and reduce bract size.
		Piccolo/Bonzi/Paczol	10 to 30 ppm spray	Use higher rates of 15 to 45 ppm in southern Florida. Applications to slower-growing cultivars in cool climates should begin when axillary shoots are 2 to 3 in. long. For vigorous growing cultivars in warm climates, applications should begin when axillary shoots are 1.5 to 3 in. long. See label for other precautions.
		Piccolo/Bonzi/Paczol/Downsize	0.237 to 0.473 mg a.i. (0.25 to 3 ppm) drench for a 6-in. pot; apply 4 fl. oz./6-in. pot	Drenches generally have less of an effect on bract size than sprays. Drench volume and mg a.i. vary with pot size. Start with lower rates.
		Concise	2.5 to 10 ppm spray	Apply when the lateral shoots are 1.5 to 2.5 in. tall (about 10 to 14 days after pinching). Test for cultivar sensitivity. Multiple applications of the lower label rate may elicit short days. For Florida only: use a foliar spray concentration between 10 to 15 ppm (2.5 to 3.8 fl. oz./gal) and do not apply after October 25.

Table 8-5. Growth Regulators for Floricultural Crops in Greenhouses

Crop	Purpose	Chemical	Rate	Precautions and Remarks
POINSETTIA (continued)	To control plant growth	Citadel/Chlormequat E-Pro /Cycocel	800 to 1,500 ppm spray	For natural season crops in N.C., don't apply Cycocel after mid-October to November 1. Late applications can reduce bract size and delay flowering.
			3,000 to 4,000 ppm drench	Drench volume varies with pot size. Consult the label for recommended volumes.
		Concise/Sumagic	2.5 to 10 ppm spray	
		Topflor	2.5 to 80 ppm spray	Use lower rates for less vigorous cultivars. SEE LABEL FOR ADDITIONAL RATE RECOMMENDATIONS.
			0.03 to 0.5 mg a.i. (0.25 to 4.2 ppm) drench for a 6-in. pot	
	To promote plant growth	Fascination	3 ppm spray	Use an early-season application during vegetative growth prior to the start of short days and flower initiation if promoting vegetative growth. SEE LABEL FOR ADDITIONAL PRECAUTIONS BEFORE USE.
		Fresco/Fascination	3 to 10 ppm spray	Use a late-season application to promote bract expansion. SEE LABEL FOR ADDITIONAL PRECAUTIONS BEFORE USE.
POINSETTIA, Tree	To control plant growth	Concise	2 to 3 ppm drench for a 6-in. pot	For use in Florida only: Apply when the lateral shoots are 1.5 to 2.5 inches tall (about 10 to 14 days after pinching). Test for cultivar sensitivity. Do not apply after October 25.
PORPHYROCOMA POHLIANA (Brazilian Fireworks)	To improve foliage color and for earlier flowering	Piccolo/Piccolo 10 XC /Bonzi/Paczol	3 to 5 ppm spray	Height control generally not needed and rates above 5 ppm can cause leaf puckering.
PORTULACA OLERACEA	To control plant growth	Abide/A-Rest	7 to 26 ppm spray	
		Concise/Sumagic	15 to 30 ppm spray	
		Piccolo/Piccolo 10 XC /Bonzi/Paczol	5 ppm drench	Apply 7 days after transplant. May replace the need to pinch.
		Topflor	30 ppm spray	Apply 7 days after transplant. Repeat 2 weeks later if needed.
	To increase lateral branching	Citadel/Cyclocel	5,000 ppm spray	Apply 5 to 6 days after pinching to improve branching of cuttings.
		Collate/Florel	300 to 500 ppm spray	Recommendations based on Michigan conditions. Defoliation can occur with rates greater than 300 ppm.
POTHOS	To control plant growth	Abide/A-Rest	25 to 132 ppm spray	
			0.25 to 0.5 mg a.i. (2 to 4 ppm) drench for a 6-in. pot; apply 4 fl. oz./6-in. pot	Drench volumes and mg a.i. vary with pot size.
		Dazide/B-Nine	2,500 to 7,500 ppm spray	
		Piccolo/Piccolo 10 XC /Bonzi/Paczol	4 to 6 mg a.i. drench for a 8-in. pot; apply 10 fl. oz./8-in. pot	
PRIMULA ACAULIS	To control plant growth	Dazide/B-Nine	1,000 to 2,500 ppm spray	PGRs usually not required.
PRIMULA OBCONICA	To control plant growth	Dazide/B-Nine	5,000 ppm spray	PGRs usually not required.
PURPLE CONEFLOWER	To control plant growth	Concise/Sumagic	30 to 40 ppm spray	
PURPLE PASSION	To control plant growth	Abide/A-Rest	26 to 132 ppm spray	
			0.25 to 0.5 mg a.i. (2 to 4 ppm) drench for a 6-in. pot; apply 4 fl. oz./6-in. pot	Drench volumes and mg a.i. vary with pot size.
RANUNCULUS	To control peduncle length	Dazide/B-Nine	2,500 to 5,000 ppm spray	Make first application after 4 weeks. Repeat at lower rate every 2 weeks if needed. Three to 4 applications may be needed. Conduct trials to determine optimal concentrations and timing.
ROSE, Pot	To control plant growth	Concise/Sumagic	0.1 to 0.2 mg a.i./pot drenches	Usually only a single application is made.
		Piccolo/Piccolo 10 XC /Bonzi/Paczol	16 to 25 ppm sprays	Begin applications after the final pinch. Make the first one in 14 to 21 days, repeat weekly if needed. Discontinue applications after visible bud.
SALVIA, Annual	To control plant growth	Abide/A-Rest	10 to 20 ppm spray	
		Dazide/B-Nine	2,500 to 5,000 ppm spray	
		Piccolo/Piccolo 10 XC /Bonzi/Paczol	20 to 60 ppm spray	
		Citadel/Chlormequat E-Pro /Cycocel	800 to 1,500 ppm spray	
		Concise/Sumagic	5 to 10 ppm spray	
		Topflor	20 to 80 ppm spray	Based on NC State University trials. Adjust rates for other locations.

PLANT GROWTH REGULATORS

Table 8-5. Growth Regulators for Floricultural Crops in Greenhouses

Crop	Purpose	Chemical	Rate	Precautions and Remarks
SALVIA PLUGS, Annual	To control plant growth	Piccolo/Piccolo 10 XC /Bonzi/Paczol	5 to 10 ppm spray	Timing of application should normally begin at the 1 to 2 true leaf stage.
SALVIA FARINACEA	To control plant growth	Citadel+Dazide/Cycocel+B-Nine	1,000 ppm + 2,500 ppm spray	Apply if growth control is needed.
		Florel/Collate	Spray	Not recommended because of flower delay.
SALVIA GUARANITICA	To control plant growth	Citadel+Dazide/Cycocel+B-Nine	1,000 to 1,500 ppm + 2,000 to 3,500 ppm spray	
SALVIA HYBRID	To control plant growth	Dazide/B-Nine	1,500 to 2,500 ppm spray	
		Piccolo/Piccolo 10 XC /Bonzi/Paczol/Downsize	0.5 to 1 ppm drench	
SALVIA LONGISPICATA x FARINACEA	To control plant growth	Dazide/B-Nine	2,500 to 3,000 ppm spray	
SALVIA PATENS	To control plant growth	Citadel+Dazide/Cycocel+B-Nine	1,000 ppm + 2,500 ppm spray	
		Piccolo/Piccolo 10 XC /Bonzi/Paczol/Downsize	1 ppm drench	Trial rate before use.
SALVIA, Perennial	To control plant growth	Piccolo/Piccolo 10 XC /Bonzi/Paczol	40 to 60 ppm spray	
SALVIA, Vegetative	To control plant growth	Dazide/B-Nine	1,000 to 2,000 ppm spray	Multiple applications may be needed to tone crop.
		Dazide + Citadel/B-Nine + Cycocel	2,000 to 3,500 ppm + 1,000 to 1,500 ppm Citadel/Cycocel applied as a tank-mix spray	
SANVITALIA	To control plant growth	Dazide/B-Nine	1,200 to 5,000 ppm spray	Use to tone plants. Cultivars' response to PGRs varies. Test a few plants to determine rate for optimal control.
SCAEVOLA AEMULA	To control plant growth	Concise/Sumagic	30 ppm spray	Based on NC State University trials, 30 ppm worked well. Adjust rates to other locations; test on a few plants to determine rate for optimal control.
			0.125 ppm drench (0.011 mg a.i.) for a 5-in. pot; apply 3 fl. oz./5-in. pot	Drench volumes vary with pot size. See label for recommended volumes. Scaevola is very responsive to Concise/Sumagic drenches. Test on a few plants to determine rate for optimal control. Recommendations based on NC State University trials.
		Dazide/B-Nine	2,500 ppm spray	
		Piccolo/Bonzi/Paczol	20 to 40 ppm spray	
			1 to 3 ppm drench (0.12 to 0.35 mg a.i.)	Drench volumes vary with pot size. See label for recommended volumes. Cultivars' response to PGRs varies. Start with lowest rate in your trials.
		Topflor	45 to 60 ppm spray	Recommendations based on NC State University trials.
			0.79 to 2.25 ppm drench (0.075 to 0.2 mg a.i.)	Drench volumes will vary with pot size. See label for recommended volumes. Scaevola is very responsive to Topflor. Test the lower rates on a few plants. Recommendations based on NC State University trials.
			2 to 4 ppm liner dip	Scaevola is very responsive to Topflor. Test the lower rates on a few plants. Recommendations based on NC State University trials.
	To increase lateral branching	Florel/Collate	300 to 500 ppm spray	Apply early, typically 2 to 3 weeks after pinching.
SCHEFFLERA	To control plant growth	Abide/A-Rest	25 to 132 ppm spray	
			0.25 to 0.5 mg a.i. (2 to 4 ppm) drench for a 6-in. pot; apply 4 fl. oz./6-in. pot	Drench volumes and mg a.i. may vary with pot size.
		Dazide/B-Nine	2,500 to 7,500 ppm spray	
	To increase lateral branching	Augeo	3,125 ppm spray	Labeled for Schefflera arboricola only.
SCHIZANTHUS	To control plant growth	Abide/A-Rest	1 to 2 ppm spray	
		Dazide/B-Nine	1,500 to 3,000 ppm spray	
SCOPARIA	To control plant growth	Dazide/B-Nine	1,000 to 2,500 ppm spray	Use to tone plants if needed.
SCUTELLARIA JAVANICA (Skullcap)	To control plant growth	Dazide + Citadel/B-Nine + Cycocel	2,500 ppm Dazide + 1,000 ppm Citadel tank mix spray	Begin applications 2 to 3 weeks after transplanting. Repeat as needed every 2 weeks.
SEMPERVIVUM spp.	To induce offsets	Configure	100 to 400 ppm spray	Based on NC State University trials when applied 2 weeks after potting. For retail sales, 400 ppm produced the most offsets. For stock plant production, 100 to 200 ppm provided a balance between an increase in offset number and a larger offset size.
SHASTA DAISY	To control plant growth	Concise/Sumagic	15 to 30 ppm spray	

PLANT GROWTH REGULATORS

409

Table 8-5. Growth Regulators for Floricultural Crops in Greenhouses

Crop	Purpose	Chemical	Rate	Precautions and Remarks
SHRIMP PLANT	To control plant growth	Abide/A-Rest	25 to 50 ppm spray	Apply after plants established.
		Dazide/B-Nine	1,000 ppm	Apply after plants established.
	To increase lateral branching	Augeo	781 to 1,562 ppm spray	
SNAPDRAGON, Seed (*ANTIRRHINUM*)	To control plant growth	Abide/A-Rest	10 to 26 ppm spray	
		Concise/Sumagic	25 to 50 ppm spray	
		Dazide + Citadel/B-Nine + Cycocel	800 to 1,000 ppm Dazide/B-Nine + 800 to 1,000 ppm Citadel/Cycocel applied as a tank-mix spray	
		Piccolo/Piccolo 10 XC /Bonzi/Paczol	30 to 90 ppm spray	Apply at an early stage of plant growth with good stem coverage, especially for vigorous varieties.
SNAPDRAGON PLUGS, Seed (*ANTIRRHINUM*)	To control plant growth	Piccolo/Piccolo 10 XC /Bonzi/Paczol	10 to 20 ppm spray	Timing of application should normally begin at the 1 to 2 true leaf stage.
SNAPDRAGON, Vegetative (*ANTIRRHINUM*)	To control plant growth	Piccolo/Bonzi/Paczol	30 to 60 ppm spray	
		Concise/Sumagic	20 to 45 ppm spray	
	To control plant growth and peduncle stretch	Dazide/B-Nine	1,500 ppm spray	Use during periods of high temperatures.
SPATHIPHYLLUM	To induce flowering	GibGro	265 ppm spray	Apply one full-coverage spray during non-seasonal bloom period (June through January). Some cultivars exhibit distorted blooms, increased petiole length and narrow leaves.
	To accelerate bloom and increase flower number	Florgib/ProGibb T&O	150 to 250 ppm spray	Use a single application approximately 9 to 12 weeks prior to expected sale date. Spray to the point of runoff and thoroughly wet all growing points.
STATICE, Cut (Limonium)	To promote plant growth and stem elongation	Florgib/ProGibb T&O	50 to 100 ppm spray	Apply when plants are 4 to 8 in. tall. Other applications can be made at 2- to 3-week intervals. See label.
	For earlier flowering and increased flowering	Florgib/ProGibb T&O	400 to 500 ppm spray	Give each plant 0.33 fl. oz. (10 ml) of solution. Use when plants are 10 in. or more in diameter (approximately 90 to 100 days after sowing). See label.
STEPHANOTIS, Pot	To tone plant growth	Dazide + Citadel/B-Nine + Cycocel	100 ppm + 100 ppm spray	Controls vine elongation and shortens days until flowering.
STOKESIA	To control plant growth	Piccolo/Piccolo 10 XC /Bonzi/Paczol	40 to 80 ppm spray	
STREPTOCARPUS	To control plant growth	Abide/A-Rest	10 to 50 ppm spray	Rate based on Loiusiana State Unversity trial.
		Dazide/B-Nine	1,500 to 2,500 ppm spray	Supplier rate recommendation.
		Topflor	5 to 20 ppm spray	
	To delay premature bloom and promote additional plant growth	Collate	250 to 1000 ppm spray	Optimal rates varied significantly by cultivar. Conduct your own trials to determine optimal rates for each Streptocarpus series and specific cultivar. Results based on Iowa State Unversity trial.
STROBILANTHES DYERIANUS (Persian Shield)	To control plant growth	Dazide/B-Nine	2,500 to 5,000 ppm spray	
		Piccolo/Piccolo 10 XC /Bonzi/Paczol	30 ppm spray	
SUNFLOWER	To control plant growth	Citadel/Chlormequat E-Pro /Cycocel	800 to 1,500 ppm spray	
		Piccolo/Bonzi/Paczol	2 to 4 mg a.i. drench; apply 4 fl. oz./6-in. pot	Optimal rate based on NC State University trials. Adjust rate for plant vigor. Drench volumes and mg a.i. vary with pot size.
		Concise/Sumagic	16 to 32 ppm sprays	Optimal rate based on NC State University trials. Adjust rate for plant vigor.
		Topflor	30 to 50 ppm spray	
			1 to 2 mg a.i. (8.45 to 16.9 ppm) drench for a 6-in. pot	
TALINUM PANICULATUM	To control plant growth	Dazide/B-Nine	2,500 to 3,500 ppm spray	For toning the crop, apply once after transplanting.
TECOMA STANS	To control plant growth	Dazide + Citadel/B-Nine + Cycocel	2,500 ppm Dazide + 1,000 ppm Citadel tank mix spray	Begin applications 2 to 3 weeks after transplanting. Repeat as needed every 2 weeks.
THUNBERGIA ALATA	To control stem elongation / plant growth	Dazide + Citadel/B-Nine + Cycocel	2,500 ppm Dazide + 1,000 ppm Citadel tank mix spray	Apply to cuttings in propagation.
TIBOUCHINA	To control plant growth	Dazide/B-Nine	2,500 ppm spray	
TOMATILLO	To control plant growth	Sumagic	2 to 10 ppm spray	See precautions listed with EGGPLANT.
TOMATO	To control plant growth	Sumagic	2 to 10 ppm spray	See precautions listed with EGGPLANT.
TORENIA FOURNIERI	To control plant growth	Concise/Sumagic	5 to 15 ppm spray	Apply if growth control is needed.
		Dazide/B-Nine	1,500 to 2,500 ppm spray	Apply if growth control is needed.

Table 8-5. Growth Regulators for Floricultural Crops in Greenhouses

Crop	Purpose	Chemical	Rate	Precautions and Remarks
TORENIA spp.	To control plant growth	Dazide/B-Nine	1,500 ppm spray	Apply if growth control is needed
		Florel/Collate	Avoid use	Florel and Collate significantly delay flowering.
TROPICAL PLANTS (Not specifically listed in this table)	To induce lateral or basal branching	Configure	50 to 500 ppm spray	The supplemental label allows legal use on greenhouse-grown plants not specifically listed on the original label. See label for trialing suggestions and precautions.
TULIP	To control plant growth	Abide/A-Rest	0.125 to 0.5 mg a.i. (1 to 4 ppm) drench for a 6-in. pot; apply 4 fl. oz./6-in. pot	Drench volumes and mg a.i. vary with pot size.
		Piccolo/Piccolo 10 XC /Bonzi/Paczol/Downsize	0.591 to 4.732 mg a.i. (5 to 40 ppm) drench for a 6-in. pot; apply 4 fl. oz./6-in. pot	Drench volumes and mg a.i. vary with pot size. Apply drenches 1 to 5 days after forcing begins.
		Piccolo/Bonzi/Paczol	2 to 5 ppm bulb soak	Soak bulbs for 1 hr. prior to planting. Ten-minute soaks of 50 ppm (1.6 oz./gal.) provided excellent results in NC State University trials. Cultivar response varied.
		Concise/Sumagic	10 ppm bulb soak	Ten minute preplant soaks provided excellent results in NC State University trials. Cultivar response varied.
		Topflor	0.5 to 1 mg a.i. (4.2 to 8.45 ppm) drench for a 6-in. pot	Based on NC State University trials. Adjust rates for other locations.
			80 to 100 ppm spray	
			10 to 40 ppm bulb soak	Ten minute preplant soaks provided excellent results in NC State University trials. Cultivar response varied.
VERBENA, Annual	To control plant growth	Dazide/B-Nine	2,500 to 5,000 ppm spray	
		Piccolo/Piccolo 10 XC /Bonzi/Paczol	15 to 30 ppm spray	
		Citadel/Chlormequat E-Pro/Cycocel	800 to 1,500 ppm spray	Begin applications 7 days after pinching. Repeat as needed every 2 weeks.
		Concise/Sumagic	15 to 30 ppm spray	
	To increase lateral branching	Augeo	521 to 1,042 ppm spray	
		Florel/Collate	500 ppm spray	Florel and Collate applications will provide some growth retardant effects and delay flowering. Read the label for restrictions on timing of applications.
VERBENA, Perennial	To control plant growth	Piccolo/Piccolo 10 XC /Bonzi/Paczol	120 to 160 ppm spray	
		Piccolo/Piccolo 10 XC /Bonzi/Paczol/Downsize	>0.36 mg a.i. (>3 ppm) drench for a 6-in. pot; apply 4 fl. oz./6-in. pot	
VERBENA PLUGS, Annual	To control plant growth	Piccolo/Piccolo 10 XC /Bonzi/Paczol	5 to 10 ppm spray	Timing of application should normally begin at the 1 to 2 true leaf stage.
VERBENA, Vegetative	To control plant growth	Dazide + Citadel/B-Nine + Cycocel	2,000 to 3,500 ppm Dazide/B-Nine + 750 to 1,000 ppm Citadel/Cycocel applied as a tank-mix spray	
		Piccolo	8 to 12 ppm liner root soak	Irrigation of the liners occured within 24 hours prior to application, which results in a moderately dry substrate (the stage the plants would be watered but not wilted). Soak for a minimum of 30 to 60 seconds. Transplant after 3-hour waiting period. Rate based on Michigan State University trials.
		Citadel/Chlormequat E-Pro /Cycocel	1,500 to 2,000 ppm spray	
		Concise/Sumagic	5 to 10 ppm spray	Apply as needed.
		Dazide/B-Nine	1,500 to 2,500 ppm spray	Do not apply within 2 weeks of a Florel or Collate application.
		Florel/Collate	250 to 300 ppm spray	Make last application 8 weeks before sale.
VERONICA	To control plant growth	Piccolo/Piccolo 10 XC /Bonzi/Paczol	20 to 40 ppm spray	
		Concise/Sumagic	20 to 40 ppm spray	
VINCA (Catharanthus)	To control plant growth	Abide/A-Rest	5 to 18 ppm spray	
		Dazide/B-Nine	2,500 to 5,000 ppm spray	
		Citadel/Chlormequat E-Pro /Cycocel	800 to 1,500 ppm spray	
		Concise/Sumagic	1 to 3 ppm spray	Apply after plants reach a height of 4 in.
		Topflor	2.5 to 7.5 ppm spray	Based on NC State University trials. Adjust rates for other locations. Vinca is very responsive to Topflor, so start trials with lower rates.
VINCA VINE (Vinca spp.)	To increase lateral branching	Florel/Collate	500 ppm spray	Florel and Collate applications will provide some growth retardant effects and delay flowering. Read the label for restrictions on timing of applications.
VIOLA	To control plant growth	Concise/Sumagic	1 to 5 ppm spray	

PLANT GROWTH REGULATORS

Table 8-5. Growth Regulators for Floricultural Crops in Greenhouses

Crop	Purpose	Chemical	Rate	Precautions and Remarks
WANDERING JEW	To control plant growth	Abide/A-Rest	26 to 132 ppm spray	
WOODY LANDSCAPE PLANTS (Not specifically listed in this table)	To control plant growth	Abide/A-Rest	50 ppm spray	Conduct trials on a small number of plants, adjusting the rates as needed for desired final plant height and duration of height control.
			0.25 mg a.i. (2 ppm) drench for a 6-in. pot; apply 4 fl. oz./ 6-in. pot	Drench volumes and mg a.i. vary with pot size. Conduct trials on a small number of plants, adjusting the rates as needed for desired final plant height and duration of height control.
		Piccolo/Piccolo 10 XC /Bonzi/Paczol	100 ppm spray	Conduct trials on a small number of plants, adjusting the rates as needed for desired final plant height and duration of height control.
		Piccolo/Piccolo 10 XC /Bonzi/Paczol/Downsize	0.47 mg a.i. (4 ppm) drench for a 6-in. pot; apply 4 fl. oz/ 6-in. pot	Conduct trials on a small number of plants, adjusting the rates as needed for desired final plant height and duration of height control.
		Concise/Sumagic	10 to 50 ppm spray	Conduct trials on a small number of plants, adjusting the rates as needed for desired final plant height and duration of height control.
			1 to 2 ppm drench	Conduct trials on a small number of plants, adjusting the rates as needed for desired final plant height and duration of height control.
ZINNIA	To control plant growth	Abide/A-Rest	7 to 26 ppm spray	
		Citadel/Chlormequat E-Pro /Cycocel	800 to 1,500 ppm spray	
		Concise/Sumagic	5 to 25 ppm spray	
		Dazide/B-Nine	2,500 to 5,000 ppm spray	Multiple applications may be required. Use higher rates for summer crops.
		Piccolo/Piccolo 10 XC /Bonzi/Paczol	15 to 45 ppm spray	
ZINNIA PLUGS	To control plant growth	Piccolo/Piccolo 10 XC /Bonzi/Paczol	4 to 10 ppm spray	Timing of application should normally begin at the 1 to 2 true leaf stage.

PLANT GROWTH REGULATORS

Growth Regulators for Woody Ornamental Crops

A. V. LeBude, Horticultural Science Department

Table 8-6. Growth Regulators for Woody Ornamental Crops

Purpose	Chemical	Rate	Precautions and Remarks
Azalea			
To produce compact plants.	succinic acid (daminozide) (B-Nine) 85% WSP	9 oz/gal water	Spray foliage to runoff. Apply during early part of July. Follow label instructions.
Woody Ornamentals			
To stimulate rooting of cuttings.	K-IAA (Rhizopon A Water Soluble Tablet) 0.0018 oz (50 mg) K-IAA	Variable	Concentration will depend on species and also on time of year rooting is to take place. Follow label instructions.
	IBA (C-mone) 1% to 2%	1% to 2%	
	K-IBA (Rhizopon AA Water Soluble Tablet) 0.0018 oz (50 mg) K-IBA	Variable	
	K-IBA (C-mone K) 1%	1%	
	K-IBA + K-NAA (C-mone K+) 1.5% L	1.5%	
	IBA + NAA (Dip 'N Grow) 1.5% L	1.5%	
	IBA (Hormex) 0.1% to 4.5% D	0.1% to 4.5%	
	IBA (Hormodin) 0.1% to 0.8% D	0.1% to 0.8%	
	IBA + thiram (Hormo-Root) 0.1% to 2% IBA + 15% thiram D	0.1% to 2% + thiram 15%	
	IBA (Rhizopon AA) 0.1% to 0.8% D	0.1 to 0.8%	
	K-NAA (Rhizopon B Water Soluble Tablet) 0.0009 oz (25 mg) K-NAA	Variable	
	NAM + thiram (Rootone) 4.24% D	4.24%	
	IBA + NAA (Wood's Rooting Compound) 1.54% L	1.54%	
To promote lateral branching and produce compact plants in various species and suppress flowering and fruit formation in various species.	dikégulac sodium (Atrimmec) 20% L; (Augeo) (Pinscher) 18.5%	Depending on species, size, vigor of plants, and specific use	Amount used and concentration will vary depending on a number of factors. Follow label directions.
To control height on a wide variety of woody landscape plants (container grown in greenhouses or shadehouses or in landscapes) using both spray or drench applications.	paclobutrazol (Bonzi) (Piccolo) 0.4% L; (Profile 2 SC) 21.8%;	variable	Amount used and concentration will vary depending on a number of factors. Follow label directions.
To reduce or eliminate undesirable fruit development on many ornamental trees and shrubs such as apples, cottonwood, crabapples, elm, flowering pear, horse chestnut, maples, oaks, pines, sour orange, sweetgum, and sycamore.	ethephon (Florel) 3.9% L; (Ethephon 2 SL) 21.7%	1 qt/10 gal water (3 oz/gal)	Timing is extremely critical. Application must be made prior to fruit set; so apply at the full bloom stage in sufficient water to wet (do not spray to run off). Follow label directions.
To retard regrowth of most trees (hickory, red oak, silver maple), shrubs (viburnum, glossy abelia), and vines (ajuga, periwinkle, English ivy).	chlorflurenol (Maintain CF 125)12.5% EC, A	0.33 pt to 8 pt/100 gal water	Concentration will depend on particular species. Apply after new flush of growth or after pruning and new leaves have fully expanded. Take care to confine the use of this material to the particular area treated.
To promote lateral shoot growth on vegetative plants of azalea, cotoneaster, juniper, taxus.	methyl decanoate/octanoate (Off-Shoot-O) 45% EC	2 to 5 oz/qt water	Amount will vary with genera. Follow label for specific conditions.
To retard regrowth of most trees, (sycamore, sweetgum, willow), shrubs (pyracantha, privet), and ivy.	maleic hydrazide (Royal Slo-Gro) 21.7% SOL	1.33 gal/100 gal water	Prune plant to desired height. After regrowth occurs (as new leaves expand), apply to drip point. Uniform coverage is important for desired results.
To reduce terminal growth on shrubs (not trees) by shortening internode length. May increase quality of plants by darkening leaf color and thickening leaves and stems.	flurprimidol (Cutless) 0.33% G; (TopFlor) 0.17% G and 0.38%	Depending on height and mass of woody stems, foliage volume, species	Prune plants to desired height. Apply any time of year to top of soil or substrate and water thoroughly.

PLANT GROWTH REGULATORS

413

Sucker Control for Flue-Cured Tobacco

L. R. Fisher, M. C. Vann, M. D. Inman, R. W. Seagroves, and D. S. Whitley, Crop and Sciences Department

Table 8-7. Sucker Control for Flue-Cured Tobacco

Type Chemical and Formulation	Purpose	Amount of Formulation Per Acre	Precautions and Remarks
Contact Type			
C8–C10 fatty alcohol (various brands) 6.01 lb/gal	Normal sucker control	2 or 2.5 gal (4% or 5%)	Apply in 48 gallons of water per acre (4% solution) to plants in button stage with second application 3 to 5 days later at any time of day, except when plants are wet or temperature exceeds 90 degrees F or plants are wilted. Use two TG-3 nozzle tips plus a TG-5 in the center or equivalents per row with approximately 20 psi operated from 12 to 16 inches above the top of the button or stalk at 2.5 to 3 mph. Rate of second application may be increased to 2.5 gallons in 47.5 gallons of water (5% solution) unless crop is tender. Will not control suckers more than 1 inch long. Excess nitrogen increases the chance of leaf drop.
C10 fatty alcohol 5.72 lb/gal		1.5 gal (3%)	Apply in 48.5 gallons water per acre (3% solution) for both applications. Follow application instructions above for C8–C10 alcohol.
C8–C10 fatty alcohol 6 .01 lb/gal	Control of late-season sucker regrowth	2.5 gal (5%)	Apply 3 to 4 weeks after MH application if suckers begin to grow. Apply in 47.5 gallons of water per acre. Follow same directions as above. Will not control suckers more than 1 inch long. Do not make more than three applications of a contact per crop per season.
Systemic Type			
Maleic hydrazide [MH]) Liquids, various brands 1.5 lb/gal 2.25 lb/gal	Normal sucker control	1.5 gal (1 qt/1,000 plants) 1 gal (1 qt/1,500 plants)	Rate varies with plant population. 1.5 gallons of the 1.5 pounds per gallons material assumes 6,000 plants per acre. For plant populations other than 6,000, adjust rate accordingly. Apply to plants 5 to 7 days after the last contact application. Apply in the morning, using 30 to 50 gallons of water per acre, two to three cone nozzle tips per row, and 40 to 60 psi. Effectiveness will be reduced if applied to wet plants or those that are drought stressed or wilted from too much rainfall or high temperatures. Do not make more than one application per season. Should wash-off occur within 6 hours, a single repeat application may be made. DO NOT APPLY AT HIGHER THAN SUGGESTED RATES OR WITHIN 7 DAYS BEFORE HARVEST IN ORDER TO MINIMIZE MH RESIDUES.
60% Water-Soluble Products Fair 80 SP or Sucker Stuff 60 WS		3.75 lb	Rate for 6,000 plants per acre. Adjust rate accordingly for other plant populations.
Royal MH-30 SG		4 to 5 lb	
Contact Local-Systemic Type			
flumetralin (Prime +, Flupro or Drexalin Plus) 1.2 lb/gal	Normal sucker control, power sprayer	2 qt	Mix in 49 gallons of water per acre and apply like a contact at elongated button to early flower stage with three nozzles per row (TG-3, TG-5, TG-3) at 20 psi. Remove suckers longer than 1 inch within 24 hours before application and remove missed suckers as observed later. Excess spray to the point of rundown on the soil increases the risk of carryover residues, which may stunt early growth of next crop, including tobacco if a dinitroaniline herbicide is also used. **Do not apply these products through any type of irrigation system, and apply only once per season.** Rainfall within 2 hours after application may reduce effectiveness. Follow WSP requirements and other precautions and restrictions listed on product labels.
flumetralin (Prime +, Flupro or Drexalin Plus) 1.2 lb/gal	Hand application	1.2 to 2.4 qt (2.5 oz/gal water)	Mix in desired amount of water at rates shown in parenthesis and apply mixture as a coarse spray or drench to top of stalk. Apply about 0.5 ounces of mixture per plant after topping and removing suckers longer than 1 inch, but do not exceed 25 to 30 gallons per acre. See remarks above for power sprayer application and follow precautions, restrictions, and WPS requirements shown on product labels.
flumetralin (Prime +, Flupro or Drexalin Plus) 1.2 lb/gal	Control of late season sucker regrowth	2 qt	Apply only if control with MH is beginning to break down. Mix in 49 gallons water per acre and apply like a contact at 20 to 25 psi 3 to 4 weeks after MH application; will not control suckers longer than 1 inch. **To reduce the risk of soil residue carryover, do not use for late-season control if used earlier in the season.**
Systemic Type + Contact Local-Systemic Type			
maleic hydrazide (MH) + flumetralin (Prime +, Flupro or Drexalin Plus)	Normal sucker control	Full rate MH + 2 qt	See precautions and remarks for MH to determine "full rate" of MH. Mix in sufficient water to total 50 gallons per acre, and apply 5 to 7 days after the last contact or when MH alone is normally applied. Apply like a contact, using three nozzles (TG-3, TG-5, TG-3) per row at approximately 20 psi. Follow precautions and restrictions on labels. **Do not apply at higher than suggested rates or within 7 days before harvest in order to minimize MH residues.**
Contact-Systemic Type			
C10 fatty alcohol + MH (FST-7 or Leven-38) 4 lb/gal	Normal sucker control	3 gal	Apply in 47 gallons water to plants in early flower stage (1 week after button) any time during the day except when plants are wet or temperatures exceed 90 degrees F or plants are wilted. Use three nozzles per row with tips that deliver a coarse spray and desired rate when operated at 20 psi. Operate sprayer at a speed of 2 1/2 to 3 miles per hour and spray 50 gallons of diluted emulsion per acre. Use a semi-coarse spray covering the top 1/3 to 1/2 of the plant and allowing the liquid to run down the stalk to the bottom of each plant. **Do not apply at higher than suggested rates or within 7 days before harvest in order to minimize MH residues.** Effectiveness will be reduced if applied to plants that are drought-stressed or wilted from too much rainfall or high temperatures.

PLANT GROWTH REGULATORS

Table 8-7. Sucker Control for Flue-Cured Tobacco

Type Chemical and Formulation	Purpose	Amount of Formulation Per Acre	Precautions and Remarks
Contact + Contact-Local Systemic			
C_8-C_{10} fatty alcohol + flumetralin (Plucker-Plus)	Normal Sucker Control	2.5 gal	Refer to the discussion above on C_8-C_{10} fatty alcohol and flumetralin. Apply in 47.5 gallons of water per acre at normal timing for flumetralin application. Remove suckers longer than 1 inch within 24 hours before application and remove missed suckers as observed later. Excess spray to the point of rundown on the soil increases the risk of carryover residues, which may stunt early growth of next crop, including tobacco if dinitroanaline herbicide is also used. Do not apply these products through any type of irrigation system. Rainfall within 2 hours after application may reduce effectiveness. Follow WPS requirements and other precautions and restrictions listed on product labels. Do not make more than two applications per season so not to exceed maximum rate of flumetralin.

Yellowing Agents for Flue-Cured Tobacco

L. R. Fisher, M. C. Vann, M. D. Inman, R. W. Seagroves, and D. S. Whitley Crop and Soil Sciences Department

Table 8-8. Yellowing Agents for Flue-Cured Tobacco

To Increase the Rate of Yellowing

Chemical	Amount of Formulation Per Acre	Pounds Active Ingredient Per Acre	Precautions and Remarks
ethephon (Prep), (Super Boll), (Mature XL), or (Ethephon 6)	1.33 to 2.67 pt	1 to 2 lb	Use after second or third priming when remaining leaves are physiologically mature. Determine if tobacco is ready to spray by treating several representative plants at several locations with test kit (or prepare test spray by mixing 1 teaspoon of product in 1 quart of water). If test leaves begin to yellow in 24 to 72 hours, apply product to tobacco in 40 to 60 gallons water per acre as a fine spray mist (40 to 60 psi). Effectiveness may be reduced by application on cool, cloudy days, poor spray coverage, or rain within 4 hours after application. Harvest leaves within 24 to 48 hours or when they reach the desired degree of yellowness; prolonged delay in harvest may result in yield and quality loss or leaf drop. Therefore, do not spray more acreage than can be harvested before major rain is anticipated. **Do not use surfactants.**
(Oskie)	2.67 to 5.33 pt	1 to 2 lb	

Sucker Control for Burley Tobacco

L. R. Fisher, M. C. Vann, M. D. Inman, R. W. Seagroves, and D. S. Whitley, Crop and Soil Sciences Department

Table 8-9. Sucker Control for Burley Tobacco

Chemicals and Formulations	Amount of Formulation Per Acre	Precautions and Remarks
Contact Type		
C8–C10 fatty alcohol (various brands) 6.01 lb/gal	1.5 to 2 gal[1] (3% to 4%)	Apply in button to early flower stage as coarse, low pressure (approximately 20 psi) spray directed downward on plant tops. Leaf burn may occur with high application rates and pressure, especially on tender or wilted plants when temperature exceeds 90 degrees F. High rates or reapplication may contribute to leaf drop. Application before dew dries will reduce effectiveness.
Systemic Type (maleic hydrazide [MH])		
Liquids, various brands 1.5 lb/gal 2.25 lb/gal	1.5 to 2 gal 1 to 1.33 gal	For all systemic products, apply to upper 1/3 to 1/4 of plant in 20 to 50 gallons water per acre at approximately 20 psi after topping to 8-inch leaf. Effectiveness reduced when applied to drought-stressed or wilted plants, or before dew has dried. Apply a single repeat application ONLY if wash-off occurs within 6 hours. For water-soluble products, see rate information below and read labels carefully for mixing instructions.
60% Water-Soluble Products Fair 80 SP or Sucker Stuff 60 WS	3.75 lb	
Royal MH-30 SG	4 to 5 lb	Rate for 6,000 plants per acre. Adjust rate accordingly for other plant populations.
Contact + Systemic Mixture		
(FST-7 or Leven-38) C10 fatty alcohol + maleic hydrazide (MH)	9 qt[1]	Apply downward on plant tops as coarse, low-pressure (approximately 20 psi) spray after topping down to 8-inch leaf. Follow precautions given above and label restrictions for both contact- and systemic-type chemicals. High rates or reapplication after wash-off may contribute to leaf drop and increase MH residues on cured tobacco.
Contact, Local-Systemic Type		
flumetralin (Prime +, Flupro or Drexalin Plus)	1 gal[1]	Apply downward on plant tops as coarse, low-pressure (approximately 20 psi) spray after topping down to 8-inch leaf. Suckers longer than 1 inch should be removed at application time and missed suckers removed when seen. Apply only once per plant per season. Excessive spray volume that causes downstalk runoff on soil increases the chance of soil residue carryover, which may stunt the growth of small grains and corn or cause early season stunting of the next tobacco crop when a dinitroanaline herbicide is also used. Rainfall within 2 hours may reduce effectiveness.
butralin (Butralin)	3 to 4 qt*	

Table 8-9. Sucker Control for Burley Tobacco

Chemicals and Formulations	Amount of Formulation Per Acre	Precautions and Remarks
Systemic + Contact, Local-Systemic		
maleic hydrazide (MH) + flumetralin (Prime +, Flupro or Drexalin Plus)	1/2 to full rate MH + 2 qt flumetralin[1]	Apply as tank mix downward on topped plants as coarse, low pressure (approximately 20 psi) spray at time recommended for MH application. Follow precautions given above and label restrictions for both systemic and contact, local-systemic-type chemicals. The 3/4 rate of MH (1.5 gallons for most products) tank mixed with Prime+ has given satisfactory sucker control on vigorous crops and/or those harvested more than 3 weeks after application.
maleic hydrazide (MH) + butralin (Butralin)	Full rate MH +2 qt Butralin	
Contact + Contact-Local Systemic		
C_8-C_{10} fatty alcohol + flumetralin (Plucker-Plus)	2.5 gal	Refer to the discussion above on C_8-C_{10} fatty alcohol and flumetralin. Apply in 47.5 gallons of water per acre at normal timing for flumetralin application. Remove suckers longer than 1 inch within 24 hours before application and remove missed suckers as observed later. Excess spray to the point of rundown on the soil increases the risk of carryover residues, which may stunt early growth of next crop, including tobacco if dinitroanaline herbicide is also used. Do not apply these products through any type of irrigation system. Rainfall within 2 hours after application may reduce effectiveness. Follow WPS requirements and other precautions and restrictions listed on product labels. Do not make more than two applications per season so not to exceed maximum rate of flumetralin.

[1] Mix in sufficient water to total 50 gallons spray per acre.

Growth Regulators for Fruiting Vegetables

C. C. Gunter, Horticultural Science Department

Table 8-10. Growth Regulators for Fruiting Vegetables

Crop	Chemical	Amount of Formulation	Precautions and Remarks
Ethephon 2 — To Stimulate Uniform Ripening for One Harvest			
Processing Tomatoes Early and midseason crops or warm conditions	Ethephon 2 (Ethrel)	1.25 to 3.25 pt/A	Apply to cover foliage and fruit uniformly when 5% to 15% of the fruit is pink and red. Apply in a minimum of 20 gallons per acre for ground application (10 gallons for aerial). Under warm temperatures (above 85 degrees F) rates as low as 1.25 pints/acre can be effective. Thorough coverage is essential. Observe treated fields closely and harvest fruit at proper maturity.
Processing Tomatoes Late season or coastal crops or cool conditions	Ethephon 2 (Ethrel)	3.25 to 6.5 pt/A	Apply to cover foliage and fruit uniformly when 5% to 15% of the fruit is pink and red. Apply in a minimum of 20 gallons per acre for ground application (10 gallons for aerial). Use the higher rate when nighttime temperatures are cool (below 65 degrees F) or vegetative growth is dense. Thorough coverage is essential. Observe treated fields closely and harvest fruit at proper maturity.
Pepper	Ethephon 2 (Ethrel)	1.25 to 4 pt/A	Apply to bell peppers when 10% have red or chocolate coloration; chili and pimento peppers when 10-30% have red to chocolate coloration. Application should not be made until enough fruit exists for sufficient yield. Product will not ripen immature, green fruit. Rates between 1.25 and 2 pints/acre should be applied in 20 gallons/acre and 3 to 4 pints/acre in 40 gallons/acre. Use of spray volumes less than 40 gallons/acre in hot dry weather may result in foliage burn. Applications should not be made at temperatures of 100 degrees F or greater. In addition, use the higher rate when nighttime temperatures are cool (below 65 degrees F) or vegetative growth is dense. Thorough coverage is essential. Observe treated fields closely and harvest fruit at proper maturity. Maximum of 4 pints/acre per year.
Sumagic — For Commercial Greenhouse, Lathhouse, and Shadehouse Use Only			
Eggplant Groundcherry Pepino Pepper Tomatillo Tomato	Sumagic (Uniconizol-P)	0.52 to 2.6 fl oz/Gal or 16 to 76 ML/Gal or 2 to 10 PPM	Apply uniformly as a foliar spray at a volume of 2 quarts/100 square feet. Initial foliar applications when 2-4 true leaves are present. Sequential applications of lower rates provide more growth control than a single high rate application. First-time user should apply the lowest recommended rate in order to determine optimal rate for individual cultivars under local environmental conditions. If additional growth regulation is required, a sequential spray application at the lowest recommended rate should be made at 7-14 days after initial application. Total uniconizol-P applied may not exceed that from a single application of a 10 ppm spray concentration at 2 quarts/100 feet. Final application may not occur later than 14 days after the 2-4 true leaf stage.

PLANT GROWTH REGULATORS

Growth Regulators for Peanuts

D. L. Jordan, Crop Science Department

Table 8-11. Growth Regulators for Peanuts

Chemical and Formulation	Amount of Formulation per Acre	Remarks
To Suppress Excessive Vegetative Growth and Reduce Pod Shed		
prohexadione calcium (Apogee) 27.5% WDG (Kudos) 27.5% WDG	7.25 + 7.25 oz	Apply Apogee or Kudos when 50% of vines from adjacent rows are touching. Follow with a repeat application 2 to 3 weeks later. Two applications are needed in most circumstances. Although a third application may be made, it is generally discouraged because of expense and possible adverse impact on peanut. Do not exceed 21.75 ounces per acre per year. The preharvest interval is 25 days. Do not apply more than 2 applications of Apogee or Kudos within 6 weeks. Apply in a minimum of 20 GPA. Always apply Apogee and Kudos with a nitrogen source. Apply one pound of spray grade ammonium sulfate for every pound of Kudos. Apply one pound of spray grade ammonium sulfate per acre or one pint of UAN (28%, 30% or 32% nitrogen solution) per acre with Apogee. Apply Apogee or Kudos with either nonionic surfactant or crop oil concentrate depending on product labels when mixing with other products.

For further information, see Extension publication *Peanut Information* (AG-331). Copies are available from your county Cooperative Extension center.

Growth Regulators for Turfgrasses

F. H. Yelverton, R. Cooper, and T. W. Gannon, Crop Science Department

Table 8-12. Growth Regulators for Turfgrasses

Brand	Amount of Formulation Per Acre	Pounds Active Ingredient Per Acre	Precautions and Remarks
Cool Season Grasses—Well-Maintained Turf: Seedhead and Foliar Suppression			
mefluidide (Embark) 0.2	5 pt/15 to 150 gal water	0.125	See Embark 2-S for low-maintenance cool-season turf. Follow label directions and precautions.
trinexapac-ethyl (Governor) 0.17 G	30 to 258 lb	0.05 to 0.44	Apply 30 to 41 pounds per acre to greens, 53 to 152 pounds per acre to fairways less than 0.5 inch cut, and 152 to 258 pounds per acre to residential and commercial turf. Do not exceed 2.5 pounds active ingredient per acre per year. These rates should provide 50% turf growth suppression for 4 weeks with minimal yellowing.
(Primo Maxx) 1 MEC or (T-Nex) 1 AQ (Primo WSB) 25 WP	6 to 44 fl oz 2.75 to 21.8 oz	0.085 to 0.34 0.085 to 0.34	Application rate varies with turfgrass species and height of cut. Apply to actively growing, nonstressed turf. More growth suppression occurs at lower mowing heights. See label for specific rate and other directions and precautions. Repeat applications can be made, but do not exceed a total of 21.4 pints per acre per year of Primo Maxx or a total of 174 ounces per acre per year of Primo WSB. Do not exceed a total of 19 pints per acre per year of T-Nex. Refer to the respective Primo label for guidelines regarding mowing prior to and following application. Mix with 0.5 to 4 gallons of water per 1,000 sq ft (20 to 174 gallons per acre). Primo can be applied to putting greens. See label for instructions.
Cool Season Grasses—Well-Maintained Turf: Foliar Suppression			
ethephon (Ethephon or Proxy) 2 SL	1.7 gal	3.4	May be applied to Kentucky bluegrass, perennial ryegrass, bentgrass, and tall and fine fescues. Apply in 22 to 174 gallons of water per acre. Do not use a surfactant. Plant growth regulator effect will not be seen until 7 to 10 days after application. May be reapplied to Kentucky bluegrass and perennial ryegrass at 7-week intervals. Repeat applications to bentgrass and tall and fine fescue may be made at 4-week intervals.
flurprimidol (Cutless 50 W) 50 WP	0.75 to 3 lb/50 to 200 gal water	0.37 to 1.5	Rates depend upon grass species and cultivar. Apply to bentgrass, Kentucky bluegrass, and perennial ryegrass in late spring-early summer and/or late summer-early fall. Time the second application to occur at least 3 months before expected winter dormancy. Do not apply to putting greens. Do not exceed 1.5 pounds per acre per application on coarse-textured soils. Treated areas should receive 0.5 inch of irrigation within 24 hours after application. Resume mowing 3 to 5 days after application.
flurprimidol + trinexapac-ethyl (Legacy) 1.51 SL	5 to 22 fl oz	0.059 to 0.26	Tolerant species include bentgrass greens and fairways, Kentucky bluegrass, and perennial ryegrass. Do not use on turf grown for sale or other commercial use as sod or seed production. Do not seed 3 weeks before or 3 weeks after application. Wait 6 to 8 weeks after sprigging or laying sod before applying. Use only 5 to 8 fluid ounces per acre on bentgrass greens. Repeat applications at 2- to 6-week intervals until 4 weeks before the onset of inactive growth.
paclobutrazol (TGR Turf Enhancer 2 SC or Trimmit 2 SC) 2 SC	1 to 2 pt /43 to 200 gal water	0.25 to 0.5	Apply in spring after greenup and after turf has been mowed once or twice. Apply at least 1 month before onset of high temperatures. In late summer-early fall, apply at least 1 month before anticipated first killing frost. Apply with 0.5 to 0.9 pound nitrogen per 1000 sq ft of a nonburning fertilizer. Apply 0.25 inch of water within 24 hours after application to remove product from foliage and onto soil surface. See label for special rates and directions for applications to bentgrass, putting greens, and overseeded bermudagrass. Repeat applications within the same growing season may be made but refer to label for instructions. Do not apply more than three times annually. Do not use on areas containing greater than 70% *poa annua*. Do not seed within 6 weeks prior to or 2 weeks after applications.
prohexadione calcium (Anuew) 27.5 WG	1.8 to 29.1 oz	0.031 to 0.5	Apply to golf course fairways, tees, greens and roughs and also athletic fields, residential and commercial lawns, sod farms, parks, cemeteries and roadsides. Apply 1.8 to 7.25 ounces per acre on bentgrass greens and tees at 1 to 2 wk intervals and 7.25 to 14.5 ounces per acre on bentgrass fairways and roughs at 2 to 4 wk intervals. Apply 14.5 to 21.8 ounces per acre on perennial bluegrass and 21.8 to 29.1 ounces per acre on perennial ryegrass at 2 to 4 wk intervals. For cool season grass sod production, apply 7.25 to 29.1 ounces per acre at 2 to 4 wk intervals. Use a spray volume of 1 to 2 gallons of water per 1000 sq ft. A nonionic surfactant may improve product performance. Do not irrigate for 4 hours after application or mow until 1 day after application.
trinexapac-ethyl (Primo Maxx) 1 MEC or (T-Nex) 1 AQ (Primo WSB) 25 WP	6 to 22 fl oz 2.75 to 10.9 oz	0.085 to 0.17 0.085 to 0.17	Application rates are for mowing heights of less than or equal to 0.5 inch Apply to actively growing, non-stressed turf. Rate varies with turfgrass species. See label for specific rate and other directions and precautions. Repeat applications can be made but do not exceed a total of 21.4 pints per acre per year of Primo Maxx or a total of 174 ounces per acre per year of Primo WSB. Do not exceed a total of 19 pints per acre per year of T-Nex. Refer to the respective Primo label for guidelines regarding mowing prior to and following applications. Mix with 0.5 to 4 gallons of water per 1,000 sq ft (20 to 174 gallons per acre). Primo can be applied to putting greens. See label for instructions.

PLANT GROWTH REGULATORS

Table 8-12. Growth Regulators for Turfgrasses

Brand	Amount of Formulation Per Acre	Pounds Active Ingredient Per Acre	Precautions and Remarks
Cool Season Grasses—Low-Maintenance Turf: Seedhead and Foliar Suppression			
chlorsulfuron (Telar DF) 75 DF + mefluidide (Embark 2-S) 2S	0.25 oz + 0.5 pt	0.012 + 0.125	For growth and seedhead suppression in fescue/bluegrass stands. Apply up until seedhead emergence. Do not apply Telar DF to turf less than 1 year old. Grass seed may be planted in treated areas 6 months after treatment but cultivation is recommended. For broadcast applications, do not exceed 0.5 ounces Telar DF per acre within a 12-month period. Telar DF alone can also be used for weed control in bahiagrass, bermudagrass, fescue, and bluegrass.
glyphosate (Touchdown Pro) 3 LC	4 to 8 fl oz/10 to 40 gal water	0.09375 to 0.1875	Touchdown Pro may be used on turf described in "GENERAL USE AREAS" section of the label. 4 to 5 ounces will suppress annual grasses, such as ryegrass, wild barley, and wild oats, growing in turf areas. 6 ounces will suppress Kentucky bluegrass and serve as a mowing substitute. 8 ounces will suppress fine fescue and tall fescue and serve as a mowing substitute. A nonionic surfactant containing at least 75% active ingredient at 0.25% v/v (1 quart per 100 gallons) or ammonium sulfate at 0.5% by weight (4.25 to 17 pounds per 100 gallons) may be added.
imazethapyr + imazapyr (Event) 1.46 lb/gal	8 to 10 fl oz	0.09 to 0.11	Apply to tall fescue, perennial ryegrass, and bluegrass only. Apply after the turf is at 100% greenup and has at least 2 inches of vertical growth. The addition of a nonionic surfactant containing at least 80% active ingredient at 0.25% v/v of the spray (2 pints per 100 gallons of spray mixture) is required. Do not use on newly established stands less than 1 year old or on highly managed turf. Do not reseed before 3 months after application. See label for herbicide tank mix options. Follow label directions and precautions.
maleic hydrazide (Retard) 2.25 lb/gal (Royal Slo-Gro) 1.5 lb/gal (Liquid Growth Retardant) 0.6 lb/gal	1.3 gal/50 gal water 2 gal/30 to 50 gal water 5 gal/45 gal water	3	Treat in the spring when the grass is actively growing but before seedhead appears. Applications made after seedhead appears will suppress subsequent seedheads. Do not apply to turf less than 3 years old, and do not reseed within 3 days after application. Treated turf may appear less dense and temporarily discolored. Optimum results may not be obtained if rainfall or overhead irrigation occurs within 12 hours following application. Remove excess grass clippings and fallen leaves before application. Do not add a surfactant. Follow label directions and precautions.
mefluidide (Embark 2-S) 2 S	1.5 to 2 pt/15 to 150 gal water	0.38 to 0.5	Apply after uniform spring greenup until approximately 2 weeks before seedheads appear. Do not apply to turf within 4 growing months after seeding, and do not reseed within 3 days after application. Treated turf may appear less dense and temporarily discolored. Optimum results may not be obtained if rainfall or overhead irrigation occurs within 8 hours following application. Remove excess clippings and fallen leaves before application. Adding 1 to 2 quarts of nonionic surfactant per 100 gallons of spray solution may enhance suppression; however, discoloration may also be increased. Follow label directions and precautions.
metsulfuron methyl (Escort XP) 60 DF	0.25 to 0.5 oz	0.009 to 0.018	Apply to well-established tall fescue and perennial bluegrass turf. Can tank mix with 0.125 to 0.25 pints per acre of Embark to improve pgr performance. Treat after 2 to 3 inches of new growth but before seed stalk formation. Temporary discoloration may occur. Do not use on stressed turf.
Warm Season Grasses— Well-Maintained Turf: Seedhead and Foliar Suppression			
trinexapac-ethyl (Governor) 0.17 G	12 to 258 lb	0.02 to 0.44	Apply 12 to 41 pounds per acre to greens, 30 to 77 pounds per acre to fairways less than 0.5 inch cut, and 41 to 258 pounds per acre to residential and commercial turf. Do not exceed 2.5 pounds active ingredient per acre per year. These rates should provide 50% turf growth suppression for 4 weeks with minimal yellowing.
(Primo Maxx) 1 MEC or (T-Nex) 1 AQ (Primo WSB) 25 WP	2.7 to 88 fl oz 1.35 to 43.6 oz	0.085 to 0.68 0.085 to 0.68	Application rate varies with turfgrass species and height of cut. Apply to actively growing, nonstressed turf. More growth suppression occurs at lower mowing heights. See label for specific rate and other directions and precautions. Repeat applications can be made but do not exceed a total of 21.4 pints per acre per year of Primo Maxx or 174 ounces per acre per year of Primo WSB. Do not exceed a total of 19 pints per acre per year of T-Nex. Refer to the respective Primo label for guidelines regarding mowing prior to and following application. Mix with 0.5 to 4 gallons of water per 1,000 sq ft (20 to 174 gallons per acre). Primo can be applied to putting greens. See label for directions.
mefluidide (Embark) 0.2	10 pt/15 to 150 gal water	0.25	For St. Augustinegrass. See Embark 2-S for low-maintenance warm season turf. Follow label directions and precautions.
Warm Season Grasses— Well-Maintained Turf: Foliar Suppression			
flurprimidol (Cutless 50 W) 50 WP	0.75 to 3 lb/50 to 200 gal water	0.37 to 1.5	Rates depend upon grass species and cultivar. Apply to Tifway, Tifgreen, common bermudagrass, or zoysiagrass. Treated areas should receive 0.5 inch of irrigation within 24 hours of application. Resume mowing. Overseed 2 to 3 weeks after fall application with a desired perennial ryegrass.
flurprimidol + trinexapac-ethyl (Legacy) 1.51 SL	8 to 15 fl oz	0.094 to 0.177	Tolerant species include Tifway and Tifsport bermudagrass, zoysiagrass, and seashore paspalum. Do not use on turf grown for sale or other commercial use as sod or seed production. Do not seed 3 weeks before or 3 weeks after application. Wait 6 to 8 weeks after sprigging or laying sod before applying. Repeat applications at 2- to 6-week intervals until 4 weeks before winter dormancy.
paclobutrazol (TGR Turf Enhancer 2 SC or Trimmit 2 SC) 2 SC	2 to 3 pt/43 to 200 gal water	0.5 to 0.75	Use any time when established hybrid bermudagrass and St. Augustinegrass are green, are actively growing, and have recovered from dormancy (filled in fully following winter). Apply with 0.5 to 0.9 pound nitrogen per 1,000 sq ft of a nonburning fertilizer. Apply 0.25 inch of water within 24 hours after application to remove product from foliage and onto soil surface. A repeat application within the same growing season may be made, but not sooner than 8 weeks following initial application. Do not apply more than 3 times annually. Do not use on areas containing greater than 70% poa annua. Refer to label to determine bermudagrass and St. Augustine cultivar response relating to sensitivity, growth, and color response. Do not seed within 6 weeks prior to or 2 weeks after application.
prohexadione calcium (Anuew) 27.5 WG	7.25 to 43.6 oz	0.125 to 0.75	Apply to golf course fairways, tees, greens and roughs and also athletic fields, residential and commercial lawns, sod farms, parks, cemeteries and roadsides. Apply 7.25 to 14.5 ounces per acre on hybrid bermudagrass greens and tees at 1 to 2 wk intervals and 29.1 to 43.6 ounces per acre on hybrid bermudagrass fairways and roughs at 2 to 4 wk intervals. For warm season grass sod production, apply 14.5 to 43.6 ounces per acre at 2 to 4 wk intervals. Use a spray volume of 1 to 2 gallons of water per 1000 sq ft. A nonionic surfactant may improve product performance. Do not irrigate for 4 hours after application or mow until 1 day after application.
trinexapac-ethyl (Primo Maxx) 1 MEC or (T-Nex) 1 AQ (Primo WSB) 25 WP	2.7 to 13 fl oz 1.35 to 6.5 oz	0.042 to 0.085 0.042 to 0.085	Application rates are for mowing heights of less than or equal to 0.5 inch Apply to actively growing, non-stressed turf. Rate varies with turfgrass species. See label for specific rate and other directions and precautions. Repeat applications can be made but do not exceed a total of 21.4 pints per acre per year of Primo Maxx or a total of 174 ounces per acre per year of Primo WSB. Do not exceed a total of 19 pints per acre per year of T-Nex. Refer to the respective Primo label for guidelines regarding mowing prior to and following applications. Mix with 0.5 to 4 gallons of water per 1,000 sq ft (20 to 174 gallons per acre). Primo can be applied to putting greens. See label for directions.

Table 8-12. Growth Regulators for Turfgrasses

Brand	Amount of Formulation Per Acre	Pounds Active Ingredient Per Acre	Precautions and Remarks
Warm Season Grasses—Low-Maintenance Turf: Seedhead and Foliar Suppression			
glyphosate (Roundup Pro) 4 lb/gal	6 fl oz/10 to 25 gal water	0.2	Apply to bahiagrass only. Apply after full greenup of the bahiagrass (about late May) and make only one application per year. Do not apply to turf less than 3 years old. Treated turf may appear less dense and temporarily discolored. Optimum results may not be obtained if rainfall or overhead irrigation occurs within 6 hours following application. This is a nonselective herbicide. If application exceeds the above recommended rates, it can result in permanent loss of turf.
(Touchdown Pro) 3 LC	0.375 to 4 pt/10 to 40 gal water	0.14 to 1.5	Touchdown Pro may be used on dormant or actively growing bermudagrass and bahiagrass turf described in "GENERAL USE AREAS" section of label. May be tank mixed with 0.25 to 2 ounces of Oust for residual weed control. Check label for correct rates. Touchdown Pro will control winter annual weeds less than 6 inches tall and also 4-to 6-leaf tall fescue in dormant turf. Use only on well-established bermudagrass. Injury may occur, but regrowth will occur under moist conditions. Bahiagrass vegetative growth and seedheads may be suppressed approximately 45 days when applied 1 to 2 wk after spring greenup and before seedhead emergence. A second application at 45 days will extend suppression to approximately 120 days.
imazapic (Plateau) 2 ASU	2 fl oz	0.031	Only government entities may buy Plateau. Used for bahiagrass seedhead suppression. Apply to bahiagrass in spring after full greenup but approximately 3 to 4 weeks prior to expected seedhead emergence or 7 to 10 days after mowing. Do not apply to wetlands. Add a surfactant according to label directions. Bahiagrass may appear less dense and discolored following application.
imazapic (Panoramic) 2 SL	2 to 3 fl oz	0.031	May be used for seedhead suppression of bahiagrass or tall fescue turf areas including industrial turf, golf courses, and non-residential areas. Apply 2-3 ounces/A for tall fescue seedhead suppression prior to seedhead emergence. Apply 2 ounces/acre after bahiagrass greenup but prior to seedhead emergence. Temporary turf discoloration may occur.
imazapic + glyphosate (Journey) 2.25 AS	11 to 32 fl oz	0.19 to 0.56	Use in noncrop areas. Temporary turf discoloration may occur. Apply 4 to 8 fluid ounces per acre on a small area first to determine rate needed for desired results. Do not use with methylated seed oil. Do not apply to drought-stressed turf. Apply after full turf greenup.
imazethapyr + imazapyr (Event) 1.46 lb/gal	8 to 10 fl oz	0.09 to 0.11	Apply to bahiagrass only. Apply after the turf is at 100% greenup and has at least 2 inches of vertical growth. The addition of a nonionic surfactant containing at least 80% active ingredient at 0.25% v/v of the spray (2 pints per 100 gallons of spray mixture) is required. Do not use on newly established stands less than 1 year old or on highly managed turf. Do not reseed before 3 months after application. See label for herbicide tank mix options. Follow label directions and precautions.
maleic hydrazide (Retard) 2.5 lb/gal (Royal Slo-Gro) 1.5 lb/gal (Liquid Growth Retardant) 0.6 lb/gal	1.3 gal/50 gal water 2 gal/30 to 50 gal water 5 gal/45 gal water	3 3 3	Apply to bahiagrass only. Apply in spring but before seedheads appear. Applications made after seedhead appearance will suppress subsequent seedheads. Do not apply to turf less than 3 years old and do not reseed within 3 days after application. Treated turf may appear less dense and temporarily discolored. Optimum results may not be obtained if rainfall or overhead irrigation occurs within 12 hours following application. Remove excess grass clippings and leaves before application. Do not add a surfactant. Follow label directions and precautions. A repeat application may be needed 6 weeks after initial application.
mefluidide (Embark 2-S) 2 S	2 qt/15 to 150 gal water	1	Apply to bermudagrass only. Apply in late spring until about 2 weeks before seedhead appearance. Do not apply to turf within 4 growing months after seeding, and do not reseed within 3 days after application. Treated turf may appear less dense and temporarily discolored. Optimum results may not be obtained if rainfall or overhead irrigation occurs within 8 hours following application. Remove excess grass clippings and leaves before application. Adding 1 to 2 quarts of a nonionic surfactant per 100 gallons of spray solution may enhance suppression; however, discoloration may also be increased. Follow label directions and precautions.
sulfometuron methyl (Oust) 75 DG	0.5 oz/30 to 50 gal water	0.02 lb	Apply to bahiagrass in late spring or early summer before seedheads appear. Do not apply to wetlands or where runoff water may flow onto agricultural lands or forests. Injury of desirable trees may result if applications are made near plants or where their roots extend or may be subjected to runoff from treated areas. Do not apply to turf less than 3 years old. Treated turf may appear less dense and temporarily discolored. Do not add a surfactant. Follow label directions and precautions.
sulfometuron methyl + chlorsulfuron (Landmark MP) 50 + 25 DG (Landmark II MP) 56.25 + 18.75 DG	0.9 oz 1.0 oz	0.042 0.047	For established bermudagrass and centipede-improved turf. Temporarily suppresses foliar and seedhead growth while controlling many grass and broadleaf weeds. Apply 30 days after breaking dormancy or either late fall or early winter. Landmark MP may discolor or cause top kill of desired turf species. Do not apply to turf less than 1 year old. Annual retreatments may reduce turf vigor.
sulfometuron methyl + metsulfuron methyl (Oust Extra) 56.25 + 15 DG	0.5 to 2 oz	0.022 to 0.088	For use on well-established, unimproved bermudagrass and centipedegrass. Apply 30 days after breaking dormancy. Can also be applied in late fall or early winter depending on weed presence. Oust Extra can be tank mixed with 3 to 4 pounds active ingredient per acre MSMA on bermudagrass during the summer. Do not add a surfactant.
Annual Bluegrass: Suppression			
flurprimidol (Cutless 50 W) 50 WP	0.25 to 0.5 lb/50 to 100 gal water	0.12 to 0.25	Apply to actively growing bentgrass putting greens in spring after third or fourth mowing or in the fall. Repeat, if necessary, at 3- to 6-week intervals, not to exceed 2 pounds per acre per growing season. Delay overseeding 2 weeks after application. Make final fall application 8 weeks before onset of winter dormancy.
	1 to 1.5 lb/50 to 200 gal water	0.5 to 0.75	Apply to bentgrass, Kentucky bluegrass, and perennial ryegrass in late spring-early summer and/or late summer-early fall. Time the second application to occur at least 3 months before expected winter dormancy. Management practices that encourage vigorous growth of perennial turfgrass following application will enhance conversion. *Poa annua* discoloration will be visible 7 to 10 days after treatment and last for 3 to 6 weeks. Do not apply to putting greens. Treated areas should receive 0.5 inch of irrigation within 24 hours after application. Resume mowing 3 to 5 days after application.
flurprimidol + trinexapac-ethyl (Legacy) 1.51 SL	5 to 30 fl oz	0.059 to 0.354	Use in cool season turfgrasses, such as bentgrass greens and fairways, Kentucky bluegrass, and perennial ryegrass. Repeat applications at 2- to 6-week intervals. Annual bluegrass suppression is gradual and could take several growing seasons. Start treatments in early spring and continue through early fall.
maleic hydrazide (Retard) 2.25 lb/gal (Royal Slo-Gro) 1.5 lb/gal (Liquid Growth Retardant) 0.6 lb/gal	1 qt/30 to 40 gal water 2 qt/30 to 40 gal water 1.25 gal/30 to 40 gal water	0.56 0.75 0.75	Treat after two normal mowings but before seedhead appears. Applications made after seedhead appears will suppress subsequent seedheads. Do not apply to golf greens. Do not apply to turf less than 3 years old, and do not reseed within 3 days after application. Treated turf may appear less dense and temporarily discolored. Optimum results may not be obtained if rainfall or overhead irrigation occurs within 12 hours following application. Remove excess grass clippings and fallen leaves before application. Do not add a surfactant. Follow label directions and precautions for use on fairways.

PLANT GROWTH REGULATORS

419

Table 8-12. Growth Regulators for Turfgrasses

Brand	Amount of Formulation Per Acre	Pounds Active Ingredient Per Acre	Precautions and Remarks
Annual Bluegrass: Suppression (continued)			
mefluidide (Embark 2-S) 2 S (Embark) 0.2	0.5 pt/15 to 150 gal water 2 to 5 pt/15 to 150 gal water	0.125 0.05 to 0.125	Apply after uniform greenup but before first appearance of seedheads. Do not apply to turf within 4 growing months after seeding, and do not reseed within 3 days after application. Treated turf may appear less dense and temporarily discolored. Optimum results may not be obtained if rainfall or overhead irrigation occurs within 8 hours following application. Remove excess grass clippings and leaves before application. Adding 1 to 2 quart of a nonionic surfactant per 100 gallons of spray solution enhances suppression; however, discoloration may also be increased. Follow label directions and precautions for use of fairways and tees.
paclobutrazol (31-3-9 Fertilizer with TGR *Poa annua* Control 0.42%)	128 lb	0.5	Apply only to bentgrass, Kentucky bluegrass, perennial ryegrass fairways, or bentgrass greens with less than a 70% *poa annua* infestation. Follow label directions and precautions. Note: This product supplies 0.9 pound N per 1,000 sq ft.
prohexadione calcium (Anuew) 27.5 WG	0.9 to 1.75 oz	0.015 to 0.03	Apply to overseeded hybrid bermudagrass during periods of active *poa annua* growth at 3 to 4 wk intervals.
(15-0-29 High K Fertilizer with TGR *Poa annua* Control 0.34%)	98 lb to 146 lb	0.33 to 0.5	Apply only to bentgrass, zoysiagrass, Kentucky bluegrass, and Kentucky bluegrass/perennial ryegrass fairways, tees, and roughs, as well as bentgrass greens with less than 70% *poa annua* infestation. Note: This product supplies 0.5 pound N per 1,000 sq ft.
(TGR Turf Enhancer 2 SC or Trimmit 2 SC) 2 SC	6.4 to 48 fl oz/43 to 200 gal water	0.1 to 0.75	Apply on hybrid bermudagrass, bentgrass, perennial ryegrass, and Kentucky bluegrass/perennial ryegrass fairways, tees, and roughs. Can also be applied to bentgrass putting greens. Apply in spring after greenup or regrowth has begun and after mowing once or twice. Apply with a nonburning fertilizer. Apply 0.25 inch of water within 24 hours after application to remove product from foliage and onto soil surface. See label for rates and other directions for applications to bentgrass putting greens and overseeded bermudagrass. Do not apply more than 3 times annually. Do not use on areas containing more than 70% *poa annua*. For bentgrass putting greens, do not apply more than 0.25 pound active ingredient per acre per application.
ethephon (Proxy) 2 SL	1.7 gal	3.4	May be used to suppress annual bluegrass seedheads and growth of other cool season turfgrasses including golf course greens, fairways, tees, and roughs. Do not use an adjuvant. Do not apply to stressed turfgrass or where excessive thatch is present. Scalping may occur on bentgrass surfaces after application. Consult label for repeat application intervals.
Overseeded Bermudagrass Turf: Foliar Suppression			
flurprimidol (Cutless 50 W) 50 WP	0.75 to 3 lb/50 to 200 gal water	0.37 to 1.5	Rates depend upon grass species and cultivar. Apply to zoysiagrass, Tifway, Tifgreen, and common bermudagrass in late spring-early summer and/or late summer-early fall. Time the second application to occur 8 to 10 weeks before expected winter dormancy. Do not apply to putting greens. Do not exceed 1.5 pound per acre per application on coarse-textured soils. Treated areas should receive 0.5 inch of irrigation within 24 hours after application. Resume mowing 3 to 5 days after application.
flurprimidol + trinexapac-ethyl (Legacy) 1.51 SL	5 to 30 fl oz	0.059 to 0.354	Use in cool season turfgrasses, such as bentgrass greens and fairways, Kentucky bluegrass, and perennial ryegrass. Repeat applications at 2- to 6-week intervals. Annual bluegrass suppression is gradual and could take several growing seasons. Start treatments in early spring and continue through early fall.
maleic hydrazide (Royal Slo-Gro) 1.5 lb/gal (Liquid Growth Retardant) 0.6 lb/gal	1.5 gal/50 gal water 3.3 gal/50 gal water	2.25	Apply in late September or early October to inhibit bermudagrass growth and allow winter overseeding to establish. Overseed no sooner than 48 hours after application. Follow label directions and precautions for use on greens and fairways.
paclobutrazol (TGR Turf Enhancer 2 SC or Trimmit 2 SC) 2 SC	6.4 to 16 fl oz/43 to 200 gal water	0.1 to 0.25	Apply any time after overseeded turf has successfully established itself. Do not apply after March 15 to avoid delay in bermudagrass green-up. Apply with 0.25 to 0.5 pound N per 1,000 sq ft of a nonburning fertilizer. Apply 0.25 inch of water within 24 hours after application to remove product from foliage and onto soil surface. Repeat applications can be made but do *not apply* more than 3 times annually. Do not use on areas containing more than 70% *poa annua*. Do not seed within 6 weeks prior to or 2 weeks after application. Do not apply to 'Tifdwarf' putting greens.
prohexadione calcium (Anuew) 27.5 WG	0.9 to 1.75 oz	0.015 to 0.03	To enhance overseeding establishment, apply to hybrid bermudagrass 3 to 5 days prior to seeding. Delay verticutting, spiking or scalping for 1 to 2 days after application.
trinexapac-ethyl (Governor) 0.17 G	129 to 165 lb	0.22 to 0.28	Apply before verticutting, scalping, or spiking the bermudagrass. Apply 1 to 5 days before overseeding. To minimize yellowing, use iron at recommended rates or available nitrogen at 0.2 to 0.5 pound per 1,000 square feet.
(Primo Maxx) 1 MEC or (T-Nex) 1 AQ (Primo WSB) 25 WP	6 to 44 fl oz 2.75 to 21.8 oz	0.08 to 0.34 0.08 to 0.34	Application rate varies with turfgrass species and height of cut. Apply to actively growing, nonstressed turf. More growth suppression occurs at lower mowing heights. See label for specific rate and other directions and precautions. Repeat applications can be made but do not exceed a total of 21.4 pints per acre per year of Primo Maxx or a total of 174 ounces per acre per year of Primo WSB. Do not exceed 19 pints per acre per year of T-Nex. Refer to the respective Primo label for guidelines regarding mowing prior to and following application. Mix with 0.5 to 4 gallons of water per 1,000 sq ft (20 to 174 gallons per acre). Primo can be applied to putting greens. See label for directions.
Lawn Edging			
maleic hydrazide (Retard) 2.25 lb/gal (Royal Slo-Gro) 1.5 lb/gal (Liquid Growth Retardant) 0.6 lb/gal	1.33 gal/100 gal water 2 gal/100 gal water 6.67 gal/100 gal water	3 3 4	Apply in spring to a 6-inch band along sidewalks. Consult instructions on applicator for delivery dosage.
mefluidide (Embark) 0.2	1.36 gal/174 gal water	0.27	For Kentucky bluegrass, tall fescue, chewings fescue, red fescue, perennial ryegrass, and St. Augustinegrass. For bermudagrass, use 5.45 gallons in 174 gallons water. Apply in 6- to 12-inch bands. Avoid overlapping.
trinexapac-ethyl (Governor) 0.17 G	100 to 259 lb	0.17 to 0.44	Do not exceed 2.5 pounds active ingredient per acre per year. These rates should provide 50% turf growth suppression for 4 weeks with minimal yellowing.
(Primo Maxx) 1 MEC (T-Nex) 1 AQ (Primo WSB) 25 WP			Apply 0.75 to 2 ounces per 1,000 linear feet of Primo Maxx or T-Nex, or 0.4 to 2 ounces per 1,000 linear ft of Primo WSB. Apply to actively growing, nonstressed turf. Apply along perimeter of lawns, sidewalks, curbs, parking lots, driveways, flower beds, or fences. Apply in an 8- to 12-inch band along the perimeter of the lawn to reduce growth of turf into adjacent areas. Application rate varies with turf species. Follow label directions for repeat applications and other precautions.

PLANT GROWTH REGULATORS

IX — ANIMAL DAMAGE CONTROL

Animal Damage Control

C. S. DePerno, Department of Forestry and Environmental Resources

General Practices

Before undertaking any control measures, read the North Carolina law regarding Wildlife Killed for Depredations or Accidentally (15 NCAC 10B.0106).

Damage caused by rodents, other mammals, and birds can be prevented or reduced by one or more of the following practices.

1. Removing or minimizing available food, water, and shelter. This includes general cleanup, rodent-proofing, and bird-proofing.
2. Using repellents.
3. Capturing offenders and releasing them away from the problem area.
4. Eliminating the offenders by trapping, shooting, or poisoning.

North Carolina legislation restricts the use of pesticides on wild birds and mammals to species that are not protected by the N.C. Wildlife Resources Commission and that have been classified as pests by the North Carolina Pesticide Board.

Currently, five rodent and eight bird species have been classified as unprotected pests, and pesticides may be used in controlling these species subject to specific restrictions of the Pesticide Board. These species are: Norway rat, black rat (roof rat), house mouse, pine vole and meadow vole, common grackle, boat-tailed grackle, brown-headed cowbird, red-winged blackbird, starling, English sparrow, pigeon, and gulls (near airports).

Federal, state, and local laws on animal control are rather strict and subject to change. Contact authorities at the N.C. Wildlife Resources Commission (1701 Mail Service Center, Raleigh, NC 27606; 919-707-0010); or State Director, USDA-APHIS, Wildlife Services (6213-E Angus Dr., Raleigh, NC 27617; 866-487-3297 or 919-786-4480) to clarify the current legal status of a particular control practice.

Report any rabid animals or suspected rabid animals to the local health department.

Traps or chemicals mentioned in this section can usually be purchased at hardware or farm supply stores. If not locally available, check with your county N.C. Cooperative Extension agent. For more information, the following publications are available at your county Cooperative Extension center:

Wildlife Damage Control: Controlling Vole Damage in Commercial Orchards and Ornamental Nurseries, AG-472-1;

Wildlife Damage Control: Snakes, AG-472-2; and

Wildlife Damage Control: Beavers, AG-472-4.

Wildlife Killed for Depredations or Accidentally, 15A N.C. Administrative Code 10B .0106

a) *Depredation Permit*

1) **Endangered or Threatened Species**. No permit shall be issued to take any endangered or threatened species of wildlife listed under 15A NCAC 10I by reason of depredations to property. An individual may take an endangered or threatened species in immediate defense of his own life or of the lives of others without a permit. Any endangered or threatened species that may constitute a demonstrable but nonimmediate threat to human safety shall be reported to a federal or state wildlife enforcement officer, who, upon verification of the report, may take or remove the specimen as provided by 15A NCAC 10I .0102.

2) **Other Wildlife Species**. Except as provided in subparagraph (1) of this paragraph, the executive director or an agent of the Wildlife Resources Commission may, upon application of a landholder and after such investigation of the circumstances as required, issue a permit to such landholder to take any species of wildlife that is or has been damaging or destroying landholder's property provided there is evidence of substantial property damage. No permit may be issued for the taking of any migratory birds and other federally protected animals unless a corresponding valid U.S. Fish and Wildlife Service depredation permit has been issued. The permit shall name the species allowed to be taken and, at the discretion of the Executive Director or an agent, may contain limitations as to age, sex or any other condition within the species so named. The permit may be used only by the landholder or another person named on the permit.

Wildlife Damage Control Agents. Upon satisfactory completion of Wildlife Resources Commission-approved training and satisfactory demonstration of a knowledge of wildlife laws and safe, humane wildlife handling techniques, an individual may apply to the Wildlife Resources Commission to become a Wildlife Damage Control Agent (WDCA). Those persons approved as agents by the commission may then issue depredation permits to landholders and list themselves as a second party to provide the control service. WDCAs may not issue depredation permits for big game animals, bats, or species listed as endangered, threatened, or of special concern under Rules 10I .0103, .0104, and .0105 of this chapter. WDCAs must report to the Wildlife Resources Commission the number and disposition of animals taken, by county, annually. Records must be available for inspection by a Wildlife Enforcement officer at any time during normal business hours. WDCA status may be revoked at any time by the executive director when there is evidence of violations of wildlife laws, failure to report, or inhumane treatment of animals by the WDCA. WDCAs may not charge for the permit but may charge for their investigations and control services. To maintain a knowledge of current laws, rules, and techniques, WDCAs must renew their agent status every three years by showing proof of having attended at least one Wildlife Commission-approved training course provided for the purpose of reviewing and updating information on wildlife laws and safe, humane wildlife handling techniques within the previous 12 months.

b) *Term of Permit*

Each depredation permit issued by the executive director or an agent shall have entered thereon a date or time of expiration after which date or time the same shall become invalid for any purpose, except as evidence of lawful possession of any wildlife that may be retained thereunder.

c) *Manner of Taking*

1) **Taking without a Permit**. Wildlife taken without a permit while committing depredations to property may, during the open season on the species, be taken by the landholder by any lawful method. During the closed season such depredating wildlife may be taken without a permit only by the use of firearms.

2) **Taking with a Permit**. Wildlife taken under a depredation permit may be taken only by the method or methods specifically authorized by the permit. When trapping is authorized, to limit the taking to the intended purpose, the permit may specify a reasonable distance from the property sought to be protected, according to the particular circumstances, within which the traps must be set. The executive director or agent may also state in a permit authorizing trapping whether or not bait may be used and the type of bait, if any, that is authorized. In addition to any trapping restrictions that may be contained in the permit, the

method of trapping must be in accordance with the requirements and restrictions imposed by G.S. 113-291.6 and other local laws passed by the N.C. General Assembly. No depredation permit shall authorize the use of poisons or pesticides in taking wildlife except in accordance with the provisions of the North Carolina Pesticide Law of 1971, the Structural Pest Control Act of 1955, and Article 22A of Chapter 113 of the General Statutes of North Carolina. No depredation permit shall authorize the taking of wildlife by any method by any landholder upon the lands of another.

3) **Intentional Wounding**. It is unlawful for any landholder, with or without a depredation permit, intentionally to wound a wild animal in a manner so as not to cause its immediate death as suddenly and humanely as the circumstances permit.

d) *Disposition of Wildlife Taken*

1) **Generally**. Except as provided by the succeeding subparagraphs of this paragraph, any wildlife killed accidentally or without a permit while committing depredations shall be buried or otherwise disposed of in a safe and sanitary manner on the property. Wildlife killed under a depredation permit may be transported to an alternate disposal site if desired. Anyone in possession of carcasses of animals being transported under a depredation permit must have the depredation permit in their possession. Except as provided by the succeeding subparagraphs of (d)(2) through (6) of this rule, all wildlife killed under a depredation permit must be buried or otherwise disposed of in a safe and sanitary manner.

2) **Deer**. The edible portions of up to five deer may be retained by the landholder for consumption but must not be transported from the property where the depredations took place without a valid depredation permit. An enforcement officer, if so requested by the permittee, shall provide the permittee a written authorization for the use by a charitable organization of the edible portions of the carcass. The nonedible portions of the carcass, including head, hide, feet, and antlers, shall be disposed of as specified in subparagraph (1) of this paragraph or turned over to a wildlife enforcement officer for disposition. When a deer is accidentally killed on a road or highway by reason of collision with a motor vehicle, the law enforcement officer who investigates the accident shall, upon request of the operator of the vehicle, provide such operator a written permit authorizing him to possess and transport the carcass of such deer for his or her personal and lawful use, including delivery of such carcass to a second person for his or her private use or the use by a charitable organization upon endorsement of such permit to such person or organization by name and when no money or other consideration of value is received for such delivery or endorsement.

3) **Fox**. Any fox killed accidentally by a dog or dogs, motor vehicle, or otherwise shall be disposed of as provided by subparagraph (1) of this paragraph. Any fox killed under a depredation permit may be disposed of in the same manner or, upon compliance with the fur tagging requirements of 15A NCAC 10B .0400, the carcass or pelt thereof may be sold to a licensed fur dealer. Any live fox taken under a depredation permit may be sold to a licensed controlled hunting preserve for fox in accordance with G.S. 113-273(g).

4) **Furbearing Animals**. The carcass or pelt of any furbearing animal killed during the open season for taking such furbearing animal either accidentally or for control of depredations to property, whether with or without a permit, may be sold to a licensed fur dealer provided that the person offering such carcass or pelt for sale has a valid hunting or trapping license, provided further that, bobcats and otters may only be sold upon compliance with any required fur tagging requirement set forth in 15A NCAC 10B .0400.

5) **Animals Taken Alive**. Wild animals in the order Carnivora and beaver shall be humanely euthanized either at the site of capture or at an appropriate facility designed to humanely handle the euthanasia or released on the property where captured. Animals transported or held for euthanasia must be euthanized within 12 hours of capture. Anyone in possession of live animals being transported for relocation or euthanasia under a depredation permit must have the depredation permit in their possession.

6) **Wild Birds or Animals Killed Accidentally with Motor Vehicles or Found Dead**. A person killing a wild bird or wild animal accidentally with a motor vehicle or finding a dead wild bird or wild animal that was killed accidentally may possess that wild bird or wild animal for a period not to exceed 10 days for the purpose of delivering it to a licensed taxidermist for preparation. The licensed taxidermist may accept the wild bird or wild animal after determining that the animal was killed accidentally. The taxidermist shall certify and record the circumstances of acquisition as determined by his or her inquiry. Licensed taxidermists shall keep accurate records of each wildlife specimen received as required by rule 10H .1003 of this chapter. Upon delivery of the finished taxidermy product to the person presenting the animal, the taxidermist shall give the person a receipt in the form required by the Wildlife Resources Commission indicating the species, date of delivery, circumstances of initial acquisition, and any other information that may be required on the form. A copy of this receipt shall be filed with the Wildlife Resources Commission within 10 days of the date of delivery of the mounted specimen. The receipt shall serve as the nontransferable permit for continued possession of the mounted specimen and shall be retained by the person for as long as the mounted specimen is kept. Mounted specimens possessed pursuant to this rule may not be sold, and if such specimens are transferred by gift or inheritance, the new owner must apply for a new permit and must submit the written receipt originally obtained from the taxidermist to document the legality of possession. This provision does not allow possession of accidentally killed raptors; migratory birds; species listed as endangered, threatened, or of special concern under Rules 10I .0103, .0104, and .0105 of this chapter; bear or wild turkey.

e) *Reporting Requirements*

Any landholder who kills a deer, bear, or wild turkey under a currently valid depredation permit shall report such kill on the form provided with the permit and mail the form immediately upon the expiration date to the Wildlife Resources Commission. The killing and method of disposition of every game animal and game bird, every furbearing animal, and every nongame animal or nongame bird for which there is no open season, when killed for committing depredations to property, without a permit, shall be reported to the Wildlife Resources Commission within 24 hours following the time of such killing, except that when the carcass or pelt of a fox killed under a depredation permit or of a furbearing animal killed with or without a permit is lawfully sold to a licensed fur dealer in this state, the fur dealer is required to report the source of acquisition, and no report is required of the seller.

If you have questions, contact:

N.C. Wildlife Resources Commission
NC State University Centennial Campus
1751 Varsity Drive
Raleigh, NC 27606
1-919-707-0010

ANIMAL DAMAGE CONTROL

Description of Potential Animal Pests

C. S. DePerno, Department of Forestry and Environmental Resources

Table 9-1. Description of Potential Animal Pests

Animal	Description	Indicators or Signs of Presence	Range and Habits	Bait or Control
Bat*	At least nine species. Mouselike body 3 to 5.5 inches long, including tail, wingspread 6 to 16 inches. Prominent ears.	Strong odor, noise, or movement, and squeaking. Brown stains (rubs) near openings or vents in roof. Bats seen entering or leaving premises at dusk or dawn.	Statewide. Roosts in caves, hollow trees, attics, and walls by day. Nocturnal, insectivorous. Some species hibernate in attics, but most are migratory.	See Animal Control Suggestions.
Beaver*	Very large rodent with brown to reddish brown fur, large orange incisors, and a large, flat, scaly tail. Weight up to record of 120 pounds, length to 48 inches, including an 11-inch scaly tail. Webbed hind feet to 7 inches long.	Girdled and felled trees, mud and stick dams, trails to terrestrial feeding sites. Dens may be lodge style of sticks and mud in middle of pond or on bank.	Statewide. Eats wide variety of herbaceous and woody plant material. Most active from dusk till dawn. Needs constant supply of water; ponded areas may cover a hundred or more acres in flat terrain. Potential for significant timber loss in bottomland drainages.	Protect gardens with electric fence, individual trees with wire mesh and pond drains with a wire cage. Remove local populations by trapping. No depredation permit needed. See N.C. Cooperative Extension publication *Wildlife Damage Control: Beavers*, AG-472-4.
Bobcat*	Brown and buff mixed with black and gray. Weight 15 to 35 pounds. Total length 24 to 40 inches. Short tail. Bobbing motion noticed while running.	Will cache food; vegetation scratched over remains of kill. Round, 2-inch footprints without claw marks. Droppings contain bone fragments and mouse and/or rabbit hair.	Statewide. Prefers thickets, river bottoms, swamps, and brushy areas. Seldom seen. Feeds primarily on rodents, rabbits, poultry, birds, amphibians, and rarely on livestock and game.	Use a No. 2 or 3 steel foot trap. Hunt or use predator call for sport. Check NCWildlife Resources Commission laws and regulations for details.
Chipmunk*	Small, striped, squirrel-like rodent; grayish or reddish brown.	1.5- to 2-inch open burrows with few signs of digging.	Piedmont and western part of state. May dig flower bulbs or seeds. Occasionally damages truck crops. Hibernates.	Use a small box-type live trap or kill trap (wooden-base rat snap) placed under a board or bucket at holes or near runways.
Cotton Rat* (field rat)	Up to 12 inches long, including 5-inch tail; dark brown, grizzled appearance. Scaly and nearly hairless tail. Hindfoot to 1.3 inches long.	Trails in grass, nests of cut grass usually above ground. Burrows with several 1.5- to 2-inch openings.	Statewide. Lives in dense cover, field edges, and wet sites. May damage alfalfa, cotton, and row crops.	Use a small box-type live trap or kill trap (wooden-base rat snap) placed under a board or bucket at holes or near runways.
Cowbird*†	Robin-sized. Males glossy black with brown head. Females plain gray with pale throat.	Damage to corn occurs when grackles expose ears.	Statewide, chiefly in eastern half. Forms large winter roosts with red-winged blackbirds and starlings.	See Animal Control Suggestions.
Coyote*	Medium-dog size, 20 to 50 pounds Gray or reddish-gray hair. Ears pointed, muzzle long and narrow, and a bushy tail. Length from tip of nose to end of tail up to 48 inches.	Hindfoot diameter is 2 to 2.75 inches with claw marks. Scat is doglike and often contains hair and bone. Dens at bases of trees in cultivated fields and pastures, brush covered slopes, sandy ridges and caves.	Statewide. All habitats. Eats vegetables, crops, rodents, fruits, poultry, game, and livestock. Will scavenge.	Shoot, trap, and animal husbandry practices.
Crow*† (American and Fish)	Abundant in the state. Solid black; about 17 inches long with long, rounded wings. Larger than blackbird.	Roosting in trees along planted fields. Uprooted seedlings; mature corn opened and shattered.	Statewide, except for high mountain peaks.	See Animal Control Suggestions.
English Sparrow	Small, reddish brown; upper parts streaked with black. Gray underparts. Males have black throat and gray cap.	Soiling of feed lots and window ledges.	Statewide. Feeds and roosts in flocks of 100 to 300 birds. Frequents feed lots; roosts on ledges of buildings and in trees, vines, and shrubs.	See Animal Control Suggestions. No depredation permit needed.
Field Mouse,* Deer Mouse, White-Footed Mouse	Up to 6 inches; tan above, light below; tail furred and bicolored. Large ears and eyes.	0.75-inch footprints and tail drag marks. Seed dug when no sprout visible. Sprouts dug from side, not pulled. Holes in bare, freshly planted fields and sparse cover.	Statewide. Often follows other rodent burrows.	Trap with small snap or mouse-size box traps or use glue boards within residences. See RODENTICIDE section.
Flying Squirrel	Head and body 6 inches with 4-inch tail. Unique folds of loose skin on sides of body allow gliding. Thick, soft fur is gray.	Seldom seen; active at night. Eats seeds, nuts, insects, and bird eggs.	Statewide. Lives in holes in trees, buildings.	Invades attics where movements and food storage usually require removal by trapping. See Animal Control Suggestions.
Grackle*† (Common and boat-tailed)	Slightly larger than robin; long, keel-shaped tail. Iridescent, glossy black.	Sprouting corn pulled up. Mature corn ears opened and shattered.	Statewide, chiefly in eastern half. Forms large winter roosts with red-winged blackbirds, cowbirds, and starlings.	See Animal Control Suggestions.
Gray Fox*	Small dog-sized, to 15 pounds and measures 32 to 45 inches long from nose to tip of tail. Bushy tail is black-tipped with a black central stripe. Grizzled gray to reddish color.	Small, dainty, doglike tracks, 1 to 2 inches wide with claw marks. Large burrows under rocks or in hollow logs or dens in piles of sawdust. Climbs trees readily.	Statewide. Prefers brush, timbered, and swampy areas. Mostly nocturnal. Opportunistic carnivore. Eats rodents, squirrels, birds, eggs, turtles, insects, fruits, poultry, and game.	Occasional rabies carrier. Report any rabid animals or suspected rabid animals to local health department. Contact N.C. Wildlife Resources Commission about depredating animals.
Gray Squirrel	Up to 18 inches long; excellent tree climber; bushy tail; grayish color. Tail 8.5 inches long.	Twigs clipped; fruit and nuts of large trees eaten; footprints 2 to 2.5 inches.	Lives close to trees and shrubs; eats nuts and fruits and occasionally girdles bark; may move into attics.	See Animal Control Suggestions.
Groundhog* (Woodchuck)	A large, grizzled brown, burrowing rodent. Weight 4 to 15 pounds. Thick bodied, short tail, short legs, coarse fur.	Large burrows with earth mounded and scattered at entrance in fence rows, hedgerows, meadows, hayfields, and edges of woods.	Mountains through piedmont. Multiple entrances to den. Mainly vegetarian. Hibernates during the winter.	Use body grip or box-type live trap. Shoot.
Gulls*†	Robust birds with long, pointed wings and webbed feet. Tails usually square.	Large flocks near dumps, cultivated fields, lakes, and rivers.	May be seen occasionally in the piedmont but more common near the coast.	See Animal Control Suggestions.

Table 9-1. Description of Potential Animal Pests

Animal	Description	Indicators or Signs of Presence	Range and Habits	Bait or Control
House Mouse	Up to 5 inches; long, nearly hairless tail; body about same color top and bottom.	Feed sacks cut; holes 1 inch and less; footprints less than 0.5 inch. Nests on finely shredded material. Black-colored droppings the size of grains of rice.	Statewide. Lives in walls, furniture, feed sacks, and other enclosures. Can live without water for a long time.	Trap with small snap or mouse-size box traps or use glue boards within residences. For poison, see RODENTICIDE section.
House Rat (Norway)	Up to 18 inches long; tail length less than head and body; dark brown color; hindfoot 1.5 inches or more.	Feed sacks cut; gnawed holes; droppings up to 0.75 inch, footprints 1.5 inches; underground burrows nearby.	Statewide. Good climber. Lives at ground level in burrows, litter, walls, and other protective enclosures.	Use environmental control. Minimize food, shelter, and water. Set snap traps and rat-size glue boards. For poison, see RODENTICIDE section. No depredation permit needed.
Meadow Vole* (Meadow Mouse)	3.5 to 5 inches long body and head; tail longer than length of hind leg; dark brown color. Tail bicolored. Lighter belly. Rather long, grizzled fur.	Small, shallow burrows and trails in grass. Trees or shrubs gnawed from ground level up under mulch or snow cover.	Statewide. Damage occurs to trees and shrubs from slightly below ground level and up. Lives in areas of dense cover.	Environmental control through close mowing or clean cultivation. Trap with mouse traps placed in runs and covered with tar paper or large flower pot. A depredation permit is required for any trapping. For poison, see RODENTICIDE section and *Wildlife Damage Control: Controlling Vole Damage in Commercial Orchards and Ornamental Nurseries, AG-472-1.*
Mole* (Eastern)	4 to 6 inches; short tail and pointed nose. Very short, paddlelike front feet; eyes and ears not visible. Very soft fur.	Extensive ridges, sometimes with mounds 1 to 2 inches high with lumpy appearance; ridges pushed up instead of dug out; no gnawings.	Statewide. Lives underground; makes tunnels, sometimes small mounds; feeds on worms, and insects, but not plants.	Use spear-type traps for best control. Ground insecticides will lessen food supply and give some control.
Mole (Star-nosed)	6-7 inches, large scaled feet and a long, thick tail with concentric rings of short, coarse hair. Has a distinctive flesh-colored ring of retractable tentacles around its nose.	Extensive ridges, sometimes with mounds 1 to 2 inches high with lumpy appearance; ridges pushed up instead of dug out; no gnawings.	Limited distribution in the central Piedmont and western mountains.	Considered "non-game" and protected. Star-nosed moles cannot be killed without requesting and receiving a permit from the North Carolina Wildlife Resources Commission (NCWRC).
Muskrat*	Dark or reddish-brown, ratlike rodent; 20 to 24 inches total length; tail 8 to 10 inches, hairless, scaly, flattened from side to side. Weight 2 to 4 pounds.	Houses of rushes, etc. or burrows in banks, with entrances below water level. Trails at water's edge.	Statewide. Eats aquatic plants; occasionally animal material and crops. Burrows into earthen dams.	See Animal Control Suggestions.
Nutria*	Large, ratlike, brownish rodent. The most noticeable characteristics are its size and a long, round tail that is scantily haired. Weight 15 to 18 pounds.	Bank burrows or platform nests anchored to marsh vegetation. Worn trails, eaten-away marsh vegetation. Trails at water's edge.	Northeastern coastal area of state. An introduced species. Feeds on rushes, grasses, seeds, roots, and coarse vegetation. Usually consumes food on shore in evening.	Use No. 1.5 foot trap, box trap, or body-grip trap.
Opossum*	Generally light gray. Average length 33 inches, including a 12-inch rat-like prehensile tail. Female has pouch for carrying young (marsupial). Weight up to 15 pounds.	Omnivorous; particularly detrimental to poultry. Badly mauled poultry typical of opossum kill.	Statewide. Prefers living by streams or swamps in wooded areas. Lives in hollow logs, rocks, crevices, etc. Frequently found in towns and cities, killed by vehicles.	Easily caught in box-style live traps.
Pigeon	Crow-sized. Slate blue background, two black wing bars. Greenish purple sheen to neck feathers.	Soiling of feed lots and window sills.	Statewide. Feed, roost, and loaf in flocks, except in breeding season.	See Animal Control Suggestions. No depredation permit needed.
Pine Vole* (Orchard Rat)	Body and head about 3 to 4.25 inches long with tail shorter than length of hind leg; dark brown with reddish tinge.	Soft, spongy ground. Extensive underground burrows. Plants severed or girdled underground or dying without visible cause. Quarter-sized holes in lawns without soil mounds.	Statewide. Lives underground. Feeds on roots and bases of plants below ground level. Damage prevalent on bulbs, ornamental shrubs, and apple trees. Often follows mole runs and other burrows.	Close mowing and clean cultivation deter voles. Set mouse traps in runs, and cover them with tar paper or large flower pot. Depredation permit required for trapping. For poison, see RODENTICIDE section and *Wildlife Damage Control: Controlling Vole Damage in Commercial Orchards and Ornamental Nurseries, AG-472-1.*
Rabbit* (Cottontail)	Up to 15 inches long, including 2-inch fluffy tail. 2.5 to 4 pounds. Long ears.	Presence indicated by round fibrous pellets. Gnawing and clipping of vegetation at 45 degree angle.	Statewide. Feeds on all parts of plants. Prefers succulent foods- young shoots and vegetables.	See Animal Control Suggestions.
Raccoon*	Small dog-size; black face mask; color rings on tail; grizzled gray color. Good climber. Weight to 35 pounds.	2- to 3-inch handlike tracks; seeds dug, leaving 2-inch holes. Ripening corn opened and eaten. May den in structures, including attics and chimneys.	Statewide. Prefers wooded areas around water; eats fish, waterfowl, poultry, eggs, grain, and fruits.	In urban settings, remove any excess pet food immediately after feeding (dog, cat, etc.) or feed pets indoors. See Animal Control Suggestions. Occasional rabies carrier. Report rabid animals or suspected rabid animals to local health department. Contact N.C. Wildlife Resources Commission about depredating animals.
Red Fox*	Medium dog-size to 15 pounds and measures 36 to 41 inches long from nose to tip of tail. Primarily reddish with white-tipped tail and white throat. However, 3-color phases (red, yellow, black) may occur.	Triangular foot prints with claw marks. Dens in burrows located in cleared or semicleared areas, under stumps, and vegetated slopes.	Statewide. Prefers fairly open country; agricultural areas and forest edges. Mostly nocturnal. Eats rodents, birds, fruits, poultry, livestock, and game.	Occasional rabies carrier. Report rabid animals or suspected rabid animals to local health department. Contact N.C. Wildlife Resources Commission about depredating animals.
Red-Winged Blackbird *†	Robin-size. Male black with red-wing patch; female brown with heavily striped underside.	Corn husks opened during soft dough stage. Soft kernels fed upon. Fungi and molds may develop on corn.	Statewide, chiefly eastern half. Forms large winter roosts with grackles, cowbirds, and starlings. Migrates north in early spring. During late summer and early fall, roosts without blackbirds and starlings.	See Animal Control Suggestions.

ANIMAL DAMAGE CONTROL

Table 9-1. Description of Potential Animal Pests

Animal	Description	Indicators or Signs of Presence	Range and Habits	Bait or Control
Roof Rat (Black or Brown)	Up to 17 inches long; tail longer than head and body; dark brown, lighter belly. Hindfoot 1.5 inches or less.	1.5-inch holes in feed sacks; droppings to 0.5 inches; large, gnawed holes; 1- to 1.5-inch tracks. Smears and swing marks.	Statewide, especially coastal area. Excellent climber; lives above ground in walls, litter, etc.	Similar to house rat. See RODENTICIDE section. No depredation permit needed.
Skunk*	House cat size; black with white markings; bushy tail; bad smell. Poor climber.	Rooting in lawns or other fertile spots that have been pushed up instead of dug; tracks usually dainty, from 1 to 3 inches	Statewide. Roots for insects; literally upends some areas; hunts mice and eggs around farm buildings. Common rabies carrier.	See Animal Control Suggestions.
Snakes (37 species in N.C.)	Beneficial for their predation on rodents. Six species are venomous. Cottonmouths, copperheads, and three species of rattlesnakes have a pit between eyes and nose with vertical eye pupil. Coral snakes have round pupils and yellow, red, and black rings encircling body. Yellow and red bands touch and the nose area is black.	Droppings with white ends similar to that of birds, shed skins.	Coral Snake (rare) found in sandhills and coastal plain.Cottonmouth Moccasin found in sandhills and coastal plain.Copperhead found statewide.Other-nonvenomous species are represented throughout state. Venomous species often nocturnal during summer months. All hibernate.	See Animal Control Suggestions and N.C. Cooperative Extension publication *Wildlife Damage Control: Snakes*, AG-472-2.
Starling	Robin-size. Large yellow beak; gold-flecked, iridescent blue-black plumage. Delta shape of wing in flight. Short tail.	Soiling of feed lots and window ledges. Nests in holes.	Statewide, chiefly in eastern half. Feeds, roosts, and loafs in flocks during nonbreeding season.	See Animal Control Suggestions. No depredation permit needed.
White-Tailed Deer*	Length 65 to 70 inches; height at shoulder 30 to 40 inches; weight 75 to 250 pounds. Reddish brown hair; slender, long-legged. Males have antlers.	Tracks are best sign. Twigs and leaves clipped off raggedly about 24 inches above ground. Raisin-sized, tear-shaped pellets in groups of 10 to 25.	Statewide. Found in all habitats from rural farming communities to urban residential subdivisions. Active any time of the day. Primarily feeds on agricultural crops, grasses, fruits, mushrooms, acorns.	For population control, see Animal Control Suggestions.
Woodpecker*†	Birds with sharply pointed beaks used for chipping and digging. Stiff tail and gripping toes.	Drumming sounds on trees or buildings. Some peck holes in siding for nest cavity. Cedar-sided houses frequently damaged.	Statewide. Undulating flight patterns.	Woodpeckers are federally protected. Prevent access by tacking netting, used to prevent bird damage to blackberries, to eves and to side of house. They may be discouraged by using frightening devices, such as rubber snakes, pie pans, metal streamers, or other materials that either flash or scare. Attaching bags of suet to trees may provide food, and thus prevent birds from drilling on siding. If the problem persists, contact State Director, USDA-APHIS-Wildlife Services Agency, 6213-E Angus Drive, Raleigh, NC 27613, (919) 786-4480.

* Covered by N.C. law regarding Wildlife Killed for Depredations or Accidentally (10B.0106).

† Covered by Migratory Bird Treaty Act (50 CFR 21).

ANIMAL DAMAGE CONTROL

Animal Control Suggestions

C. S. DePerno, Department of Forestry and Environmental Resources

Table 9-2. Animal Control Suggestions

Problem Areas	Repellent Formulation	Application Methods	Rates or Procedures	Mode of Action	Remarks
Bat — Gray bats and Indiana bats are classified as endangered species.					
Attics	Flood lights in attic. Other repellents, such as mothballs, are usually not successful.		Illumination works best on repelling colonies that have just begun.		All bats in North Carolina are beneficial insectivores. Removal or displacement of large colonies should be carried out under the supervision and guidance of federal or state agencies and licensed pest control operators. Exclusion of large natal colonies during April through August should NOT occur until after the young have left the structure.
	Ultrasound devices in attic.		Ultrasound is a new technique that has not been evaluated and may not work.		
		Bat proofing is the only permanent solution.	Seal all but a few openings with hardware cloth or steel wool for smaller openings. Wait a few days then seal remaining openings half hour after dark.		Use tongs or heavy leather gloves to pick up dead or dying bats. Bats occasionally carry rabies. If bitten or scratched, wash vigorously with soap and water and see a doctor. Collect the bat without crushing its skull, place it on ice, and take it to the county or city health department for analysis.
Cowbird — Covered by Migratory Bird Treaty Act (50 CFR 21).					
All problem areas		Scare tactics.	Propane exploders, shell crackers, and electronic sound devices.		Use combination of scare devices; vary times and locations. Be persistent.
Feedlots	4-Aminopyridine 1.0% (Avitrol)	Prebait with untreated grain, then replace with treated grain. Carefully observe cowbirds' habits to establish feeding habits and locations to prevent ingestion by desirable or non-target birds.	Rates vary with extent of bird infestation.	Blocks potassium channels resulting in increased cholinergic nervous system activity.	Restricted label. For use by or under the supervision of government agencies or pest control operators. Follow label instructions carefully.
Structures, nesting and roosting sites	4-Aminopyridine 0.5% (Avitrol)			Blocks potassium channels resulting in increased cholinergic nervous system activity.	
Standing field corn (NOT SWEET CORN)	4-Aminopyridine 0.33% (Avitrol)	Same as Feedlots, but ground and air application permitted.	Same as Feedlots.	Blocks potassium channels resulting in increased cholinergic nervous system activity.	Same as Feedlots.
Crow — Covered by Migratory Bird Treaty Act (50 CFR 21).					
Newly planted and standing crops		Scare tactics.	Propane exploders and shell crackers.		Use combination of scare devices; vary times and locations. Be persistent.
Newly planted corn fields	4-Aminopyridine 1.0% (Avitrol)	Prebait with untreated corn, then replace with treated corn. Be careful to place treated corn in the crows' feeding area to prevent ingestion by desirable or non-target birds.	Rates vary with extent of bird infestation.	Blocks potassium channels resulting in increased cholinergic nervous system activity.	Restricted label. For use by or under the supervision of government agencies or pest control operators. Follow label instructions carefully.
English Sparrow					
Structures, nesting and roosting sites		Scare tactics.	Propane exploders and shell crackers.		Use combination of scare devices; vary times and locations. Be persistent.
		Roost barrier (Nixalite).	Attach to ledges.		Long lasting way to prevent birds from roosting.
	Polybutenes 97%	Paste.	Spread on ledges.	Surface repellent that results in an unpleasant sticky sensation.	Follow label instructions carefully.
	Polybutenes 48%	Liquid spray.	Spray on shrubs and rafters.	Surface repellent that results in an unpleasant sticky sensation.	Follow label instructions carefully.
	4-Aminopyridine 0.5% (Avitrol)	Prebait with untreated grain, then replace with treated grain. Carefully observe sparrows' habits to establish feeding habits and locations to prevent ingestion by desirable or non-target birds.	Rates vary with extent of bird infestation.	Blocks potassium channels resulting in increased cholinergic nervous system activity.	Restricted label. For use by or under the supervision of government agencies or pest control operators. Follow label instructions carefully.
	Methyl anthranilate 26.4%	Liquid Spray	Spray on problem areas	Targets taste senses	Follow label instructions carefully.

ANIMAL DAMAGE CONTROL

Table 9-2. Animal Control Suggestions

Problem Areas	Repellent Formulation	Application Methods	Rates or Procedures	Mode of Action	Remarks
Grackle — Covered by Migratory Bird Treaty Act (50 CFR 21).					
All problem areas		Scare tactics	Propane exploders, shell crackers, and electronic sound devices.		Use combination of scare devices; vary times and locations. Be persistent.
Standing field corn *(NOT SWEET CORN)*	4-Aminopyridine .03% (Avitrol)	Prebait with untreated grain then replace with treated grain. Carefully observe grackles' habits to establish feeding habits and locations to prevent ingestion by desirable or non-target birds. Ground and air application permitted.	Rates vary with extent of bird infestation.	Blocks potassium channels resulting in increased cholinergic nervous system activity.	Restricted label. For use by or under the supervision of government agencies or pest control operators. Follow label instructions carefully.
Gulls — Covered by Migratory Bird Treaty Act (50 CFR 21).					
On or near airport runways, endangering air traffic; normally fall and winter flocks cause problems.		Scare tactics	Propane exploders and shell crackers.		Use combination of scare devices; vary times and locations. Be persistent.
	4-Aminopyridine 25% (Avitrol)	Prefeeding with bread is essential to deterrence. Carefully observe gulls' habits to prevent ingestion by desirable or non-target birds.	Rates vary with extent of gull problem. May need to retreat after rain.	Blocks potassium channels resulting in increased cholinergic nervous system activity.	FOR PROFESSIONAL USE ONLY. Gulls are protected by law and treaty. Use under U.S. Fish and Wildlife Service or state game commission supervision. Both federal and state permits are required to use this method of bird control.
	Polybutenes 49%	Liquid spray.	Spray on landing sites.	Surface repellent that results in an unpleasant sticky sensation.	Follow label instructions carefully.
Muskrat — Covered by 15 NCAC 10B.0106. A depredation permit is needed prior to setting a trap for this species. Where shooting is recommended, a depredation permit is also needed unless the animal is in the act of destroying property.					
Water control structures and dams		Trapping	Use 1.5 foot trap, #110 body-grip trap, single-catch box trap, or live catch muskrat colony trap. Place traps in muskrat trails and at feeding sites. Bait live-catch traps with carrot or apple.		Check traps daily.
Moles					
Burrows	Castor Oil 14.85%	Place in burrows		Surface repellent	
Pigeon					
Structures, nesting and roosting sites		Scare tactics	Propane exploders and shell crackers.		Use combination of scare devices; vary times and locations. Be persistent.
		Roost barrier (Nixalite)	Attach to ledges.		Long lasting way to prevent birds from roosting.
	Polybutenes 97%	Paste	Spread on ledges.	Surface repellent that results in an unpleasant sticky sensation.	Follow label instructions carefully.
	Polybutenes 48%	Liquid spray	Spray on shrubs and rafters.	Surface repellent that results in an unpleasant sticky sensation.	
	4-Aminopyridine 0.5% (Avitrol)	Prebait with untreated corn that looks the same as treated corn. Carefully observe pigeons' habits to establish proper feeding locations to prevent ingestion by desirable or non-target birds. Place bait in high locations and on high ledges.	Rates vary with extent of pigeon infestation.	Blocks potassium channels resulting in increased cholinergic nervous system activity.	Restricted label. For use by or under the supervision of government agencies or pest control operators. Follow label instructions carefully.
	Methyl anthranilate 26.4%	Liquid Spray	Spray on problem areas.	Targets taste senses.	Follow label instructions carefully.
In and around farm buildings; pipe yards; loading docks; building tops; inside other buildings; bridges		Shooting			Discharging firearms within city or town limits is typically illegal.

Table 9-2. Animal Control Suggestions

Problem Areas	Repellent Formulation	Application Methods	Rates or Procedures	Mode of Action	Remarks
Rabbit (cottontail) — Covered by 15 NCAC 10B.0106. A depredation permit is needed prior to setting a trap for this species. Where shooting is recommended, a depredation permit is also needed unless the animal is in the act of destroying property.					
Ornamentals, garden shrubs, vegetables, and trees	Ammonium soap 15% liquid (Hinder)	Paint or spray on plants	Rates vary according to crop or plant. See label directions.	Repellent resulting in an unpleasant odor/taste.	May be used on home gardens. Follow label instructions carefully.
	Dried blood 15% Tobacco dust 70% (Rabbit and Dog Chaser)	Ring soil around plants, trees, and shrubs with several inch-wide bands.	Use every 2 to 3 weeks.	Repellent resulting in an unpleasant odor/taste.	Follow label instructions carefully.
		Trapping	Bait box traps with apple.		Can be box trapped during hunting season with hunting license; other times, permit required.
		Fencing	2-inch mesh or smaller fencing dug in ground 2 inches, 2 feet high all around garden.		
		Shooting			Discharging firearms within city or town limits is typically illegal.
	Capsaicin 0.0001%	Spray on plants	Use every 2-3 weeks.	Repellent	Follow label instructions carefully.
Raccoon — Covered by 15 NCAC 10B.0106. A depredation permit is needed prior to setting a trap for this species. Where shooting is recommended, a depredation permit is also needed unless the animal is in the act of destroying property.					
Gardens and fields; Attics and other structures		Trapping	Live trap with box-style trap. Bait with sweet roll or corn-on-the-cob. Place bait at rear of trap to prevent stealing by raccoon. Secure trap to tree or stake.		Check traps daily. Raccoons can be vicious when trapped. Contact local state agencies (NCWRC) and licensed pest control operators for advice.
			Set a steel foot-hold trap with dirt hole.		
		Shooting			Discharging firearms within city or town limits is typically illegal.
	Capsaicin 0.256% Black Pepper, Oils Piperidine 1.48%	Spread along active areas.	Use every 2-3 weeks.	Repellent	Follow label instructions carefully.
	Meat meal 99% Red pepper 1%	Spread along active areas.	Use every 2-3 weeks.	Repellent	Follow label instructions carefully.
Chimneys		Exclusion			All chimneys should be protected from animal entry by installation of a wire chimney cap.
Red-Winged Blackbird					
All problem areas		Scare tactics	Propane exploders, shell crackers, and electronic sound devices.		Use combination of scare devices; vary times and locations. Be persistent.
Feedlots	4-Aminopyridine 1.0% (Avitrol)	Prebait with untreated grain then replace with treated grain. Carefully observe birds' habits to establish feeding habits and locations to prevent ingestion by desirable or non-target birds.	Rates vary with extent of bird infestation.	Blocks potassium channels resulting in increased cholinergic nervous system activity.	Restricted label. For use by or under the supervision of government agencies or pest control operators. Follow label instructions carefully.
Structures, nesting and roosting sites	4-Aminopyridine 0.5% (Avitrol)			Blocks potassium channels resulting in increased cholinergic nervous system activity.	Restricted label. For use by or under the supervision of government agencies or pest control operators. Follow label instructions carefully.
Standing field corn *(NOT SWEET CORN)*	4-Aminopyridine 0.03% (Avitrol)	Same as Feedlots, but ground and air application permitted.	Same as Feedlots.	Blocks potassium channels resulting in increased cholinergic nervous system activity.	Same as Feedlots.
Skunk — Covered by 15 NCAC 10B.0106. A depredation permit is needed prior to setting a trap for this species. Where shooting is recommended, a depredation permit is also needed unless the animal is in the act of destroying property.					
Houses and barns			Exclusion.		Place flour at crawl space entrance to determine if skunks have left. Once gone, seal the area. To get rid of skunk odor, use a garden sprayer to mist strong smelling areas with a solution of 1 gallon vinegar with 1 cup liquid cleanser. Bury clothes. Report rabid or suspicious animals to local health department.
	Capsaicin 0.032% Black Pepper, Oils 0.48% Piperidine 0.185%	Spread along active areas.	Use every 2-3 weeks.	Repellent	Follow label instructions carefully.

ANIMAL DAMAGE CONTROL

Table 9-2. Animal Control Suggestions

Problem Areas	Repellent Formulation	Application Methods	Rates or Procedures	Mode of Action	Remarks
Skunk (continued)					
Gardens		Trapping	Live trap with box-style trap. Bait with fish or fish-flavored cat food. Covering trap with a tarp will reduce the chance of spray.		Check traps daily and release non-target animals.
		Shooting			In agricultural settings, producers may shoot a skunk in the act of destroying property; however, it is strongly advised to contact the local Wildlife Enforcement officer or Wildlife Biologist with the N.C. Wildlife Resources Commission to obtain a Wildlife Depredation Permit. Discharging firearms within city or town limits is typically illegal.
Snakes					
Gardens and houses		Environmental	Remove debris, trash, woodpiles, thick vegetation, and food sources.		No methods are totally effective. Take sensible precautions. Learn the difference between venomous and nonvenomous snakes. Snakes are very beneficial. No chemical controls available.
		Snake-proofing	Screen or fill any openings into house. Check door sills and where pipes enter.		
		Fencing	18- to 36-inch high, 0.25 inch mesh buried several inches below surface. Keep a 2- to 3-feet strip outside fence free of vegetation.		
		Rodent glue boards	Staple several glue boards to a piece of plywood and place next to foundation under house.		Remove snakes without harming them by pouring mineral spirits or vegetable oil on the snake and glue board. Always use glue boards inside residences, never outdoors.
	Napthelene (7%) Sulfur (28%	Spread along area to protect.	Variable	Repellent	Impacts sensory receptors.
Squirrel (gray and flying) — Covered by 15 NCAC 10B.0106. A depredation permit is needed prior to setting a trap for this species. Where shooting is recommended, a depredation permit is also needed unless the animal is in the act of destroying property.					
Houses, shrubs, and trees		Trapping	Suitable size live (box-style) trap. Bait with peanuts, walnuts, peanut butter, or black oil sunflower seeds. Prebait for several days with trap doors in open position.		Check live traps daily. Release non-target animals immediately. Release squirrels at least 10 miles away. If squirrels are in the house, repair building after last squirrel is removed.
		Shooting			Discharging firearms within city or town limits is typically illegal.
Starling					
All problem areas		Scare tactics.	Propane exploders, shell crackers, and electronic sound devices.		Use combination of scare devices; vary times and locations. Be persistent.
Feedlots	4-Aminopyridine 1.0% (Avitrol)	Prebait with untreated grain, then replace with treated grain. Carefully observe starlings to establish feeding habits and locations to prevent ingestion by desirable or non-target birds.	Rates vary with extent of bird infestation.	Blocks potassium channels resulting in increased cholinergic nervous system activity.	Restricted label. For use by or under the supervision of government agencies or pest control operators. Follow label instructions carefully.
Structures, nesting and roosting sites	4-Aminopyridine 0.5% (Avitrol)	Mix with shelled corn in hopper before planting.	One packet per bushel of shelled corn.	Blocks potassium channels resulting in increased cholinergic nervous system activity.	Same as Feedlots.
	Polybutenes 97%	Paste.	Spread on ledges.	Surface repellent that results in an unpleasant sticky sensation.	Follow label instructions carefully.
	Methyl anthranilate 26.4%	Liquid spray.	Spray on problem areas.	Targets taste senses.	Follow label instructions carefully.
Standing field corn (NOT SWEET CORN)	4-Aminopyridine 0.03% (Avitrol)	Same as Feedlots, but ground and air application permitted.	Same as Feedlots.	Blocks potassium channels resulting in increased cholinergic nervous system activity.	Same as Feedlots.

ANIMAL DAMAGE CONTROL

Table 9-2. Animal Control Suggestions

Problem Areas	Repellent Formulation	Application Methods	Rates or Procedures	Mode of Action	Remarks
White-Tailed Deer — Covered by 15 NCAC 10B.0106. A depredation permit is needed prior to setting a trap for this species. Where shooting is recommended, a depredation permit is also needed unless the animal is in the act of destroying property.					
Ornamentals, gardens, field crops, and forest regeneration		Scare tactics	Propane exploders, shell crackers, and human hair hung in bags around the area.		Results may vary. Deer quickly become habituated. Many devices are expensive, require maintenance, and are dangerous.
	Soap	Hang small bars of soap from branches, using 20 pound-test fishing lines.	One bar per shrub, several per tree, replace weekly.	Repellent resulting in an unpleasant odor/taste.	Aromatic soap may be effective. However, results may vary and deer may become habituated.
	Ammonium Soap 15% Liquid (Hinder)	Painting or spraying	Rates vary according to crop or plant. See label directions.	Repellent resulting in an unpleasant odor/taste.	May be used on home gardens. Follow label instructions carefully.
	Add putrified egg solid containing the fungicide Thiram (Big Game Repellant, Deer Away, Hinder)	Spray all parts of plant within reach of animal; repeat as needed to protect new growth. Spray twigs and trunk of dormant trees to 5 feet.	Instructions for use vary depending on formulation or brand.	Repellent resulting in an unpleasant odor/taste.	***Do not*** treat vegetables, fruits, or plant parts to be eaten. Spray on dry day. Clean sprayer immediately. Follow instructions carefully. Results will vary and applications may need to be numerous and often.
		Fencing	Electric fence, high-tensile wire fence, 8 feet high woven wire, mesh fences		Must be properly maintained. Initially very expensive, however, over the long run may be cost effective.
		Shooting			In agricultural settings, producers may shoot a deer in the act of destroying property; however, it is strongly advised to contact the local Wildlife Enforcement officer or Wildlife Biologist with the N.C. Wildlife Resources Commission to obtain a Wildlife Depredation Permit. Discharging firearms within city or town limits is typically illegal.
	Capsaicin 0.0002%	Spray on plants	Use every 2-3 weeks	Repellent	Follow label instructions carefully.

Rodenticides

C. S. DePerno, Department of Forestry and Environmental Resources

Table 9-3. Rodenticides

Compound	For Use On	Formulation	Food-Handling Establishments Inside	Outside	Farm Use[1]	Precautions and Remarks
Aluminum Phosphide	Norway rats, Gophers, Mice, Moles	Commercially prepared baits 57-60%[7]	Yes	Yes	Yes	For retail sale to and use by pest control operators only. (Not for home use.)
Brodifacoum	Norway rats, Roof rats, House mice	Commercially prepared baits .005%[4]	Yes	Yes	Yes	Active against Warfarin-resistant rats. For use in homes, industrial buildings, and commercial buildings. Keep away from humans, domestic animals, and pets. For retail sale to and use by pest control operators only. For use in and around the periphery of homes, industrial, commercial, and public buildings in urban areas. Do not place bait in areas where there is a possibility of contaminating food or surfaces that come in direct contact with food.
Bromadiolone	Norway rats, Roof rats, House mice	Commercially prepared baits .005%[4]	Yes	No	Yes	For use in and around the periphery of homes, industrial, commercial, and public buildings in urban areas.
Bromethalin	Norway rats, Roof rats, House mice	Commercially prepared baits 0.1%[5]	No	No	Yes	Keep away from humans, domestic animals, pets, and wildlife.
Bromethalin	Moles, Eastern moles	Commercially prepared baits 0.025%	No	No	Yes	Keep away from humans, domestic animals, pets, and wildlife.
Cholecalciferol (Rampage)	Norway rats, House mice	Commercially prepared bait 0.075%[6]	No	No	Yes	Keep away from humans, domestic animals, pets, and wildlife.
Chlorophacinone[2]	Norway rats, Roof rats, House mice	Commercially prepared baits .005%[4]	Yes	Yes	Yes	Restricted Use; for retail sale to and use by pest control operators only. Keep away from humans, domestic animals, pets, and wildlife.
		Tracking powder 0.2%[4]	Yes	No		For use as a tracking powder in prescribed and specific operations.
	Norway rats, Roof rats, House mice, Meadow voles,[3] Pine voles[3]	Commercially prepared baits .005%[4]		Yes	Yes	Registered for home use. Follow label directions. See N.C. Cooperative Extension publication *Wildlife Damage Control: Controlling Vole Damage in Commercial Orchards and Ornamental Nurseries*, AG-472-1.

ANIMAL DAMAGE CONTROL

Table 9-3. Rodenticides

Compound	For Use On	Formulation	Food-Handling Establishments		Farm Use[1]	Precautions and Remarks
			Inside	Outside		
Diphacinone[2]	Norway rats, Norway rats, Roof rats, House mice	Commercially prepared baits .005%[4]	Yes	Yes	Yes	Keep away from humans, domestic animals, pets, and wildlife.
	Meadow voles,[3] Pine voles[3]	Commercially prepared baits .005%[4]			No	Not registered for home use. Follow label directions.
Warfarin[2]	Norway rats, Roof rats, House mice, Meadow voles,[3] Pine voles[3]	Powder concentrate 0.5%[4] Commercially prepared baits .025%[4]	Yes	Yes	Yes	Place baits in area inaccessible to children, pets, wildlife, and domestic animals or in tamper-proof bait boxes.
Zinc Phosphide	Norway rats, Roof rats, House mice, Moles, Pocket gophers	Commercially prepared baits 2%[7]	No	Yes		For retail sale to and use by pest control operators only. (Not for home use.)
		Tracking powder 10%[7]	Yes	No		For retail sale to and use by pest control operators only. (Not for home use.)
		Commercially or farm-prepared baits 2%[7]			Yes	For retail sale to and use by pest control operators only. (Not for home use.)
	Meadow voles,[3] Pine voles[3]	Commercially prepared grain baits 2%[7]			Yes	Broadcast application is ineffective against pine voles. For retail sale to and use by pest control operators only. (Not for home use.)

[1] Additional instructions on control of rodents on farms may be found in *Poultry Science and Technology Guide No. 4*, which is available at your county Cooperative Extension center.

[2] Requires multiple feedings to kill.

[3] Apply rodenticides for mole and vole control only after harvest or during the dormant period. Do not pick or use any crops after rodenticide application. Do not allow animals to graze in treated orchards.

[4] Mode of action: anticoagulant.

[5] Mode of action: oxidative phosphorylation of central nervous system.

[6] Mode of action: calcium mobilizer.

[7] Mode of action: phosphine gas.

Fish Control

C. S. DePerno, Department of Forestry and Environmental Resources

Address inquiries about fish control to the county wildlife law enforcement officer or to other representatives of the Boating and Inland Fisheries Division of the North Carolina Wildlife Resources Commission.

In overpopulated waters, reducing numbers may improve fish growth. Sometimes it is desirable to eliminate a fish population so that a more preferred mix of species can be introduced. All population control measures in public waters must, by law, be undertaken only by the North Carolina Wildlife Resources Commission. Within limits of available personnel, the Wildlife Resources Commission provides advice on materials and methods for fish control and help in determining if waters are public or private.

ANIMAL DAMAGE CONTROL

X — DISEASE CONTROL

DISEASE CONTROL

Foliar Fungicides for Wheat Leaf Disease Control

Anyone using any agricultural chemical should refer to the current chemical label, which contains information about the safe and effective use of the chemical, before using the chemical.

Table 10-1A. Foliar Fungicides for Wheat Leaf Disease Control

Disease Fungicide Type and (FRAC Code)	Fungicide[1]	Amount of Formulation Per Acre	Remarks[2]
Powdery mildew, Leaf Rust			
Triazoles (3)	metconazole (Caramba)	10 to 14 oz	For Powdery Mildew, apply fungicide only when mildew covers 5% to 10% of area of upper leaves. For leaf rust, apply fungicide only when disease covers 1% to 3% of total leaf area. Do not apply after head emergence (Feekes Growth Stage 10.5). Make no more than one application of tebuconazole per year. Apply Caramba immediately after flag leaf emergence for optimum control of diseases other than Fusarium head blight.
	propiconazole (Propimax, Tilt) 3.6 EC	4 fl oz	
	prothioconazole (Proline)	4.3 to 5.0 oz	
Combinations of Strobilurins and Triazoles (3,11)	metconazole (7.4%) + pyraclostrobin (12.0 %) (Twinline)	7 to 9 fl oz	For Powdery Mildew, apply fungicide only when mildew covers 5% to 10% of area of upper leaves. For leaf rust, apply fungicide only when disease covers 1% to 3% of total leaf area. Do not apply after head emergence (Feekes Growth Stage 10.5). Do not apply if head scab is anticipated to become a problem.
	propiconazole (11.7%) + azoxystrobin (7.0%) (Quilt)	10.5 to 14 fl oz	
	prothioconazole (10.8 %) + trifloxystrobin (32.3%) (StrategoYLD)	4.0 to 4.65 oz	
	propiconazole (11.7%) + azoxystrobin (13.5%) (QuiltXcel)	10.5 to14 fl oz	
	cyproconazoe (7.2 %) + picoxystrobin (32.3 %) (Approach Prima)	3.4 to 6.8 fl oz	
Strobilurins (11)	azoxystrobin (Quadris) 2.08 F	6.2 to 10.8 fl oz	
	pyraclostrobin (Headline) 2.09 EC	6 to 9 fl oz	
	picoxystrobin (22.5 %) Aproach	6.0 to 12.0 fl oz	
Stagonospora Leaf and Glume Blotch, Tan Spot, Helminthosporium Leaf Spot			
Multi-site action (M3)	mancozeb (various brands) 4 F 80 WP 75 DF	1.6 qt 2 lb 2 lb	If 25% of the indicator leaves have one or more lesions, then a fungicide application is indicated. Indicator leaves are: Feekes Growth Stage 6 to 8: Flag - 4 and Flag - 5 Feekes Growth Stage 8 to10: Flag - 3 Feekes Growth Stage 10 to 10.51: Flag - 2 Feekes Growth Stage 10.52 to 11: Flag - 1 Do not apply mancozeb after late heading (Feekes Growth Stage 10.5) or Tilt after flag leaf emergence (Feekes Growth Stage 8).
Strobilurins (11)	pyraclostrobin (Headline) 2.09 EC	6 to 9 fl oz	For Stagonospora, if 25% of the indicator leaves have one or more lesions, then a fungicide application is indicated. Indicator leaves are: Feekes Growth Stage 6 to 8: Flag - 4 and Flag - 5 Feekes Growth Stage 8 to 10: Flag - 3 Feekes Growth Stage 10 to 10.51: Flag - 2 Feekes Growth Stage 10.52 to 11: Flag - 1
	azoxystrobin (Quadris) 2.08 F	6.2 to 10.8 fl oz	
	picoxystrobin (22.5 %) Aproach	6.0 to 12.0 fl oz	
Combinations of Strobilurins and Triazoles (3, 11)	trifloxystrobin + prothioconazole (Stratego Yld)	4.0 to 4.65 oz	
	metconazole (7.4%) + pyraclostrobin (12.0 %) (Twinline)	7 to 9 fl oz	
	cyproconazole (7.2 %) + picoxystrobin (32.3%) Approach Prima	3.4 to 6.8 fl oz	Do not apply if head scab is anticipated to become a problem.
Head Scab			
Triazoles (3)	tebuconazole (generic brands)	4 fl oz	Specifically, forward and backward mounted nozzles, or nozzles that have two-directional spray, should be used. Spraying at 45 degrees down from horizontal has been shown to be most effective. Operate nozzles within the spray pressure directions suggested by the manufacturer. Do not make more than one application of tebuconazole per year
	Tebuconazole (19.0%) + prothioconazole (19.0%) (Prosaro 421 SC)	6.5 to 8.2 fl oz	
	metconazole (Caramba)	13.5 to 17 oz	Do not apply Caramba within 30 days of harvest.
	prothioconazole (Proline)	5.0 to 5.7 fl oz	Do not apply Proline or Prosaro within 30 days of harvest or after full flower (Feekes 10.52).

[1] Fungicides are more likely to be profitable when the yield potential is 50 bushels/acre or more.
[2] Triazole fungicides are generally more effective in control of powdery mildew, while the strobilurins are generally more effective against leaf rust and Stagonospora. Some triazoles can suppress but not eliminate head scab, whereas strobilurins should not be used if there is concern about head scab.

Further Information
Measured Crop Performance: Small Grain
Small Grain Production Guide: www.smallgrains.ncsu.edu
Copies of these publications are available from your county Cooperative Extension center.

DISEASE CONTROL

Fungicide Efficacy for Control of Wheat Diseases

The North Central Regional Committee on Management of Small Grain Diseases (NCERA-184) has developed the following information on fungicide efficacy for control of certain foliar diseases of wheat for use by the grain production industry in the U.S. Efficacy ratings for each fungicide listed in the table were determined by field testing the materials over multiple years and locations by the members of the committee. Efficacy is based on proper application timing to achieve optimum effectiveness of the fungicide as determined by labeled instructions and overall level of disease in the field at the time of application. Differences in efficacy among fungicide products were determined by direct comparisons among products in field tests and are based on a single application of the labeled rate as listed in the table. Table includes most widely marketed products and is not intended to be a list of all labeled products.

Table 10-1B. Efficacy of Fungicides for Wheat Disease Control Based on Appropriate Application Timing

Class	Active ingredient	Product	Rate/A (fl. oz)	Powdery mildew	Stagonospora leaf/glume blotch	Septoria leaf blotch	Tan spot	Stripe rust	Leaf rust	Stem rust	Head scab	Harvest Restriction
Strobilurin	Picoxystrobin 22.5%	Aproach SC	6.0 - 12	G[1]	ND	ND	ND	E	GE	ND	ND	Feekes 10.5 and 45 days
	Fluoxastrobin 40.3%	Evito 480 SC	2.0 – 4.0	G	ND	ND	VG	ND	VG	ND	NL	Feekes 10.5 and 40 days
	Pyraclostrobin 23.6%	Headline SC	6.0 - 9.0	G	VG	GE	E	E[2]	E	G	NL	Feekes 10.5
Triazole	Metconazole 8.6%	Caramba 0.75 SL	10.0 - 17.0	GE	GE	ND	GE	E	E	E	G	30 days
	Propiconazole 41.8%	Tilt 3.6 EC[3]	4.0	GE	GE	GE	GE	GE	GE	GE	P	Feekes 10.5
	Prothioconazole 41%	Proline 480 SC	5.0 - 5.7	ND	GE	GE	GE	ND	GE	GE	G	30 days
	Tebuconazole 38.7%	Folicur 3.6 F[3]	4.0	G	GE	GE	GE	E	E	E	F	30 days
	Prothioconazole 19% Tebuconazole 19%	Prosaro 421 SC	6.5 - 8.2	G	GE	GE	GE	E	E	E	G	30 days
Mixed modes of action[4]	Metconazole 7.4% Pyraclostrobin 12%	TwinLine 1.75 EC	7.0 – 9.0	G	GE	GE	E	E	E	GE	NL	Feekes 10.5
	Fluxapyroxad 14.3% Pyraclostrobin 28.6%	Priaxor	4.0 - 8.0	G	GE	GE	E	E[2]	E	GE	NL	Feekes 10.5
	Propiconazole 11.7% Azoxystrobin 7.0%	Quilt 200 SC[3]	10.5 - 14.0	GE	GE	GE	GE	E	E	GE	NL	Feekes 10.5
	Propiconazole 11.7% Azoxystrobin 13.5%	Quilt Xcel 2.2 SE	10.5 - 14.0	GE	GE	GE	GE	E	E	GE	NL	Feekes 10.5
	Prothioconazole 10.8% Trifloxystrobin 32.3%	Stratego YLD	4.0	G	GE	GE	GE	GE	GE	GE	NL	35 days
	Tebuconazole 22.6% Trifloxystrobin 22.6%	Absolute 500 SC	5.0	G	GE	GE	GE	GE	E	GE	NL	35 days

[1] Efficacy categories: NL=Not Labeled and Not Recommended; P = Poor; F = Fair; G = Good; GE =Good to Excellent (Very Good); E=Excellent; ND = Insufficient data to make statement about efficacy of this product; NL= Not labeled.

[2] Efficacy may be significantly reduced if solo strobilurin products are applied after stripe rust infection has occurred.

[3] Multiple generic products containing the same active ingredients also may be labeled in some states. Products including tebuconazole incude: Embrace, Monsoon, Muscle 3.6 F, Onset, Orius 3.6 F, Tebucon 3.6 F, Tebustar 3.6 F, Tebuzol 3.6 F, Tegrol, and Toledo. Products containing propiconazole include: Bumper 41.8 EC, Fitness, Propiconazole E-AG, and PropiMax 3.6 EC. Products containing propiconazole + azoxystrobin include: Avaris 200 SC.

[4] Products with mixed modes of action generally combine triazole (group 3) and strobilurin (group 11) active ingredients. Priaxor is an exception to this general statement and combines carboxamide (group 7) and strobilurin active ingredients.

Seed Treatment for Wheat Foliar Disease Control

Anyone using any agricultural chemical should refer to the current chemical label, which contains information about the safe and effective use of the chemical, before using the chemical.

Table 10-2. Seed Treatment for Wheat Foliar Disease Control

Disease and Fungicide	Amount of Formulation per cwt	Remarks
Seed-Borne Stagonospora Nodorum Blotch		
difenoconazole (3) metalaxyl (4) (Dividend XL)	2 fl oz	Seed treatments are moderately effective for control of seed-borne SNB.
tebuconazole + thiram (Raxil - Thiram)	3.5 to 4.6 fl oz	
tebuconazole + metalaxl (Raxil XT)	0.6 to 0.2 fl oz	
Barley Yellow Dwarf Virus		
imidacloprid (Gaucho 600 F) (Gaucho XT)	0.8 to 2.4 fl oz 3.4 fl oz	Effective
Thiametoxam (Cruiser 5FS)	0.75 to 1.33 fl oz	
Powdery Mildew		
triadimenol (Baytan 30 F) 2	1.5 fl oz	Effective

Nematode Control in Corn

Anyone using any agricultural chemical should refer to the current chemical label, which contains information about the safe and effective use of the chemical, before using the chemical.

Table 10-3. Nematode Control in Corn

Material and Formulation	Amount of Formulation Per 1,000 Feet	Amount of Formulation Per Acre (36-inch rows)	Remarks
terbufos (Counter) 20 G	5 to 6 ounces	5.0 pounds	Apply in furrow. Do not exceed 6.5 pounds per acre of Counter 20 CR.
abamectin (Avicta)	0.15 mg per seed	Seed treatment	Seed treatment

Further Information

Major Corn Diseases in North Carolina. N.C. Cooperative Extension Service, http://ipmwww.ncsu/corn/diseases/corn_diseases.html. A copy of this publication is available from your county Cooperative Extension center.

Fungicides for Control of Corn Foliar Diseases

Anyone using any agricultural chemical should refer to the current chemical label, which contains information about the safe and effective use of the chemical, before using the chemical.

Table 10-4A. Fungicides for Control of Corn Foliar Diseases (Northern and Southern Blight, Gray Leaf Spot, and Rusts)[1]

Fungicide Type and FRAC Code	Fungicide	Rate Per Acre Formulated (fluid ounces per acre)	Remarks
Strobilurins (11)	azoxystrobin (Quadris)	6.0 to 9.0 oz	See label for restrictions.
	pyraclostrobin (Headline)	6.0 to 12.0 oz	Application should be with 20 gallons of water/acre for adequate coverage with ground application, or with 5 gallons of water/acre with aerial applications.
	picoxystrobin (Aproach)	6.0 to 12.0 oz	
Triazoles (3)	propiconazole (Tilt)[1]	2.0 to 4.0 oz	See label for restrictions. Application should be with 20 gallons of water/acre for adequate coverage with ground application, or with 5 gallons of water/acre with aerial applications.
	tetraconazole (Domark)	4.0 to 6.0 oz	See label for restrictions. Application should be with 5 gallons of water/acre for adequate coverage with ground application, or with 2 gallons of water/acre with aerial applications.
	prothioconazole (Proline)	5.7 oz	Apply at R1 to R3 as a preventative or curative spray and 14 to 21 days later if disease pressure is high. Do not apply after R5 or within 21 days of harvest.
Combinations of Strobilurins and Triazoles (3,11)	azoxystrobin + propiconazole (Quilt)[1]	7.0 to 14.0 oz	See label for restrictions.
	triflozystrobin + propiconazole (Stratego)[1]	10.0 oz	Application should be with 20 gallons of water/acre for adequate coverage with ground application, or with 5 gallons of water/acre with aerial applications. Alternate with another non-Group 11 fungicide if making more than one application per season.
	prothioconazole + trifloxystrobin (Stratego YLD)	4.0 to 5.0 oz	Do not apply more than 10 ounces per year. Do not apply within 14 days of harvest.
	azoxystrobin + propiconazole (QuiltXcel)[1]	10.5 to14.0 oz	Do not apply more than 84 ounces per year. Do not apply within 30 days of harvest.
	fluoxystrobin + flutriafol (Fortix)	4.0 to 6.0 oz	Field corn only, not labeled for sweet corn.
	cyproconazoe + picoxystrobin (Aproach Prima)	3.4 to 6.8 l oz	Apply at R1 to R3 as a preventative or curative spray and 14 to 21 days later if disease pressure is high. Do not apply after R5 or within 21 days of harvest.
	tetraconazole + azoxystrobin (Affiance)	10.0 to 17.0 oz	Apply at R1 to R3 as a preventative or curative spray and 14 to 21 days later if disease pressure is high. Do not apply after R5 or within 21 days of harvest.
Combinations of dicarboximides and Strobilurins (7, 11)	Fluxapyroxad + Pyraclostrobin (Priaxor)	4.0 to 8.0 oz	Apply at R1 to R3 as a preventative or curative spray and 14 to 21 days later if disease pressure is high. Do not apply after R5 or within 21 days of harvest.

[1] Fungicides often significantly increase yields only in intensive production systems with high plant populations and adequate moisture.

Fungicide Efficacy for Control of Corn Diseases

The Corn Disease Working Group (CDWG) developed the following information on fungicide efficacy for control of major corn diseases in the United States. Efficacy ratings for each fungicide in the table were determined by field testing the materials over multiple years and locations by committee members. Efficacy ratings are based on level of disease control achieved by product, and are not necessarily reflective of yield increases obtained from product application. Efficacy depends on proper application timing, rate, and application method to achieve optimum effectiveness of the fungicide as determined by labeled instructions and overall level of disease in the field at the time of application. Differences in efficacy among fungicide products were determined by direct comparisons among products in field tests and are based on a *single application* of the labeled rate as listed in the table. **Table includes systemic fungicides available that have been tested over multiple years and locations. The table is not intended to be a list of all labeled products.**[1]

Efficacy categories: NR=Not Recommended; P=Poor; F=Fair; G=Good; GE= Good to Excellent (Very Good); E=Excellent; NL = Not Labeled for use against this disease; ND = Insufficient data to make statement about efficacy of this product for this disease.

Table 10-4B. Fungicide Efficacy for Control of Corn Diseases

Fungicide(s)				Disease						
Class	Active ingredient (%)	Product/ Trade name	Rate/A (fl oz)	Anthracnose leaf blight	Common rust	Eyespot	Gray leaf spot	Northern leaf blight	Southern rust	Harvest Restriction[2]
QoI Strobilurins Group 11	Azoxystrobin 22.9%	Quadris 2.08 SC Multiple Generics	6.0 – 15.5	GE	E	GE	E	G	G	7 days
	Pyraclostrobin 23.6%	Headline 2.09 EC/SC	6.0 – 12.0	GE	E	E	E	GE	GE	7 days
	Picoxystrobin	Aproach 2.08 SC	3.0 – 12.0	GE	GE-E	GE	F-VG	VG	G	7 days
DMI Triazoles Group 3	Propiconazole 41.8%	Tilt 3.6 EC Multiple Generics	2.0 – 4.0	NL	GE	E	G	G	F-G	30 days
	Prothioconazole 41.0%	Proline 480 SC	5.7	ND	GE	E	ND	GE	G	14 days
	Tebuconazole 38.7%	Folicur 3.6 F Multiple Generics	4.0 – 6.0	NL	ND	NL	ND	GE	F-G	36 days
	Tetraconazole 20.5%	Domark 230 ME	4.0 – 6.0	ND	ND	ND	E	ND	G	R3 (milk)
Mixed mode of action	Azoxystrobin 13.5% Propiconazole 11.7%	Quilt Xcel 2.2 SE	10.5 – 14.0	GE	GE-E	GE-E	E	GE	GE	30 days
	Bensovindiflupyr 10.27% Azoxystrobin 13.5% Propiconazole 11.7%	Trivapro A 0.83 + Trivapro B 2.2 SE	A = 4.0 B = 10.5	ND	ND	ND	E	GE	E	7 days (A) 30 days (B)
	Cyproconazole 7.17% Picoxystrobin 17.94%	Aproach Prima 2.34 SC	3.4 – 6.8	ND	ND	ND	E	GE	G-GE	30 days
	Flutriafol 19.3% Fluoxastrobin 14.84%	Fortix 3.22 SC Preemptor 3.22 SC	4.0 – 6.0	ND	ND	ND	E	GE	G-GE	30 days
	Pyraclostrobin 28.58% Fluxapyroxad 14.33%	Priaxor 4.17 SC	4.0 – 8.0	ND	GE	ND	GE	ND	G	21 days
	Pyraclostrobin 13.6% Metconazole 5.1%	Headline AMP 1.68 SC	10.0 – 14.4	ND	E	E	E	GE	G-GE	20 days
	Trifloxystrobin 32.3% Prothioconazole 10.8%	Stratego YLD 4.18 SC	4.0 – 5.0	GE	E	GE	E	GE	G-GE	14 days
	Tetraconazole 7.48% Azoxystrobin 9.35%	Affiance 1.5 SC	10.0 – 14.0	ND	ND	ND	ND	ND	G	7 days

[1] Additional fungicides are labeled for disease on corn, including contact fungicides such as chlorothalonil. Certain fungicides may be available for diseases not listed in the table, including Gibberella and Fusarium ear rot. Applications of Proline 480 SC for use on ear rots requires a FIFRA Section 2(ee) and is only approved for use in Illinois, Indiana, Iowa, Louisiana, Maryland, Michigan, Mississippi, North Dakota, Ohio, Pennsylvania, and Virginia.

[2] Harvest restrictions are listed for field corn harvested for grain. Restrictions may vary for other types of corn (sweet, seed, or popcorn, etc.), and corn for other uses such as forage or fodder.

Many products have specific use restrictions about the amount of active ingredient that can be applied within a period of time or the amount of sequential applications that can occur. Please read and follow all specific use restrictions prior to fungicide use. This information is provided only as a guide. It is the responsibility of the pesticide applicator by law to read and follow all current label directions. Reference to products in this publication is not intended to be an endorsement to the exclusion of others that may be similar. Persons using such products assume responsibility for their use in accordance with current directions of the manufacturer. Members or participants in the CDWG assume no liability resulting from the use of these products.

DISEASE CONTROL

Nematode Control on Cotton

Anyone using any agricultural chemical should refer to the current chemical label, which contains information about the safe and effective use of the chemical, before using the chemical.

Table 10-5. Nematode Control on Cotton

Nematodes Nematicide	Amount of Formulation Per Acre	Precautions and Remarks
Root-Knot, Columbia Lance, Sting, Reniform		
1,3-dichloropropene (Telone II)	3 to 6 gallons	Inject 1 to 2 weeks before planting 8 to 12 inches deep.
sodium methyldithiocarbamate (Vapam HL)	6 to 12 gallons	Inject 2 to 3 weeks before planting at 6 to 12 inches deep.
Reniform, Root-Knot		
abamectin (Avicta)	0.15 mg per seed	Seed treatment.
sodium methyldithiocarbamate (Vapam HL)	6 to 12 gallons	Inject 2 to 3 weeks before planting at 6 to 12 inches deep.
thiodicarb (Aeris) + Bacillus fermis (Poncho/Votivo)	NA	Seed treatment

Peanut Disease Control

B. B. Shew, Plant Pathology

Most peanut disease control chemicals leave residues on peanut vines that make them unsuitable for hay. Check each label before using the material if you intend to feed hay to livestock.

Note: certain products labeled on peanut in the U.S. contain active ingredients (propiconazole, phosphites, or phosphorous acid) that were not acceptable to export markets in 2016. Products containing these ingredients are not listed in the tables below. Check with county North Carolina Cooperative Extension personnel and your buyer for the latest information on the current status of these ingredients and products.

Table 10-6A. Peanut Disease Control

Disease or Diseases Controlled Pesticide Formulation (FRAC Group Number)	Amount of Formulation Per Acre	Application Schedule	Minimum Days to Harvest	Precautions and Remarks
Aspergillus Crown Rot (Aspergillus); see also seedling diseases				
azoxystrobin (Abound, various brands) [1] 2.08 F (11)	0.4 to 0.8 fl oz/1,000 ft of row	At planting	NA	Apply as in-furrow spray with 3 to 5 gallons water.
Black Root Rot (CBR) (Cylindrocladium); see also seedling diseases				
metam sodium 42% (various brands) 4.25 F	7.5 gal (36-inch rows) or 6.61 fl oz/100 ft of row	At least 2 weeks before planting or longer if cool and/or wet	NA	Inject 10 to 12 inches below the bedded soil surface. If wet and/or cold weather occurs following fumigation, the waiting period should be extended. Soil aeration helps reduce residual chemical. When in doubt use a bioassay such as the lettuce seed germination test to determine if safe to plant. Buffer zones, fumigant management plans, and other restrictions on metam sodium must be followed. See your county Extension center for details.
prothioconazole (Proline) 480 SC (3)	0.4 fl oz/1,000 ft of row	At planting or at full emergence	NA	Apply as in-furrow spray or banded at full emergence for suppression of CBR. Not a substitute for fumigation in fields with a history of more than10% CBR and rotations of less than 4 years.
prothioconazole + fluopyram (Propulse) 3.3 SC (3 + 7)	13.7 fl oz	At planting	NA	Apply as in-furrow spray for suppression of CBR and nematodes. Not a substitute for fumigation in fields with a history of more than10% CBR and rotations of less than 4 years.
Early Leafspot (Cercospora)				
cupric hydroxide (Kocide, various brands and formulations) [2] (M1)	Various; see label	Begin applications at very early pod (R3). Repeat applications every 7 to 14 days.	0	Use nozzles that give a cone-shaped spray pattern. Use 12 to 24 gallons of water for spray materials applied by ground sprayers. Use at least 5 gallons of water for materials applied by air. **Calendar program**: Five or six applications suggested. Begin applications at very early pod (R3). Repeat applications at 7- to 14-day intervals. Scout fields: if 20% or more of leaflets have spots, switch to a more effective fungicide and stay on a 14-day spray program.
basic copper sulfate (various brands and formulations) [2] (M1)	Various; see label	See above	0	See above
mancozeb and copper hydroxide (Mankocide) [2] 61.1 DF (M3)	2 to 2.6 lb	See above	14	See above
mancozeb (Manzate, Koverall, various brands) [2] (M3) M45 F45 75 WDG 80 WP	1 to 2 lb .8 to 1.6 qt 1 to 2 lb 1 to 2 lb	See above	14	See above
Sulfur (various brands and formulations) [2] (M2)	Various; see label	See above	0	See above

DISEASE CONTROL

Table 10-6A. Peanut Disease Control

Disease or Diseases Controlled Pesticide Formulation (FRAC Group Number)	Amount of Formulation Per Acre	Application Schedule	Minimum Days to Harvest	Precautions and Remarks
Early Leafspot (Cercospora); Late Leafspot (Cercosporidium); Web Blotch (Ascochyta)				
chlorothalonil (Bravo, Echo, various brands) (M5) 720, 6 F 82.5 WDG 90 DF 500	1 to1.5 pt .9 to 1.36 lb .875 to 1.25 lb 1.5 to 2.25 pt	Begin applications at very early pod (R3). Repeat applications every 14 days or according to daily weather based advisories. Begin 14-day program if web blotch is found.	14	Use nozzles that give a cone-shaped spray pattern. Use 12 to 24 gallons of water for spray materials applied by ground sprayers. Use at least 5 gallons of water for materials applied by air. **14-day program:** Five or six applications suggested. Begin applications at very early pod (R3). Repeat applications at 10- to 14-day intervals. **Advisory:** Begin applications at very early pod (R3). Repeat applications when weather conditions become favorable as determined by peanut leaf spot advisories. This schedule requires strict adherence to the program guidelines and usually results in fewer fungicide applications than the 14-day schedule. Contact your county Extension center for details. Leafspot advisories are most effective if used with long rotations, resistant varieties, and high rates of effective fungicides. Scout fields: if 20% or more of leaflets have spots, switch to a more effective fungicide and stay on a 14-day spray program. Repeated applications of chlorothalonil can make spider mites and Sclerotinia blight more difficult to control.
boscalid (Endura) [2,3] 70 WDG (7)	10 oz	Make up to 2 or 3 applications in mid-season as part of a full-season, 14-day, or advisory program	14	See above. Primarily controls web blotch. Alternate with another fungicide or mix with 0.75 to 1 pint chlorothalonil to improve leaf spot control. Also controls Sclerotinia blight; see below.
thiophanate methyl (various brands) (1) 4.5 F 70 WP 85 WDG	10 fl oz .5 lb .4 lb	14 day or advisory beginning at R3	14	See above. **Do not** apply alone. Always mix with another leaf spot fungicide.
dodine (Elast) 400 F (U12)	1.5 pt	Make no more than 3 applications as part of a full-season, 14-day, or advisory program	14	See above
cyproconazole (Alto) 100 SL (3)	5.5 fl oz	14 day or advisory beginning at R3	30	See above. Mix or alternate with another fungicide reduce the risk of fungicide resistance.
flutriafol (Topguard) 1.04 SC (3)	7 to 14 fl oz	14 day or advisory beginning at R3	14	See above. Mix or alternate with another fungicide to reduce the risk of fungicide resistance
tetraconazole (Eminent) 125 SL (3)	6 to 13 fl oz	14 day or advisory beginning at R3	14	See above. Mix or alternate with another fungicide to reduce the risk of fungicide resistance
Early Leaf Spot (Cercospora); Late Leaf Spot (Cercosporidium); Web Blotch (Ascochyta); Limb Rot (Rhizoctonia)				
tebuconazole + trifloxystrobin (Absolute) 500 SC (3 + 11)	3.5 to 7 fl oz	14 day or advisory beginning at R3	14	See above. Use higher rates for soil-borne pathogens. No more than 4 applications per season.
Early Leaf Spot (Cercospora); Late Leaf Spot (Cercosporidium); Web Blotch (Ascochyta); Limb Rot (Rhizoctonia); Stem Rot (Sclerotium rolfsii); Pod Rot (Sclerotium rolfsii, Rhizoctonia)				
prothioconazole + tebuconazole (Provost) [4] 433 SC (3+3)	7 to 10.7 fl oz	Make up to 2 to 4 applications in mid-season as part of a full-season, 14-day, or advisory program. For routine disease control, use 7 to 8 fl oz/A; also suppresses CBR at highest rate.	14	See Early Leaf Spot, Late Leaf Spot, and Web Blotch above. For best control of limb and pod rot, do not use a surfactant. Do not apply more than 3 times in a 5-spray program or after the first week in September. **Resistance management:** Site-specific fungicides (groups 3, 7, and 11) should be mixed or rotated with a fungicide from a different group to minimize the risk of fungus resistance development.
metconazole (Quash) 50 WDG (3)	2.5 to 4 oz	Make up to 2 to 4 applications in mid-season as part of a full-season, 14-day, or advisory program.	14	See above
tebuconazole + chlorothalonil (Muscle ADV) 3.84 F (3+M5)	32 fl oz	See above	14	See above. Repeated application of products containing chlorothalonil can make spider mites and Sclerotinia blight more difficult to control.
tebuconazole (various brands) (3) 3.6 F 1.67 F	7.2 fl oz 15.4 fl oz	See above	14	See Early Leaf Spot, Late Leaf Spot, and Web Blotch above. CAUTION: Not effective against many populations of leaf spot fungi. **Always** mix with chlorothalonil or another fungicide (other than group 3) that is effective against leaf spots.
prothioconazole + fluopyram (Propulse) 3.3 SC (3 + 7)	13.7 fl oz	See above	14	See Early Leaf Spot, Late Leaf Spot, and Web Blotch above. Application for leaf spot control requires a FIFRA Section 2(ee) label.
penthiopyrad (Fontelis) [3] 1.67 SC (7)	16 to 24 fl oz	Make up to 3 applications per season as part of a full-season, 14-day, or advisory program	14	See Early Leaf Spot, Late Leaf Spot, and Web Blotch above. FRAC guidelines recommend no more than 2 applications of an unmixed group 7 fungicide in a 5-spray program. Use higher rates for web blotch control. Also suppresses Sclerotinia blight; see below.
azoxystrobin (Abound; various brands) [1,4] 2.08 F (11)	12.0 to 24.6 fl oz	Make up to 2 applications per season as part of a full-season, 14-day, or advisory program. Use higher rates for limb rot and stem rot control.	14	See above. Use in mid-season for best control of soil-borne pathogens. Use no more than 2 applications in a 5 spray program. **Resistance management:** Site-specific fungicides (groups 3, 7, and 11) should be mixed or rotated with a fungicide from a different group to minimize the risk of developing fungal resistance.
Azoxystrobin + benzovindiflupyr .45 DF (11 + 7)	7.3 to 9.5 fl oz	Make up to 2 or 3 applications per season as part of a full-season, 14-day, or advisory program. Use higher rates for limb rot and stem rot control.		See above. Extended spray intervals (up to 21 days) may be possible at the highest rate. **Resistance management:** Site-specific fungicides (groups 3, 7, and 11) should be mixed or rotated with a fungicide from a different group to minimize the risk of developing fungal resistance.

439

DISEASE CONTROL

Table 10-6A. Peanut Disease Control

Disease or Diseases Controlled Pesticide Formulation (FRAC Group Number)	Amount of Formulation Per Acre	Application Schedule	Minimum Days to Harvest	Precautions and Remarks
Early Leaf Spot (Cercospora); Late Leaf Spot (Cercosporidium); Web Blotch (Ascochyta); Limb Rot (Rhizoctonia); Stem Rot (Sclerotium rolfsii); Pod Rot (Sclerotium rolfsii, Rhizoctonia) (continued)				
azoxystrobin + tebuconazole (Custodia SC; other generic) (11 + 3)	15.5 fl oz	Make up to 2 to 4 applications in mid-season as part of a full-season, 14-day, or advisory program.	14	See above
fluoxastrobin (Evito, Aftershock)[1] 480 SC (11)	5.7 fl oz	Make up to 2 applications per season as part of a full-season, 14-day, or advisory program.	14	See above **Resistance management**: Site-specific fungicides (groups 3, 7, and 11) should be mixed or rotated with a fungicide from a different group to minimize the risk of developing fungal resistance.
fluoxastrobin + tebuconazole (Evito T) (11+3)	6 to 11.2 fl oz	Make up to 2 to 4 applications in mid-season as part of a full-season, 14-day, or advisory program.	14	See above
pyraclostrobin (Headline)[1,4] 2.09 EC, 2.08 SC (11)	6 to 15 fl oz	See above	14	See above **Resistance management**: Site-specific fungicides (groups 3, 7, and 11) should be mixed or rotated with a fungicide from a different group to minimize the risk of developing fungal resistance.
fluxapyroxad + pyraclostrobin 4.17 SC (Priaxor) (7+11)	4 to 8 fl oz	Use 1 to 3 times per season a full-season, 14-day, or advisory program. Use higher rates for limb rot and stem rot control.	14	See above
Stem Rot (white mold, Southern blight, Sclerotium rolfsii); Limb Rot (Rhizoctonia); Pod Rot (Sclerotium rolfsii, Rhizoctonia)				
flutolanil (Convoy)[3] 3.8 SC (7)	10 to 16 fl oz or 20 -32 fl oz (see remarks)	Following leafspot advisories, make 1 to 3 applications in mid-season. **Does not control foliar diseases.**	40	Apply up to 16 fluid ounces per acre at 2-week intervals or up to 32 fluid ounces per acre at 3- to 4-week intervals. Do not apply more than a combined total of 64 fluid ounces in a single growing season. See label for detailed information on rates. Do not plant small grains within 5 months of last application. See label for other plant-back restrictions.
Nematodes—Fumigants				
1-3 dichloropropene 97.5% (Telone II) 93.6% (Telone EC)	Depends on application method; see label for details	At least 2 weeks before planting	NA	Inject 8 to 10 inches below the soil surface. Very effective against all nematodes. Regulations require handler training and impose buffer zones and other restrictions on fumigant use. See the label and your county Extension center for details.
1-3 dichloropropene + chloropicrin 81.2% + 16.5% (Telone C-17) 63.4% + 34.7% (Telone C-35)	Depends on application method; see label for details	At least 2 weeks before planting	NA	Inject 8 to 10 inches below the soil surface. Very effective against all nematodes. Regulations require handler training and impose buffer zones and other restrictions on fumigant use. See the label your county Extension center for details.
metam sodium 42% (various brands)[5] 4.25 F	7.5 gal	At least 2 weeks before planting	NA	Inject 8 to 10 inches below the soil surface. If wet and/or cold weather occurs following fumigation, the waiting period should be extended. Soil aeration helps reduce residual chemical. When in doubt use a bioassay such as the lettuce seed germination test to determine if safe to plant. Moderately effective against Northern root knot nematode (M. hapla). Not very effective on peanut root knot nematode (M. arenaria). Buffer zones and other restrictions on metam sodium use are required. See your county Extension center for details.
Nematodes- Nonfumigant				
fluopyram + imidacloproid (Velum Total) 3.67 SC (7 + insecticide group 4A)	18 fl oz	At planting		Apply in-furrow at planting, directed on or below the seed. Also controls thrips, leaf hoppers, and aphids (see Chapter 5). Do not exceed 0.5 pound a.i./a imidacloprid for all seed, in-furrow, and foliar applications. See label for plant-back restrictions.
Seed and Seedling Rot; Pythium Pod Rot				
mefenoxam + azoxystrobin (Uniform) 390 SE (4 + 11)	.34 fl oz/1,000 ft of row	At planting	75	Apply as an in-furrow spray at planting. Only one application per season
azoxystrobin[1] (Abound 2.08 F; various brands) (11)	.4 to .8 fl oz/1,000 ft of row	At planting	14	Apply as an in-furrow spray at planting; counts as a group 11 application for resistance management purposes
mefenoxam (Ridomil Gold GR; various brands) (4) (Ridomil Gold SL; various brands) (4)	Per 1,000 ft of row: 6.5 oz .25 pt	At planting	75	Apply in-furrow or as a 7-inch band over row at planting
mefenoxam (Ridomil Gold 2.5 GR; various brands) (4) (Ridomil Gold SL; various brands) (4)	Per 1,000 ft of row (GR): 13 oz SL per Acre: .5 to 1 pt	Early pegging	75	Apply in an 8- to 12-inch band. Do not apply to wet foliage as foliar toxicity may result. Use with other fungicides for late-season control of stem rot (Sclerotium rolfsii) and Rhizoctonia stem and pod rot (Rhizoctonia spp.).
azoxystrobin (Abound)[1] 2.08 F (11)	24 fl oz	60 and 90 days after planting	14	Use no more than 2 applications of a group 11 fungicide per season. **Resistance management**: Site-specific fungicides should be mixed or rotated with another type of fungicide to minimize risk of developing fungal resistance. Also controls stem rot, Rhizoctonia limb and pod rot, and leaf spots.

440

Table 10-6A. Peanut Disease Control

Disease or Diseases Controlled Pesticide Formulation (FRAC Group Number)	Amount of Formulation Per Acre	Application Schedule	Minimum Days to Harvest	Precautions and Remarks
Seedling Diseases—Seed Treatments				
azoxystrobin + fludioxonil + mefenoxam (Dynasty PD) [4] (11 + 12 + 4)	4 oz/100 lb seed	Seedling diseases: Apply to conditioned, untreated seed. Commercial application strongly recommended.	NA	Peanuts can be replanted immediately. Do not plant other crops within 45 days of planting treated seed.
thiamethoxam + mefenoxam +fludioxonil + azoxystrobin (CruiserMaxx Peanuts) [4] (MOA 4A + 11 + 12 + 4)	3 to 4 oz/100 lb seed	See above	NA	Peanuts can be replanted immediately. See label for additional information about plant-back restrictions. Do not make any soil or foliar application of products containing thiamethoxam to crops grown from seed treated with CruiserMaxx Peanuts. Also controls some early season insects; see Chapter 5 for more information.
Sclerotinia Blight				
fluazinam (Omega) 500 F (29)	1 to 1.5 pt	1 to 3 applications according to weather-based advisory, field history, and scouting	30	If favorable conditions persist, reapply at 21 to 30 day intervals. Do not apply more than a combined total of 4 pints in a single growing season. Contact your county Extension center for details on weather-based Sclerotinia advisories.
boscalid (Endura) [2,3] 70 WG (7)	8 to 10 oz	See above	14	If favorable conditions persist, reapply at 14- to 21-day intervals. Make no more than two consecutive applications per season. Contact your county Extension center for details on weather-based Sclerotinia advisories. Also controls or suppresses leaf spots and web blotch.
penthiopyrad (Fontelis) [3] 1.67 SC (7)	24 fl oz	1 to 3 applications according to weather-based advisory, field history, and scouting.	14	Suppression only. Apply at 2-week intervals or according to advisory. FRAC guidelines recommend no more than 2 applications of an unmixed group 7 fungicide in a 5-spray foliar disease control program. Do not apply more than 72 fluid ounces per season. Use on cultivars that have some Sclerotinia blight resistance, for example, Bailey. Also controls or suppresses leafs spots, web blotch, southern stem rot, and Rhizoctonia limb and pod rot; see above.
fluopyram + prothioconazole (Propulse) (7 + 3)	13.7 fl oz	Use 1 to 3 times per season a full-season, 14-day, or advisory program	14	Suppression only. Apply at 2-week intervals or according to advisory. Use on cultivars that have some Sclerotinia blight resistance, for example, Bailey. Also controls southern stem rot and Rhizoctonia limb rot; see above.
Iprodione 4F (Rovral, various brands) (2)	32 fl oz	Use 1 to 3 times per season on a full-season, 14-day or advisory program	10	Suppression only. Apply at 2 to 3 week intervals or according to advisory. Apply at low pressure for a spray volume of at least 40 gal per acre. Do not apply more than 72 fluid ounces per season. Use on cultivars that have some Sclerotinia blight resistance, for example, Bailey.

[1] QOI (group 11) fungicide. Do not apply group 11 fungicides more than 2 times in sequence or more than 3 times per season. See www.FRAC.org for information on fungicide resistance management.

[2] Less effective than many fungicides; more frequent application may be necessary. If using advisories, alternate or mix with more effective fungicides.

[3] Do not apply unmixed group 7 fungicides more than 2 times in sequence. FRAC guidelines: no more than 2 applications of a group 7 fungicide per 5-spray program; no more than 3 applications per ≥6-spray program. May be alternated with group 11 or group 3 fungicides. See www.FRAC.org for information on fungicide resistance management.

[4] Also suppresses CBR. See label for details.

[5] Probably not as effective as the other fumigants against nematodes.

Further Information: *2017 Peanut Information* and peanut disease control information are available at your county Cooperative Extension center.

DISEASE CONTROL

Peanut Disease Management Calendar

B. B. Shew, Plant Pathology

Table 10-6B. Peanut Disease Management Calendar

Time of Year	Disease	Threshold	Management Tactics
Spring (April–June)	Tomato spotted wilt virus (TSWV)	See TSWV risk index	Plant a resistant cultivar (Bailey, Georgia 07W, Sullivan, or Wynne); use a high seeding rate or twin rows; plant after May 5 and before May 16; apply an insecticide in furrow. Consider an additional post-emergence insecticide application.
	CBR (Cylindrocladium black rot)	1% to 10% disease in this field last time peanuts were grown	Rotate 2-4 years; avoid soybeans in rotations. Plant a resistant cultivar (Bailey or Sugg). Consider an in-furrow fungicide application.
		More than 10% disease in this field last time peanuts were grown	Rotate 3-4 years; avoid soybeans in rotations. Plant a resistant cultivar (Bailey or Sugg) and fumigate before planting.
June–Harvest	Leaf spots, Web blotch, Pepper spot	R3 (beginning pods)	Rotate at least 2 years to any crop other than peanuts. Longer rotations are preferred. Plant a partially resistant cultivar (Bailey, Sugg, Sullivan, or Wynne). Begin calendar sprays or advisory program. Use nozzles that give a cone-shaped spray pattern. Use 12 to 24 gallons of water for spray materials applied by ground sprayers. Use at least 5 gallons of water for materials applied by air. **14-day program:** Four to six applications suggested. Begin applications at very early pod (R3). The first application can be delayed 2 weeks on Bailey. Repeat applications at 14-day intervals. **Advisory:** Begin applications at very early pod (R3) or at R3 + 2 weeks on Bailey. Repeat applications when weather conditions become favorable as determined by peanut leaf spot advisories. This schedule requires strict adherence to the program guidelines and usually results in fewer fungicide applications than the 14-day schedule. Contact your county Extension center for details. Scout fields: if 20% or more of leaflets have spots, begin a 14-day spray program.
		20% leaflets with spots	Reduce intervals between sprays when over threshold. Switch to a more effective fungicide if late leaf spot, web blotch, or pepper spot becomes predominant. If using advisory, switch to a 14-day spray schedule.
	Southern stem rot	Mid-July or on demand	Plant a partially resistant cultivar (Bailey or Sugg). Avoid highly susceptible cultivars. Rotate 2-4 years to non-host crops. Use a soil fungicide or a foliar fungicide with efficacy against soil-borne pathogens at least once from Mid-July to mid-August on resistant cultivars, or up to 3 times on susceptible cultivars in fields with a history of disease, or if signs and symptoms of disease are present. See leaf spots above for application information. Using a surfactant, higher volumes of water (15-25 gallons per acre), or spraying at night may improve control.
	Sclerotinia blight	In fields with a history of disease, but less than 10% disease: early July or according to advisory	Plant a partially resistant cultivar (Bailey). Avoid highly susceptible cultivars. Rotate 4+ years with non-host crops. Scout every 2 weeks or according to advisory; begin fungicide applications if disease is seen.
		In fields with a history of greater than 10% disease: just before vines close or according to weather-based Sclerotinia advisory	Plant a partially resistant cultivar (Bailey). Avoid highly susceptible cultivars. Begin fungicide applications. Using higher volumes of water (15-25 gallons per acre) may improve control. Rotate 4-plus years with non-host crops.
September-October	CBR, Sclerotinia blight, Southern stem rot	At digging	Make disease maps to decide future rotations, use of resistant varieties, and to pinpoint areas for fumigation and fungicide application.
October–November	Nematodes, All diseases	Sample areas to be planted to peanut the following spring for nematodes and soil fertility. See the NCDA &CS website for for nematode and fertility sampling instructions.	Plan rotation and nematicide use based on recommendations. Adjust soil fertility and pH as recommended. Avoid planting in areas with high levels of Zn and/or a history of manure or litter applications.

Soybean Disease Control

L. D. Thiessen, Plant Pathology

Please refer to the current chemical label for all directions regarding safe use and the most up-to-date recommendations for application rates, timing, and harvest intervals.

Table 10-7A. Soybean Nematode Management — Nematicides

Nematodes	Nematicide and Formulation	Amount of Formulation [1]	Precautions and Remarks
Root-Knot, Columbia Lance	1,3-dichloropropene (Telone II)	3 to 6 gal/acre	Apply 10 to 14 days prior to planting. Inject at least 12 inches deep. Do not use in dry, wet, or cold soils.
Root-Knot, Columbia Lance, Lesion, Soybean Cyst, Sting	abamectin (Avicta)	0.15 mg per seed	Seed treatment.
	Bacillus fermis (Votivo)	See label	Seed treatment.

Table 10-7B. Soybean Foliar Disease Control — Frogeye Leaf Spot and Target Spot

Fungicide Type and FRAC Code	Fungicide	Rate Per Acre Formulated	Remarks
Strobilurins (11)	azoxystrobin (Quadris)	6.0 to 15.5 fl oz	Apply fungicide at R1 to R3 and make a second application 14 to 21 days later if disease pressure is high. Do not apply after R5 (small bean) or within 14 days of harvest. Higher rates provide longer residual activity and may reduce the need for a second application.
	picoxystrobin (Aproach)	6.0 to 12.0 fl oz	Apply fungicide at R1 to R3 and make a second application 14 to 21 days later if disease pressure is high. Do not apply after R5 (small bean) or within 14 days of harvest. Higher rates provide longer residual activity and may reduce the need for a second application.
	pyraclostrobin (Headline)	6.0 to 12.0 fl oz	Apply fungicide at R1 to R3 and make a second application 14 to 21 days later if disease pressure is high. Do not apply after R5 (small bean) or within 21 days of harvest. Higher rates provide longer residual activity and may reduce the need for a second application.
Nitriles (M5)	chlorothalonil (various brands)	1.5 to 2.25 pints	Apply fungicide at R1 to R3 and make a second application 14 days later if disease pressure is high. Do not apply within 48 days of harvest.
Thiophanates (1)	thiophanate-methyl (Topsin M 70WP) (Cercobin)	0.5 to 1 lb 10.9 to 21.8 fl oz	Apply fungicide at R1 to R3 and make a second application 14 to 21 days later if disease pressure is high. Do not apply after R5 (small bean). Higher rates provide longer residual activity and may reduce the need for a second application. Thiophanate-methyl is not labeled for Asiatic soybean rust.
Combinations of Strobilurins and Triazoles (3,11)	prothioconazole (10.8 %) + trifloxystrobin (32.3 %) (Stratego YLD)	4-4.65 fl oz	Apply at R1 to R3 and 14 to 21 days later if disease pressure is high. Do not apply after R5 or within 21 days of harvest.
	fluoxastrobin (14.84 %) + flutriafol (19.3 %) (Fortix)	4 to 6 fl oz	Apply at R1 to R3 and 14 to 21 days later if disease pressure is high. Do not apply after R5 or within 21 days of harvest.
	azoxystrobin (18.2 %) + difenconazole (11.4%) (Quadris Top)[1]	8 to 14 fl oz	Apply at R1 to R3 and 14 to 21 days later if rust is expected. Do not apply after R6 or within 14 days of harvest. Make no more than two applications of materials containing azoxystrobin or difenconazole per year soybean per season. Corn or wheat may be planted within 180 days of last application; do not plant other crops with 360 days of last application.
	tetraconazole (7.48) + azoxystrobin (9.35 %) (Affiance)	10.0 to 14.0 fl oz	Apply at R1 to R3 or when conditions are favorable for disease development and make a second application 15 to 21 days after application if disease pressure is high. Do not apply after R5 or within 14 days of harvest.
	cyproconazoe (7.17 %) + picoxystrobin (17.94 %) (Aproach Prima)	5.0 to 6.8 fl oz	Apply fungicide at R1 to R3 and a second application 14 to 21 days later if disease pressure is high. Do not apply after R5 or within 30 days of harvest.
Combinations of dicarboximides and Strobilurins and (7, 11)	fluxapyroxad (14.33 %) + pyraclostrobin (28.58 %) (Priaxor)	4.0 to 8.0 fl oz	Apply at R1 to R3 prior to infection and a second application 7 to 14 days later if disease pressure is high. Do not apply after R5 or within 21 days of harvest.

Further Information

Frogeye Leaf Spot of Soybean, North Carolina Cooperative Extension Service, http://www.ces.ncsu.edu/depts/pp/notes/Soybean/soy003/soy003.htm

Table 10-7C. Soybean Foliar Disease Control — Asiatic soybean rust

Fungicide type and FRAC Code	Fungicide	Rate Per Acre Formulated	Remarks
Strobilurins (11)	azoxystrobin (Quadris)	6.2 to 15.5 oz	Apply at R1 to R3 as a preventative spray and 14 to 21 days later if rust is expected. Do not apply after R5 or within 14 days of harvest. Make no more than two applications to soybean per season.
	picoxystrobin (Aproach)	6.0 to 12.0 fl oz	Apply at R1 to R3 as a preventative spray and 14 to 21 days later if disease pressure remains high. Do not apply after R5 or within 14 days of harvest. Make no more than two applications to soybean per season
	pyraclostrobin (Headline)	6.0 to 12.0 fl oz	Apply at R1 to R3 as a preventative spray and 14 to 21 days later if rust is expected. Do not apply after R5 or within 21 days of harvest. Make no more than two applications to soybean per season.

DISEASE CONTROL

Table 10-7C. Soybean Foliar Disease Control — Asiatic soybean rust

Fungicide type and FRAC Code	Fungicide	Rate Per Acre Formulated	Remarks
Triazoles (3)	cyproconazole (Alto)	2.75 to 5.5 fl oz	Apply at R1 to R3 and a second application 14 to 28 days later if rust is expected. Use higher rates if rust is present in field. Do not apply after R6 or within 21 days of harvest. Make no more than two applications of cyproconazole to soybean per season. Corn or wheat may be planted within 180 days of last application; do not plant other crops within 360 days of last application.
	flutriafol (Topguard)[1]	7 to 14 fl oz	Apply at R1 to R3 as a preventative spray and 21 to 35 days (7 fl oz) later if rust is expected. Do not apply within 21 days of harvest. Make no more than two applications of flutriafol to soybean per season. Plant-back restrictions for all crops except soybean is 120 days after last application.
	propiconazole (Tilt, Propimax, Bumper)	4.0 to 6.0 fl oz	Apply at R1 to R3 as a preventative spray and 14 to 21 days later if rust is expected. Use higher rates if rust is present in field. Do not apply after R5 or within 21 days of harvest. Make no more than two applications of propiconazole-containing materials to soybean per season.
	tetraconazole (Domark)	4.0 to 5.0 fl oz	Apply at R1 to R5 as a preventative spray if rust is expected. A second application of another fungicide may be required if disease pressure is high. Use higher rates if rust is present in field. Do not apply after R5 or within 22 days of harvest. Make no more than two applications of tetraconazole per season. Peanut, soybean and sugar beets may be planted immediately after the last application; small grains (barley, buckwheat, millet, oats, rice, rye, triticale, and wheat) and sugarcane can be planted 45 days after the last application; all other crops can be planted 120 days after the last application.
	prothioconazole (Proline)	2.5 to 3 fl oz	Apply at R1 to R3 as a preventative spray and 14 to 21 days later if rust is expected. Do not apply after R6 or within 30 days of harvest. Make no more than three applications of materials containing triazoles per year. Any crop not listed on the label may be planted within 30 days of last application.
Combinations of Strobilurins and Triazoles (3,11)	azoxystrobin (18.2 %) + difenconazole (11.4%) (Quadris Top)	8.0 to 14.0 fl oz	Apply at R1 to R3 as a preventative spray and 14 to 21 days later if rust is expected. Do not apply after R6 or within 30 days of harvest. Make no more than two applications of materials containing azoxystrobin or cyproconazole per year. Make no more than two applications of cyproconazole to soybean per season. Corn or wheat may be planted within 180 days of last application; do not plant other crops with 360 days of last application.
	azoxystrobin (7.0 %) + propiconazole (11.7 %) (Quilt)	14.0 to 20.5 fl oz	Apply at R1 to R3 as a preventative application and 14 to 21 days later if rust is expected. Do not apply after R6. Make no more than 2 applications containing propiconazole per year.
	fluoxastrobin (14.84 %) + flutriafol (19.3 %) (Fortix)	4.5 to 6 fl oz	Apply at R1 to R3 as a preventative spray and 21 to 35 days later if disease pressure is high. Do not apply after R5 or within 30 days of harvest.
	prothioconazole (10.8 %)+ trifloxystrobin (32.3 %) (Stratego YLD)	4.0 to 4.65 fl oz	Apply at R1 to R3 as a preventative spray and 14 to 21 days later if disease pressure is high. Do not apply after R5 or within 21 days of harvest. Make no more than three applications.
	trifloxystrobin (11.4 %)+ propiconazole (11.4 %) (Stratego)	10.0 fl oz	Apply at R1 to R3 as a preventative spray and 10 to 21 days later if disease pressure is high. Do not apply after R5 or within 21 days of harvest. Make no more than two applications of materials containing propiconazole per year.
	cyproconazole (7.2 %) + picoxystrobin (32.3 %) (Approach Prima)	5.0 to 6.8 fl oz	Apply at R1 to R3 as a preventative spray and 14 to 21 days later if disease pressure is high. Do not apply after R5 or within 21 days of harvest.
	tetraconazole (7.48) + azoxystrobin (9.35 %) (Affiance)	10.0 to 14.0 fl oz	Apply at R1 to R3 as a preventative spray and 14 to 21 days later if disease pressure is high. Do not apply after R5 or within 21 days of harvest.
Combinations of Dicarboximides and Strobilurins (7, 11)	fluxapyroxad (14.3 %) + pyraclostrobin (28.6 %) (Priaxor)	4.0 -8.0 fl oz	Apply at R1 to R3 prior to infection and a second application 7 to 14 days later if disease pressure is high. Do not apply after R5 or within 21 days of harvest.
Nitriles (M5)	chlorothalonil (various brands)	1.5 to 2.25 pints	Apply fungicide at R1 to R3 and make a second application 14 days later. Do not apply within 48 days of harvest.

Further Information

Soybean Rust, North Carolina Cooperative Extension Service, http://www.ces.ncsu.edu/depts/pp/notes/Soybean/soy008/soy008.htm

Using Foliar Fungicides to Manage Soybean Rust. http://www.oardc.ohio-state.edu/SoyRust/index.htm

Relative Fungicide Efficacy for Soybean Foliar Diseases

The North Central Regional Committee on Soybean Diseases (NCERA-137) has developed the following information on foliar fungicide efficacy for control of major foliar soybean diseases in the United States. Efficacy ratings for each fungicide listed in the table were determined by field-testing the materials over multiple years and locations by the members of the committee. Efficacy ratings are based upon level of disease control achieved by product, and are not necessarily reflective of yield increases obtained from product application. Efficacy depends upon proper application timing, rate, and application method to achieve optimum effectiveness of the fungicide as determined by labeled instructions and overall level of disease in the field at the time of application. **Differences in efficacy among fungicide products were determined by direct comparisons among products in field tests and are based on a single application of the labeled rate as listed in the table, unless otherwise noted. Table includes systemic fungicides available that have been tested over multiple years and locations. The table is not intended to be a list of all labeled products[1]**. Efficacy categories: NR=Not Recommended; P=Poor; F=Fair; G=Good; GE=Good to Excellent (Very Good); E=Excellent; NL=Not Labeled for use against this disease; U=Unknown efficacy or insufficient data to rank product efficacy.

Table 10-7D. Relative Fungicide Efficacy for Soybean Foliar Diseases

Class	Active ingredient (%)	Product/ Trade name	Rate/A (fl oz)	Aerial web blight	Anthracnose	Brown spot	Cercospora leaf blight[2]	Frogeye leaf spot[3]	Phomopsis/Diaporthe (Pod and stem blight)	Soybean rust	White mold[4]	Harvest restriction[5]
QoI Strobilurins Group 11	Azoxystrobin 22.9%	Quadris 2.08 SC	6.0 to 15.5	VG	VG	G	P	P	U	G-VG	P	14 days
	Fluoxastrobin 40.3%	Aftershock 480 SC Evito 480 SC	2.0 to 5.7	VG	G	G	P	P	U	U	NL	R5 (beginning seed) 30 days
	Picoxystrobin	Aproach 2.08 SC	6.0 to 12.0	VG	G	G	P	P	U	G	G-VG[9]	14 days
	Pyraclostrobin 23.6%	Headline 2.09 EC/SC	6.0 to 12.0	VG	VG	G	P	P	U	VG	NL	21 days
DMI Triazoles Group 3	Cyproconazole 8.9%	Alto 100SL	2.75 to 5.5	U	U	VG	F	F	U	VG	NL	30 days
	Flutriafol 11.8%	Topguard 1.04 SC	7.0 to 14.0	U	VG	VG	P-G	VG	U	VG-E	F	21 days
	Propiconazole 41.8%	Tilt 3.6 EC Multiple Generics[7]	2.0 to 4.0	P	VG	G	NL	F	NL	VG	NL	R5 (beginning seed)
	Prothioconazole 41.0%	Proline 480 SC[8]	2.5 to 4.3	NL	NL	NL	NL	G-VG	NL	VG	F	21 days
	Tetraconazole 20.5%	Domark 230 ME	4.0 to 5.0	NL	VG	VG	P-G	G-VG	U	VG-E	F	R5 (beginning seed)
MBC Thiophanates Group 1	Thiophanate-methyl	Topsin-M Multiple Generics	10.0 to 20.0	U	U	U	F	VG	U	G	F	21 days
SDHI Carboximides Group 7	Boscalid 70%	Endura 0.7 DF	3.5 to 11.0	U	NL	VG	U	P	NL	NL	VG	21 days

Table 10-7D. Relative Fungicide Efficacy for Soybean Foliar Diseases

Class	Active ingredient (%)	Product/ Trade name	Rate/A (fl oz)	Aerial web blight	Anthracnose	Brown spot	Cercospora leaf blight[2]	Frogeye leaf spot[3]	Phomopsis/Diaporthe (Pod and stem blight)	Soybean rust	White mold[4]	Harvest restriction[5]
Mixed mode of action	Azoxystrobin 18.2% Difenconazole 11.4%	Quadris Top 2.72 SC	8.0 to 14.0	U	U	G-VG	P-G	VG	U	VG	NL	14 days
	Azoxystrobin 19.8% Difenconazole 19.8%	Quadris Top SBX 3.76 SC	7.0 to 7.5	U	U	U	U	G-VG	U	U	U	14 days
	Azoxystrobin 7.0% Propiconazole 11.7%	Avaris 1.66 SC Quilt 1.66 SC HM-0812 1.66 SC	14.0 to 20.5	U	U	G	F	F	U	VG	NL	21 days
	Azoxystrobin 13.5% Propiconazole 11.7%	Quilt Xcel 2.2 SE	10.5 to 21.0	E	VG	G	F	F	U	VG	NL	R6
	Bensovindiflupyr 10.27% Azoxystrobin 13.5% Propiconazole 11.7%	Trivapro A 0.83 + Trivapro B 2.2 SE	A = 4.0 B = 10.5	U	U	VG	U	VG	U	U	NL	14 days R6
	Cyproconazole 7.17% Picoxystrobin 17.94%	Aproach Prima 2.34 SC	5.0 to 6.8	U	U	VG	P-G	G	U	U	NL	14 days
	Flutriafol 19.3% Fluoxastrobin 14.84%	Fortix SC Preemptor SC	4.0 to 6.0	U	U	G	U	G	U	U	U	R5
	Pyraclostrobin 28.58% Fluxapyroxad 14.33%	Priaxor 4.17 SC	4.0 to 8.0	E	VG	E	P-G	P-F	U	VG	P	21 days
	Pyraclostrobin 28.58% Fluxapyroxad 14.33% Tetraconazole 20.50%	Priaxor D 4.17 SC 1.9 SC	4.0 (each component)	U	U	VG	U	G-VG	U	U	P	21 days
	Trifloxystrobin 32.3% Prothioconazole 10.8%	Stratego YLD 4.18 SC[9]	4.0 to 4.65	VG	VG	VG	F	F	U	VG	NL	21 days
	Tetraconazole 7.48% Azoxystrobin 9.35%	Affiance 1.5 SC	10.0-14.0	U	VG	VG	F	G	U	U	U	R5 14 days

[1] Multiple fungicides are labeled for soybean rust only, powdery mildew, and Alternaria leaf spot, including tebuconazole (multiple products) and Laredo (myclobutanil). Contact fungicides such as chlorothalonil may also be labeled for use.

[2] Cercospora leaf blight efficacy relies on accurate application timing, and standard R3 application timings may not provide adequate disease control. Fungicide efficacy may improve with earlier or later applications; however, efficacy has been inconsistent with some products. Fungicides with a solo or mixed QoI or MBC mode of action may not be effective in areas where QoI or MBC resistance has been detected in the fungal population that causes Cercospora leaf blight.

[3] In areas where QoI-fungicide resistant isolates of the frogeye leaf spot pathogen are not present, QoI fungicides may be more effective than indicated in this table.

[4] White mold efficacy is based on R1-R2 application timing, and lower efficacy is obtained at R3 or later application timings, or if disease symptoms are already present at the time of application.

[5] Harvest restrictions are listed for soybean harvested for grain. Restrictions may vary for other types of soybean (edamame, etc.) and soybean for other uses such as forage or fodder.

[6] Multiple generic products containing this mode of action may also be labeled in some states.

[7] Proline has a supplemental label (2ee) for soybean, only for use on white mold in IL, IN, IA, MI, MN, NE, ND, OH, SD, WI. A separate 2ee for NY exists for white mold.

[8] Stratego YLD has a supplemental label (2ee) for white mold on soybean only in IL, IN, IA, MI, MN, NE, ND, OH, SD, WI.

[9] Rating is based on two applications of a 9 fl oz/A rate of Aproach at R1 and R3.

Many products have specific use restrictions about the amount of active ingredient that can be applied within a period of time or the amount of sequential applications that can occur. Please read and follow all specific use restrictions prior to fungicide use. This information is provided only as a guide. It is the responsibility of the pesticide applicator by law to read and follow all current label directions. Reference to products in this publication is not intended to be an endorsement to the exclusion of others that may be similar. Persons using such products assume responsibility for their use in accordance with current directions of the manufacturer. Members or participants in the NCERA-137 group assume no liability resulting from the use of these products.

DISEASE CONTROL

Tobacco Disease Control

L. D. Thiessen, Plant Pathology

Follow manufacturer's label in all cases.

Table 10-8A. Tobacco Disease Control — Nematode Control

Nematicide	Amount of Formulation Per Acre	Waiting Period Before Planting[1] (Days)	Precautions and Remarks
Fumigants dichloropropene (Telone II)	6.0 gal	21	Rates are for in-row injection. Where labeled, broadcast rates are usually 50% to 100% more than in-row rates. Apply fumigants and multi-purpose fumigants at a final depth of 12 to 14 inches. Apply only when the soil temperature is between 55 degrees F and 80 degrees F and soil is moist but not wet. Should soil become wet for an extended time following applications, a longer waiting period before transplanting may be necessary to avoid fumigant injury.
chloropicrin 98% (Chlor-O-Pic 100, Chloropicrin 100) chloropicrin 85% (Pic Plus Fumigant)	3.0 gal 4.0 gal	21	
dichloropropene + chloropicrin (Telone C-17) (Telone C-35)	10.5 gal 12.0 gal	21	

[1] Read and follow product label directions concerning worker reentry periods.

Table 10-8B. Tobacco Disease Control — Field Blue Mold, Target Spot, and Frog-eye Leafspot Control

Material (FRAC Code)[1]	Rate Per Acre (Formulated)	Method of Application
azoxystrobin (Quadris) (11)	6.0 to 12.0 fl oz	Apply on a 7- to 14-day interval with sufficient water volume for adequate coverage and canopy penetration. May be applied up to day of harvest. Do not tank mix with thiodan. **Application directions for blue mold:** Applications should begin prior to disease development or at first indication of blue mold in the area. If blue mold is present in the field, apply dimethomorph prior to Quadris applications.
Blue Mold Only (FRAC Code)[1]		
acibenzolar-S-methyl (Actigard) (21)	0.5 oz	Begin preventative applications after plants reach a height of 12 inches (Flue-Cured)/18 inches (Burley). Make up to three applications on a 10-day schedule. Apply in a minimum of 20 gallons of water per acre.
aluminum tris (O-ethyl phosphate) (Aliette WDG) (33)	2.5 to 4.0 lb	Apply immediately after transplanting and continue on a 7- to 10-day schedule. Begin with a minimum spray volume of 20 gallons per acre, and increase by 20 gallons per acre weekly to a maximum of 100 gallons per acre. The pH of spray solution should not be less than 6.0. No more than 20 pounds per season.
mandipropamid (Revus) (40)	8.0 fl oz	Apply prior to disease development and continue on a 7- to 10- day schedule. Revus can be tank mixed with another fungicide of different FRAC Code. No more than 2 consecutive applications before switching to another mode of action fungicide. No more than 32 fluid ounces per season. Do not apply within 7 days of harvest.
dimethomorph (Forum) (15)	2.0 to 8.0 oz	Increase rate and spray volume as crop size increases. MUST be used in a tank mix with another fungicide (non-Group 40) active against blue mold. Refer to the partner labeling for rates, application method, and restrictions.
mancozeb (Dithane Rainshield) (M)	2 lb	Use only if there is a threat of metalaxyl-insensitive blue mold. Mix 1.5 to 2 pounds per 100 gallons per acre. Spray weekly for complete coverage. Discontinue when threat of blue mold no longer exists. In flue-cured tobacco, do not spray after first button or within 21 days of harvest. In burley, do not spray within 30 days of harvest.
Mefenoxam (Ridomil Gold) EC, SL (Ultra Flourish) 2 E (4)	0.5 to 1 pt 1 to 2 pt	For mefenoxam-sensitive strains of the blue mold fungus, apply in a minimum of 15 gallons of water per acre. Incorporate in the top 2 to 4 inches of soil and form beds. Use highest rate for burley tobacco. For prolonged control, especially in burley, apply a supplemental soil application of either 0.5 pints per acre Ridomil Gold EC or 1 pint Ultra Flourish 2 E at layby or the last cultivation. Do not make the supplemental application if more than the highest rate was applied preplant.
Presidio (43)	4.0 fl oz	MUST be used in a tank mix with another fungicide with different mode of action active against blue mold. No more than 2 foliar applications per season.

[1] To prevent resistance in pathogens, alternate fungicides from a group with fungicides in another group. Fungicides in the "M" group are generally considered "low risk" with no signs of resistance developing to the majority of fungicides.

Table 10-8C. Tobacco Disease Control — Black Shank, Granville Wilt, and Black Root Rot Control

Material	Amount of Formulation Per Acre	Waiting Period Before Planting (Days)	Precautions and Remarks
chloropicrin 98% (Chlor-O-Pic 100, Chloropicrin 100) chloropicrin 85% (Pic Plus Fumigant)	3.0 gal 4.0 gal	21	Rates are for in-row injection. Where labeled, broadcast rates are usually 25% to 100% more than in-row rates. Apply multipurpose fumigants to a depth of 6 to 8 inches and form a high, wide bed immediately. Apply only when the soil temperature is above 55 degrees F and soil is moist but not wet. Should soil become wet for extended time following applications, a longer waiting period before transplanting may be necessary to avoid fumigant injury. Use with Ridomil for black shank control.
dichloropropene + chloropicrin (Telone C-17) (Telone C-35)	10.5 gal 12.0 gal	21	
Black Shank Only (FRAC Code)[1]			
Mefenoxam (Ridomil Gold) EC, SL (Ultra Flourish) 2 E (4)	1 to 3 pt 2 to 6 pt	0 0 0	Use in combination with crop rotation and resistant varieties where applicable. For prolonged control apply either 1 pint Ridomil Gold EC or SL or 2 pints Ultra Flourish 2 E just before transplanting followed by either 0.5 to 1 pint Gold EC or 1 to 2 pints Ultra Flourish 2 E at first cultivation and at lay-by. Also, control nematodes for best results.
Presidio (43)	4 fl oz	-	No more than 2 soil applications per season. Applications cannot be consecutive. Alternate with another fungicide of different mode of action. Apply at first cultivation or layby.

[1] To prevent resistance in pathogens, alternate fungicides within a group with fungicides in another group. Fungicides in the "M" group are generally considered "low risk" with no signs of resistance developing to the majority of fungicides.

447

DISEASE CONTROL

Table 10-8D. Tobacco Disease Control — Tobacco Seedling Disease Control

DISEASE Material (FRAC Code)[1]	Rate	Precautions and Remarks
Blue Mold		
aluminum tris (O-ethyl phosphonate) (Aliette WDG) (33)	0.5 lb/50 gal water	Apply 3 gallons of spray solution per 1,000 square feet for small plants. Increase the volume as the plants grow to a maximum of 12 gallons per 1,000 square feet. Apply preventatively or at the first sign of blue mold. Apply every 5 to 7 days, and do not exceed two applications. After application, wait 24 hours before applying any material over top. Apply insecticides that require a wash down to the soil prior to Aliette.
Blue Mold; Anthracnose; Damping-Off (Rhizoctonia); Stem Rot (Rhizoctonia); Target Spot		
mancozeb (Dithane Rainshield) (M)	1 lb/100 gal water (outdoor plant bed)	Begin sprays when seedlings are quarter size: For outdoor beds, mix 1 tablespoon per gallon water, apply 3 to 5 gallons per 900 sqare feet every 5 to 7 days.
	0.5 lb/100 gal water (greenhouse and float-bed systems)	For greenhouse and float systems, mix 1 teaspoon per gal water, apply 3 to 12 gallons per 1,000 square feet every 5 to 7 days. Use low gallonage on small plants and higher gallonage on larger plants. Do not contaminate float water with mancozeb.
Stem Rot (Rhizoctonia); Target Spot		
azoxystrobin (Quadris) (11)	0.14 oz (4ml)/1000 ft^2 (equal to 6 fl oz/acre)	Use enough water for thorough coverage (recommend 5 gallons per 1,000 square feet or more). Make ONLY ONE application prior to transplanting.
Mosaic		
milk (whole or skim) OR dry skim milk	5 gal/100 sq yd of bed 5 lb in 5 gal water/100 sq yd	Spray plants within 24 hours of pulling.
soap OR milk (skim or whole)		Wash hands with soap or dip hands every 20 minutes while pulling and transplanting to field.
dry skim milk	1 lb in 1 gal water	
Pythium Root Rot		
etridiazole (Terramaster) 4 EC (14)	1.4 oz/100 gal float water	Apply at least 2 to 3 weeks after seeding. Mix thoroughly in the float water. May be used preventively or curatively. A second application may be made, but no later than 8 weeks after seeding.
Wildfire; Angular Leafspot (Burley Tobacco); Blue Mold		
streptomycin sulfate (Agri-Mycin 17) (25)	Spray 200 ppm using 5 gal/100 sq yd, Drench 100 ppm using 10 gal/100 sq yd	Spray or drench when plants are in two-leaf stage and repeat once a week for five sprays. Prepare the solution by mixing 2 (200 ppm) or 1 (100 ppm) teaspoon of streptomycin (17% to 21%) per gallon of water.

[1] To prevent resistance in pathogens, alternate fungicides within a group with fungicides in another group. Fungicides in the "M" group are generally considered "low risk" with no signs of resistance developing to the majority of fungicides.

Table 10-8E. Tobacco Disease Control of Tomato Spotted Wilt (TSWV) — Virus suppression

Material (FRAC Code)[1]	Rate	Precautions and Remarks
acibenzolar (Actigard 50W) (21)	0.5 oz/25,000 to 50,000 plants (sprayed over the top) OR 10 to 25 ppm (added to the float water)	**Waiver of liability must be signed to obtain label.** Apply to trays or flats 5 to 7 days before transplanting. If sprayed over the top, rinse it off into potting soil. Apply only with calibrated boom sprayer to ensure no overlap. If applied to float water, ensure water is circulated uniformly to all tobacco plants. For better results, dilute Actigard in a small volume of water first, and then add this volume to the float water. Use lower rate in areas with moderate TSWV risk and highest in areas with severe TSWV risk.
	0.5 oz/acre	Up to 3 field applications in 10-day increments may be made starting 10 days after the greenhouse application. Begin applications after plants reach a height of 18 inches.
imidacloprid (Admire Pro) (4A)	1.8 oz/1,000 plants	Apply to trays IN THE GREENHOUSE 3 to 5 days prior to transplanting. Mix with water prior to application; do not add wetting agents or defoamers, and do not use in combination with other pesticides. Immediately after application, wash the material off the plants to transfer it to the potting soil. Observe worker protection standards for greenhouse application.
acibenzolar (Actigard 50W) + imidacloprid (21 + 4A)	see above	See comments above for both products. Apply Actigard first, then imidacloprid. Tank mixing has not been determined to be safe.

[1] To prevent resistance in pathogens, alternate fungicides within a group with fungicides in another group. Fungicides in the "M" group are generally considered "low risk" with no signs of resistance developing to the majority of fungicides.

DISEASE CONTROL

Turfgrass Disease Control

J. P. Kerns and E. L. Butler, Plant Pathology Extension

When more than one brand name exists for an agricultural chemical, the brand name that first came onto the market is listed first. Otherwise, brand names are listed in alphabetical order. The order in which brand names are given is not an indication of a recommendation or criticism. Products marked with an asterisk are not labeled for home lawn use.

Table 10-9. Turfgrass Disease Control

Disease	Fungicide and Formulation[1]	Amount of Formulation (oz/1,000 sq ft)[2]	Application Interval (days)[3]
Algae (Cyanobacteria)	chlorothalonil* (Daconil) 82.5 WDG (Daconil Weather Stik, Legend) 6 F (Daconil Zn) 4.16 F	1.8 to 3.25 2 to 3.6 4 to 5.5 3 to 5 6 to 8	7 to 14 7 to 14 14 7 to 14 14
	chlorothalonil + acibenzolar-S-methyl (Daconil Action) 6.1 F*	2 to 3.6 4 to 5.4	7 to 14 14
	chlorothalonil + azoxystrobin (Renown) 5.16 SC*	2.5 to 4.5	10 to 14
	chlorothalonil + fluoxastrobin (Fame C) 4.25 SC*	3 to 5.4	7 to 14
	chlorothalonil + thiophanate-methyl (Spectro) 90 WDG*	2 to 5.76	7 to 14
	chlorothalonil + triticonazole (Reserve) 4.79 SC*	3.2 to 5.4	14 to 28
	fluazinam (Secure) 4.17 SC*	0.5	14
	fluxapyroxad (Xzemplar) 2.47 SC	0.21 to 0.26	14 to 28
	mancozeb* (Fore) 80 WP (Dithane, Pentathlon) 75 DF (Pentathlon) 4 LF (Protect) 75 WP (Wingman) 75 WP	 6 6 10 6 6	 7 to 14 refer to label refer to label 7 to 14 refer to label
	mancozeb + copper hydroxide (Junction) 60 DF*	2 to 4	7 to 14
Anthracnose (Colletotrichum cereale)	azoxystrobin (Heritage, Strobe) 50 WG (Heritage) 0.8 TL (Heritage) 0.31 G (Strobe) 2 L	 0.2 to 0.4 1 to 2 2 to 4 lbs 0.38 to 0.77	 14 to 28 14 to 28 14 to 28 14 to 28
	azoxystrobin + acibenzolar-S-methyl (Heritage Action) 51 WG*	0.2 to 0.4	14 to 28
	azoxystrobin + chlorothalonil (Renown) 5.16 SC*	2.5 4.5	7 to 10 14 to 21
	azoxystrobin + difenoconazole (Briskway) 2.7 SC*	0.3 to 0.725	14
	azoxystrobin + propiconazole (Headway) 1.4 ME 1.06 G	 1.5 to 3 2 to 4 lbs	 14 to 28 14 to 28
	azoxystrobin + tebuconazole (Strobe T) 2.67 SC*	0.75 to 1.5	14 to 21
	chlorothalonil* (Daconil Ultrex) 82.5 WDG (Daconil Weather Stik, Legend) 6 F (Daconil Zn) 4.16 F (Chlorothalonil 500ZN) 4.17 F (Chlorothalonil 720SFT) 6 F (Chlorothalonil, Chlorostar) 82.5 DF (Pegasus) 6 L (Pegasus) 82.5 DF (Pegasus HPX) 6 F	2.75 to 5 3 to 3.6 3.6 to 5.5 4.4 to 5 5.3 to 8 3 to 5 7.9 2.12 to 3.5 5.5 2.8 to 3.2 3.6 to 5.5 3.25 to 5 3.6 to 5.5	7 to 14 7 to 14 14 7 to 14 14 7 to 14 14 7 to 14 14 7 to 14 7 to 14 7 to 14 7 to 14
	chlorothalonil + acibenzolar-S-methyl (Daconil Action) 6.1 F*	3 to 3.6 3.6 to 5.4	7 to 14 14
	chlorothalonil + fluoxastrobin (Fame C) 4.25 SC*	3 to 5.9	14 to 28
	chlorothalonil + iprodione + thiophanate-methyl + tebuconazole (Enclave) 5.3 F*	3 to 4 7 to 8	14 to 21 28
	chlorothalonil + propiconazole (Concert) 4.3 SC*	4.5 to 8.5	7 to 28
	chlorothalonil + propiconazole + fludioxonil (Instrata) 3.59 SC*	2.75 to 6	14 to 28
	chlorothalonil + thiophanate-methyl* (Consyst) 67 WDG (Peregrine) 67 WDG (Spectro) 90 WDG (TM/C) 67 WDG	 2 to 8 2 to 8 3.72 to 5.76 2 to 8	 7 to 14 14 7 to 14 14 to 21
	fenarimol (Rubigan) 1 AS*	1.75 to 3.5	30
	fluazinam (Secure) 4.17 SC*	0.5	14
	fludioxonil (Medallion) 50 WP	0.25 to 0.5	14

449

Table 10-9. Turfgrass Disease Control

Disease	Fungicide and Formulation[1]	Amount of Formulation (oz/1,000 sq ft)[2]	Application Interval (days)[3]
Anthracnose (continued)	fluoxastrobin (Fame) 4 SC 0.25 G	 0.18 to 0.36 2.3 to 4.6 lbs	 14 to 28 14 to 28
	fluoxastrobin + myclobutanil (Fame M) 3.9 SC	0.25 to 1	14 to 28
	fluoxastrobin + tebuconazole (Fame T) 4 SC*	0.45 to 0.9	21 to 28
	flutolanil + thiophanate-methyl (SysStar) 80 WDG	2 to 3	14 to 30
	iprodione + thiophanate-methyl* (26/36) 3.8 F (Dovetail) 3.8 F	 2 to 4 1 to 4	 14 to 21 14 to 21
	iprodione + trifloxystrobin (Interface) 2.27 SC*	4 to 7	refer to label
	metconazole (Tourney) 50 WDG	0.28 to 0.37	14 to 21
	mineral oil (Civitas) + proprietary pigment (Civitas Harmonizer)*	(8 to 32) + (1 to 4)	7 to 21
	myclobutanil (Eagle, Myclobutanil, Siskin) 20 EW	1.2	14 to 21
	penthiopyrad (Velista) 50 WG	0.3 to 0.5	14
	phosphorous acid (Jetphiter) 5.41 F	5	7
	polyoxin D (Affirm) 11.3 WDG (Endorse) 2.5 WP	 0.88 4	 7 to 14 7 to 14
	propiconazole (Banner MAXX, Kestrel, Propiconazole, Savvi, Strider) 1 ME	1 to 2	14 to 28
	pyraclostrobin (Insignia) 20 WG 2 SC	 0.5 to 0.9 0.4 to 0.7	 14 to 28 14 to 28
	pyraclostrobin + boscalid (Honor) 28 WG*	0.55 to 1.1	14 to 28
	pyraclostrobin + fluxapyroxad (Lexicon Intrinsic) 4.17 SC	0.34 to 0.47	14 to 28
	pyraclostrobin + triticonazole (Pillar) 0.81 G	3 lbs	14 to 28
	tebuconazole* (Torque) 3.6 F (Mirage) 2 SC (Skylark, Tebuconazole) 3.6 F	 0.6 to 1.1 1 to 2 0.6	 21 14 to 28 28
	thiophanate-methyl (3336) 50 WP or 4 F (3336 Plus) 2 F (SysTec 1998, T-Bird, TM) 85 WDG (3336) 2 G (SysTec 1998, T-Bird, TM) 4.5 L	 2 to 6 2 to 8 0.67 to 1.3 3 to 9 lbs 1 to 2	 14 14 to 28 14 14 14
	triadimefon (Bayleton) 50 WSP	1	30 to 45
	trifloxystrobin (Compass) 50 WDG	0.15 to 0.25	14 to 21
	trifloxystrobin + triadimefon (Armada) 50 WP (Tartan) 2 SC*	 0.6 to 1.2 1 to 2	 14 to 28 14 to 28
	triticonazole (Trinity) 1.7 SC (Triton) 70 WDG (Triton Flo) 3 F	 0.5 to 1 0.15 to 0.225 0.41 to 1.1	 14 to 28 14 to 28 14 to 28
	triticonazole + chlorothalonil (Reserve) 4.79 SC*	3.2 to 5.4	14 to 28
Brown Ring Patch (*Rhizoctonia circinata var. circinata*)	azoxystrobin (Heritage) 50 WG 0.8 TL 0.31 G	 0.2 to 0.4 1 to 2 2 to 4 lbs	 14 to 28 14 to 28 14 to 28
	azoxystrobin + acibenzolar-S-methyl (Heritage Action) 51 WG*	0.2 to 0.4	14 to 28
	azoxystrobin + difenoconazole (Briskway) 2.7 SC*	0.5 to 0.725	14 to 28
	azoxystrobin + propiconazole (Headway) 1.4 ME	1.5 to 3	14 to 28
	fluoxastrobin + myclobutanil (Fame M) 3.9 SC	0.25 to 1	14 to 28
	fluoxastrobin + tebuconazole (Fame T) 4 SC*	0.45 to 0.9	21 to 28
	penthiopyrad (Velista) 50 WG	0.5	14
	polyoxin D (Affirm) 11.3 WDG (Endorse) 2.5 WP	 0.88 4	 7 to 14 7 to 14
	pyraclostrobin (Insignia) 2 SC	0.7	14 to 28
	pyraclostrobin + fluxapyroxad (Lexicon Intrinsic) 4.17 SC	0.34 to 0.47	14 to 28
	pyraclostrobin + triticonazole (Pillar) 0.81 G	3 lbs	14 to 28
	tebuconazole* (Torque) 3.6 F (Mirage) 2 SC (Skylark, Tebuconazole) 3.6 F	 0.6 to 1.1 1 to 2 0.6	 21 14 to 28 28

Table 10-9. Turfgrass Disease Control

Disease	Fungicide and Formulation[1]	Amount of Formulation (oz/1,000 sq ft)[2]	Application Interval (days)[3]
Brown Ring Patch (continued)	triticonazole (Trinity) 1.7 SC (Triton FLO) 3 F	1 to 2 0.5 to 1.1	14 to 28 14 to 28
	triticonazole + chlorothalonil (Reserve) 4.79 SC*	3.2 to 5.4	14 to 28
Brown Patch (*Rhizoctonia solani*)	azoxystrobin (Heritage, Strobe) 50 WG (Heritage) 0.8 TL (Heritage) 0.31 G (Strobe) 2 L	0.2 to 0.4 1 to 2 2 to 4 lbs 0.38 to 0.77	14 to 28 14 to 28 14 to 28 14 to 28
	azoxystrobin + acibenzolar-S-methyl (Heritage Action) 51 WG*	0.2 to 0.4	14 to 28
	azoxystrobin + chlorothalonil (Renown) 5.16 SC*	2.5 4.5	14 14 to 21
	azoxystrobin + difenoconazole (Briskway) 2.7 SC*	0.3 to 0.725	14 to 28
	azoxystrobin + propiconazole (Headway) 1.4 ME 1.06 G	0.75 to 3 2 to 4 lbs	14 to 28 14 to 28
	azoxystrobin + tebuconazole (Strobe T) 2.67 SC*	0.75 to 1.5	14 to 21
	chloroneb (Teremec)* 65 SP 2.9 F	3 to 4 5 to 7	7 to 10 7 to 10
	chlorothalonil* (Daconil Ultrex) 82.5 WDG (Daconil Weather Stik, Legend) 6 F (Daconil Zn) 4.16 F (Chlorothalonil 500ZN) 4.17 F (Chlorothalonil 720SFT) 6 F (Chlorothalonil, Chlorostar) 82.5 DF (Pegasus) 6 L (Pegasus) 82.5 DF (Pegasus HPX) 6 F	1.8 to 3.23 3.7 to 5 2 to 3.6 4 to 5.5 3 to 5 6 to 8 3 to 5 7.9 2.12 to 3.5 5.5 1.8 to 3.2 2 to 3.6 1.82 to 3.25 2 to 3.6	7 to 14 14 7 to 14 14 7 to 14 14 7 to 14 14 7 to 14 14 7 to 14 7 to 14 7 to 14 7 to 14
	chlorothalonil + acibenzolar-S-methyl (Daconil Action) 6.1 F*	2 to 3.5 4 to 5.4	7 to 14 14
	chlorothalonil + fluoxastrobin (Fame C) 4.25 SC*	1.5 to 5.9	14 to 28
	chlorothalonil + iprodione + thiophanate-methyl + tebuconazole (Enclave) 5.3 F*	3 to 4 7 to 8	14 to 21 28
	chlorothalonil + propiconazole (Concert) 4.3 SC*	3 to 5.5 5.5 to 8.5	7 to 14 14 to 28
	chlorothalonil + propiconazole + fludioxonil (Instrata) 3.6 SC*	2.75 to 6	14 to 21
	chlorothalonil + thiophanate-methyl* (Spectro) 90 WDG (TM/C) 67 WDG	3 to 5.76 2 to 8	14 to 21 14 to 21
	fenarimol (Rubigan) 1 AS*	1.5	7 to 14
	fluazinam (Secure) 4.17 SC*	0.5	14
	fludioxonil (Medallion) 50 WP	0.2 to 0.25 0.5	7 14
	fluoxastrobin (Fame) 4 SC 0.25 G	0.09 to 0.36 1.2 to 4.6 lbs	14 to 28 14 to 28
	fluoxastrobin + myclobutanil (Fame M) 3.9 SC	0.25 to 1	14 to 28
	fluoxastrobin + tebuconazole (Fame T) 4 SC*	0.45 to 0.9	21 to 28
	fluxapyroxad (Xzemplar) 2.47 SC	0.21 to 0.26	14 to 21
	flutolanil (Prostar) 70 WP, 70 DG	1.5 to 3	14 to 21
	flutolanil + thiophanate-methyl (SysStar) 80 WDG	2 to 3	14 to 21
	iprodione 26GT, Iprodione Pro, IPro, Raven* 2 F, 2 SC, 2 SE	3 to 4	14 to 28
	iprodione + thiophanate-methyl* (26/36) 3.8 F (Dovetail) 3.8 F	2 to 4 1 to 4	14 to 21 14 to 21
	iprodione + trifloxystrobin (Interface) 2.27 SC*	3 to 5	refer to label
	mancozeb* (Fore) 80 WP (Dithane) 75 DF (Protect) 75 WP	4 4 4	7 10 7 to 14
	mancozeb + copper hydroxide (Junction) 60 DF*	2 to 4	7
	metconazole (Tourney) 50 WDG	0.28 to 0.37	14 to 21
	mineral oil (Civitas) + proprietary pigment (Civitas Harmonizer)*	(8 to 32) + (1 to 4)	7 to 21

Table 10-9. Turfgrass Disease Control

Disease	Fungicide and Formulation[1]	Amount of Formulation (oz/1,000 sq ft)[2]	Application Interval (days)[3]
Brown Patch (continued)	myclobutanil (Eagle, Myclobutanil, Siskin) 20 EW	1.2	14
	penthiopyrad (Velista) 50 WG	0.3 to 0.5	14 to 21
	phosphorous acid (Jetphiter) 5.41 F	5	7
	polyoxin D (Affirm) 11.3 WDG (Endorse) 2.5 WP	 0.88 4	 7 to 14 7 to 14
	propiconazole (Banner MAXX, Kestrel, Propiconazole, Savvi, Strider) 1 ME	1 to 2	14 to 21
	pyraclostrobin (Insignia) 20 WG 2 SC	 0.5 to 0.9 0.4 to 0.7	 14 to 28 14 to 28
	pyraclostrobin + boscalid (Honor) 28 WG*	0.55 to 1.1	14 to 28
	pyraclostrobin + fluxapyroxad (Lexicon Intrinsic) 4.17 SC	0.34 to 0.47	14 to 28
	pyraclostrobin + triticonazole (Pillar) 0.81 G	3 lbs	14 to 28
	tebuconazole* (Torque) 3.6 F (Mirage) 2 SC (Skylark, Tebuconazole) 3.6 F	 0.6 to 1.1 1 to 2 0.6	 21 14 to 28 28
	thiram (Spotrete) 4 F*	3.75 to 7.5	3 to 10
	triadimefon (Bayleton) 50 WSP, 4.15 F	0.5 to 1	15 to 30
	trifloxystrobin (Compass) 50 WDG	0.1 to 0.2 0.15 to 0.25	14 21
	trifloxystrobin + triadimefon (Armada) 50 WP (Tartan) 2 SC*	 0.6 to 1.2 1 to 2	 14 to 28 14 to 28
	triticonazole (Trinity) 1.7 SC (Triton) 70 WDG (Triton Flo) 3 F	 0.75 to 2 0.15 to 0.3 0.41 to 1.1	 14 to 28 14 to 28 14 to 28
	triticonazole + chlorothalonil (Reserve) 4.79 SC*	3.2 to 5.4	14 to 28
	vinclozolin (Curalan, Touche) 50 EG*	1	14 to 28
Copper Spot *(Gloeocercospora sorghi)*	chlorothalonil* (Daconil Ultrex) 82.5 WDG (Daconil Weather Stik, Legend) 6 F (Daconil Zn) 4.16 F (Chlorothalonil 500ZN) 6 F (Chlorothalonil 720SFT) 6 F (Chlorothalonil, Chlorostar) 82.5 DF (Pegasus) 6 L (Pegasus) 82.5 DF (Pegasus HPX) 6 F	 3.7 to 5 4 to 5.5 6 to 8 3 to 5 7.9 2.12 to 3.5 5.5 3.2 3.6 to 5.5 3.25 to 5 3.6 to 5.5	 14 14 14 7 to 10 14 7 to 10 14 7 to 10 7 to 14 7 to 14 7 to 14
	chlorothalonil + acibenzolar-S-methyl (Daconil Action) 6.1 F*	4 to 5.4	14
	chlorothalonil + azoxystrobin (Renown) 5.16 SC*	2.5	14
	chlorothalonil + fluoxastrobin (Fame C) 4.25 SC*	5.9	14
	chlorothalonil + iprodione + thiophanate-methyl + tebuconazole (Enclave) 5.3 F*	3 to 4 7 to 8	14 to 21 28
	chlorothalonil + propiconazole (Concert) 4.3 SC*	5.5 to 8.5	14
	chlorothalonil + thiophanate-methyl* (Consyst) 67 WDG (Peregrine) 67 WDG (Spectro) 90 WDG (TM/C) 67 WDG	 3 to 8 3 to 8 3 to 5.76 3 to 8	 7 to 10 14 14 14 to 21
	fenarimol (Rubigan) 1 AS*	0.75 to 1.5	10 to 28
	fluoxastrobin + myclobutanil (Fame M) 3.9 SC	0.25 to 1	14 to 21
	flutolanil + thiophanate-methyl (SysStar) 80 WDG	2 to 3	14 to 21
	iprodione + thiophanate-methyl (26/36) 3.8 F*	2 to 4	14 to 21
	mancozeb* (Fore) 80 WP (Dithane) 75 DF (Pentathlon) 4 LF (Pentathlon) 75 DF (Protect, Wingman) 75 WP	 4 to 8 4 to 8 7 to 14 4 to 8 4 to 8	 7 to 14 10 7 to 14 7 7 to 14
	mancozeb + copper hydroxide (Junction) 60 DF*	2 to 4	7 to 14
	myclobutanil (Eagle, Myclobutanil, Siskin) 20 EW	1.2	14
	tebuconazole* (Torque) 3.6 F (Skylark, Tebuconazole) 3.6 F	 0.6 to 1.1 0.6	 refer to label 28

Table 10-9. Turfgrass Disease Control

Disease	Fungicide and Formulation[1]	Amount of Formulation (oz/1,000 sq ft)[2]	Application Interval (days)[3]
Copper Spot (continued)	thiophanate-methyl (3336) 50 WP or 4 F (3336 Plus) 2 F (SysTec 1998, T-Bird, TM) 85 WDG (3336) 2 G (SysTec 1998, T-Bird, TM) 4.5 L	2 to 4 2 to 4 0.67 to 1.3 1.5 to 6 lbs 1 to 2	14 14 to 28 14 14 14
	thiram (Spotrete) 4F*	3.75 to 7.5	3 to 10
	triadimefon (Bayleton) 50 WSP, 4.15 F	0.5 to 1	15 to 30
Dead Spot *(Ophiosphaerella agrostis)*	azoxystrobin + propiconazole (Headway) 1.4 ME 1.06 G	1.5 to 3 2 to 4 lbs	14 14 to 28
	azoxystrobin + tebuconazole (Strobe T) 2.67 SC*	0.75 to 1.5	14
	boscalid* (Emerald) 70 WG	0.18	14
	chlorothalonil + thiophanate-methyl (Spectro) 90 WDG*	3.72 to 5.76	14
	fludioxonil (Medallion) 50 WP	0.3 to 0.5	14
	pyraclostrobin (Insignia) 20 WG 2 SC	0.5 to 0.9 0.4 to 0.7	14 to 28 14 to 28
	pyraclostrobin + boscalid (Honor) 28 WG*	0.55 to 1.1	14 to 28
	pyraclostrobin + fluxapyroxad (Lexicon Intrinsic) 4.17 SC	0.34 to 0.47	14 to 28
	pyraclostrobin + triticonazole (Pillar) 0.81 G	3 lbs	14 to 28
	thiophanate-methyl (3336) 50WP or 4 F (3336 Plus) 2 F (3336) 2 G	4 to 6 4 to 6 6 to 9 lbs	14 14 14
Dollar Spot *(Sclerotinia homoeocarpa)*	azoxystrobin + difenoconazole (Briskway) 2.7 SC*	0.3 to 0.725	14 to 21
	azoxystrobin + propiconazole (Headway) 1.4 ME 1.06 G	0.75 to 3 2 ot 4 lbs	7 to 28 14 to 28
	azoxystrobin + tebuconazole (Strobe T) 2.67 SC*	0.75 to 1.5	14 to 21
	boscalid* (Emerald) 70 WG	0.13 to 0.18	14 to 28
	chlorothalonil* (Daconil Ultrex) 82.5W DG (Daconil Weather Stik, Legend) 6 F (Daconil Zn) 4.16 F (Chlorothalonil 500ZN) 4.17 F (Chlorothalonil 720SFT) 6 F (Chlorothalonil, Chlorostar) 82.5 DF (Pegasus) 6 L (Pegasus) 82.5 DF (Pegasus HPX) 6 F	1 to 3.25 3.7 to 5 1 to 3.6 4 to 5.5 1.5 to 5 6 to 8 3 to 5 7.9 2.12 to 3.5 5.5 1.8 to 3.2 2 to 3.6 1.82 to 3.25 2 to 3.6	7 to 21 14 to 21 7 to 21 14 to 21 7 to 21 14 7 to 14 14 7 to 14 14 7 to 10 7 to 14 7 to 14 7 to 14
	chlorothalonil + acibenzolar-S-methyl (Daconil Action) 6.1 F*	1 to 3.5 4 to 5.4	7 to 21 14
	chlorothalonil + azoxystrobin (Renown) 5.16 SC*	2.5 to 4.5	7 to 14
	chlorothalonil + fluoxastrobin (Fame C) 4.25 SC*	3 to 5.9	14 to 21
	chlorothalonil + iprodione + thiophanate-methyl + tebuconazole (Enclave) 5.3 F*	3 to 4 7 to 8	14 to 21 28
	chlorothalonil + propiconazole (Concert) 4.3 SC*	1.5 to 3 3 to 5.5 5.5 to 8.5	7 to 10 14 to 21 14 to 28
	chlorothalonil + propiconazole + fludioxonil (Instrata) 3.6 SC*	2.75 to 6	21 to 28
	chlorothalonil + thiophanate-methyl* (Consyst) 67 WDG (Peregrine) 67 WDG (Spectro) 90 WDG (TM/C) 67 WDG	2 to 8 2 to 8 3.72 to 5.76 2 to 8	7 to 21 14 14 to 21 7 to 14
	fenarimol (Rubigan) 1 AS*	0.75 to 1.5	10 to 28
	fluazinam (Secure) 4.17 SC*	0.5	14
	fluoxastrobin (Fame) 4 SC 0.25 G	0.18 to 0.36 2.3 to 4.6 lbs	14 to 21 14 to 21
	fluoxastrobin + myclobutanil (Fame M) 3.9 SC	0.25 to 1	14 to 21
	fluoxastrobin + tebuconazole (Fame T) 4 SC*	0.45 to 0.9	21 to 28
	fluxapyroxad (Xzemplar) 2.47 SC	0.16 to 0.26	14 to 28
	flutolanil + thiophanate-methyl (SysStar) 80 WDG	2 to 3	14 to 30

453

DISEASE CONTROL

Table 10-9. Turfgrass Disease Control

Disease	Fungicide and Formulation[1]	Amount of Formulation (oz/1,000 sq ft)[2]	Application Interval (days)[3]
Dollar Spot (*Sclerotinia homoeocarpa*) (continued)	iprodione (26GT, Iprodione Pro, IPro, Raven) 2 F, 2 SC, 2 SE*	2 to 4	14 to 28
	iprodione + thiophanate-methyl* (26/36) 3.8 F (Dovetail) 3.8 F	2 to 4 / 1 to 4	14 to 21 / 14 to 21
	iprodione + trifloxystrobin (Interface) 2.27 SC*	2 to 5	refer to label
	isofetamid (Kabuto 3.33 SC)	0.4 to 0.5	14
	mancozeb* (Fore) 80 WP (Dithane) 75 DF (Pentathlon) 4 LF (Pentathlon) 75 DF (Protect, Wingman) 75 WP	6 to 8 / 6 to 8 / 10 to 14 / 6 to 8 / 6 to 8	7 to 14 / 10 / 7 to 14 / 7 / 7 to 14
	mancozeb + copper hydroxide (Junction) 60 DF*	2 to 4	7 to 14
	metconazole (Tourney) 50 WDG	0.18 to 0.37	14 to 21
	mineral oil (Civitas) + proprietary pigment (Civitas Harmonizer)*	(8 to 32) + (1 to 4)	7 to 21
	myclobutanil (Eagle, Myclobutanil, Siskin) 20 EW	0.5 to 2.4	7 to 28
	penthiopyrad (Velista) 50 WG	0.3 to 0.5	14 to 21
	propiconazole (Banner MAXX, Propiconazole, Savvi, Spectator) 1 ME	0.5 to 2	7 to 28
	pyraclostrobin (Insignia) 20 WG 2 SC	0.9 / 0.7	14 / 14
	pyraclostrobin + boscalid (Honor) 28 WG*	0.83 to 1.1	14 to 21
	pyraclostrobin + fluxapyroxad (Lexicon Intrinsic) 4.17 SC	0.34 to 0.47	14 to 28
	pyraclostrobin + triticonazole (Pillar) 0.81 G	3 lbs	14 to 28
	tebuconazole* (Torque) 3.6 F (Mirage) 2 SC (Skylark, Tebuconazole) 3.6 F	0.6 to 1.1 / 1 to 2 / 0.6	refer to label / 14 to 28 / 28
	thiophanate-methyl (3336) 50WP or 4 F (3336 Plus) 2 F (SysTec 1998, T-Bird, TM) 85 WDG (3336) 2 G (SysTec 1998, T-Bird, TM) 4.5 L	2 to 4 / 2 to 4 / 0.67 to 1.3 / 1.5 to 6 lbs / 1 to 2	14 / 14 to 28 / 14 / 14 / 14
	thiram (Spotrete) 4 F*	3.75 to 7.5	3 to 10
	triadimefon (Bayleton) 50 WSP, 4.15 F	0.25 to 1	14 to 30
	trifloxystrobin + triadimefon (Armada) 50 WP (Tartan) 2 SC*	0.6 to 1.2 / 1 to 2	14 to 28 / 14 to 28
	triticonazole (Trinity) 1.7 SC (Triton) 70 WDG (Triton FLO) 3 F	1 to 2 / 0.15 to 0.3 / 0.28 to 1.1	14 to 28 / 14 to 28 / 14 to 28
	triticonazole + chlorothalonil (Reserve) 4.79 SC*	3.2 to 4.5	14 to 28
	vinclozolin (Curalan, Touche) 50 EG*	1	21 to 28
Fairy Ring (Basidiomycetes)	azoxystrobin (Heritage, Strobe) 50 WG (Heritage) 0.8 TL (Heritage) 0.31 G	0.4 / 2 / 2 to 4 lbs	28 / 28 / 14 to 28
	azoxystrobin + acibenzolar-S-methyl (Heritage Action) 51 WG*	0.2 to 0.4	14 to 28
	azoxystrobin + difenoconazole (Briskway) 2.7 SC*	0.5 to 0.725	14 to 28
	azoxystrobin + propiconazole (Headway) 1.4ME 1.06 G	1.5 to 3 / 2 to 4 lbs	14 to 28 / 14 to 28
	azoxystrobin + tebuconazole (Strobe T) 2.67 SC*	0.75 to 1.5	28
	chlorothalonil + fluoxastrobin (Fame C) 4.25 SC*	4.5 to 5.9	21 to 28
	fluoxastrobin (Fame) 4 SC 0.25 G	0.28 to 0.36 / 2.3 to 4.6 lbs	21 to 28 / 28
	fluoxastrobin + myclobutanil (Fame M) 3.9 SC	0.5 to 1	21 to 28
	fluoxastrobin + tebuconazole (Fame T) 4 SC*	0.45 to 0.9	21 to 28
	flutolanil (Prostar) 70 WP, 70 WDG	2.2 to 4.5	21 to 30
	flutolanil + thiophanate-methyl (SysStar) 80 WDG	3 to 6.12	21 to 28
	metconazole (Tourney) 50 WDG	0.37	21
	penthiopyrad (Velista) 50 WG	0.5 to 0.7	14 to 28

Table 10-9. Turfgrass Disease Control

Disease	Fungicide and Formulation[1]	Amount of Formulation (oz/1,000 sq ft)[2]	Application Interval (days)[3]
Fairy Ring (continued)	polyoxin D (Affirm) 11.3 WDG (Endorse) 2.5 WP	1 4	7 7
	pyraclostrobin (Insignia) 20 WG 2 SC	0.9 0.7	28 28
	pyraclostrobin + boscalid (Honor) 28 WG*	1.1	28
	pyraclostrobin + fluxapyroxad (Lexicon Intrinsic) 4.17 SC	0.47	28
	pyraclostrobin + triticonazole (Pillar) 0.81 G	3 lbs	14 to 28
	tebuconazole* (Torque) 3.6 F (Mirage) 2 SC	0.6 to 1.1 1 to 2	21 28
	triadimefon (Bayleton) 50DF, 4.15 F	1 to 2	14 to 21
Gray Leaf Spot *(Pyricularia grisea)*	azoxystrobin (Heritage, Strobe) 50 WG (Heritage) 0.8 TL (Heritage) 0.31 G (Strobe) 2 L	0.2 to 0.4 1 to 2 2 to 4 lbs 0.38 to 0.77	14 to 28 14 to 28 14 to 28 14 to 28
	azoxystrobin + acibenzolar-S-methyl (Heritage Action) 51 WG*	0.2 to 0.4	14 to 28
	azoxystrobin + chlorothalonil (Renown) 5.16 SC*	2.5 to 4.5	10 to 14
	azoxystrobin + difenoconazole (Briskway) 2.7 SC*	0.5 to 0.725	14 to 21
	azoxystrobin + propiconazole (Headway) 1.4 ME 1.06 G	1.5 to 3 2 to 4 lbs	14 to 28 14 to 28
	azoxystrobin + tebuconazole (Strobe T) 2.67 SC*	0.75	1.5
	chlorothalonil* (Daconil Ultrex) 82.5 WDG (Daconil Weather Stik, Legend) 6 F (Daconil Zn) 4.16 F (Chlorothalonil 500ZN) 4.17 F (Chlorothalonil 720SFT) 6 F (Chlorothalonil, Chlorostar) 82.5 DF (Pegasus) 6 L (Pegasus) 82.5 DF (Pegasus HPX) 6 F	1.8 to 3.25 3.7 to 5 2 to 3.6 4 to 5.5 3 to 5 6 to 8 3 to 5 7.9 2.12 to 3.5 5.5 1.8 to 3.2 2 to 3.6 1.82 to 3.25 2 to 3.6	7 to 21 14 7 to 10 14 7 to 14 14 7 to 10 14 7 to 14 14 7 to 10 7 to 14 7 to 14 7 to 14
	chlorothalonil + acibenzolar-S-methyl (Daconil Action) 6.1 F*	2 to 3.5 4 to 5.4	7 to 10 14
	chlorothalonil + fluoxastrobin (Fame C) 4.25 SC*	3 to 5.9	14 to 28
	chlorothalonil + iprodione + thiophanate-methyl + tebuconazole (Enclave) 5.3 F*	3 to 4 7 to 8	14 to 21 28
	chlorothalonil + propiconazole (Concert) 4.3 SC*	3 to 5.5 5.5 to 8.5	7 to 14 14 to 21
	chlorothalonil + propiconazole + fludioxonil (Instrata) 3.6 SC*	2.75 to 6	10 to 14
	chlorothalonil + thiophanate-methyl* (Consyst) 67 WDG (Peregrine) 67 WDG (Spectro) 90 WDG (TM/C) 67 WDG	2 to 8 2 to 8 3.72 to 5.76 2 to 8	7 to 14 14 14 14 to 21
	fludioxonil (Medallion) 50 WP	0.25 to 0.5	14
	fluoxastrobin (Fame) 4 SC 0.25 G	0.18 to 0.36 2.3 to 4.6 lbs	14 to 28 14 to 28
	fluoxastrobin + myclobutanil (Fame M) 3.9 SC	0.25 to 1	14 to 28
	fluoxastrobin + tebuconazole (Fame T) 4 SC*	0.45 to 0.9	21 to 28
	flutolanil + thiophanate-methyl (SysStar) 80 WDG	2 to 3	14
	mancozeb* (Fore) 80 WP (Dithane) 75 DF (Pentathlon) 4 LF (Pentathlon) 75 DF (Wingman) 75 WP	8 6.4 to 12.8 9 to 14 8 8	14 7 to 14 5 7 7
	metconazole (Tourney) 50 WDG	0.37	14
	mineral oil (Civitas) + proprietary pigment (Civitas Harmonizer)*	(8 to 32) + (1 to 4)	7 to 21
	myclobutanil (Eagle, Siskin) 20 EW	1.2 to 2.4	14
	polyoxin D (Affirm) 11.3 WDG	0.88	7 to 14
	propiconazole (Banner MAXX, Kestrel, Propiconazole, Savvi, Strider) 1 ME	1 to 2	14

DISEASE CONTROL

Table 10-9. Turfgrass Disease Control

Disease	Fungicide and Formulation[1]	Amount of Formulation (oz/1,000 sq ft)[2]	Application Interval (days)[3]
Gray Leaf Spot (continued)	pyraclostrobin (Insignia) 20 WG 2 SC	 0.5 to 0.9 0.4 to 0.7	 14 to 28 14 to 28
	pyraclostrobin + boscalid (Honor) 28 WG*	0.55 to 1.1	14 to 28
	pyraclostrobin + fluxapyroxad (Lexicon Intrinsic) 4.17 SC	0.34 to 0.47	14 to 28
	pyraclostrobin + triticonazole (Pillar) 0.81 G	3 lbs	14 to 28
	tebuconazole* (Torque) 3.6 F (Mirage) 2 SC (Skylark, Tebuconazole) 3.6 F	 0.6 to 1.1 1 to 2 0.6	 21 14 to 28 28
	thiophanate-methyl (3336) 50 WP or 4 F (3336 Plus) 2 F (3336) 2 G (SysTec 1998, T-Bird, TM) 85 WDG (SysTec 1998, T-Bird, TM) 4.5 L	 4 to 6 4 to 8 6 to 9 lbs 2.35 to 3.53 3.5 to 5	 14 14 to 28 14 14 14
	triadimefon (Bayleton) 50 WSP, 4.15 F	0.5 to 1	14
	trifloxystrobin (Compass) 50 WDG	0.15 to 0.2 0.25	14 21
	trifloxystrobin + triadimefon (Armada) 50 WP (Tartan) 2 SC*	 0.6 to 1.2 1 to 2	 14 to 28 14 to 28
Helminthosporium Leaf Spot/Melting Out *(Bipolaris spp.; Drechslera spp.)*	azoxystrobin (Heritage, Strobe) 50 WG (Heritage) 0.8 TL (Heritage) 0.31 G (Strobe) 2 L	 0.2 to 0.4 1 to 2 2 to 4 lbs 0.38 to 0.77	 14 to 21 14 to 21 14 to 21 14 to 21
	azoxystrobin + acibenzolar-S-methyl (Heritage Action) 51 WG*	0.2 to 0.4	14 to 21
	azoxystrobin + chlorothalonil (Renown) 5.16 SC*	2.5 to 4.5	14 to 21
	azoxystrobin + difenoconazole (Briskway) 2.7 SC*	0.5 to 0.725	14 to 21
	azoxystrobin + propiconazole (Headway) 1.4 ME 1.06 G	 1.5 to 3 2 to 4 lbs	 14 to 21 14 to 21
	azoxystrobin + tebuconazole (Strobe T) 2.67 SC*	0.75 to 1.5	14 to 21
	chlorothalonil* (Daconil Ultrex) 82.5 WDG (Daconil Weather Stik, Legend) 6 F (Daconil Zn) 4.16 F (Chlorothalonil 500ZN) 4.17 F (Chlorothalonil 720SFT) 6 F (Chlorothalonil, Chlorostar) 82.5 DF (Pegasus) 6 L (Pegasus) 82.5 DF (Pegasus HPX) 6 F	 1.8 to 3.25 3.7 to 5 2 to 3.6 4 to 5.5 3 to 5 6 to 8 3 to 5 7.9 2.12 to 3.5 5.5 1.8 to 3.2 2 to 3.6 1.82 to 3.25 2 to 3.6	 7 to 21 14 to 21 7 to 21 14 7 to 21 14 7 to 10 14 7 to 10 14 7 to 10 7 to 14 7 to 14 7 to 14
	chlorothalonil + acibenzolar-S-methyl (Daconil Action) 6.1 F*	2 to 3.5 4 to 5.4	7 to 21 14
	chlorothalonil + fluoxastrobin (Fame C) 4.25 SC*	3 to 5.9	14 to 21
	chlorothalonil + propiconazole (Concert) 4.3 SC*	3 to 5.5 5.5 to 8.5	7 to 14 14 to 21
	chlorothalonil + propiconazole + fludioxonil (Instrata) 3.6 SC*	2.75 to 6	10 to 21
	chlorothalonil + thiophanate-methyl* (Consyst) 67 WDG (Peregrine) 67 WDG (Spectro) 90 WDG (TM/C) 67 WDG	 2 to 8 2 to 8 3.72 to 5.76 2 to 8	 7 to 21 14 14 14 to 21
	fluazinam (Secure) 4.17 SC*	0.5	14
	fludioxonil (Medallion) 50 WP	0.25 to 0.5	14 to 21
	fluoxastrobin (Fame) 4 SC 0.25 G	 0.18 to 0.36 2.3 to 4.6 lbs	 14 to 21 14 to 21
	fluoxastrobin + myclobutanil (Fame M) 3.9 SC	0.25 to 1	14 to 28
	fluoxastrobin + tebuconazole (Fame T) 4 SC*	0.45 to 0.9	21 to 28
	flutolanil + thiophanate-methyl (SysStar) 80 WDG	2 to 3	14
	iprodione (26GT, Iprodione Pro, IPro, Raven) 2 F, 2 SC, 2 SE*	3 to 4	14 to 28
	iprodione + thiophanate-methyl * (26/36) 3.8 F (Dovetail) 3.8 F	 2 to 4 1 to 4	 14 to 21 14 to 21

Table 10-9. Turfgrass Disease Control

Disease	Fungicide and Formulation[1]	Amount of Formulation (oz/1,000 sq ft)[2]	Application Interval (days)[3]
Helminthosporium Leaf Spot/Melting Out (continued)	iprodione + trifloxystrobin (Interface) 2.27 SC*	3 to 5	refer to label
	mancozeb* (Fore) 80 WP (Dithane) 75 DF (Pentathlon) 4 LF (Pentathlon) 75 DF (Protect, Wingman) 75 WP	4 4 5 to 14 4 4	7 to 14 10 3 to 5 7 7 to 14
	mancozeb + copper hydroxide (Junction) 60 DF*	2 to 4	7 to 14
	mineral oil (Civitas) + proprietary pigment (Civitas Harmonizer)*	(8 to 32) + (1 to 4)	7 to 21
	myclobutanil (Eagle, Myclobutanil, Siskin) 20 EW	1.2	14
	penthiopyrad (Velista) 50 WG	0.3 to 0.5	14
	polyoxin D (Affirm) 11.3 WDG (Endorse) 2.5 WP	0.88 4	7 to 14 7 to 14
	propiconazole (Banner MAXX, Kestrel, Propiconazole, Savvi, Strider) 1 ME	1 to 2	14
	pyraclostrobin (Insignia) 20 WG 2 SC	0.5 to 0.9 0.4 to 0.7	14 to 28 14 to 28
	pyraclostrobin + boscalid (Honor) 28WG*	0.55 to 1.1	14 to 28
	pyraclostrobin + fluxapyroxad (Lexicon Intrinsic) 4.17 SC	0.34 to 0.47	14 to 28
	pyraclostrobin + triticonazole (Pillar) 0.81 G	3 lbs	14 to 28
	thiophanate-methyl (3336) 50 WP or 4 F (3336 Plus) 2 F (3336) 2 G	4 to 6 4 to 8 6 to 9 lbs	14 14 to 28 14
	thiram (Spotrete) 4 F*	3.75 to 7.5	3 to 10
	trifloxystrobin (Compass) 50 WDG	0.1 to 0.15 0.15 to 0.25	14 21 to 28
	trifloxystrobin + triadimefon (Armada) 50 WP (Tartan) 2 SC*	0.6 to 1.2 1 to 2	14 to 28 14 to 28
	triticonazole (Trinity) 1.7 SC (Triton) 70 WDG	0.5 to 2 0.15 to 0.3	14 to 28 14 to 28
	triticonazole + chlorothalonil (Reserve) 4.79 SC*	3.2 to 4.5	14 to 28
	vinclozolin (Curalan, Touche) 50 EG*	1	14 to 28
Large Patch (Zoysia Patch) *(Rhizoctonia solani)*	azoxystrobin (Heritage, Strobe) 50 WG (Heritage) 0.8 TL (Heritage) 0.31 G (Strobe) 2 L	0..2 to 0.4 2 2 to 4 lbs 0.38 to 0.77	14 to 28 14 to 28 14 to 28 28
	azoxystrobin + acibenzolar-S-methyl (Heritage Action) 51 WG*	0.2 to 0.4	14 to 28
	azoxystrobin + chlorothalonil (Renown) 5.16 SC*	2.5 4.5	14 14 to 21
	azoxystrobin + difenoconazole (Briskway) 2.7 SC*	0.3 to 0.725	14 to 28
	azoxystrobin + propiconazole (Headway) 1.4 ME 1.06 G	1.5 to 3 2 to 4 lbs	14 to 28 14 to 28
	azoxystrobin + tebuconazole (Strobe T) 2.67 SC*	0.75 to 1.5	14 to 21
	chloroneb* (Teremec) 65 SP (Teremec) 2.9 F	5 9	21 to 28 21 to 28
	chlorothalonil + fluoxastrobin (Fame C) 4.25 SC*	3 to 5.9	14 to 28
	chlorothalonil + iprodione + thiophanate-methyl + tebuconazole (Enclave) 5.3 F*	3 to 4 7 to 8	14 to 21 28
	chlorothalonil + thiophanate-methyl* (Consyst) 67 WDG (Peregrine) 67 WDG	2 to 8 2 to 8	7 to 14 14
	fluoxastrobin (Fame) 4 SC 0.25	0.28 to 0.36 2.3 to 4.6 lbs	14 to 28 14 to 28
	fluoxastrobin + myclobutanil (Fame M) 3.9 SC	0.5 to 1	21 to 28
	fluxapyroxad (Xzemplar) 2.47 SC	0.21 to 0.26	14 to 28
	flutolanil (Prostar) 70 WP, 70 WDG	2.2	30
	iprodione (26GT, Iprodione Pro, Ipro, Raven) 2 F, 2 SC, 2 SE*	4	14 to 21
	iprodione + thiophanate-methyl (26/36) 3.8 F*	2 to 4	14 to 21

Table 10-9. Turfgrass Disease Control

Disease	Fungicide and Formulation[1]	Amount of Formulation (oz/1,000 sq ft)[2]	Application Interval (days)[3]
Large Patch (continued)	iprodione + trifloxystrobin (Interface) 2.27 SC*	4	14 to 21
	metconazole (Tourney) 50 WDG	0.37	14
	myclobutanil (Eagle, Myclobutanil, Siskin) 20 EW	2.4	28 (fall)
	penthiopyrad (Velista) 50 WG	0.7	14 to 28
	polyoxin D (Affirm) 11.3 WDG (Endorse) 2.5 WP	 0.88 4	 7 to 14 7 to 14
	propiconazole (Banner MAXX, Kestrel, Propiconazole, Savvi, Strider) 1 ME	3 to 4	early fall
	pyraclostrobin (Insignia) 20 WG 2 SC	 0.5 to 0.9 0.4 to 0.7	 14 to 28 14 to 28
	pyraclostrobin + boscalid (Honor) 28 WG*	1.1	14 to 28
	pyraclostrobin + fluxapyroxad (Lexicon Intrinsic) 4.17 SC	0.34 to 0.47	14 to 28
	pyraclostrobin + triticonazole (Pillar) 0.81 G	3 lbs	14 to 28
	tebuconazole* (Torque) 3.6 F (Mirage) 2 SC (Skylark, Tebuconazole) 3.6 F	 0.6 to 1.1 1 to 2 0.6	 21 28 28
	thiophanate-methyl (3336) 50WP or 4 F (3336 Plus) 2 F (SysTec 1998, T-Bird, TM) 85 WDG (3336) 2 G (SysTec 1998, T-Bird, TM) 4.5 L	 2 to 4 2 to 4 0.67 to 1.3 1.5 to 6 lbs 1 to 2	 14 14 to 28 14 14 14
	thiophanate-methyl + flutolanil (SysStar) 80 WDG	2 to 3	14 to 21
	triadimefon (Bayleton) 50 WSP, 4.15 F	1 to 2	fall and spring
	triticonazole (Trinity) 1.7 SC (Triton) 70 WDG	 1 to 2 0.15 to 0.3	 14 to 28 14 to 28
	triticonazole + chlorothalonil (Reserve) 4.79 SC*	3.2 to 5.4	14 to 28
Leaf and Sheath Spot (Rhizoctonia zeae, R. oryzae)	azoxystrobin (Heritage) 0.8 TL 0.31 G	 2 2 to 4 lbs	 14 to 28 14 to 28
	azoxystrobin + acibenzolar-S-methyl (Heritage Action) 51 WG*	0.2 to 0.4	14 to 28
	azoxystrobin + chlorothalonil (Renown) 5.16 SC*	2.5 4.5	14 14 to 21
	azoxystrobin + difenoconazole (Briskway) 2.7 SC*	0.5 to 0.725	14 to 28
	azoxystrobin + propiconazole (Headway) 1.4 ME 1.06 G	 1.5 to 3 2 to 4 lbs	 14 to 28 14 to 28
	azoxystrobin + tebuconazole (Strobe T) 2.67 SC*	0.75 to 1.5	14 to 21
	chlorothalonil + propiconazole + fludioxonil (Instrata) 3.59 SC*	2.75 to 6	14 to 21
	chlorothalonil + thiophanate-methyl (Spectro) 90 WDG*	3 to 5.76	14 to 21
	flutolanil (Prostar) 70 WDG	2.2 to 4.5	14 to 21
	penthiopyrad (Velista) 50 WG	0.3 to 0.5	14
	polyoxin D (Affirm) 11.3 WDG	0.88	7 to 14
	pyraclostrobin (Insignia) 20 WG 2 SC	 0.5 to 0.9 0.4 to 0.7	 14 to 28 14 to 28
	pyraclostrobin + boscalid (Honor) 28 WG*	1.1	14 to 28
	pyraclostrobin + fluxapyroxad (Lexicon Intrinsic) 4.17 SC	0.34 to 0.47	14 to 28
	pyraclostrobin + triticonazole (Pillar) 0.81 G	3 lbs	28
Pink Patch (Limonomyces roseipelis)	azoxystrobin (Heritage, Strobe) 50 WG (Heritage) 0.8 TL (Heritage) 0.31 G (Strobe) 2 L	 0.2 to 0.4 1 to 2 2 to 4 lbs 0.38 to 0.77	 14 to 28 14 to 28 14 to 28 14 to 28
	azoxystrobin + acibenzolar-S-methyl (Heritage Action) 51 WG*	0.2 to 0.4	14 to 28
	azoxystrobin + chlorothalonil (Renown) 5.16 SC*	2.5 to 4.5	14 to 21
	azoxystrobin + difenoconazole (Briskway) 2.7 SC*	0.5 to 0.725	14 to 28
	azoxystrobin + propiconazole (Headway) 1.4 ME 1.06 G	 1.5 to 3 2 to 4 lbs	 14 to 28 14 to 28
	azoxystrobin + tebuconazole (Strobe T) 2.67 SC*	0.75 to 1.5	14 to 21
	chlorothalonil + fluoxastrobin (Fame C) 4.25 SC*	3 to 5.9	14 to 28

DISEASE CONTROL

Table 10-9. Turfgrass Disease Control

Disease	Fungicide and Formulation[1]	Amount of Formulation (oz/1,000 sq ft)[2]	Application Interval (days)[3]
Pink Patch (continued)	chlorothalonil + propiconazole (Concert) 4.3 SC*	3 to 5.5 5.5 to 8.5	7 to 14 14 to 21
	fluazinam (Secure) 4.17 SC*	0.5	14
	fluoxastrobin (Fame) 4 SC 0.25 G	0.18 to 0.36 2.3 to 4.6 lbs	14 to 28 14 to 28
	flutolanil (Prostar) 70 WP, 70 DG	1.5	21 to 28
	flutolanil + thiophanate-methyl (SysStar) 80 WDG	2	21 to 28
	iprodione + trifloxystrobin (Interface) 2.27 SC*	3 to 4	14
	propiconazole (Banner MAXX, Kestrel, Propiconazole, Savvi, Strider) 1 ME	2	14 to 21
	pyraclostrobin (Insignia) 20 WG 2 SC	0.5 to 0.9 0.4 to 0.7	14 to 28 14 to 28
	pyraclostrobin + boscalid (Honor) 28 WG*	0.55 to 1.1	14 to 28
	pyraclostrobin + fluxapyroxad (Lexicon Intrinsic) 4.17 SC	0.34 to 0.47	14 to 28
	pyraclostrobin + triticonazole (Pillar) 0.81 G	3 lbs	14 to 28
	tebuconazole* (Torque) 3.6 F (Mirage) 2 SC (Skylark, Tebuconazole) 3.6 F	0.6 to 1.1 1 to 2 0.6	refer to label 14 to 28 28
	trifloxystrobin (Compass) 50 WDG	0.1 to 0.15 0.2 to 0.25	14 21
	trifloxystrobin + triadimefon (Armada) 50 WP (Tartan) 2 SC*	0.6 to 1.2 1 to 2	14 to 28 14 to 28
	triticonazole (Trinity) 1.7 EC	1 to 2	14 to 28
	triticonazole + chlorothalonil (Reserve) 4.79 SC*	3.2 to 4.5	refer to label
	vinclozolin (Curalan, Touche) 50 EG*	1	14 to 28
Pink Snow Mold/Microdochium Patch *(Microdochium nivale)*	azoxystrobin (Heritage, Strobe) 50 WG (Heritage) 0.8 TL (Heritage) 0.31 G (Strobe) 2L	0.2 to 0.4 0.7 2 3.5 4 lbs 7 lbs 0.77 1.35	10 to 28 1 application 10-28 1 application 10 to 28 1 application 14 1 application
	azoxystrobin + acibenzolar-S-methyl (Heritage Action) 51 WG*	0.4	refer to label
	azoxystrobin + chlorothalonil (Renown) 5.16 SC*	2.5 to 4.5	14 to 21
	azoxystrobin + difenoconazole (Briskway) 2.7 SC*	0.5 to 0.725	14 to 28
	azoxystrobin + propiconazole (Headway) 1.4 ME 1.06 G	1.5 to 3 5.25 2 to 4 lbs 5	10 to 28 1 application 14 to 28 1 application
	azoxystrobin + tebuconazole (Strobe T) 2.67 SC*	0.75 to 1.5 2.4	14 to 21 1 application
	chlorothalonil + acibenzolar-S-methyl (Daconil Action) 6.1 F*	5.4	21 to 28
	chlorothalonil + fluoxastrobin (Fame C) 4.25 SC*	3 to 5.9	28
	chlorothalonil + iprodione + thiophanate-methyl + tebuconazole (Enclave) 5.3 F*	7 to 8	28
	chlorothalonil + propiconazole (Concert) 4.3 SC*	8.5	14 to 28
	chlorothalonil + propiconazole + fludioxonil (Instrata) 3.6 SC*	5 to 11	late fall
	chlorothalonil + thiophanate-methyl* (Consyst, Peregrine, TM/C) 67 WDG (Spectro) 90 WDG	6 to 8 3.72 to 5.76	1 application 14
	fenarimol (Rubigan) 1 AS*	8 4	1 application 30 (2 applications)
	fluazinam (Secure) 4.17 SC*	0.5	late fall
	fludioxonil (Medallion) 50 WP	0.25 to 0.5	14
	fluoxastrobin (Fame) 4 SC 0.25 G	0.18 to 0.36 2.3 to 4.6 lbs	14 to 28 14 to 28
	fluoxastrobin + myclobutanil (Fame M) 3.9 SC	0.5 to 1	21 to 28
	fluoxastrobin + tebuconazole (Fame T) 4 SC*	0.45 to 0.9	30
	fluxapyroxad (Xzemplar) 2.47 SC	0.26	14 to 28
	flutolanil + thiophanate-methyl (SysStar) 80 WDG	4 to 6.12 2 to 3	1 application 14 to 21

Table 10-9. Turfgrass Disease Control

Disease	Fungicide and Formulation[1]	Amount of Formulation (oz/1,000 sq ft)[2]	Application Interval (days)[3]
Pink Snow Mold/Microdochium Patch (continued)	iprodione (26GT, Iprodione Pro, IPro, Raven) 2 F, 2 SC, 2 SE*	4 to 8	1 to 2 applications
	iprodione + thiophanate-methyl* (26/36) 3.8 F (Dovetail) 3.8 F	2 to 4 1 to 4	14 to 21 14 to 21
	iprodione + trifloxystrobin (Interface) 2.27 SC*	4 to 7	1 application
	mancozeb* (Fore) 80 WP (Dithane, Pentathlon) 75 DF (Pentathlon) 4 LF (Protect) 75 WP	6 to 8 6 to 8 10 to 14 6 to 8	14 to 42 14 to 42 14 to 42 7 to 14
	mancozeb + copper hydroxide (Junction) 60 DF*	2 to 4	14 to 42
	metconazole (Tourney) 50 WDG	0.37 to 0.44	late fall
	mineral oil (Civitas) + proprietary pigment (Civitas Harmonizer)*	(8 to 32) + (1 to 4)	7 to 21
	myclobutanil (Eagle, Myclobutanil, Siskin) 20 EW	1.2 to 2.4	prior to snow cover
	PCNB (various brands) 75 WP 10 G 4 F	3 to 8 80 to 160 12 to 16	28 to 42 prior to snowfall prior to snowfall
	polyoxin D (Affirm) 11.3 WDG (Endorse) 2.5 WP	0.88 4	7 to 14 7 to 14
	propiconazole (Banner MAXX, Kestrel, Propiconazole, Savvi, Strider) 1 ME	2 to 4	fall to early spring
	pyraclostrobin (Insignia) 20 WG 2 SC	0.5 to 0.9 0.7	14 to 28 14 to 28
	pyraclostrobin + boscalid (Honor) 28 WG*	0.55 to 1.1	14 to 28
	pyraclostrobin + fluxapyroxad (Lexicon Intrinsic) 4.17 SC	0.47	14 to 28
	pyraclostrobin + triticonazole (Pillar) 0.81 G	3 lbs	28
	tebuconazole* (Torque) 3.6 F (Mirage) 2 SC (Skylark, Tebuconazole) 3.6 F	0.6 to 1.1 1 to 2 0.6	prior to snowfall 10 to 28 prior to snowfall
	thiram (Spotrete)* 4 F 75 WDG	3 to 12 3 to 8	fall and spring fall and spring
	thiophanate-methyl (3336) 50WP or 4 F (3336 Plus) 2 F (SysTec 1998, T-Bird, TM) 85 WDG (3336) 2 G (SysTec 1998, T-Bird, TM) 4.5 L	2 to 4 2 to 4 0.67 to 1.3 1.5 to 6 lbs 1 to 2	14 14 to 28 14 14 14
	triadimefon (Bayleton) 50 WSP, 4.15 F	1 to 2	60 to 90
	trifloxystrobin (Compass) 50 WDG	0.2 to 0.25	fall to early spring
	trifloxystrobin + triadimefon (Armada) 50 WP (Tartan) 2 SC*	1.2 2	fall to early spring fall to early spring
	triticonazole (Trinity) 1.7 SC (Triton) 70 WDG (Triton Flo) 3 G	0.5 to 2 0.15 to 0.3 0.28 to 1.1	14 to 28 late fall 10 to 14
	triticonazole + chlorothalonil (Reserve) 4.79 SC*	3.2 to 4.5	14 to 28
	vinclozolin (Curalan, Touche) 50 EG*	1	10 to 21
Powdery Mildew (*Blumeria graminis*)	azoxystrobin (Heritage, Strobe) 50 WG (Heritage) 0.8 TL (Heritage) 0.31 G	0.2 to 0.4 1 to 2 2 to 4 lbs	14 to 28 14 to 28 14 to 28
	azoxystrobin + acibenzolar-S-methyl (Heritage Action) 51 WG*	0.2 to 0.4	14 to 28
	azoxystrobin + chlorothalonil (Renown) 5.16 SC*	2.5 to 4.5	14 to 21
	azoxystrobin + difenoconazole (Briskway) 2.7 SC*	0.5 to 0.725	14 to 28
	azoxystrobin + propiconazole (Headway) 1.4 ME 1.06 G	1.5 to 3 2 to 4 lbs	14 to 28 14 to 28
	azoxystrobin + tebuconazole (Strobe T) 2.67 SC*	0.75 to 1.5	14 to 21
	chlorothalonil + fluoxastrobin (Fame C) 4.25 SC*	3 to 5.9	14 to 28
	chlorothalonil + propiconazole (Concert) 4.3 SC*	4.5 to 8.5	14 to 28
	chlorothalonil + thiophanate-methyl* (Consyst) 67 WDG (Peregrine) 67 WDG (Spectro) 90 WDG (TM/C) 67 WDG	2 to 8 2 to 8 3.72 to 5.76 2 to 8	7 to 14 14 14 14 to 21

DISEASE CONTROL

Table 10-9. Turfgrass Disease Control

Disease	Fungicide and Formulation[1]	Amount of Formulation (oz/1,000 sq ft)[2]	Application Interval (days)[3]
Powdery Mildew (continued)	fenarimol (Rubigan) 1 AS*	2 to 4	1 application
	fluoxastrobin (Fame) 4 SC 0.25 G	0.18 to 0.36 2.3 to 4.6 lbs	14 to 28 14 to 28
	fluoxastrobin + myclobutanil (Fame M) 3.9 SC	0.25 to 1	14 to 28
	mancozeb + copper hydroxide (Junction) 60 DF*	2 to 4	7 to 14
	mineral oil (Civitas) + proprietary pigment (Civitas Harmonizer)*	(8 to 32) + (1 to 4)	7 to 21
	myclobutanil (Eagle, Myclobutanil, Siskin) 20 EW	1.2	14 to 28
	penthiopyrad (Velista) 50 WG	0.3 to 0.5	14
	propiconazole (Banner MAXX, Kestrel, Propiconazole, Savvi, Strider) 1 ME	1 to 2	14 to 28
	pyraclostrobin (Insignia) 20 WG 2 SC	0.5 to 0.9 0.4 to 0.7	14 to 28 14 to 28
	pyraclostrobin + boscalid (Honor) 28 WG*	0.55 to 1.1	14 to 28
	pyraclostrobin + fluxapyroxad (Lexicon Intrinsic) 4.17 SC	0.34 to 0.47	14 to 28
	pyraclostrobin + triticonazole (Pillar) 0.81 G	3 lbs	14 to 28
	tebuconazole* (Torque) 3.6 F (Skylark, Tebuconazole) 3.6 F	0.6 to 1.1 0.6	refer to label 28
	triadimefon (Bayleton) 50 WSP, 4.15 F	0.5 to 1	15 to 30
Pythium Blight (Pythium aphanidermatum)	azoxystrobin (Heritage, Strobe) 50 WG (Heritage) 0.8 TL (Heritage) 0.31 G (Strobe) 2 L	0.4 2 2 to 4 lbs 0.38 to 0.77	10 to 14 10 to 14 10 to 14 10 to 14
	azoxystrobin + acibenzolar-S-methyl (Heritage Action) 51 WG*	0.2 to 0.4	10 to 14
	azoxystrobin + propiconazole (Headway) 1.4 ME 1.06 G	3 2 to 4 lbs	10 to 14 14 to 28
	azoxystrobin + tebuconazole (Strobe T) 2.67 SC*	0.75 to 1.5	10 to 21
	chloroneb* (Teremec) 65 SP (Teremec) 2.9 F	4 7	5 to 7 5 to 7
	chlorothalonil + fluoxastrobin (Fame C) 4.25 SC*	3 to 5.9	7 to 14
	cyazofamid (Segway) 3.33 SC*	0.45 to 0.9	14 to 21
	ethazole* (Koban) 30 WP (Terrazole) 35 WP	2 to 4.5 2 to 4	10 10 to 14
	fluopicolide + propamocarb (Stellar) 5.7 SC	1.2	14
	fluoxastrobin (Fame) 4 SC 0.25 G	0.18 to 0.36 2.3 to 4.6 lbs	7 to 14 14
	fluoxastrobin + myclobutanil (Fame M) 3.9 SC	0.5 to 1	14
	fluoxastrobin + tebuconazole (Fame T) 4 SC*	0.45 to 0.9	21
	fosetyl Al (Signature, Fosetyl-Al) 80 WDG (Signature Xtra Stressgard) 60 WDG* (Autograph) 70 DF* (Viceroy) 70 DF	4 to 8 2 to 6 4.6 to 9.2 4.6 to 9.1	14 to 21 7 to 21 14 to 21 14 to 21
	mancozeb* (Fore) 80 WP (Dithane) 75 DF (Pentathlon) 4 LF (Pentathlon) 75 DF (Protect, Wingman) 75 WP	8 8 14 8 8	5 to 14 10 5 5 7 to 14
	mancozeb + copper hydroxide (Junction) 60 DF*	2 to 4	5
	mefenoxam (Subdue) 43 WSP (Subdue MAXX, Quell) 2 ME (Subdue) 1 GR (Fenox, Mefenoxam) 2 AQ, 2 EC	0.28 to 0.56 0.5 to 1 12.5 to 25 0.2 to 1	10 to 21 10 to 21 10 to 14 10 to 21
	metalaxyl (Vireo) 2 MEC	1 to 2	10 to 21
	phosphorus acid (Alude, Resyst) 3.3 F (Jetphiter) 5.41 F (Magellan) 4.3 F (Vital) 4.2 F (Vital Sign) 4.2 F	5 to 10 5 4.1 to 8.2 4 to 6 4 to 8	7 to 14 7 14 to 21 14 7 to 14
	potassium phosphite (Appear) 4.1 SC	3 to 4 4 to 6	7 to 14 14

461

Table 10-9. Turfgrass Disease Control

Disease	Fungicide and Formulation[1]	Amount of Formulation (oz/1,000 sq ft)[2]	Application Interval (days)[3]
Pythium Blight (continued)	propamocarb (Banol) 6 S*	1.3 to 4	7 to 21
	pyraclostrobin (Insignia) 20 WG 2 SC	0.9 0.7	14 to 28 10 to 14
	pyraclostrobin + boscalid (Honor) 28 WG*	1.1	10 to 14
	pyraclostrobin + fluxapyroxad (Lexicon Intrinsic) 4.17 SC	0.47	14
	pyraclostrobin + triticonazole (Pillar) 0.81 G	3 lbs	14
Pythium Root Dysfunction (*Pythium volutum*)	azoxystrobin (Heritage) 50 WG 0.8 TL	0.4 2	21 to 28 21 to 28
	azoxystrobin + acibenzolar-S-methyl (Heritage Action) 51 WG*	0.4	21 to 28
	cyazofamid (Segway) 3.33 SC*	0.45 to 0.9	14 to 21
	fluoxastrobin (Fame) 4 SC 0.25 G	0.27 to 0.36 3.6 to 4.6 lbs	14 to 28 14 to 28
	fluoxastrobin + chlorothalonil (Fame C) 4.25 SC *	4.5 to 5.9	14 to 28
	fluoxastrobin + myclobutanil (Fame M) 3.9 SC	0.5 to 1	14 to 28
	pyraclostrobin (Insignia) 20 WG 2 SC	0.9 0.7	14 to 28 14 to 28
	pyraclostrobin + boscalid (Honor) 28WG*	1.1	14 to 28
	pyraclostrobin + fluxapyroxad (Lexicon Intrinsic) 4.17 SC	0.47	14 to 28
	pyraclostrobin + triticonazole (Pillar) 0.81 G	3 lbs	14
Pythium Root Rot (*Pythium* spp.)	azoxystrobin (Heritage, Strobe) 50 WG (Heritage) 0.8 TL (Heritage) 0.31 G (Strobe) 2 L	0.4 2 2 to 4 0.38 to 0.77	10 to 14 10 to 14 10 to 14 10 to 14
	azoxystrobin + propiconazole (Headway) 1.4 ME 1.06 G	3 2 to 4 lbs	10 to 14 14 to 28
	azoxystrobin + tebuconazole (Strobe T) 2.67 SC*	0.75 to 1.5	10 to 21
	chlorothalonil + fluoxastrobin (Fame C) 4.25 SC *	3 to 5.9	7 to 10
	ethazole* (Koban) 30 WP (Terrazole) 35 WP	4.5 2 to 4	10 10 to 14
	fluoxastrobin (Fame) 4 SC 0.25 G	0.18 to 0.36 2.3 to 4.6 lbs	7 to 10 14
	fluoxastrobin + myclobutanil (Fame M) 3.9 SC	0.5 to 1	14
	fluoxastrobin + tebuconazole (Fame T) 4 SC*	0.45 to 0.9	21
	fosetyl Al (Signature, Fosetyl-Al) 80 WDG (Signature Xtra Stressgard) 60 WDG* (Autograph) 70 DF* (Viceroy) 70 DF	4 to 8 2 to 6 4.6 to 9.2 4.6 to 9.1	14 to 21 7 to 21 14 to 21 14 to 21
	phosphorous acid (Vital Sign) 2.4 F (Jetphiter) 5.41 F	6 to 8 3.5 to 5	7 to 14 7 to 28
	potassium phosphite (Appear) 4.1 SC	6 to 8	7 to 14
	propamocarb (Banol) 6 S*	1.3 to 4	7 to 21
Rapid Blight (*Labyrinthula* spp.)	iprodione + trifloxystrobin (Interface) 2.27 SC*	3 to 5	refer to label
	mancozeb (Fore) 80 WP*	8	14
	penthiopyrad (Velista) 50 WG	0.5	14
	pyraclostrobin (Insignia) 20 WG 2 SC	0.5 to 0.9 0.4 to 0.7	14 14 to 28
	pyraclostrobin + boscalid (Honor) 28WG*	0. 55 to 1.1	14 to 28
	pyraclostrobin + fluxapyroxad (Lexicon Intrinsic) 4.17 SC	0.34 to 0.47	14
	pyraclostrobin + triticonazole (Pillar) 0.81 G	3 lbs	14 to 28
	trifloxystrobin (Compass) 50 WDG	0.15 to 0.2 0.25	14 21
	trifloxystrobin + triadmefon (Armada) 50 WP	0.6 to 1.2	14 to 28

Table 10-9. Turfgrass Disease Control

Disease	Fungicide and Formulation[1]	Amount of Formulation (oz/1,000 sq ft)[2]	Application Interval (days)[3]
Red Thread (*Laetisaria fuciformis*)	azoxystrobin (Heritage, Strobe) 50 WG (Heritage) 0.8 TL (Heritage) 0.31 G (Strobe) 2 L	0.2 to 0.4 1 to 2 2 to 4 lbs 0.38 to 0.77	14 to 28 14 to 28 14 to 28 14 to 28
	azoxystrobin + acibenzolar-S-methyl (Heritage Action) 51 WG*	0.2 to 0.4	14 to 28
	azoxystrobin + chlorothalonil (Renown) 5.16 SC*	2.5 to 4.5	14 to 21
	azoxystrobin + difenoconazole (Briskway) 2.7 SC*	0.5 to 0.725	14 to 28
	azoxystrobin + propiconazole (Headway) 1.4 ME 1.06 G	1.5 to 3 2 to 4 lbs	14 to 28 14 to 28
	azoxystrobin + tebuconazole (Strobe T) 2.67 SC*	0.75 to 1.5	14 to 21
	chlorothalonil* (Daconil Ultrex) 82.5 WDG (Daconil Weather Stik, Legend) 6 F (Daconil Zn) 4.16 F (Chlorothalonil 500ZN) 4.17 F (Chlorothalonil 720SFT) 6 F (Chlorothalonil, Chlorostar) 82.5 DF (Pegasus) 6 L (Pegasus) 82.5 DF (Pegasus HPX) 6 F	1.8 to 3.25 3.25 to 5 2 to 5.5 5.5 3 to 5 5.3 to 8 3 to 5 7.9 2.12 to 3.5 5.5 1.8 to 3.2 3.6 to 5.5 3.25 to 5 3.6 to 5.5	7 to 10 14 7 to 14 14 7 to 10 14 7 to 10 14 7 to 10 14 7 to 10 7 to 14 7 to 14 7 to 14
	chlorothalonil + acibenzolar-S-methyl (Daconil Action) 6.1 F*	2 to 3.5 3.6 to 5.4	7 to 10 14
	chlorothalonil + fluoxastrobin (Fame C) 4.25 SC*	3 to 5.9	14 to 28
	chlorothalonil + iprodione + thiophanate-methyl + tebuconazole (Enclave) 5.3 F*	3 to 4 7 to 8	14 to 21 28
	chlorothalonil + propiconazole (Concert) 4.3 SC*	3 to 5.5 5.5 to 8.5	7 to 14 14 to 21
	chlorothalonil + propiconazole + fludioxonil (Instrata) 3.6 SC*	2.75 to 6	14 to 21
	chlorothalonil + thiophanate-methyl* (Consyst) 67 WDG (Peregrin) 67 WDG (Spectro) 90 WDG (TM/C) 67 WDG	3 to 8 3 to 8 3.72 to 5.76 3 to 8	7 to 10 14 14 14 to 21
	fenarimol (Rubigan) 1 AS*	8	30
	fluazinam (Secure) 4.17 SC*	0.5	14
	fluoxastrobin (Fame) 4 SC 0.25 G	0.18 to 0.36 2.3 to 4.6 lbs	14 to 28 14 to 28
	fluoxastrobin + myclobutanil (Fame M) 3.9 SC	0.25 to 1	14 to 28
	fluoxastrobin + tebuconazole (Fame T) 4 SC*	0.45 to 0.9	21 to 28
	flutolanil (Prostar) 70 WP, 70 WDG	1.5	21 to 28
	flutolanil + thiophanate-methyl (SysStar) 80 WDG	2 to 3	14 to 21
	iprodione (iprodione (26GT, Iprodione Pro, IPro, Raven) 2 F, 2 SC, 2 SE*	4	14
	iprodione + thiophanate-methyl (26/36) 3.8 F*	2 to 4	14 to 21
	iprodione + trifloxystrobin (Interface) 2.27 SC*	3 to 4	14
	mancozeb* (Fore) 80 WP (Dithane) 75 DF (Pentathlon) 4 LF (Pentathlon) 75 DF (Protect, Wingman) 75 W	4 to 8 4 to 8 7 to 14 4 to 8 4 to 8	7 to 14 10 7 to 14 7 7 to 14
	mancozeb + copper hydroxide (Junction) 60 DF*	2 to 4	7 to 14
	metconazole (Tourney) 50 WDG	0.37	14
	mineral oil (Civitas) + proprietary pigment (Civitas Harmonizer)*	(8 to 32) + (1 to 4)	7 to 21
	myclobutanil (Eagle, Myclobutanil) 20 EW	1.2	14 to 21
	penthiopyrad (Velista) 50 WG	0.3 to 0.5	14
	polyoxin D (Affirm) 11.3 WDG (Endorse) 2.5 WP	0.88 4	7 to 14 7 to 14
	Propiconazole (Banner MAXX, Kestrel, Propiconazole, Savvi, Strider) 1 ME	2	14 to 21
	pyraclostrobin (Insignia) 20 WG 2 SC	0.5 to 0.9 0.4 to 0.7	14 to 28 14 to 28

463

DISEASE CONTROL

Table 10-9. Turfgrass Disease Control

Disease	Fungicide and Formulation[1]	Amount of Formulation (oz/1,000 sq ft)[2]	Application Interval (days)[3]
Red Thread (continued)	pyraclostrobin + boscalid (Honor) 28 WG*	0.55 to 1.1	14 to 28
	pyraclostrobin + fluxapyroxad (Lexicon Intrinsic) 4.17 SC	0.34 to 0.47	14
	pyraclostrobin + triticonazole (Pillar) 0.81 G	3 lbs	14 to 28
	tebuconazole* (Torque) 3.6 F (Mirage) 2 SC (Skylark, Tebuconazole) 3.6 F	0.6 to 1.1 / 1 to 2 / 0.6	refer to label / 14 to 28 / 28
	thiophanate-methyl (3336) 50WP or 4 F (3336 Plus) 2 F (SysTec 1998, T-Bird, TM) 85 WDG (3336) 2 G (SysTec 1998, T-Bird, TM) 4.5 L	2 to 4 / 2 to 4 / 0.67 to 1.3 / 1.5 to 6 lbs / 1 to 2	14 / 14 to 28 / 14 / 14 / 14
	thiram (Spotrete) 4 F*	3.75 to 7.5	3 to 10
	triadimefon (Bayleton) 50 WSP, 4.15 F	0.5 to 1	15 to 30
	trifloxystrobin (Compass) 50 WDG	0.1 to 0.15 / 0.2 to 0.25	14 / 21
	trifloxystrobin + triadimefon (Armada) 50 WP (Tartan) 2 SC*	0.6 to 1.2 / 1 to 2	14 to 28 / 14 to 28
	triticonazole (Trinity) 1.7 SC (Triton) 70 WDG	0.5 to 1 / 0.15 to 0.3	14 to 28 / 14 to 28
	triticonazole + chlorothalonil (Reserve) 4.79 SC*	3.2 to 4.5	refer to label
	vinclozolin (Curulan, Touche) 50 EG*	1	14 to 28
Rust (*Puccinia* ssp.)	azoxystrobin (Heritage, Strobe) 50 WG (Heritage) 0.8 TL (Heritage) 0.31 G	0.2 to 0.4 / 1 to 2 / 2 to 4 lbs	14 to 28 / 14 to 28 / 14 to 28
	azoxystrobin + acibenzolar-S-methyl (Heritage Action) 51 WG*	0.2 to 0.4	14 to 28
	azoxystrobin + chlorothalonil (Renown) 5.16 SC*	2.5 to 4.5	14 to 21
	azoxystrobin + difenoconazole (Briskway) 2.7 SC*	0.5 to 0.725	14 to 28
	azoxystrobin + propiconazole (Headway) 1.4 ME 1.06 G	1.5 to 3 / 2 to 4 lbs	14 to 28 / 14 to 28
	azoxystrobin + tebuconazole (Strobe T) 2.67 SC*	0.75 to 1.5	14 to 21
	chlorothalonil* (Daconil Ultrex) 82.5 WDG (Daconil Weather Stik, Legend) 6 F (Daconil Zn) 4.16 F (Chlorothalonil 500ZN) 6 F (Chlorothalonil 720SFT) 6 F (Chlorothalonil, Chlorostar) 82.5 DF (Pegasus) 6 L (Pegasus) 82.5 DF (Pegasus HPX) 6 F	3.7 to 5 / 4.0 to 5.5 / 6 to 8 / 3 to 5 / 7.9 / 2.12 to 3.5 / 5.5 / 3.2 / 3.6 to 5.5 / 3.25 to 5 / 3.6 to 5.5	14 / 14 / 14 / 7 to 14 / 14 / 7 to10 / 14 / 7 to 14 / 7 to 14 / 7 to 14 / 7 to 14
	chlorothalonil + acibenzolar-S-methyl (Daconil Action) 6.1 F*	4 to 5.4	14
	chlorothalonil + fluoxastrobin (Fame C) 4.25 SC*	3 to 5.9	14 to 28
	chlorothalonil + propiconazole (Concert) 4.3 SC*	3 to 5.5 / 4.5 to 8.5	7 to 14 / 14 to 28
	chlorothalonil + propiconazole + fludioxonil (Instrata) 3.6 SC*	2.75 to 6	14 to 28
	chlorothalonil + thiophanate-methyl* (Consyst) 67 WDG (Peregrine) 67 WDG (Spectro) 90 WDG (TM/C) 67 WDG	3 to 8 / 3 to 8 / 3.72 to 5.76 / 3 to 8	7 to 14 / 14 / 14 / 14 to 21
	fluazinam (Secure) 4.17 SC*	0.5	14
	fluoxastrobin (Fame) 4 SC 0.25 G fluoxastrobin + myclobutanil (Fame M) 3.9 SC	0.18 to 0.36 / 2.3 to 4.6 lbs / 0.25 to 1	14 to 28 / 14 to 28 / 14 to 28
	iprodione + trifloxystrobin (Interface) 2.27 SC*	3 to 5	refer to label
	mancozeb* (Fore) 80 WP (Dithane) 75 DF (Petathlon) 4 LF (Pentathlon) 75 DF (Wingman) 75 WP	4 / 4 / 5 to 7 / 4 / 4	7 to 14 / 10 / 7 to 10 / 7 to 10 / 7 to 10
	mancozeb + copper hydroxide (Junction) 60 DF*	2 to 4	7 to 14
	metconazole (Tourney) 50 WDG	0.37	14

Table 10-9. Turfgrass Disease Control

Disease	Fungicide and Formulation[1]	Amount of Formulation (oz/1,000 sq ft)[2]	Application Interval (days)[3]
Rust (continued)	myclobutanil (Eagle, Myclobutanil, Siskin) 20 EW	1.2	14 to 28
	propiconazole (Banner Maxx, Kestrel, Propiconazole, Savvi, Strider) 1 ME	1 to 2	14 to 28
	pyraclostrobin (Insignia) 20 WG 2 SC	0.5 to 0.9 0.4 to 0.7	14 to 28 14 to 28
	pyraclostrobin + boscalid (Honor) 28WG*	0.55 to 1.1	14 to 28
	pyraclostrobin + fluxapyroxad (Lexicon Intrinsic) 4.17 SC	0.34 to 0.47	14 to 28
	pyraclostrobin + triticonazole (Pillar) 0.81 G	3 lbs	14 to 28
	tebuconazole* (Torque) 3.6 F (Mirage) 2 SC (Skylark, Tebuconazole) 3.6 F	0.6 to 1.1 1 to 2 0.6	refer to label 14 to 28 28
	thiophanate-methyl (3336) 50WP or 4 F (3336 Plus) 2 F (T-Bird) 4.5 L (SysTec 1998, T-Bird, TM) 85 WDG	4 to 6 4 to 8 3.5 to 5 2.35 to 3.53	14 14 to 28 14 14
	thiram (Spotrete) 4 F*	3.75 to 7.5	3 to 10
	triadimefon (Bayleton) 50 WSP, 4.15 F	0.5 to 1	15 to 30
	trifloxystrobin (Compass) 50 WDG	0.1 to 0.15 0.2 to 0.25	14 21
	trifloxystrobin + triadimefon (Armada) 50 WP (Tartan) 2 SC*	0.6 to 1.2 1 to 2	14 to 28 14 to 28
	triticonazole (Trinity) 1.7 SC (Triton) 70 WDG	0.5 to 1 0.15 to 0.225	14 to 28 14 to 28
	triticonazole + chlorothalonil (Reserve) 4.79 SC*	3.2 to 4.5	14 to 28
Slime Mold *(Myxomycetes spp.)*	mancozeb (Fore) 80 WP*	4 to 8	7 to 14
	mancozeb + copper hydroxide (Junction) 60 DF*	2 to 4	7 to 14
Southern Blight *(Sclerotium rolfsii)*	azoxystrobin (Heritage, Strobe) 50 WG (Heritage) 0.8 TL (Heritage) 0.31 G (Strobe) 2 L	0.2 to 0.4 1 to 2 2 to 4 lbs 0.38 to 0.77	14 to 28 14 to 28 14 to 28 14 to 28
	azoxystrobin + acibenzolar-S-methyl (Heritage Action) 51 WG*	0.2 to 0.4	14 to 28
	azoxystrobin + chlorothalonil (Renown) 5.16 SC*	2.5 to 4.5	14 to 21
	azoxystrobin + difenoconazole (Briskway) 2.7 SC*	0.5 to 0.725	14 to 28
	azoxystrobin + propiconazole (Headway) 1.4 ME 1.06 G	1.5 to 3 2 to 4 lbs	14 to 28 14 to 28
	azoxystrobin + tebuconazole (Strobe T) 2.67 SC*	0.75 to 1.5	14 to 21
	chloroneb* (Termec) 65 SP	4	5 to 7
	chlorothalonil + fluoxastrobin (Fame C) 4.25 SC*	3 to 5.9	14 to 28
	fluoxastrobin (Fame) 4 SC 0.25 G	0.18 to 0.36 2.3 to 4.6 lbs	14 to 28 14 to 28
	fluoxastrobin + myclobutanil (Fame M) 3.9 SC	0.25 to 1	14 to 28
	fluoxastrobin + tebuconazole (Fame T) 4 SC*	0.45 to 0.9	21 to 28
	flutolanil (Prostar) 70 WP, 70 WDG	1.5	21 to 28
	flutolanil + thiophanate-methyl (SysStar) 80 WDG	2	21 to 28
	triadimefon (Bayleton) 50 WSP, 4.15 F	0.5 to 2	14 to 28
	trifloxystrobin + triadimefon (Armada) 50 WP (Tartan) 2 SC*	0.6 to 1.2 1 to 2	14 14
Spring Dead Spot *(Ophiosphaerella korrae;* *O. herpotricha; O. narmari)*	azoxystrobin (Heritage, Strobe) 50 WG (Heritage) 0.8 TL (Strobe) 2L	0.4 2 0.38 to 0.77	14 to 28 14 to 28 28
	azoxystrobin + acibenzolar-S-methyl (Heritage Action) 51 WG*	0.2 to 0.4	14 to 28
	azoxystrobin + propiconazole (Headway) 1.4 ME 1.06 G	3 2 to 4 lbs	14 to 28 14 to 28
	azoxystrobin + tebuconazole (Strobe T) 2.67 SC*	1.5	14 to 21
	chlorothalonil + fluoxastrobin (Fame C) 4.25 SC*	5.9	14 to 28
	chlorothalonil + iprodione + thiophanate-methyl + tebuconazole (Enclave) 5.3 F*	3 to 4 7 to 8	14 to 21 28

465

Table 10-9. Turfgrass Disease Control

Disease	Fungicide and Formulation[1]	Amount of Formulation (oz/1,000 sq ft)[2]	Application Interval (days)[3]
Spring Dead Spot (continued)	fenarimol (Rubigan) 1 AS*	4 6	14 to 30 (2 applications) 1 application
	fluoxastrobin (Fame) 4 SC 0.25 G	 0.36 2.3 to 4.6 lbs	 14 to 28 14 to 28
	fluoxastrobin + myclobutanil (Fame M) 3.9 SC	0.5 to 1	14 to 28
	fluoxastrobin + tebuconazole (Fame T) 4 SC*	0.45 to 0.9	21 to 28
	myclobutanil (Eagle, Myclobutanil, Siskin) 20 EW	2.4	28 (fall)
	penthiopyrad (Velista) 50 WG	0.7	28
	propiconazole (Banner MAXX, Kestrel, Propiconazole, Savvi, Strider) 1 ME	4	30
	tebuconazole* (Torque) 3.6 F (Mirage) 2 SC (Skylark, Tebuconazole) 3.6 F	 0.6 to 1.1 2 0.6	 21 28 fall and spring
	thiophanate-methyl (3336) 50WP or 4 F (3336) 2 G	 4 to 6 6 to 9 lbs	 14 14
Stripe Smut *(Ustilago striiformis)*	chlorothalonil + iprodione + thiophanate-methyl + tebuconazole (Enclave) 5.3 F*	3 to 4 7 to 8	14 to 21 28
	chlorothalonil + propiconazole (Concert) 4.3 SC*	4.5 to 8.5	fall or spring
	myclobutanil (Eagle, Myclobutanil, Siskin) 20 EW	1.2	14
	propiconazole (Banner MAXX, Kestrel, Propiconazole, Savvi, Strider) 1 ME	1 to 2	fall or spring
	tebuconazole* (Torque) 3.6 F (Skylark, Tebuconazole) 3.6 F	 0.6 to 1.1 0.6	 spring spring
	thiophanate-methyl (3336) 50WP or 4 F (3336 Plus) 2 F (3336) 2 G (T-Bird) 4.5 L (SysTec 1998, T-Bird, TM) 85 WDG (SysTec 1998, T-Bird,TM) 4.5 L	 4 to 6 4 to 8 6 to 9 lbs 5 to 10 3 to 3.53 5	 14 14 to 28 14 14 to 21 14 to 21 14 to 21
	triadimefon (Bayleton) 50 WSP	1	refer to label
	trifloxystrobin + triadimefon (Armada) 50 WP (Tartan) 2 SC*	 0.6 1	 refer to label refer to label
Summer Patch *(Magnaporthe poae)*	azoxystrobin (Heritage, Strobe) 50 WG (Heritage) 0.8 TL (Heritage) 0.31 G (Strobe) 2 L	 0.2 to 0.4 1 to 2 2 to 4 lbs 0.38 to 0.77	 14 to 28 14 to 28 14 to 28 14 to 28
	azoxystrobin + acibenzolar-S-methyl (Heritage Action) 51 WG*	0.2 to 0.4	14 to 28
	azoxystrobin + difenoconazole (Briskway) 2.7 SC*	0.5 to 0.725	14 to 28
	azoxystrobin + propiconazole (Headway) 1.4 ME 1.06 G	 1.5 to 3 2 to 4 lbs	 14 to 28 14 to 28
	chlorothalonil + fluoxastrobin (Fame C) 4.25 SC*	3 to 5.9	14 to 28
	chlorothalonil + iprodione + thiophanate-methyl + tebuconazole (Enclave) 5.3 F*	3 to 4 7 to 8	14 to 21 28
	chlorothalonil + propiconazole + fludioxonil (Instrata) 3.6 SC*	6 to 11	14 to 28
	fenarimol (Rubigan) 1 AS*	2 to 4 2 4 to 8	30 (2 applications) 30 (greens) single application
	fluoxastrobin (Fame) 4 SC 0.25 G	 0.18 to 0.36 2.3 to 4.6 lbs	 14 to 28 14 to 28
	fluoxastrobin + myclobutanil (Fame M) 3.9 SC	0.25 to 1	14 to 28
	fluoxastrobin + tebuconazole (Fame T) 4 SC*	0.45 to 0.9	21 to 28
	fludioxonil (Medallion) 50 WP	0.5	14
	fluxapyroxad (Xzemplar) 2.47 SC	0.26	14 to 28
	metconazole (Tourney) 50 WDG	0.37	14
	myclobutanil (Eagle, Myclobutanil, Siskin) 20 EW	1.2 to 2.4	14 to 28
	penthiopyrad (Velista) 50 WG	0.3 to 0.5	14 to 28
	propiconazole (Banner MAXX, Kestrel, Propiconazole, Savvi, Strider) 1 ME	2 4	14 28
	pyraclostrobin (Insignia) 20 WG 2 SC	 0.5 to 0.9 0.4 to 0.7	 14 to 28 14 to 28

Table 10-9. Turfgrass Disease Control

Disease	Fungicide and Formulation[1]	Amount of Formulation (oz/1,000 sq ft)[2]	Application Interval (days)[3]
Summer Patch (continued)	pyraclostrobin + boscalid (Honor) 28 WG*	1.1	14 to 28
	pyraclostrobin + fluxapyroxad (Lexicon Intrinsic) 4.17 SC	0.34 to 0.47	14 to 28
	pyraclostrobin + triticonazole (Pillar) 0.81 G	3 lbs	28
	tebuconazole* (Torque) 3.6 F (Mirage) 2 SC (Skylark, Tebuconazole) 3.6 F	0.6 to 1.1 1 to 2 0.6	21 14 to 28 28
	thiophanate-methyl (3336) 50WP or 4 F (3336 Plus) 2 F (3336) 2 G (SysTec 1998, T-Bird, TM) 85 WDG (SysTec 1998, T-Bird, TM) 4.5 L	4 to 6 4 to 8 6 to 9 lbs 3.53 5	14 to 21 14 to 28 14 to 21 14 14
	triadimefon (Bayleton) 50 WSP, 4.15 F	1 to 2	30
	trifloxystrobin (Compass) 50 WDG	0.2 to 0.25	21 to 28
	trifloxystrobin + triadimefon (Tartan) 2 SC * (Armada) 50 WP	2 1.2	21 to 28 21 to 28
	triticonazole (Trinity) 1.7 SC (Triton) 70 WDG	1 to 2 0.3 to 0.6	14 to 28 14 to 28
	triticonazole + chlorothalonil (Reserve) 4.79 SC*	3.2 to 5.4	14 to 28
Take-All Patch *(Gaeumannomyces graminis)*	azoxystrobin (Heritage, Strobe) 50 WG (Heritage) 0.8 TL (Heritage) 0.31 G (Strobe) 2 L	0.4 2 2 to 4 lbs 0.38 to 0.77	28 28 28 28
	azoxystrobin + acibenzolar-S-methyl (Heritage Action) 51 WG*	0.2 to 0.4	28
	azoxystrobin + difenoconazole (Briskway) 2.7 SC*	0.5 to 0.725	28
	azoxystrobin + propiconazole (Headway) 1.4 ME 1.06 G	3 2 to 4 lbs	14 to 28 14 to 28
	chlorothalonil + fluoxastrobin (Fame C) 4.25 SC*	5.9	28
	fenarimol (Rubigan) 1 AS*	4 4 to 8	30 (greens) 30 (1 or 2 applications)
	fluoxastrobin (Fame) 4 SC 0.25 G	0.36 2.3 to 4.6 lbs	28 28
	fluoxastrobin + myclobutanil (Fame M) 3.9 SC	0.5 to 1	28
	fluoxastrobin + tebuconazole (Fame T) 4 SC*	0.45 to 0.9	28
	myclobutanil (Eagle, Myclobutanil, Siskin) 20 EW	2.4	28 (spring/fall)
	propiconazole (Banner MAXX, Kestrel, Propiconazole, Savvi, Strider) 1 ME	2 to 4	spring and fall
	pyraclostrobin (Insignia) 20 WG 2 SC	0.9 0.7	28 28
	pyraclostrobin + boscalid (Honor) 28 WG*	1.1	28
	pyraclostrobin + fluxapyroxad (Lexicon Intrinsic) 4.17 SC	0.47	28
	pyraclostrobin + triticonazole (Pillar) 0.81 G	3 lbs	28
	tebuconazole* (Torque) 3.6 F (Mirage) 2 SC (Skylark, Tebuconazole) 3.6 F	0.6 to 1.1 1 to 2 0.6	fall and spring 14 to 28 fall and spring
	thiophanate-methyl (3336) 50 WP or 4 F (3336 Plus) 2 F (3336) 2 G	4 to 6 4 to 8 6 to 9 lbs	14 14 to 28 14
	triadimefon (Bayleton) 50 WSP, 4.15 F	1 to 2	21 to 28
	trifloxystrobin + triadimefon (Armada) 50 WP	1.2	28
	triticonazole (Trinity) 1.7 SC (Triton) 70 WDG	1 to 2 0.15 to 0.3	14 to 28 14 to 28
	triticonazole + chlorothalonil (Reserve) 4.79 SC*	3.2 to 5.4	14 to 28
Yellow Patch *(Rhizoctonia cerealis)*	azoxystrobin (Heritage, Strobe) 50 WG (Heritage) 0.8 TL (Heritage) 0.31 G (Strobe) 2 L	0.4 2 2 to 4 lbs 0.38 to 0.77	28 28 14 to 28 28
	azoxystrobin + acibenzolar-S-methyl (Heritage Action) 51 WG*	0.2 to 0.4	14 to 28
	azoxystrobin + chlorothalonil (Renown) 5.16 SC*	2.5 to 4.5	14 to 28

Table 10-9. Turfgrass Disease Control

Disease	Fungicide and Formulation[1]	Amount of Formulation (oz/1,000 sq ft)[2]	Application Interval (days)[3]
Yellow Patch (continued)	azoxystrobin + difenoconazole (Briskway) 2.7 SC*	0.5 to 0.725	14 to 28
	azoxystrobin + propiconazole (Headway) 1.4 ME 1.06 G	3 2 to 4 lbs	28 14 to 28
	azoxystrobin + tebuconazole (Strobe T) 2.67 SC*	1.5	14 to 21
	chlorothalonil + fluoxastrobin (Fame C) 4.25 SC*	3 to 5.9	14 to 28
	chlorothalonil + propiconazole + fludioxonil (Instrata) 3.6 SC*	8 to 11	late fall
	chlorothalonil + thiophanate-methyl (Spectro) 90 WDG*	3 to 5.76	14 to 21
	fludioxonil (Medallion) 50 WP	0.5	1 application
	fluoxastrobin (Fame) 4 SC 0.25 G	0.36 2.3 to 4.6 lbs	28 14 to 28
	fluoxastrobin + myclobutanil (Fame M) 3.9 SC	0.25 to 1	28
	fluoxastrobin + tebuconazole (Fame T) 4 SC*	0.45 to 0.9	21 to 28
	flutolanil (Prostar) 70 WP, 70 WDG	1.5	21 to 28
	flutolanil + thiophanate-methyl (SysStar) 80 WDG	2	21 to 28
	metconazole (Tourney) 50 WDG	0.37 to 0.44	late fall
	polyoxin D (Affirm) 11.3 WDG (Endorse) 2.5 WP	0.88 4	7 to 14 7 to 14
	propiconazole (Banner MAXX, Kestrel, Propiconazole, Savvi, Strider) 1 ME	3 to 4	late fall
	tebuconazole (Mirage) 2 SC*	1 to 2	21 to 28
	thiophanate-methyl (3336) 50WP or 4 F (3336 Plus) 2 F (3336) 2 G	4 to 6 4 to 8 6 to 9 lbs	14 14 to 28 14
	triticonazole (Triton FLO) 3 F (Trinity) 1.75 SC	0.55 to 1.1 1 to 2	21 to 28 21 to 28
	triticonazole + chlorothalonil (Reserve) 4.79 SC*	3.2 to 5.4	21 to 28
Yellow Tuft *(Sclerophthora macrospera)*	fosetyl Al (Signature, Fosetyl-Al) 80 WDG (Signature Xtra Stressgard) 60 WDG* (Autograph) 70 DF* (Viceroy) 70 DF	4 to 8 2 to 6 4.6 to 9.2 4.6 to 9.1	14 to 21 14 to 21 14 to 21 14 to 21
	mefenoxam (Subdue WSP) 43 WSP (Subdue Maxx, Quell 2 ME (Subdue GR) 1 G (Mefenoxam, Fenox) 2 AQ, 2 EC	0.28 to 0.56 0.5 to 1 12.5 to 25 0.2 to 1	10 to 21 10 to 21 10 to 14 10 to 21
	metalaxyl (Vireo) 2 MEC	1 to 2	10 to 21
	pyraclostrobin (Insignia) 20 WG 2 SC	0.5 to 0.9 0.4 to 0.7	14 to 28 14 to 28
	pyraclostrobin + boscalid (Honor) 28 WG*	0.55 to 1.1	14 to 28
	pyraclostrobin + fluxapyroxad (Lexicon Intrinsic) 4.17 SC	0.34 to 0.47	14 to 28
	pyraclostrobin + triticonazole (Pillar) 0.81 G	3 lbs	14 to 28
Zoysia Patch	See Large Patch		

[1] Other trade names with the same active ingredients are labeled for use on turfgrasses and can be used according to label directions.

[2] Apply fungicides in 2 to 5 gallons of water per 1,000 square feet according to label directions. Use lower rates for preventive and higher rates for curative applications.

[3] Use shorter intervals when conditions are very favorable for disease.

* Products marked with an asterisk are not labeled for home lawn use.

Nematicides for Turf

J. P. Kerns and E. L. Butler, Plant Pathology Extension

Table 10-10. Nematicides for Turf

Nematicide and Formulation	Amount of Formulation Per 1,000 sq ft	Precautions and Remarks
1,3-dichloropropene (Curfew)	4 to 5	Curfew is a restricted use pesticide and may only be applied by custom applicators certified by Dow Agrosciences. Curfew should not be applied within 100 feet of an occupied structure or any well used for potable water. Do not enter treated area for 24 hours after application. Curfew may be applied throughout the year, but turfgrass should be actively growing at the time of application. Soil moisture should be adequate to support good turf growth and maintained at that level for at least 7 days after application. One-half inch of irrigation or rainfall must be applied to the area to be treated within 18 hours prior to application and as soon as possible following application. Refer to label for additional application guidelines.
abamectin (Avid) 0.15 EC	1.31	For use on golf course greens only. Only abamectin formulated as Avid can be used for nematode control in turf according to a 24(c) label. Appy Avid 0.15 EC as an early curative treatment (after appropriate nematode extraction, identification, and counts). Apply early in the morning while grass is wet with dew or irrigate the area prior to application with 0.1 inch of water. Immediately after application irrigate with 0.1 inch of water to move treatments through the thatch. Do not over irrigate. Apply 3 to 4 consecutive Avid 0.15 EC applications on a 14- to 21-day interval. Avid is labeled only for sting (*Belonolaimus longicaudatus*) and ring (*Macroposthonia* sp.) nematodes.
Bacillus firmus (Nortica)	10 to 30	Nortica is a biological agent for the protection of plant roots against plant parasitic nematodes on turf, lawns, sod farms, and golf courses. Do not mix with other chemicals or fertilizers during application without first contacting a local Bayer representative. Do not apply for a month after a formaldehyde application. Do not apply within 2 weeks of a fumigant application. Do not combine in the spray tank with pesticides, surfactants or fertilizers if there has been no previous experience or use of the combination to show it is physically compatible, effective, and non-injurious under your use conditions. Refer to product label for further information about mixing compatabilities. Nortica is suitable for application by spraying, drenching, or by drip irrigation. Optimal results are obtained by pre-plant applications (from 2 to 7 days prior to planting) and immediately irrigating after application to a minimum of 3-4 inches. If product is applied prior to planting, maintain moist soil with daily irrigation until planting. Refer to label for further information about application techniques. Make applications every 3 months as necessary and irrigate to a depth of 4 inches.
fenamiphos (Nemacur) 3 EC 10 G	 9.7 2.3 lbs	Existing stock may be used until October 2017. Nemacur 10G and 3EC are restricted use pesticides. Carefully follow label directions before and during application. Do not apply to hydrological soil group A soils that are excessively drained and predominately sand or loamy sand. Do not use on residential lawns or public recreation areas other than golf courses. Nemacur 10G may also be used on cemetery and industrial grounds turfgrass areas. Do not apply more than 200 pounds of 10G per acre per year or more than twice per year of the 3EC. Distribute product evenly over area. Irrigate area immediately using a minimum of 0.5 inch of water. Do not treat newly seeded areas. Do not apply granular nematicides with hand-held spreaders that might allow the dust to be inhaled. Do not apply where water runoff is likely to occur. Sod should not be cut or handled for 30 days after treatment. 3EC is not recommended for use on tees and greens. Do not treat more than 10 acres of turf on any golf course in a 24-hour period. Any application must be followed by a minimum 3-day interval before another application is made. Do not apply within 10 feet of any surface body of water or surface drains. Do not apply between noon and sunset from June through September.
fluensulfone (Nimitz Pro G)	1.38 to 2.75 lbs	Intended for use by a commercial applicator. For nematode control in bermudagrass, St. Augustinegrass, zoysiagrass, centipedegrass, seashore paspalum, tall fescue, and creeping bentgrass on golf courses, sports fields, commercial turfgrass areas, sod farms and residential turfgrass lawns. Nimitz must be immediately watered in after application with a minimum of ¼ inch water. Do not make applications when soil temperature is below 55 degrees F. Do not exceed 240 pounds of product per acre per calendar year. Do no allow bystanders to enter the treated area until the granules have been watered-in.
fluopyram (Indemnify)	0.195 to 0.39	Intended for use by commercial applicators. For use on golf courses, sod farms, sport fields, residential, institutional, municipal, commercial, and other turfgrass areas. Do not apply more than the maximum annual rate for each specific use from any combination of products containing fluopyram. Do not apply via aerial application. For ground application equipment, apply 2 to 5 gallons of solution per 1000 sq. ft. When using against nematodes, for optimum control irrigate within 24 hours of application to depth of the root zone to be protected. Do not apply more than 17.1 fl oz of Indemnify per acre per year. For residential turf, do not apply more than 15.5 fl oz of Indemnify per acre per year.
furfural (Multiguard Protect)	0.126 to 0.184	For terrestrial (outdoor) non-food use on established turf on golf course tees and greens, practice greens, spot treatment of fairways, roughs, and turf/sod farms. Areas to be treated must be at least 70% of field capacity before application. Apply up to 6 applications using only ground boom sprayers set to release spray at no more than 2 feet above the ground. Use the high rate at the start of the season and under high infestation and/or until acceptable control is achieved every 14 to 28 days. Then use the lower rate as a maintenance application at 14-28 day intervals.

DISEASE CONTROL

Floral, Nursery, and Landscape Diseases

Fungicides and Bactericides for Disease Control of Greenhouse Floriculture Crops

Anyone using any agricultural chemical should refer to the current chemical label, which contains information about the safe and effective use of the chemical, before using the chemical. Consult the product label to ensure that the variety of ornamental plant that you wish to treat is listed on the label. Check for phytotoxicity by making trial applications on a smaller number of plants before you treat an entire crop.

Table 10-11. Disease Control of Annual, Perennial, Bedding, and Flowering Potted Plants in Greenhouses

Disease Pesticide and Formulation	Rate of Formulation	Schedule and Remarks
Bacterial Leaf Spot *(Pseudomonas, Xanthomonas)*		
Bacillus subtilis (Rhapsody, Cease)	2 to 8 qt/100 gal	Begin applications when conditions favor disease development prior to the onset of disease. Thorough coverage is essential. Repeat at 7-day intervals.
Bacillus amyloliquefaciens strain D747 (DoubleNickel LC)	0.5 to 6 qt/100 gal	Begin preventative applications at plant emergence and repeat every 3 to 28 days depending on disease pressure. OMRI listed.
Copper hydroxide (Champ WG) (Nu-Cop) 50DF (CuPRO 2005)	0.5 lb/100 gal 1 lb/100 gal 0.75 to 2.0 lb/100 gal	Begin at first sign of disease, and repeat at 7- to 14-day intervals. Do not tank mix copper formulations with Aliette. Avoid contact with metal surfaces. Discoloration of blooms may occur on certain plant varieties- check label.
Copper hydroxide + mancozeb (Junction)	1.5 to 3.5 lb/100 gal	Begin at first sign of disease, and repeat at 7- to 14-day intervals.
Copper, metallic (Copper-Count-N)	1 qt/100 gal	Begin at first sign of disease, and repeat at 7- to 14-day intervals.
Copper octanoate (Camelot O)	0.5 to 2.0 gal/100 gal	Begin at first sign of disease, and repeat at 7- to 10-day intervals. Camelot O may cause copper toxicity on some plant species.
copper oxychloride + copper hydroxide (Badge SC) (Badge X2)	1.5-6 pt/A (high) 1.5-2 pt/A (low) 1.5 to 5 lb/A (high) 1.5 to 2 lb/A (low)	High rates should only be used during dormancy. Use lower rate (1.5 to 2 pt/A or 1.5 to 2 lb/A) when new growth is present. Repeat at 7 to 14 day intervals. Badge X2 is OMRI listed. Do not mix with Aliette.
copper sulfate pentahydrate (Phyton 27, Phyton 35)	See label	See label.
cuprous oxide (Nordox 75WG)	0.66 lb/100 gal	Begin at first sign of disease and repeat at 7 to 14 day intervals.
didecyl dimethyl ammonium chloride (KleenGrow)	0.06 to 0.38 fl. oz. per gallon of water	Apply starting at week 3 or earlier if conditions are favorable for disease. Use a watering device to drench the top and bottom of the leaves and stems, avoiding flowers in full bloom, every 14 days to prevent the spread of spores and the build-up of organic material. Remove severely infected plants and disinfect the area with 1.0 fluid ounce of KleenGrow per gallon of water.
Black root rot *(Thielaviopsis basicola)*		
etridiazole + thiophanate-methyl (Banrot) 40WP	See label	See label.
fludioxonil (Medallion)	1 to 2 oz/100 gal	Apply as a preventive drench at seeding or transplanting. Make only one application to seeding crop. If needed, re-treat transplants 21 to 28 days after initial application. Do not apply as a seed or soil drench to impatiens or New Guinea impatiens.
thiophanate-methyl (AllBan Flo) (Cleary 3336 F, or other formulations) (OHP 6672) 50WP, (T-Storm) 50WSB (SysTec 1998) FL (T-Storm Flowable), (OHP 6672 4.5L)	 7.5 to 20 oz/100 gal 8 to 16 fl oz/100 gal 12 to 16 oz/100 gal 10 to 20 fl oz/100 gal 20 fl oz/100 gal	Apply as a heavy spray or drench at the rate of 1/2 to 2 pints per sq ft. Repeat at 4- to 8-week intervals. Apply 8 oz as a drench or directed spray after seeding, or apply 12 to 16 oz after transplanting. Repeat at 21- to 28-day intervals. Apply 1 to 3 pt/sq ft after transplanting to thoroughly soak growing medium. Repeat at 21-to 28-day intervals. Apply as heavy spray or drench at a rate of 1 to 2 pints per sq ft. Repeat at 2- to 4-week intervals. Apply as heavy spray or drench at a rate of 1 to 2 pints per sq ft. Repeat at 2- to 4-week intervals.
triflumizole (Terraguard) 50W	1 to 4 oz/100 gal	Apply as soil drench at 2- to 4-week intervals. Use higher rate under heavy disease pressure.
Botrytis Blight		
azoxystrobin (Heritage)	2 to 8 oz/100 gal (see label)	Apply every 7 to 21 days prior to infection. Do not exceed 2 oz/100 gal on impatiens, pansy, or viola. Do not make more than three sequential applications before switching to a nonstrobilurin fungicide.
azoxystrobin + benzovindiflupyr (Mural)	4 to 7 oz/100 gal	Apply every 7 to 14 days.
Bacillus subtilis (Rhapsody, Cease)	2 to 8 qt/100 gal	Repeat at 7-day intervals. Thorough coverage is essential. Begin applications when conditions favor disease development prior to the onset of disease.
Bacillus amyloliquefaciens strain D747 (DoubleNickel LC)	0.5 to 6 qt/100 gal	Begin preventative applications at plant emergence and repeat every 3 to 28 days depending on disease pressure. OMRI listed.
chlorothalonil (Chlorothalonil DF) (Chlorothalonil 500ZN) (Chlorothalonil 720 SFT) (Chlorostar VI, Daconil Weather Stik, Echo 720, Manicure 6F) (Daconil Ultrex) 82.5WDG (Echo) 90DF, (Echo Ultimate) (Exotherm Termil)	 See label 1.9 pt/100 gal 1.37 pt/100 gal 1 3/8 pt/100 gal 1.4 lb/100 gal 1.25 lb/100 gal 1 can/1,000 sq ft	Repeat at 7- to 14-day intervals. Apply to foliage or flowers when plants are dry or nearly dry. Discontinue applications prior to bract formation on poinsettia. Rotate with fenhexamid, iprodione, or fludioxonil. On rose, use 1 pt/100 gal. On rose, use 1 lb/100 gal. On rose, use 0.78 lb/100 gal (Echo 90) DF or 0.9 lb/100 gal (Echo Ultimate). See label for method of application.
chlorothalonil + thiophanate-methyl (Spectro) 90WDG (Consyst WDG, TM/C WDG)	 1 to 2 lb/100 gal 0.75 to 1.5 lb/100 gal	Minimum re-treat interval is 7 days. Do not apply to green or variegated pittosporum or schefflera more than once. Repeat every 7 to 10 days as needed during disease period.

DISEASE CONTROL

Table 10-11. Disease Control of Annual, Perennial, Bedding, and Flowering Potted Plants in Greenhouses

Disease Pesticide and Formulation	Rate of Formulation	Schedule and Remarks
Botrytis Blight (continued)		
copper hydroxide (Champ WG) (Nu-Cop) 50DF (CuPRO 2005)	0.5 lb/100 gal 1 lb/100 gal 0.75 to 2 pounds/100 gal	Begin at first sign of disease, and repeat at 7- to 14-day intervals. Do not tank mix copper formulations with Aliette. Avoid contact with metal surfaces. Discoloration of blooms may occur on certain plant varieties- check label. See label for rate on Easter Lily.
copper hydroxide + mancozeb (Junction)	1.5 to 3.5 lb/100 gal	Repeat at 7- to 14-day intervals.
copper, metallic (Copper-Count-N)	1 qt/100 gal	Begin at first sign of disease, and repeat at 7- to 14-day intervals.
copper octanoate (Camelot O)	0.5-2.0 gal/100 gal	Begin at first sign of disease, and repeat at 7- to 10-day intervals. Camelot O may cause copper toxicity on some plant species.
copper sulfate pentahydrate (Phyton 35, Phyton 27)	See label	See label.
cyprodinil + fludioxonil (Palladium)	4 to 6 oz/100 gal	Spray on a 7-14 day interval while conditions are conducive to disease develepment. After two applications, alternate with another fungicide with a different MOA for two applications. Cautionary statement on label for applications to Geraniums, Impatiens and New Guinea Impatiens.
dichloran (Botran) 75W	1 lb/150 to 200 gal See label	Apply to stock cuttings or greenhouse plants. Begin when disease is anticipated or first appears. Spray foliage and flowers at 7- to 14-day intervals.
fenhexamid (Decree) 50WDG	0.75 to 1.5 lb/100 gal	RESISTANCE TO THIS CHEMICAL HAS BEEN REPORTED IN BOTRYTIS. Avoid making more than two consecutive applications of this product. Treat at 7- to 14-day intervals. Rotate with chlorothalonil, copper, mancozeb, or iprodione. Make trial application before treating poinsettia.
fludioxonil (Medallion) 50WSP (Emblem)	2 to 4 oz/100 gal 2 to 4 fl oz/100 gal	Spray to runoff. Repeat at 7- to 14-day intervals. Use no more than two consecutive applications before switching to another fungicide with a different mode of action. Do not use after bract formation on poinsettia. Foliar applications on impatiens, New Guinea impatiens, and some geranium cultivars may cause stunting or chlorosis, especially on young plants.
iprodione (Chipco 26019 N/G) (18 Plus, Iprodione Pro 2SE)	1.0 to 2.5 lb/100 gal 1 to 2.5 qt/100 gal	Spray to ensure thorough coverage. Repeat at 7- to 14-day intervals. Do not make more than four applications per year. Do not apply to Spathiphyllum. Do not apply as a soil drench on impatiens or pothos.
iprodione + thiophanate-methyl (26/36)	33 to 84 fl oz/100 gal	Spray to ensure thorough coverage. Repeat at 7 to 14 day intervals. Do not use as a drench on impatiens or pothos. Do not use on Spathiphyllum.
mancozeb (Dithane) 75DF, (Fore) 80WP, (Mancozeb DG) (Pentathlon DF) (Pentathlon LF) (Protect DF)	1.5 lb/100 gal 1 to 2 lb/100 gal 0.8 to 1.6 qt/100 gal 1 to 2 lb/100 gal 1 to 2 bags/100 gal	Do not use on French dwarf or signet-type marigolds. Begin at first sign of disease. Repeat at 7- to 10-day intervals. Most effective when applied prior to infection. Not for use on marigold. Begin at first sign of disease. Repeat at 7-to 10-day intervals. Begin at first sign of disease. Repeat at 7-to 10-day intervals. To improve performance, add 2 to 4 oz of an effective spreader-sticker. Apply at 7- to 21-day intervals.
mancozeb + myclobutanil (MANhandle)	2 lb/100 gal	Addition of Latron B-1956 will improve performance. Apply at 7- to 10-day intervals.
polyoxin D zinc salt (Endorse 2.5WP) (Affirm WDG)	1.1 to 2.2 lb/100 gal 0.25 to 0.5 lb/100 gal	Apply as a foliar spray every 7 to 10 days. Apply prior to disease development and when conditions are conducive for disease.
pyraclostrobin + boscalid (Pageant Intrinsic)	12 to 18 oz/100 gal	Apply prior to disease development. Repeat at 7- to 14-day intervals. Make no more than two sequential applications. Do not expose petunia or impatiens in flower or wintercreeper or nine bark to spray or drift as injury may occur.
thiophanate-methyl (3336 F)	12-16 fl oz/100 gal	Apply every 7 to 14 days beginning at first signs of disease.
thiophanate-methyl + mancozeb (Zyban) 79WSB	see label	Do not use on French marigold or gloxinia. Apply weekly.
trifloxystrobin (Compass O) 50WDG	2 to 4 oz/100 gal	Repeat at 7- to 14-day intervals until the threat of disease is over. Rotate to another nonstrobilurin fungicide after each application of Compass O. Make no more than four foliar applications per crop cycle or season.
triflumizole (Terraguard) 50W	4 to 8 oz/100 gal	Make initial application prior to or at first sign of disease. Repeat at 7- to 14-day intervals. Do not use on impatiens plugs.
Bulb and Corm Rots (Fusarium, Penicillium)		
iprodione (Chipco 26019 N/G) (18 Plus, Iprodione Pro 2SE)	2 lb/100 gal 2 qt/100 gal	Dip 5 minutes prior to storage.
didecyl dimethyl ammonium chloride (KleenGrow)	Immerse both crate and bulbs in a solution of 0.15 to 1.5 fl oz KleenGrow per gallon of water for 30 seconds	For control of Botrytis, Rhizoctonia, Fusarium and Penicillium bulb rots and other fungal and bacterial diseases of tulip, narcissus, gladiolus, crocus, dahlia, freesia, iris, lily, daylily, amaryllis, hyacinth, iris, scilla and ornamental onion. Remove from solution and allow to drain prior to planting or storage. Test solution regularly using QAC test strips, (La Motte #2949-BJ or equivalent). Add KleenGrow when necessary to maintain an a.i. concentration between 90 and 900 ppm.
iprodione + thiophanate-methyl (26/36)	66 fl oz/100 gal	Dip 5 minutes prior to storage.
thiabendazole (Mertect 340-F)	30 fl oz/100 gal	Clean and treat bulbs and corms within 24 to 48 hours of digging. Warm solution prior to dipping. Mix fresh solution per label guidelines. Dip bulbs 15 to 30 minutes and corms 15 minutes for *Fusarium* control, or dip bulbs 10 to 15 minutes for *Penicillium* (blue mold) control.
thiophanate-methyl (AllBan Flo) (Cleary 3336) F (OHP-6672) 50WP, (T-Storm) 50WSB (T-Storm Flowable) (SysTec 1998) FL, (OHP 6672) 4.5L	14.5 to 33 fl oz/100 gal 12 to 16 fl oz/100 gal foliar spray; 16 to 24 fl oz/100 gal dip 16 to 24 oz/100 gal 12 to 16 oz/100 gal see label	Soak clean bulbs for 15 to 30 minutes in warm (80-85 degrees F) solution. Treat bulbs within 48 hours of digging. Dry well before storing.

471

Table 10-11. Disease Control of Annual, Perennial, Bedding, and Flowering Potted Plants in Greenhouses

Disease Pesticide and Formulation	Rate of Formulation	Schedule and Remarks
Cylindrocladium Stem Canker or Root Rot		
azoxystrobin + benzovindiflupyr (Mural)	4 to 7 oz/100 gal	Apply every 7 to 14 days.
chlorothalonil (Chlorothalonil DF) (Chlorothalonil 500ZN) (Chlorothalonil 720 SFT) (Chlorostar VI, Daconil Weather Stik, Echo 720, Manicure 6F) (Daconil Ultrex, Daconil Weatherstik) 82.5WDG (Echo) 90DF, (Echo Ultimate) (Exotherm Termil)	See label 1.9 pt/100 gal 1.37 pt/100 gal 1 3/8 pt/100 gal 1.4 lb/100 gal 1.25 lb/100 gal 1 can/1,000 sq ft	Repeat at 7- to 14-day intervals. Apply to foliage or flowers when plants are dry or nearly dry. See label for method of application.
chlorothalonil + thiophanate-methyl (Spectro) 90WDG	1.0 to 2.0 lb per 100 gallons	For best results use spray mixture the same day it is prepared. Spray uniformly over the area to be treated with a properly calibrated power sprayer, apply as a full coverage spray to run-off when conditions are favorable for disease development.
fludioxonil (Medallion) 50WSP	see label	Completely drench the growing medium. Repeat at 21- to 28-day intervals. Two applications per year when conditions favor disease development are usually adequate for control.
pyraclostrobin + boscalid (Pageant Intrinsic)	12 to 18 oz/100 gal	Apply prior to disease development. Completely drench the growing medium. Repeat at 7- to 14-day intervals. Make no more than two sequential applications. Do not expose petunia or impatiens in flower or wintercreeper or nine bark to spray, as injury may occur.
thiophanate-methyl (AllBan Flo) (Cleary 3336) F (OHP-6672) 50W, (T-Storm) 50WSB (SysTec 1998) FL (T-Storm Flowable), (OHP 6672) 4.5L	7.5 to 20 fl oz/100 gal 8 to 16 fl oz/100 gal 12 to 16 oz/100 gal 10 to 20 fl oz/100 gal 20 fl oz/100 gal	Apply as drench or heavy spray at 0.5 to 2 pints per sq ft. Repeat at 4- to 8-week intervals. Apply 8 oz after seeding or sticking, or 12 to 16 oz after transplanting as a drench or directed spray at a rate that thoroughly soaks the growing media through the root zone. Repeat every 21 to 28 days. Apply as a drench or heavy spray at a rate of 1 to 3 pints per sq ft. Repeat at 21- to 28-day intervals. Apply as a drench or heavy spray at a rate of 1 to 2 pints per sq ft. Repeat at 2- to 4-week intervals. Apply as a drench or heavy spray at a rate of 1 to 2 pints per sq ft. Repeat at 2- to 4-week intervals.
triflumizole (Terraguard) SC	See label	Can be used as a cutting soak or soil drench.
Downy Mildew		
azoxystrobin (Heritage)	1 to 4 oz/100 gal 2 to 4 oz/100 gal (rose)	Apply every 7 to 14 days prior to infection. Do not apply 2 oz rate on less than 14-day intervals. May damage snapdragons; use 1 oz rate and rotate. Apply every 7 to 21 days on rose.
azoxystrobin + benzovindiflupyr (Mural)	4 to 7 oz/100 gal	Apply every 7 to 14 days.
Bacillus subtilis (Rhapsody, Cease)	2 to 8 qt/100 gal	Repeat at 3- to 10-day intervals. Thorough coverage is essential. Begin applications when conditions favor disease development, prior to the onset of disease.
Bacillus amyloliquefaciens strain D747 (DoubleNickel LC)	0.5 to 6 qt/100 gal	Begin preventative applications at plant emergence and repeat every 3 to 28 days depending on disease pressure. OMRI listed.
copper hydroxide (Champ WG) (Nu-Cop) 50DF (CuPRO 2005)	 0.5 lb/100 gal 1 lb/100 gal 0.75 to 2.0 lb/100 gal	Begin at first sign of disease, and repeat at 7- to 14-day intervals. Do not tank mix copper formulations with Aliette. Avoid contact with metal surfaces. Discoloration of blooms may occur on certain plant varieties; check label.
copper oxychloride + copper hydroxide (Badge SC) (Badge X2)	 1.5-6 pt/A (high) 1.5-2 pt/A (low) 1.5 to 5 lb/A (high) 1.5 to 2 lb/A (low)	High rates should only be used during dormancy. Use lower rate (1.5 to 2 pt/A or 1.5 to 2 lb/A) when new growth is present. Repeat at 7 to 14 day intervals. Badge X2 is OMRI listed. Do not mix with Aliette.
copper, metallic (Copper-Count-N)	1 qt/100 gal	Begin at first sign of disease, and repeat at 7- to 14-day intervals. To avoid phytotoxicity, do not use any copper compound on alyssum.
copper octanoate (Camelot O)	0.5-2.0 gal/100 gal	Begin at first sign of disease, and repeat at 7- to 10-day intervals. Camelot O may cause copper toxicity on some plant species.
copper sulfate pentahydrate (Phyton 27)	See label	See label. To avoid phytotoxicity, do not use any copper compound on alyssum.
cyazofamid (Segway)	2.1 to 3.5 fl oz/100 gal	14 to 21 day intervals using another registered fungicide with a different mode of action. Apply sufficient volume to wet all foliage until runoff (normally 50 to 100 gallons per acre).
dimethomorph (Stature SC)	6.12 to 12.25 oz/100 gal	Apply at first sign of disease. Apply to obtain complete coverage of flowers, foliage, and stems. Repeat at 10- to 14-day intervals throughout the production cycle.
dimethomorph + ametoctradin (Orvego)	11 to 14 fl oz/100 gal	Apply on 10 to 14 day intervals using another registered fungicide with a different mode of action. Apply sufficient volume to wet all foliage until runoff (normally 50 to 100 gallons per acre).
fenamidone (Fenstop)	7 to 14 fl oz/100 gal	Apply as a foliar spray until wet. Repeat as necessary on a 28-day schedule. Do not apply more than 2 applications per crop per season.
fluoxastrobin (Disarm G)	see label	Apply in advance of infection periods.
fluopicolide (Adorn)	1 to 4 fl oz/100 gal	Adorn MUST be tank mixed for resistance management with another product that is registered for use against the target disease.
fosetyl-Al (Aliette) 80WDG	1.25 to 4 lb/100 gal (bedding plants) 2.5 lb/100 gal (rose)	Systemic. Repeat as necessary, but do not make more than one application every 14 days.
mancozeb (Dithane) 75DF, (Fore) 80WSP, (Mancozeb DG) (Pentathlon DF) (Pentathlon LF) (Protect DF)	 1.5 lb/100 gal 1 to 2 lb/100 gal 0.8 to 1.6 qt/100 gal 1 to 2 lb/100 gal 1 to 2 bags/100 gal	Repeat at 7- to 10-day intervals. Begin at first sign of disease. Repeat 7- to 10-day intervals. Begin at first sign of disease. Repeat 7- to 10-day intervals. To improve performance, add 2-4 oz of an effective spreader-sticker. Repeat at 7- to 21-day intervals.
mancozeb + copper hydroxide (Junction)	1.5 to 3.5 lb/100 gal	Begin applications at first sign of disease and repeat at 7- to 14-day intervals.

Table 10-11. Disease Control of Annual, Perennial, Bedding, and Flowering Potted Plants in Greenhouses

Disease Pesticide and Formulation	Rate of Formulation	Schedule and Remarks
Downy Mildew (continued)		
mandipropamid (Micora)	4 to 8 fl oz/100 gal	This product can also be used on vegetables sold to the retail market in GH with permanent flooring. Apply prior to disease development. Repeat sprays at 7- to 14-day intervals. Make no more than two sequential applications, then rotate to another fungicide with a different MOA.
mefenoxam (Subdue Maxx, Subdue GR)	0.5 to 1 fl oz/100 gal See label for rates	Apply Subdue Maxx as a foliar spray or soil drench treatment. Apply Subdue GR as a soil surface or soil/planting media incorporation treatment.
oxathiapiprolin (Segovis)	0.6 to 2.4 fl oz/100 gal	Begin foliage applications prior to disease development and continue on 5 to 14 day interval when conditions are conducive for disease development. Do not apply more than 2 consecutive applications before switching to another non-Group U-15 fungicide.
phosphorous acid (Alude, Fosphite, Reliant)	See label for rates	Apply prior to disease development. Spray to thoroughly wet all foliage. Repeat at 14- to 21-day intervals.
polyoxin D zinc salt (Endorse 2.5WP)	1.1 to 2.2 lb/100 gal	Apply as a foliar spray every 7 to 10 days. Apply prior to disease development and when conditions are conducive for disease.
potassium bicarbonate (MilStop)	1.25 to 5 lb/100 gal	Uniform and complete coverage of foliage is essential for best results. See label for special instructions regarding poinsettia, pansy, and impatiens.
potassium phosphite (Vital)	4 pt/100 gal	Apply prior to disease onset, and repeat at 14-day intervals.
pyraclostrobin (Insignia)	4 to 8 oz/100 gal	Apply prior to onset of disease and repeat every 7 to 14 days. Do not expose flowering impatiens or flowering petunias to Insignia.
pyraclostrobin + boscalid (Pageant Intrinsic)	12 to 18 oz/100 gal	Apply prior to disease development. Repeat at 7- to 10-day intervals. Make no more than two sequential applications. Do not expose petunia or impatiens in flower or wintercreeper or nine bark to spray or drift as injury may occur.
thiophanate-methyl + mancozeb (Zyban)	1.5 lb/100 gal	Apply at first sign of disease, and repeat at 7-day intervals.
trifloxystrobin (Compass O) 50WDG	2 to 4 oz/100 gal	Apply as a foliar spray before disease is detected or when conditions are favorable for disease. Repeat at 7- to-14-day intervals until threat of disease is over.
Fungal Leaf Spots (consult label for specific fungi controlled)		
azoxystrobin (Heritage)	1 to 4 oz/100 gal	Repeat at 7- to 28-day intervals. Do not make more than three sequential applications before switching to a nonstrobilurin fungicide. Good control of *Alternaria* leaf spot.
azoxystrobin + benzovindiflupyr (Mural)	4 to 7 oz/100 gal	Apply every 7 to 21 days.
Bacillus subtilis (Rhapsody, Cease)	2 to 8 qt/100 gal	Repeat at 7-day intervals. Thorough coverage is essential. Not effective on *Alternaria*.
Bacillus amyloliquefaciens strain D747 (DoubleNickel LC)	0.5 to 6 qt/100 gal	Begin preventative applications at plant emergence and repeat every 3 to 28 days depending on disease pressure. OMRI listed.
chlorothalonil (Chlorothalonil DF) (Chlorothalonil 500ZN) (Chlorothalonil 720 SFT) (Chlorostar VI, Echo 720, Manicure 6 Flowable) (Daconil Ultrex) 82.5WDG (Echo) 90DF, (Echo Ultimate)	See label. 1.9 pt/100 gal 1.37 pt/100 gal 1 3/8 pt/100 gal 1.4 lb/100 gal 1.25 lb/100 gal	Repeat at 7- to 14-day intervals. Works well for control of *Alternaria* leaf spot. Discontinue applications prior to bract formation on poinsettias. Applications made during bloom may damage flowers. Apply to plants when both foliage and flowers are dry or nearly dry.
chlorothalonil + thiophanate-methyl (Spectro) 90WDG (Consyst WDG, TM/C WDG)	1 to 2 lb/100 gal 0.75 to 1 lb/100 gal	Apply when foliage and flowers are dry, or nearly dry. Repeat at 7-day intervals. Good control of *Colletotrichum* (anthracnose) and *Alternaria*. Not recommended for Swedish Ivy, Boston Fern, and Easter Cactus. Thorough, uniform coverage is essential for good control. Repeat in 7 to 10 days.
copper hydroxide (Champ WG) (Nu-Cop) 50DF (CuPRO 2005)	0.5 lb/100 gal 1 lb/100 gal 0.75 to 2.0 lb/100 gal	Begin at first sign of disease, and repeat at 7- to 14-day intervals. Do not tank mix copper formulations with Aliette. Avoid contact with metal surfaces. Discoloration of blooms may occur on certain plant varieties- check label.
copper oxychloride + copper hydroxide (Badge SC) (Badge X2)	1.5-6 pt/A (high) 1.5-2 pt/A (low) 1.5 to 5 lb/A (high) 1.5 to 2 lb/A (low)	High rates should only be used during dormancy. Use lower rate (1.5 to 2 pt/A or 1.5 to 2 lb/A) when new growth is present. Repeat at 7 to 14 day intervals. Badge X2 is OMRI listed. Do not mix with Aliette.
copper hydroxide + mancozeb (Junction)	1.5 to 3.5 lb/100 gal	Repeat at 7- to 21-day intervals. Discoloration of foliage and/or blooms is possible on some varieties of azalea, carnation, chrysanthemum, and rose. Do no use on French marigold.
copper, metallic (Copper-Count-N)	1 qt/100 gal	Begin at first sign of disease, and repeat at 7- to 14-day intervals
copper octanoate (Camelot O)	0.5-2.0 gal/100 gal	Begin at first sign of disease, and repeat at 7- to 10-day intervals. Camelot O may cause copper toxicity on some plant species.
copper sulfate pentahydrate (Phyton 27)	See label	See label. To avoid phytotoxicity, do not use any copper compound on alyssum.
cyprodinil + fludioxonil (Palladium)	2 to 4 oz/100 gal	Spray on a 7-14 day interval while conditions are conducive to disease develepment. After two applications, alternate with another fungicide with a different MOA for two applications. See cautionary statement on label for applications to Geraniums, Impatiens and New Guinea Impatiens.
fludioxonil (Medallion)	1 to 2 oz/100 gal	Repeat at 7- to 14-day intervals. Good control of *Alternaria* leaf spot.
iprodione (Chipco 26019 N/G) (18 Plus, Iprodione Pro 2SE)	1.0 to 2.5 lb/100 gal 1 to 2.5 qt/100 gal	Spray plants to ensure thorough coverage. Repeat at 7- to 14-day intervals. Do not make more than 4 applications per crop per year. Do not drench impatiens or pothos. Do not use on Spathiphyllum. Good control of *Alternaria* leaf spot.
iprodione + thiophanate-methyl (26/36)	33 to 84 fl oz/100 gal	Spray plants to ensure thorough coverage. Repeat at 7- to 14-day intervals. Do not make more than 4 applications per crop per year. Do not drench impatiens or pothos. Do not use on Spathiphyllum.
mancozeb (Dithane) 75DF, (Fore) 80WSP, (Mancozeb DG) (Pentathlon DF) (Pentathlon LF) (Protect DF)	1.5 lb/100 gal 1 to 2 lb/100 gal 0.8 to 1.6 qt/100 gal 1 to 2 lb/100 gal	Do not use on French dwarf or signet-type marigolds. Begin at first sign of disease. Repeat at 7- to 10-day intervals. Most effective when applied prior to infection. Begin at first sign of disease. Repeat at 7-to 10-day intervals. Begin at first sign of disease. Repeat at 7-to 10-day intervals. To improve performance, add 2-4 oz of an effective spreader-sticker. Apply at 7- to 21-day intervals.

473

Table 10-11. Disease Control of Annual, Perennial, Bedding, and Flowering Potted Plants in Greenhouses

Disease Pesticide and Formulation	Rate of Formulation	Schedule and Remarks
Fungal Leaf Spots (consult label for specific fungi controlled) (continued)		
mancozeb + myclobutanil (Clevis, MANhandle)	1 to 2 lb/100 gal	Apply at first sign of disease, and repeat at 7- to 10-day intervals. The addition of Latron B-1956 will improve performance. Good control of Alternaria.
myclolobutanil (Eagle) 20EW (Eagle) 40WP	6 to 12 fl oz/100 gal 3 to 6 oz/100 gal	Apply at 10- to 14-day intervals, not to exceed 21 days. For chrysanthemums, see label for specific rates.
pyraclostrobin (Insignia)	See label	Apply prior to onset of disease and repeat every 7 to 14 days. Do not expose flowering impatiens or flowering petunias to Insignia.
pyraclostrobin + boscalid (Pageant Intrinsic)	6 to 12 oz/100 gal	Apply prior to disease development. Repeat at 7- to 14-day intervals. Make no more than two sequential applications. Do not expose petunia or impatiens in flower or wintercreeper or nine bark to spray or drift as injury may occur.
polyoxin D zinc salt (Endorse 2.5WP)	2.2 lb/100 gal	Control of Alternaria blight, anthracnose (Colletotrichum), and Curvularia leaf spot. Apply as a foliar spray every 7 to 10 days. Apply prior to disease development and when conditions are conducive for disease.
potassium bicarbonate (MilStop)	1.25 to 5 lb/100 gal	Uniform and complete coverage of foliage is essential for best results. See label for special instructions regarding poinsettia, pansy, and impatiens.
tebuconazole (Torque)	4 to 8 fl oz/100 gal	Begin applications 14 to 21 days prior to when disease is expected, or at very first sign of disease.
thiophanate-methyl (AllBan Flo) (Cleary 3336) F, (OHP 6672) 50WP, (T-Storm) 50WSB (SysTec 1998) FL, (T-Storm Flowable)	10 to 14.5 fl oz/100 gal 12 to 16 fl oz/100 gal 20 fl oz/100 gal	Begin applications 14 to 21 days prior to when disease is expected, or at very first sign of disease. Repeat at 7- to 10-day intervals. Poor control of Alternaria leaf spot. Begin when disease first appears and repeat at 7- to 14-day intervals. Begin when disease first appears and repeat at 10- to 14-day intervals.
triadimefon (Strike) 25WDG	2 to 4 oz/100 gal	Apply as needed at first sign of disease. Good control of Alternaria leaf spot.
trifloxystrobin (Compass O) 50WDG	1 to 4 oz/100 gal	Repeat at 7- to 14-day intervals until threat of disease is over. Rotate to another nonstrobilurin fungicide after each application. Good control of Alternaria leaf spot.
triflumizole (Terraguard) 50W	4 to 8 oz/100 gal	Apply at very first sign of disease. Do not use on impatiens plugs, and do not exceed 2 oz/100 gal for impatiens transplants. Repeat at 7- to 14-day intervals.
triticonazole (Trinity) 19SC	4 to 12 fl oz/100 gal	See label as rate varies depending on fungal leaf spot pathogen. Use preventively. Begin applications when conditions favor fungal infection and before disease symptom development. Use of an adjuvant/spreader sticker can aid in control.
Fusarium Root and Crown Rot		
azoxystrobin (Heritage) azoxystrobin (Heritage)	See label See label.	Apply as a directed spray every 7 to 21 days, Apply as a soil drench at 7-to 28 day intervals.
etridiazole + thiophanate-methyl (Banrot) 40W	6 to 12 oz/100 gal	Apply in sufficient volume to saturate the soil mixture. Irrigate immediately. Repeat at 4- to 12-week intervals if necessary. See label for more details.
pyraclostrobin (Empress Intrinsic)	1 to 6 fl oz/100 gal (see label)	Apply at 1 to 3 fl oz for plants in propagation, rooted cuttings, plugs and seedlings and at 2 to 6 fl oz to all other plants. Do not apply to dry soil media. Apply preventative to disease with sequential at 7 to 28 days after the first application if needed.
fludioxonil (Medallion)	1 to 2 oz/100 gal	Apply as a drench at seeding or transplanting. Make only one application to seeding crop. If needed, retreat transplants 21 to 28 days after initial application. See label for incorporation into potting mixture.Do not apply as a seed or soil drench to impatiens or New Guinea impatiens.
fludioxonil + mefenoxam (Hurricane) 48WP	See label	Apply as a pre-potting or growing media drench per label directions. For control of Fusarium, add Medallion Fungicide at a rate of 1.0 oz/gal. Application to impatiens, New Guinea impatiens, pothos, geranium, and Easter lily may cause stunting and/or chlorosis.
pyraclostrobin + boscalid (Pageant Intrinsic)	12 to 18 oz/100 gal	Apply prior to disease development. Completely drench the growing medium. Repeat at 7- to 14-day intervals. Make no more than two sequential applications. Do not expose petunia or impatiens in flower or wintercreeper or nine bark to spray, as injury may occur.
Streptomyces griseoviridis (Mycostop)	See label	Apply inoculant as a seed dressing, soil drench spray, or transplant dip. Must be applied prior to onset of disease. See label.
thiophanate-methyl (AllBan Flo) (Cleary 3336) F (OHP-6672) 50WP, (T-Storm) 50WSB (SysTec 1998) FL (T-Storm Flowable), (OHP 6672) 4.5L	7.5 to 20 fl oz/100 gal 8 to 16 fl oz/100 gal 12 to 16 fl oz/100 gal 10 to 20 fl oz/100 gal 20 fl oz/100 gal	Apply as a heavy spray or drench at a rate of 0.5 to 2 pints per sq ft. Repeat at 4- to 8-week intervals. Apply 8 oz as a drench or directed spray after seeding or sticking or 12 to 16 oz after transplanting. Repeat at 21- to 28-day intervals. Soak growing medium after transplanting. Repeat every 21 to 28 days. Apply as a drench or heavy spray at a rate of 1 to 2 pints per sq ft. Repeat at 2- to 4-week intervals. Apply as a drench or heavy spray at a rate of 1 to 2 pints per sq ft. Repeat at 2- to 4-week intervals.
triflumizole (Terraguard) 50W	4 to 8 oz/100 gal	Apply soil drenches weekly as needed. Use higher rate under heavy disease pressure.
triticonazole (Trinity) 19SC	8 to 12 fl oz/100 gal	Use preventively. Begin applications when conditions favor fungal infection and before disease symptom development. The crown and base of the plant and the soil or potting medium surrounding the crown must be thoroughly covered.
Myrothecium Leaf Blight, Crown, or Petiole Rot		
azoxystrobin (Heritage)	1 to 4 oz/100 gal	Apply every 7 to 21 days. Do not exceed 2 oz per 100 gal on impatiens, pansy, or violas.
azoxystrobin + benzovindiflupyr (Mural)	4 to 7 oz/100 gal	Apply every 7 to 21 days.
chlorothalonil (Chlorostar VI, Daconil Weather Stik, Echo 720, Manicure 6 Flowable) (Chlorothalonil DF) (Chlorothalonil 500 ZN) (Chlorothalonil 720 SFT) (Daconil Ultrex) 82.5WDG (Echo Ultimate)	1 3/8 pt/100 gal See label 1.9 pt/100 gal 1.37 pt/100 gal 1.4 lb/100 gal 1.25 lb/100 gal	Repeat applications at 7- to 14-day intervals.

Table 10-11. Disease Control of Annual, Perennial, Bedding, and Flowering Potted Plants in Greenhouses

Disease Pesticide and Formulation	Rate of Formulation	Schedule and Remarks
Myrothecium Leaf Blight, Crown, or Petiole Rot (continued)		
cyprodinil + fludioxonil (Palladium)	2 to 4 oz/100 gal	Spray on a 7-14 day interval while conditions are conducive to disease develepment. After two applications, alternate with another fungicide with a different MOA for two applications. See cautionary statement on label for applications to Geraniums, Impatiens and New Guinea Impatiens.
fludioxonil (Medallion)	1 to 2 oz/100 gal	Spray to runoff at 7- to 14-day intervals. See label for media mix and drench applications. Drench applications to impaties or New Guinea impaties may cause stunting and/or chlorosis.
pyraclostrobin + boscalid (Pageant Intrinsic)	8 to 12 oz/100 gal	Apply prior to disease development. Repeat at 7- to 14-day intervals. Make no more than two sequential applications. Do not expose petunia or impatiens in flower or wintercreeper or nine bark to spray or drift as injury may occur.
trifloxystrobin (Compass O)	1 to 2 oz/100 gal	Repeat at 7- to 14-day intervals until threat of disease is over.
triflumizole (Terraguard) 50W	4 to 8 oz/100 gal	Apply prior to or at first sign of disease. Repeat at 7- to 14-day intervals. Use higher rates for initial application under disease pressure. Do not use on impatiens plugs. Do not exceed 2 oz per 100 gallons on impatiens transplants.
triticonazole (Trinity) 19SC	8 to 12 fl oz/100 gal	Use preventively. Begin applications when conditions favor fungal infection and before disease symptom development. Use of an adjuvant/spreader sticker can aid in control.
Phytophthora Aerial Shoot Blight		
azoxystrobin (Heritage)	1 to 2 oz/100 gal	Apply every 7 to 28 days. Do not make more than three sequential applications before switching to another effective fungicide with a different mode of action.
azoxystrobin + benzovindiflupyr (Mural)	4 to 7 oz/100 gal	Apply every 7 to 14 days.
chlorothalonil + thiophanate-methyl (Spectro) 90 WDG	1 to 2 lb/100 gal	Apply when plants are dry. Spectro has protective and curative action. Repeat at 7-day intervals.
dimethomorph (Stature SC)	12.25 fl oz/100 gal	Begin spraying at first sign of disease. Use a full-coverage spray at 10- to 14-day intervals throughout production cycle.
dimethomorph + ametoctradin (Orvego)	14 fl oz/100 gal	Apply on 10 to 14 day intervals using another registered fungicide with a different mode of action. Apply sufficient volume to wet all foliage until runoff (normally 50 to 100 gallons per acre).
fenamidone (Fenstop)	7 to 14 fl oz/100 gal	Apply as a foliar spray until wet. Repeat as necessary on a 28-day schedule. Do not apply more than 4 applications per crop per season.
fluopicolide (Adorn)	2 to 4 fl oz/100 gal	MUST ALWAYS BE TANK MIXED WITH THE LABELED RATE OF ANOTHER FUNGCIDE WITH A DIFFERENT MODE OF ACTION. Apply before disease development. Use higher rate when treating plants with high potential for disease. Reapply after 14 to 28 days. Do not apply more than 2 applications per cropping cycle.
fuoxastrobin (Disarm 480SC)	1 to 4 fl oz/100 gal	Apply as a crown spray at 7 to 28 day application interval depending on disease pressure
fosetyl-Al (Aliette) 80 WDG	1.25 to 4 lb/100 gal	Apply prior to disease development. Apply as a foliar spray. Repeat as necessary, but do not exceed one application every 14 days.
mandipropamid (Micora)	4 to 8 fl oz/100 gal	This product can also be used on vegetables sold to the retail market in GH with permanent flooring. Apply prior to disease development. Repeat at 7- to 14-day intervals. Make no more than two sequential applications before rotating to an alternate MOA.
oxathiapiprolin (Segovis)	0.6 to 2.4 fl oz/100 gal	Begin foliar applications prior to disease development and continue on 5- to 14-day interval when conditions are conducive for disease development. Do not apply more than 2 consecutive applications before switching to another non-Group U-15 fungicide.
phosphorous acid (Alude, Fosphite, Reliant)	See label for rates	Apply as a soil drench or foliar spray as a preventive.
potassium phosphite (Vital)	see label	Apply preventatively as a soil drench or foliar spray.
pyraclostrobin + boscalid (Pageant Intrinsic)	18 oz/100 gal	Apply prior to disease development. Repeat at 7- to 10-day intervals. Make no more than two sequential applications. Do not expose petunia or impatiens in flower or wintercreeper or nine bark to spray or drift as injury may occur.
Phytophthora or Pythium Root and Crown Rot		
azoxystrobin + benzovindiflupyr (Mural)	3 oz/100 gal	Apply 1 to 2 pints of drench solution per square foot surface area every 7 to 28 days.
cyazofamid (Segway)	3.0 to 6.0 fl oz/100 gal	Apply ONLY to ornamentals grown in containers in greenhouses as a soil drench. Make applications on a 14- to 21-day interval using another registered fungicide with a different mode of action. Check label for recommended maximum drench volume based on pot diameter.
dimethomorph (Stature SC)	3.06 to 6.12 oz/50 to 100 gal	Apply when plant roots are well established, or at first sign of disease on 10- to 14-day intervals throughout production cycle. Use enough solution to wet root zone. Avoid watering plants for several hours after application. See label for rates for container-grown perennials and woody ornamentals. **Not effective against Pythium root rot.**
dimethomorph + ametoctradin (Orvego)	11 to 14 fl oz/100 gal	NOT LABELED FOR PYTHIUM. Apply on 10 to 14 day intervals using another registered fungicide with a different mode of action. Apply sufficient volume to wet all foliage until runoff (normally 50 to 100 gallons per acre).
dipotassium phosphonate + dipotassium phosphate (Biophos) 1% (v/v)	See label for rates	Apply as a soil drench or foliar spray as a preventive.
etridiazole (Truban) 30WP (Terrazole) 35WP	3 to 10 oz/100 gal 3.5 to 10 oz/100 gal	Apply in sufficient volume to saturate soil. Irrigate immediately. Repeat at 4- to 12-week intervals. Drench 4-inch pot with a minimum of 2 oz and a 6-inch pot with 4 oz. Re-treat at 4- to 12-week intervals. Use higher rates for peat or other high organic potting media.
etridiazole + thiophanate-methyl (Banrot) 40W	4 to 8 oz/100 gal	Apply in sufficient volume to saturate the soil mixture. Irrigate immediately. Repeat at 4- to 12-week intervals if necessary.
fenamidone (Fenstop)	7 to 14 fl oz/50 to 100 gal/400 sq ft	Apply as a drench using 1 to 2 pints per square foot. Repeat as necessary on a 28-day application schedule. Do not apply more than 4 applications of the maximum rate per crop per season. Higher rate has shown more consistent efficacy in research trials.
fludioxonil + mefenoxam (Hurricane) 48WP	See label	Apply as a pre-potting or growing media drench per label directions. Application to impaties, New Guinea impaties, pothos, geranium, and Easter lily may cause stunting and/or chlorosis.

DISEASE CONTROL

Table 10-11. Disease Control of Annual, Perennial, Bedding, and Flowering Potted Plants in Greenhouses

Disease Pesticide and Formulation	Rate of Formulation	Schedule and Remarks
Phytophthora or Pythium Root and Crown Rot (continued)		
fluopicolide (Adorn)	1 to 4 fl oz/100 gal	MUST ALWAYS BE TANK MIXED WITH THE LABELED RATE OF ANOTHER FUNGCIDE WITH A DIFFERENT MODE OF ACTION. Apply before disease development. Use higher rate when treating plants with high potential for disease. Reapply after 14 to 28 days. Do not make more than one application per crop on poinsettia- phytotoxicity has been observed with repeat applications.
fosetyl-Al (Aliette) 80WDG	1.25 to 4 lb/100 gal	Apply as a foliar spray prior to disease development. Repeat as necessary, but do not exceed one application every 14 days.
Gliocladium virens strain GL-21	see label	Naturally occurring soil fungus. Do not use additional soil fungicides at time of incorporation. Avoid use in media totally void of organic matter. OMRI listed.
mandipropamid (Micora)	4 to 8 fl oz/100 gal	This product can also be used on vegetables sold to the retail market in GH with permanent flooring. Apply prior to disease development. Repeat at 7- to 14-day intervals. Make no more than two sequential applications.
mefenoxam (Fenox ME, Mefenoxam 2, Subdue MAXX)	See label for rates	Can be applied as a drench, soil surface spray, or incorporated into the soil mix. Consult label for specific crops and applications.
oxathiapiprolin (Segovis)	0.65 to 3.2 fl oz/100 gal	NOT LABELED FOR PYTHIUM. Apply at 5 to 14 day interval preventatively or at first sight of disease symptoms. Do not apply more than 2 consecutive applications before switching to another non-Group U-15 fungicide.
phosphorous acid (Alude, FOSPHITE, Reliant)	See label for rates	Apply as a soil drench or foliar spray as a preventive.
potassium phosphite (Vital)	1 pt/100 gal	Apply preventatively as a soil drench or foliar spray.
propamocarb hydrochloride (Banol, Proplant)	20 to 30 fl oz/100 gal	Apply at seeding or transplanting. See label. Effective for preventing Pythium infections.
pyraclostrobin (Empress Intrinsic)	1 to 3 fl oz/100 gal in propagation and 2 to 6 fl oz/100 gal for all other plants in production	Apply as a preventative drench prior to onset of disease. Can be reapplied 7 to 28 days following the intial application.
trifloxystrobin (Compass O)	1 to 2 oz/100 gal	Apply as drench to wet the upper half of growing media. Start application at time of planting and at 14 to 28 days depending on disease pressure. NOT LABELED FOR PYTHIUM.
Powdery Mildew		
azoxystrobin (Heritage)	1 to 4 oz /100 gal	Spray every 7 to 28 days as needed. To avoid fungicide resistance, do not make more than two sequential applications of Heritage before rotating with non-strobilurin products.
azoxystrobin + benzovindiflupyr (Mural)	4 to 7 oz/100 gal	Apply every 7 to 21 days. Do not make more than two sequential applications before rotating to another class of fungicide that is not Group 7 or 11.
Bacillus subtilis (Rhapsody, Cease)	2 to 4 qt/100 gal	Repeat at 7-day intervals. Thorough coverage is essential. Begin applications when conditions favor disease development prior to the onset of disease.
Bacillus amyloliquefaciens strain D747 (DoubleNickel LC)	0.5 to 6 qt/100 gal	Begin preventative applications at plant emergence and repeat every 3 to 28 days depending on disease pressure. OMRI listed.
chlorothalonil (Chlorostar VI, Daconil Weather Stik, Echo 720, Manicure 6 Flowable) (Chlorothalonil DF) (Chlorothalonil 500ZN) (Chlorothalonil 720 SFT) (Daconil Ultrex) 82.5WDG (Echo) 90DF, (Echo Ultimate)	1 3/8 pt/100 gal See label 1.9 pt/100 gal (1.6 pt/100 gal on rose) 1.37 pt/100 gal 1.4 lb/100 gal 1.25 lb/100 gal	Apply until runoff when flowers and foliage are dry. Repeat at 7- to 14-day intervals. Avoid applications during bloom where flower injury is unacceptable. Discontinue use on poinsettias prior to bract formation.
chlorothalonil + thiophanate-methyl (Spectro) 90WDG (ConSyst, TM/C) 50WDG	1 to 2 lb/100 gal 0.75 lb/100 gal	Apply when foliage and flowers are dry, or nearly dry. Re-treat at 7-day intervals. Repeat at 7- to 10-day intervals.
copper hydroxide (Champ WG) (Nu-Cop) 50DF (CuPRO 2005)	0.5 lb/100 gal 1 lb/100 gal 0.75 to 2.0 lb/100 gal	Begin at first sign of disease, and repeat at 7- to 14-day intervals. Do not tank mix copper formulations with Aliette. Avoid contact with metal surfaces. Discoloration of blooms may occur on certain plant varieties- check label.
copper octanoate (Camelot O)	0.5-2.0 gal/100 gal	Begin at first sign of disease, and repeat at 7- to 10-day intervals. Camelot O may cause copper toxicity on some plant species.
copper sulfate pentahydrate (Phyton 27)	See label	See label. To avoid phytotoxicity, do not use any copper compound on alyssum.
cyprodinil + fludioxonil (Palladium)	4 to 6 oz/100 gal	Spray on a 7-14 day interval while conditions are conducive to disease develepment. After two applications, alternate with another fungicide with a different MOA for two applications.
fluoxastrobin (Disarm 480SC)	1 to 4 fl oz/100 gal	Use for preventative applications only and apply every 7 to 28 days depending on disease pressure. Do not make more than two consecutive applications before switching to a non Group 11 fungicide.
Hydrogen peroxide + peroxyacetic acid (ZeroTol 2.0)	125 fl oz/100 gal	Apply in early morning or late evening. Apply consecutive applications until control is achieved, then begin weekly preventative program (see label).
mancozeb + myclobutanil (Clevis, MANhandle)	16 oz/100 gal	Begin at first sign of disease, and repeat at 7- to 10-day intervals. In a limited number of trials, gave excellent control of powdery mildew on rose.
myclobutanil (Eagle) 20EW (Eagle) 40WSP	6 to 12 fl oz/100 gal 3 to 6 oz/100 gal	Apply at 10- to 14-day intervals. Use caution if applying to Gerbera daisy as phytotoxicity may occur.
neem oil (Triact 70)	1 gal/100 to 200 gal	Trial first on open blooms. Retreat at 7- to 14-day intervals. Use 1:200 rate as a preventive and 1:100 rate if disease is evident.
Phosphorous acid (Rampart)	see label	Apply at two to three week intervals. Do not apply to plants that are under heat or moisture stress. Do not apply to plants treated with copper compounds at less than 20 days intervals.
piperalin (Pipron) LC	4 to 8 fl oz/100 gal	See label for precautions on hydrangea, begonia, and poinsettia. Use high rate if disease is already present.

Table 10-11. Disease Control of Annual, Perennial, Bedding, and Flowering Potted Plants in Greenhouses

Disease Pesticide and Formulation	Rate of Formulation	Schedule and Remarks
Powdery Mildew (continued)		
pyraclostrobin + boscalid (Pageant Intrinsic)	6 to 12 oz/100 gal	Apply prior to disease development. Repeat at 7- to 10-day intervals. Make no more than two sequential applications. Do not expose petunia or impatiens in flower or wintercreeper or nine bark to spray or drift as injury may occur.
potassium bicarbonate (Armicarb 100) (Kaligreen) (MilStop)	2 to 2.5 lb/100 gal 1 to 3 lb/100 gal 4 to 8 oz/100 gal	Apply every 10- to 14-days. Increase frequency to every 5 - 7 days under heavy disease pressure. Begin at first sign of disease. Repeat at 7-10 day intervals. See label for precautions for poinsettia, impatiens, and pansy. Repeat at 7- to 14-days intervals.
polyoxin D zinc salt (Endorse 2.5WP)	1.1 to 2.2 lb/100 gal	Apply as a foliar spray every 7 to 10 days. Apply prior to disease development and when conditions are conducive for disease.
tebuconazole (Torque)	4 to 8 fl oz/100 gal	Begin applications 14 to 21 days prior to when disease is expected, or at very first sign of disease.
thiophanate-methyl (AllBan Flo) (Cleary 3336) F, (OHP 6672) 50WP, (T-Storm) 50WSB (SysTec 1998) FL, (T-Storm Flowable)	10 to 20 fl oz/100 gal 12 to 24 oz/100 gal 10 fl oz/100 gal	Apply when disease first appears and repeat at 7- to 14-day intervals. Apply when disease first appears and repeat at 7- to 14-day intervals. Apply when disease first appears and repeat at 10- to 14-day intervals.
thiophanate-methyl + mancozeb (Zyban) 79W	4 bags/100 gal (24 oz/100 gal)	Repeat at weekly intervals.
triadimefon (Strike) 25WDG	2 to 4 oz/100 gal	Apply as needed at first sign of disease. Repeat at 14- to 21-day intervals. Not effective for powdery mildew control on verbena.
trifloxystrobin (Compass O) 50WDG	2 to 4 oz/100 gal	Good eradicant. Repeat at 7- to 14-day intervals. Rotate to another fungicide of nonstrobilurin chemistry after each Compass application. Use caution when applying to petunia, violets, and New Guinea impatiens due to possible phytotoxicity.
triflumizole (Terraguard SC)	4 to 16 oz/100 gal	Use 16 oz/100 gal for initial applications of existing infections. Use 4 to 8 oz/100 gal for subsequent applications and preventative sprays. Do not exceed 2 oz/100 gal for impatiens transplants. Repeat at 7- to 14-day intervals.
triticonazole (Trinity) 19SC	6 to 12 fl oz/100 gal	Use preventively. Begin applications when conditions favor fungal infection and before disease symptom development. Use of an adjuvant/spreader sticker can aid in control.
Rhizoctonia Aerial Blight		
azoxystrobin (Heritage)	1 to 4 oz/100 gal	Repeat at 7- to 28-day intervals. Do not make more than three sequential applications before switching to a nonstrobilurin fungicide.
azoxystrobin + benzovindiflupyr (Mural)	4 to 7 oz/100 gal	Apply every 7 to 14 days. Do not make more than two sequential applications before rotating to another class of fungicide that is not Group 7 or 11.
chlorothalonil (Chlorostar VI, Daconil Weather Stik, Echo 720, Manicure 6 Flowable) (Chlorothalonil DF, Chlorothalonil 720 SFT) (Chlorothalonil 500 ZN) (Daconil Ultrex) 82.5WDG (Echo Ultimate)	1 3/8 pt/100 gal See label 1.9 pt/100 gal 1.4 lb/100 gal 1.25 lb/100 gal	Apply when foliage and flowers are dry. Repeat at 7- to 14-day intervals. Apply to hydrangea foliage only. Avoid application during bloom period on plants where flower injury is unacceptable.
chlorothalonil + thiophanate-methyl (Spectro) 90WDG	1 to 2 lb/100 gal	Re-treat at a minimum of 7-day intervals.
cyprodinil + fludioxonil (Palladium)	2 to 6 oz/100 gal	Spray on a 7 to14 day interval while conditions are conducive to disease develepment. After two applications, alternate with another fungicide with a different MOA for two applications. Cautionary statement on label for applications to Geraniums, Impatiens and New Guinea Impatiens.
fludioxonil (Medallion)	1 to 2 oz/100 gal	Spray to runoff. Repeat at 7- to 14-day intervals until conditions no longer favor disease.
flutolanil (Contrast, Prostar) 70WSP	3 to 12 oz/100 gal	Apply at 14- to 21-day intervals.
iprodione (Chipco 26019 N/G) (18 Plus, Iprodione Pro 2SE)	1 to 2 lb/100 gal 1 to 2.5 qt/100 gal	Spray plants to ensure thorough coverage Repeat at 7- to 14-day intervals. Do not make more than four applications per crop per year. Do not use as a soil drench on impatiens or pothos. Do not use on Spathiphyllum.
iprodione + thiophanate-methyl (26/36)	33 to 84 fl oz/100 gal	Spray plants to ensure thorough coverage. Repeat at 7- to 14-day intervals. Do not make more than four applications per year. Do not use as a soil drench on impatiens or pothos. Do not use on Spathiphyllum.
pyraclostrobin (Empress Intrinsic)	1 to 3 fl oz/100 gal in propagation and 2 to 6 fl oz/100 gal for all other plants in production	Apply as a preventative drench prior to onset of disease. Can be reapplied 7 to 28 days following the intial application.
pyraclostrobin + boscalid (Pageant Intrinsic)	12 to 18 oz/100 gal	Apply prior to disease development. Repeat at 7- to 14 day intervals. Make no more than two sequential applications. Do not expose petunia or impatiens in flower or wintercreeper or nine bark to spray or drift as injury may occur.
polyoxin D zinc salt (Endorse 2.5WP)	1.1 to 2.2 lb/100 gal	Apply as a foliar spray every 7 to 10 days. Apply prior to disease development and when conditions are conducive for disease.
thiophanate-methyl (AllBan Flo) (Cleary 3336) F	10 to 14.5 fl oz/100 gal 12 to 16 fl oz/100 gal	Apply when disease symptoms first appear. Repeat at 7- to 14-day intervals.
thiophanate-methyl + flutolanil (SysStar WDG)	4 to 8 oz/100 gal	For best results apply before disease development.
triflumizole (Terraguard SC)	4 to 8 fl oz/100 gal	Make initial application prior to or at first sign of disease. Use the higher rate under heavy disease pressure. Repeat at 7- to 14-day intervals.

DISEASE CONTROL

Table 10-11. Disease Control of Annual, Perennial, Bedding, and Flowering Potted Plants in Greenhouses

Disease Pesticide and Formulation	Rate of Formulation	Schedule and Remarks
Rhizoctonia Stem and Root Rot		
azoxystrobin (Heritage)	1 to 4 oz/100 gal (spray) 0.9 oz/100 gal (drench)	Apply as a preventative spray or drench treatment. Repeat at 7 to 21 days. Do not exceed 2 oz/100 gal on impatiens.
azoxystrobin + benzovindiflupyr (Mural)	3 oz/100 gal	Apply 1 to 2 pints of drench solution per square foot surface area every 7 to 28 days.
Bacillus amyloliquefaciens strain D747 (DoubleNickel LC)	0.5 to 4.5 pt/100 gal	Apply as a drench or course spray to growing medium at or immediately before seeding. Repeat applications as needed. OMRI listed.
chlorothalonil + thiophanate-methyl (Spectro) 90WDG	1 to 2 lb/100 gal	Spray uniformly over area to be treated. Re-treat at 7-day intervals.
etridiazole + thiophanate-methyl (Banrot) 40W	4 to 8 oz/100 gal	Apply in sufficient volume to saturate the soil mixture. Irrigate immediately. Repeat at 4- to 12-week intervals if necessary.
fludioxonil (Medallion)	1 to 2 oz/100 gal	Apply as a drench at seeding or transplanting. Apply sufficient mix to wet the upper one-half of the growing medium. Make only one application to seedling crop. If needed, re-treat transplants 21 to 28 days after initial application. Do not apply as a seed or soil drench to impatiens or New Guinea impatiens. May cause stunting or chlorosis on some geranium cultivars.
fludioxonil + mefenoxam (Hurricane) 48 WP	See label	Apply as a pre-potting or growing media drench per label directions. Labeled for Rhizoctonia and Phytophthora/Pythium root rots and is best used when both diseases are present or suspected.
flutolanil (Contrast, Prostar) 70WSP	3 to 6 oz/100 gal	Apply drench at 2 oz per 4-inch pot. Repeat 21 to 28 days after initial application. Make no more than four applications per year.
Gliocladium virens strain GL-21	see label	Naturally occurring soil fungus. Do not use additional soil fungicides at time of incorporation. Avoid use in media totally void of organic matter. OMRI listed.
iprodione (Chipco 26019 N/G) (18 Plus, Iprodione Pro 2SE)	6.5 oz/100 gal 13 fl oz/100 gal	Apply 1 to 2 pints per sq ft at seeding or transplanting. Do not apply as a drench on impatiens or pothos. Repeat every 14 days. Do not make more than six applications per year. Do not use on Spathiphyllum.
iprodione + thiophanate-methyl (26/36)	13.5 fl oz/100 gal	Apply 1 to 2 pints per sq ft at seeding or transplanting. Do not apply as a drench on impatiens or pothos. Do not use on Spathiphyllum. Repeat every 14 days. Do not make more than 6 applications per year.
PCNB (Terraclor) 75WP	4 to 8 oz/100 gal	See label for amount to apply. One repeat application can be made 4 to 6 weeks later, if necessary.
polyoxin D zinc salt (Endorse 2.5WP) (Affrim)	1.1 to 2.2 lb/100 gal 0.25 to 0.5 lb/100 gal	Apply as a drench every 14 to 28 days.
pyraclostrobin (Empress Intrinsic)	1 to 3 fl oz/100 gal in propagation and 2 to 6 fl oz/100 gal for all other plants in production	Apply as a preventative drench prior to onset of disease. Can be reapplied 7 to 28 days following the intial application.
pyraclostrobin + boscalid (Pageant Intrinsic)	12 to 18 oz/100 gal	Apply prior to disease development. Repeat at 7- to 21-day intervals. Make no more than two sequential applications. Do not expose petunia or impatiens in flower or wintercreeper or nine bark to spray or drift as injury may occur.
thiophanate-methyl (AllBan Flo) (Cleary 3336) F (OHP 6672) 50WP, (T-Storm) 50WSB (OHP 6672) 4.5L, (T-Storm Flowable)	7.5 to 20 fl oz/100 gal 8 to 16 fl oz/100 gal 12 to 16 oz/100 gal 20 fl oz/100 gal	Apply as a heavy spray or drench at a rate of 0.5 to 2 pints per sq ft. Repeat at 4- to 8-week intervals. Apply 8 oz as a drench or directed spray after seeding or 12 to 16 oz after transplanting. Repeat at 21- to 28-day intervals. Apply as a drench at a rate of 1 to 3 pints per sq ft after transplanting. Repeat at 21- to 28-day intervals. Apply as a drench or heavy spray at a rate of 1 to 2 pints per sq ft after transplanting. Repeat at 2- to 4-week intervals.
thiophanate-methyl + flutolanil (SysStar WDG)	2 to 4 oz/100 gal	Apply according to label directions.
trifloxystrobin (Compass O) 50WDG	0.5 oz/100 gal	Apply as a drench to wet upper half of the growing media. Apply at seeding, again at transplanting, and at 21- to 28-day intervals thereafter. May injure petunia, violet, and New Guinea impatiens.
triflumizole (Terraguard) 50W	4 to 8 oz/100 gal	Apply as soil drench at 2- to 4-week intervals. Use higher rate under heavy disease pressure.
triticonazole (Trinity) 19SC	8 to 12 fl oz/100 gal	Use preventively. Begin applications when conditions favor fungal infection and before disease symptom development. The crown and base of the plant and the soil or potting medium surrounding the crown must be thoroughly covered.
Rusts		
azoxystrobin (Heritage)	1 to 4 oz/100 gal	Apply at 7- to 28-day intervals. Do not make more than 2 sequential applications of Heritage before alternating with a nonstrobilurin fungicide. Not effective for rust control on Hypericum. Rotate with mancozeb or triflumizole.
azoxystrobin + benzovindiflupyr (Mural)	4 to 7 oz/100 gal	Apply every 7 to 14 days. Do not make more than two sequential applications before rotating to another class of fungicide that is not Group 7 or 11.
Bacillus amyloliquefaciens strain D747 (DoubleNickel LC)	0.5 to 6 qt/100 gal	Begin preventative applications at plant emergence and repeat every 3 to 28 days depending on disease pressure. OMRI listed.
flutoanil (ProStar 70WP)	3 to 6 oz/100 gal	Use as foliar application and repeat every 14 to 21 days. Do not exceed 4 applications per growing season. Do not make more than two consecutive applications before switching to a non-Group 7 fungicide.
fuoxastrobin (Disarm 480SC)	1 to 4 fl oz/100 gal	Use for preventative applications only and apply every 7 to 28 days depending on disease pressure. Do not make more than two consecutive applications before switching to a non Group 11 fungicide.
chlorothalonil (Chlorostar VI, Daconil Weather Stik, Echo 720, Manicure 6 Flowable) (Chlorothalonil DF, Chlorothalonil 720 SFT) (Chlorothalonil 500 ZN) (Daconil Ultrex) 82.5WDG (Echo) 90DF, (Echo Ultimate)	1 3/8 pt/100 gal See label 1.9 pt/100 gal 1.4 lb/100 gal 1.25 lb/100 gal	Apply when foliage and flowers are dry. Repeat at 7- to 14-day intervals. Apply to hydrangea foliage only. Avoid application during bloom period on plants where flower injury is unacceptable.

Table 10-11. Disease Control of Annual, Perennial, Bedding, and Flowering Potted Plants in Greenhouses

Disease Pesticide and Formulation	Rate of Formulation	Schedule and Remarks
Rusts (continued)		
chlorothalonil + thiophanate-methyl (Spectro) 90WDG	1 to 2 lb/100 gal	Apply when foliage and flowers are dry, or nearly dry. Re-treat at a minimum of 7-day intervals.
mancozeb (Dithane) 75DF, (Fore) 80WSP, (Mancozeb DG) (Pentathlon DF) (Pentathlon LF) (Protect DF)	1.5 lb/100 gal 1 to 2 lb/100 gal 0.8 to 1.6 qt/100 gal 1 to 2 lb/100 gal	Begin at first sign of disease. Repeat at 7- to 10-day intervals. Begin application at first sign of disease. Repeat at 7- to 10-day intervals. Begin application at first sign of disease. Repeat at 7- to 10-day intervals. Apply at 7- to 21-day intervals.
mancozeb + myclobutanil (Clevis, MANhandle)	1 to 2 lb/100 gal	Apply at first sign of disease, and repeat at 7- to 10-day intervals. The addition of Latron B-1956 will improve performance. In a limited number of trials, gave very good to excellent control of rust on geranium (18 oz/100 gal) and snapdragon (1 lb/100 gal).
myclobutanil (Eagle) 20EW (Eagle) 40WP	6 to 12 fl oz/100 gal 3 to 6 oz/100 gal	Apply on a protectant application schedule at 10- to 14-day intervals. See label for rates to control white rust on chrysanthemum.
neem oil (Triact 70)	1 gal/100 to 200 gal	Apply at 7- to 14-day spray intervals. Trial first on open blooms. To control existing disease, apply on a 7-day schedule until disease pressure is eliminated. Not for impatiens, carnation, or hibiscus.
oxycarboxin (Plantvax) 75W	16 to 24 oz/100 gal	Apply at first sign of disease. Repeat at 2-week intervals for a maximum of two to four applications per season.
pyraclostrobin (Insignia)	See label	Apply prior to onset of disease and repeat every 7 to 14 days. Do not expose flowering impatiens or flowering petunias to Insignia.
pyraclostrobin + boscalid (Pageant Intrinsic)	6 to 12 oz/100 gal	Apply prior to disease development. Repeat at 7- to 10-day intervals. Make no more than two sequential applications. Do not expose petunia or impatiens in flower or wintercreeper or nine bark to spray or drift as injury may occur.
thiophanate-methyl (AllBan Flo) (Cleary 3336) F	10.75 to 20 fl oz/100 gal 12 to 16 fl oz/100 gal	Begin applications at first sign of disease. Repeat at 7- to 14-day intervals.
triadimephon (Strike) 25WDG	2 to 4 oz/100 gal	Spray to the point of drip as needed.
triflumizole (Terraguard) 50WP	2 to 8 oz/100 gal	Apply prior to, or at first sign of disease. Repeat at 7- to 14-day intervals.
Scab, Poinsettia *(Sphaceloma)*		
azoxystrobin (Heritage)	1 to 4 oz/100 gal	Apply at 7- to 21-day intervals. Test for phytotoxicity prior to treating entire crop. Do not make more than three sequential applications of Heritage before alternating with a nonstrobilurin fungicide.
azoxystrobin + benzovindiflupyr (Mural)	4 to 7 oz/100 gal	Apply every 7 to 14 days. Do not make more than two sequential applications before rotating to another class of fungicide that is not Group 7 or 11.
copper sulfate pentahydrate (Phyton 27)	2.0 to 3.5 oz/100 gal	Apply at 7-day intervals.
mancozeb (Dithane) 75DF, (Fore) 80WSP, (Mancozeb DG) (Protect DF)	1.5 lb/100 gal 1 to 2 lb/100 gal	Spray at first sign of disease. Apply at 7- to 21-day intervals. May leave a residue.
mancozeb + myclobutanil (Clevis, MANhandle)	2 lb/100 gal	Apply at first sign of disease. The addition of Latron B-1956 will improve performance.
myclobutanil (Eagle) 20EW (Eagle) 40WP	6 to 12 fl oz/100 gal 3 to 6 oz/100 gal	Retreat at 10- to 14-day intervals.
triticonazole (Trinity) 19SC	6 to 12 fl oz/100 gal	Use preventively. Begin applications when conditions favor fungal infection and before disease symptom development.
Sclerotinia Blight *(Sclerotinia sclerotiorum)*		
chlorothalonil (Chlorothalonil DF) (Chlorothalonil 500 ZN) (Chlorothalonil 720 SFT) (Chlorostar VI, Daconil Weather Stik, Echo 720, Manicure 6 Flowable) (Daconil Ultrex) 82.5WDG (Echo) 90DF, (Echo Ultimate)	See label 1.9 pt/100 gal 1.37 pt/100 gal 1 3/8 pt/100 gal 1.4 lb/100 gal 1.25 lb/100 gal	Apply at 7- to 14-day intervals. Apply when foliage and flowers are dry, or nearly dry. Applications made during bloom may damage flowers.
chlorothalonil + thiophanate-methyl (Spectro) 90WDG	1 to 2 lb/100 gal	Apply when both foliage and flowers are dry, or nearly dry. Repeat at 7-day intervals. Do not use on Swedish ivy, Boston fern, or Easter cactus.
cyprodinil + fludioxonil (Palladium)	2 to 6 oz/100 gal	Spray on a 7-14 day interval while conditions are conducive to disease develepment. After two applications, alternate with another fungicide with a different MOA for two applications.
fenhexamid (Decree) 50WDG	0.75 to 1.5 lb/100 gal	Apply at 7- to 14-day intervals. Maintain agitation during application. May cause phytotoxicity on poinsettia bracts.
fuoxastrobin (Disarm 480SC)	2 to 4 fl oz/100 gal	Use for preventative applications only and apply every 7 to 21 days depending on disease pressure. Do not make more than two consecutive applications before switching to a non Group 11 fungicide.
PCNB (Terraclor) 75WP	See label	Apply as a drench or bulb soak according to label directions.
pyraclostrobin + boscalid (Pageant Intrinsic)	12 to 18 oz/100 gal	Apply prior to disease development. Repeat at 7- to 14-day intervals. Make no more than two sequential applications. Do not expose petunia or impatiens in flower or wintercreeper or nine bark to spray or drift as injury may occur.
thiophanate-methyl (AllBan Flo) (Cleary 3336) F (OHP 6672) 50WP, (T-Storm) 50WSB (OHP 6672) 4.5L, (T-Storm Flowable) (SysTec 1998) FL	7.5 to 20 fl oz/100 gal 8 to 16 fl oz/100 gal 12 to 16 fl oz/100 gal 20 fl oz/100 gal 10 to 20 fl oz/100 gal	Repeat at 4- to 8-week intervals. Repeat at 21- to 28-day intervals. Repeat at 21- to 28-day intervals. Repeat at 2- to 4-week intervals. Repeat at 2- to 4-week intervals.

479

DISEASE CONTROL

Table 10-11. Disease Control of Annual, Perennial, Bedding, and Flowering Potted Plants in Greenhouses

Disease Pesticide and Formulation	Rate of Formulation	Schedule and Remarks
Sclerotinia Blight *(Sclerotinia sclerotiorum)* (continued)		
triticonazole (Trinity) 19SC	8 to 12 fl oz/100 gal	Use preventively. Begin applications when conditions favor fungal infection and before disease symptom development. The stem areas of the plant must be thoroughly covered using spray to runoff.
Thielaviopsis Root Rot: *See* **Black root rot.**		

Further Information — Plant Pathology Disease Notes may be accessed at: www.ces.ncsu.edu/depts/pp/notes/Ornamentals/Ornamental_contents.html. Printed copies are also available at each county Cooperative Extension Center.

Disease Control for Forest, Christmas, and Ornamental Trees

Anyone using any agricultural chemical should refer to the current chemical label, which contains information about the safe and effective use of the chemical, before using the chemical.

Table 10-12. Disease Control for Forest, Christmas, and Ornamental Trees

CROP Disease	Material	Rate	Method	Schedule	Remarks
Ash					
Anthracnose	thiophanate-methyl (AllBan Flo) (Cleary 3336)	10.75 to 20 fl oz/100 gal 12 to 16 fl oz/100 gal	foliar spray	Three to four applications at 14-day intervals.	First application at bud break.
	chlorothalonil (Daconil Ultrex) 82.5 WDG	1.4 lb/100 gal	foliar spray	Repeat in 7 to 14 days when conditions favor disease.	
	metconazole (Tourney)	1 to 4 oz/100 gal	foliar spray	Repeat in 14 to 28 days when conditions favor disease.	
	mancozeb (Dithane 75DF)	1 to 2 lb/100 gal	foliar spray	Repeat at 7 to 10 day intervals.	Begin at first sign of disease
	pyraclostrobin (Insignia)	8 to 16 oz/100gal	foliar spray	Repeat at 7 to 14 day intervals as needed	Use for preventative applications only. Do not make more than two consecutive applications before switching to a non Group 11 fungicide.
	pyraclostrobin + boscalid (Pageant Intrinsic)	18 oz./100 gal	foliar spray	Repeat applications 7 to14 days as needed	
	tebuconazole (Torque)	4 to 8 fl oz/100 gal	foliar spray	Begin applications 14 to 21 days prior to when disease is expected, or at very first sign of disease.	
Crabapple					
Fire blight	acibenzolar-s-methyl (Actigard 50WG)	see label	see label	Apply 2-3 applications between 20% bloom and petal fall.	Systemic acquired inducer. Apply in tank mixture with streptomycin.
	Bacillus amyloliquefaciens strain D747 (DoubleNickel LC)	0.5 to 6 qt/100 gal	foliar spray	Apply 2 to 3 applications between 20% bloom and petal fall.	Systemic acquired inducer. Apply in tank mixture with streptomycin or a copper product. OMRI listed.
	Bacillus subtilis (Serenade Max)	2 to 3 lb/A	Foliar spray	See label	Begin applications at 5% bloom and continue to protect open, untreated blossoms. OMRI listed.
	copper oxychloride + copper hydroxide (Badge SC) (Badge X2)	1.5 to 6 pt/A 1.5 to 5 lb/A	foliar spray	Repeat applications every 7 to 10 days until approximately 1 month after petal fall.	Can be phytotoxic after bud break. Use 4.5 to 6 pt/A (or 5 lb/A) rate one time only prior to the tight cluster stage. Use 1.5pt/A (or 1.5 to 2 lb/A) rate from after tight cluster. Badge X2 is OMRI listed.
	Reynautria sachalinensis extract	1 to 2 qt/A	foliar spray	3 to 7 days starting at petal fall if conditions favorable for shoot blight infection	For suppression of fire blight only. Use prior to predicted infection events.
	streptomycin (Agri-mycin 17)	50 to 100 ppm	foliar spray	Three to five applications starting at 20% to 30% bloom. Refer to forecasting models.	Spray every 10 to 14 days.
Rusts (cedar apple rust, quince rust, hawthorn rust)	mancozeb (Dithane 75DF)	1 to 2 lb/100 gal	foliar spray	Repeat at 7 to 10 day intervals.	Begin preventatively at tight cluster and continue through petal fall. Monitor throughout summer.
	cyprodinil + difenoconazole (Inspire Super)	12 fl oz/A	Foliar spray	Repeat at 7 to 10 day intervals.	Begin preventative treatments at tight cluster stage and continue through petal fall. Monitor throughout summer.
	flutriafol (Rhyme)	4 to 6 fl oz/A	Foliar spray	Repeat at 10 day intervals.	Begin preventative treatments at tight cluster stage and continue through petal fall. Monitor throughout summer.
	propiconazole (Topaz)	See label	Foliar spray	See label	Begin preventative treatments at tight cluster stage and continue through petal fall. Monitor throughout summer.

Table 10-12. Disease Control for Forest, Christmas, and Ornamental Trees

CROP Disease	Material	Rate	Method	Schedule	Remarks
Crabapple (continued)					
Rusts (cedar apple rust, quince rust, hawthorn rust) (continued)	tebuconazole (Toledo 45WP)	4 to 8 oz/A	foliar spray	Repeat applications 7 to 10 days as needed	Begin preventative treatments at tight cluster/pink stage and continue through petal fall. Monitor throughout summer.
	triflumazole (Procure 480SC)	8 to 16 fl oz/A	foliar spray	Every 7 to 10 days.	Begin preventative treatments at tight cluster/pink stage and continue through petal fall. Monitor throughout summer.
Apple scab, powdery mildew (rates and timings are given for apple scab, see label for powdery mildew).	benzovindiflupyr (Aprovia)	5.5 to 7 fl oz/A	foliar spray	Apply every 7 to 10 days	Begin preventative treatments starting at tight cluster and continue through petal fall. Do not make more than two consecutive applications before switching to a non Group 7 fungicide.
	flutriafol (Topguard)	13 fl oz/A	foliar spray	Apply every 7 to 10 days	Begin preventative treatments starting at tight cluster and continue through petal fall. Monitor for disease throughout summer.
	fuxapyroxad + pyraclostrobin	4 to 5.5 fl oz/A	Foliar sptray	Apply every 7 to 10 days	Begin preventative treatments starting at tight cluster and continue through petal fall. Do not make more than two consecutive applications before switching to a non Group 7 or non-Group 11 fungicide.
	Myclobutanil (Rally 40WSP)	3 oz/50 gal	foliar spray	Apply every 10 to 14 days.	Begin preventative treatments starting at tight cluster and continue through petal fall. Monitor for disease throughout summer.
	propiconazole (Banner MAXX)	2 to 8 fl oz/ 100 gal	foliar spray	Every 14 to 21 days as needed.	Begin preventative treatments starting at tight cluster and continue through petal fall. Monitor for disease throughout summer.
	chlorothalonil + propiconazole (Concert II)	9 to 35 fl oz/100 gal (see label for specific rate instructions)	foliar spray	Every 14 to 21 days	See label for additional details
	pyraclostrobin + boscalid (Pageant Intrinsic)	6 to 12 oz./100 gal	foliar spray	Repeat applications 7 to14 days as needed	
	tebuconazole (Torque)	4 to 8 fl oz/100 gal	foliar spray	Begin applications 14 to 21 days prior to when disease is expected, or at very first sign of disease.	
	trifloxystrobin (Flint)	2 to 2.5 oz/A	foliar spray	Every 7 to 14 days.	Begin preventative treatments starting at tight cluster and continue through petal fall. Do not make more than two consecutive applications before switching to a non Group 7 fungicide
Crapemyrtle					
Cercospora leaf spot	azoxystrobin (Heritage) 50 W	4 oz/100 gal	foliar spray	Spray every 7 to 28 days.	
	azoxystrobin + benzovindiflupyr (Mural)	4 to 7 oz/100 gal	Foliar spray	Apply every 7 to 21 days.	Do not make more than two sequential applications before rotating to another class of fungicide that is not Group 7 or 11.
	fluoxstrobin (Disarm 480SC)	1 to 4 fl oz/100 gal	Foliar spray	Every 7 to 28 days	
	horticultural oil (Ultra-Pure Oil)	See label	foliar spray	See label.	
	metconazole (Tourney)	1 to 4 oz/100 gal	foliar spray	Repeat in 14 to 28 days when conditions favor disease.	
	pyraclostrobin + boscalid (Pageant Intrinsic)	8 to 12 oz./100 gal	foliar spray	Repeat applications 7 to14 days as needed	
	triticonazole (Trinity) 19SC	4 to 8 fl oz/100 gal	foliar spray	Every 7 to 14 days.	Use preventively. Begin applications when conditions favor fungal infection and before disease symptom development. Use of an djuvant/spreader sticker can aid in control.
Diplodia tip blight, Lophodermium needlecast, Swiss needlecast	azoxystrobin (Heritage) 50 WG	3.2 to 8 fl oz/100 gal	foliar spray	Spray every 7 to 28 days.	To avoid fungicide resistance, do not make more than three sequential applications before rotating with nonstrobilurin products.
Powdery mildew	azoxystrobin (Heritage) 50 W	4 oz/100 gal	foliar spray	Preventative sprays at 7- to 28-day intervals.	
	chlorothalonil + propiconazole (Concert II)	22 – 35 fl oz/100 gal	foliar spray	Every 14 to 21 days	See label for details
	Fenarimol (Rubigan A.S.)	6 to 10 fl oz	Foliar spray	Apply every 10 to 14 days as needed.	
	Metconazole (Tourney)	1 to 4 oz/100 gal	foliar spray	Repeat in 14 to 28 days when conditions favor disease.	
	Myclobutanil (Eagl 20EW)	6 to 12 fl oz/100 gal	foliar spray	Apply at 10- to 14-day intervals.	
	Potassium bicarbonate (MilStop)	2 to 5 lb/A	Foliar spray	Apply every 7 to 14 days	OMRI listed

Table 10-12. Disease Control for Forest, Christmas, and Ornamental Trees

CROP Disease	Material	Rate	Method	Schedule	Remarks
Crapemyrtle (continued)					
Powdery mildew (continued)	propiconazole (Banner Maxx)	5 to 8 fl oz/100 gal	Foliar spray	See label	
	propiconazole + chlorothalonil (Concert II)	22-35 fl oz/100 gal	Foliar spray	Repeat every 14 days as needed	
	pyraclostrobin + boscalid (Pageant Intrinsic)	6 to 12 oz./100 gal	foliar spray	Repeat applications 7 to14 days as needed	
	Sulfur (Micro Sulf)	3 to 10 lb/100 gal	Foliar spray	Repeat every 5 to 10 days	Begin with onset of disease
	triadimefon (Bayleton 50)	See label.	Foliar spray	See label	
	triticonazole (Trinity) 19SC	6 to 12 fl oz/100 gal	foliar spray	Every 7 to 14 days.	Use preventively. Begin applications when conditions favor fungal infection and before disease symptom development. Use of an adjuvant/spreader sticker can aid in control.
Root rot (*Phytophthora* spp.)	cyazofamid (Segway)	3 to 6 fl oz/100 gal	soil drench	14 to 21 day intervals using another registered fungicide with a different mode of action.	
	fosetyl-AL (Aliette) WDG	2.5 lb/100 gal 2.5 to 5 lb/100 gal	dip foliar (spray to wet)	Before transplanting 30-day minimum interval	Thoroughly wet plant and root mass immediately before transplanting.
	mefenoxam (Subdue MAXX)	Seedlings: 1.25 pt/50 gal 2-0 transplants: 2.5 pt/50 gal	directed soil spray over beds	Spring and fall (every 6 months).	Apply 0.5 to 1 inch of water after application, if rain is not expected within 3 days.
	phosphorous acid (Alude, Fosphite, Reliant)	See label for rates	Apply as a soil drench or foliar spray as a preventive		
	pyraclostrobin + boscalid (Pageant Intrinsic)	12 to 18 oz/100 gal	Soil drench	Repeat applications 14 to 28 days as needed	
Dogwood					
Anthracnose (*Discula*)	azoxystrobin (Heritage) 50 WG	1 to 4 oz/100 gal	foliar spray	Spray every 7 to 28 days.	To avoid fungicide resistance, do not make more than three sequential applications before rotating with nonstrobilurin products.
	chlorothalonil (Daconil Ultrex) 82.5 WDG	1.4 lb/100 gal	foliar spray	Repeat every 7 to 14 days when conditions favor disease.	Prune out all diseased tissue. Several applications in fall before leaf drop may also be advisable.
	chlorothalonil + propiconazole (Concert II)	9 to 17 fl oz/100 gal 35 fl oz/100 gal	foliar spray	Apply every 14 days Apply every 28 days	See label for additional details.
	chlorothalonil + thiophanate-methyl (Spectro 90WDG)	1 to 2 lb/100 gal	foliar spray	Apply every 7 to 14 days	See label for additional details.
	mancozeb (Dithane, Fore, Mancozeb DG) (Protect DF)	1.5 lb/100 gal 1 to 2 lb	foliar spray	Spray every 10 to 14 days from bud break until mid-summer.	Prune out all diseased tissue. Several applications in fall before leaf drop may also be advisable.
	metconazole (Tourney)	1 to 4 oz/100 gal	foliar spray	Repeat in 14 to 28 days when conditions favor disease.	
	myclobutanil (Systhane) (Immunox) (Eagle 20EW)	4 oz/100 gal 1 fl oz/gal 6 to 12 fl oz/100 gal	foliar spray	Spray every 10 to 14 days. Spray every 14 days. Spray every 10 to 14 days	Prune out all diseased tissue. Several applications in fall before leaf drop may also be advisable.
	propiconazole (Banner MAXX, Banner MAXX II)	2 to 8 fl oz/100 gal	foliar spray	Apply every 14 to 28 days from bud break to mid-summer.	Prune out all diseased tissue. Several applications in fall before leaf drop may also be advisable. Apply 8 fl oz/100 gal rate every 28 days, or 2 to 4 fl oz/100 gal rate every 14 days.
	tebuconazole (Torque)	4 to 8 fl oz/100 gal	foliar spray	Begin applications 14 to 21 days prior to when disease is expected, or at very first sign of disease.	
	thiophanate-methyl + propiconazole (Protocol)	4 to 8 fl oz/100 gal 16 fl oz/100 gal	foliar spray	Apply every 14 days Apply every 28 days	See label for additional application instructions.
	trifloxystrobin (Compass O)	2 to 4 oz/100 gal	foliar spray	Apply every 7 to 14 days	Do not make more than two consecutive applications before switcthing to a non-Group 11 fungicide.
Powdery mildew	azoxystrobin (Heritage) 50 WG	1 to 4 oz/100 gal	foliar spray	Spray every 7 to 28 days as needed.	Do not make more than two sequential applications of Heritage before rotating with nonstrobilurin products to avoid fungicide resistance. See label.
	azoxystrobin + benzovindiflupyr (Mural)	4 to 7 oz/100 gal	foliar spray	Apply every 7 to 21 days	Do not make more than 2 sequential applications before rotating to a non-Group 7 or non-Group 11 fungicide.
	chlorothalonil + propiconazole (Concert II)	22 – 35 fl oz/100 gal	foliar spray	Every 14 to 21 days	See label for additional details
	fenarimol (Rubigan A.S.)	8 fl oz/100 gal	foliar spray	Apply every 10 to 14 days	

Table 10-12. Disease Control for Forest, Christmas, and Ornamental Trees

CROP Disease	Material	Rate	Method	Schedule	Remarks
Dogwood (continued)					
Powdery mildew (continued)	metconazole (Tourney)	1 to 4 oz/100 gal	foliar spray	Repeat in 14 to 28 days when conditions favor disease.	
	myclobutanil (Eagle) 40 WSP	3.0 oz/50 to100 gal	foliar spray	Spray every 21 days.	
	petroleum distillate: horticultural oil (Sunspray) 11 E	1 gal/100 gal	foliar spray	Spray at 14-day intervals.	
	potassium bicarbonate (MilStop)	2 to 5 lb/A	foliar spray	Apply at 7 to 14 day intervals	Begin applications at first sign of disease. See label for tank mixing considerations. OMRI listed.
	propiconazole (Banner MAXX, Banner MAXXII)	5 to 8 fl oz/100 gal	foliar spray	Spray every 21 days in spring.	
	pyraclostrobin + boscalid (Pageant Intrinsic)	6 to 12 oz/100 gal	foliar spray	Repeat applications 7 to14 days as needed	
	sulfur (Micro Sulf)	3 to 10 lb/100 gal	foliar spray	Apply every 5 to 10 days	Begin when disease first appears
	tebuconazole (Torque)	4 to 8 fl oz/100 gal	foliar spray	Begin applications 14 to 21 days prior to when disease is expected, or at very first sign of disease.	
	triadimefon (Bayleton FLO)	5.5 fl oz/275 to 550 gal	foliar spray	Repeat applications as needed	
	trifloxystrobin (Compass O)	1 to 2 oz/100 gal	foliar spray	Repeat at 7- to 14-day intervals.	Do not make more than two consecutive applications before switcthing to a non-Group 11 fungicide.
	thiophanate-methyl + chlorothalonil (Spectro) 90 WDG	2 lb/100 gal	foliar spray	Minimum repeat interval is 7 days.	Protective and curative activity.
	thiophanate-methyl + propiconazole (Protocol)	10-16 fl oz/100 gal	foliar spray	Apply every 14 days Apply every 28 days	
	triflumizol (Terraguard SC)	16 fl oz/100 gal 4 to 8 fl oz/100 gal	foliar spray	Make initial application prior to or at first sign of disease. Repeat at 7- to 14-day intervals.	Use 16 fl oz/100 gal rate only for initial application if disease is present. For subsequent and protective applications use lower rates.
	triticonazole (Trinity) 19SC	6 to 12 fl oz/100 gal	foliar spray	Spray every 7 to 14 days.	Use preventively. Begin applications when conditions favor fungal infection and before disease symptom development. Use of an adjuvant/spreader sticker can aid in control.
Root rot (*Phytophthora* spp.)	cyazofamid (Segway)	3 to 6 fl oz/100 gal	soil drench	14 to 21 day intervals using another registered fungicide with a different mode of action.	
	fosetyl-AL (Aliette) WDG	2.5 lb/100 gal 2.5 to 5 lb/100 gal	dip foliar (spray to wet)	Before transplanting 30-day minimum interval	
	mefenoxam (Fenox ME, Mefenoxam 2, Subdue MAXX)	See label.	Can be applied as a drench or soil surface spray. Consult label for specific crops and applications.	Every 2 to 3 months.	FUNGICIDE RESISTANCE IS POSSIBLE. Do not apply rates of 2 fl oz more often than every 10 weeks.
	metalaxyl (Metalazyl 2E AG)	1 to 4 fl oz/100 gal	drench	Repeat every 2 to 3 months	Apply 1 pt of solution per square foot
	oxathiapiprolin (Segovis)	See label	drench	See label.	Do not make more than two consecutive applications before switching to a fungicide with a different mode of action.
	phosphorous acid (Alude, Fosphite, Reliant)	See label for rates	Apply as a soil drench or foliar spray as a preventive.	See label	
	pyraclostrobin + boscalid (Pageant Intrinsic)	12 to 18 oz./100 gal	soil drench	Repeat applications 14 to 28 days as needed	
Spot anthracnose (*Elsinoe*) Septoria leaf spot	azoxystrobin (Heritage) 50 WG	1 to 4 oz/100 gal	foliar spray	Spray every 7 to 28 days as needed.	Do not make more than three sequential applications of Heritage before rotating with nonstrobilurin products to avoid fungicide resistance. See label.
	azoxystrobin + benzovindiflupyr (Mural)	4 to 7 oz/100 gal	foliar spray	Apply every 7 to 21 days	Do not make more than 2 sequential applications before rotating to a non-Group 7 or non-Group 11 fungicide.
	chlorothalonil (Daconil Ultrex)	1.4 lb/100 gal	foliar spray	Apply every 7 to 14 days as needed.	

483

DISEASE CONTROL

Table 10-12. Disease Control for Forest, Christmas, and Ornamental Trees

CROP Disease	Material	Rate	Method	Schedule	Remarks
Dogwood (continued)					
Spot anthracnose (*Elsinoe*) Septoria leaf spot (continued)	mancozeb (Dithane, Fore, Mancozeb DG) (Protect DF)	1.5 lb/100 gal OR 2.5 tsp/gal OR 1 to 2 lb	foliar spray	First spray as buds break in spring. Second as petals fall. Third in midsummer. Fourth when predormant (after flower buds are well formed).	See label as rate varies by product.
	myclobutanil (Immunox) (Eagle 20EW)	1 fl oz/gal 6 to 12 fl oz/100 gal	foliar spray	Spray every 14 days. Spray every 10 to 14 days.	
	potassium bicarbonate (MilStop)	2 to 5 lb/A	foliar spray	Apply at 7 to 14 day intervals	Begin applications at first sign of disease. See label for tank mixing considerations. OMRI listed.
	pyraclostrobin + boscalid (Pageant Intrinsic)	8 to 12 oz./100 gal	foliar spray	Repeat applications 7 to14 days as needed	
	tebuconazole (Torque)	4 to 8 fl oz/100 gal	foliar spray	Begin applications 14 to 21 days prior to when disease is expected, or at very first sign of disease.	
Eastern Cedar					
Annosus root rot (*Fomes annosus*)	See PINE				
Phomopsis needle blight	azoxystrobin (Heritage) 50 WG	1 to 4 oz/100 gal	foliar spray	Spray every 7 to 28 days as needed.	Do not make more than three sequential applications of Heritage before rotating with nonstrobilurin products to avoid fungicide resistance. See label.
	thiophanate methyl (AllBan Flo) (Cleary 3336) F	14.5 to 20 fl oz/100 gal 12 to 24 fl oz/100 gal	foliar spray	Beginning when disease appears or during suspected periods of disease incidence, apply every 7 to 14 days.	
Elm					
Dutch elm disease (*Ophiostoma ulmi*)	All treatments listed must be followed for effective **prevention** of disease on highly valued trees. 1. Sanitation—cut down and destroy diseased trees and dead limbs.				
	2. Elm bark beetle control.				
	3. SMDC (Vapam) - kills root grafts	1 gal SMCD/ 3 gal water 6 oz/hole	Pour in 1-in. diameter holes 15 in. deep, spaced 6 to 9 in. apart in a line between healthy and diseased trees	Apply with first appearance of disease.	Not closer than 20 ft from healthy tree. Soil temperature above 50 degrees F. Professional applicators only.
	4. Systemic chemical prevention: propiconazole (Alamo) thiabendazole (Arbortect 20S)	See label	root flare injection	See label	
	5. Therapeutic treatment: propiconazole (Alamo) thiabendazole (Arbortect)	See label		See label	
Fraser Fir					
Botrytis seedling blight	chlorothalonil (Bravo Ultrex)	1.4 to 2.5 lb/A	foliar	Apply every 7 to 14 days	Make additional applications to nursery beds when seedlings are 4 inches tall at 7- to 14-day intervals as long as favorable conditions persist.
	metconazole (Tourney)	1 to 4 oz/100 gal	foliar spray	Repeat in 14 to 28 days when conditions favor disease.	
	thiophanate methyl (Cleary 3336) F	12 to 16 fl oz/100 gal	foliar	Apply at first sign of disease.	Tank mix combination with chlorothalonil is recommended.
Damping-off, postplant	mefenoxam (Subdue MAXX) + thiophanate methyl (Cleary 3336) F	0.25 fl oz + 12 fl oz/100 gal apply 2 pt/sq ft	drench	Apply at first sign of disease.	
	metconazole (Tourney)	1 to 4 oz/100 gal	foliar spray	Repeat in 14 to 28 days when conditions favor disease.	
Diplodia tip blight, Lophodermium needlecast, Swiss needlecast	azoxystrobin (Heritage) 50 WG IF GROWN AS ORNAMENTAL	3.2 to 8 fl oz/acre	foliar spray	Spray every 7 to 28 days.	To avoid fungicide resistance, do not make more than two sequential applications before rotating with nonstrobilurin products.
	azoxystrobin ((Aframe) IF GROWN AS CHRISTMAS TREE	6 to 15.5 fl oz/acre	foliar spray	Spray every 7 to 21 days	To avoid fungicide resistance, do not make more than two sequential applications before rotating with nonstrobilurin products.

DISEASE CONTROL

Table 10-12. Disease Control for Forest, Christmas, and Ornamental Trees

CROP Disease	Material	Rate	Method	Schedule	Remarks
Fraser Fir (continued)					
Phytophthora root rot	cyazofamid (Segway)	3 to 6 fl oz/100 gal	soil drench	14 to 21 day intervals using another registered fungicide with a different mode of action.	Irrigate with at least ½ inch of water if rainfall does not occur within 24 hrs. For container plants, check label for recommended maximum drench volume based on pot diameter.
	fosetyl-AL (Aliette WDG)	2.5 lb/100 gal	dip	Before transplanting	Thoroughly wet plant and root mass immediately before transplanting.
		2.5 to 5 lb/100 gal/acre	foliar (spray to wet)	30-day minimum interval	Field-grown trees in plantations.
	mefenoxam (Subdue MAXX)	1.5 pt/acre/50 gal/ acre 2.5 pt/acre/50 gal/acre 0.63 to 1.25 gal/50 gal/acre	drench drench directed soil spray	MAXX: May and September	Do not apply to fir growing on bottomlands or poorly drained soils, or near surface water. Seed beds or plug plantings. 2-0 transplants. Field-grown trees in plantations.
	(Subdue GR)	6 to 30 lb/acre 16 to 20 lb/acre 50 to 250 lb/acre	broadcast broadcast broadcast	GR: once in spring and again in fall	Seed beds. 2-0 transplants. Field-grown trees in plantations. Apply 0.5 to 1 inch water after application if rain is not expected within 3 days.
	phosphorous acid (Alude, Fosphite, Reliant)	See label for rates	Soil drench or foliar spray.		Apply as a soil drench or foliar spray as a preventive
Hemlock					
Twig rust (*Melampsora farlowii*)	triadimefon (Bayleton) 50 WSP	5.5 oz/68.75 to 137.5 gal	foliar spray	Begin at bud break and continue every 14 days until growth stops.	
Leyland Cypress					
Needle blight (*Passalora, Cercosporidium*)	chlorothalonil (Daconil Ultrex)	1.4 lb/100 gal	foliar spray	7 to 10 days intervals	Begin scouting last year's infection sites for sporulation (tufts of olive green spores) in mid-May to mid-June. At the first sign of sporulation, make 2 applications of a systemic fungicide at 14-day intervals. Follow with 1 or 2 applications of a protectant fungicide (Daconil, Fore, Kocide, Badge, Rainshield) applied at 7-day intervals. In early August, scout trees for new infections. If sporulation is observed, re-treat with 1 application of a systemic fungicide 14 days later with 2 applications of a protectant applied at 7-day intervals.
	copper hydroxide (Kocide 2000)	0.75 lb/100 gal	foliar spray	7 to 14 day intervals	
	copper hydroxide + copper oxychloride (Badge SC)	3 to 6 pt/A	foliar spray	7 to 14 day intervals	
	mancozeb (Fore) WSP (Rainshield)	1.5 lb/100 gal	foliar spray	7 to 10 day intervals	
Diplodia tip blight, Lophodermium needlecast, Swiss needlecast	azoxystrobin (Heritage) 50 WG IF GROWN AS ORNAMENTAL	3.2 to 8 oz/acre	foliar spray	Spray every 7 to 28 days.	To avoid fungicide resistance, do not make more than two sequential applications before rotating with nonstrobilurin products.
	azoxystrobin (Aframe) IF GROWN AS CHRISTMAS TREE	6 to 15.5 fl oz/acre	foliar spray	Spray every 7 to 21 days	To avoid fungicide resistance, do not make more than two sequential applications before rotating with nonstrobilurin products. Quadris is only for Christmas tree applications.
Longleaf Pine					
Brown spot (*Scirrhia acicola*)	Bordeaux mixture (copper sulfate, lime, and water 8-8-100)	60 gal/acre	foliar spray	Spray at 10- to 14-day interval after emergence of seedlings until July 1.	
Maple					
Anthracnose	azoxystrobin (Heritage) 50 WG	1 to 4 oz/100 gal	foliar spray	Two to three applications at 14 to 28 days.	First application at bud break. Do not make more than three sequential applications of Heritage before rotating with nonstrobilurin products to avoid fungicide resistance. See label.
	chlorothalonil (Echo 720)	1.375 pt/100 gal	foliar spray	Repeat every 7 to 14 days	
	pyraclostrobin (Insignia)	8 to 16 oz/100 gal	foliar spray	Apply every 7 to 14 days	Do not make more than two sequential applications before switching to a non-Group 11 fungicides
	pyraclostrobin + boscalid (Pageant Intrinsic)	8 to 12 oz./100 gal	foliar spray	Repeat applications 7 to14 days as needed	
	thiophanate methyl (AllBan Flo) (Cleary 3336) F	10.75 to 20 fl oz/100 gal 12 to 16 fl oz/100 gal	foliar spray	Three to four applications at 10- to 14-day intervals.	Make first application at bud break.
	triticonazole (Trinity) 19SC	8 to 12 fl oz/100 gal	foliar spray	Spray every 7 to 14 days.	Use preventively. Begin applications when conditions favor fungal infection and before disease symptom development.
Oak					
Anthracnose	See MAPLE				
Leaf spot	propiconazole (Banner MAXX, Banner MAXX II)	16 fl oz/100 gal	foliar spray	Apply every 14 to 28 days.	Apply as needed.
	copper hydroxide + copper oxychloride (Badge SC)	1.5 to 2 pt/A	foliar spray	7 to 14 day intervals	See label for use of higher rates.
	metconazole (Tourney)	1 to 4 oz/100 gal	foliar spray	Repeat in 14 to 28 days when conditions favor disease.	

DISEASE CONTROL

Table 10-12. Disease Control for Forest, Christmas, and Ornamental Trees

CROP Disease	Material	Rate	Method	Schedule	Remarks
Oak (continued)					
Leaf spot (continued)	propiconazole (Bumper ES)	6 fl oz/100 gal	Foliar spray	Repeat every 14 to 28 days when conditions favor disease.	
	pyraclostrobin (Insignia)	8 to 16 oz/100 gal	foliar spray	Apply every 7 to 14 days	Do not make more than two sequential applications before switching to a non-Group 11 fungicides
	pyraclostrobin + boscalid (Pageant Intrinsic)	8 to 12 oz./100 gal	foliar spray	Repeat applications 7 to14 days as needed	
	triticonazole (Trinity) 19SC	4 to 8 fl oz/100 gal	foliar spray	Spray every 7 to 14 days.	Use preventively. Begin applications when conditions favor fungal infection and before disease symptom development. Use of an adjuvant/spreader sticker can aid in control.
Wilt (Ceratocystis)	propiconazole (Alamo)	See label	Root flare injection	See label.	
Pine					
Annosus root rot (Fomes annosus)	Borax, dry granular (sodium tetraborate decahydrate)	1 lb/50 sq ft of stump surface OR liberally cover stump surface	Sprinkle liberally on fresh-cut stump	Immediately after felling tree.	To prevent infection from freshly cut stumps.
Fusiform rust (Cronartium fusiforme)	triadimefon (Bayleton 50 DF)	4 to 16 oz/acre	foliar spray	Begin application before infection. Repeat at 2- to 3-week intervals as needed. Use higher rate in high hazard areas.	
	myclobutanil (Eagle 20EW)	12 to 18 fl oz/A	foliar spray	Apply at 2 to 3 week intervals	Begin applications before infection in early spring. Add spray adjuvant for enhanced disease control.
	prothioconazole (Proline 480SC)	5 fl oz/A	See label	14 to 21 days	
Phytophthora root rot	cyazofamid (Segway)	3 to 6 fl oz/100 gal	soil drench	14 to 21 day intervals using another registered fungicide with a different mode of action.	Irrigate with at least ½ inch of water if rainfall does not occur within 24 hrs. For container plants, check label for recommended maximum drench volume based on pot diameter.
	fosetyl-AL (Aliette WDG)	2.5 lb/100 gal 2.5 to 5 lb/acre	dip foliar (spray to wet)	Before transplanting 30-day minimum interval	Thoroughly wet plant and root mass immediately before transplanting. Field-grown trees in plantations.
	mefenoxam (Fenox ME, Mefenoxam 2, Subdue MAXX)	See label.	Can be applied as a drench or soil surface spray. Consult label for specific crops and applications.	Every 2 to 3 months.	FUNGICIDE RESISTANCE IS POSSIBLE. Do not apply rates of 2 fl oz more often than every 10 weeks.
	phosphorous acid (Alude, Fosphite, Reliant)	See label for rates	Apply as a soil drench or foliar spray as a preventive.		
Scotch and White Pine					
Needle blight (Lophoderimum pinastri)	maneb (Pentathlon DF)	1 to 2 lb/100 gal	foliar spray	Spray every 7 to 10 days August 15 to October 1.	
Sycamore					
Anthracnose	chlorothalonil (Daconil Ultrex) 82.5 WDG	1.4 lb/100 gal	foliar spray		Spray at budswell and repeat at 7- to 14-day intervals during cool, moist weather.
	copper hydroxide (Champ WG)	2 to 4 lb/100 gal	foliar spray	7 to 10 days	Make first application at bud crack and second application 7 to 10 days later at leaf expansion
	copper hydroxide + copper oxychloride (Badge SC)	0.75 to 2.5 pt/100 gal	foliar spray	7 to 10 days	Make first application at bud crack and second application 7 to 10 days later at leaf expansion
	Copper sulfate pentahydrate	0.5 to 1.0 pt/A	foliar spray	7 to 10 days	Make first application at bud crack and second application 7 to 10 days later at leaf expansion
	metconazole (Tourney)	1 to 4 oz/100 gal	foliar spray	Repeat in 14 to 28 days when conditions favor disease.	
	Propiconazole (Strider)	5 to 8.5 fl oz/100 gal	foliar spray		
	pyraclostrobin + boscalid (Pageant Intrinsic)	8 to 12 oz./100 gal	foliar spray	Repeat applications 7 to14 days as needed	
	thiophanate methyl + mancozeb (Zyban) 70 WSB	4 bags/100 gal	foliar spray		Spray at budswell and repeat at 7- to 14-day intervals during cool, moist weather.
	Thiophanate methyl (Cleary 3336) F	12 to 16 fl oz/100 gal	foliar spray		Spray at budswell and repeat at 7- to 14-day intervals during cool, moist weather.
	Triticonazole (Trinity) 19SC	8 to 12 fl oz/100 gal	foliar spray	Spray every 7 to 14 days.	Use preventively. Begin applications when conditions favor fungal infection and before disease symptom development.

Table 10-12. Disease Control for Forest, Christmas, and Ornamental Trees

CROP Disease	Material	Rate	Method	Schedule	Remarks
Sycamore (continued)					
Powdery mildew	See Crabapple				

Further Information

All the following Plant Pathology *Information Notes* can be accessed from http://www.ces.ncsu.edu/depts/pp/notes/Ornamental/ornamental_contents.html.

Diseases of Leyland Cypress, Plant Pathology Ornamental Disease Information Note No. 17.

Dutch Elm Disease. Plant Pathology Ornamental Disease Information Note No. 18.

Holly Diseases and Their Control in the Landscape, Plant Pathology Ornamental Disease Information Note No. 7.

Phytophthora Root Rot and Its Control on Established Woody Ornamentals. Plant Pathology Ornamental Disease Information Note No. 13.

Powdery Mildew of Ornamentals and Shade Trees. Plant Pathology Ornamental Disease Information Note No. 4.

Scorch Disease on Shade Trees, Plant Pathology Ornamental Disease Information Note No. 10.

Some Common Pecan Diseases and Their Control in North Carolina. Plant Pathology Ornamental Disease Information Note No. 3.

Some Common Pine Diseases in North Carolina Landscapes and Their Control. Plant Pathology Information Note No. 192.

Dogwood Diseases. Plant Pathology Information Note No. 23.

Commercial Landscape and Nursery Crops Disease Control

Anyone using any agricultural chemical should refer to the current chemical label, which contains information about the safe and effective use of the chemical, before using the chemical.

Table 10-13. Comercial Landscape and Nursery Crops Disease Control

DISEASE Pesticide and Formulation	Rate of Formulation (per 100 gallons)	Schedule and Remarks
Anthracnose (*Colletotrichum, Gleosporium, Elsinoe, Marssonina, Mycosphaerella*, and others)		
azoxystrobin (Heritage)	1 to 4 oz/100 gal	Repeat every 7 to 28 days. Apply at the first sign of disease. Should not be applied to certain plant species; see label. Do not apply to apple or flowering cherry trees. May cause phytotoxicity on certain crabapple cultivars. See label.
azoxystrobin + benzovindiflupyr (Mural)	4 to 7 oz/100 gal	Repeat every 7 to 14 days. Do not apply to apple or flowering cherry trees. May cause phytotoxicity on certain crabapple cultivars. Do not make more than two consecutive applications before switching to non-Group 7 or a non-Group 11 fungicide.
Bacillus subtilis (Rhapsody, Cease)	2 to 8 qt/100 gal	Begin applications when conditions favor disease development prior to the onset of disease. Thorough coverage is essential. Repeat at 7-day intervals.
Bacillus amyloliquefaciens strain D747 (DoubleNickel LC)	0.5 to 6 qt/100 gal	Begin preventative applications at plant emergence and repeat every 3 to 28 days depending on disease pressure. OMRI listed.
Chlorothalonil (Daconil Ultrex) 82.5WDG (Daconil WeatherStik) 54F	1.4 lb/100 gal 1.375 pt/100 gal	Reapply at 7- to 14-day intervals.
Chlorothalonil + propiconazole (Concert II)	9 to 35 fl oz/100 gal	Apply as a full coverage spray. Reapply at 14- to 21-day intervals. Refer to label for specific rate and application instructions.
Chlorothalonil + thiophanate-methyl (Spectro) 90WDG	1 to 2 lb/100 gal	Repeat applications at 7- to 21-day intervals, according to label. See label for maximum seasonal application rules.
Copper hydroxide (Champ WG) (Nu-Cop) 50DF (CuPRO 2005)	0.5 lb/100 gal 1 lb/100 gal 0.75 to 2.0 lb/100 gal	Begin at first sign of disease, and repeat at 7- to 14-day intervals. Do not tank mix copper formulations with Aliette. Avoid contact with metal surfaces. Discoloration of blooms may occur on certain plant varieties- check label.
copper oxychloride + copper hydroxide (Badge SC) (Badge X2)	1.5-6 pt/A (high) 1.5-2 pt/A (low) 1.5 to 5 lb/A (high) 1.5 to 2 lb/A (low)	High rates should only be used during dormancy. Use lower rate (1.5 to 2 pt/A or 1.5 to 2 lb/A) when new growth is present. Repeat at 7 to 14 day intervals. Badge X2 is OMRI listed. Do not mix with Aliette.
copper hydroxide + mancozeb (Junction)	1.5 to 3.5 lb/100 gal	Phytotoxicity may occur. Reapply at 7- to 14-day intervals.
Copper octanoate (Camelot O)	0.5-2.0 gal/100 gal	Begin at first sign of disease, and repeat at 7- to 10-day intervals. Camelot O may cause copper toxicity on some plant species.
Copper sulfate pentahydrate (Phyton 27)	See label	See label. To avoid phytotoxicity, do not use any copper compound on alyssum.
Cyprodinil + fludioxonil (Palladium)	2 to 4 oz/100 gal	Spray on a 7- to 14-day interval while conditions are conducive to disease 487evelopment. After two applications, alternate with another fungicide with a different MOA for two applications
mancozeb (Dithane) 75 DF, (Fore) 80 WSP, (Mancozeb DG) (Pentathlon LF) (Protect DF)	1.5 lb/100 gal 0.8 to 1.6 qt/100 gal 1 to 2 lb/100 gal	Repeat application on 7- to 10-day intervals. Use of a spreader sticker will improve performance. Repeat application on 7- to 10-day intervals. Repeat application on 7- to 21-day intervals. Use of a spreader sticker will improve performance.
Metconazole (Tourney)	1 to 4 oz/100 gal	Repeat in 14 to 28 days when conditions favor disease
polyoxin D zinc salt (Affirm)	0.25 to 0.5 lb/100 gal	Repeat every 7 to 10 days when conditions favor disease.
pyraclostrobin (Insignia)	8 to 16 oz/100 gal	Repeat every 7 to 14 days when conditions favor disease. Do not make more than two consecutive applications before switching to a non-Group 11 fungicide.

DISEASE CONTROL

Table 10-13. Comercial Landscape and Nursery Crops Disease Control

DISEASE Pesticide and Formulation	Rate of Formulation (per 100 gallons)	Schedule and Remarks
Anthracnose (Colletotrichum, Gleosporium, Elsinoe, Marssonina, Mycosphaerella, and others) (continued)		
pyraclostrobin + boscalid (Pageant Intrinsic)	8 to 12 oz/100 gal	Apply prior to disease development. Repeat at 7- to 14-day intervals. Make no more than two sequential applications. Do not expose petunia or impatiens in flower or wintercreeper or nine bark to spray or drift as injury may occur.
tebuconazole (Torque)	4 to 8 fl oz/100 gal	Begin applications 14 to 21 days prior to when disease is expected, or at very first sign of disease.
thiophanate-methyl (AllBan Flo) (Cleary 3336) F (SysTec 1998) FL	10.75 to 20 fl oz/100 gal 12 to 16 fl oz/100 gal 20 fl oz/100 gal	Apply as buds break or at first sign of disease. Repeat application on 7- to 14-day intervals.
thiophanate-methyl + mancozeb (Zyban) WSB	24 oz (4 bags) /100 gal	Apply at 7-day intervals.
triticonazole (Trinity) 19SC	8 to 12 fl oz/100 gal	Spray every 7 to 14 days. Use preventively. Begin applications when conditions favor fungal infection and before disease symptom development.
Bacterial Leaf Spot (Pseudomonas, Xanthomonas)		
Bacillus subtilis (Rhapsody, Cease)	2 to 8 qt/100 gal	Begin applications when conditions favor disease development prior to the onset of disease. Thorough coverage is essential. Repeat at 7-day intervals.
Bacillus amyloliquefaciens strain D747 (DoubleNickel LC)	0.5 to 6 qt/100 gal	Begin preventative applications at plant emergence and repeat every 3 to 28 days depending on disease pressure. OMRI listed.
copper hydroxide (Champ WG) (Nu-Cop) 50DF (CuPRO 2005)	0.5 lb/100 gal 1 lb/100 gal 0.75 to 2.0 lb/100 gal	Begin at first sign of disease, and repeat at 7- to 14-day intervals. Do not tank mix copper formulations with Aliette. Avoid contact with metal surfaces. Discoloration of blooms may occur on certain plant varieties- check label. .
copper hydroxide + mancozeb (Junction)	1.5 to 3.5 lb/100 gal	Begin at first sign of disease, and repeat at 7- to 14-day intervals.
copper oxychloride + copper hydroxide (Badge SC) (Badge X2)	1.5-6 pt/A (high) 1.5-2 pt/A (low) 1.5 to 5 lb/A (high) 1.5 to 2 lb/A (low)	High rates should only be used during dormancy. Use lower rate (1.5 to 2 pt/A or 1.5 to 2 lb/A) when new growth is present. Repeat at 7 to 14 day intervals. Badge X2 is OMRI listed. Do not mix with Aliette.
copper octanoate (Camelot O)	0.5-2.0 gal/100 gal	Begin at first sign of disease, and repeat at 7- to 10-day intervals. Camelot O may cause copper toxicity on some plant species.
cuprous oxide (Nordox 75WG)	0.66 lb/100 gal	Begin at first sign of disease and repeat at 7 to 14 day intervals.
copper sulfate pentahydrate (Phyton 27, Phyton 35)	See label	
didecyl dimethyl ammonium chloride (KleenGrow)	0.06 to 0.38 fl. oz. per gallon of water	Apply starting at week 3 or earlier if conditions are favorable for disease. Use a watering device to sprench the top and bottom of the leaves and stems, avoiding flowers in full bloom, every 14 days to prevent the spread of spores and the build-up of organic material. Remove severely infected plants and disinfect the area with 1.0 fl. oz. KleenGrow per gallon of water.
Black Root Rot (Thielaviopsis basicola)		
etridiazole + thiophanate-methyl (Banrot) 8G (Banrot) 40WP	Broadcast 8 to 12 lb/1,000 sq ft 6 to 12 oz/100 gal	After application, rake in or lightly cultivate soil. Apply in sufficient volume to saturate the soil mixture. Irrigate immediately. Repeat at 4- to 12-week intervals if necessary. Protects against *Thielaviopsis* and *Pythium* but is not as effective against *Thielaviopsis* as thiophanate-methyl-only products that have a higher concentration of active ingredient.
fludioxonil (Medallion)	1 to 2 oz/100 gal	Apply as a drench at transplanting as a preventive. If needed, re-treat transplants 21 to 28 days after initial application. Do not apply as a seed or soil drench to impatiens or New Guinea impatiens.
thiophanate-methyl (AllBan Flo) (Cleary 3336) F, (3336) F (OHP 6672) 50WP, (T-Storm) 50WSB (SysTec 1998) FL (T-Storm Flowable), OHP 6672 (4.5L)	7.5 to 20 oz/100 gal 8 to 16 fl oz/100 gal 12 to 16 oz/100 gal 10 to 20 oz/100 gal 20 fl oz/100 gal	Apply as a heavy spray or drench at the rate of 1/2 to 2 pints per sq ft. Repeat at 4- to 8-week intervals. Apply 8 oz as a drench or directed spray after seeding, or apply 12 to 16 oz after transplanting. Repeat at 21- to 28-day intervals. Apply 1 to 3 pt/sq ft after transplanting to thoroughly soak growing medium. Repeat at 21-to 28-day intervals. Apply as heavy spray or drench at a rate of 1 to 2 pints per sq ft. Repeat at 2- to 4-week intervals. Apply as heavy spray or drench at a rate of 1 to 2 pints per sq ft. Repeat at 2- to 4-week intervals.
triflumizole (Terraguard) 50W	1 to 4 oz	Apply as soil drench at 2- to 4-week intervals. Use higher rate under heavy disease pressure. For use in greenhouses, shadehouses, and nurseries.
Black Rot of Bulb Crops (Sclerotinia sclerotiorum)		
PCNB (Terraclor) (Revere) 10G	5 to 9.75 fl oz/100 gal 40 lb/sq ft	Spread evenly on soil and mix into upper 6 to 7 in. of soil.
thiophanate-methyl (AllBan Flo) (Cleary 3336) F (OHP 6672) 50WP, (T-Storm) 50WSB (SysTec 1998) FL, OHP 6672 (4.5L)	10.75 to 20 oz/100 gal 8 to 16 fl oz/100 gal 12 to 16 oz/100 gal 20 fl oz/100 gal	Apply late Spring or at first sign of disease. Repeat every 7-14 days as needed during disease period. Apply 8 oz as a drench or directed spray after seeding, or apply 12 to 16 oz after transplanting. Repeat at 21- to 28-day intervals. Apply late Spring or at first sign of disease. Repeat every 7-14 days as needed during disease period. Apply as heavy spray or drench at a rate of 1 to 2 pints per sq ft. Repeat at 2- to 4-week intervals.

Table 10-13. Comercial Landscape and Nursery Crops Disease Control

DISEASE Pesticide and Formulation	Rate of Formulation (per 100 gallons)	Schedule and Remarks
Black Spot-Rose (*Diplocarpon rosae*)		
Bacillus subtilis (Rhapsody, Cease)	2 to 8 qt/100 gal	Begin applications when conditions favor disease development, prior to the onset of disease. Thorough coverage is essential. Repeat at 7-day intervals.
Bacillus amyloliquefaciens strain D747 (DoubleNickel LC)	0.5 to 6 qt/100 gal	Begin preventative applications at plant emergence and repeat every 3 to 28 days depending on disease pressure. OMRI listed.
azoxystrobin (Heritage)	4 to 8 oz/100 gal	Apply on 7-day interval if disease conditions are favorable. If disease is severe, mix with another registerd fungicide for black spot control. Do not make more than two consecutive applications before switching to a non-Group 11 fungicide.
Captan (Captec 4L)	1 qt/100 gal	Apply at first sign of disease. Repeat 7 to 10 day intervals.
Chlorothalonil (Daconil Ultrex) 82.5 WDG (Echo) 90DF	1 lb/100 gal 0.875 lb/100 gal	Apply at bud break. Repeat applications at 7- to 14-day intervals. Knockout roses are sensitive to chlorothalonil.
mancozeb (Protect DF)	1 to 2 lb/100 gal	Apply every 7 to 21 days.
myclobutanil (Eagle 20EW)	6 to 12 fl oz	Apply on 7 to 10 day protectant schedule. Some greenhouse varieties are sensitive to myclobutanil.
parrafinic oil (Organic JMS Stylet Oil)	1 to 2 oz/gal	See label regarding application instructions and phytotoxicity warnings (with captan). OMRI listed.
propiconazole (Banner MAXX II)	5 to 8 fl oz/100 gal	Apply with contact fungicide labeled for black spot
propiconazole + cholorothalonil (Concert II)	22-35 fl oz/100 gal	Apply with contact fungicide labeled for black spot
Sulfur (Kumulus DF)	see label	Apply when disease first appears and continue at 5 to 10 day intervals.
Tebuconazole (Torque)	4 to 8 fl oz/100 gal	Begin applications 14 to 21 days prior to when disease is expected, or at very first sign of disease.
Thiophanate-methyl (AllBan Flo)	10.75 to 20 oz/100 gal	Apply late Spring or at first sign of disease. Repeat every 7-14 days as needed during disease period.
(Cleary 3336) F	8 to 16 fl oz/100 gal	Apply 8 oz as a drench or directed spray after seeding, or apply 12 to 16 oz after transplanting. Repeat at 21- to 28-day intervals.
(OHP 6672) 50WP, (T-Storm) 50WSB	12 to 16 oz/100 gal	Apply late Spring or at first sign of disease. Repeat every 7-14 days as needed during disease period.
(SysTec 1998) FL, (T-Storm Flowable), OHP 6672 (4.5L)	20 fl oz/100 gal	Apply as heavy spray or drench at a rate of 1 to 2 pints per sq ft. Repeat at 2- to 4-week intervals.
Thiophanate-methyl + chlorothalonil (Spectro 90WDG)	1 to 1.5 lb/100 gal	7 to 14 day reapplication interval.
Thiophanate-methyl + propiconazole (Protocol)	10 to 16 fl oz/100 gal	Apply in tank mixture with contact fungicide registered for black spot.
thiophanate-methyl + iprodione (26/36 Fungicide)	33 to 84 fl oz/100 gal	Do not make more than 4 applications per year.
thiophanate-methyl + mancozeb (Zyban) WSB	24 oz (4 bags)/100 gal	Apply at 7-day intervals.
trifloxystrobin (Compass O)	2 to 4 oz/100 gal	Apply at 7 to 14 day intervals. Do not make more than two consecutive applications before switching to a non-Group 11 fungicide.
Botrytis Blight (Gray Mold)		
Bacillus subtilis (Rhapsody, Cease)	2 to 8 qt/100 gal	Begin applications when conditions favor disease development prior to the onset of disease. Thorough coverage is essential. Repeat at 7-day intervals.
Bacillus amyloliquefaciens strain D747 (DoubleNickel LC)	0.5 to 6 qt/100 gal	Begin preventative applications at plant emergence and repeat every 3 to 28 days depending on disease pressure. OMRI listed.
azoxystrobin (Heritage)	4 to 8 oz/100 gal	Apply for suppression only. Repeat at 7- to 21-day intervals. Do not make more than two consecutive applications before switching to a non-Group 11 fungicid
azoxystrobin + benzovindiflupyr (Mural)	4 to 7 oz/100 gal	Apply every 7 to 14 days. Do not apply to apple or flowering cherry trees. May cause phytotoxicity on certain crabapple cultivars. Do not make more than two consecutive applications before switching to non-Group 7 or a non-Group 11 fungicide.
copper oxychloride + copper hydroxide (Badge SC) (Badge X2)	1.5-6 pt/A (high) 1.5-2 pt/A (low) 1.5 to 5 lb/A (high) 1.5 to 2 lb/A (low)	High rates should only be used during dormancy. Use lower rate (1.5 to 2 pt/A or 1.5 to 2 lb/A) when new growth is present. Repeat at 7 to 14 day intervals. May cause discoloration in some azalea varieties. Badge X2 is OMRI listed. Do not mix with Aliette.
chlorothalonil (Daconil Ultrex, Daconil Weatherstik)	1.4 lb/100 gal	Repeat applications at 7- to 14-day intervals.
cyprodinil + fludioxonil (Palladium)	4 to 6 oz/100 gal	Spray on a 7- to14-day interval while conditions are conducive to disease develepment. After two applications, alternate with another fungicide with a different MOA for two applications. See cautionary statement on label for applications to Geraniums, Impatiens and New Guinea Impatiens.
fenhexamide (Decree)	See label	Repeat applications at 7- to 14-day intervals. Do not make more than two consecutive applications of product.
fluoxastrobin (Decree 480SC)	4 to 8 fl oz/100 gal	Apply prior to infection. Repeat at 7- to 21-day intervals. Do not make more than two consecutive applications before switching to a non-Group 11 fungicide.
fludioxonil (Medallion)	2 to 4 oz/100 gal	Spray to runoff at 7- to 14-day intervals. Do not make more than two consecutive applications of Medallion before rotating to another effective product with a different mode of action.
iprodione (Chipco 26GT) (18 Plus, Iprodione Pro 2SE)	1 to 2.5 qt/100 gal 1 to 2.5 qt/100 gal	Apply at 7- to 14-day intervals. Limit total applications to a maximum of 4 per year.

489

Table 10-13. Comercial Landscape and Nursery Crops Disease Control

DISEASE Pesticide and Formulation	Rate of Formulation (per 100 gallons)	Schedule and Remarks
Botrytis Blight (Gray Mold) (continued)		
Iprodione + thiophanate methyl (26/36 Fungicide)	33 to 84 fl oz/100 gal	Do not make more than 4 applications per year.
mancozeb (Dithane, Fore, Mancozeb)	1.5 lb/100 gal	Addition of a nonionic surfactant will improve performance. Re-treat at 7- to 10-day intervals.
metconazole (Tourney)	1 to 4 oz/100 gal	Repeat in 14 to 28 days when conditions favor disease
pyraclostrobin + boscalid (Pageant Intrinsic)	8 to 12 oz/100 gal	Apply prior to disease development. Repeat at 7- to 14-day intervals. Make no more than two sequential applications. Do not expose petunia or impatiens in flower or wintercreeper or nine bark to spray or drift as injury may occur.
Cylindrocladium Stem Canker or Root Rot		
chlorothalonil (Chlorothalonil DF) (Chlorothalonil 500ZN) (Chlorothalonil 720 SFT) (Chlorostar VI, Daconil Weather Stik, Echo 720, Manicure 6F) (Daconil Ultrex) 82.5WDG (Echo) 90DF, (Echo Ultimate) (Exotherm Termil)	See label 1.9 pt/100 gal 1.37 pt/100 gal 1 3/8 pt/100 gal 1.4 lb/100 gal 1.25 lb/100 gal 1 can/1,000 sq ft	Repeat at 7- to 14-day intervals. Apply to foliage when plants are dry or nearly dry. See label for method of application.
chlorothalonil + thiophanate-methyl (Spectro) 90WDG	1.0 to 2.15 lb per 100 gallons	For best results use spray mixture the same day it is prepared. Spray uniformly over the area to be treated with a properly calibrated power sprayer, apply as a full coverage spray to run-off when conditions are favorable for disease development.
azoxystrobin + benzovindiflupyr (Mural)	5 to 7 oz/100 gal	Apply every 7 to 14 days. Do not apply to apple or flowering cherry trees. May cause phytotoxicity on certain crabapple cultivars. Do not make more than two consecutive applications before switching to non-Group 7 or a non-Group 11 fungicide.
cyprodinil + fludioxonil (Palladium)	2 to 4 oz/100 gal	For stem diseases ensure full spray coverage of all stems and inner areas of the plant to the soil/media. Apply on a 7- to 14-day interval while conditions are conducive to disease development. After two applications, alternate with another fungicide with a different MOA for two applications. No drench applications on label.
fludioxonil (Medallion) 50WSP	1 to 2 oz/100 gal	Completely drench the growing medium. Repeat at 21- to 28-day intervals. Two applications per year when conditions favor disease development are usually adequate for control.
pyraclostrobin + boscalid (Pageant Intrinsic)	8 to 12 oz/100 gal	Apply prior to disease development. Completely drench the growing medium. Repeat at 7- to 14-day intervals. Make no more than two sequential applications. Do not expose petunia or impatiens in flower or wintercreeper or nine bark to spray, as injury may occur.
thiophanate-methyl (AllBan Flo) (Cleary 3336) F (OHP-6672) 50WP, (T-Storm) 50WSB (SysTec 1998) FL (T-Storm Flowable), (OHP 6672) 4.5L	7.5 to 20 fl oz/100 gal 8 to 16 fl oz/100 gal 12 to 16 oz/100 gal 10 to 20 fl oz/100 gal 20 fl oz/100 gal	Apply as drench or heavy spray at 0.5 to 2 pints per sq ft. Repeat at 4- to 8-week intervals. Apply 8 oz after seeding or sticking, or 12 to 16 oz after transplanting as a drench or directed spray at a rate that thoroughly soaks the growing media through the root zone. Repeat every 21 to 28 days. Apply as a drench or heavy spray at a rate of 1 to 3 pints per sq ft. Repeat at 21- to 28-day intervals. Apply as a drench or heavy spray at a rate of 1 to 2 pints per sq ft. Repeat at 2- to 4-week intervals. Apply as a drench or heavy spray at a rate of 1 to 2 pints per sq ft. Repeat at 2- to 4-week intervals.
triflumizole (Terraguard) 50W	See label	Can be used as a cutting soak or soil drench.
Daylily Leaf Streak (*Aureobasidium microstictum*)		
chlorothalonil (Daconil Ultrex) 82.5WDG	1.4 lb/100 gal	Apply early in the spring as new growth emerges and before disease symptoms appear. Make three to four applications at 14-day intervals.
mancozeb (Dithane) 75 DF, (Fore) 80 WSP (Protect DF)	1.5 lb/100 gal 1 to 2 lb/100 gal	Apply early in the spring as new growth emerges and before disease symptoms appear. Make three to four applications at 14-day intervals.
thiophanate-methyl (Cleary 3336) F	12 to 16 fl oz/100 gal	Apply early in the spring as new growth emerges and before disease symptoms appear. Make three to four applications at 14-day intervals.
Daylily Rust (*Puccinia hemerocallidis*)		
Alternately apply a systemic fungicide from Category 1 with a protective fungicide from Category 2 to protect new foliage as it emerges. Re-treat at 7- to 14-day intervals.		
Category 1 Systemics		
azoxystrobin (Heritage)	1 to 4 oz/100 gal	
flutolanil (Contrast)	3 oz/100 gal	
triadimefon (Bayleton) (Strike) 25 WDG	1 PVA packet/550 to 1,100 gal 4 oz/100 gal	Bayleton cannot be used on plants being grown for sale or other commercial use. Strike is for commercial greenhouse and nursery use only.
trifloxystrobin (Compass)	2 to 4 oz/100 gal	
Category 2 Protectants		
chlorothalonil (Daconil Ultrex)	1.4 lb/100 gal	
mancozeb (Fore)	1.5 lb/100 gal	
Downy Mildew (*Bremia, Pseudoperonospora, Peronospora, Plasmopara* spp.)		
azoxystrobin (Heritage)	1 to 2 oz/100 gal (bedding plants) 2 to 4 oz/100 gal (rose)	Apply every 7 to 14 days prior to infection. Do not apply 2-oz rate on less than 14-day intervals. May damage snapdragons; use 1 oz rate and rotate. Apply every 7 to 21 days on rose during periods of active plant growth and prior to dormancy. Do not apply to apple, flowering cherry, or crabapple. See label.
azoxystrobin + benzovindiflupyr (Mural)	4 to 7 oz/100 gal	Apply every 7 to 14 days. Do not make more than two consecutive applications before switching to a non-Group 7 or non-Group 11 fungicide. Do not apply to apple, flowering cherry, or crabapple. See label.

490

Table 10-13. Comercial Landscape and Nursery Crops Disease Control

DISEASE Pesticide and Formulation	Rate of Formulation (per 100 gallons)	Schedule and Remarks
Downy Mildew *(Bremia, Pseudoperonospora, Peronospora, Plasmopara* spp.) (continued)		
Bacillus subtilis (Rhapsody, Cease)	2 to 8 qt/100 gal	Repeat at 3- to 10-day intervals. Thorough coverage is essential. Begin applications when conditions favor disease development, prior to the onset of disease.
Bacillus amyloliquefaciens strain D747 (DoubleNickel LC)	0.5 to 6 qt/100 gal	Begin preventative applications at plant emergence and repeat every 3 to 28 days depending on disease pressure. OMRI listed.
copper hydroxide (Champ WG) (Nu-Cop) 50DF (CuPRO 2005)	0.5 lb/100 gal 1 lb/100 gal 0.75 to 2.0 lb/100 gal	Begin at first sign of disease, and repeat at 7- to 14-day intervals. Do not tank mix copper formulations with Aliette. Avoid contact with metal surfaces. Discoloration of blooms may occur on certain plant varieties- check label. To avoid phytotoxicity, do not use any copper compound on alyssum.
copper oxychloride + copper hydroxide (Badge SC) (Badge X2)	1.5-2 pt/A 1.5 to 2 lb/A	Repeat at 7 to 14 day intervals. May cause discoloration in some azalea varieties.See label for other phytoxicity warnings. Badge X2 is OMRI listed. Do not mix with Aliette.
copper octanoate (Camelot O)	0.5-2.0 gal/100 gal	Begin at first sign of disease, and repeat at 7- to 10-day intervals. Camelot O may cause copper toxicity on some plant species.
copper sulfate pentahydrate (Phyton 27)	See label	See label. To avoid phytotoxicity, do not use any copper compound on alyssum.
cyazofamid (Segway)	2.1 to 3.5 fl oz/100 gal	Apply on 14- to 21-day intervals using another registered fungicide with a different mode of action. Apply sufficient volume to wet all foliage until runoff (normally 50 to 100 gallons per acre).
dimethomorph (Stature DM)	6.4 to 12.8 oz/100 gal	Apply at first sign of disease. Apply to obtain complete coverage of flowers, foliage, and stems. Repeat at 10- to 14-day intervals throughout the production cycle. For use on greenhouse and nursery-grown ornamentals.
dimethomorph + ametoctradin (Orvego)	11 to 14 fl oz/100 gal	Apply on 10- to 14-day intervals using another registered fungicide with a different mode of action. Apply sufficient volume to wet all foliage until runoff (normally 50 to 100 gallons per acre).
fosetyl-Al (Aliette) 80WDG	1.25 to 4 lb/100 gal (bedding plants) 2.5 lb/100 gal (roses)	Systemic. Apply prior to disease development. Repeat as necessary, but do not make more than one application every 14 days.
Fluoxastrobin (Disarm 480SC)	1 to 4 fl oz/100 gal	Apply every 7 to 21 days. Do not make more than two consecutive applications before switching to a non-Group 11 fungicide.
pyraclostrobin (Insignia)	4 to 8 oz/100 gal	Apply every 7 to 14 days prior to disease development. Do not make more than two consecutive applications before switching to a non-Group 11 fungicide.
Mancozeb (Dithane) 75 DF (Protect DF)	1.5 lb/100 gal 1 to 2 lb/100 gal	Repeat at 7- to 10-day intervals. Reapply in 7 to 21 days.
Mandipropamid (Micora)	4 to 8 fl oz/100 gal	This product can also be used on vegetables sold to the retail market in GH with permanent flooring. Apply prior to disease development. Repeat at 7- to 14-day intervals. Make no more than two sequential applications.
Mefenoxam (Subdue MAXX) (Subdue Gr)	0.5 to 1 fl oz/100 gal See label for rates	Apply Subdue MAXX as a foliar spray or soil drench treatment. Apply Subdue GR as a soil surface or soil/planting media incorporation treatment.
oxathiapiprolin (Segovis)	0.6 to 2.4 fl oz/100 gal	Make no more than two consecutive applications before switching to a non-Group U15 fungicide.
pyraclostrobin + boscalid (Pageant)	12 to 18 oz/100 gal	Use preventatively, prior to disease development on a 7- to 10-day schedule. Do not make more than two consecutive applications before switching to a non-Group 7 or non-Group 11 fungicide.
phosphorous acid (Alude, Reliant)	See label	Apply prior to disease development. Spray to thoroughly wet all foliage. Repeat at 14- to 21-day intervals.
potassium phosphite (Vital)	4 pt/100 gal	Apply as a foliar spray prior to disease onset, and repeat at 14-day intervals.
pyraclostrobin + boscalid (Pageant Intrinsic)	8 to 12 oz/100 gal	Apply prior to disease development. Repeat at 7- to 10-day intervals. Make no more than two sequential applications. Do not expose petunia or impatiens in flower or wintercreeper or nine bark to spray or drift as injury may occur.
trifloxystrobin (Compass O) 50WDG	2 to 4 oz/100 gal	Apply as a foliar spray before disease is detected or when conditions are favorable for disease. Repeat at 7- to-14-day intervals until threat of disease is over.
thiophanate-methyl + mancozeb (Zyban) WSB	24 oz (4 bags)/100 gal	Apply at first sign of disease, and repeat at 7-day intervals.
Entomosporium Leaf Blight		
azoxystrobin (Heritage) 50WDG	1 to 4 oz/100 gal	Spray at budbreak, and repeat at 7- to 28-day intervals as needed. To avoid fungicide resistance, make no more than three sequential applications of Heritage before rotating with nonstrobilurin products.
copper oxychloride + copper hydroxide (Badge SC) (Badge X2)	1.5-2 pt/A 1.5 to 2 lb/A	Repeat at 7 to 14 day intervals. May cause discoloration in some azalea varieties.See label for other phytoxicity warnings. Badge X2 is OMRI listed. Do not mix with Aliette.
chlorothalonil (Daconil Ultrex) 82.5 WDG	1.4 lb/100 gal	Begin applications at budbreak, and continue every 7 to 14 days.
chlorothalonil + propiconazole (Concert II)	22 to 35 fl oz/100 gal	Apply as a full coverage spray. Reapply at 14 to 21 day intervals. Refer to label for specific rate and application instructions.
chlorothalonil + thiophanate-methyl (Spectro 90)	1 to 2 lb/100 gal	Spray at a minimum of 7-day intervals. Apply when foliage is dry.
Iprodione + thiophanate methyl (26/36 Fungicide)	33 to 84 fl oz/100 gal	Do not make more than 4 applications per year.
myclobutanil (Eagle) 20EW (Eagle) 40WP	6 to 12 fl oz/100 gal 3 to 6 oz/100 gal	Spray every 10 to 14 days.

491

DISEASE CONTROL

Table 10-13. Comercial Landscape and Nursery Crops Disease Control

DISEASE Pesticide and Formulation	Rate of Formulation (per 100 gallons)	Schedule and Remarks
Entomosporium Leaf Blight (continued)		
propiconazole (Banner MAXX II)	5 to 8 fl oz/100 gal	
thiophanate-methyl (3336 F)	12 to 16 oz/100 gal	Apply when disease first appears and repeat everty 7 to 14 days.
triadimefon (Bayleton) 50WSP (Strike) 25WDG	1 PVA packet/137.5 to 275 gal 8 to 16 oz/100 gal	In early spring as growth starts, spray every 14 to 21 days until growth is fully expanded. May be phototoxic with repeated applications. Bayleton is not for plants offered for sale or other commercial use.
triticonazole (Trinity) 19SC	12 fl oz/100 gal	Spray every 7 to 14 days. Use preventively. Begin applications when conditions favor fungal infection and before disease symptom development.
Fireblight *(Erwinia amylovora)* **(or see section under fire blight control of crabapple)**		
copper hydroxide (Champ WG) (Nu-Cop) 50DF (CuPRO 2005	0.5 lb/100 gal 1 lb/100 gal . 0.75 to 2.0 lb/100 gal	Begin at first sign of disease, and repeat at 7- to 14-day intervals. Do not tank mix copper formulations with Aliette. Avoid contact with metal surfaces. Discoloration of blooms may occur on certain plant varieties- check label.
copper hydroxide + mancozeb (Junction)	1.5 to 3.5 lb/100 gal	See label for timing of application.
copper sulfate pentahydrate (Phyton 27)	See label	See label. To avoid phytotoxicity, do not use any copper compound on alyssum.
fosetyl-Al (Aliette)	2.5 lb/100 gal	See label for timing of application.
streptomycin sulfate (Agri-mycin 17)	0.5 lb/100 gal	See label.
Flower Blight. *See* Petal Blight.		
Fungal Leaf Spots *(Alternaria, Cercospora, Cylindrosporium, Phyllosticta, Septoria)* **Consult product labels for specific fungi controlled**		
azoxystrobin (Heritage)	1 to 4 oz/100 gal	Repeat every 7- to 28-days. Apply at the first sign of disease as new growth buds out. Do not apply to apple, flowering cherry, or crabapple. See label.
azoxystrobin + benzovindiflupyr (Mural)	4 to 7 oz/100 gal	Apply every 7 to 21 days.Do not make more than two consecutive applications before switching to a non-Group 7 or non-Group 11 fungicide. Do not apply to apple, flowering cherry, or crabapple. See label.
Bacillus subtilis (Rhapsody, Cease)	2 to 8 qt/100 gal	Repeat at 3- to 10-day intervals. Thorough coverage is essential. Begin applications when conditions favor disease development, prior to the onset of disease.
Bacillus amyloliquefaciens strain D747 (DoubleNickel LC)	0.5 to 6 qt/100 gal	Begin preventative applications at plant emergence and repeat every 3 to 28 days depending on disease pressure. OMRI listed.
chlorothalonil (Daconil Ultrex) (Daconil WeatherStik)	1.4 lb/100 gal 1.375 pt/100 gal	Reapply at 7- to 14-day intervals.
chlorothalonil + propiconazole (Concert II)	9 to 35 fl oz/100 gal	Apply as a full coverage spray. Reapply at 14- to 21-day intervals. Refer to label for specific rate and application instructions.
chlorothalonil + thiophanate-methyl (Spectro) 90WDG	1 to 2 lb/100 gal	Repeat applications at 7- to 21-day intervals depending on plant treated; see label.
copper hydroxide (Champ WG) (Nu-Cop) 50DF (CuPRO 2005	0.5 lb/100 gal 1 lb/100 gal 0.75 to 2.0 lb/100 gal	Begin at first sign of disease, and repeat at 7- to 14-day intervals. Do not tank mix copper formulations with Aliette. Avoid contact with metal surfaces. Discoloration of blooms may occur on certain plant varieties- check label.
copper oxychloride + copper hydroxide (Badge SC) (Badge X2)	1.5-2 pt/A 1.5 to 2 lb/A	Repeat at 7 to 14 day intervals. May cause discoloration in some azalea varieties.See label for other phytoxicity warnings. Badge X2 is OMRI listed. Do not mix with Aliette.
copper hydroxide + mancozeb (Junction)	1.5 to 3.5 lb/100 gal	Reapply at 7- to 14-day intervals.
copper octanoate (Camelot O)	0.5-2.0 gal/100 gal	Begin at first sign of disease, and repeat at 7- to 10-day intervals. Camelot O may cause copper toxicity on some plant species.
copper sulfate pentahydrate (Phyton 27)	See label	
cyprodinil + fludioxonil (Palladium)	2 to 6 oz/100 gal	Spray on a 7 to 14 day interval while conditions are conducive to disease develepment. After two applications, alternate with another fungicide with a different MOA for two applications. Cautionary statement on label for applications to Geraniums, Impatiens and New Guinea Impatiens.
mancozeb (Dithane) 75 DF, (Fore) 80 WSP, Mancozeb DG) (Pontathlon LF) (Protect DF)	1.5 lb/100 gal 1.2 qt/100 gal 1 to 2 lb/100 gal	Repeat application at 7- to 10-day intervals. Addition of a spreader sticker will improve performance.
metconazole (Tourney)	1 to 4 oz/100 gal	Repeat in 14 to 28 days when conditions favor disease
myclobutanil (Eagle) 40WP	3 to 6 oz/100 gal	Apply as a protectant every 10 to 14 days.
potassium bicarbonate (MilStop)	1.25 to 5 lb/100 gal	Uniform and complete coverage of foliage is essential for best results. See label for special instructions regarding poinsettia, pansy, and impatiens.
pyraclostrobin (Insignia)	See label	Apply prior to onset of disease and repeat every 7 to 14 days. Do not expose flowering impatiens or flowering petunias to Insignia.

Table 10-13. Comercial Landscape and Nursery Crops Disease Control

DISEASE Pesticide and Formulation	Rate of Formulation (per 100 gallons)	Schedule and Remarks
Fungal Leaf Spots (continued)		
pyraclostrobin + boscalid (Pageant Intrinsic)	8 to 12 oz/100 gal	Apply prior to disease development. Repeat at 7- to 14-day intervals. Make no more than two sequential applications. Do not expose petunia, or impatiens in flower or wintercreeper, or nine bark to spray or drift as injury may occur.
tebuconazole (Torque)	4 to 8 fl oz/100 gal	Begin applications 14 to 21 days prior to when disease is expected, or at very first sign of disease.
thiophanate-methyl (AllBan Flo) (Cleary's 3336) F (SysTec 1998) FL (Zyban) WSB	10 to 14.5 fl oz/100 gal 12 to 16 fl oz/100 gal 20 fl oz/100 gal 24 oz (4 bags) /100 gal	Apply at first sign of disease. Repeat application at 7- to 14-day intervals.
triadimefon (Bayleton, Strike)	See label	
triticonazole (Trinity) 19SC	4 to 8 fl oz/100 gal	Spray every 7 to 14 days. Use preventively. Begin applications when conditions favor fungal infection and before disease symptom development. Use of an adjuvant/spreader sticker can aid in control.
Fusarium Root and Crown Rot		
azoxystrobin (Heritage)	Directed spray: 1 to 4 oz/100 gal Drench: 0.2 to 0.9 oz/100 gal	Repeat every 7 to 21 days. Apply 1 to 2 pt of solution per sq ft every 7 to 28 days.
cyprodinil + fludioxonil (Palladium)	2 to 4 oz/100 gal	For stem diseases ensure full spray coverage of all stems and inner areas of the plant to the soil/media. Apply on a 7- to 14-day interval while conditions are conducive to disease develepment. After two applications, alternate with another fungicide with a different MOA for two applications.
fludioxonil (Medallion)	1 to 2 oz/100 gal	Wet entire medium. Reapply at 21- to 28-day intervals. May cause phytotoxicity when applied to impatiens, New Guinea impatiens, and geraniums.
thiophanate-methyl (AllBan Flo) (Cleary's 3336) G (Cleary 3336) F (SysTec 1998) FL	7.5 to 20 fl oz/100 gal 22 to 30 lb/1,000 sq ft Drench: 8 to 16 fl oz/100 gal 10 to 20 fl oz/100 gal	Apply as drench or heavy spray after transplanting. Repeat at 2- to 4-week intervals. For preventative control, incorporate into media prior to planting or as a broadcast, or make an over the top application after seeding or transplanting. For curative control, apply when disease first appears. Repeat every 21 to 28 days. Apply after seeding or transplanting at a rate to thoroughly soak growing medium. Repeat every 21 to 28 days. Apply as drench or heavy spray after transplanting. Repeat at 2- to 4-week intervals.
pyraclostrobin (Empress Intrinsic)	1 to 6 fl oz/100 gal (see label)	Apply at 1 to 3 fl oz for plants in propagation, rooted cuttings, plugs, and seedlings and at 2 to 6 fl oz to all other plants. Do not apply to dry soil media. Apply preventative to disease with sequential at 7 to 28 days after the first application if needed.
pyraclostrobin + boscalid (Pageant Intrinsic)	12 to18 oz/100 gal	Apply prior to disease development. Repeat at 7- to 14-day intervals. Make no more than two sequential applications. Do not expose petunia, or impatiens in flower or wintercreeper, or nine bark to spray or drift as injury may occur.
Streptomyces griseoviridis (Mycostop)	See label	Apply inoculant as a seed dressing, soil drench spray, or transplant dip. Must be applied prior to onset of disease. See labe l.
triflumizole (Terraguard) 50W	4 to 8 oz/100 gal	Apply soil drenches at weekly intervals. Use higher rate under heavy disease pressure.
triticonazole (Trinity) 19SC	8 to 12 fl oz/100 gal	Apply every 7 to 14 days. Use preventively. Begin applications when conditions favor fungal infection and before disease symptom development. The stem areas of the plant must be thoroughly covered using spray to runoff.
Gray Mold. *See Botrytis Blight.*		
Iris Leaf Spot *(Didymellina macrosopora/Mycosphaerella macrospora)*		
azoxystrobin (Heritage)	2 to 4 oz/100 gal	Apply every 7 to 21 days.
azoxystrobin + benzovindiflupyr (Mural)	4 to 7 oz/100 gal	Apply every 7 to 21 days. Do not make more than two consecutive applications before switching to a non-Group 7 or non-Group 11 fungicide.
chlorothalonil (Daconil Ultrex) (Daconil WeatherStik)	1.4 lb/100 gal 1.375 pt/100 gal	Apply to new growth at 7- to 14-day intervals in spring.
mancozeb (Dithane) 75 DF, (Fore) 80 WSP, Mancozeb DG) (Pentathlon LF) (Protect DF)	1.5 lb/100 gal 0.8 to 1.6 qt/100 gal 1 to 2 lb/100 gal	Addition of a spreader sticker will improve performance. Repeat applications on 7- to 10-day intervals. Apply at 7- to 21-day intervals.
myclobutanil (Eagle) 20EW (Eagle) 40WP	12 fl oz/100 gal 6 oz/100 gal	
Pyraclostrobin (Insignia)	2 to 8 oz/100 gal	Apply preventatively on a 7- to 14-day application interval. Do not make more than two consecutive applications before switching to a non-Group 7 or non-Group 11 fungicide.
Thiophanate-methyl (Incognito 85WDG)	7.2 to 9.6 oz/100 gal	Apply every 7 to 14 days prior to or immediately at the onset of disease.
Kabatina Twig Blight		
thiophanate-methyl (Cleary 3336) F	16 to 24 fl oz/100 gal	Apply at 7- to 14-day intervals. Disease not easily controlled. Fall applications may reduce disease the following year.
Leaf and Flower Gall *(Exobasidium ssp.)*		
triadimefon (Bayleton) 50WSP (Strike) 25W	1 PVA packet (11 oz/550 gal) 4 oz/100 gal	Begin applications at bud break, and apply at 10-day intervals. Bayleton cannot be used on plants grown for sale or other commercial use. Strike is for use in commercial nurseries, garden centers, and greenhouses only.
ferbam (Granuflo)	1 to 1.5 lb/100 gal	Apply to plants, flowers, and litter around plants at 3- to 4-day intervals during bloom.

493

DISEASE CONTROL

Table 10-13. Comercial Landscape and Nursery Crops Disease Control

DISEASE Pesticide and Formulation	Rate of Formulation (per 100 gallons)	Schedule and Remarks
Petal or Flower Blight of Azalea, Rhododendron, or Camellia (*Ovulinia* ssp., *Ciborinia camelliae, Sclerotinia camelliae*)		
mancozeb (Dithane) 75 DF, (Fore) 80 WSP, (Mancozeb DG) (Protect DF)	1.5 lb/100 gal 1 to 2 lb/100 gal	Beginning when flowers start to show color, spray two or three times each week during bloom. Direct spray into flowers and thoroughly spray ground under bushes.
myclobutanil (Eagle) 40WP	3 to 6 oz/100 gal	Beginning when flowers start to show color, spray every 10 to 14 days.
propiconazole (Banner MAXX, Banner MAXX II)	5 to 8 fl oz/100 gal	Spray every 21 days during bloom.
tebuconazole (Torque)	4 to 8 fl oz/100 gal	Apply 2 to 3 times per week into the flowers as they open and develop.
thiophanate-methyl (AllBan Flo) (Cleary 3336) F (OHP 6672) 50WP, (T-Storm) 50WSB (SysTec 1998) FL, (T-Storm Flowable), OHP 6672 (4.5L)	10.75 to 20 oz/100 gal 8 to 16 fl oz/100 gal 8 to 16 oz/100 gal 20 fl oz/100 gal	Apply as flowers open. Repeat every 7 to 14 days.
triadimefon (Bayleton) 50WSP (Strike) 25WDG	1 PVA packet (11 oz/137.5 to 275 gal) 8 to 16 oz/100 gal	Make one application as first flower buds show color. Spray later varieties as they show color at 7- to 14-day intervals. Bayleton cannot be used on plants being grown for sale or other commercial use. Strike is for use in commercial nurseries, garden centers, and greenhouses only.
Phomopsis Twig Blight		
azoxystrobin (Heritage)	1 to 4 oz/100 gal	Apply at the first sign of disease, as new growth buds out. Repeat every 7 to 28 days.
azoxystrobin + benzovindiflupyr (Mural)	4 to 7 oz/100 gal	Apply every 7 to 21 days. Do not make more than two consecutive applications before switching to a non-Group 7 or non-Group 11 fungicide. Do not apply to apple, flowering cherry, or crabapple. See label.
copper oxychloride + copper hydroxide (Badge SC) (Badge X2)	 1.5-2 pt/A 1.5 to 2 lb/A	Repeat at 7 to 14 day intervals. May cause discoloration in some azalea varieties. See label for other phytoxicity warnings. Badge X2 is OMRI listed. Do not mix with Aliette.
cyprodinil + fludioxonil (Palladium)	2 to 4 oz/100 gal	For stem diseases ensure full spray coverage of all stems and inner areas of the plant to the soil/media. Apply on a 7 to 14 day interval while conditions are conducive to disease develepment. After two applications, alternate with another fungicide with a different MOA for two applications.
mancozeb (Dithane) 75 DF, (Fore) 80 WSP, (Mancozeb DG) (Pentathlon LF) (Protect DF)	1.5 lb/100 gal 0.8 to 1.6 qt/100 gal 1 to 2 lb/100 gal	Addition of a spreader sticker will improve performance. Repeat application on 7- to 10-day intervals.
Propiconazole (Banner MAXX)	5 to 8 fl oz/100 gal	For junipers, make first application as soon as new growth is observed. Repeat application every 14 to 21 days during period of active plant growth.
pyraclostrobin + boscalid (Pageant Intrinsic)	8 to12 oz/100 gal	Apply prior to disease development. Repeat at 7- to 14-day intervals. Make no more than two sequential applications. Do not expose petunia, or impatiens in flower or wintercreeper, or nine bark to spray or drift as injury may occur.
thiophanate-methyl (AllBan Flo) (Cleary 3336) F (T-Storm) 50WSB (SysTec 1998) FL, (T-Storm Flowable), OHP 6672 (4.5L)	14.5 to 20 fl oz/100 gal 16 to 24 fl oz/100 gal 24 oz/100 gal 20 fl oz /100gal	Repeat at 10- to 14-day intervals. Apply when symptoms first appear. Re-treat every 7 to 14 days as needed during disease period. Apply in spring; repeat every 7 to 10 days.
thiophanate-methyl + mancozeb (Zyban)	6 bags (36 oz/75 gal)	Apply at 7- to 10-day intervals.
Phytophthora and Pythium Root Rot		
cyazofamid (Segway)	3 to 6 fl oz/100 gal	Apply at 14 to 21 day intervals using another registered fungicide with a different mode of action. Irrigate with at least ½ inch of water if rainfall does not occur within 24 hrs. For container plants, check label for recommended maximum drench volume based on pot diameter.
azoxystrobin + benzovindiflupyr (Mural)	3 oz/100 gal	Apply 1 to 2 pints of drench solution per square foot surface area every 7 to 28 days. Do not apply to apple, flowering cherry, or crabapple. See label.
dipotassium phosphonate + dipotassium phosphate (Biophos)	256 fl oz/100 gal	Apply as a soil drench or foliar spray as a preventative.
dimethomorph (Stature DM) 50W	6.4 to 12.8 oz/100 gal	Apply at 10- to 14-day intervals through production cycle. When applied as a drench, use enough solution to wet root zone of the plant. No more than two applications of Stature DM can be applied consecutively in a crop. **Not effective on Pythium root rot.**
dimethomorph + ametoctradin (Orvego)	11 to 14 fl oz	NOT LABELED FOR PYTHIUM. Apply on 10 to 14 day intervals using another registered fungicide with a different mode of action. Apply sufficient volume to wet all foliage until runoff (normally 50 to 100 gallons per acre).
etridiazole (Truban) 30WP (Terrazole) 35WP	3 to 10 oz/100 gal/400 sq ft 3.5 to 10 oz/100 gal	Apply in sufficient volume to saturate the soil mixture. Water in immediately after application. Repeat at 4- to 12-week intervals.
etridiazole + thiophanate-methyl (Banrot) 40WP (Banrot) 8G	6 to 12 oz/100 gal See label	Apply in sufficient volume to saturate the soil mixture. Irrigate immediately with additional water equal to at least half the volume of the fungicide drench. Re-treat at 4- to 12-week intervals. For use in nursery crops. See label.

Table 10-13. Comercial Landscape and Nursery Crops Disease Control

DISEASE Pesticide and Formulation	Rate of Formulation (per 100 gallons)	Schedule and Remarks
Phytophthora and Pythium Root Rot (continued)		
fluopicolide (Adorn)	1 to 4 fl oz/100 gal	MUST ALWAYS BE TANK MIXED WITH THE LABELED RATE OF ANOTHER FUNGIIDE WITH A DIFFERENT MODE OF ACTION. Apply before disease development. Use higher rate when treating plants with high potential for disease. Reapply after 14 to 28 days.
fosetyl-Al (Aliette) 80WP	See label	Can be applied as a preventative foliar or drench application. Can be incorporated into the soil for control of Phytophthora species.
mandipropamid (Micora)	4 to 8 fl oz/100 gal	This product can also be used on vegetables sold to the retail market in GH with permanent flooring. Apply prior to disease development. Repeat at 7- to 14-day intervals. Make no more than two sequential applications.
mefenoxam (Fenox ME, Mefenoxam 2, Subdue MAXX, Subdue GR)	See label	Can be applied as a drench or soil surface spray or soil incorporation treatment (granular). Consult label for specific crops and applications. Repeat at 2- to 3-month intervals. Do not apply rates of 1.25 fl oz per 100 gal more often than every 3 months.
oxathiapiprolin (Segovis)	0.65 to 3.2 fl oz/100 gal	NOT LABELED FOR PYTHIUM. Apply n 5- to 14-day interval preventatively or at first sight of disease symptoms. Do not apply more than 2 consecutive applications before switching to another non-Group U-15 fungicide.
phosphorous acid (Alude, Fosphite, Reliant)	See label for rates	Apply as a soil drench or foliar spray as a preventive.
potassium phosphite (Vital)	See label	Apply as a soil drench or foliar spray as a preventive.
propamocarb (Banol)	See label	Do not use for field-grown ornamentals.
trifloxystrobin (Compass O)	1 to 2 oz/100 gal	
Powdery Mildew		
azoxystrobin (Heritage) 50WDG	1 to 4 oz/100 gal	Apply only as a preventive. Spray every 7 to 28 days as needed. To avoid fungicide resistance, make no more than two sequential applications of Heritage before rotating with nonstrobilurin products. Do not apply to apple, flowering cherry, or crabapple. See label.
azoxystrobin + benzovindiflupyr (Mural)	4 to 7 oz/100 gal	Apply every 7 to 21 days. Do not make more than two sequential applications before rotating to another class of fungicide that is not Group 7 or 11. Do not apply to apple, flowering cherry, or crabapple. See label.
Bacillus subtilis (Rhapsody, Cease)	2 to 4 qt/100 gal	Repeat at 7-day intervals. Thorough coverage is essential. Begin applications when conditions favor disease development prior to the onset of disease.
Bacillus amyloliquefaciens strain D747 (DoubleNickel LC)	0.5 to 6 qt/100 gal	Begin preventative applications at plant emergence and repeat every 3 to 28 days depending on disease pressure. OMRI listed.
chlorothalonil (Daconil Ultrex) 82.5 WDG	1.4 lb/100 gal	Spray at 7- to 14-day intervals. Applications made during bloom may damage flowers.
chlorothalonil + propiconazole (Concert II)	22 to 35 fl oz/100 gal	Apply as a full coverage spray. Reapply at 14 to 21 day intervals. Refer to label for specific rate and application instructions.
copper oxychloride + copper hydroxide (Badge SC) (Badge X2)	1.5-2 pt/A 1.5 to 2 lb/A	Repeat at 7 to 14 day intervals. May cause discoloration in some azalea varieties. See label for other phytoxicity warnings. Badge X2 is OMRI listed. Do not mix with Aliette.
copper octanoate (Camelot O)	0.5-2.0 gal/100 gal	Begin at first sign of disease, and repeat at 7- to 10-day intervals. Camelot O may cause copper toxicity on some plant species.
copper sulfate pentahydrate (Phyton 27)	See label	
cyprodinil + fludioxonil (Palladium)	2 to 6 oz/100 gal	Spray on a 7 to 4 day interval while conditions are conducive to disease develepment. After two applications, alternate with another fungicide with a different MOA for two applications.
fuoxastrobin (Disarm 480SC)	1 to 4 fl oz/100 gal	Use for preventative applications only and apply every 7 to 28 days depending on disease pressure. Do not make more than two consecutive applications before switching to a non Group 11 fungicide.
metconazole (Tourney)	1 to 4 oz/100 gal	Repeat in 14 to 28 days when conditions favor disease
myclobutanil (Eagle) 20EW (Eagle) 40WP	8 fl oz/100 gal 3 to 6 oz/100 gal	Apply at 10- to 14-day intervals.
potassium bicarbonate (Kaligreen)	1 to 3 lb/100 gal	Apply at 7- to 10-day intervals.
propiconazole (Banner MAXX, Banner MAXX II)	5 to 12 fl oz/100 gal	See label for appropriate rate and application intervals. For application in field nurseries and landscape plantings.
pyraclostrobin (Insignia)	4 to 8 oz/100 gal	Apply at 7 to 14 day intervals. Do not make more than two consecutive applications before switching to a non Group 11 fungicide.
Pyraclostrobin + boscalid (Pageant Intrinsic)	6 to12 oz/100 gal	Apply prior to disease development. Repeat at 7- to 10-day intervals. Make no more than two sequential applications. Do not expose petunia or impatiens in flower or wintercreeper or nine bark to spray or drift as injury may occur.
Sulfur (Micro Sulf)	3 to 10 lb/100 gal	Begin when disease first appears and continue at 5 to 10 intervals. See label for phytoxicity considerations.
tebuconazole (Torque)	4 to 8 fl oz/100 gal	Apply every 14 for a total of three applications at the first sign of disease.
Thiophanate-methyl (AllBan Flo) (Cleary 3336) F (OHP 6672) 50WP, (T-Storm) 50WSB (SysTec 1998) FL (T-Storm Flowable), OHP 6672 (4.5L)	10 to 20 oz/100 gal 12 to 24 fl oz/100 gal 8 to 16 oz/100 gal 10 fl oz/100 gal 20 fl oz/100 gal	Apply when disease first appears and repeat every 7-14 days. Rotations with other effective products are recommended. Apply when disease first appears and repeat every 7-14 days. Rotations with other effective products are recommended. Apply when disease first appears and repeat every 7-14 days as needed during disease period. Apply as heavy spray or drench at a rate of 1 to 2 pints per sq ft. Repeat at 2- to 4-week intervals. Apply as heavy spray or drench at a rate of 1 to 2 pints per sq ft. Repeat at 2- to 4-week intervals.
Thiophanate-methyl + mancozeb (Zyban) WSB	24 oz (4 bags) /100 gal	Apply at first sign of disease. Repeat at 7-day intervals.

DISEASE CONTROL

Table 10-13. Comercial Landscape and Nursery Crops Disease Control

DISEASE Pesticide and Formulation	Rate of Formulation (per 100 gallons)	Schedule and Remarks
Powdery Mildew (continued)		
Triadimefon (Bayleton) 50WSP (Strike) 25WDG	1 PVA packet (11 oz/550 to 1,100 gal) 2 to 4 oz/100 gal	Spray as needed. Bayleton cannot be used on plants for sale or other commercial use. Strike is for greenhouse and nursery use only.
Trifloxystrobin (Compass O)	2 to 4 oz/100 gal	Apply to point of drip before disease is detected. Apply at 7- to 14-day intervals. Rotate to another nonstrobilurin product after each application.
Triticonazole (Trinity) 19SC	6 to 12 fl oz/100 gal	Apply every 7 to 14 days. Use preventively. Begin applications when conditions favor fungal infection and before disease symptom development. Use of an adjuvant/spreader sticker can aid in control.
Rhizoctonia Aerial Blight (Rhizoctonia solani)		
chlorothalonil (Daconil Ultrex)	1.4 lb/100 gal	Spray to runoff. Repeat at 7- to 14-day intervals until conditions no longer favor disease.
Cyprodinil + fludioxonil (Palladium)	2 to 6 oz/100 gal	Spray on a 7- to 14-day interval while conditions are conducive to disease development. After two applications, alternate with another fungicide with a different MOA for two applications.
Fludioxonil (Medallion)	1 to 2 oz/100 gal	Spray to runoff. Repeat at 7- to 14-day intervals until conditions no longer favor disease.
Flutolanil (Contrast) 70WSP (Prostar) 70WP	3 to 12 oz/100 gal	Apply at 14- to 21-day intervals.
Iprodione (Chipco 26019 N/G) (18 Plus, Iprodione Pro 2SE)	1.0 to 2.5 qt/100 gal 1.0 to 2.5 qt/100 gal	Spray plants to ensure thorough coverage Repeat at 7- to 14-day intervals. Do not make more than four applications per crop per year.
mancozeb (Dithane 75DF)	1 to 2 lb/100 gal	Apple at 7- to 10-day intervals. See label for specific instructions.
myclobutanil (Eagle 20EW)	6 to 12 fl oz/100 gal	Apply at 7- to 10-day intervals. See label for specific instructions.
polyoxin D zinc salt (Endorse 2.5 WP)	1.1 to 2.2 lb/100 gal	Apply as a foliar spray every 7 to 10 days. Apply prior to disease development and when conditions are conducive for disease.
pyraclostrobin + boscalid (Pageant Intrinsic)	12 to18 oz/100 gal	Apply prior to disease development. Repeat at 7- to 14-day intervals. Make no more than two sequential applications. Do not expose petunia or impatiens in flower or wintercreeper or nine bark to spray or drift as injury may occur.
thiophanate-methyl (Cleary 3336) F	12 to 16 fl oz/100 gal	Apply when disease symptoms first appear. Repeat at 7- to 14-day intervals during disease period.
Thiophanate-methyl + iprodione (26/36 fungicide)	33-84 fl oz/100 gal	No more than 4 applications per year.
triflumizole (Terraguard) 50W	4 to 8 oz/100 gal	Make initial application prior to or at first sign of disease. Use the higher rate under heavy disease pressure. Repeat at 7- to 14-day intervals.
Rhizoctonia Stem and Root Rot (Rhizoctonia solani)		
azoxystrobin (Heritage) 50WDG	directed spray: 1 to 4 oz/100 gal drench: 0.2 to 0.9 oz/100 gal	Apply as a directed spray every 7 to 21 days as needed. To avoid fungicide resistance, make no more than three sequential applications of Heritage before rotating with nonstrobilurin products. Apply 1 to 2 pt of solution per sq ft surface area every 7 to 28 days as a preventative drench treatment. Do not exceed 2 oz/100 gal on impatiens or pansy.
azoxystrobin + benzovindiflupyr (Mural)	3 oz/100 gal	Apply 1 to 2 pints of drench solution per square foot surface area every 7 to 28 days. Should not be applied to some tree fruit varieties. See label.
Bacillus amyloliquefaciens strain D747 (DoubleNickel LC)	0.5 to 4.5 pt/100 gal	Apply as a drench or course spray to growing medium at or immediately before seeding. Repeat applications as needed. OMRI listed.
chlorothalonil + thiophanate-methyl (Spectro) 90WDG	1 to 2 lb/100 gal	Retreat at 7-day intervals. Apply as a spray only. Do not apply more than once to green or variegated Pittosporum due to risk of phytotoxicity. Apply when foliage is dry.
cyprodinil + fludioxonil (Palladium)	2 to 4 oz/100 gal	For stem diseases ensure full spray coverage of all stems and inner areas of the plant to the soil/media. Apply on a 7 to 14 day interval while conditions are conducive to disease develepment. After two applications, alternate with another fungicide with a different MOA for two applications.
etridiazole + thiophanate-methyl (Banrot) 40W	16 to 12 oz/400 sq ft	Apply in sufficient volume to saturate the soil mixture. Irrigate immediately. Repeat at 4- to 12-week intervals if necessary.
fludioxonil (Medallion) 50W	1 to 2 oz/100 gal	Apply as a drench at seeding or transplanting. Apply sufficient mix to wet the upper one-half of the growing medium. Make only one application to seedling crop. If needed, re-treat transplants 21 to 28 days after initial application. Do not apply as a seed or soil drench to impatiens or New Guinea impatiens. May cause stunting or chlorosis on some geranium cultivars. See label for maximum amounts that can be applied per year.
flutolanil (Contrast) 70WSP (Prostar) 70WP	3 to 6 oz/100 gal	Apply drench according to label. Repeat 21 to 28 days after initial application. Make no more than four applications per year to ornamental plantings.
iprodione (Chipco 26019 N/G) (18 Plus, Iprodione Pro 2SE)	6.5 oz/100 gal 13 fl oz/100 gal	Apply 1 to 2 pints per sq ft at seeding or transplanting. Do not apply as a drench on impatiens or pothos. Repeat every 14 days. Do not make more than six applications per year. Do not use on Spathiphyllum.
metconazole (Tourney)	1 to 4 oz/100 gal	Repeat in 14 to 28 days when conditions favor disease
PCNB (Terraclor) 75WP	4 to 8 oz/100 gal	See label for amount to apply. One repeat application can be made 4 to 6 weeks later, if necessary.
polyoxin D zinc salt (Endorse 2.5 WP)	1.1 to 2.2 lb/100 gal/acre	Apply as a foliar spray every 7 to 10 days. Apply as a drench every 14 to 28 days.
pyraclostrobin + boscalid (Pageant Intrinsic)	12 to18 oz/100 gal	Apply prior to disease development. Repeat at 7- to 14-day intervals. Make no more than two sequential applications. Do not expose petunia or impatiens in flower or wintercreeper or nine bark to spray or drift as injury may occur.

Table 10-13. Comercial Landscape and Nursery Crops Disease Control

DISEASE Pesticide and Formulation	Rate of Formulation (per 100 gallons)	Schedule and Remarks
Rhizoctonia Stem and Root Rot (continued)		
thiophanate-methyl (AllBan Flo) (Cleary 3336) F (OHP-6672) 50WP	7.5 to 20 oz/100 gal 8 to 16 fl oz/100 gal 12 to 16 oz/100 gal	Repeat at 4- to 8-week intervals. Apply as a soil drench or directed spray to thoroughly soak growing media through the root zone after seeding or transplanting at 21- to 28-day interval. Apply 1 to 3 pt/sq ft after transplanting to thoroughly soak growing medium. Repeat every 21 to 28 days.
Thiophanate-methyl + iprodione (26/36 fungicide)	See label	See label.
trifloxystrobin (Compass)	0.5 oz/100 gal	Apply as a drench to wet upper half of the growing media. Apply at seeding, transplanting, and at 21- to 28-day intervals thereafter. May injure petunia, violet, and New Guinea impatiens.
triflumizole (Terraguard) 50W	4 to 8 oz/100 gal	Apply as soil drench at 2- to 4-week intervals. Use higher rate under heavy disease pressure.
Rust (also see Daylily Rust)		
azoxystrobin (Heritage)	1 to 4 oz/100 gal commercial rose production: 1.6 to 8 oz	Apply at 7- to 28-day intervals. Do not make more than three sequential applications of Heritage before alternating with a nonstrobilurin fungicide. Should not be applied to certain plant species; see label.
azoxystrobin + benzovindiflupyr (Mural)	4 to 7 oz/100 gal	Apply every 7 to 14 days. Do not make more than two sequential applications before rotating to another class of fungicide that is not Group 7 or 11.
Bacillus amyloliquefaciens strain D747 (DoubleNickel LC)	0.5 to 6 qt/100 gal	Begin preventative applications at plant emergence and repeat every 3 to 28 days depending on disease pressure. OMRI listed.
chlorothalonil (Daconil Ultrex) (Echo) 90DF	1.4 lb/100 gal; 1 lb for roses 1.4 lb/100 gal; 0.875 oz for roses	Apply when foliage and flowers are dry. Repeat at 7- to 14-day intervals. Apply to hydrangea foliage only. Avoid application during bloom period on plants where flower injury is unacceptable.
chlorothalonil + propiconazole (Concert II)	22 to 35 fl oz/100 gal	Apply as a full coverage spray. Reapply at 14 to 21 day intervals. Refer to label for specific rate and application instructions. Higher rate listed for *Melampsora occidentalis* --use 69 fl oz/100gal.
chlorothalonil + thiophanate- methyl (Spectro) 90	1 to 2 lb/100 gal	Apply when foliage and flowers are dry, or nearly dry. Re-treat at 7-day intervals. Do not exceed 50.6 lb per acre during one season for field-grown ornamentals.
flutolanil (Contrast) 70 WSP	3 to 6 oz/100 gal	Repeat at 14- to 21-day intervals.
fuoxastrobin (Disarm 480SC)	1 to 4 fl oz/100 gal	Use for preventative applications only and apply every 7 to 28 days depending on disease pressure. Do not make more than two consecutive applications before switching to a non Group 11 fungicide.
mancozeb (Dithane, Mancozeb) (Protect DF)	1.5 lb/100 gal 1 to 2 lb/100 gal	Begin at first sign of disease. Repeat at 7- to 10-day intervals. Apply at 7- to 21-day intervals. To improve performance, add 2 to 4 oz of a spreader-sticker.
metconazole (Tourney)	1 to 4 oz/100 gal	Repeat in 14 to 28 days when conditions favor disease
myclobutanil (Eagle 40WP) (Eagle 20WE)	3 to 6 oz/100 gal 6 to 12 oz/100 gal	Apply on a protectant application schedule at 10- to 14-day intervals.
neem oil (Triact 70)	1 gal/100 to 200 gal See label	Apply at 7- to 14-day spray intervals. Trial first on open blooms. To control existing disease, apply on a 7-day schedule until disease pressure is eliminated. Not for impatiens, carnation, or hibiscus.
pyraclostrobin (Insignia)	See label	Apply prior to onset of disease and repeat every 7 to 14 days. Do not expose flowering impatiens or flowering petunias to Insignia.
propiconazole (Banner MAXX)	See label	See label. Do not use in greenhouses.
pyraclostrobin + boscalid (Pageant Intrinsic)	12 to18 oz/100 gal	Apply prior to disease development. Repeat at 7- to 14-day intervals. Make no more than two sequential applications. Do not expose petunia or impatiens in flower or wintercreeper or nine bark to spray or drift as injury may occur.
tebuconazole (Torque)	4 to 8 fl oz/100 gal	Apply every 14 for a total of three applications at the first sign of disease.
thiophanate-methyl (AllBan Flo) (Cleary 3336) F (OHP 6672) 50WP, (T-Storm) 50WSB OHP 6672 (4.5L)	10.75 to 20 oz/100 gal 12 to 16 fl oz/100 gal 12 to 16 oz/100 gal 20 fl oz/100 gal	For use on crabapples. Do not use treated crabapples for food. Apply late Spring or at first sign of disease. Repeat every 7-14 days as needed during disease period. Apply 8 oz as a drench or directed spray after seeding, or apply 12 to 16 oz after transplanting. Repeat at 21- to 28-day intervals. Apply late Spring or at first sign of disease. Repeat every 7-14 days as needed during disease period. Apply as heavy spray or drench at a rate of 1 to 2 pints per sq ft. Repeat at 2- to 4-week intervals.
thiophanate-methyl + mancozeb (Zyban) WSB	See label	For use on crabapples.
triadimefon (Bayleton) (Strike)	See label	Spray to the point of drip as needed. See label for spray interval. Bayleton is not for use on plants being grown for sale.
triflumizole (Terraguard) 50W	2 to 8 oz/100 gal	Apply prior to, or at first sign of disease. Repeat at 7- to 14-day intervals.
Scab (*Cladosporium, Fusicladium, Spilocea, Venturia*) For apple scab, see disease under "crabapple" above.		
Mancozeb (Dithane, Fore, Mancozeb)	1.5 lb/100 gal	Begin spraying at first sign of disease, and repeat at 7- to 10-day intervals.
azoxystrobin (Heritage)	1 to 4 oz/100 gal	Repeat every 10 to 28 days. Do not make more than two consecutive applications before switching to a non-Group 11 fungicide. PHYTOTOXIC to some crabapple and other tree fruit cultivars. See label
azoxystrobin + benzovindiflupyr (Mural)	4 to 7 oz/100 gal	Repeat every 7 to 14 days. Do not make more than two consecutive applications before switching to a non-Group 11 fungicide. PHYTOTOXIC to some crabapple and other tree fruit cultivars. See label

Table 10-13. Comercial Landscape and Nursery Crops Disease Control

DISEASE Pesticide and Formulation	Rate of Formulation (per 100 gallons)	Schedule and Remarks
Scab (*Cladosporium, Fusicladium, Spilocea, Venturia*) (continued)		
copper oxychloride + copper hydroxide (Badge SC) (Badge X2)	1.5-2 pt/A 1.5 to 2 lb/A	Repeat at 7 to 14 day intervals. May cause discoloration in some azalea varieties. See label for other phytoxicity warnings. Badge X2 is OMRI listed. Do not mix with Aliette.
Fluoxastrobin + myclobutanil (Disarm M)	3 to 11 fl oz/100 gal	Repeat at 7 to 28 day intervals
Metconazole (Tourney)	1 to 4 oz/100 gal	Repeat in 14 to 28 days when conditions favor disease
myclobutanil (Eagle 20EW)	6 to 12 fl oz/100 gal	Spray every 10 to 14 days.
Propiconazole (Banner MAXX)	See label	See label. Do not use in greenhouses.
Pyraclostrobin (Insignia)	4 to 8 oz/100 gal	Repeat at 7- to 14-day intervals. Do not make more than 2 consecutive applications before switching to a non-Group 11 fungicide.
Pyraclostrobin + boscalid (Pageant Intrinsic)	8 to 12 oz/100 gal	Apply prior to disease development. Repeat at 7- to 10-day intervals. Make no more than two sequential applications. Do not expose petunia or impatiens in flower or wintercreeper or nine bark to spray or drift as injury may occur.
Tebuconazole (Torque)	4 to 10 fl oz/100 gal	For preventative applications, apply at least 3 times per year, 14 to 21 days apart.
Thiophanate-methyl (AllBan Flo) (Cleary 3336) F (SysTec 1998) FL	10.75 to 20 fl oz/100 gal 12 to 16 fl oz/100 gal 20 fl oz/100 gal	Spray at bud break. Repeat three to four times at 7- to 14-day intervals.
Thiophanate-methyl + iprodione (26/36 fungicide)	33 to 84 fl oz/100 gal	No more than 4 applications per year.
thiophanate-methyl + mancozeb (Zyban) WSB	24 oz (4 bags)	Repeat at 7-day intervals.
Trifloxystrobin (Compass O)	2-4 oz/100 gal	Repeat at 7- to 14-day intervals. Do not make more than 2 consecutive applications before switching to a non-Group 11 fungicide.
Triticonazole (Trinity) 19SC	6 to 12 fl oz/100 gal	Apply every 7 to 14 days. Use preventively. Begin applications when conditions favor fungal infection and before disease symptom development.
Sclerotinia Stem Rot (*Sclerotinia sclerotiorum*)		
chlorothalonil (Daconil Ultrex) 82.5 WDG	1.4 lb/100 gal	Repeat at 7- to 14-day intervals when conditions favor disease.
Cyprodinil + fludioxonil (Palladium)	2 to 4 oz/100 gal	For stem diseases ensure full spray coverage of all stems and inner areas of the plant to the soil/media. Apply on a 7- to 14-day interval while conditions are conducive to disease development. After two applications, alternate with another fungicide with a different MOA for two applications.
Metconazole (Tourney)	1 to 4 oz/100 gal	Repeat in 14 to 28 days when conditions favor disease
PCNB (Revere) 10G (Terraclor) 75W	20 lb/1,000 sq ft 6 to 12 fl oz/100 gal	Apply 1 wk prior to planting; spread on soil surface and mix into soil at a 4-in. depth. See label for amount to apply. One repeat application may be made 4 to 6 weeks later.
Pyraclostrobin + boscalid (Pageant Intrinsic)	12 to 18 oz/100 gal	Apply prior to disease development. Repeat at 7- to 10-day intervals. Make no more than two sequential applications. Do not expose petunia or impatiens in flower or wintercreeper or nine bark to spray or drift as injury may occur.
Triticonazole (Trinity) 19SC	8 to 12 fl oz/100 gal	Apply every 7 to 14 days. Use preventively. Begin applications when conditions favor fungal infection and before disease symptom development. The stem areas of the plant must be thoroughly covered using spray to runoff.
Shot Hole (*Blumeriella, Coccomyces* fungal and bacterial)		
copper hydroxide + mancozeb (Junction)	1.5 to 3.5 lb/100 gal	Begin at first sign of disease. Repeat at 7- to 14-day intervals.
Mancozeb (Pentathlon DF) (Protect DF)	1 to 2 lb/100 gal	Begin applications at first sign of disease. Apply at 7- to 10-day intervals. Begin applications at first sign of disease. Apply at 7- to 21-day intervals.
Metconazole (Tourney)	1 to 4 oz/100 gal	Apply on a 14 to 28 day interval.
thiophanate-methyl (Cleary 3336) F	12 to 16 fl oz/100 gal	Begin when disease first appears, and repeat every 7 to 14 days.
Southern Stem Blight (*Sclerotium rolfsii*)		
azoxystrobin (Heritage)	directed spray: 1 to 4 oz/100 gal drench: 0.2 to 0.9 oz/100 gal	Apply every 7 to 21 days. Can be used in outdoor nurseries, retail nurseries, residential and commercial landscape areas. Apply 1 to 2 pt solution per square foot surface area every 7 to 28 days, prior to infection. Apply to container-grown ornamentals only. Do not apply to crabapple, apple, or some tree fruit. See label.
azoxystrobin + benzovindiflupyr (Mural)	5 to 7 oz/100 gal (direct spray) 2 to 3 oz/100 gal (drench)	Repeat every 7 to 21 day. Do not make more than two consecutive applications before switching to a non-Group 11 fungicide. PHYTOTOXIC to some crabapple and other tree fruit cultivars. See label.
cyprodinil + fludioxonil (Palladium)	2 to 4 oz/100 gal	For stem diseases ensure full spray coverage of all stems and inner areas of the plant to the soil/media. Apply on a 7 to 14 day interval while conditions are conducive to disease development. After two applications, alternate with another fungicide with a different MOA for two applications.
Fluoxastrobin (Disarm 480SC)	2 to 4 fl oz/100 gal	Crown spray every 7 to 21 days.
flutolanil (Contrast) 70WSP (Prostar) 70WP	3 to 6 oz/100 gal	Drench at 21- to 28-day intervals. Uses 1 pt per sq ft or 2 pt for depths greater than 4 in. See label for container rates. For use in outdoor container and field-grown stock. Make no more than 4 applications per year to nursery ornamental plantings.
PCNB (Terraclor) 75WP	3.25 to 6.5 lb/1,000 sq ft	Apply in sufficient water to ensure uniform ground coverage prior to planting, and thoroughly incorporate to a depth of 6 to 7 in. For use in nursery and landscape plantings.

DISEASE CONTROL

Table 10-13. Comercial Landscape and Nursery Crops Disease Control

DISEASE Pesticide and Formulation	Rate of Formulation (per 100 gallons)	Schedule and Remarks
Southern Stem Blight *(Sclerotium rolfsii)* **(continued)**		
tebuconazole (Torque)	4 to 8 fl oz/100 gal	Apply every 14 for a total of three applications at the first sign of disease.
triticonazole (Trinity) 19SC	8 to 12 fl oz/100 gal	Apply every 7 to 14 days. Use preventively. Begin applications when conditions favor fungal infection and before disease symptom development. The stem areas of the plant must be thoroughly covered using spray to runoff.
Volutella Blight		
chlorothalonil (Daconil Ultrex) 82.5 WDG (Daconil WeatherStik)	1.4 lb/100 gal 1.375 pt/100 gal	Reapply at 7- to 14-day intervals.
copper hydroxide (Champ WG) (Nu-Cop) 50DF (CuPRO 2005)	0.5 lb/100 gal 1 lb/100 gal 0.75 to 2.0 lb/100 gal	Begin at first sign of disease, and repeat at 7- to 14-day intervals. Do not tank mix copper formulations with Aliette. Avoid contact with metal surfaces. Discoloration of blooms may occur on certain plant varieties- check label.
copper hydroxide + mancozeb (Junction)	1.5 to 3.5 lb/100 gal	Begin at first sign of disease, and repeat at 7- to 14-day intervals.
copper oxychloride + copper hydroxide (Badge SC) (Badge X2)	1.5-2 pt/A 1.5 to 2 lb/A	Repeat at 7 to 14 day intervals. May cause discoloration in some azalea varieties. See label for other phytoxicity warnings. Badge X2 is OMRI listed. Do not mix with Aliette.
copper sulfate pentahydrate (Phyton 27, Phyton 35)	See label	See label. To avoid phytotoxicity, do not use any copper compound on alyssum.
mancozeb (Dithane) 75 DF, (Fore) 80 WSP, Mancozeb DG) (Pentathlon LF) (Pentathlon DF) (Protect DF)	2.0 lb/50 gal/5,000 sq ft of bed 0.8 to 1.6 pt/100 gal 1 to 2 lb/100 gal 3 to 4 lb/100 gal/10,000 sq ft of bed	Start at first sign of disease, and apply at 10- to 14-day intervals. Use a drenching spray. Start at first sign of disease, and apply at 10- to 14-day intervals. Begin at first sign of disease, and repeat at 7- to 10-day intervals. Use a drenching spray. Start at first sign of disease, and apply at least 5 applications at 10- to 14-day intervals.
thiophanate-methyl + mancozeb (Zyban) WSB	24 oz (4 bags)/100 gal	Apply at 7-day intervals while disease is prevalent.

Further Information

Boxwood blight links http://plantpathology.ces.ncsu.edu/pp-ornamentals/
All Plant Pathology *Disease Notes for Ornamentals* can be accessed from
http://www.ces.ncsu.edu/depts/pp/notes/Ornamental/ornamental_contents.html.
Holly Diseases and Their Control in the Landscape. Plant Pathology Ornamental Disease Information Note 7,
http://www.ces.ncsu.edu/depts/pp/notes/oldnotes/od7.htm
Nematodes and Their Control in Woody Ornamentals in the Landscape. Plant Pathology Information Note 63,
http://www.ces.ncsu.edu/depts/pp/notes/oldnotes/no63.htm
Phytophthora Root Rot and Its Control on Established Woody Ornamentals. Plant Pathology Ornamental Disease Information Note 13,
http://www.ces.ncsu.edu/depts/pp/notes/oldnotes/odin13/od13.htm
Rose Diseases and Their Control in the Home Garden. Plant Pathology Ornamental Disease Information Note 2,
http://www.ces.ncsu.edu/depts/pp/notes/Ornamental/odin002/odin002.htm
Juniper Diseases. Plant Pathology Ornamental Disease Information Note 15, http://www.ces.ncsu.edu/depts/pp/notes/oldnotes/od15.htm
Entomosporium Leaf Spot on Redtip. Plant Pathology Ornamental Disease Information Note 11,
http://www.ces.ncsu.edu/depts/pp/notes/Ornamental/odin011/odin011.htm
Rhododendron Diseases. Plant Pathology Ornamental Disease Information Note 12, http://www.ces.ncsu.edu/depts/pp/notes/oldnotes/od12.htm
Azalea Diseases. Plant Pathology Ornamental Disease Information Note 16.

Copies of these publications are available from your local Cooperative Extension Center.

Treatments for Sanitizing Tools, Equipment, Cultivation Surfaces, Pots and Flats

Note: It was not possible to update the information in this table for the 2016 North Carolina Agricultural Chemicals Manual. The information in this table is current as of Jan. 1, 2015. Anyone using any agricultural chemical should refer to the current chemical label, which contains information about the safe and effective use of the chemical, before using the chemical.

All items should be free of organic debris before exposure to the treatments listed below. Sanitizing an entire greenhouse involves physically removing leftover debris and soil as a first step prior to disinfection, as soil and organic residues reduce the effectiveness of disinfectants. There are some commercial cleaners specifically designed for greenhouse use, e.g., Strip-It (best applied by spray, brush, or foam), which is a combination of cleaning and wetting agents formulated to remove algae, dirt, and hard water deposits. High pressure power washing with soap and water is also an option prior to disinfection as listed below.

Table 10-14. Treatments for Sanitizing Tools, Equipment, Cultivation Surfaces, and other Related Items

Material or Treatment	Trade name	Formulation	Remarks	Contact time
alcohol, ethyl and isopropyl (grain, rubbing, wood) (70-100%)	Various commercial brands; Lysol Spray (also includes quaternary ammonium)	Depends on formulation. Read label. Typically full strength for RTU (Ready To Use) formulations.	Evaporates quickly so that adequate contact time may not be achieved; high concentrations of organic matter diminish effectiveness; flammable.	10 min for equipment, pots, flats and surfaces. Tools can be dipped for 10 seconds and allowed to dry. Do not rinse.
hydrogen peroxide (hydrogen dioxide) and peroxyacetic acid mixture	ZeroTol 2.0; SaniDate 5.0; Oxidate 2.0	2.5 fl oz per gallon of water 0.5 fl oz per gallon of water 0.5 to 1.25 fl oz per gallon of water	Very corrosive; eye/skin irritant. Low odor. Use according to label. Must be stored in cool location.	1-10 min
quaternary ammonium	Physan 20;	Depends on formulation. Typically 1 tablespoon per gallon of water	Effective for non-porous surface sanitation, e.g. floors, walls, benches, pots. Low odor, irritation.	10-15 min Must remain wet for 10 min. Wipe dry with a clean cloth or sponge or allow to air dry.
	KleenGrow	For general disinfection use 0.5 to 1.0 fl oz per gallon of water	Hard, NON-POROUS surfaces use 1.0 fl oz per gal water; Tools, cutters & equipment use 0.5 fl oz per gal water. Apply solution with a cloth, mop, sponge, coarse spray device or by immersion until surfaces are wet. Prepare a fresh solution daily.	Must remain wet for 10 min. Wipe dry with a clean cloth or sponge or allow to air dry.
sodium hypochlorite (8.25%)	Clorox; Commercial bleach;	10%; or a 1:14 ratio of bleach : water	Inactivated by organic matter; fresh solutions should be prepared every 8 hr or more frequently if exposed to sunlight; corrosive to metal; irritating to eyes and skin; Exposure to sunlight reduces efficacy. Keep solution in opaque container.	10-15 min. for equipment, pots, flats and surfaces. Tools can be dipped for 10 seconds and allowed to dry. Do not rinse.
steam	NA	Cover or otherwise seal	For plastic pots and trays, heat center of steamer between 150 degrees F to 160 degrees F;	60 min.
			For less heat-sensitive objects, heat to 180 degrees F.	15 min.
solarization	NA	Place clean items on solid surface, cover tightly with CLEAR plastic	Clear plastic works much better.	140 degrees F, 4 to 8 hr/day for 7 days

Disease Control for Commercial Vegetables

L. M. Quesada-Ocampo, Inga Meadows, and Frank Louws, Plant Pathology

This section was prepared as a collaborative effort of vegetable pathology experts in the southeastern United States who yearly update the Southeastern U.S. Vegetable Crop Handbook. Contributors this year included: L. M. Quesada-Ocampo (North Carolina State University), E. Sikora (Auburn University), R. Melanson (Mississippi State University), F. Louws (North Carolina State University), E. Pfeufer (University of Kentucky), I. Meadows (North Carolina State University), and E. Eure (North Carolina State University).

Caution: At the time these tables were prepared, the entries were believed to be useful and accurate. However, labels change rapidly, and errors are possible, so the user must follow all directions on the product label. Federal tolerances for fungicides may be canceled or changed at any time.

Information in the following tables must be used in the context of a comprehensive disease control program. Many diseases are controlled by combined strategies—using resistant varieties, crop rotation, deep-turn plowing, sanitation, seed treatments, cultural practices, and fungicides. Always use top-quality seed and plants obtained from reliable sources. Seed are ordinarily treated by commercial producers for control of decay and damping off.

Nematode control chemicals and greenhouse diseases are provided in separate tables following the crop tables. Efficacy tables will help in selecting appropriate disease control materials for some vegetable crops. These tables are located at the end of each crop table.

Rates: Some foliar rates given in the tables are based on mixing a specified amount of product in 100 gallons of water and applying the finished spray for complete coverage of foliage just to the point of runoff with high pressure (over 250 psi) drop nozzle sprayers. Actual amount of product and water applied per acre will vary depending on plant size and row spacing. Typically, 25 to 75 gallons per acre of finished spray are used. Concentrate spray (air blast, aircraft, etc.) rates are based on amount of product per acre. Caution: With concentrate sprays, it is easy to apply too much product. Some fungicides are adversely affected by pH of water; adjust pH of water if specified on label.

Do not feed treated foliage to livestock unless allowed by the label. Do not reenter fields until sprays have dried; some fungicides may have a reentry requirement of one to several days. Read the label. Do not exceed maximum number of applications on the label. Do not exceed maximum limit of fungicide per acre per application or per year as stated on the label. See label for rotational crops. In all cases, follow directions on the label.

Disease Control by Crop

Asparagus

Table 10-15. Disease Control Products for Asparagus

Disease	Material	Rate of Material to Use		Minimum Days		Method, Schedule, and Remarks
		Formulation	Active Ingredient	Harvest	Reentry	
Gray mold (*Botrytis cinerea*)	fenhexamid (Elevate)	1.5 lb/A	0.75 lb/A	180	0.5	Apply at fern stage only.
Crown rot	mancozeb (various)	See label	See label	See label	See label	See label. Rates vary depending on the formulation. Soak crowns 5 minutes in burlap bag with gentle agitation, drain, and
Phytophthora crown rot, spear rot	mefenoxam (Ridomil Gold SL)	1 pt/acre	0.5 lb/A	1	2	Apply over beds after seeding or covering crowns, 30 to 60 days before first cutting, and just before harvest.
	fosetyl-AL (Alliette)	5 lb/A	4 lb a.i.	110	0.5	
Rust	myclobutanil (Rally 40W)	5 oz/A	2.0 oz a.i.	180	1	Begin applications to developing ferns after harvest has taken place. Repeat on a schedule not to exceed 14 days. Do not supply to harvestable spears.
	sulfur (various)	See label	See label	See label	See label	See label. Rates vary depending on the formulation.
	tebuconazole (Folicur) 3.6F (Orius) 3.6F (Toledo) 3.6F	4 to 6 fl oz/A	0.11 to 0.17 lb	180	0.5	Apply to developing ferns at first sign of rust and repeat on a 14-day interval; no more than 3 applications per season.
Rust, Cercospora leaf spot	chlorothalonil (various)	See label	See label	See label	See label	See label. Rates vary depending on the formulation.
	mancozeb (various)	See label	See label	See label	See label	Apply to ferns after harvest; spray first appearance, 7 to 10 day intervals. Do not exceed 8 pounds product per acre per crop.
Purple spot	azoxystrobin (Quadris)	6 to 15.5 fl oz	0.10 to 0.25 lb a.i.	100	4 hr	Do not apply more than 1 foliar application of Quadris (or other group 11 fungicide) before alternating with a fungicide that has a different mode of action.
	chlorothalonil (various)	See label	See label	See label	See label	See label. Rates vary depending on the formulation.
	trifloxystrobin (Flint 50 WDG)	3 to 4 oz/A	1.5 to 2 oz/A	180	12 hr	Make no more than one application before alternating with fungicides that have a different mode of action. Begin applications preventively when conditions are favorable for disease and continue as needed on a 7 to 14 day interval.

Table 10-16. Relative Importance of Alternative Management Practices for Disease Control in Asparagus

Scale E, excellent; G, good; F, fair; P, poor; NC, no control; ND, no data.

Management tactic	Disease				
	Rust	Cercospora blight	Stemphylium blight	Fusarium root rot	Phytophthora crown/ spear rot
Avoid overhead irrigation	F	F	F	NC	NC
Crop rotation (5 years or more)	NC	NC	NC	F	P
Clip and bury infected ferns	G	G	G	NC	NC
Destroy infected ferns	E	E	E	NC	NC
Encourage air movement/wider row spacing	P	P	G	NC	NC
Plant in well-drained soil	NC	NC	NC	F	F
Destroy volunteer asparagus	F	NC	NC	NC	NC
Pathogen-free planting material	NC	NC	NC	E	E
Resistant/tolerant cultivars	G	G	NC	G	NC

Basil

Table 10-17. Disease Control Products for Basil

Disease	Material	Rate of Material to Use		Minimum Days		Method, Schedule, and Remarks
		Formulation	Active Ingredient	Harvest	Reentry	
Damping off (*Pythium*)	mefenoxam (Ridomil Gold SL)	1.0 to 2.0 pt/acre		21	2	Limit of 2 soil applications per season.
Leaf spots, fungal (*Botrytis, Alternaria, Fusarium*), powdery mildew	cyprodinil + fludioxonil (Switch) 62.5WG	11 to 14 oz/acre	6.87 to 8.75 oz/acre	7	0.5	Limit of 56 fluid ounces per acre per season. Make no more than two consecutive applications before rotating to another effective fungicide with a different mode of action.
	fluopyram (Luna Privilege)	4.0 to 6.84 fl oz/acre		3	0.5	Limit of 13.7 fl oz per acre per season. Apply as needed on a 7- to 10-day interval. When disease pressure is severe, use the higher rates and/or shorter intervals.
	fluopyram (Luna Sensation)	5.0 to 7.6 fl oz/acre		7	0.5	Limit of 15.3 fl oz per acre per season. Apply as needed on a 7- to 10-day interval. When disease pressure is severe, use the higher rates and/or shorter intervals.
Downy mildew (*Peronospora belbahrii*)	cyazofamid (Ranman)	2.75 to 3 fl oz/acre	0.071 to 0.078 lb/acre	0	0.5	Limit of 27 fluid ounces per acre per season. Alternate with a fungicide with a different mode of action. May be applied through sprinkler irrigation system.
	manipropamid (Revus)	8 fl oz/acre	0.13 lb/acre	1	4 hr	Limit of 32 fluid ounces per acre per season. Make no more than two consecutive applications before rotating to another effective fungicide with a different mode of action. **NOTE: This is a supplemental label (EPA Reg. No. 100-1254) that expires on 12/19/2016.**
	phosphorous acid (Confine Extra, K-Phite)	1 to 3 qt/20 to 100 gal water/acre	0.53 to 1.59 qt/20 to 100 gal water/acre	0	4 hr	Do not apply at less than 3 day intervals.
	potassium phosphite (Fosphite, Fungi-phite, Helena Prophyt)	1 to 3 qt/100 gal water/acre	0.53 to 1.59 qt/100 gal water/acre	0	4 hr	Do not apply at less than 3 day intervals.
Fusarium wilt and *Pythium* and *Rhizoctonia* root rots	phosphorous acid (Confine Extra, K-Phite)	1 to 3 qt/20 to 100 gal water/acre	0.53 to 1.59 qt/20 to 100 gal water/acre	0	4 hr	Do not apply at less than 3 day intervals.
	potassium phosphite (Fosphite, Fungi-phite, Helena Prophyt)	1 to 3 qt/100 gal water/acre	0.53 to 1.59 qt/100 gal water/acre	0	4 hr	Do not apply at less than 3 day intervals.
Pythium Damping off	Metalaxyl (Ridomil Gold)	1 to 2 pt/acre	0.5 to 1 pt/acre	21	2	Apply as a basally directed spray 28 days after planting.

Bean

Table 10-18. Disease Control Products for Bean

Disease	Material	Rate of Material to Use		Minimum Days		Method, Schedule, and Remarks
		Formulation	Active Ingredients	Harvest	Reentry	
Bean, Dry (*Catjang, Chickpea, Fava, Garbanzo, Lentil, Lima, Lupine, Mung, May pea, Southernpea, Soybean*)						
Anthracnose, Alternaria leaf spot and blight, Ascochyta leaf and pod spot, rust (*Phakopsora*)	azoxystrobin (Quadris) 2.08 F	6.2 to 15.4 fl oz	0.10 to 0.25 lb	0	4 hr	Do not apply more than three sequential applications.
	azoxystrobin+propiconazole (Quilt Xcel)	10.5-14 oz/A		14	0.5	Apply when conditions are conducive for disease. Up to three applications may be made on 7- to14-day intervals.
	fluxapyroxad + pyraclostrobin (Priaxor)	4.0 to 8.0 fl oz	0.13 to 0.25 lb	21	12 hr	Begin prior to disease development and continue on a 7- to 14-day spray schedule.
Anthracnose, Botrytis gray mold, white mold (*Sclerotinia*)	boscalid (Endura) 70 WG	8 to 11 oz	5.6 to 7.7 oz	21	0.5	Use 6 oz rate for chickpeas and lentils. Many other dried and succulent beans on label.
	picoxystrobin (Approach)	6 to 12 fl oz		14	0.5	Do not apply more than three sequential applications.
Aschochita blight, Rust, white mold	prothioconazole (Proline 480 SC)	4.3 to 5.7 fl oz/acre		7	0.5	See label for specific crop restrictions.
Alternaria, Anthracnose, Ascochyta, rust, web blight	azoxystrobin+propiconazole (Quadris Opti)	10.5-14 oz/A		14	0.5	For use only on beans to be harvested dry with pods removed. Applications should begin prior to disease development. Do not apply more than two foliar applications of Quadris Opti before alternating with a fungicide with a different mode of action.
Alternaria leaf spot and blight, powdery mildew, gray mold, white mold	fluopyram (Luna Privilege)	4.11 fl oz/acre		14	0.5	Begin fungicide applications preventively. Continue as needed on a 7- to 10-day interval depending on disease pressure.

DISEASE CONTROL

Table 10-18. Disease Control Products for Bean

Disease	Material	Rate of Material to Use		Minimum Days		Method, Schedule, and Remarks
		Formulation	Active Ingredients	Harvest	Reentry	
Bean, Dry (*Catjang, Chickpea, Fava, Garbanzo, Lentil, Lima, Lupine, Mung, May pea, Southernpea, Soybean*) (continued)						
Anthracnose, Botrytis gray mold, white mold (*Sclerotinia*)	boscalid (Endura 70 WG)	8 to 11 oz	5.6 to 7.7 oz	21	0.5	Use 6-ounce rate for chickpeas and lentils. Many other dried and succulent beans on label.
	picoxystrobin (Approach)	8 to 12 fl oz		14	0.5	Do not apply more than three sequential applications. For white mold, make preventive applications beginning at bloom (see label).
	thiophanate-methyl (Topsin M) 70 WP	30-40 fl oz		28	3	For multiple applications: Make first application when 10-30% of plants have at least one open bloom and continue on 7 day intervals. Do not apply more than 80 fluid ounces per season.
	fluazinam (Omega 500)	0.5 to 0.85 pts	0.26 to 0.44	14	3	Apply at 10 to 30% bloom.
Botrytis gray mold, white mold (*Sclerotinia*)	fludioxonil (Cannonball WG)	7 oz		7	0.5	Begin before disease develops and continue on 7 day intervals until conditions no longer favor disease development. Do not apply more than 28 ounces per acre. Do not use on cowpeas.
	iprodione (Rovral) 50 WP 4 F	1.5 to 2 lb/acre	0.75 to 1 lb/acre	0	1	Do not use on cowpea. Two applications maximum per saosn with last application no late than peak bloom. Do not feed dry bean hay to livestock until 45 days after last application.
Pythium damping off	mefenoxam (Ridomil Gold) 4 SL	0.5 to 1 pt/trt acre	0.25 to 0.5 lb/trt acre	—	2	Preplant incorporate. See label for row rates.
	azoxystrobin + mefenoxam (Uniform)	0.34 fl oz/1,000 row ft	0.0071 + 0.0028 lb/acre	—		In-furrow spray. See label directions.
Rhizoctonia root rot	azoxystrobin (Quadris) 2.08 F	0.4 to 0.8 fl oz/ 1,000 row feet	0.006 to 0.013 lb	—	4 hr	Make in-furrow or banded applications shortly after plant emergence.
	azoxystrobin + mefenoxam (Uniform)	0.34 fl oz/1,000 row ft	0.0071 + 0.0028 lb/acre	—		In-furrow spray. See label directions.
	PCNB (Blocker)	2.3 to 3.3 fl oz/1000 row ft	1.0 to 1.5 lb/acre	—		In-furrow spray. See label directions.
Rust (*Uromyces*)	azoxystrobin (Quadris) 2.08 F	6.2 fl oz	0.10 lb	0	4 hr	Do not apply more than three sequential applications.
	boscalid (Endura) 70 WG	8 to 11 oz	5.6 to 7.7 oz	21	0.5	Use 6 ounce rate for chickpeas and lentils. Many other dried and succulent beans on label.
	pyraclostrobin (Headline) 2.09F	5.5 to 8 fl oz	1.5 to 2.4 lb	30	0.5	All dry beans except soybeans. Make no more than 2 applications per season
	tebuconazole (Folicur) 3.6F	4 to 6 fl oz/acre	1.5 to 2 lb/acre	14	0.5	Apply before disease appears when conditions favor rust development and repeat at 14 day intervals; maximum 12 fluid ounces per season.
Bean, Snap						
Anthracnose, Botrytis, Sclerotinia	azoxystrobin (Quadris) 2.08 F	6.2 to 15.4 fl oz	0.10 to 0.25 lb	0	4 hr	For anthracnose only. Do not apply more than three sequential applications.
	chlorothalonil (various)	See label	See label	See label	See label	See label. Rates vary depending on the formulation.
	dicloran (Botran) 75 W	2.25 to 4 lb/acre	1.7 to 3 lb/acre	2	0.5	For Sclerotinia only. Use low rate for bush varieties and high rate for pole varieties.
Anthracnose, Botrytis, Sclerotinia (continued)	thiophanate-methyl (Topsin M) 70 WP	1 to 2 lb/acre	0.7 to 1.4 lb/acre	14	1	Spray at 25% bloom, repeat at full bloom. Do not exceed 4 pounds product per season.
	fluazinam (Omega 500)	0.5 to 0.85 pts	0.26 to 0.44	14	3	Apply at 10 to 30% bloom.
	fluxapyroxad + pyraclostrobin (Priaxor)	4.0 to 8.0 fl oz	0.13 to 0.25 lb	7	12 hr	Begin prior to disease development and continue on a 7 to 14 day spray schedule.
Ascochyta blight, Botrytis gray mold, white mold	boscalid (Endura) 70 WG	8 to 11 oz	5.6 to 7.7 oz	7	0.5	Many other dried and succulent beans on label.
	penthiopyrad (Fontelis)	14 to 30 fl oz/acre	0.18 to 0.4 oz/acre	0	12 hr	Begin sprays prior to disease development.
Alternaria, Anthracnose, Ascochyta, rust, southern blight, web blight	azoxystrobin+propiconazole (Quilt) Xcel)	10.5-14 oz/A		7	0.5	Apply when conditions are conducive for disease. Up to three applications made me made on 7-14 day intervals.
Botrytis gray mold, white mold (*Sclerotinia*)	Fludioxonil (Cannonball WG)	7 oz		7	0.5	Begin before disease develops and continue on 7 day intervals until conditions no longer favor disease development. Do not apply more than 28 ounces per acre. Do not use on cowpeas.
Bacterial blights	fixed copper (various)	See label	See label	See label	See label	See label. Rates vary depending on the formulation. Spray first appearance, 10 day intervals.
Cottony leak (Pythium spp)	Reason 500 SC	5.5-8.2 fl oz	0.178 - 0.267 lb a.i./A	3	0.5	Begin applications when conditions become favorabe for disease development. Do not make more then one application before alternating to a product with a dfferent mode of action.
Downy mildew (*Phytophthora nicotianae*)	mandipropamid (Revus)	8 fl oz/acre	0.13 fl oz/acre	1	4 hr	Begin applications prior to disease development and continue throughout the season on a 7-10 day interval.

Table 10-18. Disease Control Products for Bean

Disease	Material	Rate of Material to Use		Minimum Days		Method, Schedule, and Remarks
		Formulation	Active Ingredients	Harvest	Reentry	
Bean, Snap (continued)						
Pythium damping off	mefenoxam (Ridomil Gold) 4 SL	0.5 to 1 pt/trt acre	0.25 to 0.5 lb/trt acre	—	2	Preplant incorporate. See label for row rates.
Powdery mildew	sulfur (various)	See label	See label	See label	See label	See label. Rates vary depending on the formulation. Spray at first appearance, 10 to 14 day intervals. Avoid days over 90 degrees F.
Rhizoctonia root rot	azoxystrobin (Quadris) 2.08 F	0.4 to 0.8 fl oz/ 1,000 row feet	0.006 to 0.013lb/ 1,000 row feet	—	4 hr	Apply in-furrow or banded applications shortly after plant emergence.
	myclobutanil Rally 40 WSP	4 to 5 oz/acre	1.6 to 2 oz/acre	0	1	For *Rhizoctonia* only.
	dichoropropene (Telone) C-17 C-35	10.8 to 17.1 gal/acre 13 to 20.5 gal/acre	107 to 169 lb/acre 139 to 220 lb/acre	—	5	Rate is based on soil type; see label for in-row rates.
	metam-sodium (Vapam) 42 HL	37.5 to 75 gal/trt acre	160 to 320 lb/trt acre	—	—	Rate is based on soil properties and depth of soil to be treated; apply 14 to 21 days before planting.
Rhizoctonia and Fusarium seed root and	Prothioconazole (Redigo 480)	0.16-0.32 fl oz/100 lbs seed		—	—	For seed rot and damping off caused by Rhizoctonai and Fusarium.
Rust *(Uromyces)*	azoxystrobin (Quadris) 2.08 F	6.2 to 15.4 oz/acre	0.10 to 0.25 lb/acre	0	4 hr	Make no more than three sequential applications.
	pyraclostrobin (Headline)	6.0 to 9.0 fl oz	0.09 to 0.18 lb/acre		12 hr	Make no more than two sequential applications.
	boscalid (Endura) 70 WG	8 to 11 oz/acre	5.6 to 7.7 oz/acre	7	0.5	Many other dried and succulent beans on label.
	chlorothalonil (various)	See label	See label	See label	See label	See label. Rates vary depending on the formulation.
	myclobutanil Rally 40 WSP	4 to 5 oz/acre	1.6 to 2 oz/acre	0	1	Spray at first appearance.
	sulfur (various)	See label	See label	See label	See label	See label. Rates vary depending on the formulation. Spray at 7 to 10 day intervals.
	tebuconazole (Folicur) 3.6F	4 to 6 fl oz/acre	1.5 to 2 lb/acre	7	0.5	Apply before disease appears when conditions favor rust development and repeat at 14 day intervals; maximum 24 fluid ounces per season.
White mold *(Sclerotinia)*	botran 75 W	2.5 to 4 lb/acre	1.9 to 3 lb/acre	2	0.5	Use low rate for bush varieties and high rate for pole varieties.
Bean, Lima						
Botrytis, Sclerotinia, leaf spots	azoxystrobin (Quadris) 2.08 F	6.2 to 15.4 fl oz/acre	0.10 to 0.25 lb/acre	0	4 hr	Leaf spots only; do not make more than three sequential applications.
	thiophanate-methyl (Topsin M) 70 WP	1.5 to 2 lb/acre	1.05 to 1.4 lb/acre	14	1	4 pounds limit per acre per crop.
	iprodione (Rovral) 50 WP 4 F	1.5 to 2 lb/acre 1.5 to 2 pt/acre	0.75 to 1 lb/acre	0	1	
	fluazinam (Omega 500)	0.5 to 0.85 pts	0.26 to 0.44	30	3	Apply at 10 to 30% bloom.
	penthiopyrad (Fontelis)	14 to 30 fl oz/acre	0.18 to 0.4	0	12 hr	Begin sprays prior to disease development.
	pyraclostrobin (Headline)	6.0 to 9.0 fl oz	0.09 to 0.18 lb/acre		12 hr	Make no more than two sequential applications.
	fluxapyroxad + pyraclostrobin (Priaxor)	4.0 to 8.0 fl oz	0.13 to 0.25 lb	21	12 hr	Begin prior to disease development and continue on a 7 to 14 day spray schedule.
Botrytis gray mold, white mold *(Sclerotinia)*	fludioxonil (Cannonball WG)	7 oz	3.5 oz	7	0.5	Begin before disease develops and continue on 7 day intervals until conditions no longer favor disease development. Do not apply more than 28 ounces per acre. Do not use on cowpeas.
	Coniothyrium minitans (Contans WG)	1 to 4 lb/acre	0.05 to 0.2 lb/acre	0	4 hr	OMRI listed product. Apply to soil surface and incorporate no deeper than 2 inches. Works best when applied prior to planting or transplanting. Do not apply other fungicides for 3 weeks after applying Contans.
Damping off, Pythium, Rhizoctonia	azoxystrobin (Quadris) 2.08 F	0.4 to 0.8 fl oz/ 1,000 row feet	0.006 to 0.013 lb/ 1,000 row feet	—	4 hr	*Rhizoctonia* only. Make in-furrow or banded applications shortly after plant emergence.
	mefenoxam (Ridomil Gold) 4 SL	0.5 to 2 pt/trt acre	0.25 to 0.5 lb/trt acre	—	2	For *Pythium* only. Soil incorporate. See label for row rates. Use proportionally less for band rates.
	azoxystrobin + mefenoxam (Uniform)	0.34 fl oz/1,000 row ft	0.0071 + 0.0028 lb/ acre	—		In-furrow spray. See label directions.

504

Table 10-19. Efficacy of Products for Foliar Disease Control in Beans

Scale: E, excellent; G, good; F, fair; P, poor; NC, no control; ND, no data.

Product[1]	Fungicide group[F]	Preharvest interval (Days)	Aerial Rhizoctonia	Anthracnose	Brown Spot (Pseudomonas)	Cercospora	Common Bacterial Blight	Common Rust	Downy Mildew	Gray Mold (Botrytis)	Halo Blight	Powdery Mildew	Pythium Cottony Leak	Pythium Damping off	Rhizoctonia Sore Shin	Sclerotinia Blight	Southern Blight (S. rolfsii)
azoxystrobin (Quadris)	11	14	E	G	NC	ND	NC	E	ND	P	NC	ND	F	ND	G	NC	E
azoxystrobin + mefenoxam (Uniform)	11 + 4	—	ND	ND	NC	ND	NC	ND	ND	NC	NC	NC	P	G	G	NC	NC
boscalid (Endura)	7	7 to 21	ND	ND	NC	ND	NC	ND	NC	G	NC	ND	NC	NC	ND	E	P
penthiopyrad (Fontelis)	7	0	ND	ND	NC	ND	NC	ND	NC	G	NC	ND	NC	NC	ND	E	F
dicloran (Botran)	14	—	NC	NC	NC	NC	NC	NC	NC	F	NC	NC	NC	NC	NC	F	NC
fluazinam (Omega 500)	29	30	ND	ND	NC	NC	NC	NC	ND	G	NC	NC	NC	NC	ND	G	F
chlorothalonil (various)	M	7	P	F	NC	G	NC	G	F	NC	NC	ND	NC	NC	NC	NC	NC
cyprodonil + fludioxonil (Switch)	9 + 12	7	ND	ND	NC	ND	NC	ND	NC	G	NC	ND	NC	NC	ND	E	P
fixed copper (various)	M	0	NC	P	F	P	F	P	F	P	F	P	NC	NC	NC	NC	NC
iprodione (Rovral)	2	—	P	NC	NC	NC	NC	NC	NC	G	NC	NC	NC	NC	F	G	NC
mefenoxam (Ridomil)	4	—	NC	NC	NC	NC	NC	NC	G	NC	NC	NC	F[R]	G[R]	NC	NC	NC
pyraclostrobin (Headline)	11	7 to 21	G	G	NC	G	NC	E	ND	P	NC	ND	F	ND	F	NC	F
fluxapyroxad + pyraclostrobin (Priaxor)	7 + 11	7 to 21	G	G	NC	G	NC	E	ND	G	NC	ND	F	ND	F	E	F
sulfur (various)	M	0	NC	F	NC	F	NC	F	P	P	NC	F	NC	NC	NC	NC	NC
tebuconazole (Folicur)	3	7	NC	NC	NC	F	NC	G	NC	F	NC	NC	NC	NC	P	NC	G
thiophanate-methyl (Topsin M)	1	14 to 28	P	F	NC	G	NC	ND	NC	NC	NC	ND	NC	NC	P	F	NC

[1] Efficacy ratings do not necessarily indicate a labeled use for every disease.

[F] To prevent resistance in pathogens, alternate fungicides within a group with fungicides in another group. Fungicides in the "M" group are generally considered "low risk" with no signs of resistance developing to the majority of fungicides.

[R] Resistance reported in the pathogen.

Table 10-20. Importance of Alternative Management Practices for Disease Control in Beans

Scale: E, excellent; G, good; F, fair; P, poor; NC, no control; ND, no data

Strategy	Anthracnose	Ashy stem blight	Botrytis gray mold	Cercospora	Common bacterial blight and halo blight	Fusarium root rot	Mosaic viruses	Powdery mildew	Pythium damping off	Rhizoctonia root rot	Root knot	Rust (more on pole beans)	Southern blight (Sclerotium rolfsii)	White mold (Sclerotini)
Avoid field operations when leaves are wet	E	NC	E	F	E	NC	NC	NC	NC	NC	NC	E	NC	NC
Avoid overhead irrigation	E	NC	E	E	E	NC	NC	NC	P	NC	NC	E	NC	G
Change planting date	F	F	NC	P	F	G	F	P	E	E	P	G (early)	NC	NC
Cover cropping with antagonist	NC	ND	NC	NC	NC	NC	NC	NC	NC	NC	G	NC	NC	NC
Crop rotation	G	P	F	F	G	F	P	P	F	F	G	NC	F	E
Deep plowing	E	F	E	P	E	F	NC	NC	F	F	E	NC	E	E

DISEASE CONTROL

Table 10-20. Importance of Alternative Management Practices for Disease Control in Beans

Scale: E, excellent; G, good; F, fair; P, poor; NC, no control; ND, no data

Destroy crop residue	E	F	E	F	E	NC	NC	NC	P	P	F	F	G	E
Encourage air movement	E	NC	E	F	E	NC	NC	E	P	NC	NC	F	NC	G
Increase between-plant spacing	P	NC	P	F	P	P	P	P	F	F	NC	P	F	G
Increase soil organic matter	NC	F	NC	NC	NC	F	NC	NC	NC	NC	F	NC	NC	NC
Insecticidal oils	NC	NC	NC	NC	NC	NC	F	NC	NC	NC	NC	NC	NC	NC
pH management	NC	NC	NC	NC	NC	F	NC	NC	NC	NC	NC	NC	NC	NC
Plant in well-drained soil	F	F	F	NC	F	E	NC	NC	E	E	NC	NC	P	F
Plant on raised beds	F	P	F	NC	F	E	NC	NC	E	E	NC	NC	P	F
Plastic mulch bed covers	NC	NC	NC	NC	NC	NC	NC	NC	NC	NC	NC	NC	NC	F
Postharvest temperature control	NC	NC	NC	NC	NC	NC	NC	NC	NC	NC	NC	NC	NC	E
Reflective mulch	NC	NC	NC	NC	NC	NC	G	NC	NC	NC	NC	NC	NC	P
Reduce mechanical injury	NC	NC	NC	NC	NC	F	P	NC	NC	NC	NC	NC	P	NC
Rogue diseased plants	NC	NC	P	NC	NC	NC	F	NC	NC	NC	NC	NC	P	F
Row covers	NC	NC	NC	NC	NC	F	NC	NC	NC	NC	NC	NC	NC	NC
Soil solarization	NC	NC	P	NC	NC	F	NC	NC	F	G	F	NC	F	G
Pathogen-free planting material	E	G	NC	F	E	NC	G	NC	NC	NC	NC	NC	NC	NC
Resistant cultivars	E	G	NC	E	E	G	E	E	NC	NC	NC	E	NC	F
Weed control	F	NC	F	NC	F	F	E	F	NC	NC	F	F	P	F

Brassicas (Broccoli, Brussel Sprout, Cabbage, Cauliflower)

Table 10-21. Disease Control Products for Broccoli, Brussel Sprout, Cabbage, and Cauliflower

		Rate of Material to Use		Minimum Days		
Disease	Material	Formulation	Active Ingredient	Harvest	Reentry	Method, Schedule, and Remarks
Alternaria leaf spot	azoxystrobin + difenoconazole (Quadris Top 2.72 SC)	14 fl oz/acre	0.3 lb/acre	1	0.5	Apply prior to disease, but when conditions are favorable, on 7 to 14 day schedule. Alternate to a non-QoI fungicide after 1 application. No more than 4 applications per season.
	boscalid (Endura 70 EG)	6 to 9 oz/acre	4.2 to 6.3 oz/acre	0	0.5	Begin applications prior to disease development, and continue on a 7 to 14 day interval. Make no more than 2 applications per season.
	cyprodinil + difenoconazole (Inspire Super 2.82 SC)	16 to 20 fl oz/acre	5.6 to 7.1 oz/acre	7	0.5	Begin applications prior to disease development, and continue on a 7 to 10 day interval. Make no more than 2 sequential applications before rotating to another effective fungicide with a different mode of action. Do not exceed 80 fluid ounces per season.
	cyprodinil + fludioxonil (Switch 62.5WG)	11 to 14 oz/acre	6.9 to 8.7 oz/acre	7	0.5	Apply when disease first appears, and continue on 7 to 10 day interval. Do not exceed 56 ounces of product per acre per year.
	fluxapyroxad + pyraclostrobin (Priaxor 500 SC)	6.0 to 8.2 fl oz/ acre	0.20 to 0.27 lb/acre	3	0.5	Make no more than two sequential applications before alternating with fungicides that have a different mode of action. Maximum of 3 applications. **Do not apply to turnip greens or roots.**
	triflumizole (Procure 480SC)	6 to 8 fl oz/acre	3 to 4 oz/acre	1	0.5	Apply when disease first appears and continue on 14 day interval. Do not exceed 18 fluid ounces per season.
Alternaria and gray mold	penthiopyrad (Fontelis 1.67 SC)	14 to 30 fl oz/acre	0.18 to 0.39 lb/acre	0	0.5	Do not exceed 72 fluid ounces of product per year. Make no more than 2 sequential applications per season before rotating to another effective product with a different mode of action.
Black leg	iprodione (Rovral 4F)	2 lb/acre 2 pt/acre	1 lb/acre 1 lb/acre	0	—	Apply to base of plant at 2- to 4-leaf stage. A second application may be made up to the harvest date. Do not use as a soil drench. **For broccoli only.**
	fluxapyroxad + pyraclostrobin (Priaxor 500 SC)	6.0 to 8.2 fl oz/ acre	0.20 to 0.27 lb/acre	3	0.5	Make no more than two sequential applications before alternating with fungicides that have a different mode of action. Maximum of 3 applications. **Do not apply to turnip greens or roots.**

Table 10-21. Disease Control Products for Broccoli, Brussel Sprout, Cabbage, and Cauliflower

| Disease | Material | Rate of Material to Use | | Minimum Days | | Method, Schedule, and Remarks |
		Formulation	Active Ingredient	Harvest	Reentry	
Black rot, downy mildew	acibenzolar-*S*-methyl (Actigard 50WG)	0.5 to 1 oz/acre	0.25 to 0.5 oz/acre	7	0.5	Begin applications 7 to 10 days after thinning, not to exceed 4 applications per a season.
	fixed copper (various)	See label	See label	See label	See label	See label. Rates vary depending on the formulation. Apply on 7 to 10 day intervals after transplanting or shortly after seeds have emerged. Some reddening on older broccoli leaves and flecking of cabbage wrapper leaves may occur. Check label carefully for recommended rates for each disease on each crop.
Clubroot	Cyazofamid (Ranman) 34.5 SC	*Transplant:* 12.9 to 25.75 fl oz/100 gal water *Banded:* 20 fl oz/acre	0.333 to 0.665 lb/100 gal water 0.52 lb/acre	0.5	0	Either apply immediately after transplanting with 1.7 fluid ounces of solution per transplant, or as a banded application with soil incorporation of 6 to 8 inches prior to transplanting. Do not apply more than 39.5 fluid ounces/acre/season; or 6 (1 soil + 5 foliar) applications per season. Do not make more than 3 consecutive applications without rotating to another fungicide with a different mode of action for 3 subsequent applications.
	fluazinam (Omega 500F)	*Transplant:* 6.45 fl oz/100 gal water Banded: 2.6 pts/acre	0.002 lb/gal water 1.36 lb/acre	50	50	Either apply directly as a drench to transplants or as a banded application with soil incorporation of 6 to 8 inches prior to transplanting. Use of product can delay harvest and cause some stunting without adverse effects on final yields.
Downy mildew	amectoctradin + dimethomorph (Zampro)	14 fl oz/acre	0.48 lb/acre	0	0.5	Do not make more than 2 sequential applications before alternating to a fungicide with a different mode of action. Addition of an adjuvant may improve performance (see label for specifics).
	cyazofamid (Ranman 400 SC)	2.75 fl oz/acre	0.072 lb/acre	0	0.5	Begin applications on a 7 to 10 day schedule when disease first appears or weather is conducive. Do not apply more than 39.5 fluid ounces/acre/season; or 6 (1 soil + 5 foliar) applications per season. Do not make more than 3 consecutive applications without rotating to another fungicide with a different mode of action for 3 subsequent applications.
	dimethomorph (Forum)	6 fl oz/acre		0	0.5	Use as a preventive application before infection occurs. The minimum application interval is 7 days.
	fluopicolide (Presidio 4 SC)	3 to 4 fl oz/acre	0.09 to 0.125 lb/acre	2	0.5	No more than 2 sequential applications before rotating to another effective product of a different mode of action. Limited to 4 applications, 12 fluid ounces/ acre per season.
	fosetyl-AL (Aliette 80WDG)	2 to 5 lb/acre	1.6 to 4 lb/acre	3	1	Apply when disease first appears; then repeat on 7 to 21 day intervals. Do not tank mix with copper fungicides. A maximum of 7 applications can be made per season. Also for loose-heading Chinese cabbage, kale, kohlrabi, and greens (collard, mustard, and rape).
	mandipropamid (Revus 2.08 SC)	8 fl oz/acre	0.13 lb/acre	1	0.5	Apply prior to disease development and continue throughout season at 7 to 10 day intervals; maximum 32 fluid ounces per season.
	potassium phosphite	2 to 4 pt/acre	2.1 lb/acre	0	4 hr	Apply when weather is foggy as a preventative. Do not apply to plants under water or temperature stress. Spray solution should have a pH greater than 5.5. Apply in at least 30 gallons water per acre.
Downy mildew, Alternaria leaf spot	azoxystrobin (Quadris 2.08 F)	6.2 to 15.4 fl oz	0.1 to 0.25 lb	0	4 hr	
	chlorothalonil (various)	See label	See label	See label	See label	See label. Rates vary depending on the formulation.
	cyprodinil + difenoconazole (Inspire Super 2.82 SC)	16 to 20 fl oz/acre	5.6 to 7.1 oz/acre	7	0.5	Begin applications prior to disease development, and continue on a 7 to 10 day interval. Make no more than 2 sequential applications before rotating to another effective fungicide with a different mode of action. Do not exceed 80 fluid ounces per season.
	fenamidone (Reason 500SC)	5.5 to 8.2 fl oz/acre	0.178 to 0.267 lb/acre	2	0.5	Begin applications on a 5 to 10 day schedule when disease first appears or weather is conducive. Do not apply more than 24.6 fluid ounces/acre/season. Do not make more than 1 application without rotating to another fungicide with a different mode of action.
	mancozeb (various)	See label	See label	See label	See label	See label. Rates vary depending on the formulation. Spray at first appearance of disease and continue on a 7 to 10 day interval.
	mefenoxam + chlorothalonil (Ridomil Gold/Bravo)	1.5 lb/acre	—	7	2	Begin applications when conditions favor disease but prior to symptoms. Under severe disease pressure use additional fungicides between 14 day intervals. Do not make more than four applications per crop.
Powdery mildew	azoxystrobin + difenocon-azole (Quadris Top 2.72 SC)	14 fl oz/acre	0.3 lb/acre	1	0.5	Apply prior to disease, but when conditions are favorable, on 7 to 14 day schedule. Alternate to a non-QoI fungicide after 1 application. No more than 4 applications per season.
	boscalid (Endura 70 EG)	6 to 9 oz/acre	4.2 to 6.3 oz/acre	0	0.5	Begin applications prior to disease development, and continue on a 7 to 14 day interval. Make no more than 2 applications per season; disease suppression only.

DISEASE CONTROL

Table 10-21. Disease Control Products for Broccoli, Brussel Sprout, Cabbage, and Cauliflower

Disease	Material	Rate of Material to Use		Minimum Days		Method, Schedule, and Remarks
		Formulation	Active Ingredient	Harvest	Reentry	
Powdery mildew (continued)	cyprodinil + difenoconazole (Inspire Super 2.82 SC)	16 to 20 fl oz/acre	5.6 to 7.1 oz/acre	7	0.5	Begin applications prior to disease development, and continue on a 7 to 10 day interval. Make no more than 2 sequential applications before rotating to another effective fungicide with a different mode of action. Do not exceed 80 fluid ounces per season.
	cyprodinil + fludioxonil (Switch 62.5WG)	10 to 12 oz/acre	6.25 to 7.5 oz/acre	7	0.5	Apply when disease first appears, and continue on 7 to 10 day intervals. Do not exceed 56 ounces of product per acre per year.
	fluxapyroxad + pyraclostrobin (Priaxor 500 SC)	6.0 to 8.2 fl oz/ acre	0.20 to 0.27 lb/acre	3	0.5	Make no more than two sequential applications before alter- nating with fungicides that have a different mode of action. Maximum of 3 applications. **Do not apply to turnip greens or roots.**
	penthiopyrad (Fontelis 1.67 SC)	14 to 30 fl oz/acre	0.18 to 0.39 lb/acre	0	0.5	Do not exceed 72 fluid ounces of product per year. Make no more than 2 sequential applications per season before rotating to another effective product with a different mode of action.
	sulfur (various)	See label	See label	See label	See label	See label. Rates vary depending on the formulation. Apply when disease first appears; then repeat as needed on 14 day interval. Avoid applying on days over 90 degrees F. Also for use on greens (collard, kale, and mustard), rutabaga, and turnip.
	triflumizole (Procure 480 SC)	6 to 8 fl oz/acre	3 to 4 oz/acre	1	0.5	Apply when disease first appears and continue on 14 day interval. Do not exceed 18 fluid ounces per season.
Pythium damping off, Phytophthora basal stem rot	fluopicolide (Presidio 4 F)	3 to 4 fl oz/acre	0.09 to 0.125 lb/acre	2	0.5	Apply as a soil drench at transplant. As plants enlarge, use apply directly to soil by chemigation on a 7 to 10 day schedule as conditions favor disease, but prior to disease development. No more than 2 sequential applications before rotating to another effective product of a different mode of action. Limited to 4 applications, 12 fluid ounces/ acre per season,
	mefenoxam (Ridomil Gold 4 SL)	0.25 to 2 pt/acre	0.12 to 1.0 lb/acre	—	2	Apply 1 to 2 pt per acre as a broadcast, preplant ap- plication to soil and incorporate in top 2 in. of soil. For Pythium control, use only 0.25 to 0.5 pt per acre.
	metalaxyl (MetaStar 2 E AG)	4 to 8 pt/ trt acre	0.5 to 1 lb/acre	—	2	Preplant incorporated or surface application.
Rhizoctonia bottom rot	boscalid (Endura 70 WP)	6 to 9 oz/acre	4.2 to 6.3 oz/acre	0	0.5	Begin applications prior to disease development, and continue on a 7 to 14 day interval. Make no more than 2 applications per season; disease suppression only.
Rhizoctonia stem (wire-stem) and root rot	azoxystrobin (Quadris 2.08 SC)	5.8 to 8.7 fl oz/ acre on 36-in. rows	0.1 to 0.25 lb/acre	0	4 hr	Rate is equivalent to 0.4 to 0.6 fluid ounce per 1000 row feet. Apply at planting as a directed spray to the furrow in a band 7 inches wide. See label for other row spacings.
	flutolanil (Convoy)	26 fl oz/acre		45	0.5	Apply fungicide to the row at planting as an in-furrow or di- rected spray at the base of transplants. If applied as soil drench at planting, observe 45 day PHI.
Sclerotinia stem rot (white mold)	boscalid (Endura 70 EG)	6 to 9 oz/acre	4.2 to 6.3 oz/acre	0	0.5	Begin applications prior to disease development, and continue on a 7 to 14 day interval. Make no more than 2 applications per season.
	penthiopyrad (Fontelis 1.67 SC)	16 to 30 fl oz/acre	0.21 to 0.39 lb/acre	0	0.5	Do not exceed 72 fluid ounces of product per year. Make no more than 2 sequential applications per season before rotating to another effective product with a different mode of action.
	Coniothyrium minitans (Contans WG)	1 to 4 lb/acre	0.05 to 0.2 lb/acre	0	4 hr	OMRI listed product. Apply to soil surface and incorporate no deeper than 2 inches. Works best when applied prior to planting or transplanting. Do not apply other fungicides for 3 weeks after applying Contans.

Table 10-22. Efficacy of Products for Disease Control in Brassicas

Scale: E, excellent; G, good; F, fair; P, poor; NC, no control; ND, no data.

Product[1,2]	Fungicide group[F]	Preharvest interval (Days)	Alternaria Leaf Spot	Bacterial Soft Rot	Black Rot	Black Leg	Bottom Rot (*Rhizoctonia*)	Cercospora & Cercosporella	Clubroot	Downy Mildew	Powdery Mildew	Pythium damping-off	Sclerotinia/Raisin Head	Wirestem (*Rhizoctonia*)
azoxystrobin (Amistar, Quadris)	11	—	E	NC	NC	ND	ND	F	NC	F	F	NC	NC	P
boscalid (Endura)	7	0 to 14	G	NC	NC	F	NC	NC	NC	P	P	NC	F	F
chlorothalonil (various)	M	7	F	NC	NC	NC	P	F	NC	F	F	NC	NC	NC
cyazofamid (Ranman)	21	0	NC	NC	NC	NC	NC	NC	NC	G	NC	NC	NC	NC
tebuconazole (Folicur, Tebuzol, Tegrol)	3	7	F	NC	NC	ND	NC	F	NC	NC	ND	NC	NC	NC
cyprodinil + fludioxonil (Switch)	9 + 12	7	F	NC	NC	NC	NC	F	NC	NC	F	NC	NC	NC
dimethomorph (Forum)	40	0	NC	NC	NC	NC	NC	NC	NC	F	NC	NC	NC	NC
fenamidone (Reason)	11	2	F	NC	NC	NC	NC	F	NC	E	NC	NC	NC	NC
fluopicolide (Presidio)	43	2	NC	NC	NC	NC	NC	NC	NC	E	NC	NC	NC	NC
amectoctradin + dimethomorph (Zampro)	45 + 40	0 to 7	NC	NC	NC	NC	NC	NC	NC	E	NC	NC	NC	NC
fixed copper (various)[3]	M	0	P	NC	F	NC	NC	P	NC	F	F	NC	NC	NC
fosteyl-Al[4] (Aliette)	33	3	NC	NC	NC	NC	NC	NC	NC	F	NC	NC	NC	NC
iprodione (Rovral)[5]	2	—	NC	NC	NC	F	NC	NC	NC	NC	NC	NC	P	P
difenoconazole + cyprodinil (Inspire Super)	3 + 9	7	G	NC	NC	ND	NC	G	NC	NC	F	NC	P	NC
penthiopyrad (Fontelis)	7	0	E	NC	NC	ND	NC	ND	NC	NC	G	NC	G	NC
mandipropamid (Revus)	40	1	NC	NC	NC	NC	NC	NC	NC	E	NC	NC	NC	NC
mancozeb (various)	M	7	F	NC	P	NC	NC	F	NC	F	P	NC	NC	NC
mefenoxam (Ridomil Gold EC) pre-plant	4	—	NC	NC	NC	NC	NC	NC	NC	F	NC	F[R]	NC	NC
mefenoxam + chlorothalonil (Ridomil Gold Bravo)	4 + M	7	F	NC	NC	NC	P	F	NC	F	F	NC[R]	NC	NC
fluazinam (Omega 500) [6]	29	20 to 50	NC	NC	NC	NC	NC	NC	F	NC	NC	NC	NC	NC
pyraclostrobin (Cabrio)	11	—	E	NC	NC	ND	NC	E	NC	F	F	NC	NC	NC
sulfur (various)	M	0	P	NC	NC	NC	NC	P	NC	P	F	NC	NC	NC

[1] Efficacy ratings do not necessarily indicate a labeled use for every disease.

[2] Fungicides registered specifically on Cole crops (cabbage, cauliflower, broccoli) include chlorothalonil, iprodione, maneb, and mefenoxam + chlorothalonil. Pyraclostrobin is labeled only on turnip tops. Fosteyl-Al is not labeled on turnips. Always refer to product labels prior to use.

[3] Phytotoxicity is seen when fosteyl-Al is tank-mixed with copper.

[4] When used in combination with fosteyl-Al or maneb.

[5] Applications of iprodione made for black leg may suppress *Alternaria*, *Sclerotinia*, and wirestem on broccoli.

[6] Use a 20 day PHI for Omega 500 on leafy greens and a 50 day PHI for Cole brassicas. Use a zero day PHI for Zampro on leafy greens and a 7 day PHI on Cole crops.

[F] To prevent resistance in pathogens, alternate fungicides within a group with fungicides in another group. Fungicides in the "M" group are generally considered "low risk" with no signs of resistance developing to the majority of fungicides.

[R] Resistance reported in the pathogen.

Table 10-23. Importance of Alternative Management Practices for Disease Conrol in Brassicas

Scale: E, excellent; G, good; F, fair; P, poor; NC, no control; ND, no data

Strategy	Alternaria leaf spot	Bacterial soft rot	Black rot	Black leg	Bottom rot (*Rhizoctonia*)	Cercospora	Clubroot	Downy mildew	Powdery mildew	Pythium	Sclerotinia head	Wirestem (*Rhizoctonia*)
Avoid field operations when leaves are wet	P	F	G	F	F	P	NC	P	NC	NC	NC	NC
Avoid overhead irrigation	E	E	E	E	F	E	NC	G	P	NC	NC	NC
Change planting date	P	P	NC	NC	P	NC	NC	NC	NC	P	NC	F
Cover cropping with antagonist	NC	NC	NC	NC	NC	NC	P	NC	NC	P	NC	NC
Crop rotation	F	F	G	G	P	F	NC	F	NC	NC	P	P
Deep plowing	F	F	G	G	F	F	NC	F	NC	NC	F	F
Destroy crop residue	F	F	G	G	F	F	NC	F	NC	NC	P	P
Encourage air movement	F	P	P	P	F	F	NC	F	NC	P	F	NC
Increase between-plant spacing	F	P	P	P	F	F	NC	F	NC	P	F	NC
Increase soil organic matter	NC	NC	NC	NC	P	NC	P	NC	NC	NC	NC	P
Hot water seed treatment	P	NC	E	G	NC	NC	NC	NC	NC	NC	NC	NC
pH management	NC	NC	NC	NC	NC	NC	E	NC	NC	NC	NC	NC
Plant in well-drained soil	P	F	P	P	G	P	E	P	NC	F	F	G
Plant on raised beds	NC	F	P	NC	G	NC	E	P	NC	F	F	G
Plastic mulch bed covers	P	NC	NC	NC	F	NC	NC	NC	NC	NC	NC	NC
Postharvest temperature control	NC	E	NC	NC	NC	NC	NC	NC	NC	NC	NC	NC
Reflective mulch	NC	NC	NC	NC	NC	NC	NC	NC	NC	NC	NC	NC
Reduce mechanical injury	NC	E	G	NC	NC	NC	NC	NC	NC	NC	F	P
Rogue diseased plants	P	NC	NC	F	P	NC	NC	NC	NC	NC	NC	NC
Row covers	NC	P	NC	NC	NC	NC	NC	NC	NC	NC	NC	NC
Soil solarization	NC	NC	NC	P	F	NC	NC	NC	NC	P	P	F
Pathogen-free planting material	F	NC	E	E	F	NC	G	NC	NC	NC	P	F
Resistant cultivars	NC	NC	E	NC	NC	NC	P	F	F	NC	NC	P
Weed control	F	NC	F	F	NC	F	F	F	F	NC	F	NC

Cantaloupe — See Cucurbits

Corn, Sweet

Table 10-24. Disease Control Products for Corn, Sweet

Disease	Material	Rate of Material to Use		Minimum Days		Method, Schedule, and Remarks
		Formulation	Active Ingredient	Harvest	Reentry	
Various seedling diseases	mefenoxam + PCNB + *Bacillus subtilis* GB03 (System 3)	3 oz/bu of seed		–	1	Seed treatment
	pyraclostrobin (Stamina)	1.6 fl oz/100 lbs of seed				Seed treatment
	fludioxonil (Maxim 4FS)	0.16 fl oz/100 lbs of seed			0.5	Seed treatment
Soilborne diseases	fluoxastrobin (Aftershock)	0.16 to 0.24 fl oz/ 1000 row feet	0.07 to 0.10 oz/ 1000 row feet	7	0.5	May be applied as a banded or in-furrow spray. Consult label for specifics.
Rust, Helminthosporium	chlorothalonil (various)	See label	See label	See label	See label	See label. Rates vary depending on the formulation. Spray at first appearance, 4 to 14 day intervals.

Table 10-24. Disease Control Products for Corn, Sweet

Disease	Material	Rate of Material to Use		Minimum Days		Method, Schedule, and Remarks
		Formulation	Active Ingredient	Harvest	Reentry	
Leaf blight (Northern and Southern Blight)	mancozeb (various)	See label	See label	See label	See label	See label. Rates vary depending on the formulation. Start applications when disease first appears and repeat at 4 to 7 day intervals.
Rust, Northern and Southern Blight, Gray Leaf Spot, Northern leaf spot	tebuconazole (various)	4 to 6 oz/acre	1.5 to 2.3 oz/acre	7	0.5	For optimum results use as a preventative treatment. Folicur 3.6 F must have 2 to 4 hours of drying time on foliage for the active ingredient to move systemically into plant tissue before rain or irrigation occurs.
Anthracnose, Gray leaf spot, Northern and southern corn leaf blights and spots, southern rust, rust	propiconazole (Tilt)	2 to 4 fl oz/acre		14	1	16 fluid ounces per acre per crop maximum.
	penthiopyrad (Vertisan)	10 to 24 fl oz/acre		0	0.5	No more than 2 sequential applications of the fungicide before switching to a fungicide with another mode of action.
	fluoxastrobin (Aftershock)	3.8 fl oz/acre		–	0.5	Soil and foliar treatments.
	pyraclostrobin (Headline SC & EC)	6 to 12 fl oz/acre		7	0.5	Do not exceed 2 sequential applications of this fungicide or with other group 11 fungicides.
	azoxystrobin (Quadris)	6.0 to 15.5 fl oz/acre	0.1 to 0.25 lb/acre	7	4 hr	Use lower rate for rust. Make no more than two sequential applicationsbefore alternating with fungicides that have a different mode of action. Apply no more than 2.88 quarts per crop per acre per season.
	fluoxastrobin + tebuconazole (Evito T)	4 to 9 fl oz/acre	0.052 to 0.163 lb/ arce	7	19	No more than 2 sequential applications before alternating with a different mode of action.
	pyraclostrobin + metconazole (Headline AMP)	10 to 14.4 fl oz/acre		7	0.5	No more than 2 sequential applications before alternating with a different mode of action.
	azoxystrobin + propiconazole (Quilt)	10.5 to 14 fl oz/acre	0.14 to 0.18 lb/acre	14	1	Apply when disease first appears; continue to apply on a 14 day interval if favorable conditions persist.
	azoxystrobin + propiconazole (Quilt XCEL)	10.5 to 14 fl oz/acre		14	0.5	Alternate Quilt Xcel sprays with another mode of action than a group 11 fungicide.
	azoxystrobin + propiconazole (Avaris)	7 to 14 fl oz/acre		14	0.5	Alternate Avaris sprays with another mode of action than a group 11 fungicide.
	trifloxystrobin +propiconazole (Stratego)	10 fl oz/acre		14	0.5	Apply Stratego when disease first appears and continue on a 7 to 14 day interval. Alternate applications of Stratego with another product with a different mode of action than Group 11 fungicides.
	trifloxystrobin +propiconazole (Stratego YLD)	4 to 5 fl oz/acre		14	0.5	Alternate Stratego YLD sprays with another mode of action than a group 11 fungicide.
	fluxapyroxad + pyraclostrobin (Priaxor)	4 to 8 fl oz/acre	1.7 to 3.4 oz/acre	7	0.5	Do not make more than two sequential applications of Priaxor before switching to a fungicide with a different mode of action. May be used with adjuvants (consult label for specifics).
	tebuconazole +prothioconazole (Prosaro 421SC)	6.5 fl oz/acre		7	0.5	Check label for restrictions and information about adjuvants. Effective on Eye Spot.

Cucurbits (Cucumber, Cantaloupe, Melon, Pumpkin, Squash, Watermelon)

Table 10-25. Disease Control Products for Cucurbits

Disease	Material	Rate of Material to Use		Minimum Days		Method, Schedule, and Remarks
		Formulation	Active Ingredient	Harvest	Reentry	
Angular leaf spot, Bacterial leaf spot	fixed copper (various)	See label	See label	See label	See label	See label. Rates vary depending on the formulation. Repeated use may cause leaf yellowing.
	acibenzolar-S-methyl (Actigard) 50 WP	0.5 to 1 oz/acre	0.25 to 0.5 oz/acre	0	05	Apply to healthy, actively growing plants. Do not apply to stressed plants. Apply no more than 8 ounces per acre per season.
Bacterial fruit blotch	fixed copper (various)	See label	See label	See label	See label	See label. Rates vary depending on the formulation. Start applications at first bloom; ineffective once fruit reaches full size. Repeated use may cause leaf yellowing.
	acibenzolar-S-methyl (Actigard) 50 WP	0.5 to 1 oz/acre	0.25 to 0.5 oz/acre	0	0.5	Apply to healthy, actively growing plants. Do not apply to stressed plants. Apply no more than 8 ounces per acre per season.
Bacterial wilt						See Insect Control section for Cucumber Beetles.

511

DISEASE CONTROL

Table 10-25. Disease Control Products for Cucurbits

Disease	Material	Rate of Material to Use		Minimum Days		Method, Schedule, and Remarks
		Formulation	Active Ingredient	Harvest	Reentry	
Belly (fruit) rot, *Rhizoctonia*	azoxystrobin (Quadris) 2.08 F	See label	See label	1	4 hr	Make banded application to soil surface or in-furrow application just before seed are covered.
	fluopyram + tebuconazole (Luna Experience) 3.3 F	17 fl oz/acre	0.27 to 0.44 lb/acre	7	0.5	**APPLY ONLY TO WATERMELON**. Make no more than 2 applications before alternating to a fungicide with different active ingredients. Do not rotate with tebuconazole.
	thiophanate-methyl (Topsin M) 70 WP	0.5 lb/acre	0.35 lb/acre	0	0.5	Apply in sufficient water to obtain runoff to soil surface.
Cottony leak (*Pythium*)	metalaxyl (MetaStar) 2 E	4 to 8 pt/treated acre	0.5 to 1 lb/treated acre	0	2	Soil surface application in 7-inch band.
Damping off (*Pythium*) and fruit rot	mefenoxam (Ridomil Gold) 4 SL	1 to 2 pt/treated acre	0.5 to 1 lb/acre	0	2	Preplant incorporated (broadcast or band); soil spray (broadcast or band); or injection (drip irrigation).
	metalaxyl (MetaStar) 2 E	4 to 8 pt/treated acre	0.5 to 1 lb/acre	0	2	Preplant incorporated or surface application.
	propamocarb (Previcur Flex) 6 F	12.8 fl oz/100 gal	0.6 lb/100 gal	2	0.5	Rates based on rock wool cube saturation in the greenhouse. See label for use in seed beds, drip system, and soil drench.
Downy mildew	ametoctradin + dimethomorph (Zampro) 4.38 SC	14 oz/A	0.48 lb/acre	0	0.5	Make no more than 2 applications before alternating to a fungicide with different active ingredients. Do not rotate with Forum. Maximum of 3 applications per crop per season.
	azoxystrobin (Quadris) 2.08 F	11 to 15.4 fl oz/acre	0.18 to 0.25 lb/acre	1	4 hr	Make no more than one application before alternating with a fungicide with a different mode of action. Apply no more than 2.88 quarts per crop per acre per season. Resistance reported.
	chlorothalonil (various)	See label	See label	See label	See label	See label. Rates vary depending on the formulation.
	chlorothalonil + potassium phosphite (Catamaran) 5.27 SC	6 pints/acre	3.3 pints/acre	0	0.5	Apply no more than 50 pints per crop per acre per season.
	chlorothalonil + zoxamide (Zing!)	36 fl oz/acre	1.18 lb/acre 0.20 lb/acre	0	0.5	May cause sunburn in watermelon fruit, see label for details.
	cyazofamid (Ranman) 400 SC	2.1 to 2.75 fl oz/acre	0.054 to 0.071 lb/acre	0	0.5	Do not apply more than 6 sprays per crop. Make no more than 3 consecutive applications followed by 3 applications of fungicides from a different resistance management group.
	cymoxanil (Curzate) 60 DF	3.2 oz/acre	1.9 oz/acre	3	0.5	Use only in combination with labeled rate of protectant fungicide (e.g., mancozeb or chlorothalonil).
	dimethomorph (Forum) 4.17SC	6 fl oz/acre	3.13 oz/acre	0	0.5	Must be applied as a tank mix with another fungicide with a different mode of action. Do not make more than two sequential applications.
	famoxadone + cymoxanil (Tanos) 50WP	8 oz/acre	4 oz/acre	3	0.5	Do not make more than one application before alternating with a fungicide that has a different mode of action. Must be tank-mixed with contact fungicide with a different mode of action.
	Fenamidone (Reason) 500 SC	5.5 fl oz/acre	0.178 lb/acre	14	0.5	Begin applications when conditions favor disease development, and continue on 5 to 10 day interval. Do not apply more than 22 fluid ounces per growing season. Alternate with fungicide from different resistance management group, and make no more than 4 total applications of Group 11 fungicides per season.
	fixed copper (various)	See label	See label	See label	See label	See label. Rates vary depending on the formulation. Repeated use may cause leaf yellowing.
	Fluazinam (Omega 500F)	0.75 to 1.5 pints/acre		30	0.5	**APPLY ONLY TO MELONS**. Initiate applications when conditions are favorable for disease development or when disease symptoms first appear. Repeat applications on a 7- to 10-day schedule.
	fluopicolide (Presidio) 4F	3 to 4 fl oz/acre	0.09 to 0.125 lb/acre	2	0.5	Tank mix with another downy mildew fungicide with a different mode of action.
	fosetyl-AL (Aliette) 80 WDG	2 to 5 lb/acre	1.6 to 4 lb/acre	0.5	0.5	Do not tank mix with copper-containing products. Mixing with surfactants or foliar fertilizers is not recommended.
	mandipropamid (Revus) 2.00F	8 fl oz/acre	0.13 lb/acre	1	0.5	For disease suppression only. Resistance reported.
	mancozeb (various)	See label	See label	See label	See label	See label. Rates vary depending on the formulation. Labeled on all cucurbits.
	mefenoxam + chlorothalonil (Ridomil Gold Bravo, Flouronil) 76.5 WP	2 to 3 lb/acre	1.5 lb/acre	7	2	Spray at first appearance and repeat at 14 day intervals. Apply full rate of protectant fungicide between applications. Avoid late-season application when plants reach full maturity. Resistance reported.
	oxathiapiprolin + chlorothalonil (Orondis Opti SC)	1.7 to 2.5 pt/acre		0	0.5	Limit to 10 pt per acre per year. Limit to six foliar applications per acre per year for the same crop. Do not follow soil applications of Orondis with foliar applications of Orondis. Begin foliar applications prior to disease development and continue on a 5- to 14-day interval. Use the higher rates when disease is present.

Table 10-25. Disease Control Products for Cucurbits

Disease	Material	Rate of Material to Use Formulation	Rate of Material to Use Active Ingredient	Minimum Days Harvest	Minimum Days Reentry	Method, Schedule, and Remarks
Downy mildew (continued)	oxathiapiprolin (Orondis Ultra A) mandipropamid (Orondis Ultra B)	2.0 to 4.8 fl oz/acre 8 fl oz/acre		0 0	4 hr 4 hr	Limit to 19.2 fl oz of Orondis Ultra A and 32 fl oz of Orondis Ultra B per acre per year. Limit to six foliar applications per acre per year for the same crop. Do not follow soil applications of Orondis with foliar applications of Orondis. Begin foliar applications prior to disease development and continue on a 5- to 14-day interval. Use the higher rates when disease is present.
	propamocarb (Previcur Flex) 6 F	1.2 pt/acre	0.9 lb/acre	2	0.5	Begin applications before infection; continue on a 7 to 14 day interval. Do not apply more than 6 pints per growing season. Always tank mix with another Downy mildew product.
	pyraclostrobin (Cabrio) 20 WG	8 to 12 oz/acre	1.6 to 2.4 oz/acre	0	0.5	Make no more than one application before alternating to a fungicide with a different mode of action. Resistance
	pyraclostrobin + boscalid (Pristine) 38 WG	12.5 to 18.5 oz/acre	4.8 to 7 oz/acre	0	1	Make no more than 4 applications per season. Resistance reported.
	trifloxystrobin (Flint) 50 WDG	4 oz/acre	2 oz/acre	0	0.5	Begin applications preventatively and continue as needed alternating applications of Ridomil Gold Bravo on a 7 to14 day interval. Resistance reported.
	zoxamide + mancozeb (Gavel) 75 DF	1.5 to 2 lb	1.13 to 1.5 lb	5	2	Begin applications when plants are in 2-leaf stage, and repeat at 7 to 10 day intervals. Now labeled on all cucurbits. Maximum 8 applications per season.
Fusarium wilt	prothioconazole (Proline 480 SC)	5.7 fl oz/acre		7	0.5	One soil and two foliar applications allowed by either ground or chemigation application equipment (including drip irrigation). Do not use in water used for hand transplanting. Not for use in greenhouse/transplant house.
Gummy stem blight, Black rot	prothioconazole (Proline 480 SC)	5.7 fl oz/acre		7	0.5	One soil and two foliar applications allowed by either ground or chemigation application equipment (including drip irrigation). Do not use in water used for hand transplanting. Not for use in greenhouse/transplant house.
	tebuconazole (Monsoon) 3.6 F	8 oz/acre	2 lb/acre	7	0.5	Maximum 3 applications per season. Apply as a protective spray at 10 to 14 day intervals. Add a surfactant.
Leaf spots: *Alternaria*, Anthracnose (*Colletotrichum*), *Cercospora*, Gummy stem blight (*Didymella*), Target spot (*Corynespora*)	azoxystrobin (Quadris) 2.08 F	11 to 15.4 fl oz/acre	0.18 to 0.25 lb/acre	1	4 hr	Make no more than one application before alternating with a fungicide with a different mode of action. Apply no more than 2.88 quarts per crop per acre per season. Do not use for Gummy stem blight where resistance to group 11(Qol) fungicides exists.
	azoxystrobin + difenoconazole (Quadris Top) 1.67 SC	12 to 14 fl oz/acre	3.6 to 4.2 fl oz/acre	1	0.5	Not for Target spot. Make no more than one application before alternating with fungicides that have a different mode of action. Apply no more than 56 fluid ounces per crop per acre per season.
	chlorothalonil (various)	See label	See label	See label	See label	See label. Rates vary depending on the formulation.
	chlorothalonil + potassium phosphite (Catamaran) 5.27 SC	6 pints/acre	3.3 pints/acre	0	0.5	Apply no more than 50 pints per crop per acre per season. Do not apply to watermelon fruit when stress conditions conducive to sunburn occur.
	cyprodinil + fludioxonil (Switch) 62.5 WG	11 to 14 oz/acre	6.9 to 8.9 oz/acre	1	0.5	Only for *Alternaria* and Gummy stem blight. Make no more than 2 applications before alternating to a different fungicide. Maximum of 4 to 5 applications at high and low rates.
	difenoconazole + cyprodinil (Inspire Super) 2.82 SC	16 to 20 fl oz/acre	5.3 to 7.1 fl oz/acre	7	0.5	Not for Target spot. Make no more than two sequential applications before alternating with fungicides that have a different mode of action. Apply no more than 80 fluid ounces per crop per acre per season.
	famoxadone + cymoxanil (Tanos) 50WP	8 oz/acre	4 oz/acre	3	0.5	Only for *Alternaria* and Anthracnose; do not make more than one application before alternating with a fungicide that has a different mode of action; must be tank-mixed with contact fungicide with a different mode of action
	fenamidone (Reason) 500 SC	5.5 fl oz/acre	0.178 lb/acre	14	0.5	Begin applications when conditions favor disease development, and continue on 5 to 10 day interval. Do not apply more than 22 fluid ounces per growing season. Alternate with fungicide from different resistance management group, and make no more than 4 total applications of Group 11 fungicides per season.
	fixed copper (various)	See label	See label	See label	See label	See label. Rates vary depending on the formulation. Repeated use may cause leaf yellowing.
	mancozeb (various)	See label	See label	See label	See label	See label. Rates vary depending on the formulation. Labeled on all cucurbits.
	fluopyram + tebuconazole (Luna Experience) 3.3 F	8 to 17 fl oz/acre	0.27 to 0.44 lb/acre	7	0.5	**APPLY ONLY TO WATERMELON.** Not for *Cercospora* or target spot. Make no more than 2 applications before alternating to a fungicide with different active ingredients. Do not rotate with tebuconazole.
	fluopyram + trifloxystrobin (Luna Sensation) 1.67 F	7.6 fl oz/acre	0.25 lb/acre	0	0.5	**APPLY ONLY TO WATERMELON** and only to control *Alternaria* and Anthracnose. Make no more than 2 applications before alternating to a fungicide with different active ingredients. Maximum 4 applications per season.

513

DISEASE CONTROL

Table 10-25. Disease Control Products for Cucurbits

Disease	Material	Rate of Material to Use		Minimum Days		Method, Schedule, and Remarks
		Formulation	Active Ingredient	Harvest	Reentry	
Leaf spots: *Alternaria*, Anthracnose (*Colletotrichum*), *Cercospora*, Gummy stem blight (*Didymella*), Target spot (*Corynespora*) (continued)	fluxapyroxad + pyraclostrobin (Merivon 500 SC)	4 to 5.5 fl oz/ acre	0.26 to 0.36 lb/acre	0	0.5	Make no more than two sequential applications before alternating with fungicides that have a different mode of action. Maximum of 3 applications per crop.
	polyoxin D zinc salt (OSO 5% SC)	3.75 to 13 fl oz/acre	0.21 to 0.72 oz/acre	0	4 hr	Begin preventive sprays when conditions favor disease development, and continue on a 7-14 day spray interval as needed.
	potassium phosphite + tebuconazole (Viathon)	4 pints/acre	0.21 lb	7	0.5	**APPLY ONLY TO WATERMELON.** Maximum 3 applications per crop.
	pyraclostrobin (Cabrio) 20 WG	12 to 16 oz/acre	2.4 to 3.2 oz/acre	0	0.5	Do not use for Gummy stem blight where resistance to group 11 (Qol) fungicides exists. Make no more than one application before alternating to a fungicide with a different mode of action.
	pyraclostrobin + boscalid (Pristine) 38 WG	12.5 to 18.5 oz/acre	4.8 to 7 oz/acre	0	1	Not for target spot. Do not use for gummy stem blight where resistance to group 7 and group 11 fungicides exists. Use highest rate for anthracnose. Make no more than 4 applications per season.
	thiophanate-methyl (Topsin M) 70 WP	0.5 lb/acre	0.35 lb/acre	0	0.5	Spray at first appearance and then at 7 to 10 day intervals. Resistance reported in gummy stem blight fungus.
	zoxamide + mancozeb (Gavel) 75 DF	1.5 to 2 lb	1.13 to 1.5 lb	5	2	*Cercospora* and *Alternaria* only. Begin applications when plants are in 2-leaf stage, and repeat at 7 to 10 day intervals. Now labeled on all cucurbits. Maximum 8 applications per season.
Phytophthora blight	ametoctradin + dimethomorph (Zampro) 4.38SC	14 oz/A	0.48 lb/acre	0	0.5	Make no more than 2 applications before alternating to a fungicide with different active ingredients. Do not rotate with Forum. Maximum of 3 applications per crop per season. Apply at planting as a preventive drench treatment. Addition of a spreading or penetrating adjuvant is recommended
	cyazofamid (Ranman) 400 SC	2.75 lf oz/acre	0.071 lb/acre	0	0.5	Do not apply more than 6 sprays per crop. Make no more than 3 consecutive applications followed by 3 applications of fungicides from a different resistance management group. Resistant isolates have been found.
	dimethomorph (Forum) 4.17SC	6 fl oz/acre	3.13 oz/acre	0	0.5	Must be applied as a tank mix with another fungicide with a different mode of action. Do not make more than two sequential applications.
	fluopicolide (Presidio) 4F	3 to 4 fl oz/acre	0.09 to 0.125 lb/acre	2	0.5	Tank mix with another *Phytophthora* fungicide with a different mode of action. May be applied through drip irrigation to target crown rot phase.
	mandipropamid (Revus) 2.08F	8 fl oz/acre	0.13 lb/acre	0	0.5	For disease suppression only; apply as foliar spray with copper based fungicide.
	oxathiapiprolin (Orondis Gold 200) mefenoxam (Orondis Gold B)	2.4 to 19.2 fl oz/acre 1.0 to 2.0 pt/acre		0 0	4 hr 0 hr	Limit to 38.6 fl oz of Orondis Gold 200 per acre per year. Limit to four applications per acre per year for the same crop. Do not follow soil applications of Orondis with foliar applications of Orondis. Apply at planting in furrow, by drip, or in transplant water. Use the higher rates for heavier soils, for longer application intervals, or for susceptible varieties.
	oxathiapiprolin (Orondis Ultra A) mandipropamid (Orondis Ultra B)	2.0 to 4.8 fl oz/acre 8 fl oz/acre		0 0	4 hr 4 hr	Limit to 19.2 fl oz of Orondis Ultra A and 32 fl oz of Orondis Ultra B per acre per year. Limit to six foliar applications per acre per year for the same crop. Do not follow soil applications of Orondis with foliar applications of Orondis. Begin foliar applications prior to disease development and continue on a 5- to 14-day interval. Use the higher rates when disease is present.
Plectosporium blight	azoxystrobin + difenoconazole (Quadris Top) 1.67 SC	12 to 14 fl oz/acre	3.6 to 4.2 fl oz/acre	1	0.5	Make no more than one application before alternating with fungicides that have a different mode of action. Apply no more than 56 fluid ounces per crop per acre per season.
	fluxapyroxad + pyraclostrobin (Merivon 500 SC)	4 to 5.5 fl oz/ acre	0.26 to 0.36 lb/acre	0	0.5	Make no more than two sequential applications before alternating with fungicides that have a different mode of action. Maximum of 3 applications per crop.
	trifloxystrobin (Flint) 50 WDG	1.5 to 2 oz/acre	0.75 to 1 oz/acre	0	0.5	Make no more than one application before alternating with fungicides that have a different mode of action. Begin applications preventively when conditions are favorable for disease and continue as needed on a 7 to 14 day interval.
	pyraclostrobin (Cabrio) 20WG	12 to 16 oz/acre	2.4 to 3.2 oz/acre	0	0.5	Make no more than 1 application before alternating to a fungicide with a different mode of action.
Powdery mildew	acibenzolar-S-methyl (Actigard) 50 WP	0.5 to 1 oz/acre	0.25 to 0.5 oz/acre	0	0.5	Apply to healthy, actively growing plants. Do not apply to stressed plants. Apply no more than 8 ounces per acre per season.
	azoxystrobin + difenoconazole (Quadris Top) 1.67 SC	12 to 14 fl oz/acre	3.6 to 4.2 fl oz/acre	1	0.5	Make no more than one application before alternating with fungicides that have a different mode of action. Apply no more than 56 fluid ounces per crop per acre per season.
	chlorothalonil (various)	See label	See label	See label	See label	See label. Rates vary depending on the formulation.

Table 10-25. Disease Control Products for Cucurbits

Disease	Material	Rate of Material to Use		Minimum Days		Method, Schedule, and Remarks
		Formulation	Active Ingredient	Harvest	Reentry	
Powdery mildew (continued)	chlorothalonil + potassium phosphite (Catamaran) 5.27 SC	6 pints/acre	3.3 pints/acre	0	0.5	Apply no more than 50 pints per crop per acre per season. Do not apply to watermelon fruit when stress conditions conducive to sunburn occur.
	difenoconazole + cyprodinil (Inspire Super) 2.82 SC	16 to 20 fl oz/acre	5.2 to 6.5 fl oz/acre	7	0.5	Make no more than two sequential applications before alternating with fungicides that have a different mode of action. Apply no more than 80 fl oz per crop per acre per season.
	cyprodinil + fludioxonil (Switch) 62.5 WG	11 to 14 oz/acre	6.9 to 8.9 oz/acre	1	0.5	Make no more than 2 applications before alternating to a different fungicide. Maximum of 4 to 5 applications at high and low rates. Not for target spot or anthracnose or cecospora.
	fixed copper (various)	See label	See label	See label	See label	See label. Rates vary depending on the formulation. Repeated use may cause leaf yellowing.
	cyflufenamid (Torino) 0.85 SC	3.4 oz/acre	0.02 lb/acre	0	4 hr	Do not make more than 2 applications per crop.
	fluopyram + tebuconazole (Luna Experience) 3.3F	8 to 17 fl oz/acre	0.27 to 0.44 lb/acre	7	0.5	**APPLY ONLY TO WATERMELON.** Make no more than 2 applications before alternating to a fungicide with different active ingredients. Do not rotate with tebuconazole.
	fluxapyroxad + pyraclostrobin (Merivon 500 SC)	4 to 5.5 fl oz/ acre	0.26 to 0.36 lb/acre	0	0.5	Make no more than two sequential applications before alternating with fungicides that have a different mode of action. Maximum of 3 applications per crop.
	metrafenone (Vivando)	15.4 fl oz/acre	0.3 lb/acre	0	0.5	Supplemental label expires Dec. 31, 2017. Begin applications prior to disease and continue in a 7- to 10-day
	myclobutanil (Rally) 40 WP	2.5 to 5 oz/acre	1 to 2 oz/acre	0	1	Apply no more than 1.5 pound per acre per crop. Observe a 30-day plant-back interval.
	penthiopyrad (Fontelis) 1.67 SC	12 to 16 fl oz/acre	0.16 to 0.21 lb/acre	1	0.5	Make no more than 2 sequential applications before switching to another fungicide. Do not rotate with Pristine or Luna Experience.
	pyraclostrobin + boscalid (Pristine) 38 WG	12.5 to 18.5 oz/acre	4.8 to 7 oz/acre	0	1	Make no more than 4 applications per season.
	quinoxyfen (Quintec) 2.08 SC	4 to 6 fl oz/acre	1 to 1.3 fl oz/acre	3	0.5	Make no more than 2 applications before alternating to a different fungicide. Maximum of 24 fluid ounces/acre per year. **DO NOT USE ON SUMMER SQUASH or CUCUMBER**; labeled on winter squashes, pumpkins, gourds, melon and watermelon.
	sulfur (various)	See label	See label	See label	See label	See label. Rates vary depending on the formulation. Do not use when temperature is over 90 degrees F or on sulfur-sensitive varieties.
	tebuconazole (Monsoon) 3.6F	4 to 6 fl oz/acre	1.5 to 2 lb/acre	7	0.5	Apply before disease appears when conditions favor development and repeat at 10 to 14 day intervals; max 24 fluid ounces per season.
	triflumizole (Procure) 50 WS	4 to 8 oz/acre	2 to 4 oz/acre	0	0.5	Begin applications at vining or first sign of disease, and repeat at 7 to 10 day intervals.
Scab	acibenzolar-S-methyl (Actigard) 50 WP	0.5 to 1 oz/acre	0.25 to 0.5 oz/acre	0	0.5	Apply to healthy, actively growing plants. Do not apply to stressed plants. Apply no more than 8 ounces per acre per season.
	chlorothalonil (various)	See label	See label	See label	See label	See label. Rates vary depending on the formulation.
	chlorothalonil + potassium phosphite (Catamaran) 5.27 SC	6 pints/acre	3.3 pints/acre	0	0.5	Apply no more than 50 pints per crop per acre per season. Do not apply to watermelon fruit when stress conditions conducive to sunburn occur.
Vine decline (*Monosporascus cannonballus*)	fludioxonil (Cannonball WG)	4 to 8 oz/acre		14	0.5	**APPLY ONLY TO MELONS.**

Table 10-26. Efficacy of Products for Disease Control in Cucurbits

Scale: E, excellent; G, good; F, fair; P, poor; NC, no control; ND, no data.

Product[1]	Fungicide group[F]	Preharvest interval (Days)	Alternaria Leaf Spot	Angular Leafspot	Anthracnose	Bacterial Fruit Blotch	Belly Rot	Cercospora Leaf Spot	Cottony Leak	Damping off (Pythium)	Downy Mildew	Gummy Stem Blight	Phytophthora Blight (foliage and fruit)	Phytophthora Blight (crown and root)	Plectosporium Blight	Powdery Mildew	Target Spot
acibenzolar-S-methyl (Actigard)	21	0	NC	ND	NC	F	NC	NC	ND	ND	ND	NC	ND	ND	NC	ND	NC
ametoctradin + dimethomorph (Zampro)	45 + 40	0	ND	NC	NC	NC	NC	NC	ND	ND	F	NC	G	G	NC	NC	NC
Azoxystrobin[2] (Quadris)	11	1	G	NC	G	NC	F	G	NC	NC	NC[R]	NC[R]	NC	NC	F	NC[R]	G
azoxystrobin + difenoconazole (Quadris Top)	11 + 3	1	ND	NC	G	NC	ND	ND	ND	ND	ND	F	ND	ND	F	F	ND
Chlorothalonil[9] (various)	M	0	G	NC	G	NC	NC	NC	NC	NC	F	F	NC	NC	F	F	G
cyazofamid (Ranman)	21	0	NC	NC	NC	NC	NC	NC	ND	NC	G	NC	F	NC	NC	NC	NC
cyflufenamid (Torino)	U6	0	NC	NC	NC	NC	NC	NC	NC	NC	NC	NC	NC	NC	NC	E	NC
cymoxanil (Curzate)	27	3	NC	NC	NC	NC	NC	NC	ND	ND	F	NC	F	NC	NC	NC	NC
cyprodinil + fludioxonil (Switch)	9 + 12	1	ND	NC	F	NC	ND	ND	NC	NC	NC	F	NC	NC	F	F	NC
difenoconazole + cyprodinil (Inspire Super)	3 + 9	7	ND	NC	ND	NC	NC	NC	NC	NC	NC	F	NC	NC	F	F	ND
dimethomorph (Forum)	40	0	NC	NC	NC	NC	NC	NC	NC	NC	P	NC	P	NC	NC	NC	NC
Famoxadone[2] + cymoxanil (Tanos)	11 + 27	3	ND	NC	ND	NC	ND	NC	NC	NC	F	NC	ND	NC	NC	NC	NC
fenamidone (Reason)	11	14	F	NC	NC	NC	NC	NC	ND	NC	F[R]	NC	F	NC	NC	NC	NC
fixed copper (various)[P]	M	1	P	F	P	F	NC	F	NC	NC	P	P	ND	NC	P	P	P
fluopicolide (Presidio)	43	2	NC	NC	NC	NC	NC	NC	NC	P[R]	NC	F	F	NC	NC	NC	NC
fluopyram + tebuconazole (Luna Experience)	7 + 3	7	ND	NC	NC	NC	ND	NC	NC	NC	NC	G	NC	NC	NC	G	NC
fluopyram + trifloxystrobin (Luna Sensation)	7 + 11	0	ND	NC	F	NC	NC	NC	NC	NC	NC	F	NC	NC	NC	F	NC
kresoxim-methyl (Sovran)	11	0	ND	NC	ND	NC	ND	NC	NC	NC	ND	NCR	ND	NC	ND	NCR	ND
mancozeb (various)[5]	M	5	F	NC	G	NC	NC	G	NC	NC	F	F	P	NC	F	P	G
mancozeb + fixed copper[4] (ManKocide)	M + M	5	P	F	P	F	NC	P	NC	NC	F	NC	P	NC	P	P	F
mandipropamid (Revus)	40	0	NC	NC	NC	NC	NC	NC	NC	NC	NC[R]	NC	F	P	NC	NC	NC
mefenoxam[4,3] (Ridomil Gold EC, Ultra Flourish)	4	0	NC	NC	NC	NC	NC	NC	F[R]	G[R]	NC	NC	F[R]	F[R]	NC	NC	NC
Mefenoxam[2] + chlorothalonil[5] (Ridomil Gold/Bravo, Flouronil)	4 + M	0	F	NC	NC	NC	NC	F	F[R]	F[R]	F[R]	NC	F[R]	NC	F	F	F
Mefenoxam[2] + copper[5] (Ridomil Gold/Copper)	4 + M	5	P	P	F	P	NC	P	F[R]	F[R]	F[R]	P	F[R]	NC	P	NC	P
Mefenoxam[2] + mancozeb[5] (Ridomil Gold MZ)	4 + M	5	F	NC	F	NC	NC	F	F[R]	F[R]	F[R]	F	F[R]	NC	F	NC	F
Myclobutanil[2] (Rally)	3	0	NC	NC	NC	NC	NC	NC	NC	NC	NC	NC	NC	NC	NC	F	NC
Oxathiapiprolin (Orondis Opti, Orondis Ultra, Orondis Gold)	U15		ND	NC	ND	NC	ND	ND	NC	NC	G	ND	G	G	ND	ND	ND
penthiopyrad (Fontelis)	7	1	ND	NC	F	NC	NC	NC	NC	NC	NC	NC[R]	NC	NC	NC	F	NC
Phosphonate[6] (Aliette, Agri-Fos, Phostrol, ProPhyte)	33	0.5	NC	NC	NC	NC	NC	NC	NC	NC	P	NC	NC	F	NC	NC	NC
potassium phosphite + tebuconazole (Viathon)	33 + 3	7	ND	NC	ND	NC	ND	ND	ND	ND	P	F	ND	ND	NC	F	NC
propamocarb (Previcur Flex)	28	2	NC	NC	NC	NC	NC	NC	NC	ND	F	NC	G	NC	NC	NC	NC
pyraclostrobin[2] (Cabrio)	11	0	G	NC	G	NC	NC	NC	NC	NC	NC[R]	NC[R]	P	NC	G	NC[R]	E
pyraclostrobin[2] + boscalid (Pristine)	11 + 7	0	G	NC	F	NC	ND	G	NC	NC	NC[R]	NC[R]	P	NC	F	F	E
quinoxyfen (Quintec)	13	3	NC	NC	NC	NC	NC	NC	NC	NC	NC	NC	NC	NC	NC	E	NC
Sulfur (various) [P]	M	0	NC	NC	NC	NC	NC	NC	NC	NC	NC	NC	NC	NC	NC	G	NC
tebuconazole (Monsoon)	3	7	ND	NC	NC	NC	NC	ND	NC	NC	NC	F	NC	NC	NC	F	NC
thiophanate-methyl[3] (Topsin M)	1	1	F	NC	F	NC	F	F	NC	NC	NC	NC	NC	NC	F	NC[R]	P
trifloxystrobin[3] (Flint)	11	0	G	NC	G	NC	ND	ND	NC	NC	NC[R]	NC[R]	NC	NC	G	NC[R]	G
triflumizole (Procure)	3	0	NC	NC	NC	NC	NC	NC	NC	NC	NC	NC	NC	NC	NC	F	NC
zoxamide + mancozeb (Gavel)	22 + M	5	F	NC	F	NC	NC	F	NC	NC	F	F	P	NC	F	P	F

[1] Efficacy ratings do not necessarily indicate a labeled use for every disease.
[2] Curative activity; locally systemic.
[3] Systemic.
[4] When used in combination with chlorothalonil or mancozetb, gives increased control.
[5] Contact control only; no systemic control.
[6] Check manufacturers label for compatibility with other products.
[P] Can be phytotoxic at temperatures above 90 degrees F; read the label carefully.
[F] To prevent resistance in pathogens, alternate fungicides within a group with fungicides in another group. Fungicides in the "M" group are generally considered "low risk" with no signs of resistance developing to the majority of fungicides.
[R] Resistance reported in the pathogen.

Table 10-27. Importance of Alternative Management Practices for Disease Control in Cucurbits

Scale: E, excellent; G, good; F, fair; P, poor; NC, no control; ND, no data

Strategy	Alternaria leaf blight	Angular leaf spot	Anthracnose	Bacterial fruit blotch	Bacterial wilt	Belly rot	Cercospora leaf spot	Choanephora fruit rot	Cottony leak	Downy mildew	Gummy stem blight	Mosaic virus	Phytophthora blight	Plectosporium blight	Powdery mildew	Pythium damping off	Root knot	Target spot
Avoid field operations when leaves are wet	P	F	P	F	F	NC	NC	P	NC	P	P	NC	NC	ND	NC	NC	NC	NC
Avoid overhead irrigation	F	F	F	F	P	NC	P	NC	NC	F	F	NC	F	P	P	NC	NC	P
Change planting date from Fall to Spring[1]	G	P	G	P	P	F	G	F	F	G	G	F	F	F	F	G	G	G
Cover cropping with antagonist	NC	NC	NC	NC	NC	NC	NC	NC	NC	NC	NC	NC	NC	NC	NC	NC	F	NC
Crop rotation with non-host (2 to 3 years)	F	F	F	F	NC	P	F	NC	NC	NC	F	NC	F	F	NC	P	F	F
Deep plowing	P	NC	P	NC	NC	F	P	NC	NC	NC	F	NC	P	P	NC	P	F	P
Destroy crop residue immediately	F	P	F	P	P	P	P	NC	P	F	F	F	P	P	F	NC	F	P
Encourage air movement[2]	F	P	F	P	NC	NC	F	F	F	F	F	NC	NC	P	NC	NC	NC	F
Soil organic amendments[3]	ND	NC	ND	NC	NC	P	ND	NC	F	NC	ND	NC	P	ND	NC	F	F	ND
Insecticidal/horticultural oils[4]	NC	NC	NC	NC	F	NC	NC	NC	NC	NC	NC	F	NC	NC	F	NC	NC	NC
pH management (soil)	NC	NC	NC	NC	NC	NC	NC	NC	ND	NC	NC	NC	ND	NC	NC	ND	ND	NC
Plant in well-drained soil	NC	NC	NC	NC	NC	F	NC	P	F	NC	NC	NC	F	NC	NC	F	P	NC
Plant on raised beds	NC	NC	NC	NC	NC	P	NC	P	F	NC	F	NC	F	NC	NC	F	P	NC
Plastic mulch bed covers	NC	NC	NC	NC	NC	F	NC	P	F	NC	F	NC	F	P	NC	NC	NC	NC
Postharvest temperature control (fruit)	NC	NC	F	F	NC	P	NC	F	F	NC	F	NC	F	F	NC	NC	NC	NC
Reflective mulch (additional effect over plastic mulch)	NC	NC	NC	NC	NC	NC	NC	NC	NC	NC	NC	F	NC	NC	NC	NC	NC	NC
Reduce mechanical injury	P	P	P	P	F	P	P	P	P	NC	P	P	P	P	NC	NC	NC	P
Rogue diseased plants/fruit (home garden)	F	P	P	P	P	NC	P	P	P	P	P	F	F	NC	NC	P	F	P
Row covers (insect exclusion)	NC	NC	NC	NC	G	NC	NC	NC	NC	NC	NC	G	NC	NC	NC	NC	NC	NC
Soil solarization (reduce soil inoculum)	P	NC	P	NC	NC	F	P	NC	P	NC	P	NC	P	P	NC	F	P	P
Pathogen-free planting material	P	E	F	E	NC	NC	NC	NC	NC	NC	E	NC	NC	NC	NC	F	NC	NC
Resistant cultivars[5]			E		E					E		E			E			
Destroy volunteer plants	F	F	F	F	F	NC	F	NC	NC	F	F	F	F	NC	F	NC	P	F

[1] Early planting reduces risk.

[2] Air movement can be encouraged by increasing plant spacing, orienting beds with prevailing wind direction and increasing exposure of field to prevailing wind.

[3] Soil organic amendments = cover crops; composted organic wastes.

[4] Insecticidal/Horticultural oil = Sunspray Ultra-Fine Spray Oil (Sun Company, Inc.), JMS Stylet oil; Safe-T-Side (Brandt Consolidated, Inc.); PCC 1223 (United Ag Products).

[5] Resistance available in some cucurbits.

517

DISEASE CONTROL

Example Spray Program for Foliar Disease Control in Watermelon Production

This spray program is based on research conducted at the Clemson Coastal Research and Education Center, Charleston, SC, and on a survey of watermelon fields in South Carolina in 2015 and 2016. The most common diseases in both years were gummy stem blight and powdery mildew. The spring program is designed to manage bacterial fruit blotch, bacterial leaf spots, gummy stem blight, powdery mildew, anthracnose, and downy mildew on watermelon. The fall program is designed to manage gummy stem blight, downy mildew, and anthracnose.

- Start spraying when vines start to run or no later than when the first blooms (the male ones) open.
- From vine run until mid-May, spray every 10 days. After mid-May or when powdery and downy mildew typically show up in your area, spray every week through harvest regardless of the weather. Weekly sprays are needed to protect watermelon from powdery and downy mildew. Dry weather limits gummy stem blight but promotes powdery mildew; dry weather does not limit downy mildew or anthracnose if they are already present in a field.
- **Do not stop spraying until you stop harvesting.** Downy and powdery mildew can attack a crop any time it goes more than one week without a fungicide spray. Fungicides with a 7-day PHI are not recommended during the harvest period (usually after week 5); note that mancozeb and Gavel have a 5-day PHI. If more sprays are needed after spray 8 to extend the season during harvesting, apply sprays 5 to 8 again.

If this spray schedule is used to select fungicides for other cucurbits (vine crops), note that not all fungicides in this spray schedule are labeled on other cucurbits. Luna Expe- rience and Luna Sensation are registered only on watermelon. **Luna Experience and Luna Sensation are not registered for use in Louisiana.** Quintec is not registered on cucumber. To control powdery mildew on other cucurbits or in Louisiana, use Procure in place of Luna fungicides or Quintec.

Table 10-28. Example Spray Program for Foliar Disease Control in Watermelon Production

Spray	Fungicide Program for Spring Watermelon*	Comments on Spring Program	Fungicide Program for Fall Watermelon*
1 (vine run)	mancozeb + fixed copper	For prevention of bacterial leaf spots.	chlorothalonil or Catamaran
2	tebuconazole	If fruit blotch or bacterial leaf spots are a concern, add fixed copper.	tebuconazole or Luna Sensation
3	chlorothalonil or Catamaran	Do not tank mix copper with chlorothalonil or Catamaran.	chlorothalonil + **Zampro**
4	chlorothalonil (or mancozeb)	If fruit blotch is a concern, substitute mancozeb + fixed copper.	mancozeb + Flint
5a**	mancozeb + Quintec	The protectant switches here from chlorothalonil to mancozeb to avoid injury to fruit on hot, sunny	**Gavel**
5b**	Luna Experience	Use Luna Experience if powdery mildew is present.	(same as 5a)
6	**Gavel**	Note 5-day PHI.	Quadris Top
7a**	mancozeb + Torino	Note 5-day PHI on mancozeb.	mancozeb + **Ranman**
7b**	Switch	Use Switch if gummy stem blight is present.	(same as 7a)
8	mancozeb + **Ranman**	If downy mildew is seen earlier in the season, apply Ranman as soon as possible.	chlorothalonil
9-12	If more sprays are needed after spray 8 until the last harvest, apply sprays 5 to 8 again.		

* Fungicides that control downy mildew are in bold. Fungicides that control powdery mildew are underlined.

** Option "a" is a lower cost treatment that may be less effective. Option "b" is a more expensive fungicide that also is more effective.

Eggplant

Table 10-29. Disease Control Products for Eggplant

Disease	Material	Rate of Material to Use		Minimum Days		Method, Schedule, and Remarks
		Formulation	Active Ingredient	Harvest	Reentry	
Leaf blights, fruit rots	azoxystrobin (Quadris) Flowable	6 to 15.5 fl oz/acre	0.1 to 0.25 lb/acre	0	4 hr	Apply at flowering to manage green fruit rot. Limit of 61.5 fluid ounces per acre per season. Make no more than **one** application before alternating with fungicides that have a different mode of action.
	azoxystrobin + difenoconazole (Quadris Top) 29.6SC	8 to 14 fl oz/acre	2.4 to 4.1 fl oz/acre	0	0.5	Limit of 55.3 fluid ounces per acre per season. Make no more than two consecutive applications before rotating to another effective fungicide with a different mode of action.
	boscald (Endura)	2.5 to 3.5 oz/acre	1.75 to 2.45 oz/acre	0	0.5	Limit of 21 ounces per acre per season. Make no more than two sequential applications before alternating with fungicides that have a different mode of action.
	chlorothalonil (various)	See label	See label	See label	See label	See label. Rates vary depending on the formulation.
	chlorothalonil + cymoxanil (Ariston)	2 to 2.44 pt/acre	0.84 to 1.02 pt/acre	3	0.5	Limit of 18.1 pints per acre per year.
	fixed copper (various)	See label	See label	See label	See label	See label. Rates vary depending on the formulation. Make first application 7 to 10 days after transplanting. Carefully examine field for disease to determine need for additional applications. If disease is present, make additional applications at 5-day intervals. Do not spray copper when temperatures are above 90 degrees F.

Table 10-29. Disease Control Products for Eggplant

Disease	Material	Rate of Material to Use		Minimum Days		Method, Schedule, and Remarks
		Formulation	Active Ingredient	Harvest	Reentry	
Leaf blights, fruit rots (continued)	fenamidone (Reason) 500SC	5.5 to 8.2 fl oz/acre	0.178 to 0.267 lb/ acre	14	0.5	Limit of 24.6 fluid ounces per growing season. Make no more than **one** applications before rotating to another effective fungicide with a different mode of action.
	fluoxastrobin (Aftershock, Evito 280SC)	2 to 5.7 fl oz/acre	0.81 to 2.30 fl oz/acre	3	0.5	Limit of 22.8 fluid ounces per acre per season. Make no more than **one** application before alternating with fungicides that have a different mode of action. **NOTE: Do not overhead irrigate for 24 hours following a spray application.**
	mancozeb (various)	See label	See label	See label	See label	See label. Rates vary depending on the formulation.
	mancozeb + copper (ManKocide)	2 to 3 lb/acre	1.2 to 1.8 lb/acre	7	2	Limit of 39 pounds per acre per season.
	penthiopyrad (Fontelis)	16 to 24 oz/acre	3.26 to 4.89 fl oz/acre	0	0.5	**SUPPRESSION ONLY.** Limit of 72 fluid ounces per acre per year. Make no more than two consecutive applications before rotating to another effective fungicide with a different mode of action.
	pyraclostrobin (Cabrio) EG 20%	8 to 12 oz/acre	1.6 to 2.4 oz/acre	0	4 hr	Apply at flowering to manage green fruit rot. Limit of 96 ounces per acre per season. Make no more than **one** sequential application before alternating with fungicides that have a different mode of action.
	pyraclostrobin + fluxapyroxad (Priaxor)	4.0 to 8.0 fl oz/acre	1.72 to 3.43 fl oz/acre	0	0.5	Limit of 24 fluid ounces per acre per season. Make no more than two consecutive applications before rotating to another effective fungicide with a different mode of action.
Phytophthora blight	cyazofamid (Ranman Fungicide)	2.75 fl oz/acre	0.071 lb/acre	0	0.5	Limit of 16.5 fluid ounces per acre per season. Apply to the base of the plant at transplanting or in the transplant water. Make no more than three consecutive applications followed by three consecutive applications of another effective fungicide with a different mode of action.
	fixed copper (various)	See label	See label	See label	See label	See label. Rates vary depending on the formulation. Begin applications when conditions first favor disease development and repeat at 3 to 10 day intervals if needed depending on disease severity. Use the higher rates when conditions favor disease. Do not spray copper when temperatures are above 90 degrees F.
	dimethomorph (Acrobat, Forum)	6 fl oz/acre	2.61 fl oz/acre	0	0.5	**SUPPRESSION ONLY.** Limit of 30 fluid ounces per acre per season. Make no more than two sequential before alternating with fungicides that have a different mode of action. **NOTE: Must tank mix with another fungicide with a different mode of action.**
	mancozeb (various)	See label	See label	See label	See label	See label. Rates vary depending on the formulation.
	mancozeb + copper (ManKocide)	2 to 3 lb/acre	1.2 to 1.8 lb/acre	7	2	**SUPPRESSION ONLY.** Limit of 39 pounds per acre per season.
	mefenoxam + copper hydroxide (Ridomil Gold/Copper)	2 lb/acre	1.3 lb/acre	7	2	See label for an optimal spray program. Limit of four applications per crop per year. Do not exceed 0.4 pound a.i. per acre per season of mefenoxam + metalaxyl (MetaStar).
	famoxadone + cymoxanil (Tanos)	8 to 10 oz/acre	4 to 5 oz/acre	3	0.5	**SUPPRESSION ONLY.** Make no more than **one** application before alternating with a fungicide with a different mode of action. **NOTE: Must tank mix with another fungicide with a different mode of action (i.e maneb or copper).**
	fenamidone (Reason) 500SC	8.2 fl oz/acre	0.267 lb/ acre	14	0.5	**SUPRESSION ONLY.** Limit of 24.6 fluid ounces per growing season. Make no more than **one** application before rotating to another effective fungicide with a different mode of action.
	fluazinam (Omega) 500F	1 to 1.5 pt/acre		30	0.5	Apply as a soil drench at 1.5 pints per acre. For foliar applications use 1 pt per acre. Limit of 9 pints per acre per season.
	fluopicolide (Presidio)	3 to 4 fl oz/acre	0.09 to 0.125 lb/acre	2	0.5	Limit of 12 ounces per acre per season. Make no more than two times sequentially before alternating with fungicides that have a different mode of action. **NOTE: Must be tank-mixed with another mode of action product.**
	mandipropamid (Revus, Micora)	8 fl oz/acre	0.13 lb/acre	1	0.5	**SUPPRESSION ONLY.** Limit of 32 fluid ounces per acre per season. **NOTE: Must tank mix with another fungicide with a different mode of action (i.e. copper).**
	amectoctradin + dimethomorph (Zampro)	14 fl oz/acre	0.27 lb/acre	4	0.5	Limit of 42 fluid ounces per acre per season. Make no more than two sequential applications before rotating to another effective fungicide with a different mode of action.
Powdery mildew	azoxystrobin (Quadris) Flowable	6 to 15.5 fl oz/acre	0.10 to 0.25 lb/acre	0	4 hr	Limit of 61.5 fluid ounces per acre per season. Make no more than **one** application before alternating with fungicides that have a different mode of action.
	azoxystrobin + difenoconazole (Quadris Top) 29.6SC	8 to 14 fl oz/acre	2.4 to 4.1 fl oz/acre	0	0.5	Limit of 55.3 fluid ounces per acre per season. Make no more than two consecutive applications before rotating to another effective fungicide with a different mode of action.

DISEASE CONTROL

Table 10-29. Disease Control Products for Eggplant

Disease	Material	Rate of Material to Use		Minimum Days		Method, Schedule, and Remarks
		Formulation	Active Ingredient	Harvest	Reentry	
Powdery Mildew (continued)	chlorothalonil (various)	See label	See label	See label	See label	See label. Rates vary depending on the formulation.
	chlorothalonil + cymoxanil (Ariston)	2 to 2.44 pt/acre	0.84 to 1.02 pt/acre	3	0.5	Limit of 18.1 pints per acre per year.
	pyraclostrobin (Cabrio) EG 20%	8 to 16 oz/acre	1.6 to 2.4 oz/acre	0	4 hr	Apply at flowering to manage green fruit rot. Limit of 96 ounces per acre per season. Make no more than **one** sequential application before alternating with fungicides that have a different mode of action.
	pyraclostrobin + fluxapyroxad (Priaxor)	6.0 to 8.0 fl oz/acre	2.57 to 3.43 fl oz/acre	0	0.5	Limit of 24 fluid ounces per acre per season. Make no more than two consecutive applications before rotating to another effective fungicide with a different mode of action.
	Pyraclostrobin (Cabrio) EG	8 to 16 oz/acre	1.6 to 2.4 oz/acre	0	0.5	Limit of 96 ounces per acre per season. Make no more than **one** sequential application before alternating with fungicides that have a different mode of action.
Southern blight (*Sclerotium rolfsii*)	fluoxastrobin (Aftershock, Evito 280SC)	2 to 5.7 fl oz/acre	0.81 to 2.30 fl oz/acre	3	0.5	Limit of 22.8 fluid ounces per acre per season. Make no more than **one** application before alternating with fungicides that have a different mode of action. **NOTE: Do not overhead irrigate for 24 hours following a spray application.**
	polyoxin D (OSO 5%)	6.5 to 13 fl oz/acre	0.32 TO 0.72 oz/acre	0	4 hr	**Suppression only.** Can be applied using banded or irrigation water applications. Limit of 4.2 ounces a.i. per acre per season.
	pyraclostrobin (Cabrio) EG 20%	12 to 16 oz/acre	1.6 to 2.4 oz/acre	0	4 hr	**Suppression only.** Apply at flowering to manage green fruit rot. Limit of 96 ounces per acre per season. Make no more than **one** sequential application before alternating with fungicides that have a different mode of action.
	pyraclostrobin + fluxapyroxad (Priaxor)	4.0 to 8.0 fl oz/acre	1.72 to 3.43 fl oz/acre	0	0.5	**SUPPRESSION ONLY.** Limit of 24 fluid ounces per acre per season. Make no more than two consecutive applications before rotating to another effective fungicide with a different mode of action.
Pythium root rot	mefenoxam (Ridomil Gold, Ultra Flourish)	Check label; varies between products	Check label; varies between products	—	2	**May only be applied at planting.** Apply in a 12 to 16 inches band or in 20 to 50 gallons water per acre in transplant water. Mechanical incorporation or 0.5 to 1 inch irrigation water needed for movement into root zone if rain is not expected. After initial application, two supplemental applications (1 pint per treated acre) can be applied.
	metalaxyl (MetaStar) 2E	4 to 8 pt/treated acre	0.5 to 1 lb/acre	7	2	Limit of 12 pints per acre per season. Preplant (soil incorporated), at planting (in water or liquid fertilizer), or as a basil-directed spray after planting. See label for the guidelines for supplemental applications.
Rhizoctonia seedling and root rot	azoxystrobin (Equation, Quadris, Satori) Flowable	0.4 to 0.8 fl oz/1,000 row feet	0.006 to 0.013 lb/1,000 row feet	—	4 hr	Make in-furrow or banded applications shortly after plant emergence. Under cool, wet conditions, crop injury from soil directed applications may occur.
Verticillium wilt	Polyoxin D (OSO 5%)	6.5 to 13 fl oz/acre	0.32 TO 0.72 oz/acre	0	4 hr	**SUPPRESSION ONLY.** Can be applied using banded or irrigation water applications. Limit of 4.2 ounces a.i. per acre per season.

Endive

Table 10-30. Disease Control Products for Endive

Disease	Material	Rate of Material to Use		Minimum Days		Method, Schedule, and Remarks
		Formulation	Active Ingredient	Harvest	Reentry	
Downy mildew	amectoctradin + dimethomorph (Zampro 525 SC)	14 fl oz/acre	6.6 oz/acre	0	12 hr	Do not make more than 2 sequential applications before alternating to a fungicide with a different mode of action. Addition of an adjuvant may improve performance (see label for specifics). Do not apply more than 42 fluid ounces per acre per season.
	azoxystrobin (Satori, Quadris) 2.08 F	12.3 to 15.4 fl oz/ acre	0.2 to 0.25 fl oz/ acre	0	4 hr	Make no more than two sequential applications before alternating with fungicides that have a different mode of action. Apply no more than 2.88 quarts per crop per acre per season.
	cymoxanil/famoxadone (Tanos)	8.0 oz	0.178 to 0.267 fl oz	3	0.5	See label for directions.
	dimethomorph (Acrobat, Forum) 50 WP	6.4 oz/acre	3.2 oz/acre	0	0.5	Must be applied as a tank mix with another fungicide active against downy mildew. Do not make more than two sequential applications.
	fenamidone (Reason 500 SC)	5.5 to 8.2 fl oz	0.178 to 0.267 fl oz	2	0.5	Alternate with fungicides with a different mode of action.
	fosetyl-Al (Aliette) 80 WDG	2 to 5 lb/acre	1.6 to 4 lb/acre	3	0.5	Do not mix with surfactants, foliar fertilizers, or products containing copper.
	fluopicolide (Presidio)	3 to 4 fl oz/acre	0.178 to 0.267 fl oz	2	0.5	Must be tank-mixed with a fungicide with a different mode of action. Do not exceed two sequential applications.

DISEASE CONTROL

520

Table 10-30. Disease Control Products for Endive

Disease	Material	Rate of Material to Use		Minimum Days		Method, Schedule, and Remarks
		Formulation	Active Ingredient	Harvest	Reentry	
Downy mildew (continued)	mandipropamid (Micora)	5.5 to 8 fl oz/A	0.65 to 0.9 fl oz/A	—	4 hr	Tank mix Micora with a non-Group 40 fungicide and begin applications prior to disease development. **DO NOT** apply more than two applications per crop, or in consecutive applications.
	mandipropamid (Revus) 2.08 F	8 fl oz/acre	0.13 lb/acre	1	0.5	Apply prior to disease development and continue throughout season at 7 to 10 day intervals; maximum 32 fluid ounces per season.
	Pyraclostrobin (Cabrio)	12 to 16 oz/acre	0.1 to 0.20oz/acre	0	0.5	
Decay, Seedling blight, damping-off	Fludioxonil (Spirato) 480FS (Maxim) 4FS	0.08-0.16 fl oz/ 100 lb of seed		—	12	Used to control diseases of seed such as Aspergillus, Fusarium, and Rhizoctonia among others. Does NOT control Pythium or Phytophthora.
Gray mold rot	dichloran (Botran 75-W)	2 2/3 lb/acre	1.5 to 2 lb a.i.	14	0.5	Apply 7 days following transplanting. Repeat application when plants are half mature. Do not apply to wilted plants or seedlings. Two applications may be applied per season.
	penthiopyrad (Fontelis)	14 to 24 fl oz/acre	—	0	0.5	Begin applications before disease development. **DO NOT** make more than two sequential applications before switching to a fungicide with a different mode of action.
Leaf spots	azoxystrobin (Satori, Quadris) 2.08 F	6.2 to 15.4 fl oz	0.1 to 0.25 fl oz	0	4 hr	Make no more than two sequential applications before alternating with fungicides that have a different mode of action. Apply no more than 2.88 quarts per crop per acre per season.
	Cyprodinil/fludioxonil (Switch 62.5 WDG)	11 to 14 oz/acre	6.9 to 8.8 fl oz	0	0.5	Switch also has activity against basal rot, Sclerotinia and Gray mold. Alternate with a fungicide with a different mode of action after 2 applications.
	penthiopyrad (Fontelis)	14 to 24 fl oz/acre	—	0	0.5	Begin applications before disease development. **DO NOT** make more than two sequential applications before switching to a fungicide with a different mode of action.
	Pyraclostrobin (Cabrio)	12 to 16 oz/acre	0.15 to 0.20 oz/acre	0	0.5	
Lettuce drop	penthiopyrad (Fontelis)	14 to 24 fl oz/acre	—	0	0.5	Begin applications before disease development. **DO NOT** make more than two sequential applications before switching to a fungicide with a different mode of action.
Phytophthora blight and crown rot	Oxathiapiprlin (Orondis)	2.4-19.2 oz/A		0	4	Apply at planting, in furrow, by drip, or in transplant water.
Powdery mildew	penthiopyrad (Fontelis)	14 to 24 fl oz/acre	—	0	0.5	Begin applications before disease development. **DO NOT** make more than two sequential applications before switching to a fungicide with a different mode of action.
	Quinoxyfen (Quintec)	6 fl oz	—	1	1	Alternate with fungicide with a different mode of action
	sulfur (various)	See label	See label	See label	See label	See label. Rates vary depending on the formulation. Apply at early leaf stage and repeat every 10 to 14 days or as needed. Do not apply if temperatures are expected to exceed 90 degrees F within 3 days of application due to the risk of crop injury.
	Triflumizole (Procure 480 SC)	6 to 8 fl oz/acre	3 to 6 fl oz	0	0.5	Applications should begin prior to disease development. Repeat on a 14-day schedule. Do not apply more than 18 fluid ounces per acre per season.
Pythium damping-off	mefenoxam ((Ridomil Gold) 4 SL	1 to 2 pt/trt acre	0.5 to 1 lb/trt acre	—	2	Soil-incorporate at planting. Use proportionally less for band rate. Apply preplant incorporated or surface application at planting.
	metalaxyl (MetaStar) 2 E	4 to 8 pt/trt acre	0.5 to 1 lb/trt acre	—	2	Banded over the row, preplant incorporated, or injected with liquid fertilizer.
Rust	penthiopyrad (Fontelis)	14 to 24 fl oz/acre	—	0	0.5	Begin applications before disease development. **DO NOT** make more than two sequential applications before switching to a fungicide with a different mode of action.
	sulfur (various)	See label	See label	See label	See label	See label. Rates vary depending on the formulation. Apply at early leaf stage and repeat every 10 to 14 days or as needed. Do not apply if temperatures are expected to exceed 90 degrees F within 3 days of application due to the risk of crop injury.
Septoria leaf spot	Fluxapyroxad; pyraclostrobin (Merivon)	4 fl oz/acre		1	12 hr	Make no more than two sequential applications before alternating to a non-group 7 or non-group 11 fungicide. DO NOT apply more than 33 fluid ounces per acre per season.

Garlic

Table 10-31. Disease Control Products for Garlic

Disease	Material	Rate of Material to Use		Minimum Days		Method, Schedule, and Remarks
		Formulation	Active Ingredient	Harvest	Reentry	
Botrytis blight, purple blotch, downy mildew	azoxystrobin (Amistar, Quadris) 2.08 F	6.2 to 15.4 fl oz/acre	0.10 to 0.25 lb/acre	0	4 hr	Use upper rate for downy mildew and *Botrytis*. Do not make more than two sequential applications.
	azoxystrobin + difenoconazole (Quadris Top)	14 fl oz/acre	2.38 lb/acre	7	0.5	Begin sprays prior to disease onset and spray on a 7 to 14 day schedule. Do not rotate with Group 11 fungicides.
	fluazinam (Omega 500)	1.0 pt/acre	0.52 lb/acre	7	2	Initiate sprays when conditions are favorable for disease or at disease onset. Spray on a 7 to 10 day schedule.
	boscalid (Endura) 70 WG	6.8 oz/acre	4.8 oz/acre	7	0.5	Not for downy mildew. Do not make more than 2 sequential applications or more than 6 applications per season.
	chlorothalonil (various)	See label	See label	See label	See label	See label. Rates vary depending on the formulation.
	difenoconazole + cyprodinil (Inspire Super)	16 to 20 fl oz/acre		14	0.5	Make no more than two applications before alternating with a fungicide with a different mode of action.
	fluopicolide (Presidio) 4F	3 to 4 fl oz/acre	0.09 to 0.125 lb/acre	2	0.5	For *Botrytis* and downy mildew; tank mix with another fungicide with a different mode of action.
	mefenoxam + mancozeb (Ridomil Gold MZ)	2.5 lb/acre	—	7	2	
	mefenoxam + chlorothalonil (Ridomil Gold/Bravo)	2 lb/acre	—	7	2	Spray at first appearance, 7 to 14 day intervals.
	pyraclostrobin (Cabrio) 20 WG	8 to 12 oz/acre	1.6 to 2.4 oz/acre	7	0.5	Not for *Botrytis*. Use highest rate for downy mildew. Make no more than 2 sequential applications and no more than 6 applications per season.
	pyraclostrobin + boscalid (Pristine) 38 WG	10.5 to 18.5 oz/acre	4 to 7 oz/acre	7	1	Use highest rate for suppression only on downy mildew. Make no more than 6 applications per season.
	pyrimethanil (Scala) 5 F	9 or 18 fl oz/acre	0.35 or 0.7 lb/acre	7	0.5	Not for downy mildew. Use lower rate in a tank mix with broad spectrum fungicide and higher rate when applied alone. Do not apply more than 54 fluid ounces per crop.
	tebuconazole (Folicur) 3.6F	4 to 6 fl oz/acre	1.5 to 2 lb/acre	7	0.5	Only labeled for purple blotch; apply before disease appears when conditions favor purple blotch development and repeat at 10 to 14 day intervals; maximum 32.5 fluid ounces per season.
Downy mildew	dimethomorph (Acrobat, Forum) 50 WP	6.4 oz/acre	3.2 oz/acre	0	0.5	Must be applied as a tank mix with another fungicide active against downy mildew. Do not make more than two sequential applications.
	amectoctradin + dimethomorph (Zampro)	14.0 fl oz/acre	0.47 lb/acre	0	12 hr	Tank-mix with a broad-spectrum fungicide like chlorothaloni or mancozeb.
White rot (*Sclerotium*)	azoxystrobin (Amistar, Quadris) 2.08 F	6.2 to 12.3 fl oz/acre	0.10 to 0.20 lb/acre	0	4 hr	Do not make more than two sequential applications.
	iprodione (Rovral) 50 WP	4 lb/acre	2 lb/acre	—	1	Spray cloves as they are being covered by soil (38 to 40 inches bed spacing). One application per year.
	metam-sodium (Vapam; Sectagon)	37.5 to 75 gal/acre	160 to 320 lb/acre	—	2	Rate is based on soil properties and depth of soil to be treated.
	tebuconazole (Folicur) 3.6F	20.5 fl oz/acre	8 lb/acre	7	0.5	Apply in 4 to 6 inch band over/into furrow; maybe applied in chemigation.

Hop

Table 10-32. Disease Control Products for Hop

Disease	Material	Rate of Material to Use		Minimum Days		Method, Schedule, and Remarks
		Formulation	Active Ingredient	Harvest	Reentry	
Downy mildew (*Pseudoperonospora humuli*)	Fosetyl-Al (Aliette WDG)	2.5/acre		24	0.5	Apply as a directed foliar spray. When conditions are warm and humid applications should be made as follows: (1) when shoots are 6-12 inches high; (2) after training when vines are 5-6 feet tall; (3) approximately three weeks after the second application; and (4) during bloom. Use sufficient volume of water to insure complete coverage of foliage.
	fixed copper (various)	See label	See label	See label	See label	See label. Rates vary depending on the formulation. Repeated use may cause leaf yellowing.
	dimethomorph (Forum)	6 fl oz/acre		7	0.5	Begin sprays prior to disease. Minimum interval is 10 days. Maximum 3 applications per season.
	cymoxanil (Curzate 60DF)	3.2 oz/acre		7	0.5	Tank mix with a protectant fungicide. Begin applications prior to disease and continue at 10- to 14-day intervals.
	P potassium phosphite (various)	See label	See label	See label	See label	See label. Rates vary depending on the formulation.

DISEASE CONTROL

Table 10-32. Disease Control Products for Hop

Disease	Material	Rate of Material to Use		Minimum Days		Method, Schedule, and Remarks
		Formulation	Active Ingredient	Harvest	Reentry	
Downy mildew (*Pseudoperonospora humuli*) (continued)	metalaxyl (Metastar 2E)	1 qt/acre		45	2	Apply as a soil drench (1qt/acre in 20 gals) and follow with foliar fixed copper applications. Apply as foliar spray (1 quart/acre in 50 gals) in combination with fixed copper. Do not make more than 3 applications per season.
	ametoctradin + dimethomorph (Zampro)	11 to 14 fl oz/acre		7	0.5	Begin applications prior to disease and continue at 10-day intervals. Do not make more than 3 applications per season.
	famoxadone + cymoxanil (Tanos)	8 oz/acre		7	0.5	Begin applications prior to disease and continue at 6- to 8-day intervals. Do not make more than 6 applications per season.
	mefenoxam (Ridomil Gold SL)	0.50 pt/acre	0.25 pt/acre	45	2	Can apply as soil drench or foliar spray, see label for details. Tank mix with fixed copper.
	mandipropamid (Revus)	8 fl oz/acre	0.13 lb/acre	7	4 hr	Begin applications prior to disease and continue at 7- to 10-day intervals. Do not make more than 3 applications per season.
	cyazofamid (Ranman 400 SC)	2.5 to 2.75 fl oz/acre	0.054 to 0.071 lb/acre	3	0.5	Begin applications prior to disease and continue at 7- to 10-day intervals. Do not apply more than 32 fluid ounces per season.
Powdery mildew (*Sphaerotheca humuli, S. macularis*)	tebuconazole (Folicur 3.6F)	4 to 8 fl oz/acre		14	0.5	Begin applications prior to disease and continue at 10- to 14-day intervals. Do not apply more than 16.5 fluid ounces per acre per season.
	trifloxystrobin (Flint)	See label	See label	14	0.5	Several rates available. Begin applications prior to disease and continue at 10- to 14-day intervals.
	Metrafenone (Vivando)	15.4 fl oz/acre	0.3 lb/acre	3	0.5	Begin applications prior to disease and continue at 7- to 14-day intervals. Do not make more than 2 applications per season.
	quinoxyfen (Quintec)	8.2 fl oz/acre		21	0.5	Do not make more than 4 applications per season.
	Triflumizole (Procure 400 SC)	12 fl oz/acre		7	0.5	Begin applications prior to disease and continue at 14-day intervals. Do not apply more than 36 fluid ounces per acre per season.
	pyraclostrobin + boscalid (Pristine)	See label	See label	14	0.5	Ground and aerial applications allowed, see label for details. Begin ground applications prior to disease and continue at 10- to 21-day intervals. Do not make more than 3 applications per season.

Greens, Leafy Brassica (Collard, Kale, Mustard, Rape, Turnip)

Table 10-33. Disease Control Products for Greens, Leafy Brassica

Disease	Material	Rate of Material to Use		Minimum Days		Method, Schedule, and Remarks
		Formulation	Active Ingredient	Harvest	Reentry	
For turnips harvested for roots, see Root Vegetables						
Alternaria leaf spot, Cercospora leaf spot, Anthracnose, White spot, and various foliar diseases (see specific labels)	boscalid (Endura 70 WG)	6 to 9 oz/acre	4.2 to 6.3 oz/acre	14	0.5	Begin applications prior to disease development, and continue on a 7 to 14 day interval. Make no more than 2 applications per season. **Not labeled for turnip greens or roots.**
	azoxystrobin + difenoconazole (Quadris Top 2.72 SC)	12 to 14 fl oz/acre	3.5 to 4.1 oz/acre	1	0.5	Make no more than one application before alternating to another fungicide with Group 11 mode of action (NOT Quadris or Cabrio).
	azoxystrobin (Quadris 2.08SC)	6.2 to 15.4 fl oz/ acre	0.1 to 0.25 lb/acre	0	4 hr	Make no more than two sequential applications before alternating with fungicides that have a different mode of action. May be applied to turnip grown for roots.
	fluxapyroxad + pyraclostrobin (Priaxor 500 SC)	6.0 to 8.2 fl oz/ acre	0.20 to 0.27 lb/acre	3	0.5	No more than two sequential applications before alternating with fungicides that have a different mode of action. Maximum of 3 applications. **Do not apply to turnip greens or roots.**
	pyraclostrobin (Cabrio 20 EG)	12 to 16 oz/acre 8 to 12 oz/acre (turnip)	2.4 to 3.2 oz/acre 1.6 to 2.4 oz/acre	3	0.5	Begin applications prior to disease development and continue on a 7- to 10-day interval. No more than 2 sequential applications before alternating to a fungicide with different mode of action. **Cabrio may be used for turnip harvested for leaves. Cabrio cannot be used for turnip grown for roots.**
	tebuconazole 6 F (various)	3 to 4 oz/acre	1.4 to 1.8 oz/acre	7	0.5	For optimum results use as a preventative treatment. Folicur 3.6 F must have 2 to 4 hours of drying time on foliage for the active ingredient to move systemically into plant tissue before rain or irrigation occurs. **May be used for turnip harvested for leaves. Not for use on turnip grown for roots.**
	cyprodonil + fludioxonil (Switch 62.5WG)	11 to 14 oz/acre	6.9 to 8.7 oz/acre	7	0.5	Apply when disease first appears, and continue on 7 to 10 day intervals. See label for complete list of greens.

523

DISEASE CONTROL

Table 10-33. Disease Control Products for Greens, Leafy Brassica

Disease	Material	Rate of Material to Use		Minimum Days		Method, Schedule, and Remarks
		Formulation	Active Ingredient	Harvest	Reentry	
Alternaria leaf spot, Cercospora leaf spot, Anthracnose, White spot, and various foliar diseases (see specific labels) (continued)	penthiopyrad (Fontelis 1.67 SC)	14 to 30 fl oz/acre	2.9 to 6.2 oz/acre	0	0.5	Make no more than two sequential applications before alternating with fungicides that have a different mode of action. **May be applied to turnips grown for roots.**
	difenoconazole + cyprodinil (Inspire Super 2.82SC)	16 to 20 fl oz/acre	5.2 to 6.5 oz/acre	7	0.5	Make no more than 2 sequential applications before alternating to a fungicide with a different mode of action.
Botrytis gray mold	penthiopyrad (Fontelis 1.67 SC)	14 to 30 fl oz/acre	2.9 to 6.2 oz/acre	0	0.5	Make no more than two sequential applications before alternating with fungicides that have a different mode of action. **May be applied to turnips grown for roots.**
	difenoconazole + cyprodinil (Inspire Super 2.82 SC)	16 to 20 fl oz/acre	5.2 to 6.5 oz/acre	7	0.5	Make no more than 2 sequential applications before alternating to a fungicide with a different mode of action.
Downy Mildew	pyraclostrobin (Cabrio 20 EG)	12 to 16 oz/acre	2.4 to 3.2 oz/acre	3	0.5	Begin applications prior to disease development and continue on a 7- to 10-day interval. Make no more than 2 sequential applications before alternating to a fungicide with a different mode of action.
	fluopicolide (Presidio 4 SC)	3 to 4 fl. oz/acre	0.09 to 0.125 lb/acre	2	0.5	Make applications on a 7 to 10 day schedule. Presidio must be tank mixed with another fungicide with a different mode of action. Make no more than 2 sequential applications before rotating to a fungicide with a different mode of action. Apply no more than 12 ounces per acre per season.
	cyazofamid (Ranman 400 SC)	2.75 fl. oz/acre	0.072 lb/acre	0	0.5	Make applications on a -7 to 10-day schedule. Do not apply more than 39.5 fluid ounces per acre per crop growing season.
	mandipropamid (Revus 2.08 SC)	8.0 fl oz/acre	0.13 lb/acre	1	0.5	Begin applications prior to disease development and continue on a 7- to 10-day interval. Make no more than 2 consecutive applications before switching to another effec- tive non-group 40 fungicide. **Not labeled for turnip greens or roots.**
	fenamidone (Reason 500SC)	5.5 to 8.2 oz/acre	0.178 to 0.267 lb/acre	2	0.5	Begin applications as soon as conditions become favorable for disease development. Applications should be made on a 5 to 10 day interval. Do not make more than one application of Reason 500 SC before alternating with a fungicide from a different resistance management group.
	amectoctradin + dimethomorph (Zampro 525 SC)	14 fl oz/acre	6.6 oz/acre	0	0.5	Do not make more than 2 sequential applications before alternating to a fungicide with a different mode of action. Addition of an adjuvant may improve performance (see label for specifics).
	dimethomorph (Forum 4.16 SC)	6.4 oz/acre	3.2 oz/acre	0	0.5	Must be tank-mixed with another fungicide active against Phytophthora blight. Do not make more than 2 sequential applications before alternating to another effective fungicide with a different mode of action. Do not make more than 5 applications per season. **Not labeled for turnip greens or roots.**
	fosetyl-Al (Aliette 80W DG)	2 to 5 lb/acre	1.6 to 4 lb/acre	3	1	Apply when disease first appears; then repeat on 7 to 21 day intervals. Do not tank mix with copper fungicides. A maximum of 7 applications can be made per season. **Not labeled for turnip greens or roots.**
	potassium phosphite	2 to 4 pt/acre	2.1 lb/acre	0	4 hr	Apply when weather is foggy as a preventative. Do not apply to plants under water or temperature stress. Spray solution should have a pH greater than 5.5. Apply in at least 30 gallons water per acre.
Powdery mildew	boscalid (Endura 70 WG)	6 to 9 oz/acre	4.2 to 6.3 oz/acre	14	0.5	Begin applications prior to disease development, and continue on a 7 to 14 day interval. Make no more than 2 applications per season; disease suppression only. **Not labeled for turnip greens or roots.**
	pyraclostrobin (Cabrio 20 EG)	12 to 16 oz/acre	2.4 to 3.2 oz/acre	3	0.5	Begin applications prior to disease development and continue on a 7 to 10 day interval. Make no more than 2 sequential applications before alternating to a fungicide with a different mode of action. **Cabrio may be used for turnip harvested for leaves. Cabrio cannot be used for turnip grown for roots.**
	triflumizole (Procure 480SC)	6 to 8 oz/acre	2.5 to 3.4 oz/acre	1	0.5	Make no more than two sequential applications before rotating with a fungicide with a different mode of action. Do not rotate with Rally or Nova.
	cyprodinil + fludioxonil (Switch 62.5WG)	11 to 14 oz/acre	0.9 to 0.7 oz/acre	7	0.5	Apply when disease first appears, and continue on 7 to 10 day intervals. See label for complete list of greens. **May be used on turnip where leaves only will be harvested. Not for use on turnip grown for roots.**
	penthiopyrad (Fontelis 1.67 SC)	14 to 30 fl oz/acre	2.9 to 6.2 oz/acre	0	0.5	Make no more than 2 sequential applications before alternating with fungicides that have a different mode of action. **May be applied to turnips grown for roots.**
	difenoconazole + cypro-dinil (Inspire Super)	16 to 20 fl oz/acre	5.2 to 6.5 oz/acre	7	0.5	Make no more than 2 sequential applications before alternating to a fungicide with a different mode of action.

524

Table 10-33. Disease Control Products for Greens, Leafy Brassica

Disease	Material	Rate of Material to Use		Minimum Days		Method, Schedule, and Remarks
		Formulation	Active Ingredient	Harvest	Reentry	
Powdery mildew (continued)	fluxapyroxad + pyraclostrobin (Priaxor 500 SC)	6.0 to 8.2 fl oz/ acre	0.20 to 0.27 lb/acre	3	0.5	Make no more than two sequential applications before alternating with fungicides that have a different mode of action. Maximum of 3 applications. **Do not apply to turnip greens or roots.**
	tebuconazole 6 F (various)	3 to 4 oz/acre	1.4 to 1.8 oz/acre	7	0.5	For optimum results use as a preventative treatment. Folicur 3.6 F must have 2 to 4 hours of drying time on foliage for the active ingredient to move systemically into plant tissue before rain or irrigation occurs. **May be applied to turnip grown for roots.**
Rhizoctonia bottom rot	boscalid (Endura 70 WG)	6 to 9 oz/acre	4.2 to 6.3 oz/acre	14	0.5	Begin applications prior to disease development, and continue on a 7 to 14 day interval. Make no more than 2 applications per season; disease suppression only. **Not labeled for turnip greens or roots.**
Sclerotinia stem rot (white mold)	boscalid (Endura 70 WG)	6 to 9 oz/acre	4.2 to 6.3 oz/acre	14	0.5	Begin applications prior to disease development, and continue on a 7- to 14-day interval. Make no more than 2 applications per season. **Not labeled for turnip greens or roots.**
	penthiopyrad (Fontelis 1.67 SC)	16 to 30 fl oz/acre	0.21 to 0.39 lb/acre	0	0.5	Do not exceed 72 fluid ounces of product per year. Make no more than 2 sequential applications per season before rotating to another effective product with a different mode of action.
	Coniothyrium minitans (Contans WG)	1 to 4 lb/acre	0.05 to 0.2 lb/acre	0	4 hr	OMRI listed product. Apply to soil surface and incorporate no deeper than 2 inches. Works best when applied prior to planting or transplanting. Do not apply other fungicides for 3 weeks after applying Contans.
Seedling root rot, basal stem rot (Rhizoctonia)	azoxystrobin (Quadris 2.08 SC)	0.4 to 0.8 fl oz per 1000 row feet	00.006 to 0.13lb/1000 row feet	0	4 hr	Apply at planting as a directed spray to the furrow in a band 7 inches wide.
White rust, Alternaria leaf spot, Cercospora leaf spot	azoxystrobin (Quadris 2.08 SC)	6.2 to 15.4 fl oz/ acre	0.1 to 0.25 lb/acre	0	4 hr	Make no more than 2 sequential applications.
	fenamidone (Reason 500SC)	8.2 oz/acre	0.267 lb/acre	2	0.5	Begin applications as soon as conditions become favorable for disease development. Applications should be made on a 5 to 10 day interval. Do not make more than 1 application of Reason 500 SC before alternating with a fungicide from a different resistance management group.
	fluxapyroxad + pyraclostrobin (Priaxor 500 SC)	6.0 to 8.2 fl oz/ acre	0.20 to 0.27 lb/acre	3	0.5	Make no more than two sequential applications before alternating with fungicides that have a different mode of action. Maximum of 3 applications. **Do not apply to turnip greens or roots.**

Jerusalem Artichoke

Table 10-34. Disease Control Products for Jerusalem Artichoke

Disease	Material	Rate of Material to Use		Minimum Days		Method, Schedule, and Remarks
		Formulation	Active Ingredient	Harvest	Reentry	
Pythium damping off	Mefenoxam (Ridomil Gold 4 SL)	1 to 2 pt/treated acre	0.5 to 1 lb/treated acre	1	2	Soil incorporation. See label for row rates.
	fluopicolide (Presidio 4 SC)	3 to 4 fl oz/acre	0.09 to 0.125 lb/acre	7	0.5	Do not use more than 2 times sequentially and not more than 4 times at the low rate or 3 times at the high rate per acre per season.
Southern blight	azoxystrobin (Quadris 2.08 SC)	0.4 to 0.8 fl oz/1000 row ft	0.1 to 0.2 oz/1000 row ft	14	4 hr	
Rust	azoxystrobin (Quadris 2.08 SC)	6.2 to 20.3 fl oz/ acre	1.6 to 4 oz/acre	14	4 hr	
Powdery mildew, *Cercospora* leaf spot	azoxystrobin (Quadris 2.08 SSC)	9.2 to 15.4 fl oz/ acre	2.4 to 4 oz/acre	14	4 hr	
White mold (*Sclerotinia* basal stalk rot)	boscalid (Endura 70 EG)	10 oz/acre	7 oz/acre	30	0.5	2 applications per crop per season at 10 ounces per acre.
	Coniothyrium minitans (Contans WG)	1 to 4 lb/acre	0.05 to 0.2 lb/acre	0	4 hr	OMRI listed product. Apply to soil surface and incorporate no deeper than 2 inches. Works best when applied prior to planting or transplanting. Do not apply other fungicides for 3 weeks after applying Contans.

DISEASE CONTROL

Lettuce

Table 10-35. Disease Control Products for Lettuce

| Disease | Material | Rate of Material to Use | | Minimum Days | | Method, Schedule, and Remarks |
		Formulation	Active Ingredient	Harvest	Rentry	
Bottom rot, (*Rhizoctonia*), drop (*Sclerotinia*)	azoxystrobin (Amistar, Quadris)	0.4 to 0.8 fl oz/ 1,000 row feet	0.006 to 0.013 lb	—	4 hr	*Rhizoctonia* only. Make in-furrow or banded applications shortly after plant emergence.
	boscalid (Endura)	8 to 11 oz/acre	5.6 to 7.7 oz/acre	14	0.5	Suppression only on bottom rot. Apply immediately after emergence or immediately after transplanting. Make no more than 2 applications per season.
	Coniothyrium minitans (Contans WG)	1 to 4 lb/acre	0.05 to 0.2 lb/acre	0	4 hr	OMRI listed product. Apply to soil surface and incorporate no deeper than 2 inches. Works best when applied prior to planting or transplanting. Do not apply other fungicides for 3 weeks after applying Contans.
	dicloran (Botran)	2 to 5.3 lb/acre	1.5 to 4 lb/acre	14	0.5	Rate depends on timing; 5.3 pounds per crop per season maximum.
	iprodione (Rovral)	1.5 to 2 lb/acre	0.75 to 1 lb/acre	14	1	Tank mix with another fungicide with a different mode of action.
Botrytis rot	boscalid (Endura)	8 to 11 oz/acre	5.6 to 7.7 oz/acre	14	0.5	Make no more than 2 applications per season.
	dicloran (Botran)	1.5 to 2.6 lb/acre	1.1 to 2 lb/acre	14	0.5	Apply when disease is anticipated.
Damping off (*Pythium*)	Mefenoxam (Ridomil Gold)	1 to 2 pt/trt acre	0.5 to 1 lb/trt acre	—	0.5	Preplant incorporated. Apply preplant incorporated or surface application at planting.
	metalaxyl (MetaStar)	4 to 8 pt/trt acre	0.5 to 1 lb/acre	—	2	Banded over the row, preplant incorporated, or injected with liquid fertilizer.
	propamocarb (Previcur Flex)	12.8 fl oz/100 gal water	0.6 lb/100 gal	2	0.5	Rates based on rock wool cube saturation in the greenhouse. See label for use in seed beds, drip system, and soil drench.
Downy mildew	acibenzolar-S-methyl (Actigard 50WG)	0.75 to 1 oz/acre	0.375 to 0.5 oz/ acre	7	0.5	Do not apply prior to thinning or within 5 days after transplanting. Apply preventatively every 7 to 10 days, not to exceed 4 applications (4 ounces) per a season.
	cymoxanil (Curzate)	3.2 to 5.0 oz/acre	1.92 to 3.0 oz/acre	3	0.5	Use only in combination with a protectant fungicide. Apply on a 5 to 7 day schedule, not to exceed 30 ounces/acre per a 12-month period.
	dimethomorph (Acrobat, Forum)	6.4 oz/acre	3.2 oz/acre	0	0.5	Must be applied as a tank mix with another fungicide active against downy mildew. Do not make more than two sequential applications.
	famoxadone + cymoxanil (Tanos)	8 oz/acre	4 oz/acre	3	0.5	Not for gummy stem blight or leaf lettuce. Do not make more than 1 application before alternating with a fungicide that has a different mode of action. Must be tank-mixed with contact fungicide with a different mode of action.
	fenamidone (Reason 500SC)	5.5 to 8.2 fl oz/acre	0.178 to 0.267 lb/ acre	2	0.5	Begin applications when conditions favor disease development, and continue on 5 to 10 day interval. Do not apply more than 24.6 fluid ounces per growing season. Alternate with fungicide with different resistance management group.
	fluopicolide (Presidio)	3 to 4 fl oz/acre	0.09 to 0.125 lb/ acre	2	0.5	Tank mix with another downy mildew fungicide with a different mode of action.
	mandipropamid (Revus)	8 fl oz/acre	0.13 lb/acre	1	0.5	Apply prior to disease development and continue throughout season at 7 to 10 day intervals; maximum 32 fluid ounces per season.
	oxathiapiprolin + mefenoxam (Orondis Gold 200 + Orondis Gold B)	4.8 to 19.2 fl oz/acre + 1 to 2 pt/acre		0	4 hr	**Must tank mix Orondis Gold 200 and Gold B before application.** Appy at planting, in furrow, by drip, or in transplant water. Make no more than 2 sequential applications before alternating with fungicides that have a different mode of action. Maximum of 4 applications at the low rate or 2 applications at the high rate per crop per year. **Use either soil (Orondis Gold) or foliar (Orondis Ultra) applications but not both on the same crop.**
	oxathiapiprolin + mandipropamid (Orondis Ultra + Orondis Ultra B)	2.0 to 4.8 fl oz/acre + 8 fl oz/acre		0	4 hr	**Must tank mix Orondis Ultra and Ultra B before application.** Appy prior to disease development at 3 to 14-day intervals. Make no more than 2 sequential applications before alternating with fungicides that have a different mode of action. Maximum of 6 applications at the low rate or 4 applications at the high rate per crop per year. **Use either soil (Orondis Gold) or foliar (Orondis Ultra) applications but not both on the same crop.**
	propamocarb (Previcur Flex)	2 pt/acre	1.5 lb/acre	2	0.5	Do not apply more than 8 pt per growing season; begin applications before infection, and continue on a 7 to 10 day interval.
Downy mildew, leaf spots	azoxystrobin (Amistar, Quadris)	6.2 to 15.4 fl oz/ acre	0.1 to 0.25 fl oz/ acre	7	4 hr	Use highest rate for downy mildew. Make no more than 2 sequential applications before alternating with fungicides that have a different mode of action. Apply no more than 2.88 quarts per crop per acre per season.
	fixed copper (various)	See label	See label	See label	See label	See label. Rates vary depending on the formulation.

Table 10-35. Disease Control Products for Lettuce

Disease	Material	Rate of Material to Use		Minimum Days		Method, Schedule, and Remarks
		Formulation	Active Ingredient	Harvest	Rentry	
Downy mildew, leaf spots (continued)	fluxapyroxad + pyraclostrobin (Merivon 500 SC)	4 to 11 fl oz/ acre	0.26 to 0.72 lb/acre	1	0.5	Make no more than two sequential applications before alternating with fungicides that have a different mode of action. Maximum of 3 applications per crop.
	fosetyl-Al (Aliette 80WDG)	2 to 5 lb/acre	1.6 to 4 lb/acre	3	1	Spray at first appearance of disease and continue on a 7 to 10 day interval.
	mancozeb (various)	See label	See label	See label	See label	See label. Rates vary depending on the formulation. Spray at first appearance of disease and continue on a 7 to 10 day interval.
Powdery mildew	azoxystrobin (Amistar, Quadris) 2.08 F	6.2 to 15.4 fl oz/ acre	0.1 to 0.25 fl oz/ acre	7	4 hr	Make no more 2 two sequential applications before alternating with fungicides that have a different mode of action. Apply no more than 2.88 quarts per crop per acre per season.
	fluxapyroxad + pyraclostrobin (Merivon 500 SC)	4 to 11 fl oz/ acre	0.26 to 0.72 lb/acre	1	0.5	Make no more than two sequential applications before alternating with fungicides that have a different mode of action. Maximum of 3 applications per crop.
	myclobutanil (Rally) 40 WSP	5 oz/acre	0.31 lb/acre	3	1	Apply when disease first appears and continue on a 14-day interval. Apply no more than 20 ounces (4 applications) per season.
Powdery mildew (continued)	penthiopyrad (Fontelis)	14 to 24 fl oz/acre	0.18 to 0.31 lb/acre	0	0.5	Begin applications prior to disease development on a 7 to 14 day schedule. Do not exceed 72 fluid ounces of product per year. Make no more than 2 sequential applications per season before rotating to another effective product with a different mode of action.
	quinoxyfen (Quintec)	4 to 6 fl oz/acre	1.0 to 1.6 oz/acre	1	0.5	Apply preventatively on a 10 to 14 day interval, making no more than 4 applications (24 fl oz) per a season.
	triflumizole (Procure) 480 SC	6 to 8 fl oz/acre	3 to 4 oz/acre	0	0.5	Apply when disease first appears and continue on 14-day interval. Do not exceed 18 fluid ounces per season.
	sulfur (various)	See label	See label	See label	See label	See label. Rates vary depending on the formulation.

For Melons – See Cucurbits

Okra

Table 10-36. Disease Control Products for Okra

Disease	Material	Rate of Material to Use		Minimum Days		Method, Schedule, and Remarks
		Formulation	Active Ingredient	Harvest	Reentry	
Alternaria, gray mold, powdery mildew	Cyprodinil; Fludioxonil (Switch) 62.5WG	11-14 oz/acre	—	0	0.5	Begin applications before disease development and continue on 7 to 10 day interval. Make no more than 2 consecutive applications before alternating to a fungicide with a different. Do not apply more than 56 ounces per acre per season.
Alternaria, gray mold, powdery mildew, Septoria leaf spot, target spot	penthiopyrad (Fontelis)	16 to 24 fl oz/acre	—	0	0.5	Begin applications before disease development and continue on 7 to 14 day interval. Do not exceed more than 72 ounces per acre per year.
Anthracnose, bacterial leaf spot, leaf spots, pod spots, powdery mildew	fixed copper (various)	See label	See label	See label	See label	See label. Rates vary depending on the formulation.
Anthracnose, Botrytis leaf mold, powdery mildew, Cercospora leaf spot	Chlorothalonil; cymoxanil (Ariston)	2-4.4 pints/acre	—	3	0.5	Begin applications before disease development and continue on 7 day interval.
Anthracnose, gray leaf spot, powdery mildew, Cercospora leaf spot	Difenoconazole; azoxystrobin (Quadris Top)	8-14 fl oz/acre	—	0	0.5	Begin applications before disease development and continue on 7- to 10-day interval. Make no more than 2 consecutive applications before alternating to a fungicide with a different. Do not apply more than 55 fluid ounces per acre per season.
	Difenoconazole; cyprodinil (Inspire Super)	16-20 fl oz/acre	—	0	0.5	Begin applications before disease development and continue on 7 to 10 day interval. Make no more than 2 consecutive applications before alternating to a fungicide with a different. Do not apply more that 47 fluid ounces per acre per season.
Anthracnose, gray leaf spot, powdery mildew, Cercospora leaf spot, Rhizoctonia stem rot	Difenoconazole; benzovindiflupyr	10.5-13.5 oz/A	—	0	0.5	Begin applications before disease development and continue on 7 to 10 day interval. Make no more than 2 consecutive applications before alternating to a fungicide with a different mode of action. Refer to label for informationon on addition of an adjuvant.

527

Table 10-36. Disease Control Products for Okra

Disease	Material	Rate of Material to Use — Formulation	Rate of Material to Use — Active Ingredient	Minimum Days — Harvest	Minimum Days — Reentry	Method, Schedule, and Remarks
Cercospora leaf spot	chlorothalonil (various)	See label	See label	See label	See label	See label. Rates vary depending on the formulation.
	tebuconazole (Orius 3.6F, Uppercut, Folicur)	4 to 6 fl oz/A	1.5 to 2.3 oz/A	3	0.5	**DO NOT** apply more than 24 fluid ounces per acre per season.
Downy mildew	mandipropamid (Micora)	5.5 to 8 fl oz/A	0.65 to 0.9 fl oz/A	—	4 hr	Tank mix Micora with a non-Group 40 fungicide and begin applications prior to disease development. **DO NOT** apply more than two applications per crop, or in consecutive applications.
Powdery mildew	azoxystrobin (Quadris) 2.08 F	6.2 to 15.4 fl oz/A	0.10 to 0.25 lb/A	0	4	Do not make more than two sequential applications.
	azoxystrobin (Heritage)	3.2 to 8.0 oz/A	1.125 lb/A	0	4	Do not apply more than two sequential applications before alternating with a fungicide with a different mode of action. Do not make more than 4 applications of Heritage or other strobilurin fungicide per acre per season.
	chlorothalonil (various)	See label	See label	See label	See label	See label. Rates vary depending on the formulation.
	myclobutanil (Rally 40WSP)	2.5 to 5 oz/A	0.062 to 0.125 lb/A	0	1	Do not make more than 4 applications per season. Minimum re-treatment interval: 10 to 14 days.
Phytophthora blight and crown rot	Oxathiapiprlin (Orondis)	2.4-19.2 oz/A		0	4	Apply at planting, in furrow, by drip, or in transplant water. Do not make more then 4 applications per crop.
Rhizoctonia seedling rot	azoxystrobin (Quadris) 2.08 F	0.4 to 0.8 fl oz/1,000 row feet	0.006 to 0.013 lb	—	4 hr	Make in-furrow or banded applications shortly after plant emergence.

Onion

Table 10-37. Disease Control Products for Onion

Disease	Material	Rate of Material to Use — Formulation	Rate of Material to Use — Active Ingredient	Minimum Days — Harvest	Minimum Days — Reentry	Method, Schedule, and Remarks
ONION (green)						
Damping off (*Pythium*)	mefenoxam (Ridomil Gold) 4 SL	0.5 to 1 pt/trt acre	0.25 to 0.5 lb/trt acre	—	2	See label for low rates. Also for dry onion.
	metalaxyl (MetaStar) 2 E	2 to 4 pt/trt acre	0.25 to 0.5 pt/trt acre	—	2	Preplant incorporated or soil surface spray.
Downy mildew	azoxystrobin (Amistar, Quadris) 2.08 F	9.2 to 15.4 fl oz/ acre	0.15 to 0.25 fl oz/ acre	0	4 hr	Make no more than one application before alternating with a fungicide with a different mode of action. Apply no more than 2.88 quarts per crop per acre per season.
	azoxystrobin + chlorothalonil (Quadris Opti)	2.4 to 3.7 pt/acre		14	0.5	Make no more than one application before alternating with a fungicide with a different mode of action.
	dimethomorph (Acrobat, Forum) 50 WP	6.4 oz/acre	3.2 oz/acre	0	0.5	Must be applied as a tank mix with another fungicide active against downy mildew. Do not make more than two sequential applications.
	fenamidone (Reason) 500 SC	5.5 fl oz/acre	0.178 lb/acre	7	0.5	Begin applications when conditions favor disease development, and continue on 5 to 10 day interval. Do not apply more than 22 fluid ounces per growing season. Alternate with fungicide from different resistance group.
	mandipropamid (Revus) 2.08F	8 fl oz/acre	0.13 lb/acre	7	0.5	Apply prior to disease development and continue throughout season at 7 to 10 day intervals; maximum 24 fluid ounces per season.
	amectoctradin + dimethomorph (Zampro)	14.0 fl oz/acre	0.27 lb/acre	0	12 hr	Begin applications prior to disease development and continue on a 5- to 7- day spray interval.
	mefenoxam + chlorothalonil (Ridomil Gold/Bravo)	2 lb/acre	—	21	2	
	pyraclostrobin (Cabrio) 20 WG	8 to 12 oz/acre	1.6 to 2.4 oz/acre	7	0.5	Make no more than 2 sequential applications and no more than 6 applications per season.
	pyraclostrobin + boscalid (Pristine) 38WG	18.5 oz/acre	4 to 7 oz/acre	7	1	For suppression only. Make a maximum of 6 applications per season.
Leaf blight (*Botrytis*)	azoxystrobin (Amistar, Quadris) 2.08 F	6.2 to 15.4 fl oz/ acre	0.1 to 0.25 fl oz/acre	7	4 hr	Make no more than two sequential applications before alternating with fungicides that have a different mode of action. Apply no more than 2.88 quarts per crop per acre per season.
	azoxystrobin + chlorothalonil (Quadris Opti)	1.6 to 3.2 pt/acre		14	0.5	Make no more than one application before alternating with a fungicide with a different mode of action.
	boscalid (Endura) 70 WG	6.8 oz/acre	4.8 oz/acre	7	0.5	Do not make more than 2 sequential applications or more than 6 applications per season.
	penthiopyrad (Fontelis)	16 to 24 fl oz/acre	0.21 to 0.31 fl oz/ acre	3	12 hr	Begin sprays prior to disease development and continue on a 7 to 14 day schedule.
	chlorothalonil (various)	See label	See label	See label	See label	See label. Rates vary depending on the formulation.

Table 10-37. Disease Control Products for Onion

Disease	Material	Rate of Material to Use		Minimum Days		Method, Schedule, and Remarks
		Formulation	Active Ingredient	Harvest	Reentry	
ONION (green) (continued)						
Leaf blight *(Botrytis)* (continued)	cyprodinil + fludioxonil (Switch) 62.5 WG	11 to 14 oz/acre	6.9 to 8.8 oz/acre	7	0.5	Do not plant rotational crops other than onions or strawberries for 12 months following the last application.
	dicloran (Botran) 75 W	1.5 to 2.7 lb/acre	1.1 to 2 lb/acre	14	0.5	
	difenoconazole + cyprodinil (Inspire Super)	16 to 20 fl oz/acre		7 to 14	0.5	Make no more than two applications before alternating with a fungicide with a different mode of action.
	pyraclostrobin + boscalid (Pristine) 38 WG	14.5 to 18.5 oz/acre	5.5 to 7 oz/acre	7	1	Make a maximum of 6 applications per season.
	pyrimethanil (Scala) 5 F	9 or 18 fl oz/acre	0.35 or 0.7 lb/acre	7	0.5	Use lower rate in a tank mix with broad-spectrum fungicide and higher rate when applied alone. Do not apply more than 54 fluid ounces per crop.
	azoxystrobin + chlorothalonil (Quadris Opti)	1.6 to 3.6 pts/acre	1.1 to 2.5 lb/acre	7	2	Applications should begin prior to disease onset and subsequent applications should be made on a 7 to 14 day interval.
Purple blotch	azoxystrobin (Amistar, Quadris) 2.08 F	6.2 to 12.3 fl oz/ acre	0.1 to 0.2 fl oz/acre	7	4 hr	Make no more than two sequential applications before alternating with a fungicide with a different mode of action. Apply no more than 2.88 quarts per crop per acre per
	azoxystrobin + chlorothalonil (Quadris Opti)	1.6 to 3.2 pt/acre		14	0.5	Make no more than one application before alternating with a fungicide with a different mode of action.
	boscalid (Endura) 70WG	6.8 oz/acre	4.8 oz/acre	7	0.5	Do not make more than 2 sequential applications or more than 6 applications per season.
	penthiopyrad (Fontelis)	16 to 24 fl oz/acre	0.21 to 0.31 fl oz/ acre	3	12 hr	Begin sprays prior to disease development and continue on a 7-14 day schedule.
	cyprodinil + fludioxonil (Switch) 62.5 WG	11 to 14 oz/acre	6.9 to 8.8 oz/acre	7	0.5	Do not plant rotational crops other than onions or strawberries for 12 months following the last application.
	difenoconazole + cyprodinil (Inspire Super)	16 to 20 fl oz/acre		7-14	0.5	Make no more than two applications before alternating with a fungicide with a different mode of action.
	fenamidone (Reason) 500 SC	5.5 to 8.2 fl oz	0.178 lb/acre	7	0.5	Begin applications when conditions favor disease development, and continue on 5 to 10 day interval. Do not apply more than 22 fluid ounces per growing season. Alternate with fungicide from different resistance management group.
	pyraclostrobin (Cabrio) 20 WG	8 to 12 oz/acre	1.6 to 2.4 oz/acre	7	0.5	Make no more than 2 sequential applications and no more than 6 applications per season.
	pyraclostrobin + boscalid (Pristine) 38 WG	10.5 to 18.5 oz/acre	4 to 7 oz/acre	7	1	Make a maximum of 6 applications per season.
	pyrimethanil (Scala) 5 F	9 or 18 fl oz/acre	0.35 or 0.7lb/acre	7	0.5	Use lower rate in a tank mix with broad spectrum fungicide and higher rate when applied alone. Do not apply more than 54 fluid ounces per crop.
	tebuconazole (Folicur) 3.6F	4 to 6 fl oz/acre	1.5 to 2 lb/acre	7	0.5	Apply before disease appears when conditions favor purple blotch development and repeat at 10 to 14 day intervals; maximum 24 fluid ounces per season.
Stemphylium leaf blight	pyraclostrobin + boscalid (Pristine) 38 WG	10.5 to 18.5 oz/acre	4 to 7 oz/acre	7	1	Make no more than 6 applications per season.
ONION (dry)						
Damping off *(Pythium)*	mefenoxam (Ridomil Gold) 4 SL	0.5 to 1 pt/trt acre	0.25 to 0.5 lb/trt acre	—	2	See label for row rates. Also for green onion.
	metalaxyl (Metatar) 2 E	2 to 4 pt/trt acre	0.25 to 0.5 pt/trt acre	—	2	Preplant incorporated or soil surface spray.
Downy mildew	azoxystrobin (Amistar, Quadris) 2.08 F	9.2 to 15.4 fl oz/ acre	0.15 to 0.25 fl oz/ acre	0	4 hr	Make no more than one application before alternating with a fungicide with a different mode of action. Apply no more than 2.88 quarts per crop per acre per season.
	amectoctradin + dimethomorph (Zampro)	14.0 fl oz/acre	0.27 lb/acre	0	12 hr	Begin applications prior to disease development and continue on a 5- to 7-day spray interval.
	fenamidone (Reason)	5.5 fl oz/acre	0.178 lb/acre	7	12 hr	Use as soon as environmental conditions are favorable.
	azoxystrobin + chlorothalonil (Quadris Opti)	2.4 to 3.2 pt/acre		14	0.5	Make no more than one application before alternating with a fungicide with a different mode of action.
	dimethomorph (Acrobat, Forum) 50 WP	6.4 oz/acre	3.2 oz/acre	0	0.5	Must be applied as a tank mix with another fungicide active against downy mildew. Do not make more than two sequential applications.
	famoxadone + cymoxanil (Tanos)	8.0 oz/acre	0.25 lb/acre	3	0.5	Apply preventively on a 5- to 7-day schedule and do not rotate with group 11 fungicides.
	fluazinam (Omega 500)	1.0 pt/acre	0.52 lb/acre	7	2	Initiate sprays when conditions are favorable for disease or at disease onset. Spray on a 7- to 10-day schedule.
	mandipropamid (Revus) 2.08 F	8 fl oz/acre	0.13 lb/acre	1	0.5	Apply prior to disease development and continue throughout season at 7 to 10 day intervals; maximum 32 fluid ounces per season.
	mefenoxam + mancozeb (Ridomil Gold MZ)	2.5 lb/trt acre	—	7	2	
	pyraclostrobin (Cabrio) 20 WG	12 oz/acre	2.4 oz/acre	7	0.5	Make no more than 2 sequential applications and no more than 6 applications per season.

Table 10-37. Disease Control Products for Onion

Disease	Material	Rate of Material to Use		Minimum Days		Method, Schedule, and Remarks
		Formulation	Active Ingredient	Harvest	Reentry	
ONION (dry) (continued)						
Downy mildew (continued)	pyraclostrobin + boscalid (Pristine) 38 WG	18.5 oz/acre	4 to 7 oz/acre	7	1	Suppression only. Make no more than 6 applications per season.
Leaf blight (*Botrytis*)	azoxystrobin (Amistar, Quadris) 2.08 F	6.2 to 15.4 fl oz/ acre	0.1 to 0.25 fl oz/acre	7	4 hr	Make no more than two sequential applications before alternating with fungicides with different mode of action. Apply no more than 2.88 quarts per crop per acre per season.
	azoxystrobin + chlorothalonil (Quadris Opti)	1.6 to 3.2 pt/acre		14	0.5	Make no more than one application before alternating with a fungicide with a different mode of action.
	boscalid (Endura) 70 WG	6.8 oz/acre	4.8 oz/acre	7	0.5	Do not make more than 2 sequential applications or more than 6 applications per season.
	penthiopyrad (Fontelis)	16 to 24 fl oz/acre	0.21 to 0.31 fl oz/ acre	3	12 hr	Begin sprays prior to disease development and continue on a 7- to 14-day schedule.
	cyprodinil + fludioxonil (Switch) 62.5WG	11 to 14 oz/acre	6.9 to 8.8 oz/acre	7	0.5	Do not plant rotational crops other than onions or strawberries for 12 months following the last application.
	dicloran (Botran) 75 W	1.5 to 2.7 lb/acre	1.1 to 2 lb/acre	14	0.5	Use lower rate in a tank mix with broad-spectrum fungicide and higher rate when applied alone. Do not apply more than 54 fluid ounces per crop.
	difenoconazole + cyprodinil (Inspire Super)	16 to 20 fl oz/acre		7 to 14	0.5	Make no more than two applications before alternating with a fungicide with a different mode of action.
	fixed copper (various)	See label	See label	See label	See label	See label. Rates vary depending on the formulation. Spray at first appearance, 7 to 10 day intervals. Do not apply to exposed bulbs.
	pyraclostrobin + boscalid (Pristine) 38 WG	14.5 to 18.5 oz/acre	5.5 to 7 oz/acre	7	1	Make no more than 6 applications per season.
	pyrimethanil (Scala) 5 F	9 or 18 fl oz/acre	0.35 or 0.7 lb/acre	7	0.5	Use lower rate in a tank mix with broad-spectrum fungicide and higher rate when applied alone. Do not apply more than 54 fluid ounces per crop.
	azoxystrobin + difenoconazole (Quadris Top)	14 fl oz/acre	2.38 lb/acre	7	0.5	Begin sprays prior to disease onset and spray on a 7- to 14-day schedule. Do not rotate with Group 11 fungicides.
	fluazinam (Omega 500)	1.0 pt/acre	0.52 lb/acre	7	2	Initiate sprays when conditions are favorable for disease or at disease onset. Spray on a 7- to 10-day schedule.
Neck rot (*Botrytis*), purple blotch (*Alternaria*), downy mildew	azoxystrobin + chlorothalonil (Quadris Opti)	1.6 to 3.2 pt/acre		14	0.5	Make no more than one application before alternating with a fungicide with a different mode of action.
	penthiopyrad (Fontelis)	16 to 24 fl oz/acre	0.21 to 0.31 fl oz/ acre	3	12 hr	Begin sprays prior to disease development and continue on a 7- to 14-day schedule.
	chlorothalonil (various)	See label	See label	See label	See label	See label. Rates vary depending on the formulation.
	fixed copper (various)	See label	See label	See label	See label	See label. Rates vary depending on the formulation. May reduce bacterial rots.
	fosetyl-Al (Aliette) 80 WDG	2 to 3 lb/acre	1.6 to 2.4 lb/acre	7	0.5	Do not mix with surfactants, foliar fertilizers, or products containing copper; will not control neck rot.
	iprodione (Rovral) 50 WP	1.5 lb/acre	0.75 lb/acre	7	0.5	Apply when conditions are favorable; 14-day intervals. Rovral is not for downy mildew.
	mancozeb (various)	See label	See label	See label	See label	See label. Rates vary depending on the formulation.
	mefenoxam + chlorothalonil (Ridomil Gold/Bravo)	2 lb/acre	—	7	2	
	vinclozolin (Ronilan) 50 DF	See label	0.75 to 1 lb/acre	18	1	Three treatments minimum.
Purple blotch, leaf blight	azoxystrobin (Amistar, Quadris) 2.08 F	6.2 to 15.4 fl oz/ acre	0.1 to 0.25 fl oz/acre	7	4 hr	Make no more than two sequential applications before alternating with fungicides with a different mode of action. Apply no more than 2.88 quarts per crop per acre per season.
	azoxystrobin + chlorothalonil (Quadris Opti)	1.6 to 3.2 pt/acre		14	0.5	Make no more than one application before alternating with a fungicide with a different mode of action.
	famoxadone + cymoxanil (Tanos)	8.0 oz/acre	0.25 lb/acre	3	0.5	Apply preventively on a 5 to 7 day schedule and do not rotate with group 11 fungicides.
	boscalid (Endura) 70 WG	6.8 oz/acre	4.8 oz/acre	7	0.5	Do not make more than 2 sequential applications or more than 6 applications per season.
	penthiopyrad (Fontelis)	16 to 24 fl oz/acre	0.21 to 0.31 fl oz/ acre	3	12 hr	Begin sprays prior to disease development and continue on a 7 to 14 day schedule.
	cyprodinil + fludioxonil (Switch) 62.5 WG	11 to 14 oz/acre	6.9 to 8.8 oz/acre	7	0.5	Do not plant rotational crops other than onions or strawberries for 12 months following the last application.
	difenoconazole + cyprodinil (Inspire Super)	16 to 20 fl oz/acre		7 to 14	0.5	Make no more than two applications before alternating with a fungicide with a different mode of action.
	fenamidone (Reason) 500 SC	5.5 fl oz/acre	0.178 lb/acre	7	0.5	Begin applications when conditions favor disease development, and continue on 5 to 10 day interval. Do not apply more than 22 fluid ounces per growing season. Alternate with fungicide from different resistance group.
	iprodione (Rovral) 50 WP	1.5 lb/acre 50 to 100 gal/acre	0.75 lb/acre 50 to 100 gal/acre	7	0	Start 7-day foliar sprays at first appearance of favorable conditions.

530

Table 10-37. Disease Control Products for Onion

Disease	Material	Rate of Material to Use		Minimum Days		Method, Schedule, and Remarks
		Formulation	Active Ingredient	Harvest	Reentry	
ONION (dry) (continued)						
Purple blotch, leaf blight (continued)	pyraclostrobin (Cabrio) 20 WG	8 to 12 oz/acre	1.6 to 2.4 oz/acre	7	0.5	Make no more than 2 sequential applications and no more than 6 applications per season.
	pyraclostrobin + boscalid (Pristine) 38 WG	10.5 to 18.5 oz/acre	4 to 7 oz/acre	7	1	Make no more than 6 applications per season.
	pyrimethanil (Scala) 5 F	9 or 18 fl oz/acre	0.35 or 0.7 lb/acre	7	0.5	Use lower rate in a tank mix with broad-spectrum fungicide and higher rate when applied alone. Do not apply more than 54 fluid ounces per crop.
	tebuconazole (Folicur) 3.6 F	4 to 6 fl oz/acre	—	7	0.5	Do not apply more than 12 fluid ounces per acre per
	vinclozolin (Ronilan) 50 DF	See label		18	1	Two treatments Minimum. Effective on purple leaf blotch when disease pressure is low.
	azoxystrobin + difenoconazole (Quadris Top)	14 fl oz/acre	2.38 lb/acre	7	0.5	Begin sprays prior to disease onset and spray on a 7 to 14 day schedule. Do not rotate with Group 11 fungicides.
	fluazinam (Omega 500)	1.0 pt/acre	0.52 lb/acre	7	2	Initiate sprays when conditions are favorable for disease or at disease onset. Spray on a 7 to 10 day schedule.
Smut	mancozeb (various)	See label	See label	See label	See label	See label. Rates vary depending on the formulation.
Stemphylium leaf blight	difenoconazole + cyprodinil (Inspire Super)	16 to 20 fl oz/acre		7 to 14	0.5	Make no more than two applications before alternating with a fungicide with a different mode of action.
	pyraclostrobin + boscalid (Pristine) 38 WG	10.5 to 18.5 oz/acre	4 to 7 oz/acre	7	1	Make no more than 6 applications per season.
	azoxystrobin + difenoconazole (Quadris Top)	14 fl oz/acre	2.38 lb/acre	7	0.5	Begin sprays prior to disease onset and spray on a 7 to 14 day schedule. Do not rotate with Group 11 fungicides.
	fluazinam (Omega 500)	1.0 pt/acre	0.52 lb/acre	7	2	Initiate sprays when conditions are favorable for disease or at disease onset. Spray on a 7 to 10 day schedule.
	penthiopyrad (Fontelis)	16 to 24 fl oz/acre	0.21 to 0.31 fl oz/acre	3	12 hr	Begin sprays prior to disease development and continue on a 7 to 14 day schedule.
White rot *(Sclerotium)*	azoxystrobin + chlorothalonil (Quadris Opti)	1.6 to 3.2 pt/acre		14	0.5	Make no more than one application before alternating with a fungicide with a different mode of action.
	penthiopyrad (Fontelis)	16 to 24 fl oz/acre	0.21 to 0.31 fl oz/acre	3	12 hr	Begin sprays prior to disease development and continue on a 7 to 14 day schedule.
White rot *(Sclerotium)* (continued)	dicloran (Botran) 75 W	5.3 lb/acre	4 lb/acre	14	0.5	Apply 5-inch band over seed row and incorporate in top 1.5 to 3 inches of soil, 1 to 2 weeks before seeding.
	thiophanate-methyl (Topsin M) 70 WP	See label		—	—	Spray into open furrow at time of seeding or planting in row.
	vinclozolin (Ronilan) 50 DF	See label		18	0.5	Three treatments minimum.

Table 10-38. Efficacy of Products for Disease Control in Onion

Scale: E, excellent; G, good; F, fair; P, poor; NC, no control; ND, no data.

Product[1]	Fungicide group[F]	Preharvest interval (Days)	Bacterial Streak (Pseudomonas viridiflava)	Black Mold (Aspergillus niger)	Botrytis Leaf Blight (B. squamosa)	Botrytis Neck Rot (B. allii)	Damping off (Pythium spp.)	Downy Mildew (P. destructor)	Fusarium Basal Rot (F. oxysporum)	Onion Smut (Urocystis colchici)	Center Rot (Pantoea ananatis)	Pink Root (Phoma terrestris)	Purple Blotch (Alternaria porri)	Stemphylium Leaf Blight and Stalk Rot	White Rot (Sclerotium cepivorum)
amectoctradin + dimethomporph (Zampro)	40 + 45	0	NC	NC	NC	NC	NC	G	NC	NC	NC	NC	NC	NC	NC
azoxystrobin (Quadris)	11	7	NC	ND	F	NC	NC	ND	NC	ND	NC	NC	G	G	ND
azoxystobin + difenoconazole (Quadris Top)	11+ 3	1	NC	NC	F	NC	NC	ND	NC	NC	NC	NC	G	G	NC
Chlorothalonil (various)	M	14	NC	NC	F	NC	NC	P	NC	NC	NC	NC	F	F	NC
cyprodinil + fludioxonil (Switch)	9 + 12	7	NC	NC	G	ND	NC	NC	NC	NC	NC	NC	F	F	NC
dichloropropene + chloropicrin, fumigant (Telone C-17)	—	—	NC	NC	NC	NC	P	NC	F	NC	NC	F	NC	NC	F
dimethomorph (Forum)	40	0	NC	NC	NC	NC	NC	F	NC	NC	NC	NC	NC	NC	NC
fenamidone (Reason)	11	7	NC	NC	P	NC	NC	G	NC	NC	NC	NC	P	P	NC
famoxadone/cymoxanil (Tanos)	11 + 27	3	NC	NC	F	NC	NC	P	NC	NC	NC	NC	F	F	NC
fixed copper (various)	M	1	F	NC	F	NC	NC	F	NC	NC	F	NC	F	NC	NC
fluazinam (Omega 500)	29	2	NC	NC	G	NC	NC	G	NC	NC	NC	NC	E	E	NC
fosetyl-Al (Aliette)	33	7	NC	NC	NC	NC	NC	NC	F	NC	NC	NC	NC	NC	NC
iprodione (Rovral)	2	7	NC	NC	E	P	NC	NC	NC	NC	NC	NC	E	F	F
mancozeb (various)	M	7	NC	NC	F	NC	NC	F	NC	E	NC	NC	F	F	NC
mancozeb + copper (ManKocide)	M + M	7	F	NC	F	NC	NC	F	NC	F	F	NC	F	F	NC
mandipropamid (Revus)	40	7	NC	NC	NC	ND	F	NC	NC	NC	NC	NC	NC	NC	NC
mefenoxam (Ridomil Gold EC)	4	7	NC	NC	NC	NC	F	ND	NC	NC	NC	NC	NC	NC	NC
mefenoxam + chlorothalonil (Ridomil Gold Bravo)	4 + M	14	NC	NC	F	NC	P	F	NC	NC	NC	NC	F	F	NC
mefenoxam + copper (Ridomil Gold/ Copper)	4 + M	7	F	NC	NC	NC	P	F	NC	NC	F	NC	NC	NC	NC
mefenoxam + mancozeb (Ridomil Gold MZ)	4 + M	7	NC	NC	F	NC	P	F	NC	F	NC	NC	F	F	NC
metam sodium, fumigant (Vapam)	—	—	NC	NC	NC	NC	F	NC	F	NC	NC	E	NC	NC	F
pyraclostrobin (Cabrio)	11	7	NC	ND	F	NC	NC	F	NC	ND	NC	NC	G	G	ND
pyraclostrobin + boscalid (Pristine)	11 + 7	7	NC	ND	G	F	NC	F	NC	ND	NC	NC	E	E	ND
pyrimethanil (Scala)	9	7	NC	ND	F	NC	NC	NC	NC	ND	NC	NC	F	F	NC
tebuconazole (Folicur)	3	7	NC	ND	NC	NC	NC	NC	ND	ND	NC	ND	F	F·	F

[1] Efficacy ratings do not necessarily indicate a labeled use for every disease.

[F] To prevent resistance in pathogens, alternate fungicides within a group with fungicides in another group. Fungicides in the "M" group are generally considered "low risk" with no signs of resistance developing to the majority of fungicides.

[R] Resistance reported in the pathogen.

Parsley

Table 10-39. Disease Control Products for Parsley

| Disease | Material | Rate of Material to Use | | Minimum Days | | Method, Schedule, and Remarks |
		Formulation	Active Ingredient	Harvest	Reentry	
Cercospora leaf spot (Early blight)	azoxystrobin (Quadris) 2.08 F	2 to 5 oz/acre	0.1 to 0.25 fl oz/acre	0	4 hr	Make no more than two sequential applications before alternating with fungicides that have a different mode of action. Apply no more than 1.88 pounds per crop per acre per season.
	fluxapyroxad + pyraclostrobin (Merivon 500 SC)	4 to 11 fl oz/ acre	0.26 to 0.72 lb/acre	3	0.5	Make no more than two sequential applications before alternating with fungicides that have a different mode of action. Maximum of 3 applications per crop.
	penthiopyrad (Fontelis)	14 to 24 fl oz	0.18 to 0.31 lb/acre	3	0.5	Do not make more than two sequential applications. Maximum of 72 fluid ounces per acre per year.
	pyraclostrobin (Cabrio) EG	12 to 16 oz/acre	2.4 to 3.2 oz/acre	0	0.5	Make no more than two sequential applications before alternating with fungicides that have a different mode of action. Apply no more than 64 ounces per crop per acre per season.
Damping off and root rot (*Pythium, Phytophthora*)	mefenoxam (Ridomil Gold) 4 SL	1 to 2 pt/treated acre	0.5 to 1 lb/treated acre	0	0.5	Apply preplant incorporated or surface application at planting.
	metalaxyl (MetaStar) 2 E	2 to 8 pt/treated acre	0.5 to 1 lb/treated acre	0	2	Banded over the row, preplant incorporated, or injected with liquid fertilizer.
Powdery mildew, *Septoria* leaf spot (late blight)	azoxystrobin (Quadris) 2.08 F	2 to 5 oz/acre	0.1 to 0.25 fl oz/acre	0	4 hr	Make no more than two sequential applications before alternating with fungicides that have a different mode of action. Apply no more than 1.88 pound per crop per acre per season.
	cyprodinil + fludioxonil (Switch) 62.5 WG	11 to 14 oz/acre	6.9 to 8.8 oz/acre	0	0.5	Make no more than two sequential applications before alternating with fungicides that have a different mode of action for two applications. Apply no more than 56 ounces per crop per acre per season.
	fixed copper (various)	See label	See label	See label	See label	See label. Rates vary depending on the formulation. Spray at first disease appearance, 7- to 10-day intervals.
	fluxapyroxad + pyraclostrobin (Merivon 500 SC)	4 to 11 fl oz/ acre	0.26 to 0.72 lb/acre	3	0.5	Make no more than two sequential applications before alternating with fungicides that have a different mode of action. Maximum of 3 applications per crop.
	penthiopyrad (Fontelis) 1.67 F	14 to 24 fl oz	0.18 to 0.31 lb/acre	3	0.5	Do not make more than two sequential applications. Maximum of 72 fluid ounces per acre per year.
	pyraclostrobin (Cabrio) EG	12 to 16 oz/acre	2.4 to 3.2 oz/acre	0	0.5	Make no more than two sequential applications before alternating with fungicides that have a different mode of action. Apply no more than 64 ounces per crop per acre per season.
Web blight and root rot (*Rhizoctonia*)	azoxystrobin (Quadris) 2.08 F	0.125 to 0.25 oz/ 1000 row ft	0.10 to 0.20 oz/1000 row ft	0	4 hr	Apply as banded spray to the lower stems and soil surface. Make no more than two sequential applications. Apply no more than 1.88 pounds per crop per acre per season. Soil applications are included in this maximum.
White mold	cyprodinil + fludioxonil (Switch) 62.5 WG	11 to 14 oz/acre	6.9 to 8.8 oz/acre	0	0.5	Make no more than two sequential applications before alternating with fungicides that have a different mode of action for two applications. Apply no more than 56 ounces per crop per acre per season. First application at thinning and second application 2 weeks later.
	penthiopyrad (Fontelis) 1.67 F	16 to 30 fl oz	0.21 to 0.39 lb/acre	3	0.5	Do not make more than two sequential applications. Maximum of 72 fluid ounces per acre per year.
	Coniothyrium minitans (Contans WG)	1 to 4 lb/acre	0.05 to 0.2 lb/acre	0	4 hr	OMRI listed product. Apply to soil surface and incorporate no deeper than 2 inches. Works best when applied prior to planting or transplanting. Do not apply other fungicides for 3 weeks after applying Contans.

DISEASE CONTROL

Table 10-40. Importance of Alternative Management Practices for Disease Control in Parsley

Scale: E, excellent; G, good; F, fair; P, poor; NC, no control; ND, no data.

Strategy	Powdery mildew	*Pythium* damping off and root rot	*Rhizoctonia* damping off and root rot	Root knot (nematode)	*Sclerotinia* white mold	*Septoria* blight
Avoid field operations when leaves are wet	NC	NC	NC	NC	P	G
Avoid overhead irrigation	NC	NC	NC	NC	E	E
Biofungicide	F	NC	NC	NC	F	NC
Change planting date	NC	NC	E (early)	E (early)	E (late)	NC
Suppressive cover crops	NC	NC	NC	F	NC	NC
Crop rotation with non-host	NC	F	P	P	F	E
Deep plowing	P	NC	F	P	F	G
Destroy crop residue	P	NC	P	P	F	G
Encourage air movement	P	P	NC	NC	E	G
Flooding (where feasible)	NC	NC	F	G	G	NC
Increase soil organic matter	F	P	P	F	NC	P
Hot water seed treatment	NC	NC	NC	NC	NC	E
Plant in well-drained soil	NC	E	G	NC	F	P
Plant on raised beds	NC	E	G	NC	F	NC
Plastic mulch bed covers	F	NC	F	NC	P	NC
Postharvest temperature control	NC	NC	NC	NC	E	NC
Reduce mechanical injury	NC	NC	P	NC	G	NC
Soil solarization	NC	P	F	F	P	F
Pathogen-free planting material	P	NC	NC	NC	P	E
Resistant/tolerant cultivars	NC	NC	P	NC	NC	F
Weed control	F	NC	NC	F	F	NC

Pea

Table 10-41. Disease Control Products for Pea

Disease	Material	Rate of Material to Use		Minimum Days		Method, Schedule, and Remarks
		Formulation	Active Ingredient	Harvest	Reentry	
PEA (English)						
Anthracnose	azoxystrobin (Quadris) 2.08 F	6.2 to 15.4 fl oz/ acre	0.10 to 0.25 lb/acre	0	4 hr	Do not make more than two sequential applications.
	penthiopyrad (Fontelis) 1.67 F	14 to 30 fl oz	0.18 to 0.39 lb/acre	0	0.5	Do not make more than two sequential applications. Maximum of 72 fluid ounces per acre per crop.
	pyraclostrobin + fluxapyroxad (Priaxor) 500 SC	4.0 to 8.0 fl oz/acre	0.13 to 0.26 lb/acre	7	12 hr	Do not make more than two sequential applications. Maximum of 16 fluid ounces per acre per crop.
Ascochyta leaf spot and blight	Azoxystrobin (Quadris) 2.08 F	6.2 to 15.4 fl oz/ acre	0.10 to 0.25 lb/acre	0	4 hr	Do not make more than two sequential applications.
	boscalid (Endura) 70 WG	8 to 11 oz/ acre	5.6 to 7.7 oz/acre	7	0.5	Maximum of 2 applications per crop.
	penthiopyrad (Fontelis) 1.67 F	14 to 30 fl oz	0.18 to 0.39 lb/acre	0	0.5	Do not make more than two sequential applications. Maximum of 72 fluid ounces per acre per crop.
	pyraclostrobin + fluxapyroxad (Priaxor) 500 SC	4.0 to 8.0 fl oz/acre	0.13 to 0.26 lb/acre	7	12 hr	Do not make more than two sequential applications. Maximum of 16 fluid ounces per acre per crop.
Gray mold (*Botrytis*), White mold (*Sclerotinia*)	boscalid (Endura) 70 WG	8 to 11 oz/ acre	5.6 to 7.7 oz/acre	7	0.5	Maximum of 2 applications per crop.
	penthiopyrad (Fontelis) 1.67 F	14 to 30 fl oz	0.18 to 0.39 lb/acre	0	0.5	Do not make more than two sequential applications. Maximum of 72 fluid ounces per acre per year.
	pyraclostrobin + fluxapyroxad (Priaxor) 500 SC	4.0 to 8.0 fl oz/acre	0.13 to 0.26 lb/acre	7	12 hr	Do not make more than two sequential applications. Maximum of 16 fluid ounces per acre per crop.
White mold (*Sclerotinia*)	*Coniothyrium minitans* (Contans WG)	1 to 4 lb/acre	0.05 to 0.2 lb/acre	0	4 hr	OMRI listed product. Apply to soil surface and incorporate no deeper than 2 inches. Works best when applied prior to planting or transplanting. Do not apply other fungicides for 3 weeks after applying Contans.
Powdery mildew	boscalid (Endura) 70 WG	8 to 11 oz/ acre	5.6 to 7.7 oz/acre	7	0.5	Maximum of 2 applications per crop.
	fixed copper (various)	See label	See label	See label	See label	See label. Rates vary depending on the formulation.
	penthiopyrad (Fontelis) 1.67 F	14 to 30 fl oz	0.18 to 0.39 lb/acre	0	0.5	Do not make more than two sequential applications. Maximum of 72 fluid ounces per acre per year.

Table 10-41. Disease Control Products for Pea

Disease	Material	Rate of Material to Use		Minimum Days		Method, Schedule, and Remarks
		Formulation	Active Ingredient	Harvest	Reentry	
PEA (English) (continued)						
Powdery mildew (continued)	pyraclostrobin + fluxapyroxad (Priaxor) 500 SC	4.0 to 8.0 fl oz/acre	0.13 to 0.26 lb/acre	7	12 hr	Do not make more than two sequential applications. Maximum of 16 fluid ounces per acre per crop.
	sulfur (various)	See label	See label	See label	See label	See label. Rates vary depending on the formulation. Spray at first appearance, 10 to 14 day intervals. Do not use sulfur on wet plants or on hot days (in excess of 90 degrees F).
Pythium damping off	mefenoxam (Ridomil Gold) 4 EC	0.5 to 1 pt/trt acre	0.25 to 0.5 lb/trt acre	—	2	Incorporate in soil. See label for row rates.
Rhizoctonia root rot	azoxystrobin (Quadris) 2.08 F	0.4 to 0.8 fl oz/ 1,000 sq ft	0.0006 to 0.13 lb/ 1,000 sq ft	—	4 hr	Make in-furrow or banded applications shortly after plant emergence.
Rust *(Uromyces)*	azoxystrobin (Quadris) 2.08 F	6.2 fl oz/acre	0.10 lb/acre	0	4 hr	Do not make more than two sequential applications.
	penthiopyrad (Fontelis) 1.67 F	14 to 30 fl oz	0.18 to 0.39 lb/acre	0	0.5	Do not make more than two sequential applications. Maximum of 72 fluid ounces/acre per year.
	pyraclostrobin + fluxapyroxad (Priaxor) 500 SC	4.0 to 8.0 fl oz/acre	0.13 to 0.26 lb/acre	7	12 hr	Do not make more than two sequential applications. Maximum of 16 fluid ounces/acre per crop.
PEA (Southern)						
Anthracnose	thiophanate-methyl (Topsin M) 70 WP	1 to 1.5 lb/acre	0.7 to 1.1 lb/acre	28	0.5	Use no more than 4 lb (2.8 lb a.i.) per acre per year.
Anthracnose, Rust	azoxystrobin (Quadris) 2.08 F	2 to 5 oz/acre	0.1 to 0.25 lb/acre	14 (dry) 0 (succulent)	4 hr	Make no more than 2 sequential applications before alternating with a fungicide with a different mode of action. Use no more than 1.5 pounds a.i. per acre per season.
Ascochyta blight, Gray mold, White mold	boscalid (Endura) 70 WG	8 to 11 oz/acre	5.6 to 7.7 oz/acre	21 (dried) or 7 (succulent)	0.5	Maximum of 2 applications per season.
Ascochyta blight, Rust, white mold	prothioconazole (Proline) 480 SC	5.7 fl oz /acre	2.85 oz /acre	7	0.5	Maximum of 3 applications per year. Use no more than 17.1 fluid ounces per acre per year.
Downy mildew, Bacterial blights	fixed copper (various)	See label	See label	See label	See label	See label. Rates vary depending on the formulation.
Downy mildew, *Cercospora*, Anthracnose. Rust	chlorothalonil (various)	See label	See label	See label	See label	See label. Rates vary depending on the formulation.
Downy mildew, *Cercospora*, Anthracnose, Rust, Powdery mildew	Pyraclostrobin (Headline) 2.09 EC	6 to 9 fl oz/acre	0.1 to 0.15 lb/acre	21	0.5	Make no more than 2 sequential applications before alternating with a fungicide with a different mode of action. Use no more than 18 fluid ounces per acre per season.
	sulfur (various)	See label	See label	See label	See label	See label. Rates vary depending on the formulation. Spray at first appearance; 7 to 10 day interval.
Pythium damping off	mefenoxam (Ridomil Gold) 4 SL	0.5 to 1 pt/ treated acre	0.25 to 0.5 lb/ treated acre	—	0.5	Broadcast or banded over the row as a soil spray at planting or preplant incorporation into the top 2 inches of soil.
	metalaxyl (MetaStar) 2 E	2 to 4 pt/treated acre	0.5 to 1 lb/treated acre	—	2	Broadcast or banded over the row as a soil spray at planting or preplant incorporation into the top 2 inches of soil.
Rhizoctonia root rot	azoxystrobin (Quadris) 2.08 F	0.4 to 0.8 fl oz/1,000 row feet	0.006 to 0.013 lb/1,000 row feet	—	4 hr	Make in-furrow or banded application shortly after plant emergence.
White mold *(Sclerotinia)*	*Coniothyrium minitans* (Contans WG)	1 to 4 lb/acre	0.05 to 0.2 lb/acre	0	4 hr	OMRI listed product. Apply to soil surface and incorporate no deeper than 2 inches. Works best when applied prior to planting or transplanting. Do not apply other fungicides for 3 weeks after applying Contans.
	fludioxonil (Cannonball) 50 WP	7 oz/acre	3.5 oz/acre	7	0.5	Make no more than 2 sequential applications before alternating with a fungicide with a different mode of action for 2 applications. Use no more than 28 oz/ acre per year.

Pepper

Table 10-42. Disease Control Products for Pepper

Disease	Material	Rate of Material to Use		Minimum Days		Method, Schedule, and Remarks
		Formulation	Active Ingredient	Harvest	Reentry	
Aphid-transmitted viruses: PVY, TEV,	JMS Stylet-Oil	3 qt/100 gal water	—	0	Dry	Use in 50 to 200 gallons per acre depending on plant size. Spray weekly when winged aphids first appear.

535

Table 10-42. Disease Control Products for Pepper

Disease	Material	Rate of Material to Use		Minimum Days		Method, Schedule, and Remarks
		Formulation	Active Ingredient	Harvest	Reentry	
Anthracnose fruit rot	azoxystrobin (Quadris) Flowable	6 to 15.5 fl oz/acre	0.1 to 0.25 lb/acre	0	4 hr	Apply at flowering to manage green fruit rot. Limit of 61.5 fluid ounces per acre per season. Make no more than **one** application before alternating with fungicides that have a different mode of action.
	azoxystrobin + difenoconazole (Quadris Top) 29.6 SC	8 to 14 fl oz/acre	2.4 to 4.1 fl oz/acre	0	0.5	Limit of 55.3 fluid ounces per acre per season. Make no more than two consecutive applications before rotating to another effective fungicide with a different mode of action.
	chlorothalonil (various)	See label	See label	See label	See label	See label. Rates vary depending on the formulation.
	chlorothalonil + cymoxanil (Ariston)	2 to 2.44 pt/acre	0.84 to 1.02 pt/acre	3	0.5	Limit of 18.1 pints per acre per year.
	famoxadone + cymoxanil (Tanos)	8 to 10 oz/acre	4 to 5 oz/acre	3	0.5	Make no more than **one** application before alternating with a fungicide with a different mode of action. **NOTE: Must tank mix with another fungicide with a different mode of action (i.e. maneb or copper).**
	fenamidone (Reason) 500SC	5.5 to 8.2 fl oz/acre	0.178 to 0.267 lb/ acre	14	0.5	Limit of 24.6 fluid ounces per growing season. Make no more than **one** applications before rotating to another effective fungicide with a different mode of action.
	mancozeb (various)	See label	See label	See label	See label	See label. Rates vary depending on the formulation.
	mancozeb + copper (ManKocide)	2 to 3 lb/acre	1.2 to 1.8 lb/acre	7	2	Limit of 39 pounds per acre per season.
	pyraclostrobin (Cabrio) EG 20%	8 to 12 oz/acre	1.6 to 2.4 oz/acre	0	4 hr	Apply at flowering to manage green fruit rot. Limit of 96 ounces per acre per season. Make no more than **one** sequential application before alternating with fungicides that have a different mode of action.
	pyraclostrobin + fluxapyroxad (Priaxor)	4.0 to 8.0 fl oz/acre	1.72 to 3.43 fl oz/acre	0	0.5	**RIPE ROT ONLY.** Limit of 24 fluid ounces per acre per season. Make no more than two consecutive applications before rotating to another effective fungicide with a different mode of action.
	penthiopyrad (Fontelis)	24 fl oz/acre	4.89 fl oz/acre	0	0.5	**SUPPRESSION ONLY.** Limit of 72 fluid ounces per acre per year. Make no more than two consecutive applications before rotating to another effective fungicide with a different mode of action.
	trifloxystrobin (Flint)	3 to 4 oz/acre	1.5 to 2 oz/acre	3	0.5	**SUPPRESSION ONLY.** Limit of 16 ounces per acre per year. Make no more than **one** application before alternating with fungicides that have a different mode of action.
Bacterial soft rot	famoxadone + cymoxanil (Tanos)	8 to 10 oz/acre	4 to 5 oz/acre	3	0.5	**SUPPRESSION ONLY.** Make no more than **one** application before alternating with a fungicide with a different mode of action. **NOTE: Must tank mix with another fungicide with a different mode of action (i.e maneb or copper).**
Bacterial spot (field)	acibenzolar-S-methyl (Actigard) 50 WG	0.33 oz to 0.75 oz/ acre	0.17 to 0.38 oz/ acre	14	0.5	**FOR CHILI PEPPERS ONLY EXCECPT IN THE STATE OF GEORGIA.** Begin applications within one week of transplanting or emergence. Make up to six weekly, consecutive applications.
	fixed copper (various)	See label	See label	See label	See label	See label. Rates vary depending on the formulation. Make first application 7 to 10 days after transplanting. Carefully examine field for disease to dwhere mine need for additional applications. If disease is present, make additional applications at 5-day intervals. Applying mancozeb with copper significantly enhances bacterial spot control. Do not spray copper when temperatures are above 90 degrees F.
	famoxadone + cymoxanil (Tanos)	8 to 10 oz/acre	4 to 5 oz/acre	3	0.5	**SUPPRESSION ONLY.** Make no more than **one** application before alternating with a fungicide with a different mode of action. **NOTE: Must tank mix with another fungicide with a different mode of action (i.e. maneb or copper).**
	mancozeb (various)	See label	See label	See label	See label	See label. Rates vary depending on the formulation.
	mancozeb + copper (ManKocide)	2 to 3 lb/acre	1.2 to 1.8 lb/acre	7	2	Limit of 39 pound per acre per season.
	quinoxyfen (Quintec)	6.0 oz/acre	1.35 oz/acre	3	0.5	Use 6 ounces of product per acre in no less than 30 gallons of water per acre. **NOTE: May only be used to manage bacterial spot in Georgia, Florida, North Carolina, and South Carolina (Section 2(ee)).**
Bacterial spot (transplants)	streptomycin sulfate (Agri-Mycin 17, Firewall, Streptrol)	1 lb/100 gal	0.8 oz/100 gal	—	1	**MAY ONLY BE APPLIED TO TRANSPLANTS.** Spray when seedlings are in the 2-leaf stage and continue at 5-day intervals until transplanted into field. **NOTE: Some pathogen strains are resistant to streptomycin sulfate.**
	fixed copper (various)	See label	See label	See label	See label	See label. Rates vary depending on the formulation. Begin applications when conditions first favor disease development and repeat at 3 to 10 day intervals if needed depending on disease severity. Use the higher rates when conditions favor disease. Do not spray copper when temperatures are above 90 degrees F.

536

Table 10-42. Disease Control Products for Pepper

Disease	Material	Rate of Material to Use		Minimum Days		Method, Schedule, and Remarks
		Formulation	Active Ingredient	Harvest	Reentry	
Bacterial spot (seed)	sodium hypochlorite (Clorox 5.25%, regular formulation)	1 pt + 4 pt water	—	—	—	Add 1 tablespoon of surfactant (Tween-20 or 80, Silwet) to improve coverage on the seed.
Cercospora leaf spot	azoxystrobin + difenoconazole (Quadris Top) 29.6 SC	8 to 14 fl oz/acre	2.4 to 4.1 fl oz/acre	0	0.5	Limit of 55.3 fluid ounces per acre per season. Make no more than two consecutive applications before rotating to another effective fungicide with a different mode of action.
	fixed copper (various)	See label	See label	See label	See label	See label. Rates vary depending on the formulation. Begin applications when conditions first favor disease development and repeat at 3 to 10 day intervals if needed depending on disease severity. Use the higher rates when conditions favor disease. Do not spray copper when temperatures are above 90 degrees F.
	mancozeb (various)	See label	See label	See label	See label	See label. Rates vary depending on the formulation.
	mancozeb + copper (ManKocide)	2 to 3 lb/acre	1.2 to 1.8 lb/acre	7	2	Limit of 39 pound per acre per season.
Phytophthora foliar blight and fruit rot (*Phytophthora capsici*)	cyazofamid (Ranman Fungicide)	2.75 fl oz/acre	0.071 lb/acre	0	0.5	Limit of 16.5 fluid ounces per acre per season. Apply to the base of the plant at transplanting or in the transplant water. Make no more than three consecutive applications followed by three consecutive applications of another effective fungicide with a different mode of action.
	fixed copper (various)	See label	See label	See label	See label	See label. Rates vary depending on the formulation. Begin applications when conditions first favor disease development and repeat at 3 to 10 day intervals if needed depending on disease severity. Use the higher rates when conditions favor disease. Do not spray copper when temperatures are above 90 degrees F.
	dimethomorph (Acrobat, Forum)	6 fl oz/acre	2.61 fl oz/acre	0	0.5	**SUPPRESSION ONLY.** Limit of 30 fluid ounces per acre per season. Make no more than two sequential before alternating with fungicides that have a different mode of action. **NOTE: Must tank mix with another fungicide with a different mode of action.**
	mancozeb (various)	See label	See label	See label	See label	See label. Rates vary depending on the formulation.
	mancozeb + copper (ManKocide)	2 to 3 lb/acre	1.2 to 1.8 lb/acre	7	2	**SUPPRESSION ONLY.** Limit of 39 pound per acre per season.
	mefenoxam + copper hydroxide (Ridomil Gold/Copper)	2 lb/acre	1.3 lb/acre	7	2	See label for an optimal spray program. Limit of four applications per crop per year. Do not exceed 0.4 pound a.i. per acre per season of mefenoxam + metalaxyl (MetaStar).
	famoxadone + cymoxanil (Tanos)	8 to 10 oz/acre	4 to 5 oz/acre	3	0.5	**SUPPRESSION ONLY.** Make no more than **one** application before alternating with a fungicide with a different mode of action. **NOTE: Must tank mix with another fungicide with a different mode of action (i.e maneb or copper).**
	fenamidone (Reason) 500SC	8.2 fl oz/acre	0.267 lb/ acre	14	0.5	**SUPRESSION ONLY.** Limit of 24.6 fluid ounces per growing season. Make no more than **one** applications before rotating to another effective fungicide with a different mode of action.
	fluazinam (Omega) 500F	1 to 1.5 pt/acre		30	0.5	Apply as a soil drench at 1.5 pints per acre. For foliar applications use 1 pint per acre. Limit of 9 pints per acre per season.
	fluopicolide (Presidio)	3 to 4 fl oz/acre	0.09 to 0.125 lb/acre	2	0.5	Limit of 12 fluid ounces per acre per season. Make no more than 2 times sequentially before alternating with fungicides that have a different mode of action. **NOTE: Must be tank-mixed with another mode of action product.**
	mandipropamid (Revus, Micora)	8 fl oz/acre	0.13 lb/acre	1	0.5	**SUPPRESSION ONLY.** Limit of 32 fluid ounces per acre per season. **NOTE: Must tank mix with another fungicide with a different mode of action (i.e. copper).**
	amectoctradin + dimethomorph (Zampro)	14 fl oz/acre	0.27 lb/acre	4	0.5	Limit of 42 fluid ounces per acre per season. Make no more than two sequential applications before rotating to another effective fungicide with a different mode of action.
Phytophthora or *Pythium* root rot (field)	mefenoxam (Ridomil Gold, Ultra Flourish)	Check label; varies between products	Check label; varies between products	—	2	**MAY ONLY BE APPLIED AT PLANTING.** Apply in a 12 to 16 inch band or in 20 to 50 gallons water per acre in transplant water. Mechanical incorporation or 0.5 to 1 inches irrigation water is needed for movement into root zone if rain is not expected. After initial application, two supplemental applications (1 pint per treated acre) can be applied. **NOTE: Strains of *Phytophthora capsici* insensitive to Ridomil Gold have been detected in some North Carolina and Louisiana pepper fields.**
	metalaxyl (MetaStar) 2E	4 to 8 pt/treated acre	0.5 to 1 lb/acre	7	2	Limit of 12 pints per acre per season. Preplant (soil incorporated), at planting (in water or liquid fertilizer), or as a basil-directed spray after planting. See label for the guidelines for supplemental applications.

Table 10-42. Disease Control Products for Pepper

Disease	Material	Rate of Material to Use		Minimum Days		Method, Schedule, and Remarks
		Formulation	Active Ingredient	Harvest	Reentry	
Powdery mildew	azoxystrobin (Quadris) Flowable	6 to 15.5 fl oz/acre	0.1 to 0.25 lb/acre	0	4 hr	Limit of 61.5 fluid ounces per acre per season. Make no more than **one** application before alternating with fungicides that have a different mode of action.
	azoxystrobin + difenoconazole (Quadris Top) 29.6 SC	8 to 14 fl oz/acre	2.4 to 4.1 fl oz/acre	0	0.5	Limit of 55.3 fluid ounces per acre per season. Make no more than two consecutive applications before rotating to another effective fungicide with a different mode of action.
	chlorothalonil + cymoxanil (Ariston)	2 to 2.44 pt/acre	0.84 to 1.02 pt/acre	3	0.5	Limit of 18.1 pints per acre per year.
	penthiopyrad (Fontelis)	16 to 24 fl oz/acre	4.89 fl oz/acre	0	0.5	Limit of 72 fluid ounces per acre per year. Make no more than two consecutive applications before rotating to another effective fungicide with a different mode of action.
	pyraclostrobin + fluxapyroxad (Priaxor)	6.0 to 8.0 fl oz/acre	2.57 to 3.43 fl oz/acre	0	0.5	Limit of 24 fluid ounces per acre per season. Make no more than two consecutive applications before rotating to another effective fungicide with a different mode of action.
	quinoxyfen (Quintec)	4.0 to 6.0 fl oz/acre	0.90 to 1.35 fl oz/acre	3	0.5	Limit of 24 fluid ounces per acre per year. Make no more than two consecutive applications before alternating with fungicides that have a different mode of action. **NOTE: Under certain environmental conditions leaf spotting or chlorosis may occur after application; discontinue use if symptoms occur.**
	sulfur (various)	See label	See label	See label	See label	See label. Rates vary depending on the formulation. Apply at first appearance and repeat at 14-day intervals as needed.
	trifloxystrobin (Flint)	1.5 to 2 oz/acre	0.75 to 1 oz/acre	3	0.5	Limit of 16 ounces per acre per year. Make no more than **one** application before alternating with fungicides that have a different mode of action.
Southern blight (*Sclerotium rolfsii*)	fluoxastrobin (Aftershock, Evito 280SC)	2 to 5.7 fl oz/acre	0.81 to 2.30 fl oz/acre	3	0.5	Limit of 22.8 fluid ounces per acre per season. Make no more than **one** application before alternating with fungicides that have a different mode of action. **NOTE: Do not overhead irrigate for 24 hours following a spray application.**
	pyraclostrobin (Cabrio) EG 20%	12 to 16 oz/acre	1.6 to 2.4 oz/acre	0	4 hr	**SUPPRESSION ONLY.** Apply at flowering to manage green fruit rot. Limit of 96 ounces per acre per season. Make no more than **one** sequential application before alternating with fungicides that have a different mode of action.
	pyraclostrobin + fluxapyroxad (Priaxor)	4.0 to 8.0 fl oz/acre	1.72 to 3.43 fl oz/acre	0	0.5	**SUPPRESSION ONLY.** Limit of 24 fluid ounces per acre per season. Make no more than two consecutive applications before rotating to another effective fungicide with a different mode of action.
Target spot (*Corynespora cassiicola*)	boscald (Endura)	3.5 oz/acre	2.45 oz/acre	0	0.5	Limit of 21 ounces per acre per season. Make no more than two sequential applications before alternating with fungicides that have a different mode of action.
	cyprodinil + difenoconazole (Inspire Super)	16 to 20 fl oz/acre	0.60 to 0.75 lb/acre	0	0.5	Limit of 47 fluid ounces per acre per season.
	fluoxastrobin (Aftershock, Evito 480SC)	2 to 5.7 fl oz/acre	0.81 to 2.30 fl oz/acre	3	0.5	Limit of 22.8 fluid ounces per acre per season. Make no more than **one** application before alternating with fungicides that have a different mode of action. **NOTE: Do not overhead irrigate for 24 hours following a spray application.**
	penthiopyrad (Fontelis)	16 to 24 fl oz/acre	3.26 to 4.89 fl oz/acre	0	0.5	**SUPPRESSION ONLY.** Limit of 72 fluid ounces per acre per year. Make no more than two consecutive applications before rotating to another effective fungicide with a different mode of action.
	pyraclostrobin (Cabrio) EG 20%	8 to 12 oz/acre	1.6 to 2.4 oz/acre	0	4 hr	Apply at flowering to manage green fruit rot. Limit of 96 ounces per acre per season. Make no more than **one** sequential application before alternating with fungicides that have a different mode of action.
	pyraclostrobin + fluxapyroxad (Priaxor)	4.0 to 8.0 fl oz/acre	1.72 to 3.43 fl oz/acre	0	0.5	Limit of 24 fluid ounces per acre per season. Make no more than two consecutive applications before rotating to another effective fungicide with a different mode of action.

Table 10-43. Relative Effectiveness of Various Chemicals for Pepper Disease Control

Scale: E, excellent; G, good; F, fair; P, poor; NC, no control; ND, no data.

Product[1]	Fungicide Group[1]	Preharvest Interval (Days)	Anthracnose (immature fruit rot)	Bacterial Spot	Phytophthora Blight (root and crown)	Phytophthora Blight (fruit and foliage)	Pythium Damping off	Southern Blight (Basal Stem Rot)
azoxystrobin (Quadris)	11	0	F	NC	NC	NC	NC	ND
Chlorothalonil (various)	M	3	P	NC	NC	P	NC	NC
cyazofamid (Ranman)	21	0	NC	NC	F	G	NC	NC
dimethomorph (Acrobat, Forum)	40	4	NC	NC	NC	P	NC	NC
dimethomorph + amectoctradin (Zampro)	40 + 45	4	NC	NC	F	G	ND	NC
Famoxadone + cymoxanil (Tanos)	11 + 27	3	P	NC	NC	P	NC	ND
fixed copper (various)	M	Check label	P	F	NC	F	NC	NC
fluopicolide (Presidio)	43	2	NC	NC	F	G	NC	NC
Floxystrobin (Evito)	11	3	NC	NC	NC	NC	NC	ND
fluxapyroxad + pyraclostrobin (Priaxor)	11 + 7	7	F	NC	NC	NC	NC	ND
Mancozeb[2] (various)	M	5	F	P	P	P	NC	NC
mandipropamid (Revus)	40	1	NC	NC	F	G	NC	NC
mefenoxam[R] (Ridomil Gold EC, Ultra Flourish)	4	0	NC	NC	E	NA	G	NC
mefenoxam[R] + copper (Ridomil Gold + copper)	4 + M	14	P	F	NA	G	NC	NC
penthiopyrad (Fontelis)	7	0	ND	NC	NC	NC	NC	ND
proamocarb (Previcur Flex)	28	5	NC	NC	NC	NC	F	NC
pyraclostrobin (Cabrio)	11	0	G	NC	NC	NC	NC	ND
quinoxyfen (Quintec)	13	3	NC	P	NC	NC	NC	NC
streptomycin sulfate[3] (Agri-mycin, Stretrol, Firewall)	25	Not for field use	NC	F	NC	NC	NC	NC
sulfur (various)	M	0	NC	NC	NC	NC	NC	NC

[1] Efficacy ratings do not necessarily indicate a labeled use for every disease.
[2] Copper tank-mixed with mancozeb enhances the efficacy against bacterial spot.
[3] Streptomycin may only be used on transplants; not registered for field use.
[R] Resistance reported in the pathogen.

DISEASE CONTROL

Table 10-44. Importance of Alternative Management Practices for Disease Control in Pepper

Scale: E, excellent; G, good; F, fair; P, poor; NC, no control; ND, no data.

Strategy	Anthracnose (immature fruit)	Aphid-transmitted viruses (PVX, CMV, TEV, AMV, PVY)	Bacterial soft rot of fruit	Bacterial spot	Blossom-end rot	Phytophthora blight (fruit and foliage)	Phytophthora blight (root and crown)	Pythium damping off	Root-knot nematode	Southern blight	Tomato Spotted Wilt Virus
Avoid field operations when foliage is wet	F	NC	NC	G	NC	F	P	NC	NC	NC	NC
Avoid overhead irrigation	G	NC	F	G	NC	G	G	P	NC	NC	NC
Change planting date within a season	NC	F (early)	NC	F (early)	NC	NC	NC	P (late)	F (early)	P (early)	Varies
Cover cropping with antagonist	NC	NC	NC	NC	NC	NC	NC	NC	F	NC	NC
Rotation with non-host (2 to 3 years)	G	NC	NC	NC	NC	P	P	NC	F	P	NC
Deep plowing	F	NC	NC	NC	NC	NC	NC	NC	P	F	NC
Prompt destruction of crop residue	F	F	NC	P	NC	P	P	NC	F	P	NC
Promote air movement	P	NC	NC	F	NC	P	P	NC	NC	NC	NC
Use of soil organic amendments	NC	NC	NC	NC	NC	P	P	P	F	P	NC
Application of insecticidal/horticultural oils	NC	F	NC	NC	NC	NC	NC	NC	NC	NC	NC
pH management (soil)	NC	NC	NC	NC	F	NC	NC	NC	F	NC	NC
Plant in well-drained soil/raised beds	NC	NC	NC	NC	NC	NC	G	G	NC	NC	NC
Eliminate standing water/saturated areas	NC	NC	NC	NC	NC	NC	G	G	NC	NC	NC
Postharvest temp control (fruit)	NC	NC	G	NC	NC	NC	NC	NC	NC	NC	NC
Use of reflective mulch	NC	F	NC	NC	NC	NC	NC	NC	NC	NC	G
Reduce mechanical injury	NC	NC	NC	NC	NC	NC	NC	NC	NC	NC	NC
Rogue diseased plants/fruit	NC	NC	NC	NC	NC	F	F	NC	NC	NC	NC
Soil solarization	NC	NC	NC	NC	NC	NC	P	NC	F	NC	NC
Use of pathogen-free planting stock	F	NC	NC	G	NC	NC	NC	NC	NC	NC	NC
Use of resistant cultivars	NC	NC	NC	G	F	F	F	NC	G	NC	G
Weed management	P	F	NC	NC	NC	P	P	NC	F	NC	P

Potato, Irish

Table 10-45. Disease Control Products for Potato, Irish

Diseasse	Material	Rate of Material to Use		Minimum Days		Method, Schedule, and Remarks
		Formulation	Active Ingredient	Harvest	Reentry	
Black and silver scurf	azoxystrobin (various)	See label	See label	see label	see label	Apply in furrow at planting according to label directions.
	fluopyram (Luna Privilege)	5.47 fl oz/acre (ground); 2.82 oz/acre (aerial)	0.1 to 0.2 lb/acre (ground); 0.1 lb/acre (aerial)	7	0.5	Use on a 5 to 7-day interval. Do not apply more than 10.95 ounces per acre per season for ground application and no more than 8346 ounces per acre per season for aerial application. Do not make more than 2 applications before alternating with a fungicide with a different mode of action. Labeled for silver scurf only.
	fluoxastrobin (Aftershock, Evito 480 SC)	0.16 to 0.24 fl oz/1,000 ft of row	0.06 to 0.1 fl oz/1,000 ft	7	0.5	Apply in furrow at planting according to label directions. Do not apply more than 22.8 fluid ounes of product per acre per year including seed treatment use. Alternate with fungicide from different resistance management group.
	flutolanil (Moncut) 70DF	0.71 to 1.1 lb/acre	0.50 to 0.77 lb/acre	—	0.5	For black scurf only.
	penthiopyrad (Vertisan)	0.7 to 1.6 fl oz/1,000 ft of row	0.1 to 0.3 fl oz/acre	7	0.5	Maximum rate is 24 fluid ounces. Labeled for black scurf only.

DISEASE CONTROL

Table 10-45. Disease Control Products for Potato, Irish

| Diseasse | Material | Rate of Material to Use | | Minimum Days | | Method, Schedule, and Remarks |
		Formulation	Active Ingredient	Harvest	Reentry	
Black and silver scurf (continued)	polyoxin (Ph-D) (Oso 5%SC)	6.2 fl oz/acre 3.75 to 13.0 fl oz/acre	0.7 fl oz/acre 0.2 to 0.65 fl oz/acre	0	0.2	Apply at 7 to 14 day intervals. Both products are registered for black scurf only.
	thiophanate-methyl + mancozeb + cymoxanil (Evolve)	0.75 lb/100 lb seed	—	—	—	
Fusarium seedpiece decay, Rhizoctonia stem canker, Streptomyces common scab	azoxystrobin + mefenoxam 2.08 F + 4 F	0.82 fl oz/1,000 ft of row	—	—	0	Apply as an in-furrow spray in 3 to 5 gallons of water per acre at planting. Not for Fusarium or Scab.
	fludioxonil (Maxim 4FS) (Maxim PSP)	0.08 to 0.16 oz/100 lb seed 0.5 lb/100 lb seed 0.5	--	—	0.5	NOT labeled for Streptomyces. Do not use for food or feed. See label for mixing instructions.
	fludioxinil + mancozeb (Maxim MZ)	0.5 lb/100 lb seed	--	--	1	NOT labeled for Streptomyces. Do not use for food or feed. See label for mixing instructions.
	mancozeb (various)	See label	See label	See label	See label	See label. Rates vary depending on the formulation.
	maneb (Manex 4F)	0.8 qt/10 gal water	0.02 lb/gal water	—	—	
	penthiopyrad (Vertisan)	0.7 to 1.6 oz/1,000 ft of row	0.1 to 0.3 oz/acre	7	0.5	Maximum rate is 24 fluid ounces. Labeled for Rhizoctonia stem canker only.
	polyoxin (Ph-D)	6.2 fl oz/acre	0.7 fl oz/acre	0	0.2	Apply at 7 to 14 day intervals. Ph-D is registered for Rhizoctonia stem canker only.
	thiophanate-methyl + mancozeb (Tops MZ)	0.75 lb/100 lb seed	—	—	—	
	thiophanate-methyl + mancozeb + cymoxanil (Evolve)	0.75 lb/100 lb seed	—	—	—	
Early blight, white mold	azoxystrobin + difenoconazole (Quadris Top)	8 to 14 fl oz/acre	2.4 to 4.1 fl oz/acre	14	0.5	Apply at 7- to 14-day-intervals. Apply no more than 2 sequential applications without alternating with a fungicide with a different mode of action. Do not apply more than 55.3 lb product per acre per year. Do not apply more than 0.46 lb a.i./acre /year of difenoconazole-containing products; do not apply more than 2.0 pounds a.i./acre /year of azoxystrobin-containing products. Labeled for early blight only.
	boscalid (Endura)	2.5 to 10 oz/acre	0.11 to 0.44 lb/acre	30	0.5	For control of Sclerotinia White mold, use 5.5 to 10 ounce rate and begin applications prior to row closure or at the onset of disease. Make a second application 14 days later if conditions continue to favor disease development. For Early blight control, use 2.5 to 4.5 ounces rate. **DO NOT** apply more than 20.5 ounces of product per acre per season.
	fluopyram (Luna Privilege)	4.0 to 5.47 oz/acre (ground); 2.82 oz/acre (aerial)	0.1 to 0.2 lb/acre (ground); 0.1 lb/acre (aerial)	7	0.5	Use on a 5 to 7-day interval. Do not apply more than 10.95 ounces per acre per season for ground application and no more than 8346 ounces per acre per season for aerial application. Do not make more than 2 applications before alternating with a fungicide with a different mode of action.
	fluopyram + pyrimethanil (Luna Tranquility)	11.2 oz/acre	—	7	0.5	Apply at 7 to 14 day intervals. Do not make more than 2 sequential applications without switching to a fungicide outside of Group 7 or Group 9.
	fluxapyroxad + pyraclostrobin (Priaxor Xemium)	4 to 8 oz/acre	—	7	0.5	Apply at 7 to 14 day intervals. Do not apply more than 24 ounces per acre per season including infurrow and foliar uses. Use 6 to 8 ounces per acre for control of white mold.
	iprodione (various)	See label	See label	See label	See label	Do not exceed 4 applications/acre/season.
	metiram + pyraclostrobin (Cabrio Plus)	2.0 to 2.9 lb/acre	—	3 (FL); 14	—	Apply at 7 to 14 day-intervals. Do not apply more than 17.4 pounds per acre product per season. Do not apply more than 2 sequential applications before alternating with a fungicide with a different mode of action. Use at 2.9 pounds peracre for white mold.
	penthiopyrad (Vertisan)	10 to 24 oz/acre	0.1 to 0.3 oz/acre	7	0.5	Apply at 7-14 day intervals. Make no more than 2 applications before alternating with a fungicide with a different mode of action. For white mold control, use at 14 to 24 ounces per acre. Do not exceed 72 ounces per acre per year. Do not apply more than 11.25 ounces a.i. per acre per year in total from any combination of seed, soil, or foliar appliactions.
	pyraclostrobin (Headline; Headline SC)	6 to 12 fl oz/acre	1.4 to 2.8 fl oz/acre	3	1	**DO NOT** exceed more than six foliar applications or 72 total ounces of product per acre per season. For early blight, use 6- to 9-ounces rate; for white mold, use 6- to 12-ounce rate, depending on weather conditions and disease pressure. Do not apply more than one time before alternating with a fungicide with a different mode of action.
	pyrimethanil (Scala SC)	7 oz/acre	3.8 oz/acre	7	0.5	Apply at 7 to 14 day intervals. Do not apply more than 35 ounces per acre per season. For control of early blight only.
	thiophanate-methyl (various)	See label	See label	See label	See label	Rates may vary depending on the product.

541

Table 10-45. Disease Control Products for Potato, Irish

Diseasse	Material	Rate of Material to Use		Minimum Days		Method, Schedule, and Remarks
		Formulation	Active Ingredient	Harvest	Reentry	
Late blight, white mold	fluazinam (Omega) 500 F	5.5 to 8 oz/acre	0.16 to 0.26 oz/acre	14	1	Begin applications when plants are 6 to 8 inches tall or when conditions favor disease development. Repeat applications at 7- to 10-day intervals. When white mold pressure is low to moderate, use 5.5 fluid ounces. When conditions favor moderate to high white mold pressure, increase the rate to 8 fluid ounces. **DO NOT** apply more than 3.5 pints per acre during each growing season.
Early blight, late blight	azoxystrobin (various)	See label	See label	See label	See label	Rates may vary depending on the product. Alternate with protectant fungicide (chlorothalonil or mancozeb). **DO NOT** apply more than 1.88 pounds per acre per season.
	azoxystrobin + chlorothalonil (Quadris Opti)	1.6 pt/acre	—	14	0.5	Apply at 5 to 7 day-intervals. Do not apply more than 1.5 pounds a.i. per acre per year of azoxystrobin; do not apply more than 11.25 pounds a.i. per acre per year of chlorothalonil.
	chlorothalonil (various)	See label	See label	See label	See label	See label. Rates vary depending on the formulation.
	chlorothalonil + zoxamide (Zing!)	32 to 34 fl oz/acre	—	7	0.5	Apply at 5 to 7 day intervals. Do not make more than 2 sequential applications before alternating with a fungicide that has a different mode of action. Do not make more than 8 applications or apply more than 1.52 pounds zoxamide and 8.88 lb chlorothalonil per acre per season.
	cymoxanil + famoxadone (Tanos)	6 to 8 fl oz/acre	—	14	0.5	Use rate of 6 fluid ounces only for early blight. Do not apply more than 48 fluid ounces per acre per crop season and no more than 72 fluid ounces per acre per 12 months. Do not make more than one application before alternating with a fungicide with a different mode of action. Do not make more than 6 applications of any fungicide in Group 11 per cropping cycle.
	fixed copper (various)	See label	See label	See label	See label	See label. Rates vary depending on the formulation.
	famoxadone + cymoxanil (Tanos)	6 to 8 oz/acre	—	14	1	Begin applications when conditions favor disease development or when disease is present in area. Should be tank mixed with a protectant fungicide (chlorothalonil or mancozeb). **DO NOT** apply more than 48 ounces per acre per season.
	fenamidone (Reason) 500 SC	5.5 to 8.2 fl oz/acre	0.178 to 0.267 lb/acre	14	0.5	Begin applications when conditions favor disease development, and continue on 5 to 10 day interval. Do not apply more than 24.6 fluid ounces per growing season. Alternate with fungicide from different resistance management group.
	fluoxastrobin (Aftershock, Evito) 480 SC	2 to 3.8 fl oz/acre	0.16 to 0.12 lb/acre	7	0.5	Begin applications when conditions favor disease development, on 7 to 10 day intervals. Do not apply more than once before alternating with fungicides that have a different mode of action. Do not apply more than 22.8 fluid ounces per acre per season. For late blight, apply at full label rate.
	fluxapyroxad + pyraclostrobin (Priaxor)	4 to 8 fl oz/acre	—	7	0.5	Apply at 7 to 14 day intervals. Do not apply more than 24 ounces per acre per season, including infurrow and foliar uses.
	mancozeb (various)	See label	See label	See label	See label	See label. Rates vary depending on the formulation. Apply when conditions favor disease but before symptoms appear. Continue applications on a 5 to 10 day schedule depending on disease pressure. Under high disease pressure, shorten spray interval and use maximum rate.
	mancozeb + chlorothalonil (Elixir)	1.2-2.0 lb/acre	—	7 (FL): 14 (NC, VA)	1	Do not apply more than 18.0 pounds per acre per crop. Product is only registered in FL, NC, and VA in the southeastern US.
	mandipropamid + difenoconazole (Revus Top)	5.5 to 7 oz/acre	—	1	0.5	Begin applications when conditions favor disease development, on 7 to 10 day intervals. Do not apply more than twice before alternating with fungicides that have a different mode of action. Do not apply more than 28 fluid ounces acre per season.
	maneb (Manex) 4F	0.8 to 1.6 qt/acre	0.8 to 1.6 lb/acre	14	1	
	mefenoxam+ chlorothalonil (Ridomil Gold/Bravo, Flouronil)	2 lb/acre	2.5 pints/acre	14	2	See label for limits on application limits per season and application interval.
	mefenoxam+ mancozeb (Ridomil Gold MZ)	2.5 lb/acre	—	14	2	Apply at 14-day intervals for up to 3 applications.
	metiram (Polyram) 80 DF	1.5 to 2 lb/acre	1.2 to 1.6 lb/acre	14	1	
	metiram + pyraclostrobin (Cabrio Plus)	2.0 to 2.9 lb/acre	1.2 to 1.74 lb/acre	3 (FL); 14	1	Do not apply more than 17.4 pounds per acre product per season. Do not apply more than 2 sequential applications before alternating with a fungicide with a different mode of action.
	polyoxin (Ph-D) (Oso 5%SC)	6.2 fl oz/acre 3.75 to 13.0 fl oz/acre	0.7 fl oz/acre 0.2 to 0.65 fl oz/acre	0	0.2	Labeled for Early Blight only.

DISEASE CONTROL

Table 10-45. Disease Control Products for Potato, Irish

| Diseasse | Material | Rate of Material to Use | | Minimum Days | | Method, Schedule, and Remarks |
		Formulation	Active Ingredient	Harvest	Reentry	
Early blight, late blight (continued)	propamocarb hydrochloride (Previcur Flex) 6 F	0.7 to 1.2 pt/acre	0.5 to 0.9 lb/acre	14	0.5	Do not apply more than 6 pints per growing season. Use lower rates when conditions favor disease but disease is not present. Increase rate as risk of disease development increases.
	pyraclostrobin (Headline, Headline SC) 2.08 F	6 to 12 oz/acre	1.4 to 2.8 oz/acre	3	1	**DO NOT** exceed more than six foliar applications or 72 total ounces of product per acre per season. For early blight, use 6- to 9-ounce rate; for late blight, use 6- to 12-ounce rate, depending on weather conditions and disease pressure.
	pyraclostrobin + chlorothalonil (Cabrio Plus)	2.0 to 2.9 lb/acre	--	14	1	Do not apply more than 2 applications before switching to a different mode of action. Do not exceed 17.4 lbs/acre/season.
	pyrimethanil (Scala) 5 F	7 fl oz/acre	0.27 lb/acre	7	0.5	Early blight only. Use only in a tank mix with another Early blight fungicide. Do not apply more than 35 fluid ounces per crop.
	trifloxystrobin (Gem) 500 SC	6 to 8 oz/acre	1.5 to 2 oz/acre	7	0.5	Begin applications preventively and continue as needed on a 7 to 10 day schedule. Alternate every other application with a protectant fungicide. **DO NOT** apply more than 48 ounces per acre per season.
	triphenyltin hydroxide (Super Tin) 80 WP	2.5 to 3.75 oz/acre	0.2 to 0.3 lb/acre	21	1	Add to 3 to 15 gallons of water depending on method of application. Do not exceed 15 ounces of product per acre per season.
	zoxamide + mancozeb (Gavel 75DF)	1.5 to 2.0 lb/acre	—	14	2	Apply at 5 to 10 day intervals except shorten to 5 to 7 day intervals when late blight is present. Do not make more than 6 applications or apply more than 12 pounds (8.0 pounds mancozeb and 1.0 zoxamide) per acre per season crop.
Late blight	ametoctradin + dimethomorph (Zampro)	11 to 14 fl oz/acre	—	4	0.5	Apply at 5 to 7 day intervals. Do not exceed 42 oz per acre per season. Do not make more than 2 sequential appliactions without alternating to a fungicide with a different mode of action and no more than 3 total applications in a season.
	cyazofamid (Ranman) 400 SC	1.4 to 2.75 fl oz/acre	0.036 to 0.071 lb/acre	7	0.5	**Late blight only.** Do not apply more than 10 sprays per crop. Make no more than 3 consecutive applications followed by 3 applications of fungicides from a different resistance management group.
	cymoxanil (Curzate) 60 DF	3.2 oz/acre	0.1 lb/acre	14	1	Use Curzate or Acrobat in combination with a protectant fungicide (chlorothalonil, mancozeb, metiram, or triphenyltin hydroxide). **DO NOT** exceed 32 ounces per acre per season.
	dimethomorph (Acrobat, Forum) 50 WP	4 to 6.4 oz/acre	2 to 3.2 oz/acre	4	0.5	For late blight control only. Apply on a 5 to 10 day schedule depending on disease pressure. Check label for rotational crop guidelines. Use in combination with a protectant fungicide. Do not exceed 30 ounces per acre per season. **DO NOT** tank mix with metalaxyl or mefenoxam.
	mefenoxam + copper hydroxide (Ridomil Gold/ Copper)	2 lb/acre	--	14	2	Apply at 14-day intervals for up to 3 applications. Alternate with a protectant fungicide.
	mono- and dipotassium salts of phosphorous acid (Alude;Phostrol)	2.5 to 10 pints/acre	--	0	4h	Mix with a fungicide labeled for control of late blight. See label for in-furrow application or foliar application rates.
	oxathiapiprolin + chlorothalonil (Orondis Opti)	See label	--	See label	See label	Orondis is sold in co-pack with chlorothalonil; see label.
	oxathiapiprolin + mandipropamid (Orondis Ultra)	See label	--	See label	See label	Orondis sold in co-pack with mandipropamid; see label.
Pink rot, *Pythium* leak, tuber rot	azoxystrobin + mefenoxam (Quadris Ridomil Gold) 2.08 F + 4 F	0.82 fl oz/1,000 fl of row	—	—	0	Apply as an in-furrow spray in 3 to 15 gallons of water per acre at planting.
	cyazofamid (Ranman) 400 SC	1.4 to 2.75 fl oz/acre	0.036 to 0.071 lb/acre	7	0.5	For pink rot and *Pythium* leak, apply at the high rate. Do not apply more than 10 sprays per crop. Make no more than 3 consecutive applications followed by 3 applications from a different resistance management group.
	mefenoxam (Ridomil Gold SL) (Ultraflourish)	0.42 fl oz/1,000 ft of row 0.84 fl oz/1,000 ft of row	0.013 lb/1,000 ft of row 0.21 fl oz/1,000 ft of row	14	2	See labels for maximum amount of product allowable per season. PHI is based on foliar application for Ultraflourish.
	mefenoxam + chlorothalonil (Ridomil Gold/Bravo. Flouronil)	2 lb/acre	—	14	2	Apply at flowering and then continue on a 14-day interval. Do not exceed more than four applications per crop.
	mefenoxam + copper hydroxide (Ridomil Gold/ Copper)	2 lb/acre	—	14	2	Apply at 14-day intervals for up to 3 applications. Alternate with a protectant fungicide.
	mefenoxam + mancozeb (Ridomil Gold MZ)	2.5 lb/acre	—	14	2	Apply at 14-day intervals for up to 3 applications.

543

Table 10-45. Disease Control Products for Potato, Irish

Diseasse	Material	Rate of Material to Use		Minimum Days		Method, Schedule, and Remarks
		Formulation	Active Ingredient	Harvest	Reentry	
Pink rot, *Pythium* leak, tuber rot (continued)	metalaxyl (MetaStar) 2 E	4 to 8 pt/trt acre	0.5 to 1 lb/acre	7	2	Preplant incorporated or soil surface spray
	mono- and di- potassium salts of phosphorous acid (Alude)	1.25 qt/acre in 90-375 gal water	1.62 lb/acre	0	0.16	Apply at 5-14 days depending on disease incidence.
Powdery mildew	azoxystrobin (various)	See label	See label	See label	See label	See label. Rates may vary depending on the product. Apply in furrow at planting according to label direction. Do not apply more than one application without alternating away from fungicides in Group 11.
	azoxystrobin + chlorothalonil (Quadris Opti)	1.6 pt/acre	—	14	0.5	Apply at 5 to 7 day intervals. Do not apply more than 1.5 pounds a.i./acre/year of azoxystrobin; do not apply more than 11.25 pounds a.i./acre /year of chlorothalonil. Do not make more than 1 application before alternating with a fungicide with a different mode of action. Do not apply this product or other fungicides in Group 11 more than 6 times in a season.
	azoxystrobin + difenoconazole (Quadris Top)	8 to 14 fl oz/acre	—	14	0.5	Apply at 7 to 14 day intervals. Apply no more than 2 sequential applications without alternating with a fungicide with a different mode of action. Do not apply more than 55.3 pounds product per acre per year. Do not apply more than 0.46 pound a.i./acre /year of difenoconazole-containing products; do not apply more than 2.0 pounds a.i./acre /year of azoxystrobin-containing products.
	chlorothalonil + sulfur (Bravo S)	4.3 pt/acre	—	7	2	Apply when disease first appears; then repeat as needed. Check label for application intervals. Avoid applying sulfur on days over 90 degrees F.
	fluopyram + pyrimethanil (Luna Tranquility)	11.2 oz/acre	—	7	0.5	Apply at 7 to 14 day intervals. Do not make more than 2 sequential applications without switching to a fungicide outside of Group 7 or Group 9.
	fluxapyroxad + pyraclostrobin (Priaxor Xemium)	4 to 8 oz/acre	—	7	0.5	Apply at 7 to 14 day intervals. Do not apply more than 24 ounces per acre per season including infurrow and foliar uses.
	mandipropamid + difenoconazole (Revus Top)	5.5 to 7 fl oz/acre	—	14	0.5	Begin applications when conditions favor disease development, on 7 to 10 day intervals. Do not apply more than twice before alternating with fungicides that have a different mode of action. Do not apply more than 28 fluid ounces per acre per season.
	metiram + pyraclostrobin (Cabrio Plus)	2.9 lb/acre	—	3 (FL); 14	1	Apply at 7 to 14 day-intervals. Do not apply more than 17.4 pounds per acre product per season. Do not apply more than 2 sequential applications before alternating with a fungicide with a different mode of action.
	penthiopyrad (Vertisan)	10 to 24 oz/acre	0.1 to 0.3 oz/acre	7	0.5	Apply at 7-14 day intervals. Make no more than 2 applications before alternating with a fungicide with a different mode of action. Do not exceed 72 ounces per acre per year. Do not apply more than 11.25 ounces a.i. per acre per year in total from any combination of seed, soil, or foliar appliactions.
	pyraclostrobin (Headline; Headline SC)	6 to 12 fl oz/acre	1.4 to 2.8 fl oz/acre	3	1	**DO NOT** exceed six foliar applications or 72 total ounces of product per acre per season. Do not apply more than one time before alternating with a fungicide with a different mode of action.
	sulfur (various)	See label	See label	See label	See label	See label. Rates vary depending on the formulation.

For Pumpkin, Winter Squash, and Summer Squash – See Cucurbits
Radish – See Root Vegetables
Scallion - See Onion, Green Shallot – See Onion, Dry

DISEASE CONTROL

Root Vegetables (Except Sugar Beet) Beet, Carrot, Parsnip, Radish, Turnip

Table 10-46. Disease Control Products for Root Vegetables (Except Sugar Beet) — Beet (red, garden or table), Carrot, Parsnip, Radish, Turnip – Harvested for roots only

Disease	Material	Rate of Material to Use		Minimum Days		Method, Schedule, and Remarks
		Formulation	Active Ingredient	Harvest	Reentry	
Alternaria blight, black crown rot	iprodione (Rovral) 4 F	1 to 2 pt/acre	0.5 to 1 lb/acre	0	1	**For use on carrot s only**. Make no more than 4 applications per season.
Alternaria leaf blight, Cercospora leaf spot	azoxystrobin (Quadris) 2.08 F	9.0 to 15.5 fl oz/acre	0.15 to 0.25 lb/acre	0	4 hr	No more than 1 application before alternating with a fungicide with a different mode of action. Make no more than 123 fluid ounces per acre per year.
	azoxystrobin + propiconazole (Quilt) 1.66SC	14 fl oz	0.18 lb/acre	14	0.5	**For use on carrot s only**. No more than 1 application before alternating with a non-Group 11 fungicide. Make no more than 55 fluid ounces per acre per year.
	boscalid (Endura) 70WG	4.5 oz/acre	0.2 lb/acre	0	0.5	**For use on carrot s only**. Not for *Cercospora*. Do not make more than 2 consecutive applications or more than 5 applications per season.
	chlorothalonil (various)	See label	See label	See label	See label	See label. Rates vary depending on the formulation.
	cyprodinil + fludioxonil (Switch) 62.5 WG	11 to 14 oz/acre	6.9 to 8.7 oz/acre	7	0.5	**Not for *Cercospora***. Apply when disease first appears, and continue on 7 to 10 day intervals if conditions remain favorable for disease development. Do not exceed 56 ounces of product per acre per year.
	fixed copper (various)	See label	See label	See label	See label	See label. Rates vary depending on the formulation.
	penthiopyrad (Fontelis) 1.67 SC	16 to 30 fl oz/acre	3.26 to 6.78 fl oz/acre	0	0.5	Make no more than 2 consecutive applications before alternating with a fungicide with a different mode of action. Apply no more than 61 fluid ounces/acre per year.
	pyraclostrobin (Cabrio) 20 WG	8 to 12 oz/acre	1.6 to 2.4 oz/acre	0	0.5	Alternate with a fungicide with a different mode of action.
	pyraclostrobin + boscalid (Pristine) 38 WG	8 to 10.5 oz/acre	3 to 4 oz/acre	0	0.5	**FOR USE ON CARROTS ONLY.** Make no more than 2 consecutive applications before alternating with a different mode of action. Use no more than 63 ounces or make no more than 6 applications per season.
	trifloxystrobin (Flint) 50 WDG	2 to 3 oz/acre *For radish:* 2 to 4 oz/acre	1 to 1.5 oz/acre *For radish:* 1 to 2 oz/acre	7	0.5	Make no more than 1 application before alternating with a fungicide with another mode of action.
	trifloxystrobin (Gem) 500 SC	1.9 to 2.9 fl oz/acre	0.81 to 1.24 fl oz/acre			Make no more than 4 applications per season.
Cercospora leaf spot or blight, powdery mildew	propiconazole (Tilt) 3.6 F	3 to 4 oz/acre	0.084 to 0.112 lb/acre	14	0.5	**FOR USE ON GARDEN BEETS AND CARROTS ONLY.** Use higher rate for carrots. Make no more than 2 consecutive applications before alternating with a fungicide with a different mode of action. Apply no more than 16 fluid ounces per acre per season.
	tebuconazole (Folicur) 3.6EC	4 to 7.2 fl oz/acre	0.113 to 0.203 lb/acre	7	0.5	**FOR USE ON TURNIP ONLY.** Repeat applications at 12 to 14 day intervals. Apply no more than 28 fluid ounces/acre/season.
Downy mildew	fixed copper (various)	See label	See label	See label	See label	See label. Rates vary depending on the formulation. Spray or dust at first appearance, 7 to 10 day intervals.
	Actigard 50WG	0.5 to 1 oz/acre	0.25 to 0.5 oz/acre	7	0.5	**FOR RADISH, RUTABAGA, AND TURNIP ONLY.** Make up to 4 applications at 7-day intervals.
Leaf spots	fixed copper (various)	See label	See label	See label	See label	See label. Rates vary depending on the formulation. Spray or dust at first appearance, 7- t o 10-day intervals.
Phytophthora basal stem rot	mefenoxam (Ridomil Gold) 4 SL	1 to 2 pt/trt acre	0.5 to 1 lb/trt acre	—	2	Apply as a preplant application to soil and incorporate into top 2 inches of soil. Broadcast or band.
	metalaxyl (MetaStar) 2E	4 to 8 pt/trt acre	0.5 to 1 lb/trt/acre	—	2	
Pythium root rot, root dieback, cavity spot	mefenoxam (Ridomil Gold) 4 SL (Ultra Flourish) 2 EC	1 to 2 pt/trt acre 2 to 4 pt/trt acre	0.5 to 1 lb/trt acre	—	2	Apply preplant incorporated into top 2 inches, as a soil spray at planting, or in drip irrigation at planting.
	metalaxyl (MetaStar) 2 E	4 to 8 pt/trt acre	0.9 to 1.8 pt/trt acre	—	2	Apply preplant incorporated into top 2 inches of soil or as a post-plant soil spray at planting followed by 0.5 to 1 inch of water.
	fenamidone (Reason) 500 SC	8.2 fl oz/acre	0.27 lb/acre	14	0.5	Make no more than 1 application before alternating with a mefenoxam-containing fungicide. Apply no more than 24.6 fluid ounces per growing season. Applied with sprayer or in sprinkler irrigation.
	cyazofamid (Ranman) 400 SC	6 fl oz/acre	0.156 lb/acre	14	0.5	**FOR USE ON CARROTS ONLY.** May be applied preplant incorporated, as a pre-emergent surface band, or in sprinkler irrigation. Applications can be repeated at 14- day intervals, but must alternate with a Pythium fungicide with a different mode of action.
	fluopicolide (Presidio) 4 F	3 to 4 fl oz/acre	0.09 to 0.125 lb/acre	7	0.5	Can be applied with a sprayer or in sprinkler irrigation. Regardless of method, must be applied in combination with a fungicide with a different mode of action and labeled for that method. No more than 2 consecutive applications before alternating with a Pythium fungicide with a different mode of action. Maximum of 12 fluid ounces per acre per year. For carrots only, may also be applied preplant incorporated.

DISEASE CONTROL

Table 10-46. Disease Control Products for Root Vegetables (Except Sugar Beet) — Beet (red, garden or table), Carrot, Parsnip, Radish, Turnip – Harvested for roots only

Disease	Material	Rate of Material to Use		Minimum Days		Method, Schedule, and Remarks
		Formulation	Active Ingredient	Harvest	Reentry	
Powdery mildew	azoxystrobin (Quadris) 2.08 F	9.0 to 15.5 fl oz/acre	0.15 to 0.25 lb/acre	0	4 hr	No more than 1 application before alternating with a fungicide with a different mode of action. Apply no more than 123 fluid ounces per acre per season.
	azoxystrobin + propiconazole (Quilt) 1.66SC	14 fl oz	0.18 lb/acre	14	0.5	**For use on carrot s only**. No more than 1 application before alternating with a non-Group11 fungicide. Apply no more than 55 fluid ounces per acre per year.
	boscalid (Endura) 70 WG	See label	0.2 lb/acre	0	0.5	**For use on carrot s only**. Make more than 2 consecutive applications before alternating with a fungicide with a different mode of action. Make no more than 5 applications per season.
	cyprodinil + fludioxonil (Switch) 62.5 WG	11 to 14 oz/acre	0.430 to 0.547 lb/acre	7	0.5	Make no more than 2 applications per crop. For radish, apply no more than 28 ounces per crop or 56 ounces per plot of land per year.
	fluxapyroxad + pyraclostrobin (Merivon) 4.18 SC	4 to 5.5 fl oz/acre	0.13 to 0.18 lb/acre	7	0.5	Do not make more than 2 consecutive applications before rotating to a labeled non-Group 7 or non-Group 11 fungicide. Make no more than 3 applications per season.
	pyraclostrobin (Cabrio) 20 EG	8 to 12 oz/acre	1.6 to 2.4 oz/acre	0	0.5	Alternate with a fungicide with a different mode of action.
	pyraclostrobin + boscalid (Pristine) 38 WG	8 to 10.5 oz/acre	3 to 4 oz/acre	0	0.5	**FOR USE ON CARROTS ONLY.** Make no more than 2 consecutive applications before alternating with a different mode of action. Use no more than 63 ounces or make no more than 6 applications per season.
	penthiopyrad (Fontelis) 1.67 SC	16 to 30 fl oz/acre	3.26 to 6.78 fl oz/acre	0	0.5	Make no more than 2 consecutive applications before alternating with a fungicide with a different mode of action. Apply no more than 61 fluid ounces per acre per year.
	sulfur (various)	See label	See label	See label	See label	See label. Rates vary depending on the formulation. Spray at first appearance. Avoid applying on days over 90 degrees F.
Rhizoctonia root canker	azoxystrobin (Quadris) 2.08 F	0.4 to 0.8 fl oz/1000 row ft	0.0065 to 0.013 lb/1000 row ft	0	4 hr	Make one application, applied either in-furrow at planting, in a 7-inch band over the row prior to or shortly after planting, or in drip irrigation.
Rust	penthiopyrad (Fontelis) 1.67 SC	16 to 30 fl oz/acre	3.26 to 6.78 fl oz/acre	0	0.5	Make no more than 2 sequential applications before alternating with a fungicide with a different mode of action. Apply no more than 61 fluid ounces per acre per year.
	sulfur (various)	See label	See label	See label	See label	See label. Rates vary depending on the formulation.
Sclerotinia, Botrytis	boscalid (Endura) 70 WG	7.8 oz	0.34 lb/acre	0	0.5	**FOR USE ON CARROTS ONLY.** No more than 2 applications before alternating with a fungicide with a different mode of action. Limit of 3 applications per season.
	penthiopyrad (Fontelis) 1.67 SC	16 to 30 fl oz/acre	3.26 to 6.78 fl oz/acre	0	0.5	Make no more than 2 sequential applications before alternating with a fungicide with a different mode of action. Apply no more than 61 fluid ounces per acre per year.
Sclerotinia, Botrytis	thiabendazole (Mertect 340-F) 4.1 F	41 fl oz/100 gal	1.31 lb/100 gal	—	0.5	Dip harvested roots 5 to 10 seconds. Do not rinse.
Southern blight	dichloropropene (Telone) C-17 C-35	10.8 to 17.1 gal/acre 13 to 20.5 gal/acre	107 to 169 lb/acre 139 to 220 lb/acre	—	5	Fumigate soil in-the-row 3 to 6 weeks before seeding. Rate is based on soil type; see label for in-row rates.
	pyraclostrobin + boscalid (Pristine) 38 WG	8 to 10.5 oz/acre	3 to 4 oz/acre	0	0.5	**FOR USE ON CARROTS ONLY.** Suppression only. Make no more than 6 applications per season.
	azoxystrobin (Quadris) 2.08 F	0.4 to 0.8 fl oz/1000 row ft	0.0065 to 0.013 lb/1000 row ft	0	4 hr	Make one application, applied either in-furrow at planting, in a 7-inch band over the row prior to or shortly after planting, or in drip irrigation.
	penthiopyrad (Fontelis) 1.67 SC	16 to 30 fl oz/acre	3.26 to 6.78 fl oz/acre	0	0.5	Make no more than 2 sequential applications before alternating with a fungicide with a different mode of action. Apply no more than 61 fluid ounces per acre per year.
White rust	azoxystrobin (Quadris) 2.08 F	6.0 to 15.5 fl oz/acre	0.1 to 0.25 lb/acre	0	4 hr	No more than 1 application before alternating with a fungicide with a different mode of action. Apply no more than 123 fl oz per acre per season.
	pyraclostrobin (Cabrio) 20 EG	8 to 16 oz/acre	1.6 to 3.2 oz/acre	0	0.5	Alternate with a fungicide with a different mode of action. Apply no more than 48 ounces/acre/season.
	mefenoxam + copper hydroxide (Ridomil Gold/ Copper)	2 lb/acre	—	7	1	Spray leaves. Use with preplant Ridomil 2E soil applications. Make 2 to 4 applications if needed on 14-day intervals.

Table 10-47. Importance of Alternative Management Practices for Disease Control in Carrot

Scale: E, excellent; G, good; F, fair; P, poor; NC, no control; ND, no data.

Strategy	Alternaria blight	Cercospora blight	Powdery mildew	Pythium cavity spot	Pythium damping off	Southern blight	Rhizoctonia cavity spot	Sclerotinia postharvest	Botrytis postharvest	Bacterial leaf blight	Root-knot nematode
Avoid field operations when leaves are wet	P	P	NC	NC	NC	NC	NC	NC	NC	F	NC
Avoid overhead irrigation	F	F	NC	NC	NC	NC	NC	F	NC	F	NC
Change planting date	P	P	NC	F	F	F	NC	NC	NC	NC	F
Cover cropping with antagonist	NC	NC	NC	NC	NC	NC	NC	NC	NC	NC	F
Crop rotation	F	F	NC	P	P	P	P	P	NC	F	P
Deep plowing	G	G	P	NC	NC	F	F	F	P	G	NC
Destroy crop residue	E	E	P	NC	NC	NC	P	NC	P	E	P
Encourage air movement	F	F	NC	NC	NC	NC	NC	F	NC	NC	NC
Plant in well-drained soil	NC	NC	NC	G	G	P	F	F	NC	NC	NC
Plant on raised beds	NC	NC	NC	F	F	NC	F	P	NC	NC	NC
Postharvest temperature control	NC	NC	NC	NC	NC	NC	NC	E	E	NC	NC
Reduce mechanical injury	NC	NC	NC	NC	NC	NC	NC	F	G	NC	NC
Destroy volunteer carrots	F	F	P	NC	NC	NC	NC	NC	NC	NC	NC
Pathogen-free planting material	E	E	NC	NC	NC	NC	NC	NC	NC	E	NC
Resistant cultivars	G	G	F	NC	NC	NC	NC	NC	NC	NC	NC

Spinach

Table 10-48. Disease Control Products for Spinach

Disease	Material	Rate of Material to Use — Formulation	Rate of Material to Use — Active Ingredients	Minimum Days — Harvest	Minimum Days — Reentry	Method, Schedule, and Remarks
Damping off (*Pythium*)	mefenoxam (Ridomil Gold) 4 SL (Ultra Flourish) 2 EC	1 to 2 pt/trt acre 2 to 4 pt/trt acre	0.5 to 1 lb/trt acre	21	2	Broadcast or banded over the row as a soil spray or preplant incorporation into the top 2 inches of soil.
	metalaxyl (MetaStar) 2 E	4 to 8 pt/trt acre	1 to 2 lb/trt acre	21	2	Broadcast or banded over the row as a soil spray or preplant incorporation into the top 2 inches of soil.
Seedling blight (*Rhizoctonia*), damping off (*Pythium*)	azoxystrobin + mefenoxam (Uniform)	0.34 fl oz/ 1000 ft of row	0.13 fl oz/ 1000 ft of row	—	0	Apply as an in furrow spray in 5 gallons of water per acre prior to covering seed. Make only 1 application per season.
Downy mildew	cymoxanil (Curzate) 60DF	5 oz /acre	0.1875 lb/acre	1	0.5	Apply with a protectant fungicide. Apply no more than 30 ounces per acre in a 12-month period.
	mandipropamid (Revus) 2.08 F	8 fl oz/acre	0.13 lb/acre	1	4 hr	Make no more than 2 consecutive applications before alternating with a fungicide with a different mode of action. Apply no more than 32 fluid ounces/acre/season.
	pyraclostrobin (Cabrio) 20 EG	12 to 16 oz/acre	0.15 to 0.2 lb/acre	0	0.5	Make no more than 2 consecutive applications before alternating with a fungicide with a different mode of action. Apply no more than 64 ounces per acre per growing season.
Downy mildew, white rust	acibenzolar-*S*-methyl (Actigard) 50 WG	0.5 to 0.75 oz/acre	0.25 to 0.37 oz/acre	7	0.5	Do not apply to young seedlings or plants stressed due to drought, excessive moisture, cold weather, or herbicide injury.
	famoxadone + cymoxanil (Tanos) DF	8 to 10 oz/acre	—	1	0.5	Must be tank-mixed with a contact downy mildew fungicide with a different mode of action. Make no more than 1 application before alternating with a fungicide with a different mode of action. Apply no more than 84 ounces/acre per cropping season.
	fluopicolide (Presidio) 4F	3 to 4 fl oz/acre	0.09 to 0.125 lb/acre	2	0.5	Tank mix with another downy mildew fungicide with a different mode of action.

DISEASE CONTROL

Table 10-48. Disease Control Products for Spinach

Disease	Material	Rate of Material to Use		Minimum Days		Method, Schedule, and Remarks
		Formulation	Active Ingredients	Harvest	Reentry	
Downy mildew, white rust (continued)	fluxapyroxad + pyraclostrobin (Merivon) 4.18 SC	4 to 11 fl oz/acre	0.13 to 0.36 lb/acre	1	0.5	**Do not tank-mix Merivon with any pesticides, adjuvants, fertilizers, nutrients, or any other additives.** Do not make more than 2 consecutive applications before rotating to a labeled non-Group 7 or non-Group 11 fungicide. Make no more than 3 applications per season.
	fixed copper (various)	See label	See label	See label	See label	See label. Rates vary depending on the formulation. Some formulations of copper may cause leaf flecking.
	fosetly-Al (Aliette) 80 WDG	2 to 5 lb/acre	1.4 to 4 lb/acre	3	0.5	Do not mix with surfactants, foliar fertilizers, or products containing copper.
	mefenoxam (Ridomil) 4 SL	0.25 pt/acre	0.125 lb/acre	21	2	Apply preplant as above for damping off. Shank in 21 days after planting or after first cutting. Another application may be shanked in after the next cutting. A total of 2 shank applications may be made on a 21-day interval. Do not use post-emergence if a preplant application was not made.
	mefenoxam + copper hydroxide (Ridomil Gold/Copper)	2.5 lb/acre	—	21	2	Spray to foliage. Use with preplant Ridomil Gold soil application.
	metalaxyl (MetaStar) 2 E	1 pt/trt acre	0.125 lb/acre	21	2	Shank in 21 days after planting. Apply no more than 2 shanked applications on 21-day intervals.
White rust	cyazofamid (Ranman) 400 SC	2.1 to 2.75 fl oz/ acre	0.054 to 0.071 lb/acre	0	0.5	No more than 5 applications per crop. No more than 3 consecutive applications followed by at least three applications of a fungicide with a different mode of action. Do not apply more than 13.75 fluid ounces per acre per crop per growing season.
	fenamidone (Reason) 500 SC	5.5 to 8.2 fl oz/acre	0.18 to 0.27 lb/acre	2	0.5	Make no more than 1 application before alternating with a fungicide with a different mode of action. Apply no more than 24.6 fluid ounces/acre per growing season.
	pyraclostrobin (Cabrio) 20 EG	8 to 12 oz/acre	0.1 to 0.15 lb/acre	0	0.5	Make no more than 2 sequential applications before alternating with a fungicide with a different mode of action. Apply no more than 64 ounces per acre per growing season.
Downy mildew, leaf spot	pyraclostrobin (Cabrio) 20 EG	12 to 16 oz/acre	0.15 to 0.2 lb/acre	0	0.5	Make no more than 2 sequential applications before alternating with a fungicide with a different mode of action. Apply no more than 64 ounces per acre per growing season.
Downy mildew, leaf spot, white	fixed copper (various)	See label	See label	See label	See label	See label. Rates vary depending on the formulation. Some formulations of copper may cause flecking on the leaves.
Leaf spot, white rust	azoxystrobin (Quadris) 2.08 F	6 to 15.5 fl oz/acre	0.10 to 0.25 lb/acre	0	4 hr	Make no more than 2 sequential applications before alternating with a fungicide with a different mode of action. Apply no more than 92.3 fluid ounces per acre per season.
Leaf spots	penthiopyrad (Fontelis) SC	14 to 24 fl oz/acre	2.86 to 4.90 fl oz/acre	3	0.5	Make no more than 2 sequential applications before alternating with a fungicide with a different mode of action. Apply no more than 72 fluid ounces per acre per year.

For Winter Squash and Summer Squash – See Cucurbits

Sweetpotato

Table 10-49. Disease Control Products for Sweetpotato

Disease	Material	Rate of Material to Use		Minimum Days		Method, Schedule, and Remarks
		Formulation	Active Ingredient	Harvest	Reentry	
Black rot (*Ceratocystis fimbriata*), scurf (*Monilochaetes infuscans*). and foot rot	thiabendazole (Mertect 340 F)	107 fl oz/100 gal	4 lb/100 gal	—	—	**Dip seed roots** 1 to 2 minutes and plant immediately; not for *Sclerotial* blight.
Black rot (*Ceratocystis fimbriata*)	thiabendazole (Mertect 340 F)	3	0.42 fl oz per 2,000 lb of roots or 0.42 fl oz/gal	0.5	0.5	**SECTION 18 LABEL ONLY IN NORTH CAROLINA (until December 31, 2016)** Postharvest treatment of sweetpotato for control of black rot. Limit to one application during packing. Mist washed roots on a conveyor line, with tumbling action, before packing with 0.42 fl oz of Mertect to each 2,000 lb of roots in sufficient water for complete coverage. Or dip the roots for 20 seconds in 0.42 fl oz of Mertect per gal of water. Ensure roots are dry before packing.
Circular spot, Sclerotial blight, *Rhizoctonia* stem canker, *Pythium* root rot	azoxystrobin (Quadris) 2.08 F	0.4 to 0.8 fl oz/1,000 row feet	0.006 to 0.013 lb	—	4 hr	Make in-furrow or banded applications shortly after transplanting.

548

Table 10-49. Disease Control Products for Sweetpotato

Disease	Material	Rate of Material to Use		Minimum Days		Method, Schedule, and Remarks
		Formulation	Active Ingredient	Harvest	Reentry	
Seed-borne and soilborne fungi that cause decay, damping off or seedling blight	azoxystrobin (Dynasty) 0.83 F	0.19 to 0.38 fl oz per 100 lb of propagating roots	0.009 to 0.036 fl oz per 100 lb of propagating roots	—	4 hr	Apply uniformly to seed roots as a water-based slurry.
	fludioxonil (Maxim 4 FS)	0.08 to 0.16 fl oz per 100 lb of propagating roots	0.032 to 0.064 fl oz per 100 lb of propagating roots	—	0.5	Apply uniformly to seed roots as a water-based slurry.
Damping off (*Pythium*)	cyazofamid (Ranman 400 SC)	6/1 fl oz/acre		7	0.5	Apply at planting. Refer to label for details.
	mefenoxam (Ridomil Gold) 4 SL	1 to 2 pt/treated acre	0.5 to 1 lb/treated acre	—	2	Incorporate in soil. See label for row rate.
	metalaxyl (MetaStar) 2 E	4 to 8 pt/treated acre	0.5 to 1 lb/acre	7	2	Preplant incorporated or soil surface spray.
Foliar diseases (*Alternaria* and powdery mildew)	azoxystrobin + difenoconazole (Quadris Top)	8 to 14 fl oz/ acre		14	0.5	Begin foliar applications prior to disease and continue on a 7- to 14-day interval.
	cyprodinil + fludioxonil (Switch 62.5WG)	11 to 14 oz/acre		7	0.5	Begin foliar applications prior to disease and continue on a 7- to 10-day interval.
	fenamidone (Reason 500 SC)	5.5 to 8.2 oz/acre	0.178 to 0.267 fl oz/acre	14	0.5	Begin foliar applications prior to disease and continue on a 5- to 10-day interval.
	pyrimethanil (Scala SC)	7 fl oz/acre	0.27 lb/acre	7	0.5	Begin foliar applications prior to disease and continue on a 7- to 14-day interval.
Foliar (*Alternaria*) and soilborne (*Rhizoctonia, Sclerotinia*) fungal diseases	azoxystrobin (Aframe, various)	6 to 15.5 fl oz/acre	0.1 to 0.25 lb/acre	0	4 hr	Limit to 123 fl oz per acre per season. For soilborne disease control refer to label. Begin foliar applications prior to disease and continue on a 5- to 7-day interval.
	boscalid (Endura)	2.5 to 10 oz/acre		10	0.5	For soilborne disease control refer to label. Begin foliar applications prior to disease and continue on a 7- to 10-day interval.
	fluopyram + pyrimethanil (Luna Tranquility)	7 + 9	11.2 fl oz/acre	7	0.5	Limit to 54.7 fl oz per acre per year. Apply at the critical timings for disease control Continue as needed on a 7- to 14-day interval. When disease pressure is severe, use the shorter intervals.
	fluoxastrobin (Aftershock)	2 to 3.8 fl oz/acre	0.06 to 0.12 lb/acre	7	0.5	Limit to 22.8 fl oz per acre per year. For soilborne disease control refer to label. Begin foliar applications prior to disease and continue on a 7- to 10-day interval.
	metconazole (Quash)	2.5 to 4 oz/acre	0.078 to 0.125 lb/acre	1	0.5	Begin foliar applications prior to disease and continue on a 7- to 10-day interval.
	penthiopyrad (Vertisan)	0.7 to 24 fl oz/acre		7	0.5	For soilborne disease control refer to label. Begin foliar applications prior to disease and continue on a 7- to 14-day interval.
Mottle necrosis (*Pythium* postharvest)	potassium phosphite (Alude)	1 ¼ quarts/acre		0	4 hr	Foliar spray at 5 to 14 day intervals depending on disease incidence.
Postharvest sanitation	calcium hypochlorite 65%	3 to 10 oz/100 gal	150 to 500 ppm	—	—	Dip or spray 2 to 5 minutes. Monitor chlorine concentration and add chlorine or change solution as needed.
Postharvest Rhizopus soft rot	dicloran (Botran) 75 W	1 lb/100 gal	0.75 lb/100 gal	—	—	Spray or dip. Dip for 5 to 10 seconds in well-agitated suspension. Add 1/2 pound Botran to 100 gallons of treating suspension after 500 bushels treated. Do not rinse.
	fludioxonil (Scholar 1.9 SC)	16 to 32 fl oz/100 gal	3.3 to 6.5 fl oz/100 gal	—	—	Dip for approximately 30 seconds in well-agitated solution and allow sweetpotatoes to drain. Add 8 fl oz to 100 gals after 500 bushels are treated. OR mix 16 fluid ounces in 7 to 25 gallons of water, wax/emulsion, or aqueous dilution of wax/ oil emulsion. Can also be used to disinfest tanks, refer to label.
Scurf (*Monilochaetes infuscans*) and Sclerotial blight (*Sclerotium rolfsii*)	dicloran (Botran) 75 W	1 lb/100 gal	0.75 lb/100 gal	—	—	Seed dip: Dip seed sweetpotatoes 10 to 15 seconds in a well-agitated fungicide suspension. Drain sweetpotatoes and bed promptly. Prepare fresh fungicide suspension daily. Plant bed spray: Spray or sprinkle over bedded sweetpotatoes before covering them with soil.
White rust	azoxystrobin (Quadris) 2.08 F	6.2 to 15.4 fl oz/ acre	0.1 to 0.25 fl oz/acre	7	4 hr	Make no more than two sequential applications before alternating with fungicides that have a different mode of action. Apply no more than 2.88 quarts per crop per acre per season.
	fenamidone (Reason 500 SC)	5.5 to 8.2 fl oz/acre	0.178 to 0.267 lb/ acre	14	0.5	Begin applications when conditions favor disease development, and continue on 5 to 10 day interval. Do not apply more than 16.4 fluid ounces per growing season. Alternate with a fungicide from different resistance management group.
	pyraclostrobin (Cabrio) 20 WG	8 to 16 oz/acre	1.6 to 3.2 oz/acre	0	0.5	Do not apply more than 48 ounces per acre per season. Alternate with a fungicide with a different mode of action after each use.

DISEASE CONTROL

Table 10-50. Efficacy of Products for Disease Control in Sweetpotato

Scale: E, excellent; G, good; F, fair; P, poor; NC, no control; ND, no data.

Product	Nematicide (N) or Fungicide (F)	Bacterial stem and root rot (*E. chrysanthemi*)	Black rot (*C. fimbriata*)	Foot rot (*P. destruens*)	Fusarium root rot and stem canker (*F. solani*)	Fusarium surface rot (*F. oxysporum*)	Fusarium wilt (*F. oxysporum* f. sp. *batatas*)	Java black rot (*D. gossypina*)	Root-knot and Reniform nematodes (*Meloidogyne & Rotylenchus* spp.)	Rhizopus soft rot (*R. stolonifer*)	Sclerotial blight/Circular spot (*S. rolfsii*)	Scurf (*M. infuscans*)	Soil rot/Pox (*S. ipomoea*)	Sweetpotato feathery mottle virus
aldicarb (Temik)	N	ND	ND	ND	ND	ND	ND	ND	G	ND	ND	ND	ND	ND
Pseudomonas syringae (Bio-Save)	F	ND	ND	ND	ND	ND	ND	ND	ND	P	ND	ND	ND	ND
chlorine	F	F	F	P	ND	ND	ND	P	ND	F	ND	P	NC	ND
chloropicrin	N, F	ND	P	P	F	P	F	F	F	ND	F	ND	F	ND
dicloran (Botran 75W)	F	NC	P	ND	ND	F	ND	P	ND	F	P	F	NC	ND
ethoprop (Mocap)	N	ND	ND	NC	ND	ND	ND	ND	P	ND	ND	ND	ND	ND
fludioxonil (Scholar)	F	NC	F	ND	ND	ND	ND	ND	NC	F	NC	ND	NC	ND
metam sodium (Vapam)	N	ND	P	P	F	P	F	ND	F	ND	ND	ND	ND	ND
oxamyl (Vydate)	F	ND	ND	NC	ND	ND	ND	ND	F	ND	ND	ND	ND	ND
dichloropropene (Telone II)	N	ND	ND	NC	P	P	P	ND	G	ND	ND	ND	ND	ND
thiabendazole (Mertect 340-F)	F	NC	E	F	P	P	ND	F	ND	E	F	P	NC	ND

Table 10-51. Importance of Alternative Management Practices for Disease Control in Sweetpotato

Scale: E, excellent; G, good; F, fair; P, poor; NC, no control; ND, no data.

Strategy	Bacterial stem and root rot (*E. chrysanthemi*)	Black rot (*C. fimbriata*)	Foot rot (*P. destruens*)	Fusarium root rot & stem canker (*F. solani*)	Fusarium surface rot (*F. oxysporum*)	Fusarium wilt (*F. oxysporum* f. sp. *batatas*)	Java black rot (*D. gossypina*)	Root-knot & Reniform nematodes (*Meloidogyne and Rotylenchus*)	Rhizopus soft rot (*R. stolonifer*)	Sclerotial blight/Circular spot (*S. rolfsii*)	Scurf (*M. infuscans*)	Soil rot/Pox (*S. ipomoea*)	Sweepotato Feathery Mottle Virus
Crop rotation (3 to 4 years)	P	F	P	G	P	F	F	F	NC	F	P	F	NC
Disease-free planting stock	G	E	G	E	P	E	E	F	NC	P	E	P	G
Resistant cultivars	F	NC	P	F	P	G	F	F[s]	F	F	P	G	F
Careful handling to reduce mechanical injury	P	F	NC	F	E	NC	NC	NC	E	NC	NC	NC	NC
Cutting plants (in beds) above soil line	P	E	G	F	NC	F	F	NC	NC	NC	G	P	NC
Soil sample for nematode analysis	NC	NC	NC	NC	NC	NC	NC	G	NC	NC	NC	NC	NC
Sanitation (equipment, fields, storage houses)	F	E	F	P	F	NC	F	NC	E	NC	P	NC	NC
Manage insects that transmit pathogens	NC	E	NC	NC	NC	NC	NC	NC	NC	NC	NC	NC	NC
Sulfur added to soil to reduce pH	NC	NC	NC	NC	NC	NC	NC	NC	NC	NC	NC	F	NC
Prompt curing and proper storage conditions	F	F	NC	F	E	NC	F	NC	E	NC	NC	NC	NC
Site selection (drainage)	P	NC	NC	F	F	F	F	F	F	P	NC	P	NC
Manage insects that cause feeding injuries to roots	P	E	NC	NC	NC	NC	P	NC	P	NC	NC	NC	NC
Avoid harvesting when soils are wet	F	F	NC	NC	E	NC	F	NC	F	NC	NC	NC	NC

Note: Resistant cultivars for root knot nematode are susceptible to reniform nematode

Sweetpotato Storage House Sanitation – See Sanitation

Tomatillo

Table 10-52. Disease Control Products for Tomatillo

Disease	Material	Rate of Material to Use		Minimum Dasy		Method, Schedule, and Remarks
		Formulation	Active Ingredients	Harvest	Reentry	
Early Blight	azoxystrobin (Quadris, Equation)	5 to 6.2 fl oz/acre	0.08 to 0.1 fl oz/acre	0	4 hr	Limit of 37 fl ounces per crop per acre per season. Make no more than **one** application before alternating with fungicides that have a different mode of action.
	azoxystrobin + difenoconazole (Quadris Top)	8 fl oz/acre	2.4 fl oz/acre	0	0.5	Limit of 47 fluid ounces per acre per season. Do not apply until 21 days after transplanting or 35 days after seeding.
	boscald (Endura)	2.5 to 3.5 oz/acre	1.75 to 2.45 oz/acre	0	0.5	Limit of 21 ounces per acre per season. Make no more than two sequential applications before alternating with fungicides that have a different mode of action.
	cyprodinil + difenoconazole (Inspire Super)	16 to 20 fl oz/acre	0.60 to 0.75 lb/acre	0	0.5	Limit of 47 fluid ounces per acre per season.
	cyprodinil + fludioxonil (Switch) 62.5 WG	11 to 14 oz/acre	0.43 to 0.55 lb/acre	0	0.5	Limit of 56 ounces per acre per year. After two applications, rotate to another fungicide with a different mode of action for two applications.
	difenoconozole + mandipropamid (Revus Top)	5.5 to 7 fl oz/acre	2.41 to 3.1 fl oz/acre	1	0.5	Limit of 28 fluid ounces per acre per season. Make no more than two consecutive applications per season before alternating with fungicides that have a different mode of action.
	fenamidone (Reason) 500SC	5.5 to 8.2 fl oz/acre	0.178 to 0.267 lb/acre	14	0.5	Limit of 24.6 fluid ounces per growing season. Make no more than **one** applications before rotating to another effective fungicide with a different mode of action.
	fluoxastrobin (Aftershock, Evito 480SC)	2.0 to 5.7 fl oz/acre	0.81 to 2.3 fl oz/A	3	0.5	Limit of 22.8 fluid ounces per acre per season. Make no more than **one** application before alternating with fungicides that have a different mode of action. **NOTE: Do not overhead irrigate for 24 hours following a spray application.**
	penthiopyrad (Fontelis)	10 to 24 fl oz/acre	0.13 to 0.31 lb/acre	0	0.5	Do not exceed 72 fluid ounces of product per year. Make no more than two sequential applications per season before alternating with fungicides that have a different mode of action.
	polyoxin D (Ph-D)	6.2 oz/acre	0.7 oz/acre	0	4 hr	Limit of five applications per season. Make no more than **one** sequential applications before alternating with fungicides that have a different mode of action.
	pyraclostrobin (Cabrio) 20% EG	8 to 16 oz/acre	1.6 to 3.2 oz/acre	0	4 hr	Limit of 96 oz per acre per season. Make no more than **one** sequential application before alternating with fungicides that have a different mode of action.
	Trifloxystrobin (Flint)	2 to 3 oz/acre	1 to 3.5 oz/acre	3	0.5	Limit of 16 ounces per acre per year. Make no more than **one** sequential applications before alternating with fungicides that have a different mode of action.
Powdery mildew	azoxystrobin (Quadris) Flowable	5 to 6.2 fl oz/acre	0.08 to 0.1 fl oz/acre	0	4 hr	Limit of 37 fluid ounces per crop per acre per season. Make no more than **one** application before alternating with fungicides that have a different mode of action.
	azoxystrobin + difenoconazole (Quadris Top)	8 fl oz/acre	2.4 fl oz/acre	0	0.5	Limit of 47 fluid ounces per acre per season. Do not apply until 21 days after transplanting or 35 days after seeding.
	chlorothalonil (various)	See label	See label	See label	See label	See label. Rates vary depending on the formulation.
	chlorothalonil + cymoxanil (Ariston)	2 to 2.44 pt/acre	0.84 to 1.02 pt/acre	3	0.5	Limit of 18.1 pints per acre per year.
	cyprodinil + difenoconazole (Inspire Super)	16 to 20 fl oz/acre	0.60 to 0.75 lb/acre	0	0.5	Limit of 47 fluid ounces per acre per season.
	cyprodinil + fludioxonil (Switch) 62.5 WG	11 to 14 oz/acre	0.43 to 0.55 lb/acre	0	0.5	Limit 56 ounces per acre per year. After two applications, rotate to another fungicide with a different mode of action for two applications.
	penthiopyrad (Fontelis)	10 to 24 fl oz/acre	0.13 to 0.31 lb/acre	0	0.5	Do not exceed 72 fluid ounces of product per year. Make no more than two consecutive applications per season before alternating with fungicides that have a different mode of action.
	polyoxin D (Ph-D)	6.2 oz/acre	0.7 oz/acre	0	4 hr	Limit five applications per season. Make no more than **one** application before alternating with fungicides that have a different mode of action.
	pyraclostrobin (Cabrio) 20% EG	8 to 16 oz/acre	1.6 to 3.2 oz/acre	0	4 hr	Limit of 96 ounces per acre per season. Make no more than **one** sequential application before alternating with fungicides that have a different mode of action.
	mandipropamid + difenoconazole (Revus Top)	5.5 to 7 fl oz/acre	0.089 to 0.114 lb/acre	0	0.5	Limit of 28 fluid ounces per acre per season. Make no more than two consecutive applications before alternating with fungicides that have a different mode of action.

DISEASE CONTROL

Tomato
I. Meadows, Entomology and Plant Pathology

Table 10-53. Disease Control Products for Tomato

Disease	Material	Rate of Material to Use		Minimum Days		Method, Schedule, and Remarks
		Formulation	Acive Ingredients	Harvest	Reentry	
Tomato (grown for transplant and retail sale to consumers)						
Late blight	mandipropamid (Micora)	5.5 to 8.0 fl oz/acre	0.65 to 0.9 fl oz/acre	NA	4 hr	Apply no more than 2 applications containing Micora per crop.
Tomato (transplants) Treating seed to eliminate plant pathogens on or within the seed is recommended. For a list of seed treatments that are compatible with raw (naked) seed see seed treatment table.						
Bacterial canker, Bacterial spot, Bacterial speck	sodium hypochlorite (Clorox 5.25%)	1 qt + 4 qt water	same	NA	0	Wash seed for 40 minutes in solution with continuous agitation; air dry promptly. Use 1 gallon of solution per 1 pound seed.
	streptomycin sulfate (Agri-Mycin 17 WP) (Streptrol 21.3%)	1 lb/100 gal	--	NA	0	Begin application at first true leaf stage, repeat weekly until transplanting. For plant bed use only.
Bacterial spot, Bacterial speck	bacteriophage (AgriPhage)	3 to 8 oz/9,600 sq ft	--	NA	0	Works best if applied during or after last watering of the day.
	fixed copper (various)	See label	See label	See label	See label	See label. Rates vary depending on the formulation. Begin application at first true leaf stage, repeat at 3 to day intervals until transplanting. Alternating with streptomycin sulfate is recommended.
	mancozeb (various)	See label	See label	See label	See label	See label. Rates vary depending on the formulation. **NOTE: Use a full rate of fixed copper in combination with mancozeb.** Mancozeb alone does not control bacteria.
	streptomycin sulfate (Agri-Mycin 17, Ag Streptomycin, Bac-Master, FireWall)	1 lb/100 gal	0.17 lb/100 gal	NA	0	Begin application at first true leaf stage, repeat weekly until transplanting. For plant bed use only.
Botrytis (gray mold), Botrytis stem canker	chlorothalonil + potassium phosphite (Catamaran)	4.5 to 5.5 pts/acre	2.5 to 3.1 pt/acre	NA	0	Limit of 50 pints/acre/season. NOTE: Phytotoxicity may occur on greenhouse transplants.
	dicloran (Botran 75W)	1 lb/100 gal	0.75 lb/100 gal	NA	0	Limit of 5.3 pounds product per acre per year. Do not make more than 4 applications per crop. Apply as a foliar spray to seedlings.
Early blight, Late blight	chlorothalonil + potassium phosphite (Catamaran)	4.5 to 5.5 pts/acre	2.5 to 3.1 pt/acre	NA	0	Limit of 50 pints/acre/season. NOTE: Phytotoxicity may occur on greenhouse transplants.
	mancozeb (various)	See label	See label	See label	See label	See label. Rates vary depending on the formulation. **NOTE: Use a full rate of fixed copper in combination with mancozeb if bacteria control is also required.**
Pythium damping off	cyazofamid (Ranman)	3.0 fl oz/100 gallons water	0.078 lb	NA	0.5	Make a single **DRENCH** application to seedling tray at the time of planting or at any time thereafter up until one week before transplanting.
	propamocarb (Previcur Flex)	1.5 pt/acre	1 pt/acre	5	0.5	Limit of 7.5 pints/acre/season. Do not apply more than once before alternating with fungicides that have a different mode of action.
Southern Blight	penthiopyrad (Fontelis 1.67 SC)	1 to 1.6 fl oz/1000 row ft	0.2 to 0.3 fl oz/1000 row ft	NA	0.5	Apply as a soil drench to seedling tray or at time of transplant.
Tomato (field)						
Anthracnose	azoxystrobin (various)	5 to 6.2 fl oz/acre	0.08 to 0.1 lb/acre	0	4 hr	Limit of 37 fluid ounces/acre/season. Do not make more than one application before alternating to a fungicide with a different mode of action. **NOTE: Under high temperatures Satori in combination with some additives or adjuvants may cause crop injury.**
	chlorothalonil (various)	See label	See label	See label	See label	See label. Rates vary depending on the formulation.
	azoxystrobin + benzovindiflupyr (Mural)	0.6 oz/5,000 ft^2	--	0	0.5	Make no more than 2 applications before switching to a different mode of action. Mixing with some adjuvants may cause phytotoxicity; see label. Limits of a.i.'s apply; see label.
	azoxystrobin + chlorothalonil (Quadris Opti)	1.6 pt/acre	--	0	4 hr	Must alternate with a non-FRAC code 11; use of an adjuvant may cause phytotoxicity.
	fixed copper (various)	See label	See label	See label	See label	See label. Rates vary depending on the formulation.
	difenoconazole + benzovindiflupyr (Aprovia Top)	10.5 to 13.5 fl oz/acre	--	0	0.5	Do not make more than 2 applications before switching to a non-Group 7 fungicide. See label for application invervals and limits per season.
	difenoconazole + cyprodinil (Inspire Super)	16 to 20 fl oz	--	0	0.5	Limit of 80 fluid ounces/acre/season. Do not make more than two consecutive applications before alternating with fungicides that have a different mode of action.

Table 10-53. Disease Control Products for Tomato

Disease	Material	Rate of Material to Use		Minimum Days		Method, Schedule, and Remarks
		Formulation	Acive Ingredients	Harvest	Reentry	
Tomato (field) (continued)						
Anthracnose (continued)	fluopyram + trifloxystrobin (Luna Sensation)	7.6 fl oz/acre	--	3	0.5	Disease suppression ONLY. Do not exceed 5 applications or 27.1 fl oz product/acre/season. Do not make more than 2 applications without switching to a different mode of action.
	flutriafol (Rhyme)	5 to 7 fl oz/acre	--	0	0.5	Do not exceed more than 4 applications or 28 fl oz product/acre/season.
	fluxapyroxad + pyraclostrobin (Priaxor)	4 to 8 fl oz/acre	1.7 to 3.4 fl oz/acre	0	0.5	Limit of 24 fluid ounces/acre/season. Do not make more than two consecutive applications before alternating with fungicides that have a different mode of action.
	famoxadone + cymoxanil (Tanos)	8 oz/acre	--	3	0.5	Limit of 72 fluid ounces/acre/season (12 month cycle). Do not make more than one application before alternating to a fungicide with a different mode of action. **NOTE: Must be tanked mixed with a contact fungicide that has a different mode of action.**
	cymoxanil + chlorothalonil (Ariston)	1.9 to 3 pt/acre	--	3	0.5	Check copper labels for specific precautions and limitations for mixing with this product.
	mancozeb (various)	See label	See label	See label	See label	See label. Rates vary depending on the formulation.
	mandipropamid + difenoconazole (Revus Top)	5.5 to 7 oz/acre	--	1	0.5	Limit of 28 fl oz product/acre/season. Do not make more than two consecutive applications before alternating with fungicides with different mode of action.
	penthiopyrad (Fontelis 1.67 SC)	24 fl oz/acre	4.8 fl oz/acre	0	0.5	Disease suppression only. Limit of 72 fluid ounces/acre/season. Do not make more than two consecutive applications before alternating with fungicides that have a different mode of action.
	polyoxin D zinc salt (Ph-D) (Oso 5% SC)	6.2 oz/acre 3.75 to 13.0 fl oz/acre	0.7 oz/acre	0	4 hr	Alternate with non-FRAC code 19 fungicide.
	pyraclostrobin (Cabrio)	8 to 12 oz/acre	1.6 to 3.2 oz/acre	0	0.5	Limit of 96 fluid ounces/acre/season. Do not make more than two applications before alternating to a fungicide with a different mode of action.
	trifloxystrobin (Flint 50WDG) (Gem 500 SC)	2 to 4 oz/acre 3 to3.8 oz/acre	1.5 to 2 oz	3	0.5	**DISEASE SUPPRESSION ONLY.** Limit of 16 fluid ounces/acre/season. Do not make more than one application before alternating to a fungicide with a different mode of action.
	zinc dimethyldithiocarbamate (Ziram 76 DF)	3 to 4 lb/acre	2.3 to 3 lb/acre	7	2	Limit of 24 pounds/acre/season. **NOTE: Do not use on cherry tomatoes.**
Bacterial speck, Bacterial spot	acibenzolar-S-methyl (Actigard 50 WG)	0.33 to 0.75 oz/acre	0.165 to 0.375 oz/ acre	14	0	Begin within one week of transplanting, spray at 7- to 10-day intervals for up to six applications. High rates may cause plant stunting.
Bacterial speck, Bacterial spot, Bacterial canker	famoxadone + cymoxanil (Tanos)	8 oz/acre	--	3	0.5	**DISEASE SUPPRESSION ONLY.** Limit of 72 fluid ounces/acre/season (12 month cycle). Do not make more than one application before alternating to a fungicide with a different mode of action. **NOTE: Must be tanked mixed with a contact fungicide that has a different mode of action.**
	fixed copper (various)	See label	See label	See label	See label	See label. Rates vary depending on the formulation. A fixed copper spray can suppress spot and foliar canker. Most strains of spot and speck are copper resistant. Spot control is improved if product is tanked mix with mancozeb + fixed copper, see above. REI varies with copper formulation, check label.
Botrytis (gray mold)	boscalid (Endura)	9 to 12.5 oz/acre	6.3 to 8.75 oz/acre	0	0.5	Limit of 25 ounces/acre/season. Make no more than 2 sequential applications and no more than 2 per crop year.
	chlorothalonil (various)	See label	See label	See label	See label	See label. Rates vary depending on the formulation.
	chlorothalonil + cymoxanil (Ariston)	1.9 pt/acre	--	3	0.5	Limit of 32 pints/acre/season.
	cyprodinil + fludioxonil (Switch 62.5 WG)	11 to 14 oz/acre	--	0	0.5	Limit of 56 ounces/acre/season. After two applications alternate with another fungicide with a different mode of action for two applications.
	difenoconazole + cyprodinil (Inspire Super)	16 to 20 fl oz	--	0	0.5	Limit of 47 fluid ounces/acre/season. Do not make more than 2 consecutive applications before alternating with fungicides with different mode of action.
	fluopyram + trifloxystrobin (Luna Sensation)	7.6 fl oz/acre	--	3	0.5	Do not exceed 5 applications or 27.1 fl oz/acre/season. Do not make more than 2 applications without switching to a different mode of action.
	fludioxonil (Scholar SC)	16 to 32 fl oz/100 gal	3.3 to 4.1 fl oz/100 gal	NA	NA	Use as a postharvest dip only. **NOTE: Do not use on processing tomato varieties.**

553

Table 10-53. Disease Control Products for Tomato

Disease	Material	Rate of Material to Use		Minimum Days		Method, Schedule, and Remarks
		Formulation	Acive Ingredients	Harvest	Reentry	
Tomato (field) (continued)						
Botrytis (gray mold) (continued)	fluxapyroxad + pyraclostrobin (Priaxor)	4 to 8 fl oz/acre	--	0	0.5	Limit of 24 fluid ounces/acre/season. Do not make more than two consecutive applications before alternating with fungicides with different mode of action.
	penthiopyrad (Fontelis 1.67 SC)	24 fl oz/acre	4.8 fl oz/acre	0	0.5	**DISEASE SUPPRESSION ONLY. Limit of 72 fluid ounces/acre/season.** Do not make more than two consecutive applications before alternating with fungicides that have a different mode of action.
	pyraclostrobin (Cabrio)	12 to 16 oz/acre	--	0	0.5	**Disease suppression only.** No more than 2 applications allowed before switching to a different mode of action. Do not exceed 96 oz/acre/season.
	pyrimethanil (Scala) SC	7 fl oz/acre	0.27 lb/acre	1	0.5	Limit of 35 fluid ounces/acre/season.
Buckeye rot	azoxystrobin (various)	See label	See Label	See label	See Label	Do not make more than one application before alternating to a fungicide with a different mode of action. See label for application restrictions.
	azoxystrobin + chlorothalonil (Quadris Opti)	1.6 pt/acre	--	0	0.5	Limit of five applications. Do not make more than one application before alternating to a fungicide with a different mode of action.
	famoxadone + cymoxanil (Tanos)	8 oz/acre	--	3	0.5	**DISEASE SUPPRESSION ONLY.** Limit of 72 fluid ounces/acre/season. Do not make more than one application before alternating to a fungicide with a different mode of action. **NOTE: Must be tanked mixed with a contact fungicide that has a different mode of action.**
	mancozeb + zoxamide (Gavel 75DF)	1.5 to 2 lb/acre	--	5	2	Limit of 8 pounds/acre/season West of the Mississippi River. Limit of 16 pound/acre/season East of the Mississippi River.
	mefenoxam + copper hydroxide (Ridomil Gold/ Copper)	2 lb/acre	--	14	2	Tank mix with 0.8 pound a.i. of either maneb or mancozeb.
	oxathiapiprolin + chlorothalonil (Orondis Opti)	See label	See label	See label	See label	Co-pack with chlorothalonil; see label for rates and restrictions.
	oxathiapiprolin + mandipropamid (Orondis Ultra)	See label	See label	See label	See label	Co-pack with mandipropamid; see label for rates and restrictions.
Cristulariella leaf spot, Powdery mildew	myclobutanil (Rally 40 WSP)	2.5 to 4 oz/acre	1 to 1.6 oz/acre	1	1	Spray weekly beginning at first sign of disease. Do not apply more than 1.25 lb/acre.
Damping off *(Pythium)*, Root and fruit rots *(Phytophthora)*	fosetyl-Al (Aliette 80 WDG)	2.5 to 5 lb/acre	2 to 4 lb/acre	14	0.5	Start sprays at 2 to 4 leaf stage or at transplanting on a 7 to 14 day schedule. Not for *Phytophthora* fruit rot. Check label for specific counties in each state where use is prohibited.
	mefenoxam (Ridomil Gold GR) (Ridomil Gold 4 SL) Ultra Flourish 2 EC)	20 lb/treated acre 1 to 2 pt/treated A 1 to 2 qt/treated A	0.5 lb/treated acre 0.5 to 1 lb/treated A 0.5 to 1 lb/treated A	7 28 28	2	Apply uniformly to soil at time of planting. Incorporate mechanically if rainfall is not expected before seeds germinate. A second application may be made up to 4 weeks before harvest. Do not exceed 3 pints of Ridomil Gold EC or 40 lb of Ridomil Gold GR per acre.
	propamocarb (Previcur Flex)	1.5 pt/acre	1 pt/acre	5	0.5	Limit of 7.5 pints/acre/season. Do not apply more than once before alternating with fungicides that have a different mode of action. For Pythium only.
Gray Leaf Spot *(Stemphylium* spp.)	azoxystrobin + difenoconazole (Quadris Top 2.72 SC)	8 fl oz/acre	--	0	0.5	Do not apply until 21 days after transplanting or 35 days after seeding. Limit of 47 fl oz/acre/season.. Make no more than 2 consecutive applications before rotating to another effective fungicide with a different mode of action. **NOTE: Do not use on tomato varieties that bear mature fruit less than 2 inches in diameter.**
	azoxystrobin + benzovindiflupyr (Mural)	0.6 oz/5,000 ft^2	--	0	0.5	Make no more than 2 applications before switching to a different mode of action. Mixing with some adjuvants may cause phytotoxicity; see label. Limits of a.i.'s apply; see label.
	chlorothalonil (various)	See label	See label	See label	See label	See label. Rates vary depending on the formulation.
	difenoconazole + benzovindiflupyr (Aprovia Top)	10.5 to 13.5 fl oz/acre	--	0	0.5	Do not make more than 2 applications before switching to a non-Group 7 fungicide. See label for application invervals and limits per season.
	difenoconazole + cyprodinil (Inspire Super)	16 to 20 fl oz	--	0	0.5	Limit of 80 fluid ounces/acre/season. Do not make more than two consecutive applications before alternating with fungicides that have a different mode of action.
	fluopyram + trifloxystrobin (Luna Sensation)	7.6 fl oz/acre	--	3	0.5	Do not exceed 5 applications or 27.1 fl oz/acre/season. Do not make more than 2 applications without switching to a different mode of action.

Table 10-53. Disease Control Products for Tomato

| Disease | Material | Rate of Material to Use | | Minimum Days | | Method, Schedule, and Remarks |
		Formulation	Acive Ingredients	Harvest	Reentry	
Tomato (field) (continued)						
Gray Leaf Spot (*Stemphylium* spp.) (continued)	fluopyram + pyrimethanil (Luna Tranquility)	11.2 fl oz/acre	--	1	0.5	See label for limits on application amounts per season. Do not make more than 2 applications of Group 7 or 9 fungicides without switching to a different mode of action.
	mancozeb (various)	See label	See label	See label	See label	See label. Rates vary depending on the formulation.
	mancozeb + copper (ManKocide)	1 to 3 lb/acre	--	5	2	Limit of 42.7 pounds/acre/season West of the Mississippi River. Limit of 58 pounds/acre/season East of the Mississippi River.
	mancozeb + zoxamide (Gavel 75DF)	1.5 to 2 lb/acre	--	5	2	Limit of 8 pounds/acre/season West of the Mississippi River. Limit of 16 pounds/acre/season East of the Mississippi River.
	mandipropamid + difenoconazole (Revus Top)	5.5 to 7 oz/acre	--	1	0.5	Limit of 28 fluid ounces/acre/season. Do not make more than two consecutive applications before alternating with fungicides that have a different mode of action.
	trifloxystrobin (Flint)	3 to 4 oz	1.5 to 2 oz	3	0.5	Disease suppression only. Limit of 16 fluid ounces/acre/season. Do not make more than one application before alternating to a fungicide with a different mode of action.
Early blight, *Septoria* leaf spot and Target spot	azoxystrobin (various)	See label	See label	See label	See label	Limit of 37 fluid ounces/acre/season. Do not make more than one application before alternating to a fungicide with a different mode of action. **NOTE: Under high temperatures Satori in combination with some additives or adjuvants may cause crop injury.**
	azoxystrobin + benzovindiflupyr (Mural)	0.6 oz/5,000 ft²	--	0	0.5	Make no more than 2 applications before switching to a different mode of action. Mixing with some adjuvants may cause phytotoxicity; see label. Limits of a.i.'s apply; see label.
	azoxystrobin + chlorothalonil (Quadris Opti)	1.6 pt/acre	--	0	0.5	Must alternate with a non-FRAC code 11; use of an adjuvant may cause phytotoxicity. Do not make more than 5 applications of a Group 11 fungicide/acre/season.
	azoxystrobin + difenoconzole (Quadris Top)	8 fl oz/acre	--	0	0.5	Do not apply until 21 days after transplanting or 35 days after seeding. Limit of 47 fluid ounces/acre/season. Make no more than two consecutive applications before rotating to another effective fungicide with a different mode of action. **NOTE: Do not use on tomato varieties that bear mature fruit less than 2 inches in diameter.**
	chlorothalonil (various)	See label	See label	See label	See label	Refer to label for rates and restrictions.
	azoxystrobin + chlorothalonil (Quadris Opti)	1.6 pt/acre	0.8 pt/acre	0	4 hr	Must alternate with a non-FRAC code 11; use of an adjuvant may cause phytotoxicity.
	boscalid (Endura)	2.5 to 3.5 oz/acre	1.75 to 2.45 oz/acre	0	0.5	Limit of 21 ounces/acre/season. Do not make more than two applications of the high rate per season and no more than six application of the low rate per season. **NOTE: Endura is not labeled for *Septoria* leaf spot management.**
	cyprodinil + fludioxonil (Switch 62.5 WG)	11 to 14 oz/acre	--	0	0.5	Limit of 56 ounces/acre/season. After two applications alternate with another fungicide with a different mode of action for two applications. **NOTE: Switch 62.5 WG is not labeled for *Septoria* leaf spot or Target spot management.**
	difenoconazole + benzovindiflupyr (Aprovia Top)	10.5 to 13.5 fl oz/acre	--	0	0.5	Do not make more than 2 applications before switching to a non-Group 7 fungicide. See label for application invervals and limits per season.
	difenoconazole + cyprodinil (Inspire Super)	16 to 20 fl oz	--	0	0.5	Limit of 80 fluid ounces/acre/season. Do not make more than two consecutive applications before alternating with fungicides that have a different mode of action.
	famoxadone + cymoxanil (Tanos)	6 to 8 oz/acre	--	3	0.5	Limit of 72 fluid ounces/acre/season. Do not make more than one application before alternating to a fungicide with a different mode of action. **NOTE: Must be tanked mixed with a contact fungicide that has a different mode of action. For *Septoria* leaf spot and target spot use 8 ounces per acre.**
	cymoxanil + chlorothalonil (Ariston)	1.9 to 3.0 pt/acre	--	3	0.5	Check copper labels for specific precautions and limitations for mixing with this product.
	fenamidone (Reason) 500 SC	5.5 to 8.2 fl oz/acre	0.18 to 0.27 fl oz/acre	14	4 hr	Limit of 24.6 fluid ounces/acre/season. Do not apply more than once before alternating with fungicides that have a different mode of action. **NOTE: Reason 500 SC is not labeled for Target spot management.**

Table 10-53. Disease Control Products for Tomato

Disease	Material	Rate of Material to Use		Minimum Days		Method, Schedule, and Remarks
		Formulation	Acive Ingredients	Harvest	Reentry	
Tomato (field) (continued)						
Early blight, *Septoria* leaf spot and Target spot (continued)	fluopyram + trifloxystrobin (Luna Sensation)	7.6 fl oz/acre	--	3	0.5	Do not exceed 5 applications or 27.1 fl oz/acre/season. Do not make more than 2 applications without switching to a different mode of action.
	fluopyram + pyrimethanil (Luna Tranquility)	11.2 fl oz/acre	--	1	0.5	See label for limits on application amounts per season. Do not make more than 2 applications of Group 7 or 9 fungicides without switching to a different mode of action.
	fluoxastrobin (Aftershock, Evito 480 SC)	2.0 to 5.7 fl oz/acre	0.06 to 0.18 lb/acre	3	0.5	Limit of 22.8 fluid ounces/acre/season. Do not apply more than once before alternating with fungicides that have a different mode of action.
	flutriafol (Rhyme)	5 to 7 fl oz/acre	--	0	0.5	Do not exceed more than 4 applications or 28 fl oz product/acre/season. NOT LABELED FOR SEPTORIA.
	fluxapyroxad + yraclostrobin (Priaxor)	4 to 8 fl oz/acre	--	0	0.5	Limit of 24 fluid ounces/acre/season. Do not make more than two consecutive applications before alternating with fungicides that have a different mode of action.
	mancozeb (various)	See label	See label	See label	See label	See label. Rates vary depending on the formulation.
	mancozeb + zoxamide (Gavel 75DF)	1.5 to 2 lb/acre	--	5	2	Limit of 8 pounds/acre/season west of the Mississippi River. Limit of 16 pounds/acre/season east of the Mississippi River.
	mandipropamid + difenoconazole (Revus Top)	5.5 to 7 oz/acre	--	1	0.5	Limit of 28 fluid ounces/acre/season. Do not apply more than two consecutive applications before alternating with a fungicide that have a different mode of action.
	penthiopyrad (Fontelis 1.67 SC)	16 to 24 fl oz	--	0	0.5	Do not exceed 72 fl oz/acre/season. Do not make more than 2 applications before switching to a different mode of action.
	propamocarb (Previcur Flex)	0.7 to 1.5 pt/acre	0.47 to 1 pt/acre	5	0.5	Limit of 7.5 pints/acre/season. Do not apply more than once before alternating with fungicides that have a different mode of action. Tank mix with a compatible fungicide for optimal Early blight control. **NOTE: Previcur Flex is not labeled for *Septoria* leaf spot or Target spot management.**
	polyoxin D zinc salt (Ph-D) (Oso 5% SC)	6.2 oz/acre 3.75 to 13.0 fl	0.7 oz/acre	0	4 hr	Alternate with non-FRAC code 19 fungicide.
	pyraclostrobin (Cabrio)	8 to 12 oz/acre	1.6 to 3.2 oz/acre	0	0.5	Limit of 96 fluid ounces/acre/season. Do not make more than two applications before alternating to a fungicide with a different mode of action.
	pyrimethanil (Scala) SC	7 fl oz/acre	0.27 lb/acre	1	0.5	Limit of 35 fluid ounces/acre/season. Use only in a tank mix with another fungicide recommended for Early blight. **NOTE: Scala SC is not labeled for *Septoria* leaf spot management. Scala SC is only labeled for Target spot in Florida.**
	trifloxystrobin (Flint)	3 to 4 oz/acre	1.5 to 2 oz/acre	3	0.5	Limit of 16 fluid ounces/acre/season. Do not make more than one application before alternating to a fungicide with a different mode of action. **NOTE: Flint is not labeled for Early blight or Target spot**
	zinc dimethyldithiocarbamate (Ziram) 76 DF	3 to 4 lb/acre	2.3 to 3 lb/acre	7	2	Limit of 24 pounds/acre/season. **NOTE: Do not use on cherry tomatoes. Ziram 76DF is not labeled for Target spot management.**
	zoxamide + chlorothalonil (Zing)	36 fl oz/acre	—	5	0.5	Do not use more than 2 sequential applications before alternating to a fungicide with a different mode of action. **Not labeled for Target spot management.**
Powdery mildew	azoxystrobin (various)	5 to 6.2 fl oz/acre	0.08 to 0.1 fl oz/acre	0	4 hr	Limit of 37 fluid ounces per crop/acre/season. Make no more than **one** application before alternating with fungicides that have a different mode of action.
	azoxystrobin + benzovindiflupyr (Mural)	0.6 oz/5,000 ft²	--	0	0.5	Make no more than 2 applications before switching to a different mode of action. Mixing with some adjuvants may cause phytotoxicity; see label. Limits of a.i.'s apply; see label.
	azoxystrobin + chlorothalonil (Quadris Opti)	1.6 pt/acre	--	0	0.5	Must alternate with a non-FRAC code 11; use of an adjuvant may cause phytotoxicity. Do not make more than 5 applications of a Group 11 fungicide/acre/season.
	azoxystrobin + difenoconazole (Quadris Top)	8 fl oz/acre	2.4 fl oz/acre	0	0.5	Limit of 47 fluid ounces/acre/season. Do not apply until 21 days after transplanting or 35 days after seeding.

DISEASE CONTROL

Table 10-53. Disease Control Products for Tomato

Disease	Material	Rate of Material to Use		Minimum Days		Method, Schedule, and Remarks
		Formulation	Acive Ingredients	Harvest	Reentry	
Tomato (field) (continued)						
Powdery mildew (continued)	chlorothalonil + cymoxanil (Ariston)	2 to 2.44 pt/acre	--	3	0.5	Limit of 18.1 pints per acre per year.
	cyprodinil + fludioxonil (Switch 62.5 WG)	11 to 14 oz/acre	--	0	0.5	Limit of 56 ounces/acre/season. After two applications alternate with another fungicide with a different mode of action for two applications.
	cyprodinil + difenoconazole (Inspire Super)	16 to 20 fl oz/acre	--	0	0.5	Limit of 80 fluid ounces/acre/season.
	difenoconazole + benzovindiflupyr (Aprovia Top)	10.5 to 13.5 fl oz/acre	--	0	0.5	Do not make more than 2 applications before switching to a non-Group 7 fungicide. See label for application invervals and limits per season.
	fluopyram + trifloxystrobin (Luna Sensation)	5 to 7.6 fl oz/acre	--	3	0.5	Do not exceed 5 applications or 27.1 fl oz/acre/season. Do not make more than 2 applications without switching to a different mode of action.
	mandipropamid + difenoconazole (Revus Top)	5.5 to 7 fl oz/acre	--	0	0.5	Limit of 28 fluid ounces/acre/season. Make no more than two consecutive applications before alternating with fungicides that have a different mode of action.
	myclobutanil (Rally 40 WSP) (Sonoma 40WSP) (Sonoma 20EW)	2.5 to 4 oz/acre 2.5 to 4 oz/acre 4.75-7.6 fl oz/acre	1 to 1.6 oz/acre	1	1	Limit of 1.25 pounds product/acre/season. Do not apply more than 38 fl oz Sonoma 20EW/acre/season.
	penthiopyrad (Fontelis)	16 to 24 fl oz/acre	0.13 to 0.31 lb/acre	0	0.5	Do not exceed 72 fluid ounces of product per year. Make no more than two consecutive applications per
	polyoxin D zinc salt (Ph-D) (Oso 5% SC)	6.2 oz/acre 3.75 to 13.0 fl oz/acre	0.7 oz/acre	0	4 hr	Alternate with non-FRAC code 19 fungicide.
	pyraclostrobin (Cabrio 20% EG)	8 to 16 oz/acre	1.6 to 3.2 oz/acre	0	4 hr	Limit of 96 ounces/acre/season. Make no more than **one** sequential application before alternating with
	sulphur (various)	See labels	See labels	1	1	Follow labels; may cause leaf burn if used under high temperatures.
	trifloxystrobin (Gem 500 SC) (Flint 50WDG)	2.0 fl oz/acre 3 to 4 oz/acre	--	3	0.5	Do not apply more than 16 oz product/acre/season. Alternate with a fungicide outside of Group 11. Do not exceed 5 applications/acre/season.
Late blight	azoxystrobin (various)	See label	See label	See label	See label	Limit of 37 fluid ounces/acre/season. Do not make more than one application before alternating to a fungicide with a different mode of action. **NOTE: Apply at 5 to 7 day intervals for effective late blight management.**
	chlorothalonil (various)	See label	See label	See label	See label	See label. Rates vary depending on the formulation.
	azoxystrobin + chlorothalonil (Quadris Opti)	1.6 pt/acre	--	0	4 hr	Must alternate with a non-FRAC code 11; use of an adjuvant may cause phytotoxicity.
	cymoxanil + chlorothalonil (Ariston)	1.9 to 3.0 pt/acre	--	3	0.5	Check copper labels for specific precautions and limitations for mixing with this product.
	cyazofamid (Ranman 400 SC)	2.1 to 2.8 fl oz/ acre	--	0	0.5	Limit of 16.5 fluid ounces/acre/season. Do not make more than one application before alternating to a fungicide with a different mode of action. **NOTE: Ranman must be tank mixed with an organosilicate**
	cymoxanil (Curzate 60 DF)	3.2 to 5 oz/acre	1.9 to 3.0 oz	3	0.5	Limit of 30 ounces per 12 month period. Use only in combination with a labeled rate of a protectant fungicide. If late blight is present use 5 ounces per acre on a 5 day schedule.
	dimethomorph (Acrobat 50 WP) (Forum 4.18 F)	6.4 oz/acre 6 oz/acre	3.2 oz/acre 2.6 oz/acre	4 4	0.5 0.5	Limit of 30 ounces/acre/season. Performance is improved if tanked mix with another fungicide with a different mode of action. Follow label restrictions carefully.
	dimethomorph + ametoctradin (Zampro)	14 fl oz/acre	--	4	0.5	Limit of 42 fluid ounces/acre/season. Do not make more than two consecutive applications before alternating to a fungicide with a different mode of action. The addition of a spreading or penetrating adjuvant is recommended to improve product performance.
	fenamidone (Reason 500 SC)	5.5 to 8.2 fl oz/acre	0.18 to 0.27 fl oz/ acre	14	4 hr	Limit of 24.6 fluid ounces/acre/season. Do not apply more than once before alternating with fungicides that have a different mode of action.
	fluopicolide (Presidio 4F)	3 to 4 fl oz/acre	0.09 to 0.13 lb/acre	2	0.5	Limit of 12 fluid ounces/acre/season or four applications. Do not make more than two consecutive applications before alternating to a fungicide with a different mode of action. Use only in combination with a labeled rate of another fungicide product with a different mode of action.

Table 10-53. Disease Control Products for Tomato

Disease	Material	Rate of Material to Use		Minimum Days		Method, Schedule, and Remarks
		Formulation	Acive Ingredients	Harvest	Reentry	
Tomato (field) (continued)						
Late Blight (continued)	fluoxastrobin (Aftershock or Evito 480 SC)	5.7 fl oz/acre	0.18 lb/acre	3	0.5	**DISEASE SUPPRESSION ONLY.** Limit of 22.8 fluid ounces/acre/season. Do not apply more than once before alternating with fungicides that have a different mode of action.
	fluxapyroxad + pyraclostrobin (Priaxor)	8 fl oz/acre	--	0	0.5	**DISEASE SUPPRESSION ONLY.** Limit of 24 fluid ounces/acre/season. Do not make more than two consecutive applications before alternating with fungicides that have a different mode of action.
	mancozeb (various)	See label	See label	See label	See label	See label. Rates vary depending on the formulation.
	mancozeb + copper hydroxide	1 to 3 lb/acre	--	5	1	Apply at 7- to 10-day intervals. Do not mix with other fungicide products.
	mancozeb + zoxamide (Gavel 75DF)	1.5 to 2 lb/acre	1.1 to 1.5 lb/acre	5	2	Limit of 8 pounds/acre/season West of the Mississippi River. Limit of 16 pounds/acre/season East of the Mississippi River.
	mandipropamid + difenoconazole (Revus Top)	5.5 to 7 oz/acre	--	1	0.5	Limit of 28 fluid ounces/acre/season. Do not apply more than two consecutive applications before alternating with a fungicide that have a different mode of action.
	mefenoxam + chlorothalonil (Ridomil Gold Bravo)	2.5 to 3 lb/acre	--	14	2	Do not make more than 3 applications of Ridomil Gold/Bravo or Ridomil Gold MZ per crop. Tank mix Bravo MZ58 with 0.45 lb mancozeb 80W.
	mefenoxam + mancozeb (Ridomil Gold MZ)	2.5 lb/acre	--	5	2	Do not make more than 3 applications of Ridomil Gold/Bravo or Ridomil Gold MZ per crop. Tank mix Bravo MZ58 with 0.45 pound mancozeb 80W.
	mono- and di-potassium salts of phosphorous acid (Alude)	1.5 to 2 qt/acre	--	3	4 hr	
	oxathiapiprolin + chlorothalonil (Orondis Opti)	See label	See label	See label	See label	Co-pack with chlorothalonil; see label for rates and restrictions.
	oxathiapiprolin + mandipropamid (Orondis Ultra)	See label	See label	See label	See label	Co-pack with mandipropamid; see label for rates and restrictions.
	polyoxin D zinc salt (Ph-D) (Oso 5% SC)	6.2 oz/acre 3.75 to 13.0 fl oz/acre	0.7 oz/acre	0	4 hr	Alternate with non-FRAC code 19 fungicide.
	propamocarb (Previcur Flex)	0.7 to 1.5 pt/acre	0.47 to 1 pt/acre	5	0.5	Limit of 7.5 pints/acre/season. Do not apply more than once before alternating with fungicides that have a different mode of action.
	pyraclostrobin (Cabrio)	8 to 16 oz/acre	--	0	0.5	No more than 2 applications allowed before switching to a different mode of action. Do not exceed 96 oz/acre/season.
	trifloxystrobin (Flint)	4 oz/acre	2 oz/acre	3	0.5	Limit of 16 fluid ounces/acre/season. Do not make more than one application before alternating with a protectant fungicide. NOTE: Apply Flint with 75% of the labeled rate of a protectant fungicide.
	zoxamide + chlorothalonil (Zing)	36 fl oz	--	5	0.5	Do not use more than 2 sequential applications before alternating to a fungicide with a different mode of action. Do not tank mix with another fungicide if the target pest is only late blight. Tank mix only if a partner is required to control other diseases.
Leaf mold (*Fulvia fulva* = *Passlora fulva*)	azoxystrobin + difenoconzole (Quadris Top)	8 fl oz/acre	--	0	0.5	Do not apply until 21 days after transplanting or 35 days after seeding. Limit of 47 fluid ounces/acre/season. Make no more than two consecutive applications before rotating to another effective fungicide with a different mode of action. **NOTE: Do not use on tomato varieties that bear mature fruit less than 2 inches in diameter.**
	difenoconazole + benzovindiflupyr (Aprovia Top)	10.5 to 13.5 fl oz/acre	--	0	0.5	Do not make more than 2 applications before switching to a non-Group 7 fungicide. See label for application invervals and limits per season.
	difenoconazole + cyprodinil (Inspire Super)	16 to 20 fl oz/acre	--	0	0.5	Limit of 80 fluid ounces/acre/season. Do not make more than two consecutive applications before alternating with fungicides that have a different mode of action.
	famoxadone + cymoxanil (Tanos)	8 oz/acre	--	3	0.5	Limit of 72 fluid ounces/acre/season (12 month cycle). Do not make more than one application before alternating to a fungicide with a different mode of action. **NOTE: Must be tanked mixed with a contact fungicide that has a different mode of action.**
	fluopyram + trifloxystrobin (Luna Sensation)	7.6 fl oz/acre	--	3	0.5	Do not exceed 5 applications or 27.1 fl oz/acre/season. Do not make more than 2 applications without switching to a different mode of action.
	mancozeb (various)	See label	See label	See label	See label	See label. Rates vary depending on the formulation.

Table 10-53. Disease Control Products for Tomato

Disease	Material	Rate of Material to Use		Minimum Days		Method, Schedule, and Remarks
		Formulation	Acive Ingredients	Harvest	Reentry	
Tomato (field) (continued)						
Leaf mold (*Fulvia fulva* = *Passlora fulva*) (continued)	mancozeb + copper hydroxide (ManKocide)	1 to 3 lb/acre	--	5	1	Apply at 7- to 10-day intervals. Do not mix with other fungicide products.
	mancozeb + zoxamide (Gavel 75DF)	1.5 to 2 lb/acre	--	5	2	Limit of 8 pounds/acre/season West of the Mississippi River. Limit of 16 pounds/acre/season East of the Mississippi River.
	polyoxin D zinc salt (Ph-D) (Oso 5% SC)	6.2 oz/acre 3.75 to 13.0 fl oz	0.7 oz/acre	0	4 hr	Alternate with non-FRAC code 19 fungicide.
Sour rot (*Geotrichum candidum*)	propiconazole (Mentor)	See label	See label	0	0	Use Mentor as a post-harvest dip, drench or high volume spray to control certain post-harvest rots. See label for details.
Southern blight	difenoconazole + benzovindiflupyr (Aprovia Top)	10.5 to 13.5 fl oz/acre	10.5 to 13.5 fl oz/acre	0	0.5	Do not make more than 2 applications before switching to a non-Group 7 fungicide. See label for application invervals and limits per season.
	fluoxastrobin (Aftershock, Evito 480 SC)	2.0 to 5.7 fl oz/acre	0.06 to 0.18 lb/acre	3	0.5	Begin applications when conditions favor disease development, on 7- to 10-day intervals. Do not apply more than once before alternating with fungicides that have a different mode of action. Do not apply more than 22.8 fluid ounces/acre/season.
	fluxapyroxad + pyraclostrobin (Priaxor)	4 to 8 fl oz/acre	--	0	0.5	**DISEASE SUPPRESSION ONLY.** Limit of 24 fluid ounces/acre/season. Do not make more than two consecutive applications before alternating with fungicides that have a different mode of action.
	PCNB (Blocker 4F) (transplanting)	4.5 to 7.5 pt/100 gal; use 0.5 pt of solution per plant.	1.8 to 3 pt/100 gal	NA	0.5	Transplanting: Apply at the time of transplanting for Southern blight suppression. The solution should be agitated often to maintain a uniform mixture to assure proper dosage. Limit of 7.5 pounds a.i./acre/season.
	PCNB (Blocker 4F) (in furrow)	1.2 to 1.9 gal; apply 10.6 to 16.7 fl oz product per 1000 ft of row	4.2 to 6.7 fl oz per 1000 ft of row	NA	0.5	In furrow: Apply in 8 to 10 gallons of water per acre based on 36-inch row spacing. Apply as in-furrow sprays to the open "V" trench just prior to planting. When cultivating, set plows as flat as possible to avoid getting non-treated soil against stems or plants. Limit of 7.5 pounds a.i./acre/season.
	penthiopyrad (Fontelis)	1 to 1.6 fl oz/1000 row ft	0.2 to 0.3 fl oz/1000 row ft	NA	0.5	Apply as a soil drench to seedling tray or at the time of transplant.
	pyraclostrobin (Cabrio)	12 to 16 oz/acre	2.4 to 3.2 oz/acre	0	4 hr	**DISEASE SUPPRESSION ONLY.** Limit of 96 fluid ounces/acre/season. Do not make more than two applications before alternating to a fungicide with a different mode of action.
Timber rot , white mold or Sclerotinia stem rot	fluxapyroxad + pyraclostrobin (Priaxor)	4 to 8 fl oz/acre	--	0	0.5	**DISEASE SUPPRESSION ONLY.** Limit of 24 fluid ounces/acre/season. Do not make more than two consecutive applications before alternating with fungicides that have a different mode of action.
	pyraclostrobin (Cabrio 20% EG)	12 to 16 oz/acre	2.4 to 3.2 oz/acre	0	4 hr	**DISEASE SUPPRESSION ONLY.** Limit of 96 fluid ounces/acre/season. Do not make more than two applications before alternating to a fungicide with a different mode of action.

559

DISEASE CONTROL

Table 10-54. Importance of Alternative Management Practices for Disease Control in Tomato

Scale: E, excellent; G, good; F, fair; P, poor; NC, no control; ND, no data.

Strategy	Bacterial canker*	Bacterial speck	Bacterial spot	Botrytis	Buckeye rot	Early blight	Late blight	Leaf Mold (greenhouse or open field)	Powdery mildew	Septoria leaf spot	Target Spot (greenhouse or open field)	Tomato spotted wilt virus**
Use of resistant cultivars	NR	P	F	NR	NR	F	F	P	F	NR	P	G
Use of disease free seed or transplants	G	G	G	NC	NC	NC	NC	F	NC	P	F	NA
Use of seed treatments	G	G	G	NC	NC	P	P	F	NA	P	ND	NA
Use of sanitation practices at the transplant stage	G	G	G	G	NC	NC	NC	F	NC	NC	F	NC
Crop rotation (3-4 years)	F	P	P	NC	F	F	NC	F	NC	P	P	NC
Control of solanaceous weeds	F	NC	NC	F	F	F	F	F	F	F	F	F
Fertility	NC	NC	NC	F	NC	F	NC	ND	NC	NC	ND	NC
Use of cover crops	NC	NC	NC	NC	F	P	NC	ND	NC	NC	ND	NC
Destroy crop residue	F	NC	NC	NC	NC	P	NC	F	NC	F	F	ND
Rogue plants	F	NC	NC	NC	NC	NC	NC	NC	NC	NC	NC	NC
Promote air movement	F	F	F	F	P	P	F	F	P	F	F	NA
Use of plastic or reflective mulches	NC	NC	NC	NC	F	F	NC	NC	NC	F	NC	G
Do not handle plants when wet	G	G	G	NC	NC	P	P	F	NC	P	F	NC
Use of drip irrigation (avoiding overhead irrigation)	F	F	F	F	P	F	F	F	NC	F	F	NC
Use of biological control or biorational products	P	P	F	P	NC	P	P	P	P	NC	P	NC
Use of foliar fungicides/bactericides	F	P	F	F	F	G	G	F	G	G	F	NA
Use of insecticides	NC	NC	NC	NC	NC	NC	NC	NC	NC	NC	NC	F
Soil fumigation	NC	NC	NC	NC	F	P	NC	NC	NC	NC	NC	NC

Table 10-55. Efficacy of Products for Disease Control in Tomato

Scale: E, excellent; G, good; F, fair; P, poor; NC, no control; ND, no data.

Product[1]	Fungicide Group[F]	Preharvest Interval (Days)	Bacterial Canker (foliar)	Bacterial Speck	Bacterial Spot	Botrytis Graymold	Buckeye Rot	Early Blight	Late Blight	Leaf mold (Fulvia fulva)	Powdery Mildew	Septoria Leaf Spot	Target Spot
azoxystrobin[2] (Quadris)	11	1	NC	NC	NC	NC	ND	E[R]	F	G	E	G	F[R]
famoxadone + cymoxanil (Tanos)	11 + 27	3	NC	NC	NC	NC	P	F	F	F	ND	F	F[R]
bacteriophage[3] (AgriPhage)	NA	0	NC	P	P	NC	NC	NC	NC	NC	NC	NC	NC
acibenzolar-S-methyl9 (Actigard)	21	14	ND	F	F	NC	NC	NC	NC	NC	NC	NC	F
boscalid (Endura)	7	0	NC	NC	NC	F	NC	G	NC	ND	ND	ND	F
chlorothalonil (various)	M	0	NC	NC	NC	F	P	F	G	G	P	G	F
chlorothalonil + cymoxanil (Ariston)	M + 27	3	NC	NC	NC	F	P	F	G	F	P	F	F
cyazofamid (Ranman)	21	0	NC	NC	NC	NC	NC	F	F	NC	NC	NC	NC
cymoxanil (Curzate)	27	3	NC	NC	NC	NC	ND	NC	F	ND	NC	ND	NC
cyprodinil + fludioxonil (Switch)	9 + 12	0	NC	NC	NC	F	NC	F	NC	NC	F	NC	NC
dimethomorph (Acrobat, Forum)	40	4	NC	NC	NC	NC	F	NC	F	NC	NC	NC	NC
difenoconazole + cyprodinil (Inspire Super)	3 + 9	0	NC	NC	NC	NC	NC	F	NC	F	G	F	F
dimethomorph + ametoctradin (Zampro)	40 + 45	4	NC	NC	NC	NC	NC	NC	G	NC	NC	NC	NC
fenamidone (Reason)	11	14	NC	NC	NC	NC	F	F	F	P	ND	P	F[R]
fixed copper (various)	M	0	F	F[R]	F[R]	NC	P	F	F	F	P	F	NC
fluopicolide (Presidio)	43	2	NC	NC	NC	NC	P	NC	G	NC	NC	NC	NC
fluxapyroxad + pyraclostrobin (Priaxor)	11 + 7	0	NC	NC	NC	F	NC	G	NC	ND	ND	ND	F
mancozeb (various)	M	5	NC	NC	NC	NC	P	F	F	F	NC	F	F
mancozeb + fixed copper (ManKocide)	M + M	5	F	F	F	NC	P	F	F	F	P	F	P
mancozeb + zoxamide (Gavel)	M + 22	3	NC	NC	NC	NC	P	F	F	F	NC	F	NC
mandipropamid + difenoconazole (Revus Top)	40 + 3	1	NC	NC	NC	NC	P	F	G	ND	NC	ND	G
mefenoxam[8] + chlorothalonil (Ridomil Gold Bravo, Flouronil)	4 + M	14	NC	NC	NC	P	E	P	E	F	NC	F	F
mefenoxam + copper (Ridomil Gold/Copper)	4 + M	14	F	F	F	NC	E	F	E[R]	F	P	F	NC
mefenoxam + mancozeb (Ridomil Gold MZ)	4 + M	5	NC	NC	NC	NC	E	F	G[R]	F	NC	F	P
myclobutanil (Rally)	3	1	NC	NC	NC	NC	NC	NC	NC	NC	G	NC	ND
penthiopyrad (Fontelis)	7	0	NC	NC	NC	F	NC	G	NC	ND	ND	ND	F
polyoxin D zinc salt (Ph-D; Oso 5% SC)	19	0	ND	ND	ND	F	ND	F	ND	ND	F	ND	F
propamocarb (Previcur Flex)	28	5	NC	NC	NC	NC	P	NC	F	NC	NC	NC	NC
pyraclostrobin (Cabrio)	11	0	NC	NC	NC	NC	P	E[R]	F	G	E	G	F[R]
pyrimethanil (Scala)	9	1	NC	NC	NC	F	NC	F	NC	ND	ND	ND	F
streptomycin[4] (Agri-Mycin, Streptrol)	25	0	F	F	F	NC	NC	NC	NC	NC	NC	NC	NC
sulfur (various)[5]	M	0	NC	NC	NC	NC	NC	NC	NC	NC	F	NC	NC
zinc dimethyldithiocarbamate[10] (Ziram)	M	7	NC	NC	NC	NC	NC	F	ND	F	ND	F	ND

[1] Efficacy ratings do not necessarily indicate a labeled use for every disease.

[2] Contact control only; not systemic.

[3] Biological control product consisting of a virus that attacks pathogenic bacteria.

[4] Streptomycin may only be used on transplants; not registered for field use.

[5] Sulfur may be phytotoxic; follow label carefully.

[7] Curative activity; not systemic.

[8] Curative activity; systemic.

[9] Systemic activated resistance.

[10] Do not use on cherry tomatoes.

[F] To prevent resistance in pathogens, alternate fungicides within a group with fungicides in another group. Fungicides in the "M" group are generally considered "low risk" with no signs of resistance developing to the majority of fungicides.

[R] Resistance reported in the pathogen.

561

DISEASE CONTROL

Table 10-56. Example Spray Program for Foliar Disease Control in Fresh Market Tomato Production

Week	Chemical[1]	Number of Applications of Chemical Per Season
Before Harvest (weeks 1 to 8)		
1	mancozeb + copper + Actigard[1]	mancozeb, 1; copper 1; Actigard, 1
2	mancozeb + copper	mancozeb, 2; copper 2
3	mancozeb + Fontelis OR Endura[,2] + Actigard	mancozeb, 3; Fontelis, 1; Endura, 1; Actigard, 2
4	mancozeb+ copper	mancozeb, 4; copper 3
5	mancozeb + Inspire Super + Actigard	mancozeb, 5; Inspire Super, 1; Actigard, 3
6	mancozeb + copper	mancozeb, 6; copper 4
7	mancozeb + Fontelis OR Endura[2] + Actigard	mancozeb, 7; Fontelis, 2; Endura, 2; Actigard, 4
8	mancozeb + copper	mancozeb, 8; copper 5
During Harvest (weeks 9 to 15) (Note: For fields in with low or no danger of Late Blight, Revus Top, Ranman and Presidio may not be essential tools; the protectant, chlorothalonil may suffice. Contact your local Extension office to learn if Late Blight is present in your region).		
9	chlorothalonil + Inspire Super	chlorothalonil, 1; Inspire Super, 2
10	Revus Top OR Presidio OR Ranman	Revus Top, 1; Presidio, 1; Ranman, 1
11	chlorothalonil + Fontelis OR Endura[2]	chlorothalonil, 2; Fontelis, 3; Endura, 3;
12	Revus Top OR Presidio OR Ranman	Revus Top, 2; Presidio, 2; Ranman, 2
13	chlorothalonil + Inspire Super	chlorothalonil, 3; Inspire Super, 3
14	Revus Top OR Presidio OR Ranman	Revus Top, 3; Presidio, 3; Ranman, 3
15	chlorothalonil + (Fontelis[3]) OR Endura[2]	chlorothalonil, 4; Fontelis, 3; Endura, 4
	Finish season with chlorothalonil	

Refer to the label for rates to use in volume-based spraying.

[1] For most products, the total number of applications or amount of product applied per season is restricted by the label.

[2] Strobilurins were removed from this guide due to widespread resistance in the pathogen that causes early blight. However, a strobilurin may be substituted for Fontelis OR Endura if resistance doesn't occur in your area.

[3] This application exceeds the limit on amount applied per season if the higher rate has been used. Be sure to check label and apply on if within the allowable amount.

Turnip Greens – see Greens and Leafy Brassicas

Turnip Roots – see Root Vegetables

For Watermelons – see Cucurbits

Nematode Control in Vegetable Crops

Crop losses due to nematodes can be avoided or reduced by using the following management tactics.

- Practice crop rotation.
- Plow out and expose roots immediately after the last harvest.
- Plow or disk the field two to four times before planting.
- Use nematode-free planting material.
- Sample soil and have it assayed for nematodes, preferably in the fall. There is a fee for each sample. Ship sample via DHL, FedEx, or UPS to: North Carolina Department of Agriculture and Consumer Services.
- Where warranted, fumigate or use other nematicides according to guidelines listed on the label. (Soil should be warm, well worked, and free from undecomposed plant debris and have adequate moisture for seed germination.)
- For in-row application, insert chisel 6 to 8 inches deep and throw a high, wide bed up over it; do not rework rows after fumigating.
- For broadcast treatments, insert chisels 6 to 8 inches deep, and space chisels 12 inches apart for most fumigants; use 5-inch spacing for Vapam.

Row rates in this section are stated for rows on 40-inch spacing. For other row spacings, multiply the stated acre rate by the appropriate conversion factor to determine the amount of material applied per acre (Do not alter stated amount per 100-foot row). This will be a guide to the amount of material to purchase for the acreage you want to treat.

For example, if 10 gallons per acre are used on 40-inch rows, for 36-inch rows, it will take 11.1 gallons to treat an acre.

CAUTION: Read labels carefully. Some products have restrictive crop rotations.

Your Row Spacing (inches)	Conversion Factor
24	1.67
26	1.54
28	1.43
30	1.33
32	1.25
34	1.18
36	1.11
38	1.05
40	1.00
42	0.952
44	0.909
46	0.870
48	0.833
5 ft	0.667
6 ft	0.556
7 ft	0.476
8 ft	0.417

Relative Efficacy of Currently Registered Fumigants or Fumigant Combinations for Managing Soilborne Nematodes, Diseases, and Weeds

Table 57A. Relative Efficacy of Currently Registered Fumigants or Fumigant Combinations for Managing Soilborne Nematodes, Diseases, and Weeds[1]

Product	Rate per Treated Acre[2]		Nematodes	Disease	Nutsedge	Weeds: Annual
	Volume (gal)	Weight (lb)				
Telone II (1,3-D)	15-27	153-275	+++++	+	-	-
Telone EC[3]	9-24[4]	91-242[4]	+++++	+	-	-
Telone C17 (1,3-D + chloropicrin)	32.4-42	343-445	+++++	+++	+	+
Telone C35 (1,3-D + chloropicrin)	39–50	437-560	+++++	+++++	+	++
InLine (1,3-D + chloropicrin)[3]	29–57.6 (See Label)	325-645 (See Label)	+++++	+++++	+	+++
Pic-Clor 60 (chloropicrin + 1,3-D)	48.6	588	+++++	+++++	+	+++
Pic-Clor 60 EC[3]	42.6	503	+++++	+++++	+	+++
Metam potassium[5]	30-62	318-657	++	+++	+	++++
Metam sodium[5] (MS)	37.5-75	379-758	++	+++	+	++++
Chloropicrin + MS[5]	19.5–31.5 + 37.5-75	275-444 + 379-758	++	+++++	++	++++
Chloropicrin	48.6	150-350	+	+++++	—	—
Tri-Pic 100EC[3]	8-24	100-300	+	+++++	—	—
Paladin (dimethyl disulphide)[6]	35.0–51.3	310-455	++++	++++	++++	+++[5]
Paladin EC[3,6]	37.0-54.2	326-479	++++	++++	++++	+++[5]
Dominus (allyl isothiocyanate)[7]	25-40[4]	212-340[4]	++	+++	+	+++

[1] Fumigants with lower efficacy against weeds may require a complementary herbicide or hand-weeding program, although use of virtually impermeable film (VIF) or totally impermeable film (TIF) may increase weed control, particularly with Telone C35 or Paladin. Refer to the Herbicide Recommendation section of this guide for directions pertaining to herbicide applications. Telone can persist more than 21 days under cool or wet soil conditions.

[2] Rates can sometimes be reduced if products are applied with VIF or TIF.

[3] Product is formulated for application through drip lines under a plastic mulch; efficacy is dependent on good distribution of the product in the bed profile.

[4] Labelled rates are per *broadcast-equivalent* acre, NOT per treated acre.

[5] Metam potassium can be Metam KLR, K-Pam, Sectagon K54 or other registered formulations, and should be used in soils with high sodium content. Metam sodium can be Vapam, Sectagon 42, Metam CLR or other registered formulations.

[6] Paladin should be applied with 21% chloropicrin and VIF or TIF to enhance disease control, and has low efficacy on certain small seeded broadleaf weeds and grasses. Paladin may not be registered in all States.

[7] Dominus is registered but there is limited experience with the product through University or independent trials in our region; growers may want to consider this on an experimental basis. Planting interval is 10 days. The active ingredient allyl isothiocyanate is similar to the active ingredient in metam sodium products (methyl isothiocyanate) and is likely to behave in a similar manner with a similar pest control profile.

DISEASE CONTROL

Management of Soilborne Nematodes with Non-Fumigant Nematicides

Nematodes are best managed through an integrated program (IPM). Key management options may include securing advisory/predictive soil samples, crop rotation, fallow periods, host resistance, soil amendments, flooding, soil solarization, suppressive cover crops and other options. For more details see https://edis.ifas.ufl.edu/cv112.

Table 10-57B. Management of Soilborne Nematodes with Non-Fumigant Nematicides

Commodity	Material	Application Method for Given Soils	Formulated Rate per Acre	Formulated Rate Per 100 Sq Ft Or 100 Ft Row	Schedule and Remarks
Most vegetables	Ditera DF Biological nematicide	Banded or low pressure drip irrigation	See label	See label	See label.
Bean (snap and lima)	ethoprop (Mocap) various formulations	Broadcast or banded	See label	See label	Incorporate 2 to 4 inches deep. See label.
Cabbage	ethoprop (Mocap) various formulations	Broadcast or banded	See label	See label	Restricted use pesticide. Incorporate 2 to 4 inches deep.
Carrot	oxamyl (Vydate) L	Preplant broadcast	2 gal in 20 gal	1 gal in 20 gal	Apply (in furrow) within 1 wk of planting, and thoroughly incorporate into soil 4 to 6 inches deep. May be applied via chemigation or In-Furrow – see label. 14 days to harvest restriction.
		At planting seed furrow	1 to 2 gal in 20 gal	0.3 to 0.6 fl oz in 5.9 fl oz	
Cucumber	ethoprop (Mocap) various formulations	Banded only	See label	See label	Incorporate 2 to 4 inches deep. See label.
Cucurbits (cucumber, squash, cantaloupe, watermelon, honeydew, pumpkin)	oxamyl (Vydate) L	Preplant broadcast	1 to 2 gal	—	Incorporate 2 to 4 inches into soil.
		Foliar	2 - 4 pt	—	Apply by air or ground with the first spray 2-4 weeks after planting; apply second spray 2-3 weeks after first spray.
		Drip chemigation or Soil injection	2 to 4 pt	0.07 to 0.15 fl oz	Treatments should be initiated at the time of seedling emergence or transplanting, or within 14 days of seedling emergence or transplanting. Sequential applications should be made on 14 to 21 day intervals. Do not treat within one day of harvest.
Cucurbits (Group 9) including Cucumber, squash, honeydew, watermelon, and many others (see label). Fruiting vegetables (Group 8) including tomato, pepper and eggplant and many others (see label).	Fluensulfone (Nimitz) NOTE: Nimitz received federal registration Sept 2014 – check registration status in your State.	3.5 to 5 pints/ treated acre (56 to 80 fl. oz./treated acre)	—	—	Applications must be broadcast incorporated, banded and incorporated, or by drip irrigation. Apply at a minimum of 7 days before transplanting. Do not apply more than one application per crop, and no more than 112 fl oz of product per acre, per year (365 days).
Cucurbits (Group 9) see label for list Fruiting vegetables (Group 8) see label for list	heat-killed *Burkholderia* spp. strain A396	4 – 8 qt	—	—	Majestene is a biological nematicide approved for organic vegetable production. It has not been extensively field-tested in the Southeast and Mid-Atlantic states, but research to date suggests useful activity against root-knot, lesion, sting, stunt, ring, and reniform nematodes. Can be applied as a pre-plant incorporated, in-furrow or banded spray as long as spray volume is sufficient to thoroughly soak the root zone. However, Majestene can also be drip-applied prior to planting, at planting or shortly thereafter, and again later in the season. Higher rates are likely more effective, and repeated applications also increase the extent and duration of nematode control. If nematode populations are high, another product may also be necessary for control.
Eggplant (Also see fruiting vegetables above)	fenamiphos (Nemacur)	Banded	5.9 fl oz/1000 ft		On narrow rows do not let bands overlap. See label.
	oxamyl (Vydate) L	Banded on soil	—	1 gal	See label.
		Foliar	4 pt	—	Apply twice by ground equipment at 1 to 2 week intervals 2 to 4 weeks after the second soil treatment.
		Drip chemigation or Soil injection	2 to 4 pt	0.07 to .15 fl oz	Treatments should be initiated either at the time of transplanting, or within 14 days of transplanting. Sequential applications should be made on 10 to 14 day intervals. Do not treat within 7 days of harvest.

Table 10-57B. Management of Soilborne Nematodes with Non-Fumigant Nematicides

Commodity	Material	Application Method for Given Soils	Formulated Rate per Acre	Formulated Rate Per 100 Sq Ft Or 100 Ft Row	Schedule and Remarks
Pepper (bell & non-bell) (Also see fruiting vegetables above)	oxamyl (Vydate) L	Transplant water	2 pt in 200 gal	—	
		Foliar	2 to 4 pt	—	Apply 14 days after transplant repeat 1 to 2 week intervals. Do not treat within 7 days of harvest.
		Drip chemigation or Soil injection	2 to 4 pt in	—	Treatments should be initiated either at the time of transplanting, or within 14 days of transplanting. Sequential applications should be made on 10 to 14 day intervals. Do not treat within 7 days of harvest.
Potato	ethoprop (Mocap) various formulations	Broadcast or banded	See label	See label	Incorporate 2 to 4 in. deep. See label.
	oxamyl (Vydate) L	Furrow	1 to 2 gal in 20 gal	—	Apply to seed furrow at planting; begin foliar sprays when early season control has diminished.
		Foliar	2 to 4 pt		
Sweetpotato	aldicarb (Temik) 15 G	40-in. rows	10 to 20 lb	0.4 to 0.7 oz	Apply in a 12- to 15-inch band and incorporate 4 to 8 in. deep, plant. Do not harvest within 120 days of treatment. Do not use vines. **NOTE: For use ONLY in LA or MS.**
	ethoprop (Mocap) various formulations	Banded only	See label	See label	Incorporate 2 to 4 inches deep. See label.
	oxamyl (Vydate) L	Preplant broadcast	2 to 3 gal in 20 gal	—	Thoroughly incorporate into soil 4 to 6 in. deep and plant.
		Transplant water	1 to 2 gal in 200 gal		
Sweet corn	ethoprop (Mocap) various formulations	Broadcast or banded	See label	See label	Incorporate 2 to 4 in. deep. See label.
Sweet corn, Popcorn	terbufos (Counter) various formulations (see label)	Row, 30-in. min.	See label	See label	Apply in furrow
Tomato	oxamyl (Vydate) L	Foliar	2 to 4 pt	—	Spray when plants are established. Repeat 1- to 2-week intervals.
		Drip irrigation	2 to 8 pt	—	Apply at first irrigation to field. Repeat every 1 to 2 weeks while plants are small. As plants enlarge, increase dosage progressively to 8 pints.

Greenhouse Disease Control

R. A. Melanson, Extension Plant Pathologist, Mississippi State University; F. Louws, Plant Pathologist, NCState University; M. L. Lewis Ivey, Plant Pathologist, The Ohio State University; A. Keinath, Plant Pathologist, Clemson University

Note: Follow manufacturer's directions on label in all cases.

Caution: At the time this table was prepared, the entries were believed to be useful and accurate. However, labels change rapidly and errors are possible, so the user must follow all directions on the pesticide container. See product labels for application limits per crop/season.

Information in the following table must be used in the context of a total disease control program. For example, many diseases are controlled by the use of resistant varieties, crop rotation, sanitation, seed treatment, and cultural practices. Always use top-quality seed or plants obtained from reliable sources. Seeds are ordinarily treated by the seed producer for the control of seed decay and damping-off.

Most foliar diseases can be reduced or controlled by maintaining relative humidity under 90 percent, by keeping the air circulating in the house with a large overhead polytube, and by avoiding water on the leaves.

Caution: The risk of pesticide exposure in the greenhouse is high. Use protective clothing laundered daily or after each exposure. Ventilate during application and use appropriate personal protective equipment (PPE).

Table 10-58A. Greenhouse Disease Control for Various Vegetable Crops

Disease/Location	Product[1]	FRAC Group	Rate of Formulation	Minimum Days Harvest	Minimum Days Reentry	Schedule and Remarks
Greenhouse						
Sanitation	Solarization		140 degrees F, 4 to 8 hours for 7 days	—	—	Close greenhouse during hottest and sunniest part of summer for at least 1 week. Greenhouse must reach at least 140 degrees F each day. Remove debris and heat sensitive materials and keep greenhouse and contents moist. Will not control pests 0.5 inches or deeper in soil. Not effective against TMV.

DISEASE CONTROL

Table 10-58A. Greenhouse Disease Control for Various Vegetable Crops

Disease/Location	Product[1]	FRAC Group	Rate of Formulation	Minimum Days Harvest	Minimum Days Reentry	Schedule and Remarks
	Added heat		180 degrees F for 30 minutes	—	—	Remove all debris and heat-sensitive materials. Keep house and contents warm.
Basil						
Alternaria leaf spot (*Alternaria* spp.) Botrytis leaf blight (*Botrytis* spp.), Fusarium blight (*Fusarium* spp.)	cyprodinil + fludioxonil (Switch 62.5WG)	9 + 12	11 to 14 oz/acre	7	0.5	After two applications, alternate with another fungicide with a different mode of action for two applications. See label for application limits.
Downy mildew (*Peronospora belbahrii*)	cyazofamid (Ranman, Ranman SC400)	21	2.75 to 3.0 fl oz/acre	0	0.5	Do not exceed 27 fl oz per acre per growing season. See label for surfactant recommendations. Alternate applications with fungicides that have a different mode of action. Do not make more than three consecutive applications before switching to products that have a different mode of action for three applications before returning to Ranman/Ranman 400SC.
	mandipropamid (Micora, Revus)	40	8.0 fl oz/acre	1	4 hr	**Micora: For basil grown for transplants and retail sale to consumers only.** **Revus:** Do not exceed 32 fluid ounces of product per acre per season. See label for additional limits when producing multiple croppings per year.
Downy mildew, Fusarium wilt, *Pythium* and *Rhizoctonia* root rots	phosphorous acid (various)	33	See label	See label	See label	See label for restrictions.
	potassium phosphite	33	See label	See label	See label	See label for restrictions.
Cucurbits						
Alternaria leaf blight and spot (*Alternaria cucumerina* and *A. alternata*), anthracnose (*Colletotrichum orbiculare*), Cercospora leaf spot (*Cercospora citrullina*), downy mildew (*Pseudoperonospora cubensis*), gummy stem blight (*Didymella bryoniae*), powdery mildew (*Sphaerotheca fuliginea, Erysiphe cichoracearum*), Septoria leaf spot (*Septoria cucurbitacearum*)	azoxystrobin + difenoconazole (Quadris Top)	11 + 9	12 to 14 fl oz/acre	1	0.5	Do not make more than one application of a QoI-containing fungicide before switching to a fungicide with a different mode of action. Do not exceed 56 fl oz of product per acre per season. See product label for other diseases labeled.
Angular leaf spot, downy mildew	copper, fixed (various)	M1	See label	See label	See label	**Some products are OMRI-listed.** See product label for complete application instructions, specific crop and disease labels, and greenhouse usage.
Powdery mildew	cyflufenamid (Torino)	U6	3.4 oz/acre	0	4 hours	Do not make more than two applications per year.
Downy mildew (*Pseudoperonospora cubensis*)	cymoxanil (Curzate 60DF)	27	3.2 to 5.0 oz/acre	3	12 hr	Always apply in a tank mix with the labeled rate of a protectant fungicide. Do not exceed 30 ounces of product per 12-month period.
Alternaria leaf blight and spot (*A. cucumerina* and *A. alternata*), gummy stem blight (*Didymella*), powdery mildew (*S. fuliginea, E. cichoracearum*)	cyprodinil + fludioxonil (Switch 62.5WG)	9 + 12	11 to 14 oz/acre	1	0.5	After two applications, alternate with another fungicide with a different mode of action for two applications.See label for application limits.
Alternaria leaf blight and spot (*A. cucumerina* and *A. alternata*), anthracnose (*C. orbiculare*), Cercospora leaf spot (*C. citrullina*), gummy stem blight (*Didymella*), powdery mildew (*S. fuliginea, E. cichoracearum*), Septoria leaf blight (*S. cucurbitacearum*)	difenoconazole + cyprodinil (Inspire Super)	3 + 9	16 to 20 fl oz/acre	7	0.5	Do not apply more than 80 fluid ounces of product per acre per season. Make no more than two consecutive applications per season before alternating with fungicides that have a different mode of action.
Alternaria leaf blight (*A. cucumerina*), anthracnose (*Colletotrichum* spp.), downy mildew (*P. cubensis*) **Suppression:** Bacterial fruit blotch (*Acidovoraz avenae* subsp. *citrulli*), Phytophthora blight (*Phytophthora capsici*)	famoxadone + cymoxanil (Tanos)	11 + 27	8 oz/acre 8 to 10 oz/acre (for diseases listed under suppression)	3	0.5	Do not exceed 32 oz/acre of product per crop crop cycle or 72 oz/acre per 12 month period. Do not make more than one application of product before alternating with a fungicide that has a different mode of action. See label for tank mixing instructions. For suppression of foliar and fruit phases ONLY of Phytophthora blight.
Botrytis gray mold	fenhexamid (Decree 50 WDG)	17	1.5 lbs/acre (stand alone) 1.0 to 1.5 lbs/acre (tank-mix)	0	0.5	**Labeled for cucumbers ONLY.** For use in transplant production and greenhouse production. Do not make more than two consecutive applications. See labels for additional tank-mixing instructions. Do not exceed 6.0 lbs product per acre (transplants) or per acre per season (greenhouse production).
Alternaria leaf spot, anthracnose, Cercospora leaf spot, downy mildew, gummy stem blight, scab	mancozeb (various)	M3	See label	See label	See label	See product labels for complete application instructions, specific crop and disease labels, and greenhouse usage.
Alternaria leaf spot, Cercospora leaf spot, downy mildew, Phytophthora rot (*Phytopthora capsici*)	mancozeb + zoxamide (Gavel 75SF)	M3 + 22	1.4 to 2.0 lb/acre	5	2	Do not exceed 16 lb per acre per season.

Table 10-58A. Greenhouse Disease Control for Various Vegetable Crops

Disease/Location	Product[1]	FRAC Group	Rate of Formulation	Minimum Days Harvest	Reentry	Schedule and Remarks
Cucurbits (continued)						
Suppression: Downy mildew (*P. cubensis*), Phytophthora blight (*P. capsici*)	mandipropamid (Revus)	40	8.0 fl oz/acre (Revus)	0	4 hr	Do not exceed 32 fl oz of product per acre oer season. See label for application instructions specific to each disease.
Alternaria leaf spot and blight, Botrytis gray mold, gummy stem blight (*Didymella*), powdery mildew, Sclerotinia stem rot	penthiopyrad (Fontelis)	7	0.375 to 0.5 fl oz/ gal to treat 1,360 sq ft	1	0.5	Do not exceed 67 fluid ounces of product per year. Make no more than two consecutive applications per season before alternating with fungicides that have a different mode of action. See label for cucurbit restrictions.
Botrytis gray mold, Corynespora leaf spot (*Corynespora cassicola*), early blight (*Alternaria* sp.), gummy stem blight (*Didymella*), powdery mildew (*Spthaerotheca* sp.), scab (*Cladosporium* sp.)	polyoxin D zinc salt (Affirm WDG, OSO 5%SC; Ph-D)	19	6.2 oz/acre (Affirm) 3.75 to 13.0 fl oz/acre (OSO) 6.2 oz/acre (Ph-D)	0	4 hr	Check products labels for maximum limits of product per season. Alternate with fungicides that have a different mode of action. Check product labels for other tomato diseases also on the label.
Alternaira leaf spot, anthracnose, *Botrytis*, Cercospora leaf spot, powdery mildew	potassium bicarbonate (Milstop)	NC	1.25 to 5.0 lbs/100 gal	0	1 hr	**OMRI-listed.** Do not exceed 0.5 lbs of product per 4,350 sq ft or 1.15 lbs product per 10,000 sq ft per application. Do not store unused spray solution. See label for additional diseases labeled.
Alternaria leaf spot, anthracnose, Cercospora leaf spot, downy mildew, gummy stem blight (GSB), scab, target spot	potassium phosphite + chlorothalonil (Catamaran)	33 + M5	4 pt/acre (anthracnose, downy mildew, target spot) 6 pt/acre (Alternaria and Cercospora leaf spot, GSB, scab)	1	0.5	Do not exceed 50 pt per acre per season. Phytotoxicity potential. Do not combine with other pesticides, serfactants or fertilizers. Do not apply on the day of harvest.
Damping off and root rots (*Phytophthora* spp, *Pythium* spp.)	promocarb hydrochloride (Previcur Flex)	28	See label	2	0.5	Product applied through a drip system or as a soil drench. Do not apply more than four applications of product after transplanting per crop cycle. Do not mix with other products. Phytotoxicity may occur if applied to dry growing media.
Powdery mildew	sulfur (various)	M2	See label	See label	See label	**Some products are OMRI-listed.** See product labels for complete application instructions, specific crop and disease labels, and greenhouse usage. Not all products are registered for use in all states.
Powdery mildew	triflumizole (Procure 480SC, Terraguard SC)	3	4 to 8 fl oz/acre (Procure) 2 to 4 fl oz/100 gal (Terraguard)	1	0.5	**Procure:** Do not exceed 40 fl oz of product per acre per season. Label specifices the following powdery mildew species: *Erysiphe cichoracearum, Podosphaera xanthii*). **Terraguard SC:** For use only as a foliar spray. For use in commercial greenhouse production only. Can be used on greenhouse transplants. Do not exceed 40 fl oz of product per acre per cropping system. See label for additional application instructions. **Procure 480SC and Terraguard SC not registered for use in all states.**
Lettuce						
Downy mildew (*Bremia lactucae*)	cymoxanil (Curzate 60DF)	27	3,2 to 5.0 oz/acre (head lettuce) 5.0 oz/acre (leafy lettuce)	3 (head) 1 (leafy)	12 hr	For use on head and leaf lettuce. Use with the labeled rate of a protectant fungicide. Do not exceed 30 oz of product per 12-month period.
Alternaria leaf spot, *Botrytis* gray mold, Sclerotinia rot, basal rot (*Phoma*), Septoria leaf spot **Suppression:** Powdery mildew	cyprodinil + fludioxonil (Switch 62.5WG)	9 + 12	11 to 14 oz/acre	0	0.5	For use on head and leaf lettuce. After two applications, alternate with another fungicide with a different mode of action for two applicaions. See label for application limits.
Botrytis gray mold rot, drop rot, *Sclerotinia minor*, watery soft rot (*Sclerotinia sclerotiorum*)	dicloran (Botran 5F)	14	0.6 qt/acre (at planting) 0.6 to 1.8 qt/acre (prethinning) 1.8 to 3.2 qt/acre (post-thinning [direct seeded] and established transplants)	14	0.5	For use on head and leaf lettuce. See label for detailed instructions. Do not exceed 3.2 qt of product per acre per year or within 14 days of harvest. **Not registered for use in SC, TN, or VA.**
Downy mildew (*B. lactucae*), white rust (*Albugo occidentalis*)	famoxadone + cymoxanil (Tanos)	11 + 27	8 to 10 oz/acre	1	0.5	For use on head and leaf lettuce. Do not exceed 48 oz/acre of product per crop season. Do not make more than one application of product before alternating with a fungicide that has a different mode of action. See label for tank mixing instructions.
Botrytis gray mold	fenhexamid (Decree 50 WDG)	17	1.5 lbs/acre (stand-alone) 1.0 to 1.5 lbs/acre (tank-mix)	3	0.5	For use in transplant production and greenhouse production. Do not make more than two consective applications. See label for additional tank-mixing instructions. Do not exceed 3.0 lbs product per acre (transplants) or per acre per crop (greenhouse production).
Anthracnose, downy mildew	mancozeb (various)	M3	See label	See label	See label	See product labels for complete application instructions, specific crop and disease labels, and greenhouse usage.

567

Table 10-58A. Greenhouse Disease Control for Various Vegetable Crops

Disease/Location	Product[1]	FRAC Group	Rate of Formulation	Minimum Days Harvest	Reentry	Schedule and Remarks
Lettuce (continued)						
Downy mildew	mandipropamid (Micora, Revus)	40	5.5 to 8.0 fl oz/acre (Micora) 8.0 fl oz/acre (Revus)	-- 1	4 hr	For use on head and leaf lettuce. **Micora: For lettuce grown for transplants and retail sale to consumers only.** Do not make more than two applications per crop. After making an application of Micora, alternate with a fungicide with a different mode of action. Apply in a tank-mix with another downy mildew fungicide with a different mode of action. **Revus:** Do not exceed 32 fl oz of product per acre oer season. Do not make more than 2 applications before switching to a fungicide with a different mode of action.
Alternaria leaf spot (*Alternaria sonchi*), Botrytis gray mold, Cercospora leaf spot (*Cercospora* spp.), powdery mildew (*Erysiphe cichoracearum*), Septoria leaf spot (*Septoria* spp.)	penthiopyrad (Fontelis)	7	14 to 24 fl oz/acre	3	0.5	For use on head and leaf lettuce. Do not exceed 72 fl oz of product per year. Make no more than two consecutive applications per season before alternating with fungicides that have a different mode of action. See label for additional diseases that may be controlled.
Alternaira leaf spot, anthracnose, *Botrytis*, Cercospora leaf spot, powdery mildew	potassium bicarbonate (Milstop)	NC	1.25 to 5.0 lbs/100 gal	0	1 hr	**OMRI-listed.** Do not exceed 0.5 lbs of product per 4,350 sq ft or 1.15 lbs product per 10,000 sq ft per application. Do not store unused spray solution. See label for additional diseases labeled.
Damping off and root rots (*Phytophthora* spp, *Pythium* spp.)	promocarb hydrochloride (Previcur Flex)	28	See label	2	0.5	**For use on leaf lettuce only.** Product applied as a foliar treatment. Do not apply more than two applications of product per crop cycle. Do not mix with other products. Phytotoxicity may occur if applied to dry growing media.
Powdery mildew	sulfur (various)	M2	See label	See label	See label	**Some products are OMRI-listed.** See product labels for complete application instructions, specific crop and disease labels, and greenhouse usage. Not all products are registered for use in all states.
Alternaria leaf spot/black spot (*Alternaria* spp.), powdery mildew (*Erysiphe* spp.)	triflumizole (Procure 480SC)	3	6 to 8 fl oz/acre	0	0.5	For use on head and leaf lettuce. Do not exceed 18 fl oz of product per acre per season. **Product not registered for use in all states.**
Tomato (transplant production)						
Damping off (*Pythium* spp.)	cyazofamid (Ranman, Ranman 400SC)	21	3 fl oz/100 gal	--	0.5	**For transplant production only.** Apply as a soil drench. Do not use a surfactant. One fungicide application can be made to the seedling tray at planting or any time afterwards until one week before transplanting.
Late blight (*Phytophthora infestans*)	mandipropamid (Micora)	40	5.5 to 8.0 fl oz/acre	--	4 hr	**For tomatoes grown for transplants and retail sale to consumers only.** Do not make more than two applications per crop. Do not make more than 2 consecutive applications before switching to a fugnide from a different FRAC group.
Damping off and root rots (*Phytophthora* spp, *Pythium* spp.)	promocarb hydrochloride (Previcur Flex)	28	See label.	--	0.5	**For preseeding and/or seedling treatment (before transplanting).** Do not mix with other products. See label for specific use directions. Phytotoxicity may occur if applied to dry growing media.
Bacterial canker, speck, and/or spot	*Streptomycin* sulfate (Agri-Mycin 17, Ag Streptomycin, Firewall 17 WP, Firewall 50 WP, Harbour)	25	See labels	--	0.5	**For transplant production only.** Begin applications at the first true leaf stage. Repeat at 4 to 5 day intervals until transplanting in the field. Check product labels for specific diseases labeled and application rates. Firewall 50 WP has a maximum of 6 applications per year.
Crown and basal rot (*Fusarium* spp., *Rhizoctonia solani*, *Sclerotinia* spp.), damping off (*Pythium* spp., *Rhizoctonia*), downy mildew (*Peronospora* spp., *Plasmopara* spp.), spots and blights (*Alternaria* spp., *Cercospora* spp., *Phoma* spp., *Septoria* spp.), Phytophthora blight, powdery mildew (*Leveilula* spp. and *Oidiopsis* spp.), rots and blights (*Botrytis*)	pyraclostrobin + boscalid (Pageant Intrinsic)	7 + 11	See label	--	0.5	For use on transplants intended for the home consumer market. **Do not use product for transplants that are intended for agricultural production fields.** Do not tank mix with adjuvants or other agricultural products. See label for application rates specific to a particular disease and for application limits (growing cycle, consecutive crops, production structure).
Tomato (after transplanting in greenhouse)						
Anthracnose (*Colletotrichum* spp.), black mold (*Alternaria alternata*), early blight (*Alternaria solani*), gray leaf spot (*Stemphylium botryosum*), powdery mildew (*Leveilula taurica*), Septoria leaf spot (*Septoria lycopersici*), target spot (*Corynespora cassiicola*)	azoxystrobin + difenoconazole (Quadris Top)	11 + 9	8 fl oz/acre	0	0.5	Do not use for transplant production. Do not make more than two consecutive applications before switching to a fungicide with a different mode of action. Do not exceed 47 fl oz of product per acre per season. Do not apply until 21 days after transplanting or 35 days after seeding. Do not use with adjuvants or tank mix with any EC product on fresh market tomatoes. Plant injury may occur with the use of adjuvants. See label for specifics.
Botrytis gray mold, powdery mildew	Banda de Lupinus albus doce (BLAD) (Fracture)	NG	24.4 to 36.6 fl oz/acre	1	4 hr	Do not make more than two sequential applications before alternating to a fungicide with a different mode of action.

Table 10-58A. Greenhouse Disease Control for Various Vegetable Crops

Disease/Location	Product[1]	FRAC Group	Rate of Formulation	Minimum Days Harvest	Minimum Days Reentry	Schedule and Remarks
Tomato (after transplanting in greenhouse) (continued)						
Anthracnose, bacterial speck and spot, early blight, gray leaf mold, late blight, Septoria leaf spot	copper, fixed (various)	M1	See label	See label	See label	**Some products are OMRI-listed.** See product labels for complete application instructions, specific crop and disease labels, and greenhouse usage.
Late blight (*P. infestans*), Phytophthora blight (*P. capsici*)	cyazofamid (Ranman 400SC)	21	2.1 to 2.75 fl oz/acre (late blight) 2.75 fl oz/acre (Phytophthora blight)	0	0.5	Do not exceed 16.5 fl oz per acre per year. See label for surfactant recommendations. Alternate applications with fungicides that have a different mode of action. Do not make more than three consecutive applications before switching to products that have a different mode of action for three applications before returning to Ranman 400SC. See label for application instructions specific to target disease.
Late blight (*Phytophthora infestans*)	cymoxanil (Curzate 60DF)	27	3.2 to 5.0 oz/acre	3	12 hr	Use with the labeled rate of a protectant fungicide. Do not exceed 30 oz of product per 12-month period.
Early blight (*A.solani*), Botrytis gray mold, powdery mildew (*L. taurica*)	cyprodinil + fludioxonil (Switch 62.5WG)	9 + 12	11 to 14 oz/acre	0	0.5	After two applications, alternate with another fungicide with a different mode of action for two applicaions. See label for application limits. Do not apply to small tomatoes such as cherry or grape-type tomatoes in the greenhouse.
Anthracnose (*Colletotrichum* spp.), black mold (*A. alternata*), Botrytis gray mold, early blight (*A. solani*), gray leaf spot (*S. botryosum*), leaf mold (*Fulvia fulva*), powdery mildew (*L. taurica*), Septoria leaf spot, target spot (*C. cassiicola*)	difenoconazole + cyprodinil (Inspire Super)	3 + 9	16 to 20 fl oz/acre	0	0.5	Do not apply more than 47 fl oz of product per acre per season. Make no more than two consecutive applications per season before alternating with fungicides that have a different mode of action.
Phytophthora and Pythium root rots	etridiazole (Terramaster 4EC)	14	6 to 7 fl oz/acre	3	0.5	For application by drip irrigation. Apply as a 0.01% solution (6.5 fl oz/500 gal water) no sooner than 3 weeks after transplanting or a previous application. Do not exceed 27.4 fl oz of product per acre per crop season. Additional indoor restrictions regarding REI are provided on the label. **Product has a Section 24c registration for this use on greenhouse toamtoes in FL, KY, MS, OK, TN, TX, and VA. Product is not registered for this use in AL, GA, LA, NC, or SC.**
Anthracnose (*Colletotrichum* spp.), early blight (*A. solani*) late blight (*P. infestans*), leaf mold (*Cladosporium fulvum*), Septoria leaf spot (*S. lycopersici*), target spot (*C. cassiicola*) **Suppression:** Bacterial canker, bacterial speck, bacterial spot, buckeye rot (*Phytophthora* spp.)	famoxadone + cymoxanil (Tanos)	11 + 27	6 to 8 oz/acre (early blight) 8 oz/acre (other labeled diseases)	3	0.5	Do not exceed more than 72 oz/acre per crop cycle or 12 month period. Do not make more than one application of product before alternating with a fungicide that has a different mode of action. See label for tank mixing instructions.
Botrytis gray mold	fenhexamid (Decree 50 WDG)	17	1.5 lbs/acre (stand-alone) 1.0 to 1.5 lbs/acre (tank-mix)	0	0.5	Do not make more than two consective applications. See label for additional tank-mixing instructions. Do not exceed 6.0 lbs product per acre per season for greenhouse production.
Anthracnose, bacterial speck and spot, early blight, gray leaf spot, late blight, leaf mold, Septoria leaf spot	mancozeb (various)	M3	See label	See label	See label	See product labels for complete application instructions, specific crop and disease labels, and greenhouse usage.
Anthracnose, bacterial speck and spot, buckeye rot, early blight, gray leaf spot, late blight, leaf mold, Septoria leaf spot	mancozeb + zoxamide (Gavel 75DF)	M3 + 22	1.5 to 2.0 lb/acre	5	2	Do not exceed 8 lb per acre per season (west of the Mississippi River) or 16 lb per acre per season (east of the Mississippi River). For bacterial speck and spot, apply the full rate of product in a tank mix with a full rate of a fixed copper. See label for other application limits. **Product has a 2ee registration for anthracnose.**
Anthracnose (*Colletotrichum* spp.), black mold (*A. alternata*), early blight, (*A. solani*), gray leaf spot (*S. botryosum*), late blight (*P. infestans*), leaf mold (*F. fulva*), powdery mildew (*L. taurica*), Septoria leaf spot, target spot (*C. cassiicola*)	mandipropamid + difenoconazole (Revus Top)	3 + 40	5.5 to 7.0 fl oz/acre	1	0.5	Do not make more than two consecutive applications per season before alternating with fungicides that have a different mode of action. Do not exceed 28 fl oz of product per acre per season.
Alternaria blights and leaf spots, black mold (*A. alternata*), early blight, Botrytis gray mold, powdery mildew (*L. taurica*), basal stem rot (*Sclerotium rolfsii*), Septoria leaf spot, target spot (*C. cassiicola*) **Suppression:** Anthracnose	penthiopyrad (Fontelis)	7	0.5 to 0.75 fl oz/gal per 1,360 sq ft	0	0.5	Do not exceed 72 fl oz of product per year. Do not make more than two consecutive applications per season before alternating with fungicides that have a different mode of action. See label for specific instructions for basal stem rot.
Botrytis rot, early blight, powdery mildew (*L. taurica, Oidiopsis sipula*) **Suppression:** Anthracnose	polyoxin D zinc salt (Affirm WDG, OSO 5%SC; Ph-D)	19	6.2 oz/acre (Affirm) 3.75 to 13.0 fl oz/acre (OSO) 6.2 oz/acre (Ph-D)	0	4 hr	Check products labels for maximum limits of product per season. Alternate with fungicides that have a different mode of action. Check product labels for other tomato diseases also on the label. **OSO 5%SC is not registered for use in AR, KY, MS, or OK.**

DISEASE CONTROL

Table 10-58A. Greenhouse Disease Control for Various Vegetable Crops

Disease/Location	Product[1]	FRAC Group	Rate of Formulation	Minimum Days Harvest	Minimum Days Reentry	Schedule and Remarks
Tomato (after transplanting in greenhouse) (continued)						
Alternaira leaf spot, anthracnose, *Botrytis*, Cercospora leaf spot, powdery mildew, Septoria leaf spot	potassium bicarbonate (Milstop)	NC	1.25 to 5.0 lbs/100 gal	0	1 hr	**OMRI-listed.** Do not exceed 0.5 lbs of product per 4,350 sq ft or 1.15 lbs product per 10,000 sq ft per application. Do not store unused spray solution. See label for additional diseases labeled.
On foliage: Early blight, gray leaf mold, gray leaf spot, late blight, Septoria leaf spot, target spot *On fruit:* Anthracnose, Alternaria fruit rot, *Botrytis*, late blight rot, Rhizoctonia rot	potassium phosphite + chlorothalonil (Catamaran)	33 + M5	4.5 to 5.5 pt/acre (foliage diseases) 7 pt/acre (fruit diseases)	0	0.5	Do not exceed 50 pt per acre per season. Phytotoxicity potential. Do not combine with other pesticides, surfactants or fertilizers. May be applied on the day of harvest.
Damping off and root rots (*Phytophthora* spp., *Pythium* spp.)	promocarb hydrochloride (Previcur Flex)	28	See label	5	0.5	Product applied through a drip system or as a soil drench. Do not apply more than four applications of product after transplanting per crop cycle. Do not mix with other products. Phytotoxicity may occur if applied to dry growing media.
Botrytis gray mold	pyraclostrobin + boscalid (Pageant Intrinsic)	7 + 11	23 oz/acre	0	0.5	Do not tank mix with adjuvants or other agricultural products. Do not exceed 69 oz/acre of product per crop cycle. Do not make more than one application of product before switching to a fungicide with a different mode of action.
Botrytis gray mold, early blight	pyrimethanil (Scala SC)	9	7 fl oz/acre	1	0.5	Plant injury may occur in non-ventilated houses; ventilation for at least 2 hours after application of product. Use only in a tank mix with another fungicide for early blight.Do not exceed 35 fl oz/acre per crop.
Powdery mildew	sulfur (various)	M2	See label	See label	See label	**Some products are OMRI-listed.** See product labels for complete application instructions, specific crop and disease labels, and greenhouse usage. Not all products are registered for use in all states.
Powdery mildew	triflumizole (Terraguard SC)	3	2 to 4 fl oz/100 gal	1	0.5	For use only as a foliar spray. For use in commercial greenhouse production only. Can be used on greenhouse transplants. Do not exceed 40 fl oz of product per acre per cropping system. See label for additional application instructions. **Products not registered for use in all states.**

Biorational and Biological Products for Greenhouse Disease Management

Table 10-58B. Biorational and Biological Products for Greenhouse Disease Management

Material	Disease	Rate of Formulation	Rate of Active Ingredient	Minimum Days Harvest	Minimum Days Reentry	Schedule and Remarks
Bacillus subtilis strain QST 713 (Serenade, CEASE)	Target spot, Early blight, Bacterial spot and speck, Powdery mildew, *Botrytis* gray mold	3 to 6 qt/acre	0.04 to 0.08 qt/100 gal	0	4 hr	**OMRI-listed.** Begin applications prior to disease development, and repeat at 5 to 10 day intervals depending on the formulation.
Bacillus amyloliquefaciens strain D 747	Bacterial spot and speck, gray mold, powdery mildew, early blight, ;ate blight, various soilborne	0.25 to 3 lb/acre		0	4 hr	**OMRI-listed.**
Coniothyrium minitans (Contans WG)	Timber rot or White mold (*Sclerotinia*)	0.75 to 1.5 oz/1,000 sq ft applied to growing medium	0.04 to 0.08 oz/1,000 sq ft	0	4 hr	**OMRI-listed.**
Extract of *Reynoutria sachalinensis* (Regalia)	*Botrytis* gray mold Suppression of bacterial canker	0.5-1.0% v/v solution	—	0	4 hr	Apply preventatively in 25 to 100 gal per acre. Repeat applications in 7 to 14 day intervals depending on crop growth and disease pressure.
Milk (skim)	Viruses[2] (ToMV, TMV)	5 lb/100 gal/100 sq yards	minimum 3.5% protein	Until spray dries	0	Spray plants until run-off. **NOTE: Sooty mold may develop on treated plants.** Dip hands every 5 minutes while handling plants. Dip tools for 1 minute. Do not rinse.
Neem oil	Leaf mold, Powdery mildew, *Botrytis* gray mold	1:200 up to 1:100 solution	—	0	Until spray dries	**OMRI-listed.** Apply to just before runoff. **NOTE: Toxic to honeybees.**
Streptomyces griseoviridis (Mycostop)	Pythium, Rhizoctonia, Phytophthora suppression	1 to 2 g/100 sq ft of area enough to water to cover root zone	0.04 to 0.08 oz/1,000 sq ft	0	4 hr	**OMRI-listed.**

DISEASE CONTROL

Table 10-58B. Biorational and Biological Products for Greenhouse Disease Management

Material	Disease	Rate of Formulation	Rate of Active Ingredient	Minimum Days Harvest	Minimum Days Reentry	Schedule and Remarks
sulfur (various)	Powdery mildew	See label	See label	See label	See label	See label. Rates vary depending on the formulation. **OMRI-listed.** Do not apply if temperatures will exceed 90 degrees F for 3 days following application as plant injury may occur.
Trichoderma harzianum (Plant Shield)	Pythium root rot, *Botrytis* gray mold	See Remarks	See Remarks	0	0	Suspend in 100 gallons water. Apply 50 to 100 gallons per 800 square feet area as a drench for *Pythium*. Use 0.5 to 1 ounces per gallon water for *Botrytis* as a foliar spray.

[1] Other formulations may be available.
[2] Use in combination with seed treatments and sanaition practices.

Efficacy of Products for Greenhouse Disease Control

F. Louws, Plant Pathologist, NCState University; M. Paret, Plant Pathologist, University of Florida; M. L. Lewis Ivey, Plant Pathologist, The Ohio State University; A. Keinath, Plant Pathologist, Clemson University; R. A. Melanson, Extension Plant Pathologist, Mississippi State University

Table 10-59. Efficacy of Products for Greenhouse Disease Control

Scale: E, excellent; G, good; F, fair; P, poor; NC, no control; ND, no data

Product[1]	Fungicide Group F	Preharvest Interval (Days)	Anthracnose (*Colletotrichum coccodes*)	Bacterial Soft Rot (*Erwinia* sp.)	Bacterial Canker (*Clavibacter michiganensis* subsp. *michiganensis*)	Botrytis Gray Mold (*Botrytis cinerea*)	Early Blight (*Alternaria solani*)	Leaf Mold (*Passalora ulva*[3])	Powdery Mildew (*Leveillula taurica*)	Phytophthora Root Rot (*Phytophthora* sp.)	Pythium Root Rot (*Pythium myriotylum*)	Rhizoctonia Root Rot (*Rhizoctonia solani*)	Septoria Leaf Spot (*Septoria lycopersici*)	Target Spot (*Corynespora cassicola*)	Timber Rot (White Mold) (*Sclerotinia sclerotiorum*)
acibenzolar-*S*-methyl (Actigard)	P	14	NC	F	F	NC	NC	NC	NC	NC	NC	NC	NC	NC	NC
bacteriophage (AgriPhage)	NC	0	NC	P	F	NC	NC	NC	NC	NC	NC	NC	NC	NC	NC
Bacillus subtilis (Rhapsody)	NC	0	ND	ND	NC	P	P	ND	P	NC	NC	NC	ND	P	ND
BLAD (Fracture)	NC	1	ND	NC	NC	F	ND	ND	F	NC	NC	ND	ND	ND	ND
boscalid (Endura)	7	0	ND	NC	NC	F	F	ND	NC	NC	NC	NC	ND	G	F
Coniothyrium minitans (Contans WG)	NC	0	NC	NC	NC	NC	NC	NC	NC	NC	NC	NC	NC	NC	P
cyprodinil + fludioxonil (Switch 62.5WG)	9 + 12	0	G	NC	NC	E	F	ND	G	NC	NC	F	F	ND	ND
difenoconazol + cyprodinil (Inspire Super)	3 + 9	0	G	NC	NC	E	G	ND	G	NC	NC	NC	ND	G	F
etridiazole (Terramaster)[2]	14	3	NC	NC	NC	NC	NC	NC	NC	F	F	ND	NC	NC	NC
famoxadone + cymoxanil (Tanos)	11 + 27	3	F	ND	P	ND	F	F	P	NC	NC	NC	F	F	ND
fenhexamide (Decree 50 WDG)	17	1	NC	NC	NC	NC	NC	NC	F	NC	NC	NC	NC	NC	NC
fixed copper	M1	0	P	F	F	P	F	P	P	NC	NC	NC	P	P	NC
hydrogen dioxide (Oxidate)	NC	0	ND	ND	NC	ND	ND	ND	P	P	P	ND	ND	ND	ND
mancozeb (Dithane 75 DF)	M3	5	G	NC	NC	F	G	F	ND	NC	NC	NC	F	F	ND
neem oil (Trilogy, Triact 70)	NC	0	ND	ND	ND	ND	ND	ND	ND	ND	ND	ND	ND	ND	ND
penthiopyrad (Fontelis)	7	0	P	NC	NC	F	G	G	G	NC	NC	NC	F	G	F
polyoxin D zinc salt (Ph-D; OSO 5% SC)	19	0	F	NC	NC	F	F	ND	F	ND	ND	ND	ND	F	ND
propamocarb hydrochloride (Previcur Flex)	28	5	NC	NC	NC	NC	NC	NC	NC	F	F	NC	NC	NC	NC

Table 10-59. Efficacy of Products for Greenhouse Disease Control

Scale: E, excellent; G, good; F, fair; P, poor; NC, no control; ND, no data

pyrimethanil (Scala)	9	1	NC	NC	NC	FR	NC	NC	NC	NC	NC	NC	NC	NC	NC
Streptomyces griseoviridis (Mycostop)	NC	0	NC	NC	NC	NC	NC	NC	NC	NC	F	F	NC	NC	NC
streptomycin sulfate (Agri-Mycin 17)[2]	18	0	NC	F	F	NC	NC	NC	NC	NC	NC	NC	NC	NC	NC
sulfur[P] (Microthiol Disperss)	M2	0	P	NC	NC	NC	NC	NC	F	NC	NC	NC	NC	NC	NC
Trichoderma harzianum (Plant Shield)	NG	0	NC	NC	NC	NC	NC	NC	NC	NC	P	F	NC	NC	NC

[1] Efficacy ratings do not necessarily indicate a labeled use for every disease.

[2] For use on transplants only.

[P] Sulfur may be phytotoxic; follow label carefully.

[F] To prevent resistance in pathogens, alternate fungicides within a group with fungicides in another group. Fungicides in the "M" groups are generally considered "low risk" with no signs of resistance developing to the majority of fungicides. "NC" indicates that the product has not been classified into a fungicide group.

Seed Treatments

Seed sanitation to eradicate bacterial or viral plant pathogens: When treating vegetable seeds, **it is critical to follow the directions exactly**, because germination can be reduced by the treatment and/or the pathogen may not be completely eliminated. The effect of a treatment on germination should be determined on a small lot of seeds prior to treating large amounts of seed. Treatments should not be applied to: 1) pelleted seed, 2) previously treated seed, or 3) old or poor quality seed. A protective fungicide treatment (see below) can be applied to the seed following treatment for bacterial pathogens.

Seed treatments to prevent damping off diseases: Most commercially available vegetable seeds come treated with at least one fungicide and/or insecticide. Vegetable producers who would like to apply their own seed treatment should purchase non-treated seed. While many fungicides are labeled for use on vegetable seed, most fungicides are restricted to commercial treatment only and should not be applied by producers. Labeled fungicides can be applied to seed following treatment for bacterial pathogens (see above). **Do not use fungicide treated seed for food or feed.**

HOT WATER TREATMENT

By soaking seed in hot water, seedborne fungi and bacteria can be reduced, if not eradicated, from the seed coat. Hot water soaking will not kill pathogens associated with the embryo nor will it remove seedborne plant viruses from the seed surface.

1. Place seed loosely in a weighted cheesecloth or nylon bag.
2. Warm the seed by soaking it for 10 minutes in 100 degree Fahrenheit (37 Celsius) water.
3. Transfer the warmed seed into a water bath already heated to the temperature recommended for the vegetable seed being treated. The seeds should be completely emerged in the water for the recommended amount of time (see table bellow). Agitation of the water during the treatment process will help to maintain a uniform temperature in the water bath.
4. Transfer the hot water treated seed into a cold-water bath for five minutes to stop the heating action.
5. Remove seed from the cheesecloth or nylon bag and spread them evenly on clean paper towel or a sanitized drying screen to dry. Do not dry seed in areas where fungicides, pesticides or other chemicals are located.
6. Seed can be treated with a labeled fungicide to protect against damping off pathogens.

CHLORINE BLEACH TREATMENT

Treating seeds with a solution of chlorine bleach can effectively remove bacterial pathogens and some viruses (i.e. Tobacco Mosaic Virus) that are borne on the surface of seeds.

1. Add 1 quart (946 ml) of Clorox® bleach to 5 quarts (4.7 L) of potable water.
2. Add a drop or two of liquid dish detergent or a commercial surfactant such as Activator 90 or Silwet to the disinfectant solution. Add seed to the disinfectant solution (1 pound of seed per 4 quarts of disinfectant solution) and agitate for 1 minute.
3. Prepare fresh disinfectant solution for each batch of seeds to be treated.
4. Rinse the seed in a cold water bath for 5 minutes to remove residual disinfectant.
5. Spread seeds evenly on clean paper towel or a sanitized drying screen to dry. Do not dry seed in area where fungicides, pesticides, or other chemicals are located.
6. Seed can be treated with a labeled fungicide to protect against damping off pathogens.

HYDROCHLORIC ACID TREATMENT

Tomato seed can be treated with a dilute solution of hydrochloric acid (HCl) solution to eliminate seedborne bacterial pathogens such as *Xanthomonas* spp. (Bacterial leaf spot), *Pseudomonas syringae* pv. *tomato* (Bacterial speck) and *Clavibacter michiganensis* subsp. *michiganensis* (Bacterial canker). Hydrochloric acid can also be used to remove TMV from the surface of tomato seed. **Do not use HCL-treated seed for food or animal feed.**

1. Prepare a 5% solution of HCl by adding one part acid to 19 parts potable water. Prepare the acid solution in a well-ventilated area and avoid direct skin contact with the acid.
2. Soak seeds for 6 hours with gentle agitation.
3. Carefully drain the acid off of the seed and rinse seed under running potable water for 30 minutes. Alternatively, rinse the seeds 10 to 12 times with potable water to remove residual acid.
4. Spread seeds evenly on clean paper towel or a sanitized drying screen to dry. Do not dry seed in area where fungicides, pesticides, or

other chemicals are located.

5. Seed can be treated with a labeled fungicide to protect against damping off pathogens.

TRISODIUM PHOSPHATE TREATMENT

Tomato seed can be treated with trisodium phosphate (TSP) to eradicate seed-transmitted TMV. **Do not use TSP-treated seed for food or animal feed.**

1. Prepare a 10% solution of TSP (1 part TSP in 9 parts potable water). Trisodium phosphate is available at most home supply or paint stores. Avoid direct skin contact with the TSP solution.
2. Soak seed for 15 minutes in the disinfectant solution.
3. Rinse the seed in a cold water bath for 5 minutes to remove residual disinfectant.
4. Spread seeds evenly on clean paper towel or a sanitized drying screen to dry. Do not dry seed in area where fungicides, pesticides, or other chemicals are located.
5. Seed can be treated with a labeled fungicide to protect against damping off pathogens.

TESTING SEED GERMINATION AFTER SEED TREATMENTS

1. Randomly select 100 seeds from each seed lot.
2. Treat 50 seeds using one of the sanitizers described above.
3. After the treated seed has dried and before application of a protectant fungicide, plant the treated and non-treated seed separately in flats containing planting mix according to standard practice. Label each group as treated or non-treated.
4. Allow the seeds to germinate and grow until the first true leaf appears (to allow for differences in germination rates to be observed).
5. Count seedlings in each group separately.
6. Determine the percent germination for each group: # seedlings emerged ÷ # seeds planted x 100.
7. Compare percent germination between the treated and non-treated groups. Percent germination should be within 5% of each other.

Table 10-60. Recommended Temperatures and Treatment Times for Hot Water Disinfestation of Vegetable Seed

Vegetable Crop	Water Temperature (°F/°C)	Soaking Time (Minutes)
Broccoli	122/50	20 to 25
Brussels sprout	122/50	25
Cabbage	122/50	25
Carrot	122/50	15 to 20
Cauliflower	122/50	20
Celery	122/50	25
Chinese cabbage	122/50	20
Collard	122/50	20
Cucumber[1]	122/50	20
Eggplant	122/50	25
Garlic	120/49	20
Kale, Kohlrabi	122/50	20
Lettuce	118/48	30
Mint	112/44	10
Mustard, Cress, Radish	122/50	15
Onion	115/46	60
Pepper	125/51	30
Rape, Rutabaga	122/50	20
Shallot	115/46	60
Spinach	122/50	25
Tomato	122/50	25
Turnip	122/50	20

[1] Cucurbits other than cucumbers can be severely damaged by hot water treatment and should be disinfested using chlorine bleach.

DISEASE CONTROL

Table 10-61. Products for Seed Treatment

	42-S Thiram (thiram)	Allegiance (metalaxyl)	Acquire (metalaxyl)	Apron (mefenoxam)	Belmont 2.7FS (metalaxyl)	Botran 75W (dicloran)	Captan 400 (captan)	Cruiser Maxx (thiamethoxam + mefenoxam + fludioxonil)	Cruiser Maxx Potato (thiamethoxam + fludioxonil)	Dyna-Shield (fludioxonil)	Dividend Extreme (difenoconazole + mefenoxam)	Dynasty (azoxystrobin)	Emesto Silver (penflufen + prothioconazole)	Maxim 4FS (fludioxonil)	Maxim MZ (fludioxonil + mancozeb)	MetaStar 2E AG (metalaxyl)	MonCoat MZ (flutolanil + mancozeb)	Sebring 318S (metalaxyl)	Sebring 480S (metalaxyl)	Spirato 480 FS (fludioxonil)	Vitaflo-280 (carboxin + thiram)	Tops MZ-Gaucho (thiophanate-methyl + mancozeb)	Trilex (trifloxystrobin + metalaxyl)
Fungicide Group[F]	M3	4	4	4	4	14	M3	4+12	12	12	3+4	11	3	12	12+M3	4	7+M3	4	4	M3	7+M3	1+M3	11+ 4
Beans, Snap	X	X	X	X	X		X	X		X		X		X		X		X	X	X	X		X
Beans, Lima	X	X	X	X	X		X	X		X		X		X		X		X	X	X	X		X
Beets	X	X	X	X	X		X			X		X		X		X		X	X	X			
Broccoli	X		X	X	X		X			X		X		X				X	X	X			
Carrots	X	X	X	X	X					X		X		X		X		X		X			
Celery				X										X						X			
Chinese Cabbage	X		X	X	X							X		X						X			
Cole Crops	X		X	X	X		X			X		X		X						X			
Cucumbers	X	X	X	X	X		X			X		X		X		X		X	X	X			X
Eggplants	X		X	X	X					X		X		X				X		X			X
Garlic				X						X		X		X						X			
Greens, Mustard	X		X	X	X		X			X		X		X				X		X			
Greens, Turnip	X		X	X	X		X			X		X		X						X			
Horseradish			X	X	X					X		X		X						X			
Leeks				X						X		X		X									
Lettuce	X		X	X	X					X		X		X				X		X			
Muskmelons	X		X	X	X		X			X		X		X				X		X			
Okra	X		X		X					X		X		X		X		X	X				
Onions, Dry	X		X	X	X					X		X		X						X			
Onions, Green	X		X	X	X					X		X		X						X			
Parsley			X	X	X					X		X		X				X		X			
Parsnips			X	X	X					X		X		X				X		X			
Peas	X	X	X	X	X		X	X		X				X				X	X	X			X
Peppers	X		X	X	X		X			X		X		X				X		X			X
Pumpkins	X		X	X	X		X			X		X		X				X		X			X
Radish	X		X	X	X		X			X		X		X				X		X			
Spinach	X		X	X	X		X			X		X		X		X		X	X	X			
Squash, Summer	X		X	X	X		X			X		X		X				X		X			X
Squash, Winter	X		X	X	X		X			X		X		X				X		X			X
Sweet Corn	X	X		X			X			X	X	X		X				X		X	X		
Sweet Potatoes				X		X				X		X		X				X		X			
Tomatoes	X	X	X	X	X					X		X		X				X		X			X
Watermelon	X	X	X	X	X		X			X		X		X				X		X			X
White Potatoes (Irish)							X		X[2]	X		X	X[3]	X	X		X[3]			X		X[4]	X

[F] To prevent resistance in pathogens, alternate fungicides within a group with fungicides in another group. Fungicides in the "M" group are generally considered "low risk" with no signs of resistance developing to the majority of fungicides.

Table 10-62. Biocontrol Agents and Disinfestants Registered for Seed Treatment

Crop	Actinovate STP (Streptomyces lydicus WYEC 108)	Kodiak HB (Bacillus subtilis GB03)	Mycostop (Streptomyces griseoviridis K61)	T-22 Planter Box (Trichoderma harzianum)	Yield Shield (Bacillus pumilus GB34)	Clorox	Hot water	Hydrochloric acid (HCl)	Trisodium phosphate (TSP)
Beans, Snap	X	X	X	X	X				
Beans, Lima	X	X	X	X	X				
Beets	X	X	X	X					
Broccoli	X	X					X		
Carrots	X	X	X	X			X		
Celery	X	X					X		
Chinese Cabbage	X	X					X		
Cole Crops	X	X	X				X		
Cucumbers	X	X		X		X	X		
Eggplants	X	X	X			X	X		
Garlic	X	X		X			X		
Greens, Mustard	X	X				X	X		
Greens, Turnip	X	X				X	X		
Horseradish	X	X							
Leeks	X	X	X						
Lettuce	X	X	X			X	X		
Muskmelons	X	X		X		X			
Okra	X	X							
Onions, Dry	X	X	X	X			X		
Onions, Green	X	X	X	X			X		
Parsley	X	X	X			X			
Parsnips	X	X		X					
Peas	X	X	X	X	X				
Peppers	X	X	X			X	X		
Pumpkins/Winter	X	X		X		X			
Radish	X	X	X	X			X		
Spinach	X	X	X				X		
Squash, Summer	X	X		X		X			
Sweet Corn	X	X		X					
Sweet Potatoes	X			X					
Tomatoes	X	X	X			X	X	X	X
Watermelon	X	X		X		X			
White Potatoes (Irish)				X					

DISEASE CONTROL

Sanitizing

Follow manufacturers label in all cases. The objective of treating soil in greenhouses and plant beds is to reduce to acceptable levels weeds, nematodes, insects, fungi, and bacteria in the soil that cause damage and disease in plants. This is critical for the successful production of greenhouse crops and healthy plants for field use.

Follow general procedures for successful soil fumigation as outlined in commercial literature. Do not overtreat soil. Treated soils might develop nutrient imbalance and be especially susceptible to secondary infestation of disease-causing agents. Always follow directions on the label on the pesticide containers.

CAUTION: Some products or treatments cannot be used for crops.

Table 10-63. Sanitizing Greenhouses and Plant Beds (Soils and Beds)

Scale: E, excellent; G, good; F, fair; P, poor; NC, no control; ND, no data.

Material	Rate to Use		Weeds	Nematodes	Insects	Bacteria	Fungi	Viruses	Use and Remarks
	Formulation	Application							
dry heat	180 degrees F for 30 min	Place small quantities in oven.	G	G	G	G	G	P	
Metam-sodium (Metam CLR, Sectagon, Vapam)	11 to 22 oz/100 sq ft (37.5 to 75 gal/acre)	Inject 4 to 6 in. deep, space chisels 6 in. apart. Cover with plastic 7 to 15 days. Aerate 2 to 4 weeks.	F	G	G	F	G	P	All crops, all pests. Long waiting period after fumigating in cold soil (under 60 degrees F).
	11 to 22 oz/100 sq ft (37.5 to 75 gal/acre)	Dilute with sufficient water and sprinkle uniformly to penetrate 2 in. into soil. A hose proportioner may be used. Cover with plastic 7 to 15 days. Aerate 2 to 4 weeks.							
solarization	140 degrees F, 4 to 8 hr/day for 7 days	Pots, benches, tools, shallow soil, structures	G	F	G	F	G	P	Close greenhouse during hot, sunny days in summer for at least 1 week. Greenhouse must reach 140 degrees F or higher each day. Remove debris and heat-sensitive materials and keep greenhouse and contents moist. Will not control TMV or pests 0.5 inch or deeper in soil.
steam	Heat soil from 180 degrees to 200 degrees F (30 min) 6 in.	Perforated pipes on or in soil, cover with tarp	G	G	G	G	G	F	All crops, most pests.
Telone C-17 + herbicide	10.3 to 17.1 gal/acre See label for herbicide rates	Inject 10 to 12 in. deep with chisels spaced 12 in. apart. Seal the soil by packing, wetting, or covering with plastic mulch.	G	G	F	F	G	P	Use higher rates for heavy soils. See label for organic soils.
Telone II + herbicide	7 to 16 oz/100 sq ft (25 to 54 gal/acre) See label for herbicide rates	Inject 4 to 6 in. deep, space chisels 12 in. apart. Cover 1 week, aerate 3 weeks.	G	G	F	P	P	P	Not for greenhouse use.

NOTE: *Follow manufacturer's label in all cases*

Table 10-64. Sweetpotato Storage House Sanitation

Material	Rate per 1,000 Cubic Feet of Space	Methods and Remarks
Heat	140 degrees F 4 to 8 hr/day for 7 days or 180 degrees F for 30 minutes	See remarks under sanitizing greenhouses. The storage house, ventilation system, and equipment must be very clean and moist during the procedure. *Caution:* rot-causing organisms inside a drain will probably not be exposed to lethal temperature.

Table 10-65. Water, Produce, and Equipment Sanitation

Medium	Sanitizer	Contact Time (minutes)	Rate of Material to Use		Method, Schedule, and Remarks
			Target Rate (ppm)	Formulation	
Wash water, dump tank water, or vegetable wash water*	calcium hypochlorite (Aquafit)	2	25	1 oz/200 gal	
	chlorine dioxide (Harvest Wash, ProOxine, Anthium Dioxide, Adox 750)	1 to 10	3 to 5	Varies between products; see product labels.	Maintain water pH between 6.0 and 10. Restricted to large operations. Requires automated and controlled injection systems. **NOTE: Chlorine dioxide is explosive.**
	chlorine gas (99.9%)	—	Contact supplier for rates.		Restricted to very large operations. Requires automated and controlled injection systems. Regulated by both the EPA (water) and FDA (food contact surfaces).
	hydrogen dioxide or peroxide (StorOx)	—	Varies based on method of application.		
	hydrogen peroxide + peroxyacetic acid (BioSide HS) (Keystone Fruit and Vegetable Wash) (PAA Sanitizer FP) (Perasan A) (SaniDate 5.0) Tsunami 100 (Victory) (VigorOx 15 F&V)	1 1.5 0.75 — 0.75 — 1.5 —	80 peroxyacetic acid 80 peroxyacetic acid 88 to 130 (peroxyacetic acid, non-porous surfaces) 25 peroxyacetic acid 24 to 85 peroxyacetic acid 30 to 80 80 5 to 85 peroxyacetic acid (postharvest pathogens) 45 peroxyacetic acid (foodborne pathogens)	1 fl oz/16.4 gal 1 fl oz/16.4 gal 3 to 3.5 fl oz/16 gal 1 oz/20 gal 5.9 to 20.9 fl oz/100 gal 2.5 to 6.7 fl oz/100 gal 1 fl oz/16.4 gal 0.1 to 1 fl oz/16 gal 0.54 fl oz/16 gal	Contact times vary depending on the governing sanitary code. Post-sanitation rinse is not necessary.
	sodium hypochlorite (5.25%) (12.75%)	— —	150 150	2.9 ml/L 1.18 ml/L	Maintain water pH between 6.0 and 7.5. Noxious chlorine gas can be released when the pH drops below 6.0. **NOTE: Household bleach is NOT registered for use with fresh produce.**
	sodium hypochlorite (Agclor 310) (Dibac) (Dynachlor) (Extract-2) (JP Optimum CRS) (Maxxum 700) (Zep FS Formula 4665)	— 2 2 2 - 2 2	65 to 400 25 25 25 25 25 25	0.5 to 3 gal/1000 gal 1 oz/20 gal 5 oz/200gal 5 oz/200gal 0.75 oz/10 gal 8 oz/200gal 5 oz/200gal	Monitor residual chlorine or change solution when it is visibly dirty. Rinse produce with potable water prior to packing.
Equipment** (conveyors, scrubbers, plastic harvest containers, peelers, field equipment, etc.)	calcium hypochlorite (Aquafit)	2	600 ppm (porous surfaces)	3 oz/20 gal	Do not rinse or soak equipment overnight.
	chlorine dioxide (ProOxine, Sanogene, Anthium Dioxide, Adox 750)	1 to 10	10 to 20 (porous or non-porous surfaces) 500 (ceilings, floors and walls)	Varies between products; see product labels.	
	hydrogen dioxide or peroxide (StorOx)	—	1 to 3 ppm (1:300 to 1:100) (non-porous surfaces)	0.5 to 1.25 fl oz/gal	Apply until run-off. Requires a thorough post-sanitation rinse with potable water.
	hydrogen peroxide + peroxyacetic acid (BioSide HS) (PAA Sanitizer FP) (Perasan A) (SaniDate 5.0) (VigorOx 15 F&V)	1 or more 1 or more 1 1 1	93-500 peroxyacetic acid 88 to 130 peroxyacetic acid (non-porous surfaces) 82 to 500 peroxyacetic acid ~128 85 peroxyacetic acid	0.7 to 3.8 fl oz/10 gal 1 to 1.5 fl oz/5 gal 1 to 6.1 oz/6 gal 1.6 fl oz/5 gal 3.1 fl oz/50 gal	Contact time varies depending on the governing sanitary code. Consult labels as some products require a post-disinfection rinse with potable water.
	sodium hypochlorite (5.25%)	2 2 —	100 to 200 (non-porous surfaces) 600 (porous surfaces) 1000 to 2000 (floors and walls)	1.9 to 3.8 ml/L 11.4 ml/L 1900 to 3800 ml/L	Noxious chlorine gas can be released when the pH drops below 6.0. Porous surfaces require a thorough post-disinfection rinse with potable water. Allow all surface types to air dry prior to re-use.
	sodium hypochlorite (12.75%)	2 2 —	100 to 200 (non-porous surfaces) 600 (porous surfaces) 1000 to 2000 (ceilings, floors and walls)	0.78 to 1.56 ml/L 4.68 ml/L 780 to 1560 ml/L	
	sodium hypochlorite (Agclor 310, Dibac, Dynachlor, Extract-2, JP Optimum CRS, Maxxum 700, Zep FS Formula 4665)	1	Varies based on method of application.		
	quaternary ammonia (DDAC) (KleenGrow)	10	—	1 fl oz/gal	Allow surfaces to air dry. If treated surfaces will contact food, thoroughly rinse surfaces with potable water.

* Recommendations are for potable water only. Recommended rates are not effective in reducing pathogen populations in non-potable water (i.e. surface or ground water).

** Recommendations are for potable water only. Always wash off organic debris and soil with water prior to sanitizing. Rates and contact time are dependent on surface type.

DISEASE CONTROL

Generic and Alternative Fungicides

Table 10-66. Generic and Alternative Fungicides for Use on Vegetable Crops

Common Name	Trade Name(s)
chlorothalonil	Bravo Ultrex *(Syngenta)*
	Bravo Weather Stik *(Syngenta)*
	Bravo Zn *(Syngenta)*
	Chloronil 720 *(Syngenta)* Chlorothalonil 720 SC *(Arysta)* Echo 720 *(SipcamAdvan)* Echo 90DF *(SipcamAdvan)* Echo Zn *(SipcamAdvan)* Equus 500 Zn *(MANA)*
	Equus 720 SST *(MANA)*
	Equus DF *(MANA)*
	Initiate 720 *(Loveland Products)*
	Initiate ZN *(Loveland Products)*
copper hydroxide	Champ DP Dry Prill *(Nufarm)*
	Champ Formula 2 Flowable *(Nufarm)*
	Champ WG *(Nufarm)*
	Champion Wettable Powder *(Nufarm)*
	Kentan DF *(Isagro USA)*
	Kocide 2000 *(DuPont)*
	Kocide 3000 *(DuPont)* Nu Cop 3L *(Albaugh)* Nu Cop 50DF *(Albaugh)*
	NU Cop 50WP *(Albaugh)*
	Nu Cop HB *(Albaugh)*
copper sulfate (basic)	Basic Copper 53 *(Albaugh)*
	Copper Z 4/4 *(Helena)*
	Cuprofix Ultra 40 Disperss *(UPI)*
	Cuproxat *(NuFarm)*
fosetyl-Al	Aliette WDG Fungicide *(Bayer)*
	Linebacker WDG *(NovaSource)*
fludioxonil	Mamim 4FS *(Syngenta)*
	Spirato 480 FS *(Nufarm)*
iprodione	Enclosure 4 *(Devgen)* Iprodione 4L AG *(Arysta)* Meteor *(UPI)*
	Nevado 4F *(MANA)*
	Rovral 4 Flowable Fungicide *(Bayer)*
	Rovral 4 Flowable-*(FMC)*
mancozeb	Dithane F-45 Rainshield *(Dow)* Dithane M-45 *(Dow)*
	Koverall *(Cheminova)* Manzate Flowable *(UPI)* Manzate Max *(UPI)* Manzate Pro-Stick *(UPI)* Penncozeb 4FL *(UPI)* Penncozeb 75DF *(UPI)* Penncozeb 80WP *(UPI)*
	Roper DF Rainshield *(Loveland Products)*
mefenoxam	Ridomil Gold GR *(Syngenta)* Ridomil Gold SL *(Syngenta)* Ultra Flourish *(Nufarm)*
myclobutanil	Rally 40WSP *(Dow)*
	Sonoma 25EW AG *(Albaugh)*
	Sonoma 40WSP *(Albaugh)*
pentachloronitrobenzene	Blocker 4F
phosphite, potassium	Helena Prophyt *(Helena)*
	Confine Extra *(Winfield Solutions)*
	Reveille *(Helena)*
phosphite (mono- and dibasic salts)	Phostrol *(Nufarm)*
phosphorous acid (mono- and dipotassium salts)	Alude *(Cleary)*
	Fosphite Fungicide *(JK Biotech)*
	Fungi-Phite *(Plant Protectants)*
	K-Phite 7LP AG *(Plant Food Systems)*
	Rampart *(Loveland Products)*
propamocarb hydrochloride	Previcur Flex *(Bayer)*
	Promess *(Agriphar)*
propiconazole	AmTide Propiconazole 41.8% EC *(AmTide)*
	Bumper 41.8 EC *(MANA)*
	Bumper ES *(MANA)*

Table 10-66. Generic and Alternative Fungicides for Use on Vegetable Crops

Common Name	Trade Name(s)
Propiconazole (continued)	Fitness *(Loveland Products)*
	Propi-star EC *(Albaugh)*
	Propicure 3.6F *(Direct Ag Source)*
	Propimax EC *(Dow AgroSciences)*
	Shar-Shield PPZ *(Sharda USA)*
	Tilt *(Syngenta)*
	Topaz *(Winfield Solutions)*
	Willowood Propicon 3.6EC *(Willowood USA)*
sulfur	Cosavet-DF *(Sulphur Mills Limited)*
	CSC 80% Thiosperse *(Martin Resources)* CSC Dusting Sulfur *(Martin Resources)* CSC Thioben 90 *(Martin Resources)* CSC Wettable Sulfur *(Martin Resources)*
	Dusting Sulfur *(Loveland Products; Wilbur-Ellis)* First Choice Dusting Sulfur *(Loveland Products)* IAP Dusting Sulfur *(Independent Agribusiness Professionals)*
	InteGro Magic Sulfur Dust *(InteGro Inc.)*
	Kumulus DF *(Arysta)* Liquid Sulfur Six *(Helena)* Micro Sulf *(Nufarm)*
	Microfine Sulfur *(Loveland Products)*
	Microthiol Disperss *(UPI)*
	Special Electric Sulfur *(Wilbur-Ellis)*
	Spray Sulfur *(Wilbur-Ellis)*
	Sulfur 6L *(Arysta)* Sulfur 90W *(Drexel)* Sulfur DF *(Wilbur-Ellis)*
	That Flowable Sulfur *(Stoller Enterprises)*
	Thiolux *(Loveland Products)*
	Wettable Sulfur *(Helena)*
	Yellow Jacket Dusting Sulfur *(Georgia Gulf Sulfur)*
	Yellow Jacket Wettable Sulfur *(Georgia Gulf Sulfur)*
tebuconazole	AmTide TEBU 3.6F *(AmTIde)*
	Barrier *(Real Farm Technologies)*
	Folicur *(Bayer)*
	Monsoon *(Loveland Products)* Onset 3.6L *(Winfield Solutions)* Orius 3.6 F *(MANA)*
	Solera Tebuconazole 3.6F *(Solera)*
	Tebu-Crop 3.6F *(Sharda USA)*
	Tebucon 3.6F *(Repar Corp.)*
	TebuStar 3.6L *(Albaugh)* Tebuzol 3.6F *(UPI)* Toledo 3.6F *(Rotam)*
thiophanate-methyl	Incognito 4.5 F *(MANA)*
	Incognito 85 WDG *(MANA)* Thiophanate Methyl 85 WDG *(MANA)* T-Methyl 4.5F *(Nufarm)*
	T-Methyl 70W WSB *(Nufarm)*
	Topsin 4.5FL *(UPI)*
	Topsin M 70 WDG *(UPI)*

DISEASE CONTROL

Table 10-67. Biopesticides and Fungicide Alternatives for Vegetables

Active Ingredient	Product	Crops	Target Diseases/Pests	Greenhouse Use	OMRI-Listed	Comments
Acibenzolar-*S*-methyl	Actigard	chili pepper, cucurbits, lettuce, onion, spinach, tomato	bacterial blights, downy mildew, powdery mildew (crop dependent, see label)	No	No	Do not apply to plants stressed by heat, cold, or moisture extremes.
Bacillus pumilus QST2808	Ballad Plus, Sonata	Cole crops, cucurbits, legumes, bulb vegetables, root crops, pepper, tomato, sweet corn	early blight, late blight, downy mildew, powdery mildew, leaf blights, rust	Yes	Yes	Ballad Plus can be used on sweet corn only.
Bacillus subtilis GB03	Companion	most vegetables (see label)	root diseases	Yes	Yes	
Bacillus subtilis MBI 600	Subtilex NG	cucurbits, eggplant, pepper, tomato	root diseases, powdery mildew	Yes	No	Apply to soil or potting medium; use as a foliar spray
Bacillus subtilis QST713	Cease, Serenade Max	Cole crops, leafy vegetables, legumes, cucurbits, pepper, tomato	downy mildew, powdery mildew, leaf blights	Yes	Yes	
Coniothyrium minitans	Contans WG	most vegetables (see label)	*Sclerotinia sclerotiorum* (white mold, timber rot, drop)	Yes	Yes	Apply to soil or potting medium.
Gliocladium cantenulatum	PreStop Biofungicide	most vegetables (see label)	seed rots, root diseases, *Botrytis* stem canker	Yes		
Gliocladium virens GL-21	SoilGard 12G	most vegetables (see label)	seed rots, root diseases	Yes	Yes	Do not apply in conjunction with chemical fungicides.
Hydrogen peroxide (Oxidate)	Oxidate, Terracide	most vegetables (see label)	root diseases, leaf blights	Yes	Yes	
Myrothecium verrucaria	DiTera DF	Cole crops, cucurbits, eggplant, leafy vegetables, legumes, pepper, root and tuber vegetables, tomato	nematodes	Yes	Yes	
Neem oil	Trilogy	vegetables	foliar diseases	Yes	Yes	May cause leaf burn; test a small number of plants before spraying entire crop.
Oils from cottonseed, corn, and garlic	Mildew Cure	tomato, cucurbits	powdery mildew	Yes		May cause leaf burn; test a small number of plants before spraying entire crop.
Oils from clove, rosemary, and thyme	Sporatec	most vegetables (see label)	powdery mildew, fungal leaf blights	Yes	Yes	Addition of a spray adjuvant (spreader or penetrant) is recommended.
Oil from soybean	Oleotrol-M	most vegetables (see label)	*Botrytis* gray bold, downy mildew, powdery mildew	Yes	Yes	Tank-mix with a spreader-sticker.
Paecilomyces lilacinus	MeloCon WG	most vegetables (see label)	nematodes	Yes	Yes	
Phage	AgriPhage	most vegetables (see label)	bacterial spot, speck	Yes	No	
Phosphorous compounds	Alude, Fosphite					
Fungi-Phite, Phostrol, ProPhyt, Rampart	most vegetables (see label)	downy mildew, powdery mildew, leaf blights	Yes	No		
Potassium bicarbonate	Armicarb, Kaligreen, Milstop	most vegetables (see label)	powdery mildew, fungal leaf blights	Yes	Yes (Kaligreen, Milstop)	pH of spray solution should not be below 7.0.
Potassium salts of fatty acids	M-Pede	most vegetables (see label)	powdery mildew	Yes		Do not mix with surfactants or apply to stressed plants to avoid plant injury.
Potassium silicate	Sil-MATRIX	most vegetables (see label)	powdery mildew, *Botrytis* gray mold	Yes	Yes	Tank-mix with a non-ionic surfactant for best results.
Pseudomonas chloroaphis	Ateze	most vegetables (see label)	stem, root diseases	Yes	No	Greenhouse use only.
Reynoutria sachalinensis	Regalia	most vegetables (see label)	powdery mildew, fungal leaf blights	Yes	Yes	First application should be made before symptoms appear.
Streptomyces griseoviridis	Mycostop	most vegetables (see label)	seedling, root, and stem rots	Yes	Yes	Can be added to potting mix or applied in-furrow to field soil.
Streptomyces lydicus	Actinovate AG	most vegetables (see label)	seedling, root, and stem rots; foliar blights	Yes	Yes	Can be used as a soil or foliar treatment
Streptomyces lydicus + iron, molybdenum, and humic acid	Actino-Iron	most vegetables (see label)	seedling, root, and stem rots	Yes	Yes	Soil treatment
Trichoderma harzianum	T-22, RootShield, PlantShield	Cole crops, eggplant, leafy vegetables, pepper, tomato	seedling, root, and stem rots	Yes	Yes	Can be added to potting mix or applied in-furrow to field soil.
Trichoderma viride	Binab	most vegetables	seedling, root, and stem rots	Yes	No	

Fungicide Resistance Management

Fungicides are organized according to FRAC groups, chemical structure and Mode of Action (MoA). Fungicides within a given FRAC group control fungi in a similar manner and share the same risk for fungicide resistance development. Some fungicides are referred to as high- or at-risk fungicides because of their specific MoAs and therefore have a high risk for resistance development. Groups of fungicides, such as the QoI's (FRAC group 11) or Phenylamides (FRAC group 4) are prone to resistance development due to very specific MoAs. Fungicides in high- or at-risk groups should be rotated and/or tank-mixed with broad spectrum protectant fungicides (FRAC group M3 or M5) to delay the development of resistant strains of fungi. For more information on fungicide resistance management see: http://www.frac.info/

Table 10-68. Fungicide Modes of Action for Fungicide Resistance Management

FRAC Code	Fungicide Resistance Risk	Group Name	Example Active ingredients	Example Products
P1	Unknown	Benzo-thiadiazole (BTH)	Acibenzolar-S-methyl	Actigard
M1	Low	Inorganic copper	Fixed copper	Copper (various)
M2	Low	Inorganic sulfur	Sulfur	Sulfur (various)
M3	Low	Dithiocarbamates	Mancozeb	Mancozeb (various)
M5	Low	Chloronitriles	Chlorothalonil	Chlorothalonil (various)
1	High	Methyl benzimidazole carbamates (MBC)	Thiophanate-methyl	Topsin M
2	Medium to high	Dicarboximides	Iprodione	Rovral
3	Medium	Demethylation inhibitors (DMI)	Triflumizole Myclobutanil	Procure Rally
4	High	Phenylamide	Mefenoxam	Ridomil Gold
7	Medium to high	Succinate dehydrogenase inhibitors (SDHI)	Boscalid Penthiopyrad	Endura Fontelis
9	Medium	Anilino-pyrimidines (AP)	Pyrimethanil	Scala
11	High	Quinone outside inhibitors (QoI)	Pyraclostrobin Trifloxystrobin Azoxystrobin	Cabrio Flint Quadris
12	Low to medium	Phenylpyrroles (PP)	Fludioxinil	Maxim
13	Medium	Aza-naphthalenes	Quinoxyfen	Quintec
14	Low to medium	Aromatic hydrocarbons (AH)	Dichloran	Botran
21	Medium to high	Quinone inside Inhibitors (QiI)	Cyazofamid	Ranman
22	Low to medium	Benzamides (toluamides) Thiazole carboxamide	Zoxamide	Gavel (contains zoxamide and mancozeb)
27	Low to medium	Cyanoacetamide-oximes	Cymoxanil	Curzate
28	Low to medium	Carbamates	Propamocarb	Presidio
29	Unknown	Dinitroanilines	Fluazinam	Omega
33	Low	Phosphonates	Fosetyl A!	Aliette
40	Low to medium	Carboxylic acid amides (CAA)	Dimethomorph Mandipropamid	Forum Revus
43	High	Benzamides	Fluopicolide	Presidio

INDEX

ABBREVIATIONS

a.i.	active ingredient
A	aerosol, acre
AE	acid equivalent
AS	aqueous solution
B	bait
bu	bushel
cc	cubic centimeters
CEC	cation exchange capacity
conc	concentration
cu ft	cubic foot (feet)
D	dust
DF	dry flowable
EC or E	emulsifiable concentrate
ES	emulsifiable solution
F	Fahrenheit
F	flowable, liquid
FME	flowable microencapsulated
g	gram(s)
G	granule
gal	gallon(s)
gpa	gallons per acre
gpm	gallons per minute
hr	hour(s)
IE	invert emulsion
in	inch(es)
IU	international units
kg	kilogram(s)
L	liquid, flowable
LC	liquid concentrate
LD	lethal dose (e.g. LD 50)
lb	pound(s)
lb/gal	pounds per gallon
liq	liquid
LS	liquid solution
LV	low volume
ml	milliliter(s)
mm	millimeter(s)
mg	milligram(s)
min	minutes
mmd	mass median diameter
No	number
OC	oil concentration
OD	oil dispersible
OS	oil solution
oz	ounce(s)
P, PS	pellets
PO	pour on
ppm	parts per million
psi	pounds per square inch
pt	pint(s)

qt	quart(s)
RS	resin strip
S	sprayable, solution
SC	spray concentrate
SL	soluble liquid
SOL	solution
SP	soluble powder
sq ft	square foot (feet)
tbsp	tablespoon(s)
TM	trade mark
trt	treated
tsp	teaspoon(s)
ULV	ultra low volume
WDG	water dispersible granule
WDL	water dispersible liquid
WP or W	wettable powder
WS	water soluble
WSG	water soluble granule
WSL	water soluble liquid
WSP	water soluble powder
XLR	extra long residual
<	less than
>	more than

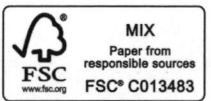